GPCRs as Therapeutic Targets

GPCRs as Therapeutic Targets

Volume 1

Edited by

Annette Gilchrist
College of Pharmacy-Downers Grove, Midwestern University,
Downers Grove, IL, USA

WILEY

This edition first published 2023

Registered Office
John Wiley & Sons, Inc., 111 River Street, Hoboken, NJ 07030, USA

Editorial Office
John Wiley & Sons, Inc., 111 River Street, Hoboken, NJ 07030, USA

For details of our global editorial offices, customer services, and more information about Wiley products visit us at www.wiley.com.

Wiley also publishes its books in a variety of electronic formats and by print-on-demand. Some content that appears in standard print versions of this book may not be available in other formats.

Library of Congress Cataloging-in-Publication Data

Names: Gilchrist, Annette, editor.
Title: GPCRs as therapeutic targets / edited by Annette Gilchrist.
Other titles: G-protein-coupled receptors as therapeutic targets
Description: Hoboken, NJ : Wiley, 2023. | Includes bibliographical references and index.
Identifiers: LCCN 2022016266 (print) | LCCN 2022016267 (ebook) | ISBN 9781119564744 (cloth) | ISBN 9781119564799 (adobe pdf) | ISBN 9781119564720 (epub)
Subjects: MESH: Receptors, G-Protein-Coupled | Drug Delivery Systems
Classification: LCC RS199.5 (print) | LCC RS199.5 (ebook) | NLM QU 55.7 | DDC 615/.6–dc23/eng/20220608
LC record available at https://lccn.loc.gov/2022016266
LC ebook record available at https://lccn.loc.gov/2022016267

Cover image: © Jon Burkhart; acrylic (Media)
Cover design by Wiley

Set in 9.5/12.5pt STIXTwoText by Straive, Chennai, India

Contents

Preface

G protein-coupled receptors (GPCRs) are the largest group of cell surface receptors. They regulate nearly all known human physiological processes from the sensory modalities of vision, taste, and smell to hormones that control our growth and development to neurotransmitters that govern behavior. Given their role in normal homeostasis and a broad array of pathological conditions including cancer, diabetes, cardiovascular disease, and asthma to name just a few, they serve as the targets for hundreds of drugs and more recently biologics such as monoclonal antibodies. Yet our understanding of GPCRs continues to evolve and texts that discuss these receptors must constantly be revisited. For example, we are just beginning to appreciate the importance of genetic variation in targeted GPCRs.

In the 30 years I have studied GPCRs many novel pharmacological concepts have been advanced. We have seen the emergence of constitutively active GPCRs and inverse agonism, arrestin signaling and functional selectivity, receptor dimerization, and signaling through subcellular receptors. We have seen many GPCRs without known endogenous ligands (orphans) undergo deorphanization, and accepted that some orphan receptors may only function constitutively in a ligand-independent manner. Advances in our understanding of their basal activity, their ability to bind a diverse array of ligands, how they communicate a signal across the cell membrane or within the cells, when they dimerize, their crosstalk with other receptors, and that their genetic variation can lead to disease or differences in drug response has expanded our appreciation for GPCRs. In addition, the depth of our understanding of GPCR pharmacology has in turn altered the drug discovery process itself, expanding the ways in which they are screened for compounds that modulate their signaling.

This two volume book set is organized into 26 chapters and will serve as a resource for any scientists investigating GPCRs, be it in academia or industry. The first volume provides in-depth information about the molecular pharmacology of this important target class and presents up-to-date material on GPCR structures and structure based drug design. There are eight chapters on the evolving pharmacology for GPCRs, including chapters discussing allosteric modulation, receptor dimerization, deorphanization, ubiquitination, intracellular trafficking, and subcellular GPCR signaling. The next six chapters discuss the rapidly growing field of GPCR structures and structure based drug design. Included in this section are chapters on the structural basis of G protein selectivity, as well as rational drug design for not only GPCRs but downstream signaling molecules such as phospholipase C. The second volume includes information on the role of GPCRs in disease and novel approaches for studying this receptor family. There are seven chapters addressing how GPCRs play a role in a wide range of pathological states including cancer, substance use disorders, cerebrovascular disease, and metabolic disease. The final five chapters present recent approaches employed to study GPCRs including mass spectrometry, bioluminescence, single molecule microscopy, and optogenetics. Together, the two volume book set provides a thorough overview of GPCRs in terms of their structure, pharmacology, function, and role in disease states, and provides information on novel approaches to measure GPCR activity.

Annette Gilchrist
Midwestern University

List of Contributors

John A. Allen
Department of Pharmacology and Toxicology
University of Texas Medical Branch
Galveston, TX
USA

and

Department of Neuroscience and Cell Biology
University of Texas Medical Branch
Galveston, TX
USA

and

Center for Addiction Research
University of Texas Medical Branch
Galveston, TX
USA

Salete J. Baptista
Data-Driven Molecular Design Group
CNC – Center for Neuroscience and Cell Biology
University of Coimbra
Coimbra
Portugal

and

Centro de Ciências e Tecnologias Nucleares, Instituto Superior Técnico
Universidade de Lisboa
Bobadela LRS
Portugal

Carlos A.V. Barreto
Data-Driven Molecular Design Group
CNC – Center for Neuroscience and Cell Biology
University of Coimbra
Coimbra
Portugal

and

Institute for Interdisciplinary Research, University of Coimbra
PhD Programme in Experimental Biology and Biomedicine
Coimbra
Portugal

Eric Bell
Department of Computational Medicine and Bioinformatics
University of Michigan
Ann Arbor, MI
USA

Kyla Bourque
Department of Pharmacology and Therapeutics
McGill University
Montréal, Québec
Canada

Mariaelana Brenner
Department of Medicine, Division of Endocrinology, Diabetes and Metabolism
University of Illinois College of Medicine
Chicago, IL
USA

Seema Briyal
Pharmaceutical Sciences
Midwestern University
Downers Grove, IL
USA

Michael R. Bruchas
Center of Excellence in the Neurobiology of Addiction, Pain, and Emotion, Department of Anesthesiology
University of Washington
Seattle, WA
USA

and

Department of Pharmacology
University of Washington
Seattle, WA
USA

Beatriz Bueschbell
Department of Pharmaceutical and Medicinal Chemistry
Pharmaceutical Institute, University of Bonn
Bonn
Germany

Davide Calebiro
College of Medical and Dental Sciences
Institute of Metabolism and Systems Research, University of Birmingham
Birmingham
UK

and

Centre of Membrane Proteins and Receptors (COMPARE)
Universities of Nottingham and Birmingham
UK

Wallace Chan
Department of Computational Medicine and Bioinformatics
University of Michigan
Ann Arbor, MI
USA

and

Department of Biological Chemistry
University of Michigan
Ann Arbor, MI
USA

and

Department of Pharmacology
University of Michigan
Ann Arbor, MI
USA

Naincy R. Chandan
Department of Pharmacology
University of Michigan Medical School
Ann Arbor, MI
USA

Sun Choi
College of Pharmacy and Graduate School of Pharmaceutical Sciences
Ewha Womans University
Seoul
Republic of Korea

Irene Coin
Faculty of Life Sciences
Institute of Biochemistry, Leipzig University
Leipzig
Germany

Bryan A. Copits
Pain Center, Department of Anesthesiology
Washington University School of Medicine
St. Louis, MO
USA

Natasha C. Dale
Molecular Endocrinology and Pharmacology
Harry Perkins Institute of Medical Research, QEII Medical Centre
Nedlands, Western Australia
Australia

and

Centre for Medical Research
The University of Western Australia
Crawley, Western Australia
Australia

and

National Centre
Australian Research Council Centre for Personalised Therapeutics Technologies
Australia

Michael R. Dores
Department of Biology
Hofstra University
Hempstead, NY
USA

Stefan Ernicke
Faculty of Life Sciences, Institute of Biochemistry
Leipzig University
Leipzig
Germany

Daniel E. Felsing
Department of Pharmacology and Toxicology
University of Texas Medical Branch
Galveston, TX
USA

and

Center for Addiction Research
University of Texas Medical Branch
Galveston, TX
USA

Geordi Frere
Department of Chemistry, Chemical and Physical Sciences
University of Toronto
Mississauga, ON
Canada

Elyssa Frohlich
Department of Pharmacology and Therapeutics
McGill University
Montréal, Québec
Canada

Annette Gilchrist
Department of Pharmaceutical Sciences, College of Pharmacy-Downers Grove
Midwestern University
Downers Grove, IL
USA

Sophie Gough
Department of Medicine, Division of Endocrinology, Diabetes and Metabolism
University of Illinois College of Medicine
Chicago, IL
USA

Jerome Gould
Department of Chemistry, Chemical and Physical Sciences
University of Toronto
Mississauga, ON
Canada

Karen J. Gregory
Drug Discovery Biology
Monash Institute of Pharmaceutical Science, Monash University
Parkville
Australia

Jak Grimes
College of Medical and Dental Sciences
Institute of Metabolism and Systems Research, University of Birmingham
Birmingham
UK

and

Centre of Membrane Proteins and Receptors (COMPARE)
Universities of Nottingham and Birmingham
UK

Anil Gulati
Pharmaceutical Sciences
Midwestern University
Downers Grove, IL
USA

and

Pharmazz, Inc.
Willowbrook, IL
USA

Patrick T. Gunning
Department of Chemistry, Chemical and Physical Sciences
University of Toronto
Mississauga, ON
Canada

Michael A. Hanson
SB SciTech
San Marcos, CA
USA

Steven K. Harmon
Department of Neuroscience
Washington University School of Medicine
Saint Louis, MO
USA

Advait Hasabnis
Department of Chemistry, Chemical and Physical Sciences
University of Toronto
Mississauga, ON
Canada

Terence E. Hébert
Department of Pharmacology and Therapeutics
McGill University
Montréal, Québec
Canada

Laura H. Heitman
Division of Drug Discovery and Safety
LACDR
Leiden University
The Netherlands

and

Oncode Institute
Leiden
The Netherlands

Carole Le Henaff
Department of Molecular Pathobiology
New York University College of Dentistry
New York, NY
USA

Iara C. Ibay
Department of Pharmaceutical Sciences, College of Pharmacy-Downers Grove
Midwestern University
Downers Grove, IL
USA

Adriaan P. IJzerman
Division of Drug Discovery and Safety, LACDR
Leiden University
The Netherlands

Roshanak Irannejad
Department of Biochemistry and Biophysics
Cardiovascular Research Institute
University of California
San Francisco, CA
USA

Ralf Jockers
Institut Cochin, CNRS, INSERM
Université de Paris
Paris
France

Elizabeth K.M. Johnstone
Molecular Endocrinology and Pharmacology
Harry Perkins Institute of Medical Research, QEII Medical Centre
Nedlands, Western Australia
Australia

and

Centre for Medical Research
The University of Western Australia
Crawley, Western Australia
Australia

and

National Centre
Australian Research Council Centre for Personalised Therapeutics Technologies
Australia

Yuh-Jiin I. Jong
Department of Neuroscience
Washington University School of Medicine
Saint Louis, MO
USA

Tracy M. Josephs
Drug Discovery Biology
Monash Institute of Pharmaceutical Science, Monash University
Parkville
Australia

Andrew N. Keller
Drug Discovery Biology
Monash Institute of Pharmaceutical Science, Monash University
Parkville
Australia

Terry Kenakin
Department of Pharmacology
University of North Carolina School of Medicine
Chapel Hill, NC
USA

Brian T. Layden
Department of Medicine, Division of Endocrinology, Diabetes and Metabolism
University of Illinois College of Medicine
Chicago, IL
USA

and

Jesse Brown Veterans Affairs Medical Center
Department of Medicine, Section of Endocrinology
Chicago, IL
USA

Raudah Lazim
College of Pharmacy and Graduate School of Pharmaceutical Sciences
Ewha Womans University
Seoul
Republic of Korea

Kristen R. Lednovich
Department of Medicine, Division of Endocrinology, Diabetes and Metabolism
University of Illinois College of Medicine
Chicago, IL
USA

Yoonji Lee
College of Pharmacy
Chung-Ang University
Seoul
Republic of Korea

Katie Leach
Drug Discovery Biology
Monash Institute of Pharmaceutical Science, Monash University
Parkville
Australia

Agostinho Lemos
Data-Driven Molecular Design Group
CNC – Center for Neuroscience and Cell Biology
University of Coimbra
Coimbra
Portugal

Braden T. Lobingier
Department of Chemical Physiology and Biochemistry
Oregon Health and Sciences University
Portland, OR
USA

Yan Lu
iHuman Institute, ShanghaiTech University
Shanghai
China

and

School of Life Science and Technology
ShanghaiTech University
Shanghai
China

Miguel Machuqueiro
Departmento de Química e Bioquímica, Faculdade de Ciências
Universidade de Lisboa
BioISI-Biosystems and Integrative Sciences Institute
Lisboa
Portugal

Pedro R. Magalhães
Departmento de Química e Bioquímica, Faculdade de Ciências
Universidade de Lisboa
BioISI-Biosystems and Integrative Sciences Institute
Lisboa
Portugal

Rita Melo
Data-Driven Molecular Design Group
CNC – Center for Neuroscience and Cell Biology
University of Coimbra
Coimbra
Portugal

and

Centro de Ciências e Tecnologias Nucleares, Instituto Superior Técnico
Universidade de Lisboa
Bobadela LRS
Portugal

Ravi Mistry
College of Medical and Dental Sciences
Institute of Metabolism and Systems Research, University of Birmingham
Birmingham
UK

and

Centre of Membrane Proteins and Receptors (COMPARE)
Universities of Nottingham and Birmingham
UK

Irina S. Moreira
Data-Driven Molecular Design Group
CNC – Center for Neuroscience and Cell Biology
University of Coimbra
Coimbra
Portugal

and

Department of Life Sciences, Faculty of Science and Technology
University of Coimbra
Coimbra
Portugal

and

CIBB – Center for Innovative Biomedicine and Biotechnology
University of Coimbra
Coimbra
Portugal

Pratanphorn Nakliang
College of Pharmacy and Graduate School of Pharmaceutical Sciences
Ewha Womans University
Seoul
Republic of Korea

Ashley N. Nilson
Department of Neuroscience and Cell Biology
University of Texas Medical Branch
Galveston, TX
USA

and

Center for Addiction Research
University of Texas Medical Branch
Galveston, TX
USA

Atsuro Oishi
Institut Cochin, CNRS, INSERM
Université de Paris
Paris
France

and

Department of Anatomy
Kyorin University Faculty of Medicine
Tokyo
Japan

and

Cancer RNA Research Unit
National Cancer Center Research Institute
Tokyo
Japan

Karen L. O'Malley
Department of Neuroscience
Washington University School of Medicine
Saint Louis, MO
USA

Patrick R. O'Neill
Hatos Center for Neuropharmacology
Department of Psychiatry and Biobehavioral Sciences
University of California Los Angeles
Los Angeles, CA
USA

Maria C. Orencia
GPCR Consortium
San Marcos, CA
USA

Aditya Pandey
Department of Chemistry, Chemical and Physical Sciences
University of Toronto
Mississauga, ON
Canada

and

Department of Biochemistry
University of Toronto
Toronto, ON
Canada

Nicola C. Partridge
Department of Molecular Pathobiology
New York University College of Dentistry
New York, NY
USA

Frank Park
Department of Pharmaceutical Sciences
University of Tennessee Health Science Center
Memphis, TN
USA

Kevin D.G. Pfleger
Molecular Endocrinology and Pharmacology
Harry Perkins Institute of Medical Research
QEII Medical Centre
Nedlands, Western Australia
Australia

and

Centre for Medical Research
The University of Western Australia
Crawley, Western Australia
Australia

and

National Centre
Australian Research Council Centre for Personalised Therapeutics Technologies
Australia

and

Dimerix Limited
Nedlands, Western Australia
Australia

Hoa T.N. Phan
Department of Pharmacology
University of Michigan Medical School
Ann Arbor, MI
USA

António J. Preto
Data-Driven Molecular Design Group
CNC – Center for Neuroscience and Cell Biology
University of Coimbra
Coimbra
Portugal

and

Institute for Interdisciplinary
Research, University of Coimbra
PhD Programme in Experimental
Biology and Biomedicine
Coimbra
Portugal

Robert S. Prosser
Department of Chemistry, Chemical
and Physical Sciences
University of Toronto
Mississauga, ON
Canada

and

Department of Biochemistry
University of Toronto
Toronto, ON
Canada

Talha Qadri
Department of Medicine, Division of
Endocrinology, Diabetes and
Metabolism
University of Illinois College of
Medicine
Chicago, IL
USA

Shanshan Qin
iHuman Institute, ShanghaiTech
University
Shanghai
China

Amaresh Ranjan
Pharmaceutical Sciences
Midwestern University
Downers Grove, IL
USA

Nícia Rosário-Ferreira
Data-Driven Molecular Design Group
CNC – Center for Neuroscience and
Cell Biology
University of Coimbra
Coimbra
Portugal

and

Chemistry Department, Faculty of
Science and Technology, Coimbra
Chemistry Center
University of Coimbra
Coimbra
Portugal

Fredrik Sadler
Biochemistry, Molecular Biology and
Biophysics Graduate Program
University of Minnesota
Minneapolis, MN
USA

Anke C. Schiedel
Department of Pharmaceutical and
Medicinal Chemistry, Pharmaceutical
Institute
University of Bonn
Bonn
Germany

Wenqing Shui
iHuman Institute
ShanghaiTech University
Shanghai
China

and

School of Life Science and Technology
ShanghaiTech University
Shanghai
China

Sivaraj Sivaramakrishnan
Biochemistry, Molecular Biology and
Biophysics Graduate Program
University of Minnesota
Minneapolis, MN
USA

and

Department of Genetics
Cell Biology, and Development
University of Minnesota
Minneapolis, MN
USA

Alan V. Smrcka
Department of Pharmacology
University of Michigan Medical School
Ann Arbor, MI
USA

Emma Tripp
College of Medical and Dental
Sciences
Institute of Metabolism and Systems
Research, University of Birmingham
Birmingham
UK

and

Centre of Membrane Proteins and
Receptors (COMPARE)
Universities of Nottingham and
Birmingham
UK

Nagarajan Vaidehi
Department of Computational and
Quantitative Medicine, City of Hope
Cancer Center
Beckman Research Institute of the
City of Hope
Duarte, CA
USA

Xuesong Wang
Division of Drug Discovery and Safety
LACDR
Leiden University
The Netherlands

Gerard J. P. van Westen
Division of Drug Discovery and Safety
ALCDR
Leiden University
The Netherlands

Carl W. White
Molecular Endocrinology and
Pharmacology
Harry Perkins Institute of Medical
Research, QEII Medical Centre
Nedlands, Western Australia
Australia

and

Centre for Medical Research
The University of Western Australia
Crawley, Western Australia
Australia

and

National Centre
Australian Research Council Centre
for Personalised Therapeutics
Technologies
Australia

Jiansheng Wu
Department of Computational Medicine and Bioinformatics
University of Michigan
Ann Arbor, MI
USA

and

School of Geographic and Biological Information
Nanjing University of Posts and Telecommunications
Nanjing
China

Yang Zhang
Department of Computational Medicine and Bioinformatics
University of Michigan
Ann Arbor, MI
USA

and

Department of Biological Chemistry
University of Michigan
Ann Arbor, MI
USA

Part I

GPCR Pharmacology/Signaling

1

An Overview of G Protein Coupled Receptors and Their Signaling Partners

Iara C. Ibay and Annette Gilchrist

Department of Pharmaceutical Sciences, College of Pharmacy-Downers Grove, Midwestern University, Downers Grove, IL, USA

1.1 Overview of GPCR Superfamily

G protein coupled receptors (GPCRs) encompass a large and diverse protein superfamily with over 800 members identified in the human genome. Of these 390 are odorant receptors, 33 are taste receptors, 10 are visual receptors, and 5 are pheromone receptors with the remaining receptors having non-sensory mechanistic properties. Of these, ~120 remain "orphan" receptors whose endogenous agonist is unknown [1]. A recent publication reported that 134 GPCRs are targets for some 700 medications approved in the United States or European Union accounting for ~35% of the current drugs on the market [2].

The structures of GPCRs have been widely studied. All GPCR members are comprised of seven transmembrane (TM) domains with an extracellular amino (N)-terminus that is highly varied, three extracellular loops, three intracellular loops, and an intracellular carboxyl (C)-terminus. Detailed crystal structures of GPCRs have been utilized to understand their molecular and mechanistic properties. The Nomenclature and Standards Committee of the International Union of Basic and Clinical Pharmacology (NC-IUPHAR) classification divides GCPRs into six classes (Classes A–F) depending on their amino acid sequence and functional similarities (designed fingerprints) of the seven hydrophobic domains [3–5]. Class A GPCRs, also known as the rhodopsin-like family, include the vast majority of receptors accounting for nearly 85% of GPCRs. Like all GPCRs, Class A receptors have seven TM helices, but there is an eighth helix, formed by a palmitoylated cysteine in the C-terminal tail. For members of this class the orthosteric binding site is deep in the transmembrane helices [6]. There are around

GPCRs as Therapeutic Targets, Volume 1, First Edition. Edited by Annette Gilchrist.

70 members of Class B GPCRs, and multiple members belong to the secretin receptor family. Many members have a large N-terminal domain of around 120 residues stabilized by disulfide bonds that serves as the orthosteric ligand binding side and is often activated by peptides [7]. Structures have been determined for several Class B receptors including glucagon receptor [8, 9], glucagon like peptide 1 (GLP-1) receptor [10, 11], and parathyroid hormone receptor-1 (PTH1) [12]. The adhesion family of GPCRs is phylogenetically related to class B receptors. They differ by possessing large extracellular N-termini that are proteolytically cleaved at a conserved site within a larger autoproteolysis-inducing domain. Class C contains 22 members, including the eight metabotropic glutamate receptors, γ-aminobutyric acid ($GABA_B$) receptors, calcium sensing receptors, taste receptors, and retinoic acid-inducible orphan GPCRs. In addition to having a characteristically large extracellular domain to which ligands bind, many of the receptors are obligatory dimers [13]. In 2014, the Ray Stevens group provided the structure for metabotropic glutamate receptor 1 ($mGlu_1$) [14]. This was followed in 2021 by structural information for the calcium-sensing receptor [15, 16]. Class D GPCRs are found exclusively in fungi where they regulate survival and reproduction. Within Class D, the fungal GPCRs are further categorized into 10 classes on the basis of sequence homology. Chris Tate and colleagues recently provided the first structure for a Class D GPCR, namely that of Ste2 coupled to a heterotrimeric G protein and bound to α-factor [17]. Class E GPCRs constitute cyclic adenosine monophosphate (cAMP) receptors from a protozoan amoeba (*Dictyostelium discoideum*) that are involved in chemotaxis. While the biochemical aspects of these receptors are well characterized, less is known about their structure [18, 19]. Class F GPCRs include the frizzled or smoothened receptors that are fundamental for mediating hedgehog signaling and Wnt binding. The Class F GPCRs ligands vary in size from small molecules and peptides to large proteins [7]. In 2019, Xiaochun Li's group published the structure of Smoothened bound to 24(*S*),25-epoxycholesterol and coupled to a heterotrimeric G_i protein [20]. As Class A, B, C, D, and F structures have been published, the only GPCR family without an atomic level structure is Class E GPCRs.

With the advancement of technologies such as X-ray crystallography, cryogenic electron microscopy (cryo-EM), nuclear magnetic resonance (NMR), electron paramagnetic resonance (EPR) spectroscopy techniques such as double electron–electron resonance (DEER), and molecular dynamics (MD) simulation many high resolution GPCR structures have been determined experimentally allowing us to understand some of the differences between individual receptors. In recent years, computational biology methods have utilized homology modeling and machine learning with programs such as MODELLER, RoseTTAFold, AlphaFold, and GPCR Dock to expand our ability to accurately predict GPCR structures [21–23]. To date, high resolution structures have been published for

107 different GPCRs with over 120 GPCR-G protein complexes [24]. Structures for GPCRs have been determined with the receptor in active, inactive, as well as intermediate states with arrestin or heterotrimeric G proteins present [25]. A common feature of GPCRs that have 3D structures for both an antagonist bound state and agonist bound ternary complex is the large differences in receptor conformation between these states [26]. Early studies with Class A GPCRs indicated there was movement in the TM6 domain when an extracellular signal is bound [27]. Rasmussen et al. showed that there were changes in the cytoplasm facing side of the receptor, which included TM5 and TM6 moving outwards and TM7 and TM3 moving inwards [28]. Similar movements in TM3, TM6, and TM7 were observed with adenosine A_{2A} receptors when bound to an agonist [29]. The outward movement of TM6 and inward movement of TM7 described for class A receptors, were also observed with Class B GPCRs including calcitonin, and GLP-1R, although there were differences in helix 8 [25]. The 3D structures from X-ray crystallography and cryo-EM have provided in-depth information about the orthosteric and allosteric binding sites of GPCRs. They have also provided novel insights to receptor dimerization [30]. Structural advances have been complemented by studies using MD simulation [31], deep mutational scanning [32], genome sequencing [33], and signal protein profiling [34, 35].

1.2 GPCR Signaling

GPCRs share functional likenesses, serving as biosensors; however, they exhibit versatility in the mechanisms by which they communicate extracellular signals to the cells. To this end, binding of an endogenous ligand produces a conformational change in the receptor that allows the recruitment of other proteins such as heterotrimeric G-proteins, β-arrestins, and G-protein coupled receptor kinases (GRKs). GPCRs show considerable promiscuity in their ability to bind these proteins. For example, many GPCRs can bind to several different G proteins resulting in a wide array of downstream signaling processes. In addition, GPCRs can act independent of G-proteins, sending signals through β-arrestin. β-arrestins have the capacity to bind the receptor and lead to (i) desensitization, after which the cell does not respond to further stimuli for a period of time; (ii) internalization, whereby receptors undergo endocytosis; and (iii) downregulation, with resultant decrease in the number of receptors displayed at the cell surface. At the same time, β-arrestin binding to a GPCR means the receptor is unable to interact with G-proteins thus terminating G protein effector signaling. GPCR signaling can be altered by receptor phosphorylation brought about by kinases such as GRKs with the phosphorylation pattern serving as a barcode [36]. The versatility of GPCRs has been demonstrated by their ability to interact with other membrane proteins. As an example, homo- and hetero-dimerization of GPCRs can lead to their activation [37]. Furthermore, transactivation by receptor tyrosine kinases can trigger GPCR downstream signaling events [38]. Notably, although historically GPCRs were considered cell surface receptors a large number of GPCRs have been found at intracellular sites such as the endosomes, golgi apparatus, and nuclear membrane [39]. The functional effects observed with activation of the intracellular GPCRs are often different from those seen with the same GPCR at the plasma membrane, even when identical G protein(s) and downstream second messengers are activated [40]. These traits provide GPCRs with the ability to regulate a wide array of physiological processes.

1.2.1 GPCR Signaling and G-Proteins

Typically, the extracellular and/or transmembrane portions of the GPCR possess the portion that bind to activating ligands, a vital step in the GPCR being able to transduce signals intracellularly. This leads to a conformational change in the receptor such that the intracellular ends of transmembrane helices H5 and H6 move outward to form a cleft in the receptor where the C-terminus of the Gα-subunit binds. In its inactive state, the Gα subunit is bound to guanosine diphosphate (GDP) and the Gβγ subunit to make a heterotrimeric structure. The Gβγ subunit increases the affinity Gα has for GDP by more than 100-fold [41]. Binding of the G protein to the GPCR leads to a conformational change in the receptor that results in concomitant closure of the loops over the orthosteric binding site, resulting in both an increase in agonist affinity [42] and a decrease in its off-rate [43]. This subsequently results in release of GDP from Gα. Thus, activated GPCRs function as guanine-nucleotide exchange factors (GEFs), to promote the release of GDP from Gα. Recent work by Su et al. suggests the role of the GPCR is to deform the GDP-binding pocket and accelerate GDP release from the G protein [44]. As guanosine triphosphate (GTP) is at a much higher level in cells it binds to the empty Gα. This event changes the conformation in the heterotrimeric G-protein causing a dissociation of it from the GPCR with the Gα and Gβγ proteins each going on to elicit independent downstream effects [45]. The intrinsic GTP hydrolysis of the Gα subunit returns the protein to the GDP-bound state allowing it to interact with Gβγ [46]. GPCR-G protein signaling can be modified via regulator of G-protein signaling (RGS) proteins that accelerate the rate of GTP hydrolysis by Gα [47].

Gα proteins are divided into four subfamilies: $G_{i/o}$, G_s, $G_{q/11}$, and $G_{12/13}$. Each subfamily can interact with different effectors resulting in distinct downstream signaling events. The activation of the Gαs usually leads to an increase of the catalytic activity of adenylyl cyclase and an increase in second messenger cAMP.

The increase in cAMP activates protein kinase A, which leads to initiation of other signaling pathways. Meanwhile, signaling through Gα_i usually leads to an inhibition of adenylyl cyclase, and decreased cAMP levels. The G$\alpha_{q/11}$ subfamily can active phospholipase C (PLC), resulting in hydrolyzation of phosphatidylinositol 4,5-bisphosphate. As a result, two second messengers, diacylglycerol and inositol phosphate, are produced. The production of diacylglycerol leads to activation of protein kinase C and inositol triphosphate which subsequently induce the release of Ca^{2+} in the cell. G_{12} family members (G$\alpha_{12/13}$) usually act via the small GTPase RhoA. Most Gα subunits are localized at the membrane by palmitoylation at residues near their N-termini. Members of the Gα_i subfamily are also myristoylated at residues near their N-termini but rather than promoting membrane interaction this post-translational event increases affinity for adenylyl cyclase, G$\beta\gamma$ subunits, and the GEF/chaperone Ric-8A [46].

As GPCRs bind to G-proteins at a distant site from the nucleotide binding pocket, it has been long recognized that activation must be achieved through allosteric mechanisms. This transmission of information also enables G proteins to allosterically influence ligand affinity for GPCRs [43]. As with GPCRs, high-resolution structures, molecular dynamics simulation, and a variety of biophysical approaches have expanded our understanding of the receptor-G protein interface, and conformational changes in Gα that occur upon receptor binding [28, 48, 49]. While conformational differences between the Gα subunit when bound to either GDP and G$\beta\gamma$ or GTP are relatively small, GPCR binding causes global conformational changes in G proteins with a striking conformational rearrangement resulting in a large separation of the Ras-like and helical domains of Gα. In fact, a 60° rotation of the α5 helix is present in all the published cryo-EM structures of Gα [50]. The interdomain separation is not the trigger for nucleotide release, but rather, a sequel to allosteric mechanisms that trigger the release, facilitating escape of the released nucleotide [49]. The interruption of the contacts between H1 and H5 is the key step for GDP release with H1 serving as the molecular switch for GDP release, and H5 being the distal trigger that is "pulled" on receptor binding [51]. Comparison of $G_{i/o}$ and G_s heterotrimers has shown that there are G protein family specific variations in the activation process [26].

1.2.2 GPRC Signaling and Arrestins

Arrestins were first discovered in photoreceptor cells and named for their ability to "arrest" the signaling of rhodopsin. There are four mammalian arrestins: arrestin-1 (S-antigen, 48 kDa protein, visual or rod arrestin), arrestin-2 (β-arrestin or β-arrestin1), arrestin-3 (β-arrestin2 or hTHY-ARRX), and arrestin-4 (cone or X-arrestin), and they are classified as visual (arrestin-1 and arrestin-4) or non-visual (arrestin-2 and arrestin-3) [6]. As noted earlier, arrestins terminate

GPCR signaling by competing with heterotrimeric G proteins for binding at the receptor. In addition, GPCRs can mediate downstream responses through arrestin-dependent signaling, in a manner that is independent of G protein activation. That GPCRs can signal through both G proteins and arrestins has led to the recognition of biased signaling by some agonists yielding activation of only one pathway [52]. Arrestins also play a critical role in GPCR internalization and intracellular trafficking [53]. In addition, arrestins can serve as scaffolds for proteins like mitogen-activated protein kinase (MAPK), c-Jun N-terminal kinase (JNK), AKT, SRC, and phosphoinositide 3-kinase that are all important in downstream signaling [54]. Another piece in the GPCR signaling puzzle is that arrestins play a pivotal role in desensitization of GPCRs [55]. Some GPCRs are coupled to arrestins for only a short amount of time (acute) and dissociate once the receptor is internalized with examples including β_2AR, dopamine, and endothelin A receptors. Other GPCRs are coupled to arrestins for a longer period of time (prolonged), dissociating slowly due to a strong, stable bond with angiotensin type 1A receptor, and thyrotropin-releasing hormone receptor as examples [56]. There are also distinct ways in which desensitization can be induced. Heterologous desensitization refers to receptors that are phosphorylated in the absence of an agonist, while homologous desensitization is agonist induced. GPCR desensitization is largely regulated by specific kinases (GPCR kinases or GRKs) that phosphorylate the receptor with subsequent binding of arrestins [57]. In addition to playing a role in receptor internalization, and desensitization, arrestins are also involved in GPCR trafficking, recycling, ubiquitination, and degradation [58].

1.2.3 GPRC Signaling and GRKs

The first step of GPCR interaction with arrestins is the phosphorylation of the receptor by GRKs. Both the number and arrangement of phosphates can vary with different phosphorylation patterns (aka barcodes) triggering different arrestin-mediated effects [59]. Currently, there are seven known GRKs. They have a similar structure with a domain where the protein kinase resides and a variable C-terminus [55]. GRK1, GRK2, GRK3, and GRK7 are activated exclusively by binding to active GPCRs. GRK4 constitutively phosphorylates the dopamine D1 receptor, while GRK5 and GRK6 can phosphorylate inactive GPCRs. Notably, GRKs can phosphorylate and thus regulate many non-GPCR substrates. In addition to regulating GPCR signaling via phosphorylation, GRKs can control signaling in a phosphorylation-independent manner via direct protein–protein interaction. GRKs possess an RGS homology (RH) domain and the RH domains of GRK2 and GRK3 bind active G$\alpha_{q/11}$ and by doing so reduce $G_{q/11}$-mediated signaling through sequestration. Although all GRK isoforms are equipped with the RH domain, only GRK2 and GRK3 function in this manner, as other GRKs are missing key binding residues. In addition, GRK2 and GRK3 possess a pleckstrin homology (PH) domain in their C-terminus capable of binding Gβγ. As a result, GRK2 and GRK3 can regulate Gβγ-dependent signaling through sequestration in the same manner as the RH domain regulates $G_{q/11}$-mediated signaling. Finally, some GRK functions are mediated by GRK interacting proteins (GITs), which are large multidomain scaffolding proteins that interact with multiple partners and are involved in numerous cellular processes [60].

1.2.4 GPRC Signaling and RGS Proteins

The RGS family of proteins was initially identified as GTPase-accelerating proteins (GAPs) capable of acting on Gα subunits that were also subsequently found to modulate G-protein effector interactions [61]. By acting catalytically to increase the rate of Gα-catalyzed GTP hydrolysis with subsequent termination of signaling RGS proteins can control the intensity and the time of the cellular responses by GPCRs and the loss of RGS-mediated control can lead to a range of pathologies. Unlike Ras-family GAPS, RGS proteins do not participate in the chemistry of GTP hydrolysis. Rather, their binding stabilize the transition state for GTP hydrolysis resulting in the release of interactions between the γ-phosphate of GTP and the N-terminus of switch II on Gα allowing new interactions with the Gβγ interface [46]. Interestingly, RGS proteins and Gα effectors occupy distinct and non-overlapping binding sites on Gα [62].

To date, 20 mammalian RGS proteins (RGS1–14 and RGS16–21) have been shown to serve as GAPs and several RGS protein families have been established, named after their prototypical members: A/RZ, B/R4, C/R7, and D/R12 [63]. One of the unique features to RGS proteins is that there is specificity of different RGS proteins for distinct Gα subunits. Recent comprehensive work by Masuho et al. mapped nearly all of the 300 theoretically possible Gα-RGS pairings for the preferences that RGS proteins have for Gα substrates [64]. RGS proteins are regulated by their local protein concentration (at the site of signaling), and this can be controlled by regulating subcellular localization, protein stability, transcription, or through epigenetic regulation [65].

GAP activity in RGS proteins is conveyed by a ~120 residue α-helical domain referred to as the RH domain [66]. The smallest members of the RGS family (RGS1–5, RGs8, RGS13, RGS16, RGS18, RGS21) vary in size from 20 to 30 kDa [67] and possess a simple structure consisting of a conserved RH domain and short C-termini. Other RGS proteins also contain GAP-independent domains and/or regulatory elements that target or regulate the RGS domain GAP functionality, or that possess non-canonical signaling properties. An example is found in the R7 family of RGS proteins that includes RGS6, RGS7, RGS9, and RGS11 [68]. This family contains a functional region known as G protein γ-like (GGL) that has been implicated in protein–protein interactions with Gβ5 as well as GPCRs. A second disheveled/EGL-1/Plextril (DEP) domain has also been demonstrated to enhance GPCR interactions. In addition, R7BP (for R7 binding protein) and R9AP (RGS9 associated protein, specifically in photoreceptor cells) are membrane tethered proteins that enhance proximity of R7 family RGS proteins to the plasma membrane [69]. Another example of a GAP-independent domain that mediates RGS function is the $G\alpha_{i/o}$-Loco (GoLoco) motif present in R12 family members RGS12 and 14. The GoLoco motif can (i) bind Gα and inhibits GTP exchange, thereby preventing G protein activation; and (ii) block the association of Gα with Gβγ, potentially leading to prolonged Gβγ signaling [70]. RGS12 and RGS14 also have Ras binding domain(s) that may play a role in integrating GPCR activation with the Ras/MAPK signaling pathway [71]. Through the phosphotyrosine binding (PTB) domain, RGS12 can interact with, and modulate the activity of, N-type calcium channels in a phosphorylation-dependent manner [72]. However, the presence of additional domains is not necessary for an RGS protein to exert GAP-independent protein–protein interactions as RGS2 can directly interact with adenylate cyclase [73], eukaryotic initiation factor 2Bε [74], and several GPCRs, including α_{1A} adrenoceptor [75] and M_1 muscarinic receptor [76]. In fact, members from all four RGS families have been shown to interact with GPCRs.

One of the pitfalls of studying RGS function is that their GAP activity can only be measured when the RGS protein is bound to the Gα subunit, thus providing a difficult platform for studying its function. As a result, many studies focus on the downstream effects of RGS RH domain-Gα interactions to shed light on the cellular functions of RGS proteins. Yet even in light of the challenges, drug discovery efforts aimed at RGS are underway given that they may provide a means by which one can target pathological GPCR signaling [67].

1.3 GPCR Pharmacology

GPCR are a popular target for drug discovery efforts. In fact, 107 different GPCRs have been successfully targeted by approximately 481 drugs, or 34% of FDA approved drugs [77]. This number continues to rise as more discoveries are made including novel biological agents such as peptide agonists and monoclonal antibodies (mAbs), allosteric modulators, and biased agonists/antagonists. Investigators continue to be surprised by the complex nature and variety of signaling mechanisms that GPCRs employ. Some of the new therapeutics for GPCRs include the mAb Erenumab (Aimovig) that targets calcitonin gene-related peptide receptor (CGRPR), Lemborexant (Dayvigo) a dual orexin receptor antagonist, and difelikefalin (Korsuva) a κ opioid receptor agonist. Yet, there are over 200 GPCRs that have not been exploited as targets [77].

GPCR ligands include light, ions, small molecules (odorants, vitamins, neurotransmitters), hormones (estrogen, growth hormone-releasing hormone), lipids (sterols, fatty acids, sphingolipids), peptides (bradykinin, somatostatin), and proteins (Wnt, chemokines). Ligands are classified based on their ability to elicit a signal and include full agonists, partial agonists, inverse agonists, and neutral antagonists. Full agonists produce the most GPCR activity. Partial agonists can act as antagonists to the endogenous ligand as well as function as activating modulators when there are excessive amounts of receptors and a lack of agonists. Inverse agonists turn off constitutive signaling between GPCRs and their cognate G proteins. Lastly, neutral antagonists disable the action of the agonists while having no effect on receptor signaling. Full agonists stabilize the active state, while inverse agonists stabilize the inactive state. Neutral antagonists and partial agonists can bind receptors in both active and inactive states. To reach the fully active state GPCRs need both a full agonist and a G-protein. Thus, GPCR signal transduction through G proteins is an allosteric process that involves communication between sites that are conformationally linked [30].

With few exceptions, nearly all of the approved drugs for GPCRs target the orthosteric binding site. For Class A GPCRs, the orthosteric site is between TM6 and extracellular loop 2 (ECL2), positioned near the extracellular portion of the receptor. The site is easily reached, such that ligands can bind without having to cross the membrane. These highly conserved binding sites vary in shape and chemical properties that are often dependent on the nature of the ligand. For example, a lipophilic ligand may enter from the membrane while peptide ligands may require their binding sites face the extracellular space due to their large size.

Although the main focus for GPCR drug discovery efforts has been on orthosteric binding sites, interest in allosteric binding sites is on the rise [78]. Allosteric modulators bind somewhere other than the orthosteric site. These allosteric sites have been found throughout the GPCR [79] including ECL2, TM3. Negative allosteric modulators (NAMs) decrease GPCR activity while positive allosteric modulators (PAMs) increase GPCR activity. Although, NAMs and PAMs do not interact with the orthosteric site they can control the function of the orthosteric ligands. There are currently only a few FDA approved allosteric modulators for GPCR targets including cinacalcet (calcium sensing receptor), maraviroc (CCR5 chemokine receptor), and sonedigib and vismodegib (smoothened receptor). In addition to NAMs and PAMs, which require the presence of the endogenous ligand to work, there is another class of drug known as allosteric agonists that have the capacity to activate GPCRs but do so by binding to a site distinct from the orthosteric site.

Rather than target orthosteric sites or allosteric sites, some investigators have explored bitopic ligands that can occupy both sites simultaneously to provide more target specificity [80–82]. Because of their dual steric nature, these ligands can interrupt the conformational flexibility of the GPCR and introduce a bias in signaling. Furthermore, some investigators have also looked at bivalent ligands for GPCR dimers [83, 84].

Ligands can also play a role in the stabilization of specific GPCR conformations enabling them to elicit biased-agonism. An example of biased signaling can be found in the angiotensin II type 1 receptor (AT1R), which can signal through both $G_{q/11}$ protein and arrestin. The AT1R ligand TRV120027 is biased towards β-arrestin, resulting in no activation of $G_{q/11}$, while being able to produce heart contractility via β-arrestin-coupling [85]. However, some ligands thought to be biased agonists have since been shown to be partial agonists. For example, when Gillis et al. used an approach that minimized the effects of system bias and receptor reserve many opioids previously described as biased agonists were found to be partial agonists without significant bias [86]. Yet, identification of biased ligands remains a promising direction for future drug discovery efforts [87].

1.4 Forging Ahead

One of the challenges that remain with GPCRs involves receptor dimerization. Dimers are formed when two GPCRs interact creating one signaling entity, a phenomenon that adds to the complexity of GPCR signaling. There is irrefutable evidence that GPCRs can form dimers and higher order oligomers. However, even though the dysfunction of GPCR dimers has been associated with multiple diseases, for example with angiotensin II type 1 and bradykinin receptor B2 heterodimer complexes with pre-eclampsia [88], controversy remains as to whether the mechanism is necessary for all GPCRs [89]. An indisputable and widely accepted example of GPCR dimerization is the Class C GABA receptors where dimerization is obligatory for function [90]. When $GABA_{B2}$ receptors are not present, $GABA_{B1}$ receptors are unable to localize to the cell surface due to the endoplasmic reticulum retention signal that is attached to $GABA_{B1}$ receptor. When heterodimerization occurs, the retention signal on $GABA_{B1}$ receptors is blocked by $GABA_{B2}$ receptors. Heterodimerization is necessary for signal transduction with $GABA_{B1}$ receptors being responsible for binding the agonist while $GABA_{B2}$ receptors activate the G protein. Cryo-EM structures of the $GABA_B$ receptors resulting from heterodimerization of $GABA_{B1}$ and $GABA_{B2}$ further demonstrated dimerization [91]. Similarly, crystal structure of $mGlu_1$ indicated parallel dimer formation, with the extracellular domains mediating receptor homo- and hetero-dimerization [29]. Recently, cryo-EM structures of $mGlu_2$ and $mGlu_4$ receptors bound to heterotrimeric G_i revealed a G-protein-binding site formed by three intracellular loops and helices III and IV, that is distinct from other GPCR structures [92]. Recent work with Class A GPCR dimerization suggested that only about 20% form dimers [93]. Yet, the phenomenon of oligomerization can exert a significant impact on ligand binding, downstream signaling, crosstalk, internalization, and trafficking. Further studies are needed in order to gain a better understanding of GPCR dimerization and resulting physiological effects. Given that dimerization of GPCRs has been implicated in several diseases, a strong interest remains for identifying compounds that can alter this interaction.

One new area for GPCR exploration has been with cholesterol. GPCRs are integral membrane proteins whose function, organization, and dynamics are regulated by membrane lipids, such as cholesterol. Cholesterol is an abundant member of plasma membranes. Some GPCRs have been shown to have cholesterol interaction motifs, putative interaction sites that may facilitate cholesterol-sensitive functions [94]. While a general idea of cholesterol interaction sites has emerged, no consensus model has been reached. Cholesterol has been implicated as an allosteric modulator of GPCR function, shifting the receptor to a high-affinity state for both agonist and antagonist binding with evidence that it can positively modulate a number of receptors including glutamate [14], cannabinoid [95], smoothened [96], oxytocin [97], and chemokine receptors [98]. Molecular dynamics simulations have explored the effect of cholesterol on GPCRs and found they affect not only receptor structure but also organization (oligomerization), which in turn can influence receptor cross-talk and drug efficacy [99]. Further studies are needed to understand the physiological relevance of the link between GPCRs and cholesterol, and to capitalize on the information for drug discovery efforts.

References

1 Laschet, C., Dupuis, N., and Hanson, J. (2018). The G protein-coupled receptors deorphanization landscape. *Biochem. Pharmacol.* 153: 62–74.

2 Sriram, K. and Insel, P.A. (2018). G protein-coupled receptors as targets for approved drugs: how many targets and how many drugs? *Mol. Pharmacol.* 93 (4): 251–258.

3 Attwood, T.K. and Findlay, J.B. (1994). Fingerprinting G-protein-coupled receptors. *Protein Eng.* 7 (2): 195–203.

4 Kolakowski, L.F. Jr., (1994). GCRDb: a G-protein-coupled receptor database. *Recept Channels* 2 (1): 1–7.

5 Hu, G.M., Mai, T.L., and Chen, C.M. (2017). Visualizing the GPCR network: classification and evolution. *Sci. Rep.* 7 (1): 15495.

6 Gurevich, V.V. and Gurevich, E.V. (2019). GPCR signaling regulation: the role of GRKs and arrestins. *Front. Pharmacol.* 10: 125.

7 Alexander, S.P.H., Christopoulos, A., Davenport, A.P. et al. (2019). THE CONCISE GUIDE TO PHARMACOLOGY 2019/20: G protein-coupled receptors. *Br. J. Pharmacol.* 176 Suppl 1 (Suppl 1): S21–s141.

8 Chang, R., Zhang, X., Qiao, A. et al. (2020). Cryo-electron microscopy structure of the glucagon receptor with a dual-agonist peptide. *J. Biol. Chem.* 295 (28): 9313–9325.

9 Qiao, A., Han, S., Li, X. et al. (2020). Structural basis of G_s and G_i recognition by the human glucagon receptor. *Science* 367 (6484): 1346–1352.

10 Liang, Y.L., Khoshouei, M., Glukhova, A. et al. (2018). Phase-plate cryo-EM structure of a biased agonist-bound human GLP-1 receptor-G_s complex. *Nature* 555 (7694): 121–125.

11 Wu, F., Yang, L., Hang, K. et al. (2020). Full-length human GLP-1 receptor structure without orthosteric ligands. *Nat. Commun.* 11 (1): 1272.

12 Zhao, L.H., Ma, S., Sutkeviciute, I. et al. (2019). Structure and dynamics of the active human parathyroid hormone receptor-1. *Science* 364 (6436): 148–153.

13 Møller, T.C., Moreno-Delgado, D., Pin, J.P. et al. (2017). Class C G protein-coupled receptors: reviving old couples with new partners. *Biophys. Rep.* 3 (4): 57–63.

14 Wu, H., Wang, C., Gregory, K.J. et al. (2014). Structure of a Class C GPCR metabotropic glutamate receptor 1 bound to an allosteric modulator. *Science* 344 (6179): 58–64.

15 Wen, T., Wang, Z., Chen, X. et al. (2021). Structural basis for activation and allosteric modulation of full-length calcium-sensing receptor. *Sci. Adv.* 7 (23): eabg1483.

16 Chen, X., Wang, L., Cui, Q. et al. (2021). Structural insights into the activation of human calcium-sensing receptor. *Elife* 10: e68578.

17 Velazhahan, V., Ma, N., Pándy-Szekeres, G. et al. (2021). Structure of the Class D GPCR Ste2 dimer coupled to two G proteins. *Nature* 589 (7840): 148–153.

18 Kamimura, Y. and Ueda, M. (2021). GPCR signaling regulation in sictyostelium chemotaxis. *Methods Mol. Biol.* 2274: 317–336.

19 Greenhalgh, J.C., Chandran, A., Harper, M.T. et al. (2020). Proposed model of the dictyostelium cAMP receptors bound to cAMP. *J. Mol. Graphics Modell.* 100: 107662.

20 Qi, X., Liu, H., Thompson, B. et al. (2019). Cryo-EM structure of oxysterol-bound human smoothened coupled to a heterotrimeric G_i. *Nature* 571 (7764): 279–283.

21 Ballante, F., Kooistra, A.J., Kampen, S. et al. (2021). Structure-based virtual screening for ligands of G protein-coupled receptors: what can molecular docking do for you? *Pharmacol. Rev.* 73 (4): 527–565.

22 Tiss, A., Ben Boubaker, R., Henrion, D. et al. (2021). Homology modeling of Class A G-protein-coupled receptors in the age of the structure boom. *Methods Mol. Biol.* 2315: 73–97.

23 Chan, W.K.B. and Zhang, Y. (2020). Virtual screening of human Class-A GPCRs using ligand profiles built on multiple ligand-receptor interactions. *J. Mol. Biol.* 432 (17): 4872–4890.

24 Pándy-Szekeres, G., Esguerra, M., Hauser, A.S. et al. (2022). The G protein database, GproteinDb. *Nucleic Acids Res.* 50 (D1): D518–d525.

25 Weis, W.I. and Kobilka, B.K. (2018). The molecular basis of G protein-coupled receptor activation. *Annu. Rev. Biochem.* 87: 897–919.

26 Glukhova, A., Draper-Joyce, C.J., Sunahara, R.K. et al. (2018). Rules of engagement: GPCRs and G proteins. *ACS Pharmacol. Transl. Sci.* 1 (2): 73–83.

27 Rosenbaum, D.M., Cherezov, V., Hanson, M.A. et al. (2007). GPCR engineering yields high-resolution structural insights into β_2-adrenergic receptor function. *Science* 318 (5854): 1266–1273.

28 Rasmussen, S.G., DeVree, B.T., Zou, Y. et al. (2011). Crystal structure of the β_2 adrenergic receptor-G_s protein complex. *Nature* 477 (7366): 549–555.

29 Xu, F., Wu, H., Katritch, V. et al. (2011). Structure of an agonist-bound human A_{2A} adenosine receptor. *Science* 332 (6027): 322–327.

30 Thal, D.M., Glukhova, A., Sexton, P.M. et al. (2018). Structural insights into G-protein-coupled receptor allostery. *Nature* 559 (7712): 45–53.

31 Kapla, J., Rodríguez-Espigares, I., Ballante, F. et al. (2021). Can molecular dynamics simulations improve the structural accuracy and virtual screening performance of GPCR models? *PLoS Comput. Biol.* 17 (5): e1008936.

32 Jones, E.M., Lubock, N.B., Venkatakrishnan, A.J. et al. (2020). Structural and functional characterization of G protein-coupled receptors with deep mutational scanning. *Elife* 9: e54895.

33 Karczewski, K.J., Francioli, L.C., Tiao, G. et al. (2020). The mutational constraint spectrum quantified from variation in 141,456 humans. *Nature* 581 (7809): 434–443.

34 Inoue, A., Raimondi, F., Kadji, F.M.N. et al. (2019). Illuminating G-protein-coupling selectivity of GPCRs. *Cell* 177 (7): 1933–1947.e25.

35 Hauser, A.S., Kooistra, A.J., Munk, C. et al. (2021). GPCR activation mechanisms across classes and macro/microscales. *Nat. Struct. Mol. Biol.* 28 (11): 879–888.

36 Sente, A., Peer, R., Srivastava, A. et al. (2018). Molecular mechanism of modulating arrestin conformation by GPCR phosphorylation. *Nat. Struct. Mol. Biol.* 25 (6): 538–545.

37 Pin, J.P., Kniazeff, J., Prézeau, L. et al. (2019). GPCR interaction as a possible way for allosteric control between receptors. *Mol. Cell. Endocrinol.* 486: 89–95.

38 Kilpatrick, L.E. and Hill, S.J. (2021). Transactivation of G protein-coupled receptors (GPCRs) and receptor tyrosine kinases (RTKs): recent insights using luminescence and fluorescence technologies. *Curr. Opin. Endocr. Metab. Res.* 16: 102–112.

39 Crilly, S.E. and Puthenveedu, M.A. (2021). Compartmentalized GPCR signaling from intracellular membranes. *J. Membr. Biol.* 254 (3): 259–271.

40 Mohammad Nezhady, M.A., Rivera, J.C., and Chemtob, S. (2020). Location bias as emerging paradigm in GPCR biology and drug discovery. *iScience* 23 (10): 101643.

41 Higashijima, T., Ferguson, K.M., Sternweis, P.C. et al. (1987). Effects of Mg^{2+} and the beta gamma-subunit complex on the interactions of guanine nucleotides with G proteins. *J. Biol. Chem.* 262 (2): 762–766.

42 Warne, T., Edwards, P.C., Doré, A.S. et al. (2019). Molecular basis for high-affinity agonist binding in GPCRs. *Science* 364 (6442): 775–778.

43 DeVree, B.T., Mahoney, J.P., Vélez-Ruiz, G.A. et al. (2016). Allosteric coupling from G protein to the agonist-binding pocket in GPCRs. *Nature* 535 (7610): 182–186.

44 Su, M., Zhu, L., Zhang, Y. et al. (2020). Structural basis of the activation of heterotrimeric G_s-protein by isoproterenol-bound β_1-adrenergic receptor. *Mol. Cell* 80 (1): 59–71.e4.

45 Hamm, H.E. and Gilchrist, A. (1996). Heterotrimeric G proteins. *Curr. Opin. Cell Biol.* 8 (2): 189–196.

46 Sprang, S.R. (2016). Invited review: activation of G proteins by GTP and the mechanism of Gα-catalyzed GTP hydrolysis. *Biopolymers* 105 (8): 449–462.

47 Almutairi, F., Lee, J.K., and Rada, B. (2020). Regulator of G protein signaling 10: structure, expression and functions in cellular physiology and diseases. *Cell. Signalling* 75: 109765.

48 Chung, K.Y., Rasmussen, S.G., Liu, T. et al. (2011). Conformational changes in the G protein G_s induced by the β_2 adrenergic receptor. *Nature* 477 (7366): 611–615.

49 Dror, R.O., Mildorf, T.J., Hilger, D. et al. (2015). SIGNAL TRANSDUCTION. Structural basis for nucleotide exchange in heterotrimeric G proteins. *Science* 348 (6241): 1361–1365.

50 Draper-Joyce, C. and Furness, S.G.B. (2019). Conformational transitions and the activation of heterotrimeric G proteins by G protein-coupled receptors. *ACS Pharmacol. Transl. Sci.* 2 (4): 285–290.

51 Flock, T., Ravarani, C.N.J., Sun, D. et al. (2015). Universal allosteric mechanism for Gα activation by GPCRs. *Nature* 524 (7564): 173–179.

52 Smith, J.S., Lefkowitz, R.J., and Rajagopal, S. (2018). Biased signalling: from simple switches to allosteric microprocessors. *Nat. Rev. Drug Discovery* 17 (4): 243–260.

53 Eichel, K. and von Zastrow, M. (2018). Subcellular organization of GPCR signaling. *Trends Pharmacol. Sci.* 39 (2): 200–208.

54 Laporte, S.A. and Scott, M.G.H. (2019). β-Arrestins: multitask scaffolds orchestrating the where and when in cell signalling. *Methods Mol. Biol.* 1957: 9–55.

55 Tian, X., Kang, D.S., and Benovic, J.L. (2014). β-Arrestins and G protein-coupled receptor trafficking. *Handb. Exp. Pharmacol.* 219: 173–186.

56 Rajagopal, S. and Shenoy, S.K. (2018). GPCR desensitization: acute and prolonged phases. *Cell. Signalling* 41: 9–16.

57 Carmona-Rosas, G., Alcántara-Hernández, R., and Hernández-Espinosa, D.A. (2019). The role of β-arrestins in G protein-coupled receptor heterologous desensitization: a brief story. *Methods Cell Biol.* 149: 195–204.

58 Jean-Charles, P.Y., Freedman, N.J., and Shenoy, S.K. (2016). Cellular roles of β-arrestins as substrates and adaptors of ubiquitination and deubiquitination. *Prog. Mol. Biol. Transl. Sci.* 141: 339–369.

59 Nobles, K.N., Xiao, K., Ahn, S. et al. (2011). Distinct phosphorylation sites on the β_2-adrenergic receptor establish a barcode that encodes differential functions of β-arrestin. *Sci. Signal.* 4 (185): ra51.

60 Premont, R.T., Claing, A., Vitale, N. et al. (1998). β_2-Adrenergic receptor regulation by GIT1, a G protein-coupled receptor kinase-associated ADP ribosylation factor GTPase-activating protein. *Proc. Natl. Acad. Sci. U.S.A.* 95 (24): 14082–14087.

61 Stewart, A. and Fisher, R.A. (2015). Introduction: G protein-coupled receptors and RGS proteins. *Prog. Mol. Biol. Transl. Sci.* 133: 1–11.

62 Sprang, S.R., Chen, Z., and Du, X. (2007). Structural basis of effector regulation and signal termination in heterotrimeric Gα proteins. *Adv. Protein Chem.* 74: 1–65.

63 Abramow-Newerly, M., Roy, A.A., Nunn, C. et al. (2006). RGS proteins have a signalling complex: interactions between RGS proteins and GPCRs, effectors, and auxiliary proteins. *Cell. Signalling* 18 (5): 579–591.

64 Masuho, I., Balaji, S., Muntean, B.S. et al. (2020). A global map of G protein signaling regulation by RGS proteins. *Cell* 183 (2): 503–521.e19.

65 Alqinyah, M. and Hooks, S.B. (2018). Regulating the regulators: epigenetic, transcriptional, and post-translational regulation of RGS proteins. *Cell. Signalling* 42: 77–87.

66 Tesmer, J.J., Berman, D.M., Gilman, A.G. et al. (1997). Structure of RGS4 bound to AlF4 – activated $G_{i\alpha 1}$: stabilization of the transition state for GTP hydrolysis. *Cell* 89 (2): 251–261.

67 O'Brien, J.B., Wilkinson, J.C., and Roman, D.L. (2019). Regulator of G-protein signaling (RGS) proteins as drug targets: progress and future potentials. *J. Biol. Chem.* 294 (49): 18571–18585.

68 Sjögren, B. (2017). The evolution of regulators of G protein signalling proteins as drug targets – 20 years in the making: IUPHAR review 21. *Br. J. Pharmacol.* 174 (6): 427–437.

69 Drenan, R.M., Doupnik, C.A., Jayaraman, M. et al. (2006). R7BP augments the function of RGS7*Gβ5 complexes by a plasma membrane-targeting mechanism. *J. Biol. Chem.* 281 (38): 28222–28231.

70 Traver, S., Splingard, A., Gaudriault, G. et al. (2004). The RGS (regulator of G-protein signalling) and GoLoco domains of RGS14 co-operate to regulate G_i-mediated signalling. *Biochem. J* 379 (Pt 3): 627–632.

71 Shu, F.J., Ramineni, S., and Hepler, J.R. (2010). RGS14 is a multifunctional scaffold that integrates G protein and Ras/Raf MAPkinase signalling pathways. *Cell. Signalling* 22 (3): 366–376.

72 Schiff, M.L., Siderovski, D.P., Jordan, J.D. et al. (2000). Tyrosine-kinase-dependent recruitment of RGS12 to the N-type calcium channel. *Nature* 408 (6813): 723–727.

73 Salim, S., Sinnarajah, S., Kehrl, J.H. et al. (2003). Identification of RGS2 and type V adenylyl cyclase interaction sites. *J. Biol. Chem.* 278 (18): 15842–15849.

74 Nguyen, C.H., Ming, H., Zhao, P. et al. (2009). Translational control by RGS2. *J. Cell Biol.* 186 (5): 755–765.

75 Hague, C., Bernstein, L.S., Ramineni, S. et al. (2005). Selective inhibition of α_{1A}-adrenergic receptor signaling by RGS2 association with the receptor third intracellular loop. *J. Biol. Chem.* 280 (29): 27289–27295.

76 Bernstein, L.S., Ramineni, S., Hague, C. et al. (2004). RGS2 binds directly and selectively to the M_1 muscarinic acetylcholine receptor third intracellular loop to modulate $G_{q/11\alpha}$ signaling. *J. Biol. Chem.* 279 (20): 21248–21256.

77 Hauser, A.S., Attwood, M.M., Rask-Andersen, M. et al. (2017). Trends in GPCR drug discovery: new agents, targets and indications. *Nat. Rev. Drug Discovery* 16 (12): 829–842.

78 Slosky, L.M., Caron, M.G., and Barak, L.S. (2021). Biased allosteric modulators: new frontiers in GPCR drug discovery. *Trends Pharmacol. Sci.* 42 (4): 283–299.

79 Chan, H.C.S., Li, Y., Dahoun, T. et al. (2019). New binding sites, new opportunities for GPCR drug discovery. *Trends Biochem. Sci* 44 (4): 312–330.

80 Reinecke, B.A., Wang, H., and Zhang, Y. (2019). Recent advances in the drug discovery and development of dualsteric/bitopic activators of G protein-coupled receptors. *Curr. Top. Med. Chem.* 19 (26): 2378–2392.

81 Fronik, P., Gaiser, B.I., and Sejer Pedersen, D. (2017). Bitopic ligands and metastable binding sites: opportunities for G protein-coupled receptor (GPCR) medicinal chemistry. *J. Med. Chem.* 60 (10): 4126–4134.

82 Schrage, R. and Kostenis, E. (2017). Functional selectivity and dualsteric/bitopic GPCR targeting. *Curr. Opin. Pharmacol.* 32: 85–90.

83 Huang, B., St. Onge, C.M., Ma, H. et al. (2021). Design of bivalent ligands targeting putative GPCR dimers. *Drug Discovery Today* 26 (1): 189–199.

84 Shonberg, J., Scammells, P.J., and Capuano, B. (2011). Design strategies for bivalent ligands targeting GPCRs. *ChemMedChem* 6 (6): 963–974.

85 Ikeda, Y., Kumagai, H., Motozawa, Y. et al. (2015). Biased agonism of the angiotensin II type I receptor. *Int. Heart J.* 56 (5): 485–488.

86 Gillis, A., Gondin, A.B., Kliewer, A. et al. (2020). Low intrinsic efficacy for G protein activation can explain the improved side effect profiles of new opioid agonists. *Sci. Signal.* 13 (625).

87 Fernandez, T.J., De Maria, M., and Lobingier, B.T. (2020). A cellular perspective of bias at G protein-coupled receptors. *Protein Sci.* 29 (6): 1345–1354.

88 AbdAlla, S., Lother, H., el Massiery, A. et al. (2001). Increased AT_1 receptor heterodimers in preeclampsia mediate enhanced angiotensin II responsiveness. *Nat. Med.* 7 (9): 1003–1009.

89 Bouvier, M. and Hébert, T.E. (2014). CrossTalk proposal: weighing the evidence for Class A GPCR dimers, the evidence favours dimers. *J. Physiol.* 592 (12): 2439–2441.

90 Kniazeff, J., Prézeau, L., Rondard, P. et al. (2011). Dimers and beyond: the functional puzzles of Class C GPCRs. *Pharmacol. Ther.* 130 (1): 9–25.

91 Kim, Y., Jeong, E., Jeong, J.H. et al. (2020). Structural basis for activation of the heterodimeric $GABA_B$ receptor. *J. Mol. Biol.* 432 (22): 5966–5984.

92 Lin, S., Han, S., Cai, X. et al. (2021). Structures of G_i-bound metabotropic glutamate receptors $mGlu_2$ and $mGlu_4$. *Nature* 594 (7864): 583–588.

93 Felce, J.H., Latty, S.L., Knox, R.G. et al. (2017). Receptor quaternary organization explains G protein-coupled receptor family structure. *Cell Rep.* 20 (11): 2654–2665.

94 Sarkar, P. and Chattopadhyay, A. (2020). Cholesterol interaction motifs in G protein-coupled receptors: slippery hot spots? *Wiley Interdiscip. Rev. Syst. Biol. Med.* 12 (4): e1481.

95 Yeliseev, A., Iyer, M.R., Joseph, T.T. et al. (2021). Cholesterol as a modulator of cannabinoid receptor CB_2 signaling. *Sci. Rep.* 11 (1): 3706.

96 Huang, P., Nedelcu, D., Watanabe, M. et al. (2016). Cellular cholesterol directly activates smoothened in hedgehog signaling. *Cell* 166 (5): 1176–1187.e14.

97 Waltenspühl, Y., Schöppe, J., Ehrenmann, J. et al. (2020). Crystal structure of the human oxytocin receptor. *Sci. Adv.* 6 (29): eabb5419.

98 van Aalst, E. and Wylie, B.J. (2021). Cholesterol is a dose-dependent positive allosteric modulator of CCR_3 ligand affinity and G protein coupling. *Front. Mol. Biosci.* 8: 724603.

99 Sengupta, D. and Chattopadhyay, A. (2015). Molecular dynamics simulations of GPCR–cholesterol interaction: an emerging paradigm. *Biochim. Biophys. Acta* 1848 (9): 1775–1782.

2

Recent Advances in Orphan GPCRs Research and Therapeutic Potential

Atsuro Oishi[1,2,3] *and Ralf Jockers*[1]

[1] *Institut Cochin, CNRS, INSERM, Université de Paris, Paris, France*
[2] *Department of Anatomy, Kyorin University Faculty of Medicine, Tokyo, Japan*
[3] *Cancer RNA Research Unit, National Cancer Center Research Institute, Tokyo, Japan*

2.1 Introduction

The G protein-coupled receptor (GPCR) subfamily is the largest membrane protein family, encoded by approximately 800 genes in the human genome, which can be roughly divided into 400 odorant and 400 non-odorant receptors. Among the non-odorant GPCRs, 100–150 are currently targeted by clinically relevant drugs representing 30–40% of all commercially available drugs [1, 2]. The other 250–300 GPCRs are waiting for therapeutic applications, among those approximately 100 orphan GPCRs that are not matched with any endogenous ligand as of the beginning of 2020. Currently, none of the orphan GPCRs is targeted by clinically relevant drugs. This can be explained not only by the absence of knowledge about their cognate ligands but also by the limited knowledge about their role in physiology and/or pathological circumstances. Progress on both aspects is expected to provide valuable information on future therapeutic applications and drug development.

In this review, we will begin by briefly looking back on the concise history of GPCR deorphanization, then the strategy for deorphanization and recent technical advances in the field of orphan GPCRs will be reviewed. As another important point of interest, ligand-independent features of orphan GPCRs will be presented and their therapeutic relevance considered. For the years of 2013 or earlier, the reader is referred to other complementary expert reviews [3] including phylogenetic aspects [4] and the web site of the International Union of Basic and Clinical Pharmacology (IUPHAR) [5] that is regularly updated (https://www.guidetopharmacology.org/latestPairings.jsp). The current review focuses on deorphanizations from 2013. For further complementary information within this time frame see also the recent review from Laschet et al. [6].

GPCRs as Therapeutic Targets, Volume 1, First Edition. Edited by Annette Gilchrist.

2.2 Concise History of Orphan GPCRs Research

In 1987 the gene of the β2-adrenergic receptor (β2AR) was cloned by Lefkowitz and coworkers [7]. He looked back to this moment as "In what was at the time a remarkable surprise we observed sequence homology, ..., with the visual light-sensing protein rhodopsin" [8], which had been cloned in 1983 [9]. This sequence homology evoked the idea of a 7TM receptor superfamily [8]. Indeed, subsequently cloned adrenergic receptor genes confirmed this feature, followed by many other 7TM receptor genes fully establishing the theory of a 7TM receptor superfamily [10]. Along these lines for most GPCRs the cloning of the gene preceded the isolation of the receptor protein. Until the mid-1990s, known ligands were matched with unknown genes suspected to code for GPCRs based on their sequence homology [11, 12]. At the same time, homology cloning discovered new putative GPCR genes, which did not respond to the ligand of interest and for which the cognate ligands had to be subsequently identified [13]. These genes were then expressed in heterologous systems and tested in functional assays against libraries of molecules of known function or therapeutic value to identify ligand–receptor pairs. The first examples of successful "deorphanization" by this so-called reverse pharmacology strategy (from receptor to the ligand) were the dopamine D2 receptor [14] and serotonin (5HT) 1A receptor [15]. The strategy was expanded toward collections of unlisted ligands of synthetic or natural origin and was common practice in the late 1990s and early 2000s with the successful deorphanization of orexin receptors [16] and the ghrelin receptor [17]. Since the late 2000s the speed of deorphanization slowed down reaching a mean deorphanization rate of 0–5 GPCRs/year in the last 15 years with currently around 100 non-odorant GPCRs left without cognate ligand [6].

2.3 Current Deorphanization Strategies

Reverse pharmacology with tissue extracts successfully contributed to the deorphanization of many GPCRs. For the remaining orphan GPCRs, several reasons might explain the incapability of the tested extracts to deorphanize them: (i) the cognate ligand does not exist in the targeted tissue extracts or was not present at the time of tissue collection, (ii) the cognate ligand exists in the targeted tissue, but lost its activity during the extraction process or the functional assay, (iii) the cognate ligand exists in the tissue extract in its active form, but the detection system (assay choice) is not appropriate or not sensitive enough to detect existing concentrations. Apart from further optimizing the preparation of tissue extracts, several other techniques and strategies have been developed in the last few years to deorphanize the remaining orphan GPCRs as detailed in the following text.

2.3.1 Criteria of Deorphanization

The IUPHAR [3] established two key requirements for the successful deorphanization of GPCRs: (i) reproducibility; more than two independent groups should have demonstrated the activity of a ligand on an orphan receptor; (ii) its presence in tissues (the ligand should be detectable in tissues or body fluids *in vivo*). The current state of deorphanization of GPCRs is continuously updated and revised by IUPHAR [5] including recently proposed receptor–ligand pairings. This includes cases that comply to the aforementioned criteria but also those with only a single report. Contradictory cases have been summarized recently [6].

2.3.2 Cognate and Surrogate Ligands

According to the definition given by IUPHAR, deorphanization translates into the identification of the natural, endogenous ligand(s), called the cognate ligand. Apart from this specific goal, the identification of non-endogenous or synthesized molecules that bind to orphan GPCRs, so-called surrogate ligands, is an alternative goal that can lead to the elucidation of the function of orphan receptors and can present the first step toward drug development. The development of surrogate ligands with inverse agonistic activity could be of therapeutic interest for those orphan GPCRs with significant spontaneous receptor activity (see Section 2.4.1). For orphan GPCRs that engage with other GPCRs or proteins in heteromeric complexes (see Section 2.4.7), synthetic agonists might be of interest to modulate the function of the partner protein. This is illustrated by the surrogate ligand of the orphan GPR139, which diminishes opioid-induced analgesia in a mouse model of drug abuse due to its inhibitory effect on μ-opioid receptor (MOR) activity in heteromeric complexes [18] (see also Section 2.4.7.3). For the orphan GPR27, two surrogate ligands were identified that induce β-arrestin2 recruitment [19]. Neither ligand was able to promote G protein signaling in the TGFα shedding assay (see Section TGFα Shedding Assay), suggesting that either GPR27 or the two surrogate ligands are biased toward β-arrestin2. A complete list of surrogate ligands for orphan GPCRs is found at the IUPHAR website [5].

2.3.3 Techniques and Strategies for Deorphanization

2.3.3.1 Recent Advances in Signaling Assays

Many classical GPCR assays are based on the determination of second messengers such as cAMP and inositol phosphate, on the analysis of the phosphorylation state of proteins downstream of GPCR activation such as ERK or CREB, or the expression levels of transcription factors such as c-fos. Whereas the repertoire of activated G protein pathways can vary widely between different GPCRs, recruitment of β-arrestins seems to be a general trait for most of them. Therefore, the β-arrestin recruitment assays became popular for deorphanization after 2000 [20]. Notably, some GPCRs such as CXCR7 were shown not to activate G proteins but to recruit β-arrestins [21, 22]. Additional assays have been developed and applied to the deorphanization of other orphan GPCRs as detailed in the following text.

PRESTO-Tango Assay Several quantitative GPCR-β-arrestin recruitment assays were developed with different techniques including BRET (Bioluminescence Resonance Energy Transfer) [23], enzyme complementation [24, 25], and Tango assay [26]. The Tango assay is a luciferase reporter assay based on the β-arrestin recruitment of a β-arrestin2-Tobacco Etch Virus (TEV) protease fusion protein to the GPCR-V2Ctail-TEV cleavage site-transcriptional activator (tTA) fusion protein. In this approach, the GPCR contains the carboxyl-terminal tail of the vasopressin V2 receptor (V2Ctail), which is known to be an efficient β-arrestin recruiter and the tTA after a TEV cleavage site. Following recruitment of β-arrestin2-TEV protease fusion protein cleaves the tTA from the receptor, and tTA translocates into the nucleus where it activates luciferase reporter gene transcription (Figure 2.1a). Roth and coworkers implemented this assay into a platform containing 315 non-odorant GPCRs, including many orphans named PRESTO (parallel receptorome expression and screening via transcriptional output)-Tango [27]. Key features of the platform are (i) high sensitivity due to signal amplification, (ii) availability of 384-well format; (iii) high responsiveness to ligand stimulation of non-orphan GPCRs, especially class A GPCRs; (iv) detectability of ligand-independent spontaneous β-arrestin2 recruitment. With this platform, 446 FDA-approved drugs were screened against 91 orphans or understudied GPCRs and nateglinide, a drug approved for diabetes, was identified as an agonist for MRGPRX4 [27]. The whole PRESTO-Tango assay and single components of it are available through Addgene. The PRESTO-Tango assay has been used for the functional validation of predicted candidate ligands thus contributing to the pairing of several peptides with GPR1, BB3, GPR15, GPR55, and GPR68 (for more details see Section 2.3.3.3) [28].

TGFα Shedding Assay Although many GPCRs are known to activate G12/G13 proteins, high-throughput assays are not established yet for this pathway. Current low-throughput methods to measure the activation of this pathway are based on monitoring the activity of downstream effectors such as the small G proteins RhoA, Rac1, and Cdc42. FRET (Förster Resonance Energy Transfer) [29, 30] or pull-down assays [31] are both used to monitor the active, GTP-bound form of small G proteins. In addition, translocation of small G proteins to the plasma membrane [32] or stress fiber formation [33] is commonly used.

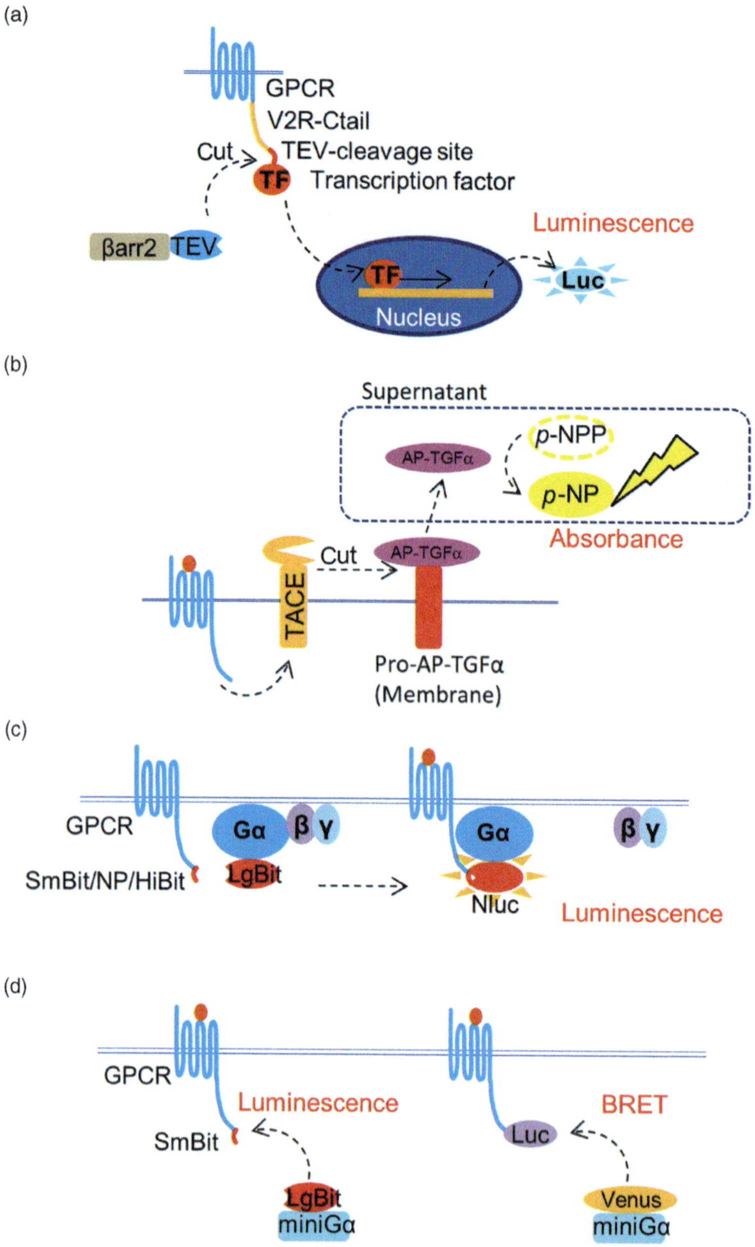

Figure 2.1 Recently developed assays to monitor GPCRs activation. (a) PREST-Tango assay. βarr2; β-arrestin2, TEV, Tobacco Etch Virus protease; TF, transcription factor; Luc, luciferase. (b) TGFα shedding assay. TACE, TNFα-converting enzyme; AP-TGFα, alkaline phosphatase-tagged TGFα; *p*-NPP, *para*-nitrophenylphosphate; *p*-NP, *para*-nitrophenol. (c) GPCR–G protein interaction assay. NP, natural peptide. (d) GPCR-mini Gα protein interaction assay.

In 2012, a TGFα shedding assay was established allowing the detection of G12/G13 and Gq/G11 protein activation in a high-throughput-compatible format [34]. In this assay, Gq/G11- and/or G12/G13-coupled GPCRs activate the TNFα-converting enzyme (TACE) metalloprotease, also known as ADAM17, which cleaves off TGFα from its membrane-bound precursor by ectodomain shedding. The amount of released TGFα is quantified by measuring the activity of alkaline phosphatase to which TGFα is fused [34, 35] (Figure 2.1b). By applying this technique to orphan GPCRs, three orphans, P2Y10, A630033H20, and GPR174 were found to couple to G12/G13 proteins in response to lysophosphatidylserine [34]. Later on, this assay also contributed to the discovery of new pairings such as bile acids with MRGPRX4 receptors [36]. Originally this assay was restricted to Gq/G11- and G12/G13-coupled GPCRs but by replacing the six last amino acids of Gαq with those of its Gαs,i,o,z counterparts or by using the promiscuous Gα16, the assay becomes responsive to GPCRs coupled to these G proteins [34]. The assay was used to determine the G protein coupling profile for the two surrogate ligands of GPR27, initially identified based on their capacity to recruit β-arrestin2 to GPR27. The absence of any response suggested that the two surrogate ligands or GPR27 itself are biased toward β-arrestin2 [19]. Notably this assay detects also spontaneous, ligand-independent GPCR activity [34] and thus could be considered for the study of ligand-independent features of orphan GPCRs. In conclusion, high sensitivity, application to a broad range of GPCRs, easy implementation and high-throughput compatibility are the main features of the TGFα shedding assay.

G Protein Recruitment Assay and Mini-G Proteins Classical assays to measure GPCR activation rely on downstream events (i.e. effector activities), rather than on the direct measurement of G protein activation. Later, BRET assays were established to measure G protein activation by means of the ligand-induced rearrangements between Gα and Gβγ subunits [37]. Recently, several groups developed optimized assay systems to monitor the coupling of Gα proteins to GPCRs by BRET [38] or Nanoluciferase (Nanoluc)-complementation assays [39, 40]. Among these, the Hanson lab generated sensors for all major Gα subunits (Gαs, Gαo, Gαi1, Gαi2, Gαi3, Gαq, Gα11, Gα12, Gα13) by conjugating them with the LargeBit fragment of Nanoluc. Nanoluc complementation is achieved when Gα-LargeBit is recruited to GPCRs conjugated with the SmallBit fragment of Nanoluc. The assay has a high signal-to-noise ratio and low variability ($Z' > 0.5$) and can be optimized by varying the affinity of the 13 amino acid, long SmallBit fragment for the LargeBit fragment [39] (Figure 2.1c).

Another recent assay is based on the recruitment of so-called mini-G proteins (mG) to GPCRs by BRET or Nanoluc complementation assay [41–43]. Mini G proteins are engineered G proteins consisting essentially of the GTPase domain of Gα subunits. The mG were originally developed for structural studies but have turned out to be useful G protein surrogates to study GPCR activation. They are located in the cytosol in their basal state and translocate to activated GPCRs (Figure 2.1d). Currently available sensors are mGs and mG12 sensors and chimeric mGs/i and mGs/q sensors as Gi and Gq protein surrogates, respectively. The assay has a high signal-to-noise ratio and is applicable to high-throughput measurements. These sensors are likely to become important tools for deorphanization studies that rely on the direct detection of ligand-induced Gα protein recruitment, in particular for Gα12/13 and Gz proteins for which only a limited number of dedicated signaling assays are available.

2.3.3.2 Recent Advances in Ligand Binding Assays

In recent years, classical radioligand binding assays were complemented with non-radioactive detection methods based on BRET [44–46] or homogenous time-resolved fluorescence (HTRF®) [47]. Common features of these assays are: (i) ligand labeling with fluorophores; (ii) use of fusion proteins with attachment of Nanoluc (for BRET) or Snap-tags (for HTRF) to the N-terminal extracellular receptor domain; (iii) interaction between labeled ligands and receptors relies on their molecular proximity measured as energy transfer between both. For the moment, these techniques don't directly assist GPCR deorphanization but it can be easily envisioned that they can be used to identify fluorescent tracer compounds from a library of fluorescently labeled synthetic compounds that could then be used to screen for natural ligands.

2.3.3.3 Bioinformatics-Assisted Approaches

The availability of whole-genome sequence information for many species as well as a wealth of structural information for GPCRs has provided researchers with a vast amount of new information that are potentially helpful for the deorphanization of orphan GPCRs. For structure-based strategies, GPCRs deorphanization or virtual screening of ligands had limited success until recently [48]. To refine the sequence comparison algorithm and identify pharmacological neighbors for orphan GPCRs, the algorithm "GPCR contact-informed neighboring pocket" (GPCR–CoINPocket) was designed focusing on persistence and the strength of residues important for ligand binding based on 27 crystallographically characterized GPCRs [49]. Then the algorithm was applied to the analysis of the transmembrane domains of orphan GPCRs [49]. GPCR–CoINPocket correctly identified "pharmacological similarity" between two phylogenetically distant GPCRs, CXCR4 and the recently deorphanized ACKR3, which share the same endogenous ligand, CXCL12 [3, 5, 50]. GPCR–CoINPocket also proposed a new classification of orphan GPCRs based on pharmacological similarity. The bombesin, orexin, pyroglutamylated RFamide peptide (QRFP), and neuropeptide

S receptors were identified as pharmacological neighbors of GPR37L1. This pharmacological proximity helped in the discovery of the first surrogate ligands for GPR37L1 with a 30% success rate [49].

In another recent bioinformatics analysis, 21 orphan GPCRs were selected based on shared characteristics of known peptide-activated GPCRs in terms of sequence signatures and 3D-structure. In parallel, a library of 218 peptides was generated using a proteome-wide machine-learning approach including multiple sequence alignments of more than 300 species, evolutionary trace analysis and cleavage site predictions of precursors. Experimental validation of predicted receptor–peptide pairs was performed with three assays, mass redistribution, receptor internalization, and β-arrestin recruitment (PRESTO-Tango assay) [27]. At the end of this process, 17 peptides could be experimentally paired with 5 orphan GPCRs (BB3, GPR1, GPR15, GPR55, and GPR68). This large-scale analysis also confirmed the redundancy of peptide receptors to not only bind their primary ligand but in many cases several "secondary" ligands, increasing the chance for receptor–ligand pairing but also illustrating the difficulty to identify the primary ligand [28].

2.3.3.4 Crystal Structures of Orphan GPCRs

The elucidation of the crystal structure of several non-orphan GPCR has contributed largely to recent advances in the understanding of these receptors. In February 2020, the first structure of on orphan GPCR, GPR52, was published and provided important information [51]. The structure revealed the structural basis for the known constitutive activity of GPR52 toward the Gs/cAMP pathway. The authors show that the extracellular loop 2 (ECL2) folds into a lid to close the suspected ligand binding pocket and in addition acts as a tethered ligand. This together with the absence of a sodium binding pocket, known to be important for damping constitutive activity of GPCRs, is likely to explain the high spontaneous activity of GPR52. This hypothesis was further confirmed by the cryo-electron microscopy structure of GPR52 with mGs (Section G Protein Recruitment Assay and Mini-G Proteins) that represents the fully active state of the receptor. The binding mode of a surrogate ligand with agonistic activity was investigated in cocrystals and revealed an allosteric binding pocket adjacent to the suspected orthosteric binding pocket. This study illustrates the multitude of novel insights that can be obtained from the structure of orphan GPCR in terms of their activity state, the shape of the suspected orthosteric binding pocket, the nature of the natural ligand, and the binding mode of surrogate ligands.

2.4 Analysis of Orphan GPCR Function and Expression Profiles

The absence of any endogenous ligand for orphan GPCRs is a major handicap for their functional characterization. However, a battery of experimental strategies are now available that can be applied to orphan GPCRs to provide further information on their function. In terms of receptor signaling studies, surrogate ligands or chimeric proteins as well as ligand-independent, spontaneous receptor activity can be very insightful. Furthermore, orphan receptors may allosterically regulate the function of other proteins. At the organism level, determination of the expression profile of orphan GPCRs at different developmental stages and under physiological and pathological conditions as well as the phenotype of knockout animals may provide important hints. In the following we will discuss these different aspects apart from surrogate ligands, which were described earlier.

2.4.1 Spontaneous Activity of Orphan GPCRs

Spontaneous activity of GPCRs is considered as the ligand-independent, intrinsic GPCR activity that induces intracellular signaling, due to the high spontaneous tendency of the receptor to reach active conformations. Depending on the specific conformation stabilized in the absence of ligand, spontaneous receptor activity may be measured at one or several G protein- and/or β-arrestin-dependent pathways [52–54]. Whereas for non-orphan GPCRs the spontaneous activity can be quantified relative to the ligand-induced activity, for orphan GPCRs with spontaneous activity the quantification of this activity and the relevance is more difficult to judge.

Many studies have shown that the spontaneous activity depends on several parameters. First, it is proportional to the amount of receptor expressed [55]. Most studies measure spontaneous activity of orphans by expressing them ectopically in heterologous expression systems such as HEK293, HeLa, or CHO cells [56]. By comparing the activation level of mock-transfected with receptor-transfected cells, the spontaneous activity of an orphan receptor can be evaluated. Second, spontaneous activity may depend on the specific level of a signaling pathway at which it is monitored, i.e. the G protein level, the second messenger level, or the reporter gene expression level. The lack of a consensus on the cutoff line that defines whether an orphan GPCR is judged to have spontaneous activity, or not, introduces further heterogeneity and renders the comparison of different studies difficult. In conclusion, experiments to evaluate the spontaneous activity of orphan GPCRs have to be carefully controlled. Several well-controlled studies have shown that GPR3, GPR6, GPR12, GPR26, GPR52, GPR61, GPR62, GPR78, GPR119, and GPR135 have significant spontaneous activity, at least in heterologous cell systems [55, 57–62]. For Gs-coupled GPR52, the molecular basis for spontaneous receptor activity has been revealed (see Section 2.3.3.4 for more details) [51] but it is currently unknown whether a similar mechanism applies to the other orphan GPCRs. Furthermore several surrogate ligands with inverse agonistic activity have been identified in screening campaigns for GPR3 [63] and GPR6 [64].

The appreciation of the presence of a spontaneous activity in a given tissue that can be attributed to an orphan or non-orphan GPCR is not trivial. *In vivo* evidence for the existence of spontaneous receptor activity has been obtained for histamine H3, serotonin 5-HT2_A and 5-HT$_{2C}$, cannabinoid CB1, MOR and ghrelin receptors [65–68] by using synthetic inverse agonists. The existence of natural antagonists or inverse agonists like liver-expressed antimicrobial peptide 2 (LEAP2) for the ghrelin receptor [69] and the agouti-related protein AgRP (83–192) for the melanocortin-4 receptor (MC4R) [70] further support the *in vivo* relevance of spontaneous GPCR activity. Similar data are lacking for orphan GPCRs as only few adequate inverse agonists have been described and because the presence of endogenous ligands is always difficult to rule out. Collectively, defining the spontaneous activity of orphan GPCRs is still in its infancy but *in vitro* studies indicate that it may be an important feature of some orphans. Knockdown and knockout strategies together with the identification of surrogate inverse agonists will be appropriate tools to tackle this question in the future.

2.4.2 Elucidation of Signaling Properties of Orphan GPCR by Chimeric Receptors

An important element of the deorphanization process is the choice of an appropriate readout system. Signaling pathways activated by surrogate ligands or spontaneously activated by orphan GPCRs may provide valuable hints although it is likely they don't reflect the full spectrum of signaling properties. Engineered chimeric GPCRs were recently introduced providing new insights for several orphan GPCRs [71]. These chimeric GPCRs are composed of the ligand binding domain of rhodopsin and the intracellular loops and the C-tail of the GPCR of interest (Figure 2.2). These chimera respond to light stimulation like rhodopsin but evoke, depending on the origin of the intracellular domains, unique signaling profiles on cAMP, Ca^{2+}, MAPK/ERK, and Rho-dependent pathways. This approach was applied to 63 class A orphan or understudied (for which the initial ligand proposal has still to be confirmed) GPCRs and succeeded in identifying intracellular signaling profiles for GPR1, GPR3, GPR4, GPR21, GPR32, GPR61, GPR63, GPR68, GPR78, GPR85, GPR88, GPR135, and GPR150 [71]. The resulting information will help to design appropriate assay readouts for future deorphanization studies.

2.4.3 Expression Profile of Orphan GPCRs

Collecting information on the expression of orphan GPCRs at the tissue and cellular level constitute a very important element in the study of orphan GPCRs. In 2007, the Coughlin lab provided a comprehensive mRNA expression map of

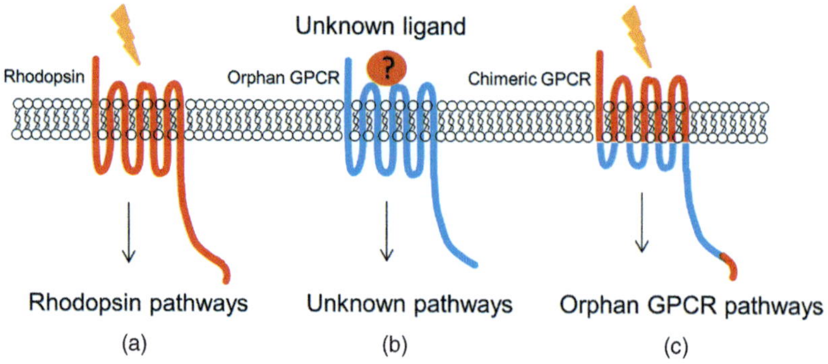

Figure 2.2 Light-sensitive chimeric GPCRs. (a) Light activation of rhodopsin triggers intracellular signaling pathways. (b) Unknown ligand can bind orphan GPCR but activated pathways are unknown. (c) Chimeric GPCRs consisting of extracellular domains, transmembrane domains, and end of C-tail of rhodopsin (red) along with the intracellular domain of orphan GPCR (blue). Chimeric GPCRs are responsive to light and the repertoire of activated signaling pathways depends on the intracellular domains of orphan GPCRs.

353 GPCRs including most of the orphan GPCRs in 41 adult mouse tissues [72]. In 2018, Ehrlich et al. provided a new brain atlas (Images available at: http://ogpcr-neuromap.douglas.qc.ca.) dedicated to the expression profile of murine and human orphan GPCR mRNA detected by *in situ* hybridization [73]. These general sources constitute an excellent starting point that will have to be complemented, depending on each receptor, with other relevant information like developmental expression patterns and the expression under different environmental conditions. A recent study found that mRNA levels of several orphan GPCRs changed upon 24 hours cold-exposure of mice: in the brown adipose tissue GPR120 (up), GPR109a (down), GPR81 (down), GPR89 (up); in the white adipose tissue GPR85 (up), GPR152 (down); in the liver GPR135 (up), GPRC5b (up) [74]. This knowledge will help to orient future research on these receptors in these target tissues that will have to be extended to the protein level.

In a study focusing on neuropathic pain, transcriptomic analysis of orphan GPCR expression in the spinal cord identified 61 receptors, among them GPR160, with increased expression following traumatic nerve injury [75]. By comparing tissue expression profiles of GPR160 and endogenously expressed orphan ligands, a high correlation between GPR160 and the cocaine- and amphetamine-regulated transcript peptide (CARTp) was observed. The identity of CARTp as a likely endogenous ligand for GPR160 is supported by the fact that CARTp induced ERK activation and c-fos expression *in vitro* and that CAPTp evoked painful hypersensitivity that was sensitive to siRNA molecules and neutralizing antibody for GPR160. This example illustrates the great potential of comparing expression profiles of orphan GPCR and orphan ligands in deorphanization.

2.4.4 Expression Map of Orphan GPCRs in Cancer

Several GPCRs are overexpressed under pathological conditions. However, a comprehensive database assembling this information has not been established yet. Insel and coworkers determined GPCR expression patterns (including orphans) in several cancer cell lines (leukemia cells, pancreatic ductal carcinoma) and cancer cells from patients [76]. This study revealed that each cancer cell type expresses a unique repertoire of GPCRs that is different from the corresponding non-cancer cells. In addition, the highly expressed GPCRs are different between cancers suggesting that the GPCR expression profile contributes to the specificity of each cancer type. Some orphan GPCRs like GPRC5A and GPR68 have been only detected in cancer cells but not in the corresponding non-malignant cells [76]. Other orphan GPCRs have been associated with several cancers from various sources ranging from cancer cell lines and human and mouse cancer tissues (Table 2.1). These receptors might contribute to the malignant phenotype of these cancer cells and they constitute interesting biomarker candidates with diagnostic or therapeutic value.

2.4.5 Evolutionary Conservation of Orphan GPCRs

Evolutionary conservation of an orphan GPCR is another important argument supporting its functional relevance. This argument can be extended to GPCR ligands assuming that ligand–receptor pairs often co-evolve [28]. Accordingly, ligand–receptor pairings established in one species combined with the existence of the same ligand in another species can encourage the search for the cognate receptor among the orphan GPCRs of this species. Extended phylogenetic analysis may also provide new insights on the classification of orphan GPCRs, which might facilitate its deorphanization based on the assumption that phylogenetically close GPCRs are likely to respond to a similar type of endogenous ligand. The classification of some orphan GPCRs was indeed revisited regularly between 2005 and 2015 with the availability of sequence information from an increasing number of different species [4, 118, 119]. Illustrative examples of this re-classification are GPR61, GPR62, and GPR135. These orphans, initially scattered within the phylogenetic GPCR tree, were proposed to cluster with the melatonin receptor subfamily based on the inclusion of sequence information from human, mouse, and rat. However, dedicated ligand binding experiments showed that melatonin is not the ligand for these receptors, and they remain orphans [55]. This may be due to the relatively low sequence homology of 20–25% with melatonin receptors and the absence of distinctive features of this subfamily. These results illustrate the limits of phylogenetic predictions and the need for experimental validation.

Table 2.1 Endogenous expression of orphan and understudied GPCRs in cancer.

Official name/ previous name	Type of cancer	PMID	References
GPR17	Rodent and human oligodendrocyte differentiation	29706593	[77]
GPR18	Metastatic melanoma	20880198	[78]
	Glioblastoma multiforme cell lines	27018161	[79]
GPR19	Breast cancer cells	28476646	[80]
	Human lung cancer cells	22912338	[81]
GPR34	Non-Hodgkin lymphoma, marginal zone lymphomas	22966169	[82]
GPR37	Hepatocellular carcinoma	25169131	[83]
LGR4/GPR48	Prostate cancer	25636507	[84]
	Lewis lung carcinoma cells	17178856	[85]
LGR5/GPR49	Basal cell carcinoma	18688030	[86]
	Human colon and ovarian primary tumors	16575208	[87]
GPR50	Lung cancer	29572483	[88]
GPR55	MDA-MB-231 breast cancer cell line	20590578	[89]
	22 cell lines out of 24 examined cancer cell lines	20818416	[90]
	Human breast cancers, pancreatic ductal carcinoma, glioblastoma		
	Prostate and ovarian cancer cell lines	20838378	[91]
	Human squamous cell carcinomas	22751111	[92]
ADGRG1/GPR56	Colorectal cancer cells	30066935	[93]
	Ovarian serous carcinoma/epithelial ovarian cancer cells	27881002	[94]
	Gliomas and functions in tumor cell adhesion	15674329	[95]

Table 2.1 (Continued)

Official name/ previous name	Type of cancer	PMID	References
ADGRG2/GPR64	Parathyroid tumors	27760455	[96]
GPR68	Neuroendocrine tumors, pheochromocytomas, cervical adenocarcinomas, endometrial cancer, paragangliomas, medullary thyroid carcinomas, gastrointestinal stromal tumors, and pancreatic adenocarcinomas	31652823	[97]
	Pancreatic ductal adenocarcinoma, cancer associated fibroblasts	29872392	[76]
		29092903	[98]
GPR78	Lung cancer	27697106	[99]
GPR87	Squamous cells of the lung	18057535	[100]
	Large cell and adenocarcinomas of the lung and transitional cell carcinomas of the urinary bladder	18183596	[101]
ADGRG3/GPR97	Lymphatic endothelial cells	24178298	[102]
GPR107	Gastric tumor cell line KATOIII	22933024	[103]
ADGRF/GPR110	Glioma	28728843	[104]
ADGRG4/GPR112	Gastrointestinal neuroendocrine carcinomas	18953328	[105]
ADGRF3/GPR113	Gastrointestinal neuroendocrine tumors	23158174	[106]
GPR132	Integrated stress response in acute myeloid leukemia	31127149	[107]
ADGRD1/GPR133	Glioblastoma	27775701	[108]
GPR137	Hepatoma	25490967	[109]
	Pancreatic cancer	25471990	[110]
GPR160	Prostate cancer	26871479	[111]
GPR161	Breast cancer cells	24599592	[112]
GPR182	Sinusoidal endothelial differentiation	29408502	[113]
	Adenoma	28094771	[114]
GPRC5A	Prostate cancer	31276604	[115]
	PDAC and colon cancer cells	29872392	[76]
ML-18	Lung cancer	25554218	[116]
ADGRL4/ELTD1	Hepatocellular carcinoma	30026838	[117]

2.4.6 Targeted Gene Deletion and Genetic Mutations of Orphan GPCRs

In the absence of knowledge about the identity of cognate ligands, phenotyping of animals models, in particular mouse models with a targeted deletion of the corresponding gene, is one of the most efficient ways to obtain insights on the physiological function of orphan GPCRs *in vivo* (Table 2.2). These information are useful not only to learn more about the physiological function of orphan GPCRs but also to select those of therapeutic interest for which surrogate ligands could be developed. That the complete phenotype might be only revealed in particular disease models or particular experimental conditions is frequently observed. The recent example of GPR139 knockout mice illustrates this aspect, as no obvious phenotype was initially observed but a more focused examination revealed a hypersensitivity to opioids [18] (see also Section 2.4.7.3). In the case of GPR88, knockout mice were first reported in 2009 and an altered striatal phenotype was observed [147]. Only later on, more detailed phenotypic features were discovered like impaired motor- and cue-based behaviors, motor deficit, improved spatial learning, low anxiety, and alcohol use disorders [148–150] (Section 2.4.7.4). Furthermore, tissue-specific gene deletions might uncover hidden phenotypes. For GPRC5B, ubiquitous gene deletion resulted in neurological [177, 178] and metabolic phenotypes [179], whereas smooth-muscle cell-specific gene deletion showed enhanced prostacyclin receptor signaling associated with facilitation of vascular relaxation, thus revealing the cardiovascular function of GPRC5B in smooth muscle contractility [181].

The identification of natural mutations of poorly examined proteins may provide useful information on their potential function. Recently, rare variants of 85 orphan or understudied GPCRs were analyzed with the sequence kernel association tests (SKATs) in an unselected cohort of 51 289 individuals [185], and 28 GPCRs were proposed to have relevance to congruent diseases. This analysis generated interesting hypothesis about their physiological function that now need to be confirmed experimentally.

2.4.7 Orphan GPCRs as Regulators of Other Proteins

Non-orphan GPCRs were shown to expand their functional diversity by forming heteromers with new heteromer-specific functional properties [186–192]. This paradigm has also been shown for orphan GPCRs that can form heteromeric complexes with non-orphan GPCRs [193]. An increasing number of orphan GPCRs have been shown to allosterically modulate the function of the non-orphan GPCR through these complex. Other scenarios like a more "passive" scaffolding functions have also been reported.

Table 2.2 Phenotype of orphan and understudied GPCRs gene deleted mice.

Official name/ previous name	Phenotype of gene deletion	PMID	References
GPR3	Premature ovarian aging	15956199	[120]
	Reduced β-amyloid deposition	19213921	[121]
	Hypersensitivity to thermal pain	21352831	[122]
GPR4	Increased perinatal mortality, impaired vascular proliferation	17145776	[123]
GPR6	Selected alterations in instrumental conditioning	17934457	[124]
GPR12	Dyslipidemia and obesity	16887097	[125]
GPR15	Tissue damage, inflammatory cytokine expression	23661644	[126]
GPR20	Hyperactivity	US patent	[127]
GPR21	Resistant to diet-induced obesity, increase in glucose tolerance and insulin sensitivity, modest lean phenotype	22653059	[128]
GPR22	Increased severity of functional decompensation following aortic banding	18539757	[129]
GPR26	Increased levels of anxiety, depression-like behaviors	21924326	[130]
GPR27	Endogenous mouse insulin promotor activity and glucose-stimulated insulin secretion	22253604	[131]
GPR34	Enhanced immune response	21097509	[132]
GPR37	Altered striatal signaling	21372109	[133]
GPR39	Obesity and altered adipocyte metabolism	21784784	[134]
LGR4/GPR48	Multiple developmental disorders	18487371	[135]
		19605502	[136]
		18955481	[137]
		18424556	[138]
GPR50	Abnormal thermoregulation, fasting-induced torpor	22197240	[139]
	Rapid tumor progression	29572483	[88]

(Continued)

Table 2.2 (Continued)

Official name/ previous name	Phenotype of gene deletion	PMID	References
GPR52	Psychosis-related behaviors	24587241	[140]
GPR61	Obesity associated with hyperphagia	21971119	[141]
ADGRG2/GPR64	Obstructive infertility	15367682	[142]
GPR82	Lower body weight and body fat content associated with reduced food intake, decreased serum triglyceride levels, higher insulin sensitivity and glucose tolerance	22216272	[143]
GPR83	Protected from obesity and glucose intolerance with HFD	23744028	[144]
GPR85	Significant brain weight increase with a trend of enhanced memory	18413613	[145]
	Enhanced ability to discriminate spatial relationships	22697179	[146]
GPR88	Altered striatal signaling	19796684	[147]
	Increased alcohol seeking and drinking	29580570	[148]
	Striatal deficits, better in spatial tasks, reduced anxiety	26188600	[149]
	Hyperactivity, poor motor coordination and impaired cue-based learning	23064379	[150]
	Striatal dysfunctions and impaired learning	28729439	[151]
GPR97	Macrophages invasion in the liver and kidney with HFD	27089991	[152]
	Ameliorate renal dysfunction, histologic lesions, and inflammatory responses in AKI	29531097	[153]
GPR98	Lower femoral bone mineral density associated with increased Rankl expression in osteoblasts	22419726	[154]
ADGRF5/GPR116	Chronic airway inflammation	30654796	[155]

Table 2.2 (Continued)

Official name/ previous name	Phenotype of gene deletion	PMID	References
ADGRA2/GPR124	Embryonic lethal by CNS-specific angiogenesis arrest in forebrain and neural tube	21071672	[156]
	Embryonic lethal with abnormal angiogenesis of the forebrain and spinal cord	21421844	[157]
ADGRG6/GPR126	Delayed axonal sorting by Schwann cells, Schwann cells arrest at the promyelinating stage	21613327	[158]
	Hypotrabeculation and affected mitochondrial function	24082093	[159]
GPR139	Enhanced opioid-induced inhibition of neuronal firing	31416932	[18]
GPR149	Increased fertility and enhanced ovulation	19887567	[160]
GPR161	Midgestation lethality and increased Sonic hedgehog signaling in the neural tube	23332756	[161]
MRGPRE/GPR167	Delayed allodynia onset accompanied with affected development of neuropathic pain	18197975	[162]
GPR176	Shorter circadian period	26882873	[163]
GPR183	Reduced early antibody response to a T-dependent antigen, failure of B cells to move to the outer follicle	19597478	[164]
ADGRB2/BAI2	Antidepressant-like behavior	21110148	[165]
CLESR1/ADGRC1	Defects in the orientation of hair outgrowth in the adult mice	19357712	[166]
	Defects in hindbrain neuron migration	20631168	[167]
CELSR2/ADGRC2	Defects in the migration of hindbrain neurons.	20631168	[167]
	Reduced quiescence of hematopoietic stem cells	22817897	[168]
	Defects in CSF dynamics and hydrocephalus	20473291	[169]

(Continued)

Table 2.2 (Continued)

Official name/ previous name	Phenotype of gene deletion	PMID	References
CELSR3/ADGRC3	Defects in anterior–posterior axon tract organization in the brainstem	21106844	[170]
	Defects in tangential cortical interneuron migration	19332558	[171]
	Defects in pancreatic β-cell differentiation	23177622	[172]
ADGRE1/EMR1	Failure in peripheral immune tolerance	15883173	[173]
ADGRV1	Loss-of-function mutations are associated with Usher syndrome, a sensory deficit disorder	18463160	[174]
GPRC5A	Developed more lung tumors	18000218	[175]
	Severe inflammatory conditions associated with abnormally activated nuclear factor-κB	20354164	[176]
GPRC5B	Altered spontaneous activity pattern and decreased response to a new environment	21840300	[177]
	Altered cortical neurogenesis	24089469	[178]
	Reduced obesity and obesity-associated inflammation	23169819	[179]
	Impaired long-term motor learning	29481883	[180]
	Facilitation of vascular relaxation and prevention of smooth muscle dedifferentiation	31941358	[181]
GPRC5C	Altered acid–base homeostasis as lower blood pH and higher urine pH	29196502	[182]
GPRC6A	Lower response in bone marrow and testis by testosterone stimulation	20947496	[183]

HFD, high fat diet; AKI, acute kidney injury.
Source: Adapted and updated from Alexander et al. [5] and Ahmed et al. [184].

2.4.7.1 GPR50

GPR50 is an orphan receptor, which due to its high sequence homology to melatonin MT_1 and MT_2 receptors belongs to the melatonin receptor subfamily. Despite this classification, GPR50 does not bind to melatonin. Phylogenetic studies suggest that the ortholog of GPR50 in lower vertebrates binds to melatonin indicating that GPR50 lost its capacity to bind to melatonin during evolution [194]. Evolutionary conservation of GPR50 expression might be thus due to other, ligand-independent functions of GPR50 [193]. One of these functions could be the modulation of the function of MT_1 in heterodimeric complexes [195]. When engaged in these complexes, GPR50 behaves as a negative allosteric modulator of MT_1 function by preventing melatonin binding and signal transduction of MT_1. The physiological significance of this complex is still elusive but might be relevant in the basal hypothalamic region expressing both receptors [196].

More recently, the property of GPR50 to form heteromeric complexes has been expanded toward the transforming growth factor (TGF) β type I receptor (TβRI), a well-known molecule in cancer pathophysiology [88]. GPR50 stabilizes TβRI in its active state by substituting for the canonical dimerization partner TβRII, and rendering TβRI constitutively active. Consequently, GPR50 expression results in anti-proliferative effects in cancer cells and inhibited tumor growth in a xenograft mouse model in a TβRI-dependent manner. Conversely, deletion of the GPR50 gene in a mouse model of spontaneous mammary cancer development showed a reduction of overall survival of mice indicating that GPR50 has antitumor-properties in wild type mice. A similar relation was observed in human breast cancer databases that show an association between poor relapse-free survival ratios and low GPR50 mRNA expression levels. Whether activation of GPR50 by a natural or surrogate ligand is able to modulate the activation state of TβRI or the interaction with TβRI and further downstream events remains to be determined once such ligands are identified.

2.4.7.2 GPR61, GPR62, and GPR135

Three poorly related orphan GPCRs, GPR61, GPR62, and GPR135 were proposed to belong to the melatonin receptor subfamily based on phylogenetic criteria [4, 119]. Binding of melatonin to these receptors was not confirmed experimentally. The relatively low sequence homology of these receptors with the three current members of this subfamily argues against the classification of GPR61, GPR62, and GPR135 into this subfamily. Complementary *in vitro* experimental data indicate however a possible modulatory role of these three orphans on MT_2 function [55]. All three orphans inhibit melatonin-induced β-arrestin recruitment to MT_2 while maintaining G protein-dependent signaling of MT_2 in transfected HEK293 cells, possibly through common heteromeric complexes [55]. Although the existence of such complexes in tissues remains to be demonstrated, this case illustrates the capacity of orphan GPCRs to function as allosteric modulators of other receptors in heteromeric complexes.

2.4.7.3 GPR139

An *in vivo* screening campaign for MOR-interacting genes using a *Caenorhabditis elegans* platform for unbiased discovery of genes influencing opioid responsiveness, picked up GPR139. Interestingly, GPR139 was shown to physically interact with MOR, to inhibit MOR function, and to affect opioid withdrawal [18]. These results were replicated in a mammalian setting, as GPR139 knockout mice showed enhanced opioid sensitivity. Furthermore, a surrogate agonist for GPR139, JNJ-63533054, enhanced the GPR139 action on opioid sensitivity *in vivo*. This pioneering work classifies GPR139 as a new target for the opioid abuse, and demonstrates the usefulness of the *C. elegans in vivo* platform to screen for new orphan GPCRs functions.

2.4.7.4 GPR88

GPR88 is one of the most examined GPCRs among the remaining orphans. Genetic association studies and the discovery of a deleterious mutation in the *GPR88* gene indicated a role of GPR88 in neurodevelopment [197] and psychiatric disorders such as bipolar disorders and schizophrenia [198]. Phenotyping of GPR88 knockout mice showed several neuronal abnormalities [147–150] (Table 2.1). Interestingly, the impaired motor- and cue-dependent behaviors of GPR88 deficient mice were normalized by opioid receptor (OR) antagonists [149]. *In vitro* studies showed the formation of heteromers between GPR88 and all three ORs and an overall inhibitory action of GPR88 on OR function in HEK293 cells. This is consistent with facilitated morphine-induced locomotor sensitization, withdrawal, and supra-spinal analgesia that were observed in GPR88 deficient mice [199]. Activation of GPR88 with a synthetic agonist further potentiated the inhibitory effect of GPR88 on MOR signaling in wild-type mice. Collectively, the different features of GPR88 make this orphan receptor an attractive therapeutic candidate.

2.4.7.5 GPR158 and GPR179

Regulators of G protein signaling (RGS) modulate the speed of G protein signaling. This usually occurs by interacting directly with G proteins but may also occur through direct interaction with GPCRs [200]. During the quest for proteins regulating RGS7, two orphan GPCRs, GPR158 and GPR179, belonging to the class C GPCR family, were identified in a proteomic screen [201]. The scaffolding function of GPR158/179 for RGS7 was demonstrated in retinas of GPR179 knockout mice in which targeting of RGS7 to the postsynaptic compartment of bipolar neurons was prevented [201]. Recruitment of RGS7 to these membranes controls the inactivation speed of another GPCR, the metabotropic glutamate receptor (mGluR6), and its post-synaptic response to light [202]. A second proteomic screen, focused on binding partners of the large extracellular domain of GPR179, revealed extracellular matrix proteins, in particular of the heparan sulfate proteoglycan (HSPG) family [202]. In the photoreceptor synapse, the HSPG family member pikachurin interacts with the post-synaptic GPR179/RGS7 complex and the pre-synaptic dystrophin-dystroglycan complex. The integrity of this complex is essential for proper synaptic transmission of photoreceptor signals [202]. Collectively, GPR158 and GPR179 appear to serve a dual scaffolding function with intracellular and extracellular components in synapsis. None of the discovered GPR158/GPR179 interactors appear to show ligand properties but act rather as allosteric regulators.

2.4.7.6 GPR31

KRAS, a major oncogene with frequent mutations in various cancers, requires plasma membrane localization through farnesylation of its C-terminal CAAX motif [203]. Farnesyltransferase inhibitors intended to inhibit KRAS plasma membrane translocation failed in clinical trials as anticancer drug, because of compensatory prenylation of KRAS by geranylgeranyl transferase [204]. To overcome this important problem, an unbiased genome-wide siRNA screen for genes required for plasma membrane localization of KRAS was performed [205]. The orphan GPR31 was one of the hits that facilitated membrane association of KRAS through its chaperone activity in the ER. Consistently, GPR31 silencing in cancer cells has shown that this orphan receptor is necessary for KRAS-promoted oncogenic signaling, cancer cell proliferation, and micropinocytosis. Thus, silencing of GPR31 has the potential to counteract the oncogenic properties of KRAS.

2.5 Conclusion and Perspectives

Here we reviewed the status quo of the orphan GPCRs field including current deorphanization strategies and the search for surrogate ligands. Deorphanization became more difficult in the last 10 years as testified by the high number of remaining orphan GPCRs, but it still remains a major goal in GPCR research. Recent technological advances in terms of readouts and bioinformatics data mining opened new perspectives that hopefully will have a long-lasting impact on the GPCR deorphanization rate. As several GPCRs show spontaneous, ligand-independent activity or have modulatory roles on the function of other GPCRs or proteins, the obligatory existence of a cognate ligand for every orphan GPCRs has been questioned [193, 206]. The possible co-existence of ligand-dependent and ligand-independent GPCR functions is increasingly recognized [207], and GPCRs with known ligands indicating that both functions are not mutually exclusive. An increasing number of surrogate ligands have been discovered for orphan GPCRs that are of importance, not only to understand their physiological function but also to target a specific aspect of receptor function that might be of particular value for future therapeutic applications. Antibodies and small seize nanobodies are likely to become innovate tools for orphan GPCRs in the future. Finally, further knockout mice models and refined phenotyping of existing models are likely to continue to make important contributions to decipher the function of orphan GPCRs, in parallel to the other approaches intended to modulate their function by natural or surrogate ligands.

Acknowledgments

We acknowledge the members of the Jockers lab (Institut Cochin, France), the department of anatomy (Kyorin University, Japan), and the Yoshimi lab (National Cancer Center Research Institute, Japan) for the kind support. This work was supported by grants from the Vehicle Racing Commemorative Foundation, Daiwa Securities Health Foundation, SENSHIN Medical Research Foundation (to A.O.) and the Fondation pour la Recherche Médicale (Equipe FRM 2019, EQU201903008055 to R.J.), Agence Nationale de la Recherche (ANR-15-CE14-0025-02, ANR-19-CE16 (Mito-GPCR) to R.J.), La Ligue Contre le Cancer N/Ref: RS19/75-127 and the "Who am I?" laboratory of excellence No. ANR-11-LABX-0071 funded by the French Government through its "Investments for the Future" program operated by The French National Research Agency under Grant no. ANR-11-IDEX-0005-01 (to R.J.), Inserm and CNRS.

References

1 Sriram, K. and Insel, P.A. (2018). G protein-coupled receptors as targets for approved drugs: how many targets and how many drugs? *Mol. Pharmacol.* 93 (4): 251–258. Epub 2018/01/05.

2 Hauser, A.S., Attwood, M.M., Rask-Andersen, M. et al. (2017). Trends in GPCR drug discovery: new agents, targets and indications. *Nat. Rev. Drug Discovery* 16 (12): 829–842. Epub 2017/10/28.

3 Davenport, A.P., Alexander, S.P., Sharman, J.L. et al. (2013). International Union of Basic and Clinical Pharmacology. LXXXVIII. G protein-coupled receptor list: recommendations for new pairings with cognate ligands. *Pharmacol. Rev.* 65 (3): 967–986. Epub 2013/05/21.

4 Civelli, O., Reinscheid, R.K., Zhang, Y. et al. (2013). G protein-coupled receptor deorphanizations. *Annu. Rev. Pharmacol. Toxicol.* 53: 127–146. Epub 2012/10/02.

5 Alexander, S.P.H., Christopoulos, A., Davenport, A.P. et al. (2019). THE CONCISE GUIDE TO PHARMACOLOGY 2019/20: G protein-coupled receptors. *Br. J. Pharmacol.* 176 (Suppl 1): S21–S141. Epub 2019/11/12.

6 Laschet, C., Dupuis, N., and Hanson, J. (2018). The G protein-coupled receptors deorphanization landscape. *Biochem. Pharmacol.* 153: 62–74. Epub 2018/02/20.

7 Dixon, R.A., Kobilka, B.K., Strader, D.J. et al. (1986). Cloning of the gene and cDNA for mammalian β-adrenergic receptor and homology with rhodopsin. *Nature* 321 (6065): 75–79. Epub 1986/05/01.

8 Lefkowitz, R.J. (2013). A brief history of G-protein coupled receptors (Nobel lecture). *Angew. Chem. Int. Ed. Engl.* 52 (25): 6366–6378. Epub 2013/05/08.

9 Nathans, J. and Hogness, D.S. (1983). Isolation, sequence analysis, and intron-exon arrangement of the gene encoding bovine rhodopsin. *Cell* 34 (3): 807–814. Epub 1983/10/01.

10 Dohlman, H.G., Caron, M.G., and Lefkowitz, R.J. (1987). A family of receptors coupled to guanine nucleotide regulatory proteins. *Biochemistry* 26 (10): 2657–2664. Epub 1987/05/19.

11 Loosfelt, H., Misrahi, M., Atger, M. et al. (1989). Cloning and sequencing of porcine LH-hCG receptor cDNA: variants lacking transmembrane domain. *Science* 245 (4917): 525–528. Epub 1989/08/04.

12 Ebisawa, T., Karne, S., Lerner, M.R. et al. (1994). Expression cloning of a high-affinity melatonin receptor from *Xenopus* dermal melanophores. *Proc. Natl. Acad. Sci. U.S.A.* 91 (13): 6133–6137. Epub 1994/06/21.

13 Libert, F., Parmentier, M., Lefort, A. et al. (1989). Selective amplification and cloning of four new members of the G protein-coupled receptor family. *Science* 244 (4904): 569–572. Epub 1989/05/05.

14 Bunzow, J.R., Van Tol, H.H., Grandy, D.K. et al. (1988). Cloning and expression of a rat D_2 dopamine receptor cDNA. *Nature* 336 (6201): 783–787. Epub 1988/12/22.

15 Fargin, A., Raymond, J.R., Lohse, M.J. et al. (1988). The genomic clone G-21 which resembles a β-adrenergic receptor sequence encodes the 5-HT_{1A} receptor. *Nature* 335 (6188): 358–360. Epub 1988/09/22.

16 Sakurai, T., Amemiya, A., Ishii, M. et al. (1998). Orexins and orexin receptors: a family of hypothalamic neuropeptides and G protein-coupled receptors that regulate feeding behavior. *Cell* 92 (4): 573–585. Epub 1998/03/10.

17 Kojima, M., Hosoda, H., Date, Y. et al. (1999). Ghrelin is a growth-hormone-releasing acylated peptide from stomach. *Nature* 402 (6762): 656–660. Epub 1999/12/22.

18 Wang, D., Stoveken, H.M., Zucca, S. et al. (2019). Genetic behavioral screen identifies an orphan anti-opioid system. *Science* 365 (6459): 1267–1273. Epub 2019/08/17.

19 Dupuis, N., Laschet, C., Franssen, D. et al. (2017). Activation of the orphan G protein-coupled receptor GPR27 by surrogate ligands promotes β-arrestin 2 recruitment. *Mol. Pharmacol.* 91 (6): 595–608. Epub 2017/03/21.

20 Yin, H., Chu, A., Li, W. et al. (2009). Lipid G protein-coupled receptor ligand identification using β-arrestin PathHunter assay. *J. Biol. Chem.* 284 (18): 12328–12338. Epub 2009/03/17.

21 Kalatskaya, I., Berchiche, Y.A., Gravel, S. et al. (2009). AMD3100 is a CXCR7 ligand with allosteric agonist properties. *Mol. Pharmacol.* 75 (5): 1240–1247. Epub 2009/03/04.

22 Rajagopal, S., Kim, J., Ahn, S. et al. (2010). β-Arrestin- but not G protein-mediated signaling by the "decoy" receptor CXCR7. *Proc. Natl. Acad. Sci. U.S.A.* 107 (2): 628–632. Epub 2009/12/19.

23 Angers, S., Salahpour, A., Joly, E. et al. (2000). Detection of β_2-adrenergic receptor dimerization in living cells using bioluminescence resonance energy transfer (BRET). *Proc. Natl. Acad. Sci. U.S.A.* 97 (7): 3684–3689. Epub 2000/03/22.

24 Dixon, A.S., Schwinn, M.K., Hall, M.P. et al. (2016). NanoLuc complementation reporter optimized for accurate measurement of protein interactions in cells. *ACS Chem. Biol.* 11 (2): 400–408. Epub 2015/11/17.

25 Yan, Y.X., Boldt-Houle, D.M., Tillotson, B.P. et al. (2002). Cell-based high-throughput screening assay system for monitoring G protein-coupled receptor activation using β-galactosidase enzyme complementation technology. *J. Biomol. Screening* 7 (5): 451–459. Epub 2003/11/06.

26 Barnea, G., Strapps, W., Herrada, G. et al. (2008). The genetic design of signaling cascades to record receptor activation. *Proc. Natl. Acad. Sci. U.S.A.* 105 (1): 64–69. Epub 2008/01/01.

27 Kroeze, W.K., Sassano, M.F., Huang, X.P. et al. (2015). PRESTO-Tango as an open-source resource for interrogation of the druggable human GPCRome. *Nat. Struct. Mol. Biol.* 22 (5): 362–369. Epub 2015/04/22.

28 Foster, S.R., Hauser, A.S., Vedel, L. et al. (2019). Discovery of human signaling systems: pairing peptides to G protein-coupled receptors. *Cell* 179 (4): 895–908.e21. Epub 2019/11/02.

29 Itoh, R.E., Kurokawa, K., Ohba, Y. et al. (2002). Activation of rac and cdc42 video imaged by fluorescent resonance energy transfer-based single-molecule probes in the membrane of living cells. *Mol. Cell. Biol.* 22 (18): 6582–6591. Epub 2002/08/23.

30 Yoshizaki, H., Ohba, Y., Kurokawa, K. et al. (2003). Activity of Rho-family GTPases during cell division as visualized with FRET-based probes. *J. Cell Biol.* 162 (2): 223–232. Epub 2003/07/16.

31 Kimura, K., Tsuji, T., Takada, Y. et al. (2000). Accumulation of GTP-bound RhoA during cytokinesis and a critical role of ECT2 in this accumulation. *J. Biol. Chem.* 275 (23): 17233–17236. Epub 2000/06/06.

32 Oishi, A., Makita, N., Sato, J. et al. (2012). Regulation of RhoA signaling by the cAMP-dependent phosphorylation of RhoGDIα. *J. Biol. Chem.* 287 (46): 38705–38715. Epub 2012/09/27.

33 Ridley, A.J. and Hall, A. (1992). The small GTP-binding protein rho regulates the assembly of focal adhesions and actin stress fibers in response to growth factors. *Cell* 70 (3): 389–399. Epub 1992/08/07.

34 Inoue, A., Ishiguro, J., Kitamura, H. et al. (2012). TGFα shedding assay: an accurate and versatile method for detecting GPCR activation. *Nat. Methods* 9 (10): 1021–1029. Epub 2012/09/18.

35 Inoue, A., Arima, N., Ishiguro, J. et al. (2011). LPA-producing enzyme PA-$PLA_{1\alpha}$ regulates hair follicle development by modulating EGFR signalling. *EMBO J.* 30 (20): 4248–4260. Epub 2011/08/23.

36 Yu, H., Zhao, T., Liu, S. et al. (2019). MRGPRX4 is a bile acid receptor for human cholestatic itch. *eLife* 8: e48431. Epub 2019/09/11.

37 Gales, C., Van Durm, J.J., Schaak, S. et al. (2006). Probing the activation-promoted structural rearrangements in preassembled receptor-G protein complexes. *Nat. Struct. Mol. Biol.* 13 (9): 778–786. Epub 2006/08/15.

38 Yano, H., Provasi, D., Cai, N.S. et al. (2017). Development of novel biosensors to study receptor-mediated activation of the G-protein α subunits Gs and Golf. *J. Biol. Chem.* 292 (49): 19989–19998. Epub 2017/10/19.

39 Laschet, C., Dupuis, N., and Hanson, J. (2018). A dynamic and screening-compatible nanoluciferase-based complementation assay enables profiling of individual GPCR–G protein interactions. *J. Biol. Chem.* Epub 2018/12/30.

40 Yano, H., Cai, N.S., Javitch, J.A. et al. (2018). Luciferase complementation based-detection of G-protein-coupled receptor activity. *Biotechniques* 65 (1): 9–14. Epub 2018/07/18.

41 Wan, Q., Okashah, N., Inoue, A. et al. (2018). Mini G protein probes for active G protein-coupled receptors (GPCRs) in live cells. *J. Biol. Chem.* 293 (19): 7466–7473. Epub 2018/03/11.

42 Carpenter, B. and Tate, C.G. (2016). Engineering a minimal G protein to facilitate crystallisation of G protein-coupled receptors in their active conformation. *Protein Eng. Des. Sel.* 29 (12): 583–594. Epub 2016/09/28.

43 Nehme, R., Carpenter, B., Singhal, A. et al. (2017). Mini-G proteins: novel tools for studying GPCRs in their active conformation. *PLoS One* 12 (4): e0175642. Epub 2017/04/21.

44 Bouzo-Lorenzo, M., Stoddart, L.A., Xia, L. et al. (2019). A live cell NanoBRET binding assay allows the study of ligand-binding kinetics to the adenosine A_3 receptor. *Purinergic Signal.* 15 (2): 139–153. Epub 2019/03/29.

45 Peach, C.J., Kilpatrick, L.E., Friedman-Ohana, R. et al. (2018). Real-time ligand binding of fluorescent VEGF-A isoforms that discriminate between VEGFR2 and NRP1 in living cells. *Cell Chem. Biol.* 25 (10): 1208–18 e5. Epub 2018/07/31.

46 Stoddart, L.A., Johnstone, E.K.M., Wheal, A.J. et al. (2015). Application of BRET to monitor ligand binding to GPCRs. *Nat. Methods* 12 (7): 661–663. Epub 2015/06/02.

47 Auriau, J., Roujeau, C., Belaid Choucair, Z. et al. (2018). Gain of affinity for VEGF165 binding within the VEGFR2/NRP1 cellular complex detected by an HTRF-based binding assay. *Biochem. Pharmacol.* 158: 45–59. Epub 2018/09/22.

48 Ngo, T., Kufareva, I., Coleman, J. et al. (2016). Identifying ligands at orphan GPCRs: current status using structure-based approaches. *Br. J. Pharmacol.* 173 (20): 2934–2951. Epub 2016/02/03.

49 Ngo, T., Ilatovskiy, A.V., Stewart, A.G. et al. (2017). Orphan receptor ligand discovery by pickpocketing pharmacological neighbors. *Nat. Chem. Biol.* 13 (2): 235–242. Epub 2016/12/20.

50 Thelen, M. and Thelen, S. (2008). CXCR7, CXCR4 and CXCL12: an eccentric trio? *J. Neuroimmunol.* 198 (1, 2): 9–13. Epub 2008/06/06.

51 Lin, X., Li, M., Wang, N. et al. (2020). Structural basis of ligand recognition and self-activation of orphan GPR52. *Nature* 579 (7797): 152–157.

52 Bond, R.A. and Ijzerman, A.P. (2006). Recent developments in constitutive receptor activity and inverse agonism, and their potential for GPCR drug discovery. *Trends Pharmacol. Sci.* 27 (2): 92–96. Epub 2006/01/13.

53 Gilliland, C.T., Salanga, C.L., Kawamura, T. et al. (2013). The chemokine receptor CCR1 is constitutively active, which leads to G protein-independent, β-arrestin-mediated internalization. *The Journal of biological chemistry* 288 (45): 32194–32210. Epub 2013/09/24.

54 Zhang, B., Albaker, A., Plouffe, B. et al. (2014). Constitutive activities and inverse agonism in dopamine receptors. *Adv. Pharmacol.* 70: 175–214. Epub 2014/06/17.

55 Oishi, A., Karamitri, A., Gerbier, R. et al. (2017). Orphan GPR61, GPR62 and GPR135 receptors and the melatonin MT_2 receptor reciprocally modulate their signaling functions. *Sci. Rep.* 7 (1): 8990. Epub 2017/08/23.

56 Milligan, G. (2003). Constitutive activity and inverse agonists of G protein-coupled receptors: a current perspective. *Mol. Pharmacol.* 64 (6): 1271–1276. Epub 2003/12/04.

57 Muroi, T., Matsushima, Y., Kanamori, R. et al. (2017). GPR62 constitutively activates cAMP signaling but is dispensable for male fertility in mice. *Reproduction* 154 (6): 755–764. Epub 2017/09/16.

58 Engelstoft, M.S., Norn, C., Hauge, M. et al. (2014). Structural basis for constitutive activity and agonist-induced activation of the enteroendocrine fat sensor GPR119. *Br. J. Pharmacol.* 171 (24): 5774–5789. Epub 2014/08/15.

59 Jones, P.G., Nawoschik, S.P., Sreekumar, K. et al. (2007). Tissue distribution and functional analyses of the constitutively active orphan G protein coupled receptors, GPR26 and GPR78. *Biochim. Biophys. Acta* 1770 (6): 890–901. Epub 2007/03/17.

60 Oishi, A., Dam, J., and Jockers, R. (2020). β-Arrestin-2 BRET biosensors detect different β-arrestin-2 conformations in interaction with GPCRs. *ACS Sens.* 5 (1): 57–64. Epub 2019/12/19.

61 Tanaka, S., Ishii, K., Kasai, K. et al. (2007). Neural expression of G protein-coupled receptors GPR3, GPR6, and GPR12 up-regulates cyclic AMP levels and promotes neurite outgrowth. *J. Biol. Chem.* 282 (14): 10506–10515. Epub 2007/02/08.

62 Martin, A.L., Steurer, M.A., and Aronstam, R.S. (2015). Constitutive activity among orphan class-A G protein coupled receptors. *PLoS One* 10 (9): e0138463. Epub 2015/09/19.

63 Ayukawa, K., Suzuki, C., Ogasawara, H. et al. (2020). Development of a high-throughput screening-compatible assay for discovery of GPR3 inverse agonists using a cAMP biosensor. *SLAS Discov.* 25 (3): 287–298. Epub 2019/09/14.

64 Laun, A.S., Shrader, S.H., and Song, Z.H. (2018). Novel inverse agonists for the orphan G protein-coupled receptor 6. *Heliyon* 4 (11): e00933. Epub 2018/11/28.

65 Bilsky, E.J., Giuvelis, D., Osborn, M.D. et al. (2010). In vitro and in vivo assessment of μ opioid receptor constitutive activity. *Methods Enzymol.* 484: 413–443. Epub 2010/11/03.

66 Petersen, P.S., Woldbye, D.P., Madsen, A.N. et al. (2009). In vivo characterization of high basal signaling from the ghrelin receptor. *Endocrinology* 150 (11): 4920–4930. Epub 2009/10/13.

67 Aloyo, V.J., Berg, K.A., Clarke, W.P. et al. (2010). Inverse agonism at serotonin and cannabinoid receptors. *Prog. Mol. Biol. Transl. Sci.* 91: 1–40. Epub 2010/08/10.

68 Morisset, S., Rouleau, A., Ligneau, X. et al. (2000). High constitutive activity of native H_3 receptors regulates histamine neurons in brain. *Nature* 408 (6814): 860–864. Epub 2000/12/29.

69 Ge, X., Yang, H., Bednarek, M.A. et al. (2018). LEAP2 is an endogenous antagonist of the ghrelin receptor. *Cell Metab.* 27 (2): 461–9 e6. Epub 2017/12/14.

70 Nijenhuis, W.A., Oosterom, J., and Adan, R.A. (2001). AgRP(83–132) acts as an inverse agonist on the human-melanocortin-4 receptor. *Mol. Endocrinol.* 15 (1): 164–171. Epub 2001/01/06.

71 Morri, M., Sanchez-Romero, I., Tichy, A.M. et al. (2018). Optical functionalization of human Class A orphan G-protein-coupled receptors. *Nat. Commun.* 9 (1): 1950. Epub 2018/05/18.

72 Regard, J.B., Sato, I.T., and Coughlin, S.R. (2008). Anatomical profiling of G protein-coupled receptor expression. *Cell* 135 (3): 561–571. Epub 2008/11/06.

73 Ehrlich, A.T., Maroteaux, G., Robe, A. et al. (2018). Expression map of 78 brain-expressed mouse orphan GPCRs provides a translational resource for neuropsychiatric research. *Commun. Biol.* 1: 102. Epub 2018/10/03.

74 Shore, A.M., Karamitri, A., Kemp, P. et al. (2013). Cold-induced changes in gene expression in brown adipose tissue, white adipose tissue and liver. *PLoS One* 8 (7): e68933. Epub 2013/07/31.

75 Yosten, G.L., Harada, C.M., Haddock, C.J. et al. (2020). GPR160 de-orphanization reveals critical roles in neuropathic pain in rodents. *J. Clin. Invest.* 130 (5): 2587–2592. Epub 2020/01/31.

76 Insel, P.A., Sriram, K., Wiley, S.Z. et al. (2018). GPCRomics: GPCR expression in cancer cells and tumors identifies new, potential biomarkers and therapeutic targets. *Front. Pharmacol.* 9: 431. Epub 2018/06/07.

77 Merten, N., Fischer, J., Simon, K. et al. (2018). Repurposing HAMI3379 to block GPR17 and promote rodent and human oligodendrocyte differentiation. *Cell Chem. Biol.* 25 (6): 775–86.e5. Epub 2018/05/01.

78 Qin, Y., Verdegaal, E.M., Siderius, M. et al. (2011). Quantitative expression profiling of G-protein-coupled receptors (GPCRs) in metastatic melanoma: the constitutively active orphan GPCR GPR18 as novel drug target. *Pigment Cell Melanoma Res.* 24 (1): 207–218. Epub 2010/10/01.

79 Finlay, D.B., Joseph, W.R., Grimsey, N.L. et al. (2016). GPR18 undergoes a high degree of constitutive trafficking but is unresponsive to *N*-arachidonoyl glycine. *PeerJ* 4: e1835. Epub 2016/03/29.

80 Rao, A. and Herr, D.R. (2017). G protein-coupled receptor GPR19 regulates E-cadherin expression and invasion of breast cancer cells. *Biochim. Biophys. Acta, Mol. Cell. Res.* 1864 (7): 1318–1327. Epub 2017/05/10.

81 Kastner, S., Voss, T., Keuerleber, S. et al. (2012). Expression of G protein-coupled receptor 19 in human lung cancer cells is triggered by entry into S-phase and supports G_2-M cell-cycle progression. *Mol. Cancer Res.* 10 (10): 1343–1358. Epub 2012/08/23.

82 Ansell, S.M., Akasaka, T., McPhail, E. et al. (2012). t(X;14)(p11;q32) in MALT lymphoma involving GPR34 reveals a role for GPR34 in tumor cell growth. *Blood* 120 (19): 3949–3957. Epub 2012/09/12.

83 Liu, F., Zhu, C., Huang, X. et al. (2014). A low level of GPR37 is associated with human hepatocellular carcinoma progression and poor patient survival. *Pathol. Res. Pract.* 210 (12): 885–892. Epub 2014/08/30.

84 Liang, F., Yue, J., Wang, J. et al. (2015). GPCR48/LGR4 promotes tumorigenesis of prostate cancer via PI3K/Akt signaling pathway. *Med. Oncol.* 32 (3): 49. Epub 2015/02/01.

85 Gao, Y., Kitagawa, K., Hiramatsu, Y. et al. (2006). Up-regulation of GPR48 induced by down-regulation of $p27^{Kip1}$ enhances carcinoma cell invasiveness and metastasis. *Cancer Res.* 66 (24): 11623–11631. Epub 2006/12/21.

86 Tanese, K., Fukuma, M., Yamada, T. et al. (2008). G-protein-coupled receptor GPR49 is up-regulated in basal cell carcinoma and promotes cell proliferation and tumor formation. *Am. J. Pathol.* 173 (3): 835–843. Epub 2008/08/09.

87 McClanahan, T., Koseoglu, S., Smith, K. et al. (2006). Identification of overexpression of orphan G protein-coupled receptor GPR49 in human colon and ovarian primary tumors. *Cancer Biol. Ther.* 5 (4): 419–426. Epub 2006/04/01.

88 Wojciech, S., Ahmad, R., Belaid-Choucair, Z. et al. (2018). The orphan GPR50 receptor promotes constitutive TGFβ receptor signaling and protects against cancer development. *Nat. Commun.* 9 (1): 1216. Epub 2018/03/25.

89 Ford, L.A., Roelofs, A.J., Anavi-Goffer, S. et al. (2010). A role for L-α-lysophosphatidylinositol and GPR55 in the modulation of migration, orientation and polarization of human breast cancer cells. *Br. J. Pharmacol.* 160 (3): 762–771. Epub 2010/07/02.

90 Andradas, C., Caffarel, M.M., Perez-Gomez, E. et al. (2011). The orphan G protein-coupled receptor GPR55 promotes cancer cell proliferation via ERK. *Oncogene* 30 (2): 245–252. Epub 2010/09/08.

91 Pineiro, R., Maffucci, T., and Falasca, M. (2011). The putative cannabinoid receptor GPR55 defines a novel autocrine loop in cancer cell proliferation. *Oncogene* 30 (2): 142–152. Epub 2010/09/15.

92 Perez-Gomez, E., Andradas, C., Flores, J.M. et al. (2013). The orphan receptor GPR55 drives skin carcinogenesis and is upregulated in human squamous cell carcinomas. *Oncogene* 32 (20): 2534–2542. Epub 2012/07/04.

93 Ji, B., Feng, Y., Sun, Y. et al. (2018). GPR56 promotes proliferation of colorectal cancer cells and enhances metastasis via epithelial mesenchymal transition through PI3K/AKT signaling activation. *Oncol. Rep.* 40 (4): 1885–1896. Epub 2018/08/02.

94 Liu, Z., Huang, Z., Yang, W. et al. (2017). Expression of orphan GPR56 correlates with tumor progression in human epithelial ovarian cancer. *Neoplasma* 64 (1): 32–39. Epub 2016/11/25.

95 Shashidhar, S., Lorente, G., Nagavarapu, U. et al. (2005). GPR56 is a GPCR that is overexpressed in gliomas and functions in tumor cell adhesion. *Oncogene* 24 (10): 1673–1682. Epub 2005/01/28.

96 Balenga, N., Azimzadeh, P., Hogue, J.A. et al. (2017). Orphan adhesion GPCR GPR64/ADGRG2 is overexpressed in parathyroid tumors and attenuates calcium-sensing receptor-mediated signaling. *J. Bone Miner. Res.* 32 (3): 654–666. Epub 2016/10/30.

97 Herzig, M., Dasgupta, P., Kaemmerer, D. et al. (2019). Comprehensive assessment of GPR68 expression in normal and neoplastic human tissues using a novel rabbit monoclonal antibody. *Int. J. Mol. Sci.* 20 (21): 5261. Epub 2019/10/28.

98 Wiley, S.Z., Sriram, K., Liang, W. et al. (2018). GPR68, a proton-sensing GPCR, mediates interaction of cancer-associated fibroblasts and cancer cells. *FASEB J.* 32 (3): 1170–1183. Epub 2017/11/03.

99 Dong, D.D., Zhou, H., and Li, G. (2016). GPR78 promotes lung cancer cell migration and metastasis by activation of Gαq-Rho GTPase pathway. *BMB Rep.* 49 (11): 623–628. Epub 2016/10/05.

100 Gugger, M., White, R., Song, S. et al. (2008). GPR87 is an overexpressed G-protein coupled receptor in squamous cell carcinoma of the lung. *Dis. Markers* 24 (1): 41–50. Epub 2007/12/07.

101 Glatt, S., Halbauer, D., Heindl, S. et al. (2008). hGPR87 contributes to viability of human tumor cells. *Int. J. Cancer* 122 (9): 2008–2016. Epub 2008/01/10.

102 Valtcheva, N., Primorac, A., Jurisic, G. et al. (2013). The orphan adhesion G protein-coupled receptor GPR97 regulates migration of lymphatic endothelial cells via the small GTPases RhoA and Cdc42. *J. Biol. Chem.* 288 (50): 35736–35748. Epub 2013/11/02.

103 Yosten, G.L., Redlinger, L.J., and Samson, W.K. (2012). Evidence for an interaction of neuronostatin with the orphan G protein-coupled receptor, GPR107. *Am. J. Physiol. Regul. Integr. Comp. Physiol.* 303 (9): R941–R949. Epub 2012/08/31.

104 Shi, H. and Zhang, S. (2017). Expression and prognostic role of orphan receptor GPR110 in glioma. *Biochem. Biophys. Res. Commun.* 491 (2): 349–354. Epub 2017/07/22.

105 Leja, J., Essaghir, A., Essand, M. et al. (2009). Novel markers for enterochromaffin cells and gastrointestinal neuroendocrine carcinomas. *Mod. Pathol.* 22 (2): 261–272. Epub 2008/10/28.

106 Carr, J.C., Boese, E.A., Spanheimer, P.M. et al. (2012). Differentiation of small bowel and pancreatic neuroendocrine tumors by gene-expression profiling. *Surgery* 152 (6): 998–1007. Epub 2012/11/20.

107 Nii, T., Prabhu, V.V., Ruvolo, V. et al. (2019). Imipridone ONC212 activates orphan G protein-coupled receptor GPR132 and integrated stress response in acute myeloid leukemia. *Leukemia* 33 (12): 2805–2816. Epub 2019/05/28.

108 Bayin, N.S., Frenster, J.D., Kane, J.R. et al. (2016). GPR133 (ADGRD1), an adhesion G-protein-coupled receptor, is necessary for glioblastoma growth. *Oncogenesis* 5 (10): e263. Epub 2016/10/25.

109 Shao, X., Liu, Y., Huang, H. et al. (2015). Down-regulation of G protein-coupled receptor 137 by RNA interference inhibits cell growth of two hepatoma cell lines. *Cell Biol. Int.* 39 (4): 418–426. Epub 2014/12/11.

110 Cui, X., Liu, Y., Wang, B. et al. (2015). Knockdown of GPR137 by RNAi inhibits pancreatic cancer cell growth and induces apoptosis. *Biotechnol. Appl. Biochem.* 62 (6): 861–867. Epub 2014/12/05.

111 Zhou, C., Dai, X., Chen, Y. et al. (2016). G protein-coupled receptor GPR160 is associated with apoptosis and cell cycle arrest of prostate cancer cells. *Oncotarget* 7 (11): 12823–12839. Epub 2016/02/13.

112 Feigin, M.E., Xue, B., Hammell, M.C. et al. (2014). G-protein-coupled receptor GPR161 is overexpressed in breast cancer and is a promoter of cell proliferation and invasion. *Proc. Natl. Acad. Sci. U.S.A.* 111 (11): 4191–4196. Epub 2014/03/07.

113 Schmid, C.D., Schledzewski, K., Mogler, C. et al. (2018). GPR182 is a novel marker for sinusoidal endothelial differentiation with distinct GPCR signaling activity in vitro. *Biochem. Biophys. Res. Commun.* 497 (1): 32–38. Epub 2018/02/07.

114 Kechele, D.O., Blue, R.E., Zwarycz, B. et al. (2017). Orphan Gpr182 suppresses ERK-mediated intestinal proliferation during regeneration and adenoma formation. *J. Clin. Invest.* 127 (2): 593–607. Epub 2017/01/18.

115 Sawada, Y., Kikugawa, T., Iio, H. et al. (2020). GPRC5A facilitates cell proliferation through cell cycle regulation and correlates with bone metastasis in prostate cancer. *Int. J. Cancer* 146 (5): 1369–1382. Epub 2019/07/06.

116 Moody, T.W., Mantey, S.A., Moreno, P. et al. (2015). ML-18 is a non-peptide bombesin receptor subtype-3 antagonist which inhibits lung cancer growth. *Peptides* 64: 55–61. Epub 2015/01/03.

117 Kan, A., Le, Y., Zhang, Y.F. et al. (2018). ELTD1 function in hepatocellular carcinoma is carcinoma-associated fibroblast-dependent. *J. Cancer* 9 (14): 2415–2427. Epub 2018/07/22.

118 Bjarnadottir, T.K., Gloriam, D.E., Hellstrand, S.H. et al. (2006). Comprehensive repertoire and phylogenetic analysis of the G protein-coupled receptors in human and mouse. *Genomics* 88 (3): 263–273. Epub 2006/06/07.

119 Gloriam, D.E., Fredriksson, R., and Schioth, H.B. (2007). The G protein-coupled receptor subset of the rat genome. *BMC Genomics* 8: 338. Epub 2007/09/26.

120 Ledent, C., Demeestere, I., Blum, D. et al. (2005). Premature ovarian aging in mice deficient for Gpr3. *Proc. Natl. Acad. Sci. U.S.A.* 102 (25): 8922–8926. Epub 2005/06/16.

121 Thathiah, A., Spittaels, K., Hoffmann, M. et al. (2009). The orphan G protein-coupled receptor 3 modulates amyloid-β peptide generation in neurons. *Science* 323 (5916): 946–951. Epub 2009/02/14.

122 Ruiz-Medina, J., Ledent, C., and Valverde, O. (2011). GPR3 orphan receptor is involved in neuropathic pain after peripheral nerve injury and regulates morphine-induced antinociception. *Neuropharmacology* 61 (1, 2): 43–50. Epub 2011/03/01.

123 Yang, L.V., Radu, C.G., Roy, M. et al. (2007). Vascular abnormalities in mice deficient for the G protein-coupled receptor GPR4 that functions as a pH sensor. *Mol. Cell. Biol.* 27 (4): 1334–1347. Epub 2006/12/06.

124 Lobo, M.K., Cui, Y., Ostlund, S.B. et al. (2007). Genetic control of instrumental conditioning by striatopallidal neuron-specific S1P receptor Gpr6. *Nat. Neurosci.* 10 (11): 1395–1397. Epub 2007/10/16.

125 Bjursell, M., Gerdin, A.K., Jonsson, M. et al. (2006). G protein-coupled receptor 12 deficiency results in dyslipidemia and obesity in mice. *Biochem. Biophys. Res. Commun.* 348 (2): 359–366. Epub 2006/08/05.

126 Kim, S.V., Xiang, W.V., Kwak, C. et al. (2013). GPR15-mediated homing controls immune homeostasis in the large intestine mucosa. *Science* 340 (6139): 1456–1459. Epub 2013/05/11.

127 Brennan TJ, Moore M. and Matthews W. (2002) Transgenic mice containing GPCR5-1 gene disruptions. International Patent Number: WO2002079440A2. https://patentimages.storage.googleapis.com/1b/28/6c/e04e112ab48c32/WO2002079440A2.pdf.

128 Osborn, O., Oh, D.Y., McNelis, J. et al. (2012). G protein-coupled receptor 21 deletion improves insulin sensitivity in diet-induced obese mice. *J. Clin. Invest.* 122 (7): 2444–2453. Epub 2012/06/02.

129 Adams, J.W., Wang, J., Davis, J.R. et al. (2008). Myocardial expression, signaling, and function of GPR22: a protective role for an orphan G protein-coupled receptor. *Am. J. Physiol. Heart Circ. Physiol.* 295 (2): H509–H521. Epub 2008/06/10.

130 Zhang, L.L., Wang, J.J., Liu, Y. et al. (2011). GPR26-deficient mice display increased anxiety- and depression-like behaviors accompanied by reduced phosphorylated cyclic AMP responsive element-binding protein level in central amygdala. *Neuroscience* 196: 203–214. Epub 2011/09/20.

131 Ku, G.M., Pappalardo, Z., Luo, C.C. et al. (2012). An siRNA screen in pancreatic beta cells reveals a role for Gpr27 in insulin production. *PLoS Genet.* 8 (1): e1002449. Epub 2012/01/19.

132 Liebscher, I., Muller, U., Teupser, D. et al. (2011). Altered immune response in mice deficient for the G protein-coupled receptor GPR34. *J. Biol. Chem.* 286 (3): 2101–2110. Epub 2010/11/26.

133 Marazziti, D., Di Pietro, C., Mandillo, S. et al. (2011). Absence of the GPR37/PAEL receptor impairs striatal Akt and ERK2 phosphorylation, DeltaFosB expression, and conditioned place preference to amphetamine and cocaine. *FASEB J.* 25 (6): 2071–2081. Epub 2011/03/05.

134 Petersen, P.S., Jin, C., Madsen, A.N. et al. (2011). Deficiency of the GPR39 receptor is associated with obesity and altered adipocyte metabolism. *FASEB J.* 25 (11): 3803–3814. Epub 2011/07/26.

135 Jin, C., Yin, F., Lin, M. et al. (2008). GPR48 regulates epithelial cell proliferation and migration by activating EGFR during eyelid development. *Invest. Ophthalmol. Vis. Sci.* 49 (10): 4245–4253. Epub 2008/05/20.

136 Luo, J., Zhou, W., Zhou, X. et al. (2009). Regulation of bone formation and remodeling by G-protein-coupled receptor 48. *Development* 136 (16): 2747–2756. Epub 2009/07/17.

137 Song, H., Luo, J., Luo, W. et al. (2008). Inactivation of G-protein-coupled receptor 48 (Gpr48/Lgr4) impairs definitive erythropoiesis at midgestation through down-regulation of the ATF4 signaling pathway. *J. Biol. Chem.* 283 (52): 36687–36697. Epub 2008/10/29.

138 Weng, J., Luo, J., Cheng, X. et al. (2008). Deletion of G protein-coupled receptor 48 leads to ocular anterior segment dysgenesis (ASD) through down-regulation of Pitx2. *Proc. Natl. Acad. Sci. U.S.A.* 105 (16): 6081–6086. Epub 2008/04/22.

139 Bechtold, D.A., Sidibe, A., Saer, B.R. et al. (2012). A role for the melatonin-related receptor GPR50 in leptin signaling, adaptive thermogenesis, and torpor. *Curr. Biol.* 22 (1): 70–77. Epub 2011/12/27.

140 Komatsu, H., Maruyama, M., Yao, S. et al. (2014). Anatomical transcriptome of G protein-coupled receptors leads to the identification of a novel therapeutic candidate GPR52 for psychiatric disorders. *PLoS One* 9 (2): e90134. Epub 2014/03/04.

141 Nambu, H., Fukushima, M., Hikichi, H. et al. (2011). Characterization of metabolic phenotypes of mice lacking GPR61, an orphan G-protein coupled receptor. *Life Sci.* 89 (21, 22): 765–772. Epub 2011/10/06.

142 Davies, B., Baumann, C., Kirchhoff, C. et al. (2004). Targeted deletion of the epididymal receptor HE6 results in fluid dysregulation and male infertility. *Mol. Cell. Biol.* 24 (19): 8642–8648. Epub 2004/09/16.

143 Engel, K.M., Schrock, K., Teupser, D. et al. (2011). Reduced food intake and body weight in mice deficient for the G protein-coupled receptor GPR82. *PLoS One* 6 (12): e29400. Epub 2012/01/05.

144 Muller, T.D., Muller, A., Yi, C.X. et al. (2013). The orphan receptor Gpr83 regulates systemic energy metabolism via ghrelin-dependent and ghrelin-independent mechanisms. *Nat. Commun.* 4: 1968. Epub 2013/06/08.

145 Matsumoto, M., Straub, R.E., Marenco, S. et al. (2008). The evolutionarily conserved G protein-coupled receptor SREB2/GPR85 influences brain size, behavior, and vulnerability to schizophrenia. *Proc. Natl. Acad. Sci. U.S.A.* 105 (16): 6133–6138. Epub 2008/04/17.

146 Chen, Q., Kogan, J.H., Gross, A.K. et al. (2012). SREB2/GPR85, a schizophrenia risk factor, negatively regulates hippocampal adult neurogenesis and neurogenesis-dependent learning and memory. *Eur. J. Neurosci.* 36 (5): 2597–2608. Epub 2012/06/16.

147 Logue, S.F., Grauer, S.M., Paulsen, J. et al. (2009). The orphan GPCR, GPR88, modulates function of the striatal dopamine system: a possible therapeutic target for psychiatric disorders? *Mol. Cell. Neurosci.* 42 (4): 438–447. Epub 2009/10/03.

148 Ben Hamida, S., Mendonca-Netto, S., Arefin, T.M. et al. (2018). Increased alcohol seeking in mice lacking Gpr88 involves dysfunctional mesocorticolimbic networks. *Biol. Psychiatry* 84 (3): 202–212. Epub 2018/03/28.

149 Meirsman, A.C., Le Merrer, J., Pellissier, L.P. et al. (2016). Mice lacking GPR88 show motor deficit, improved spatial learning, and low anxiety reversed by delta opioid antagonist. *Biol. Psychiatry* 79 (11): 917–927. Epub 2015/07/21.

150 Quintana, A., Sanz, E., Wang, W. et al. (2012). Lack of GPR88 enhances medium spiny neuron activity and alters motor- and cue-dependent behaviors. *Nat. Neurosci.* 15 (11): 1547–1555. Epub 2012/10/16.

151 Rainwater, A., Sanz, E., Palmiter, R.D. et al. (2017). Striatal GPR88 modulates foraging efficiency. *J. Neurosci.* 37 (33): 7939–7947. Epub 2017/07/22.

152 Shi, J., Zhang, X., Wang, S. et al. (2016). Gpr97 is dispensable for metabolic syndrome but is involved in macrophage inflammation in high-fat diet-induced obesity in mice. *Sci. Rep.* 6: 24649. Epub 2016/04/20.

153 Fang, W., Wang, Z., Li, Q. et al. (2018). Gpr97 exacerbates AKI by mediating Sema3A signaling. *J. Am. Soc. Nephrol.* 29 (5): 1475–1489. Epub 2018/03/14.

154 Urano, T., Shiraki, M., Yagi, H. et al. (2012). GPR98/Gpr98 gene is involved in the regulation of human and mouse bone mineral density. *J. Clin. Endocrinol. Metab.* 97 (4): E565–E574. Epub 2012/03/16.

155 Kubo, F., Ariestanti, D.M., Oki, S. et al. (2019). Loss of the adhesion G-protein coupled receptor ADGRF5 in mice induces airway inflammation and the expression of CCL2 in lung endothelial cells. *Respir. Res.* 20 (1): 11. Epub 2019/01/19.

156 Kuhnert, F., Mancuso, M.R., Shamloo, A. et al. (2010). Essential regulation of CNS angiogenesis by the orphan G protein-coupled receptor GPR124. *Science* 330 (6006): 985–989. Epub 2010/11/13.

157 Cullen, M., Elzarrad, M.K., Seaman, S. et al. (2011). GPR124, an orphan G protein-coupled receptor, is required for CNS-specific vascularization and establishment of the blood-brain barrier. *Proc. Natl. Acad. Sci. U.S.A.* 108 (14): 5759–5764. Epub 2011/03/23.

158 Monk, K.R., Oshima, K., Jors, S. et al. (2011). Gpr126 is essential for peripheral nerve development and myelination in mammals. *Development* 138 (13): 2673–2680. Epub 2011/05/27.

159 Patra, C., van Amerongen, M.J., Ghosh, S. et al. (2013). Organ-specific function of adhesion G protein-coupled receptor GPR126 is domain-dependent. *Proc. Natl. Acad. Sci. U.S.A.* 110 (42): 16898–16903. Epub 2013/10/02.

160 Edson, M.A., Lin, Y.N., and Matzuk, M.M. (2010). Deletion of the novel oocyte-enriched gene, Gpr149, leads to increased fertility in mice. *Endocrinology* 151 (1): 358–368. Epub 2009/11/06.

161 Mukhopadhyay, S., Wen, X., Ratti, N. et al. (2013). The ciliary G-protein-coupled receptor Gpr161 negatively regulates the Sonic hedgehog pathway via cAMP signaling. *Cell* 152 (1, 2): 210–223. Epub 2013/01/22.

162 Cox, P.J., Pitcher, T., Trim, S.A. et al. (2008). The effect of deletion of the orphan G-protein coupled receptor (GPCR) gene MrgE on pain-like behaviours in mice. *Mol. Pain* 4: 2. Epub 2008/01/17.

163 Doi, M., Murai, I., Kunisue, S. et al. (2016). Gpr176 is a Gz-linked orphan G-protein-coupled receptor that sets the pace of circadian behaviour. *Nat. Commun.* 7: 10583. Epub 2016/02/18.

164 Pereira, J.P., Kelly, L.M., Xu, Y. et al. (2009). EBI2 mediates B cell segregation between the outer and centre follicle. *Nature* 460 (7259): 1122–1126. Epub 2009/07/15.

165 Okajima, D., Kudo, G., and Yokota, H. (2011). Antidepressant-like behavior in brain-specific angiogenesis inhibitor 2-deficient mice. *J. Physiol. Sci.* 61 (1): 47–54. Epub 2010/11/27.

166 Ravni, A., Qu, Y., Goffinet, A.M. et al. (2009). Planar cell polarity cadherin Celsr1 regulates skin hair patterning in the mouse. *J. Invest. Dermatol.* 129 (10): 2507–2509. Epub 2009/04/10.

167 Qu, Y., Glasco, D.M., Zhou, L. et al. (2010). Atypical cadherins Celsr1-3 differentially regulate migration of facial branchiomotor neurons in mice. *J. Neurosci.* 30 (28): 9392–9401. Epub 2010/07/16.

168 Sugimura, R., He, X.C., Venkatraman, A. et al. (2012). Noncanonical Wnt signaling maintains hematopoietic stem cells in the niche. *Cell* 150 (2): 351–365. Epub 2012/07/24.

169 Tissir, F., Qu, Y., Montcouquiol, M. et al. (2010). Lack of cadherins Celsr2 and Celsr3 impairs ependymal ciliogenesis, leading to fatal hydrocephalus. *Nat. Neurosci.* 13 (6): 700–707. Epub 2010/05/18.

170 Fenstermaker, A.G., Prasad, A.A., Bechara, A. et al. (2010). Wnt/planar cell polarity signaling controls the anterior–posterior organization of monoaminergic axons in the brainstem. *J. Neurosci.* 30 (47): 16053–16064. Epub 2010/11/26.

171 Ying, G., Wu, S., Hou, R. et al. (2009). The protocadherin gene Celsr3 is required for interneuron migration in the mouse forebrain. *Mol. Cell. Biol.* 29 (11): 3045–3061. Epub 2009/04/01.

172 Cortijo, C., Gouzi, M., Tissir, F. et al. (2012). Planar cell polarity controls pancreatic beta cell differentiation and glucose homeostasis. *Cell Rep.* 2 (6): 1593–1606. Epub 2012/11/28.

173 Lin, H.H., Faunce, D.E., Stacey, M. et al. (2005). The macrophage F4/80 receptor is required for the induction of antigen-specific efferent regulatory T cells in peripheral tolerance. *J. Exp. Med.* 201 (10): 1615–1625. Epub 2005/05/11.

174 Jacobson, S.G., Cideciyan, A.V., Aleman, T.S. et al. (2008). Usher syndromes due to MYO7A, PCDH15, USH2A or GPR98 mutations share retinal disease mechanism. *Hum. Mol. Genet.* 17 (15): 2405–2415. Epub 2008/05/09.

175 Tao, Q., Fujimoto, J., Men, T. et al. (2007). Identification of the retinoic acid-inducible Gprc5a as a new lung tumor suppressor gene. *J. Natl. Cancer Inst.* 99 (22): 1668–1682. Epub 2007/11/15.

176 Deng, J., Fujimoto, J., Ye, X.F. et al. (2010). Knockout of the tumor suppressor gene Gprc5a in mice leads to NF-κB activation in airway epithelium and promotes lung inflammation and tumorigenesis. *Cancer Prev. Res.* 3 (4): 424–437. Epub 2010/04/01.

177 Sano, T., Kim, Y.J., Oshima, E. et al. (2011). Comparative characterization of GPRC5B and GPRC5CLacZ knockin mice; behavioral abnormalities in GPRC5B-deficient mice. *Biochem. Biophys. Res. Commun.* 412 (3): 460–465. Epub 2011/08/16.

178 Kurabayashi, N., Nguyen, M.D., and Sanada, K. (2013). The G protein-coupled receptor GPRC5B contributes to neurogenesis in the developing mouse neocortex. *Development* 140 (21): 4335–4346. Epub 2013/10/04.

179 Kim, Y.J., Sano, T., Nabetani, T. et al. (2012). GPRC5B activates obesity-associated inflammatory signaling in adipocytes. *Sci. Signal.* 5 (251): ra85. Epub 2012/11/22.

180 Sano, T., Kohyama-Koganeya, A., Kinoshita, M.O. et al. (2018). Loss of GPRC5B impairs synapse formation of Purkinje cells with cerebellar nuclear neurons and disrupts cerebellar synaptic plasticity and motor learning. *Neurosci. Res.* 136: 33–47. Epub 2018/02/27.

181 Carvalho, J., Chennupati, R., Li, R. et al. (2020). Orphan G-protein-coupled receptor GPRC5B controls smooth muscle contractility and differentiation by inhibiting prostacyclin receptor signaling. *Circulation* 141 (14): 1168–1183. Epub 2020/01/17.

182 Rajkumar, P., Cha, B., Yin, J. et al. (2018). Identifying the localization and exploring a functional role for Gprc5c in the kidney. *FASEB J.* 32 (4): 2046–2059. Epub 2017/12/03.

183 Pi, M., Parrill, A.L., and Quarles, L.D. (2010). GPRC6A mediates the non-genomic effects of steroids. *J. Biol. Chem.* 285 (51): 39953–39964. Epub 2010/10/16.

184 Ahmad, R., Wojciech, S., and Jockers, R. (2015). Hunting for the function of orphan GPCRs – beyond the search for the endogenous ligand. *Br. J. Pharmacol.* 172 (13): 3212–3228. Epub 2014/09/19.

185 Dershem, R., Metpally, R.P.R., Jeffreys, K. et al. (2019). Rare-variant pathogenicity triage and inclusion of synonymous variants improves analysis of disease associations of orphan G protein-coupled receptors. *J. Biol. Chem.* 294 (48): 18109–18121. Epub 2019/10/20.

186 Nelson, G., Chandrashekar, J., Hoon, M.A. et al. (2002). An amino-acid taste receptor. *Nature* 416 (6877): 199–202. Epub 2002/03/15.

187 Nelson, G., Hoon, M.A., Chandrashekar, J. et al. (2001). Mammalian sweet taste receptors. *Cell* 106 (3): 381–390. Epub 2001/08/18.

188 Jones, K.A., Borowsky, B., Tamm, J.A. et al. (1998). $GABA_B$ receptors function as a heteromeric assembly of the subunits $GABA_BR1$ and $GABA_BR2$. *Nature* 396 (6712): 674–679. Epub 1999/01/01.

189 White, J.H., Wise, A., Main, M.J. et al. (1998). Heterodimerization is required for the formation of a functional $GABA_B$ receptor. *Nature* 396 (6712): 679–682. Epub 1999/01/01.

190 AbdAlla, S., Lother, H., el Massiery, A. et al. (2001). Increased AT_1 receptor heterodimers in preeclampsia mediate enhanced angiotensin II responsiveness. *Nat. Med.* 7 (9): 1003–1009. Epub 2001/09/05.

191 Ayoub, M.A., Couturier, C., Lucas-Meunier, E. et al. (2002). Monitoring of ligand-independent dimerization and ligand-induced conformational changes of melatonin receptors in living cells by bioluminescence resonance energy transfer. *J. Biol. Chem.* 277 (24): 21522–21528. Epub 2002/04/10.

192 Oishi, A., Cecon, E., and Jockers, R. (2018). Melatonin receptor signaling: impact of receptor oligomerization on receptor function. *Int. Rev. Cell Mol. Biol.* 338: 59–77. Epub 2018/04/28.

193 Levoye, A., Dam, J., Ayoub, M.A. et al. (2006). Do orphan G-protein-coupled receptors have ligand-independent functions? New insights from receptor heterodimers. *EMBO Rep.* 7 (11): 1094–1098. Epub 2006/11/02.

194 Dufourny, L., Levasseur, A., Migaud, M. et al. (2008). GPR50 is the mammalian ortholog of Mel1c: evidence of rapid evolution in mammals. *BMC Evol. Biol.* 8: 105. Epub 2008/04/11.

195 Levoye, A., Dam, J., Ayoub, M.A. et al. (2006). The orphan GPR50 receptor specifically inhibits MT_1 melatonin receptor function through heterodimerization. *EMBO J.* 25 (13): 3012–3023. Epub 2006/06/17.

196 Sidibe, A., Mullier, A., Chen, P. et al. (2010). Expression of the orphan GPR50 protein in rodent and human dorsomedial hypothalamus, tanycytes and median eminence. *J. Pineal Res.* 48 (3): 263–269. Epub 2010/03/10.

197 Alkufri, F., Shaag, A., Abu-Libdeh, B. et al. (2016). Deleterious mutation in *GPR88* is associated with chorea, speech delay, and learning disabilities. *Neurol. Genet.* 2 (3): e64. Epub 2016/04/29.

198 Del Zompo, M., Deleuze, J.F., Chillotti, C. et al. (2014). Association study in three different populations between the GPR88 gene and major psychoses. *Mol. Genet. Genomic Med.* 2 (2): 152–159. Epub 2014/04/02.

199 Laboute, T., Gandia, J., Pellissier, L.P. et al. (2020). The orphan receptor GPR88 blunts the signaling of opioid receptors and multiple striatal GPCRs. *eLife* 9: e50519. Epub 2020/02/01.

200 Maurice, P., Daulat, A.M., Turecek, R. et al. (2010). Molecular organization and dynamics of the melatonin MT_1 receptor/RGS20/G(i) protein complex reveal asymmetry of receptor dimers for RGS and G(i) coupling. *EMBO J.* 29 (21): 3646–3659. Epub 2010/09/23.

201 Orlandi, C., Posokhova, E., Masuho, I. et al. (2012). GPR158/179 regulate G protein signaling by controlling localization and activity of the RGS7 complexes. *J. Cell Biol.* 197 (6): 711–719. Epub 2012/06/13.

202 Orlandi, C., Omori, Y., Wang, Y. et al. (2018). Transsynaptic binding of orphan receptor GPR179 to dystroglycan–Pikachurin complex is essential for the synaptic organization of photoreceptors. *Cell Rep.* 25 (1): 130–45 e5. Epub 2018/10/04.

203 Wright, L.P. and Philips, M.R. (2006). Thematic review series: lipid posttranslational modifications. CAAX modification and membrane targeting of Ras. *J. Lipid Res.* 47 (5): 883–891. Epub 2006/03/18.

204 Cox, A.D., Der, C.J., and Philips, M.R. (2015). Targeting RAS membrane association: back to the future for anti-RAS drug discovery? *Clin. Cancer Res.* 21 (8): 1819–1827. Epub 2015/04/17.

205 Fehrenbacher, N., Tojal da Silva, I., Ramirez, C. et al. (2017). The G protein-coupled receptor GPR31 promotes membrane association of KRAS. *J. Cell Biol.* 216 (8): 2329–2338. Epub 2017/06/18.

206 Clement, N., Renault, N., Guillaume, J.L. et al. (2018). Importance of the second extracellular loop for melatonin MT_1 receptor function and absence of melatonin binding in GPR50. *Br. J. Pharmacol.* 175 (16): 3281–3297. Epub 2017/09/13.

207 Kern, A., Albarran-Zeckler, R., Walsh, H.E. et al. (2012). Apo-ghrelin receptor forms heteromers with DRD2 in hypothalamic neurons and is essential for anorexigenic effects of DRD2 agonism. *Neuron* 73 (2): 317–332. Epub 2012/01/31.

3

The Evolution of Our Understanding of "GPCRs"

Terry Kenakin

Department of Pharmacology, University of North Carolina School of Medicine, Chapel Hill, NC, USA

3.1 Introduction

Pharmacology emerged from physiology some 150–180 years ago when physiologists, who had used chemicals to probe physiological systems to explore how they functioned, became more interested in the chemical probes than in the systems. Pharmacology has always functioned in a mechanistic vacuum in that pharmacologists most often do not fully understand the systems they used to study drugs. This is because the technology available to probe the workings of cells and organs oftentimes is insufficient to provide detailed mechanistic knowledge of pharmacological systems. Given this, the law of parsimony was the standard providing the most simple mechanisms and explanations possible in lieu of detailed information. Within this context, the main pharmacological tool is the null method where the effects of a drug are observed in a system, an intervention is performed on the system (i.e. addition of an antagonist, etc.), and then the effects of the same drug examined again in the presence of the intervention. The assumption here is that any changes seen are due to the intervention since the procedure before and after the intervention is the same. Approaches based on this principle provided some of the most important tools in pharmacology for use in the discovery and classification of new drugs, i.e. the agonist "potency ratio," *vide infra*.

Taken in isolation, physiology appears complex and seemingly without order as the same chemical can have a diverse array of physiological effects. However, the concept of the "receptor" was introduced into pharmacology to provide order and allow a control point for pharmacologists and medicinal chemists to modify the desired therapeutic effect of drugs. Thus, medicinal chemists had a control point where modification of activity (receptor activity) could change a broad array of physiological effects. Receptors at that point in time were only a concept [1] and

GPCRs as Therapeutic Targets, Volume 1, First Edition. Edited by Annette Gilchrist.

not a biochemically defined entity, but they did serve as a starting point for the mathematical models devised in early pharmacological literature that were used to identify drug universal activity constants. This was, in turn, extremely important in view of the fact that drugs produce different effects in different systems while activating the same receptor. Under these circumstances, the observed effects in one test system could not be automatically applied to other systems as would be required to predict therapeutic value. Drugs are rarely if ever tested in the direct therapeutic system; therefore, a scale is needed to make predictions in all systems; pharmacological models provided those scales.

A very important component of the pharmacological systems needed to classify drugs is a means to show drug effect; until the 1980s, these were uniformly animal isolated tissues. These were used to induce a visible response with the assumption that the tissue simply amplifies this initial signal (defined by Stephenson as "stimulus" [2]) through a biochemical cascade referred to as the "stimulus-response" mechanism. The assumption with these types of systems is that the cells provide a uniform amplification of a uniform receptor stimulus; therefore, ratios of activity accurately reflected ratios of pharmacological effect at the receptor. This reasoning led to valuable pharmacological tools such as the agonist potency ratio (PR) which can quantify relative agonist activity in test systems with ratios that would then allow prediction of similar activity in all systems. It should be noted that before the mid-1980s, the assumption in this process was that stimulus–response cascades were monotonic (only one "y," tissue response, for every "x," drug concentration). This assumption is required for PR values to be accurate predictors of agonist effect in all tissues.

With regard to isolated tissue systems, the two obvious weaknesses were, first, the fact that animal receptors and cells are different from human receptors and cells and thus errors in activity translation would occur; and second, the processing of receptor stimulus by cellular stimulus–response systems is much more complex than previously appreciated. It will be seen that a perfect storm of ideas entered the field in the mid-to-late 1980s and upended this second assumption.

3.2 The "Perfect Storm" of Ideas Enters Pharmacology

In the late 1980s, two general ideas, already prevalent in other realms, began to infiltrate pharmacological literature. The first was that the receptor is not as simple as previously thought. Specifically, the parsimonious models previously put forward suggested that receptors could exist in an "inactive" and an "active" state and that agonism was initiated when a ligand converted a receptor inactive state to an active state through the process of conformational selection [3]. However, this soon was seen to not accommodate experimental observations as this simplistic scheme was not in agreement with protein biochemistry and molecular dynamic theories describing the conformation of proteins (including receptors). These minimal models describe an inactive ($[R_i]$) and active ($[R_a]$) receptor active state as unliganded species interacting with a single signaling protein to form a ternary complex species that mediates cellular response due to a ligand ($[A]$) [4]:

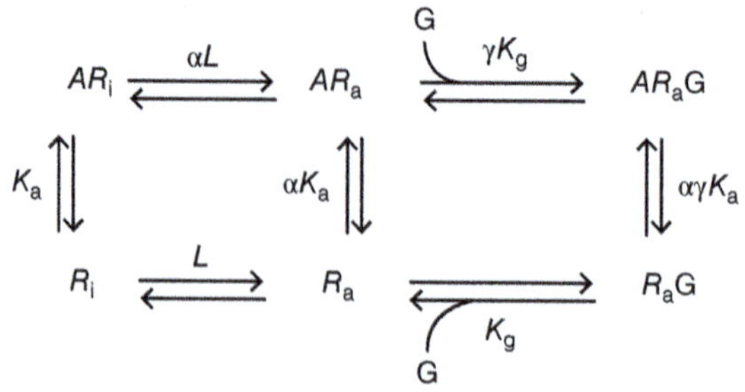

(3.1)

Ligand $[A]$ has an equilibrium association constant for the receptor–ligand complex of K_a for R_i and αK_a for R_a and the equilibrium between R_i and R_a is given by L, where $L = [R_a]/[R_i]$. The signaling protein (in this case denoted as G) has an equilibrium association constant for the unliganded receptor of K_g and for the ligand-bound receptor of γK_g. Response is produced by the activated receptor complex coupling to the signaling protein, and it appears from this scheme that the only two receptor species present appear to be the inactive state ($[R_i]$) and the active state ($[R_a]$), i.e. the system operates on the basis of only two receptor states, R_i and R_a. However, this is an illusion since the binding of an agonist to the receptor leads to a possibly infinite number of receptor states as predicted by variation of the γ term with different agonists. Thus, the value of γ determines agonism which is not specifically linked to R_a, but rather can be different for different agonists [5]. The seven transmembrane receptors are allosteric proteins which bind ligands and signaling proteins at different sites. Therefore, the principles of allosteric interaction govern receptor function. One of those principles is probe dependence whereby different bodies binding to the receptor can confer different effects, in this case, different agonists confer unique affinities of the receptor for G proteins; characterized by the value of γ. Given unique ligand-associated values of γ, it can be seen that although the initial condition of the system contains only two receptor species (R_i and R_a), upon agonist binding, a potentially infinite number of ARaG species can be formed.

The models of receptor function up to this time and including that shown in Figure 3.1 are limited also in that they must predefine the protein species present and also define a linear mechanism for the formation of new species. These are referred to as "linkage models" in that the various protein species are defined and then linked together in ways that conserve energy, i.e. transition between species

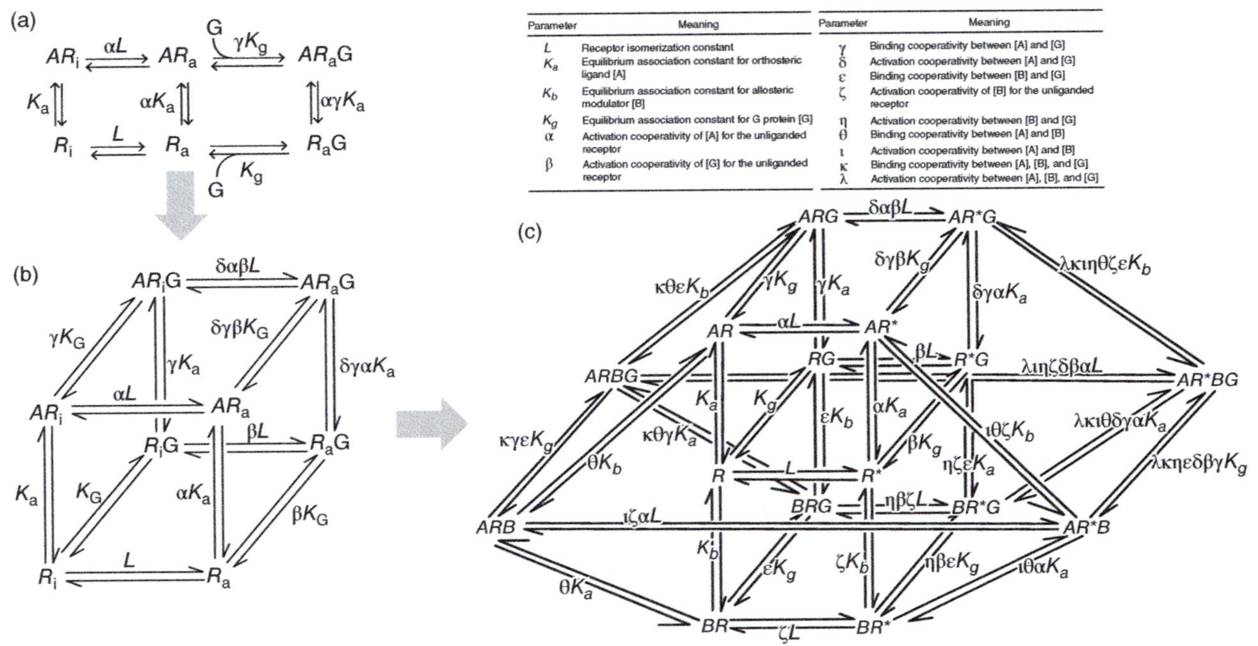

Parameter	Meaning	Parameter	Meaning
L	Receptor isomerization constant	γ	Binding cooperativity between [A] and [G]
K_a	Equilibrium association constant for orthosteric ligand [A]	δ	Activation cooperativity between [A] and [G]
K_b	Equilibrium association constant for allosteric modulator [B]	ε	Binding cooperativity between [B] and [G]
K_g	Equilibrium association constant for G protein [G]	ζ	Activation cooperativity of [B] for the unliganded receptor
α	Activation cooperativity of [A] for the unliganded receptor	η	Activation cooperativity between [B] and [G]
β	Activation cooperativity of [G] for the unliganded receptor	θ	Binding cooperativity between [A] and [B]
		ι	Activation cooperativity between [A] and [B]
		κ	Binding cooperativity between [A], [B], and [G]
		λ	Activation cooperativity between [A], [B], and [G]

Figure 3.1 Linkage models for receptor systems. (a) Extended ternary complex model for two state receptors interacting with G protein. Source: Redrawn from Samama et al. [4]. (b) Cubic ternary complex model allowing the inactive state to couple to the G protein as well. Source: Redrawn from Refs [6–8]. (c) Binding model for two receptor states interacting with G protein and an allosteric modulator. Source: Christopoulos and Kenakin [9]/ASPET.

by all routes are energy neutral. The linearity is imposed through requisite precursor species that must be created before a defined other species can be created. There a number of limitations to these types of models. For instance, if there are limited points of detection of change in the system, then the involvement of intermediate species cannot be verified. This was especially true for isolated tissue experiments where the only point of detection for changes in the system was tissue response. Therefore, if the phenomenon of receptor internalization was the point of the study, the only indication of receptor internalization was disappearance of tissue response. The assumption therefore was that receptor activation had to precede receptor internalization. The problems associated with this limitation became evident as more vantage points into the system became available. With the advent of imaging techniques, receptor internalization can be viewed directly, and with these tools, it can be seen that receptor activation does not need to precede receptor internalization, i.e. many receptor antagonists actively internalize receptors without producing response [10, 11]. These and other data dispelled any notion of linearity between efficacies, and with the introduction of multiple assays for different agonist activity came the realization that efficacy is collateral, i.e. a given ligand can create subsets of efficacies mediated by a given receptor.

A second shortcoming of linkage models is the need to predefine the various protein species in the system. This is a random process governed by parsimony in light of the unknown conformations possibly stabilized by signaling proteins and/or ligands. Figure 3.1 shows how linkage systems can quickly become unwieldy as more receptor behaviors are incorporated. Thus, Figure 3.1a shows the extended ternary complex model and defines a two state receptor system whereby one of the states binds to a G protein; Figure 3.1b shows the cubic ternary complex model [6–8], where both states can bind to the G protein; and Figure 3.1c [9] shows where all states bind to G proteins with the introduction of an allosteric modulator as well. The problem with such models, aside from an exponential increase in complexity, is the inability to verify microconstants within the model needed to fit data.

The need to predefine protein species was eliminated by the introduction of molecular dynamics into pharmacology. As described by Onaran and Costa [12, 13], the probability of ligands stabilizing subsets of an array of interchangeable and coexisting tertiary receptor conformations (termed a protein ensemble) formed the basis for agonism. This constituted one of the two major drivers for the pharmacological perfect storm as it paved the way toward ideas describing allosteric drug function and biased receptor signaling.

Molecular dynamics brings two important ideas to receptor pharmacology. The first is that receptors (in fact all proteins) can create a myriad of tertiary conformations of similar free energy and that ligands select these through selective affinities [14–19]. Thus, dynamic systems are formed from these collections of states (ensembles) [20–22] that combine with signaling systems through a full range of allosteric linkages [23, 24]. These ensembles have been described in terms of oscillating dynamic systems of multiple conformations [24, 25] giving rise to "fluctuating networks" operating on a real time scale of microseconds [26].

Agonism is thus described in terms of the probability of changing a receptor state; this is the discerning property of an agonist in terms of producing cellular response [12, 13]. The second important idea brought to receptor pharmacology by molecular dynamics is that the inherent linearity of the linkage models previously used to describe signaling is not required, but rather the existence of a given receptor active conformation is governed by a free movement of the receptor on an "energy landscape" [27–36]. Thus, there is no need to transition through one conformation to get to another – see Figure 3.2. Under these circumstances, agonist efficacy becomes "collateral," and the agonist-bound receptor can select distinct conformations with no regard to precursor conformations [37].

The other component of the perfect storm is the increased number of functional assays that have entered pharmacology; these now allow examination of the particular efficacy associated with each these multiple conformations. The availability

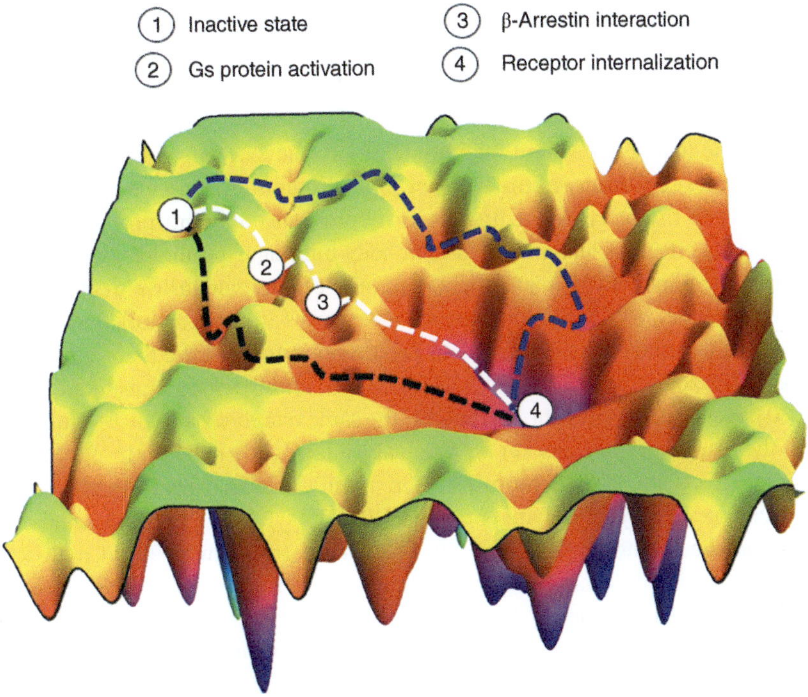

Figure 3.2 Energy landscape with some identified energy states associated with function. Paths show three alternative routes for the receptor arriving at receptor internalization only one of which involves receptor activation.

of multiple functional readouts of agonist response is the single most important driver for the discovery of biased signaling. One of the most important influences related to the entry of new assays in pharmacology is the change from an emphasis on binding studies to functional studies. Binding was adopted on a large scale in the 1970s because it is amenable to high-throughput screening and also was considered simple. This latter idea is a fallacy (note Figure 3.1c which actually is a binding assay scheme) as binding can be extremely complex. The main advantage to functional assays is that they give the ability to see cellular function, detect allosteric effects, and provide a wider range of drug-effect detection. With regard to this latter idea, radioligand binding detects drug effect that interferes with a single receptor probe (the radioligand), whereas a functional assay utilizes a much wider range of receptor probes in the form of the signaling proteins that interrogate the receptor conformation from the cell cytosol [38]. This change in emphasis also ushered in the capability to detect and characterize new types of chemical entities, namely allosteric modulators [39]. Further, diversity was created by the introduction of recombinant receptor systems whereby receptors could be transfected into host cells to be made to interact with a range of different signaling and accessory proteins. This diversity was caused by the allosteric nature of seven transmembrane receptors. Specifically, the provision of more cytosolic interactants for receptors resulted in systems that could readily display the variable behaviors possible for these proteins.

Just as the ligand-dependent magnitude of γ in the extended ternary complex model dictates the affinity of the receptor for G proteins, this same effect dictates the affinity of the receptor for ligands as it couples to different proteins in the cytosol. This effect is clearly seen in nanobody binding experiments, where it is seen that the nanobody αNb80 enhances affinity of norepinephrine 1000-fold for β_2-adrenoceptors, while the nanobody αNb60 reduces affinity 100-fold. This effect transcends to antagonists as seen with β_2-adrenoceptors for the β-adrenoceptor blockers carazolol, carvedilol, and ICI118,551, the affinity of which is reduced by αNb80 by respective factors of 3, 5, and 11; the same nanobody enhances the affinity of the β-adrenoceptor blocker alprenolol by a factor of 6.48 [40].

These effects are also demonstrated in natural physiological systems with the receptor activity modifying proteins (RAMPs). RAMPs produce different effects on the calcitonin/CGRP family of peptides (calcitonin, α and β CGRP, amylin, adrenomedullin, and adrenomedullin 2/intermedin) to confer selectivity of effect [41]. The impact of these effects pharmacologically is shown by the selectively high affinity for amylin by cellular co-expression of RAMP3; this produces completely different pharmacology compared to cells not containing RAMP3 leading to a selective change in antagonist potency of peptide antagonists such as AC66 51 [42]. Specifically, without RAMP3, AC66 has a K_B (equilibrium dissociation constant of the antagonist-receptor complex indicating affinity) for blockade of responses to amylin and calcitonin of 0.25 nM, but with co-expressed RAMP3, there is a selective sevenfold decrease in potency of AC66 for blockade of amylin responses (potency = 1.8 nM) and no concomitant change in potency for calcitonin responses [42].

In general, the perfect storm period in pharmacology (1989–2000) provided new viewpoints for drug response (new pharmacological assays, recombinant assays providing new pairings of receptor, and cytosolic proteins) and new ideas for linking experimental observations with protein (receptor) behaviors. This confluence provided the ideal atmosphere for the emergence of an important concept in receptor behavior to follow, namely biased receptor signaling.

3.3 Biased Receptor Signaling

Experimentation during this time provided data to support the notion that agonists stabilize specific tertiary pleiotropically coupled receptor conformations that go on to activate certain selected signaling pathways at the expense of others, a phenomenon given the name biased receptor signaling. This phenomenon was discovered through data that did not conform to the previous models of receptor behavior. Specifically, the prevailing view was that receptors produced stimulus to the cell which is then amplified to yield observable cell response through a biochemical cascade linked to the receptor and thus is presumably utilized by all agonists for that receptor. Receptor theory describes agonist potency as being the ratio of agonist affinity (expressed as the equilibrium dissociation constant of the agonist-receptor complex denoted by K_A) and efficacy (the ability of the agonist to induce an active state receptor related to an intrinsic property of the agonist and the sensitivity of the functional systems used to make the measurement) [43]. If ratios of potency are compared and the sensitivity of the functional system is assumed to be the same for both agonists, then the resulting potency ratios will be system independent measure of agonism based on affinity and efficacy. The fact that these ratios are viewed through a common amplifier (cellular stimulus–response cascade) should not affect the value of the ratio thus potency ratios were considered to be predictive and system independent measures of agonism that should not vary between tissues and with differences in cellular receptor density.

Beginning in the 1980s, literature reports began to cite instances where the constancy of potency ratios did not adhere to this scheme [44–53]. As multiple assays for different pleiotropic signaling pathways from receptors became available, pharmacologists could measure the potency ratio of agonists acting on the same receptor through two different stimulus–response readouts. It is at this point that clear differences began to emerge as the potency ratio for two agonists for one pathway was found to be different from the potency ratio of the same two agonists acting at the same receptor through a different signaling pathway. For instance, when cyclic AMP was measured from agonism at the PACAP receptor, it was found that the relative order of potency for the agonists $PACAP_{1-27}$ and $PACAP_{1-38}$ is $PACAP_{1-27} > PACAP_{1-38}$; however, when inositol phosphate metabolism was measured for the same two agonists via the same receptor, the relative order of potency reversed to $PACAP_{1-38} > PACAP_{1-27}$ [54]. These types of data could not be reconciled with a single stimulus–response mechanism processing the relative receptor signals. In 1989, a molecular mechanism for these effects was presented stating that agonists producing different receptor active states will produce differences in agonism when these difference states interact with different signaling mechanisms, i.e. signaling will be biased [55]. This mechanism was formally presented as the reason for biased signaling (denoted at that time as "stimulus trafficking") and variable potency ratios in 1995 [56], although numerous other groups published similar data under difference nomenclatures (functional dissociation [57], biased agonism [58], biased inhibition [59], differential engagement [60], functional selectivity [50, 61, 62], and ligand directed signaling [63]).

In this period, biased signaling (as the phenomenon has generally been termed) was treated as a unique function of receptors, but now it is recognized as simply being the outcome of standard allosteric behavior of receptor proteins. Specifically, the unique signaling to different cytosolic proteins by an agonist is actually a simple probe-dependent facilitation of the normal receptor-signaling protein interaction by a modulator (in this case, the agonist). Therefore, biased signaling is not in any way special, but rather reflects normal allosteric receptor behavior, i.e. the allosteric effect resulting in agonism is different for one probe-guest pair such as receptor-G proteins over another guest-receptor probe such as receptor/β-arrestin.

The phenomenon of biased signaling became established through many studies of single agonist-receptor pairs producing variant agonism in two signaling pathways, the most prevalent being G protein versus β-arrestin signaling. However, as the number of vantage points for viewing receptor signaling increased, it became clear that texture in agonism can be found along numerous steps in the cellular stimulus–response cascade. Considering a gradient of biochemical reactions beginning with the ligand–receptor interaction and ending with cellular response, it is a general rule that the farther away from the receptor binding step the signaling measurement is made, the more diverse the bias will be seen between agonists. For example, the simple *in vitro* bias profiles of parathyroid hormone (PTH) and the analog [D-Trp12,Tyr34]-bPTH(7–34) does not predict the differences in the *in vivo* transcriptomes produced by these agonists. This leads to the conclusion that a therapeutically useful prediction of bone mass effects of the biased agonist cannot be made from the simple two pathway analysis [64–66]. In fact, such complex arrays of responses in the form of transcriptome analysis from genetically modified animals have been proposed as being optimal to gauge the translation of biased signaling to the therapeutic arena [66]. For example, in bone remodeling studies in mice, the biased PTH receptor agonist [D-Trp12,Tyr34]-bPTH(7–34) produces a transcriptome with remarkably little similarity to the one produced by the natural agonist PTH, leading to the conclusion that the *in vivo* bone mass effects of the biased agonist could not have been predicted by its *in vitro* bias profile [64–69]. An even more effective use of whole animals and complex readouts of biased response is transcriptome analysis in control versus genetically modified animals [66].

There are two important messages emerging from the perfect storm period in pharmacology: (i) ligands acting on the same receptor can produce very different overall signaling effects (i.e. not all ligands are created equal) and (ii) the way these complex signaling profiles interact with cells makes translation of *in vitro* agonist data to *in vivo* systems much more variable than previously realized. The first point has opened up vast new areas in pharmacologic drug discovery as it becomes clear that new synthetic ligands need not simply be copies of natural neurotransmitters and hormones or molecules that prevent their binding. A wealth of new ideas entered the field as biased signaling was seen to be a means to (i) emphasize beneficial signaling pathways, i.e. PTH-mediated bone building for osteoporosis [64, 70]; (ii) de-emphasize harmful signaling pathways, i.e. respiratory depression for opioid analgesics [71–73]; (iii) de-emphasize harmful pathways and prevent the natural agonist from activating these pathways, i.e. biased angiotensin blockers for heart failure [74, 75]; and (iv) allow pursuit of previously forbidden drug targets due to side effects, i.e. κ-opioid receptor analgesics [76, 77]. This has, in turn, revitalized drug research focusing on seven transmembrane receptors [78].

Biased signaling is seen in natural systems which once appeared to be redundant (i.e. the multiple crossover of activities of chemokines among a limited number of chemokine receptors). Specifically, bias causes fine-tuning of receptor signaling as in the case of the chemokines CCL19 and CCL21. Both are natural agonists for the CCR7 receptor, but one (CCL21) activates only G proteins, while the other (CCL19) activates G proteins and receptor agonist-dependent phosphorylation and recruitment of β-arrestin to terminate the G protein stimulus [79–81]. There are different methods of visualizing these diverse signaling patterns such as radar plots [82, 83] (see Figure 3.3a) or statistically analyzed clusters from heatmaps [78, 86, 87] (see Figure 3.3b).

In addition to proactive approaches to planned bias for better drugs, a more exploratory application of bias has also entered drug discovery, namely determining bias in new molecules even when it is not clear if it will be beneficial or not. Thus, libraries are cross screened in multiple assays to identify and classify new molecular entities which are then carried forward into more complex assays to determine possible advantage. This approach is in keeping with the basic tenet

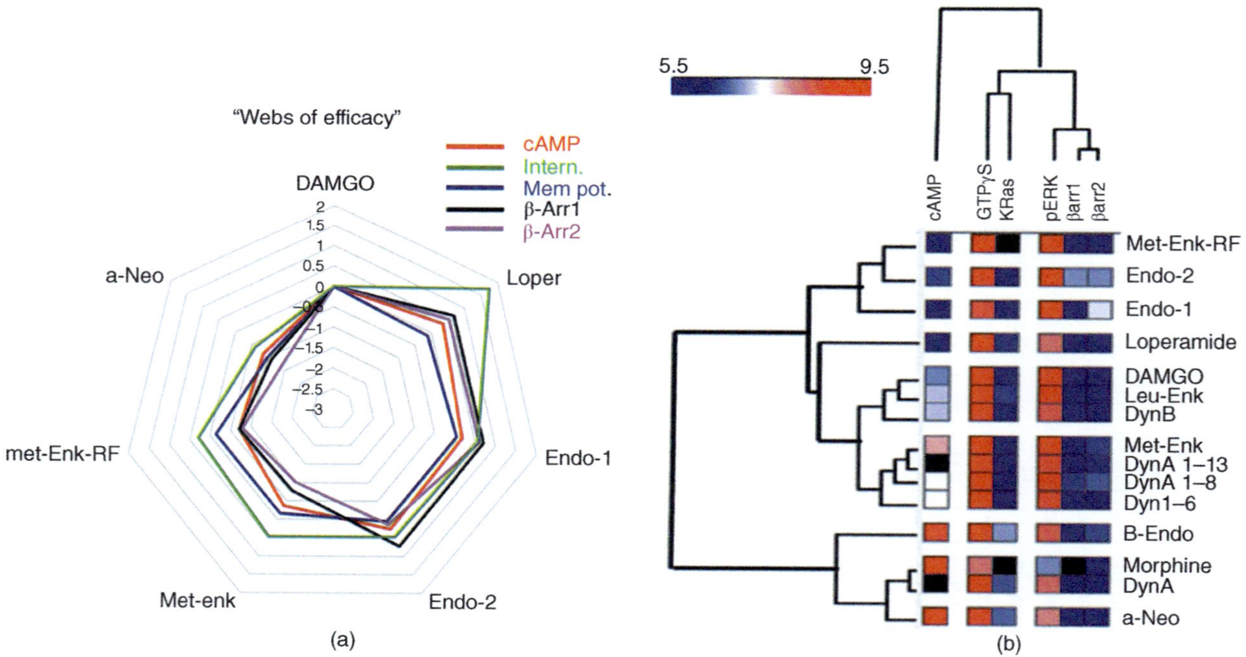

Figure 3.3 Two methods of depicting biased signaling in multiple ligands for multiple pathways. (a) Biased signaling for six opioid receptor agonists in five signaling pathways as expressed in a radar plot showing ΔLog(max/EC_{50}) values (with DAMGO as the reference agonist). Values >0 indicate bias toward that pathway, while values <0 indicate the reverse. Intern. = receptor internalization, Mem Pot. = membrane potential Source: Data taken from Thompson et al. [84] redrawn as radar plot. (b) Cluster analysis of 15 opioid agonists on the basis of signaling (Log(max/EC_{50}) values) in six functional assays. Source: Data from Thompson et al. [85]; analysis redrawn from Kenakin [78].

that complex assays detect subtle differences in activity to a greater extent than simple assays. Thus, molecules that appear to be similar, can actually be shown to be quite different in more complex systems (note the previous example with PTH and [D-Trp12,Tyr34]-bPTH(7–34)). In addition, the clustering similarities in ligand activity through signaling pathways may assist medicinal chemists in structure–activity relationships toward improved profiles (i.e. see Figure 3.3b).

The discovery of biased signaling also brought with it a realization that the previously conceived pharmacology of agonists whereby cells only amplify stimulus through a monotonic function is flawed. Specifically, agonists that produce different receptor active states will activate multiple signaling pathways in cells in different ways and the link between stimulus and response ceases to be monotonic (i.e. agonists may utilize different stimulus–response cascades in the same cell). Under these circumstances, agonist potency ratios become cell-dependent and fail to be universal predictors of relative agonist activity in all systems [88]. For example, the potency ratios for three agonists (porcine calcitonin [PCal], human calcitonin [hCal], human CGRP [hCGRP]) for the human calcitonin receptor expressed in COS cells is pCal > hCal > hCGRP with EC_{50} ratios of 1/11/64, while for the same agonists and the same receptor expressed in CHO cells, the ratios for pCal > hCal > hCGRP are $1/10^3/10^5$ [89].

In terms of drug discovery, the introduction of molecular dynamics, new assay technology, and recombinant systems also introduced new classes of chemical targets for potential drugs. Historically, most chemical targets were confined to copies of neurotransmitters or hormones as agonists or competitive blockers as antagonists, but with newly introduced assays and the appreciation of receptors as allosteric proteins, new types of molecules in the form of negative and positive allosteric modulators (NAMs and PAMs) entered the field. Since allosteric modulators bind to their own site on the receptor, their effects are saturable, and this allows them to modify natural signaling without completely converting the signal. Allosteric molecules work with endogenous molecules *in vivo* to modify natural response, unlike orthosteric ligands which impose their own efficacy on the system. In addition, allosteric effects are probe-dependent (a modulator may change the response to one agonist/radioligand without affected the response to another) to customize the resulting signaling. Such probe-dependence may be useful therapeutically as in the blockade of HIV-1 infection via CCR5 receptors. Specifically, there is evidence to suggest that the natural chemokine function of CCR5 is protective after HIV-1 infection with respect to progression to AIDS [90]; therefore, blockade of HIV-1/receptor interaction that allows otherwise normal receptor function toward natural chemokines, i.e. direct blockade of HIV-1/receptor binding and sparing of natural chemokine CCR5-mediated protection [91]. Such an effect is evident for the allosteric HIV-1 inhibitor TAK 652 [91].

An especially unique class of therapeutic molecules emerging from these approaches are potentiators of physiological response (positive allosteric modulators referred to as PAMs). These effects can be achieved through increasing the affinity of the receptor for the agonist (described as an α-effect) or through increasing the efficacy of the agonist (through what is alternately described as a β-effect) or both; PAMs can thus rejuvenate failing systems in disease. As well as modifying natural agonism, PAMs and NAMs can produce direct agonism. When this occurs, a broader range of modifying effects on natural agonism can occur (potentiation, no effect, or inhibition). This is in contrast to normal orthosteric partial agonist which block natural agonist response. An added advantage of allosteric ligands is that drugs for intractable receptors such as Family B peptide receptors which are notoriously resistant to binding small drug-like ligands can be found [92].

3.4 Continuing Influences

The last 15 years have revolutionized pharmacological drug discovery but new technology and innovations have continued to modify the landscape. Thus, the introduction of structural biology with the solving of receptor X-ray crystal structures has provided new approaches for medicinal chemistry to design new drugs. Once thought to be too flexible to allow crystal structures [93], receptor states have now been revealed to allow docking of real and virtual molecules for optimization of drug binding [94, 95]. This technology also has enabled virtual screening to provide an important starting point for new drug discovery and development. An example of this approach is the discovery of PAM ligand ogerin for the orphan receptor GPR68. Thus, 3307 homology models of the receptor were created which led to an intensive campaign of virtual docking of 3.1 million lead-like molecules to yield 3.3 trillion calculated complexes. This lead to the identification of ZINC67740571 (subsequently called ogerin [96]), a potent PAM of GPR68 hydrogen sensing activity.

In addition, the impact of the elucidated human genome has been immense providing recombinant assays, genetic knockout techniques (i.e. the reduction in morphine-mediated respiratory depression in β-arrestin knockout mice [71–73]) and the reduction in the bone-building effects of PTH in β-arrestin knockout mice [64, 70] and genetic knock-in approaches (i.e. phosphorylation-deficient muscarinic M_3 receptors have been used to study receptor internalization and phospholipase coupling [97–100] to allow the study of effect of various components of physiological systems). Genetically modified receptors that respond only to synthetic agonists, a technique pioneered through the production of Receptor Activated Solely by a Synthetic Ligands (RASSLs [101]) and then improved with the production of Designer Receptors Exclusively Activated by Designer Drugs

(DREADDs) [102–105] have been instrumental in determining the physiological significance of various receptor-signaling systems for therapeutic exploitation.

3.5 Conclusion

While pharmacologists still work in semidefined systems, tremendous advances have been in the understanding of the consequences of receptor activation and blockade have been made in recent years. These have defined new receptor modes of action and new pharmacological entities that are now being applied to therapeutic problems. As new advances build on these achievements, there is every reason to suppose that this trend will accelerate and hopefully result in increased molecules for the drug treatment of disease.

References

1 Rang, H.P. (2009). The receptor concept: pharmacology's big idea. *Br. J. Pharmacol.* 147: S9–S16.

2 Burgen, A.S. (1981). Conformational changes and drug action. *Fed. Proc.* 40: 2723–2728.

3 Stephenson, R.P. (1956). A modification of receptor theory. *Br. J. Pharmacol. Chemother.* 11: 379–393.

4 Samama, P., Cotecchia, S., Costa, T., and Lefkowitz, R.J. (1993). A mutation-induced activated state of the β_2-adrenergic receptor: extending the ternary complex model. *J. Biol. Chem.* 268: 4625–4636.

5 Kenakin, T.P. (2012). Biased signalling and allosteric machines: new vistas and challenges for drug discovery. *Br. J. Pharmacol.* 165: 1659–1166.

6 Weiss, J.M., Morgan, P.H., Lutz, M.W., and Kenakin, T.P. (1996). The cubic ternary complex receptor-occupancy model. I Model description. *J. Theroet. Biol.* 178: 151–167.

7 Weis, J.M., Morgan, P.H., Lutz, M.W., and Kenakin, T.P. (1996). The cubic ternary complex receptor-occupancy model. II. Understanding apparent affinity. *J. Theroet. Biol.* 178: 169–182.

8 Weiss, J.M., Morgan, P.H., Lutz, M.W., and Kenakin, T.P. (1996). The cubic ternary complex receptor-occupancy model. III. Resurrecting efficacy. *J. Theoret. Biol.* 181: 381–397.

9 Christopoulos, A. and Kenakin, T. (2002). G protein-coupled receptor allosterism and complexing. *Pharmacol. Rev.* 54: 323–374.

10 Roettger, B.F., Ghanekar, D., Rao, R. et al. (1997). Antagonist-stimulated internalization of the G protein-coupled cholecystokinin receptor. *Mol. Pharmacol.* 51: 357–362.

11 Willins, D.L., Berry, S.A., Alsayegh, L. et al. (1999). Clozapine and other 5-hydroxytryptamine-$_{2A}$ receptor antagonists alter the subcellular distribution of 5-hydroxytryptamine-$_{2A}$ receptors in vitro and in vivo. *Neuroscience* 91: 599–606.

12 Onaran, H.O. and Costa, T. (1997). Agonist efficacy and allosteric models of receptor action. *Ann. N.Y. Acad. Sci.* 812: 98–115.

13 Onaran, H.O., Scheer, A., Cotecchia, S., and Costa, T. (2000). A look at receptor efficacy. From the signaling network of the cell to the intramolecular motion of the receptor. In: *The Pharmacology of Functional, Biochemical, and Recombinant Systems*, Handbook of Experimental Pharmacology, vol. 148 (ed. T.P. Kenakin and J.A. Angus), 217–280. Heidelberg: Springer.

14 Park, P.S. (2012). Ensemble of G protein-coupled receptor active states. *Curr. Med. Chem.* 19: 1146–1154.

15 Nygaard, R., Zou, Y., Dror, R.O. et al. (2013). The dynamic process of β_2-adrenergic receptor activation. *Cell* 152: 532–542.

16 Motlagh, H.N., Wrabl, J.O., and Hilser, V.J. (2014). The ensemble nature of allostery. *Nature* 508: 331–339.

17 Boehr, D.D., Nussinov, R., and Wright, P.E. (2009). The role of dynamic conformational ensembles in biomolecular recognition. *Nat. Chem. Biol.* 5: 789–796.

18 Dror, R.O., Jensen, M.O., Borhani, D.W., and Shaw, D.E. (2010). Exploring atomic resolution physiology on a femtosecond to millisecond timescale using molecular dynamics simulations. *J. Gen. Physiol.* 135: 555–562.

19 Dror, R.O., Arlow, D.H., Shaw, D.E. et al. (2011). Activation mechanism of the β_2-adrenergic receptor. *Proc. Natl. Acad. Sci. U.S.A.* 108: 18684–18689.

20 Vardy, E. and Roth, B.L. (2013). Conformational ensembles in GPCR activation. *Cell* 152: 385–386.

21 Manglik, A. and Kobilka, B. (2014). The role of protein dynamics in GPCR function: insights from the β_2AR and rhodopsin. *Curr. Opin. Cell Biol.* 27: 136–143.

22 Manglik, A., Kim, T.H., Masureel, M. et al. (2015). Structural insights into the dynamic process of β_2 adrenergic receptor signaling. *Cell* 161: 1101–1111.

23 Monod, J., Wyman, J., and Changeux, J.-P. (1965). On the nature of allosteric transitions; a plausible model. *J. Mol. Biol.* 12: 88–118.

24 Changeux, J.P. and Edelstein, S. (2011). Conformational selection or induced fit? 50 years of debate resolved. *F1000 Biol. Rep.* 3: 19.

25 Cui, Q. and Karplus, M. (2008). Allostery and cooperativity revisited. *Protein Sci.* 17: 1295–1307.

26 Ichikawa, O., Fujimoto, K., Yamada, A. et al. (2016). G-protein/β-arrestin-linked fluctuating network of G-protein-coupled receptors for predicting drug efficacy and bias using short-term molecular dynamics simulation. *PLoS One* 11: e0155516.

27 Frauenfelder, H., Parak, F., and Young, R.D. (1988). Conformational substates in proteins. *Annu. Rev. Biophys. Biophys. Chem.* 17: 451–479.

28 Frauenfelder, H., Sligar, S.G., and Wolynes, P.G. (1991). The energy landscapes and motions of proteins. *Science* 254: 1598–1603.

29 Woodward, C. (1993). Is the slow exchange core the protein folding core? *Trends Biochem. Sci.* 18: 359–360.

30 Dill, K.A. and Chan, H.S. (1997). From Levinthal to pathways to funnels. *Nat. Struct. Biol.* 4: 10–19.

31 Hilser, J. and Freire, E. (1997). Predicting the equilibrium protein folding pathway: structure-based analysis of staphylococcal nuclease. *Protein Struct. Funct. Genet.* 27: 171–183.

32 Miller, D.W. and Dill, K.A. (1997). Ligand binding to proteins: the binding landscape model. *Protein Sci.* 6: 2166–2179.

33 Hilser, V.J., Dowdy, D., Oas, T.G., and Freire, E. (1998). The structural distribution of cooperative interactions in proteins: analysis of the native state ensemble. *Proc. Natl. Acad. Sci. U.S.A.* 95: 9903–9908.

34 Hilser, V.J., García-Moreno, E.B., Oas, T.G. et al. (2006). A statistical thermodynamic model of the protein ensemble. *Chem. Rev.* 106: 1545–1558.

35 Freire, E. (1998). Statistical thermodynamic linkage between conformational and binding equilibria. *Adv. Protein Chem.* 51: 255–279.

36 Ma, B., Shatsky, M., Wolfson, H.J., and Nussinov, R. (2002). Multiple diverse ligands binding at a single protein site: a matter of pre-existing conformations. *Protein Sci.* 11: 184–197.

37 Kenakin, T. (2002). Drug efficacy at G protein-coupled receptors. *Annu. Rev. Pharmacol. Toxicol.* 42: 349–379.

38 Kenakin, T.P. (2009). Cellular assays as portals to seven-transmembrane receptor-based drug discovery. *Nature Rev. Drug Disc.* 8: 617–626.

39 Rees, S., Morrow, D., and Kenakin, T. (2002). GPCR drug discovery through the exploitation of allosteric drug binding sites. *Recept. Channels* 8: 261–268.

40 Staus, D.P., Strachan, R.T., Manglik, A. et al. (2016). Allosteric nanobodies reveal the dynamic range and diverse mechanisms of G-protein-coupled receptor activation. *Nature* 535: 448–452.

41 Hay, D.L., Garelja, M.L., Poyner, D.R., and Walker, C.S. (2018). Update on the pharmacology of calcitonin/CGRP family of peptides: IUPHAR review. *Br. J. Pharmacol.* 175: 3–17.

42 Armour, S.L., Foord, S., Kenakin, T., and Chen, W.J. (1999). Pharmacological characterization of receptor-activity-modifying proteins (RAMPs) and the human calcitonin receptor. *J. Pharmacol. Toxicol. Methods* 42: 217–224.

43 Black, J.W., Leff, P., Shankley, N.P., and Wood, J. (1985). An operational model of pharmacological agonism: the effect of E/[A] curve shape on agonist dissociation constant estimation. *Br. J. Pharmacol.* 84: 561–571.

44 Roth, B.L. and Chuang, D.M. (1987). Multiple mechanisms of serotonergic signal transduction. *Life Sci.* 41: 1051–1064.

45 Mottola, D.M., Cook, L.L., Jones, S.R. et al. (1991). Dihydrexidine, a selective dopamine receptor agonist that may discriminate postsynaptic D_2 receptors. *Soc. Neurosci. Abstr.* 17: 818.

46 Roerig, S.C., Loh, H.H., and Law, P.Y. (1992). Identification of three separate guanine nucleotide- binding proteins that interact with the δ-opioid receptor in NG108-15 X glioma hybrid cells. *Mol. Pharmacol.* 41: 822–831.

47 Fisher, A., Heldman, E., Gurwitz, D. et al. (1993). Selective signaling via unique M_1 muscarinic agonists. *Ann. N.Y. Acad. Sci.* 695: 300–303.

48 Gurwitz, D., Haring, R., Heldman, E. et al. (1994). Discrete activation of transduction pathways associated with acetylcholine M_1 receptor by several muscarinic ligands. *Eur. J. Pharmacol.* 267: 21–31.

49 Lawler, C.O., Watts, V.J., Booth, R.G. et al. (1994). Discrete functional selectivity of drugs: OPC-14597, a selective antagonist for post-synaptic dopamine D_2 receptors (Abstract). *Soc. Neurosci. Abstr.* 20: 525.

50 Lawler, C.P., Prioleau, C., Lewis, M.M. et al. (1999). Interactions of the novel antipsychotic aripiprazole (OPC-14597) with dopamine and serotonin receptor subtypes. *Neuropsychopharmacology* 20: 612–627.

51 Ward, J.S., Merrit, L., Calligaro, D.O. et al. (1995). Functionally selective M_1 muscarinic agonists. 3. Side chains and azacycles contributing to functional muscarinic selectivity among pyrazacycles. *J. Med. Chem.* 38: 3469–3481.

52 Heldman, E., Barg, J., Fisher, A. et al. (1996). Pharmacological basis for functional selectivity of partial muscarinic receptor agonists. *Eur. J. Pharmacol.* 297: 283–291.

53 Mailman, R.B., Lawler, C.P., Lewis, M.M., and Blake, B. (1998). Functional effects of novel dopamine ligands: dihydrexidine and Parkinson's disease as a first step. In: *Dopamine Receptor Subtypes: From Basic Science to Clinical Application* (ed. P. Jenner and R. Demirdamar), 112–119. Fairfox, CA: IOS Press.

54 Spengler, D., Waeber, C., Pantaloni, C. et al. (1993). Differential signal transduction by five splice variants of the PACAP receptor. *Nature* 365: 170–175.

55 Kenakin, T.P. and Morgan, P.H. (1989). Theoretical effects of single and multiple transducer receptor coupling proteins on estimates of the relative potency of agonists. *Mol. Pharmacol.* 35: 214–222.

56 Kenakin, T. (1995). Agonist-receptor efficacy. II. Agonist trafficking of receptor signals. *Trends Pharmacol. Sci.* 16: 232–238.

57 Whistler, J.L., Enquist, J., Marley, A. et al. (2002). Modulation of postendocytic sorting of G protein-coupled receptors. *Science* 297: 615–620.

58 Jarpe, M.B., Knall, C., Mitchell, F.M. et al. (1998). [D-Arg1,D-Phe5,D-Trp7,9,Leu11]Substance P acts as a biased agonist toward neuropeptide and chemokine receptors. *J. Biol. Chem* 273: 3097–3104.

59 Kudlacek, O., Waldhoer, M., Kassack, M.U. et al. (2002). Biased inhibition by a suramin analogue of A_1-adenosine receptor/G protein coupling in fused receptor/G protein tandems: the A_1-adenosine receptor is predominantly coupled to Go_{α} in human brain. *Naunyn Schmiedebergs Arch. Pharmacol.* 365: 8–16.

60 Manning, D.R. (2002). Measures of efficacy using G proteins as endpoints: differential engagement of G proteins through single receptors. *Mol. Pharmacol.* 62: 451–452.

61 Kilts, J.D., Connery, H.S., Arrington, E.G. et al. (2002). Functional selectivity of dopamine receptor agonists. II. Actions of dihydrexidine in D_{2L} receptor-transfected MN9D cells and pituitary lactotrophs. *J. Pharmacol. Exp. Ther.* 301: 1179–1189.

62 Shapiro, D.A., Renock, S., Arrington, E. et al. (2003). Aripiprazole, a novel atypical antipsychotic drug with a unique and robust pharmacology. *Neuropsychopharmacology* 28: 1400–1411.

63 Michel, M.C. and Alewijnse, A.E. (2007). Ligand-directed signaling: 50 ways to find a lover. *Mol. Pharmacol.* 72: 1097–1099.

64 Gesty-Palmer, D., Yuan, L., Martin, B. et al. (2013). β-Arrestin-selective G protein coupled receptor agonists engender unique biological efficacy in vivo. *Mol. Endocrinol.* 27: 296–314.

65 Appleton, K.M., Lee, M.-H., Alele, C. et al. (2013). Biasing the parathyroid hormone receptor: relating in vitro ligand efficacy to in vivo biological activity. *Methods Enzymol.* 522: 230–259.

66 Luttrell, L.M., Maudsley, S., and Gesty-Palmer, D. (2018). Translating in vitro ligand bias into in vivo efficacy. *Cell. Signalling* 41: 46–55.

67 Maudsley, S., Martin, B., Gesty-Palmer, D. et al. (2015). Delineation of a conserved arrestin-biased signaling repertoire in vivo. *Mol. Pharmacol.* 87: 706–717.

68 Maudsley, S., Martin, B., Janssens, J. et al. (2016). Informatic deconvolution of biased GPCR signaling mechanisms from in vivo pharmacological experimentation. *Methods* 92: 51–63.

69 Bradley, S.J. and Tobin, A.B. (2016). Design of next-generation G protein-coupled receptor drugs: linking novel pharmacology and in vivo animal models. *Annu. Rev. Pharmacol. Toxicol.* 56: 535–559.

70 Gesty-Palmer, D. and Luttrell, L.M. (2011). 'Biasing' the parathyroid hormone receptor: a novel anabolic approach to increasing bone mass? *Br. J. Pharmacol.* 164: 59–67.

71 Raehal, K.M., Walker, J.K., and Bohn, L.M. (2005). Morphine side effects in β-arrestin 2 knockout mice. *J. Pharmacol. Exp. Ther.* 314: 1195–1201.

72 Kelly, E. (2013). Efficacy and ligand bias at the μ-opioid receptor. *Br. J. Pharmacol.* 169: 1430–1446.

73 Koblish, M., Carr, R. III, Siuda, E.R. et al. (2017). TRV0109101, a G protein-biased agonist of the m-opioid receptor, does not promote opioid-induced mechanical allodynia following chronic administration. *J. Pharmacol. Exp. Ther.* 362: 254–262.

74 Violin, J.D., DeWire, S.M., Barnes, W.G., and Lefkowitz, R.J. (2006). G protein-coupled receptor kinase and β-arrestin-mediated desensitization of the angiotensin II type 1A receptor elucidated by diacylglycerol dynamics. *J. Biol. Chem.* 281: 36411–36419.

75 Violin, J.D., DeWire, S.M., Yamashita, D. et al. (2010). Selectively engaging β-arrestins at the angiotensin II type 1 receptor reduces blood pressure and increases cardiac performance. *J. Pharmacol. Exp. Ther.* 335: 572–579.

76 White, K.L., Scopton, A.P., Rives, M.L. et al. (2014). Identification of novel functionally selective κ-opioid receptor scaffolds. *Mol. Pharmacol.* 85: 83–90.

77 Brust, T.F., Morgenweck, J., Kim, S.A. et al. (2016). Biased agonists of the kappa opioid receptor suppress pain and itch without causing sedation or dysphoria. *Sci. Signal.* 9: ra11.

78 Kenakin, T. (2015). New lives for seven transmembrane receptors as drug targets. *Trends Pharmacol. Sci.* 36: 705–706.

79 Kohout, T.A., Nicholas, S.L., Perry, S.J. et al. (2004). Differential desensitization, receptor phosphorylation, β-arrestin recruitment, and ERK1/2 activation by the two endogenous ligands for the CC chemokine receptor 7. *J. Biol. Chem.* 279: 23214–23222.

80 Byers, M.A., Calloway, P.A., Shannon, L. et al. (2008). Arrestin 3 mediates endocytosis of CCR7 following ligation of CCL19 but not CCL21. *J. Immunol.* 181: 4723–4732.

81 Hauser, M.A. and Legler, D.F. (2016). Common and biased signaling pathways of the chemokine receptor CCR7 elicited by its ligands CCL19 and CCL21 in leukocytes. *J. Leukocyte Biol.* 99: 869–882.

82 Evans, B.A., Sato, M., Sarwar, M. et al. (2010). Ligand-directed signalling at β-adrenoceptors. *Br. J. Pharmacol.* 159: 1022–1038.

83 Zhou, L., Lovell, K.M., Frankowski, K.J. et al. (2013). Development of functionally selective, small molecule agonists at κ opioid receptors. *J. Biol. Chem.* 288: 36703–36716.

84 Thompson, G.L., Lane, J.R., Coudrat, T. et al. (2015). Biased agonism of endogenous opioid peptides at the μ-opioid receptor. *Mol. Pharmacol.* 88: 335–346.

85 Thompson, G.L., Lane, J.R., Coudrat, T. et al. (2016). Systematic analysis of factors influencing observations of biased agonism at the μ-opioid receptor. *Biochem. Pharmacol.* 113: 70–87.

86 Huang, X.P., Setola, V., Yadav, P.N. et al. (2009). Parallel functional activity profiling reveals valvulopathogens are potent 5-hydroxytryptamine$_{2B}$ receptor agonists: implications for drug safety assessment. *Mol. Pharmacol.* 76: 710–722.

87 Namkung, Y., LeGouill, C., Kumar, S. et al. (2018). Functional selectivity profiling of the angiotensin II type 1 receptor using pathway-wide BRET signaling sensors. *Sci. Signal.* 11 (559): eaat1631.

88 Kenakin, T.P. (2016). Synoptic pharmacology: Detecting and assessing the pharmacological significance of ligands for orphan receptors. *Pharmacol. Res.* 114: 284–290.

89 Christmanson, L., Westermark, P., and Betsholtz, C. (1994). Islet amyloid polypeptide stimulates cyclic AMP accumulation via the porcine calcitonin receptor. *Biochem. Biophys. Res. Commun.* 205: 1226–1235.

90 Gonzalez, E., Kulkarni, H., Bolivar, H. et al. (2005). The influence of CCL3L1 gene-containing segmental duplications on HIV-1/AIDS susceptibility. *Science* 307: 1434–1440.

91 Muniz-Medina, V.M., Jones, S., Maglich, J.M. et al. (2009). The relative activity of "function sparing" HIV-1 entry inhibitors on viral entry and CCR internalization: is allosteric functional selectivity a valuable therapeutic property? *Mol. Pharmacol.* 75: 490–501.

92 Kenakin, T.P. (2010). Ligand detection in the allosteric world. *J. Biomol. Screening* 15 (2): 119–130.

93 Liapakis, G., Cordomi, A., and Pardo, L. (2012). The G protein-coupled receptor family: actors with many faces. *Curr. Pharm. Des.* 18: 175–185.

94 Hauser, A.S., Attwood, M.M., Rask-Andersen, M. et al. (2017). Trends in GPCR drug discovery: new agents, targets and indications. *Nature Rev. Drug Disc.* 16: 829–842.

95 Wang, S., Wacker, D., Levit, A. et al. (2017). D_4 dopamine receptor high-resolution structures enable the discovery of selective agonists. *Science* 358: 381–386.

96 Huang, X.P., Karpiak, J., Kroeze, W.K. et al. (2015). Allosteric ligands for the pharmacologically dark receptors GPR68 and GPR65. *Nature* 527: 477–483.

97 Kong, K.C., Butcher, A.J., McWilliams, P. et al. (2010). M_3-muscarinic receptor promotes insulin release via receptor phosphorylation/arrestin-dependent activation of protein kinase D_1. *Proc. Natl. Acad. Sci. U.S.A.* 107: 21181–21186.

98 Poulin, B., Butcher, A., McWilliams, P. et al. (2010). The M_3-muscarinic receptor regulates learning and memory in a receptor phosphorylation/arrestin-dependent manner. *Proc. Natl. Acad. Sci. U.S.A.* 107: 9440–9445.

99 Torrecilla, I., Spragg, F.J., Poulin, B. et al. (2007). Phosphorylation and regulation of a G protein-coupled receptor by protein kinase CK2. *J. Cell Biol.* 177: 127–137.

100 Budd, D.C., Willars, G.B., McDonald, J.E., and Tobin, A.B. (2001). Phosphorylation of the $G_{q/11}$-coupled M_3 muscarinic receptor is involved in receptor activation of the ERK1/2 mitogen-activated protein kinase pathway. *J. Biol. Chem.* 276: 4581–4587.

101 Coward, P., Wada, H.G., Falk, M.S. et al. (1998). Controlling signaling with a specifically designed G_i-coupled receptor in transgenic mice. *Nat. Biotechnol.* 17: 165–169.

102 Urban, D.J. and Roth, B.L. (2015). DREADDs (designer receptors exclusively activated by designer drugs): chemogenetic tools with therapeutic utility. *Annu. Rev. Pharmacol. Toxicol.* 55: 399–417.

103 Conklin, B.R., Hsiao, E.C., Claeysen, S. et al. (2008). Engineering GPCR signaling pathways with RASSLs. *Nat. Methods* 5: 673–667.

104 Armbruster, B.N., Li, X., Pausch, M.H. et al. (2007). Evolving the lock to fit the key to create a family of G protein–coupled receptors potently activated by an inert ligand. *Proc. Natl. Acad. Sci. U.S.A.* 104: 5163–5166.

105 Giguere, P.M., Kroeze, W.K., and Roth, B.L. (2014). Tuning up the right signal: chemical and genetic approaches to study GPCR function. *Curr. Opin. Cell Biol.* 27: 51–55.

4

Approaching GPCR Dimers and Higher-Order Oligomers Using Biophysical, Biochemical, and Proteomic Methods

Kyla Bourque, Elyssa Frohlich, and Terence E. Hébert

Department of Pharmacology and Therapeutics, McGill University, Montréal, Québec, Canada

4.1 Introduction

G protein-coupled receptors (GPCRs) comprise a superfamily of seven transmembrane domain-containing receptors whose signal transduction pathways drive distinct biochemical events via activation of multiple effectors such as heterotrimeric G proteins or β-arrestins [1, 2]. Ligand occupancy at the orthosteric site triggers cellular responses that can ultimately integrate at the level of gene transcription, modulating biological processes such as cell proliferation, differentiation, and survival. Accordingly, GPCR signaling is tightly interconnected with the homeostasis and support of numerous organs in the circulatory, respiratory, nervous, and immune systems. Perhaps unsurprisingly, dysregulated GPCR signaling leads to various human diseases. In this regard, GPCRs are considered as attractive drug targets and it is estimated that approximately one-third of all current pharmaceuticals target this receptor family. However, GPCR biology and attendant pharmacology are recognized as increasingly complex and our incomplete understanding of canonical versus disease-relevant signal transduction mechanisms impedes development of new GPCR-targeting drugs.

Evolving past the simplistic "lock and key," or "on–off" mode of activation, GPCRs can be allosterically controlled by a plethora of different modulators binding at sites distinct from the canonical orthosteric binding pocket [3–5]. Occupancy of the orthosteric site by different ligands can also result in biased signaling, resulting in distinct receptor conformations and further parsing GPCR activation and function into distinct outcomes [3, 6, 7]. Furthermore, as membrane-bound proteins, GPCRs are highly accessible to the extracellular and intracellular spaces and can engage in numerous protein–protein interactions within the plane of the lipid bilayer further influencing receptor activity [8]. The

GPCRs as Therapeutic Targets, Volume 1, First Edition. Edited by Annette Gilchrist.

molecular complexities associated with GPCR signal transduction combined with the numerous regulatory mechanisms operating on them make these receptors challenging drug targets. Class A or rhodopsin-like GPCRs are particularly interesting pharmacological targets as their activity can be modulated not only by the aforementioned mechanisms but also by their ability to function as monomers, homodimers, heterodimers, and higher oligomeric complexes. Multiple GPCRs can thus engage in intermolecular interactions creating distinct signaling entities. These interactions can result in a number of functional consequences, altering GPCR trafficking, internalization, G protein, and effector coupling and signaling [9–12]. These multimeric receptor complexes can thus be functionally unique from their respective receptor monomers and can display distinct pharmacological properties.

Despite the fact that distinct GPCR dimers can have very different lifetimes, larger metastable GPCR complexes are increasingly important potential therapeutic targets for pharmacological regulation as they play important roles in disease progression. For instance, the angiotensin II type I receptor (AT_1R) can form functional heterodimers with the bradykinin type 2 receptor (B_2R) and has been suggested to be involved in the pathogenesis of preeclampsia [13, 14]. The AT_1R/B_2R heterodimer is thought to cause harm by coupling to $G\alpha_q$ leading to angiotensin hypersensitivity and aberrant calcium signaling, worsening preeclampsia symptoms [14, 15]. In this case, downregulating or disrupting the heterodimer could have therapeutic benefit [14]. In contrast, the occurrence of dimeric species is believed to be protective in HIV since the Δ32 mutation in the C–C chemokine receptor type 5 or CCR5 encodes a nonfunctional receptor unable to be expressed at the cell surface thus reducing viral entry into the cell. Individuals heterozygous for the mutation are thought to be protected as the Δ32/CCR5 dimer prevents the receptor complex from reaching the cell surface where it is instead retained in the endoplasmic reticulum [16]. On the other hand, there are other important dimer pairs that could be taken advantage of in the pathogenesis of Parkinson's disease such as the D_1/D_3 dopamine dimer or the adenosine $A_{2A}R/D_2R$ dimer. Since agonist occupancy at the $A_{2A}R$ reduces D_2R signaling, it has been speculated that $A_{2A}R$ antagonists may help alleviate Parkinson's disease (PD) symptoms by increasing D_2 dopamine signaling [17]. In addition, D_2R can also form homodimers and expression of this dimer has been reported to be elevated in postmortem tissue of patients suffering from schizophrenia or major depression, possibly representing a new target for these diseases [18, 19].

Despite substantial evidence pointing toward the existence of class A GPCRs dimers, and how they are differentially regulated in health and disease, they remain a challenging and controversial subject. Technical limitations hinder gathering of empirical evidence creating a debate in the field regarding their relevance *in vivo* as most studies are conducted *in vitro* and relatively few in native tissues. Here, we review various methodologies that have been used to study GPCR dimers and oligomers and discuss how new technologies may help move the field forward.

4.2 Biochemical or Antibody-Based Methods to Study GPCR Dimers and Higher-Order Oligomers

4.2.1 Co-immunoprecipitation

When attempting to discern whether two proteins interact, co-immunoprecipitation experiments are commonly performed (see Figure 4.1a). Co-immunoprecipitation (co-IP) was used to demonstrate the interactions between numerous GPCR heterodimer partners such as the Prostanoid F and angiotensin II type I (FP/AT_1R) [21], bradykinin type 2 and thromboxane TP (B_2R/TP) [22], β_2-adrenergic and oxytocin (B_2AR/OTR) [23], dopamine D_1R/D_2R [24], and MT_1/MT_2 melatonin receptors [25] among many examples. These experiments require the solubilization of membrane-bound proteins, extracting them from their native lipid environment usually with the use of detergents. Putative dimer partners can be immunoprecipitated out of cell lysates using antibodies harboring affinity for an antigen-tagged or endogenous receptor. Since antibodies against native GPCRs are notoriously difficult to obtain, epitope tags such as FLAG or HA are often used [26]. If two GPCRs interact, the interacting protomer can subsequently be observed via SDS-PAGE. Sample preparation typically includes strong ionic detergents used to disrupt noncovalent interactions causing proteins or multimeric protein aggregates to dissociate into linearized polypeptide chains. This may disrupt GPCR complexes, yet membrane proteins may also aggregate when boiled in reducing and denaturing sample buffer causing artifacts to appear.

Epitope-tagged β_2AR was first shown to form homodimers by co-immunoprecipitation in SDS, under nonreducing conditions, and the interaction was shown to be disrupted using a peptide derived from the sixth transmembrane domain of the β_2AR [27]. Nonreducing conditions were key to preserving protein–protein interactions as the addition of reducing agents could disrupt covalent disulfide bonds. Careful consideration must be taken when designing and analyzing co-immunoprecipitation experiments as oligomers may or may not be observed based on the biochemical properties of the reagents used [28]. In some cases, native or nondenaturing gels should be used. The absence of an interaction can also be a consequence of the transient nature of interactions between protomers. All things considered, interactions revealed by immunoprecipitation are considered indirect and should be confirmed using other methods. Ironically,

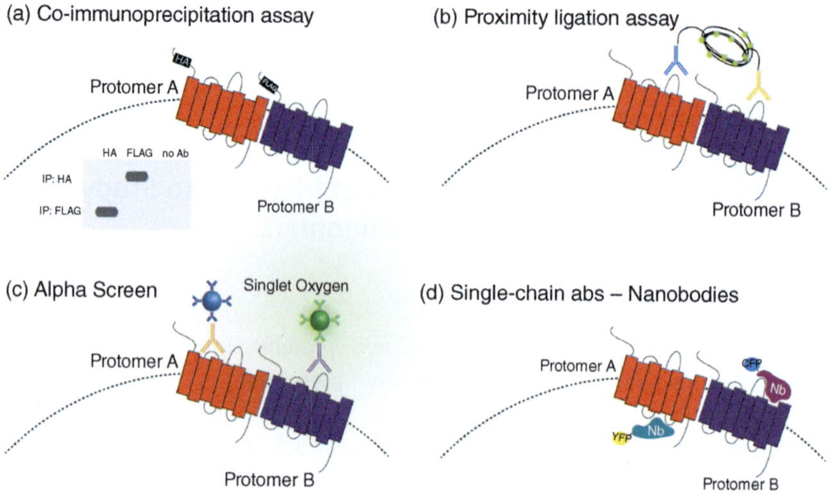

Figure 4.1 Biochemical approaches to studying GPCR dimers. (a) Co-immunoprecipitation-based experiments use antibodies to immunoprecipitate putative partner protomers from cell lysates. Immuno-detection can use antibodies against either epitope-tagged (for example HA or FLAG tags) or endogenous receptors. If two GPCRs interact, the interacting protomer can subsequently be observed via western blot. (b) Proximity ligation technology uses primary or secondary antibodies conjugated to oligonucleotides or PLA-proximity probes. If two protomers are in close enough proximity, a complementary connector oligonucleotide can bind, ligating the two probes together. This newly ligated proximity-dependent DNA molecule is amplified by a polymerase when fluorescently labeled oligonucleotides are added to the reaction allowing visualization of the protein–protein interaction via microscopy (c) AlphaScreen technology or luminescent oxygen channeling immunoassay uses latex nanobeads conjugated to primary antibodies against the receptors under study. If two protomers interact, excitation of the donor beads leads to the release of a singlet oxygen causing a reaction that results in the measured chemiluminescence of the acceptor. Source: Based on Fernandez-Duenas et al. [20]. (d) Single-chain antibodies or nanobodies can be used as alternatives to conventional antibodies as their small size facilitates access to crevices and may be able to distinguish monomers versus oligomeric species. Nanobodies can be raised against extracellular or intracellular epitopes and can be tagged with fluorescent proteins to permit the visualization of dimeric complexes via microscopy.

although biophysical techniques described below were in part developed because of this concern, most journal reviewers now demand both types of approaches.

4.2.2 Proximity Ligation Assays

With better antibodies against *some* native GPCRs, the use of proximity ligation assays (PLAs) has become a promising technique in the study of GPCR oligomerization [29–31]. PLA technology features the use of primary or secondary antibodies conjugated to oligonucleotides or PLA-proximity probes (see Figure 4.1b). If two protomers are in close enough proximity, a complementary connector oligonucleotide can bind, ligating the two probes together [29]. This newly ligated proximity-dependent DNA molecule can be amplified by a polymerase when fluorescently labeled oligonucleotides are added to the reaction allowing visualization of the protein–protein interaction via microscopy [30]. PLA combined with confocal laser scanning microscopy revealed the existence of $D_2R/A_{2A}R$ dimers in mouse striatal brain slices [32]. In that study, the distance requirements for ligation of the secondary antibodies was determined to be 16 nm, marginally larger than the 10 nm requirement for resonance energy transfer (RET)-based biophysical methods (discussed below) [32]. In addition, *in situ* PLA provided evidence for the formation of D_1R/D_2R heterodimers in the nucleus accumbens of rats and monkeys [24, 33]. The clear advantage of this technique is that it can be performed *ex vivo* in intact fixed slices. This technology is thus amenable to human-derived samples and the ligation of both PLA probes followed by DNA amplification allows detection of low levels of endogenous proteins. To date, PLA has been used to demonstrate the existence of several GPCR heterodimers such as the dopamine D2 and adenosine$_{2A}$ ($D_2R/A_{2A}R$) [32, 34], angiotensin II type I and dopamine D2 (AT_1R/D_2R) [35], μ-opioid- and adenosine 2A receptors ($\mu OR/A_{2A}R$) [36], cannabinoid CB_1R/CB_2R [37], serotonin and CB2 cannabinoid receptors ($5HT_{1A}R/CB_2R$) [38], and free fatty acid receptors 2 and 3 (FFAR2/FFAR3) [39]. Of course, as stated for co-IP experiments, PLA conditions must also be optimized and stringent controls must be incorporated in the experimental design to ensure the interaction is specific and not artifactual [40].

4.2.3 Luminescent Oxygen Channeling Immunoassay

Recently, a new technique has emerged, similar in principle to PLA as a means to study GPCR oligomerization. The technique, based on AlphaScreen technology, is a luminescent oxygen channeling immunoassay that uses latex nanobeads conjugated to primary antibodies against the receptors under study, illustrated in Figure 4.1c [20]. AlphaScreen was applied to the $A_{2A}R/D_2R$ dimer where excitation of the donor beads led to the release of a singlet oxygen causing a reaction that resulted in chemiluminescence of the acceptor [20]. These nanobeads were conjugated to secondary antibodies which recognized primary antibodies against the $A_{2A}R/D_2R$ pair. This method reinforced results observed in mice and revealed the existence of the $A_{2A}R/D_2R$ dimer in postmortem human caudate tissue [20]. Taking into consideration the potential relevance of this dimer in Parkinson's disease, the AlphaScreen technology further assessed the dimerization index in healthy versus Parkinsonian tissue samples. Interestingly, the proportion of dimers was shown to be increased in PD samples, an observation that contrasts results extrapolated from some animal models [20]. Even if this work requires further validation and a larger sample number to confirm this striking result, it highlights the importance of developing tools to study dimers that are amenable to use in human samples as disease progression modeled in animals may manifest differently than in human patients.

4.2.4 Heavy Chain Antibodies or Nanobodies

Nanobodies or single chain antibodies emerged as crystallization chaperones for GPCRs and their use resulted in the structure of several Class A GPCRs [41–44], reviewed in [45]. Originally derived from camelids, nanobodies are monomeric heavy chain-only antibodies whose small size (15 kDa) affords them access to small crevices and epitopes not accessible to conventional antibodies [46]. To date, nanobodies generated in camelids or identified using synthetic libraries or by phage-display have been reported to act as G protein mimetics stabilizing a spectrum of distinct GPCR conformational states [45, 47]. Metabotropic glutamate receptors are Class C GPCRs, classically known to form obligate dimers [48–50]. Recently, nanobodies recognizing mGlu2 but not the other mGluR isoforms have been isolated. Two of these, called DN10 and DN13, were found to bind to the active form of mGlu2, acting as positive allosteric modulators [51]. Using fluorescently labeled nanobodies, time-resolved FRET (TR-FRET) demonstrated that two nanobodies could bind mGlu2 homodimers as predicted [51]. DN10 and DN13 were also shown to be selective for mGlu2 homodimers as opposed to mGlu2/3 or mGlu2/4-containing heterodimers [51]. This study highlights the potential of nanobodies as powerful tools to further explore activation and conformation-specific properties of GPCR dimers in their native environments. Theoretically, engineering nanobodies to recognize dimeric species could help stabilize these protein–protein interactions allowing to study dimer specific signaling properties and perhaps even permit the visualization of these structures via X-ray crystallography or by cryogenic electron microscopy [52, 53]. The linking of nanobodies to form bivalent or multivalent equivalents may also prove useful in understanding the impact and extent of GPCR oligomerization [54]. Fluorescently tagged nanobodies, as shown in Figure 4.1d, likely have a "bright" future in such studies as they will allow the investigation of endogenous protein interactions in native tissues with functional relevance [55].

4.3 Biophysical Approaches to Study GPCR Dimers and Higher-Order Oligomers

Resonance energy transfer (RET)-based techniques are well-established tools for the study of GPCRs [53, 56, 57]. This method, represented in Figure 4.2a, relies

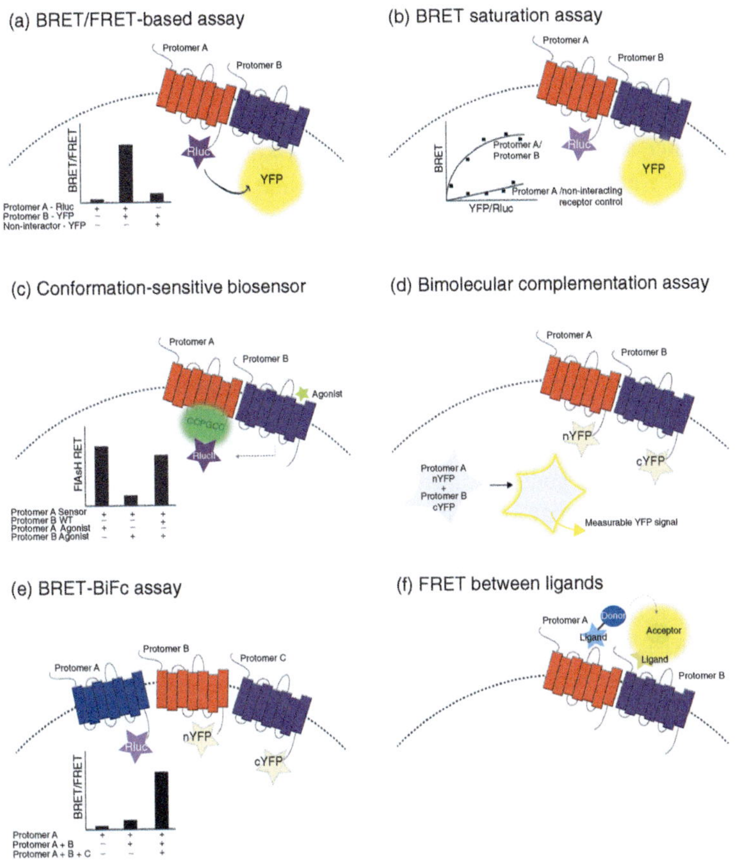

Figure 4.2 Resonance energy transfer-based approaches to studying GPCR dimers. (a) BRET/FRET based experiments rely on the nonradiative transfer of resonance energy between a donor and acceptor chromophore. If the two proteins of interest are within 10 nm, a BRET/FRET signal can be measured. (b) RET saturation curves are typically used to assess the specificity between putative interactors. A constant amount of the Rluc-tagged protomer A is co-transfected with increasing amounts of the YFP-tagged protomer B. Once all the donors are occupied with an acceptor, the RET signal saturates indicative of a specific interaction. Nonspecific interactions result in a quasilinear relationship. (c) Bimolecular complementation assays can be used to assess whether two GPCRs interact. Putative interacting proteins can be fused to a split fluorophore such as the N-terminal and C-terminal fragments of YFP. If receptors do not interact, no fluorescence signal is detected. However, interactions between dimeric species result in the reconstitution of the split fluorophore and the subsequent emission of light at a characteristic wavelength. (d) BRET-BiFC assays can be combined since when the split YFP reconstitutes itself, it becomes an ideal acceptor for the Rluc donor moiety. This method is amenable to studies of higher-order oligomers. (e) FlAsH RET-based conformation-sensitive biosensors can be used to assess how allosteric information is transferred between protomers via conformational rearrangements. Physical interaction with another protomer can affect the conformation of the sensor component. (f) FRET can be measured at the level of ligand binding. Fluorescently labeled ligands (agonists/antagonists) can help deduce whether two proteins interact by the detection of a proximity-dependent FRET signal.

on the exchange of resonance energy from a donor molecule to a compatible acceptor chromophore. Compatibility is determined by the energy spectra of the paired chromophores where the donor's emission spectrum must overlap with the acceptor's excitation wavelength. For the transfer of resonance energy to occur, the distance and orientation requirement necessitate that the RET pair be within 10 nm. The practical applications of this method are considerable, and of course, highly amenable to interrogating protein–protein interactions. Using this experimental design, GPCR oligomerization can be assessed in intact live cells, with receptors expressed in their native membrane environment. A RET donor can be a bioluminescent enzyme, as in BRET, or a fluorophore, as in FRET [58]. Due to the nature of the donor moiety, BRET has a higher signal-to-noise ratio as it does not require external excitation with subsequent bleed-through [59]. Using a FRET-compatible fluorescent acceptor, the use of a laser for donor excitation can inadvertently excite the acceptor resulting in higher fluorescent backgrounds [59]. However, since BRET is generally associated with lower emitted light levels, the higher fluorophore signal intensity associated with FRET is more amenable to microscopic studies. As a prototypical GPCR, the β_2AR was shown to form homodimers by co-immunoprecipitation, further confirmed through BRET-based assays. Using tagged receptors fused with either Renilla luciferase (Rluc) or YFP, a detectable BRET signal was detected upon substrate addition [60]. Agonist treatment increased the BRET signal suggesting that agonist occupancy increased the formation of dimers or altered the conformation of the existing dimers [60].

4.3.1 RET Saturation Curves

When assessing whether two proteins interact, appropriate controls must be incorporated into the experimental design to be able to distinguish *specific* versus *non-specific* interactions. To this end, saturation assays are commonly used as they provide the ability to determine the specificity between putative dimer partners. In the most common design as depicted in Figure 4.2b, a constant amount of the donor tagged protomer A is co-transfected with increasing amounts of the acceptor tagged protomer B. Once all the donors are occupied with an acceptor, then the RET signal saturates and the RET signal can no longer increase. This provides evidence that the dimer pair is not a consequence of random collisions, but rather a result of specific interactions. If two protomers do not interact, a quasi-linear curve is observed. Saturation curves are key experiments since high concentrations of any pair of overexpressed receptors can result in "dimer artifacts," due to random collisions. Thus, RET saturation assays should be conducted to assess the specificity of the observed homo- or heteromeric complexes. In the first instance where such an analysis was performed, saturation was reached using β_2AR tagged with either GFP or Rluc while β_2AR-Rluc transfected alongside soluble GFP resulted in a linear relationship [61]. β_2AR homodimerization has also been recapitulated by other groups [62]. Demonstrating specificity in dimer studies should not be underestimated, as it is further validates protein–protein interactions. In addition, competition assays can also be performed by tagging the two putative dimer partners with compatible chromophores and competing the interactions with an untagged receptor. If the untagged receptor is able to reduce the RET signal in a dose-dependent manner, then the dimer is considered more likely.

4.3.2 RET-Based Conformational Profiling of GPCR Heterodimers

When two receptors form a dimer, it can be imagined that an allosteric connection spans both receptors subsequently impacting the overall activity of the complex. Ligand occupancy of one protomer can thus transmit allosteric information to the partner protomer via conformational rearrangements. Exploring conformation as an experimental output can provide a mechanism explaining how the dimer functions as a unit. This has been reported for the prostanoid F and angiotensin II type I receptor (FP/AT_1R) heterodimer and revealed that the transfer of information between FP and AT_1R is asymmetric [63]. Basically, as depicted in Figure 4.2c, a conformation-sensitive, BRET-based biosensor was engineered into FP by inserting the fluorescein biarsenical hairpin derivative binding (FlAsH) tetracysteine tag into various positions in the third intracellular loop along with RlucII at the distal C-terminus [63, 64]. The FlAsH reagent has high affinity for thiols in the "CCPGCC" sites introduced into the FP sequence resulting in the emission of green fluorescence [65]. After FP labeling with FlAsH, prostaglandin F2α (PGF2α) binding at the orthosteric site led to a conformational change in FP measured by BRET. Co-transfection of this FP conformational sensor with wildtype AT_1R followed by angiotensin II (Ang II) treatment also resulted in a conformational change in FP [63]. This showed that Ang II binding of AT_1R led to a conformational rearrangement that was allosterically transmitted to FP as a consequence of dimerization. Using clustered regularly interspaced short palindromic repeats (CRISPR) engineered knockout cells, it was further demonstrated that $G\alpha_{q/11}$ was necessary for the exchange of allosteric information between the two receptors suggesting that the heterodimer organized into a complex with a shared $G\alpha_{q/11}$. Interestingly, allosteric information was not transmitted in the opposite configuration, since co-transfecting AT_1R conformational sensors with wildtype FP followed by PGF2α stimulation [63, 66] did not result in a change in AT_1R conformation, showing that the information transfer was asymmetric.

A similar experimental design has been applied for the α_2-adrenergic and μ-opioid receptor (α_2AR/μOR) heterodimer [67]. FlAsH-FRET conformational biosensors were engineered in the α_2AR and the addition of norepinephrine resulted in a decrease in FRET; typical of a conformational change associated with receptor activation. When this biosensor-tagged receptor was cotransfected with μOR, costimulation with norepinephrine and morphine affected the conformation of the α_2AR observed as a decrease in FRET [67]. Based on this data, morphine occupancy appeared to transmit an allosterically driven effect observed as a conformational change in the α_2AR. It was further explored whether this altered α_2AR conformational state affected G protein-mediated signaling. In dual receptor expressing cells, a G protein activation sensor revealed that morphine impaired norepinephrine-induced Gα_i activation, which was mirrored by a decrease in norepinephrine-mediated extracellular signal-regulated kinase (ERK) phosphorylation [67]. The primary advantage of the FlAsH system is its ability to elucidate underlying allosteric effects occurring at the level of the GPCR dimers in question. Heterodimer-specific effects likely occur through conformational rearrangements, and the FlAsH-RET approach has the potential to demonstrate this in screening campaigns.

4.3.3 RET and Higher-Order Oligomers

When describing GPCR dimers, the nomenclature can be complicated as the metastable complex of receptors and their associated signaling partners can become quite large. Theoretically, the minimal protein units necessary to form a dimer involves interactions between two receptors. However, other proteins can also participate thus modulating signaling events mediated by such complexes. In view of this, BRET can be combined with other techniques to determine the stoichiometry of receptor oligomers. Bimolecular fluorescence complementation (BiFC) is another technique used to demonstrate dimerization between two GPCRs [68]. Briefly, putative interacting proteins under investigation can be fused to a split fluorophore such as the N-terminal and C-terminal fragments of YFP. If receptors do not interact, no fluorescence signal is detected. However, interactions between dimeric species result in the reconstitution of the split fluorophore and the subsequent emission of light at a characteristic wavelength (see Figure 4.2d). This technique was used to demonstrate homodimerization between the β_2AR [62], D_2R [69], and heterodimerization of the D_1R/D_3R [70] and FP/AT_1R [21]. Since fluorophores can potentially interact in a nonspecific manner, appropriate controls should always be included [71]. A relatively easy control is the use of a GPCR known not to interact with the receptor under study demonstrating that the fluorophore pair does not undergo spontaneous reconstitution in the cellular environment in which it is expressed. Developments within these systems no longer suffer from the inherent irreversibility of the fluorophore reconstitution. Innovative systems such as split Nanoluc are reversible, providing the added benefit of exploring interaction kinetics. As such, the nanoluciferase complementation assay (NanoBiT) was used to demonstrate that D_2R homodimerization was sensitive to the binding of a D_2R antagonist [69].

BiFC can be combined with RET (FRET or BRET) to study higher-order oligomers, represented in Figure 4.2e. This has been done for adenosine receptor heterotetramers as three fusion proteins were generated by tagging the $A_{2A}R$ with Rluc, N-YFP, or C-YFP [72]. Since the split YFP constructs retain the same spectral properties as its full counterpart, it remains a compatible BRET acceptor for Rluc. Accordingly, BRET saturation curves revealed the presence of the dimer and a trimeric higher-order species, as this design was able to capture interactions between three $A_{2A}R$ equivalents [72]. This method may underestimate the "oligomeric" state of adenosine receptors since if two $A_{2A}R$-Rluc or two $A_{2A}R$-YFP protomers would interact, this population would not be detected.

Over time, several versions of RET have been developed and further refined to study higher-order oligomers. In a recent publication, authors described the use of a new experimental design; protomer A was tagged with GFP and Rluc while putative protomer B could be fused to FK506 binding $FKBP_1$ or $FKBP_3$ proteins that can be chemically induced to dimerize [73]. In the basal state, and in absence of any significant interactions, the reported BRET signal is expected to be low. Upon the addition of AP20187, a ligand that led to the induced association of protomer B, if protomer A and protomer B assemble as heterodimers, an increase in the BRET efficiency should be observed. Using fusion constructs, this method was validated with the obligate GPCR dimer pair $GABA_{BR1}$ and $GABA_{BR2}$ but was not able to reproduce findings of β_2AR homodimerization [73]. Enhanced BRET efficiency was also observed between the chemokine receptors CXCR4 and CCR5 when tagged with GFP/Rluc or FKBP suggesting a heterodimer organization. This is an interesting example of how different biological tools can be combined to create new systems to interrogate GPCR dimerization.

Another technique used to study higher-order oligomers is sequential BRET/FRET or SRET [74]. Fluorophore pairs that are compatible with this technique are Rluc-GFP^2-YFP or Rluc-YFP-DsRed as their spectral properties are RET-compatible. In principle, this approach relies on two energy transfer events, or sequential events such as those occurring between Rluc-YFP and YFP-DsRed. This technique was validated using tagged versions of the heterotrimeric G proteins as positive controls; $G\alpha$-GFP^2, $G\beta$-YFP, and $G\gamma$-Rluc [74]. Since these proteins assemble as heterotrimers, substrate addition led to the emission of $G\gamma$-Rluc compatible with the spectrum of $G\alpha$-GFP^2 and $G\beta$-YFP. Thus, if these three proteins interact a YFP signal should be recorded. Since FRET and co-immunoprecipitation experiments revealed that $D_2R/A_{2A}R$, $A_{2A}R/CB_1R$, and D_2R/CB_1R form dimers, the authors sought to understand if there could be an interplay between all three receptors. They thus performed FRET experiments with D_2R-GFP and CB_1R-YFP and attempted to compete the interaction with the

$A_{2A}R$ [74]. Instead of the expected decrease, they observed an increase in FRET. It was speculated that the presence of the $A_{2A}R$ likely led to a conformational change in the FRET pair increasing the efficiency of energy transfer. Since the $A_{2A}R$ does not compete with this receptor pair, it was suggested that they interacted in an oligomeric organization. SRET was thus used to detect the occurrence of a hetero-oligomer containing the $A_{2A}R$, D_1R, and CB_1R receptors [74]. SRET saturation curves demonstrated the presence of the three receptors, whereas saturation curves using D_4R or $GABA_{B2}$ as the donor or 5-HT_{2B}-YFP yielded linear curves [74], used as negative controls.

Three-color FRET has also been used to demonstrate the presence of higher-order oligomers for the ghrelin receptor and the D_2R in liposomes [75]. This method depends on a donor, fluorescein-tagged receptor, a transmitter, Cy3-tagged receptor, and an acceptor Cy5-tagged receptor. Emission of the transmitter fluorophore would indicate a ghrelin receptor dimer, whereas emission of Cy3 and Cy5 revealed the likelihood of three protomers in an oligomer. The propensity of the ghrelin receptor to form oligomers was also tested by adding unlabeled receptor with labeled receptor in liposomes [75]. A decrease in two- and three-color FRET was observed indicating that the unlabeled receptor was competing by forming oligomers with the labeled version [75]. This was also tested with the D_2R which is known to form dimers with the ghrelin receptor. Here, only two-color FRET was observed, consistent with a dimeric configuration.

To investigate the stoichiometry within a higher-order complex, single particle photo-bleaching can be used and has demonstrated the tetrameric organization of the muscarinic M2 receptor [76]. The number of steps or drops in fluorescence intensity can be correlated to the number of protomers in a complex. Following this design, the M_2R was fused to eGFP and immobilized on coverslips. Using photobleaching, the M_2R was shown to organize in tetramers and such results could be confirmed by western blot using an antibody specific to GFP [76]. The presence of carbachol did not affect the signal suggesting that M_2R must be present in part as constitutive higher order species [76]. It was also demonstrated that these four M_2R were in complex with four heterotrimeric $G\alpha_{i1}$ proteins. Activation by carbachol or GTPγS affected $G\alpha_{i1}$ organization but not that of the receptors revealed by a reduction in the $G\alpha_{i1}$ oligomeric state from tetramers to dimers or monomers [76, 77]. As a consequence of this altered oligomeric G protein state, the M_2R tetramer may be accessed by other effector molecules altering the signal transduction pathways engaged by the activated complex.

4.3.4 GPCR Dimerization at the Single Molecule Level

Single molecule pull-downs combined with single molecule photobleaching has been able to determine the stoichiometry of β_2AR complexes in individual cells [78]. Briefly, biotin-coated coverslips were used in order to capture neutravidin-labeled antibodies against proteins of interest. Cell extracts were subsequently added to the cover slip allowing the antibody to bind a bait protein. Proteins thought to interact with the bait can then be visualized via microscopy, by conjugating fluorescent antibodies to putative interactors, by immunofluorescence [79, 80]. This technique further revealed that β_2AR monomers are preferentially phosphorylated by GPCR kinases (GRKs) while dimers are phosphorylated by Protein kinase A (PKA) [78]. The oligomeric state of β_2ARs also affected their spatial segregation since upon agonist stimulation, β_2AR monomers internalized while the dimers remained at the membrane. In hippocampal neurons, GRK-phosphorylated monomers where found to be in the soma while the dimers were located in dendrites and the two receptor populations modulated the activity of calcium channels in distinct ways [78].

Further, the transient nature of D_2R homodimers has been demonstrated using single fluorescence molecule tracking [81]. Experiments were performed in CHO cells expressing ACP-tagged D_2R, which were covalently linked to the fluorescent molecule ATTO594. Co-localization of two D_2R was observed when two ATTO594 molecules were within 220 nm distance [81]. The latter represents a large distance, however, it was estimated that if two molecules interact for long periods of time, longer than 19 seconds for instance, then this would be considered an interaction rather than random collisions as the latter should be transient and short lived. In addition, the authors showed that GPCR dimers were not static entities but are rather dynamic in nature. D_2R homodimerization was also shown to be modulated by the addition of agonists. The lifetime was prolonged by 1.5-fold in presence of agonist and was unchanged by a neutral antagonist [81]. Transient homodimerization has been detected for other GPCRs, *N*-formyl peptide [82], β_1-adrenergic [83], M1 muscarinic receptors [84] and also revealed a temperature dependence for such interactions.

4.4 Engineering Ligands as Tools to Study GPCR Dimers

Time-resolved FRET (TR-FRET) is another technique that can be used to demonstrate receptor dimerization. TR-FRET is particularly interesting as the long fluorescence lifespan of the donor allows a time delay before reading the emission from the acceptor reducing background signals caused by direct excitation of the acceptor. Most commonly, FRET-compatible fluorophores are tagged to the receptors under study. However, in some cases, two ligands can also be fluorescently tagged [85]. In this arrangement, ligand-binding FRET is indicative of dimerization (see Figure 4.2f). This allows the design of dimer-specific tools as this type of FRET can be monitored between ligands in native systems. FRET occurred between antagonists for the vasopressin V2 receptor and oxytocin receptors suggestive of a dimeric organization [85]. This approach was applied to mammary gland membrane preparations and FRET was detected between fluorescently labeled antagonists in this native environment [85]. The measured FRET signal was shown to be specific as it was sensitive to the addition of unlabeled oxytocin. Since these membrane preparations may have originated from the endoplasmic reticulum (ER) or Golgi, FRET was performed in intact mammary gland organ patches. An increase in FRET further confirmed the binding of two antagonists to receptor dimers in this system unlike brain patches which express lower amounts of OTR [85]. However, fluorescently labeled agonists revealed a low FRET signal suggesting that upon the first agonist binding event, the second agonist binding is less favorable, indicating negative cooperativity in multiple agonist binding [85].

There are several functional consequences associated with the activation of GPCR homo- and heterooligomers. For one, receptor pharmacology can be altered as the physical association between both receptors can reshape GPCR topology and corresponding ligand binding properties. The development of ligands that selectively recognize higher-order species can thus aid in understanding dimer-specific signaling events. Bivalent ligands containing two pharmacophores separated by a polyethylene glycol-based spacer of various lengths offer an excellent strategy to interrogate GPCR heterodimers and can exclusively target a heterodimer pair. Numerous hetero-bivalent ligands have been designed targeting μOR/δOR [86, 87], δOR/κOR [88], μOR/D_2R [89], μOR/mGluR5 [90], MT_1/MT_2 [91], among others. Likewise, the development of biased agonists tailored to the selective activation of either G proteins versus β-arrestins has been an ongoing pursuit for years. One particularly interesting advance has been the development of *b*iased *u*n*m*atched *b*ivalent *l*igands (BUmBL) [92]. The generation of biased bivalent ligands constructed by fusing a melanocortin agonist and antagonist showed that melanocortin homodimers exhibit asymmetrical signaling properties. The rationale behind this was the prediction that antagonizing one signaling pathway while activating the other, would create a signaling bias. This strategy led to the successful development of a particular BUmBL that was a potent activator of cyclic adenosine monophosphate (cAMP) signaling and a weak recruiter of β-arrestin [92]. Such approaches can be applied to several other asymmetric GPCR dimers to fine tune signaling responses via system bias by swapping the ligands for other receptors. The μOR, for example engages in dimers with several GPCRs, thus developing biased bivalent ligands that favor G protein activation may help alleviate the side effects associated with current μOR targeting drugs. G protein biased bivalent ligands have also been synthesized for the D_2R [93]. To address whether D_2R homodimer-targeting bivalent ligands were indeed specific, cell penetrating dimer disrupting peptides were used [94]. The bivalent ligand exhibited enhanced binding affinity to the D_2R compared to monovalent ligands and was shown to have decreased binding affinity in presence of a dimer disrupting peptide [94]. In sum, bivalent ligands will help dissect dimer-specific signaling signatures with the potential of exploiting the biased nature of GPCRs.

4.5 Proteomic Approaches to Study GPCR Dimers and Higher-Order Oligomers

It is well established that the interaction between protomers results in changes in signal transduction events, manifested as distinct biological outcomes in health or disease. It is likely that these functional consequences are directly related to changes in the complement of proteins and molecules interacting with such higher-order complexes. The proteome associated with dimers, referred to as homodimer-omes or heterodimer-omes can be characterized thanks to advances in proteomics (see Figure 4.3). Tandem affinity purification (TAP) has been used to identify GPCR-interacting proteins [95]. In this system, fusion constructs are generated by tagging the GPCR of interest with the TAP tag composed of two IgG binding units of protein A, a TEV protease cleavage site and a calmodulin binding domain. Newer versions have been developed using HA and FLAG epitope tags. Based on the original protocols, the bait protein, or GPCR of interest can be purified along with any interacting proteins using two affinity purification steps, the first using an IgG resin or agarose beads. The TEV site then gets cleaved by a protease releasing the first eluted complex bound on calmodulin-coated beads. The addition of a calcium chelator, EGTA, results in the elution of the GPCR of interest along with interacting proteins. The purified protein complex can then be prepared for mass spectroscopy analyses. With two purification steps, this technique results in a "cleaner" sample preparation and less background signals compared to co-immunoprecipitation, although this technique can also be used.

The TAP methods have been used to identify proteins that interact with the melatonin 1 and 2 (MT_1, MT_2) receptors [96]. The TAP tag was introduced at the C-terminal part of the receptors and validated for cell surface localization and functionality considering the large size of the tag. Mitochondrial and ribosomal proteins were abundant in all samples including nontransfected cells, and these were considered as nonspecific interactors [96]. All other proteins not found to be expressed in the nontransfected cells were deemed as specific melatonin receptor interactors. All three $G\alpha_i$ isoforms were identified as interactors consistent with the $G\alpha_i$ coupling of the MT_1 and MT_2. Several other interacting proteins were identified such as the small GTPase Rac1 for the MT_1 and p120 for MT_2 [96]. Since GPCRs encompass a large family of receptors, it is likely that several GPCRs share

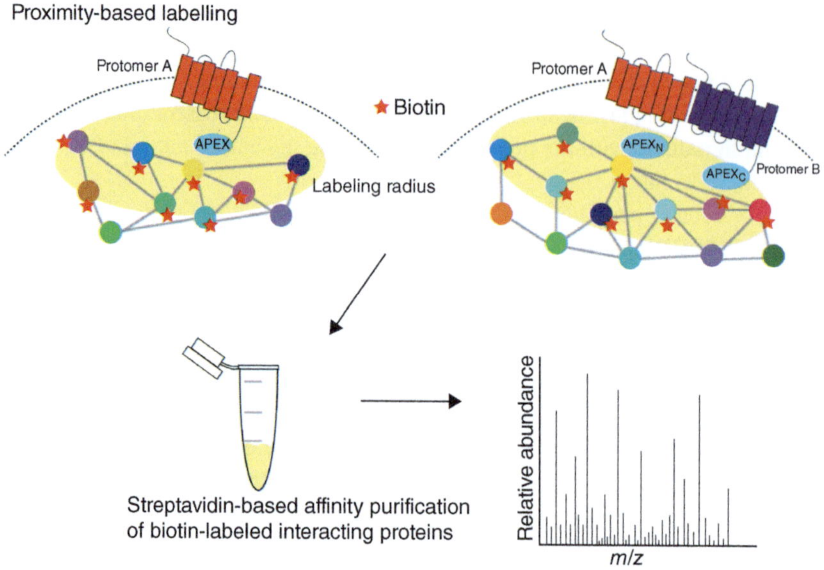

Figure 4.3 Proteomic approaches to studying GPCR dimers. Enzymes such as APEX2 or BirA can be fused to GPCRs of interest to catalyze biochemical reactions that result in the biotinylation of nearby interacting proteins. Split-APEX, requiring the dimerization-dependent reconstitution of two fragments, $APEX_N$ and $APEX_C$, can be used to detect monomer versus dimer-specific proteomes. Neighboring proteins are then enriched for with streptavidin-based affinity purification of biotin-labeled interacting proteins. The identity of these interacting proteins can be revealed by mass spectroscopy.

a similar pool of interactors, estimated at 35% for MT_1 and 45% for MT_2 [96]. Transient interactors may be underrepresented using this technique as the majority of proteins purified by the TAP method represent stable interactors. The use of chemical crosslinking may help stabilize such interactions. Although to address this limitation, other proximity-based labeling approaches have been developed and are discussed below.

4.5.1 Proximity-Based Labeling

A recent unbiased approach used to detect protein–protein interactions in live cells is proximity-based labeling. One such method, BioID, utilizes a biotin ligase, BirA, fused to a protein of interest that can result in biotinylation of nearby interacting proteins. Neighboring proteins are then enriched for using streptavidin-coated beads and subsequently identified by mass spectroscopy. Compared to other biochemical methods, BioID has the advantage of being able to capture transient protein interactions. However, the labeling time of BioID requires approximately 12–24 hours, thus it is not well suited to the signal transduction timescale of GPCRs ranging on the order of seconds to minutes [97, 98]. Under these conditions, other biotinylating enzymes can be used such as APEX2, an engineered ascorbate peroxidase, as it offers higher temporal resolution as the enzymatic reaction only requires 30–60 seconds for biotinylation [99, 100]. Instead of a biotin ligase, APEX2 utilizes an ascorbate peroxidase and upon the addition of hydrogen peroxide, APEX2 catalyzes the addition of biotin phenol to nearby proteins (see Figure 4.3). Since the labeling reaction for BioID and APEX2 occurs at 37 °C, it is amenable for studies using mammalian cells. In addition, TurboID has recently been developed and since the biochemical reaction occurs at 30 °C, it is applicable to yeast (30 °C), flies (25 °C), and worms (20 °C) as model systems [101].

Proximity-based labeling with APEX2 has been accomplished by tagging the C-terminus of the δOR. In this instance, the researchers were able to capture spatial and temporal protein interactions. With a 20 nm radius, APEX2 can label interacting partners, but does not distinguish proteins known to be specific to a subcellular location; otherwise, referred to as "bystanders." Using plasma membrane, cytosolic, or endosomal markers tagged with APEX2, bystander proteins can be identified and filtered-out from subsequent analyses. Through the identification of the δOR interactome, researchers were able to determine that WWP2 and TOM1 were required for the trafficking of δOR to lysosomes, and knocking down either of these proteins impaired δOR degradation [97]. Another group tagged AT_1R and β_2AR with APEX2 [98]. For both receptors, multiple Gα subtypes were strongly enriched in the absence of ligand treatment, whereas β-arrestin2 was not. Interestingly, β_2AR was already associated with adenylyl cyclase, as they showed its enrichment in the absence of ligand. A time course experiment was then able to identify interacting proteins over 30 minutes. This allowed tracking of G protein enrichment and endocytosis of the receptor. AT_1R and β_2AR displayed differences in regard to their internalization kinetics; interactions with the clathrin machinery and endosomal markers peaked at 180 seconds for AT_1R, but was only detected at 600 seconds for β_2AR. In addition, the β_2AR was found to interact with 16 different GPCRs, most of which remained constant even through internalization [98], highlighting the promise of such techniques to track dimer partners.

TAP, BioID, or APEX2 are highly applicable for the mapping of the interactome of single proteins or monomeric GPCRs. However, in order to map the interactome specific to GPCR homodimers and heterodimers, these approaches require further development. When applied to GPCR dimerization, proximity-based labeling has the ability to identify novel GPCR and effector interactors in a vast array of different conditions. For instance, it could allow us to probe for how protein interactomes change after the dual stimulation of agonists specific for each receptor in the heterodimer. APEX2 can track changes in the complement of interacting nearby effectors with spatial and temporal resolution. The latter was shown for the β_2AR as internalization of GPCRs and their complexes was tracked from the plasma membrane to the endosome [98]. Interacting proteins identified using this technique can further be analyzed by knockdown experiments to provide biological significance to the proteomic observations.

In the same vein, split-TAP tag technology has been modified to tag two proteins of interest expanding its utility for dimer-based studies. One protein is tagged with the IgG-binding peptide (TAP-A), while the other is fused to a calmodulin-binding protein (TAP-C), thus allowing purification of proteins that interact within a complex composed of two proteins of interest [102]. Likewise, APEX2 labeling has evolved and can reveal how interactomes are modulated in response to GPCR dimerization. Split APEX2 is tailored for such applications, where the N-terminus and C-terminus could be tagged onto different GPCRs and used to investigate the interactome of homo- and heterodimers [103]. With the split APEX2 enzymes, the interactomes could be monitored to understand how the dimer signals and how the protein interactomes may be different compared to their respective monomer components. With a labeling radius prerequisite for the biotinylation to occur, if the receptor pair exhibits different internalization kinetics in response to complex activation, nearby proteins may no longer be biotinylated. There is still much to be learned when applying split-APEX to GPCR dimerization studies. Dynamic changes occurring at the protein level such as in the repertoire of G proteins or β-arrestins and their recruitment parameters may help elucidate mechanisms that underlie changes in signal transduction events as a consequence of dimerization.

4.5.2 Phosphoproteomics

In addition to investigating the interactomes of GPCR monomers or dimers, the phosphorylation patterns of receptors can also significantly guide our understanding of GPCR signaling and subsequent internalization. Through mass spectrometry, distinct phosphorylation sites can be determined, thus phosphorylation patterns of GPCRs under different conditions can be studied. Phosphoproteomics has been used to study GPCR phosphorylation "barcodes" as means to investigate how different protein kinases play a role in receptor desensitization and how this may be receptor-specific. For example, using a phosphoproteomic approach, one group was able to identify the β_2AR phosphorylation site (of 13 putative sites) preferentially phosphorylated by GRK2 or GRK6, as well as agonist-specific phosphorylation patterns [104]. The potential combination of proximity-based labeling and phosphoproteomics should be noted here, as both GRK2 and GRK6 were identified as interactors of β_2AR [98]. Although β_2AR is very well studied as a model GPCR, phosphorylation patterns of GPCRs in the dimeric context are less clear. Phosphorylation patterns may tell us a lot about individual receptors but could also be applied in the study of dimers. Phosphoproteomic approaches could be used to study how internalization parameters are changed through the physical interaction between two or more GPCRs. Agonist stimulation of one protomer may affect the phosphorylation status of the second protomer altering the internalization kinetics of the metastable complex. By combining proximity-based labeling and phosphoproteomic tools, we can begin to design strategies to understand receptor dimers. While phosphoproteomics could be used to elucidate phosphorylation patterns or barcodes for different receptors, proximity-based labeling would allow to determine the kinases that are responsible.

While not the focus of this article, broad approaches to study potential functional asymmetries in signaling modes for higher-order GPCR complexes using different combinations of agonists and antagonists for constituent receptors needs to be done. BRET-based biosensor approaches designed to interrogate large sections of the signaling phenotype or signature of a given GPCR can also be applied to the study of dimers [105]. Different orthosteric ligand pairs, in different orders of application could be tested using such assays to capture information on potential allosteric interactions between the two receptors as well as identify any functional asymmetries with respect to the function of these dimers. As controls, dual ligand stimulation should also be performed when single receptors are expressed alone. Where potential interactions are identified, conventional signaling assays should be used to verify results.

4.6 Perspectives

With the first report of GPCR oligomers more than two decades ago, several techniques have since emerged to study the complexities associated with these new signaling entities. However, there are recurrent concerns revolving around the physiological relevance of both dimers and higher-order oligomers. The root of this debate often stems from the use of artificial systems such as HEK 293 cells and other immortalized cell lines as vehicles for receptor expression. Such heterologous cell systems were invaluable tools in many GPCR studies as they allowed the controlled expression of the receptor partners under study. GPCR dimers have been observed in several cellular systems including primary cell cultures and in native tissues. However, with all the collected experimental evidence, there is still a sense of controversy associated with the physiological relevance of GPCR dimers. It is surprising that CRISPR-mediated gene editing has not yet been used more extensively in studies of GPCR oligomerization as a means to introduce an epitope tag in frame into endogenously expressed GPCRs. Tagging putative GPCR partners with different tags would allow the study of GPCR dimerization at an endogenous level bypassing the need to overexpress the protomers under study and avoiding concerns about antibody specificity. Cellular transfection may result in the accumulation of proteins at levels that are not physiologically representative thus creating overexpression artifacts that may inadvertently generate apparent dimers by random collisions. However, it can be imagined that epitope tags driven by endogenous promoters may result in expression too low for meaningful conclusions to be drawn. As hinted above, the use of fluorescently tagged nanobodies present themselves as attractive new tools for dimer studies. For example, Nb80 recognizes the activated conformational state of the β_2AR, while Nb60 binds the inactive inverse agonist-bound receptor [41, 106, 107]. Upon ligand stimulation, these nanobodies may be able to reveal the activation status of the receptor or perhaps even that of β_2AR homodimers or heterodimers. Nanobodies thus have the potential to reveal asymmetries within partner receptors.

Finally, the translation of biological data for clinical use is hampered by the limited physiological relevance to human disease of heterologous cell systems and animal models. With the advent of inducible pluripotent stem cells (iPSCs), studies on GPCR oligomers can now transition toward the use of physiological and disease relevant cellular models. The techniques discussed in this review are manifestly amenable for the study of GPCR oligomerization in human inducible pluripotent stem cells and their differentiated derivatives. iPSC-derived neurons or cardiomyocytes will be able to address cell type specific signaling effects as these cells should express all relevant GPCRs and effectors removing the need for overexpression systems. Using iPSCs as a model system sets the stage for deciphering disease-relevant signaling phenotypes driven by changes in GPCR dimers that may be down- or upregulated in disease states. The context dependence of GPCR dimers thus increases the requirement of using physiologically relevant cells as a model system. In our view, iPSCs will facilitate studies critical to expanding the GPCR drug development pipeline by increasing the translatability of experimental findings. The unbiased nature of proteomic approaches may also be able to provide molecular explanations for signaling changes occurring at the level of GPCR dimers as the interacting protein pool, or "interactomes" may be different compared to monomers. Thus, changes in G protein versus β-arrestin coupling or internalization patterns may be rationalized by proteomics-based methods. In conclusion, with recent advances in various methodologies to probe for GPCR dimerization combined with the promise of iPSCs, mining the physiological relevance of GPCR dimers will be greatly facilitated.

References

1 Marinissen, M.J. and Gutkind, J.S. (2001). G-protein-coupled receptors and signaling networks: emerging paradigms. *Trends Pharmacol. Sci.* 22 (7): 368–376.

2 Wootten, D., Christopoulos, A., Marti-Solano, M. et al. (2018). Mechanisms of signalling and biased agonism in G protein-coupled receptors. *Nat. Rev. Mol. Cell Biol.* 19 (10): 638–653.

3 Smith, J.S., Lefkowitz, R.J., and Rajagopal, S. (2018). Biased signalling: from simple switches to allosteric microprocessors. *Nat. Rev. Drug Discovery* 17 (4): 243–260.

4 Pin, J.P., Kniazeff, J., Prezeau, L. et al. (2019). GPCR interaction as a possible way for allosteric control between receptors. *Mol. Cell. Endocrinol.* 486: 89–95.

5 Wootten, D., Christopoulos, A., and Sexton, P.M. (2013). Emerging paradigms in GPCR allostery: implications for drug discovery. *Nat. Rev. Drug Discovery* 12 (8): 630–644.

6 Kenakin, T. (2011). Functional selectivity and biased receptor signaling. *J. Pharmacol. Exp. Ther.* 336 (2): 296–302.

7 Kenakin, T.P. (2012). Biased signalling and allosteric machines: new vistas and challenges for drug discovery. *Br. J. Pharmacol.* 165 (6): 1659–1669.

8 Paila, Y.D. and Chattopadhyay, A. (2010). Membrane cholesterol in the function and organization of G-protein coupled receptors. In: *Cholesterol Binding and Cholesterol Transport Proteins: Structure and Function in Health and Disease* (ed. J.R. Harris), 439–466. Dordrecht: Springer Netherlands.

9 Bulenger, S., Marullo, S., and Bouvier, M. (2005). Emerging role of homo- and heterodimerization in G-protein-coupled receptor biosynthesis and maturation. *Trends Pharmacol. Sci.* 26 (3): 131–137.

10 Milligan, G. (2010). The role of dimerisation in the cellular trafficking of G-protein-coupled receptors. *Curr. Opin. Pharmacol.* 10 (1): 23–29.

11 Milligan, G. and Smith, N.J. (2007). Allosteric modulation of heterodimeric G-protein-coupled receptors. *Trends Pharmacol. Sci.* 28 (12): 615–620.

12 Milligan, G. (2004). G protein-coupled receptor dimerization: function and ligand pharmacology. *Mol. Pharmacol.* 66 (1): 1–7.

13 AbdAlla, S., Lother, H., el Massiery, A. et al. (2001). Increased AT_1 receptor heterodimers in preeclampsia mediate enhanced angiotensin II responsiveness. *Nat. Med.* 7 (9): 1003–1009.

14 Quitterer, U., Fu, X., Pohl, A. et al. (2019). β-Arrestin1 prevents preeclampsia by downregulation of mechanosensitive AT1-B2 receptor heteromers. *Cell* 176 (1, 2): 318–33.e19.

15 Quitterer, U. and AbdAlla, S. (2019). Discovery of pathologic GPCR aggregation. *Front. Med.* 6: 9.

16 Benkirane, M., Jin, D.Y., Chun, R.F. et al. (1997). Mechanism of transdominant inhibition of CCR5-mediated HIV-1 infection by ccr5Δ32. *J. Biol. Chem.* 272 (49): 30603–30606.

17 Pinna, A., di Chiara, G., Wardas, J. et al. (1996). Blockade of A2a adenosine receptors positively modulates turning behaviour and c-Fos expression induced by D1 agonists in dopamine-denervated rats. *Eur. J. Neurosci.* 8 (6): 1176–1181.

18 Wang, M., Pei, L., Fletcher, P.J. et al. (2010). Schizophrenia, amphetamine-induced sensitized state and acute amphetamine exposure all show a common alteration: increased dopamine D2 receptor dimerization. *Mol. Brain* 3: 25.

19 Pei, L., Li, S., Wang, M. et al. (2010). Uncoupling the dopamine D1–D2 receptor complex exerts antidepressant-like effects. *Nat. Med.* 16 (12): 1393–1395.

20 Fernández-Dueñas, V., Gómez-Soler, M., Valle-León, M. et al. (2019). Revealing adenosine A_2A-dopamine D_2 receptor heteromers in Parkinson's disease post-mortem brain through a new AlphaScreen-based assay. *Int. J. Mol. Sci.* 20 (14): 3600.

21 Goupil, E., Fillion, D., Clement, S. et al. (2015). Angiotensin II type I and prostaglandin F2α receptors cooperatively modulate signaling in vascular smooth muscle cells. *J. Biol. Chem.* 290 (5): 3137–3148.

22 Dagher, O.K., Jaffa, M.A., Habib, A. et al. (2019). Heteromerization fingerprints between bradykinin B2 and thromboxane TP receptors in native cells. *PLoS One* 14 (5): e0216908.

23 Wrzal, P.K., Devost, D., Petrin, D. et al. (2012). Allosteric interactions between the oxytocin receptor and the β_2-adrenergic receptor in the modulation of ERK1/2 activation are mediated by heterodimerization. *Cell. Signal.* 24 (1): 342–350.

24 Hasbi, A., Perreault, M.L., Shen, M.Y.F. et al. (2017). Activation of dopamine D1–D2 receptor complex attenuates cocaine reward and reinstatement of cocaine-seeking through inhibition of DARPP-32, ERK, and DeltaFosB. *Front. Pharmacol.* 8: 924.

25 Ayoub, M.A., Levoye, A., Delagrange, P. et al. (2004). Preferential formation of MT_1/MT_2 melatonin receptor heterodimers with distinct ligand interaction properties compared with MT_2 homodimers. *Mol. Pharmacol.* 66 (2): 312–321.

26 Michel, M.C., Wieland, T., and Tsujimoto, G. (2009). How reliable are G-protein-coupled receptor antibodies? *Naunyn-Schmiedeberg's Arch. Pharmacol.* 379 (4): 385–388.

27 Hébert, T.E., Moffett, S., Morello, J.P. et al. (1996). A peptide derived from a β_2-adrenergic receptor transmembrane domain inhibits both receptor dimerization and activation. *J. Biol. Chem.* 271 (27): 16384–16392.

28 Nimchinsky, E.A., Hof, P.R., Janssen, W.G. et al. (1997). Expression of dopamine D_3 receptor dimers and tetramers in brain and in transfected cells. *J. Biol. Chem.* 272 (46): 29229–29237.

29 Fredriksson, S., Gullberg, M., Jarvius, J. et al. (2002). Protein detection using proximity-dependent DNA ligation assays. *Nat. Biotechnol.* 20 (5): 473–477.

30 Leuchowius, K.J., Weibrecht, I., and Söderberg, O. (2011). In situ proximity ligation assay for microscopy and flow cytometry. *Curr. Protoc. Cytom.* 56 (1): 9–36.

31 Gomes, I., Sierra, S., and Devi, L.A. (2016). Detection of receptor heteromerization using in situ proximity ligation assay. *Curr. Protoc. Pharmacol.* 75: 2.16.1–2.16.31.

32 Trifilieff, P., Rives, M.L., Urizar, E. et al. (2011). Detection of antigen interactions ex vivo by proximity ligation assay: endogenous dopamine D_2-adenosine A_2A receptor complexes in the striatum. *Biotechniques* 51 (2): 111–118.

33 Rico, A.J., Dopeso-Reyes, I.G., Martinez-Pinilla, E. et al. (2017). Neurochemical evidence supporting dopamine D_1—D_2 receptor heteromers in the striatum of the long-tailed macaque: changes following dopaminergic manipulation. *Brain Struct. Funct.* 222 (4): 1767–1784.

34 Zhu, Y., Mészáros, J., Walle, R. et al. (2019). Detecting GPCR complexes in postmortem human brain with proximity ligation assay and a Bayesian classifier. *Biotechniques* 68: 122–128.

35 Martinez-Pinilla, E., Rodriguez-Perez, A.I., Navarro, G. et al. (2015). Dopamine D_2 and angiotensin II type 1 receptors form functional heteromers in rat striatum. *Biochem. Pharmacol.* 96 (2): 131–142.

36 Sun, G.C., Ho, W.Y., Chen, B.R. et al. (2015). GPCR dimerization in brainstem nuclei contributes to the development of hypertension. *Br. J. Pharmacol.* 172 (10): 2507–2518.

37 Navarro, G., Borroto-Escuela, D., Angelats, E. et al. (2018). Receptor–heteromer mediated regulation of endocannabinoid signaling in activated microglia. Role of CB_1 and CB_2 receptors and relevance for Alzheimer's disease and levodopa-induced dyskinesia. *Brain Behav. Immun.* 67: 139–151.

38 Franco, R., Villa, M., Morales, P. et al. (2019). Increased expression of cannabinoid CB_2 and serotonin 5-HT_{1A} heteroreceptor complexes in a model of newborn hypoxic–ischemic brain damage. *Neuropharmacology* 152: 58–66.

39 Ang, Z., Xiong, D., Wu, M. et al. (2018). $FFAR_2$—$FFAR_3$ receptor heteromerization modulates short-chain fatty acid sensing. *FASEB J.* 32 (1): 289–303.

40 Alsemarz, A., Lasko, P., and Fagotto, F. (2018). Limited significance of the in situ proximity ligation assay. *BioRxiv* https://doi.org/10.1101/411355.

41 Rasmussen, S.G., DeVree, B.T., Zou, Y. et al. (2011). Crystal structure of the β_2-adrenergic receptor-Gs protein complex. *Nature* 477 (7366): 549–555.
42 Kruse, A.C., Ring, A.M., Manglik, A. et al. (2013). Activation and allosteric modulation of a muscarinic acetylcholine receptor. *Nature* 504 (7478): 101–106.
43 Huang, W., Manglik, A., Venkatakrishnan, A.J. et al. (2015). Structural insights into μ-opioid receptor activation. *Nature* 524 (7565): 315–321.
44 Che, T., Majumdar, S., Zaidi, S.A. et al. (2018). Structure of the nanobody-stabilized active state of the κ-opioid receptor. *Cell* 172 (1, 2): 55–67.e15.
45 Manglik, A., Kobilka, B.K., and Steyaert, J. (2017). Nanobodies to study G protein-coupled receptor structure and function. *Annu. Rev. Pharmacol. Toxicol.* 57: 19–37.
46 Muyldermans, S. (2013). Nanobodies: natural single-domain antibodies. *Ann. Rev. Biochem.* 82 (1): 775–797.
47 Steyaert, J. and Kobilka, B.K. (2011). Nanobody stabilization of G protein-coupled receptor conformational states. *Curr. Opin. Struct. Biol.* 21 (4): 567–572.
48 Romano, C., Yang, W.L., and O'Malley, K.L. (1996). Metabotropic glutamate receptor 5 is a disulfide-linked dimer. *J. Biol. Chem.* 271 (45): 28612–28616.
49 Kniazeff, J., Prezeau, L., Rondard, P. et al. (2011). Dimers and beyond: the functional puzzles of class C GPCRs. *Pharmacol. Ther.* 130 (1): 9–25.
50 Niswender, C.M. and Conn, P.J. (2010). Metabotropic glutamate receptors: physiology, pharmacology, and disease. *Annu. Rev. Pharmacol. Toxicol.* 50: 295–322.
51 Scholler, P., Nevoltris, D., de Bundel, D. et al. (2017). Allosteric nanobodies uncover a role of hippocampal mGlu2 receptor homodimers in contextual fear consolidation. *Nat. Commun.* 8 (1): 1967.
52 Domanska, K., Vanderhaegen, S., Srinivasan, V. et al. (2011). Atomic structure of a nanobody-trapped domain-swapped dimer of an amyloidogenic β_2-microglobulin variant. *Proc. Natl. Acad. Sci. U.S.A.* 108 (4): 1314–1319.
53 El Khamlichi, C., Reverchon-Assadi, F., Hervouet-Coste, N. et al. (2019). Bioluminescence resonance energy transfer as a method to study protein-protein interactions: application to G protein coupled receptor biology. *Molecules* 24 (3): 537.
54 Mujic-Delic, A., de Wit, R.H., Verkaar, F. et al. (2014). GPCR-targeting nanobodies: attractive research tools, diagnostics, and therapeutics. *Trends Pharmacol. Sci.* 35 (5): 247–255.
55 Farrants, H., Gutzeit, V.A., Acosta-Ruiz, A. et al. (2018). SNAP-tagged nanobodies enable reversible optical control of a G protein-coupled receptor via a remotely tethered photoswitchable ligand. *ACS Chem. Biol.* 13 (9): 2682–2688.

56 Pfleger, K.D.G. and Eidne, K.A. (2006). Illuminating insights into protein–protein interactions using bioluminescence resonance energy transfer (BRET). *Nat. Methods* 3 (3): 165–174.

57 Cui, Y., Zhang, X., Yu, M. et al. (2019). Techniques for detecting protein–protein interactions in living cells: principles, limitations, and recent progress. *Sci. China Life Sci.* 62 (5): 619–632.

58 Harekrushna, S. (2011). Förster resonance energy transfer – a spectroscopic nanoruler: principle and applications. *J. Photochem. Photobiol., C* 12 (1): 20–30.

59 Boute, N., Jockers, R., and Issad, T. (2002). The use of resonance energy transfer in high-throughput screening: BRET versus FRET. *Trends Pharmacol. Sci.* 23 (8): 351–354.

60 Angers, S., Salahpour, A., Joly, E. et al. (2000). Detection of β_2-adrenergic receptor dimerization in living cells using bioluminescence resonance energy transfer (BRET). *Proc. Natl. Acad. Sci. U.S.A.* 97 (7): 3684–3689.

61 Mercier, J.F., Salahpour, A., Angers, S. et al. (2002). Quantitative assessment of β_1- and β_2-adrenergic receptor homo- and heterodimerization by bioluminescence resonance energy transfer. *J. Biol. Chem.* 277 (47): 44925–44931.

62 Parmar, V.K., Grinde, E., Mazurkiewicz, J.E. et al. (2017). β_2-Adrenergic receptor homodimers: role of transmembrane domain 1 and helix 8 in dimerization and cell surface expression. *Biochim. Biophys. Acta Biomembr.* 1859 (9 Pt A): 1445–1455.

63 Sleno, R., Devost, D., Pétrin, D. et al. (2017). Conformational biosensors reveal allosteric interactions between heterodimeric AT1 angiotensin and prostaglandin F2α receptors. *J. Biol. Chem.* 292 (29): 12139–12152.

64 Sleno, R., Pétrin, D., Devost, D. et al. (2016). Designing BRET-based conformational biosensors for G protein-coupled receptors. *Methods* 92: 11–18.

65 Griffin, B.A., Adams, S.R., and Tsien, R.Y. (1998). Specific covalent labeling of recombinant protein molecules inside live cells. *Science* 281 (5374): 269–272.

66 Devost, D., Sleno, R., Pétrin, D. et al. (2017). Conformational profiling of the AT_1 angiotensin II receptor reflects biased agonism, G protein coupling, and cellular context. *J. Biol. Chem.* 292 (13): 5443–5456.

67 Vilardaga, J.-P., Nikolaev, V.O., Lorenz, K. et al. (2008). Conformational cross-talk between α_{2A}-adrenergic and μ-opioid receptors controls cell signaling. *Nat. Chem. Biol.* 4 (2): 126–131.

68 Wouters, E., Vasudevan, L., Crans, R.A.J. et al. (2019). Luminescence- and fluorescence-based complementation assays to screen for GPCR oligomerization: current state of the art. *Int. J. Mol. Sci.* 20 (12): 2958.

69 Wouters, E., Marin, A.R., Dalton, J.A.R. et al. (2019). Distinct dopamine D_2 receptor antagonists differentially impact D_2 receptor oligomerization. *Int. J. Mol. Sci.* 20 (7): 1686.

70 Guitart, X., Navarro, G., Moreno, E. et al. (2014). Functional selectivity of allosteric interactions within G protein-coupled receptor oligomers: the dopamine D_1–D_3 receptor heterotetramer. *Mol. Pharmacol.* 86 (4): 417–429.

71 Kerppola, T.K. (2006). Design and implementation of bimolecular fluorescence complementation (BiFC) assays for the visualization of protein interactions in living cells. *Nat. Protoc.* 1 (3): 1278–1286.

72 Gandia, J., Galino, J., Amaral, O.B. et al. (2008). Detection of higher-order G protein-coupled receptor oligomers by a combined BRET-BiFC technique. *FEBS Lett.* 582 (20): 2979–2984.

73 Felce, J.H., MacRae, A., and Davis, S.J. (2019). Constraints on GPCR heterodimerization revealed by the type-4 induced-association BRET assay. *Biophys. J.* 116 (1): 31–41.

74 Carriba, P., Navarro, G., Ciruela, F. et al. (2008). Detection of heteromerization of more than two proteins by sequential BRET–FRET. *Nat. Methods* 5 (8): 727–733.

75 Damian, M., Pons, V., Renault, P. et al. (2018). GHSR-D2R heteromerization modulates dopamine signaling through an effect on G protein conformation. *Proc. Natl. Acad. Sci. U.S.A.* 115 (17): 4501–4506.

76 Shivnaraine, R.V., Fernandes, D.D., Ji, H. et al. (2016). Single-molecule analysis of the supramolecular organization of the M_2 muscarinic receptor and the $G\alpha_{i1}$ protein. *J. Am. Chem. Soc.* 138 (36): 11583–11598.

77 Li, Y., Shivnaraine, R.V., Huang, F. et al. (2018). Ligand-induced coupling between oligomers of the M_2 receptor and the G_{i1} protein in live cells. *Biophys. J.* 115 (5): 881–895.

78 Shen, A., Nieves-Cintron, M., Deng, Y. et al. (2018). Functionally distinct and selectively phosphorylated GPCR subpopulations co-exist in a single cell. *Nat. Commun.* 9 (1): 1050.

79 Jain, A., Liu, R., Ramani, B. et al. (2011). Probing cellular protein complexes using single-molecule pull-down. *Nature* 473 (7348): 484–488.

80 Jain, A., Liu, R., Xiang, Y.K. et al. (2012). Single-molecule pull-down for studying protein interactions. *Nat. Protoc.* 7 (3): 445–452.

81 Kasai, R.S., Ito, S.V., Awane, R.M. et al. (2018). The class-A GPCR dopamine D_2 receptor forms transient dimers stabilized by agonists: detection by single-molecule tracking. *Cell Biochem. Biophys.* 76 (1, 2): 29–37.

82 Kasai, R.S., Suzuki, K.G., Prossnitz, E.R. et al. (2011). Full characterization of GPCR monomer–dimer dynamic equilibrium by single molecule imaging. *J. Cell Biol.* 192 (3): 463–480.

83 Calebiro, D., Rieken, F., Wagner, J. et al. (2013). Single-molecule analysis of fluorescently labeled G-protein-coupled receptors reveals complexes with distinct dynamics and organization. *Proc. Natl. Acad. Sci. U.S.A.* 110 (2): 743–748.

84 Hern, J.A., Baig, A.H., Mashanov, G.I. et al. (2010). Formation and dissociation of M_1 muscarinic receptor dimers seen by total internal reflection fluorescence imaging of single molecules. *Proc. Natl. Acad. Sci. U.S.A.* 107 (6): 2693–2698.

85 Albizu, L., Cottet, M., Kralikova, M. et al. (2010). Time-resolved FRET between GPCR ligands reveals oligomers in native tissues. *Nat. Chem. Biol.* 6 (8): 587–594.

86 Daniels, D.J., Lenard, N.R., Etienne, C.L. et al. (2005). Opioid-induced tolerance and dependence in mice is modulated by the distance between pharmacophores in a bivalent ligand series. *Proc. Natl. Acad. Sci. U.S.A.* 102 (52): 19208–19213.

87 Lenard, N.R., Daniels, D.J., Portoghese, P.S. et al. (2007). Absence of conditioned place preference or reinstatement with bivalent ligands containing μ-opioid receptor agonist and δ-opioid receptor antagonist pharmacophores. *Eur. J. Pharmacol.* 566 (1–3): 75–82.

88 Daniels, D.J., Kulkarni, A., Xie, Z. et al. (2005). A bivalent ligand (KDAN-18) containing δ-antagonist and κ-agonist pharmacophores bridges δ_2 and κ_1 opioid receptor phenotypes. *J. Med. Chem.* 48 (6): 1713–1716.

89 Qian, M., Vasudevan, L., Huysentruyt, J. et al. (2018). Design, synthesis, and biological evaluation of bivalent ligands targeting dopamine D_2-like receptors and the μ-opioid receptor. *ChemMedChem* 13 (9): 944–956.

90 Peterson, C.D., Kitto, K.F., Akgun, E. et al. (2017). Bivalent ligand that activates μ-opioid receptor and antagonizes mGluR5 receptor reduces neuropathic pain in mice. *Pain* 158 (12): 2431–2441.

91 Karamitri, A., Sadek, M.S., Journe, A.S. et al. (2019). *O*-Linked melatonin dimers as bivalent ligands targeting dimeric melatonin receptors. *Bioorg. Chem.* 85: 349–356.

92 Lensing, C.J., Freeman, K.T., Schnell, S.M. et al. (2019). Developing a biased unmatched bivalent ligand (BUmBL) design strategy to target the GPCR homodimer allosteric signaling (cAMP over β-arrestin 2 recruitment) within the melanocortin receptors. *J. Med. Chem.* 62 (1): 144–158.

93 Bonifazi, A., Yano, H., Ellenberger, M.P. et al. (2017). Novel bivalent ligands based on the sumanirole pharmacophore reveal dopamine D_2 receptor (D_2R) biased agonism. *J. Med. Chem.* 60 (7): 2890–2907.

94 Pulido, D., Casado-Anguera, V., Perez-Benito, L. et al. (2018). Design of a true bivalent ligand with picomolar binding affinity for a G protein-coupled receptor homodimer. *J. Med. Chem.* 61 (20): 9335–9346.

95 Daulat, A.M., Maurice, P., and Jockers, R. (2011). Tandem affinity purification and identification of GPCR-associated protein complexes. *Methods Mol. Biol.* 746: 399–409.

96 Daulat, A.M., Maurice, P., Froment, C. et al. (2007). Purification and identification of G protein-coupled receptor protein complexes under native conditions. *Mol. Cell. Proteomics* 6 (5): 835–844.

97 Lobingier, B.T., Huttenhain, R., Eichel, K. et al. (2017). An approach to spatiotemporally resolve protein interaction networks in living cells. *Cell* 169 (2): 350–60.e12.

98 Paek, J., Kalocsay, M., Staus, D.P. et al. (2017). Multidimensional tracking of GPCR signaling via peroxidase-catalyzed proximity labeling. *Cell* 169 (2): 338–49.e11.

99 Gingras, A.C., Abe, K.T., and Raught, B. (2019). Getting to know the neighborhood: using proximity-dependent biotinylation to characterize protein complexes and map organelles. *Curr. Opin. Chem. Biol.* 48: 44–54.

100 Lam, S.S., Martell, J.D., Kamer, K.J. et al. (2015). Directed evolution of APEX2 for electron microscopy and proximity labeling. *Nat. Methods* 12 (1): 51–54.

101 Branon, T.C., Bosch, J.A., Sanchez, A.D. et al. (2018). Efficient proximity labeling in living cells and organisms with TurboID. *Nat. Biotechnol.* 36 (9): 880–887.

102 Caspary, F., Shevchenko, A., Wilm, M. et al. (1999). Partial purification of the yeast U2 snRNP reveals a novel yeast pre-mRNA splicing factor required for pre-spliceosome assembly. *EMBO J.* 18 (12): 3463–3474.

103 Han, Y., Branon, T.C., Martell, J.D. et al. (2019). Directed evolution of split APEX2 peroxidase. *ACS Chem. Biol.* 14 (4): 619–635.

104 Nobles, K.N., Xiao, K., Ahn, S. et al. (2011). Distinct phosphorylation sites on the β_2-adrenergic receptor establish a barcode that encodes differential functions of β-arrestin. *Sci. Signal.* 4 (185): ra51.

105 Namkung, Y., LeGouill, C., Kumar, S. et al. (2018). Functional selectivity profiling of the angiotensin II type 1 receptor using pathway-wide BRET signaling sensors. *Science signaling* 11 (559): eaat1631.

106 Staus, D.P., Strachan, R.T., Manglik, A. et al. (2016). Allosteric nanobodies reveal the dynamic range and diverse mechanisms of G-protein-coupled receptor activation. *Nature* 535: 448–452.

107 Staus, D.P., Wingler, L.M., Strachan, R.T. et al. (2014). Regulation of β_2-adrenergic receptor function by conformationally selective single-domain intrabodies. *Mol. Pharmacol.* 85 (3): 472–481.

5

Arrestin and G Protein Interactions with GPCRs: A Structural Perspective

Carlos A.V. Barreto[1,2], Salete J. Baptista[1,3], Beatriz Bueschbell[4], Pedro R. Magalhães[5], António J. Preto[1,2], Agostinho Lemos[1], Nícia Rosário-Ferreira[1,6], Anke C. Schiedel[4], Miguel Machuqueiro[5], Rita Melo[1,3], and Irina S. Moreira[1,7,8]

[1] *Data-Driven Molecular Design Group, CNC – Center for Neuroscience and Cell Biology, University of Coimbra, Coimbra, Portugal*
[2] *Institute for Interdisciplinary Research, University of Coimbra, PhD Programme in Experimental Biology and Biomedicine, Coimbra, Portugal*
[3] *Centro de Ciências e Tecnologias Nucleares, Instituto Superior Técnico, Universidade de Lisboa, Bobadela LRS, Portugal*
[4] *Department of Pharmaceutical and Medicinal Chemistry, Pharmaceutical Institute, University of Bonn, Bonn, Germany*
[5] *Departmento de Química e Bioquímica, Faculdade de Ciências, Universidade de Lisboa, BioISI-Biosystems and Integrative Sciences Institute, Lisboa, Portugal*
[6] *Chemistry Department, Faculty of Science and Technology, Coimbra Chemistry Center, University of Coimbra, Coimbra, Portugal*
[7] *Department of Life Sciences, Faculty of Science and Technology, University of Coimbra, Coimbra, Portugal*
[8] *CIBB – Center for Innovative Biomedicine and Biotechnology, University of Coimbra, Coimbra, Portugal*

5.1 Overview of GPCR Biology

5.1.1 GPCRs

G protein-coupled receptors (GPCRs), well-known seven-transmembrane (7TM) spanning receptors, represent the largest family of cell-surface receptors of the human genome. From a structural point of view, GPCRs consist of a single polypeptide, characterized by an extracellular NH_2-terminal domain, followed by seven hydrophobic transmembrane α-helices (TM1–TM7) connected by three extracellular (ECL1–ECL3) and three intracellular loops (ICL1–ICL3), and an intracellular COOH-terminal domain [1–3]. To date, over 800 human GPCRs have been identified and categorized into six families on the basis of their amino acid sequence similarity and phylogenetic analysis: Class A (rhodopsin-like receptors), Class B (the secretin family), Class C (metabotropic glutamate

GPCRs as Therapeutic Targets, Volume 1, First Edition. Edited by Annette Gilchrist.

receptors [mGluRs]), Class D (fungal mating pheromone receptors), Class E (cyclic adenosine monophosphate [cAMP] receptors), and Class F (Frizzled and Smoothened receptors) [4, 5]. The rhodopsin-like receptors can be further subdivided into "rhodopsin receptors" and "nonrhodopsin receptors." The members of this large superfamily mediate several downstream signaling pathways triggered by a spectrum of structurally diverse ligands, including endogenous (biogenic amines, glycoproteins, hormones, ions, lipids, neurotransmitters, nucleotides, and peptides) and exogenous ligands (odorants, photons, tastants, and therapeutic agents) [6, 7]. There is still a considerable number of orphan GPCRs for which the endogenous ligands remain unknown [8, 9]. Remarkably, GPCRs are one of the most thoroughly investigated therapeutic targets in drug discovery, mostly due to their substantial involvement in numerous pathologies, including cardiovascular diseases [10, 11], cancer [12, 13], central nervous system (CNS) disorders [14–17], and immune system diseases [18–20]. Approximately, 30–40% of the marketed drugs approved by US Food and Drug Administration (FDA) represent GPCR-targeted therapeutic agents [21].

5.1.2 Signaling Pathways

GPCRs mediate several signaling pathways through a general mechanism that involves the activation of the GPCR, initiating a chain of events that lead to the release of second messengers, the molecules responsible to carry the signal inside the cell [22]. The complexity of GPCR-mediated physiological functions is determined by the association between the activated receptors and the heterotrimeric guanine nucleotide-binding proteins (G proteins). Heterotrimeric G proteins comprise α (G_α), β (G_β), and γ (G_γ) subunits, which are involved in signal transduction. G_α signals by itself, whereas G_β and G_γ form heterodimers ($G_{\beta\gamma}$) [23–25]. Both G_α and G_γ are linked to the plasma membrane by lipids [26]. Once activated through the binding of a first messenger, a GPCR undergoes conformational changes and, due to an allosteric effect, acts as a guanine nucleotide exchange factor (GEF), catalyzing the guanosine diphosphate (GDP) release and the binding of guanosine triphosphate (GTP) to the G_α. Once formed, the GTP–G_α complex binds to the GPCR, while the $G_{\beta\gamma}$ dimer disconnects. As such, both G_α and $G_{\beta\gamma}$ become available for other interactions [27]. It was also reported that the $G_{\beta\gamma}$ can be pivotal to the exchange of GDP for GTP by aiding the G_α to reposition itself toward a more suitable conformation for the exchange, once it detaches itself [28]. This signal transduction pathway, which is started by first messengers (ligands) binding to GPCRs, contributes to the interaction with other proteins through G_α and the $G_{\beta\gamma}$, which later promotes the second messenger production [22]. The transduction pathway ends when GTP is hydrolyzed via the intrinsic GTPase activity of G_α that is promoted by GTPase-activating G proteins, such as regulators of G protein signaling (RGS) [29]. RGS act in intermediate states between G_α–GDP and G_α–GTP bound states. This happens by lowering the energy required for GTP hydrolysis, thus favoring it which, in turn, supports the inactive conformation where the GPCR reconnects to a heterotrimeric G protein [30].

There are several effectors of GPCRs that can be up- and/or downregulated depending on the ligand and the receptor [29]. Currently, 20 G_α subunits, 6 G_β subunits, and 11 G_γ subunits have been described [31]. Regarding G_α subunits, four different classes have been reported: $G_{\alpha s}$, $G_{\alpha i/o}$, $G_{\alpha q/11}$, and $G_{\alpha 12/13}$ [32]. The first G_α subunit to be studied was $G_{\alpha s}$ [33], which upregulates adenylate cyclase, increasing the concentration of cAMP in the cell from ATP's conversion. This activates the cAMP-dependent protein kinase (PKA). Through phosphorylation of Ser and Thr residues in a broad spectrum of target proteins, PKA modulates downstream metabolic signaling networks. One example of the PKA's role is the transcription regulation as PKA is responsible for the phosphorylation of cAMP-response element-binding protein, which will be needed for other cofactors [34]. While $G_{\alpha s}$ is responsible for the activation of the cAMP-dependent pathway, $G_{\alpha i/o}$ downregulates PKA activation by inhibiting cAMP production [33].

Another subunit, $G_{\alpha q/11}$, activates the membrane-bound protein phospholipase C (PLC), which upregulates the phosphatidylinositol signaling pathway, responsible for the synthesis of phosphatidylinositol (PI) and lipid signaling [29]. In the plasma membrane, PLC splits phosphatidylinositol 4,5-bisphosphate into inositol triphosphate (IP3) and diacylglycerol (DAG). IP3 is released into the cytosol and activates receptors in the smooth endoplasmic reticulum, leading to the opening of calcium channels and increasing the concentration of Ca^{2+} in the cytosol. Unlike IP3, DAG remains in the membrane where it activates protein kinase C (PKC) [35].

The mechanisms of action of the $G_{\alpha 12/13}$ family are still unclear. Therefore, further studies are needed for better understanding and confirmation of previous results, since disruptions of their regulation are linked to numerous pathologies [36]. While $G_{\alpha 12}$ seems to interact with several proteins such as phospholipase D, c-Src, and PKC, $G_{\alpha 13}$ seems to regulate Rho by interacting with the small GTPase Rho and p115RhoGEF, and modulate the phosphatidylinositol-4,5-bisphosphate 3-kinase pathway through protein-tyrosine kinase 2 activation [31]. It is also suggested that the mitogen-activated protein kinase pathway is modulated through GPCR signaling. However, additional research should be conducted toward a full understanding [37].

Nevertheless, signaling regulation is not exclusive to G_α. The $G_{\beta\gamma}$ dimer also has multiple signaling regulating abilities, both directly, with many effectors reported and indirectly, by modulating G_α activity [38]. The portfolio of $G_{\beta\gamma}$ interactions includes some isoforms of adenylate cyclase and phospholipase Cβ, as well as Kir3 and voltage-gated calcium channels, which exert their action in different cellular compartments allowing the dimer to broaden its influence to other effectors [39]. As previously mentioned, the two best-described signaling pathways in which GPCRs have active roles are cAMP-dependent and phosphatidylinositol pathways [31], which are shown in Figure 5.1.

The long-term or repeated stimulation of GPCRs by their respective ligands induces a number of processes that can progressively attenuate the receptor responsiveness, in a process of desensitization [40, 41]. Second messenger-dependent protein kinases, PKA and PKC, and G protein-coupled receptor kinases (GPCRKs) are the two families of second messenger kinases that participate in receptor desensitization [40, 41]. The second-messenger-dependent protein kinases PKA and PKC induce the phosphorylation of multiple GPCRs, suppressing the responsiveness of these receptors [42]. Hence, these proteins are involved in a process of heterologous desensitization. The GPCRK family members specifically phosphorylate Ser/Thr residues of agonist-occupied GPCRs (homologous desensitization) [40]. The GPCRK-induced phosphorylation promotes the recruitment of a class of intracellular scaffolding proteins, β-arrestins, to the receptor complex [43–45]. High affinity interaction between β-arrestins and the activated GPCRs inhibits further G protein binding by steric hindrance. This leads to the attenuation of receptor signaling and, ultimately, to endocytosis of the GPCR, predominantly mediated by clathrin [43–45]. Once an activated receptor is internalized, distinct intracellular trafficking routes may occur. Desensitized GPCRs can be sorted to lysosomal degradation or dephosphorylated in the endocytic compartment for recycling to the cell surface. Although the clathrin-mediated endocytosis is the major pathway of GPCR internalization, the caveolae-dependent and clathrin/caveolae-independent pathways constitute alternative mechanisms of GPCR endocytosis [43–45].

5.1.3 Biased Signaling and Application in Drug Discovery

Drug discovery research and development would benefit from a better understanding of the biology and mechanism of agonists and antagonists toward the desired outcome from a signaling pathway. Thus, the goal of using biased signaling to improve drug discovery is a subject of great interest. Over the last 25 years, the concept of biased signaling, also called functional selectivity, has challenged the drug discovery process and redefined the definition of target coverage. Biased signaling describes the ability of ligands to stabilize different receptor conformations [46]. Subsequently, the efficacy of a particular ligand/drug becomes more complex since there are multiple efficacies that translate into multiple observed activities [46]. Moreover, this signaling event does not only comprise different conformational states of a receptor (conformational bias) but may also result in different downstream signaling routes (G protein bias or arrestin bias). The ability to select which

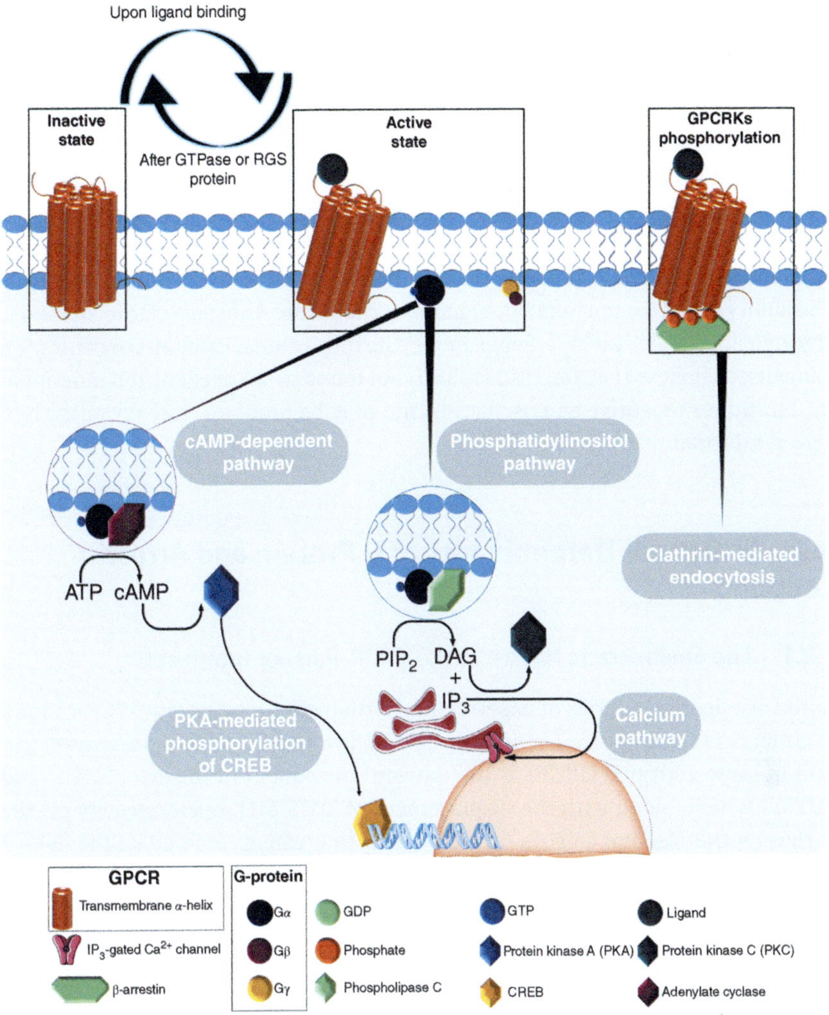

Figure 5.1 Schematic diagram of GPCR signaling pathways highlighting the shift from inactivated to activated GPCR and GPCR kinases (GPCRKs) phosphorylation. Depiction of the main pathways in which activated GPCRs are involved: (i) modulation of adenylyl cyclase affecting the cAMP-dependent pathway (ii) activation of phospholipase C affecting the phosphatidylinositol pathway downstream. Source: Calcium channel, endoplastic reticulum and DNA images by Servier Medical Art by Servier (available at http://smart.servier.com/), licensed under CC BY 3.0.

signaling pathway is activated opens a new platform for the development of safer and more selective drugs [47, 48]. Grundmann and Kostenis [47] have described another bias that complicates ligand–receptor interaction events even more. Their research shows that, signaling can be subdivided into at least three components: (i) quality (which substances take part in the event), (ii) space (where the event takes place), and (iii) time (when the events take place) [47]. The dynamic arrangement between those three dimensions is deeply integrated in GPCR signal transduction. Temporal bias is distinct from conformational bias, where ligands induce different receptor conformations that are linked to distinct signaling routes [47]. The ability of cells to time-encode signals for the secure transport of information is physiologically relevant [47]. For example, the intracellular calcium concentration following a single extracellular stimulus is not found to be transient and monophasic, but rather repetitive and oscillating and may be proof for time-encoding biological information [47].

5.2 Structural Determinants of G Protein and Arrestin Coupling

5.2.1 The Electrostatic Nature of the GPCR-Partner Interfaces

In the last decade, a series of experimental structures for activated GPCRs bound to either G proteins [49–56] or arrestins [57–60] were determined. A common feature in these activated GPCRs is the outward movement of the cytoplasmic end of TM6 which, along with the displacement of TM5 [61], opens a cavity on the surface of the receptor (Figure 5.2, left panel) that can accommodate the helix 5 (H5) of G proteins or the finger loop of arrestins. It was observed that binding to this pocket triggers G protein activation [66]. There is often a low sequence identity between GPCRs, but the majority of the residues involved in direct interactions with their partners are conserved, including the ones in the cytoplasmic cavity [67]. This may explain how so many different receptors are able to couple to the same G protein and how a limited number of arrestins (4 in humans) are able to modulate the function of so many GPCRs (~800 in the human genome) [68]. In activated receptors, the cytoplasmic cavity is significantly cationic at the surface (Figure 5.2, right panel) due to the presence of several Lys and Arg residues, most of which are conserved. Interestingly, deeper into the helix bundle core, there are only a few acidic residues, like Asp79 of the adrenergic β_2 receptor (β_2AR), whose protonation state change has been linked to the receptor activation and may help to explain its experimentally observed pH-dependent behavior [69].

The cationic nature of the cytoplasmic cavity has pointed several authors to the possible role of electrostatics in the binding recognition between the

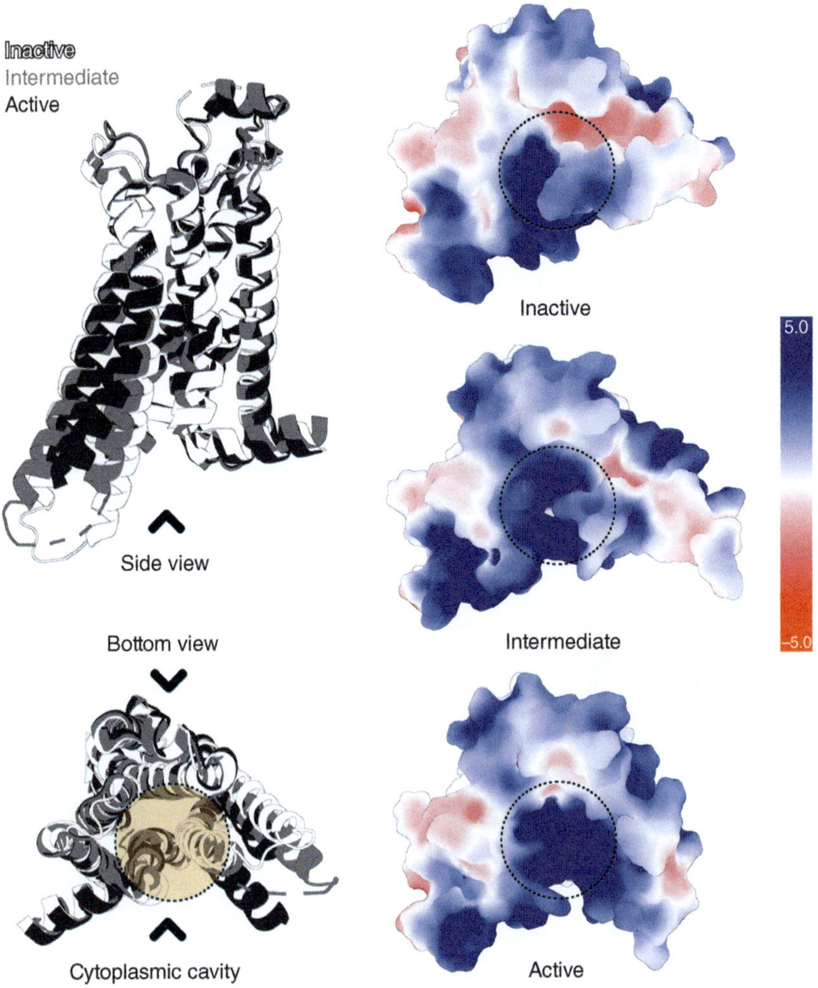

Figure 5.2 Structural features of the adenosine A_{2A} receptor in the inactive (PDB ID: 3REY [62]), intermediate (PDB ID: 2YDV [63]), and active (PDB ID: 5G53 [50]) states. On the left side, a cartoon representation highlighting the opening of the cytoplasmic cavity upon activation and, on the right side, the electrostatic potential mapped on the receptors' surface illustrating the cationic nature of the cavity being exposed. The electrostatic potential was calculated using the default parameters of APBS [64], within the PyMOL software [65]. The unit of the electrostatic potential is *kT/e*. Source: Based on Refs [50, 62–65].

receptors and their partners [67, 68, 70]. In particular, it has been proposed that charge–charge interactions could be the explanation for the promiscuity among arrestins [67, 68]. However, a detailed analysis of the electrostatic potential of several G proteins and arrestins shows that the H5 of $G_{\alpha s}$ in G proteins (Figure 5.3, top panel) is far more negatively charged than the finger loop helix in arrestins (Figure 5.3, bottom panel). In fact, it is not surprising that arrestins, which are more promiscuous, slightly shift the GPCR recognition from electrostatic factors, which favor selectivity to more hydrophobic interactions, which provide affinity. Song and coworkers [70] used atomistic molecular dynamics simulations and observed that the rhodopsin–arrestin complex is stabilized mainly by hydrophobic contacts and a complex network of long-lived hydrogen bonds that naturally emerge at the interface [70]. The authors looked specifically for charge–charge

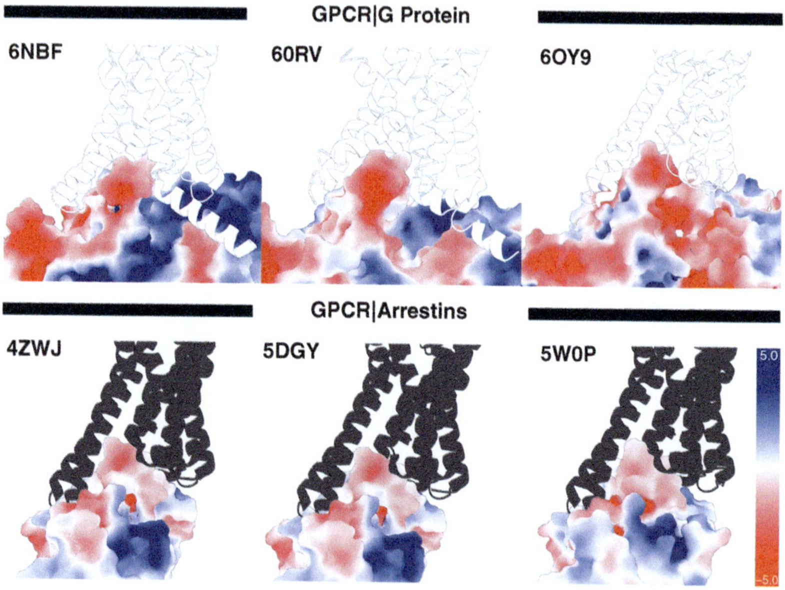

Figure 5.3 Electrostatic potential surfaces of the different partners bound to GPCRs (represented in cartoon). On the top panel, we selected three recent G protein structures with PDB IDs: 6NBF, which is a cryoelectron microscopy (Cryo-EM) structure of parathyroid hormone receptor type 1 in complex with a G protein [54]; 6ORV, which is the glucagon-like peptide-1 receptor bound to a nonpeptide agonist and G protein [56]; and 6OY9, which is a structure of the rhodopsin–transducin complex [55]. On the bottom panel, three different arrestin structures with PDB IDs: 4ZWJ, which is the femtosecond X-ray laser crystal structure of rhodopsin bound to arrestin [57]; 5DGY which is the X-ray laser diffraction structure of the rhodopsin–arrestin complex [58]; and 5W0P which is the X-ray free electron laser crystal structure of rhodopsin bound to visual arrestin [59]. The unit of the electrostatic potential is kT/e. Source: Based on Refs [54–59].

interactions in their trajectories and were only able to identify one water-mediated interaction between Asp72 of arrestin and Lys66 of rhodopsin, which is outside of the cytoplasmic cavity.

Dror and coworkers [71] observed that the receptor structural changes that give rise to arrestin activation are less in the cytoplasmic cavity than in the remaining regions interfacing arrestin [71]. This is significantly different from G protein activation and may support the future design of biased ligands for selective destabilization of the complexes [72, 73]. More recently, these authors even suggested that β-arrestin can interact cooperatively with the phospholipid bilayer which leads to enhanced stability and increased functional versatility [74]. Furthermore, it was also observed that receptor phosphorylation can influence the receptor–arrestin binding and subsequent arrestin-mediated signaling, especially in rhodopsin-like receptors [59, 75, 76]. The phosphorylation occurs on the C-terminal tail of the GPCR receptor, leading to conformational and electrostatic changes that are critical for the first and key step of arrestin binding [75].

5.2.2 Conformational Differences Between GPCR-G Protein and GPCR–Arrestin Complexes

As stated above, both G proteins and arrestins are able to couple to GPCRs, triggering different signaling pathways. Positional changes involving TM5, TM6, and TM7 are reported upon GPCR activation, particularly the inward movement of TM7 and the outward movement of TM6 [67, 77]. These changes create a cavity in the cytoplasmic surface of the receptor, allowing the binding of different intracellular partners, namely G proteins and arrestins. A characteristic rearrangement of the three highly conserved residues Arg3.50, Tyr5.58, and Tyr7.53 was observed upon receptor activation, independent of the binding partner. Furthermore, one of the most significant differences between G protein and arrestin coupling to GPCRs is the extension of TM6 movement, ranging from 8 Å [78] to 17 Å [50], being the largest shift associated with $G_{\alpha s}$. Comparing G protein and arrestin coupling, higher TM3–TM6 and lower TM3–TM7 interhelical distances are reported for the $G_{\alpha s}$ protein [49, 79, 80]. Additionally, Dror and coworkers [81], in an extensive molecular dynamics (MD) study with angiotensin II type I receptor (AT1R), reported alternative intracellular conformations that showed different coupling profiles. Apart from the canonical active-form able to couple with G protein and arrestins, an alternative conformation was observed, that accommodates β-arrestins but not G proteins. In this alternative conformation, TM7 shifts toward TM3 which rotates the side-chains from Tyr7.53 and Arg3.50 to the intracellular side, which causes a structural clash with H5 of $G_{\alpha q}$. Two other conformations were identified that might hinder β-arrestin coupling by creating a larger intracellular cavity, thus presenting some selectivity for G protein coupling [81]. Noteworthy is the fact that arrestins (particularly visual arrestins) tend to create a larger interaction interface with GPCRs when compared to G proteins, despite the high level of shape complementarity between both partners at the receptor interface. In fact, TM3, ICL2, TM5, and TM6 are common structural motifs involved in interactions with both G proteins and arrestins [67]. Interestingly, Xu and coworkers [60] recently reported that β-arrestin 1 (arrestin-2), when complexed to the nonvisual GPCR neurotensin receptor 1, presents a rotation of 90° in relation to the visual arrestin–rhodopsin complex, leading to different binding interfaces [60]. However, a newly solved Cryo-EM complex of arrestin-2 with the M2 muscarinic receptor revealed an identical orientation of 7TMs when compared to the rhodopsin-visual arrestin crystal complex [74]. This suggests that arrestin binding is, after all, dependent on which GPCR subfamily member is coupled. Further studies including new structures of different GPCRs complexed with different G proteins and arrestin members are needed to investigate all the conformational changes involved in selective partner coupling and the subsequent signaling of different biological responses.

5.3 Selectivity Between G Proteins

As previously reported, once activated, GPCRs can interact with different GPCR effectors at the cytoplasmic receptor interface. Even though the GPCR coupling selectivity is still unclear, the advances in structural studies, such as Cryo-EM and X-ray free-electron lasers not only increased in the number of GPCR complexes with different partners available but also knowledge about structural and dynamic determinants that underlie GPCR coupling [82, 83]. Although several GPCRs reveal a preference toward a specific G protein family, a GPCR member is often able to couple to more than one G protein [84]. For instance, β_2AR primarily not only couples to $G_{\alpha s}$, resulting in smooth muscle relaxation but can also couple to $G_{\alpha i}$, inhibiting this response [85]. In fact, it is reported that promiscuous GPCRs display hotspot residues in the G protein-binding interface that allow the binding of different G proteins [86]. Contrarily, different GPCRs are able to couple to the same G_{α} protein [87]. For example the β_1-adrenergic receptor [88] and 5HT$_6$ receptor [89] are able to activate $G_{\alpha s}$, leading to heart muscle contraction and excitatory neurotransmission, respectively [90].

The C-terminus of the G_{α} subunit was reported as the domain primarily responsible for coupling selectivity [86, 91], as it has been shown conclusively that the switch of the last three H5 C-terminal amino acids between $G_{\alpha i}$ and $G_{\alpha q}$ induces promiscuous signaling in HEK293 cells [86]. Nevertheless, Lambert and coworkers [91] reported that, for some receptors, the G_{α} subunit core is more relevant for coupling selectivity than the G_{α} C-terminus. The analysis of the coupling of 14 representative GPCR members to the 16 G_α subunits, namely to a representative G_α of the 4 G_α subunits and 12 G_α chimeras (displaying the C-terminus of one representative partner and the Gα core of another), showed that different GPCRs were able to couple to the same G protein by the recognition of different conserved features of the partners' family. They have also stated that, $G_{\alpha i}$-coupled receptors usually tend to be more selective than $G_{\alpha s}$ and $G_{\alpha q}$-coupled receptors. Moreover, in a previous study using G protein sequence analysis, Babu and coworkers [87] reported a selectivity barcode on each human G protein that generally includes residues at s2s3 and h4s6 loops, β-sheet S5 and helix 4 (H4), as well as the H5 C-terminus as the most involved structural motif in selective coupling to different GPCRs. This unique amino acid interaction pattern at the GPCR–G protein interface involving each of the 16 G_α is surrounded by conserved amino acid positions that might define the similar binding orientation of each GPCR–G protein complex, besides the G_α subtype coupled. A rotation and translation of H5 was also observed for each receptor–G protein interface, exposing other residues that might also be engaged in G_α subtype selective coupling. Nevertheless, besides the physicochemical differences at the binding interface, one must consider other factors such as levels of G_α, precoupling, posttranslational modifications and different membrane composition that affect G protein binding selectivity [87].

5.3.1 Patterns for $G_{\alpha s}$

Crystal and Cryo-EM structures of different GPCRs complexed with $G_{\alpha s}$ were solved, revealing information about $G_{\alpha s}$ coupling selectivity. In fact, both crystal structures of class A β_2AR (PDB ID: 3SN6 [49]) and adenosine A_{2A} (A_{2A}R) (PDB ID: 5G53 [50]) receptors complexed with $G_{\alpha s}$ and an engineered mini-$G_{\alpha s}$, respectively, as well as the Cryo-EM structures of class B GLP-1 (PDB ID: 5VAI [52]) and calcitonin receptors (PDB ID: 5UZ7 [92]) complexed with $G_{\alpha s}$, display a quite similar interaction profile at the binding interface [93]. In the case of β_2AR–$G_{\alpha s}$ and A_{2A}–mini$G_{\alpha s}$ complexes, some residues included in the $G_{\alpha s}$ selectivity barcode, particularly the two H5 C-terminus residues His387$^{H5.19}$ and Tyr391$^{H5.23}$, are able to establish strong interactions with both receptors [49, 50]. Even though these receptors generally interact with similar positions at $G_{\alpha s}$ and receptor, different residues might be involved in the interaction. By contrast, β_2AR and GLP-1 create an identical cavity able to recognize and interact with the H5 C-terminus residues by nonpolar interactions. In this case, some equivalent residues are engaged in these interactions, such as X3.50, X3.53, X5.65, and X6.36. Even though the molecular details involved in the recognition interaction pattern on the receptor are distinct between the two GPCRs, H5 is highly flexible and able to establish different interactions by using the same set of C-terminus amino acids [52]. This means that receptors belonging to distinct subfamilies might read the $G_{\alpha s}$ selectivity barcode distinctively [94]. Recently, the structure of β_2AR complexed with the C-terminus of $G_{\alpha s}$ (14 amino acids) and GDP-bound $G_{\alpha s}$ heterotrimer (PDB ID: 6EG8 [94]), which represents an intermediate state in the β_2AR-$G_{\alpha s}$ activation process, was solved. This complex highlights the key interactions of two H5 residues involved in conserved interactions within class A GPCRs, Arg389$^{H5.21}$ and Glu392$^{H5.24}$. Despite not being identified in β_2AR–$G_{\alpha s}$ crystal structure [49], these residues are reported to be engaged in key interactions with the β_2AR. The larger outward movement of TM6, required for the orientation of the G_s C-terminus might suggest the involvement of this intermediate structure in G protein coupling specificity [94].

5.3.2 Patterns for $G_{\alpha i/o}$

A few Cryo-EM structures of $G_{\alpha i/o}$ complexed with class A GPCRs have been solved. The structures of μOR–$G_{\alpha i}$ (PDB ID: 6DDE [95]), 5-HT$_{1B}$R–$G_{\alpha o}$ (PDB ID: 6G79 [96]), A_1R–$G_{\alpha i}$ (PDB ID: 6D9H [80]), and rhodopsin–$G_{\alpha i}$ (PDB ID: 6CMO [97]), give further insights into G protein coupling specificity [94]. Comparing the previously solved β_2AR–$G_{\alpha s}$ and rhodopsin–$G_{\alpha i}$ model, some interesting differences were recognized. In good agreement with the work of Vaidehi and coworkers [86], the tilt angle of the H5 C-terminus is different depending on the bound G protein subtype. In fact, while the tilt angle of H5 C-terminus is distinct between β_2AR–$G_{\alpha s}$ and rhodopsin–$G_{\alpha i}$ complexes, it is identical when comparing the rhodopsin–$G_{\alpha i}$ and metarhodopsin II-11-amino acid C-terminal fragment derived from $G_{\alpha t}$ complexes. As $G_{\alpha t}$ belongs to the $G_{\alpha i}$ subfamily, this indicates a clear difference regarding the interaction of the C-terminal of $G_{\alpha i}$ versus $G_{\alpha s}$ subfamilies [93]. By inspecting the dynamic interactions of the C-termini of $G_{\alpha s}$, $G_{\alpha i}$, and $G_{\alpha q}$ proteins complexed with seven GPCRs (β_2AR, β_3AR, D_1R, α_{2A}AR, CB_1R, α_{1A}AR, and V_1AR), Vaidehi and coworkers [86] highlighted that the insertion angle of the H5 C-terminus with GPCR tends to be greater for $G_{\alpha i}$-coupled receptors (α_{2A}AR, CB_1R (PDB ID: 5TGZ [98])) in comparison with $G_{\alpha s}$- (receptors β_2AR, β_3AR, and D_1R) and $G_{\alpha q}$-coupled receptors (α_{1A}AR and V_1AR). Furthermore, they found that the extreme C-terminal region is directed toward TM6 and TM7 in $G_{\alpha s}$-coupled receptors, while it is directed toward TM2, ICL1, and ICL2 in $G_{\alpha q}$ and $G_{\alpha i}$-coupled receptors. By combining MD simulations and fluorescence resonance energy transfer sensor measurements, this study also defined some selectivity hotspot residues, such as Glu392$^{H5.24}$ for $G_{\alpha s}$, Leu349$^{H5.16}$, Glu355$^{H5.22}$, and Asn357$^{H5.24}$ $G_{\alpha q}$, and Gly352$^{H5.24}$ for $G_{\alpha i}$. This is in agreement with the selectivity barcode reported by Babu and coworkers [87]. Actually, H5.16, H5.22, and H5.24 are shown to be crucial for the binding of all three G_α subtypes, $G_{\alpha s}$, $G_{\alpha q}$, and $G_{\alpha i}$. Besides TM3, TM5, and TM6, all three structural motifs, mainly engaged in interactions with the C-terminal H5 of $G_{\alpha s}$, $G_{\alpha q}$, and $G_{\alpha i}$, ICL1 and ICL2, are also involved in important contacts with $G_{\alpha q}$, and $G_{\alpha i}$, but not with $G_{\alpha s}$ [86]. Furthermore, Hubbell and coworkers [93] stated the crucial role of Lys349$^{H5.21}$, Asp350$^{H5.22}$, Cys351$^{H5.23}$, Gly352$^{H5.24}$, and Phe354$^{H5.26}$ at the H5 C-terminus, alongside Asp193$^{s2s3.02}$ and Leu194$^{S3.01}$ at the s2s3 loop and S3 sheet, respectively, for $G_{\alpha i}$ protein coupling selectivity. The type of C-terminal residues influences the insertion angle of H5, and the consequent interaction with TM6, whose movement is generally lower in $G_{\alpha i}$–protein coupling (in comparison with $G_{\alpha s}$), is also reported as a determinant of G protein selectivity [93].

5.3.3 Other Families

Interesting studies were published regarding the coupling selectivity of $G_{\alpha q/11}$ and $G_{\alpha 12/13}$ subfamilies. As stated before, a selectivity barcode primarily involving mostly the extreme of C-terminus is also identified for $G_{\alpha q}$ and $G_{\alpha 12/13}$ subfamilies, besides the most studied $G_{\alpha s}$ and $G_{\alpha i}$ proteins [87]. A compelling study about the widely investigated muscarinic acetylcholine receptors used MD simulations alongside thermodynamic analyses of both cognate (M_1-$G_{\alpha q}$ and M_2-$G_{\alpha i}$) and noncognate (M_1-$G_{\alpha i}$ and M_2-$G_{\alpha q}$) complexes. This study revealed that only two pairs of interactions are specific to each of cognate receptor–G protein complexes (M_1-$G_{\alpha q}$ and M_2-$G_{\alpha i}$), involving residues at hns1.2 position for both complexes and H5.13 and H5.25 for M_2-$G_{\alpha i}$ and M_1-$G_{\alpha q}$, respectively [99]. Notably, none of these residues are part of the selectivity barcode described by Babu and coworkers [87], but specific interactions involving $G_{\alpha i}$ in both cognate and noncognate complexes were observed, engaging some residues of the selectivity barcode previously reported, such as Ile345$^{H5.16}$ and Asn348$^{H5.19}$. Specific interactions only involving $G_{\alpha q}$ complexes were not observed.

Although the $G_{\alpha 12/13}$ subfamily is the least studied among G proteins, mostly due to the lack of selective inhibitors and the difficulty in obtaining assays for specific evaluation of their activity, they are known to be activated by different GPCRs. For example, the orphan receptor GPR35, which has been emerging as an interesting target in several diseases [79, 100], is reported to primarily couple to $G_{\alpha 13}$ and less efficiently to $G_{\alpha 12}$. Milligan and coworkers investigated the G protein coupling selectivity to GPR35, by using bioluminescence resonance energy transfer assay sensors and $G_{\alpha 12/13}$ chimeras, in which single residue mutations were applied to determine the coupling selectivity. They showed that the coupling selectivity between $G_{\alpha 12}$ and $G_{\alpha 13}$ was mostly dependent on the single Leu$^{H5.23}$Val mutation. Indeed, the exchange of leucine by other amino acids led to the almost complete loss of $G_{\alpha 13}$ coupling. Moreover, the introduction of leucine at H5.23 of $G_{\alpha q}$ promotes the coupling of this subunit to GPR35, reinforcing the key role of leucine at this position in order to allow an efficient coupling, and therefore, the higher selectivity of $G_{\alpha 13}$ in comparison with $G_{\alpha 12}$ to couple to GPR35 [79]. It is important to note that receptor–G protein coupling selectivity is strictly dependent on the specific determinants of the GPCR analyzed, besides the G protein subunit coupled. A combination of residue interactions seems to be key for G protein selective coupling, rather than a selective residue pattern.

5.4 Arrestins-Binding Differences

All known arrestins have high sequence conservation with the same structural arrangement: two domains organized in a seven-strand β-sandwich (N-domain and C-domain) connected by an interdomain region [76]. GPCRs interact with the concave sides of the arrestins [76] in a two-step process. First, the polar core in the N-domain acts as phosphorylation sensor and binds to phosphorylated residues in the receptor. Arrestin then undergoes several structural changes, such as the disruption of the polar core and consequent release of the C-tail by the three-element region [76, 101]. Second, these motions expose structural elements engaged in a high-affinity recognition that are located in the interdomain surface (finger loop, C-loop, the lariat loop, and the middle loop) and that are part of the binding interface with the receptor. Additionally, a small rotation of the C-domain relative to the N-domain was also observed. Although this is the general mechanism proposed, the binding affinity of arrestin varies not only with the receptor type but also with the receptor phosphorylation state. The search for a phosphorylation selectivity footprint is still an important focus in GPCR research. Veprintsev and coworkers [102] systematically evaluated the phosphorylation footprint of rhodopsin not only for arrestin affinity but also arrestin activation and arrestin conformation. Three types of functional motifs were identified: "key sites" for arrestin binding and activation, an "inhibitory site" that revokes arrestin binding, and "modulator sites" that adjust the global arrestin conformation. pThr340 and pSer343, and pThr360 and pSer363 were identified as "key sites" in rhodopsin and the vasopressin receptor 2 (V_2R), respectively. Upstream, a negatively charged region was also described as a contributor to arrestin binding both receptors. Between these two regions, a group of four phosphorylation sites was found to modulate the global conformation and flexibility of arrestin-1. Additionally, residues Thr342 on rhodopsin and Ser362 on V_2R, when phosphorylated, appear to play a neutral to inhibitory role depending on the level of phosphorylation of the receptor, i.e. the inhibitory action seems stronger in low-phosphorylated receptors [102]. Visual arrestins show high affinity for phosphorylated active receptors, whereas arrestin-3 exhibits a higher preference for non-phosphorylated receptors [103]. Multiple reports explaining this difference in selectivity have been described: β-arrestins possess a weaker hydrogen bond network in the N–C-domain interface [104]; the C-terminal tail from arrestin-3 can populate multiple conformations [105]; and strand XIV of arrestin-3 has a different conformation when compared to other subtypes [106].

In regards to the nonvisual arrestin interactions, a classification was proposed for receptors: Class A receptors bind to arrestin-3 with higher-affinity than arrestin-2 and do not interact with visual arrestins, whereas class B receptors bind both nonvisual arrestins with similar high affinity and also interact with visual arrestins [107]. An explanation for this behaviour has been described by Veprintsev and coworkers [102]. They aligned the C-termini of various class A and class B receptors and showed that both classes have Thr and Ser residues in the "key sites" region, but for class B the group of negatively charged residues, which have been linked to being necessary for stable binding of arrestin-2 is missing [102].

5.5 Modulating GPCR Signaling

The mechanistic and structural basis of ligand binding and the subsequent signaling events of GPCRs are far from being completely understood. Besides the widely explored mechanisms of agonism and antagonism and their second messenger signaling, other ways of modulating GPCR signaling were reported. According to Kenakin [46] drugs are characterized by three molecular properties: affinity, efficacy, and target coverage [25, 46]. They further state that affinity and efficacy are useful scales to determine the activity of a particular drug–receptor pair and may predict the potential therapeutic usage [46, 108]. Today, GPCRs are described as "allosteric microprocessors" as they can generate a vast number of conformations which are also (life)time-dependent according to the chemical properties of the ligand [25, 48]. Furthermore, this kind of conformational dynamics was reported to be linked to the level of constitutive activity of a receptor and modulated by the strength of interactions within its transmembrane core [25]. The full spectrum of GPCR signaling includes not only ligands and cells but also complex probe-dependent cooperative interactions between co-binding ligands (so-called "allosteric modulators"), oligomerization and biased cell signaling [48, 109]. The concepts of one single active receptor conformation (known as the active state) or "two-state models" become obsolete here [109, 110]. Meanwhile, it is proven that a GPCR can adopt multiple active and inactive conformations, even in the apo-state [25]. From the therapeutic point of view, the understanding of biased signaling and allosteric modulation of GPCR-directed therapeutics leads to the involvement of the desired pathways over those not involved in the therapeutic effect [47]. Ideally, this would reduce or even abolish unwanted and adverse effects [47, 111].

5.5.1 Biased Ligands

Functional selectivity/biased ligands may offer enormous potential as drug targets with advantages over orthosteric ligands [112]. This implies that only a subset of the possible signaling repertoire is recruited to the receptor at the relative expense of the other pathways [113]. Until very recently, G protein signaling and the generation of second messengers was believed to occur exclusively at the plasma membrane. Different studies could demonstrate that some GPCRs may sustain $G_{\alpha s}$ protein signaling and cAMP production in endosomes after internalization [114–118]. Since endosomal G protein signaling is not only temporally but also spatially distinct from the usual pathway activation it may also be considered as another concept of biased signaling. For example, $G_{\alpha i}$ coupling-dependent endosomal signaling was reported for the sphingosine-1-phosphate receptor [119]. The V_2R [120] is also an example where two structurally similar ligands (nonapeptides, differing only in two residues) can temporally and spatially differentially activate the same downstream pathways [114]. Vasopressin leads to persistent cAMP production from early endosomes, whereas oxytocin-triggered cAMP signaling is restricted to the plasma membrane. These findings can partially explain the different physiological effects of vasopressin and oxytocin. It was also reported that two different β-arrestin-biased agonists were able to activate different downstream signaling pathways, which is probably due to the fact that each agonist stabilizes a completely different receptor conformation [114].

However, it is still unclear which ligand properties lead to the promotion of either pathway. According to Glass and coworkers [121], it is important to consider that different agonists activating different pathways are a consequence of kinetics; for example, very slow dissociating agonists may allow receptor conformations which favor low-affinity interactions for a GPCR/signaling molecule-pair to persist long enough for downstream signaling [121]. Moreover, the concept of signaling bias outdates traditional parameters such as EC_{50} and efficacy as some pathways may achieve maximum response at lower receptor occupancy leading to a higher potency, which can then be described as "pathway bias" [121]. Lastly, the performance of ligands also strongly depends on the *in vitro* or the *in vivo* assay system. Sexton and coworkers [25] summarized key features for the best-studied GPCR classes on structural details that determine functional selectivity. For class A GPCRs, it was determined that signaling bias results from the formation of a hydrogen-bond network that stabilizes the active G protein signaling state through novel interactions with the highly conserved NPxxY motif [122]. In a spectroscopy study with β_2AR [25, 123], G protein signaling was promoted through conformational shifts on TM6 and TM7 [25, 122]. For ligands, which displayed limited G protein downstream activity and therefore favored β-arrestin-mediated signaling, only changes on TM7 were observed [25].

Similarly, differential β-arrestin recruitment with changes on TM7 was observed for opioid receptors for 3-iodobenzoyl naltrexamine, $5HT_{1B}$ and $5HT_{2B}$ for ergotamine [25]. Moreover, the phosphorylation state of the C-terminus of GPCRs was also shown to result in different arrestin arrangements [25]. This was explained by Sexton et al., by the fact that β-arrestin exists in two major conformations: a tail and a core conformation which differ in different GPCR–β-arrestin-pairs [25]. It still remains unclear to which extent the C-terminus undergoes structural reorganization upon phosphorylation by GPCRKs to influence receptor and arrestin conformations [25].

The current model for Class B GPCRs activation for peptide agonists comprises the reorganization of the receptor's ECLs, which accommodate the peptide binding to the receptor core and causes conformational changes to it [25]. Toward the intracellular face of the receptor, hydrophobic residues stabilize G protein-complexed (active and inactive) states which are mostly localized in TM6 [25]. By comparing the structures of the GLP-1R, bound to its physiological ligand GLP-1 [24, 124] or the biased agonist exendin-p5 [25, 92], key differences within the ECL3 and the top of TM1 were observed [25].

The structural hallmark of class C GPCRs are the very large extracellular domains, the conserved "venus fly trap" (VFT) module containing the endogenous ligand-binding site [25]. Orthosteric agonist binding to the VFT leads to a shift of this module from an open to a closed state which alters the orientations of the TMs [25]. In addition, the receptors always exist as dimers. Until now, no structures of full-length receptors are available, and the nature of the dimers and which TMs are relevant in dimer-formation remain unclear. There is evidence, that the cysteine disulfide bridging of the TM4–TM5 interface in the $mGlu_2R$ dimer subunits prevents agonist-mediated activation, while TM6–TM6 crosslinking leads to a constitutively active receptor [25, 125]. Despite the limited information about this class C receptor activation mode, many allosteric modulators were discovered for mGluRs which bind within the TMs (Table 5.1) [25, 185]. According to Sexton and coworkers [25], most positive allosteric modulators (PAMs) lack intrinsic efficacy, although they could activate N-terminally truncated receptors lacking the extracellular domain, which can be interpreted as that this domain contributes to the maintenance of an inactive receptor state [25].

More recently, bitopic ligands have emerged as a new approach to further investigate biased GPCR signaling [155]. Bitopic ligands comprise two pharmacophores, where one binds to the orthosteric site of a receptor, while the other binds to the allosteric site [155]. By simultaneously targeting the ortho- and allosteric site of a receptor, these ligands show increased subtype selectivity and high affinity associated with well-defined pharmacophores [155]. This approach has been successful in targeting D_3R [186]. Consequently, bitopic ligands are able to stabilize a specific receptor conformation thereby facilitating or inhibiting

Table 5.1 Representatives for allosteric modulators of G protein-coupled receptors.

GPCR/Ligand	Allosteric binding mode	Biased signaling	Reference
Class A			
5-Hydroxytryptamine 1A receptor ($5HT_{1A}R$)			
Anandamide	Endogenous modulator		[113]
Cholesterol	Endogenous modulator		[113]
5-Hydroxytryptamine 1B receptor ($5HT_{1B}R$)			
5HT-moduline	Endogenous modulator		[113]
Sodium	Endogenous modulator		[113, 126]
5-Hydroxytryptamine 1D receptor ($5HT_{1D}R$)			
5HT-moduline	Endogenous modulator		[113]
5-Hydroxytryptamine 2A receptor ($5HT_{2A}R$)			
Anandamide	Endogenous modulator		[113]
Oleamide	Endogenous modulator		[113]
Sodium	Endogenous modulator		[113]
5-Hydroxytryptamine 2C receptor ($5HT_{2C}R$)			
CYD-1–79 cis-4-Undecyl-piperidine-2-carboxylic acid (2,3-dihydroxypropyl)amide	PAM		[127]
PNU-69176E	PAM, Ago-PAM		[127]
VA-024	PAM		[127]
Anandamide	Endogenous modulator		[113]
Oleamide	Endogenous modulator		[113]
5-Hydroxytryptamine 4 receptor ($5TH_4R$)			

IgG	Endogenous modulator		[113]
5-Hydroxytryptamine 7 receptor ($5TH_7R$)			
Anandamide	Endogenous modulator		[113]
Adenosine A_1 receptor (A_1R)			
Amilorides	NAM		[108, 128]
LUF5484 2-Amino-3-benzoylthiophene (2A3BT) derivatives	PAM	Inhibits G protein signaling pathway	[113, 129]
LUF6258	Ago-PAM; hybrid ortho/allosteric ligand		[108]
"PD" series in Brun et al. (1990) → PD81723; PD 71,605; PD 81,723; PD 117,975; PD81723 2A3BT derivatives	PAMs		[108, 113, 127, 130, 131]
SCH-202676 *N*-(2,3-diphenyl-1,2,4-thiadiazol-5-(2*H*)-ylidene)methanamine	PAM and NAM	PAM and NAM activity dependent of bound ligand	[127, 132]
T62 2A3BT derivatives	PAM	Inhibit G protein signaling pathway	[127, 129]
VCP333 2A3BT derivatives	PAM	cAMP > pERK1/2	[127, 129]
VCP520 2A3BT derivatives	PAM	cAMP > pERK1/2	[110, 127, 129]
VCP746	Allosteric and bitopic ligand	Yes	[133]
Sodium	Endogenous modulator		[113, 128]
Adenosine A_{2A} receptor ($A_{2A}R$)			

(continued)

Table 5.1 (Continued)

GPCR/Ligand	Allosteric binding mode	Biased signaling	Reference
1-[4-(9-Benzyl-2-phenyl-9*H*-purin-6-ylamino)-phenyl]-3-phenyl-urea derivatives	PAMs		[132]
1-[4-(9-Benzyl-2-phenyl-9H-8-azapurin-6-ylamino)-phenyl]-3-phenyl-urea derivatives	PAMs		[132]
Amilorides	NAM		[108, 127, 128, 131]
HMA 5-(*N*,*N*-hexamethylene)amiloride	NAM		[127]
Sodium	Endogenous modulator		[113, 128]
Adensoine A_{2B} receptor ($A_{2B}R$)			
1-Benzyl-3-ketoindole derivatives	PAMs and NAMs		[127, 132]
Amilorides	NAM		[108]
Adenosine A_3 receptor (A_3R)			
3-(2-Pyridinyl)isoquinoline derivatives	NAMs		[132]
Amilorides	NAM		[108]
DU124183 1*H*-Imidazo-[4,5-*c*]quinolin-4-amines	NAM		[113, 134]
LUF6000	PAM		[127, 132, 135]
LUF6096	PAM		[132, 135]
MRS542	Ago-PAM	Partial agonist only in β-arrestin translocation	[135]
MRS1760	Ago-PAM	Partial agonist only in β-arrestin translocation	[135]

VU5455Z 3-(2-Pyridinyl)isoquinoline	PAM		[132, 134]
VU8507Z	PAM		[108, 136]
Xanthine 7-riboside (DBXRM)	Ago-PAM	Partial agonist only in β-arrestin translocation	[135]
2-Arachidonylglycerol	Endogenous modulator		[113]
Sodium	Endogenous modulator		[113]
Adrenergic α_1 receptor (α_{1A}AR)			
9-Aminoacridine	NAM		[137, 138]
ρ-TIA (conotoxin)	NAM		[113, 137]
Amilorides	NAM		[113, 128]
Diazepam	PAM		[113, 137]
Midazolam	PAM		[113, 137]
Lorazepam	PAM		[113, 137]
IgG	Endogenous modulator		[113]
Sodium	Endogenous modulator		[113, 128]
Adrenergic α_{2A} receptor (α_{2A}AR)			
Amilorides	NAM		[113, 128]
Cholesterol	Endogenous modulator		[113]
Sodium	Endogenous modulator		[113, 128]
Adrenergic α_{2B} receptor (α_{2B}AR)			

(*continued*)

Table 5.1 (Continued)

GPCR/Ligand	Allosteric binding mode	Biased signaling	Reference
Amilorides	NAM		[113, 128]
Sodium	Endogenous modulator		[113, 128]
Adrenergic α_{2D} receptor (α_{2D}AR)			
Agmatine	PAM, endogenous modulator		[113, 139]
Sodium	Endogenous modulator		[113, 128]
Adrenergic β_1 receptor (β_1AR)			
Magnesium	Endogenous modulator		[113]
Manganese	Endogenous modulator		[113]
IgG	Endogenous modulator		[113]
Adrenergic β_2 receptor (β_2AR)			
Compound-15	NAM		[127]
Cholesterol	Endogenous modulator		[113]
IgG	Endogenous modulator		[113]
Zinc	Endogenous modulator		[113, 136]
Angiotensin II receptor type 1 (AT_1R)			
TRV027	NAM	β-Arrestin	[114]
IgG	Endogenous modulator		[113]
Sodium	Endogenous modulator		[113]
Cannabinoid receptor 1 (CB_1)			

Cannabidiol	NAM	Alters the signaling of CB1/CB2 heteromers; CB1: $G\alpha s > G\alpha i/o > \beta arr1$, $G\alpha q$, $G\beta\gamma$	[127, 140]
GAT211	PAM		[127]
GAT228	PAM		[127]
GAT229	PAM		[127]
Fenofibrate	NAM		[127]
Lipoxin A4	PAM		[127]
ORG27569	PAM, NAM	β-Arrestin, inhibits G protein signaling → pathway- and time-specific NAM for Gαi (CP55940); allosteric agonist for ERK1/2 phosphorylation; PAM for CP55940 affinity	[112, 113, 127]
ORG29647	NAM	β-Arrestin inhibits G protein signaling	[112, 113, 127]
Pregnenolone	NAM	Yes Inhibits pERK1/2 but not cAMP signaling	[127, 140]

(continued)

Table 5.1 (Continued)

GPCR/Ligand	Allosteric binding mode	Biased signaling	Reference
PSNCBAM-1	NAM	β-Arrestin inhibits G protein signaling	[112, 113, 127, 140, 144]
RTI-371	PAM		[112, 113, 127]
ZCZ011	PAM		[127]
Pepcan-12	Endogenous modulator		[113]
Pregnenolone	Endogenous modulator		[113]
Sodium	Endogenous modulator		[113]
C–C chemokine receptor type 1 (CCR1)			
BX-471	—		[113]
CP-481715	—		[113]
RVSM Arg-Val-Ser-Met	—		[136]
ASLW (Ala-Ser-Leu-Trp)	—		[136]
Trichosanthin	—		[136]
C–C chemokine receptor type 2 (CCR2)			
CCR2-RA-[R] (2*R*)-3-Acetyl-1-(4-chloro-2-fluorophenyl)-2-cyclohexyl-4-hydroxy-2*H*-pyrrol-5-one	NAM		[113, 127]
JNJ-27141491	—		[113]
S.D.-24	—		[113]

C–C chemokine receptor type 3 (CCR3)		
BX-471	—	[136]
CP-481-715	—	[136]
TAK779	—	[113]
UCB35625	—	[113]
C–C chemokine receptor type 4 (CCR4)		
Pyrazinyl sulfonamides	—	[113]
C–C chemokine receptor type 5 (CCR5)		
AK530	—	[113, 136]
Ancriviroc	NAM	[113, 136]
Aplaviroc	NAM	[109, 113, 136]
AK602 or GSK-873140		
Maraviroc	NAM	[126, 127]
SCH 351125	NAM	[109, 113]
TAK-K220	NAM	[113, 136, 145]
TAK779	NAM	[109, 113, 136]
Vicriviroc	NAM	[113, 136, 145]
Trichosanthin	NAM	[113, 136]
C–C chemokine receptor type 9 (CCR9)		
Vercirnon	NAM	[127]

(continued)

Table 5.1 (Continued)

GPCR/Ligand	Allosteric binding mode	Biased signaling	Reference
C–X–C chemokine receptor type 1 ($CXCR_1$)			
Reparixin	NAM		[113, 127, 136]
AZ compounds A&B	—		[113]
Navarixin SCH 527123	NAM		[113, 136]
C–X–C chemokine receptor type 1 and 2 ($CXCR_1$/$CXCR_2$)			
Ladarixin	NAM		[127]
C–X–C chemokine receptor type 2 ($CXCR_2$)			
AZ compounds A&B	—		[113]
Navarixin SCH 527123	NAM		[113, 136]
Elubrixin SB 656933	NAM		[113, 136]
C–X–C chemokine receptor type 3 ($CXCR_3$)			
8-Azaquinazolinone derivatives	PAMs	Inhibits CXCL11 effects	[127]
IP-10 CXCL10	Possible PAM	β-Arrestin > G protein	[113, 136, 146]
I-TAC CXCL11	Possible NAM	G protein > β-arrestin	[113, 136, 146]
VUF10661	PAM		[130]
C–X–C chemokine receptor type 4 ($CXCR_4$)			
ASLW (Ala-Ser-Leu-Trp)	PAM	Substantial activation > internalization	[110, 113]

ATI-2341 (pepducin)	Ago-PAM	Gαi > Gα13, but does not induce β-arrestin recruitment	[124, 147, 148]
Plerixafor AMD3100	NAM		[127, 130]
RVSM Arg-Val-Ser-Met	—		[113, 130]
Trichosanthin	NAM		[113, 136]
C–X–C chemokine receptor type 7 ($CXCR_7$)			
GSLW Gly-Ser-Leu-Trp	—		[113]
Cholecystokinin A receptor (CCK_1)			
Benzodiazepines	—		[113]
Devazepide	NAM	Inhibitor of CCK binding	[113, 149]
T-0632 1*H*-Indole-3-propanoic acid	Allosteric antagonist		[113]
GI181771X	NAM		[113, 150]
SR146131	PAM-agonist		[150]
Cholecystokinin B receptor (CCK_2)			
GI181771X	NAM		[113, 150]
Dopamine receptor D_1 (D_1R)			
DETQ	PAM		[151, 152]
[2-(2,6-dichlorophenyl)-1-((1*S*,3*R*)-3-(hydroxy-methyl)-5-(2-hydroxypropan-2-yl)-1-methyl-3,4dihydroiso-quinolin-2(1*H*)-yl)ethan-1-one]			

(*continued*)

Table 5.1 (Continued)

GPCR/Ligand	Allosteric binding mode	Biased signaling	Reference
LY3154207	PAM		[151]
BMS compound4	PAM		[151]
[1-((rel-1*S*,3*R*,6*R*)-6-(benzo[*d*][1,3]dioxol-5-yl) bicyclo[4.1.0]heptan-3-yl)-4-(2-bromo-5-chlorobenzyl) piperazine]			
BMS compound5	PAM		[151]
[rel-(9*R*,10*R*,12*S*)-*N*-(2,6-dichloro-3-methylphenyl)-12-methyl-9,10-dihydro-9,10-eth-anoanthracene-12-carboxamide]			
MLS6585	PAM		[151, 153]
MLS1082	PAM		[151, 153]
Compound B	PAM		[151]
Compound 6 (Astellas)	PAM		[151]
ASP4345	PAM		[151, 154]
Compound 7 (UCB)	PAM		[151]
Zinc	Endogenous modulator		[113]
Dopamine receptor D2 (D_2R)			
AB02-12	PAM DRD2, DRD3		[155, 156]
PLG (Pro-Leu-Gly); melanocyte-stimulating hormone release inhibiting factor (MIF-1)	PAM DRD2 and DRD4		[127, 147]
PAOPA (3(*R*)-[(2(*S*)-pyrrolidinyl carbonyl) amino]-2-oxo-1-pyrrolidine acetamide)	DRD2-selective PAM		[127, 157]
SB269652	NAM DRD2 and DRD3	Bitopic ligand, only NAM activity in DRD2 homodimers	[113]

IgG	Endogenous modulator		[113]
Melanotropin release inhibiting factor 1	Endogenous modulator		[113]
Zinc	Endogenous modulator		[113]
Dopamine receptor D3 (D_3R)			
AB02-12	PAM DRD2, DRD3		[155, 156]
SB269652	NAM DRD2 and DRD3	Bitopic ligand, only NAM activity in DRD2 homodimers	[127, 158, 159]
Dopamine receptor D4 (D_4R)			
PLG (Pro-Leu-Gly); melanocyte-stimulating hormone release inhibiting factor (MIF-1)	PAM DRD2 and DRD4		[127, 147]
Endothelin A receptor (ET_AR)			
Acetylsalicylic acid	NAM		[113, 160]
Sodium salicylate	NAM		[113, 160]
IgG	Endogenous modulator		[113]
Sodium	Endogenous modulator		[113]
Free fatty acid receptor 1 (FFA1R)			
TAK 875	PAM		[113, 127]
AM 8182	PAM		[127]
AM 1638	PAM		[113, 127]
AMG 837	PAM		[127]
MK8666	PAM		[127]
AP8	PAM, Ago-PAM		[127]

(continued)

Table 5.1 (Continued)

GPCR/Ligand	Allosteric binding mode	Biased signaling	Reference
TAK-875	Ago-PAM		[127]
Docosahexaenoic acid	Endogenous modulator		[113]
Free fatty acid receptor 2 (FFA2R)			
4-CMTB 4-Chloro-α-(1-methylethyl)-*N*-2-thiazolylbenzeneacetamide	PAM		[127]
AZ1729	PAM		[127]
Phenylacetamide 1	Ago-PAM		[113, 147]
Phenylacetamide 2	Ago-PAM		[113, 147]
Free fatty acid receptor 3 (FFA3R)			
Hexahydroquinolone-3-carboxamide compound 1	—		[113, 127]
Hexahydroquinolone-3-carboxamide compound 109	Ago-PAM		[127]
Follicle-stimulating hormone receptor (FSHR)			
ADX61623	NAM	Inhibits cAMP and progesterone but not estradiol production; increases FSH binding	[161]
ADX68692	NAM	Inhibits cAMP and progesterone and estradiol production	[161]
ADX68693	NAM	Inhibits cAMP and progesterone but not estradiol production	[161]
BMS compounds 2–7	—		[113]
Org 24444-0	PAM		[161]

Tetrahydroquinolines	NAMs	Inhibition of FSHR-induced cAMP production without inhibiting FSH binding	[161]
Gonadotropin-releasing hormone receptor (GnRHR)			
Furan-1	—		[113]
HMA 5-(*N*,*N*-hexamethyl-ene)amiloride	—		[113]
Growth hormone secretagogue receptor (GHSR)			
L-692,429	—		[113]
GH-releasing peptide 6	—		[113]
Luteinizing hormone receptor (LHR)			
Org 41841	—		[113]
Org42599	—		[130]
[3*H*]-Org 43553	—		[113]
Muscarinic acetylcholine receptor M_1 (M_1)			
77-LH-28-1	PAM, Ago-PAM		[130, 135]
AC260584	PAM		[130]
AC-42	PAM, Ago-PAM		[130, 135]
Brucine	PAM		[113]
Benzyl quinolone carboxylic acid (BQCA)	PAM		[113, 127, 162, 163]

(*continued*)

Table 5.1 (Continued)

GPCR/Ligand	Allosteric binding mode	Biased signaling	Reference
Clozapine	Ago-PAM		[135]
MK7622	—		[113]
ML169	—		[113]
MT3	—		[113]
MT7	—		[113]
N-Desmethylclozapine	Ago-PAM		[135]
Staurosporine	—		[113]
Tacrine	—		[113]
TBPB	PAM		[130]
VU0029767	PAM	$G\alpha q > G\alpha_{12}$ PAM for intracellular Ca^{2+} mobilization (*ACh*); NAL for PLD activation (*ACh*); weak PAM for PI hydrolysis (*ACh*)	[113, 158, 164]
VU0090157	PAM		[164]
VU0108370	PAM		[127]
VU0448350	PAM		[127]
VU0119498	—		[113]
Arachidonic acid	Endogenous modulator		[113]
IgG	Endogenous modulator		[113]

Sodium	Endogenous modulator		[113]
Muscarinic acetylcholine receptor M_2 (M_2)			
Alcuronium	NAM, PAM		[113, 145]
C7/3-phth	NAM		[113]
DUO3	—		[113]
Gallamine	NAM		[113, 130, 145]
LY2033298	PAM	PAM for ERK1/2 phosphorylation (*ACh*, *Oxo*, *Oxo-M*, *TMA*, *or McN-343*); NAM for ERK1/2 phosphorylation (*Pilocarpine or Xanomeline*)	[109, 110, 113]
LY2119620	PAM		[113, 127]
Tacrine	—		[113]
W84	—		[113]
Arachidonic acid	Endogenous modulator		[113]
Dynorphin-A	Endogenous modulator		[113]
IgG	Endogenous modulator		[113]
Myelin basic protein	Endogenous modulator		[113]
Major basic protein	Endogenous modulator		[113]

(continued)

Table 5.1 (Continued)

GPCR/Ligand	Allosteric binding mode	Biased signaling	Reference
Protamine	Endogenous modulator		[113]
Sodium	Endogenous modulator		[113]
Muscarinic acetylcholine receptor M_3 (M_3)			
Brucine	PAM	Only in the mutant M_3K7.32E → Allosteric PAM agonist for Gαq (*CCh*); PAM for Gα12 (*CCh*); NAL for Gαi (*CCh*)	[113, 135]
CNO	PAM	β-Arrestin	[135]
N-Chloromethyl	—		[113]
WIN62577	—		[113]
VU0119498	—		[113]
Arachidonic acid	Endogenous modulator		[113]
IgG	Endogenous modulator		[113]
Sodium	Endogenous modulator		[113]
Muscarinic acetylcholine receptor M_4 (M_4)			
Alcuronium	—		[113]
LY2033298	PAM	Functional cooperativity factors	[113, 127, 130, 158, 165, 166]
LY2119620	—		[113]
ML108	PAM		[127]

ML173	PAM	[127]
ML293	PAM	[127]
MT3	—	[113]
Thiochrome	—	[113]
VU0010010	—	[113]
VU0152099	PAM	[113, 127]
VU0152100	—	[113]
VU0464090	PAM	[127]
VU10010	PAM	[127, 158]
VU6000918	PAM	[127]
Arachidonic acid	Endogenous modulator	[113]
Sodium	Endogenous modulator	[113]
Muscarinic acetylcholine receptor M_5 (M_5)		
ML129 (VU0238429)	PAM	[127]
ML326	—	[113]
ML375	NAM	[113, 127]
ML380	—	[113]
ML381	—	[113]
VU0119498	Pan-PAM	[127]

(continued)

Table 5.1 (Continued)

GPCR/Ligand	Allosteric binding mode	Biased signaling	Reference
VU6000181	NAM		[127]
Arachidonic acid	Endogenous modulator		[113]
Mu (μ) opioid receptor (MOR)			
BMS-986121	PAM		[127]
BMS-986122	PAM		[127]
BMS-986124	NAL		[127]
Cannabidiol	—		[113]
Ignavine	PAM		[127]
TRV130	—	G protein > β-arrestin	[114]
Magnesium	Endogenous modulator		[113]
Manganese	Endogenous modulator		[113]
Sodium	Endogenous modulator		[113]
Delta (δ) opioid receptor (DOR)			
BMS-986187	PAM		[113, 127]
BMS-986188	—		[113]
Cannabidiol	—		[113]
Manganese	Endogenous modulator		[113]
Sodium	Endogenous modulator		[113]
Delta/kappa (δ/κ) opioid receptors (DOR/KOR)			
BMS-986187	PAM		[127, 167]
Kappa (κ) opioid receptor (KOR)			

Magnesium	Endogenous modulator		[113]
Manganese	Endogenous modulator		[113]
Sodium	Endogenous modulator		[113]
Melanin-concentrating hormone receptor 1 (MCH_1R)			
MQ1	NAM		[127]
Neurotensin receptor 1 (NTSR1)			
ML301	PAM		[168]
ML314	PAM	β-Arrestin biased	[168]
SBI-553	PAM		[168]
Neurotensin receptor 2 (NTSR2)			
ML301	PAM		[168]
Neuropeptide Y receptor Y4 (PPY4R)			
tBPC	PAM		[127]
N-Methyl D-aspartate receptor (NMDAR)			
Ifenprodil	NAM		[145]
Oxoeicosanoid receptor 1 (OXER1)			
Gue1654	NAM	selectively inhibits Gβγ but not Gαi signaling induced by the OXE-R agonist, 5-oxo-ETE	[169]
Oxytocin receptor (OXTR)			

(*continued*)

Table 5.1 (Continued)

GPCR/Ligand	Allosteric binding mode	Biased signaling	Reference
Cholesterol	Endogenous modulator		[113]
Progesterone (rat)	Endogenous modulator		[113]
5-Dihydroprogesterone (human)	Endogenous modulator		[113]
Prostaglandin DP_2 receptor (DP_2)			
1-(4-Ethoxyphenyl)-5-methoxy-2-methylindole-3-carboxylic acid	NAM, PAM	NAMs for arrestin recruitment (*prostaglandin D2*) PAM for G protein signaling	[110, 113, 170]
N_{α}-tosyltryptophan	PAM		[170]
Prostaglandin $F_{2\alpha}$ receptor ($PGF_{2\alpha}$)			
PDC113.824	PAM, NAM	NAMP for Gα 12 (*PGF2αP*); PAM for Gαq (*PGF2α*)	[113, 147]
Protease-activated receptor 2 (PAR2)			
AZ3451	NAM		[127]
P2Y purinoreceptor 1 (P2Y1R)			
BPTU *N*-[2-[2-(1,1-Dimethylethyl)phenoxy]-3-pyridinyl]-*N′*-[4-(trifluoromethoxy)phenyl]urea	PAM		[127]
P2Y purinoreceptor 2 (P2Y2R)			
2,2′-Pyridylsatogen tosylate	—		[113]
Compound **89**	PAM		[127]
Rhodopsin			
Cholesterol	Endogenous modulator		[113]

Relaxin receptor 1 (RXFP1)			
ML290	—		[113]
Relaxin receptor 3 (RXFP3)			
135PAM1	—		[113]
Sphingosine 1-phosphate receptor 2 (S1PR2)			
CYM-5520	—		[113]
Sphingosine 1-phosphate receptor 3 (S1PR3)			
CYM-5541	—		[113]
Tachykinin receptor 1 (NK1)			
Heparin	—		[113]
Tachykinin receptor 2 (NK2)			
[*N*-Benzyl, *N*-(2-naphthylmethyl)-amino]-acetonitrile	—		[113]
Compound **184**	NAM	Yes	[127]
LPI805	NAM, PAM	NAM for cAMP production (*Neurokinin A*) NAL for intracellular Ca^{2+} mobilization (*Neurokinin A*) PAM: G_s > Gq	[113, 145]
Thyrotropin receptor (TSHR)			
IgG	Endogenous modulator		[113]

(*continued*)

Table 5.1 (Continued)

GPCR/Ligand	Allosteric binding mode	Biased signaling	Reference
Class B			
Calcitonin receptor (CT)			
Pyrazolopyrimidines 2d, 2e, 2f, 2g	—		[113]
Calcitonin gene-related peptide receptor (CGRPR)			
BIBN4096BS	—		[113]
Corticotropin-releasing hormone receptor 1 (CRF_1R)			
Antalarmin	—		[113]
CP-376395	—		[113]
DMP696	—		[113]
N_α-tosyltryptophan	PAM	G_i + β-arrestin > G_i	[145]
NBI 27914	—		[113]
NBI 35965	—		[113]
Glucagon receptor (GCGR)			
Antibody mAb7	NAM		[171]
Bay27–9955	—		[113]
L-168,049	—		[113]
Glucagon-like peptide-1 receptor (GLP1R)			
1,3-bis[[4-tert-butoxy-carbonylamino)benzoyl]amino]-2,4-bis[3-methoxy-4-(thiophene-2-carbonyloxy)-phenyl] cyclobutane-1,3-dicarboxylic acid (Boc5); substituted cyclobuthanes	PAM		[171]

4-(3-Benzyloxyphenyl)-2-(ethylsulfinyl)-6-(trifluoromethyl) pyrimidine (BETP)	PAM	Potentiates oxyntomodulin-dependant increase of cAMP	[113, 171]
6,7-Dichloro-2-methylsulfonyhl-3-*tert*-butylaminoquinoxaline (NOVO2, compound 2); Novo Nordisk Compounds 1–6 (quinoxalines)	PAM	Potentiates oxyntomodulin-dependant increase of cAMP > Ca^{2+}	[109, 113, 130, 171]
TT-15	PAM		[171]
TTP273	PAM		[171]
Class C			
Calcium sensing receptor (CaSR)			
Calhex 231	—		[113]
Cinacalcet	PAM	Different kinetics of ERK1/2 phosphorylation, depending on the levels of Ca_o^{2++} / PAM for intracellular Ca^{2+} mobilization (Ca^{2++}) NAL for ERK1/2 phosphorylation (Ca^{2+})	[113, 126, 135, 165]
Fendeline	—		[113]
NPS 467	—		[113]
NPS R-568 *N*-(3-[2-chlorophenyl]propyl)-(*R*)-α-methyl-3-methoxybenzylamine	PAM		[113, 130, 172]

(*continued*)

Table 5.1 (Continued)

GPCR/Ligand	Allosteric binding mode	Biased signaling	Reference
NPS-R467	PAM	Different kinetics of ERK1/2 phosphorylation, depending on the levels of Ca_o^{2+}	[165, 173]
NPS-R568	PAM		[148]
NPS-R2143 2-Chloro-6-[(2*R*)-2-hydroxy-3-[(2-methyl-1-naphthalen-2-ylpropan-2-yl)amino] propoxy]benzonitrile	NAM	Different kinetics of ERK1/2 phosphorylation, depending on the levels of Ca_o^{2+}	[113, 148, 165, 172]
L-Phe	Endogenous modulator		[113]
L-Trp	Endogenous modulator		[113]
L-Tyr	Endogenous modulator		[113]
Glutathione	Endogenous modulator		[113]
IgG	Endogenous modulator		[113]
γ-Aminobutyric $acid_B$ receptor ($GABA_B$ R)			
CGP7930	PAM		[113, 174]
CGP13501	PAM		[113, 174]
GS39783	PAM		[113, 174]
L-Leu	Endogenous modulator		[113]
L-Ile	Endogenous modulator		[113]

L-Phe	Endogenous modulator		[113]
IgG	Endogenous modulator		[113]
Metabotropic glutamate receptor type 1 (mGluR1)			
(−)-CPCCOEt	NAM		[113, 143, 175]
(−)-PHCCC	NAM	Equipotent PAM at mGluR4	[176]
[^{3}H]R214127	—		[113]
Bay36–7620	NAM		[113, 176]
CFMTI	NAM		[175]
cis-64a	—		[113]
EM-TBPC	—		[113]
FITM	NAM		[176]
Gadolinium (Gd^{3+})	PAM, NAM	PAM on agonist-mediated Gαq; NAM on Gαs pathway	[147]
JNJ-16259685	NAM		[113, 176]
NPS2390	—		[113]
Ro 01–6128	PAM		[113, 176]
Ro 67–4853	PAM		[176]
Ro 67–7476	PAM		[113, 176]

(*continued*)

Table 5.1 (Continued)

GPCR/Ligand	Allosteric binding mode	Biased signaling	Reference
Ro 07–11401	PAM		[113, 176]
VU-71	PAM		[176]
VU0405623	PAM		[176]
VU0465334	NAM		[176]
VU0469650	NAM		[175, 176]
VU0470300	NAM		[176]
VU0483737	PAM		[176]
VU0483605	PAM		[176]
YM-298198	—		[113]
IgG	Endogenous modulator		[113]
Metabotropic glutamate receptor type 2 (mGluR2)			
ADX71149/JNJ-40411813	PAM		[158, 177, 178]
AZD8529	PAM		[158, 177]
BINA	—		[113, 177]
CBiPES	—		[113]
DN10 nanobody	PAM, Ago-PAM		[179]
DN13 nanobody	PAM		
JNJ-40068782	—		[113, 177]
JNJ-46281222	PAM		[177]

LY181837	—	[113]
LY2119620	PAM	[131]
LY2607540	—	[113]
LY487379	—	[113]
MNI-137	—	[113]
Ro-4491533	NAM	[113, 177]
Ro-4995819/Decoglurant	NAM	[177]
Ro-4988546	—	[113]
Ro-5488608	—	[113]
Ro-676221	NAM	[177]
SAR218645	PAM	[175]
VU6001966	NAM	[175]
mGluR2/mGluR4 heteromer complex		
VU0155041	PAM	[158]
Lu AF21934	PAM	[158]
Metabotropic glutamate receptor type 3 (mGluR3)		
AZD8529	—	[113]
MNI-137	—	[113]
RO4491533	—	[113]

(*continued*)

Table 5.1 (Continued)

GPCR/Ligand	Allosteric binding mode	Biased signaling	Reference
RO4988546	—		[113]
RO5488608	—		[113]
VU0650786	NAM		[175]
Metabotropic glutamate receptor type 4 (mGluR4)			
(−)-PHCCC	PAM	Equipotent NAM at mGluR1	[176]
Histamine	PAM	Gq > Gαi/o	[135]
LY2033298	NAL		[131]
MPEP	—		[113]
ML182	PAM		[176]
PTX002331	PAM		[175]
SIB-1893	—		[113]
Valiglurax	PAM		[175]
VU0080421	—		[113]
VU0155041	—		[113]
VU0155094	—		[113]
VU0359516	PAM		[176]
VU0400195	PAM		[176]
VU0405623	PAM		[176]
VU0422288	—		[113]

VU0652957	PAM		[175]
mGluR4/mGluR4 homodimer complex			
PHCCC	PAM		[158]
VU0155041	PAM		[158]
Metabotropic glutamate receptor type 5 (mGluR5)			
5-MPEP 5-Methyl-6-(phenylethynyl)pyridine	PAM		[113, 180]
5PAM523	—		[113]
ADX-47273	—		[113]
Basimglurant	NAM		[175]
CDPPB (3-Cyano-*N*-(1,3-diphenyl-1*H*-pyrazol-5-yl)benzamide)	PAM	Yes	[111, 113, 130, 175]
CPPHA *N*-(4-Chloro-2-[(1,3-dioxo-1,3-dihydro-2*H*-isoindol-2-yl)methyl]phenyl)-2-hydroxybenzamide	PAM, NAM	PAM for intracellular Ca^{2+} mobilization (*Glutamate or DHPG*); NAM for ERK1/2 phosphorylation (*Glutamate or DHPG, low concentrations*); NAM for ERK1/2 phosphorylation (*Glutamate or DHPG, high concentrations*)	[113, 135, 181]

(*continued*)

Table 5.1 (Continued)

GPCR/Ligand	Allosteric binding mode	Biased signaling	Reference
CTEP	NAM		[175]
DFB	—		[113]
DMeOB	—		[113]
DCB	—		[113]
Dipraglurant	NAM		[182]
DPFE (1-(4-(2,4-Difluorophenyl) piperazin-1-yl)-2-((4-fluorobenzyl)oxy)ethanone)	PAM	yes	[111, 113]
EMMCA	NAM	β-Arrestin > G protein	[135]
Fenobam	NAM		[113, 182]
Mavoglurant	NAM		[113, 175]
M-5MPEP	NAM	NAM for intracellular Ca^{2+} mobilization (*quisqualate or DHPG*)	[113, 182]
MPEP 2-Methyl-6-(phenylethynyl)pyridine	NAM	Ca^{2+} > Inositol	[113, 143]
MTEP	NAM		[113, 182]
SIB-1757	NAM		[175]
SIB-1893	NAM		[175]
VU-29	—		[113]
VU-0357121	—		[113]

VU-0360172 (*N*-Cyclobutyl-6-((3-fluorophenyl) ethynyl)picolinamide)	PAM	yes	[111, 113]
VU-0366058	NAM		[182]
VU-0366248	NAM		[182]
VU-0409106	NAM		[182]
VU0409551, ((4-fluorophenyl)(2-(phenoxymethyl)-6,7-dihydrooxazolo[5,4-c]pyridin-5(4H)-yl)methanone)	PAM	Yes	[111, 158, 175, 183]
VU0424465, (R)-5-((3-fluo-rophenyl)ethynyl)-N-(3-hydroxy-3-methylbutan-2-yl)picolinamide	PAM	Yes	[111]
IgG	Endogenous modulator		[113]
Metabotropic glutamate receptor type 6 (mGluR6)			
PHCCC	—		[113]
VU-0155041	—		[113]
VU-0422288	—		[113]
Metabotropic glutamate receptor type 7 (mGluR7)			
ADX-71743	NAM		[113, 175]
AMN-082	—		[113, 130]
MMPIP 6-(4-Methoxyphenyl)-5-methyl-3-(4-pyridinyl)-isoxazolo[4,5-*c*]pyridin-4(5*H*)-one	NAM		[113, 184]
VU-0155041	—		[113]

(*continued*)

Table 5.1 (Continued)

GPCR/Ligand	Allosteric binding mode	Biased signaling	Reference
VU-0155094	—		[113]
VU-0422288	—		[113]
Metabotropic glutamate receptor type 8 (mGluR8)			
AZ-12216052	PAM		[113, 175]
VU-0155041	—		[113]
VU-0155094	—		[113]
VU-0422288	—		[113]
Taste receptor type 1 member 1 (T1R1)			
IMP	—		[113]
S-807	—		[113]
Taste receptor type 1 member 2 (T1R2)			
S-819	—		[113]
SE-2	—		[113]
SE-3	—		[113]
Senomyx	—		[113]
Taste receptor type 1 member 3 (T1R3)			
Cyclamate	—		[113]
Lactisole	—		[113]

Class F		
Smoothened receptor		
Sant1	—	[113]
Sant2	—	[113]
Sonedigib	—	[126]
nat-20(*S*)-OHC	—	[113]
Oxysterols	Endogenous modulator	[113]
Vismodegib	—	[126]
Frizzled4 receptor		
FzM1	—	[113]
Orphan GPCRs		
GPR68		
Ogerin	PAM	[126]

Endogenous modulators: Note that some of these examples are more appropriately considered putative endogenous allosteric modulators.
Abbreviations: Ago-PAM (agonistic-positive allosteric modulator), AB (AbbVie), ADX (Addex), AMD (Sanofi Aventis), AZ/AZD (AstraZeneca), BCQA (benzyl quinolone carboxylic acid), BMS (Bristol-Myers Squibb), BAY (Bayer AG), CP (C.P. Pharmaceuticals), CGP (Ciba Geigy), DMP (DuPont Merck Pharmaceuticals), GSK (GlaxoSmithKline), HMA (5-(*N,N*-hexamethylene)amiloride), JNJ (Johnson & Johnson), L (Labaz Group), LY (Eli Lilly), MK (Merck & Co), MT (Micromet), NBI (Neurocrine Biosciences), NAM (negative allosteric modulator), ORG (Organon), Oxo (oxotremorine (1-(4-pyrrolidin-1-ylbut-2-yn-1-yl)pyrrolidin-2-one)), Oxo-M (oxotremorine methiodide (*N,N,N*-trimethyl-4-(2-oxo-1-pyrolidinyl)-2-butyn-1-ammonium iodide)), PAM (positive allosteric modulator), PD/PDC (Parke–Davis), PNU (Pharmacia & Upjohn), PLD (phospholipase D), RO (Roche), SB/SBI (SmithKline Beecham), SCH (Schering–Plough), TAK (Takeda), TMA (tetramethylammonium), TRV (Trevena Inc.), UCB (UCB Pharma).
Source: Based on Refs [25, 108, 147, 185].

the recruitment of either G proteins or β-arrestins and therefore diminishing functional selectivity [155]. In conclusion, quantifying biased signaling involves determining at least two or more cellular responses initiated by two or more agonists and comparing them [121].

5.5.2 Allosteric Modulation

Allostery is a phenomenon that describes the ability of a ligand to bind to a spatially distinct binding site from the orthosteric one [113]. From a theoretical perspective, allosteric phenomena of enzymes (for which allostery was first described) were expressed in models such as the Monod–Wyman–Changeux (MWC) and in the Koshland–Nemethy–Filmer (KNF) models [113, 127, 187, 188]. While the MWC model states allostery as a concerted process (i.e. conformational selection), the KNF model describes the phenomenon as a sequential process (i.e. conformational induction) [113]. When applying these models to GPCRs, it can be stated that each model reflects valid key aspects of ligand–receptor-binding events such as ligand-mediated shifts in the population of preexisting conformational ensembles of a receptor followed by changes in the interactive properties of new ensembles [113]. Other models, such as the ternary-complex-mass-action model describes in more detail the allosteric relationship between two ligands simultaneously binding to separate sites on the same receptor [175]. Since the mode of action of allosteric modulators is by definition to potentiate or inhibit the signaling of the bound orthosteric ligand, the model expresses this as "cooperativity factor" [175]. This framework provides a theoretical basis for the direction and magnitude of functional response of the orthosteric ligand when combined with an allosteric ligand [175]. Notably, the intrinsic GPCR signal transduction is allosteric as it involves an extracellular stimulus and subsequent propagation of a signal to the topologically distinct intracellular site which is occupied by β-arrestin and/or G proteins [113].

From a mechanistic point of view, allosteric modulators can either increase the functional response of an orthosteric ligand, which is defined as a PAM or decrease the functional response of an orthosteric ligand, which is defined as a negative allosteric modulator (NAM) [108, 127]. Besides these types of modulators, neutral/silent allosteric ligands (NALs/SALs) have been described [108, 127], which do not modulate the receptor function itself but rather compete with other allosteric ligands at the allosteric binding site [127]. Allosteric modulation activities in principle require the presence of an orthosteric ligand [108]. However, known exceptions are the so-called "Ago-PAMs" (or allosteric agonists), which potentiate the agonist's response as well as display intrinsic efficacy without the presence of an orthosteric agonist [127]. Notably, NAMs may act directly via allosteric antagonism of an orthosteric agonist's stimulation or they may assert antagonism by binding to an allosteric site and inducing changes to the receptor's structure or "life cycle" that prohibit eventual receptor activation by an orthosteric agonist [127].

Allosteric modulators display a unique range of activities compared to orthosteric molecules in particular antagonists and partial agonists [109]. The orthosteric ligands all preclude the access of other ligands (such as agonists) resulting in a preemptive system with a maximal result [109]. While antagonists lead to the full inactivation of a receptor, partial agonists cause only partial activation of the receptor [109]. The level of saturation of these effects is not necessarily conducted by allosteric modulation. According to Kenakin [109] allosteric modulators uniquely: (i) alter the interaction of very large proteins; (ii) possess the potential to modulate but not completely activate/inhibit receptor function; (iii) preserve physiological patterns; (iv) yield therapies with reduced side effects; (v) negative modulation can produce texture in antagonism (vi); have separate effects on agonist affinity and efficacy; (vii) show extraordinary selectivity for receptor types and (viii) exercise agonist dependence – since allosterism changes the shape of a protein it holds the possibility that such a change although catastrophic for one agonist, may have no effect at all for another [109].

Allosteric effects may be quantified *in vitro* by comparing dose–response curves, obtained in the presence of different concentrations of the allosteric ligand together with the agonist. In a review by Kenakin [109], they also mathematically approached how to calculate the allosteric effect in models. In terms of binding, the Stockton–Ehlert allosteric binding model is a standard to define the allosterism of a candidate which then can be used in the Black–Leff operational model of agonism [189]. The chemistry of allosteric modulators ranges from cations, small molecules, peptides, lipids, autoantibodies up to receptor dimerization (either homo- or heterodimers) (Table 5.1) [108]. Furthermore, very subtle degrees of positive or negative cooperativity (which may already be sufficient for certain GPCRs and pathophysiological conditions) were reported to result in an allosteric "effect ceiling" [113]. This mechanism increases the likelihood of on-target safety in overdose situations [113]. The ability to achieve unprecedented modes of on-target selectivity and/or fine-tune endogenous responses as a consequence of pure PAM or NAM activity proves how tight physiological regulation is vital, which is particularly relevant for neurodegeneration, schizophrenia, diabetes, and endocrine disorders [113].

Recently, allosteric modulation of GPCRs has emerged as a promising new area for the development of selective and potent ligands, especially for CNS disorders [127]. Two characteristics render the design and development of allosteric ligands interesting: the potential to exhibit greater receptor subtype selectivity, for two reasons: (i) a decreased evolutionary pressure as the allosteric binding sites are less conserved among receptor subtypes compared to the orthosteric ones, which are usually identical for a GPCR receptor family [127]; (ii) and/or selective cooperativity with an orthosteric site at one receptor subtype while exhibiting neutral cooperativity at the other subtypes of this receptor family [108, 113]. Second, the nature of cooperativity between orthosteric and allosteric sites on GPCRs, which has practical and therapeutic implications [113]. Although most studies on allosteric ligands of GPCRs are focused on exogenous modulators since these are therapeutically relevant, there is also evidence that endogenous ligands may act as allosteric modulators [113, 190]. The best-studied example for allosteric modulation at GPCRs is the positive cooperativity between the G protein-binding site and the orthosteric ligand-binding site [113]. Recently, the interaction of GPCRs with β-arrestins underwent a paradigm shift. Originally, it was believed to only be relevant for the termination of receptor activation, internalization, and recycling. Now, there is evidence that this protein scaffold can be involved in G protein-independent signal transduction [113]. However, this is under debate as already mentioned above. This implies that unique signaling properties result due to specific receptor conformations [113]. Other examples for GPCRs association with other proteins such as RAMPS (receptor activity-modifying G proteins) or MRAPS (melanocortin receptor accessory proteins), and the formation of homo- and hetero-GPCR-dimers display the variability to alter pharmacological properties of GPCR signaling [190]. Besides the ubiquitous allosteric sites at GPCRs utilized by β-arrestin and G proteins, it was shown for class A GPCRs that sodium is crucial for stabilizing the inactive state of many receptors via an allosteric site centered on the highly conserved Asp2.50 [113, 190]. In class C GPCRs, such as mGluRs, aromatic amino acids (L-Phe, L-Trp, L-Tyr) bind to a spatially distinct site next to the VFT [113]. For the calcium-sensing receptor, this leads to a potentiation of actions of extracellular calcium at a number of intracellular signaling pathways [113, 191–193].

Another concept of modulation has been introduced into GPCR signaling: functional selectivity for allosteric modulators [147]. According to Al-Qattan and Mordi, the classification of PAMs and NAMs is not precise enough, since the same allosteric ligand may cause different receptor downstream signaling effects, i.e. acts as PAM on certain pathways and as NAM on others which would be then described as biased signaling [108]. Such allosteric modulators, which include ions, ligands, small and large molecules (e.g. antibodies), and receptor-oligomers were reported to modulate hormone binding and/or intracellular coupling through differential cooperative effects on the binding of the orthosteric ligand (positive and negative; [147]). Furthermore, there are numerous endogenous allosteric modulators described (Table 5.1), which were reported to play crucial physiological roles in keeping diverse biological functions mediated [147]. Such endogenous modulators are ions, antibodies, amino acids, and other natural ligands (Table 5.1) [147]. In comparison to pure allosteric modulators and allosteric agonists/antagonists, bitopic ligands are capable of binding to and bridging both allosteric and orthosteric binding sites [108]. They are described as hybrid molecules, which possess orthosteric and allosteric pharmacophores with the intention to mediate a completely new pharmacology of a GPCR [113]. Some allosteric agonists have therefore been reclassified as bitopic ligands [113].

Taken together, it can be summarized that a deeper characterization of allosteric and biased ligands is necessary, although both represent a new generation of drugs. However, these binding mechanisms may actually comprise a more complex and highly dynamic pattern of functional selectivities than it could just be explained by either G protein or β-arrestin-dependent signaling pathways. To facilitate novel allosteric drug design, two requirements must be addressed. First, the full signaling behavior should be studied across multiple pathways in order to increase the likelihood that it results from different receptor conformations [165]. Second, the requirement of quantification of the affinity of the GPCR is of interest for allosteric ligands as well as the effects they mediate [165].

5.6 Case Study: Dopamine Receptor Family

In Section 5.2, it has been made clear that besides the number of GPCRs that are involved in biological processes and pathways, the complexity of the interactions they establish with intracellular partners (G proteins and arrestins) further extends the network of different effects that can emerge from GPCR-related activity.

Focusing on the five known human dopamine receptors (D_1R–D_5R) alone and their documented interactions with intracellular partners, 10 different G proteins and 2 arrestins and 40 dimeric complexes have been modelled. To calculate these complexes, of which most of the single chains' structures were unknown, the sequences were subjected to a homology modeling protocol, taking as templates structures with high sequence similarity and biological functional proximity. Furthermore, the DRs' structures were subjected to in-membrane MD energy minimization, followed by dimer interface structure refinement. Upon analyzing the complexes, structural and dynamic information were used to convey a phylogenetic approach. The latter shows clear distinctions between D_1-like (D_1R and D_5R) and D_2-like (D_2R, D_3R, and D_4R) complexes. Furthermore, it clearly groups the different intracellular partners upon complexation into three different classes: arrestins, G_i and G_s. The groupings sorted by the protocol are consistent with the literature referring to similar GPCRs and partners. When analyzing specific interaction motifs and interface amino acid content, complex-specific differences start to emerge, indicating partner specificity dependence on a residue level. In both, the G protein and arrestin systems, a high degree of structural complementarity was visible, although these two partner groups have significant differences. When looking at DRs interface, the ICLs, the perimembrane helix (HX8) and the intracellular side of the TMs, in particular TM3, TM5, and TM6, are key on the interaction with the partner. For the arrestins, the finger loop and the C-loop tend to be the most relevant regions for interaction. Regarding G proteins, the substructures H4, h4s6, and H5 play a large role in the interaction. Furthermore, specific interactions between DR-arrestins and DR-G protein were also found. For instance, in complexes with arrestins, HX8 tends to have more interacting residues with D_1-like receptors than with D_2-like receptors. Similarly, ICL3 seems to play an important part when considering DR–G protein interactions. These interactions seem to be a distinctive factor between arrestins and G protein complexes and could serve as a lead for drug discovery studies. The distances between TM3–TM6 and TM3–TM7 were also analyzed, as they are indicative of the activation state. For the DRs, it was noticed that G_s induces a larger TM3–TM6 and shorter TM3–TM7 distances than all the other analyzed partners [194].

Overall, this study is in agreement with the literature, when considering DRs are similar in function and structure to other GPCRs. Furthermore, it highlights structural and sequence-specific motifs that emerge in the GPCR–arrestin or GPCR–G protein complex interaction. By doing so, it underlines the large array of degrees of complexity that can emerge from only very few protein sequences.

General Considerations

GPCRs are the most representative family of membrane receptors, key players in several biological processes and, consequently, involved in a vast number of diseases. Among the different intracellular partners that directly interact with GPCRs, arrestins and G proteins are two major effectors, triggering different signaling pathways. This work gives an overview about recent highlights of the known structural determinants involved in the GPCR coupling to their intracellular partners, and the main factors engaged in the coupling selectivity, namely biased signaling, allosteric modulation, and electrostatic nature of GPCRs–partners interface.

It is highlighted that different structural and/or electrostatic hallmarks underlie the ligand binding and the intracellular partner coupling of the main classes of GPCRs. The electrostatic nature of GPCRs–partner interface was associated with the highest promiscuity involving GPCR–arrestin coupling. On the other hand, bitopic ligands, able to bind both orthosteric or allosteric sites, have emerged as an approach to study biased ligand signaling and, consequently, the GPCR selective coupling and signaling. It is worth mentioning that specific structural and sequence motifs were reported to be involved in the binding of different arrestins or

G proteins subfamilies, namely G_s, G_i, and G_q. Interestingly, an alternative intracellular coupling conformation was reported for the coupling of β-arrestin with AT_1R, revealing the complexity and the broad structural coverage implications engaged in the GPCR coupling.

Altogether, more efforts are required in order to better characterize the arrestin and G protein interactions with GPCRs, and how these interactions trigger distinct biological pathways.

Funding

This work was funded by the European Regional Development Fund (ERDF), through the Centro 2020 Regional Operational Programme under project CENTRO-01-0145-FEDER-000008: BrainHealth 2020, and through the COMPETE 2020 – Operational Programme for Competitiveness and Internationalisation and Portuguese national funds via FCT – Fundação para a Ciência e a Tecnologia, under projects POCI-01-0145-FEDER-031356, PTDC/QUI-OUT/32243/2017, and UIDB/04539/2020. António J. Preto, Carlos A.V. Barreto, B. Bueschbell and N. Rosário-Ferreira were also supported by FCT through PhD scholarship SFRH/BD/144966/2019, SFRH/BD/145457/2019, SFRH/BD/149709/2019 and PD/BD/135179/2017, respectively. Authors would also like to acknowledge ERNEST – European research Network on Signal Transduction, CA18133. M. Machuqueiro and P. Magalhães acknowledge the financial support from FCT (CEECIND/02300/2017, PTDC/BIA-BFS/28419/2017, and UID/MULTI/04046/2019).

References

1 Kobilka, B.K. (2007). G protein coupled receptor structure and activation. *Biochim. Biophys. Acta* 1768 (4): 794–807.

2 Moreira, I.S. (2014). Structural features of the G-protein/GPCR interactions. *Biochim. Biophys. Acta* 1840 (1): 16–33.

3 Rosenbaum, D.M., Rasmussen, S.G.F., and Kobilka, B.K. (2009). The structure and function of G-protein-coupled receptors. *Nature* 459 (7245): 356–363.

4 Lee, S.-M., Booe, J.M., and Pioszak, A.A. (2015). Structural insights into ligand recognition and selectivity for classes A, B, and C GPCRs. *Eur. J. Pharmacol.* 763 (Pt B): 196–205.

5 Peng, W.-T., Sun, W.-Y., Li, X.-R. et al. (2018). Emerging roles of G protein-coupled receptors in hepatocellular carcinoma. *Int. J. Mol. Sci.* 19 (5): 1366.

6 Hilger, D., Masureel, M., and Kobilka, B.K. (2018). Structure and dynamics of GPCR signaling complexes. *Nat. Struct. Mol. Biol.* 25 (1): 4–12.

7 Marinissen, M.J. and Gutkind, J.S. (2001). G-protein-coupled receptors and signaling networks: emerging paradigms. *Trends Pharmacol. Sci.* 22 (7): 368–376.

8 Tang, X., Wang, Y., Li, D. et al. (2012). Orphan G protein-coupled receptors (GPCRs): biological functions and potential drug targets. *Acta Pharmacol. Sin.* 33 (3): 363–371.

9 Civelli, O., Reinscheid, R.K., Zhang, Y. et al. (2013). G protein-coupled receptor deorphanizations. *Annu. Rev. Pharmacol. Toxicol.* 53 (1): 127–146.

10 Foster, S.R., Roura, E., Molenaar, P., and Thomas, W.G. (2015). G protein-coupled receptors in cardiac biology: old and new receptors. *Biophys. Rev.* 7 (1): 77–89.

11 Jialu, W., Clarice, G., and Rockman, H.A. (2018). G-protein–coupled receptors in heart disease. *Circ. Res.* 123 (6): 716–735.

12 Arakaki, A.K.S., Pan, W.-A., and Trejo, J. (2018). GPCRs in cancer: protease-activated receptors, endocytic adaptors and signaling. *Int. J. Mol. Sci.* 19 (7): 1886.

13 Insel, P.A., Sriram, K., Wiley, S.Z. et al. (2018). GPCRomics: GPCR expression in cancer cells and tumors identifies new, potential biomarkers and therapeutic targets. *Front. Pharmacol.* 9: 431.

14 Guerram, M., Zhang, L.-Y., and Jiang, Z.-Z. (2016). G-protein coupled receptors as therapeutic targets for neurodegenerative and cerebrovascular diseases. *Neurochem. Int.* 101: 1–14.

15 Huang, Y., Todd, N., and Thathiah, A. (2017). The role of GPCRs in neurodegenerative diseases: avenues for therapeutic intervention. *Curr. Opin. Pharmacol.* 32: 96–110.

16 Lemos, A., Melo, R., Moreira, I.S., and Cordeiro, M.N.D.S. (2018). Computer-aided drug design approaches to study key therapeutic targets in Alzheimer's disease. In: *Computational Modeling of Drugs Against Alzheimer's Disease*, Neuromethods, vol. 132 (ed. K. Roy), 61–106. New York, NY: Humana Press.

17 Lemos, A., Melo, R., Preto, A.J. et al. (2018). In silico studies targeting G-protein coupled receptors for drug research against Parkinson's disease. *Curr. Neuropharmacol.* 16 (6): 786–848.

18 Neumann, E., Khawaja, K., and Müller-Ladner, U. (2014). G protein-coupled receptors in rheumatology. *Nat. Rev. Rheumatol.* 10 (7): 429–436.

19 Unutmaz, D., KewalRamani, V.N., and Littman, D.R. (1998). G protein-coupled receptors in HIV and SIV entry: new perspectives on lentivirus–host interactions and on the utility of animal models. *Semin. Immunol.* 10 (3): 225–236.

20 Zhang, L. and Shi, G. (2016). Gq-coupled receptors in autoimmunity. *J. Immunol. Res.* 2016: 3969023.

21 Hauser, A.S., Attwood, M.M., Rask-Andersen, M. et al. (2017). Trends in GPCR drug discovery: new agents, targets and indications. *Nat. Rev. Drug Discovery* 16 (12): 829–842.

22 Tuteja, N. (2009). Signaling through G protein coupled receptors. *Plant Signal. Behav.* 4 (10): 942–947.

23 Gilman, A.G. (1987). G proteins: transducers of receptor-generated signals. *Annu. Rev. Biochem.* 56 (1): 615–649.

24 Pierce, K.L., Premont, R.T., and Lefkowitz, R.J. (2002). Seven-transmembrane receptors. *Nat. Rev. Mol. Cell Biol.* 3 (9): 639–650.

25 Wootten, D., Christopoulos, A., Marti-Solano, M. et al. (2018). Mechanisms of signalling and biased agonism in G protein-coupled receptors. *Nat. Rev. Mol. Cell Biol.* 19 (10): 638–653.

26 Mystek, P., Rysiewicz, B., Gregrowicz, J. et al. (2019). Gγ and Gα identity dictate a G-protein heterotrimer plasma membrane targeting. *Cells* 8 (10): 1246.

27 Hanlon, C.D. and Andrew, D.J. (2015). Outside-in signaling – a brief review of GPCR signaling with a focus on the *Drosophila* GPCR family. *J. Cell Sci.* 128 (19): 3533–3542.

28 Mahoney, J.P. and Sunahara, R.K. (2016). Mechanistic insights into GPCR–G protein interactions. *Curr. Opin. Struct. Biol.* 41: 247–254.

29 Syrovatkina, V., Alegre, K.O., Dey, R., and Huang, X.-Y. (2016). Regulation, signaling, and physiological functions of G-proteins. *J. Mol. Biol.* 428 (19): 3850–3868.

30 Druey, K.M. (2017). Emerging roles of regulators of G protein signaling (RGS) proteins in the immune system. In: *Advances in Immunology*, G Protein-Coupled Receptors in Immune Response and Regulation, vol. 136 (ed. A.K. Shukla), 315–351. Academic Press.

31 Neves, S.R., Ram, P.T., and Iyengar, R. (2002). G protein pathways. *Science* 296 (5573): 1636–1639.

32 Jean-Charles, P.-Y., Kaur, S., and Shenoy, S. (2017). G protein-coupled receptor signaling through β-arrestin–dependent mechanisms. *J. Cardiovasc. Pharmacol.* 70 (3): 142–158.

33 Isobe, K., Jung, H.J., Yang, C.-R. et al. (2017). Systems-level identification of PKA-dependent signaling in epithelial cells. *Proc. Natl. Acad. Sci. U.S.A.* 114 (42): E8875–E8884.

34 Sassone-Corsi, P. (2012). The cyclic AMP pathway. *Cold Spring Harbor Perspect. Biol.* 4 (12): a011148.

35 Epand, R.M. (2017). Features of the phosphatidylinositol cycle and its role in signal transduction. *J. Membr. Biol.* 250 (4): 353–366.

36 Suzuki, N., Hajicek, N., and Kozasa, T. (2009). Regulation and physiological functions of $G_{12/13}$-mediated signaling pathways. *Neurosignals* 17 (1): 55–70.
37 Jain, R., Watson, U., Vasudevan, L., and Saini, D.K. (2018). ERK activation pathways downstream of GPCRs. In: *International Review of Cell and Molecular Biology*, G Protein-Coupled Receptors: Emerging Paradigms in Activation, Signaling and Regulation Part A, vol. 338 (ed. A.K. Shukla), 79–109. Academic Press.
38 Senarath, K., Kankanamge, D., Samaradivakara, S. et al. (2018). Regulation of G protein βγ signaling. In: *International Review of Cell and Molecular Biology (G Protein-Coupled Receptors: Emerging Paradigms in Activation, Signaling and Regulation Part B)*, vol. 339 (ed. A.K. Shukla), 133–191. Academic Press.
39 Dupré, D.J., Robitaille, M., Rebois, R.V., and Hébert, T.E. (2009). The role of Gβγ subunits in the organization, assembly, and function of GPCR signaling complexes. *Annu. Rev. Pharmacol. Toxicol.* 49 (1): 31–56.
40 Kelly, E., Bailey, C.P., and Henderson, G. (2008). Agonist-selective mechanisms of GPCR desensitization. *Br. J. Pharmacol.* 153 (S1): S379–S388.
41 Rajagopal, S. and Shenoy, S.K. (2018). GPCR desensitization: acute and prolonged phases. *Cell Signal.* 41: 9–16.
42 Ghadessy, R.S. and Kelly, E. (2002). Second messenger-dependent protein kinases and protein synthesis regulate endogenous secretin receptor responsiveness. *Br. J. Pharmacol.* 135 (8): 2020–2028.
43 Gurevich, V.V. and Gurevich, E.V. (2019). GPCR signaling regulation: the role of GRKs and arrestins. *Front. Pharmacol.* 10: 125.
44 Lefkowitz, R.J. (1998). G protein-coupled receptors III. New roles for receptor kinases and β-arrestins in receptor signaling and desensitization. *J. Biol. Chem.* 273 (30): 18677–18680.
45 Moore, C.A.C., Milano, S.K., and Benovic, J.L. (2007). Regulation of receptor trafficking by GRKs and arrestins. *Annu. Rev. Physiol.* 69 (1): 451–482.
46 Kenakin, T. (2011). Functional selectivity and biased receptor signaling. *J. Pharmacol. Exp. Ther.* 336 (2): 296–302.
47 Grundmann, M. and Kostenis, E. (2017). Temporal bias: time-encoded dynamic GPCR signaling. *Trends Pharmacol. Sci.* 38 (12): 1110–1124.
48 Ilter, M., Mansoor, S., and Sensoy, O. (2019). Utilization of biased G protein-coupled receptor signaling towards development of safer and personalized therapeutics. *Molecules* 24 (11): 2052.
49 Rasmussen, S.G.F., DeVree, B.T., Zou, Y. et al. (2011). Crystal structure of the β_2 adrenergic receptor–Gs protein complex. *Nature* 477 (7366): 549–555.
50 Carpenter, B., Nehmé, R., Warne, T. et al. (2016). Structure of the adenosine A_{2A} receptor bound to an engineered G protein. *Nature* 536 (7614): 104–107.
51 Liang, Y.-L., Khoshouei, M., Radjainia, M. et al. (2017). Phase-plate cryo-EM structure of a class B GPCR–G-protein complex. *Nature* 546 (7656): 118–123.

52 Zhang, Y., Sun, B., Feng, D. et al. (2017). Cryo-EM structure of the activated GLP-1 receptor in complex with a G protein. *Nature* 546 (7657): 248–253.

53 dal Maso, E., Glukhova, A., Zhu, Y. et al. (2019). The molecular control of calcitonin receptor signaling. *ACS Pharmacol. Transl. Sci.* 2 (1): 31–51.

54 Zhao, L.-H., Ma, S., Sutkeviciute, I. et al. (2019). Structure and dynamics of the active human parathyroid hormone receptor-1. *Science* 364 (6436): 148–153.

55 Gao, Y., Hu, H., Ramachandran, S. et al. (2019). Structures of the rhodopsin-transducin complex: insights into G-protein activation. *Mol. Cell* 75 (4): 781–790.e3.

56 Zhao, P., Liang, Y.-L., Belousoff, M.J. et al. (2020). Activation of the GLP-1 receptor by a non-peptidic agonist. *Nature* 577 (7790): 432–436.

57 Kang, Y., Zhou, X.E., Gao, X. et al. (2015). Crystal structure of rhodopsin bound to arrestin by femtosecond X-ray laser. *Nature* 523 (7562): 561–567.

58 Zhou, X.E., Gao, X., Barty, A. et al. (2016). X-ray laser diffraction for structure determination of the rhodopsin–arrestin complex. *Sci. Data* 3 (1): 160021.

59 Zhou, X.E., He, Y., de Waal, P.W. et al. (2017). Identification of phosphorylation codes for arrestin recruitment by G protein-coupled receptors. *Cell* 170 (3): 457–469.e13.

60 Yin, W., Li, Z., Jin, M. et al. (2019 Dec). A complex structure of arrestin-2 bound to a G protein-coupled receptor. *Cell Res.* 29 (12): 971–983.

61 Filipek, S. (2019). Molecular switches in GPCRs. *Curr. Opin. Struct. Biol.* 55: 114–120.

62 Doré, A.S., Robertson, N., Errey, J.C. et al. (2011). Structure of the adenosine A_{2A} receptor in complex with ZM241385 and the xanthines XAC and caffeine. *Structure* 19 (9): 1283–1293.

63 Lebon, G., Warne, T., Edwards, P.C. et al. (2011). Agonist-bound adenosine A_{2A} receptor structures reveal common features of GPCR activation. *Nature* 474 (7352): 521–525.

64 Baker, N.A., Sept, D., Joseph, S. et al. (2001). Electrostatics of nanosystems: application to microtubules and the ribosome. *Proc. Natl. Acad. Sci. U.S.A.* 98 (18): 10037–10041.

65 The PyMOL Molecular Graphics System, Version 2.0 Schrödinger, LLC.

66 Dror, R.O., Mildorf, T.J., Hilger, D. et al. (2015). Structural basis for nucleotide exchange in heterotrimeric G proteins. *Science* 348 (6241): 1361–1365.

67 Carpenter, B. and Tate, C.G. (2017). Active state structures of G protein-coupled receptors highlight the similarities and differences in the G protein and arrestin coupling interfaces. *Curr. Opin. Struct. Biol.* 45: 124–132.

68 Kang, Y., Gao, X., Zhou, X.E. et al. (2016). A structural snapshot of the rhodopsin-arrestin complex. *FEBS J.* 283 (5): 816–821.

69 Ranganathan, A., Dror, R.O., and Carlsson, J. (2014). Insights into the role of Asp79$^{2.50}$ in β_2 adrenergic receptor activation from molecular dynamics simulations. *Biochemistry* 53 (46): 7283–7296.

70 Wang, D., Yu, H., Liu, X. et al. (2017). The orientation and stability of the GPCR–arrestin complex in a lipid bilayer. *Sci. Rep.* 7 (1): 16985.

71 Latorraca, N.R., Wang, J.K., Bauer, B. et al. (2018). Molecular mechanism of GPCR-mediated arrestin activation. *Nature* 557 (7705): 452–456.

72 McCorvy, J.D., Butler, K.V., Kelly, B. et al. (2018). Structure-inspired design of β-arrestin-biased ligands for aminergic GPCRs. *Nat. Chem. Biol.* 14 (2): 126–134.

73 Wingler, L.M., Elgeti, M., Hilger, D. et al. (2019 Jan). Angiotensin analogs with divergent bias stabilize distinct receptor conformations. *Cell* 176 (3): 468–478.e11.

74 Staus, D.P., Hu, H., Robertson, M.J. et al. (2020). Structure of the M2 muscarinic receptor–β-arrestin complex in a lipid nanodisc. *Nature* 579: 1–10.

75 Zhou, X.E., Melcher, K., and Xu, H.E. (2017). Understanding the GPCR biased signaling through G protein and arrestin complex structures. *Curr. Opin. Struct. Biol.* 45: 150–159.

76 Gurevich, V.V. and Gurevich, E.V. (2013). Structural determinants of arrestin functions. Chapter 3. In: *Progress in Molecular Biology and Translational Science*, The Molecular Biology of Arrestins, vol. 118 (ed. L.M. Luttrell), 57–92. Academic Press.

77 Weis, W.I. and Kobilka, B.K. (2018). The molecular basis of G protein-coupled receptor activation. *Annu. Rev. Biochem.* 87 (1): 897–919.

78 Scheerer, P., Park, J.H., Hildebrand, P.W. et al. (2008). Crystal structure of opsin in its G-protein-interacting conformation. *Nature* 455 (7212): 497–502.

79 Mackenzie, A.E., Quon, T., Lin, L.-C. et al. (2019). Receptor selectivity between the G proteins $G\alpha_{12}$ and $G\alpha_{13}$ is defined by a single leucine-to-isoleucine variation. *FASEB J.* 33 (4): 5005–5017.

80 Draper-Joyce, C.J., Khoshouei, M., Thal, D.M. et al. (2018). Structure of the adenosine-bound human adenosine A_1 receptor–G_i complex. *Nature* 558 (7711): 559–563.

81 Suomivuori, C.-M., Latorraca, N.R., Wingler, L.M. et al. (2020). Molecular mechanism of biased signaling in a prototypical G protein-coupled receptor. *Science* 367 (6480): 881–887.

82 Du, Y., Duc, N.M., Rasmussen, S.G.F. et al. (2019). Assembly of a GPCR-G protein complex. *Cell* 177 (5): 1232–1242.e11.

83 Ishchenko, A., Gati, C., and Cherezov, V. (2018). Structural biology of G protein-coupled receptors: new opportunities from XFELs and cryoEM. *Curr. Opin. Struct. Biol.* 51: 44–52.

84 Harding, S.D., Sharman, J.L., Faccenda, E. et al. (2018). The IUPHAR/BPS Guide to PHARMACOLOGY in 2018: updates and expansion to encompass the new guide to IMMUNOPHARMACOLOGY. *Nucleic Acids Res.* 46 (D1): D1091–D1106.

85 Li, F., Godoy, M.D., and Rattan, S. (2004). Role of adenylate and guanylate cyclases in β_1-, β_2-, and β_3-adrenoceptor-mediated relaxation of internal anal sphincter smooth muscle. *J. Pharmacol. Exp. Ther.* 308 (3): 1111–1120.

86 Sandhu, M., Touma, A.M., Dysthe, M. et al. (2019). Conformational plasticity of the intracellular cavity of GPCR–G-protein complexes leads to G-protein promiscuity and selectivity. *Proc. Natl. Acad. Sci. U.S.A.* 116 (24): 11956–11965.

87 Flock, T., Hauser, A.S., Lund, N. et al. (2017). Selectivity determinants of GPCR–G-protein binding. *Nature* 545 (7654): 317–322.

88 Frielle, T., Collins, S., Daniel, K.W. et al. (1987). Cloning of the cDNA for the human β_1-adrenergic receptor. *Proc. Natl. Acad. Sci. U.S.A.* 84 (22): 7920–7924.

89 Ruat, M., Traiffort, E., Arrang, J.M. et al. (1993). A novel rat serotonin (5-HT6) receptor: molecular cloning, localization and stimulation of cAMP accumulation. *Biochem. Biophys. Res. Commun.* 193 (1): 268–276.

90 Southan, C., Sharman, J.L., Benson, H.E. et al. (2016). The IUPHAR/BPS Guide to PHARMACOLOGY in 2016: towards curated quantitative interactions between 1300 protein targets and 6000 ligands. *Nucleic Acids Res.* 44 (D1): D1054–D1068.

91 Okashah, N., Wan, Q., Ghosh, S. et al. (2019). Variable G protein determinants of GPCR coupling selectivity. *Proc. Natl. Acad. Sci. U.S.A.* 116 (24): 12054–12059.

92 Liang, Y.-L., Khoshouei, M., Glukhova, A. et al. (2018). Phase-plate cryo-EM structure of a biased agonist-bound human GLP-1 receptor–Gs complex. *Nature* 555 (7694): 121–125.

93 Eps, N.V., Altenbach, C., Caro, L.N. et al. (2018). G_i- and G_s-coupled GPCRs show different modes of G-protein binding. *Proc. Natl. Acad. Sci. U.S.A.* 115 (10): 2383–2388.

94 Liu, X., Xu, X., Hilger, D. et al. (2019). Structural insights into the process of GPCR–G protein complex formation. *Cell* 177 (5): 1243–1251.e12.

95 Koehl, A., Hu, H., Maeda, S. et al. (2018). Structure of the μ-opioid receptor–G_i protein complex. *Nature* 558 (7711): 547–552.

96 García-Nafría, J., Nehmé, R., Edwards, P.C., and Tate, C.G. (2018). Cryo-EM structure of the serotonin 5-HT_{1B} receptor coupled to heterotrimeric G_o. *Nature* 558 (7711): 620–623.

97 Kang, Y., Kuybeda, O., de Waal, P.W. et al. (2018). Cryo-EM structure of human rhodopsin bound to an inhibitory G protein. *Nature* 558 (7711): 553–558.

98 Hua, T., Vemuri, K., Pu, M. et al. (2016). Crystal structure of the human cannabinoid receptor CB_1. *Cell* 167 (3): 750–762.e14.

99 Santiago, L.J. and Abrol, R. (2019). Understanding G protein selectivity of muscarinic acetylcholine receptors using computational methods. *Int. J. Mol. Sci.* 20 (21): 5290.

100 Divorty, N., Mackenzie, A.E., Nicklin, S.A., and Milligan, G. (2015). G protein-coupled receptor 35: an emerging target in inflammatory and cardiovascular disease. *Front. Pharmacol.* 6: 41.

101 Ostermaier, M.K., Schertler, G.F., and Standfuss, J. (2014). Molecular mechanism of phosphorylation-dependent arrestin activation. *Curr. Opin. Struct. Biol.* 29: 143–151.

102 Mayer, D., Damberger, F.F., Samarasimhareddy, M. et al. (2019). Distinct G protein-coupled receptor phosphorylation motifs modulate arrestin affinity and activation and global conformation. *Nat. Commun.* 10 (1): 1–14.

103 Sensoy, O., Moreira, I.S., and Morra, G. (2016). Understanding the differential selectivity of arrestins toward the phosphorylation state of the receptor. *ACS Chem. Neurosci.* 7 (9): 1212–1224.

104 Kim, Y.J., Hofmann, K.P., Ernst, O.P. et al. (2013). Crystal structure of pre-activated arrestin p44. *Nature* 497 (7447): 142–146.

105 Zhuo, Y., Vishnivetskiy, S.A., Zhan, X. et al. (2014). Identification of receptor binding-induced conformational changes in non-visual arrestins. *J. Biol. Chem.* 289 (30): 20991–21002.

106 Zhan, X., Gimenez, L.E., Gurevich, V.V., and Spiller, B.W. (2011). Crystal structure of arrestin-3 reveals the basis of the difference in receptor binding between two non-visual subtypes. *J. Mol. Biol.* 406 (3): 467–478.

107 Oakley, R.H., Laporte, S.A., Holt, J.A. et al. (2000). Differential affinities of visual arrestin, βarrestin1, and βarrestin2 for G protein-coupled receptors delineate two major classes of receptors. *J. Biol. Chem.* 275 (22): 17201–17210.

108 Al-Qattan, M.N.M. and Mordi, M.N. (2019). Molecular basis of modulating adenosine receptors activities. *Curr. Pharm. Des.* 25 (7): 817–831.

109 Kenakin, T.P. (2012). Biased signalling and allosteric machines: new vistas and challenges for drug discovery. *Br. J. Pharmacol.* 165 (6): 1659–1669.

110 Kenakin, T. and Christopoulos, A. (2013). Signalling bias in new drug discovery: detection, quantification and therapeutic impact. *Nat. Rev. Drug Discovery* 12 (3): 205–216.

111 Hellyer, S.D., Albold, S., Sengmany, K. et al. (2019). Metabotropic glutamate receptor 5 (mGlu5)-positive allosteric modulators differentially induce or potentiate desensitization of mGlu5 signaling in recombinant cells and neurons. *J. Neurochem.* 151 (3): 301–315.

112 Ahn, K.H., Mahmoud, M.M., Shim, J.-Y., and Kendall, D.A. (2013). Distinct roles of β-arrestin 1 and β-arrestin 2 in ORG27569-induced biased signaling and internalization of the cannabinoid receptor 1 (CB1). *J. Biol. Chem.* 288 (14): 9790–9800.

113 Gentry, P.R., Sexton, P.M., and Christopoulos, A. (2015). Novel allosteric modulators of G protein-coupled receptors. *J. Biol. Chem.* 290 (32): 19478–19488.

114 Pupo, A.S., Duarte, D.A., Lima, V. et al. (2016). Recent updates on GPCR biased agonism. *Pharmacol. Res.* 112: 49–57.

115 Calebiro, D., Nikolaev, V.O., Gagliani, M.C. et al. (2009). Persistent cAMP-signals triggered by internalized G-protein–coupled receptors. *PLoS Biol.* 7 (8): e1000172.

116 Ferrandon, S., Feinstein, T.N., Castro, M. et al. (2009). Sustained cyclic AMP production by parathyroid hormone receptor endocytosis. *Nat. Chem. Biol.* 5 (10): 734–742.

117 Kotowski, S.J., Hopf, F.W., Seif, T. et al. (2011). Endocytosis promotes rapid dopaminergic signaling. *Neuron* 71 (2): 278–290.

118 Irannejad, R., Tomshine, J.C., Tomshine, J.R. et al. (2013). Conformational biosensors reveal GPCR signalling from endosomes. *Nature* 495 (7442): 534–538.

119 Mullershausen, F., Zecri, F., Cetin, C. et al. (2009). Persistent signaling induced by FTY720-phosphate is mediated by internalized S1P1 receptors. *Nat. Chem. Biol.* 5 (6): 428–434.

120 Feinstein, T.N., Yui, N., Webber, M.J. et al. (2013). Noncanonical control of vasopressin receptor type 2 signaling by retromer and arrestin. *J. Biol. Chem.* 288 (39): 27849–27860.

121 Ibsen, M.S., Connor, M., and Glass, M. (2017). Cannabinoid CB_1 and CB_2 receptor signaling and bias. *Cannabis Cannabinoid. Res.* 2 (1): 48–60.

122 Ceraudo, E., Horioka, M., Mattheisen, J.M. et al. (2019). Uveal melanoma oncogene CYSLTR2 encodes a constitutively active GPCR highly biased toward Gq signaling. *bioRxiv* 663153.

123 Liu, J.J., Horst, R., Katritch, V. et al. (2012). Biased signaling pathways in β_2-adrenergic receptor characterized by 19F-NMR. *Science* 335 (6072): 1106–1110.

124 Zweemer, A.J.M., Toraskar, J., Heitman, L.H., and IJzerman, A.P. (2014). Bias in chemokine receptor signalling. *Trends Immunol.* 35 (6): 243–252.

125 Geng, Y., Mosyak, L., Kurinov, I. et al. (2016). Structural mechanism of ligand activation in human calcium-sensing receptor. (ed. E.Y. Isacoff). *eLife* 5: e13662.

126 Wacker, D., Stevens, R.C., and Roth, B.L. (2017). How ligands illuminate GPCR molecular pharmacology. *Cell* 170 (3): 414–427.

127 Wold, E.A., Chen, J., Cunningham, K.A., and Zhou, J. (2019). Allosteric modulation of class A GPCRs: targets, agents, and emerging concepts. *J. Med. Chem.* 62 (1): 88–127.

128 Katritch, V., Fenalti, G., Abola, E.E. et al. (2014). Allosteric sodium in class A GPCR signaling. *Trends Biochem. Sci* 39 (5): 233–244.

129 Valant, C., Aurelio, L., Urmaliya, V.B. et al. (2010). Delineating the mode of action of adenosine A_1 receptor allosteric modulators. *Mol. Pharmacol.* 78 (3): 444–455.

130 Lane, J.R., Abdul-Ridha, A., and Canals, M. (2013). Regulation of G protein-coupled receptors by allosteric ligands. *ACS Chem. Neurosci.* 4 (4): 527–534.

131 Lane, J.R., May, L.T., Parton, R.G. et al. (2017). A kinetic view of GPCR allostery and biased agonism. *Nat. Chem. Biol.* 13 (9): 929–937.

132 Vecchio, E.A., Baltos, J.-A., Nguyen, A.T.N. et al. (2018). New paradigms in adenosine receptor pharmacology: allostery, oligomerization and biased agonism. *Br. J. Pharmacol.* 175 (21): 4036–4046.

133 Baltos, J.-A., Gregory, K.J., White, P.J. et al. (2016). Quantification of adenosine A_1 receptor biased agonism: implications for drug discovery. *Biochem. Pharmacol.* 99: 101–112.

134 Gao, Z.-G., Kim, S.-K., IJzerman, A., and Jacobson, K. (2005). Allosteric modulation of the adenosine family of receptors. *Mini-Rev. Med. Chem.* 5 (6): 545–553.

135 Gao, Z.-G. and Jacobson, K.A. (2013). Allosteric modulation and functional selectivity of G protein-coupled receptors. *Drug Discovery Today Technol.* 10 (2): e237–e243.

136 Conn, P.J., Christopoulos, A., and Lindsley, C.W. (2009). Allosteric modulators of GPCRs: a novel approach for the treatment of CNS disorders. *Nat. Rev. Drug Discovery* 8 (1): 41–54.

137 Williams, L.M., He, X., Vaid, T.M. et al. (2019). Diazepam is not a direct allosteric modulator of α_1-adrenoceptors, but modulates receptor signaling by inhibiting phosphodiesterase-4. *Pharmacol. Res. Perspect.* 7 (1): e00455.

138 Campbell, A.P., Wakelin, L.P.G., Denny, W.A., and Finch, A.M. (2017). Homobivalent conjugation increases the allosteric effect of 9-aminoacridine at the α_1-adrenergic receptors. *Mol. Pharmacol.* 91 (2): 135–144.

139 Molderings, G.J., Menzel, S., Kathmann, M. et al. (2000). Dual interaction of agmatine with the rat α_{2D}-adrenoceptor: competitive antagonism and allosteric activation. *Br. J. Pharmacol.* 130 (7): 1706–1712.

140 Al-Zoubi, R., Morales, P., and Reggio, P.H. (2019). Structural insights into CB_1 receptor biased signaling. *Int. J. Mol. Sci.* 20 (8): 1837.

141 Khajehali, E., Malone, D.T., Glass, M. et al. (2015). Biased agonism and biased allosteric modulation at the CB_1 cannabinoid receptor. *Mol. Pharmacol.* 88 (2): 368–379.

142 Baillie, G.L., Horswill, J.G., Anavi-Goffer, S. et al. (2013). CB_1 receptor allosteric modulators display both agonist and signaling pathway specificity. *Mol. Pharmacol.* 83 (2): 322–338.

143 Langmead, C.J. and Christopoulos, A. (2014). Functional and structural perspectives on allosteric modulation of GPCRs. *Curr. Opin. Cell Biol.* 27: 94–101.

144 Khurana, L., Fu, B.-Q., Duddupudi, A.L. et al. (2017). Pyrimidinyl biphenylureas: identification of new lead compounds as allosteric modulators of the cannabinoid receptor CB1. *J. Med. Chem.* 60 (3): 1089–1104.

145 Luttrell, L.M. and Kenakin, T.P. (2011 [cited 2020 Feb 3]). Refining efficacy: allosterism and bias in G protein-coupled receptor signaling. In: *Signal Transduction Protocols* [Internet], Methods in Molecular Biology, vol. 756 (ed. L.M. Luttrell and S.S.G. Ferguson), 3–35. Totowa, NJ: Humana Press.

146 Nedjai, B., Li, H., Stroke, I.L. et al. (2012). Small molecule chemokine mimetics suggest a molecular basis for the observation that CXCL10 and CXCL11 are allosteric ligands of CXCR3. *Br. J. Pharmacol.* 166 (3): 912–923.

147 Khoury, E., Clément, S., and Laporte, S.A. (2014). Allosteric and biased G protein-coupled receptor signaling regulation: potentials for new therapeutics. *Front. Endocrinol.* 5: 68.

148 Violin, J.D., Crombie, A.L., Soergel, D.G., and Lark, M.W. (2014). Biased ligands at G-protein-coupled receptors: promise and progress. *Trends Pharmacol. Sci.* 35 (7): 308–316.

149 Gao, F., Sexton, P.M., Christopoulos, A., and Miller, L.J. (2008). Benzodiazepine ligands can act as allosteric modulators of the Type 1 cholecystokinin receptor. *Bioorg. Med. Chem. Lett.* 18 (15): 4401–4404.

150 Desai, A.J., Mechin, I., Nagarajan, K. et al. (2019). Molecular basis of action of a small-molecule positive allosteric modulator agonist at the type 1 cholecystokinin holoreceptor. *Mol. Pharmacol.* 95 (3): 245–259.

151 Felsing, D.E., Jain, M.K., and Allen, J.A. (2019). Advances in dopamine D_1 receptor ligands for neurotherapeutics. *Curr. Top. Med. Chem.* 19 (16): 1365–1380.

152 Svensson, K.A., Heinz, B.A., Schaus, J.M. et al. (2017). An allosteric potentiator of the dopamine D_1 receptor increases locomotor activity in human

D_1 knock-in mice without causing stereotypy or tachyphylaxis. *J. Pharmacol. Exp. Ther.* 360 (1): 117–128.

153 Luderman, K.D., Conroy, J.L., Free, R.B. et al. (2018). Identification of positive allosteric modulators of the D_1 dopamine receptor that act at diverse binding sites. *Mol. Pharmacol.* 94 (4): 1197–1209.

154 Hall, A., Provins, L., and Valade, A. (2019). Novel strategies to activate the dopamine D_1 receptor: recent advances in orthosteric agonism and positive allosteric modulation. *J. Med. Chem.* 62 (1): 128–140.

155 Bonifazi, A., Yano, H., Guerrero, A.M. et al. (2019). Novel and potent dopamine D_2 receptor Go-protein biased agonists. *ACS Pharmacol. Transl. Sci.* 2 (1): 52–65.

156 Wood, M., Ates, A., Andre, V.M. et al. (2016). In vitro and in vivo identification of novel positive allosteric modulators of the human dopamine D_2 and D_3 receptor. *Mol. Pharmacol.* 89 (2): 303–312.

157 Klein, M.O., Battagello, D.S., Cardoso, A.R. et al. (2019). Dopamine: functions, signaling, and association with neurological diseases. *Cell Mol. Neurobiol.* 39 (1): 31–59.

158 Foster, D.J. and Conn, P.J. (2017). Allosteric modulation of GPCRs: new insights and potential utility for treatment of schizophrenia and other CNS disorders. *Neuron* 94 (3): 431–446.

159 Reilly, S.W., Riad, A.A., Hsieh, C.-J. et al. (2019). Leveraging a low-affinity diazaspiro orthosteric fragment to reduce dopamine D_3 receptor (D_3R) ligand promiscuity across highly conserved aminergic G-protein-coupled receptors (GPCRs). *J. Med. Chem.* 62 (10): 5132–5147.

160 Talbodec, A., Berkane, N., Blandin, V. et al. (2000). Aspirin and sodium salicylate inhibit endothelin ET_A receptors by an allosteric type of mechanism. *Mol. Pharmacol.* 57 (4): 797–804.

161 Landomiel, F., De Pascali, F., Raynaud, P. et al. (2019). Biased signaling and allosteric modulation at the FSHR. *Front. Endocrinol.* 10: 148.

162 Canals, M., Lane, J.R., Wen, A. et al. (2012). A Monod–Wyman–Changeux mechanism can explain G protein-coupled receptor (GPCR) allosteric modulation. *J. Biol. Chem.* 287 (1): 650–659.

163 Chambon, C., Jatzke, C., Wegener, N. et al. (2012). Using cholinergic M1 receptor positive allosteric modulators to improve memory via enhancement of brain cholinergic communication. *Eur. J. Pharmacol.* 697 (1): 73–80.

164 Marlo, J.E., Niswender, C.M., Days, E.L. et al. (2009). Discovery and characterization of novel allosteric potentiators of M_1 muscarinic receptors reveals multiple modes of activity. *Mol. Pharmacol.* 75 (3): 577–588.

165 Davey, A.E., Leach, K., Valant, C. et al. (2012). Positive and negative allosteric modulators promote biased signaling at the calcium-sensing receptor. *Endocrinology* 153 (3): 1232–1241.

166 Leach, K., Loiacono, R.E., Felder, C.C. et al. (2010). Molecular mechanisms of action and in vivo validation of an M_4 muscarinic acetylcholine receptor allosteric modulator with potential antipsychotic properties. *Neuropsychopharmacology* 35 (4): 855–869.

167 Stanczyk, M.A., Livingston, K.E., Chang, L. et al. (2019). The δ-opioid receptor positive allosteric modulator BMS 986187 is a G-protein-biased allosteric agonist. *Br. J. Pharmacol.* 176 (11): 1649–1663.

168 Pinkerton, A.B., Peddibhotla, S., Yamamoto, F. et al. (2019). Discovery of β-arrestin biased, orally bioavailable, and CNS penetrant neurotensin receptor 1 (NTR1) allosteric modulators. *J. Med. Chem.* 62 (17): 8357–8363.

169 Wisler, J.W., Xiao, K., Thomsen, A.R., and Lefkowitz, R.J. (2014). Recent developments in biased agonism. *Curr. Opin. Cell Biol.* 27: 18–24.

170 Mathiesen, J.M., Ulven, T., Martini, L. et al. (2005). Identification of indole derivatives exclusively interfering with a G protein-independent signaling pathway of the prostaglandin D_2 receptor $CRTH_2$. *Mol. Pharmacol.* 68 (2): 393–402.

171 Wootten, D. and Miller, L.J. (2020). Structural basis for allosteric modulation of class B G protein-coupled receptors. *Annu. Rev. Pharmacol. Toxicol.* 60 (1): 89–107.

172 Mos, I., Jacobsen, S.E., Foster, S.R., and Bräuner-Osborne, H. (2019). Calcium-sensing receptor internalization is β-arrestin–dependent and modulated by allosteric ligands. *Mol. Pharmacol.* 96 (4): 463–474.

173 Holstein, D.M., Berg, K.A., Leeb-Lundberg, L.M.F. et al. (2004). Calcium-sensing receptor-mediated $ERK_{1/2}$ activation requires $G\alpha_{i2}$ coupling and dynamin-independent receptor internalization. *J. Biol. Chem.* 279 (11): 10060–10069.

174 Pin, J.-P. and Prezeau, L. (2007). Allosteric modulators of GABAB receptors: mechanism of action and therapeutic perspective. *Curr. Neuropharmacol.* 5 (3): 195–201.

175 Stansley, B.J. and Conn, P.J. (2019). Neuropharmacological insight from allosteric modulation of mGlu receptors. *Trends Pharmacol. Sci.* 40 (4): 240–252.

176 Cho, H.P., Garcia-Barrantes, P.M., Brogan, J.T. et al. (2014). Chemical modulation of mutant mGlu1 receptors derived from deleterious GRM1 mutations found in schizophrenics. *ACS Chem. Biol.* 9, 10: 2334–2346.

177 Pérez-Benito, L., Doornbos, M.L.J., Cordomí, A. et al. (2017). Molecular switches of allosteric modulation of the metabotropic glutamate 2 receptor. *Structure* 25 (7): 1153–1162.e4.

178 Hopkins, C.R. (2013). Is there a path forward for mGlu2 positive allosteric modulators for the treatment of schizophrenia? *ACS Chem. Neurosci.* 4 (2): 211–213.

179 De Groof, T.W.M., Bobkov, V., Heukers, R., and Smit, M.J. (2019). Nanobodies: new avenues for imaging, stabilizing and modulating GPCRs. *Mol. Cell. Endocrinol.* 484: 15–24.

180 Noetzel, M.J., Gregory, K.J., Vinson, P.N. et al. (2013). A novel metabotropic glutamate receptor 5 positive allosteric modulator acts at a unique site and confers stimulus bias to mGlu5 signaling. *Mol. Pharmacol.* 83 (4): 835–847.

181 Zhang, Y., Rodriguez, A.L., and Conn, P.J. (2005). Allosteric potentiators of metabotropic glutamate receptor subtype 5 have differential effects on different signaling pathways in cortical astrocytes. *J. Pharmacol. Exp. Ther.* 315 (3): 1212–1219.

182 Sengmany, K., Hellyer, S.D., Albold, S. et al. (2019). Kinetic and system bias as drivers of metabotropic glutamate receptor 5 allosteric modulator pharmacology. *Neuropharmacology* 149: 83–96.

183 Rook, J.M., Xiang, Z., Lv, X. et al. (2015). Biased mGlu5-positive allosteric modulators provide in vivo efficacy without potentiating mGlu5 modulation of NMDAR currents. *Neuron* 86 (4): 1029–1040.

184 Niswender, C.M., Johnson, K.A., Miller, N.R. et al. (2010). Context-dependent pharmacology exhibited by negative allosteric modulators of metabotropic glutamate receptor 7. *Mol. Pharmacol.* 77 (3): 459–468.

185 Leach, K. and Gregory, K.J. (2017). Molecular insights into allosteric modulation of Class C G protein-coupled receptors. *Pharmacol. Res.* 116: 105–118.

186 Hayatshahi, H.S., Xu, K., Griffin, S.A. et al. (2018). Analogues of arylamide phenylpiperazine ligands to investigate the factors influencing D_3 dopamine receptor bitropic binding and receptor subtype selectivity. *ACS Chem. Neurosci.* 9 (12): 2972–2983.

187 Monod, J., Wyman, J., and Changeux, J.-P. (1965). On the nature of allosteric transitions: a plausible model. *J. Mol. Biol.* 12 (1): 88–118.

188 Koshland, D.E., Némethy, G., and Filmer, D. (1966). Comparison of experimental binding data and theoretical models in proteins containing subunits. *Biochemistry* 5 (1): 365–385.

189 Kenakin, T. (2017). A scale of agonism and allosteric modulation for assessment of selectivity, bias, and receptor mutation. *Mol. Pharmacol.* 92 (4): 414–424.

190 van der Westhuizen, E.T., Valant, C., Sexton, P.M., and Christopoulos, A. (2015). Endogenous allosteric modulators of G protein-coupled receptors. *J. Pharmacol. Exp. Ther.* 353 (2): 246–260.

191 Conigrave, A.D., Quinn, S.J., and Brown, E.M. (2000). L-Amino acid sensing by the extracellular Ca^{2+}-sensing receptor. *Proc. Natl. Acad. Sci. U.S.A.* 97 (9): 4814–4819.

192 Zhang, Z., Qiu, W., Quinn, S.J. et al. (2002). Three adjacent serines in the extracellular domains of the CaR are required for L-amino acid-mediated potentiation of receptor function. *J. Biol. Chem.* 277 (37): 33727–33735.

193 Mun, H.-C., Culverston, E.L., Franks, A.H. et al. (2005). A double mutation in the extracellular Ca^{2+}-sensing receptor's venus flytrap domain that selectively disables L-amino acid sensing. *J. Biol. Chem.* 280 (32): 29067–29072.

194 Preto, A.J., Barreto, C.A.V., Baptista, S.J. et al. (2020). Understanding the binding specificity of G-protein coupled receptors towards G-proteins and arrestins: application to the dopamine receptor family. *J. Chem. Inf. Model.* 60 (8): 3969–3984.

6

GPCRs at Endosomes: Sorting, Signaling, and Recycling

Roshanak Irannejad[1] and Braden T. Lobingier[2]

[1] *Department of Biochemistry and Biophysics, Cardiovascular Research Institute, University of California, San Francisco, CA, USA*
[2] *Department of Chemical Physiology and Biochemistry, Oregon Health and Sciences University, Portland, OR, USA*

6.1 Recycling Pathways for GPCRs at Endosomes

6.1.1 Roads Split After Endocytosis

The majority of GPCRs which undergo endocytosis are funneled into the early endosome [1, 2]. From the early endosome, GPCRs can be sorted back to the plasma membrane, to the Golgi, or to the lysosome [3]. Active endosomal sorting to these first two destinations is thought to protect membrane proteins from proteolysis at the lysosome [4]. In the upcoming section, our focus will be on pathways and mechanisms for GPCR recycling to the plasma membrane – either directly or via the Golgi. We point the reader to excellent reviews which discuss the mechanisms by which GPCRs, and membrane proteins generally, are trafficked to the lysosome [3, 5].

6.2 Sequence-Directed GPCR Recycling

What determines if a GPCR is sorted to recycling or degradative pathways? As we will review below, the prevailing model is that cis-acting sequences in the carboxy-terminal tail of the GPCR direct the receptor into recycling pathways; in the absence of such a signal, the receptor is routed to the lysosome. Additionally, recycling sequences appear to be the dominant endosomal sorting signal, as a lysosomal-targeted GPCR can be rerouted to the plasma membrane by "transplanting" a recycling motif onto it.

GPCRs as Therapeutic Targets, Volume 1, First Edition. Edited by Annette Gilchrist.

6.2.1 Class I PDZ Binding Motifs

One of the best understood sequences for GPCR recycling is the class I PDZ binding motif (PDZbm), which is bound by PDZ domain containing proteins [6]. The minimum consensus sequence for a class I PDZbm is found in the last three C-terminal residues: Ser/Thr-X-Φ-COOH (X: any amino acid; Φ: hydrophobic amino acids including Val/Leu/Ile/Phe/Met). The role of a class I PDZbm in GPCR recycling was first recognized for the beta 2 adrenergic receptor (B2AR) [7]. Three important observations were made in these early studies: (1) Appending a single alanine to the C-terminus of a B2AR blocks the function of the PDZbm; (2) Disrupting PDZbm function, and thus recycling, results in rerouting of B2AR to the lysosome; (3) The PDZbm is necessary and sufficient to produce recycling in lysosome-targeted GPCRs [7, 8]. The PDZ mechanism is not exclusive to B2AR. Multiple GPCRs – including B1 adrenergic receptor (B1AR), parathyroid hormone 1 receptor (PTH1R), serotonin receptor 4A (5HT4A), somatostatin receptor 5 (SSTR5) – have class I PDZbm which have been shown to be important in endosomal sorting of these receptors [6]. Later in this chapter, we will discuss the mechanisms for recycling of GPCRs with a class I PDZbm via Sorting Nexin 27 (SNX27) and Retromer.

It is important to note, however, that only a minority of GPCRs have a class I PDZbm. Bioinformatic searches have identified 30 GPCRs with a class I PDZbm (~4% of all GPCRs) [3, 6]. These observations raise the question: how do GPCRs without a PDZbm recycle?

6.2.2 Nonconsensus Recycling Motifs

This question points toward a knowledge gap in the field. Here we will highlight multiple examples of necessary and sufficient GPCR recycling sequences which do not resemble a class I PDZbm. As such, we will refer to them as "nonconsensus" to underscore the lack of similarity to other known endosomal recycling motifs [9]. *Mu opioid receptor (MOR).* A sequence of seven amino acids was identified in the C-terminus of MOR (384-RTNHQ*LENLEAE*TAPLP-400) that is necessary and sufficient for recycling [10]. Interestingly, splice-variants of MOR have been identified lacking this recycling sequence, and study of these variants as recombinant transgenes verified that they are targeted to the lysosome [11]. *Formyl-peptide receptor type 2 (FPR2).* The last five residues of the FPR2 C-terminus (343-PAET*ELQAM*-351) were shown to be necessary and sufficient for GPCR recycling [12]. Further analysis demonstrated that all five residues were important for receptor recycling. In contrast to a class I PDZbm, appending an alanine at the C-terminus of FPR2 did not disrupt recycling [12]. *Somatostatin receptor 2 (SSTR2).* Seven amino acids in the C-terminus (360-*LLNGDLQ*TSI-369) are necessary and sufficient to mediate GPCR recycling [13]. Intriguingly, SSTR2 also contains a class I-like PDZbm (TSI). While TSI is not part of the necessary and sufficient recycling motif, further study suggests that it plays a role in directing SSTR2 recycling between Rab4 and Rab11 pathways [13]. *D1 dopamine receptor (D1R).* Truncation of the majority of the C-tail of D1R (last 87 amino acids) strongly reduced receptor recycling [14]. Further mapping identified the necessary and sufficient sequence for receptor recycling between residues 360–382 of D1R. In contrast to the general model of GPCR sorting described above, truncation of the D1R C-tail did not result in rerouting the receptor to the lysosome even though recycling was strongly reduced [14]. *Luteinizing hormone receptor (LHR).* Human LHR (hLHR) was among the first GPCRs in which a sufficient recycling motif was identified that lacked the consensus sequence of a PDZbm. This sequence was found through comparison of the trafficking behavior of LHR from different species [15]. While LHR from rat, mouse, and pig was sorted into the lysosome, hLHR was recycled efficiently [15]. Sequence alignments identified a series of amino acids (GTALL) present only in the C-terminus of hLHR (683-LHCQ***GTALL***DKTRYTEC-699) which was sufficient for recycling. The GTALL sequence was sufficient to mediate recycling as a transplantable motif that could reroute lysosomal-targeted GPCRs to the recycling pathway. Subsequent studies further highlighted the important role of the glycine and threonine in GTALL function [16]. It is intriguing to note that hLHR contains a class I-like PDZbm (TEC) [17]. While the terminal residue of hLHR is a cysteine and thus does not resemble a classical class I PDZbm, studies have shown that this sequence binds to the PDZ domain of the GAIP-interacting protein C terminus (GIPC) [17]. GIPC is important in sorting/retaining hLHR to APPL1-positive endosomes and that APPL1 function is important in hLHR recycling [17, 18]. We emphasize that the GPCRs discussed above do not represent a comprehensive list of receptors which recycle without a class I PDZbm. We focused on GPCRs in which a necessary and sufficient recycling sequence had been identified as they represent the minimal components of the membrane protein-sorting complex we will describe in more detail in Section 6.4.

Together, these data point toward a significant knowledge gap in the GPCR field: what are the sorting adaptors and sorting complexes which bind these nonconsensus recycling motifs? Do the pathways by which nonconsensus GPCRs recycle differ from those using a class I PDZbm? Recent technological advances, such as genome-wide CRISPR screening and high-resolution proximity labeling within living cells, offer possible methods by which this knowledge gap can be bridged [19–22]. With the recognition that many GPCRs can initiate G protein signaling from endosomes, identifying the mechanisms of "nonconsensus" motif recycling will be paramount in understanding both acute and long-term control of cellular GPCR signaling [23].

6.3 Endosomes as a Platform for Sorting into Recycling Pathways: Structure and Function

6.3.1 Endosomal Tubules

An important step in the recycling of membrane proteins from the early endosome is sorting into endosomal tubules [9, 24]. Tubules, cytoplasmic-facing membrane projections from the endosome, are the defining structural intermediate of the recycling process much the same way that the clathrin-coated pit is the defining structure of one endocytic pathway. Much is still unknown about sorting into endosomal tubules and the subsequent resolution of these structures into transport vesicles. First, the molecular composition of endosomal tubules is not fully clear. Polymerized actin exists on some, but not all tubules [25]. Endosomal tubules have been seen to be decorated with endosomal sorting complexes, and in some cases, sorting complexes have been observed to be restricted to different endosomal tubules [26]. However, it is not clear if one type of sorting complex per tubule is the exception or the rule. To underscore their heterogeneity, endosomal tubules have distinct lifetimes [25]. Additionally, it is not fully clear how scission between the tubules and endosome occurs. Multiple models, which are not mutually exclusive, have been put forward including the cooperative action of microtubular motors, actin polymerization, SNX-BAR proteins, and/or EHD-family proteins [4, 9, 27]. Together, these data highlight the importance of the endosomal tubule as a dynamic structure as well as the many unresolved questions that surround them.

6.3.2 Endosomal Actin

Branched actin exists on the endosome in small patches and appears, by light and electron microscopy, to sit at the base as well as along the length of a subset of endosomal tubules [25]. Endosomal actin has multiple roles. Chemical or genetic interference with endosomal actin result in perturbations to the entire endo-lysosomal pathway including changes in shape and subcellular location of both endosomes and lysosomes [28]. It is also clear that endosomal actin is required for the recycling of some membrane proteins such as the SNX27/retromer-dependent B2AR, and that B2AR enters actin-decorated endosomal tubules [7, 25]. How endosomal actin affects trafficking of other membrane proteins, such as the recycling of TFR or lysosomal targeting of EGFR, is somewhat in debate. While not conclusive, the weight of evidence at present suggests that inhibition of endosomal actin perturbs the trafficking of EGFR or TFR but does not stop these from reaching their respective destinations [7, 29–31].

6.3.3 Associated Protein Complexes

There are multiple protein complexes associated with the endosome that contribute to recycling of membrane proteins while not acting as direct sorting complexes. The first of these is the WASH (Wiskott–Aldrich syndrome protein and SCAR homolog) complex. WASH consists of five proteins – WASH1, SWIP, Strumpellin, CCDC53, and FAM21 – and has multiple roles. One role is to direct the postendosome trafficking of tubule-derived transport vesicles through regulating PI(4)P levels at the Golgi [32]. Another role of the WASH complex is to act as a scaffold. The WASH complex recruits the ARP2/3 complex to endosomes [29, 33]. ARP2/3 is localized to the base of endosomal tubules and nucleates endosomal actin [25, 28]. The WASH complex also binds sorting complexes such as Retromer and Retriever (discussed below) [9]. Finally, the WASH complex also binds to a large protein complex called CCC for its subunits (COMMD1, CCDC22, CCDC93, and additional subunits of the COMMD family of proteins) [34, 35]. The molecular identity and function of the CCC complex is still under investigation. Recent studies have provided insight into at least one function of the CCC complex [35]. Disruption of CCC complex activity increases PI(3)P levels at the endosome and, consequently, increases endosomal WASH localization and actin polymerization. The consequence of these CCC defects was a failure of Retriever and Retromer cargos to recycle [35]. While it is unclear why WASH needs PI(3)P for recruitment, these data suggest a WASH-CCC complex axis controls two necessary components for endosomal sorting: branched actin and inositol phosphates.

6.4 Endosomal Recycling Complexes

6.4.1 Retromer

The first identified endosomal recycling complex was called Retromer. The Retromer core complex in mammals is composed of three genes named for the vacuole protein sorting screen (VPS): VPS26, VPS29, and VPS35. A combination of structural efforts – including crystallography and cryo electron microscopy using the yeast version of Retromer – has led to a model in which VPS35 forms arches rising up from the membrane and decorated on the distal end (away from the bilayer) by VPS29 and on the proximal end (near the bilayer) by VPS26 [9, 36, 37]. Evidence suggests that Retromer is recruited to endosomes through interactions with activated Rab7 (GTP-bound) or through association with SNX proteins [38, 39]. Functionally, Retromer interacts with cargos (membrane proteins), cargo adaptors (SXN27 or SNX3), and actin polymerization machinery (WASH complex; discussed above) [9].

6.4.2 SNX27/Retromer

SNX27 belongs to the large family of PX-domain PI(3)P-binding proteins. SNX27 contains an N-terminal PDZ-domain which binds to class I PDZbms [40–43]. As discussed in section I, multiple GPCRs have been shown to be sorted by SNX27. Closer analysis of the SNX27 PDZ domain revealed a small 13 amino acid insertion not found in other PDZs [41, 44]. This 13 amino acid loop binds to the Retromer subunit VPS26 [44, 45]. Of particular note, the affinity of SNX27 for a PDZbm increases by an order of magnitude when bound to VPS26 [45]. Thus, SNX27 acts as an adaptor linking membrane proteins with a class I PDZbm to Retromer. Excellent biochemical and structural studies of SNX27 have identified a strong preference for negative charge in the amino acids upstream of the PDZbm [46]. Specifically, with the class I PDZ bm consensus sequence S/T-X-Φ as the −2, −1, and 0 positions respectively, SNX27 has preference for E/D or phosphorylated S/T in the −6, −5, and −3 positions. It is not yet clear how this *in vitro* preference operates in the cell. The last-four amino acids of B2AR (DSLL) – thus lacking the E/D or S/T in the −6 and −5 positions – are sufficient to mediate GPCR recycling in cultured cells, suggesting that the negative charge in or around the PDZbm is not an obligate requirement for SNX27 function [8]. Lastly, SNX27 also contains a C-terminal FERM-like domain which, *in vitro*, has been shown to find to Φ-X-N-P-X-pY sorting motifs (X: any amino acid; Φ: hydrophobic amino acids; pY: phosphotyrosine) [47]. It is not currently clear if the SNX27-FERM-like domain participates in sorting and recycling of membrane proteins in a manner analogous to its PDZ domain.

6.4.3 SNX3/Retromer

Another example of the adaptor/Retromer architecture is the SNX3/Retromer complex and its consensus sorting motif Φ-X-[L/M/V] motif [9]. The structure of the recycling sequence from DMT1-II bound to SNX3/Retromer was solved by crystallography, and the resulting model showed the EL*YLL* sequence from DMT1-II at the interface between VPS26 and SNX3 [37]. Thus, with SNX27/Retromer the recycling motif only contacts SNX27 while in SNX3/Retromer a hybrid interface of VPS26/SNX3 is used. Additionally, the *in vitro* affinity of DMT1-II recycling sequence for SNX3/Retromer (~127 μM) was about an order of magnitude weaker than many of those observed for the class I PBMs and SNX27/VPS26 [37, 45, 46]. It should be noted that to our knowledge, no GPCR has been shown to be direct cargo for SNX3/Retromer.

SNX3 has also been shown to be involved in recruiting Retromer to endosomal membranes, and SNX3 depletion results in loss of Retromer localized to endosomes [48, 49]. As a result, it is important to differentiate cargo which is directly bound by SNX3 from those which require Retromer function and thus, indirectly, are dependent on SNX3. Studies of another SNX3/Retromer cargo, Wntless, have identified dual Φ-X-[L/M/V] motifs [50]. While it has not been proven that Wntless binds to SNX3/Retromer in a manner similar to DMT1-II, loss of SNX3 or Retromer function leads to Wntless missorting [49, 51]. High-resolution live cell imaging has provided insights into differences in sorting between SNX27/Retromer and SNX3/Retromer. Both B2AR (a SNX27/Retromer cargo) and Wntless accumulate in the same endosomal tubules, Wntless shows a much stronger steady-state enrichment in these domains [50]. Despite colocalization into the same endosomal subdomains, Wntless and B2AR have different postendosome trafficking itineraries: Wntless transits to the trans Golgi network (TGN) while B2AR is routed to the plasma membrane. Curiously, these distinct destinations are not encoded by their recycling motifs (Φ-X-[L/M/V] or a class I PDZbm): chimeric constructs which swap these motifs change the enrichment of two proteins in endosomal tubules without changing the final destination [50]. While FAM21 has been shown to negatively regulate B2AR trafficking to the TGN, it is not clear how this system could occur concurrently with Wntless – localized to the same endosomal tubules – trafficking to the Golgi [32]. One possibility is for subsequent sorting at the posttubulation transport vesicle [50]. An alternative possibility is for differential retention of cargos at the Golgi before trafficking onto the plasma membrane. Regardless, it is clear that differential trafficking of membrane proteins – including rates of endosomal exit and postendosome destination—has the potential to significantly shape both intracellular and total GPCR signaling [23].

6.4.4 SNX17/Retriever

SNX17 binds to Φ-X-N-P-X-[F/Y] or Φ-X-N-X-X-[F/Y] motifs via its FERM-like domain [9, 47, 52, 53]. SNX17 function has been linked to trafficking of multiple cargos including LRP, B1-integrin, amyloid precursor protein (APP), and P-selectin [54–60]. A crystal structure of SNX17 in complex with the sorting motif from P-selectin (NAAY) was solved and showed the sorting motif from P-selectin bound to the F3 submodule of SNX17 FERM-like domain [47]. *In vitro* measurements using purified peptides corresponding to the sorting motifs in P-selectin and APP estimated affinities of 2.7 and 22 μM, respectively, for SNX17 [47].

Analogous to the adaptor/Retromer model observed with SNX27/Retromer and SNX3/Retromer, SNX17 can function as a cargo adaptor for a Retromer-like complex called Retriever [53]. Retriever is composed of three subunits: DSCR3, C16orf62, and VPS29. DSCR3 has also been called VPS26C due to its sequence similarity with the Retromer subunits VPS26A/B [53]. Additionally, modeling of the C-terminus of C16orf62 shows a similar fold to VPS35, suggesting broad similarities between Retriever and Retromer [53]. However, there are several key and intriguing differences between these complexes. VPS26C/DSCR3 has a loop insertion not found in VPS26A/B and a shortened C-terminus. Retriever is not known to directly associate with SNX3 or Rab7, the two known mechanisms for Retromer recruitment to endosomes. Perhaps relatedly, only Retriever appears to associate directly with the CCC complex, although CCC function is necessary for Retromer trafficking [35, 53]. Additionally, while both Retriever and Retromer associated with the WASH complex, Retromer does so directly via VPS35 and FAM21 while Retriever does so indirectly via the CCC complex [34, 61, 62]. To our knowledge, it is currently unclear if any GPCRs use the SNX17/Retriever pathway. A peptide array suggested potential specific binding between SNX17 and Φ-X-N-P-X-[F/Y] or Φ-X-N-X-X-[F/Y] motifs in GPR37, GRM7, and OX2R [47].

6.4.5 SNX-BAR

In yeast, the SNX-BAR proteins VPS5 and VPS17 form a heterodimer and stably associate with the Retromer complex [63]. In higher eukaryotes, SNX-BAR proteins are not tightly bound and, thus, are considered functionally rather than physically associated [9, 26, 64]. SNX-BAR proteins, such as SNX1, SNX2, SNX5, and SNX6 have N-terminal PX domains and C-terminal BAR domains [65]. At low concentrations, BAR domains bind ("detect") curved membranes while at high concentrations BAR domains can directly bend the membrane [66]. As such, SNX-BARs are thought to play key roles in the endosomal tubulation process.

Recently, it has been recognized that SNX-BAR heterodimers of SNX1/2 and SNX5/6 are cargo binding molecules in their own right, and retrograde trafficking on CI-MPR can be disrupted by loss of SNX-BAR function [26, 64]. Analysis of the SNX-BAR proteins reveal a 38 amino acid loop present in the PX domain of SNX5/6 not seen in other SNX-BAR proteins, and this loop was necessary (SNX5) and sufficient (if transplanted to SNX1) for CI-MPR binding (Kd ~25 μM) [67, 68]. A crystal structure of the SNX5 with the portion of the cytoplasmic tail of CIMPR has been solved, and the model revealed a bipartite sorting motif in CI-MPR: VSYKYSK and E*WLM*EEI. This bipartite motif is intriguing in at least three different ways: (i) the WLM sequence has been previously recognized as sorting motif bound by Retromer; (ii) the VSYKYSK motif plays the major role in SNX5 binding; (iii) the bipartite motif strongly overlaps with known AP-1/AP-2 binding sites [68, 69]. Recent work reconciled these data and demonstrated that CI-MPR can be independently sorted by either SNX3/Retromer or SNX-BAR complex [70]. One key finding from this study is that the SNX3/Retromer or SNX-BAR complex require different Golgi-localized tethers – GCC88 or Golgin-245, respectively – for retrograde trafficking of CI-MPR [70]. Additional cargo proteins carrying the SNX-BAR consensus motif Φ-X-Ω-X-Φ-(X)n-Φ, such as SEMA4C and IGF1R, have been shown to bind SNX5 and require SNX5 function for proper trafficking [68]. Importantly, imaging studies suggest that the SNX-BAR complex resides on distinct tubules from Retromer, suggesting a potential spatio-functional separation of tubules with distinct molecular compositions [64].

6.5 GPCR Signaling from Endosomes

Emerging studies in the past decade have shown that GPCR signaling is not limited to the plasma membrane and that receptors can also be activated and mediate downstream signaling pathways from internal membrane compartments. In this review, we will focus on the endosome, but we would be remiss not to mention that GPCR signaling has been linked to many other organelles [71, 72]. One of the first examples of endosomal signaling came from studies on pheromone-activated GPCR (encoded by the Ste2 gene) in budding yeast. Slessareva et al. identified that Ste2 can activate Gα subunit on endosomes through its association with the endosome-associated PI3 kinase encoded by Vps34 [73]. The first few examples of such signaling in mammalian system came from studies on thyroid stimulating hormone (TSH) and parathyroid hormone (PTH)/PTH-related peptide GPCRs, where TSHR and PTHR were shown to activate Gs-mediated cAMP response for a long period of time after internalization to the early endosomes [74, 75]. These types of sustained signaling from endosomes have now been demonstrated for a number of other GPCRs including: vasopressin 2 receptors, Neurokinin 1 receptors, and calcitonin receptor-like receptors [76–78]. Two hypotheses have been proposed to explain sustained signaling from the endosomes. In one model, it was thought that sustained signaling is the consequence of reduced destruction of cAMP by phosphodiesterase, through activation of G protein-independent signaling mediated by arrestin and activation of MAP kinase pathways on the endosomes [79]. In another model, it was shown that sustained signaling is unique to GPCRs that have prolonged interaction with β-arrestin on the endosomes and GPCR-G- β-arrestin form a complex on the endosomes to produce sustained signaling [77, 78, 80, 81].

However, GPCR signaling from endosomes can also occur in response to acute agonist stimulation. These types of acute responses on the endosomes are mediated by GPCR activation of Gs proteins, independent of β-arrestin, and can only occur after receptors are dissociated from β-arrestin [82]. Indeed, these two phases of acute Gα-mediated cAMP response from the plasma membrane and the endosomes were directly visualized by biosensors derived from single-domain antibodies (nanobodies) that probed the activity of both GPCR and Gα protein in living cells [83, 84].

As it has been discussed extensively in previous reviews [23, 85–88], there are now a number of studies demonstrating β-arrestin-dependent,

G protein-dependent, or β-arrestin/G protein dependent signaling on the endosomes. If removal from the plasma membrane is not the mechanism for receptor inactivation, and receptors can initiate signaling from the endosomes, a key question now is to determine the mechanisms by which GPCR-mediated signaling from endosomes are turned off.

As described in the previous sections, once internalized, many GPCRs either recycle back to the cell surface to start another round of signaling or are sorted to lysosomes for degradation. Recent evidence suggests that there are multiple mechanisms that regulate receptor trafficking from the endosome and can consequently manipulate GPCR signaling from the endosomes. These include (i) reducing the receptor resident time on the endosome by regulating receptor interaction with the retromer machinery which, in turn, promotes its exit from the endosomes and disrupts GPCR-Gα protein coupling [50, 79, 89], (ii) prolonging the endosomal occupancy time by interacting with arrestin domain-containing protein 3 (ARRDC3), a member of the β-arrestin family of visual/β-arrestins [90], and (iii) activating v-ATPases that regulate endosomal acidification which will ultimately promote dissociation of agonist-GPCR complex [91]. Whether there are other proteins that interact with any of these regulatory machineries that will, in turn, regulate their interaction with receptors, is unclear and needs to be further investigated.

6.6 Conclusion

In light of these data discussed above, it is clear that recycling of membrane proteins – including GPCRs – is an intricately regulated process. A divergence of sorting sequences, and sorting complexes, open the possibility for fine tuning of the kinetics and destinations of the endosomal sorting and recycling processes. In parallel with our growing understanding of endosomal sorting and recycling, it is now understood that many GPCRs can initiate signaling events from the endosome [23, 92]. Thus, the rapid interplay of GPCR trafficking and signaling – classically thought to occur exclusively at the cell surface – now must be understood in the chemical environment of the endosome.

References

1 Irannejad, R., Tsvetanova, N.G., Lobingier, B.T., and von Zastrow, M. (2015). Effects of endocytosis on receptor-mediated signaling. *Curr. Opin. Cell Biol.* 35: 137–143.

2 Hanyaloglu, A.C. (2018). Advances in membrane trafficking and endosomal signaling of G protein-coupled receptors. *Int. Rev. Cell. Mol. Biol.* 339: 93–131.

3 Marchese, A., Paing, M.M., Temple, B.R.S., and Trejo, J. (2008). G protein-coupled receptor sorting to endosomes and lysosomes. *Annu. Rev. Pharmacol. Toxicol.* 48: 601–629.

4 Naslavsky, N. and Caplan, S. (2018). The enigmatic endosome - sorting the ins and outs of endocytic trafficking. *J. Cell Sci.* 131 (13): http://dx.doi.org/10.1242/jcs.216499.

5 Raiborg, C. and Stenmark, H. (2009). The ESCRT machinery in endosomal sorting of ubiquitylated membrane proteins. *Nature* 458 (7237): 445–452.

6 Romero, G., von Zastrow, M., and Friedman, P.A. (2011). Role of PDZ proteins in regulating trafficking, signaling, and function of GPCRs: means, motif, and opportunity. *Adv. Pharmacol.* 62: 279–314.

7 Cao, T.T., Deacon, H.W., Reczek, D. et al. (1999). A kinase-regulated PDZ-domain interaction controls endocytic sorting of the beta2-adrenergic receptor. *Nature* 401 (6750): 286–290.

8 Gage, R.M., Kim, K.A., Cao, T.T., and von Zastrow, M. (2001). A transplantable sorting signal that is sufficient to mediate rapid recycling of G protein-coupled receptors. *J. Biol. Chem.* 276 (48): 44712–44720.

9 Cullen, P.J. and Steinberg, F. (2018). To degrade or not to degrade: mechanisms and significance of endocytic recycling. *Nat. Rev. Mol. Cell Biol.* 19 (11): 679–696.

10 Tanowitz, M. and von Zastrow, M. (2003). A novel endocytic recycling signal that distinguishes the membrane trafficking of naturally occurring opioid receptors. *J. Biol. Chem.* 278 (46): 45978–45986.

11 Tanowitz, M., Hislop, J.N., and von Zastrow, M. (2008). Alternative splicing determines the post-endocytic sorting fate of G-protein-coupled receptors. *J. Biol. Chem.* 283 (51): 35614–35621.

12 Thompson, D., McArthur, S., Hislop, J.N. et al. (2014). Identification of a novel recycling sequence in the C-tail of FPR2/ALX receptor: association with cell protection from apoptosis. *J. Biol. Chem.* 289 (52): 36166–36178.

13 Olsen, C., Memarzadeh, K., Ulu, A. et al. (2019). Regulation of somatostatin receptor 2 trafficking by C-tail motifs and the retromer. *Endocrinology* 160 (5): 1031–1043.

14 Vargas, G.A. and Von Zastrow, M. (2004). Identification of a novel endocytic recycling signal in the D1 dopamine receptor. *J. Biol. Chem.* 279 (36): 37461–37469.

15 Kishi, M., Liu, X., Hirakawa, T. et al. (2001). Identification of two distinct structural motifs that, when added to the C-terminal tail of the rat LH receptor, redirect the internalized hormone-receptor complex from a degradation to a recycling pathway. *Mol. Endocrinol.* 15 (9): 1624–1635.

16 Galet, C., Min, L., Narayanan, R. et al. (2003). Identification of a transferable two-amino-acid motif (GT) present in the C-terminal tail of the human

lutropin receptor that redirects internalized G protein-coupled receptors from a degradation to a recycling pathway. *Mol. Endocrinol.* 17 (3): 411–422.

17 Hirakawa, T., Galet, C., Kishi, M., and Ascoli, M. (2003). GIPC binds to the human lutropin receptor (hLHR) through an unusual PDZ domain binding motif, and it regulates the sorting of the internalized human choriogonadotropin and the density of cell surface hLHR. *J. Biol. Chem.* 278 (49): 49348–49357.

18 Sposini, S., Jean-Alphonse, F.G., Ayoub, M.A. et al. (2017). Integration of GPCR signaling and sorting from very early endosomes via opposing APPL1 mechanisms. *Cell. Rep.* 21 (10): 2855–2867.

19 Kim, D.I. and Roux, K.J. (2016). Filling the void: proximity-based labeling of proteins in living cells. *Trends Cell Biol.* 26 (11): 804–817.

20 Han, S., Li, J., and Ting, A.Y. (2018). Proximity labeling: spatially resolved proteomic mapping for neurobiology. *Curr. Opin. Neurobiol.* 50: 17–23.

21 Lobingier, B.T., Hüttenhain, R., Eichel, K. et al. (2017). An approach to spatiotemporally resolve protein interaction networks in living cells. *Cell* 169 (2): 350–360.e12.

22 Paek, J., Kalocsay, M., Staus, D.P. et al. (2017). Multidimensional tracking of GPCR signaling via peroxidase-catalyzed proximity labeling. *Cell* 169 (2): 338–349.e11.

23 Lobingier, B.T. and von Zastrow, M. (2019). When trafficking and signaling mix: How subcellular location shapes G protein-coupled receptor activation of heterotrimeric G proteins. *Traffic* 20 (2): 130–136.

24 Klumperman, J. and Raposo, G. (2014). The complex ultrastructure of the endolysosomal system. *Cold Spring Harbor Perspect. Biol.* 6 (10): a016857.

25 Puthenveedu, M.A., Lauffer, B., Temkin, P. et al. (2010). Sequence-dependent sorting of recycling proteins by actin-stabilized endosomal microdomains. *Cell* 143 (5): 761–773.

26 Kvainickas, A., Jimenez-Orgaz, A., Nägele, H. et al. (2017). Cargo-selective SNX-BAR proteins mediate retromer trimer independent retrograde transport. *J. Cell. Biol.* 216 (11): 3677–3693.

27 Wang, J., Fedoseienko, A., Chen, B. et al. (2018). Endosomal receptor trafficking: retromer and beyond. *Traffic* 19 (8): 578–590.

28 Seaman, M.N.J., Gautreau, A., and Billadeau, D.D. (2013). Retromer-mediated endosomal protein sorting: all WASHed up! *Trends Cell Biol.* 23 (11): 522–528.

29 Derivery, E., Sousa, C., Gautier, J.J. et al. (2009). The Arp2/3 activator WASH controls the fission of endosomes through a large multiprotein complex. *Dev. Cell* 17 (5): 712–723.

30 Gomez, T.S., Gorman, J.A., de Narvajas, A.A.-M. et al. (2012). Trafficking defects in WASH-knockout fibroblasts originate from collapsed endosomal and lysosomal networks. *Mol. Biol. Cell* 23 (16): 3215–3228.

31 Duleh, S.N. and Welch, M.D. (2010). WASH and the Arp2/3 complex regulate endosome shape and trafficking. *Cytoskeleton* 67 (3): 193–206.
32 Lee, S., Chang, J., and Blackstone, C. (2016). FAM21 directs SNX27–retromer cargoes to the plasma membrane by preventing transport to the Golgi apparatus. *Nat. Commun.* 7: http://dx.doi.org/10.1038/ncomms10939.
33 Gomez, T.S. and Billadeau, D.D. (2009). A FAM21-containing WASH complex regulates retromer-dependent sorting. *Dev. Cell* 17 (5): 699–711.
34 Phillips-Krawczak, C.A., Singla, A. et al. (2015). COMMD1 is linked to the WASH complex and regulates endosomal trafficking of the copper transporter ATP7A. *Mol. Biol. Cell* 26 (1): 91–103.
35 Singla, A., Fedoseienko, A., Giridharan, S.S.P. et al. (2019). Endosomal PI(3)P regulation by the COMMD/CCDC22/CCDC93 (CCC) complex controls membrane protein recycling. *Nat. Commun.* 10 (1): 4271.
36 Kovtun, O., Leneva, N., Bykov, Y.S. et al. (2018). Structure of the membrane-assembled retromer coat determined by cryo-electron tomography. *Nature* 561 (7724): 561–564.
37 Lucas, M., Gershlick, D.C., Vidaurrazaga, A. et al. (2016). Structural mechanism for cargo recognition by the retromer complex. *Cell* 167 (6): 1623–1635.e14.
38 Rojas, R., Kametaka, S., Haft, C.R., and Bonifacino, J.S. (2007). Interchangeable but essential functions of SNX1 and SNX2 in the association of retromer with endosomes and the trafficking of mannose 6-phosphate receptors. *Mol. Cell. Biol.* 27 (3): 1112–1124.
39 Rojas, R., van Vlijmen, T., Mardones, G.A. et al. (2008). Regulation of retromer recruitment to endosomes by sequential action of Rab5 and Rab7. *J. Cell. Biol.* 183 (3): 513–526.
40 Temkin, P., Lauffer, B., Jäger, S. et al. (2011). SNX27 mediates retromer tubule entry and endosome-to-plasma membrane trafficking of signalling receptors. *Nat. Cell Biol.* 13 (6): 715–721.
41 Lauffer, B.E.L., Melero, C., Temkin, P. et al. (2010). SNX27 mediates PDZ-directed sorting from endosomes to the plasma membrane. *J. Cell. Biol.* 190 (4): 565–574.
42 Joubert, L., Hanson, B., Barthet, G. et al. (2004). New sorting nexin (SNX27) and NHERF specifically interact with the 5-HT4a receptor splice variant: roles in receptor targeting. *J. Cell Sci.* 117 (Pt 22): 5367–5379.
43 Lunn, M.-L., Nassirpour, R., Arrabit, C. et al. (2007). A unique sorting nexin regulates trafficking of potassium channels via a PDZ domain interaction. *Nat. Neurosci.* 10 (10): 1249–1259.
44 Steinberg, F., Gallon, M., Winfield, M. et al. (2013). A global analysis of SNX27-retromer assembly and cargo specificity reveals a function in glucose and metal ion transport. *Nat. Cell Biol.* 15 (5): 461–471.

45 Gallon, M., Clairfeuille, T., Steinberg, F. et al. (2014). A unique PDZ domain and arrestin-like fold interaction reveals mechanistic details of endocytic recycling by SNX27-retromer. *Proc. Natl. Acad. Sci. U.S.A.* 111 (35): E3604–E3613.

46 Clairfeuille, T., Mas, C., Chan, A.S.M. et al. (2016). A molecular code for endosomal recycling of phosphorylated cargos by the SNX27-retromer complex. *Nat. Struct. Mol. Biol.* 23 (10): 921–932.

47 Ghai, R., Bugarcic, A., Liu, H. et al. (2013). Structural basis for endosomal trafficking of diverse transmembrane cargos by PX-FERM proteins. *Proc. Natl. Acad. Sci. U.S.A.* 110 (8): E643–E652.

48 Harrison, M.S., Hung, C.-S., Liu, T.-T. et al. (2014). A mechanism for retromer endosomal coat complex assembly with cargo. *Proc. Natl. Acad. Sci. U.S.A.* 111 (1): 267–272.

49 Harterink, M., Port, F., Lorenowicz, M.J. et al. (2011). A SNX3-dependent retromer pathway mediates retrograde transport of the Wnt sorting receptor Wntless and is required for Wnt secretion. *Nat. Cell Biol.* 13 (8): 914–923.

50 Varandas, K.C., Irannejad, R., and von Zastrow, M. (2016). Retromer endosome exit domains serve multiple trafficking destinations and regulate local G protein activation by GPCRs. *Curr. Biol.* 26 (23): 3129–3142.

51 Zhang, P., Wu, Y., Belenkaya, T.Y., and Lin, X. (2011). SNX3 controls Wingless/Wnt secretion through regulating retromer-dependent recycling of Wntless. *Cell Res.* 21 (12): 1677–1690.

52 Ghai, R., Mobli, M., Norwood, S.J. et al. (2011). Phox homology band 4.1/ezrin/radixin/moesin-like proteins function as molecular scaffolds that interact with cargo receptors and Ras GTPases. *Proc. Natl. Acad. Sci. U.S.A.* 108 (19): 7763–7768.

53 McNally, K.E., Faulkner, R., Steinberg, F. et al. (2017). Retriever is a multiprotein complex for retromer-independent endosomal cargo recycling. *Nat. Cell Biol.* 19 (10): 1214–1225.

54 Burden, J.J., Sun, X.-M., ABG, G., and Soutar, A.K. (2004). Sorting motifs in the intracellular domain of the low density lipoprotein receptor interact with a novel domain of sorting Nexin-17. *J. Biol. Chem.* 279: 16237–16245. http://dx.doi.org/10.1074/jbc.m313689200.

55 van Kerkhof, P., Lee, J., McCormick, L. et al. (2005). Sorting nexin 17 facilitates LRP recycling in the early endosome. *EMBO J.* 24 (16): 2851–2861.

56 Lee, J., Retamal, C., Cuitiño, L. et al. (2008). Adaptor protein sorting nexin 17 regulates amyloid precursor protein trafficking and processing in the early endosomes. *J. Biol. Chem.* 283 (17): 11501–11508.

57 Florian, V., Schlüter, T., and Bohnensack, R. (2001). A new member of the sorting nexin family interacts with the C-terminus of P-selectin. *Biochem. Biophys. Res. Commun.* 281 (4): 1045–1050.

58 Knauth, P., Schlüter, T., Czubayko, M. et al. (2005). Functions of sorting nexin 17 domains and recognition motif for P-selectin trafficking. *J. Mol. Biol.* 347 (4): 813–825.

59 Steinberg, F., Heesom, K.J., Bass, M.D., and Cullen, P.J. (2012). SNX17 protects integrins from degradation by sorting between lysosomal and recycling pathways. *J. Cell. Biol.* 197 (2): 219–230.

60 Böttcher, R.T., Stremmel, C., Meves, A. et al. (2012). Sorting nexin 17 prevents lysosomal degradation of β1 integrins by binding to the β1-integrin tail. *Nat. Cell Biol.* 14: 584–592. http://dx.doi.org/10.1038/ncb2501.

61 Harbour, M.E., Breusegem, S.Y., and Seaman, M.N.J. (2012). Recruitment of the endosomal WASH complex is mediated by the extended "tail" of Fam21 binding to the retromer protein Vps35. *Biochem. J.* 442 (1): 209–220.

62 Jia, D., Gomez, T.S., Billadeau, D.D., and Rosen, M.K. (2012). Multiple repeat elements within the FAM21 tail link the WASH actin regulatory complex to the retromer. *Mol. Biol. Cell* 23 (12): 2352–2361.

63 Seaman, M.N., McCaffery, J.M., and Emr, S.D. (1998). A membrane coat complex essential for endosome-to-Golgi retrograde transport in yeast. *J. Cell. Biol.* 142 (3): 665–681.

64 Simonetti, B., Danson, C.M., Heesom, K.J., and Cullen, P.J. (2017). Sequence-dependent cargo recognition by SNX-BARs mediates retromer-independent transport of CI-MPR. *J. Cell. Biol.* 216 (11): 3695–3712.

65 Cullen, P.J. (2008). Endosomal sorting and signalling: an emerging role for sorting nexins. *Nat. Rev. Mol. Cell Biol.* 9 (7): 574–582.

66 Peter, B.J., Kent, H.M., Mills, I.G. et al. (2004). BAR domains as sensors of membrane curvature: the amphiphysin BAR structure. *Science* 303 (5657): 495–499.

67 Teasdale, R.D., Loci, D., Houghton, F. et al. (2001). A large family of endosome-localized proteins related to sorting nexin 1. *Biochem. J.* 358 (Pt 1): 7–16.

68 Simonetti, B., Paul, B., Chaudhari, K. et al. (2019). Molecular identification of a BAR domain-containing coat complex for endosomal recycling of transmembrane proteins. *Nat. Cell Biol.* 21 (10): 1219–1233.

69 Seaman, M.N.J. (2007). Identification of a novel conserved sorting motif required for retromer-mediated endosome-to-TGN retrieval. *J. Cell Sci.* 120 (Pt 14): 2378–2389.

70 Cui, Y., Carosi, J.M., Yang, Z. et al. (2019). Retromer has a selective function in cargo sorting via endosome transport carriers. *J. Cell Biol.* 218: 615–631. http://dx.doi.org/10.1083/jcb.201806153.

71 Tadevosyan, A., Vaniotis, G., Allen, B.G. et al. (2012). G protein-coupled receptor signalling in the cardiac nuclear membrane: evidence and possible roles in physiological and pathophysiological function. *J. Phys.* 590 (6): 1313–1330.

72 Bénard, G., Massa, F., Puente, N. et al. (2012). Mitochondrial CB_1 receptors regulate neuronal energy metabolism. *Nat. Neurosci.* 15 (4): 558–564.

73 Slessareva, J.E., Routt, S.M., Temple, B. et al. (2006). Activation of the phosphatidylinositol 3-kinase Vps34 by a G protein alpha subunit at the endosome. *Cell* 126 (1): 191–203.

74 Calebiro, D., Nikolaev, V.O., Gagliani, M.C. et al. (2009). Persistent cAMP-signals triggered by internalized G-protein-coupled receptors. *PLoS Biol.* 7 (8): e1000172.

75 Ferrandon, S., Feinstein, T.N., Castro, M. et al. (2009). Sustained cyclic AMP production by parathyroid hormone receptor endocytosis. *Nat. Chem. Biol.* 5 (10): 734–742.

76 Jensen, D.D., Lieu, T., Halls, M.L. et al. (2017). Neurokinin 1 receptor signaling in endosomes mediates sustained nociception and is a viable therapeutic target for prolonged pain relief. *Sci Transl Med.* 9 (392): http://dx.doi.org/10.1126/scitranslmed.aal3447.

77 Feinstein, T.N., Yui, N., Webber, M.J. et al. (2013). Noncanonical control of vasopressin receptor type 2 signaling by retromer and arrestin. *J. Biol. Chem.* 288 (39): 27849–27860.

78 Yarwood, R.E., Imlach, W.L., Lieu, T. et al. (2017). Endosomal signaling of the receptor for calcitonin gene-related peptide mediates pain transmission. *Proc. Natl. Acad. Sci. U.S.A.* 114 (46): 12309–12314, 14.

79 Feinstein, T.N., Wehbi, V.L., Ardura, J.A. et al. (2011). Retromer terminates the generation of cAMP by internalized PTH receptors. *Nat. Chem. Biol.* 7 (5): 278–284.

80 Wehbi, V.L., Stevenson, H.P., Feinstein, T.N. et al. (2013). Noncanonical GPCR signaling arising from a PTH receptor-arrestin-Gβγ complex. *Proc. Natl. Acad. Sci. U.S.A.* 110 (4): 1530–1535.

81 Thomsen, A.R.B., Plouffe, B., Cahill, T.J. 3rd, et al. (2016). GPCR-G protein-β-arrestin super-complex mediates sustained G protein signaling. *Cell* 166 (4): 907–919.

82 Kotowski, S.J., Hopf, F.W., Seif, T. et al. (2011). Endocytosis promotes rapid dopaminergic signaling. *Neuron* 71 (2): 278–290.

83 Irannejad, R., Tomshine, J.C., Tomshine, J.R. et al. (2013). Conformational biosensors reveal GPCR signalling from endosomes. *Nature* 495 (7442): 534–538.

84 Stoeber, M., Jullié, D., Lobingier, B.T. et al. (2018). A genetically encoded biosensor reveals location bias of opioid drug action. *Neuron* 98 (5): 963–976.e5.

85 Eichel, K. and von Zastrow, M. (2018). Subcellular organization of GPCR signaling. *Trends Pharmacol. Sci.* 39 (2): 200–208.

86 Pavlos, N.J. and Friedman, P.A. (2017). GPCR signaling and trafficking: the long and short of it. *Trends Endocrinol. Metab.* 28 (3): 213–226.

87 Tsvetanova, N.G., Irannejad, R., and von Zastrow, M. (2015). G protein-coupled receptor (GPCR) signaling via heterotrimeric G proteins from endosomes. *J. Biol. Chem.* 290 (11): 6689–6696.

88 Irannejad, R. and von Zastrow, M. (2014). GPCR signaling along the endocytic pathway. *Curr. Opin. Cell Biol.* 27: 109–116.

89 Xiong, L., Xia, W.-F., Tang, F.-L. et al. (2016). Retromer in osteoblasts interacts with protein phosphatase 1 regulator subunit 14C, terminates parathyroid hormone's signaling, and promotes its catabolic response. *EBioMedicine* 9: 45–60.

90 Tian, X., Irannejad, R., Bowman, S.L. et al. (2016). The α-arrestin ARRDC3 regulates the endosomal residence time and intracellular signaling of the β2-adrenergic receptor. *J. Biol. Chem.* 291 (28): 14510–14525.

91 Gidon, A., Al-Bataineh, M.M., Jean-Alphonse, F.G. et al. (2014). Endosomal GPCR signaling turned off by negative feedback actions of PKA and v-ATPase. *Nat. Chem. Biol.* 10 (9): 707–709.

92 Thomsen, A.R.B., Jensen, D.D., Hicks, G.A., and Bunnett, N.W. (2018). Therapeutic targeting of endosomal G-protein-coupled receptors. *Trends Pharmacol. Sci.* 39 (10): 879–891.

7

Posttranslational Control of GPCR Signaling

Michael R. Dores

Department of Biology, Hofstra University, Hempstead, NY, USA

7.1 Introduction

Posttranslational modification of G protein-coupled receptors (GPCRs) alters the way receptors receive and transfer incoming signals to the cytoplasm. There are three major ways in which posttranslational modification contributes to GPCR signaling–regulating surface expression of GPCRs, altering GPCR conformational change, and recruiting signaling effectors. Each of these expands the variety of cellular responses to GPCR signaling beyond the level of ligand and receptor diversity, leading to cell- and tissue-specific signaling effects.

GPCRs are modulated on multiple levels at the plasma membrane and within the cell, leading to a diverse array of tissue- and context-dependent signaling outcomes. Posttranslational modifications are a critical mechanism for directing GPCR signaling. The three most common posttranslational modifications, glycosylation, phosphorylation, and ubiquitination, fundamentally shape how GPCRs interact with their ligands, signaling effectors, and adaptor proteins (APs). This chapter will describe how GPCRs are posttranslationally modified, and the role of posttranslational modification in regulating receptor signaling through altering receptor conformation and the recruitment of accessory proteins.

7.2 Posttranslational Modifications

7.2.1 Glycosylation

For GPCRs, glycosylation serves both as a quality control measure during protein synthesis and as a mechanism for altering ligand binding and signaling bias. Glycosylation of proteins involves the covalent attachment of sugar molecules to

GPCRs as Therapeutic Targets, Volume 1, First Edition. Edited by Annette Gilchrist.

asparagine (Asn), serine (Ser), or threonine (Thr) residues within the lumen of the rough endoplasmic reticulum (ER). The ability of carbohydrates to polymerize through glycosidic linkages to form both linear and branching structures consisting of multiple types of sugar monomers adds to the diversity of glycosylation and is the basis for the array of functions glycosylation has in regulating GPCRs.

7.2.2 N-Linked Glycosylation

During N-linked glycosylation, the side-chain nitrogen atom of an asparagine residue is covalently modified with a preassembled polysaccharide consisting of an *N*-acetylglucosamine disaccharide stalk with a branching structure of nine mannose and three glucose molecules (Figure 7.1). An oligosaccharide transferase enzyme complex within the lumen of the ER glycosylates specific Asn residues by targeting the consensus sequence Asn–X–Ser/Thr (where X is any amino acid except proline) [1]. In GPCRs, this consensus site has been identified within the N-terminus as well as the three extracellular loops (Figure 7.1). Once translation of the peptide is complete, the terminal glucose molecules are trimmed as part of a mechanism that controls for proper folding prior to sorting to the *cis*-golgi. The mannose tree is cleaved further within the *cis*-golgi, allowing receptors to sort to the *medial*- and *trans*-golgi where the carbohydrate tree can be modified further through the addition of sugars.

Although N-linked glycosylation serves a general role in quality control during sorting within the ER and golgi, N-linked glycosylation can also have significant effects on GPCR signaling. The extracellular domains of GPCRs are critical for ligand binding while controlling receptor conformation to impact intracellular signaling. N-Linked glycosylation affects ligand affinity by impacting receptor shape, as is the case for the Calcitonin receptor [2, 3]. Glycosylation can also influence the way a receptor changes once its ligand is bound. When glycosylated, the second extracellular loop of Protease-activated Receptor 1 (PAR1) regulates signaling bias by stabilizing the receptor in an active conformation that leads to preferential coupling to $G\alpha_{12/13}$ over $G\alpha q$ [4]. In addition, glycosylation affects the sorting of some GPCRs at the plasma membrane, as seen by the sequestration of D2 and D3 dopaminergic receptors into cholesterol-rich lipid rafts [5]. In addition, glycosylation is critical for sorting of these receptors into caveolae where they are internalized from the cell surface. N-Linked glycosylation also mediates the endocytosis and desensitization of the purinergic receptor $P2Y_2$ [6]. These examples illustrate the diverse roles that N-linked glycosylation can play in regulating GPCR signaling.

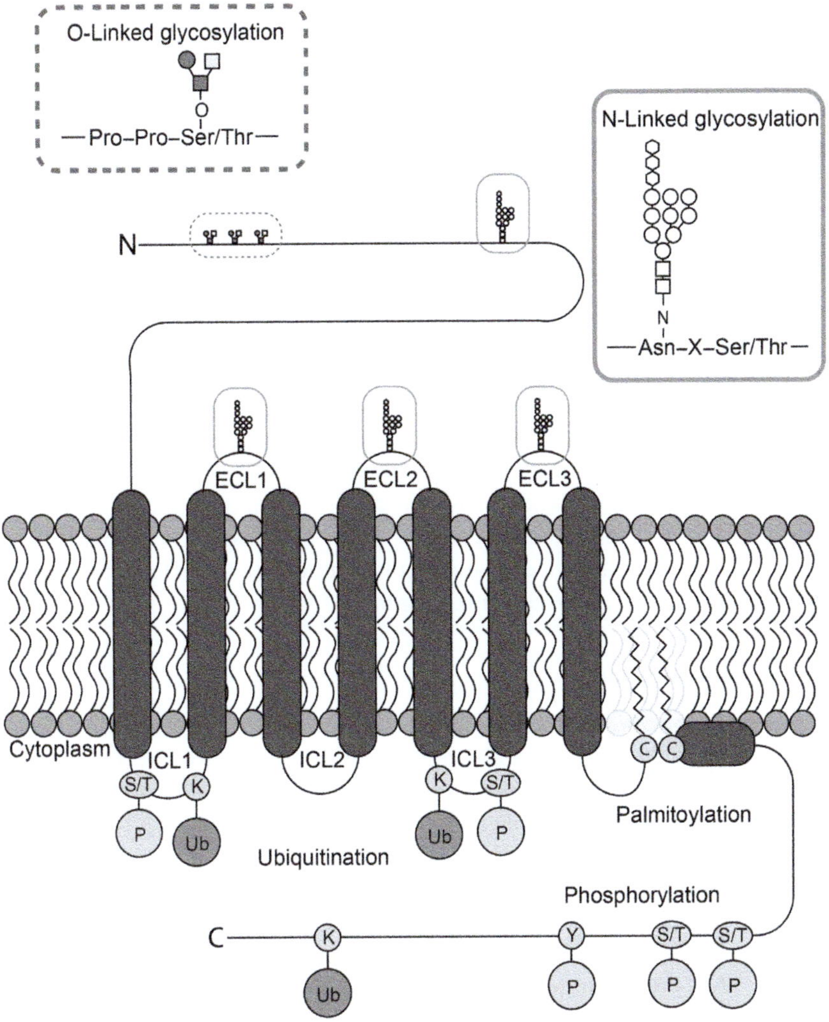

Figure 7.1 A diagram of GPCR posttranslational modifications. N-Linked glycosylation (solid inset) occurs on the N-terminus and extracellular loops (ECLs) of GPCRs at the consensus sequence Asn–X–Ser/Thr (where X is any amino acid except proline). O-Linked glycosylation (dotted insert) has been observed on the N-terminus of GPCRs. Open squares: *N*-acetylglucosamine; gray squares: *N*-acetylgalactosamine; open circles: mannose, gray circles: galactose; and hexagons: glucose. Palmitoylation of cysteine residues within the C-terminal tail of GPCRs anchors specific regions to the membrane. Phosphorylation of GPCRs can occur on serine, threonine, or tyrosine residues within intracellular loop 1 (ICL1), ICL3, and the C-terminal tail. Ubiquitination of lysine residues has been observed within ICL1, ICL3, and the C-terminal tail.

7.2.3 O-Linked Glycosylation

O-Linked glycosylation is the process in which polysaccharides are attached to the oxygen atom of serine or threonine amino acids. *N*-Acetylglucosamine and *N*-acetylgalactosamine are core monosaccharides for O-linked modifications, which are attached to target residues by a family of transferases expressed within the ER and Golgi [7, 8]. Although a strict consensus sequence has not been identified, O-linked glycosylation occurs on serine or threonine residues within proline-rich regions (Figure 7.1). O-Linked glycosylation prevents extracellular proteases from damaging exposed regions of transmembrane proteins. Proteolytic cleavage of GPCRs like the β_1-adrenergic receptor (β_1AR) by matrix metalloproteinases leads to receptor internalization. Glycosylation of serine residues within the β_1AR N-terminus prevents cleavage, resulting in enhanced signaling [9]. Among these glycosylation sites is Ser-49, and a common polymorphism (S49G) correlates with altered β_1AR signaling and the development of heart failure [10, 11]. In addition, O-linked glycosylation increases affinity of the CC-chemokine receptor CCR5 for the ligand MIP1 (macrophage inflammatory protein 1) as well as for the exterior envelope glycoproteins of HIV and other similar viruses [12]. O-Linked glycosylation represents a mechanism for heterogeneity in signaling response and receptor regulation.

7.2.4 Palmitoylation

Palmitoylation is the reversible covalent attachment of the 16-carbon saturated lipid palmitic acid (aka palmitate) to substrate cysteine residues. Protein acyl transferases (PATs) that harbor a catalytic aspartate–histidine–histidine–cysteine (DHHC) core mediate the formation of a thioester bond between the substrate cysteine sulfur atom and the palmitate fatty acid. The palmitoylation of GPCRs occurs within the C-terminal tail, facilitating critical changes in the conformation of the cytoplasmic regions of the receptor that is responsible for receptor signaling and endocytic trafficking. Many GPCRs and G-proteins are palmitoylated by specific PATs [13]. In cardiovascular tissue, palmitoylation of the α1D adrenergic receptor is essential for receptor signaling, and mutations to the tissue-specific PAT zDHHC21 lead to tachycardia and hypotension [14]. For many Class A GPCRs, the palmitate anchor helps stabilize the association of the structurally conserved eighth α-helix with the membrane. Molecular dynamics studies using the dopamine D2 receptor show that depalmitoylation releases the eighth helix into the cytoplasm [15], and this conformational change may affect G-protein coupling and receptor desensitization. In fact, palmitoylation of PAR1 regulates how endocytic adaptor proteins access the C-terminal tail [16], which suggests that depalmitoylation of the receptor is critical for endo-lysosomal sorting and degradation of activated receptors.

Cytosolic proteins can become depalmitoylated via nonenzymatic hydrolysis, or through the activity of acyl-protein thioesterases (APTs) that catalyze the hydrolysis of the thioester bond between palmitate and substrate cysteines. Humans express three APT homologues – APT1, APT2, and APTL1. While APT1 has been directly linked to the cycling of palmitoylation on small GTPases [17], including Gα subunits [18], its role in regulating GPCR palmitoylation has not been defined. However, APT2 regulates the depalmitoylation and desensitization of the melanocortin receptor MC1R. Individuals who express the MC1R RHC (red hair color) variant are more susceptible to the development of melanoma, in part because of poor MC1R signaling related to the depalmitoylated state of the receptor [19]. Pharmacological inhibition of APT2 restores MC1R RHC palmitoylation, and consequently rescues receptor signaling and reduces ultra violet B (UVB)-induced melanoma formation [20]. Given the role of dynamic palmitoylation in regulating GPCR signaling and trafficking, PATs and APTs represent a promising avenue for pharmacological development.

7.2.5 Phosphorylation

Phosphorylation is the addition of a phosphate (PO_4^-) to the side chains of serine, threonine, and tyrosine amino acids. This posttranslational modification is reversible, allowing phosphorylation to conditionally regulate modified receptors. By adding a negative charge to intracellular regions of a GPCR (Figure 7.1), phosphorylation causes changes in protein structure and can serve as a binding site for cytosolic regulatory proteins. G protein-coupled receptor kinases (GRKs) mediate Ser/Thr phosphorylation of GPCRs in response to receptor activation (Figure 7.2). Originally identified as kinases for rhodopsin (GRK1) [25, 26] and the β_2-adrenergic receptor (GRK2) [27], the human GRK kinase family is now thought to include seven members. In addition to GRKs, multiple signaling kinases have been shown to phosphorylate GPCRs. Protein Kinase A (PKA) and Protein Kinase C (PKC) are both activated by the production of specific second messengers following GPCR activation. Phosphorylation of the Gαs-coupled β_2-adrenergic receptor (PKA) [28] and the Gαq-coupled CXCR4 (PKC) [29] acts as a feedback mechanism to regulate signaling.

Kinases activated by other signaling receptors also target GPCRs and form the basis for crosstalk between different signaling pathways. A classic example of this is the interaction between adrenergic signaling and insulin signaling. Adrenaline stimulates the release of glucose from glycogen stores through the activation of the β_2-adrenergic receptor in response to low blood glucose levels or stress. The circulating hormone insulin controls the opposite physiological process – the uptake of blood glucose through the activation of a receptor-tyrosine kinase known as the insulin receptor. Stimulated insulin receptor activates Protein Kinase B (PKB), also

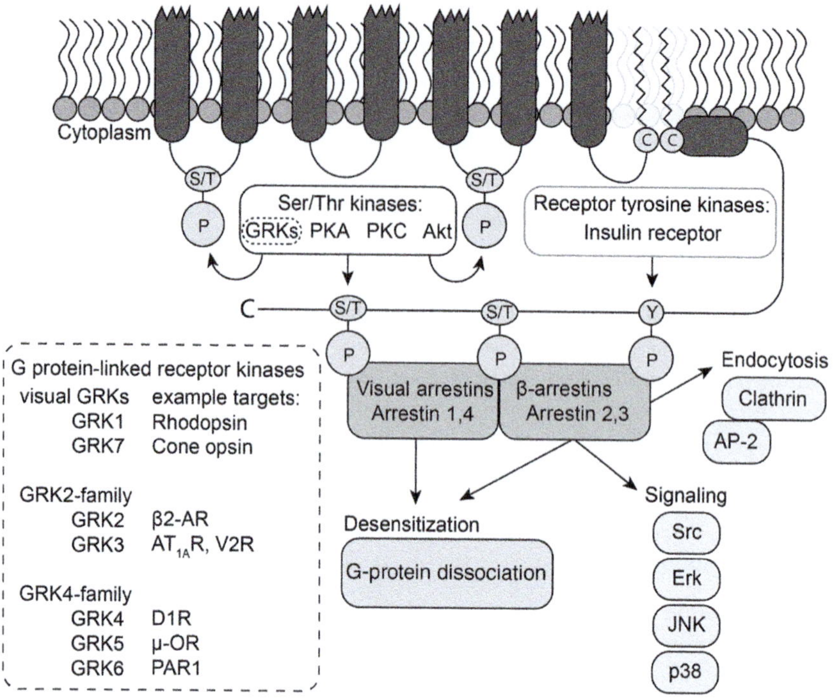

Figure 7.2 A model of GPCR phosphorylation. GPCRs are phosphorylated on intracellular loop 1 (ICL1), ICL3, and the C-terminal tail by serine/threonine kinases: GRKs (see inset), PKA, PKC, and Akt and receptor tyrosine kinases. Arrestins bind phosphorylated GPCRs and mediate desensitization, receptor endocytosis, and G-protein independent signaling. Source: Based on Refs [21–24].

known as Akt, upon binding to insulin. Akt/PKB phosphorylates the β_2-adrenergic receptor, leading to its desensitization and internalization [30]. In addition, insulin receptor itself can directly phosphorylate tyrosine residues within the C-terminal tail of β_2-adrenergic receptor [31], leading to a shift from Gαs signaling to signaling via the mitogen-activated protein kinase (MAPK) Erk and Src kinase recruitment [32]. Additionally, the act of insulin secretion from the pancreas is regulated by crosstalk between the casein kinase (CK) family of serine/threonine kinases and GPCRs. Acetylcholine stimulates muscarinic M_3 receptors on pancreatic β-cells and initiates the release of insulin. The M_3 receptor is a substrate of both CK1α and CK2α [33, 34], and phosphorylation leads to desensitization of the receptor. Inhibition of CK2α increases insulin release and represents a promising target for the treatment of both type 1 and type 2 diabetes mellitus [35].

Phosphorylation regulates GPCR signaling by recruiting a family of adaptor proteins known as arrestins. Upon binding GPCRs, arrestins inhibit G-protein coupling and can act as scaffolds for G-protein independent signaling pathways. There are four highly conserved arrestin isoforms (arrestin-1–4) which bind to phosphorylated Ser/Thr residues on the GPCR via the polar core of the protein. The visual arrestins (arrestin-1 and arrestin-4) are expressed in photoreceptors within the retina, and their primary function is to desensitize receptors by inhibiting G-protein coupling [36]. Similarly, the ubiquitously expressed β-arrestins (arrestin-2 and 3) function to inhibit G-protein binding, but can also serve as signaling scaffolds that recruit G-protein independent signaling factors like Src kinase [37]. As a signaling platform, β-arrestins can mediate the activation of MAPK pathways, including c-Jun N-terminal kinase (JNK), extracellular signal regulated kinase (ERK), and p38 [21–24] (Figure 7.2), and can control p53 expression through the activation of the ubiquitin ligase Mdm2 [38]. Given these abilities, β-arrestin signaling has been implicated in cancer proliferation and metastasis. By creating a binding site for arrestins, GPCR phosphorylation acts as a switch that changes GPCR signaling from G-protein-dependent second messenger production to a signaling kinase scaffold, altering the types of cellular responses a receptor can initiate.

Phosphorylation also mediates GPCR desensitization by targeting activated receptors for clathrin-mediated endocytosis. Internalization via clathrin-coated pits requires coupling of the transmembrane cargo with the clathrin lattice. β-Arrestins direct GPCRs like the β_2-adrenergic receptor and dopamine receptors into clathrin-coated pits by serving as an adaptor protein between the receptor and the clathrin coat [39, 40]. Additionally, agonist-induced phosphorylation of the PAR1 C-terminal tail leads to the recruitment of the AP-2 clathrin adaptor complex [41]. AP-2 is part of a family of heterotetrameric clathrin adaptors, and consists of four subunits (α, β_2, μ, and $\sigma 2$). While the α and β_2 subunits are known for their ability to bind clathrin, the α subunit of AP-2 also harbors a patch of basic amino acids that can serve as a binding site for negatively charged endocytic sorting signals like phosphorylated Ser/Thr [42]. Through β-arrestin and the AP-2 complex, GPCRs are removed from the cell surface via clathrin-coated vesicles, functionally desensitizing the cell from agonist stimulation.

A key feature of phosphorylation as a posttranslational modification is that it is reversible by the hydrolysis, a reaction catalyzed by phosphatases. Dephosphorylation is a critical step in regulating the signaling of many GPCRs, including the β_2-adrenergic receptor and the D1 dopamine receptor [43, 44]. Some candidate phosphatases include Protein Phosphatase 1 and Protein Phosphatase 2 (PP1 and PP2) [45, 46], both of which harbor multiple isoforms. PP1β, for example, is required for attenuation of β-arrestin signaling from the somatostatin receptor sst_{2A} [45]. In addition, many GPCRs are recycled back to the plasma membrane upon endocytosis as a way for the cell to rapidly resensitize to agonist stimulation. PP2A localizes to endosomes and facilitates the dephosphorylation of receptors like the neurokinin receptor NK_1R, resetting the receptor to an unphosphorylated state prior to its return back to the plasma membrane [47].

7.2.6 Ubiquitination

Ubiquitin is a 76 amino acid polypeptide that can be attached to exposed lysine residues on target substrates. The process of ubiquitination requires a cascade of three enzymes (E1, E2, and E3) to chemically modify ubiquitin and catalyze the formation of a covalent bond with lysine residues within the substrate. The E1-activating enzyme utilizes ATP to form a thioester bond with the C-terminus of ubiquitin, which is then transferred to the catalytic cysteine of an E2 enzyme. E2 enzymes then couple with an E3 ubiquitin ligase that determines substrate specificity. The addition of a single ubiquitin moiety is known as monoubiquitination; however, polymers of ubiquitin can be built off the seven lysine residues within ubiquitin itself (K6, K11, K27, K29, K33, K48, and K63). Interestingly, only monoubiquitin or K48 and K63 polyubiquitin chains are known to regulate GPCRs. The type of ubiquitination determines what ubiquitin-binding adaptor proteins bind receptors, and by extension tunes the regulation of modified receptors. GPCR ubiquitination can occur on cytosolic lysine residues within any of the intracellular loops or the C-terminal tail (Figure 7.1). Like phosphorylation, ubiquitination has a diverse array of roles in regulating GPCR signaling, from controlling receptor surface expression, endocytosis, lysosomal degradation, and modulation of G-protein independent signaling.

E3 ubiquitin ligases can be separated into three structurally distinct families, RING (really interesting new gene) domain ubiquitin ligases, the RBR (RING in between RING) ligases, and HECT (homologues to E6-AP carboxy terminus) E3 ligases. While all E3 ligases can bind to specific substrates, RING/RBR E3 ligases remain in a functional complex with their E2 enzymes in order to transfer ubiquitin to the substrate, whereas HECT E3 ligases catalyze the transfer independently. Of the 28 HECT E3 ubiquitin ligases in the human genome, the NEDD4 (neural precursor cell expressed, developmentally downregulated-4) family of E3-ligases stand out as common regulators of GPCR signaling. The nine NEDD4 ligases share highly conserved functional domains, including an N-terminal C2-domain, multiple WW domains, and the catalytic HECT domain. The WW domains bind target substrates on P/LPXY motifs (where X is any amino acid except proline) and via phosphorylated Ser/Thr residues, connecting GPCR phosphorylation to modification with ubiquitin. RING, RBR, and HECT E3 ligases all have roles in mediating both unstimulated GPCR ubiquitination as well as agonist-induced GPCR ubiquitination.

Ubiquitination of unstimulated GPCRs occurs during the process of biogenesis, either at the ER or the golgi, and can even occur once receptors have reached the plasma membrane. The basal expression of multiple opioid receptors is controlled at the ER and involves sorting of newly synthesized receptors into the endoplasmic reticulum-associated degradation (ERAD) pathway [48, 49], which terminates in translocation into the cytoplasm and proteasomal degradation. The RBR ligase Parkin controls the basal expression of the orphan GPCR GPR37 via ERAD as well [50]. Although there are no known ligands for GPR37, the receptor is a chaperone for the Wnt co-receptor LRP6, and Wnt signaling is important in the development of Parkinson's disease. Ubiquitin provides a new cytosolic binding surface, and the proteasome recognizes K48-linked polyubiquitin chains [51]. In contrast, basal ubiquitination can also serve to occlude access of adaptor proteins to their binding motifs within GPCR cytosolic domains. For example, the AP-2 clathrin adaptor mediates a high constitutive endocytic turnover of PAR1 receptors at the cell surface; however, ubiquitination of the AP-2 binding motif slows PAR1 turnover and increases the number of receptors available for signaling [52].

Ubiquitination of stimulated GPCRs provides new binding sites for the recruitment of endocytic adaptors and protein scaffolds involved in the attenuation of receptor signaling. Epsin 1 and Epsin 2 are ubiquitin-binding clathrin adaptors that sort stimulated GPCRs like PAR1 to clathrin-coated pits [41]. These adaptor proteins often work in concert with other endocytic mechanisms, like arrestins or AP-2. In many cases, receptors are ubiquitinated at the cell surface, but the ubiquitin modification instead participates in receptor sorting within the endosomal system. GPCRs internalized to early endosomes are either recycled back to the plasma membrane or sorted for degradation within the lumen of a lysosome. Ubiquitin-mediated lysosomal degradation has been well defined for the β_2-adrenergic receptor, chemokine receptor CXCR4 [53], PAR2 [54], μ and δ-opioid receptors [55, 56] and many others. Interestingly, each receptor is ubiquitinated by different E3 ligases (Figure 7.3). To reach the lumen of the lysosome, GPCRs must first be sequestered from recycled cargo, and as early endosomes mature into late endosomes, these receptors must be packaged into intraluminal vesicles. This process is mediated by a series of four ESCRT (Endosomal Sorting Complexes Required for Transport) complexes. ESCRT-0 first binds ubiquitinated receptors on the limiting membrane of early endosomes [57]. As the endosome matures, ESCRT-0 passes ubiquitinated receptors to the ubiquitin-binding components of the ESCRT-I complex [58]. The ESCRT-II complex forces the endosomal membrane inward, forming intraluminal buds while also binding ubiquitinated receptors [59]. Once packaged into mature intraluminal buds, receptors are deubiquitinated. The ESCRT-III complex then assembles to seal the vesicle and release it into the lumen of the late endosome, forming a structure known as the multivesicular endosome [60]. The multivesicular endosome then

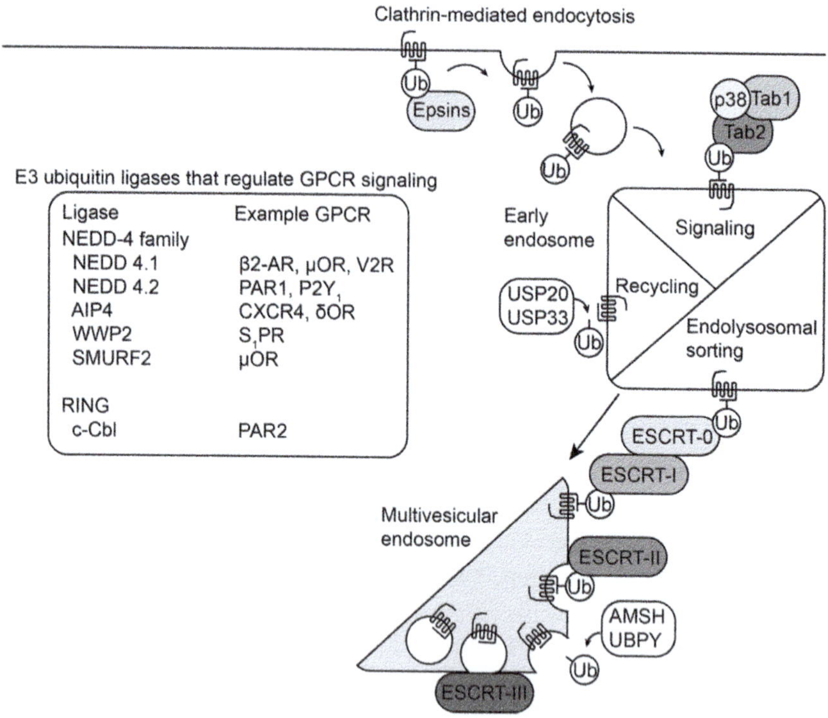

Figure 7.3 Ubiquitination regulates GPCR signaling and endocytic trafficking. Ubiquitin can serve as an endocytic sorting signal that can be bound by Epsin clathrin adaptors at the plasma membrane. At the early endosome, ubiquitin can also nucleate p38 signaling through the recruitment of the adaptor Tab2. Many ubiquitinated GPCRs are bound by the ESCRT complexes and packaged into multivesicular endosomes. For these GPCRs, deubiquitination is critical for recycling back to the plasma membrane.

fuses with lysosomes containing lipases and proteases. GPCR degradation represents the most complete way to attenuate receptor signaling while preventing rapid resensitization to ligand stimulation.

In addition to direct sorting into lysosomes, GPCR expression can be regulated through autophagy. Autophagosomes are specialized membrane structures that form in response to cellular stress. They surround damaged organelles and mediate the recycling of lipids, amino acids, and carbohydrate building blocks for use as metabolic substrates. Signaling by the β_2-adrenergic receptor induces autophagosome formation [61], and persistent adrenergic receptor signaling, combined with metabolic stress leads to enhanced receptor ubiquitination and degradation within autophagosomes [62]. This process does not require the ESCRT complexes, and highlights the role of dynamic, context-dependent ubiquitination of GPCRs in the regulation of important cellular processes.

GPCR ubiquitination can also serve to modulate receptor-signaling pathways by recruiting specific signaling scaffold proteins to activated receptors. A good example of this is PAR1 and the purinergic receptor $P2Y_1$, which are ubiquitinated upon receptor activation. Unlike many GPCRs, lysosomal sorting of activated PAR1 and $P2Y_1$ does not require ubiquitination. Instead, the HECT E3 ligase NEDD4-2 modifies both receptors with K63-linked polyubiquitin in order to recruit TAB2, a signaling scaffold protein that harbors an ubiquitin-binding domain. TAB2 couples to its homolog TAB1 and recruits the MAP kinase p38 [63]. Interestingly, the formation of this signaling scaffold occurs after PAR1 and $P2Y_1$ are internalized, indicating that ubiquitin-dependent p38 signaling occurs from endosomes.

Like phosphorylation, ubiquitination is a reversible posttranslational modification. Ubiquitin-specific proteases (USPs) mediate the removal of ubiquitin from target substrates, and highlight the role of dynamic ubiquitination in the endocytic trafficking and regulation of GPCRs. USP20 and USP33 deubiquitinate the β_2-adrenergic receptor on endosomes, allowing the receptor to recycle to the plasma membrane [62]. Likewise, USP8 prevents the lysosomal degradation of the human Frizzled receptor (FZR4), leading to extended Wnt signaling [64]. In contrast, deubiquitination is also a critical step during receptor sorting into multivesicular endosomes. The ESCRT-0 complex recruits both ubiquitin-specific protease Y (UBPY) and AMSH (associated molecule with the SH3 domain of STAM) to deubiquitinate GPCRs during packaging into multivesicular endosomes. Inhibition of AMSH and UBPY enhances ubiquitination of PAR2, but as a consequence, the receptor is stuck in an early endosomal compartment [54]. Similarly, USP14 binds and deubiquitinates CXCR4 and the metabotropic γ-aminobutyric acid $(GABA)_{b(b1)}$ receptor, processes necessary for lysosomal degradation and downregulation of these receptors [65, 66]. GPCRs have the ability to cycle between ubiquitinated and deubiquitinated states throughout their lifetime, creating a flexible mechanism for fine-tuning the cell's response to GPCR signaling.

7.3 Cytosolic Peptide Motifs and Their Accessory Proteins

7.3.1 Tyrosine and Dileucine Motifs

GPCRs harbor multiple intrinsic motifs for coupling to cytosolic adaptor proteins that regulate receptor desensitization and membrane trafficking. Endocytic sorting of multiple GPCRs is regulated by the Adaptor Protein (AP) family of clathrin-binding proteins. There are five AP-complexes (AP-1, AP-2, AP-3,

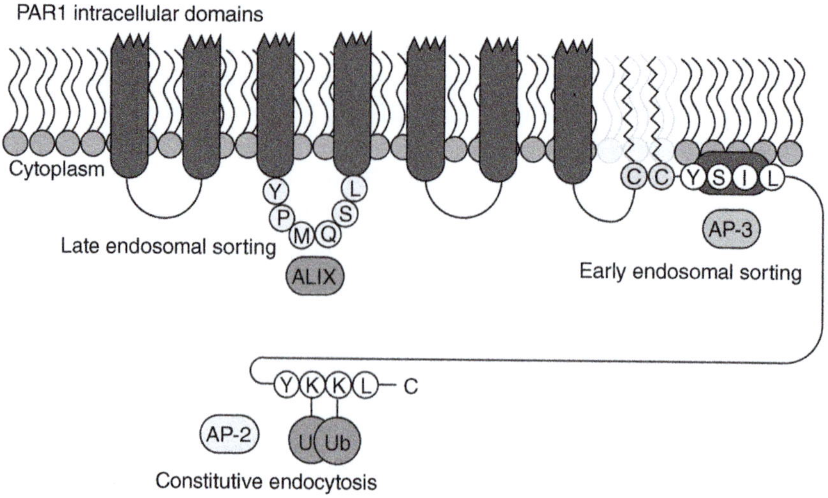

Figure 7.4 The cytosolic peptide motifs of protease-activated receptor 1. PAR1 harbors multiple, highly conserved intrinsic sorting motifs. Each plays a different role in PAR1 regulation. The YPMQSL motif within ICL2 is bound by the late endosomal adaptor protein ALIX, which facilitates lysosomal degradation of PAR1. The proximal YSIL motif within the C-terminal tail recruits the AP-3 complex, which mediates PAR1 sorting at the early endosome. The distal YKKL motif regulates constitutive internalization of PAR1 by recruiting the AP-2 clathrin adaptor complex. Source: Based on Dores et al. [68, 69].

AP-4, and AP-5) and all contain structurally conserved δ- and μ-subunits which bind the consensus motifs Asp/Glu–X–X–X–Leu–Leu/Ile, Tyr–X–X–ϕ, and Tyr–X–X–X–ϕ (where X is any amino acid except Pro, and ϕ represents any bulky hydrophobic amino acid) [67]. These AP-binding motifs are found on multiple GPCRs, however, only AP-2 and AP-3 have been shown to participate in surface expression and endo-lysosomal sorting of GPCRs. Posttranslational modifications can play key roles in regulating the accessibility of these motifs, as alluded to in Sections 7.2.5 and 7.2.6. An example of this is PAR1, whose C-terminal tail harbors two tyrosine motifs separated by the eighth helix (Figure 7.4). At the plasma membrane, palmitoylation of PAR1 keeps the proximal Y–S–I–L motif occluded [16], while leaving the distal Y–K–K–L motif accessible to the AP-2 complex for constitutive internalization. Ubiquitination of the Lys residues within the distal motif blocks AP-2 association and slows constitutive internalization of PAR1 [52], leading to greater signaling response. These findings demonstrate how intrinsic peptide motifs can be dynamically regulated during GPCR signaling and trafficking.

7.3.2 Endosomal Sorting Motifs

Intrinsic motifs and their conjugate adaptor proteins also control the packaging of some GPCRs into multivesicular endosomes. The GASP (G protein-coupled receptor associated sorting protein) family of adaptor proteins binds to the C-terminal tails of multiple GPCRs, including δ- and κ-opioid receptors, and β_2-adrenergic receptor [70–72]. GASP-1 and GASP-2 couple these receptors to the ESCRT machinery, re-routing GPCRs to the lysosome in the absence of receptor ubiquitination. Similarly, the adaptor protein ALIX sorts receptors like PAR1 and $P2Y_1$ to lysosomes by linking them to the ESCRT-III complex. The second intracellular loop of both receptors contains an ALIX-binding motif, a highly conserved Tyr–Pro–X_3–Leu (where X is any amino acid) peptide sequence [68, 69] (Figure 7.4). This mechanism allows both PAR1 and $P2Y_1$ to utilize receptor ubiquitination for G-protein independent signaling from endosomes.

7.4 Summary

Posttranslational modification of GPCRs provides a mechanism for fine-tuning receptor signaling by regulating surface expression, ligand affinity, receptor conformation, and coupling to intracellular signaling effectors. Extracellular modifications like N- and O-linked glycosylation are important for mediating receptor sorting to the plasma membrane and can influence how extracellular regulatory proteins and ligands access receptors. Intracellular modifications like palmitoylation change the way receptors interface with the plasma membrane, altering the shape of the receptor before and after ligand binding. Phosphorylation and ubiquitination represent rapidly reversible modifications that alter receptor binding to signaling effectors and endocytic trafficking proteins. Posttranslational modifications add a layer of diversity to GPCR signaling, which contributes to cell- and tissue-specific responses to GPCR ligands.

References

1 Mohorko, E., Glockshuber, R., and Aebi, M. (2011). Oligosaccharyltransferase: the central enzyme of N-linked protein glycosylation. *J. Inherit. Metab. Dis.* 34 (4): 869–878.

2 Lee, S.-M., Booe, J.M., Gingell, J.J. et al. (2017). N-Glycosylation of asparagine 130 in the extracellular domain of the human calcitonin receptor significantly increases peptide hormone affinity. *Biochemistry* 56 (26): 3380–3393.

3 Lee, S.-M., Jeong, Y., Simms, J. et al. (2020). Calcitonin receptor N-glycosylation enhances peptide hormone affinity by controlling receptor dynamics. *J. Mol. Biol.* 432 (7): 1996–2014.

4 Soto, A.G., Smith, T.H., Chen, B. et al. (2015). N-Linked glycosylation of protease-activated receptor-1 at extracellular loop 2 regulates G-protein signaling bias. *Proc. Natl. Acad. Sci. U.S.A.* 112 (27): E3600–E3608.

5 Min, C., Zheng, M., Zhang, X. et al. (2015). N-Linked glycosylation on the N-terminus of the dopamine D_2 and D_3 receptors determines receptor association with specific microdomains in the plasma membrane. *Biochim. Biophys. Acta* 1853 (1): 41–51.

6 Nakagawa, T., Takahashi, C., Matsuzaki, H. et al. (2017). *N*-Glycan-dependent cell-surface expression of the $P2Y_2$ receptor and *N*-glycan-independent distribution to lipid rafts. *Biochem. Biophys. Res. Commun.* 485 (2): 427–431.

7 Strous, G.J. (1979). Initial glycosylation of proteins with acetylgalactosaminylserine linkages. *Proc. Natl. Acad. Sci. U.S.A.* 76 (6): 2694–2698.

8 Röttger, S., White, J., Wandall, H.H. et al. (1998). Localization of three human polypeptide GalNAc-transferases in HeLa cells suggests initiation of O-linked glycosylation throughout the Golgi apparatus. *J. Cell Sci.* 111 (Pt 1): 45–60.

9 Goth, C.K., Tuhkanen, H.E., Khan, H. et al. (2017). Site-specific O-glycosylation by polypeptide *N*-acetylgalactosaminyltransferase 2 (GalNAc-transferase T_2) co-regulates β_1-adrenergic receptor N-terminal cleavage. *J. Biol. Chem.* 292 (11): 4714–4726.

10 Börjesson, M., Magnusson, Y., Hjalmarson, A., and Andersson, B. (2000). A novel polymorphism in the gene coding for the β^1-adrenergic receptor associated with survival in patients with heart failure. *Eur. Heart J.* 21 (22): 1853–1858.

11 Zhang, F. and Steinberg, S.F. (2013). S49G and R389G polymorphisms of the β^1-adrenergic receptor influence signaling via the cAMP-PKA and ERK pathways. *Physiol. Genomics* 45 (23): 1186–1192.

12 Bannert, N., Craig, S., Farzan, M. et al. (2001). Sialylated *O*-glycans and sulfated tyrosines in the NH_2-terminal domain of CC chemokine receptor 5 contribute to high affinity binding of chemokines. *J. Exp. Med.* 194 (11): 1661–1674.

13 Naumenko, V.S. and Ponimaskin, E. (2018). Palmitoylation as a functional regulator of neurotransmitter receptors. *Neural Plast.* [Internet]. 2018: 5701348. Available from: https://www.ncbi.nlm.nih.gov/pmc/articles/PMC5903346/.

14 Marin, E.P., Jozsef, L., Di Lorenzo, A. et al. (2016). The protein acyl transferase $ZDHHC_{21}$ modulates α1 adrenergic receptor function and regulates hemodynamics. *Arterioscler. Thromb. Vasc. Biol.* 36 (2): 370–379.

15 Sensoy, O. and Weinstein, H. (2015). A mechanistic role of Helix 8 in GPCRs: computational modeling of the dopamine D_2 receptor interaction with the GIPC1–PDZ-domain. *Biochim. Biophys. Acta* 1848 (4): 976–983.

16 Canto, I. and Trejo, J. (2013). Palmitoylation of protease-activated receptor-1 regulates adaptor protein complex-2 and -3 interaction with tyrosine-based motifs and endocytic sorting. *J. Biol. Chem.* 288 (22): 15900–15912.

17 Dekker, F.J., Rocks, O., Vartak, N. et al. (2010). Small-molecule inhibition of APT1 affects Ras localization and signaling. *Nat. Chem. Biol.* 6 (6): 449–456.

18 Siegel, G., Obernosterer, G., Fiore, R. et al. (2009). A functional screen implicates microRNA-138-dependent regulation of the depalmitoylation enzyme APT1 in dendritic spine morphogenesis. *Nat. Cell Biol.* 11 (6): 705–716.

19 Chen, S., Zhu, B., Yin, C. et al. (2017). Palmitoylation-dependent activation of MC1R prevents melanomagenesis. *Nature* 549 (7672): 399–403.

20 Chen, S., Han, C., Miao, X. et al. (2019). Targeting MC1R depalmitoylation to prevent melanomagenesis in redheads. *Nat. Commun.* 10 (1): 877.

21 Zhan, X., Kook, S., Gurevich, E.V., and Gurevich, V.V. (2014). Arrestin-dependent activation of JNK family kinases. *Handb. Exp. Pharmacol.* 219: 259–280.

22 DeFea, K.A., Zalevsky, J., Thoma, M.S. et al. (2000). β-Arrestin–dependent endocytosis of proteinase-activated receptor 2 is required for intracellular targeting of activated Erk1/2. *J. Cell Biol.* 148 (6): 1267–1282.

23 Luttrell, L.M., Roudabush, F.L., Choy, E.W. et al. (2001). Activation and targeting of extracellular signal-regulated kinases by β-arrestin scaffolds. *Proc. Natl. Acad. Sci. U.S.A.* 98 (5): 2449–2454.

24 Gong, K., Li, Z., Xu, M. et al. (2008). A novel protein kinase A-independent, β-arrestin-1-dependent signaling pathway for p38 mitogen-activated protein kinase activation by β_2-adrenergic receptors. *J. Biol. Chem.* 283 (43): 29028–29036.

25 Kuehn, H. (1978). Light-regulated binding of rhodopsin kinase and other proteins to cattle photoreceptor membranes. *Biochemistry* 17 (21): 4389–4395.

26 Kühn, H. (1974). Light-dependent phosphorylation of rhodopsin in living frogs. *Nature* 250 (5467): 588–590.

27 Benovic, J.L., Kühn, H., Weyand, I. et al. (1987). Functional desensitization of the isolated β-adrenergic receptor by the β-adrenergic receptor kinase: potential role of an analog of the retinal protein arrestin (48-kDa protein). *Proc. Natl. Acad. Sci. U.S.A.* 84 (24): 8879–8882.

28 Zamah, A.M., Delahunty, M., Luttrell, L.M., and Lefkowitz, R.J. (2002). Protein kinase A-mediated phosphorylation of the β_2-adrenergic receptor regulates its coupling to Gs and Gi. Demonstration in a reconstituted system. *J. Biol. Chem.* 277 (34): 31249–31256.

29 Luo, J., Busillo, J.M., Stumm, R., and Benovic, J.L. (2017). G protein-coupled receptor kinase 3 and protein kinase C phosphorylate the distal C-terminal tail of the chemokine receptor CXCR4 and mediate recruitment of β-arrestin. *Mol. Pharmacol.* 91 (6): 554–566.

30 Doronin, S., Shumay, E., Wang, H., and Malbon, C.C. (2002). Akt mediates sequestration of the β_2-adrenergic receptor in response to insulin. *J. Biol. Chem.* 277 (17): 15124–15131.

31 Karoor, V., Wang, L., Wang, H.Y., and Malbon, C.C. (1998). Insulin stimulates sequestration of β-adrenergic receptors and enhanced association of β-adrenergic receptors with Grb2 via tyrosine 350. *J. Biol. Chem.* 273 (49): 33035–33041.

32 Fan, G., Shumay, E., Malbon, C.C., and Wang, H. (2001). c-Src tyrosine kinase binds the β_2-adrenergic receptor via phospho-Tyr-350, phosphorylates G-protein-linked receptor kinase 2, and mediates agonist-induced receptor desensitization. *J. Biol. Chem.* 276 (16): 13240–13247.

33 Budd, D.C., McDonald, J.E., and Tobin, A.B. (2000). Phosphorylation and regulation of a $G_{q/11}$-coupled receptor by casein kinase 1α. *J. Biol. Chem.* 275 (26): 19667–19675.

34 Foley, J.F. (2007). The M_3-muscarinic receptor is phosphorylated and regulated by casein kinase 2. *Sci. Signal.* 2007 (382): tw134–tw134.

35 Ampofo, E., Nalbach, L., Menger, M.D. et al. (2019). Protein kinase CK2—a putative target for the therapy of diabetes mellitus? *Int. J. Mol. Sci.* 20 (18): 4398.

36 Peterhans, C., Lally, C.C.M., Ostermaier, M.K. et al. (2016). Functional map of arrestin binding to phosphorylated opsin, with and without agonist. *Sci. Rep.* 6 (1): 28686.

37 Strungs, E.G. and Luttrell, L.M. (2014). Arrestin-dependent activation of ERK and Src family kinases. *Handb. Exp. Pharmacol.* 219: 225–257.

38 Boularan, C., Scott, M.G.H., Bourougaa, K. et al. (2007). β-Arrestin 2 oligomerization controls the Mdm2-dependent inhibition of p53. *Proc. Natl. Acad. Sci. U.S.A.* 104 (46): 18061–18066.

39 Zhang, J., Barak, L.S., Winkler, K.E. et al. (1997). A central role for β-arrestins and clathrin-coated vesicle-mediated endocytosis in β_2-adrenergic receptor resensitization. Differential regulation of receptor resensitization in two distinct cell types. *J. Biol. Chem.* 272 (43): 27005–27014.

40 Ferguson, S.S.G., Downey, W.E., Colapietro, A.-M. et al. (1996). Role of β-arrestin in mediating agonist-promoted G protein-coupled receptor internalization. *Science* 271 (5247): 363–366.

41 Chen, B., Dores, M.R., Grimsey, N. et al. (2011). Adaptor protein complex-2 (AP-2) and epsin-1 mediate protease-activated receptor-1 internalization via phosphorylation- and ubiquitination-dependent sorting signals. *J. Biol. Chem.* 286 (47): 40760–40770.

42 Lindwasser, O.W., Smith, W.J., Chaudhuri, R. et al. (2008). A diacidic motif in human immunodeficiency virus type 1 Nef is a novel determinant of binding to AP-2. *J. Virol.* 82 (3): 1166–1174.

43 Gardner, B., Liu, Z.F., Jiang, D., and Sibley, D.R. (2001). The role of phosphorylation/dephosphorylation in agonist-induced desensitization of D_1 dopamine receptor function: evidence for a novel pathway for receptor dephosphorylation. *Mol. Pharmacol.* 59 (2): 310–321.

44 Tran, T.M., Friedman, J., Baameur, F. et al. (2007). Characterization of β_2-adrenergic receptor dephosphorylation: comparison with the rate of resensitization. *Mol. Pharmacol.* 71 (1): 47–60.

45 Pöll, F., Doll, C., and Schulz, S. (2011). Rapid dephosphorylation of G protein-coupled receptors by protein phosphatase 1β is required for termination of β-arrestin-dependent signaling. *J. Biol. Chem.* 286 (38): 32931–32936.

46 Pitcher, J.A., Payne, E.S., Csortos, C. et al. (1995). The G-protein-coupled receptor phosphatase: a protein phosphatase type 2A with a distinct subcellular distribution and substrate specificity. *Proc. Natl. Acad. Sci. U.S.A.* 92 (18): 8343–8347.

47 Murphy, J.E., Roosterman, D., Cottrell, G.S. et al. (2011). Protein phosphatase 2A mediates resensitization of the neurokinin 1 receptor. *Am. J. Physiol. Cell Physiol.* 301 (4): C780–C791.

48 Petaja-Repo, U.E., Hogue, M., Laperriere, A. et al. (2001). Newly synthesized human δ opioid receptors retained in the endoplasmic reticulum are retrotranslocated to the cytosol, deglycosylated, ubiquitinated, and degraded by the proteasome. *J. Biol. Chem.* 276 (6): 4416–4423.

49 Chaturvedi, K., Bandari, P., Chinen, N., and Howells, R.D. (2001). Proteasome involvement in agonist-induced down-regulation of μ and δ opioid receptors. *J. Biol. Chem.* 276 (15): 12345–12355.

50 Berger, B.S., Acebron, S.P., Herbst, J. et al. (2017). Parkinson's disease-associated receptor GPR37 is an ER chaperone for LRP6. *EMBO Rep.* 18 (5): 712–725.

51 Hershko, A. and Ciechanover, A. (1998). The ubiquitin system. *Annu. Rev. Biochem.* 67: 425–479.

52 Wolfe, B.L., Marchese, A., and Trejo, J. (2007). Ubiquitination differentially regulates clathrin-dependent internalization of protease-activated receptor-1. *J. Cell Biol.* 177 (5): 905–916.

53 Marchese, A., Raiborg, C., Santini, F. et al. (2003). The E3 ubiquitin ligase AIP4 mediates ubiquitination and sorting of the G protein-coupled receptor CXCR4. *Dev. Cell* 5 (5): 709–722.

54 Hasdemir, B., Murphy, J.E., Cottrell, G.S., and Bunnett, N.W. (2009). Endosomal deubiquitinating enzymes control ubiquitination and down-regulation of protease-activated receptor 2. *J. Biol. Chem.* 14: M109.025692.

55 Henry, A.G., White, I.J., Marsh, M. et al. (2011). The role of ubiquitination in lysosomal trafficking of δ-opioid receptors. *Traffic* 12 (2): 170–184.

56 Hislop, J.N., Henry, A.G., and von Zastrow, M. (2011). Ubiquitination in the first cytoplasmic loop of μ-opioid receptors reveals a hierarchical mechanism of lysosomal down-regulation. *J. Biol. Chem.* 286 (46): 40193–40204.

57 Katzmann, D.J., Stefan, C.J., Babst, M., and Emr, S.D. (2003). Vps27 recruits ESCRT machinery to endosomes during MVB sorting. *J. Cell Biol.* 162 (3): 413–423.

58 Bilodeau, P.S., Winistorfer, S.C., Kearney, W.R. et al. (2003). Vps27-Hse1 and ESCRT-I complexes cooperate to increase efficiency of sorting ubiquitinated proteins at the endosome. *J. Cell Biol.* 163 (2): 237–243.

59 Hierro, A., Sun, J., Rusnak, A.S. et al. (2004). Structure of the ESCRT-II endosomal trafficking complex. *Nature* 431 (7005): 221–225.

60 Babst, M., Katzmann, D.J., Estepa-Sabal, E.J. et al. (2002). ESCRT-III: an endosome-associated heterooligomeric protein complex required for MVB sorting. *Dev. Cell* 3 (2): 271–282.

61 Aránguiz-Urroz, P., Canales, J., Copaja, M. et al. (2011). Beta(2)-adrenergic receptor regulates cardiac fibroblast autophagy and collagen degradation. *Biochim. Biophys. Acta* 1812 (1): 23–31.

62 Kommaddi, R.P., Jean-Charles, P.-Y., and Shenoy, S.K. (2015). Phosphorylation of the deubiquitinase USP_{20} by protein kinase A regulates post-endocytic trafficking of β_2 adrenergic receptors to autophagosomes during physiological stress. *J. Biol. Chem.* 290 (14): 8888–8903.

63 Grimsey, N.J., Aguilar, B., Smith, T.H. et al. (2015). Ubiquitin plays an atypical role in GPCR-induced p38 MAP kinase activation on endosomes. *J. Cell Biol.* 210 (7): 1117–1131.

64 Mukai, A., Yamamoto-Hino, M., Awano, W. et al. (2010). Balanced ubiquitylation and deubiquitylation of Frizzled regulate cellular responsiveness to Wg/Wnt. *EMBO J.* 29 (13): 2114–2125.

65 Lahaie, N., Kralikova, M., Prézeau, L. et al. (2016). Post-endocytotic deubiquitination and degradation of the metabotropic γ-aminobutyric acid receptor by the ubiquitin-specific protease 14. *J. Biol. Chem.* 291 (13): 7156–7170.

66 Mines, M.A., Goodwin, J.S., Limbird, L.E. et al. (2009). Deubiquitination of CXCR4 by USP14 is critical for both CXCL12-induced CXCR4 degradation and chemotaxis but not ERK activation. *J. Biol. Chem.* 284 (9): 5742–5752.

67 Sanger, A., Hirst, J., Davies, A.K., and Robinson, M.S. (2019). Adaptor protein complexes and disease at a glance. *J. Cell Sci.* 132 (20): jcs222992. Available from: https://jcs.biologists.org/content/132/20/jcs222992.

68 Dores, M.R., Chen, B., Lin, H. et al. (2012). ALIX binds a YPX3L motif of the GPCR PAR1 and mediates ubiquitin-independent ESCRT-III/MVB sorting. *J. Cell Biol.* 197 (3): 407–419.

69 Dores, M.R., Grimsey, N.J., Mendez, F., and Trejo, J. (2016). ALIX regulates the ubiquitin-independent lysosomal sorting of the $P2Y_1$ purinergic receptor via a YPX_3L motif. *PLoS One* 11 (6): e0157587.

70 Cho, D.I., Zheng, M., Min, C. et al. (2013). ARF6 and GASP-1 are post-endocytic sorting proteins selectively involved in the intracellular trafficking of dopamine D_2 receptors mediated by GRK and PKC in transfected cells. *Br. J. Pharmacol.* 168 (6): 1355–1374.

71 Tschische, P., Moser, E., Thompson, D. et al. (2010). The G-protein coupled receptor associated sorting protein GASP-1 regulates the signalling and trafficking of the viral chemokine receptor US28. *Traffic* 11 (5): 660–674.

72 Simonin, F., Karcher, P., Boeuf, J.J.M. et al. (2004). Identification of a novel family of G protein-coupled receptor associated sorting proteins. *J. Neurochem.* 89 (3): 766–775.

8

GPCR Signaling from Intracellular Membranes

Yuh-Jiin I. Jong, Steven K. Harmon, and Karen L. O'Malley

Department of Neuroscience, Washington University School of Medicine, Saint Louis, MO, USA

8.1 Introduction

G protein-coupled receptors (GPCRs) constitute the largest group of membrane proteins in eukaryotes, transforming a myriad of stimuli (e.g. photons, odorants, pheromones, hormones, neurotransmitters, peptides, nucleotides, fatty acids, and even mechanical stimuli) into signaling pathways regulating growth, survival, differentiation, migration, metabolism, reproduction, stress, sleep, mood, and many other processes. Traditionally, the stimuli that GPCRs detect are signals that come from the extracellular environment; hence, these receptors have primarily been considered cell surface proteins. This concept began to change however when immunogold electron microscopy as well as subcellular fractionation and functional assays revealed intracellular organelles such as nuclei with functional GPCRs [1–4]. The notion of intracellular GPCR signaling got a further boost with the development of targeted fluorescence resonance energy transfer (FRET) and bioluminescence resonance energy transfer (BRET) biosensors capable of revealing discrete GPCR signaling from subcellular compartments such as endosomes, the *trans*-Golgi network (TGN), and Golgi [5–11]. These important technological advances are now altering the GPCR landscape such that the notion of intracellular GPCR signaling is no longer met with skepticism but rather is becoming one prospect among many. Knowledge that a GPCR can potentially signal from many intracellular locales is, in turn, recalibrating the way we think about drugging these receptors. From biased ligands to compartment-specific drugs, where a receptor signals may be just as important as the G protein(s) it binds to and the second messengers it activates. The challenge is in understanding the downstream consequences of spatially restricted GPCR signaling. Gaining knowledge of the physiological and pathological consequences of GPCR

GPCRs as Therapeutic Targets, Volume 1, First Edition. Edited by Annette Gilchrist.

activation or inhibition on an intracellular membrane versus the cell surface is critical in the development of new drugs necessary to block one such function but not another.

Although nuclear GPCRs were among the first to have a function ascribed to them [1, 3, 4], work over the last decade on GPCR desensitization, internalization, and endocytic sorting has elegantly demonstrated GPCR signaling from a multitude of endosomal subtypes [3, 4, 12]. Besides nuclei and endosomes, functional GPCRs are also turning up on and in mitochondria, lysosomes, the TGN, the Golgi, as well as the endoplasmic reticulum (ER) itself. As master regulators of cellular function, it makes sense that GPCRs maintain and regulate the complex spatial and temporal interactions among the various organelles as well as the dynamic multi-directional transport of vesicles, proteins, and lipids [13]. While most of the details of these processes remain to be discovered, recent discoveries offer tantalizing clues as to how these intracellular GPCRs are trafficked to their final destination, how they are activated, how they signal, and what the physiological and pathological significance of compartmentalized signaling might be.

Table 8.1 summarizes ~120 GPCRs that, based on past and current reports, exhibit compartmentalized signaling. Given that there are just over 800 GPCRs in the human genome [231] and that ~half of these are olfactory receptors [232], our current list of GPCRs that can function from within the cell (Table 8.1) suggests that at least 30% of the nonolfactory GPCRs are capable of intracellular signaling. All of the major GPCR family classes are represented with Family A being almost 90% of the current list, Family B, 7%, and Family C, 5%.

Selected receptors for which adequate literature exists to ascribe an intracellular location and function have been further categorized into their main signaling compartment and receptor ligand type (Table 8.2). To date, 50 GPCRs have been well described as having nuclear functions, of which >50% are GPCRs which bind peptides, ~20% are receptors activated by neurotransmitters such as catecholamines, glutamate or gamma aminobutyric acid (GABA), and 24% are receptors binding small lipophilic molecules such as leukotrienes, prostanoids, or lysophospholipids. The second most common intracellular signaling platform is the endosome. This is a category that is sure to grow as more information becomes available. In fact, it may be that every arrestin-mediated GPCR internalization leads to a second *de novo* signaling paradigm. Of the 53 GPCRs for which existing data shows that endosomal signaling occurs and is different from those pathways triggered at the cell surface, ~50% are peptide receptors followed by ~30% neurotransmitter GPCRs, and 13% small lipophilic molecule receptors (Table 8.2). Other GPCRs on membranes such as the ER, Golgi, mitochondria, and lysosomes have only recently been explored and currently only a handful of GPCRs have been observed as signaling from these organelles (Table 8.2). Although described

Table 8.1 Current list of intracellular GPCRs with strong evidence for location and function.

Class	Family name	IUPHAR receptor name	Human gene	Receptor	Subcellular compartment	Tissue	Evidence	Effects	References
A	5-Hydroxy-tryptamine receptors	5-HT_{1A} receptor	HTR1A	Serotonin 5-HT1A receptor	Endosomes	CHO-K1 cells	IF, FACS, FA	Ca^{2+}/CAM dependence of ERK activation	[14]
A	5-Hydroxy-tryptamine receptors	5-HT_{1B} receptor	HTR1B	Serotonin 5-HT2A receptor	Endosomes	Cortical neurons, MEFs in vivo	β-arr KOs, IF, WB, FA	ERK activation, behavioral assays	[15]
A	5-Hydroxy-tryptamine receptors	5-HT_{2C} receptor	HTR2C	Serotonin 5-HT2C (INI) receptor	Endosomes	HK cells, neurons	CM, ELISA, PhA, DN	ERK activation	[16, 17]
A	5-Hydroxy-tryptamine receptors	5-HT_4 receptor	HTR4	Serotonin 5-HT4 receptor	Mitochondria	Cardiomyocytes	SCF, WB, IF, FA	Stress and O_2 tension-mediated homeostasis of [Ca^{2+}], ROS, and adenosine triphosphate (ATP) generation	[18]
A	Acetylcholine receptors (muscarinic)	M_1 receptor	CHRM1	Muscarinic (M1) mAChRs	ER, Golgi, trans-golgi network, nuclei	Corneal epithelial, endothelial cells, brain	PhA, BA, ICC, FA	↑ DNA Pol II activity, ↑ nuclear cGMP, ↑ MAPK, ↑ synaptic plasticity	[19–21]

A	Acetylcholine receptors (muscarinic)	M_2 receptor	CHRM2	Muscarinic (M2) mAChRs	Endosomes	HEK cells	IF, WB, FA, FRET-based β-arr2 biosensors	β-arr mediated downstream signalling via Src, ERK, and JNK	[22]
A	Acetylcholine receptors (muscarinic)	M_3 receptor	CHRM3	Muscarinic (M3) mAChRs	ER, endosomes	Rat reticular thalamic nucleus, mouse, and human airway smooth muscle	EM, WB, BA, Ca^{2+} mobilization, FA	KO and KD of β-arr1 impairs M_3 airway resistance	[23, 24]
A	Adenosine receptors	A1 receptor	ADORA1	Adenosine A1	Endosomes	Smooth muscle cells	IF, β-arr KD	MAPK activation	[25]
A	Adenosine receptors	A3 receptor	ADORA3	Adenosine A3	Endosomes	HEK	β-arr FA, G_i activation, cAMP accumulation, Ca^{2+} mobilization, survival, ERK activation	Cell survival signaling, Ca^{2+} mobilization, inhibition of cAMP accumulation	[26, 27]
Ad	Adhesion Class GPCRs	ADGRE2	ADGRE2	Myeloid cell-restricted EMR-2; EMR2 (EGF-like module-containing, mucin-like, hormone receptor-like 2)	Nuclei	Breast cancer cells	IF	Survival	[28]

(continued)

Table 8.1 (Continued)

Class	Family name	IUPHAR receptor name	Human gene	Receptor	Subcellular compartment	Tissue	Evidence	Effects	References
A	Adrenoceptors	α_{1A} adrenoceptor	ADRA1A	α_{1A}-adreno-ceptors	Nuclei	Cardiac myocytes	BA, SCF, FA	G_q/G_{11}, PKC-δ activation, ERK activation, proliferation, survival, troponin phos-phorylation, sarcomere shortening	[29–32]
A	Adrenoceptors	α_{1B} adrenoceptor	ADRA1B	α_{1B}-adreno-ceptors	Nuclei	Cardiac myocytes	SCF, WB, BA, ICC, reconstitution of KO	ERK activation, receptor oligomerization	[33, 34]
A	Adrenoceptors	β_1 adrenoceptor	ADRB1	β_1-adrenergic receptor (adrenocep-tors)	Nuclei, Golgi	Cardiomyocytes, HeLa cells	BA, WB, IF, conformation-specific nanobody	G_s-mediated cAMP ↑, ↑transcription, PKA activation, ↑ ICER,	[1, 6, 35, 36]
A	Adrenoceptors	β_2 adrenoceptor	ADRB2	β_2-adreno ceptors	Endosomes	HEK cells, cardiac myocytes	Nanobodies, single particle EM analysis, cryo-EM, FA	Sustained cAMP effects, PKA activation, transcriptional responses	[5, 6, 36–38]
A	Adrenoceptors	β_3 adrenoceptor	ADRB3	β_3-adrenergic receptor	Nuclear	Cardiomyocyte	BA, ICC, WB	↑ eNOS, ERK, PI3K, transcription	[35, 39]

A	Angiotensin receptors	AT_1 receptor	AGTR1	Angiotensin Ang-II type 1 receptors AT1R	Nuclei, mitochondria, endosomes	Heterologous cells, atrial fibroblasts, squamous cell cancer cells	SCF, WB, in situ uncaging of ligand	MAPK activation, Ca^{2+} mobilization, ↑transcription, ↑cell proliferation, ↑collagen production and secretion	[40–43]
A	Angiotensin receptors	AT_2 receptor	AGTR2	Angiotensin Ang II type 2 receptors (AT2R)	Nuclei, mitochondria	Myocytes, hepatocytes, atrial fibroblasts, squamous cell cancer cells	BA, FA	NO mobilization, ↑ transcription, ↑cell proliferation, ↑collagen production and secretion, ↓apoptosis, ↑proliferation and invasion in squamous cancer cells	[41–43]
A	Apelin receptor	Apelin receptor	APLNR	Apelin	Nucleoplasm	HEK cells, COS7 cells, cerebellum-derived D283 cells, cerebellum, hypothalamus; pulmonary epithelium	ICC, IF, BA, FA	ND	[40, 44]

(continued)

Table 8.1 (Continued)

Class	Family name	IUPHAR receptor name	Human gene	Receptor	Subcellular compartment	Tissue	Evidence	Effects	References
A	Bile acid receptor	GPBA receptor	GPBAR1	Takeda GPCR, TGR5 (GPBAR1)	Nuclei, ER, MVB, vesicles	Cholangiocytes; monocytic THP1, HEK, BMDM cells	IG-EM, IF co-localization with organelle markers; TGR5-mediated KD and/or KO inhibition of viral response; GRK-β-arr-SRC assays	Cholangiocyte functional response to bile acid signaling; intracellular TGR5-mediated activation of GRK-β-arr-SRC axis to promote antiviral response	[45, 46]
A	Bombesin receptors	BB_1 receptor	NMBR	NMB-R $\vert$ BB_1 $\vert$neuromedin B receptor	Endosomes	Macrophages, RAW 264.7 macrophage cells	BB1 KD, PH, Co-IP, Co-localization endosome markers, cytokine microarray profiling	BB1 tethered to TLR-7 and -9 forms novel GPCR signaling platform on late endosomes in macrophages and RAW cells essential for ligand activation of TLRs and subsequent pro-inflammatory cell responses.	[47]

A	Bombesin receptors	BB_3 receptor	BRS3	BB_3 | bombesin receptor subtype-3	Endosome	CHO cells, HEK 293 cells	BA, enzyme fragment complementation, Ca^{2+} mobilization, β-arr recruitment; mass internalization assays, receptor internalization, β-arr recruitment (Tango, Pathfinder)	Peptide identification, constitutive activation	[48, 49]
A	Bradykinin receptors	B_1 receptor	BDKRB1	Bradykinin, B_1 receptor (B_1)	Nuclei	Cancer cells, breast cancer	EM, BA, WB, FA	↑Proliferation and colony formation	[50]
A	Bradykinin receptors	B_2 receptor	BDKRB2	Bradykinin, kinin receptor, B2	Nuclei	Cancer cells, HEK cells, COS7 cells, cerebellum-derived D283 cells	EM, BA, WB, FA⁻	Proliferation and survival maintenance	[40, 51]

(*continued*)

Table 8.1 (Continued)

Class	Family name	IUPHAR receptor name	Human gene	Receptor	Subcellular compartment	Tissue	Evidence	Effects	References
B	Calcitonin receptors	Calcitonin receptor-like receptor	CALCLR	Calcitonin receptor-like receptor (CLR)	Endosomes	HEK cells, spinal neurons	BRET-mediated changes in clathrin-mediated endocytosis, plasma membrane (KRas), early endosomes (Rab5a), and recycling endosomes (Rab11)	Sustained cAMP, PKC, cytosolic and nuclear ERK; nociception	[52]
C	Calcium-sensing receptor	CaS receptor	CASR	Calcium-sensing receptor (CASR)	Endosomes	HEK cells	↓ internalization-dependent signaling, genetic mutations in pathway	Sustained MAPK signaling	[53]
A	Cannabinoid receptors	CB_1 receptor	CNR1	Cannabinoid CB_1 receptors	Mitochondria	Brain	EM, IHC, IF, FA	G_i-mediated inhibition of soluble adenylate cyclase, ↓ cAMP	[54–61]
A	Cannabinoid receptors	CB_2 receptor	CNR2	Cannabinoid CB_2 receptor	Endolysosomes, nuclei	Prefrontal cortex, cardiomyocytes	SCF, WB, BA, ePhys	Ca^{2+} release from IP_3-sensitive- and acidic-like Ca^{2+} stores	[62–64]

A	Chemerin receptors	$Chemerin_1$	CMKLR1	Chemerin receptor 1, CMKLR1	Endosomes	CHO cells, MEFs	IF, G protein BRET, β-arr BRET, FA	β-arr dependent activation of ERK	[65]
A	Chemerin receptors	$Chemerin_2$	GPR1	Chemerin receptor 2, GPR1	Endosomes	CHO cells, MEFs	IF, G protein BRET, β-arr BRET, FA	β-arr dependent activation of ERK	[65]
A	Chemokine receptors	CCR2	CCR2	Chemokine receptor (CCR2)	Nuclei	HEK cells	SCF, WB, IP	ND	[66]
A	Chemokine receptors	CXCR4	CXCR4	cxc chemokine receptor type 4 (cxcr4)	Nuclear membrane, endosomes	Cancer cells (breast, colorectal, lung, gastric)	IHC, Flow Cyt	Survival, sustained AKT signaling, sustained FOX01/O3a signaling, inhibition of cell death	[67–73]
A	Chemokine receptors	CXCR5	CXCR5	cxc chemokine receptor type 5 (cxcr5)	Endosomes	HeLa cells	WB, suppression of internalization	Survival, sustained AKT signaling, sustained FOX01/O3a signaling, inhibition of cell death	[73]

(*continued*)

Table 8.1 (Continued)

Class	Family name	IUPHAR receptor name	Human gene	Receptor	Subcellular compartment	Tissue	Evidence	Effects	References
A	Cholecy-stokinin receptors	CCK_1 receptor	CCKAR	Cholecy-stokinin-1 receptor (CCK1R)	Endosomes	Pancreatic beta cells	Si-RNA KD, arrestin KO, SCF, WB, FA	β-arr dependent sustained ERK activation, ERK-pRSK90-Bad-Caspase-3 anti-apoptosis	[74, 75]
A	Cholecy-stokinin receptors	CCK_2 receptor	CCKBR	Cholecy-stokinin-2 receptor (CCK2R)	Endosomes	HEK cells	CI-Live with IF, β-arr-BRET assays	β-arr recruitment	[76]
A	Class A Orphans	GPR17	GPR17	GPR17 P2Y-like receptor	Endosomes	Oligodendrocyte-precursor cells	IHC, IP, FA	Ligand-induced differential GRK recruitment, ligand-specific activation, ligand-specific nuclear CREB activation	[77]
A	Class A Orphans	GPR3	GPR3	GPR3	Endosomes, Golgi	Cerebellar granular neurons	IF, CI-Live, co-localization with organellar markers	ERK1/2 and AKT activation; cell survival	[78, 79]
A	Class A Orphans	GPR37	GPR37	GPR37, Parkin-associated endothelin-like receptor (PAEL-R)	ER, Golgi, Trans-Golgi Network	HEK, H1703 cells, neurons, neural precursor cells	IP, IF, co-localization of organellar markers	↑Maturation of LRP6, thereby ↑ Wnt signaling; protects LRP from ERAD; ↑ neuronal fate	[80, 81]

A	Class A Orphans	GPR55	GPR55	GPR55	Lysosomes	Cardiomyocytes	Ca^{2+}, voltage imaging; microinjection of GPR55 ligands	Ca^{2+} release from NAADP-sensitive two-pore channels; hyper-polarization of the sarcolemma	[82]
A	Class A Orphans	GPR6	GPR6	GPR6: con-stitutively active orphan receptor; homologous to GPR3 and GPR12;	Endosomes	CHO cells	WB, FA	ERK activation; anti-apoptosis	[79]
A	Class A Orphans	GPR88	GPR88	GPR88 striatum-specific G-protein coupled receptor	Nuclei	Cortex, amygdala, hypothalamus	IF, co-localization with nuclear markers, SCF, IG-EM	ND	[83]
A	Class A Orphans	LGR4	LGR4	Leucine-rich repeat-containing GPCR (LGR4)	Nuclei	Gastric carcinoma	IHC	Neoplasia	[84]

(*continued*)

Table 8.1 (Continued)

Class	Family name	IUPHAR receptor name	Human gene	Receptor	Subcellular compartment	Tissue	Evidence	Effects	References
A	Class A Orphans	LGR5	LGR5	Leucine-rich G protein-coupled receptor-5 (LGR5)	*trans*-Golgi network	HEK cells	IF, on-cell ELISA	↑Wnt/β-catenin signaling	[85, 86]
A	Class A Orphans	MAS1	MAS1	Ang (1–7) receptor (Mas oncogene receptor)	Nuclear	Renal cells	BA, SCF, agonist induction of NO in isolated nuclei	NO production, ROS production, ↑ transcription	[87]
C	Class C Orphans	GPR158	GPR158	GPR158	Nucleus	Trabecular meshwork cells from retina	ICC	↑Cell proliferation, ↓ cellular permeability	[88]
C	Class C Orphans	GPRC6 receptor	GPRC6A	G protein-coupled receptor family C group 6 member A (GPRC6 A^{ICL3_KGKY})	Endosomes	CRISPR-modified prostate cancer cells	KO cells, WB, real time imaging,	ERK, AKT activation, mTORC1 activation, cell proliferation, ↓autophagy	[89]

B	Corticotropin-releasing factor receptors	CRF_1 receptor	CRHR1	Corticotropin-releasing hormone receptor 1 (CRHR)	Endosomes	Hippocampal cell line, hippocampal neurons	cAMP and Ca^{2+} FRET biosensors, CI-Live, Flow Cyt, IG-EM; Sex-dependent β-arr internalization	Sustained cAMP, sex-dependent β-arr endosomal signaling	[90–94]
A	Dopamine receptors	D_1 receptor	DRD1	Dopamine D_1 receptor	Endosomes	HEK cells, striatal spiny neurons	TIRF microscopy, IF	cAMP accumulation	[95]
A	Dopamine receptors	D_2 receptor	DRD2	Dopamine D_2 receptor	Endosomes; golgi	Striatum, *in vivo*	BRET, β-arr KO, WB, IHC, IF, FA	Sustained dephosphorylation/inactivation of AKT; dopamine stimulated D2LR- and PDGFRβ-containing EVs transport to Golgi, activation of G_{i3} and ERK; dendritic spine formation.	[96–99]

(*continued*)

Table 8.1 (Continued)

Class	Family name	IUPHAR receptor name	Human gene	Receptor	Subcellular compartment	Tissue	Evidence	Effects	References
A	Endothelin receptors	ET_A receptor	EDNRA	Endothelin-1 (ET_A) receptor	Nuclear	Myocytes	IF, BA, WB	Nuclear Ca^{2+} mobilization	[100]
A	Endothelin receptors	ET_B receptor	EDNRB	Endothelin-1 (ET_B) receptor	Nuclei	Cardiac myocytes, liver, cochlear neurons	BA, SCF, IF	Nuclear Ca^{2+} mobilization, nuclear NO production	[101, 102]
A	Formylpeptide receptors	FPR2/ALX	FPR2	Formyl peptide receptor 2	Nuclei	Human lung carcinoma CaLu-6 and human gastric adenocarcinoma AGS cell lines	SCF, IF, BA, FA	ERK activation	[103]
A	Free fatty acid receptors	FFA1	FFAR1	Free fatty acid receptor 1 (FFAR1) GPR40	Endosomes of cell lines, nuclei and/or perikarya of neurons	Insulinoma 1 clone 832/13 (INS-1 832/13) cell line, HEK cells, primate brain, spinal cord	KD and WB of signaling molecules, IF, IHC	Continuous G protein-dependent signaling, ↓ mTORC2 activity	[104–107]
A	Free fatty acid receptors	FFA2	FFAR2	Free Fatty Acid Receptor 2, FFA2R, GPR43	Endosomes	HEK, HeLa cells	β-arr2-dependent bimolecular luminescence complementation, WB ,IP	FFA2R-β-arr2 dependent inhibition of NF-κB	[108]

A	Free fatty acid receptors	FFA3	FFAR3	Free Fatty Acid Receptor 3, FFA3R, GPR41	Endosomes	HEK cells	(FRET)-based β-arr2 biosensors, IF, WB. IP	FFAR2/3 heterodimer complex ↑ β-arr2 recruitment; ↑ p38 phosphorylation	[109]
A	Free fatty acid receptors	FFA4	FFAR4	Free Fatty Acid Receptor, FFA4R, GPR120	Endosomes	HEK cells	(FRET)-based β-arr2 biosensors, IF, WB, FA	β-arr mediated downstream signalling via ERK and JNK; continuous G protein-dependent signaling	[22, 105]
A	G protein-coupled estrogen receptor	GPER	GPER	GPR30, G-protein-coupled estrogen receptor, GER	ER, golgi	Cancer, brain and peripheral cell types	IHC, fluorescent BA and localization, FA	$G_{i/o}$, G_s mediated effects including PI3K/AKT, ERK1/2, adenylyl cyclase, Ca^{2+} mobilization, eNOS	[110–113]

(*continued*)

Table 8.1 (Continued)

Class	Family name	IUPHAR receptor name	Human gene	Receptor	Subcellular compartment	Tissue	Evidence	Effects	References
C	$GABA_B$ receptors	$GABA_{B1}$	GABBR1	γ-Amino butyric acid (GABA) B receptor 1	ER, nuclei	HEK cells, glial-derived DI-TNC1 cell line, rat visual cortex	Reporter gene assays, surface biotinylation, WB, IF, IG-EM, BRET detection of GABAB-R1 homodimers, FA	$G_{i/o}$/MEK activation of ERK	[114, 115]
A	Galanin receptors	GAL_2 receptor	GALR2	GAL2 receptor, galanin receptor 2	Endosomes	HEK cells	FLASH/BRET-based biosensors of β-arr2 combined with NanoBit technology to measure β-arr2-Galr2 interactions in real-time living systems	β-arr1/2 dependent-pERK	[116]
A	Ghrelin receptor	Ghrelin receptor	GHSR	Growth hormone secretagogue receptor, GHSR1a	Endosomes, nuclei	HEK cells, MEFs	Mutational analysis, WB, real-time imaging, KDs, KOs, BRET monitoring of β-arr	ERK, AKT activation, proliferation, adipogenesis, RhoA activation and stress fiber formation	[117–121]

B	Glucagon receptors	GIP receptor	GIPR	Glucose-dependent Insulinotropic Peptide receptor, GIPR (gastric inhibitory polypeptide receptor)	Endosomes	HEK cells	Suppression of internalization-dependent cAMP BRET; endosomal cAMP FRET sensor; G^*_s nanobody; PKA FRET sensor	Sustained cAMP, PKA production	[122]
B	Glucagon receptors	GLP-1 receptor	GLP1R	Glucagon-like peptide-1 receptor, GLP-1R	Endosomes	CHO cells, pancreatic beta cells	ICC, FA, PhA, IG-EM	cAMP accumulation, sustained pERK signaling	[123, 124]
A	Glycoprotein hormone receptors	FSH receptor	FSHR	Follicle-stimulating hormone receptor (FSHR)	Very early endosomes	HEK cells, Sertoli cells	CI-live, TIRF, IF, FA	Sustained ERK activation	[125, 126]
A	Glycoprotein hormone receptors	LH receptor	LHCGR	Luteinizing hormone receptor LHR	Endosomes, very early endosomes	HEK cells, ovarian follicles	Pharmacological suppression of internalization, FRET microscopy; BA	Sustained cAMP production; sustained ERK activation, meiosis resumption	[127, 128]

(*continued*)

Table 8.1 (Continued)

Class	Family name	IUPHAR receptor name	Human gene	Receptor	Subcellular compartment	Tissue	Evidence	Effects	References
A	Glycoprotein hormone receptors	TSH receptor	TSHR	Thyroid-stimulating hormone TSH receptor, TSHR	Golgi, trans golgi network	Thyroid follicles	IF, FA	Sustained cAMP production, G_s-protein signaling, PKA II activation, gene transcription	[129, 130]
A	Gonado-trophin-releasing hormone receptors	$GnRH_1$ receptor	GNRHR	GnRH receptor	Nucleus	brain	BA	Histone H3 acetylation and phosphorylation	[131]
A	Kisspeptin receptor	kisspeptin receptor	KISS1R	Kisspeptin G-protein-coupled receptor (GPR54/KISS1R)	Endosomes	HEK cells, GT1-7 GnRH hypothalamic neuronal cell line, breast cancer cells	IF, Co-IP, KDs, FA	β-arr mediated ERK activation; differential regulation of gene expression; EGFR transactivation	[132–134]
A	Leukotriene receptors	$CysLT_1$ receptor	CYSLTR1	Leukotriene receptor CysLT1	Nuclei	Cancer cells, colon tissue, epithelial cells, fibroblasts and pre-keratinocytes	IF, SCF, WB, FA, co-localization of marker proteins	Nuclear Ca^{2+} mobilization, ERK activation, ↑ proliferation, translocation of NOX4, generation of ROS, oxidative nuclear DNA damage, apoptosis and necrosis	[135, 136]

A	Leukotriene receptors	$CysLT_2$ receptor	CYSLTR2	Leukotriene receptor CysLT2	Nuclei	Epithelial cells, fibroblasts and pre-keratinocytes	IF, WB, FA, co-localization of marker proteins	Translocation of NOX4, generation of ROS, oxidative nuclear DNA damage, apoptosis and necrosis	[136]
A	Lysophospholipid (LPA) receptors	LPA_1 receptor	LPAR1	Lysophosphatidic acid type 1 receptor (LPA1R)	Nuclear membrane	Cerebral microvascular endothelial cells	BA, IG-EM, SCF, WB	PTX toxin sensitive nuclear Ca^{2+} transients, ↑ iNOS expression	[69, 70, 137, 138]
A	Lysophospholipid (S1P) receptors	$S1P_1$ receptor	S1PR1	Sphingosine 1-phosphate receptor (S1P1)	Golgi, trans-Golgi network, nucleus	CHO cells, primary endothelial cells (HUVECs), CD4 T cells	ICC, WB, FA	G_i activation, sustained ↓ cAMP, ↑ERK activation, HUVEC migration, Cyr61/CTGF transcription	[139–142]
A	Melanocortin receptors	MC_2 receptor	MC2R	Melanocortin type 2 receptor (MC2R)	Nuclei	Adrenal cells	Co-IP with pore proteins, SCF; FA	ND	[143]

(*continued*)

Table 8.1 (Continued)

Class	Family name	IUPHAR receptor name	Human gene	Receptor	Subcellular compartment	Tissue	Evidence	Effects	References
A	Melano-cortin receptors	MC2 receptor	MC3R	Melano-cortin 3 receptor (MC3R)	Endosomes	CAD brain stem cells	ICC, WB, proliferation assays	↑ MAPK and AKT pathways, ↑ proliferation	[144]
A	Melano-cortin receptors	MC4 receptor	MC4R	Melano-cortin type 4 receptor (MC4R)	Endosomes, ER	HEK, N2A, hypothalamic cells	ICC, SCF, IP, WB, BA	Sustained cAMP and AMPK signaling	[145, 146]
A	Melatonin receptors	MT_1 receptor	MTNR1A	Melatonin MT_1 receptor	Mitochondria	Brain	SCF, WB	G_i protein activation, ↓ cAMP, ↓cytochrome c release, neuroprotection	[147–149]
A	Melatonin receptors	MT_2 receptor	MTNR1B	Melatonin MT_2 receptor	Mitochondria, nuclei	Choriocar-cinoma cells, gastric endothelial cells	IF, FA	↑Mitochondrial membrane potential, ATP synthesis and angiogenisis	[150, 151]
C	Metabotropic glutamate receptors	$mGlu_1$ receptor	GRM1	Metabotropic glutamate, $mGlu_1$	Nuclei, ER	Brain	IHC, IF, BA, FA	$G_q/G_{11}/PLC/IP_3$ activation, ↑intracellular Ca^{2+}	[152]
C	Metabotropic glutamate receptors	$mGlu_5$ receptor	GRM5	Metabotropic glutamate, $mGlu_5$	Nuclei, ER	Brain, spinal cord dorsal horn	EM, Ab stain, BA, FA	$G_q/G_{11}/PLC/IP_3$ activation, ↑intracellular Ca^{2+}	[153–159]

A	Neuropeptide Y receptors	Y_1 receptor	NPY1R	Neuropeptide Y Receptor (Y1)	Nuclear	Cardiac endothelial cells, pituitary cells,	IF, BA, WB	Nuclear Ca^{2+} mobilization, regulation of excitation-secretion coupling	[160]
A	Neurotensin receptors	NTS_1 receptor	NTSR1	Neurotensin receptor 1 (NTR1)	Endosomes, nuclei	NCM460 human colonic epithelial cells	IF, real time imaging, WB, gene silencing, FA	MAP kinase activation, NF-κB activation, proinflamma-tory actions of NT	[161]
A	Opioid receptors	NOP receptor	OPRL1	Nociceptin/orphanin FQ (N/OFQ) peptide (NOP)	Endosomes	HEK cells	Mutational analyses, BRET β-arr analyses; KDs, whole cell recording, behavioral assays	JNK activation, anxiolytic-like effects	[162–164]
A	Opioid receptors	δ receptor	OPRD1	Opioid receptor, delta DOR	Endosomes, Golgi, Golgi outposts, nuclei	HEK cells, striatal spiny neurons, NG108-15 cells	TIRF, conformation sensitive nanobody, photobleaching	Sustained ↓ of cAMP	[165, 166]
A	Opioid receptors	κ receptor	OPRK1	Opioid receptor, kappa KOR	Nuclei, endosomes	Myocardial cells, AtT-20 cells, primarty astrocytes, striatal neurons	FA, mutational anaylsis, KOs, KDs, BA, real time imaging	↑ Opioid peptide gene transcription, p38 activation	[167, 168]

(*continued*)

Table 8.1 (Continued)

Class	Family name	IUPHAR receptor name	Human gene	Receptor	Subcellular compartment	Tissue	Evidence	Effects	References
A	Opioid receptors	μ receptor	OPRM1	Opioid receptor, mu, MOR	Endosomes, Golgi, Golgi outposts, nuclei	HEK cells, striatal spiny neurons, mesothelial cell line	TIRF, conformation sensitive nanobody, photobleaching	Sustained ↓ of cAMP	[165, 169]
A	Orexin receptors	OX_1 receptor	HCRTR1	Orexin receptor 1 (OXR1)	Endosomes	HEK, MEFS	IF, real time imaging, Co-IP, FA	Sustained ERK activation	[170]
A	Orexin receptors	OX_2 receptor	HCRTR2	Orexin receptor 2 (OXR2)	endosomes	HEK cells	BRET, ELISA, FA	Sustained ERK and receptor-β-arr-ubiquitin complex formation	[171]
7 TM	Other 7TM proteins	GPR107	GPR107	GPR107	Golgi	HEK cells	IF, WB	Retrograde trafficking of cholera toxin	[172]
7 TM	Other 7TM proteins	GPR12	GPR12	GPR12: constitutively active orphan receptor; homologous to GPR3 and GPR6	Non ligand-induced β-arrestin2 recruitment to endosomes	PC12 cells	ICC, WB, FA	ERK activation; neurite induction	[79]

7 TM	Other 7TM proteins	GPR137	GPR137	GPR137, GPR137A | TM7SF1L1 | TM7SF1	Lysosomes	HeLa cells	FA	Rag-dependent change in lysosomal morphology; adaptor for RagA-dependent mTORC1 translocalization	[173]
7 TM	Other 7TM proteins	GPR137B	GPR137B	GPR137B	Lysosomes	HeLa cells, MEFs, cardiomyocytes	FA	Rag-dependent change in lysosomal morphology; adaptor for RagA-dependent mTORC1 translocalization	[173]
7 TM	Other 7TM proteins	GPR137C	GPR137C	GPR137C	Lysosomes	HeLa cells	FA	Rag-dependent change in lysosomal morphology; adaptor for RagA-dependent mTORC1 translocalization	[173]
7 TM	Other 7TM proteins	GPR143	GPR143	GPR143, ocular albinism 1, OA1	Melanosomes, lysosomes	Pirmented, non-pigmented cells	ICC, WB, IP, FA, KO and KD systems	G_{i3} activation, melanosomal biogenesis	[174, 175]

(continued)

Table 8.1 (Continued)

Class	Family name	IUPHAR receptor name	Human gene	Receptor	Subcellular compartment	Tissue	Evidence	Effects	References
A	P2Y receptors	$P2Y_1$ receptor	P2RY1	Purine P2Y1 receptor	Mitochondria	Isolated brain, heart, lung, liver mitochondria	SCF, WB, IG-EM, FA	Cytosolic [ATP] governs mitochondrial ATP production through regulation of mCa^{2+} uptake	[176]
A	P2Y receptors	$P2Y_{12}$ receptor	P2RY12	Purine P2Y12 receptor	Mitochondria; endosomes	Astrocytes; COS7 cells	CM; het-erodimerization and co-internalization and localization with β-arr2 on endosomes.	Recruitment and activation of AKT on endosomes	[177, 178]
A	P2Y receptors	$P2Y_2$ receptor	P2RY2	Purine P2Y1 receptor	Mitochondria	Isolated brain, heart, lung, liver mitochondria	SCF, WB, IG-EM, FA	Mitochondrial Ca^{2+} uptake	[176, 179]
B	Parathyroid hormone receptors	PTH1 receptor	PTH1R	Parathyroid hormone receptor (PTH1R)	Nuclei, endosomes	Liver, kidney, uterus, gut, ovary and osteoblast-like cells	IF, WB	Sustained cAMP generation, DNA synthesis, mitosis	[180–184]
A	Platelet-activating factor receptor	PAF receptor	PTAFR	Platelet-activating factor receptor (Ptafr) PAF	Nuclei	Retinal microvascular endothelial cells	EM, BA, IF, WB, FA	G_i-mediated ↑ NOS3, VEGFA	[185, 186]

A	Prostanoid receptors	EP_1 receptor	PTGER1	Prostaglandin receptor EP1	Nuclei	Endothelial	EM, IF, BA	Nuclear Ca^{2+} mobilization	[187]
A	Prostanoid receptors	EP_2 receptor	PTGER2	Prostaglandin receptor EP2	Nuclei	Endothelial cells, brain, liver, myometrium	BA, EM, IF, FA	Nuclear Ca^{2+} mobilization, ERK, PKB activation, eNOS activation, ↑ inflammation	[187–189]
A	Prostanoid receptors	EP_3 receptor	PTGER3	Prostaglandin receptor EP3	Nuclei	Brain endothelial cells and neurons	EM, IF, BA	Nuclear Ca^{2+} mobilization, ↑ nuclear phospho-ERK and eNOS	[190]
A	Prostanoid receptors	EP_4 receptor	PTGER4	Prostaglandin receptor EP4	Nuclei, golgi	Brain endothelial cells and neurons	EM, IF, BA, SCF	ND	[188]
A	Prostanoid receptors	FP receptor	PTGFR	Prostaglandin F2α receptor (PTGFR)	Perinuclear, nuclear	Luteinizing granulosa cells	IF, co-localization with nuclear markers, SCF, WB	Regulation of progesterone production via PLC/PKC pathway	[191]
A	Prostanoid receptors	TP receptor	TBXA2R	Thromboxane A2 receptor (TPR)	Nuclei	Oligodendrocytes	SCF, WB, IF, FA	Oligodendrocyte maturation, ↑myelin basic protein expression	[192, 193]

(*continued*)

Table 8.1 (Continued)

Class	Family name	IUPHAR receptor name	Human gene	Receptor	Subcellular compartment	Tissue	Evidence	Effects	References
A	Proteinase-activated receptors	PAR1	F2R	Protease-activated receptor PAR1	Endosomes	HeLa cells, HUVECs	IF, WB, IP	p38 activation	[194, 195]
A	Proteinase-activated receptors	PAR2	F2RL1	Protease-activated receptor PAR2, F2rl1	Endosomes, nuclei	Rat kidney cell line (KNRK), enterocytes, DRG neurons, colonic nociceptors, retina ganglion cells	Suppression of endocytosis, PhA, IF, WB, IP, Biosensors, BRET assays	RAF1 recruitment, MAPK/ERK activation, nociception, hyperexciability of DRG neruons; nuclear recruitment of Sp1 to trigger Vegfa expression, neo-vascularization.	[196–199]
A	Proteinase-activated receptors	PAR4	F2RL3	Protease-activated receptor PAR4	Endosomes	COS-7 cells, Dami megakaryocytic cells	Activated PAR4 $P2Y_{12}$ receptor heterodimerization/internalization and β-arr recruitment; BRET assays, IF, WB	Sustained p-AKT activation	[178]

A	Somatostatin receptors	SST_1 receptor	SSTR1	Somatostatin type 1 (sst1) receptor	Endosome, ER, Golgi	CHO-K1 cells, hypothalamic neurons	IF, IG-EM	Regulation of growth hormone ultradian rhythm in hypothalamus	[200, 201]
A	Somatostatin receptors	SST_2 receptor	SSTR2	Somatostatin type 2 (sst2) receptor	Endosomes, Golgi	Hippocampal neurons, dendatre molecular layer, granule cell layer	ICC, IG-EM	β-arr1 translocation to the nucleus	[202, 203]
A	Somatostatin receptors	SST_5 receptor	SSTR5	Somatostatin type 5 (sst5) receptor	Golgi, nucleus	HEK cells; PC-3, DU-145, LNCaP cells	CI-live, IF, IP; SCF, WB	ND	[204, 205]
A	Tachykinin receptors	NK_1 receptor	TACR1	Neurokinin NK1 receptors	Endosomes, nuclei	Spinal cord, DRG, prefrontal cortex, VTA	ICC, IG-EM	G_q signaling, ↑ cAMP, PKC, nuclear ERK, persistent pain	[206–208]
A	Tachykinin receptors	NK_3 receptor	TACR3	Neurokinin NK3 receptors	Endosomes, nuclei	Prefrontal cortex, VTA, paraventricular nucleus, BNST, caudate, SON, amgdala	ICC, IG-EM, SCF, WB	Stress-induced nuclear trafficking	[208–211]

(continued)

Table 8.1 (Continued)

Class	Family name	IUPHAR receptor name	Human gene	Receptor	Subcellular compartment	Tissue	Evidence	Effects	References
A	Trace amine receptor	TA_1 receptor	TAAR1	Trace amine associated receptor 1 (TAAR1)	Nuclei, ER, intracellular membranes	Breast cancer cells, astrocytes, neurons	IF, colocatization with nuclear markers, FACS, FRET imaging, biotinylation	G_s and G_{13} activation, RhoA activation, PKA activation	[212–215]
A	Urotensin receptor	UT receptor	UTS2R	Urotensin II receptor, UT receptor	Buclear, endosomes	Heart, brain, HEK cells	SCF, WB, IF, BA, BRET biosensors, FRET IP_1, cAMP, ERK assays, FA	↑ Transcription, ERK activation, NFκB phosphorylation	[216–219]
A	Vasopressin and oxytocin receptors	OT receptor	OXTR	Oxytocin receptor OXTR	Nuclei	Primary osteoblasts, and MC3t3.E1 preosteoclastic cells	Real-time fluorescent imaging, IG-EM, SCF,	Activation of ERK; up-regulation of osterix, Atf4, bone sialoprotein, and osteocalcin	[220]
A	Vasopressin and oxytocin receptors	V_{1A} receptor	AVPR1A	Vasopressin V1A	Nucleoplasts	Primary osteoblasts	IF, SCF, WB, MALDI-TOF of nuclear extracts	ND	[221]
A	Vasopressin and oxytocin receptors	V_2 receptor	AVPR2	Vasopressin type 2 receptor (V2R)	Endosomes	HEK cells, mouse renal collecting duct cells mpkCCDC14, MDCK	IF, TIRF, FA	Sustained cAMP and PKA activation	[222]

B	VIP and PACAP receptors	PAC_1 receptor	ADCYA-P1R1	PAC1 receptor	Endosomes	HEK cells, cardiac neurons	Suppression of membrane trafficking, IF, WB FA	pERK activation, ↑ccardiac neuron excitability	[223–225]
B	VIP and PACAP receptors	$VPAC_1$ receptor	VIPR1	Vasoactive intestinal peptide receptor ($VPAC_1R$)	Nuclei	Breast cancer cells, T cells, astrocytomas	ICC, BA, WB	cAMP production, adhesion and migration, survival and differentiation, anti-apoptosis	[226–230]
B	VIP and PACAP receptors	$VPAC_2$ receptor	VIPR2	Vasoactive intestinal peptide receptor ($VPAC_2$)	Nuclei	Glioblastomas	IF, SCF, WB	ND	[226]

Above receptors are listed alphabetically according to family name. Only those receptors for which firm documentation exists as to location and function are included.

Abbreviations include β-arr, β-arrestin; BA, ligand binding assays; BRET, bioluminescence resonance energy transfer; CI-Live, Live cell imaging; CF, cellular fractionation; Co-IP, co-immunoprecipitation; CM, confocal microscopy; DN, dominant negative; EM, electron microscopy; ePhys, electrophysiology; ELISA, Enzyme-linked immunosorbent assay; ERK, extracellular-signal-regulated kinase; ERK1/2; FACS, fluorescence-activated cell sorting; FRET, Fluorescence Resonance Energy Transfer; FA, functional assay/effects/response; ICC, immunocytochemistry; IF, immunofluorescence; IG-EM, Immunogold EM; IP, immunoprecipitation; IUPHAR, International Union of Basic and Clinical Pharmacology; KD, knockdown; KO, knockout; MAPK, mitogen-activated protein kinase; ND, not determined; PhA, pharmacological assessment; RIA, radioimmunassay; SCF, subcellular fractionation; TIRF, Total Internal Reflectance Fluorescence; WB, western blot, immunoblot.

Table 8.2 Primary subcellular destinations of GPCRs.

Receptor ligand types	No.	Examples	Function
		Nuclear GPCRs	
Neurotransmitters	9	Adrenergic, $mGlu_{1/5}$, $GABA_{B1}$, muscarinic	G_q, G_i; ↑ $[Ca^{2+}]$; ERK activation; ↑ transcription; ↓ NFκB
Peptides	27	Angiotensin, bradykinin, chemokine, endothelin, neurokinin, opioid	G_q, G_i; ↑ $[Ca^{2+}]$; ↑ROS; ↑ Transcription; ↑ Proliferation; anti-apoptotic
Lipophilic	12	Leukotriene, prostanoid, lysophospholipid	↑ $[Ca^{2+}]$; ERK activation; ↑ NOS; ↑ Proliferation; cell migration
Other	2	GPR88, GPR158	↑ Proliferation; chromatin remodeling; transcription
		Endosomal GPCRs	
Neurotransmitters	16	Adrenergic, dopaminergic, serotonergic	cAMP accumulation, ERK activation
Peptides	27	Opioid, NK_1, OXR_1	G_q: ↑ cAMP, ↑ transcription; G_i: ↓ cAMP; ERK activation;
Lipophilic	7	FFA, TPR	↓NFκB
Other	3	GPR3, GPR12, GPR17	ERK activation; cell survival
		Mitochondrial GPCRs	
Neurotransmitter	5	$P2Y_{1/12/2}$, 5-HT_4	Regulation of mCa^{2+} uptake leading to changes in ATP production
Peptides	2	$AT_{1/2}$	NO formation, ↓ respiration, ↓ Ca^{2+} uptake, ↓ cAMP
Lipophilic	2	CB_1, MT_1, MT_2	↓ cAMP; ↓ Cytc release; ↓ apoptosis
		Lysosomal GPCRs	
Small molecule receptors	2	CB_2, GPR55,	Ca^{2+} release from IP_3-sensitive- and acidic-like Ca^{2+} stores
Lipophilic	3	GPR137, $GPR137_B$, $GPR137_C$	mTORC1 regulation

Table 8.2 (Continued)

Receptor ligand types	No.	Examples	Function
		Golgi GPCRs	
Neurotransmitters	3	M_1, β_1, D_2	Internal G_s-cAMP signal; ERK activation
Peptides	8	GPR37; TSH, Opioid	Internal G_s-cAMP signal; ERK activation; G_i: ↓ cAMP
Lipophilic	4	GPR3, GPER, $S1P_1$, EP_4	ERK and AKT activation

Above receptors are grouped according to their activating ligand type. Other refers to orphan receptors for which the ligand has yet to be determined. Abbreviations include ERK, extracellular-signal-regulated kinase; ERK1/2; ROS, reactive oxygen species; NOS, nitric oxide synthase; mCa^{2+}, mitochondrial Ca^{2+}; Cytc, cytochrome c.

in more detail below, several common themes emerge when examining the functional consequences of receptor activation. Specifically, intracellular GPCRs frequently mediate extracellular signal-regulated kinase 1/2 (ERK1/2) activation as well as increased transcription regardless of whether the receptor signals pass through G proteins or arrestin pathways. Increased cellular proliferation and migration are additional common outcomes following nuclear GPCR activation (Table 8.2). Mitochondrial GPCRs play compelling roles in respiratory function (increasing or decreasing respiration), regulating soluble adenylate cyclase and influencing apoptosis.

Inasmuch as the data in Table 8.1 largely represent endogenous receptors expressed in many different cell types from human and/or rodent tissues, it is clear that GPCR signaling is far more complex than originally thought. The current list of intracellular GPCRs may simply represent the tip of the iceberg and doubtless additional receptors will be defined. Importantly, in many cases, unique physiological and/or pathological functions have been attributed to intracellular GPCRs underscoring the need for more selective drugs capable of targeting location-dependent receptors. Thus, understanding how GPCRs get to a given intracellular membrane and determining what their intracellular function might be is critical to fully understand a receptor's cellular repertoire and possibly to manipulate it. With that in mind, the focus of this review is to provide a broad overview of the many intracellular GPCRs that have been described, highlighting, where possible, how intracellular GPCRs get to a given destination, how they might be activated in that location, what their signaling effectors might be, and what the physiological significance of its compartmentalized signaling might be. Because many excellent reviews have recently been published on endosomal GPCR signaling [216, 233–235], here we largely focus on the commonalities of intracellular signaling versus specifics of every receptor in every locale. More details for individual receptors can be found in Table 8.1 and the references cited for noted GPCRs.

8.2 How Do GPCRs Get to a Given Location?

8.2.1 Getting to the Cell Surface (Inside Out)

A large literature describes how plasma membrane proteins including GPCRs are synthesized, folded, and assembled in the ER before moving to the Golgi and the TGN. Most transmembrane proteins are tagged with *N*-linked oligosaccharides as they are translocated through the ER membrane. These glycosylation groups are further modified as proteins move from the ER through the Golgi, and thus differential glycosylation can be used to monitor protein trafficking through these compartments. As with many other cellular processes, GPCR maturation and exit from the ER is a highly regulated process, one with many escort proteins, sorting molecules, and signaling proteins along the way to ensure that properly folded receptors get to the right membrane. This process also ensures that immature and/or improperly folded receptors do not make it out of the ER. Instead, they pass through a quality control system with misfolded proteins being targeted to the proteosomal degradation pathway [236, 237].

Many GPCRs destined for the cell surface exit the ER in ER-derived coat protein complex II (COPII) vesicles. As master regulators of vesicular trafficking, Rab GTPases, are involved in almost every step of COPII-vesicular trafficking from the ER through the Golgi [238]. For example, transport of angiotensin II receptor type 1 (AT_1), α_{1A}-adrenoceptor (α_{1A}AR), α_{1B}-adrenoceptor (α_{1B}AR), β_2-adrenoceptor (β_2AR), and Ca^{2+}-sensing (CaS) receptors are all dependent upon Rab1 for transport through these compartments [239–241]. Other critical proteins are so-called "Tre-2/Bub2/Cdc16 (TBC) domain-containing proteins" which serve as specific GTPase activating proteins (GAPs) for Rab GTPases [242]. Recently, six different TBC proteins were described as crucial regulators for α_{2B}AR cell surface transport from the ER through the Golgi via TBC-mediated inactivation of a given Rab GTPase [243]. The newly identified TBC proteins significantly inhibited the cell surface expression of α_{2A}AR, β_2AR, and AT_1 receptors as well. Thus, GPCR trafficking is controlled by specific Rabs which are in turn tightly controlled by specific TBC proteins [243].

Cell surface trafficking of GPCRs is also controlled by export motifs embedded within the receptors [244–246]. For example, a number of studies suggest that a diacidic motif (DXE), a diphenylalanine (FF) motif as well as FXXXFXXXF, FNXXLLXXXL, and FXXXXXXLL motifs all play important roles in GPCR export trafficking [247]. Deletion or mutation of these export motifs results in either a delayed exit or retention of the protein in the ER resulting in decreased cell-surface expression [247]. Various GPCRs also require a family of chaperone proteins known as receptor-activity-modifying proteins or RAMPs [248]. RAMPs serve as accessory proteins that help certain GPCRs bypass ER retention and thus be trafficked to the Golgi. RAMPs also serve a critical role in regulating the expression and pharmacology of calcitonin (CT) receptor and calcitonin receptor-like receptor (CALCLR; [249]).

8.2.2 And Back Again...

Inasmuch as there are many GPCRs on the outer (ONM) or inner nuclear membranes (INMs), the question arises as to how they get there. Because the ONM is contiguous with the ER, newly synthesized transmembrane proteins do not necessarily have to go out to the cell surface and come back. Thus, a diffusion–retention model has been proposed for many INM proteins (e.g. lamin B receptor (LBR) in which proteins synthesized in the ER rapidly diffuse laterally in the ONM, pass through peripheral channels existing between the nuclear pore complex (NPC) and the pore membrane, and then become tethered in the INM via nucleoplasmic interactions with nuclear lamins or chromatin (Figure 8.1; [250–252]). Because most transmembrane proteins are tagged with particular glycosylation groups as they move from the ER through the Golgi, their differential glycosylation can be used to monitor protein movement through these compartments. For example, a GPCR that was synthesized in the ER and then immediately trafficked back to the nucleus would be sensitive to Endo H digestion, whereas Endo H resistance suggests a protein has been trafficked at least to the Golgi where PNGase F can remove all N-linked glycosyl groups. Using such an assay, Merlen et al. [255] showed that the endothelin B receptor (ET_BR) traffics directly to the nuclear membrane after biosynthesis by passing the Golgi apparatus. In contrast, >90% of nuclear metabotropic glutamate receptor, $mGlu_5$, is processed through the Golgi before trafficking to the nucleus [256] ruling out lateral diffusion directly to the INM model. To date, not enough nuclear-bound GPCRs have been examined for clues determining their final destination to predict which membrane(s) a receptor might end up on. An overview model representing some of the ways that GPCRs are trafficked to various intracellular destinations is shown in Figure 8.1.

Various GPCRs use canonical nuclear localization signals (NLSs) for nuclear import (Figure 8.2). Typically, an NLS consists of short stretches of basic amino

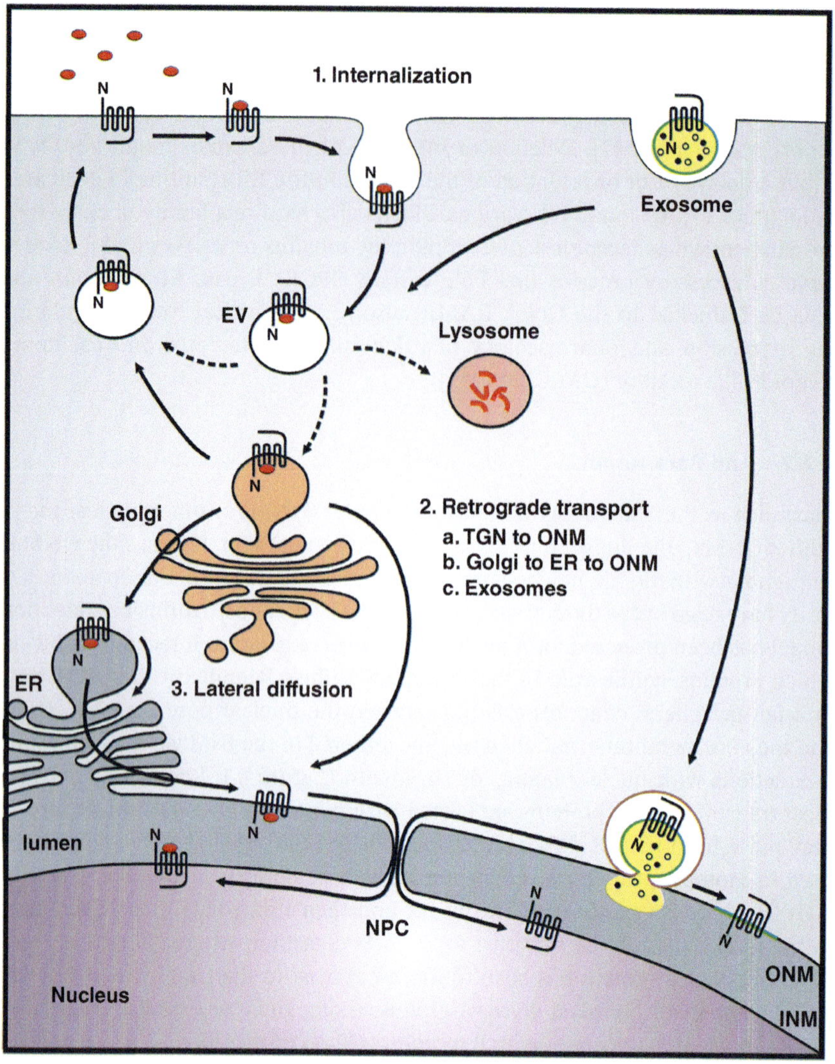

Figure 8.1 Overview of pathways used by various GPCRs to reach intracellular destinations. (1) Many GPCRs are internalized in endosomal vesicles (EVs) and subsequently redirected for degradation, recycling to the plasma membrane, and/or transport to other intracellular destinations. (2) Late EVs can undergo retrograde transport to (a) the TGN and then to the outer nuclear membrane (ONM), or (b) to the Golgi, ER, and ONM. Exosomes, containing receptors and/or ligands generated elsewhere, can also be endocytosed and either trafficked directly to another intracellular membrane such as the ONM or to a multivesicular body for similar re-direction (c). (3) Many GPCRs may be synthesized and processed in the ER and then simply undergo lateral diffusion to reach the ONM and inner nuclear membrane (INM). In other cases, a given GPCR may be processed and trafficked to at least the medial Golgi body before undergoing retrograde transport. NPC represents the nuclear pore complex shown in more detail in Figure 8.2. Source: Based on Refs [250–254].

acids that are recognized by specific members of the karyopherin superfamily [33, 40, 101, 259]. Most karyopherins such as importinβ (Impβ) bind directly to their cargo proteins in order to translocate through the NPC [260, 261]. For example, after ligand-mediated receptor internalization, the Proteinase-activated type 2 receptor (PAR2) also called the F2R-like trypsin receptor 1 (F2RL1) is recognized by importin β1 (KPNB1) which binds to its two NLS motifs in order to translocate it from the cell surface to the ONM via sorting nexin11 and dynein transport along microtubules [196]. In contrast, the oxytocin receptor is transported to the nucleus via Transportin-1 (TPNO1) following β-arrestin-mediated internalization [220] as are the chemokine receptors type 2 (CCR2) and CXC chemokine receptor 4 (CXCR4; [66, 262]). Very recently, unique NLS sequences recognized by Transportin 1 have been described that are so-called "RG-rich sequences or RSY motifs" [263]. It remains to be determined whether such motifs are used by the oxytocin receptor or CCR2 and CXCR4. The platelet-activating factor (PAF) receptor, however, is trafficked through the TGN before being re-directed to the nucleus via a process involving Rab11a and importin-5 (IPO5) [185, 264]. Some NLS sequences are also bound by the karyopherin adaptor protein Importin-α (Impα) [260, 261]. Impα binds directly to both the NLS in a cargo protein and to Impβ [265]. Subsequently, Impβ interacts with the nuclear pore complex to carry the Impβ–Impα–cargo complex into the nucleus. Parathyroid hormone receptor (PTH1R) is trafficked to the nucleus via such an association [180]. It is worth noting that despite a receptor having a putative NLS sequence, in certain cases, it does not seem to be utilized. For example, the bradykinin 2 (B_2) receptor has an NLS at its C-terminus; however, deleting or mutating it had no effect on receptor trafficking to the INM [266]. $mGlu_5$ is also anchored on the INM by direct interactions with the chromatin via a 20 amino acid sequence in its C-terminal domain (Figure 8.2; [256]). Taken together, multiple different transport pathways and signaling motifs are responsible for getting GPCRs to nuclear membranes.

8.2.3 Subnuclear Targeting

Certain GPCRs are not only associated with the ONM and INM but are also found within the nucleoplasm itself. Since in these cases, the entire receptor seems to be present in this hydrophilic milieu, the mechanism by which seven hydrophobic domains maintain their structure and function outside of a membrane is puzzling. By analogy, numerous receptor tyrosine kinases (RTKs) have also been found within the nucleoplasm having undergone endocytosis, retrograde trafficking to the Golgi, the ER, the INM, and finally the nucleoplasm itself where they appear as subnuclear particles. One such RTK, the epidermal growth factor receptor (EGFR) acts as a transcriptional regulator, signal transducer, and protein

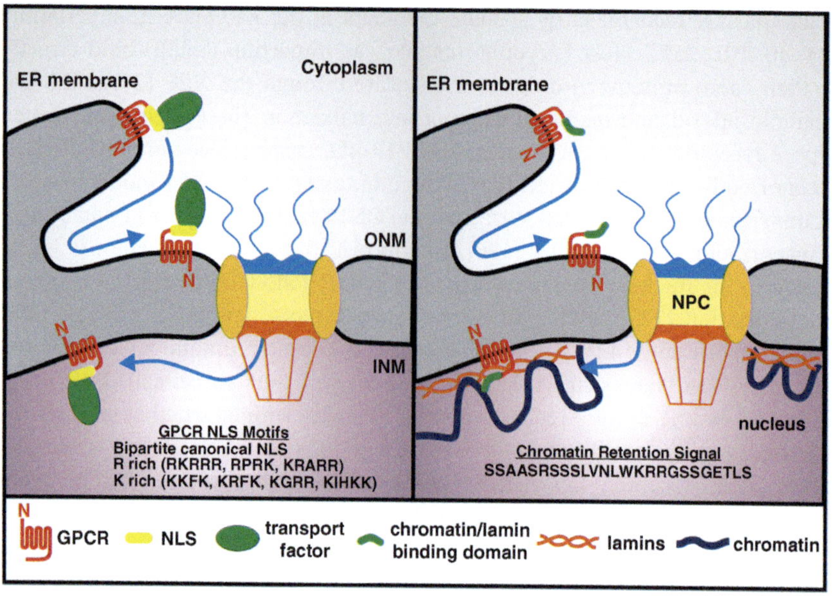

Figure 8.2 Proposed diffusion–retention model targeting GPCRs to the inner nuclear membrane (INM) through lateral channels of the nuclear pore complex (NPC). The model suggests that a given GPCR diffuses freely from the ER to the outer nuclear membrane (ONM), and from there to the INM using peripheral channels existing between the NPC and the pore membrane. Left Panel, many GPCRs destined for the INM contain NLS sequences that facilitate the protein's interaction with a karyopherin in the cytoplasm. The karyopherin then shuttles the protein through the peripheral channel/NPC prior to release in the INM. Note identified GPCR NLS motifs. Right Panel, GPCRs become tethered in the INM via nucleoplasmic interactions with nuclear lamins or chromatin [250, 257, 258]. One such chromatin retention signal is shown for the GPCR, $mGlu_5$ [256]. Source: Based on Refs [250, 256–258].

modulator within the nucleus resulting in cell proliferation, DNA replication, DNA repair, and tumor progression [267]. Two mechanisms have been proposed to account for EGFR distribution within the nucleoplasm. First, studies indicate that EGFR associates with the ER translocon Sec61β which is required for the translocation of the EGFR to the nucleoplasm [268, 269]. Second, because the inner nuclear membrane or in some cases both the inner and the outer nuclear membranes can invaginate into the nucleus forming a so-called "nucleoplasmic reticulum" (Figure 8.3; [270, 271]), it is thought that RTKs including the EGFR are associated with the nuclear reticulum which also stores and releases Ca^{2+} in an IP_3-sensitive fashion [270]. In support of this model, it has recently been shown that the EGFR can initiate Ca^{2+} release and regulate transcription from these structures [272]. By analogy with the RTK pathway, subnuclear GPCRs might also appear to be within the nucleoplasm via association with the nucleoplasmic reticulum. If so, this model would suggest that subnuclear GPCRs are never out of a membranous environment (Figure 8.3). Inasmuch as many GPCRs described in the nucleoplasm are associated with proliferation in cancerous cell types, understanding the mechanisms underlying nucleoplasmic translocation and compartmentalization, and how these processes vary in cells with different proliferative and oncogenic capacity might aid in designing new anti-cancer therapies.

8.2.4 Subcellular Targeting: Endosomes

Traditionally, GPCRs were only thought to transmit extracellular signals to the cytoplasm from their position on the cell surface. Following ligand binding and subsequent G protein-dependent signal transduction, activated receptors were phosphorylated at specific sites via GPCR kinases (GRKs) leading to the recruitment of β-arrestins which would bind to the receptor and prevent further G protein-dependent signaling. In many cases, the phosphorylated receptor with bound β-arrestin subsequently served as a target for the adaptor protein 2 (AP2) and clathrin leading to the clustering of GPCRs in clathrin-coated pits. Clathrin-coated as well as nonclathrin-coated GPCRs are internalized and are delivered to early endosomes. From the early endosome, receptors are differentially sorted (i) to recycle back to the plasma membrane, (ii) be sent to the lysosome for proteolytic degradation, or (iii) be trafficked to the Golgi apparatus [253] (Figure 8.1).

As indicated in Table 8.1 and as detailed elsewhere in this volume, although the β_2AR has been the most studied endosomal GPCR, many others have now been described (>50, Table 8.1). These include other neurotransmitter GPCRs such as dopamine D_1, D_2, D_3 receptors [95], the muscarinic M_2 receptor [22], the serotonin receptor 5-HT_{2A} [15] as well as peptide hormone receptors such as PTH1R [181], thyroid-stimulating hormone receptor (TSH) [129], glucagon-like peptide 1 (GLP-1) receptor [273], the pituitary adenylate cyclase activating polypeptide type 1 (PAC_1) receptor [223], and even small lipophilic molecule receptors such as the free fatty acid (FFA) receptors [104], and others (Table 8.1). Besides desensitizing a given GPCR response, studies show that β-arrestins also act as scaffolds to recruit additional proteins that can mediate G-protein-independent, endosomal signaling [274, 275]. Intriguingly, many internalized GPCRs not only use different effector molecules to generate a set of actions but endosomal GPCRs also induce a prolonged signal versus a rapid, transient response as seen on the cell surface [274]. Thus, plasma membrane-mediated signaling is often linked to brief, short-term signals, whereas the second signaling phase detected on endosomes can last long after cell surface signals have diminished [276].

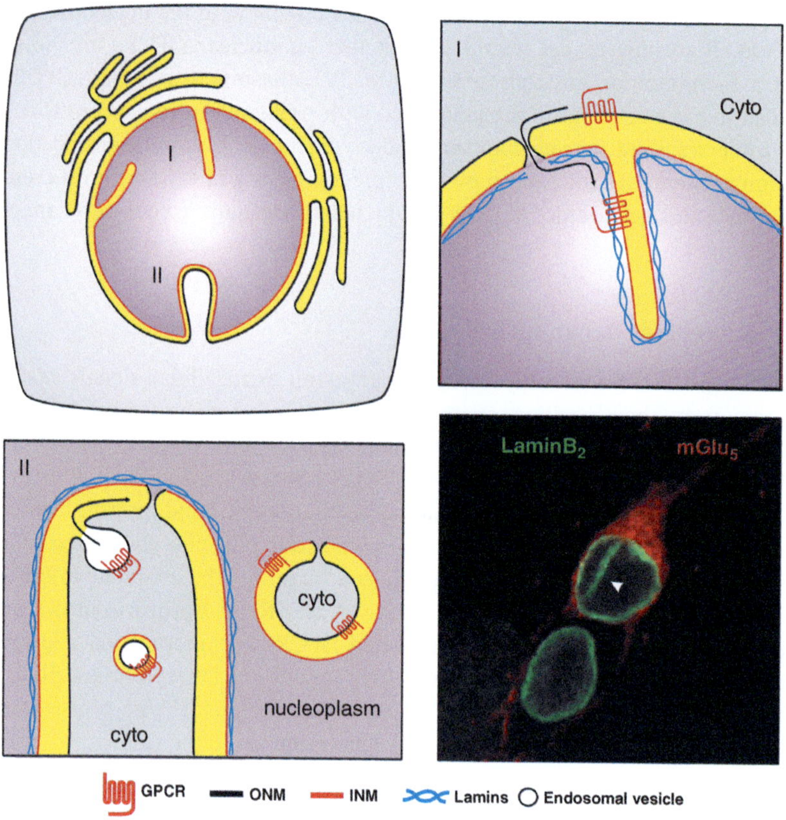

Figure 8.3 Schematic representations of the nuclear reticulum. Left upper panel, whole cell schematic emphasizing two types of membrane invaginations into the nucleoplasm. Type I in which only the INM extends into the nucleoplasm and Type II in which the invaginations into the nucleoplasm consists of INM and ONM providing a cytoplasmic conduit into the nucleus. Right, panel I shows an enlarged view of the Type I invagination, whereas the lower left, panel II shows the complex topology of the type II reticulum with its cytoplasmic interior that might contain endosomal vesicles with trafficked GPCRs or intracrine modulators. Lower right, image of rat striatal neurons expressing the GPCR $mGlu_5$ maintained in culture for eight days. Neurons were fixed, stained, and analyzed by confocal microscopy to detect $mGlu_5$ (red) or the inner nuclear lamin B_2 (green). Images represent single optical section of 0.4 μm merged such that yellow indicates colocalization of the specific antigens. Arrowhead points to a nuclear reticular structure within the nucleus. Source: Based on Refs [256, 258, 270, 271].

One long-term consequence of sustained signaling might be distinctive transcriptional responses. For example, in the case of the β_2AR, the sustained endosomal response induces unique 3′–5′-cyclic adenosine monophosphate (cAMP)-generated transcriptional responses compared to those generated by the cell surface receptor [276]. Although initially arrestin-mediated signaling focused on the mitogen-activated protein kinase/extracellular signal-regulated kinase 1/2 (MAPK/ERK1/2) cascade, many other signaling moieties can interact with the receptor/G-protein/β-arrestin complex such as protein kinase B (Akt), p38MAPK, c-Jun N-terminal kinases (JNKs), and signal transducers and activators of transcription (STATs) [277]. In turn, these proteins mediate downstream functions such as growth, cell survival, apoptosis, contractility, cell migration, and cytoskeletal reorganization [277].

8.2.5 TGN/Golgi

Endosomal GPCRs can also redistribute via retrograde trafficking to the TGN and Golgi apparatus [278], although the physiological consequences and *in vivo* relevance of receptors signaling at the Golgi are just beginning to emerge. Compartmentalized Golgi signaling is being explored using many of the same tools used to verify endosomal signaling. In particular, genetically encoded FRET biosensors have proved invaluable for visualizing GPCR signaling dynamics in real-time [279]. For example, biosensors that detect changes in cAMP as well as kinase levels have revealed compartmentalized signaling in various areas of the cell. Many of these biosensors have been further modified by the addition of targeting tags to allow visualization of GPCR signaling in discrete subcellular locations. Thus, FRET biosensors can monitor plasma membrane and cytoplasmic changes in cAMP (exchange protein 2 activated by cyclic-AMP, Epac2-camps), cytoplasmic and nuclear changes in ERK activity (ERK activity reporter, EKAR), and cytoplasmic and plasma membrane protein kinase C (PKC) activation (C kinase activity reporter, CKAR) [7]. In addition, biosensors derived from conformation-specific nanobodies that are capable of detecting ligand-induced activation are also proving invaluable to achieve spatial and temporal resolution of GPCR activation within the cell without destroying cellular function [11]. Tools like these have taken the lead in demonstrating compartmentalized GPCR signaling. As described above, the latter approach led to the early demonstration by Irannejad et al. [5] of an active β_2AR-G_s complex on endosomes. Steady-state Golgi localization and signaling has been described for the somatostatin receptor (SST_5; [204]), the delta (δ) opioid receptor [280], and the β_1AR [6]. More recently, the von Zastrow lab has demonstrated δ and mu (μ) opioid receptor activation in the Golgi and dendritic Golgi elements in neurons [165]. Irannejad et al. [6] also used nanobody sensors to describe β_1AR signaling on the Golgi apparatus in heterologous cells. These observations were extended in a more physiological setting when Nash et al. [281] established that endogenous β_1ARs could generate cAMP at the Golgi apparatus in cardiac myocytes. As with other spatially restricted receptors, future work to develop therapeutic strategies that spatially target receptor modulation may offer new treatment options.

8.3 Activation of Intracellular GPCRs

Just as there are many different transport mechanisms by which GPCRs are targeted to different subcellular compartments, there are also a variety of ways in which they can be activated. These include ligands being endocytosed along with their receptors, ligands being endocytosed by themselves or within an exosome, ligands for which a specific transporter is available, permeable ligands diffusing across cellular membranes, or ligands made *in situ* via localized biosynthetic machinery (Figure 8.4). In certain cases, no ligand is necessary since the GPCR undergoes a conformational change in which it becomes constitutively active. For example, the sphingosine 1-phosphate 2 ($S1P_2$) receptor is shed in exosomes from breast cancer cells and is then taken up by fibroblasts, where its N-terminus is processed to a shorter form that is constitutively active promoting ERK1/2 signaling and proliferation [283]. Alternative strategies to generate GPCR constitutive activity include close association with other proteins. For instance, Homer1a can lead to agonist-independent $mGlu_5$ receptor activation [284] and agonist-independent activation of the PAC_1 receptor occurs via proximity with the insulin-like growth factor 1 receptor and subsequent transactivation by tyrosine kinase Src [285].

8.3.1 Peptide Ligands

Many peptide-activated GPCRs are internalized with their receptor-bound ligands. Although such ligands are frequently destroyed by the acidic environment of the endosome, several factors help determine the stability and prolonged interaction of agonist-bound receptor interactions. These might include pH, affinity, and peptidase interactions that affect the strength and length of time the agonist remains bound and hence continued sequestration and endosomal signaling by a given receptor [235]. For instance, various lengths of the PTH bind with different affinities to PTH1R. PTH(1–34) is associated with the receptor for long periods of time leading to prolonged endosomal signaling, whereas PTH(1–36) rapidly dissociates exhibiting only transient endosomal effects [181, 286]. Peptides that are resistant to the peptidase endothelin-converting-enzyme-1 can also exhibit sustained endosomal signaling, whereas peptides such as Substance P, calcitonin gene-related peptide (CGRP), and somatostatin are rapidly destroyed by this enzyme and thus

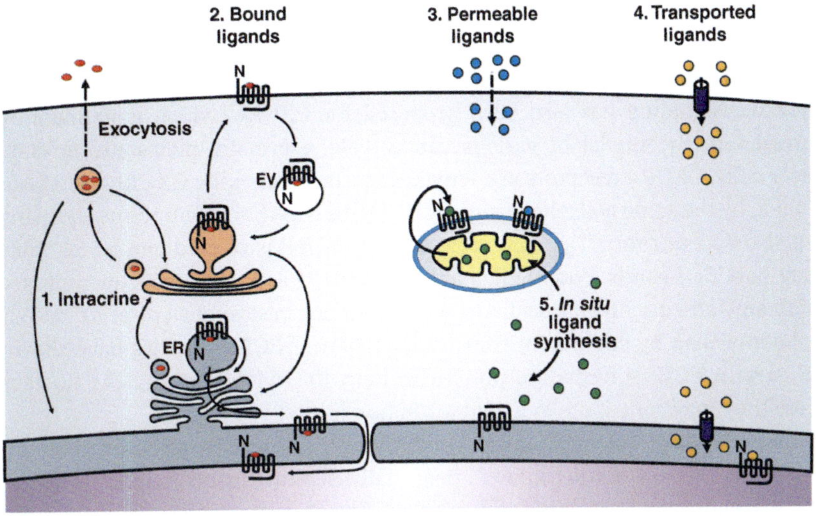

Figure 8.4 Ligand activation of intracellular GPCRs. Several different strategies have been described by which intracellular GPCRs are activated. These include the production of peptide ligands that can not only be released via exocytosis to act extracellularly but also serve as intracrine signals. In the latter case, vesicles containing processed peptides can be trafficked to intracellular organelles prior to fusion and peptide release (1). Peptides can also be released via exosomes (not shown). In addition, intracellular GPCRs can be activated via receptor-bound ligands (2) that can be internalized with a given GPCR and trafficked elsewhere as shown. Alternatively, channels, transporters or exchangers (3) can transport specific ligands across the plasma membrane as well as intracellular membranes to activate receptors whose ligand-binding domain faces the ER or nuclear lumen. Permeable ligands such as endocannabinoids can freely diffuse across cell membranes to activate their corresponding receptor (4). Ligands can also be synthesized within the cell (5) and either diffuse or be trafficked to a given cellular compartment. Source: Based on Refs [1, 4, 147, 166, 282].

have shorter peptide–receptor interactions and shorter, less sustained endosomal signals [235].

As described above, certain peptide receptors are not only sequestered in endosomes in a ligand-bound state but also subsequently transported back to the nucleus via karyopherin recognition of NLS sequences within the receptor [3, 4]. For example, PAR2 and PAF receptors are internalized and retrogradely transported to the nucleus in the ligand-bound state [185, 196] as is the oxytocin receptor [220, 287]. Alternatively, peptide ligands can also activate endogenous GPCRs present within the nucleus or on nuclear membranes (NMs). For example, radiolabeled gonadotropin-releasing hormone (GnRH) can be found in the nucleus along with the nuclear GnRH receptor where it can trigger histone H3 acetylation and phosphorylation following activation [131]. It is not clear what the source of the ligand is, but conceivably, secretory vesicles filled with processed peptides might be redirected from the secretory pathway such that intravesicular peptides are released inside the cell in an intracrine fashion (Figure 8.4; [282]). Intracrine signaling has also been proposed for CXCR4, which is abundantly expressed in the nuclei of various cancer cells especially metastatic prostate cancer cells. CXCR4 receptors are activated by the chemokine CXCL12α that can be made by the same cells that express CXCR4 on NMs [262]. Intracrine signaling may also account for ET_B receptor activation on INMs of cardiomyocytes. Like many peptide ligands, endothelins are synthesized as large precursor proteins, which are subsequently cleaved into active peptides in several steps prior to their final conversion by endothelin converting enzymes (ECEs). Studies have shown that certain ECE isoforms are present in both the cytoplasm and the nucleus where they may contribute to ligand formation [39].

Although the exact mechanisms underlying peptidergic activation of intracellular GPCRs are not yet known, potentially peptide ligands might also enter cells via endocytosis of extracellular vesicles such as exosomes. Exosomes are small membrane-bound structures which serve as intercellular carriers of many different cargos including proteins, lipids, and nucleic acids [254]. Internalized exosomes are typically transported in structures with hallmarks of late endosomes (Rab7+) to lysosomes, the ER and/or the ONM (Figure 8.1; [254]). Very recent data has revealed the complex manner in which both soluble and membranous contents of exosomes can be extracted and transferred into the nucleoplasmic reticulum [288, 289]. Since both GPCRs and cognate ligands can serve as exosomal cargos, conceivably an exosome might serve as one type of carrier for ligand delivery to nuclear GPCRs. If so, how membranous components of exosomes are extracted from the late endosomal membrane upon fusion and what the transport mechanism through nuclear pores might be remains to be discovered.

8.3.2 Transporters

Other means of ligand entry include active transport via specific transporters, channels, or exchangers (Figure 8.4). Given that most ligand-binding sites would be within the vesicle or the luminal region of the ER or nucleus, extracellular ligands would have to cross both the plasma membrane as well as the intracellular membrane to activate intracellular GPCRs [153, 154]. For example, Irannejad et al. demonstrated that Golgi-localized β_2AR could be activated by epinephrine, an endogenous ligand, by facilitated transport via the organic cation transporter 3 (Oct3) [6]. Oct3 performs a similar function in cardiac myocytes where this transporter brings norepinephrine (NE) into the cell to activate β_1ARs on cardiac myocytes. In fact, blocking Oct3 prevents NE ligand-dependent cardiac myocyte hypertrophy [281]. Because Oct3 can transport ligand in both directions, it can facilitate ligand transport across not just the plasma membrane but also the outer nuclear membrane to activate βARs or αARs [29]. For example, NE activation of the α_1-adrenergic receptor (α_1AR) localized at cardiac nuclear membranes, is also mediated via Oct3 leading to cardioprotective functions in the heart [30]. The GPCR, mGlu$_5$, is also activated by ligand entry via an excitatory amino acid transporter (EAAT3) and/or via the cysteine – glutamate exchanger [152, 154]. Conditions that block either transporter reduce glutamate uptake in central nervous system (CNS) and peripheral nervous system (PNS) neurons [154–156]. Direct demonstration of intracellular glutamate effects was achieved using microinjection of caged glutamate into neurons endogenously expressing mGlu$_5$, followed by uncaging via restricted photoactivation [290]. Only photoactivated neurons expressing caged glutamate exhibited mGlu$_5$-mediated Ca^{2+} responses, showcasing the spatial and temporal resolution of intracellular GPCRs [290].

8.3.3 Lipophilic Ligands

Certain GPCRs can be activated by ligands that are lipophilic and thus able to pass easily through membranes. Lipids such as prostanoids (prostaglandins), leukotrienes, lysophospholipids (platelet-activating factor, lysophosphatidic acid, sphingosine-1-phosphate, and ceramide-1-phosphate), and endocannabinoids, initiate signal transduction via specific receptors both on the cell surface and on intracellular membranes leading to modulation of cellular function. For example, fatty acids and their derivatives, are now recognized not just as critical components of cellular membranes and as energy carriers but also as important signaling molecules affecting a broad range of biological functions such as metabolic and immune processes [291, 292]. In the FFA receptors, FFA1 and FFA4 bind medium- to long-chain fatty acids, whereas FFA2 and FFA3 bind short-chain fatty acid receptors [293]. Interestingly, all four FFA receptors can signal from endosomes either via β-arrestin pathways or G-protein-mediated endosomal signaling [293]. For example, *in vivo* studies have revealed that FFA4 receptor activation via a synthetic agonist or docosahexaenoic acid (DHA), eicosapentaenoic acid (EPA), and polylactic acid (PLA) can block osteoclast formation through inhibition of nuclear factor kappa B (NF-κB) and MAPK signaling pathways thereby stimulating bone formation. Knocking down FFA4 receptors or β-arrestin prevented the inhibitory effects of these fatty acids on NF-κB and MAPK signaling pathways [294]. These results suggest that fatty acids already present or endogenously produced in osteoclasts may contribute to the FFA4-β-arrestin2 pathway effects [294]. Other lipophilic compounds such as thromboxanes also interact with GPCRs; one such is present in oligodendrocyte nuclei, thromboxane A_2 receptor (TPR). Mir and Le Breton [192] reported that stimulation of the nuclear TPR pool increased nuclear cAMP levels, activated cAMP response element-binding protein (CREB), and promoted oligodendrocyte survival [192]. Thromboxane itself appeared to be synthesized *in situ* since in response to added arachidonic acid, isolated nuclei had the requisite enzymes necessary for thromboxane production [192]. This study also emphasized that outside of the nucleus, the TPR coupled to G_q and G_{13}, whereas nuclear TPR coupled to G_s in oligodendrocytes. Another lipophilic compound, the steroid hormone, estrogen, mediates important physiological effects via classical nuclear estrogen receptors as well as the G-protein-coupled estrogen receptor, GPER, which is largely localized to the ER and Golgi apparatus in many different cell types (nervous, reproductive, digestive, and muscle tissue; [110]). Activated GPER can couple to $G_{i/o}$ and G_s proteins that together with associated $G_{\beta\gamma}$ subunits regulate many downstream effectors including phosphatidylinositol-4,5-bisphosphate 3-kinase (PI3K)/Akt, ERK1/2, adenylyl cyclase, Ca^{2+} mobilization, and nitric oxide synthase (NOS; [111, 295]).

Other lipid mediators include endocannabinoids, anandamide, and 2-arachidonoylglycerol (2-AG). Due to their lipophilicity, endocannabinoids are considered retrograde messengers in neurons since they are produced postsynaptically yet diffuse back across the synaptic cleft to bind the presynaptic GPCR, cannabinoid 1 (CB_1) receptor, which limits further transmitter release. CB_1 receptors have also been localized to mitochondrial, endosomal, and lysosomal compartments [54–56, 62, 296, 297]. Activation of mitochondrial CB_1 receptors suppresses respiration, whereas blockade is associated with increased mitochondrial biogenesis, β-oxidation, and energy production [54, 55]. The cannabinoid 2 (CB_2) receptor is also present in the brain where it influences excitatory synaptic transmission, plasticity, and long-term potentiation [298–300]. Located intracellularly, *in situ* 2-AG activation results in inositol triphosphate 3 receptor (IP_3R)-dependent opening of Ca^{2+}-activated chloride channels and decreased neuronal excitability [62–64]. Thus, both CB_1 and CB_2 receptors play a role inside the cell that contributes to aspects of learning and memory.

8.3.4 *In situ* Ligand Production

As described above, many ligands, especially lipid mediators such as the FFAs, thromboxane, and endocannabinoids can also be made *in situ* via localized biosynthetic machinery. In addition, prostaglandin, platelet-activating factor, and lysophosphatidic acid receptors, whose ligands are also lipid mediators derived from membrane hydrolysis, are located on nuclear membranes where they can easily activate their cognate receptors [301]. Melatonin, another small, lipophilic ligand, is found in high levels in mitochondria, where it can activate melatonin receptors 1 (MT_1). The key enzymes in melatonin synthesis are located in the mitochondria matrix, and using its precursor, deuterated-serotonin, along with mass spectrometry, d4-melatonin products were detected [147]. Once synthesized, mitochondrial melatonin can activate MT_1 receptors on the outer mitochondrial membrane. Those receptors in turn activate G_i and block adenylate cyclase activity, leading to the inhibition of stress-induced cytochrome c release and caspase activation (Figure 8.4; [147]). As further evidence of the importance of mitochondrial MT_1, its targeted overexpression inhibited neuronal death resulting from hypoxic/ischemic injury [147]. Thus, as long as a ligand is either made *in situ* or transported to the site of action, an intracellular receptor can be activated [1, 4, 166] (Figure 8.4).

8.4 Signaling of Intracellular GPCRs

Over two decades ago, it was recognized that many components of G protein signaling pathways were also found in the nucleus or associated with nuclear membranes. These included phospholipase C (PLC) isozymes, nuclear inositol phosphates, diacylglycerol (DAG), PKC isozymes, adenylate cyclase, as well as heterotrimeric G proteins themselves [157, 302–306]. Those observations as well as subsequent direct evidence of nuclear receptor G protein signaling such as was shown for the prostaglandin E_2 receptors (EP_3, EP_4; [188, 190]) and $mGlu_5$ [153] represent some of the earliest experiments showing that GPCRs signaling from intracellular membranes represented a versatile, independent system by which intracellular function is regulated. The simplicity of the isolated nuclear system belied the importance of these early results: GPCRs could signal from inside the cell and such signaling was typically prolonged.

Besides the observation of intracellular GPCR functionality, early evidence of prolonged signaling in the nucleus came from the same studies on the EP_3 nuclear receptors [188]. EP_3 receptor-selective agonists increased Ca^{2+} concentrations in isolated liver nuclei in a dose-dependent fashion; representative traces from all doses exhibited prolonged responses lasting many minutes (>15; [188]). Prolonged Ca^{2+} responses were also observed when endogenous $mGlu_5$ or $mGlu_1$ were activated in isolated neuronal nuclei [152, 154]. Inasmuch as the EP_3 receptor response was blocked by pertussis toxin implying G_i or G_o involvement, whereas $mGlu_5$ and $mGlu_1$ responses are coupled to G_q [157]; sustained signaling is generated by a G-protein-dependent process involving at least two different G_α proteins. Despite these early studies, few reports showing nuclear GPCRs have actually looked at time courses of receptor-mediated Ca^{2+} responses so the generalizability of these observations is still unknown.

Sustained endosomal signaling was reported initially for GPCRs that had been internalized following β-arrestin binding and subsequent activation of mitogen-activated protein kinase cascades [307]. G protein-dependent sustained endosomal signaling is also seen. For example, sustained G_i-mediated inhibition of cAMP was reported following activation of $S1P_1$ receptors [139], whereas PTH(1–34) activation of PTH1R led to sustained G_s activation via endosomal PTH1R [181]. Prolonged TSH receptor signaling via G_s was also observed for receptors and G proteins colocalized in Golgi compartments after internalization [129]. More recently, Jensen et al. [206] used a variety of biophysical techniques to show that following ligand stimulation neurokinin 1 (NK_1) receptors are colocalized in early endosomes along with G_q. G_q inhibition blocked NK_1 receptor endosomal signals which included activation of nuclear ERK, cytosolic PKC, and cytosolic cAMP leading to nociception [206]. Similarly, members of this same group also found that CGRP receptors could also signal from endosomes using G_q and G_s proteins to activate cytosolic PKC and nuclear ERK respectively [52]. These data reinforce the notion that GPCRs can signal from many different platforms using many different effector molecules to generate a plethora of downstream sequelae with unique spatio-temporal profiles.

8.5 Functional Roles of Intracellular GPCRs

8.5.1 Nuclear GPCRs Control Transcription, Proliferation, and Survival

Recent observations suggest that nuclear GPCRs regulate important physiological and pathophysiological processes in intact tissues and animals. When activated, many such receptors lead to increased nucleoplasmic Ca^{2+}, which in turn can trigger increased transcription (Table 8.2). As described above, activation of intracellular $mGlu_5$ with its transported agonist, quisqualate led to sustained Ca^{2+} responses followed by a cascade of molecular events starting with the phosphorylation of ERK1/2 and Elk-1 followed by the enhanced expression of synaptic plasticity genes such as c-*fos*, early growth response 1 (*egr-1*), fos-related proteins (*Fras*), and *FosB* [154, 158]. Using an unbiased bioinformatics approach, later studies showed that the very first genes affected by nuclear $mGlu_5$ activation were primarily stimuli–responsive transcription factors involved in neuronal survival and growth (Atf3, Nr4a1, Trib1, cyclic AMP element modulator [CREM], JunB, and Arid5a) as well as effector proteins such as activity-regulated cytoskeleton-associated protein (Arc), which is involved in gene regulation and synaptic plasticity [308]. Interestingly, many of the affected genes are downstream of intracellular $mGlu_5$-activated ERK1/2 as well as activated Elk-1 [159]. Elk-1 can partner with the serum response factor (SRF) transcription factor to activate promoters carrying serum response elements or function independently to upregulate transcription machinery as a whole [250]. Thus, one initial outcome of intracellular $mGlu_5$ activation is to promote SRF-mediated immediate early genes. One such gene is Arc, which is critical for long-term memory and synaptic plasticity [309]. For instance, increased Arc expression in slices and cultures decreases α-amino-3-hydroxy-5-methyl-4-isoxazolepropionic acid (AMPA; GluA) receptor-mediated synaptic currents presumably via its ability to regulate GluA receptor trafficking on the synaptic membrane [310–312]. Arc-mediated GluA receptor endocytosis is thought to contribute to long-term depression (LTD), a key component of synaptic plasticity [313, 314]. Collectively, these findings point to a critical role for intracellular $mGlu_5$ in shaping synaptic processes.

In many cases, nuclear GPCRs also play a role in proliferation. As indicated in Table 8.1, several GPCRs have been found on cardiac nuclear membranes including both angiotensin receptors, AT_1 and AT_2 [166]. Similar to $mGlu_5$, angiotensin-II (Ang-II) activation of nuclear AT_1 receptors leads to nuclear Ca^{2+} release via IP_3Rs followed by increased transcription, whereas activation of AT_2 receptors increases transcription via NOS activation [41]. Interestingly, extracellular addition of Ang-II alone resulted in modest proliferation of cardiac fibroblasts but not collagen gene expression, whereas intracellular uncaging of caged Ang-II significantly increased fibroblast proliferation, increased collagen gene expression and collagen secretion. These results suggest that intracellular AT_1 and AT_2 receptors play a greater role in shaping cardiac function than cell surface angiotensin receptors. Identification of functional intracellular Ang-II signaling in cardiac fibroblasts has exciting implications for the development of novel pharmacological interventions in cardiac fibrosis and associated heart conditions such as congestive heart failure.

An early example of nuclear GPCRs playing a role in survival is that of the vasoactive intestinal peptide type 1 ($VPAC_1$) receptor. The latter is a class B GPCR that can be activated by pituitary adenylate cyclase activating polypeptide (PACAP) and vasoactive intestinal polypeptide (VIP) with equal binding affinity [315]. Widely expressed in the CNS and many other tissues, several reports have noted the nuclear localization of $VPAC_1$ after its activation via VIP1 [226]. In particular, >50% of human glioblastoma cells have strong $VPAC_1$ receptor staining in the nucleus, where it is positively associated with glioma grade [226]. Nuclear localization is also found in a colonic adenocarcinoma cell line [316] as well as estrogen-dependent and independent human breast cancer cell lines [227]. Given that nuclear translocation of $VPAC_1$ is also associated with anti-apoptotic activity of $VPAC_1$, this may account for cells with nuclear $VPAC_1$ exhibiting increased viability [228]. Collectively, nuclear GPCRs are functional, often leading to prolonged signaling responses that result in increased transcription, increased proliferation, and increased survival of cells.

8.5.2 Endosomal GPCRs Control Physiological Processes

The discovery of biased ligands that can induce either G-protein or β-arrestin-dependent GPCR signaling opened the door for the selection of desirable responses versus those with adverse drug reactions. An early attractive target for drug design was the μ opioid receptor whose agonist, morphine, is an effective pain reliever but which also leads to respiratory suppression, constipation, tolerance, and drug dependence [317]. Subsequent studies using β-arrestin knockout mice revealed that it was receptor-β-arrestin internalization and endosomal signaling that produced the undesirable drug responses (constipation, tolerance, and respiratory suppression). In contrast, pain relief was generated via receptor-G protein-dependent signaling [318, 319]. Further *in vitro* and *in vivo* studies have led to the development of many new agonists targeting receptor domains biased toward G-protein signaling and thus high analgesic activity versus unwanted effects such as respiratory suppression [320]. Although it will take actual clinical trials to reveal the true utility of a given compound, exploring response-selective signaling is opening the door to tailoring desired versus adverse receptor responses.

Another peptide receptor, the tachykinin receptor, NK_1, also redistributes from the cell surface to endosomes where it signals via prolonged G_q-mediated ERK1/2 activation which underlies inflammation and chronic pain [206]. Therefore, in the case of NK_1, its endosomal signaling is a possible target for pain relief. Using membrane-permeant peptides as well as nanoparticles to deliver antagonists, Jensen et al. [206], were able to block NK_1 receptor-mediated nociception providing direct support for the role of endosomal NK_1 in pain. The Burnett lab has also shown that the CALCLR, is co-expressed on spinal neurons with NK_1. CALCLR also undergoes endocytosis followed by sustained endosomal signaling [52]. Like NK_1, inhibitors of dynamin, ERK, and PKC also prevent CALCLR from being internalized and/or from activating effector molecules leading to pain transmission. CALCLR is currently emerging as a novel target for the treatment of migraine [321]. Specifically, inhibitors of either the ligand or the receptor can prevent migraine effects. However, as described, CALCLR not only signals from the cell membrane but also from endosomes [321]. Thus, targeting endosomal NK_1 or CALCLR may be the best therapeutic strategy for ameliorating chronic pain.

Other examples of endosomal GPCR signaling affecting physiological processes include the binding of FFAs to the FFA receptors. Specifically, FFA1, also known as GPR40, is primarily coupled to $G_{q/11}$. When activated via endogenous fatty acids, FFA1 triggers a signaling cascade leading to the activation of protein kinase D_1 and insulin release in mouse islet cells [322]. FFA1 also undergoes desensitization and sequestration, a process mediated by β-arrestin. Functional selectivity was subsequently documented using synthetic agonists and BRET biosensors to reveal that the synthetic ligand, TAK875, recruited β-arrestin with higher efficacy than natural ligands, whereas the latter promoted G-protein-dependent signaling. Importantly, TAK875 dramatically induced insulin secretion compared to FFAs [322]. Moreover, FFA1/β-arrestin-dependent signaling also promoted anti-inflammatory and anti-apoptotic effects [323]. In this and many other examples, it is clear that a specific physiological process can be selectively activated by targeting one subcellular location and one predominant function using biased tools.

8.5.3 Mitochondrial GPCRs Control Respiration and Apoptosis

As with nuclear and endosomal GPCRs, there is a growing list of mitochondrial GPCRs (Table 8.2). These include the purine, $P2Y_1$ and $P2Y_2$ receptors [176], serotonin 5-HT_4 receptor [18], AT_1 and AT_2 receptors [42, 324], melatonin MT_1 receptors [148], and cannabinoid CB_1 receptors [54–60]. Activation and/or inhibition of mitochondrial GPCRs appears to affect processes such as mitochondrial respiration, the production of reactive oxygen species, and even apoptosis. For example, a large body of data supports the existence of an independent intracellular renin–angiotensin system in many tissues [324]. Thus, angiotensin peptides can be synthesized *in situ* where they can activate AT_1 and AT_2 receptors on nuclear membranes (Table 8.1) as well as in mitochondria in CNS neurons, cardiomyocytes, vascular endothelium, hepatocytes and proximal tubules, and distal nephrons [324, 325]. Interestingly, as animals age, there is a shift in the balance of AT receptors. For example, levels of AT_1 are low in young animals, but they significantly increase as the animal ages, whereas the opposite occurs with AT_2. Activation of AT_2 receptors on the mitochondrial inner membrane is coupled to increased nitric oxide (NO) production and suppression of respiration which can be reversed by an AT_2 specific antagonist or by an inhibitor of NO synthesis [42].

The lipophilic compound, melatonin, is also found in high levels in mitochondria, where it can activate MT_1 receptors. Recent work shows that mitochondria can synthesize melatonin within the matrix where its release activates MT_1 receptors on the outer mitochondrial membrane [147]. Mitochondrial MT_1 signal transduction activates G_i and blocks adenylate cyclase activity, leading to the inhibition of stress-induced cytochrome c release and caspase activation [147]. As further evidence of the importance of mitochondrial MT_1, its targeted overexpression inhibited neuronal death resulting from hypoxic/ischemic injury [147]. Taken together, there is increasing evidence that mitochondrial GPCRs play important roles in many processes previously thought to be mediated by plasma membrane receptors, including respiration and apoptosis.

8.5.4 ER, Golgi, Lysosomal GPCRs Control Metabolic Functions: ER

Numerous GPCRs have also been described on other intracellular membranes including the ER, the Golgi and lysosomes. One of the best-described ER GPCRs is GPER, described above, a novel estrogen receptor that mediates signaling in multiple cell types. Since estrogen is a lipophilic compound it can easily permeate membranes to activate receptors throughout the cell. Depending upon the tissue, activated GPER can couple to $G_{i/o}$ and G_s proteins that regulate many downstream effectors including phosphatidylinositol-4,5-bisphosphate 3-kinase (PI3K)/Akt, ERK1/2, adenylyl cyclase, Ca^{2+} mobilization, and NOS [111]. GPER is widely expressed in various cancer cell types and tumors, where it promotes proliferation, migration, and invasion of cancer cells. GPER also influences lipid and glucose metabolism, inflammation, and even further estrogen synthesis, actions that enhance tumor growth and metastasis [111]. Another ER-localized GPCR is $mGlu_5$ which is >90% localized on nuclear and ER membranes [326]. Using selective uncaging of glutamate as well as extensive ultrastructural localization, $mGlu_5$ was localized on ER membranes especially neuronal dendrites. As with the INM or ONM, activation of $mGlu_5$ on ER membranes was linked to intracellular Ca^{2+} release as well as synaptic plasticity in the form of LTD [155]. These studies further emphasize the diversity of intracellular GPCR responses and the wide-range of roles such receptors play.

8.5.5 TGN

Recent work by Godbole et al. [130] demonstrated that following stimulation, the TSH receptor is internalized in mouse primary thyroid cells and then retrogradely trafficked to the TGN where it can locally produce cAMP and activated protein kinase A. Godbole et al. suggested that the TGN serves as a central hub for GPCR trafficking and signaling close to the nucleus since blocking receptor internalization or retrograde trafficking impairs nuclear CREB signaling in response to TSH. In contrast to TSH receptor trafficking to the TGN, Irannejad et al. [6] found that β_1ARs are endogenously expressed at the Golgi in HeLa cells where they can induce effector responses when stimulated by a cell permeant agonist, or the endogenous ligand, epinephrine. The latter accesses the receptor via the organic cation transporter, Oct3 [6]. Subsequently, Nash et al. [281] showed that in cardiac myocytes, endogenous β_1ARs at the Golgi stimulate G_s to produce cAMP which, in turn, activates the Epac/phospholipase Cε (PLCε)/muscle-specific A kinase anchoring protein β (mAKAPβ) complex leading to DAG production. Inasmuch as chronic stimulation of β_1ARs is associated with heart failure, blocking Golgi-restricted receptors may be a new, more selective approach for this disorder [281]. Collectively, these examples of TGN/Golgi GPCRs reflect similar mechanisms of GPCRs that are endogenously expressed at nuclear or ER membranes as well as those that are trafficked there.

8.5.6 Lysosomes

Lysosomes are often thought of as the organelle in which proteins, membranes, and macromolecules are deconstructed. However, so-called "endolysosomes," Rab7+, acidic organelles, exhibit many physical and functional interactions with other intracellular organelles including the plasma membrane, ER, mitochondria, and peroxisomes suggesting a possible role in cellular homeostasis. GPCRs found on endolysosomal membranes include AT_1 [327], ET_B [328], CB_1, and CB_2 receptors [61] as well as GPR55 [82]. An example of an endolysosomal GPCR function includes the versatile AT_1 receptor. In hypothalamic neurons, Deliu et al. [327] showed that microinjected Ang-II can be taken up by micro-autophagy to activate AT_1 receptors located on Rab7+ lysosomes. This, in turn, stimulated PLC and IP_3 production, the release of Ca^{2+} from intracellular stores, cation influx via transient receptor potential (TRP) channels ending in membrane depolarization. Using similar techniques, this same group showed that the CB_2 receptor was also localized to Rab7+ lysosomes [62]. Microinjection of 2-AG activated G_q-coupled endolysosomal CB_2 receptors leading to the mobilization of nicotinic acid adenine dinucleotide phosphate (NAADP)-sensitive Ca^{2+} stores and IP_3R release of intracellular Ca^{2+}. Yet another GPCR which resides on Rab7+ lysosomes is the L-α-lysophosphatidylinositol (LPI)-sensitive receptor, GPR55, which is expressed on the plasma membrane and intracellularly in cardiomyocytes [82]. Using Ca^{2+} and voltage imaging as well as the microinjection of GPR55 ligands, it was shown that activation of GPR55 at the plasma membrane mobilizes Ca^{2+} from the sarcoplasmic reticulum, whereas activation of GPR55 on endolysosomes elicited Ca^{2+} release from acidic-like Ca^{2+} stores via the endolysosomal NAADP-sensitive two-pore channels. The former produced Ca^{2+}-independent membrane depolarization while activation of intracellular GPR55 resulted in hyperpolarization of the sarcolemma. In yet another example of endolysosomal signaling, the lysosome-localized orphan GPCR, GPR137B, was recently shown to function by interacting with Rag proteins, thereby increasing their concentration at lysosomes [173]. This process, in turn, causes the recruitment of mammalian target of rapamycin complex 1 (mTORC1), resulting in increased autophagy in RagA/B-knockout mouse embryonic fibroblasts and cardiomyocytes. Thus, GPR137B controls the dynamic exchange of active Rags at lysosomes and hence regulates mTORC1 activity [173]. Close homologs to GPR137B, GPR137 and GPR137C, probably have a similar role in regulating mTORC1 as these homologous also increased mTOR translocation to lysosomes upon overexpression [173]. Together, these results suggest that GPR137B is part of a family of lysosome-localized GPCRs that regulate mTORC1 translocation to lysosomes. Collectively, these studies emphasize the importance of yet another intracellular membrane, the endolysosome, as a GPCR platform for signaling distinct from other cellular sites.

8.6 GPCR Localization Plays an Important Role in Disease Processes

Inasmuch as GPCRs play a central role in every aspect of cellular physiology, it is not surprising that they play important roles in tumorigenesis, metastasis, angiogenesis, and treatment resistance [329]. As such, they represent attractive therapeutic targets for anticancer therapies [329]. Intriguingly, immunological profiling of clinical biopsy specimens from cancer patients revealed many GPCRs on intracellular membranes including nuclear membranes. Indeed, >25% of nuclear GPCRs are greatly increased in cancers including breast, gastric, colo-rectal, prostate, and lung (Table 8.1). For example, nuclear GPCRs implicated in breast cancer include CXCR4 [67], GPR30 [330, 331], bradykinin receptors, B_1, and B_2 [50, 51]. In the case of CXCR4, nuclear receptors have been shown to be ligand-responsive. Their activation leads to increased nucleoplasmic Ca^{2+} levels, which appear to enhance and extend signals initiated in the cytoplasm. Since increased amplitude and/or duration of nuclear Ca^{2+} is known to differentially activate unique transcription factors, nuclear chemokine receptors may be critical in regulating the process of tumorigenesis in breast cancer. If so, impermeable or nontransported antagonists targeting cell surface receptors may be ineffectual for nuclear receptors. Human GPER is also significantly expressed in triple-negative breast cancer cells, including MB-468 and MDA-MB-436 [331]. Thus, GPER likely plays a significant role in breast cancer biology via an estrogen-independent pathway. Recent data using quantitative immunogold-EM examining bradykinin B_2 receptors at both the plasma membrane and the nuclear envelop indicates that this receptor is four times higher on cancerous nuclear membranes than on normal nuclei [51]. Using permeable versus nonpermeable B_2 receptor antagonists, these studies showed that permeable antagonists exhibited important anticancer activities such as growth arrest and apoptosis versus nonpermeable, cell surface only drugs [51]. Bradykinin B_1 receptors play a similar role in triple negative breast cancer cells where high levels of receptor expression are predominantly located on the INM. Akin to B_2 receptor-expressing cells, antagonists designed to be permeable were able to block cancer activities, whereas impermeant drugs could not [50]. These data reinforce the notion that GPCR subcellular localization contributes to receptor function and even disease pathology. Therapeutic strategies must take receptor localization into account to be the most efficacious.

Intracellular GPCRs also play a role in congestive heart failure. In particular, evidence shows that cardiac cells can synthesize their own biologically active Ang-II [[166]]. Not only can intracellular Ang-II levels be manipulated by exposing cardiomyocytes to pharmacological manipulations but Ang-II can also be microinjected such that cardiac physiological properties such as cell volume, cell communication, and ion–channel function are altered possibly via activation of nuclear AT_1 and AT_2 receptors [332]. To directly test this hypothesis, Tadevosyan et al. [41] used a photo-releasable caged-Ang-II derivative, [Tyr(DMNB)^4]Ang-II [333] to demonstrate that activation of nuclear AT_1 and AT_2 receptors led to increased nuclear Ca^{2+} and NO mobilization along with associated transcriptional responses in atrial fibroblasts independently of plasma membrane AT_1 receptors. These short-term effects led, in turn, to the control of fibroblast proliferation and collagen-1 secretion as well as alterations in the glycosylation of nuclear AT_2 in congestive heart failure [41]. These findings suggest that intracellular Ang-II may participate in structural remodeling by controlling fibroblast function through activation of nuclear AT_1 and AT_2 receptors.

Another nuclear GPCR playing a role in a disease process is $mGlu_5$. In the spinal cord, the $mGlu_5$ receptor is a key mediator of neuroplasticity underlying persistent pain. Biochemical, pharmacological, and ultrastructural studies all support the notion that increased $mGlu_5$ receptor levels are observed on spinal cord dorsal horn nuclei following neuropathic pain injury, whereas decreased numbers of receptors are present on the cell surface or intracellular membranes [156]. Interestingly, pharmacological blockade of spinal nuclear $mGlu_5$ receptors inhibits pain behaviors, whereas blockade of cell surface $mGlu_5$ receptors has little effect. Moreover, inhibition of a glutamate transporter mimics the effects of intracellular $mGlu_5$ antagonism by preventing intracellular uptake of the ligand. These studies reinforce the idea that drugs designed to block intracellular $mGlu_5$ versus cell surface receptors would be the most efficacious.

A different intracellular receptor involved in nociception is the tachykinin NK_1 receptor [206]. As described above, this receptor is internalized following activation but continues to signal from endosomes. Jensen et al. [206] discovered that NK_1 receptor-mediated nociception is mediated by the internalized, endosomal NK_1 receptors, not cell surface receptors. Inhibition of endocytosis (dynamin, clathrin, or β-arrestin inhibitors) prevented pain signaling and promoted antinociception as did targeting NK_1 receptor antagonists to the endosomes [206]. These findings reinforce the notion that location-dependent signaling plays a critical role in whether a given ligand can modulate its receptor. In this case, conventional NK_1 receptor antagonists are clinically ineffective for the treatment of chronic pain because the receptor has been internalized [206]. However, they are now widely used in the treatment of chemotherapy-induced nausea and vomiting [334]. Taken together, GPCR localization clearly contributes to receptor function and even disease pathology, underscoring the importance of designing drugs to block or enhance a given biological output in the context of the receptor's cellular location.

8.7 Looking Forward

The wealth of new data demonstrating GPCRs on almost every type of intracellular membrane (nuclei, ER, mitochondria, lysosomes, and endosomes) represents a paradigm shift in GPCR research and opens the door for a host of new translational applications. For the most part, drug discovery efforts have focused on the identification of agonists or antagonists of GPCRs on the cell surface, not receptors primarily localized on intracellular membranes. Thus, drugs with a desirable pharmacokinetic outcome might be further optimized for a desirable cell surface and/or intracellular response. In the case of intracellular receptors, the same key parameters associated with drug development for cell surface receptors such as efficacy, potency, and specificity are still essential for intracellular GPCR drug design. In addition, targeting the intracellular receptor might also require that the drug accessing the cell's interior do so without perturbing its cell surface counterpart. Many *in vitro* studies have used caged ligands in which the biologically active molecule is protected by a functional group, which, upon cellular uptake, can be activated by intracellular enzymes, pH, light, and so forth. In fact, caged compounds have been shown to activate intracellular receptors *in vitro* [290, 333]. However, caged compounds are less useful for *in vivo* applications. Fortunately, a wave of new techniques drawn from chemistry, materials science, and nanotechnology are creating a plethora of novel delivery strategies [335]. As described for the NK_1 receptor [206], it is possible to engineer a targeted agonist that preferentially interacts with its cognate receptor within the acidic endosomal environment. However, it remains to be determined whether this is the case for other endosomal GPCRs and other ligands. Other new techniques include polymer-based nanocarriers, which can be tailored to display a given charge or combined with other biomolecules such as drugs, antibodies, proteins, and oligonucleotides in order to more effectively deliver the desired entity to the particular intracellular location [336–338]. Next-generation technologies include delivering ligands via nanoparticles or exosomes, which can enter cells by clathrin-dependent endocytosis allowing an unprecedented ability to deliver therapeutics via the cell's own machinery [339]. Taken together, these new tools and new strategies will allow an unprecedented ability to deliver therapeutics to every part of the cell.

Although pharmacological isolation is a useful strategy in defining a given receptor's *in vivo* physiological role, the development of genetically isolated animals in which receptors are targeted or excluded from a given intracellular membrane would reinforce and/or uncover the role of a given intracellular receptor. For example, Joyal et al. used a genetic approach to distinguish plasma membrane from nuclear PAR2 functions *in vivo* [196, 282]. In these studies [196, 282], intravitreally injected viral constructs were targeted to either the cell surface or the nuclei of retinal ganglion neurons of PAR2 knockout mice. Plasma membrane-localized PAR2 retinas exhibited increased Ang1, an angiogenic factor associated with vascular remodeling and maturation. In contrast, nuclear-localized PAR2 showed increased vascular endothelial growth factor (Vegfa) expression, which is associated with neovascularization [196, 282]. Injection of the native PAR2 increased the expression of both angiogenic factors [196, 282]. Similarly, we have used clustered regularly interspaced short palindromic repeats (CRISPR) techniques to add tags to the $mGlu_5$ receptor to isolate it either entirely on the cell surface or inside the cell. The latter animal, the first engineered, has recapitulated findings observed using pharmacological isolation (O'Malley, unpublished observations). Animals such as these can potentially serve as model systems for the development of drugs optimized for a desirable cell surface and/or intracellular response.

In summary, current efforts to develop allosteric modulators and biased ligands for GPCRs might be further enhanced by recognizing that intracellular GPCRs are functional. Given the "drugability" of GPCRs, the need to selectively tailor agonists and/or antagonists to both intracellular and cell surface receptors is critical. This would be highly significant for translational applications, since developing the most highly effective and minimally toxic drugs is the long-term goal for any effective treatment.

Funding

This work was supported, in whole or in part, by National Institutes of Health Grants MH119197, NS102783, and IDDRC Grant P50 HD103525.

Abbreviations

AR	adrenergic receptor
Akt	protein kinase B
AMPA	α-amino-3-hydroxy-5-methyl-4-isoxazolepropionic acid
Ang-II	angiotensin-II
2-AG	2-arachidonoylglycerol
Arc	activity-regulated cytoskeleton-associated protein
AT_1	angiotensin II receptor type 1

AT_2	angiotensin II receptor type 2
B_2	bradykinin B_2 receptors
cAMP	3′–5′-cyclic adenosine monophosphate
BRET	bioluminescence resonance energy transfer
CB_1	cannabinoid 1 receptor
CB_2	cannabinoid 2 receptor
CGRP	calcitonin gene-related peptide
CNS	central nervous system
COPII	coat protein complex II
CREB	cAMP response element-binding protein
CXCR4	CXC chemokine receptor 4
DAG	diacylglycerol
ECEs	endothelin converting enzymes
EGFR	epidermal growth factor receptor
EP_3	Prostaglandin E_2 receptor type 3
EP_4	Prostaglandin E_2 receptor type 4
Epac	exchange factor activated by cAMP
ER	endoplasmic reticulum
ERK1/2	extracellular-signal-regulated kinase1/2
ET_B	endothelin B receptor
EVs	extracellular vesicles
FFA	Free Fatty Acid receptor
FRET	fluorescence resonance energy transfer
PAR2	Proteinase-activated receptor type 2 (F2rl1)
GAPs	GTPase activating proteins
GPCR	G protein-coupled receptors
GPER	G-protein-coupled estrogen receptor
GnRH	gonadotropin-releasing hormone
Impα	Importin-α
Impβ	importin β
INM	inner nuclear membranes
IP_3R	inositol triphosphate 3 receptor
LBR	lamin B receptor
LTD	long-term depression
MAPK	mitogen-activated protein kinase
$mGlu_5$	metabotropic glutamate receptor type 5
MT_1	melatonin receptor 1
mTOR	mammalian target of rapamycin
mTORC1	mTOR complex 1
NAADP	nicotinic acid adenine dinucleotide phosphate
NK_1	neurokinin 1 receptors

NLSs	nuclear localization signals
NE	norepinephrine
NF-κB	nuclear factor kappa B
NOS	nitric oxide synthase
NPC	nuclear pore complex
Oct3	organic cation transporter 3
ONM	outer nuclear membrane
PAF	platelet-activating factor
PKC	protein kinase C
PLC	phospholipase C
PTH	parathyroid hormone
PTH1R	parathyroid hormone receptor type1
RTKs	receptor tyrosine kinases
RAMPs	receptor-activity-modifying proteins
TBC	Tre-2/Bub2/Cdc16
TGN	*trans*-Golgi network
TPR	thromboxane A_2 receptor
SRF	serum response factor
STATs	signal transducers and activators of transcription
Vegfa	vascular endothelial growth factor
VIP	vasoactive intestinal polypeptide
$VPAC_1$	vasoactive intestinal peptide receptor type 1

References

1 Vaniotis, G., Allen, B.G., and Hébert, T.E. (2011). Nuclear GPCRs in cardiomyocytes: an insider's view of β-adrenergic receptor signaling. *Am. J. Physiol. Heart Circ. Physiol.* 301 (5): H1754–H1764. Review.

2 Bhosle, V.K., Gobeil, F. Jr., Rivera, J.C. et al. (2015). High resolution imaging and function of nuclear G protein-coupled receptors (GPCRs). In: *Nuclear G-Protein Coupled Receptors*, Methods in Molecular Biology (Methods and Protocols), vol. 1234 (ed. B. Allen and T. Hébert), 81–97. New York, NY: Humana Press.

3 Jong, Y.I., Harmon, S.K., and O'Malley, K.L. (2018). Intracellular GPCRs play key roles in synaptic plasticity. *ACS Chem. Neurosci.* 9 (9): 2162–2172. Review.

4 Jong, Y.I., Harmon, S.K., and O'Malley, K.L. (2018). GPCR signalling from within the cell. *Br. J. Pharmacol.* 175 (21): 4026–4035. Review.

5 Irannejad, R., Tomshine, J.C., Tomshine, J.R. et al. (2013). Conformational biosensors reveal GPCR signalling from endosomes. *Nature* 495 (7442): 534–538.

6 Irannejad, R., Pessino, V., Mika, D. et al. (2017). Functional selectivity of GPCR-directed drug action through location bias. *Nat. Chem. Biol.* 13: 799–806.
7 Halls, M.L. and Canals, M. (2018). Genetically encoded FRET biosensors to illuminate compartmentalised GPCR signalling. *Trends Pharmacol. Sci.* 39 (2): 148–157. Review.
8 Halls, M.L. (2019). Localised GPCR signalling as revealed by FRET biosensors. *Curr. Opin. Cell Biol.* 57: 48–56. Review.
9 De Groof, T.W.M., Bobkov, V., Heukers, R. et al. (2019). Nanobodies: new avenues for imaging, stabilizing and modulating GPCRs. *Mol. Cell. Endocrinol.* 484: 15–24. Review.
10 Heukers, R., De Groof, T.W.M., Smit, M.J. et al. (2019). Nanobodies detecting and modulating GPCRs outside in and inside out. *Curr. Opin. Cell Biol.* 57: 115–122. Review.
11 Lobingier, B.T. and von Zastrow, M. (2019). When trafficking and signaling mix: how subcellular location shapes G protein-coupled receptor activation of heterotrimeric G proteins. *Traffic* 20 (2): 130–136. Review.
12 Eichel, K. and von Zastrow, M. (2018). Subcellular organization of GPCR signaling. *Trends Pharmacol. Sci.* 39 (2): 200–208. Review.
13 Cohen, S., Valm, A.M., and Lippincott-Schwartz, J. (2018). Interacting organelles. *Curr. Opin. Cell Biol.* 53: 84–91. Review.
14 Della Rocca, G.J., Mukhin, Y.V., Garnovskaya, M.N. et al. (1999). Serotonin 5-HT1A receptor-mediated Erk activation requires calcium/calmodulin-dependent receptor endocytosis. *J. Biol. Chem.* 274: 4749–4753.
15 Schmid, C.L., Raehal, K.M., and Bohn, L.M. (2008). Agonist-directed signaling of the serotonin 2A receptor depends on β-arrestin-2 interactions *in vivo*. *Proc. Natl. Acad. Sci. U.S.A.* 105 (3): 1079–1084.
16 Chanrion, B., Mannoury la Cour, C., Gavarini, S. et al. (2008). Inverse agonist and neutral antagonist actions of antidepressants at recombinant and native 5-hydroxytryptamine 2C receptors: differential modulation of cell surface expression and signal transduction. *Mol. Pharmacol.* 73: 748–757.
17 Labasque, M., Reiter, E., Becamel, C. et al. (2008). Physical interaction of calmodulin with the 5-hydroxytryptamine 2C receptor C-terminus is essential for G protein-independent, arrestin-dependent receptor signaling. *Mol. Biol. Cell* 19 (11): 4640–4650.
18 Wang, Q., Zhang, H., Xu, H. et al. (2016). 5-HTR_3 and 5-HTR_4 located on the mitochondrial membrane and functionally regulated mitochondrial functions. *Sci. Rep.* 6: 37336.
19 Yamasaki, M., Matsui, M., and Watanabe, M. (2010). Preferential localization of muscarinic M_1 receptor on dendritic shaft and spine of cortical pyramidal

cells and its anatomical evidence for volume transmission. *J. Neurosci.* 30 (12): 4408–4418.

20 Anisuzzaman, A.S., Uwada, J., Masuoka, T. et al. (2013). Novel contribution of cell surface and intracellular M_1-muscarinic acetylcholine receptors to synaptic plasticity in hippocampus. *J. Neurochem.* 126 (3): 360–371.

21 Uwada, J., Anisuzzaman, A.S., Nishimune, A. et al. (2011). Intracellular distribution of functional M_1-muscarinic acetylcholine receptors in N1E-115 neuroblastoma cells. *J. Neurochem.* 118 (6): 958–967.

22 Nuber, S., Zabel, U., Lorenz, K. et al. (2016). β-Arrestin-biosensors reveal a rapid, receptor-dependent activation/deactivation cycle. *Nature* 531 (7596): 661–664.

23 Oda, S., Sato, F., Okada, A. et al. (2007). Immunolocalization of muscarinic receptor subtypes in the reticular thalamic nucleus of rats. *Brain Res. Bull.* 74: 376–384.

24 Pera, T., Hegde, A., Deshpande, D.A. et al. (2015). Specificity of arrestin subtypes in regulating airway smooth muscle G protein-coupled receptor signaling and function. *FASEB J.* 10: 4227–4235.

25 Jajoo, S., Mukherjea, D., Kumar, S. et al. (2010). Role of β-arrestin1/ERK MAP kinase pathway in regulating adenosine A_1 receptor desensitization and recovery. *Am. J. Physiol. Cell Physiol.* 298 (1): C56–C65.

26 Baltos, J.A., Paoletta, S., Nguyen, A.T. et al. (2016). Structure–activity analysis of biased agonism at the human adenosine A_3 receptor. *Mol. Pharmacol.* 90 (1): 12–22.

27 Pottie, E., Tosh, D.K., Gao, Z.G. et al. (2020). Assessment of biased agonism at the A_3 adenosine receptor using β-arrestin and miniGα_i recruitment assays. *Biochem. Pharmacol.* 177: 113934.

28 Davies, J.Q., Lin, H.H., Stacey, M. et al. (2011). Leukocyte adhesion-GPCR EMR2 is aberrantly expressed in human breast carcinomas and is associated with patient survival. *Oncol. Rep.* 25 (3): 619–627.

29 Dahl, E.F., Wu, S.C., Healy, C.L. et al. (2019). ERK mediated survival signaling is dependent on the G_q-G-protein coupled receptor type and subcellular localization in adult cardiac myocytes. *J. Mol. Cell Cardiol.* 127: 67–73.

30 Wu, S.C. and O'Connell, T.D. (2015). Nuclear compartmentalization of α_1-adrenergic receptor signaling in adult cardiac myocytes. *J. Cardiovasc. Pharmacol.* 65 (2): 91–100. Review.

31 Huang, Y., Wright, C.D., Merkwan, C.L. et al. (2007). An α_1A-adrenergic-extracellular signal-regulated kinase survival signaling pathway in cardiac myocytes. *Circulation* 115 (6): 763–772.

32 Wu, S.C., Dahl, E.F., Wright, C.D. et al. (2014). Nuclear localization of a_1A-adrenergic receptors is required for signaling in cardiac myocytes: an "inside-out" a_1-AR signaling pathway. *J. Am. Heart Assoc.* 3 (2): e000145.

33 Wright, C.D., Wu, S.C., Dahl, E.F. et al. (2012). Nuclear localization drives α_1-adrenergic receptor oligomerization and signaling in cardiac myocytes. *Cell. Signalling* 24: 794–802.

34 Wright, C.D., Chen, Q., Baye, N.L. et al. (2008). Nuclear α_1-adrenergic receptors signal activated ERK localization to caveolae in adult cardiac myocytes. *Circ. Res.* 103 (9): 992–1000.

35 Boivin, B., Lavoie, C., Vaniotis, G. et al. (2006). Functional β-adrenergic receptor signalling on nuclear membranes in adult rat and mouse ventricular cardiomyocytes. *Cardiovasc. Res.* 71 (1): 69–78.

36 Bedioune, I., Lefebvre, F., Lechêne, P. et al. (2018). PDE4 and mAKAPβ are nodal organizers of β_2-ARs nuclear PKA signalling in cardiac myocytes. *Cardiovasc. Res.* 114 (11): 1499–1511.

37 Shukla, A.K., Westfield, G.H., Xiao, K. et al. (2014). Visualization of arrestin recruitment by a G-protein-coupled receptor. *Nature* 512 (7513): 218–222.

38 Thomsen, A.R.B., Plouffe, B., Cahill, T.J. 3rd. et al. (2016). GPCR-G protein-β-arrestin super-complex mediates sustained G protein signaling. *Cell* 166 (4): 907–919.

39 Audet, N., Dabouz, R., Allen, B.G. et al. (2018). Nucleoligands-repurposing G protein-coupled receptor ligands to modulate nuclear-localized G protein-coupled receptors in the cardiovascular system. *J. Cardiovasc. Pharmacol.* 71 (4): 193–204. Review.

40 Lee, D.K., Lança, A.J., Cheng, R. et al. (2004). Agonist-independent nuclear localization of the Apelin, angiotensin AT_1, and bradykinin B_2 receptors. *J. Biol. Chem.* 279: 7901–7908.

41 Tadevosyan, A., Xiao, J., Surinkaew, S. et al. (2017). Intracellular angiotensin-II interacts with nuclear angiotensin receptors in cardiac fibroblasts and regulates RNA synthesis, cell proliferation, and collagen secretion. *J. Am. Heart Assoc.* 6 (4): e004965.

42 Abadir, P.M., Foster, D.B., Crow, M. et al. (2011). Identification and characterization of a functional mitochondrial angiotensin system. *Proc. Natl. Acad. Sci. U.S.A.* 108 (36): 14849–14854.

43 Escobales, N., Nuñez, R.E., and Javadov, S. (2019). Mitochondrial angiotensin receptors and cardioprotective pathways. *Am. J. Physiol. Heart Circ. Physiol.* 316 (6): H1426–H1438.

44 Piairo, P., Moura, R.S., Nogueira-Silva, C. et al. (2011). The apelinergic system in the developing lung: expression and signaling. *Peptides* 32 (12): 2474–2483.

45 Masyuk, A.I., Huang, B.Q., Radtke, B.N. et al. (2013). Ciliary subcellular localization of TGR5 determines the cholangiocyte functional response to bile acid signaling. *Am. J. Physiol. Gastrointest. Liver Physiol.* 304 (11): G1013–G1024.

46 Hu, M.M., He, W.R., Gao, P. et al. (2019). Virus-induced accumulation of intracellular bile acids activates the TGR5-β-arrestin-SRC axis to enable innate antiviral immunity. *Cell Res.* 29 (3): 193–205.

47 Abdulkhalek, S. and Szewczuk, M.R. (2013). Neu1 sialidase and matrix metalloproteinase-9 cross-talk regulates nucleic acid-induced endosomal TOLL-like receptor-7 and -9 activation, cellular signaling and pro-inflammatory responses. *Cell. Signalling* 25 (11): 2093–2105.

48 Ikeda, Y., Kumagai, H., Okazaki, H. et al. (2015). Monitoring β-arrestin recruitment via β-lactamase enzyme fragment complementation: purification of peptide E as a low-affinity ligand for mammalian bombesin receptors. *PLoS One* 10 (6): e0127445.

49 Foster, S.R., Hauser, A.S., Vedel, L. et al. (2019). Discovery of human signaling systems: pairing peptides to G protein-coupled receptors. *Cell* 179 (4): 895–908.e21.

50 Dubuc, C., Savard, M., Bovenzi, V. et al. (2019). Antitumor activity of cell-penetrant kinin B_1 receptor antagonists in human triple-negative breast cancer cells. *J. Cell. Physiol.* 234 (3): 2851–2865.

51 Dubuc, C., Savard, M., Bovenzi, V. et al. (2018). Targeting intracellular B_2 receptors using novel cell-penetrating antagonists to arrest growth and induce apoptosis in human triple-negative breast cancer. *Oncotarget* 9 (11): 9885–9906.

52 Yarwood, R.E., Imlach, W.L., Lieu, T. et al. (2017). Endosomal signaling of the receptor for calcitonin gene-related peptide mediates pain transmission. *Proc. Natl. Acad. Sci. U.S.A.* 114 (46): 12309–12314.

53 Gorvin, C.M., Babinsky, V.N., Malinauskas, T. et al. (2018). A calcium-sensing receptor mutation causing hypocalcemia disrupts a transmembrane salt bridge to activate β-arrestin-biased signaling. *Sci. Signal.* 11 (518): eaan3714.

54 Hebert-Chatelain, E., Reguero, L., Puente, N. et al. (2014). Studying mitochondrial CB_1 receptors: yes we can. *Mol. Metab.* 3 (4): 339.

55 Hebert-Chatelain, E., Reguero, L., Puente, N. et al. (2014). Cannabinoid control of brain bioenergetics: exploring the subcellular localization of the CB_1 receptor. *Mol. Metab.* 3 (4): 495–504.

56 Hebert-Chatelain, E., Desprez, T., Serrat, R. et al. (2016). A cannabinoid link between mitochondria and memory. *Nature* 539 (7630): 555–559.

57 Bénard, G., Massa, F., Puente, N. et al. (2012). Mitochondrial CB_1 receptors regulate neuronal energy metabolism. *Nat. Neurosci.* 15 (4): 558–564.

58 Ma, L., Jia, J., Niu, W. et al. (2015). Mitochondrial CB_1 receptor is involved in ACEA-induced protective effects on neurons and mitochondrial functions. *Sci. Rep.* 5: 12440.

59 Xu, Z., Lv, X.A., Dai, Q. et al. (2016). Acute upregulation of neuronal mitochondrial type-1 cannabinoid receptor and it's role in metabolic defects and neuronal apoptosis after TBI. *Mol. Brain* 9 (1): 75.

60 Melser, S., Pagano Zottola, A.C., Serrat, R. et al. (2017). Functional analysis of mitochondrial CB_1 cannabinoid receptors ($mtCB_1$) in the brain. *Methods Enzymol.* 593: 143–174.

61 Brailoiu, G.C., Oprea, T.I., Zhao, P. et al. (2011). Intracellular cannabinoid type 1 (CB_1) receptors are activated by anandamide. *J. Biol. Chem.* 286 (33): 29166–29174.

62 Brailoiu, G.C., Deliu, E., Marcu, J. et al. (2014). Differential activation of intracellular versus plasmalemmal CB_2 cannabinoid receptors. *Biochemistry* 53: 4990–4999.

63 Currie, S., Rainbow, R.D., Ewart, M.A. et al. (2008). IP_3R-mediated Ca^{2+} release is modulated by anandamide in isolated cardiac nuclei. *J. Mol. Cell Cardiol.* 45: 804–811.

64 den Boon, F.S., Chameau, P., Schaafsma-Zhao, Q. et al. (2012). Excitability of prefrontal cortical pyramidal neurons is modulated by activation of intracellular type-2 cannabinoid receptors. *Proc. Natl. Acad. Sci. U.S.A.* 109: 3534–3539.

65 De Henau, O., Degroot, G.N., Imbault, V. et al. (2016). Signaling properties of chemerin receptors CMKLR1, GPR1 and CCRL2. *PLoS One* 11 (10): e0164179.

66 Favre, N., Camps, M., Arod, C. et al. (2008). Chemokine receptor CCR2 undergoes transportin 1-dependent nuclear translocation. *Proteomics* 8: 4560–4576.

67 Werner, R.A., Kirche, r.S., Higuchi, T. et al. (2019). CXCR4-directed imaging in solid tumors. *Front. Oncol.* 9: 770.

68 Shibuta, K., Mori, M., Shimoda, K. et al. (2002). Regional expression of CXCL12/CXCR4 in liver and hepatocellular carcinoma and cell-cycle variation during *in vitro* differentiation. *Jpn. J. Cancer Res.* 93 (7): 789–797.

69 Gobeil, F. Jr., Zhu, T., Brault, S. et al. (2006). Nitric oxide signaling via nuclearized endothelial nitric-oxide synthase modulates expression of the immediate early genes iNOS and mPGES-1. *J. Biol. Chem.* 281 (23): 16058–16067.

70 Gobeil, F., Fortier, A., Zhu, T. et al. (2006). G-protein-coupled receptors signalling at the cell nucleus: an emerging paradigm. *Can. J. Physiol. Pharmacol.* 84: 287–297. Review.

71 Spano, J.P., Andre, F., Morat, L. et al. (2004). Chemokine receptor CXCR4 and early-stage non-small cell lung cancer: pattern of expression and correlation with outcome. *Ann. Oncol.* 15 (4): 613–617.

72 Yoshitake, N., Fukui, H., Yamagishi, H. et al. (2008). Expression of SDF-1 α and nuclear CXCR4 predicts lymph node metastasis in colorectal cancer. *Br. J. Cancer* 98 (10): 1682–1689.

73 English, E.J., Mahn, S.A., and Marchese, A. (2018). Endocytosis is required for CXC chemokine receptor type 4 (CXCR4)-mediated Akt activation and antiapoptotic signaling. *J. Biol. Chem.* 293 (29): 11470–11480.

74 Roettger, B.F., Rentsch, R.U., Pinon, D. et al. (1995). Dual pathways of internalization of the cholecystokinin receptor. *J. Cell Biol.* 128 (6): 1029–1041.

75 Ning, S.L., Zheng, W.S., Su, J. et al. (2015). Different downstream signalling of CCK1 receptors regulates distinct functions of CCK in pancreatic beta cells. *Br. J. Pharmacol.* 172 (21): 5050–5067.

76 Magnan, R., Escrieut, C., Gigoux, V. et al. (2013). Distinct CCK-2 receptor conformations associated with β-arrestin-2 recruitment or phospholipase-C activation revealed by a biased antagonist. *J. Am. Chem. Soc.* 135 (7): 2560–2573.

77 Daniele, S., Trincavelli, M.L., Fumagalli, M. et al. (2014). Does GRK-β arrestin machinery work as a "switch on" for GPR17-mediated activation of intracellular signaling pathways? *Cell. Signalling* 26 (6): 1310–1325.

78 Miyagi, T., Tanaka, S., Hide, I. et al. (2016). The subcellular dynamics of the Gs-linked receptor GPR3 contribute to the local activation of PKA in cerebellar granular neurons. *PLoS One* 11 (1): e0147466.

79 Laun, A.S., Shrader, S.H., Brown, K.J. et al. (2019). GPR3, GPR6, and GPR12 as novel molecular targets: their biological functions and interaction with cannabidiol. *Acta Pharmacol. Sin.* 40 (3): 300–308.

80 Berger, B.S., Acebron, S.P., Herbst, J. et al. (2017). Parkinson's disease-associated receptor GPR37 is an ER chaperone for LRP6. *EMBO Rep.* 18 (5): 712–725.

81 Imai, Y., Soda, M., Inoue, H. et al. (2001). An unfolded putative transmembrane polypeptide, which can lead to endoplasmic reticulum stress, is a substrate of Parkin. *Cell* 105: 891–902.

82 Yu, J., Deliu, E., Zhang, X.Q. et al. (2013). Differential activation of cultured neonatal cardiomyocytes by plasmalemmal versus intracellular G protein-coupled receptor 55. *J. Biol. Chem.* 288 (31): 22481–22492.

83 Massart, R., Mignon, V., Stanic, J. et al. (2016). Developmental and adult expression patterns of the G-protein-coupled receptor GPR88 in the rat: establishment of a dual nuclear-cytoplasmic localization. *J. Comp. Neurol.* 524 (14): 2776–2802.

84 Steffen, J.S., Simon, E., Warneke, V. et al. (2012). LGR4 and LGR6 are differentially expressed and of putative tumor biological significance in gastric carcinoma. *Virchows Arch.* 461 (4): 355–365.

85 Snyder, J.C., Rochelle, L.K., Lyerly, H.K. et al. (2013). Constitutive internalization of the leucine-rich G protein-coupled receptor-5 (LGR_5) to the *trans*-Golgi network. *J. Biol. Chem.* 288 (15): 10286–10297.

86 Carmon, K.S., Lin, Q., Gong, X. et al. (2012). GR_5 interacts and cointernalizes with Wnt receptors to modulate Wnt/β-catenin signaling. *Mol. Cell. Biol.* 32 (11): 2054–2064.

87 Gwathmey, T.M., Westwood, B.M., Pirro, N.T. et al. (2010). Nuclear angiotensin-(1–7) receptor is functionally coupled to the formation of nitric oxide. *Am. J. Physiol. Renal. Physiol* 299 (5): F983–F990.

88 Patel, N., Itakura, T., Gonzalez, J.M. Jr. et al. (2013). GPR158, an orphan member of G protein-coupled receptor Family C: glucocorticoid-stimulated expression and novel nuclear role. *PLoS One* 8 (2): e57843.

89 Ye, R., Pi, M., Nooh, M.M. et al. (2019). Human GPRC6A mediates testosterone-induced mitogen-activated protein kinases and mTORC1 signaling in prostate cancer cells. *Mol. Pharmacol.* 95 (5): 563–572.

90 Punn, A., Levine, M.A., and Grammatopoulos, D.K. (2006). Identification of signaling molecules mediating corticotropin-releasing hormone-R1α-mitogen-activated protein kinase (MAPK) interactions: the critical role of phosphatidylinositol 3-kinase in regulating ERK1/2 but not p38 MAPK activation. *Mol. Endocrinol.* 20: 3179–3195.

91 Reyes, B.A., Valentino, R.J., and Van Bockstaele, E.J. (2008). Stress-induced intracellular trafficking of corticotropin-releasing factor receptors in rat locus coeruleus neurons. *Endocrinology* 149 (1): 122–130.

92 Valentino, R.J., Van Bockstaele, E., and Bangasser, D. (2013). Sex-specific cell signaling: the corticotropin-releasing factor receptor model. *Trends Pharmacol. Sci.* 34: 437–444.

93 Inda, C., Dos Santos Claro, P.A., Bonfiglio, J.J. et al. (2016). Different cAMP sources are critically involved in G protein-coupled receptor CRHR1 signaling. *J. Cell Biol.* 214 (2): 181–195.

94 Inda, C., Armando, N.G., Dos Santos Claro, P.A. et al. (2017). Endocrinology and the brain: corticotropin-releasing hormone signaling. *Endocr. Connect.* 6 (6): R99–R120. Review.

95 Kotowski, S.J., Hopf, F.W., Seif, T. et al. (2011). Endocytosis promotes rapid dopaminergic signaling. *Neuron* 71 (2): 278–290.

96 Beaulieu, J.M., Espinoza, S., and Gainetdinov, R.R. (2015). Dopamine receptors – IUPHAR Review 13. *Br. J. Pharmacol.* 172 (1): 1–23.

97 Peterson, S.M., Pack, T.F., and Caron, M.G. (2015). Receptor, ligand and transducer contributions to dopamine D_2 receptor functional selectivity. *PLoS One* 10 (10): e0141637.

98 Peterson, S.M., Pack, T.F., Wilkins, A.D. et al. (2015). Elucidation of G-protein and β-arrestin functional selectivity at the dopamine D_2 receptor. *Proc. Natl. Acad. Sci. U.S.A.* 112 (22): 7097–7102.

99 Shioda, N., Yabuk, i.Y., Wang, Y. et al. (2017). Endocytosis following dopamine D_2 receptor activation is critical for neuronal activity and dendritic spine formation via Rabex-5/PDGFRβ signaling in striatopallidal medium spiny neurons. *Mol. Psychiatry* 22 (8): 1205–1222.

100 Boivin, B., Chevalier, D., Villeneuve, L.R. et al. (2003). Functional endothelin receptors are present on nuclei in cardiac ventricular myocytes. *J. Biol. Chem.* 278 (31): 29153–29163.

101 Branco, A.F. and Allen, B.G. (2015). G protein-coupled receptor signaling in cardiac nuclear membranes. *J. Cardiovasc. Pharmacol.* 65 (2): 101–109.

102 Vaniotis, G., Glazkova, I., Merlen, C. et al. (2013). Regulation of cardiac nitric oxide signaling by nuclear β-adrenergic and endothelin receptors. *J. Mol. Cell Cardiol.* 62: 58–68.

103 Cattaneo, F., Parisi, M., Fioretti, T. et al. (2016). Nuclear localization of formyl-peptide receptor 2 in human cancer cells. *Arch. Biochem. Biophys.* 603: 10–19.

104 Marafie, S.K., Al-Shawaf, E.M., Abubaker, J. et al. (2019). Palmitic acid-induced lipotoxicity promotes a novel interplay between Akt-mTOR, IRS-1, and FFAR1 signaling in pancreatic β-cells. *Biol. Res.* 52 (1): 44.

105 Sosa-Alvarado, C., Hernández-Méndez, A., Romero-Ávila, M.T. et al. (2015). Agonists and protein kinase C-activation induce phosphorylation and internalization of FFA1 receptors. *Eur. J. Pharmacol.* 768: 108–115.

106 Qian, J., Wu, C., Chen, X. et al. (2014). Differential requirements of arrestin-3 and clathrin for ligand-dependent and -independent internalization of human G protein-coupled receptor 40. *Cell. Signalling* 26 (11): 2412–2423.

107 Ma, D., Tao, B., Warashina, S. et al. (2007). Expression of free fatty acid receptor GPR40 in the central nervous system of adult monkeys. *Neurosci. Res.* 58 (4): 394–401.

108 Lee, S.U., In, H.J., Kwon, M.S. et al. (2013). β-Arrestin 2 mediates G protein-coupled receptor 43 signals to nuclear factor-κB. *Biol. Pharm. Bull.* 36 (11): 1754–1759.

109 Ang, Z., Xiong, D., Wu, M. et al. (2018). FFAR2-FFAR3 receptor heteromerization modulates short-chain fatty acid sensing. *FASEB J.* 32 (1): 289–303.

110 Lu, C.L. and Herndon, C. (2017). New roles for neuronal estrogen receptors. *Neurogastroenterol. Motil.* 29 (7). Review: 1–7.

111 Filardo, E.J. (2018). A role for G-protein coupled estrogen receptor (GPER) in estrogen-induced carcinogenesis: dysregulated glandular homeostasis, survival and metastasis. *J. Steroid Biochem. Mol. Biol.* 176: 38–48. Review.

112 Revankar, C.M., Cimino, D.F., Sklar, L.A. et al. (2005). A transmembrane intracellular estrogen receptor mediates rapid cell signaling. *Science* 307: 1625–1630.

113 Waters, E.M., Thompson, L.I., Patel, P. et al. (2015). G-protein-coupled estrogen receptor 1 is anatomically positioned to modulate synaptic plasticity in the mouse hippocampus. *J. Neurosci.* 35: 2384–2397.

114 Gonchar, Y., Pang, L., Malitschek, B. et al. (2001). Subcellular localization of $GABA_B$ receptor subunits in rat visual cortex. *J. Comp. Neurol.* 431 (2): 182–197.

115 Richer, M., David, M., Villeneuve, L.R. et al. (2009). GABA-B_1 receptors are coupled to the ERK1/2 MAP kinase pathway in the absence of GABA-B_2 subunits. *J. Mol. Neurosci.* 38 (1): 67–79.

116 Reyes-Alcaraz, A., Lee, Y.N., Yun, S. et al. (2018). Conformational signatures in β-arrestin2 reveal natural biased agonism at a G-protein-coupled receptor. *Commun. Biol.* 1: 128.

117 Camiña, J.P., Lodeiro, M., Ischenko, O. et al. (2007). Stimulation by ghrelin of p42/p44 mitogen-activated protein kinase through the GHS-R1a receptor: role of G-proteins and beta-arrestins. *J. Cell. Physiol.* 213 (1): 187–200.

118 Lodeiro, M., Theodoropoulo, u.M., Pardo, M. et al. (2009). c-Src regulates Akt signaling in response to ghrelin via beta-arrestin signaling-independent and -dependent mechanisms. *PLoS One* 4 (3): e4686.

119 Lodeiro, M., Alén, B.O., Mosteiro, C.S. et al. (2011). The SHP-1 protein tyrosine phosphatase negatively modulates Akt signaling in the ghrelin/GHSR1a system. *Mol. Biol. Cell* 22 (21): 4182–4191.

120 Evron, T., Peterson, S.M., Urs, N.M. et al. (2014). G protein and β-arrestin signaling bias at the ghrelin receptor. *J. Biol. Chem.* 289 (48): 33442–33455.

121 Bouzo-Lorenzo, M., Santo-Zas, I., Lodeiro, M. et al. (2016). Distinct phosphorylation sites on the ghrelin receptor, GHSR1a, establish a code that determines the functions of ß-arrestins. *Sci. Rep.* 6: 22495.

122 Ismail, S., Gherardi, M.J., Froese, A. et al. (2016). Internalized receptor for glucose-dependent insulinotropic peptide stimulates adenylyl cyclase on early endosomes. *Biochem. Pharmacol.* 120: 33–45.

123 Fletcher, M.M., Halls, M.L., Zhao, P. et al. (2018). Glucagon-like peptide-1 receptor internalisation controls spatiotemporal signalling mediated by biased agonists. *Biochem. Pharmacol.* 156: 406–419.

124 Buenaventura, T., Kanda, N., Douzenis, P.C. et al. (2018). A targeted RNAi screen identifies endocytic trafficking factors that control GLP-1 receptor signaling in pancreatic β-cells. *Diabetes* 67 (3): 385–399.

125 Jean-Alphonse, F., Bowersox, S., Chen, S. et al. (2014). Spatially restricted G protein-coupled receptor activity via divergent endocytic compartments. *J. Biol. Chem.* 289 (7): 3960–3977.

126 De Pascali, F., Tréfier, A., Landomiel, F. et al. (2018). Follicle-stimulating hormone receptor: advances and remaining challenges. *Int. Rev. Cell Mol. Biol.* 338: 1–58. Review.

127 Lyga, S., Volpe, S., Werthmann, R.C. et al. (2016). Persistent cAMP signaling by internalized LH receptors in ovarian follicles. *Endocrinology* 157 (4): 1613–1621.

128 Sposini, S., Jean-Alphonse, F.G., Ayoub, M.A. et al. (2017). Integration of GPCR signaling and sorting from very early endosomes via opposing APPL1 mechanisms. *Cell Rep.* 21 (10): 2855–2867.

129 Calebiro, D., Nikolaev, V.O., Gagliani, M.C. et al. (2009). Persistent cAMP-signals triggered by internalized G-protein-coupled receptors. *PLoS Biol.* 7 (8): e1000172.

130 Godbole, A., Lyga, S., Lohse, M.J. et al. (2017). Internalized TSH receptors en route to the TGN induce local G_s-protein signaling and gene transcription. *Nat. Commun.* 8 (1): 443.

131 Re, M., Pampillo, M., Savard, M. et al. (2010). The human gonadotropin releasing hormone type I receptor is a functional intracellular GPCR expressed on the nuclear membrane. *PLoS One* 5 (7): e11489.

132 Pampillo, M., Camuso, N., Taylor, J.E. et al. (2009). Regulation of GPR54 signaling by GRK2 and β-arrestin. *Mol. Endocrinol.* 23 (12): 2060–2074.

133 Szereszewski, J.M., Pampillo, M., Ahow, M.R. et al. (2010). GPR54 regulates ERK1/2 activity and hypothalamic gene expression in a $G\alpha_{q/11}$ and β-arrestin-dependent manner. *PLoS One* 5 (9): e12964.

134 Zajac, M., Law, J., Cvetkovic, D.D. et al. (2011). GPR54 (KISS1R) transactivates EGFR to promote breast cancer cell invasiveness. *PLoS One* 6 (6): e21599.

135 Nielsen, C.K., Campbell, J.I., Ohd, J.F. et al. (2005). A novel localization of the G-protein-coupled CysLT1 receptor in the nucleus of colorectal adenocarcinoma cells. *Cancer Res.* 65 (3): 732–742.

136 Dvash, E., Har-Tal, M., Barak, S. et al. (2015). Leukotriene C4 is the major trigger of stress-induced oxidative DNA damage. *Nat. Commun.* 6: 10112.

137 Gobeil, F. Jr., Bernier, S.G., Vazquez-Tello, A. et al. (2003). Modulation of pro-inflammatory gene expression by nuclear lysophosphatidic acid receptor type-1. *J. Biol. Chem.* 278 (40): 38875–38883.

138 Waters, C.M., Saatian, B., Moughal, N.A. et al. (2006). Integrin signalling regulates the nuclear localization and function of the lysophosphatidic acid receptor-1 (LPA1) in mammalian cells. *Biochem. J* 398 (1): 55–62.

139 Mullershausen, F., Zecri, F., Cetin, C. et al. (2009). Persistent signaling induced by FTY720-phosphate is mediated by internalized $S1P_1$ receptors. *Nat. Chem. Biol.* 5 (6): 428–434.

140 Liao, J.J., Huang, M.C., Graler, M. et al. (2006). Distinctive T cell-suppressive signals from nuclearized type 1 sphingosine 1-phosphate G protein-coupled receptors. *J. Biol. Chem.* 282 (3): 1964–1972.

141 Estrada, R., Wang, L., Jala, V.R. et al. (2009). Ligand-induced nuclear translocation of $S1P_1$ receptors mediates Cyr61 and CTGF transcription in endothelial cells. *Histochem. Cell Biol.* 131 (2): 239–249.

142 Wang, C., Mao, J., Redfield, S. et al. (2014). Systemic distribution, subcellular localization and differential expression of sphingosine-1-phosphate receptors in benign and malignant human tissues. *Exp. Mol. Pathol.* 97 (2): 259–265.

143 Doufexis, M., Storr, H.L., King, P.J. et al. (2007). Interaction of the melanocortin 2 receptor with nucleoporin 50: evidence for a novel pathway between a G-protein-coupled receptor and the nucleus. *FASEB J.* 21 (14): 4095–4100.

144 Nyan, D.C., Anbazhagan, R., Hughes-Darden, C.A. et al. (2008). Endosomal colocalization of melanocortin-3 receptor and beta-arrestins in CAD cells with altered modification of AKT/PKB. *Neuropeptides* 42 (3): 355–366.

145 Gao, Z., Lei, D., Welch, J. et al. (2003). Agonist-dependent internalization of the human melanocortin-4 receptors in human embryonic kidney 293 cells. *J. Pharmacol. Exp. Ther.* 307 (3): 870–877.

146 Granell, S., Molden, B.M., and Baldini, G. (2013). Exposure of MC4R to agonist in the endoplasmic reticulum stabilizes an active conformation of the receptor that does not desensitize. *Proc. Natl. Acad. Sci. U.S.A.* 110 (49): E4733–E4742.

147 Suofu, Y., Li, W., Jean-Alphonse, F.G. et al. (2017). Dual role of mitochondria in producing melatonin and driving GPCR signaling to block cytochrome c release. *Proc. Natl. Acad. Sci. U.S.A.* 114 (38): E7997–E8006.

148 Gbahou, F., Cecon, E., Viault, G. et al. (2017). Design and validation of the first cell-impermeant melatonin receptor agonist. *Br. J. Pharmacol.* 174 (14): 2409–2421.

149 Wang, X., Sirianni, A., Pei, Z. et al. (2011). The melatonin MT1 receptor axis modulates mutant Huntingtin-mediated toxicity. *J. Neurosci.* 31 (41): 14496–14507.

150 Lanoix, D., Ouellette, R., and Vaillancourt, C. (2006). Expression of melatoninergic receptors in human placental choriocarcinoma cell lines. *Hum. Reprod.* 21 (8): 1981–1989.

151 Ahluwalia, A., Brzozowska, I.M., Hoa, N. et al. (2018). Melatonin signaling in mitochondria extends beyond neurons and neuroprotection: Implications for angiogenesis and cardio/gastroprotection. *Proc. Natl. Acad. Sci. U.S.A.* 115 (9): E1942–E1943.

152 Jong, Y.J., Schwetye, K.E., and O'Malley, K.L. (2007). Nuclear localization of functional metabotropic glutamate receptor $mGlu_1$ in HEK293 cells and cortical neurons: role in nuclear calcium mobilization and development. *J. Neurochem.* 101 (2): 458–469.

153 O'Malley, K.L., Jong, Y.J., Gonchar, Y. et al. (2003). Activation of metabotropic glutamate receptor on nuclear membranes mediates intranuclear Ca^{2+} changes in heterologous cell types and neurons. *J. Biol. Chem.* 278 (30): 28210–28219.

154 Jong, Y.J., Kumar, V., Kingston, A.E. et al. (2005). Functional metabotropic glutamate receptors on nuclei from brain and primary cultured striatal neurons. Role of transporters in delivering ligand. *J. Biol. Chem.* 280 (34): 30469–30480.

155 Purgert, C.A., Izumi, Y., Jong, Y.J. et al. (2014). Intracellular $mGluR_5$ can mediate synaptic plasticity in the hippocampus. *J. Neurosci.* 34 (13): 4589–4598.

156 Vincent, K., Cornea, V.M., Jong, Y.J. et al. (2016). Intracellular $mGluR_5$ plays a critical role in neuropathic pain. *Nat. Commun.* 7: 10604.

157 Kumar, V., Jong, Y.J., and O'Malley, K.L. (2008). Activated nuclear metabotropic glutamate receptor $mGlu_5$ couples to nuclear $G_{q/11}$ proteins to generate inositol 1,4,5-trisphosphate-mediated nuclear Ca^{2+} release. *J. Biol. Chem.* 283 (20): 14072–14083.

158 Jong, Y.J., Kumar, V., and O'Malley, K.L. (2009). Intracellular metabotropic glutamate receptor 5 ($mGluR_5$) activates signaling cascades distinct from cell surface counterparts. *J. Biol. Chem.* 284 (51): 35827–35838.

159 Kumar, V., Fahey, P.G., Jong, Y.J. et al. (2012). Activation of intracellular metabotropic glutamate receptor 5 in striatal neurons leads to up-regulation of genes associated with sustained synaptic transmission including Arc/Arg3.1 protein. *J. Biol. Chem.* 287 (8): 5412–5425.

160 Jacques, D., Sader, S., Perreault, C. et al. (2006). Roles of nuclear NPY and NPY receptors in the regulation of the endocardial endothelium and heart function. *Can. J. Physiol. Pharmacol.* 84 (7): 695–705. Review.

161 Law, I.K., Murphy, J.E., Bakirtzi, K. et al. (2012). Neurotensin-induced proinflammatory signaling in human colonocytes is regulated by β-arrestins and endothelin-converting enzyme-1-dependent endocytosis and resensitization of neurotensin receptor 1. *J. Biol. Chem.* 287 (18): 15066–15075.

162 Zhang, N.R., Planer, W., Siuda, E.R. et al. (2012). Serine 363 is required for nociceptin/orphanin FQ opioid receptor (NOPR) desensitization, internalization, and arrestin signaling. *J. Biol. Chem.* 287 (50): 42019–42030.

163 Asth, L., Ruzza, C., Malfacini, D. et al. (2016). Beta-arrestin 2 rather than G protein efficacy determines the anxiolytic-versus antidepressant-like effects of nociceptin/orphanin FQ receptor ligands. *Neuropharmacology* 105: 434–442.

164 Donica, C.L., Awwad, H.O., Thakker, D.R. et al. (2013). Cellular mechanisms of nociceptin/orphanin FQ (N/OFQ) peptide (NOP) receptor regulation and heterologous regulation by N/OFQ. *Mol. Pharmacol.* 83: 907–918.

165 Stoeber, M., Jullié, D., Lobingier, B.T. et al. (2018). A genetically encoded biosensor reveals location bias of opioid drug action. *Neuron* 98 (5): 963–976.e5.

166 Tadevosyan, A., Vaniotis, G., Allen, B.G. et al. (2012). G protein-coupled receptor signalling in the cardiac nuclear membrane: evidence and possible roles in physiological and pathophysiological function. *J. Phys.* 590 (6): 1313–1330. Review.

167 Ventura, C., Maioli, M., Pintus, G. et al. (1998). Nuclear opioid receptors activate opioid peptide gene transcription in isolated myocardial nuclei. *J. Biol. Chem.* 273 (22): 13383–13386.

168 Bruchas, M.R., Macey, T.A., Lowe, J.D. et al. (2006). Kappa opioid receptor activation of p38 MAPK is GRK3- and arrestin-dependent in neurons and astrocytes. *J. Biol. Chem.* 281 (26): 18081–18089.

169 Khorram-Manesh, A., Nordlander, S., Novotny, A. et al. (2009). Nuclear expression of μ -opioid receptors in a human mesothelial cell line. *Auton. Autacoid. Pharmacol.* 29 (4): 165–170.

170 Milasta, S., Evans, N.A., Ormiston, L. et al. (2005). The sustainability of interactions between the orexin-1 receptor and β-arrestin-2 is defined by a single C-terminal cluster of hydroxy amino acids and modulates the kinetics of ERK MAPK regulation. *Biochem. J.* 387 (Pt 3): 573–584.

171 Dalrymple, M.B., Jaeger, W.C., Eidne, K.A. et al. (2011). Temporal profiling of orexin receptor-arrestin-ubiquitin complexes reveals differences between receptor subtypes. *J. Biol. Chem.* 286 (19): 16726–16733.

172 Tafesse, F.G., Guimaraes, C.P., Maruyama, T. et al. (2014). GPR107, a G-protein-coupled receptor essential for intoxication by *Pseudomonas aeruginosa* exotoxin A, localizes to the Golgi and is cleaved by furin. *J. Biol. Chem.* 289 (35): 24005–24018.

173 Gan, L., Seki, A., Shen, K. et al. (2019). The lysosomal GPCR-like protein GPR137B regulates Rag and $mTORC_1$ localization and activity. *Nat. Cell Biol.* 21 (5): 614–626.

174 Schiaffino, M.V. (2010). Signaling pathways in melanosome biogenesis and pathology. *Int. J. Biochem. Cell Biol.* 42 (7): 1094–1104. Review.

175 Young, A., Jiang, M., Wang, Y. et al. (2011). Specific interaction of $G\alpha_{i3}$ with the Oa1 G-protein coupled receptor controls the size and density of melanosomes in retinal pigment epithelium. *PLoS One* 6 (9): e24376.

176 Belous, A., Wakata, A., Knox, C.D. et al. (2004). Mitochondrial P2Y-Like receptors link cytosolic adenosine nucleotides to mitochondrial calcium uptake. *J. Cell. Biochem.* 92 (5): 1062–1073.

177 Krzeminski, P., Misiewicz, I., Pomorsk, i.P. et al. (2007). Mitochondrial localization of P2Y1, P2Y2 and P2Y12 receptors in rat astrocytes and glioma C6 cells. *Brain Res. Bull.* 71 (6): 587–592.

178 Smith, T.H., Li, J.G., Dores, M.R. et al. (2017). Protease-activated receptor-4 and purinergic receptor P2Y12 dimerize, co-internalize, and activate Akt signaling via endosomal recruitment of β-arrestin. *J. Biol. Chem.* 292 (33): 13867–13878.

179 Belous, A.E., Jones, C.M., Wakata, A. et al. (2006). Mitochondrial calcium transport is regulated by P2Y1- and P2Y2-like mitochondrial receptors. *J. Cell. Biochem.* 99 (4): 1165–1174.

180 Pickard, B.W., Hodsman, A.B., Fraher, L.J. et al. (2006). Type 1 parathyroid hormone receptor (PTH1R) nuclear trafficking: association of PTH1R with importin α_1 and β. *Endocrinology* 147 (7): 3326–3332.

181 Ferrandon, S., Feinstein, T.N., Castro, M. et al. (2009). Sustained cyclic AMP production by parathyroid hormone receptor endocytosis. *Nat. Chem. Biol.* 5 (10): 734–742.

182 Watson, P.H., Fraher, L.J., Hendy, G.N. et al. (2000). Nuclear localization of the type 1 PTH/PTHrP receptor in rat tissues. *J. Bone Miner. Res.* 15 (6): 1033–1044.

183 Watson, P.H., Fraher, L.J., Natale, B.V. et al. (2000). Nuclear localization of the type 1 parathyroid hormone/parathyroid hormone-related peptide receptor in MC3T3-E1 cells: association with serum-induced cell proliferation. *Bone* 26 (3): 221–225.

184 Faucheux, C., Horton, M.A., Price, J.S. et al. (2002). Nuclear localization of type I parathyroid hormone/parathyroid hormone-related protein receptors in deer antler osteoclasts: evidence for parathyroid hormone-related protein and receptor activator of NF-κB-dependent effects on osteoclast formation in regenerating mammalian bone. *J. Bone Miner. Res.* 17 (3): 455–464.

185 Bhosle, V.K., Rivera, J.C., Zhou, T.E. et al. (2016). Nuclear localization of platelet-activating factor receptor controls retinal neovascularization. *Cell Discov.* 2: 16017.

186 Marrache, A.M., Gobeil, F. Jr., Bernier, S.G. et al. (2002). Proinflammatory gene induction by platelet-activating factor mediated via its cognate nuclear receptor. *J. Immunol.* 169 (11): 6474–6481.

187 Bhattacharya, M., Peri, K.G., Almazan, G. et al. (1998). Nuclear localization of prostaglandin E_2 receptors. *Proc. Natl. Acad. Sci. U.S.A.* 95 (26): 15792–15797.

188 Bhattacharya, M., Peri, K., Ribeiro-da-Silva, A. et al. (1999). Localization of functional prostaglandin E_2 receptors EP_3 and EP_4 in the nuclear envelope. *J. Biol. Chem.* 274: 15719–15724.

189 Provost, C., Choufani, F., Avedanian, L. et al. (2010). Nitric oxide and reactive oxygen species in the nucleus revisited. *Can. J. Physiol. Pharmacol.* 88 (3): 296–304. Review.

190 Gobeil, F. Jr., Dumont, I., Marrache, A.M. et al. (2002). Regulation of eNOS expression in brain endothelial cells by perinuclear EP_3 receptors. *Circ. Res.* 90 (6): 682–689.

191 Kim, J. and Shim, M. (2015). Prostaglandin F2α receptor (FP) signaling regulates Bmp signaling and promotes chondrocyte differentiation. *Biochim. Biophys. Acta* 1853 (2): 500–512.

192 Mir, F. and Le Breton, G.C. (2008). A novel nuclear signaling pathway for thromboxane A_2 receptors in oligodendrocytes: evidence for signaling compartmentalization during differentiation. *Mol. Cell. Biol.* 28 (20): 6329–6341.

193 Ramamurthy, S., Mir, F., Gould, R.M. et al. (2006). Characterization of thromboxane A_2 receptor signaling in developing rat oligodendrocytes: nuclear receptor localization and stimulation of myelin basic protein expression. *J. Neurosci. Res.* 84 (7): 1402–1414.

194 Dores, M.R., Chen, B., Lin, H. et al. (2012). ALIX binds a YPX(3)L motif of the GPCR PAR_1 and mediates ubiquitin-independent ESCRT-III/MVB sorting. *J. Cell Biol.* 197 (3): 407–419.

195 Grimsey, N., Lin, H., and Trejo, J. (2014). Endosomal signaling by protease-activated receptors. *Methods Enzymol.* 535: 389–401. Review.

196 Joyal, J.S., Nim, S., Zhu, T. et al. (2014). Subcellular localization of coagulation factor II receptor-like 1 in neurons governs angiogenesis. *Nat. Med.* 20: 1165–1173.

197 DeFea, K.A., Zalevsky, J., Thoma, M.S. et al. (2000). Beta-arrestin-dependent endocytosis of proteinase-activated receptor 2 is required for intracellular targeting of activated ERK1/2. *J. Cell Biol.* 148 (6): 1267–1281.

198 Stalheim, L., Ding, Y., Gullapalli, A. et al. (2005). Multiple independent functions of arrestins in the regulation of protease-activated receptor-2 signaling and trafficking. *Mol. Pharmacol.* 67 (1): 78–87.

199 Jimenez-Vargas, N.N., Pattison, L.A., Zhao, P. et al. (2018). Protease-activated receptor-2 in endosomes signals persistent pain of irritable bowel syndrome. *Proc. Natl. Acad. Sci. U.S.A.* 115 (31): E7438–E7447.

200 Kumar, U., Sasi, R., Suresh, S. et al. (1999). Subtype-selective expression of the five somatostatin receptors (hSSTR1-5) in human pancreatic islet cells: a quantitative double-label immunohistochemical analysis. *Diabetes* 48 (1): 77–85.

201 Stroh, T., Sarret, P., Tannenbaum, G.S. et al. (2006). Immunohistochemical distribution and subcellular localization of the somatostatin receptor subtype 1 (sst1) in the rat hypothalamus. *Neurochem. Res.* 31 (2): 247–257.

202 Csaba, Z., Lelouvier, B., Viollet, C. et al. (2007). Activated somatostatin type 2 receptors traffic *in vivo* in central neurons from dendrites to the *trans* Golgi before recycling. *Traffic* 8 (7): 820–834.

203 Lelouvier, B., Tamagno, G., Kaindl, A.M. et al. (2008). Dynamics of somatostatin type 2A receptor cargoes in living hippocampal neurons. *J. Neurosci.* 28 (17): 4336–4349.

204 Bauch, C., Koliwer, J., Buck, F. et al. (2014). Subcellular sorting of the G-protein coupled mouse somatostatin receptor 5 by a network of PDZ-domain containing proteins. *PLoS One* 9 (2): e88529.

205 Ruscica, M., Magni, P., Steffani, L. et al. (2014). Characterization and sub-cellular localization of SS1R, SS2R, and SS5R in human late-stage prostate cancer cells: effect of mono- and bi-specific somatostatin analogs on cell growth. *Mol. Cell. Endocrinol.* 382 (2): 860–870.

206 Jensen, D.D., Lieu, T., Halls, M.L. et al. (2017). Neurokinin 1 receptor signaling in endosomes mediates sustained nociception and is a viable therapeutic target for prolonged pain relief. *Sci. Transl. Med.* 9 (392): eaal3447.

207 Boer, P.A. and Gontijo, J.A. (2006). Nuclear localization of SP, CGRP, and NK_1R in a subpopulation of dorsal root ganglia subpopulation cells in rats. *Cell Mol. Neurobiol.* 26 (2): 191–207.

208 Lessard, A., Savard, M., and Gobeil, F. Jr. (2009). The neurokinin-3 (NK_3) and the neurokinin-1 (NK_1) receptors are differentially targeted to mesocortical and mesolimbic projection neurons and to neuronal nuclei in the rat ventral tegmental area. *Synapse* 63 (6): 484–501.

209 Hether, S., Misono, K., and Lessard, A. (2013). The neurokinin-3 receptor (NK_3R) antagonist SB222200 prevents the apomorphine-evoked surface but not nuclear NK_3R redistribution in dopaminergic neurons of the rat ventral tegmental area. *Neuroscience* 247: 12–24.

210 Sladek, F.M. (2011). What are nuclear receptor ligands? *Mol. Cell. Endocrinol.* 334 (1, 2): 3–13. Review.

211 Miklos, Z., Flynn, F.W., and Lessard, A. (2014). Stress-induced dendritic internalization and nuclear translocation of the neurokinin-3 (NK_3) receptor in vasopressinergic profiles of the rat paraventricular nucleus of the hypothalamus. *Brain Res.* 1590: 31–44.

212 Cisneros, I.E. and Ghorpade, A. (2014). Methamphetamine and HIV-1-induced neurotoxicity: role of trace amine associated receptor 1 cAMP signaling in astrocytes. *Neuropharmacology* 85: 499–507.

213 Pitts, M.S., McShane, J.N., Hoener, M.C. et al. (2019). TAAR1 levels and sub-cellular distribution are cell line but not breast cancer subtype specific. *Histochem. Cell Biol.* 152 (2): 155–166.

214 Gainetdinov, R.R., Hoener, M.C., and Berry, M.D. (2018). Trace amines and their receptors. *Pharmacol. Rev.* 70 (3): 549–620. Review.

215 Underhill, S.M., Hullihen, P.D., Chen, J. et al. (2019). Amphetamines signal through intracellular TAAR1 receptors coupled to $G_{\alpha 13}$ and $G_{\alpha}S$ in discrete subcellular domains. *Mol. Psychiatry* https://doi.org/10.1038/s41380-019-0469-2.

216 Nguyen, A.H., Thomsen, A.R.B., and Cahill, T.J. 3rd. (2019). Structure of an endosomal signaling GPCR-G protein-β-arrestin megacomplex. *Nat. Struct. Mol. Biol.* 26 (12): 1123–1131.

217 Doan, N.D., Nguyen, T.T., Létourneau, M. et al. (2012). Biochemical and pharmacological characterization of nuclear urotensin-II binding sites in rat heart. *Br. J. Pharmacol.* 166 (1): 243–257.

218 Nguyen, T.T., Létourneau, M., Chatenet, D. et al. (2012). Presence of urotensin-II receptors at the cell nucleus: specific tissue distribution and hypoxia-induced modulation. *Int. J. Biochem. Cell Biol.* 44 (4): 639–647.

219 Brulé, C., Perzo, N., Joubert, J.E. et al. (2014). Biased signaling regulates the pleiotropic effects of the urotensin II receptor to modulate its cellular behaviors. *FASEB J.* 28 (12): 5148–5162.

220 Di Benedetto, A., Sun, L., Zambonin, C.G. et al. (2014). Osteoblast regulation via ligand-activated nuclear trafficking of the oxytocin receptor. *Proc. Natl. Acad. Sci. U.S.A.* 111 (46): 16502–16507.

221 Sun, L., Tamma, R., Yuen, T. et al. (2016). Functions of vasopressin and oxytocin in bone mass regulation. *Proc. Natl. Acad. Sci. U.S.A.* 113 (1): 164–169.

222 Feinstein, T.N., Yu, i.N., Webber, M.J. et al. (2013). Noncanonical control of vasopressin receptor type 2 signaling by retromer and arrestin. *J. Biol. Chem.* 288 (39): 27849–27860.

223 Merriam, L.A., Baran, C.N., Girard, B.M. et al. (2013). Pituitary adenylate cyclase 1 receptor internalization and endosomal signaling mediate the pituitary adenylate cyclase activating polypeptide-induced increase in guinea pig cardiac neuron excitability. *J. Neurosci.* 33 (10): 4614–4622.

224 Tompkins, J.D., Clason, T.A., Hardwick, J.C. et al. (2016). Activation of MEK/ERK signaling contributes to the PACAP-induced increase in guinea pig cardiac neuron excitability. *Am. J. Physiol. Cell Physiol.* 311 (4): C643–C651.

225 Tompkins, J.D., Clason, T.A., Buttolph, T.R. et al. (2018). Src family kinase inhibitors blunt PACAP-induced PAC_1 receptor endocytosis, phosphorylation of ERK, and the increase in cardiac neuron excitability. *Am. J. Physiol. Cell Physiol.* 314 (2): C233–C241.

226 Barbarin, A., Séité, P., Godet, J. et al. (2014). Atypical nuclear localization of VIP receptors in glioma cell lines and patients. *Biochem. Biophys. Res. Commun.* 454 (4): 524–530.

227 Valdehita, A., Bajo, A.M., Fernández-Martínez, A.B. et al. (2010). Nuclear localization of vasoactive intestinal peptide (VIP) receptors in human breast cancer. *Peptides* 31 (11): 2035–2045.

228 Yu, R., Liu, H., Peng, X. et al. (2017). The palmitoylation of the N-terminal extracellular Cys37 mediates the nuclear translocation of $VPAC_1$ contributing to its anti-apoptotic activity. *Oncotarget* 8 (26): 42728–42741.

229 Goetzl, E.J. (2006). Hypothesis: VPAC G protein-coupled receptors for vasoactive intestinal peptide constitute a dynamic system for signaling T cells from plasma membrane and nuclear membrane complexes. *Regul. Pept.* 137 (1-2): 75–78.

230 Cochaud, S., Chevrier, L., Meunier, A.C. et al. (2010). The vasoactive intestinal peptide-receptor system is involved in human glioblastoma cell migration. *Neuropeptides* 44 (5): 373–383.

231 Lagerström, M.C. and Schiöth, H.B. (2008). Structural diversity of G protein-coupled receptors and significance for drug discovery. *Nat. Rev. Drug Discovery* 7 (4): 339–357.

232 Maßberg, D. and Hatt, H. (2018). Human olfactory receptors: novel cellular functions outside of the nose. *Physiol. Rev.* 98 (3): 1739–1763.

233 Sposini, S. and Hanyaloglu, A.C. (2017). Spatial encryption of G protein-coupled receptor signaling in endosomes: mechanisms and applications. *Biochem. Pharmacol.* 143: 1–9. Review.

234 Hanyaloglu, A.C. (2018). Advances in membrane trafficking and endosomal signaling of G protein-coupled receptors. *Int. Rev. Cell Mol. Biol.* 339: 93–131. Review.

235 Thomsen, A.R.B., Jensen, D.D., Hicks, G.A. et al. (2018). Therapeutic targeting of endosomal G-protein-coupled receptors. *Trends Pharmacol. Sci.* 39 (10): 879–891. Review.

236 Ulloa-Aguirre, A. and Conn, P.M. (2009). Targeting of G protein-coupled receptors to the plasma membrane in health and disease. *Front. Biosci. (Landmark Ed).* 14: 973–994.

237 Wu, G. (2012). Regulation of post-Golgi traffic of G protein-coupled receptors. *Subcell Biochem.* 63: 83–95.

238 Pfeffer, S.R. (2017). Rab GTPases: master regulators that establish the secretory and endocytic pathways. *Mol. Biol. Cell* 28 (6): 712–715.

239 Zhuang, X., Chowdhury, S., Northup, J.K. et al. (2010). Sar-1 dependent trafficking of human calcium receptor to the cell surface. *Biochem. Biophys. Res. Commun.* 396 (4): 874–880.

240 Wu, G., Zhao, G., and He, Y. (2003). Distinct pathways for the trafficking of angiotensin II and adrenergic receptors from the endoplasmic reticulum to the cell surface: Rab1-independent transport of a G protein-coupled receptor. *J. Biol. Chem.* 278 (47): 47062–47069.

241 Ray, K. (2015). Calcium-sensing receptor: trafficking, endocytosis, recycling, and importance of interacting proteins. *Prog. Mol. Biol. Transl. Sci.* 132: 127–150.

242 Frasa, M.A., Koessmeier, K.T., Ahmadian, M.R. et al. (2012). Illuminating the functional and structural repertoire of human TBC/RABGAPs. *Nat. Rev. Mol. Cell Biol.* 13 (2): 67–73.

243 Wei, Z., Zhang, M., Li, C. et al. (2019). Specific TBC domain-containing proteins control the ER-golgi-plasma membrane trafficking of GPCRs. *Cell Rep.* 28 (2): 554–566.e4.

244 Dong, C., Filipeanu, C.M., Duvernay, M.T. et al. (2007). Regulation of G protein-coupled receptor export trafficking. *Biochim. Biophys. Acta* 1768 (4): 853–870.

245 Duvernay, M.T., Dong, C., Zhang, X. et al. (2009). Anterograde trafficking of G protein-coupled receptors: function of the C-terminal F(X)6LL motif in export from the endoplasmic reticulum. *Mol. Pharmacol.* 75 (4): 751–761.

246 Wang, G., Wei, Z., and Wu, G. (2018). Role of Rab GTPases in the export trafficking of G protein-coupled receptors. *Small GTPases* 9 (1, 2): 130–135.

247 Barlowe, C. (2003). Signals for COPII-dependent export from the ER: what's the ticket out? *Trends Cell Biol.* 13 (6): 295–299.

248 Bomberger, J.M., Parameswaran, N., and Spielman, W.S. (2012). Regulation of GPCR trafficking by RAMPs. *Adv. Exp. Med. Biol.* 744: 25–37.

249 McLatchie, L.M., Fraser, N.J., Main, M.J. et al. (1998). RAMPs regulate the transport and ligand specificity of the calcitonin-receptor-like receptor. *Nature* 393 (6683): 333–339.

250 Lusk, C.P., Blobel, G., and King, M.C. (2007). Highway to the inner nuclear membrane: rules for the road. *Nat. Rev. Mol. Cell Biol.* 8 (5): 414–420. Review.

251 Zuleger, N., Kerr, A.R., and Schirmer, E.C. (2012). Many mechanisms, one entrance: membrane protein translocation into the nucleus. *Cell. Mol. Life Sci.* 69 (13): 2205–2216. Review.

252 Katta, S.S., Smoyer, C.J., and Jaspersen, S.L. (2014). Destination: inner nuclear membrane. *Trends Cell Biol.* 24: 221–229. Review.

253 Naslavsky, N. and Caplan, S. (2018). The enigmatic endosome – sorting the ins and outs of endocytic trafficking. *J. Cell Sci.* 131 (13): jcs216499. Review.

254 Mathieu, M., Martin-Jaular, L., Lavieu, G. et al. (2019). Specificities of secretion and uptake of exosomes and other extracellular vesicles for cell-to-cell communication. *Nat. Cell Biol.* 21 (1): 9–17.

255 Merlen, C., Farhat, N., Luo, X. et al. (2013). Intracrine endothelin signaling evokes IP_3-dependent increases in nucleoplasmic Ca^{2+} in adult cardiac myocytes. *J. Mol. Cell Cardiol.* 62: 189–202.

256 Sergin, I., Jong, Y.I., Harmon, S.K. et al. (2017). Sequences within the C terminus of the metabotropic glutamate receptor 5 ($mGluR_5$) are responsible for inner nuclear membrane localization. *J. Biol. Chem.* 292 (9): 3637–3655.

257 Hinshaw, J.E., Carragher, B.O., and Milligan, R.A. (1992). Architecture and design of the nuclear pore complex. *Cell* 69 (7): 1133–1141.

258 Zuleger, N., Korfali, N., and Schirmer, E.C. (2008). Inner nuclear membrane protein transport is mediated by multiple mechanisms. *Biochem. Soc. Trans.* 36 (6): 1373–1377.

259 Morinelli, T.A., Raymond, J.R., Baldys, A. et al. (2007). Identification of a putative nuclear localization sequence within ANG II AT(1A) receptor associated with nuclear activation. *Am. J. Physiol. Cell Physiol.* 292: C1398–C1408.

260 Soniat, M. and Chook, Y.M. (2015). Nuclear localization signals for four distinct karyopherin-β nuclear import systems. *Biochem. J* 468 (3): 353–362.

261 Çağatay, T. and Chook, Y.M. (2018). Karyopherins in cancer. *Curr. Opin. Cell Biol.* 52: 30–42.

262 Don-Salu-Hewage, A.S., Chan, S.Y., McAndrews, K.M. et al. (2013). Cysteine (C)-x-C receptor 4 undergoes transportin 1-dependent nuclear localization and remains functional at the nucleus of metastatic prostate cancer cells. *PLoS One* 8: e57194.

263 Bourgeois, B., Hutten, S., Gottschalk, B. et al. (2020). Nonclassical nuclear localization signals mediate nuclear import of CIRBP. *Proc. Natl. Acad. Sci. U.S.A.* 117 (15): 8503–8514.

264 Bhosle, V.K., Rivera, J.C., and Chemtob, S. (2019). New insights into mechanisms of nuclear translocation of G-protein coupled receptors. *Small GTPases* 10 (4): 254–263.

265 Moroianu, J., Blobel, G., and Radu, A. (1995). Previously identified protein of uncertain function is karyopherin alpha and together with karyopherin beta docks import substrate at nuclear pore complexes. *Proc. Natl. Acad. Sci. U.S.A.* 92 (6): 2008–2011.

266 Takano, M., Kanoh, A., Amako, K. et al. (2014). Nuclear localization of bradykinin B_2 receptors reflects binding to the nuclear envelope protein lamin C. *Eur. J. Pharmacol.* 723: 507–514.

267 Lee, H.H., Wang, Y.N., and Hung, M.C. (2015). Non-canonical signaling mode of the epidermal growth factor receptor family. *Am. J. Cancer Res.* 5 (10): 2944–2958.

268 Wang, Y.N., Yamaguchi, H., Hsu, J.M. et al. (2010). Nuclear trafficking of the epidermal growth factor receptor family membrane proteins. *Oncogene* 29: 3997–4006.

269 Wang, Y.N., Lee, H.H., Lee, H.J. et al. (2012). Membrane-bound trafficking regulates nuclear transport of integral epidermal growth factor receptor (EGFR) and ErbB-2. *J. Biol. Chem.* 287 (20): 16869–16879.

270 Malhas, A., Goulbourne, C., and Vaux, D.J. (2011). The nucleoplasmic reticulum: form and function. *Trends Cell Biol.* 21 (6): 362–373.

271 Jorgens, D.M., Inman, J.L., Wojcik, M. et al. (2017). Deep nuclear invaginations are linked to cytoskeletal filaments – integrated bioimaging of epithelial cells in 3D culture. *J. Cell Sci.* 130 (1): 177–189.

272 de Miranda, M.C., Rodrigues, M.A., de Angelis Campos, A.C. et al. (2019). Epidermal growth factor (EGF) triggers nuclear calcium signaling through the intranuclear phospholipase C delta-4 (PLCδ4). *J. Biol. Chem.* 294 (45): 16650–16662.

273 Kuna, R.S., Girada, S.B., Asalla, S. et al. (2013). Glucagon-like peptide-1 receptor-mediated endosomal cAMP generation promotes glucose-stimulated insulin secretion in pancreatic β-cells. *Am. J. Physiol. Endocrinol. Metab.* 305 (2): E161–E170.

274 Shenoy, S.K. and Lefkowitz, R.J. (2011). β-Arrestin-mediated receptor trafficking and signal transduction. *Trends Pharmacol. Sci.* 32 (9): 521–533.

275 Bahouth, S.W. and Nooh, M.M. (2017). Barcoding of GPCR trafficking and signaling through the various trafficking roadmaps by compartmentalized signaling networks. *Cell. Signalling* 36: 42–55.

276 Tsvetanova, N.G. and von Zastrow, M. (2014). Spatial encoding of cyclic AMP signaling specificity by GPCR endocytosis. *Nat. Chem. Biol.* 10 (12): 1061–1065.

277 Reiter, E., Ahn, S., Shukla, A.K. et al. (2012). Molecular mechanism of beta-arrestin-biased agonism at seven-transmembrane receptors. *Annu. Rev. Pharmacol. Toxicol.* 52: 179–197. Review.

278 Lu, L. and Hong, W. (2014). From endosomes to the *trans*-Golgi network. *Semin. Cell Dev. Biol.* 31: 30–39. Review.

279 Calebiro, D., Sungkaworn, T., and Maiellaro, I. (2014). Real-time monitoring of GPCR/cAMP signalling by FRET and single-molecule microscopy. *Horm. Metab. Res.* 46 (12): 827–832.

280 Weinberg, Z.Y., Crilly, S.E., and Puthenveedu, M.A. (2019). Patial encoding of GPCR signaling in the nervous system. *Curr. Opin. Cell Biol.* 57: 83–89.

281 Nash, C.A., Wei, W., Irannejad, R. et al. (2019). Golgi localized β_1-adrenergic receptors stimulate Golgi PI4P hydrolysis by PLCε to regulate cardiac hypertrophy. *Elife* 8: e48167.

282 Joyal, J.S., Bhosle, V.K., and Chemtob, S. (2015). Subcellular G-protein coupled receptor signaling hints at greater therapeutic selectivity. *Expert Opin. Ther. Targets* 19: 717–721.

283 El Buri, A., Adams, D.R., Smith, D. et al. (2018). The sphingosine 1-phosphate receptor 2 is shed in exosomes from breast cancer cells and is N-terminally processed to a short constitutively active form that promotes extracellular signal regulated kinase activation and DNA synthesis in fibroblasts. *Oncotarget* 9 (50): 29453–29467.

284 Ango, F., Prézeau, L., Muller, T. et al. (2001). Agonist-independent activation of metabotropic glutamate receptors by the intracellular protein Homer. *Nature* 411: 962–965.

285 Delcourt, N., Thouvenot, E., Chanrion, B. et al. (2007). PACAP type I receptor transactivation is essential for IGF-1 receptor signalling and antiapoptotic activity in neurons. *EMBO J.* 26: 1542–1551.

286 Dean, T., Vilardaga, J.P., Potts, J.T. Jr. et al. (2008). Altered selectivity of parathyroid hormone (PTH) and PTH-related protein (PTHrP) for distinct conformations of the PTH/PTHrP Receptor. *Mol. Endocrinol.* 22 (1): 156–166.

287 Kinsey, C.G., Bussolati, G., Bosco, M. et al. (2007). Constitutive and ligand-induced nuclear localization of oxytocin receptor. *J. Cell. Mol. Med.* 11: 96–110.

288 Santos, M.F., Rappa, G., Karbanová, J. et al. (2018). VAMP-associated protein-A and oxysterol-binding protein-related protein 3 promote the entry of late endosomes into the nucleoplasmic reticulum. *J. Biol. Chem.* 293 (36): 13834–13848.

289 Santos, M.F., Rappa, G., Karbanová, J. et al. (2019). Anti-human CD9 antibody Fab fragment impairs the internalization of extracellular vesicles and the nuclear transfer of their cargo proteins. *J. Cell. Mol. Med.* 23 (6): 4408–4421.

290 Jong, Y.I. and O'Malley, K.L. (2017). Mechanisms associated with activation of intracellular metabotropic glutamate receptor, $mGluR_5$. *Neurochem. Res.* 42: 166–172.

291 Casares, D., Escribá, P.V., and Rosselló, C.A. (2019). Membrane lipid composition: effect on membrane and organelle structure, function and compartmentalization and therapeutic avenues. *Int. J. Mol. Sci.* 20 (9): 2167. Review.

292 Falomir-Lockhart, L.J., Cavazzutti, G.F., Giménez, E. et al. (2019). Fatty acid signaling mechanisms in neural cells: fatty acid receptors. *Front Cell Neurosci.* 13: 162. Review.

293 Hara, T. (2016). Ligands at free fatty acid receptor 1 (GPR40). In: *Free Fatty Acid Receptors*, Handbook of Experimental Pharmacology, vol. 236 (ed. G. Milligan and I. Kimura), 1–16. Cham: Springer.

294 Kasonga, A.E., Kruger, M.C., and Coetzee, M. (2019). Free fatty acid receptor 4-β-arrestin 2 pathway mediates the effects of different classes of unsaturated fatty acids in osteoclasts and osteoblasts. *Biochim. Biophys. Acta, Mol. Cell. Biol. Lipids* 1864 (3): 281–289.

295 Leung, C.C.Y. and Wong, Y.H. (2017). Role of G protein-coupled receptors in the regulation of structural plasticity and cognitive function. *Molecules* 22 (7): 1239. Review.

296 Rozenfeld, R. and Devi, L.A. (2008). Regulation of CB_1 cannabinoid receptor trafficking by the adaptor protein AP-3. *FASEB J.* 22 (7): 2311–2322.

297 Koch, M., Varela, L., Kim, J.G. et al. (2015). Hypothalamic POMC neurons promote cannabinoid-induced feeding. *Nature* 519 (7541): 45–50.

298 Kim, J. and Li, Y. (2015). Chronic activation of CB_2 cannabinoid receptors in the hippocampus increases excitatory synaptic transmission. *J. Phys.* 593 (4): 871–886.

299 Li, Y. and Kim, J. (2016). Deletion of CB_2 cannabinoid receptors reduces synaptic transmission and long-term potentiation in the mouse hippocampus. *Hippocampus* 26 (3): 275–281.

300 Stempel, A.V., Stumpf, A., Zhang, H.Y. et al. (2016). Cannabinoid type 2 receptors mediate a cell type-specific plasticity in the hippocampus. *Neuron* 90 (4): 795–809.

301 Zhu, T., Gobeil, F., Vazquez-Tello, A. et al. (2006). Intracrine signaling through lipid mediators and their cognate nuclear G-protein-coupled receptors: a paradigm based on PGE_2, PAF, and LPA_1 receptors. *Can. J. Physiol. Pharmacol.* 84 (3, 4): 377–391.

302 Martelli, A.M., Evangelisti, C., Nyakern, M. et al. (2006). Nuclear protein kinase C. *Biochim. Biophys. Acta* 1761 (5, 6): 542–551. Review.

303 Visnjic, D. and Banfic, H. (2007). Nuclear phospholipid signaling: phosphatidylinositol-specific phospholipase C and phosphoinositide 3-kinase. *Pflugers Arch.* 455 (1): 19–30. Review.

304 Tresguerres, M., Levin, L.R., and Buck, J. (2011). Intracellular cAMP signaling by soluble adenylyl cyclase. *Kidney Int.* 79 (12): 1277–1288. Review.

305 Campden, R., Audet, N., and Hébert, T.E. (2015). Nuclear G protein signaling: new tricks for old dogs. *J. Cardiovasc. Pharmacol.* 65 (2): 110–122. Review.

306 Romanauska, A. and Köhler, A. (2018). The inner nuclear membrane is a metabolically active territory that generates nuclear lipid droplets. *Cell* 174 (3): 700–715.e18.

307 Peterson, Y.K. and Luttrell, L.M. (2017). The diverse roles of arrestin scaffolds in G protein-coupled receptor signaling. *Pharmacol. Rev.* 69 (3): 256–297.

308 López de Maturana, R. and Sánchez-Pernaute, R. (2010). Regulation of corticostriatal synaptic plasticity by G protein-coupled receptors. *CNS Neurol. Disord. Drug Targets.* 9 (5): 601–615.

309 Bramham, C.R., Worley, P.F., Moore, M.J. et al. (2008). The immediate early gene Arc/Arg3.1. Regulation, mechanisms, and function. *J. Neurosci.* 28 (46): 11760–11767.

310 Chowdhury, S., Shepherd, J.D., Okuno, H. et al. (2006). Arc/Arg3.1 interacts with the endocytic machinery to regulate AMPA receptor trafficking. *Neuron* 52 (3): 445–459.

311 Rial Verde, E.M., Lee-Osbourne, J., Worley, P.F. et al. (2006). Increased expression of the immediate-early gene Arc/Arg3.1 reduces AMPA receptor-mediated synaptic transmission. *Neuron* 52 (3): 461–474.

312 Shepherd, J.D., Rumbaugh, G., Wu, J. et al. (2006). Arc/Arg3.1 mediates homeostatic synaptic scaling of AMPA receptors. *Neuron* 52 (3): 475–484.

313 Park, S., Park, J.M., Kim, S. et al. (2008). Elongation factor 2 and fragile X mental retardation protein control the dynamic translation of Arc/Arg3.1 essential for mGluR-LTD. *Neuron* 59 (1): 70–83.

314 Waung, M.W., Pfeiffer, B.E., Nosyreva, E.D. et al. (2008). Rapid translation of Arc/Arg3.1 selectively mediates mGluR-dependent long term depression (LTD) through persistent increases in AMPAR endocytosis rate. *Neuron* 59 (1): 84–97.

315 Vaudry, D., Falluel-More, I.A., Bourgault, S. et al. (2009). Pituitary adenylate cyclase-activating polypeptide and its receptors: 20 years after the discovery. *Pharmacol. Rev.* 61 (3): 283–357.

316 Omary, M.B. and Kagnoff, M.F. (1987). Identification of nuclear receptors for VIP on a human colonic adenocarcinoma cell line. *Science* 238 (4833): 1578–1581.

317 Schmid, C.L., Kennedy, N.M., Ross, N.C. et al. (2017). Bias factor and therapeutic window correlate to predict safer opioid analgesics. *Cell* 171 (5): 1165–1175.e13.

318 Bohn, L.M., Lefkowitz, R.J., Gainetdinov, R.R. et al. (1999). Enhanced morphine analgesia in mice lacking β-arrestin2. *Science* 286 (5449): 2495–2498.

319 Bohn, L.M., Gainetdinov, R.R., Lin, F.T. et al. (2000). Mu-opioid receptor desensitization by β-arrestin2 determines morphine tolerance but not dependence. *Nature* 408 (6813): 720–723.

320 Grim, T.W., Acevedo-Canabal, A., and Bohn, L.M. (2020). Toward directing opioid receptor signaling to refine opioid therapeutics. *Biol. Psychiatry* 87 (1): 15–21.

321 Hendrikse, E.R., Liew, L.P., Bower, R.L. et al. (2020). Identification of small-molecule positive modulators of calcitonin-like receptor-based receptors. *ACS Pharmacol. Transl. Sci.* 3 (2): 305–320.

322 Mancini, A.D., Bertrand, G., Vivot, K. et al. (2015). β-arrestin recruitment and biased agonism at free fatty acid receptor 1. *J. Biol. Chem.* 290 (34): 21131–21140.

323 Verma, M.K., Sadasivuni, M.K., Yateesh, A.N. et al. (2014). Activation of GPR40 attenuates chronic inflammation induced impact on pancreatic β-cells health and function. *BMC Cell Biol.* 15: 24.

324 Abadir, P.M., Walston, J.D., and Carey, R.M. (2012). Subcellular characteristics of functional intracellular renin-angiotensin systems. *Peptides* 38: 437–445.

325 Ellis, B., Li, X.C., Miguel-Qin, E. et al. (2012). Evidence for a functional intracellular angiotensin system in the proximal tubule of the kidney. *Am. J. Physiol. Regul. Integr. Comp. Physiol.* 302 (5): R494–R509. Review.

326 Jong, Y.I., Harmon, S.K., and O'Malley, K.L. (2019). Location and cell-type-specific bias of metabotropic glutamate receptor, $mGlu_5$, negative allosteric modulators. *ACS Chem Neurosci.* 10 (11): 4558–4570.

327 Deliu, E., Brailoiu, G.C., Eguchi, S. et al. (2014). Direct evidence of intracrine angiotensin II signaling in neurons. *Am. J. Physiol. Cell Physiol.* 306 (8): C736–C744.

328 Deliu, E., Brailoiu, G.C., Mallilankaraman, K. et al. (2012). Intracellular endothelin type B receptor-driven Ca^{2+} signal elicits nitric oxide production in endothelial cells. *J. Biol. Chem.* 287 (49): 41023–41031.

329 Lappano, R. and Maggiolini, M. (2011). G protein-coupled receptors: novel targets for drug discovery in cancer. *Nat. Rev. Drug Discovery* 10 (1): 47–60.

330 Goshima, Y., Nakamura, F., Masukawa, D. et al. (2014). Cardiovascular actions of DOPA mediated by the gene product of ocular albinism 1. *J. Pharmacol. Sci.* 126 (1): 14–20.

331 De Filippo, E., Schiedel, A.C., and Manga, P. (2017). Interaction between G protein-coupled receptor 143 and tyrosinase: implications for understanding ocular albinism type 1. *J. Invest. Dermatol.* 137 (2): 457–465.

332 De Mello, W.C. (2014). Beyond the circulating renin-angiotensin aldosterone system. *Front. Endocrinol. (Lausanne).* 5: 104.

333 Tadevosyan, A., Villeneuve, L.R., Fournier, A. et al. (2016). Caged ligands to study the role of intracellular GPCRs. *Methods* 92: 72–77.

334 Navari, R.M. and Schwartzberg, L.S. (2018). Evolving role of neurokinin 1-receptor antagonists for chemotherapy-induced nausea and vomiting. *Onco. Targets Ther.* 11: 6459–6478.

335 Stewart, M.P., Sharei, A., Ding, X. et al. (2016). *In vitro* and ex vivo strategies for intracellular delivery. *Nature* 538 (7624): 183–192.

336 Cohen, O. and Granek, R. (2014). Nucleus-targeted drug delivery: theoretical optimization of nanoparticles decoration for enhanced intracellular active transport. *Nano Lett.* 14 (5): 2515–2521.

337 Wang, F., Wang, Y., Zhang, X. et al. (2014). Recent progress of cell-penetrating peptides as new carriers for intracellular cargo delivery. *J. Controlled Release* 174: 126–136.

338 Ye, J., Liu, E., Yu, Z. et al. (2016). CPP-assisted intracellular drug delivery, what is next? *Int. J. Mol. Sci.* 17 (11): 1892.

339 Sun, J., Liu, Y., Ge, M. et al. (2017). A distinct endocytic mechanism of functionalized-silica nanoparticles in breast cancer stem cells. *Sci. Rep.* 7 (1): 16236–16249.

Part II

Structures and Structure-Based Drug Design

9

Ten Years of GPCR Structures

Michael A. Hanson[1] and Maria C. Orencia[2]

[1] *SB SciTech, San Marcos, CA, USA*
[2] *GPCR Consortium, San Marcos, CA, USA*

9.1 Introduction

G-protein coupled receptors (GPCRs) are a fascinating class of targets. They are the most abundant protein family in the human genome and represent over 35% of the drugged targets in the modern-day pharmacopeia [1]. In just 40 years, the status of GPCRs has progressed from institutionalized doubt regarding their existence [2] to a multi-faceted field with sub-specialties in biophysical analysis, pharmacology, medicinal chemistry, structural biology and cell biology [3, 4]. They are the subjects of extensive research and funding efforts and are prized targets for pharmaceutical intervention [1].

In 1985, the first high-resolution crystal structure of a membrane protein, the photosynthetic reaction center, broke new ground in structural biology and now serves as the foundation of membrane protein structural biology [5]. While additional milestones in membrane protein structural biology were to follow [6] structural interrogation of the GPCR family was also getting underway [7]. While not a GPCR, bacteriorhodopsin does have a seven-transmembrane helical bundle and was used as a structural and methodological template for GPCRs. Initially solved at low resolution using electron crystallography and then at higher resolution using X-ray diffraction [8, 9], the efforts on this bacterial protein often presage techniques applicable for GPCRs. The first success with GPCRs came with the structure of bovine rhodopsin [10] which brought the possibility of structure-function analysis on the GPCR family to the brink of realization. However, it would not be straightforward applying the techniques used on rhodopsin and other membrane proteins to the GPCR family in general, and another seven years would elapse before additional structural details would emerge. Developed

GPCRs as Therapeutic Targets, Volume 1, First Edition. Edited by Annette Gilchrist.

during this time, technological advancements in crystallization, diffraction, and protein engineering combined, culminating in the "Golden Age" of GPCR structural biology that we currently enjoy [11] (Figure 9.1).

Now, twenty years after the first rhodopsin structure, a generalized toolbox is in place, and years of accumulated knowledge on handling membrane proteins are available. Researchers around the globe have utilized this set of techniques as a foundation for evolving new structure determination strategies. The result is a catalog of structures of important GPCR drug targets available to aid in the discovery of novel biology and superior drugs. New pharmacological paradigms are employed to reevaluate well-established pathways. Furthermore, previously undruggable receptors can now be accessed with unique chemistry based on structural insights. In this review, we will focus on the following topics:

(1) Origins of GPCR structure
(2) Early platform development to expand beyond Rhodopsin
(3) The first wave of GPCR Structural Biology
(4) Broadening the scope to agonists and complexes
(5) The impact of the GPCR Consortium
(6) A unified analysis across families
(7) The next 10 years

9.2 Origins of GPCR Structure

Membrane protein structural biology is hard. This statement is true for many reasons, but among them, two are at the forefront and are a constant in any membrane protein structural biology project:

1. Membrane proteins are generally not highly expressed in sufficient quantities needed for biophysical analysis and require extensive optimization to produce enough starting material.
2. Membrane proteins are not very stable once extracted from the lipid membrane during the purification process and therefore require a specialized lipid and detergent environment to maintain both fold and function.

In the case of GPCRs, compared with the early successes in membrane protein structural biology, there were additional difficulties that required a strong foundation of past work and over three decades to overcome.

9.2.1 Bacteriorhodopsin

The field of GPCR structural biology owes much to a distantly related, bacterial light-driven proton pump, bacteriorhodopsin [12]. Discovered in a microorganism

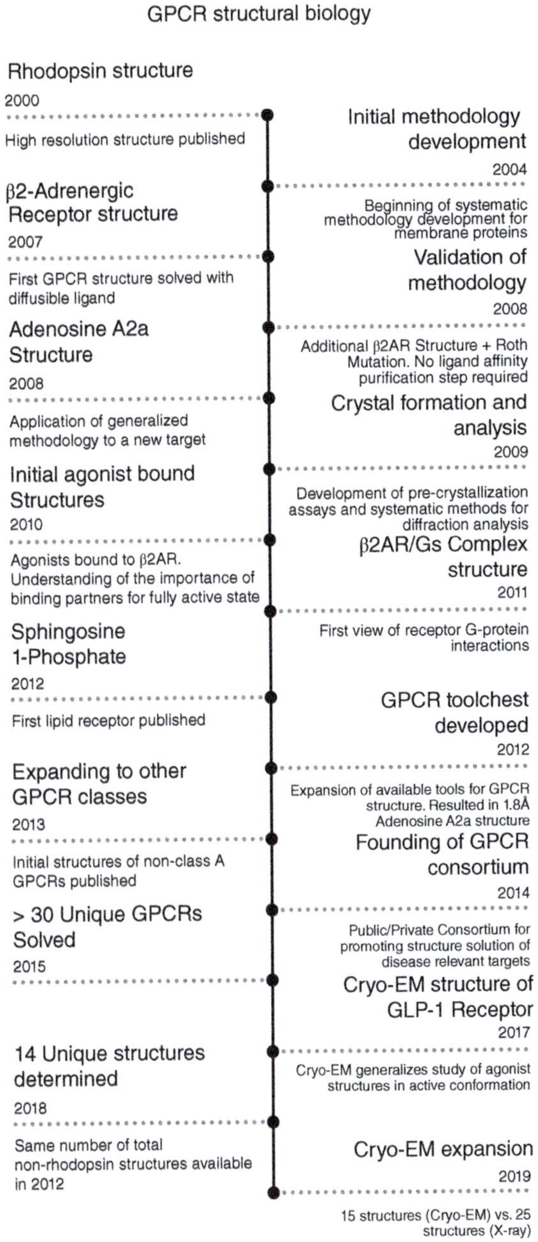

Figure 9.1 Timeline representing important events in the evolution of GPCR structural biology.

that grows under conditions of high salinity, bacteriorhodopsin possesses a striking red color and, in large amounts, is visible to the naked eye. Scholars have suggested that large numbers of these microorganisms present during the biblical event of the first plague of Egypt led to a misinterpretation of water turning to blood [13].

The abundance of the protein available in nature and the visual red coloration were essential in probing the structural characteristics of bacteriorhodopsin. While the knowledge of biochemistry and structural biology derived from these studies were critical in furthering our understanding [14]. Arguably, one of the most important outcomes from this early work on membrane protein structural biology was the development of lipidic cubic phase (LCP) crystallization [9, 15]. Modifications to this critical technology have directly led to many of the current successes in membrane protein research.

9.2.2 Rhodopsin

Researchers linked the phenomenon of vision to rhodopsin in the year 1877 when they noticed the tendency of the color of retinal membranes containing rhodopsin to bleach when exposed to light that subsequently regenerated when in the dark [16]. Initially, the primary sequence of rhodopsin provided hints about its seven-transmembrane helical architecture [17, 18]. Later, the low-resolution electron crystallography projection map revealed this feature in more detail [19]. The crystal structure of bovine rhodopsin at atomic resolution showed more specific aspects of the GPCR seven-transmembrane bundle, enabling unambiguous assignment of side-chain interactions, loop conformation, and helical orientations [10]. This achievement was genuinely groundbreaking. However, subsequent work on GPCRs made it apparent that rhodopsin occupies a highly specialized niche within the family [20], and extensive work on platform technologies was needed before additional successes in GPCR structural biology would be seen.

9.3 Early Platform Development to Expand Beyond Rhodopsin

The structure of bovine rhodopsin initiated a period of intense research on GPCR structural biology with multiple groups attempting to replicate the success. However, the critical properties of rhodopsin that are lacking in other family members prevented a direct translation of the techniques. Rhodopsin can be isolated in large quantities from native tissue, specifically, bovine eyes, which are easily accessible from the local butcher or meat packing facility [21]. After purification from the outer rod segment of the bovine retina, the protein is stable in detergent micelles, having evolved as a type of light switch to be in one of two conformations [21, 22]. This combination of factors enabled the use of techniques employed on other membrane protein structures directly for the manipulation of rhodopsin.

Non-rhodopsin GPCRs, on the other hand, have evolved much more complicated molecular pharmacology profiles. Most GPCRs sample multiple conformational states of variable intrinsic stability, allowing them to perform their cellular functions [23]. The combination of lower expression levels, lower stability, and multiple heterogeneous conformations supplied significant roadblocks for structural analysis of this target class. Four key issues first needed to be addressed to advance the field:

1. **Expression**: Low expression levels due to improper folding, post-translational modifications, or incomplete trafficking [24, 25]
2. **Purification**: Low stability due to the presence of multiple conformational states
3. **Receptor Integrity**: Monitoring continued integrity of the protein throughout the purification and crystallization process [26, 27]
4. **Crystallization**: Lack of crystals due to inferior packing interfaces [28]

It would require seven years and development of new techniques for membrane protein production, stabilization, and crystallization before the next GPCR structure would be solved. This investment of time and resources was ultimately worthwhile as it "primed the pumps" for an ensuing flood of information that has increased in magnitude every year since 2007.

9.3.1 Expression

Unlike bacteriorhodopsin and rhodopsin, the majority of GPCRs are not found naturally in sufficient quantities for biochemical analysis, and a suitable protein over-expression system was needed. At the time, most efforts focused on bacterial systems such as *Escherichia coli*, with generally disappointing results [29–31]. It was evident that the comparatively simple bacterial expression systems were not capable of reliably producing adequate levels of mammalian membrane proteins. There was more success with yeast systems for a handful of GPCRs [32]. However, yields were typically not high enough for biochemical studies. Also, the mammalian expression systems used to develop GPCR based assays were limited in scale, expensive, and not readily expandable to enable sufficient quantities of protein [25, 33–35]. These limitations occurred for a variety of reasons, including improper folding, post-translational modifications, and incomplete trafficking. While advanced protein engineering techniques can now solve many of these issues, at the time, utilization of an expression system that balanced simplicity with scalability was required. Ultimately, eukaryotic cells were necessary for proper protein trafficking and post-translational modifications [36].

A breakthrough in the field occurred with the deployment of higher eukaryotic expression systems such as insect cell expression mediated by baculovirus. The development of suitable expression vectors for baculovirus along with a general understanding of the factors involved in proper translation and trafficking of GPCRs to the cell membrane facilitated initial successes in over-expression and characterization of the protein material. Miniaturization techniques developed for bacterial expression from structural genomics initiatives were then adapted, generating GPCR expression screening platforms for higher throughput analysis of multiple constructs [26].

A streamlined process employing benchtop flow cytometers was used for rapid feedback of expression success, allowing optimization of constructs at a higher pace [26, 27]. Incorporation of expression tags, green fluorescent protein, and well-developed antibody-epitope sequences allowed differentiation between total protein expression and the amount of protein trafficking to the cell surface, a straightforward metric for assessing construct modifications [26]. The ratio between cell surface trafficked protein and total expressed material also turned out to be a reasonably good predictor of GPCR functional integrity and, by extension, relative stability during subsequent purification and crystallization experiments [27].

9.3.2 Purification

For years, the gold standard used to evaluate the functional integrity of a given receptor after purification was receptor binding to commercially available radiolabeled ligands. For this analysis to be meaningful, though, the protein needs to be purified in the absence of ligand. Many efforts, therefore, focused on stepwise purification of the material without ligand, and the remaining functional protein was measured [37–40]. The core problem with this approach is that the ligand forms an integral component of the receptor, and most GPCRs are not sufficiently stable in the absence of ligand to withstand detergent extraction and purification. The result was preparations without enough recovery of functional protein to be useful in structural studies.

Monodispersity can be a useful metric for calculating the percentage of functionally folded material in a given sample [41]. For soluble proteins, dynamic light scattering and size exclusion chromatography are commonly used [42]. For membrane proteins, however, the detergent micelles interfere with dynamic light scattering analysis, and the presence of empty micelles at a size range of 30–80 kDa quickly overwhelm any meaningful signal [43]. Size exclusion chromatography also has limitations, such as the large sample volumes often requiring an entire membrane preparation and a high degree of non-specific interactions between the protein and column material. Switching to an analytical HPLC column allows for only a small volume of sample from a protein preparation to be tested while minimizing non-specific binding interactions by washing the column with bovine serum albumin before sample injection [27]. These changes allow for routine monitoring throughout the purification process, and assays based on this technique provide a quick understanding of the stabilizing effect of different ligands and mutations on any target [44].

Benchtop flow cytometry, combined with analytical SEC techniques, enabled researchers to routinely quantify functional material throughout the purification procedure, effectively eliminating the need for radiolabeled ligand binding. Ligand could, therefore, be present at all times, and work on the β_2-adrenergic receptor showed that using it throughout the purification helped maintain the yield and integrity of the receptor [27, 45]. Efforts to optimize the detergent environment revealed that the addition of cholesterol hemisuccinate during purification further increased protein stability. This strategy of using ligand and a cholesterol analog in all steps of the purification initially tested on the β_2-adrenergic receptor [45] and confirmed to be generally applicable in work done on the adenosine A_{2a} receptor. It is now the standard in almost every protocol utilized to purify and crystallize other members of the GPCR family.

9.3.3 Receptor Integrity

Membrane protein purification requires inducing the membrane protein into a soluble state through extraction from the plasma membrane in the presence of detergent micelles. This extraction introduces an unfavorable environment and often leads to denaturation [46]. Further, these techniques require the minimization of the micelle to promote crystal packing interactions. Unfortunately, detergents with small micelles do not adequately stabilize the membrane protein. Also, chances for success increase if the membrane protein has a sizeable soluble region, which can promote crystal packing even in the presence of detergent micelles [47]. Therefore, detergent-based methods are most effective for those membrane proteins that are highly resistant to denaturation and have large soluble regions.

For the majority of GPCRs, however, the process is not straightforward. The family represents an almost worst-case scenario in the sense that they have evolved for both conformational flexibility and low intrinsic stability to perform their primary function in the cells [48, 49]. Furthermore, they often do not have significant polar surface area for supplying crystallization contacts [50]. As a result, detergents that stabilize a GPCR sufficiently for crystallization often do not allow enough contact potential for productive crystal formation [51].

A significant component of the purification optimization process during this period was an exhaustive exploration of different detergent molecules. Factors such as the size of the detergent micelle, the properties of the detergent, e.g. the critical micelle concentration, mixtures of different detergents, and the addition of phospholipids during the purification were all examined extensively [52]. The goal was to balance stabilizing the core of the protein so that it would not unfold while introducing a favorable environment for the soluble portions of the target to participate in crystal packing interactions [53]. The obstacle of finding a suitable detergent mixture remained problematic and ultimately was solved using LCP crystallography as described in the following text.

9.3.4 Crystallization

The final piece of the puzzle for a robust structural biology platform was developing a crystallization protocol. The high conformational flexibility associated with the typical GPCRs and overcoming the minimal soluble crystal packing interface were the two remaining problems to circumvent. The solution to both issues converged in the deployment of LCP crystallography while solving the structure of the β_2AR. Together with the expression, purification, and analytical methodology established over seven years, these methods became the backbone of a structure determination platform widely used for addressing the entire family of GPCRs and extended to many other classes of membrane proteins as well.

9.4 The First Wave of GPCR Structural Biology

As the tools for the platform were in development, progress was slow due to a lack of positive feedback to determine what was and what was not working. The first non-rhodopsin GPCR structure was a landmark in its own right, but it had the added importance of providing controls for future work. Each subsequent structure during this early period of platform development contributed significantly to our understanding of the process of GPCR expression, purification, and structure solution. In light of this, we will describe three of the first structures in the context of the tools utilized, highlighting the importance of each as a guidepost for efforts on new targets.

9.4.1 β2-Adrenergic Receptor

Solving the structure of β_2AR has been called a tour-de-force in protein engineering, protein chemistry, and the result of intense innovations in crystallization and diffraction analysis [54]. Protein engineering efforts attempted to solve some of the issues with GPCR crystallization, building on the idea of utilizing soluble fusion proteins as tools to increase the solubility, stability, and crystal packing surface area of a different class of membrane protein, lactose permease [55]. A turning point in the project occurred with the successful engineering of a full-length soluble protein, T4-lysozyme, into the third intracellular loop of β_2AR [56]. This modification stabilized the receptor by eliminating a large flexible loop while simultaneously increasing the crystal packing potential of the mostly hydrophobic protein. The critical insight, in the case of β_2AR, was that the junctions between the soluble protein and GPCR could be optimized to balance receptor flexibility with a loss of function as the distance between the two proteins is minimized [56].

9.4.1.1 Crystallization

Simultaneous with protein engineering, efforts toward finding an alternative crystallization platform were underway. A relatively new technique termed LCP crystallization had shown promise for test proteins but was cumbersome and not widely utilized in the field. First developed to solve the structure of bacteriorhodopsin, LCP protocols relied on protein properties of high abundance, stability, and vibrant color [9, 15]. Adaptation of protocols to less abundant, non-colored samples with varying degrees of stability would require persistence, specialized equipment, and eventually, development of a robotics system to miniaturize the process [57–61]. The successful utilization of LCP for the β_2AR structure set the stage for the increase in the availability of GPCR structures now being enjoyed in the field [62].

One of the main benefits of utilizing LCP crystallization, particularly for membrane proteins with minimal soluble surface area, is that with it the transmembrane hydrophobic regions can participate in crystal packing interactions facilitated by the LCP lipid bilayer [63]. A critical insight related to GPCR crystallization, in particular, was that this process is considerably more favorable if the lipid matrix contains a small percentage (10%) of cholesterol. Cholesterol alters the properties of LCP slightly, providing a component of native membranes that is missing and, in some cases, is necessary to promote crystal packing interfaces. Depending on the receptor, this is occasionally also important for occupying non-annular cholesterol binding sites on the GPCR surface [64].

9.4.1.2 Diffraction Analysis

The process of LCP based crystallization, however, did serve up additional technical challenges. For example, harvesting crystals required tools for delicately breaking the glass sandwich plates and a steady hand for retrieving the samples without damage. Another issue involved the lipid most commonly used for LCP crystallization. When frozen, this lipid turns completely opaque, making locating the crystal for standard cryogenic diffraction analysis a matter of guessing where it might be within the monolithic frozen mass of lipid [65]. Initially, the location and centering of the crystal for data collection was found by tediously probing for diffraction with a highly collimated and attenuated X-ray beam. The rapid deterioration of the small crystalline samples due to radiation decay allowed the collection of only partial datasets from each sample [65, 66]. Multiple samples had to be located and analyzed in this way to collect enough data for structure solution. Eventually, algorithms were developed and designed for automation of this crystal location procedure, allowing the crystallographer to merely push a button to locate the crystals in the loop [67]. Finally, stitching together all of the multiple datasets, each containing various degrees of damage due to radiative decay, required a painstaking manual triaging of diffraction images frame by frame. Replacement of this manual procedure occurred with the development of automated algorithms for data processing [68].

9.4.1.3 Structure Solution

Initially published as a relatively low resolution, partially resolved structure, β_2AR bound to the antagonist ligand carazolol was the first glimpse of a non-rhodopsin GPCR [69]. Using the LCP based crystallization method, the structure was published again one month later at a significantly higher resolution [62]. The contrast between the two structures highlights the advantages of using the LCP based crystallization method underscoring the promise of the technique for many membrane proteins. Beyond the technological advancements in structure solution and protein engineering, this structure also provided insights into the biology associated both with GPCRs in general and the β_2-adrenergic system in particular.

9.4.1.4 Structural Details

The first complete structure of the β_2AR was published in back to back Science reports [56, 62]. The structural description in the two papers focuses on the analysis of the topological details of the receptor. The researchers also observed the importance of cholesterol for facilitating crystal packing interactions and speculated on its pharmacological importance. The analysis also detailed the comparison to rhodopsin, the only other GPCR with a high-resolution structure determined at the time.

With rhodopsin as a frame of reference, the initial β_2AR structure represents a paradigm shift in how we think about the structure-function of GPCRs in general. For example, the β_2AR structure revealed for the first time that the ionic lock mechanism at the base of TM-III and TM-VI was not a common feature of inactive receptors [70, 71]. Indeed, subsequent inactive structures of other members of the family routinely fail to reproduce the ionic lock interactions postulated to be present throughout the GPCR family [72–74].

Furthermore, the β_2AR structure highlights the involvement of lipid molecules in the pharmacology of GPCRs and membrane proteins in general. In the initial β_2AR structure, the cholesterol molecule affects the crystal packing interactions and appears to mediate a parallel dimeric interface between two symmetry-related monomers. A subsequent structure of β_2AR revealed a new crystal packing lattice that was not dependent on cholesterol, establishing the pharmacological relevance of this interaction. Notably, this second structure still retains the cholesterol molecule bound to the same pocket as in the original structure allowing for the possibility of a more prominent role of cholesterol in the physiology of the receptor [45].

9.4.2 Adenosine A_{2A} Receptor

Throughout the process of generating the β_2AR structure, there were many false paths and blind alleys related to how the protein should be processed. Perhaps most importantly, was the question of whether a ligand affinity purification step was necessary [30, 51].

9.4.2.1 Generalized Purification Methodology

The logic around the requirement for a ligand affinity purification step was sound, but the implications were daunting for other GPCRs. Generally, the idea centered around protein expression as a mixture of misfolded and correctly folded material. The only way to isolate the correctly folded material for use in generating crystals was to pass the mixture over an immobilized ligand, specific for that receptor's binding pocket. The issue was that in favorable circumstances, it would require a good deal of time and optimization to develop a ligand affinity purification step for each receptor of interest; in unfavorable conditions, it would not be possible at all.

Fortunately for the field, premises underlying this logic were not accurate. In many cases, the GPCR targets were capable of expressing and trafficking to the plasma membrane in a functional state. It was the act of extracting the protein from the cell membrane with various detergents that caused the mixture of unfolded and functional protein. This realization shifted the focus from developing a ligand affinity purification step to extracting a functionally folded sample. Utilization of two primary methods accomplished this shift:

1. Extraction of the protein only after incubating with a high-affinity ligand
2. Optimization of the detergent lipid system for stability, not extraction yield or the ability to facilitate crystallization

The standardized methodology translated to other receptor systems as demonstrated with the second structure of β_2AR bound to a different antagonist [45];

more convincingly with the solution of the adenosine A2a receptor; and to significant effect with numerous structures over the next 10 years.

9.4.2.2 Structural Details

The second human GPCR target to be determined to atomic resolution provided a new view of the family and how structural adaptations within the context of the overall core fold could give sufficient flexibility to bind distinct small-molecule ligands. Specifically, a comparison of the adenosine A_{2a} receptor to β_2AR and rhodopsin revealed a structurally different extracellular region that defied modeling attempts using β_2AR or rhodopsin as a template [75]. Further, the subtle divergence of helical positions relative to β_2AR and rhodopsin demonstrated the malleability of the GPCR family's binding pocket. Combining these structural changes with the sequence differences in the binding pocket generated an orientation of the antagonist ligand perpendicular to the plane of the membrane that surprised almost all practitioners in the field at the time [76]. The idea that there was no general, family conserved, receptor-binding pocket was introduced and set the stage for understanding the wealth of diverse binding interactions driven by subtle helical shifts combined with different primary sequences across the family of GPCRs.

9.4.3 Sphingosine 1-Phosphate Receptor 1

With two successful projects completed, it was time to test the structure platform in a more commercial setting. To demonstrate the utility of the approach for more relevant drug targets, a compelling drug discovery candidate that addressed practical issues was needed. Sphingosine-1-phosphate receptor type 1 ($S1P_1$ receptor) was chosen for this purpose as it was a commercially viable target with on-going development efforts, and it had known ligands with suboptimal drug properties upon which structural insights could lead to improvements. At the time, this target had a great deal of interest in the pharmaceutical community due to Novartis' compound FTY720 entering into Phase II clinical trials for multiple sclerosis [77]. FTY720 was first isolated from natural sources and discovered as a component in traditional Chinese medicine preparations [78]. However, the compound is very lipid-like and suffers from poor pharmaceutics properties.

Protocols developed for adenosine A_{2a} receptor served as a starting point for structural studies with a few modifications. Initially, the only ligand available for purchase was a tool compound [79] with a relatively weak affinity for the receptor. Despite these unfavorable properties of the ligand, the protein readily crystallized within six months of initiating the project. The structure followed soon after, and we were ready for analysis and application to the drug discovery program at the newly founded company, Receptos.

9.4.3.1 Structural Details

The structure of the $S1P_1$ receptor was the first of a lipid-binding GPCR with an extracellular region that showed substantial structural divergence compared with the previously determined structures [68]. The extracellular organization acts to encapsulate the binding pocket, protecting it from exposure to the aqueous milieu. The packing of the N-terminus with the extracellular loops was sufficiently close that it appeared to require a novel mode of entry for the ligand through the membrane itself. In this case, the path of entry for the lipid signaling molecule is through a flexible gap between TM-I and TM-VII. Previously there had been speculation regarding this mode of entry into the binding pocket, which was now more firmly established based on structural information [80–82]. What was starting to emerge in analysis of the field was the idea that the family has adapted to bind and recognize a diverse set of ligands, partially through subtle modifications to the transmembrane region, but importantly in the adaptation of the extracellular region as well [83]. This region has a good deal more flexibility in presenting structural features that naturally complement the type of ligand.

9.4.4 Better Tools, Harder Targets

The ever-increasing rate of structure solution was a revolution in the early stages of GPCR structural analysis [84–87]. More groups were getting in on the excitement as students passed through the leading laboratories, learned the techniques, and then went on to form groups of their own, many who would later participate in The GPCR Consortium described in the following text. The projects began with targets that were easy to express and had known ligands purchased through established vendors. Through iterative cycles of breakthroughs combined with refinements and improvements to the technologies [88], the successes grew to include representatives from many of the classes and groups of the GPCR family (Table 9.1). Each effort had its own unique set of difficulties to overcome and highlighted which techniques could be improved to obtain more structures in the large GPCR family. It became apparent, though, that additional tools were needed to advance the field.

9.4.4.1 Improved Protein Engineering

The idea of utilizing T4-lysozyme as a tool for increasing crystallizable surface area had now been in use for GPCRs for four years. When it worked, it was a well-established means of generating crystal contacts between planes of the type I crystals grown in the LCP environment [112]. Unfortunately, it did not always work and seemed to be receptor-dependent in terms of its ability to stabilize and promote crystallization and high-resolution diffraction. In response to this

Table 9.1 GPCR structures determined before formation of the GPCR consortium.

Date	UniProt | family	Significance/role of receptor	Key citations	PDBid (Res. Å)
August, 2000	RHO (bovine) Class A Sensory	• First GPCR structure determined • No drugs approved or in development	[10]	1F88 (2.8)
November, 2007	ADRB2 Class A Aminergic	• First druggable GPCR structure determined • First GPCR structure determined with a diffusible ligand • The target for 49 unique drugs in 198 associated diseases • Agonists utilized in asthma, COPD and emphysema	[62] [56]	2RH1 (2.4)
July, 2008	ADRB1 (turkey) Class A Aminergic	• First GPCR to be solved with STAR technology (later utilized at Heptares) • Antagonists such as pindolol are beta-blockers used to treat hypertension and angina pectoris • The target for 31 unique drugs in 197 associated diseases • Antagonists utilized mainly in cardiovascular indications	[89]	2VT4 (2.7)
November, 2008	ADORA2a Class A Nucleotide	• The first demonstration of generalized protocols for GPCR family • Mediates important physiological role of endogenous adenosine • The target for 16 unique drugs in 125 associated diseases • Antagonists such as caffeine, theophylline utilized for migraine and asthma	[76]	3EML (2.6)
November, 2010	CXCR4 Class A Peptide	• First peptide GPCR structure • Associated with 23 types of cancers promoting metastasis, angiogenesis and tumor survival • The target for six unique drugs in 36 associated diseases • Mozobil (Plerixafor) approved drug used for hematopoietic stem cell mobilization	[90]	3ODU (2.5) 3OE6 (3.2) 3OE8 (3.1) 3OE9 (3.1) 3OE0 (2.9)

Table 9.1 (Continued)

Date	UniProt | family	Significance/role of receptor	Key citations	PDBid (Res. Å)
November, 2010	DRD3 Class A Aminergic	• Selectivity analysis among aminergic GPCRs • The target for 25 unique drugs in 91 associated diseases • Agonists utilized for Parkinson's, infertility and hyperprolactinemia • Antagonists used for schizophrenia, bipolar disorder, and unipolar depression	[91]	3PBL (2.9)
July, 2011	HRH1 Class A Aminergic	• Determination of receptor binding interactions for doxepin, a marketed antidepressant • The target for 46 unique drugs with 122 associated diseases • Antagonists utilized for allergy-related indications and cough	[32]	3RZE (3.1)
September, 2011	ADRB2 Class A Aminergic	• The first structure of a GPCR in complex with G-proteins	[92]	3SN6 (3.2)
February, 2012	S1PR1 Class A Lipid	• The first lipid activated GPCR structure • The target for six unique drugs in 13 therapeutic areas • Agonists approved for use in multiple sclerosis	[68]	3V2W (3.4) 3V2Y (2.8)
February, 2012	CHRM2 Class A Aminergic	• Back-to-back publication of the first muscarinic receptor structures • The target of 14 unique drugs with 51 associated diseases • Antagonists utilized to treat overactive bladder	[93]	3UON (3.0)
February, 2012	CHRM3 (rat) Class A Aminergic	• Back-to-back publication of the first muscarinic receptor structures • The target of 30 unique drugs with 82 associated diseases • Antagonists utilized to treat overactive bladder, hyperhidrosis, bronchitis, COPD, and emphysema	[94]	4DAJ (3.4)

(Continued)

Table 9.1 (Continued)

Date	UniProt | family	Significance/role of receptor	Key citations	PDBid (Res. Å)
March, 2012	OPRK1 Class A Peptide	• The first structure of the target class for opioid pain medications • The target of 15 unique drugs with 78 associated diseases • Utilized to treat pain and migraine	[95]	4DJH (2.9)
May, 2012	OPRM1 (mouse) Class A Peptide	• Structure analysis of the target for synthetic opioids such as morphine • The target of 43 unique drugs with 194 associated diseases • Utilized in the treatment of pain, migraine, constipation, diarrhea, and cancer	[96]	4DKL (2.8)
May, 2012	OPRL1 Class A Peptide	• Expanded analysis of opioid receptor selectivity | first novel GPCR solved with bRIL • The target of 2 unique drugs with six associated diseases • Agonists completed Phase II clinical trials for the treatment of pain, osteoarthritis and diabetic neuropathy • Antagonists completed Phase II clinical trials for the treatment of alcohol dependence and unipolar depression	[97]	4EA3 (3.0)
May, 2012	OPRD1 (mouse) Class A Peptide	• Representatives for all opioid sub-types now determined • The target of 11 unique drugs with 72 associated diseases • Utilized in the treatment of pain, migraine, constipation, diarrhea, and cancer	[98]	4EJ4 (3.4)
October, 2012	NTSR1 (rat) Class A Peptide	• First structure with endogenous peptide bound • The target of one unique drug with one associated disease • Antagonist completed Phase II for treatment of small cell lung carcinoma	[99]	4GRV (2.8)

Table 9.1 (Continued)

Date	UniProt | family	Significance/role of receptor	Key citations	PDBid (Res. Å)
November, 2012	CXCR1 Class A Peptide	• First NMR structure of a GPCR • The target of three unique drugs with eight associated diseases • Completed Phase III for treatment of type I diabetes mellitus	[100]	2LNL – NMR
December, 2012	PAR1 Class A Peptide	• Tight binding antagonist to counteract tethered ligand • The target of 3 unique drugs with 23 associated diseases • Antagonists utilized in the treatment of stroke, myocardial infarction, and peripheral arterial disease	[101]	3VW7 (2.2)
July, 2013	CRHR1 Class B Peptide	• The first class B GPCR structure • The target of 5 unique drugs with seven associated diseases • Antagonists completed Phase II clinical trials for PTSD, depression, alcohol dependence and irritable bowel syndrome	[102]	4K5Y (3.0)
May, 2013	HTR1B Class A Aminergic	• Two sub-types for serotonin receptor class bound to the same ligand enhances understanding of selectivity • The target of 13 unique drugs with 21 associated diseases • Agonists utilized in the treatment of migraine	[103]	4IAQ (2.8) 4IAR (2.7)
	HTR2B Class A Aminergic	• Two sub-types for serotonin receptor class bound to the same ligand enhances understanding of selectivity • The target of 7 unique drugs with 15 associated diseases • Partial agonists utilized in the treatment of hemorrhage		4IB4 (2.7)

(Continued)

Table 9.1 (Continued)

Date	UniProt | family	Significance/role of receptor	Key citations	PDBid (Res. Å)
May, 2013	SMO Class F Protein	• The first structure of Frizzled class • The target of 8 unique drugs with 33 associated diseases • Antagonists utilized in the treatment of basal cell carcinoma and acute myeloid leukemia	[104]	4JKV (2.5)
July, 2013	GCGR Class B Peptide	• The initial structure of an essential class of incretin peptide receptor • The target of seven unique drugs with nine associated diseases • Agonists approved for the treatment of type I diabetes mellitus	[105]	4L6R (3.3)
September, 2013	CCR5 Class A Peptide	• A new class of chemokine receptor; structure with marketed HIV drug Maraviroc bound • The target of 10 unique drugs with 27 associated diseases • Antagonists approved for the treatment of HIV infection	[106]	4MBS (2.7)
February, 2014	OPRD1 Class A Peptide	• Human OPRD1 at significantly higher resolution implicates sodium ions as modulators of function	[107]	4N6H (1.8)
April, 2014	GRM1 Class C Amino Acid	• The first class C GPCR • No drugs currently available	[108]	4OR2 (2.8)
April, 2014	$P2RY_{12}$ Class A Nucleotide	• The first structure of P2Y sub-type; in complex with three ligands • The target of seven unique drugs with 65 associated diseases • Antagonists utilized in the treatment of acute coronary syndrome	[109]	4NTJ (2.6) 4PXZ (2.5) 4PY0 (3.1)

Table 9.1 (Continued)

Date	UniProt | family	Significance/role of receptor	Key citations	PDBid (Res. Å)
September, 2014	GPR40 Class A Lipid	• Second lipid receptor structure; unexpected binding pocket • The target of two unique drugs with three associated diseases • Positive allosteric modulators have completed Phase III clinical trials for the treatment of type II diabetes mellitus	[110]	4PHU (2.3)

Note: Summary table for structures determined by the field before initiation of the GPCR Consortium. Entries that are not shaded represent structures determined by groups that would later become part of the GPCR consortium effort. Data compiled from literature and the Open Targets database populates the column designating the significance and role of the receptor [111].

shortcoming and to expand the number of accessible GPCR targets, researchers performed a systematic search for additional fusion proteins [113].

The protein databank was screened for structures of proteins to test as potential replacements. Two established receptor systems, β_2AR and adenosine A_{2a} receptor, were used in the experiments. Five new fusion proteins were engineered into the third intracellular loop of each receptor, followed by adjustments of the junctions between the fusion protein and receptor to optimize the sites. Select constructs exhibited enhanced expression and stability and these were tested in crystallization and diffraction studies. While the new fusion proteins for β_2AR were able to generate crystals and diffraction, the resolution was never equal to that found in the original structure, and the T4L fusion was indeed superior in that case.

On the other hand, the adenosine A_{2a} system preferred an alternate fusion protein, the cytochrome b_{562}RIL (bRIL). This fusion was superior in stabilizing the receptor fold and improved the resolution from 2.6 Å with the T4L fusion to 1.8 Å with the bRIL [114, 115]. Indeed, this bRIL fusion has since gone on to represent the majority of the new structures determined after its original discovery in 2011 [116, 117]. What is clear from these experiments is that screening multiple candidate fusion proteins is necessary to ensure continued success as the field progresses to harder targets.

9.4.4.2 Emergence of Mutations

Early on, mutation strategies were particularly useful for engineering targets for higher throughput structure-based drug design efforts as exemplified by the company Heptares (now Sosei-Heptares) [89, 118–122]. If high affinity stabilizing ligands are available, however, it is not as necessary to engage in extensive mutagenesis campaigns. Instead, selective mutations are utilized in conjunction with protein engineering techniques to modestly stabilize the GPCR fold, placing the receptor in a stability range sufficient for structural studies. Indeed, because of the fold similarity in the GPCR helical bundle, there are a few almost universal mutation sites that can frequently have an impact on the stability of most GPCRs.

A sequence comparison between β_2AR and bovine rhodopsin revealed one such site [27]. A glutamate residue located in the middle of a transmembrane helix predicted to be facing the plasma membrane. Energetically speaking, this is highly unfavorable and will result in a significant penalty for burying a charged residue in a hydrophobic environment. Inspection of other GPCR sequences revealed that a charged residue in this location was unique to β_2AR, and most of the time, an aromatic or small hydrophobic residue occupies the site. In bovine rhodopsin, one of the most stable inactive GPCRs known, the amino acid occupying this site is a tryptophan. Several amino acid replacements were evaluated for β_2AR in this position and tested for expression, stability, and retention of ligand binding. Replacing the glutamate residue with tryptophan resulted in the most stable mutant construct, and the use of this construct generated the second structure of β_2AR now bound to the much less stabilizing ligand timolol [45].

The general observation that a tryptophan residue at this position in the GPCR helical architecture (now known as the Roth mutation) will generally stabilize the inactive state of GPCRs has been utilized and transferred to many members of the family. The discovery of this one critical position opened up multiple branches of the phylogenetic tree for structural interrogation as it provided the needed stability for crystallizing the different classes of GPCRs [91, 123–128]. Additional stabilizing sites identified since this initial discovery provide data for analysis of sequences in the context of stability and expression data. Moreover, this is a fruitful area of research for applying machine learning approaches to protein engineering [117, 129–131].

9.4.5 Pushing the Boundaries

9.4.5.1 Agonists and Active Conformations

It has been observed since the origins of the structural biology platform for GPCRs, that the antagonist bound inactive conformation of the receptor is more tractable. This observation is likely attributable to the inactive state of the receptor being more stable while not in complex with downstream binding partners. Occasionally, agonist compounds or peptides co-crystal structures were successful. Invariably the receptor conformation remained in the inactive state or some intermediate meta-stable state reminiscent of the full agonist bound conformation [99, 132–135].

9.4.5.2 Signaling Complexes

In 2011, the Kobilka lab published a report on the β_2AR in complex with G-protein signaling machinery [92]. This structure opened a new chapter in GPCR structural biology. It engendered a new understanding of how GPCRs fit in with the rest of the signaling apparatus of the cell. Duplication of this effort on other receptor systems has taken a few years to materialize. Further success has mainly been dependent on moving away from an analysis of crystals of the complex using X-ray diffraction in favor of cryo-EM for resolution of GPCRs in complex with their cognate signaling partners [136–138].

To date, there have been 15 GPCRs determined in complex with various signaling partners using cryo-EM at an average resolution of 3.5 Å and as high as 2.0 Å. This resolution is generally sufficient for mapping out the secondary structure. When combined with higher resolution X-ray studies and molecular modeling techniques, the mechanism by which GPCRs bind agonists and transmit that binding event to the intracellular signaling apparatus has become more evident with each passing month. Further, the resolution and data collection process have been improving rapidly enabling routine sub 3.0 Å structures, placing the technique of cryo-EM in the range of utilization for structure-based drug design efforts. For the first time, we can start gathering snapshots of agonists binding in the context of various signaling complexes opening the door to understanding the structural details of biased ligand signaling [139].

9.5 The Impact of the GPCR Consortium: Widening Access to Ligands

One constant issue throughout the progression of GPCR structural biology is the underlying need for access to ligands. Early on, this was reasonably straightforward, as the selection of the target was dependent on the commercial availability of the ligands. It grew increasingly difficult to find new projects that met this requirement as the structures of the well-characterized receptors started to come out. It was common to screen at least twenty high affinity, diverse ligands to find those that not only bound to the protein with high affinity but also stabilized it in a unique fold amenable for crystallization [140]. In most cases, the one or two ligands available for a given target of interest are not enough for crystallization trials and are still a rate-limiting step in projects today.

In 2014, the GPCR Consortium, a public–private collaboration, was founded to alleviate this bottleneck. Its mission is to bring together pharmaceutical companies and leading academic researchers to generate additional structural information for medically important targets while further developing technology to facilitate the process of GPCR structure generation. Pharmaceutical members provide the receptor-specific compounds necessary to stabilize a receptor in various conformations amplifying the ability for the academic groups to solve the structures of highest interest. Initially planned for eight members, the high interest in these objectives expanded membership to include nine pharmaceutical companies and six academic sites in the first year. Members shared ligands with academic groups, and the resulting pre-competitive data was shared equally among all members and published without restrictions. At the time the consortium began, there were 26 known structures (Table 9.1). Over the next five years, this collaboration has resulted in over 14 published structures for targets requested by industry members, with an additional 16 from participating GPCR Consortium labs (Table 9.2). With many more structures determined and in various stages of the publication process, the next few years promise to be exciting.

9.6 A Unified Analysis Across GPCR Families

The great success of everyone involved in structural analysis of the GPCR family over the last thirteen years of research presents an opportunity for a greater understanding of the target class through an examination of the structural similarities and differences and how they vary among the different sub-classes. While an in-depth analysis is out of scope for this review, we describe a technique for parametrizing and comparing helices [167, 168]. Parameterization of the helices allows one to calculate and analyze the helical axes and positions along those axes that correspond to GPCR index positions taken from the GPCRdb [169]. By using a unique index position set at the helical centroid, we avoid sequence alignment as a prerequisite for structural alignment, and helical torsional deviations affecting the analysis are eliminated (Figure 9.2).

To initiate the alignment, we choose a consistent set of index positions on each helix that minimizes the deviation across the set of all GPCRs. The seven index positions utilized for this alignment are class and helix specific (Table 9.3). This methodology introduces a rigorous method for the alignment of sequences across classes based on the optimal structural alignment and serves as an anchor for the analysis of structural differences. In this way, we can directly compare modifications of angles, helical lengths, torsions, and translations associated with structural changes across the family. As this type of analysis will provide a rigorous measurement of structural modifications, we can begin to understand

Table 9.2 GPCR structures determined by GPCR consortium labs.

Date	UniProt | family	Significance/role of receptor	Citations	PDBid (Resolution Å)
April, 2015	P2RY1	• Both $P2Y_1R$ and $P2Y_{12}R$ are activated by adenosine 5′diphosphate (ADP) to induce platelet activation, which plays a pivotal role in thrombosis formation. • The blockade of either receptor significantly decreases ADP-induced platelet aggregation. Most available drugs target $P2Y_{12}R$. Targeting $P2Y_1R$ has been suggested as a new target, possibly offering safety advantages. • This paper reports $P2Y_1R$ in complex with a non-nucleotide antagonist BPTU and a nucleotide antagonist MRS2500.	[141]	4XNV (2.2) 4XNW (2.7)
May, 2015	AGTR1	• Angiotensin II type 1 receptor (AT_1R) serves as a primary regulator for blood pressure maintenance. • Although several anti-hypertension drugs have been developed at AT_1R blockers, the structural basis for ligand binding and regulation has been elusive. • This paper reports AT_1R in complex with a selective antagonist ZD7155. The use of XFEL allowed room temperature determination bypassing the issues obtaining high-quality crystals needed for synchrotron radiation.	[142]	4YAY (2.9)
March, 2015	CXCR4	• Chemokines and their receptors control migration during development, immune system responses in numerous diseases, including inflammation and cancer. • This paper reports CXCR4 in complex with the viral chemokine antagonist vMIP-II. The structure helped rationalize a large body of mutagenesis data and modeling of endogenous ligand CXCL12.	[125]	4RWS (3.1)

(Continued)

Table 9.2 (Continued)

Date	UniProt | family	Significance/role of receptor	Citations	PDBid (Resolution Å)
June, 2015	LPAR1	• The effects of lysophosphatidic acid (LPA) are mediated by six GPCRs, LPA_1–LPA_6. Since LPA is present in nearly all cells, tissues, and fluids of the body, targeting deletion of LPA receptors has revealed effects on every organ system examined thus far. • Disease indications include hydrocephalus, infertility, fibrosis, pain, and cancer. • This paper presents structures of the $LPAR_1$ in complex with three different antagonists.	[44]	4Z35 (2.9) 4Z36 (2.9) 4Z34 (3.0)
October, 2016	CNR1	• The cannabinoid receptor 1 (CB1) is the principal target of the psychoactive constituent of marijuana, the partial agonist THC. • This paper reports an antagonist bound crystal structure of human CB_1.	[143]	5TGZ (2.8)
April, 2017	AGTR2	• The angiotensin II receptors AT_1R and AT_2R serve as critical components of the renin-angiotensin-aldosterone system. • AT_1R has a central role in the regulation of blood pressure, but the function of AT_2R is less clear and has a variety of reported effects. • This paper reports human AT_2R bound to a selective ligand and a dual ligand capturing the receptor in an active-like conformation.	[144]	5UNG (2.8) 5UNF (2.8) 5UNH (2.9)
5/17/2017	SMO	• The smoothened receptor (SMO) belongs to the class Frizzled of the GPCR superfamily, constituting a vital component of the Hedgehog signaling pathway. • This paper reports the multi-domain human SMO bound and stabilized by a designed tool ligand TC114.	[145]	5V57 (3.0) 5V56 (2.9)

Table 9.2 (Continued)

Date	UniProt | family	Significance/role of receptor	Citations	PDBid (Resolution Å)
June, 2017	APLNR	• Apelin receptor (APJR) is a crucial regulator of human cardiovascular function and is activated by two different endogenous peptide ligands, apelin, and Elabela, each with different isoforms diversified by length and amino acid sequence. • This paper reports a crystal structure of human APJR in complex with a designed 17 amino acid apelin mimetic peptide agonist.	[146]	5VBL (2.6)
June, 2017	GLP1R	• The glucagon-like peptide-1 receptor (GLP-1R) and the glucagon receptor (GCGR) are members of the secretin-like class B family of GPCRs. They have opposing physiological roles in insulin release and glucose homeostasis. • The treatment of T2 diabetes requires positive modulation of GLP-1R to inhibit glucagon secretion and stimulate insulin secretion in a glucose-dependent manner. • This paper reports the human GLP-1R TM domain in complex with two different negative allosteric modulators.	[147]	5VEW (2.7) 5VEX (3.0)
June, 2017	GCGR	• The human glucagon receptor, GCGR, belongs to the class B GPCR family and plays a crucial role in glucose hemostasis and the pathology of type 2 diabetes. • This paper reports the full-length GCGR containing both the extracellular domain and the TM domain in an inactive conformation.	[148]	5XEZ (3.0) 5XF1 (3.2)

(Continued)

Table 9.2 (Continued)

Date	UniProt | family	Significance/role of receptor	Citations	PDBid (Resolution Å)
July, 2017	HTR2B	• Monoclonal antibodies provide an attractive alternative to small molecule therapies for a wide range of diseases. • Given the importance of GPCRs as pharmaceutical targets, there has been an immense interest in developing therapeutic monoclonal antibodies that act on GPCRs. • This paper presents the 5-HT_{2B} receptor and an antibody Fab fragment bound to the extracellular side of the receptor.	[149]	5TUD (3.0)
July, 2017	CNR1	• The cannabinoid receptor 1 (CB1) is the principal target of the psychoactive constituent of marijuana, the partial agonist THC. • This paper reports two agonist bound crystal structures.	[150]	5XRA (2.8) 5XR8 (3.0)
January, 2018	GCGR	• Activation of GCGR by its endogenous ligand glucagon triggers the release of glucose from the liver during fasting, and thus has an essential role in glucose homeostasis and is a potential target for type 2 diabetes. • This paper reports a full-length human GCGR in complex with a glucagon analog and partial agonist.	[151]	5YQZ (3.0)
February, 2018	HTR2C	• Currently, drugs targeting several serotonin receptors, including the 5-HT_{2C} receptor, are useful for treating obesity, drug abuse, and schizophrenia. • The competing challenges of developing selective 5-HT_{2C} receptor ligands or creating drugs with a defined polypharmacological profile remains extremely difficult. • This paper describes 5-HT_{2C} with the promiscuous agonist ritanserin and ergotamine.	[152]	6BQH (2.7) 6BQG (3.0)

Table 9.2 (Continued)

Date	UniProt | family	Significance/role of receptor	Citations	PDBid (Resolution Å)
April, 2018	NPY1R	• Neuropeptide Y (NPY) receptors have essential roles in food intake, anxiety, and cancer biology. The NPY-Y receptor system has emerged as one of the most complex networks with three peptide ligands (NPY, peptide YY, and pancreatic polypeptide) binding in 4 receptors, Y_1, Y_2, Y_4, and Y_5. • This paper reports the human Y_1R bound to 2 selective antagonists UR-MK299 and BMS-193885.	[128]	5ZBQ (2.7) 5ZBH (3.0)
May, 2018	PTAFR	• Platelet-activating factor receptor (PAFR) responds to platelet-activating factor (PAF), a phospholipid mediator of cell-to-cell communication. It is considered an important drug target for treating asthma, inflammation, and cardiovascular diseases. • This paper reports the human PAFR in complex with antagonist SR27417 and inverse agonist ABT-491.	[153]	5ZKQ (2.9) 5ZKP (2.8)
August, 2018	FZD4	• Frizzled receptors (FZDs) are class-F GPCRs that function in Wnt signaling and are essential for developing and adult organism. • As central mediators in this complex signaling pathway, FZDs serve as gatekeeping proteins both for drug intervention and for the development of probes in basic and in therapeutic research. • This paper presents the FZD4 transmembrane domain in the absence of a bound ligand.	[154]	6BD4 (2.4)
January, 2019	PTGER3	• Misoprostol is a life-saving drug in many developing countries for women at risk of postpartum hemorrhaging owing to its affordability, stability, ease of administration, and clinical efficacy. However, misoprostol lacks receptor and tissue selectivity, and thus its use has several serious side effects. • This paper presents the structure of misoprostol acid free form bound to EP3.	[155]	6M9T (2.5)

(Continued)

Table 9.2 (Continued)

Date	UniProt | family	Significance/role of receptor	Citations	PDBid (Resolution Å)
January, 2019	TBXA2R	• Thromboxane A_2 receptor (TP) plays a pivotal role in cardiovascular homeostasis and is considered an important drug target for cardiovascular disease. • This paper reports crystal structures of the human TP bound to two nonprostanoid antagonists, ramatroban and daltroban.	[156]	6IIU (2.5) 6IIV (3.0)
January, 2019	CNR2	• The cannabinoid receptor CB2 is predominantly expressed in the immune system. Selective modulation of CB2 without the psychoactivity of CB1 has therapeutic potential in inflammatory, fibrotic, and neurodegenerative diseases. • This paper reports the human CB2 receptor in complex with a rationally designed antagonist, AM10257.	[157]	5ZTY (2.8)
February, 2019	TACR1	• Neurokinin 1 receptor (NK1R) has essential regulating functions in the central and peripheral nervous systems. • NK1R antagonists such as aprepitant have been approved for treating chemotherapy-induced nausea and vomiting. However, the lack of data on NK1R structure and biochemistry has limited further drug development targeting this receptor. • This paper combines NMR and X-ray crystallography to provide a dynamic and static characterization of the binding mode of aprepitant in complex with human NK1R variants.	[158]	6J21 (3.2) 6J20 (2.7)
April, 2019	MTNR1A	• Melatonin is a neurohormone that maintains circadian rhythms by synchronization to environmental cues and is involved in diverse physiological processes such as the regulation of blood pressure and core body temperature, oncogenesis, and immune function. • This paper presents XFEL structures of MT_1 in complex with four agonists: 2-PMT, ramelteon, agomelatine, and 2-iodomelatonin.	[159]	6ME3 (2.9) 6ME2 (2.8) 6ME5 (3.2) 6ME4 (3.2)

Table 9.2 (Continued)

Date	UniProt | family	Significance/role of receptor	Citations	PDBid (Resolution Å)
April, 2019	MTNR1B	• The human MT_1 and MT_2 melatonin receptors help regulate circadian rhythm and sleep patterns. Drug development has targeted both receptors for the treatment of insomnia, circadian rhythm and mood disorders, and cancer. MT_2 has also been implicated in type 2 diabetes. • This paper reports XFEL MT_2 receptors in complex with the agonists melatonin analog 2-PMT and insomnia drug ramelteon.	[160]	6ME7 (3.2) 6ME6 (2.8) 6ME9 (3.3) 6ME8 (3.1)
November, 2019	OPRD1	• Selective activation of the δ-opioid receptor (DOP) has excellent potential for the treatment of chronic pain, benefiting from ancillary anxiolytic, and antidepressant-like effects. Moreover, DOP agonists show reduced adverse effects as compared with mu-opioid receptor (MOP) agonists that are in the spotlight of the current "opioid crisis." • This paper reports the first crystal structures of the DOP in an activated state, in complex with two relevant and structurally diverse agonists.	[161]	6PT3 (3.3) 6PT2 (2.8)
December, 2019	CYSLTR2	• Cysteinyl leukotriene G protein-coupled receptors $CysLT_1$ and $CysLT_2$ regulate pro-inflammatory responses associated with allergic disorders. While selective inhibition of $CysLT_1R$ has been used for treating asthma and associated diseases for over two decades, $CysLT_2R$ has recently started to emerge as a potential drug target against atopic asthma, brain injury, and CNS disorders, as well as several types of cancer. • This paper describes four structures of $CysLT_2R$ in complex with dual antagonists.	[162]	6RZ7 (2.4) 6RZ6 (2.4) 6RZ9 (2.7) 6RZ8 (2.7)

(Continued)

Table 9.2 (Continued)

Date	UniProt | family	Significance/role of receptor	Citations	PDBid (Resolution Å)
February, 2020	GPR52	• Highly expressed in the brain and represents a promising therapeutic target for the treatment of Huntington's disease and several psychiatric disorders. • This paper presents the structure of human GPR52 in three states: ligand-free, G_s coupled self-activation state, and a potential allosteric ligand-bound state.	[163]	6LI0 (2.2) 6LI2 (2.8) 6LI1 (2.9) 6LI3 (3.3)
February, 2020	CNR2	• Human endocannabinoid systems modulate multiple physiological processes, mainly through the activation of cannabinoid receptors CB1 and CB2. Missing structural information has significantly held back the development of promising CB2 selective agonists for treating inflammatory and neuropathic pain without the psycho-activity of CB1. • This paper describes structures of both CB1 and CB2 bound to ligands.	[164]	6KPC (3.2) 6KPF (2.9) 6KPG (3.0)
March, 2020	GCGR	• Glucagon plays a central role in the regulation of blood glucose levels and glucose homeostasis. • This paper uses cryo-EM to determine the structures of the human glucagon receptor (GCGR) bound to glucagon and distinct classes of heterotrimeric G proteins, G_s or G_i	[139]	6LML (3.9) 6LMK (3.7)
March, 2020	CHRM4	• Human muscarinic receptor M4 has emerged as an attractive drug target for the treatment of Alzheimer's, schizophrenia, and Parkinson's. • This paper reports a mutation-induced inactivated M4 receptor. Virtual screening of a focused library suggests the inactive M4 prefers antagonists more than agonists.	[165]	6KP6 (3.0)

Table 9.2 (Continued)

Date	UniProt | family	Significance/role of receptor	Citations	PDBid (Resolution Å)
April, 2020	MC4R	• The melanocortin-4 receptor is involved in energy homeostasis and is an important drug target for syndromic obesity. • This paper reports MC4R bound to the antagonist SHU9119 and identifies Ca2+ as a co-factor	[166]	6W25 (2.8)

Note: Summary table for structures of targets that were determined by GPCR Consortium collaborators after formation. Shaded entries represent targets selected by Consortium industry members. The "Significance and Role of Receptor" column describes highlights of the GPCR target and important findings from the paper as judged by the authors of this review.

how sequence affects the overall structural properties. For now, we will be content to represent key members of the family aligned in this way as a gallery of structural images (Figure 9.3).

9.7 The Next 10 Years of Discovery

The original title for this review was "10 Years of GPCR Structures"; however, the analysis grew to ensure that we acknowledge work associated with rhodopsin and bacteriorhodopsin within the context of this chapter given its appropriate place in a historical view of the progression of GPCR structural biology research (Figure 9.1). We incorporated technical aspects that enlightened our search for a uniform method of solving complicated membrane protein structures in an attempt to capture the process of making the roadmap from the author's perspective and not merely summarizing the collection of structures. At the same time, we believe there is much to learn from a quantitative analysis of structural information at a scale that is sufficiently granular to notice essential details but expansive enough to include the entire family. We are now in the fortunate situation of having too many structures of GPCRs to include in the course of a single review [72].

We are poised over the next 10 years, to see GPCR structural biology expanding in scope both horizontally and vertically. The horizontal expansion will

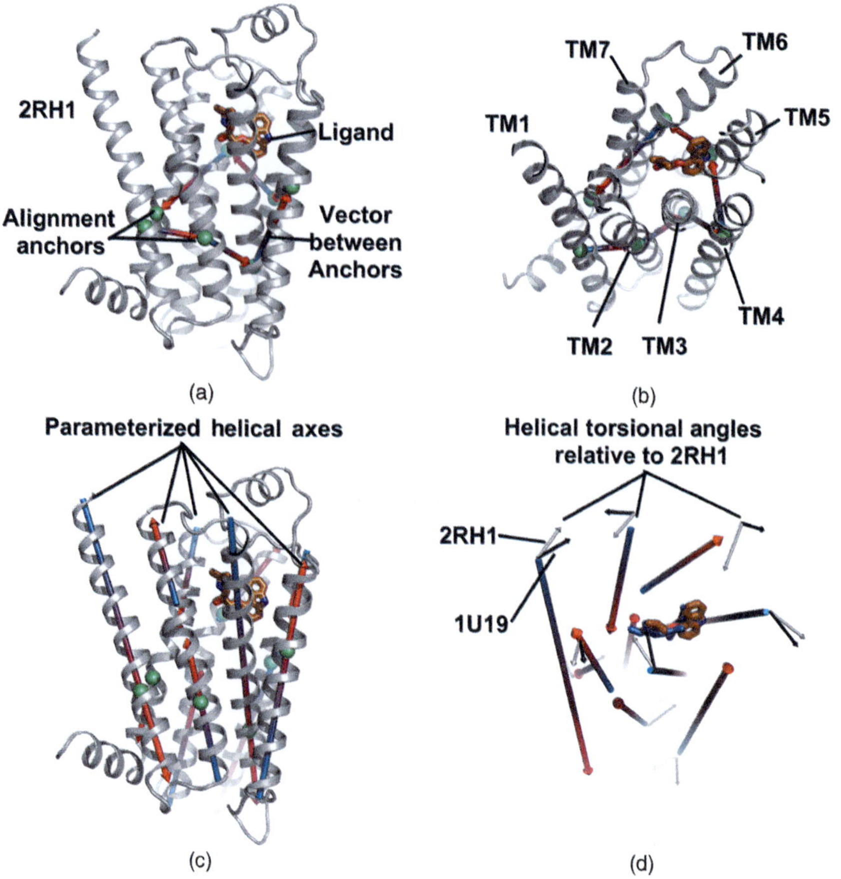

Figure 9.2 Schematic of the methodology utilized for overlaying structures across the GPCR family. (a) A unique position located on the axis of each helix was selected to minimize the RMSD during the superposition of these sites. Each class of GPCRs has a unique set of index positions that are structurally equivalent across the family (Table 9.3). (b) Top view of the anchor positions as represented by β_2AR (pdbid: 2RH1). The positions represent the most structurally invariant sites determined through an exhaustive search algorithm and based on minimal RMSD. (c) The helical axis is calculated for each receptor based on optimized parameters. The axes are represented by vectors with origins located at the beginning and end of each helical segment. Helical positions can then be represented numerically with spherical coordinates and origin shifts. (d) Additional information regarding the torsional offset of the helix relative to a standard frame of reference can also be calculated and used in this analysis. Here we show the angular offset of rhodopsin (pdbid: 1U19) relative to β_2AR (pdbid: 2RH1).

Table 9.3 Class specific alignment indices.

Class	Reference member	TM-I Ind | Res	TM-II Ind | Res	TM-III Ind | Res	TM-IV Ind | Res	TM-V Ind | Res	TM-VI Ind | Res	TM-VII Ind | Res
A	β2AR	50 | 51	46 | 75	46 | 127	54 | 162	50 | 211	50 | 288	48 | 321
B	CRFR1	55 | 135	53 | 158	52 | 211	54 | 240	50 | 283	52 | 326	52 | 358
C	GRM1	56 | 609	42 | 634	51 | 679	44 | 718	48 | 761	51 | 799	44 | 827
F	SMO	47 | 245	48 | 271	45 | 334	56 | 371	50 | 407	50 | 469	49 | 529

Note: Structurally analogous helical centroid positions and corresponding index positions across classes. A common framework for aligning structures across the GPCR family can be derived from this analysis (i.e. Class B index position 55 for TM-I is structurally equivalent to Class A index position 50).

include technological advancements that lower the barriers to entry for academic and industry groups who wish to utilize structural information as a tool. This expansion is already occurring within the confines of multiple academic and industry groups and will spread as the process becomes more robust. As a result, we will start to uncover a more detailed analysis of the molecular mechanics of signaling through specific point mutations, and we will further enable orthogonal biophysical methods to capture the dynamics of the family and how they respond to different stimuli in real-time. Finally, we will begin to assemble all of this information into a framework that will facilitate a more robust understanding of GPCR biology, such as allosteric modulation, biased ligand signaling, and the effect of the composition of lipid membrane on the signaling properties of the receptors. This framework will further our knowledge of the class and facilitate an expansion of their utility for our efforts at the discovery and development of novel pharmaceuticals to alleviate human suffering. At the same time, our knowledge will expand vertically, taking advantage of our ability as a field to visualize signaling complexes. We will generate an understanding of the molecular details of GPCR signaling not just along the canonical G-protein pathways but with other binding partners as well. Finally, we may begin to understand the role the cellular organization has on GPCR pharmacology bridging the gap between molecular and systems biology, bringing our molecular view of cellular signal processing into a broader scope.

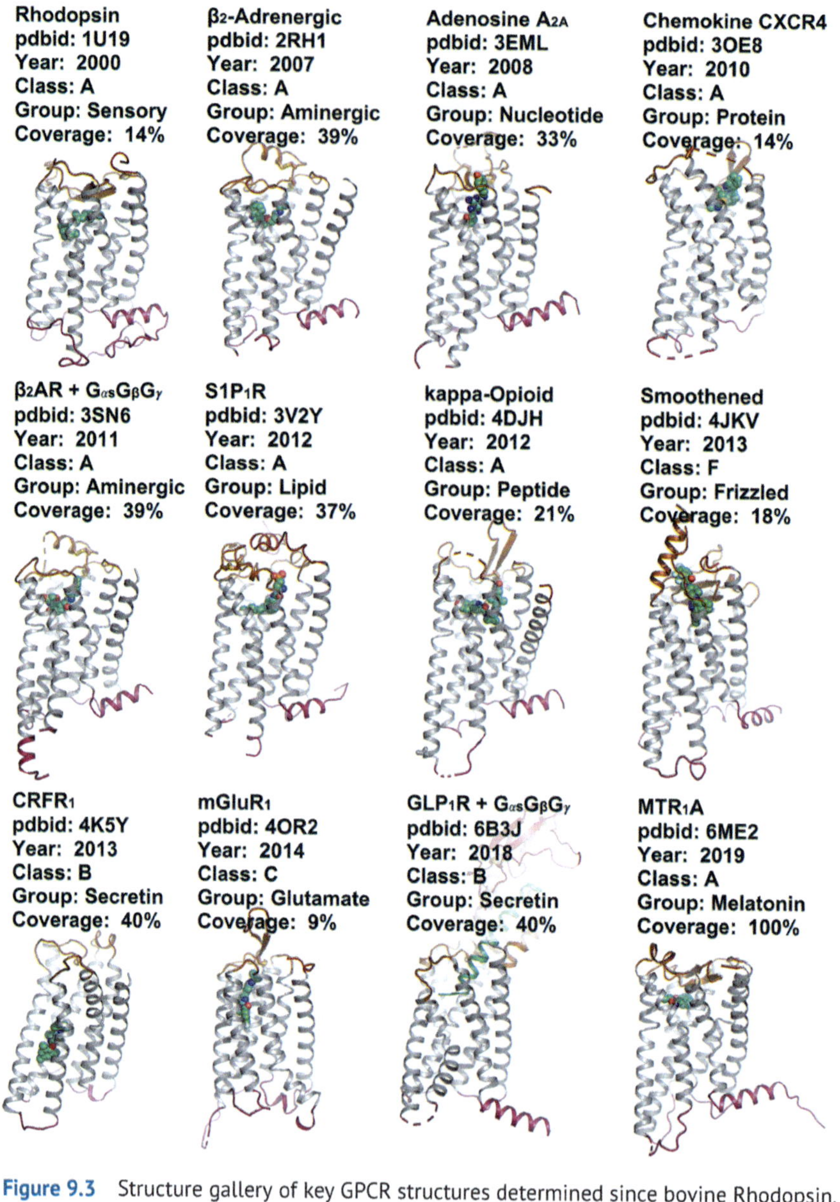

Figure 9.3 Structure gallery of key GPCR structures determined since bovine Rhodopsin. Each target has supporting information associated with it, including the percentage of structural coverage related to its group within the GPCR family. Notably, all of the receptors within the melatonin group have been determined, and 40% of Class B secretin receptor group has been solved. Groups that do not yet have any structural representation include Class A: Steroid and Carboxylic acid; Class B: Adhesion and the Taste receptors. Of note, the olfactory receptors are not included as structures have not yet been solved for this class. Finally a recent determination of the GPR52 receptor represents the first orphan structure determined. Source: Adapted from Lin et al. [163].

References

1 Sriram, K. and Insel, P.A. (2018). GPCRs as targets for approved drugs: how many targets and how many drugs? *Mol. Pharmacol.* 93: 251–258.

2 Benovic, J.L. (2012). G-protein-coupled receptors signal victory. *Cell* 151: 1148–1150.

3 Calebiro, D. and Grimes, J. (2019). G protein–coupled receptor pharmacology at the single-molecule level. *Annu. Rev. Pharmacol. Toxicol.* 60: 1–15.

4 Weis, W.I. and Kobilka, B.K. (2014). The molecular basis of G protein–coupled receptor activation. *Annu. Rev. Biochem.* 87: 897–919.

5 Deisenhofer, J., Epp, O., Miki, K. et al. (1985). Structure of the protein subunits in the photosynthetic reaction centre of *Rhodopseudomonas viridis* at 3Å resolution. *Nature* 318: 618–624.

6 Moraes, I., Evans, G., Sanchez-Weatherby, J. et al. (2014). Membrane protein structure determination – The next generation. *Biochim. Biophys. Acta* 1838: 78–87.

7 Dixon, R.A.F., Kobilka, B.K., Strader, D.J. et al. (1986). Cloning of the gene and cDNA for mammalian β-adrenergic receptor and homology with rhodopsin. *Nature* 321: 75–79.

8 Henderson, R., Baldwin, J.M., Ceska, T.A. et al. (1990). Model for the structure of bacteriorhodopsin based on high-resolution electron cryo-microscopy. *J. Mol. Biol.* 213: 899–929.

9 Pebay-Peyroula, E., Rummel, G., Rosenbusch, J.P. et al. (1997). X-ray structure of bacteriorhodopsin at 2.5 angstroms from microcrystals grown in lipidic cubic phases. *Science* 277: 1676–1681.

10 Palczewski, K., Kumasaka, T., Hori, T. et al. (2000). Crystal structure of rhodopsin: A G protein-coupled receptor. *Science* 289: 739–745.

11 Kolb, P. and Klebe, G. (2011). The golden age of GPCR structural biology: any impact on drug design? *Angew. Chem. Int. Ed.* 50: 11573–11575.

12 Oesterhelt, D. and Stoeckenius, W. (1973). Functions of a new photoreceptor membrane. *Proc. Natl. Acad. Sci. U.S.A.* 70: 2853–2857.

13 Oren, A. (2002). *Halophilic Microorganisms and their Environments*, 1e. Dordrecht: Springer.

14 Henderson, R. and Unwin, P. (1975). Three-dimensional model of purple membrane obtained by electron microscopy. *Nature* 257: 28–32.

15 Landau, E.M. and Rosenbusch, J.P. (1996). Lipidic cubic phases: a novel concept for the crystallization of membrane proteins. *Proc. Natl. Acad. Sci. U.S.A.* 93: 14532–14535.

16 Boll, F. (1877). Zur anatomie und physiologie der retina. *Arch. Anat. Physiol.* 4: 783–787.

17 Hargrave, P.A., McDowell, J.H., Curtis, D.R. et al. (1983). The structure of bovine rhodopsin. *Biophys. Struct. Mech.* 9: 235–244.

18 Ovchinnikov, L.A., Abdulaev, N.G., Feigina, M.L. et al. (1983). Visual rhodopsin. III. Complete amino acid sequence and topography in a membrane. *Bioorg. Khim* 9 (10): 1331–1340.

19 Schertler, G.F.X., Villa, C., and Henderson, R. (1993). Projection structure of rhodopsin. *Nature* 362: 770–772.

20 Costanzi, S., Siegel, J., Tikhonova, I. et al. (2009). Rhodopsin and the others: a historical perspective on structural studies of G protein-coupled receptors. *Curr. Pharm. Des.* 15: 3994–4002.

21 Stuart, J.A. and Birge, R.R. (1996). Characterization of the primary photochemical events in bacteriorhodopsin and rhodopsin. In: *Rhodopsin and G-protein Linked Receptors*, vol. 2 (ed. A.G. Lee), 33–139.

22 Palczewski, K. (2006). G protein–coupled receptor rhodopsin. *Annu. Rev. Biochem.* 75: 743–767.

23 Kenakin, T. (2003). Ligand-selective receptor conformations revisited: the promise and the problem. *Trends Pharmacol. Sci.* 24: 346–354.

24 Midgett, C.R. and Madden, D.R. (2007). Breaking the bottleneck: eukaryotic membrane protein expression for high-resolution structural studies. *J. Struct. Biol.* 160: 265–274.

25 Lundstrom, K. (2003). Semliki Forest virus vectors for rapid and high-level expression of integral membrane proteins. *Biochim. Biophys. Acta, Biomembr.* 1610: 90–96.

26 Hanson, M.A., Brooun, A., Baker, K.A. et al. (2007). Profiling of membrane protein variants in a baculovirus system by coupling cell-surface detection with small-scale parallel expression. *Protein Expression Purif.* 56: 85–92.

27 Roth, C.B., Hanson, M.A., and Stevens, R.C. (2008). Stabilization of the human β2-adrenergic receptor TM4-TM3-TM5 helix interface by mutagenesis of Glu122(3.41), a critical residue in GPCR structure. *J. Mol. Biol.* 376: 1305–1319.

28 Blois, T.M. and Bowie, J.U. (2009). G-protein-coupled receptor structures were not built in a day. *Protein Sci.* 18: 1335–1342.

29 Grisshammer, R., White, J.F., Trinh, L.B. et al. (2005). Large-scale expression and purification of a G-protein-coupled receptor for structure determination – an overview. *J. Struct. Funct. Genomics* 6: 159–163.

30 Lundstrom, K. (2005). Structural genomics of GPCRs. *Trends Biotechnol.* 23: 103–108.

31 McCusker, E.C., Bane, S.E., O'Malley, M.A. et al. (2007). Heterologous GPCR expression: a bottleneck to obtaining crystal structures. *Biotechnol. Progr.* 23: 540–547.

32 Shimamura, T., Shiroishi, M., Weyand, S. et al. (2011). Structure of the human histamine H1 receptor complex with doxepin. *Nature* 475: 65–70.

33 Akermoun, M. et al. (2005). Characterization of 16 human G protein-coupled receptors expressed in baculovirus-infected insect cells. *Protein Expr. Purif.* 44: 65–74.

34 Hassaine, G. et al. (2006). Semliki forest virus vectors for overexpression of 101 G protein-coupled receptors in mammalian host cells. *Protein Expr. Purif.* 45: 343–351.

35 Massotte, D. (2003). G protein-coupled receptor overexpression with the baculovirus–insect cell system: a tool for structural and functional studies. *Biochim. Biophys. Acta Biomembr.* 1610: 77–89.

36 Lundstrom, K., Wagner, R., Reinhart, C. et al. (2006). Structural genomics on membrane proteins: comparison of more than 100 GPCRs in 3 expression systems. *J. Struct. Funct. Genomics* 7: 77–91.

37 Harding, P.J., Attrill, H., Ross, S. et al. (2007). Neurotensin receptor type 1: *Escherichia coli* expression, purification, characterization and biophysical study. *Biochem. Soc. Trans.* 35: 760–763.

38 Lee, B.-K., Jung, K.-S., Son, C. et al. (2007). Affinity purification and characterization of a G-protein coupled receptor, *Saccharomyces cerevisiae* Ste2p. *Protein Expression Purif.* 56: 62–71.

39 Warne, T., Chirnside, J., and Schertler, G.F.X. (2003). Expression and purification of truncated, non-glycosylated turkey beta-adrenergic receptors for crystallization. *Biochim. Biophys. Acta, Biomembr.* 1610: 133–140.

40 Yao, Z. and Kobilka, B. (2005). Using synthetic lipids to stabilize purified β2 adrenoceptor in detergent micelles. *Anal. Biochem.* 343: 344–346.

41 Kawate, T. and Gouaux, E. (2006). Fluorescence-detection size-exclusion chromatography for precrystallization screening of integral membrane proteins. *Structure* 14: 673–681.

42 Wilson, W.W. (2003). Light scattering as a diagnostic for protein crystal growth – a practical approach. *J. Struct. Biol.* 142: 56–65.

43 Wiener, M.C. (2004). A pedestrian guide to membrane protein crystallization. *Methods* 34: 364–372.

44 Chrencik, J.E., Roth, C.B., Terakado, M. et al. (2015). Crystal structure of antagonist bound human lysophosphatidic acid receptor 1. *Cell* 161: 1633–1643.

45 Hanson, M.A., Cherezov, V., Griffith, M.T. et al. (2008). A specific cholesterol binding site is established by the 2.8 A structure of the human beta2-adrenergic receptor. *Structure* 16: 897–905.

46 Rosenbusch, J.P. (2001). Stability of membrane proteins: relevance for the selection of appropriate methods for high-resolution structure determinations. *J. Struct. Biol.* 136: 144–157.

47 Michel, H. (1983). Crystallization of membrane proteins. *Trends Biochem. Sci* 8: 56–59.

48 Bockaert, J. and Pin, J.P. (1999). Molecular tinkering of G protein-coupled receptors: an evolutionary success. *EMBO J.* 18: 1723–1729.

49 Preininger, A.M., Meiler, J., and Hamm, H.E. (2013). Conformational flexibility and structural dynamics in GPCR-mediated G protein activation: a perspective. *J. Mol. Biol.* 425: 2288–2298.

50 Ghosh, E., Kumari, P., Jaiman, D. et al. (2015). Methodological advances: the unsung heroes of the GPCR structural revolution. *Nat. Rev. Mol. Cell Biol.* 16: 69–81.

51 Grisshammer, R. (2009). Purification of recombinant G-protein-coupled receptors. *Methods Enzymol.* 463: 631–645.

52 Birch, J., Axford, D., Foadi, J. et al. (2018). The fine art of integral membrane protein crystallisation. *Methods* 147: 150–162.

53 Qutub, Y., Reviakine, I., Maxwell, C. et al. (2004). Crystallization of transmembrane proteins in cubo: mechanisms of crystal growth and defect formation. *J. Mol. Biol.* 343: 1243–1254.

54 Blumer, K.J. and Thorner, J. (2007). An adrenaline (and gold?) rush for the GPCR community. *ACS Chem. Biol.* 2: 783–786.

55 Privé, G.G., Verner, G.E., Weitzman, C. et al. (1994). Fusion proteins as tools for crystallization: the lactose permease from *Escherichia coli*. *Acta Crystallogr. Sect. D Biol. Crystallogr.* 50: 375–379.

56 Rosenbaum, D.M., Cherezov, V., Hanson, M.A. et al. (2007). GPCR engineering yields high-resolution structural insights into β2-adrenergic receptor function. *Science* 318: 1266–1273.

57 Cherezov, V., Fersi, H., and Caffrey, M. (2001). Crystallization screens: compatibility with the lipidic cubic phase for in meso crystallization of membrane proteins. *Biophys. J.* 81: 225–242.

58 Cherezov, V., Clogston, J., Misquitta, Y. et al. (2002). Membrane protein crystallization in meso: lipid type-tailoring of the cubic phase. *Biophys. J.* 83: 3393–3407.

59 Cherezov, V., Peddi, A., Muthusubramaniam, L. et al. (2004). A robotic system for crystallizing membrane and soluble proteins in lipidic mesophases. *Acta Crystallogr, Sect. D: Biol. Crystallogr.* 60: 1795–1807.

60 Misquitta, Y., Cherezov, V., Havas, F. et al. (2004). Rational design of lipid for membrane protein crystallization. *J. Struct. Biol.* 148: 169–175.

61 Cherezov, V., Clogston, J., Papiz, M.Z. et al. (2006). Room to move: crystallizing membrane proteins in swollen lipidic mesophases. *J. Mol. Biol.* 357: 1605–1618.

62 Cherezov, V., Rosenbaum, D.M., Hanson, M.A. et al. (2007). High resolution crystal structure of an engineered human β2-adrenergic G protein-coupled receptor. *Science* 318: 1258–1265.

63 Wöhri, A.B., Johansson, L.C., Wadsten-Hindrichsen, P. et al. (2008). A lipidic-sponge phase screen for membrane protein crystallization. *Structure* 16: 1003–1009.

64 Stevens, R.C., Hanson, M.A., Cherezov, V., et al. (2009). *Methods and compositions for obtaining high-resolution crystals of membrane proteins,* Patent number: WO2009055512A3.

65 Cherezov, V., Hanson, M.A., Griffith, M.T. et al. (2009). Rastering strategy for screening and centring of microcrystal samples of human membrane proteins with a sub-10 μm size X-ray synchrotron beam. *J. R. Soc. Interface* 6: S587–S597.

66 Sanishvili, R., Yoder, D.W., Pothineni, S.B. et al. (2011). Radiation damage in protein crystals is reduced with a micron-sized X-ray beam. *Proc. Natl. Acad. Sci. U.S.A.* 108: 6127–6132.

67 Hilgart, M.C., Sanishvili, R., Ogata, C.M. et al. (2011). Automated sample-scanning methods for radiation damage mitigation and diffraction-based centering of macromolecular crystals. *J. Synchrotron. Radiat.* 18: 717–722.

68 Hanson, M.A., Roth, C.B., Jo, E. et al. (2012). Crystal structure of a lipid G protein-coupled receptor. *Science* 335: 851–855.

69 Rasmussen, S.G.F., Choi, H.-J., Rosenbaum, D.M. et al. (2007). Crystal structure of the human β2 adrenergic G-protein-coupled receptor. *Nature* 450: 383–387.

70 Shi, L., Liapakis, G., Xu, R. et al. (2002). β2 adrenergic receptor activation. *J. Biol. Chem.* 277: 40989–40996.

71 Ballesteros, J.A., Jensen, A.D., Liapakis, G. et al. (2001). Activation of the β2-adrenergic receptor involves disruption of an ionic lock between the cytoplasmic ends of transmembrane segments 3 and 6. *J. Biol. Chem.* 276: 29171–29177.

72 Hanson, M.A. and Stevens, R.C. (2009). Discovery of new GPCR biology: one receptor structure at a time. *Structure* 17: 8–14.

73 Lodowski, D.T., Angel, T.E., and Palczewski, K. (2009). Comparative analysis of GPCR crystal structures. *Photochem. Photobiol.* 85: 425–430.

74 Schneider, E.H., Schnell, D., Strasser, A. et al. (2010). Impact of the DRY motif and the missing “ionic lock” on constitutive activity and G-protein coupling of the human histamine h4 receptor. *J. Pharmacol. Exp. Ther.* 333: 382–392.

75 Participants GD 2008, Michino, M., Abola, E. et al. (2009). Community-wide assessment of GPCR structure modelling and ligand docking: GPCR Dock 2008. *Nat. Rev. Drug Discovery* 8: 455–463.

76 Jaakola, V.-P., Griffith, M.T., Hanson, M.A. et al. (2008). The 2.6 angstrom crystal structure of a human A2A adenosine receptor bound to an antagonist. *Science* 322: 1211.

77 Napoli, K.L. (2000). The FTY720 story. *Ther. Drug. Monit.* 22: 47–51.

78 Adachi, K. and Chiba, K. (2008). FTY720 Story. Its discovery and the following accelerated development of sphingosine 1-phosphate receptor agonists as immunomodulators based on reverse pharmacology. *Perspect. Med. Chem.* 1: 1177391X0700100002.

79 Sanna, M.G., Wang, S.-K., Gonzalez-Cabrera, P.J. et al. (2006). Enhancement of capillary leakage and restoration of lymphocyte egress by a chiral S1P1 antagonist in vivo. *Nat. Chem. Biol.* 2: 434–441.

80 Hurst, D.P., Grossfield, A., Lynch, D.L. et al. (2010). A lipid pathway for ligand binding is necessary for a cannabinoid G protein-coupled receptor. *J. Biol. Chem.* 285: 17954–17964.

81 Schadel, S.A., Heck, M., Maretzki, D. et al. (2003). Ligand channeling within a G-protein-coupled receptor. The entry and exit of retinals in native opsin. *J. Biol. Chem.* 278: 24896–24903.

82 Filipek, S., Stenkamp, R.E., Teller, D.C. et al. (2003). G protein-coupled receptor rhodopsin: a prospectus. *Annu. Rev. Physiol.* 65: 851–879.

83 Venkatakrishnan, A.J., Deupi, X., Lebon, G. et al. (2013). Molecular signatures of G-protein-coupled receptors. *Nature* 494: 185–194.

84 Katritch, V., Cherezov, V., and Stevens, R.C. (2013). Structure-function of the G protein–coupled receptor superfamily. *Annu. Rev. Pharmacol. Toxicol.* 53: 531–556.

85 Katritch, V., Cherezov, V., and Stevens, R.C. (2012). Diversity and modularity of G protein-coupled receptor structures. *Trends Pharmacol. Sci.* 33: 17–27.

86 Hulme, E.C. (2013). GPCR activation: a mutagenic spotlight on crystal structures. *Trends Pharmacol. Sci.* 34: 67–84.

87 Lee, S.-M., Booe, J.M., and Pioszak, A.A. (2015). Structural insights into ligand recognition and selectivity for classes A, B, and C GPCRs. *Eur. J. Pharmacol.* 763: 196–205.

88 Lv, X., Liu, J., Shi, Q. et al. (2016). In vitro expression and analysis of the 826 human G protein-coupled receptors. *Protein Cell* 7: 325–337.

89 Warne, T., Serrano-Vega, M.J., Baker, J.G. et al. (2008). Structure of a β1-adrenergic G-protein-coupled receptor. *Nature* 454: 486–491.

90 Wu, B., Chien, E.Y.T., Mol, C.D. et al. (2010). Structures of the CXCR4 chemokine GPCR with small-molecule and cyclic peptide antagonists. *Science* 330: 1066–1071.

91 Chien, E.Y.T., Liu, W., Zhao, Q. et al. (2010). Structure of the human dopamine D3 receptor in complex with a D2/D3 selective antagonist. *Science* 330: 1091–1095.

92 Rasmussen, S.G.F., DeVree, B.T., Zou, Y. et al. (2011). Crystal structure of the β2 adrenergic receptor-Gs protein complex. *Nature* 477: 549–555.

93 Haga, K., Kruse, A.C., Asada, H. et al. (2012). Structure of the human M2 muscarinic acetylcholine receptor bound to an antagonist. *Nature* 482: 547–551.

94 Kruse, A.C., Hu, J., Pan, A.C. et al. (2012). Structure and dynamics of the M3 muscarinic acetylcholine receptor. *Nature* 482: 552–556.

95 Wu, H., Wacker, D., Mileni, M. et al. (2012). Structure of the human κ-opioid receptor in complex with JDTic. *Nature* 485: 327–332.

96 Manglik, A., Kruse, A.C., Kobilka, T.S. et al. (2012). Crystal structure of the μ-opioid receptor bound to a morphinan antagonist. *Nature* 485: 321–326.

97 Thompson, A.A., Liu, W., Chun, E. et al. (2012). Structure of the nociceptin/orphanin FQ receptor in complex with a peptide mimetic. *Nature* 485: 395–399.

98 Granier, S., Manglik, A., Kruse, A.C. et al. (2012). Structure of the δ-opioid receptor bound to naltrindole. *Nature* 485: 400–404.

99 White, J.F., Noinaj, N., Shibata, Y. et al. (2012). Structure of the agonist-bound neurotensin receptor. *Nature* 490: 508–513.

100 Park, S.H., Das, B.B., Casagrande, F. et al. (2012). Structure of the chemokine receptor CXCR1 in phospholipid bilayers. *Nature* 491: 779–783.

101 Zhang, C., Srinivasan, Y., Arlow, D.H. et al. (2012). High-resolution crystal structure of human protease-activated receptor 1. *Nature* 492: 387–392.

102 Hollenstein, K., Kean, J., Bortolato, A. et al. (2013). Structure of class B GPCR corticotropin-releasing factor receptor 1. *Nature* 499: 438–443.

103 Wang, C., Jiang, Y., Ma, J. et al. (2013). Structural basis for molecular recognition at serotonin receptors. *Science* 340: 610–614.

104 Wang, C., Wu, H., Katritch, V. et al. (2013). Structure of the human smoothened receptor bound to an antitumour agent. *Nature* 497: 338–343.

105 Siu, F.Y., He, M., de Graaf, C. et al. (2013). Structure of the human glucagon class B G-protein-coupled receptor. *Nature* 499: 444–449.

106 Tan, Q., Zhu, Y., Chen, Z. et al. (2013). Structure of the CCR5 chemokine receptor-HIV entry inhibitor maraviroc complex. *Science* 341: 1387–1390.

107 Fenalti, G., Giguere, P.M., Katritch, V. et al. (2014). Molecular control of δ-opioid receptor signalling. *Nature* 1–18.

108 Wu, H., Wang, C., Gregory, K.J. et al. (2014). Structure of a class C GPCR metabotropic glutamate receptor 1 bound to an allosteric modulator. *Science* 344: 58–64.

109 Zhang, K., Zhang, J., Gao, Z.-G. et al. (2014). Structure of the human P2Y12 receptor in complex with an antithrombotic drug. *Nature* 509: 1–17.

110 Srivastava, A., Yano, J., Hirozane, Y. et al. (2014). High-resolution structure of the human GPR40 receptor bound to allosteric agonist TAK-875. *Nature* 513: 124–127.

111 Carvalho-Silva, D., Pierleoni, A., Pignatelli, M. et al. (2019). Open targets platform: new developments and updates two years on. *Nucleic Acids Res.* 47: D1056–D1065.

112 Caffrey, M. (2000). A lipid's eye view of membrane protein crystallization in mesophases. *Curr. Opin. Struct. Biol.* 10: 486–497.

113 Chun, E., Thompson, A.A., Liu, W. et al. (2012). Fusion partner toolchest for the stabilization and crystallization of G protein-coupled receptors. *Structure* 20: 967–976.

114 Liu, W., Chun, E., Thompson, A.A. et al. (2012). Structural basis for allosteric regulation of GPCRs by sodium ions. *Science* 337: 232–236.

115 Chu, R., Takei, J., Knowlton, J.R. et al. (2002). Redesign of a four-helix bundle protein by phage display coupled with proteolysis and structural characterization by NMR and X-ray crystallography. *J. Mol. Biol.* 323: 253–262.

116 Pándy-Szekeres, G., Munk, C., Tsonkov, T.M. et al. (2017). GPCRdb in 2018: adding GPCR structure models and ligands. *Nucleic Acids Res.* 46: D440–D446.

117 Munk, C., Mutt, E., Isberg, V. et al. (2019). An online resource for GPCR structure determination and analysis. *Nat. Methods* 16: 1–12.

118 Standfuss, J., Xie, G., Edwards, P.C. et al. (2007). Crystal structure of a thermally stable rhodopsin mutant. *J. Mol. Biol.* 372: 1179–1188.

119 Warne, T., Serrano-Vega, M.J., Tate, C.G. et al. (2009). Development and crystallization of a minimal thermostabilised G protein-coupled receptor. *Protein Expression Purif.* 65: 204–213.

120 Warne, T., Moukhametzianov, R., Baker, J.G. et al. (2011). The structural basis for agonist and partial agonist action on a β1-adrenergic receptor. *Nature* 469: 241–244.

121 Congreve, M., Rich, R.L., Myszka, D.G. et al. (2011). Fragment screening of stabilized G-protein-coupled receptors using biophysical methods. *Methods Enzymol.* 493: 115–136.

122 Doré, A.S., Robertson, N., Errey, J.C. et al. (2011). Structure of the adenosine A2A receptor in complex with ZM241385 and the xanthines XAC and caffeine. *Structure* 19: 1283–1293.

123 Wacker, D., Fenalti, G., Brown, M.A. et al. (2010). Conserved binding mode of human β2 adrenergic receptor inverse agonists and antagonist revealed by X-ray crystallography. *J. Am. Chem. Soc.* 132: 11443–11445.

124 Sato, T., Baker, J., Warne, T. et al. (2015). Pharmacological analysis and structure determination of 7-methylcyanopindolol–bound β1-adrenergic receptor. *Mol. Pharmacol.* 88: 1024–1034.

125 Qin, L., Kufareva, I., Holden, L.G. et al. (2015). Crystal structure of the chemokine receptor CXCR4 in complex with a viral chemokine. *Science* 347: 1117–1122.

126 Yin, W., Zhou, X.E., Yang, D. et al. (2018). Crystal structure of the human 5-HT1B serotonin receptor bound to an inverse agonist. *Cell Discov.* 4: 12.

127 McCorvy, J.D., Wacker, D., Wang, S. et al. (2018). Structural determinants of 5-HT2B receptor activation and biased agonism. *Nat. Struct. Mol. Biol.* 25: 787–796.

128 Yang, Z., Han, S., Keller, M. et al. (2018). Structural basis of ligand binding modes at the neuropeptide Y Y1 receptor. *Nature* 556: 1–22.

129 Popov, P., Kozlovskii, I., and Katritch, V. (2019). Computational design for thermostabilization of GPCRs. *Curr. Opin. Struct. Biol.* 55: 25–33.

130 Muk, S., Ghosh, S., Achuthan, S. et al. (2019). Machine learning for prioritization of thermostabilizing mutations for G-protein coupled receptors. *Biophys. J.* 117: 2228–2239.

131 Jana, S., Ghosh, S., Muk, S. et al. (2019). Prediction of conformation specific thermostabilizing mutations for class A G protein-coupled receptors. *J. Chem. Inf. Model.* 59: 3744–3754.

132 Rasmussen, S.G.F., Choi, H.-J., Fung, J.J. et al. (2011). Structure of a nanobody-stabilized active state of the β2 adrenoceptor. *Nature* 469: 175–180.

133 Rosenbaum, D.M., Zhang, C., Lyons, J.A. et al. (2011). Structure and function of an irreversible agonist-β2 adrenoceptor complex. *Nature* 469: 236–240.

134 Xu, F., Wu, H., Katritch, V. et al. (2011). Structure of an agonist-bound human A2A adenosine receptor. *Science* 332: 322–327.

135 Wacker, D., Wang, C., Katritch, V. et al. (2013). Structural features for functional selectivity at serotonin receptors. *Science* 340: 615–619.

136 Liang, Y.-L., Khoshouei, M., Radjainia, M. et al. (2017). Phase-plate cryo-EM structure of a class B GPCR–G-protein complex. *Nature* 159: 1–18.

137 Liang, Y.-L., Khoshouei, M., Deganutti, G. et al. (2018). Cryo-EM structure of the active, Gs-protein complexed, human CGRP receptor. *Nature* 561: 1–24.

138 Liang, Y.-L., Khoshouei, M., Glukhova, A. et al. (2018). Phase-plate cryo-EM structure of a biased agonist-bound human GLP-1 receptor–Gs complex. *Nature* 555: 1–20.

139 Qiao, A., Han, S., Li, X. et al. (2020). Structural basis of Gs and Gi recognition by the human glucagon receptor. *Science* 367: 1346–1352.

140 Zhang, X., Stevens, R.C., and Xu, F. (2015). The importance of ligands for G protein-coupled receptor stability. *Trends Biochem. Sci* 40: 79–87.

141 Zhang, D., Gao, Z.-G., Zhang, K. et al. (2015). Two disparate ligand-binding sites in the human P2Y1 receptor. *Nature* 520: 317–321.

142 Zhang, H., Unal, H., Gati, C. et al. (2015). Structure of the angiotensin receptor revealed by serial femtosecond crystallography. *Cell* 161: 1–13.

143 Hua, T., Vemuri, K., Pu, M. et al. (2016). Crystal structure of the human cannabinoid receptor CB1. *Cell* 167: 750–755.e14.

144 Zhang, H., Han, G.W., Batyuk, A. et al. (2017). Structural basis for selectivity and diversity in angiotensin II receptors. *Nature* 544: 327–332.

145 Zhang, X., Zhao, F., Wu, Y. et al. (2017). Crystal structure of a multi-domain human smoothened receptor in complex with a super stabilizing ligand. *Nat. Commun.* 8: 15383.

146 Ma, Y., Yue, Y., Ma, Y. et al. (2017). Structural basis for apelin control of the human apelin receptor. *Structure* 25: 858–866.e4.

147 Song, G., Yang, D., Wang, Y. et al. (2017). Human GLP-1 receptor transmembrane domain structure in complex with allosteric modulators. *Nature* 546: 312–315.

148 Zhang, H., Qiao, A., Yang, D. et al. (2017). Structure of the full-length glucagon class B G-protein-coupled receptor. *Nature* 546: 259–264.

149 Ishchenko, A., Wacker, D., Kapoor, M. et al. (2017). Structural insights into the extracellular recognition of the human serotonin 2B receptor by an antibody. *Proc. Natl. Acad. Sci. U.S.A.* 114: 8223–8228.

150 Hua, T., Vemuri, K., Nikas, S.P. et al. (2017). Crystal structures of agonist-bound human cannabinoid receptor CB1. *Nature* 547: 468–471.

151 Zhang, H., Qiao, A., Yang, L. et al. (2018). Structure of the glucagon receptor in complex with a glucagon analogue. *Nature* 553: 106–110.

152 Peng, Y., McCorvy, J.D., Harpsøe, K. et al. (2019). 5-HT2C receptor structures reveal the structural basis of GPCR polypharmacology. *Cell* 172: 719–730.e14.

153 Cao, C., Tan, Q., Xu, C. et al. (2018). Structural basis for signal recognition and transduction by platelet-activating-factor receptor. *Nat. Struct. Mol. Biol.* 25: 488–495.

154 Yang, S., Wu, Y., Xu, T.-H. et al. (2018). Crystal structure of the Frizzled 4 receptor in a ligand-free state. *Nature* 560: 666–670.

155 Audet, M., White, K.L., Breton, B. et al. (2019). Crystal structure of misoprostol bound to the labor inducer prostaglandin E2 receptor. *Nat. Chem. Biol.* 15: 11–17.

156 Fan, H., Chen, S., Yuan, X. et al. (2019). Structural basis for ligand recognition of the human thromboxane A2 receptor. *Nat. Chem. Biol.* 15: 27–33.

157 Li, X., Hua, T., Vemuri, K. et al. (2019). Crystal structure of the human cannabinoid receptor CB2. *Cell* 176: 459–467.e13.

158 Chen, S., Lu, M., Liu, D. et al. (2019). Human substance P receptor binding mode of the antagonist drug aprepitant by NMR and crystallography. *Nat. Commun.* 10: 638.

159 Stauch, B., Johansson, L.C., McCorvy, J.D. et al. (2019). Structural basis of ligand recognition at the human MT1 melatonin receptor. *Nature* 569: 1–21.

160 Johansson, L.C., Stauch, B., McCorvy, J.D. et al. (2019). XFEL structures of the human MT2 melatonin receptor reveal the basis of subtype selectivity. *Nature* 569: 1–19.

161 Claff, T., Yu, J., Blais, V. et al. (2019). Elucidating the active δ-opioid receptor crystal structure with peptide and small-molecule agonists. *Sci. Adv.* 5: eaax9115.

162 Gusach, A., Luginina, A., Marin, E. et al. (2019). Structural basis of ligand selectivity and disease mutations in cysteinyl leukotriene receptors. *Nat. Commun.* 10: 5573.

163 Lin, X., Li, M., Wang, N. et al. (2020). Structural basis of ligand recognition and self-activation of orphan GPR52. *Nature* 1–6.

164 Hua, T., Li, X., Wu, L. et al. (2020). Activation and signaling mechanism revealed by cannabinoid receptor-Gi complex structures. *Cell* 180: 655–665.e18.

165 Wang, J., Wu, M., Wu, L. et al. (2020). The structural study of mutation-induced inactivation of human muscarinic receptor M4. *IUCrJ* 7: 294–305.

166 Yu, J., Gimenez, L.E., Hernandez, C.C. et al. (2020). Determination of the melanocortin-4 receptor structure identifies Ca2+ as a cofactor for ligand binding. *Science* 368: 428–433.

167 Ren, Z., Ren, P.X., Balusu, R. et al. (2016). Transmembrane helices tilt, bend, slide, torque, and unwind between functional states of rhodopsin. *Sci. Rep.* 6: 1–13.

168 Ren, Z. (2013). Reverse engineering the cooperative machinery of human hemoglobin. *PLoS One* 8: e77363–e77314.

169 Isberg, V., de Graaf, C., Bortolato, A. et al. (2015). Generic GPCR residue numbers – aligning topology maps while minding the gaps. *Trends Pharmacol. Sci.* 36: 22–31.

10

Activation of Class B GPCR by Peptide Ligands: General Structural Aspects

Stefan Ernicke and Irene Coin

Faculty of Life Sciences, Institute of Biochemistry, Leipzig University, Leipzig, Germany

10.1 Structure Determination of Class B GPCRs

G protein-coupled receptors (GPCRs) of the class B are a small family of 15 highly homologous receptors structurally related to the prototypical secretin receptor (secretin-like GPCRs) [1]. Class B GPCRs regulate vital functions in mammals, including glucose and calcium homeostasis, pain transmission, gastrointestinal and metabolic regulation, and stress response. They represent highly relevant targets for pharmacological intervention for the treatment of widespread pathologies, such as type 2 diabetes and obesity [2], migraine [3], osteoporosis [4], and other bone diseases [5]. Indeed, a number of therapeutic drugs targeting class B GPCRs are already on the market.

Like all GPCRs, class B receptors feature a core domain made up of seven helices than span the cell membrane (7TM bundle or transmembrane domain [TMD]) and are connected by three extracellular loops (ECLs) and three intracellular loops (ICLs) (Figure 10.1a). The N-terminus and the C-terminus face the extracellular space and the cytosol, respectively. Secretin-like receptors are defined by a large, complex glycosylated N-terminal domain (NTD), which features ~120–160 residues in length. They are activated by long peptide hormones, which bear distinct functional sites at the two ends. The C-terminal portion of the peptide forms selective high-affinity interactions with the NTD and drives selective receptor binding, while the N-terminal segment interacts with the TMD and triggers receptor activation [6] (Figure 10.1b). Thus, the NTD of class B receptors can be referred as the "binding domain," whereas the 7TM represents the "activation domain." This so-called two-domain model for peptide binding to class B receptors was formulated 20 years ago on the basis of extensive structure–activity relationship (SAR) studies on the ligands [8–11] and

GPCRs as Therapeutic Targets, Volume 1, First Edition. Edited by Annette Gilchrist.

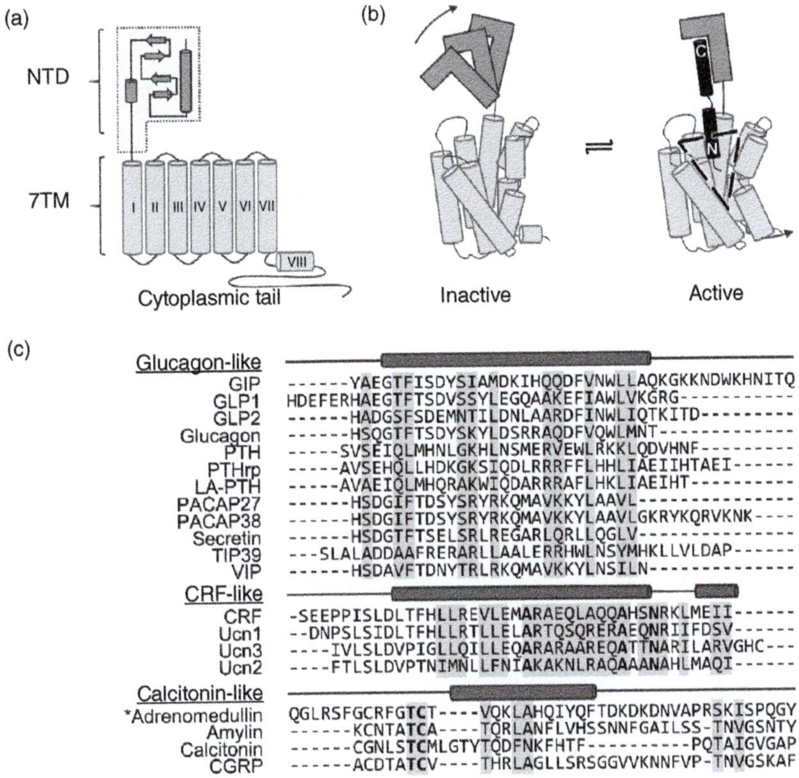

Figure 10.1 Structural elements of class B GPCRs and their ligands. (a) Schematic representation of a class B GPCR highlighting the N-terminal extracellular domain (NTD), the transmembrane core (7TM), and the C-terminal intracellular helix VIII. The NTD comprises an N-terminal α-helix followed by two pairs of antiparallel β-sheets. (b) Ligand-induced conformational changes of class B GPCRs. In the inactive state, the NTD oscillates between an "open" and a "closed" conformation. Ligand binding stabilizes the active "open" state. The peptide C-terminus (C) docks to the NTD, whereas the peptide N-terminus (N) penetrates the 7TM (two-domain model). The wide orthosteric pocket in the TMD of class B GPCRs features a characteristic V-shape, marked in the figure with a dashed triangle. (c) Sequence alignment of class B GPCR ligands according to the three subfamilies: glucagon-like, corticotropin releasing factor-like (CRF-like), and calcitonin-like. Conserved regions are shown on a grey background, with identical residues printed in bold letters. *Residues 1–8 of adrenomedullin are not shown. Source: (b) Based on Pal et al. [6] and Yang et al. [7]. (c) Modified from Pal et al. [6].

studies with receptor and peptide chimeras [12–15]. However, due to the lack of structural data, the molecular details of the activation of class B receptors had long remained elusive.

How the C-termini of peptide ligands bind the N-termini of class B receptors has been known since the mid-2000s, when the first structures of class B NTDs in complex with their ligands were solved via NMR and crystallography [6, 16]. Despite the low sequence homology, class B NTDs were found to share a common architecture, which consists of a N-terminal α-helix and a short consensus repeat (SRC), or Sushi domain (two pairs of antiparallel β-sheets) [17], stabilized by three disulfide bonds. This folding of the NTD generates a central hydrophobic groove. The C-termini of class B peptides fold as a rigid amphipathic α-helix and dock to this groove, whereby they resemble a "hot dog in the bun" [18].

Deciphering the molecular interaction of the N-terminus of class B ligands with the TM bundle of the receptors has been very challenging. The first crystal structures of class B 7TMs, the glucagon receptor (GCGR) and the corticotropin releasing factor receptor type 1 (CRF_1R), lack the NTD and represent the inactive receptors in complex with small molecule antagonists [19–21]. Interestingly, both antagonists in these structures bind to allosteric sites that are likely inaccessible to the natural peptide ligands: CP-376395 sits in a pocket of CRF_1R which is located between helices III, V, and VI in the cytoplasmic half of the 7TM [19], whereas both NNC0640 and MK-0893 bind outside the 7TM of GCGR, in a site between helices VI and VII deep into the membrane [21]. The same extra-helical allosteric pocket has been observed also in the more recent structure of the isolated 7TM of the glucagon-like peptide 1 receptor (GLP-1R) bound to negative allosteric modulators (NAMs) [22].

The real breakthrough into understanding class B activation has come with a series of structures of full-length class B GPCRs that started in 2017 and are reviewed here up to June 2019 (Table 10.1). Four of these structures were solved by crystallography: GLP-1R bound to an artificial short peptide agonist [25], the GCGR in complex with a small-molecule NAM [23], and peptide NNC1702 (analog to glucagon) [24], the parathyroid hormone receptor (PTH1R) bound to a PTH analog [30]. Other five structures, all including the G protein complex, were achieved via single-particle cryo-electron microscopy (cryo-EM): the calcitonin receptor (CTR) bound to salmon calcitonin (sCT) [28], the GLP-1R bound to GLP-1 [26] and to the biased agonist Exendin-P5 (ExP5) [27], the calcitonin gene-related peptide (CGRP) receptor bound to CGRP [29], the PTH1R receptor in complex with a long-acting PTH analog (LA-PTH) [31].

The flourishing of cryo-EM for the determination of GPCR structures may be surprising because the method has been classically applied to investigate large complexes at low resolution (e.g. ribosomes [32]). Recent technological advances, including the introduction of the direct-electron detectors, the application of new phase contrast methods (Volta phase plate), combined to enhanced vitrification techniques and improved procedures of data processing, have dramatically increased the resolution of cryo-EM ("the resolution revolution" [33]), making of it an extremely powerful technique for the structural investigation in atomic detail even of membrane proteins and relatively small complexes (<200 kDa) [34, 35].

Table 10.1 List of structures of class B GPCRs that include the TMD, updated to June 2019.

Receptor	Ligand	Type of ligand	G protein	State	PDB	EMD	Date Year/ month	References
GCGR*	NNC0640	Antagonist	No	Inactive	4L6R		2013/07	[20]
GCGR*	MK-0893	Antagonist	No	Inactive	5EE7		2016/05	[21]
GCGR	NNC0640	NAM	No	Inactive	5XEZ 5XF1		2017/06	[23]
GCGR	NNC1702	Partial agonist	No		5YQZ		2018/1	[24]
GLP-1*	PF-06372222 NNC0640	NAM NAM	No	Inactive	5VEW 5VEX		2017/06	[22]
GLP-1	Peptide 5	Agonist (short)	No	Active	5NX2		2017/06	[25]
GLP-1	GLP-1	Agonist	Yes	Active	5VAI	8653	2017/06	[26]
GLP-1	ExP5	Biased Agonist	Yes	Active	6B3J	7039	2018/03	[27]
CTR	sCT	Agonist	Yes	Active	5UZ7	8623	2017/06	[28]
CGRPR	CGRP	Agonist	Yes	Active	6E3Y	8978	2018/09	[29]
PTHR	engineered PTH (ePTH)	Agonist	No	Transition	6FJ3		2018/12	[30]
PTHR	LA-PTH	Agonist	Yes	Active 1 Active 2 Active 3	6NBF 6NBH 6NBI	0410 0411 0412	2019/04	[31]
CRF1R*	CP-376395		No	Inactive	4K5Y		2013/07	[19]

When not otherwise stated, the structures depict the full-length receptors. The * marks structures lacking the NTD (TMD only). Accession numbers to the atomic coordinates deposited in the Protein Data Bank (PDB) are reported for all structures. The Electron Microscopy Database (EMD) entry IDs are indicated for cryo-EM structures.

High-resolution cryo-EM gives some advantages in respect to crystallography for structure determination of GPCR complexes. In the first place, EM analysis does not require large amounts of a perfectly homogeneous sample, but is carried out on microgram quantities of analyte. Secondly, the analyte particles are not packed in a regular crystal but remain isolated in the vitreous sections. The particles are imaged one at a time, so that different conformational states can coexist in the same probe. The latter aspect is particularly beneficial for class B GPCRs, as the presence of the long NTD adds a further degree of flexibility to the intrinsically wide conformational landscape of GPCRs in general. Moreover, the larger tolerance toward conformational inhomogeneity obviates the need of performing a bunch of manipulations of the receptor sequence, which are usually required to obtain GPCR crystals (thermostabilizing mutations, protein fusions). In this way, images of wild type receptors can be achieved.

We provide here a general description of the new findings about peptide binding and activation of class B receptors that have come from the EM structures of peptide-receptor-G protein complexes and additional crystallography data (Table 10.1). Class B ligands can be grouped by sequence similarity into three subfamilies: glucagon-like, calcitonin-like, CRF-like, with CRF indicating the corticotropin releasing factor (Figure 10.1c) [6]. Unfortunately, while 3D data are available for the first two subfamilies, no full-length structure has been solved for a CRF receptor so far. Therefore, we will include in the discussion the most recent models for binding of the natural agonists CRF and Urocortin 1 (Ucn1) to the CRF_1R [36, 37]. The models are built on the existing structures of isolated domains, are based on extensive cross-linking studies using genetically incorporated cross-linkers [38, 39] and are further validated by *in silico* experiments. Although such models cannot offer the precision of structures based on large datasets of biophysical data, they still represent the best available approximation [40] and add useful terms of comparison for a general overview.

10.2 The Receptor-Bound Class B Peptides

All receptor-bound class B peptide ligands form extensive interactions both with the NTD and the 7TM of their receptors. In all available structures, the NTD extends out of the lipid layer that contains the TM bundle. The position of the C-terminus of the peptide on the NTD is identical or near-to-identical to the position observed in the previously solved structures of the ligands in complex with the isolated NTDs. At the other end of the peptide, the N-terminus interacts with the receptor core. Without a doubt, the recent structures have fully validated the two-domain model for peptide binding to class B GPCRs.

When isolated in aqueous solvents, all class B peptides show a high propensity to adopt a α-helical conformation. The structures of the peptide–receptors complexes show that in the bound state the ligands also adopt a mostly helical conformation. These helices are amphipathic, and face with the hydrophobic side toward the hydrophobic regions of the receptor, i.e. the binding groove in the NTD and the hydrophobic environment of the TMD [6, 28]. However, there are important differences between the subfamilies. Glucagon-like peptides, including the GCG analog, GLP-1, ExP5, and PTH dock to the receptors as rigid helices and penetrate the TMD at a steep angle, which is quite variable between the peptides, with PTH showing the most vertical inclination (Figure 10.2a). These peptides

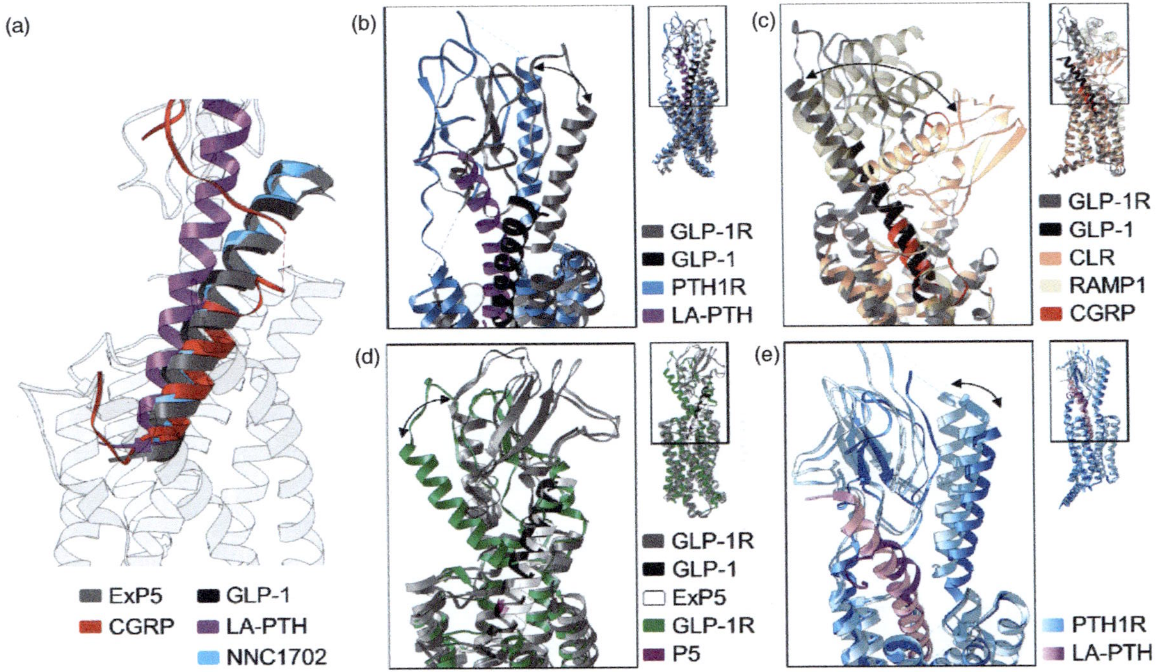

Figure 10.2 Ligand binding at class B GPCRs. (a) Ribbon representation of the class B peptide agonists ExP5 (PDB 6B3J), GLP-1 (PDB 5VAI), CGRP (PDB 3E3Y), LA-PTH (PDB 6NBF), and the GCGR-agonist NNC1702 (PDB 5YQZ) at the entrance of the TMD of their receptors (transparent). The image was generated by structural superimposition of the receptors. For the sake of clarity, only the GLP-1R (5VAI) is shown. The helical ligands show a vertical inclination at different angles. Note that the C-terminus of CGRP is unstructured. (b–e). Distinct orientation of the NTD relative to the TMD. The structures of (b) LA-PTH-bound PTH1R (PDB 6NBH) and (c) CGRP-bound CLR-RAMP1 complex (PDB 6E3Y) were superimposed to the structure of GLP-1-bound GLP-1R (PDB 5VAI). (d) GLP-1R bound either to GLP-1 (PDB 5VAI), or the short peptide agonist P5 (PDB 5NX2), or the biased agonist ExP5 (PDB 6B3J). Note that the orientation of the NTD in the GLP-1- and ExP5-bound state is very similar. Arrows indicate different positions of the N-terminal α-helix of the NTD. (e) Coexisting conformations in the EM-structures of LA-PTH-bound PTH1R (PDB 6NBF/H/I). The arrows highlight the shift in the positioning of the αN helix, which is taken reference for the whole NTD.

maintain the helical structure until either the N-terminal or penultimate residue at the N-terminus (GCG) [24, 26, 31]. Instead, CT and CGRP show only partial helical segments [28, 29]. Strikingly, the whole C-terminal segment of CGRP that docks to the NTD is unstructured. Both peptides are helical at the entrance of the TMD, which they approach at a slightly milder angle compared to GCG and in a position closer to helix I. As they penetrate the TMD, both CT and CGRP form a kink and unwrap the helix toward their N-terminus, with the N-terminal amino acids (5 and 7 residues for CT and CGRP, respectively) forming a loop. In the models for peptide binding to the CRF_1R [36, 37], cross-linking patterns suggest that CRF agonists approach the TMD in a helical conformation, but form a kink at the entrance of the TMD binding pocket. After the kink, the peptides resume the helical conformation for two helical turns, and finally adopt an extended conformation in the N-terminal stretch, which is similar to the CGRP conformation. The very N-terminus of CRF peptides, which is longer than that of all other class B ligands, exits the TMD and is exposed to the solvent. Overall, the structures have shown that, although peptide ligands bind to the NTDs of class B receptors in a similar fashion, they adopt different conformations in the TMD pocket, with different binding paths being followed by shorter peptides (glucagon like, <30 residues) or longer ones.

At the cytosolic side, a notable peculiarity of class B GCPRs is the existence of a long helix VIII, which was observed in all structures, although in minor extent in the CGRPR structure. This helix is longer than observed in class A structure and has an amphipathic character. The hydrophobic side contains conserved bulky aromatic residues which are buried in the detergent the receptor is embedded in. In the cell, helix VIII is expected to be in close contact with the membrane and may stabilize the receptor. Indeed, truncation of helix VIII at the CTR leads to reduced cell-surface expression [28]. Instead, the hydrophilic side of the helix VIII displays polar residues and interacts with the G protein. A distal tail of further amino acids downstream from helix VIII is in general not resolved in the structures, which suggests a high flexibility and lack of interactions with the G protein.

10.3 The NTD and Hinge Region

An intriguing issue for class B GPCRs is the orientation of the NTD in respect to the membrane plane and the TMD. Based on molecular dynamics (MDs) simulations, hydrogen/deuterium exchange (HDX), and disulfide cross-linking studies at the GCG receptor, it has been proposed that in the apo-state the receptor oscillates between an "open" conformational state, in which the NTD is almost perpendicular to the membrane plane, and a "closed" conformational state, in which the NTD covers the extracellular surface of the TMD [7] (Figure 10.1b). This equilibrium is essential for the initial recruitment of the peptide ligand, which stabilizes the open conformation upon complete binding. The structural data strongly support this hypothesis. In the inactive conformation of the GCGR bound to the small-molecule NAM [23] the NTD sits on top of the TMD, whereas it assumes an "open" conformation when the GCGR is bound to a glucagon analog [24].

Experimental findings at the GCGR, along with MD simulations, support the notion that the closed conformation is stabilized by interactions of the NTD with ECL1 and ECL3. A tight spatial proximity between the NTD and both ECL1 and ECL3 in the apo-GCGR is suggested by the occurrence of disulfide cross-linking between the NTD and both the ECL1 [23] and the ECL3 [7] which was confirmed by independent disulfide trapping experiments. Moreover, in a study with GLP-1R/GCGR chimeras, mutations have been identified both in the NTD and in ELC3, which can independently increase the basal activity of the GCGR, possibly by destabilizing the closed state [41]. Likewise, mutations in ECL3 of the CGRPR have been identified that increase the basal and ligand-induced activity of this receptor [42].

All structures of peptide-bound class B receptors feature an "open" or "extended" conformation, which applies both for the structures with peptide ligands of natural length and the crystal of GLP-1R bound to the short 11-mer agonist "Peptide5" (P5) [25]. However, the precise orientation of the NTD in respect to the TMD is not the same for all receptors (Figure 10.2b,c) nor for the same receptor bound to different ligands (Figure 10.2d). The degree of mobility of the NTD also differs between the receptors. In the case of the CTR and the PTHR, the structures suggest that even in the peptide-bound state, the orientation of the NTD remains flexible. In particular, the variability in the conformation of the NTD relative to the transmembrane core in the CTR structure limited the maximal achievable resolution of the structure in this region [28]. In the structure of the PTHR bound to PTH, three distinct positions of the NTD were clearly identified (Figure 10.2e). In all three conformational states, the NTD is almost perpendicular to the membrane plane, but it is rotated differently with respect to the TMD [31]. This rotation does not influence the position of the PTH agonist in the TMD, which remains the same, so that the NTD twists around the ligand. Interestingly, in one state the PTH C-terminus loses contact with the extracellular domain (ECD) and forms a kink, protruding toward the solvent. A predominant position of the NTD was observed in all structures of the GLP-1R, which features a lower mobility compared to the CTR and PTHR.

A special case is represented by the CGRPR, which is a complex of the calcitonin receptor-like receptor (CLR) and the receptor activity-modifying protein 1 (RAMP1) [29]. While the CTR, which has high homology with the CLR, showed a large variability in the orientation of the NTD, the ECD of the ligand-bound CTR-RAMP1 complex features a single predominant orientation. In the membrane, the RAMP is stably anchored to the receptor by strong interactions with helices III, IV, and V. The NTD of the RAMP is connected to the TM helix by a linker that is unstructured, but is tightly connected via hydrogen bonding both with the ECL2 and the NTD of the CLR. In the soluble extracellular region, the NTD of the RAMP is positioned almost as a clam at the side of the CLR NTD. In this way, the NTD of the CLR is kept stably in place by the ligand on one side and by the RAMP at the other side, showing overall a very limited mobility.

Distinct positioning of the NTD is not only observed between different receptors but also when the same receptor is bound to different ligands, suggesting that the ligand influences the orientation of the NTD (Figure 10.2e). When GLP-1R is bound to the short 11-mer agonist "Peptide 5" the NTD positioning is even not compatible with the binding of ligands of natural length [25]. This conformation is probably induced by the interaction of the N-terminal helix of the NTD with the short ligand, interaction that does not exist in the GLP-1 bound structure. Only subtle differences are instead observed when the receptor is bound to GLP-1 and ExP5, which is a G protein biased agonist of the same length as GLP-1 [27]. Hints that the bound ligand influences the orientation of the NTD come also by the peptide-CRF_1R models [36].

NTD and TMD are connected by a flexible region upstream of helix I, indicated as the "hinge" or "stalk." By comparing the NAM-bound and the peptide-bound glucagon structure, a major conformational rearrangement of the stalk accompanies the repositioning of the NTD. In the structure of the inactive GCGR bound to a NAM [23], the stalk forms a short β-strand that runs parallel to the membrane plane across the TMD from the top of helix III toward helix I (Figure 10.3a). At its side, ECL1 forms a β-hairpin that strongly interacts with it via hydrogen bonding, thus building a compact three-strand β-sheet that partially occludes the binding site for the peptide ligand. The β-sheet structure is not observed in the peptide-bound GCGR structure [24]. Instead, when the peptide is bound, the hinge adopts a helical conformation that extends for three turns above the plane of the other helices and supports the "open" positioning of the NTD (Figure 10.3a, middle panel). The same conformation of helix I had been observed in the crystal of the isolated TMD of GCGR, where the NTD was replaced by the globular protein BRIL [20]. In the peptide-bound structure, the stalk does not interact with ECL1, but rather engages extensive interaction with the mid-region of the peptide ligand. Overall, it is likely that the β-sheet observed in the NAM-bound GCGR stabilizes the inactive, closed conformation of the receptor, whereas ligand binding promotes the formation of the helical stalk and stabilizes it in the open conformation.

Due to the lack of full-length structures for other inactive class B receptors, it is unclear whether the β-sheet formed by ECL1 and the stalk is a common feature for all receptors of this class. It may be worth remembering that the structure of the inactive NAM–GCGR complex includes an antibody (mAb1) that stabilizes the

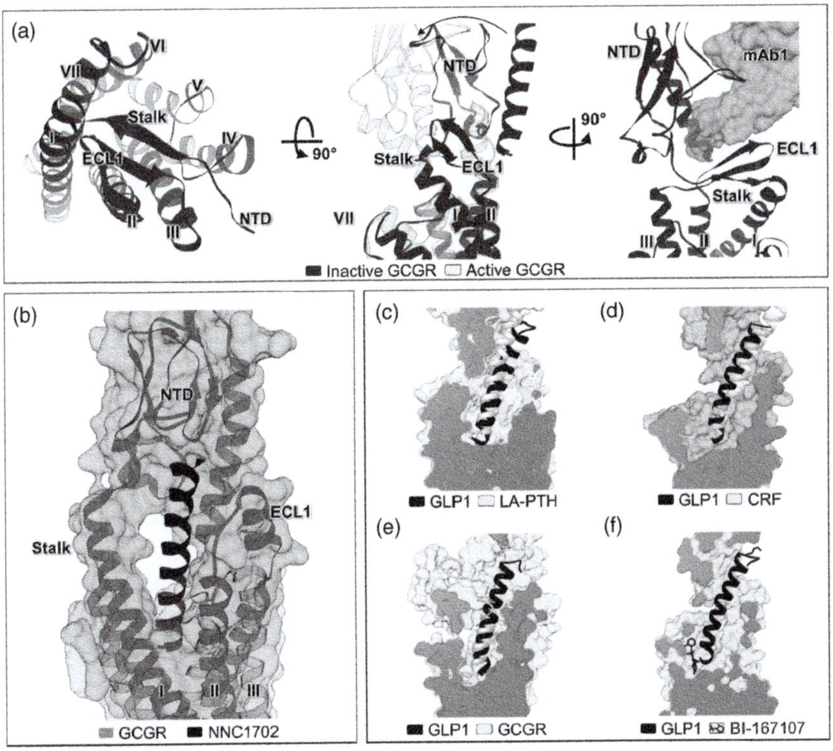

Figure 10.3 Stalk and extracellular loops of inactive and active class B receptor conformations. (a) The stalk and the ECL1 form a compact three-strand β-sheet that partially occludes the ligand-binding pocket in the inactive structure of GCGR (PDB 5XEZ). This interaction changes the orientation of the NTD in comparison to an agonist-bound receptor structure (middle panel; PDB 5YQZ, transparent). The structure of the inactive state of the GCGR includes an antibody (mAb1) that interacts both with the NTD and ECL1. The arrows highlight the shift in the positioning of the αN helix. (b) Binding mode of the peptide agonist NNC1702 to the GCGR (PDB 5YGZ): in the peptide-bound conformation, ECL1 features a short helical structure. The position of the peptide is stabilized by interactions with the ECL1 on one side and the stalk on the other side. (c–f) Class B peptides penetrate the TMDs of their cognate receptor to a similar depth. Pairwise comparison of the GLP-1 in the TMD binding pocket of GLP-1R (PDB 5VAI) with (c) LA-PTH (PDB 6NBF), (d) CRF in the most recent molecular model of the CRF-CRF$_1$R complex, (e) CGRP (PDB 6E3Y), and (f) the highly potent small-molecule agonist of the β_2-AR BI-167107 (PDB 3SN6).

NTD (Figure 10.3a, right panel). This antibody interacts with both the NTD and ECL1, and it cannot be excluded that it exerts an influence on the NTD orientation observed in this crystal. Also the helical conformation of the hinge region in the active conformation has not been observed in all structures. For instance, while helix I extends above the plane of the other helices in the CT and CGRP receptors [29], the hinge region is unstructured in the GLP-1R and PTHR structures.

Finally, a role in constraining the mobility of the NTD in the "open" conformation of class B peptide-receptor complexes may be played by ECL1. The ECL1 of class B GPCRs is in general longer compared to most class A GPCRs [20]. The conformation of ECL1 differs between those receptors that show a predominant stable position of the NTD (GLP-1R and GCGR) and those that feature higher NTD flexibility (CTR, PTHR). In the structures of GLP-1R bound to GLP-1 and ExP5, ECL1 features a short helical segment which packs against the middle segment of the ligand (Figure 10.3b). In this way, the peptide is kept tightly in place by ECL1 on one side and the stalk on the other side, which stabilize the whole complex. In addition, a helical extension of the extracellular tip of helix II elevates the position of ECL1 above the membrane plane, so that the loop is suitably positioned to directly engage in intramolecular interactions with the NTD. Helix II and ECL1 adopt a similar conformation in the complex of GLP-1R with the 11-mer agonist, where ECL1 interacts both with the ligand and the NTD [25]. Likewise, an extension of helix II and a helical folding of ECL1 are observed in the peptide-bound GCGR crystal [24]. Similar to the GLP-1R structures, ECL1 packs against the bound glucagon analog and contacts the NTD. Instead, ECL1 is unstructured in the CTR and in both the crystal and EM structure of the PTHR, showing no contacts with the ligands or the NTDs. The propensity of ECL1 to form a helical segment is likely receptor-specific and not necessarily dependent on the ligand, as the loop can feature a helical segment also in the absence of peptide (e.g. in the isolated TMD of CRF_1R) [19].

10.4 The Activated Class B TMD

Activation of class B GPCRs is triggered by the interaction of the N-terminus of the peptide ligand with the TMD of the receptor. The structures show that class B TMDs feature wide binding pockets, ranging from 3300 to 3700 $Å^3$ in size. Such large pockets are suited to accommodate large peptide ligands, although the actual pocket occupancy ranges between 30% and 50% [31]. Besides this general shape, each TMD pocket features a unique amino acid composition that is tailored to the sequence of the N-terminus of the cognate peptide. In this way, the TMD contributes a further degree of specificity for the recognition of the cognate ligand in addition to the highly specific recognition of the peptide C-terminus by the NTD [6].

The comparison between active peptide-bound structures and the inactive structures shows major rearrangements in the conformation of the TMD (Figure 10.4a). On the extracellular side, active structures feature an outward movement of the extracellular halves of helix VI and helix VII around the conserved $Gly^{6.50}$ and $Gly^{7.50}$ flexible hinges (number in superscript refer to Wootten numbering for class B GPCRs [43]), which takes along a shift of the position of ECL3. An inherent

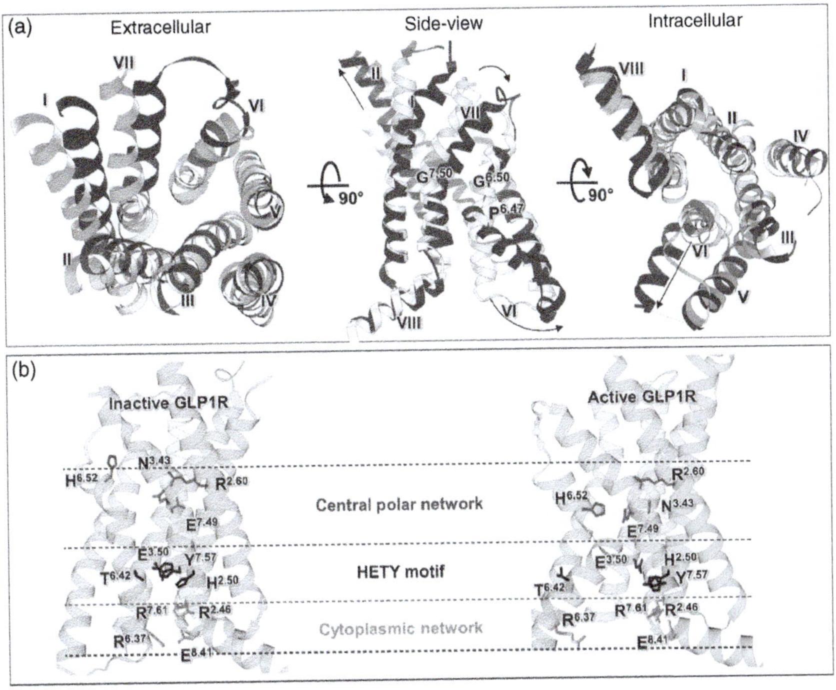

Figure 10.4 Ligand-induced TMD conformational changes at class B GPCRs. (a) Superimposition of the inactive (light grey; PDB 5VEW) and active (dark grey; PDB 5VAI) structures of GLP-1R. A hinge around the conserved $Gly^{6.50}$ and $Gly^{7.50}$ enables a pronounced outward movement of the extracellular side of helices VI and VII and a simultaneous shift of the intracellular tip of helix VI. Elongation of helix II at the extracellular side supports the helical structure of ECL1. (b) Polar networks identified in the TMD of the GLP-1R. The networks rearrange in the transition from the inactive (left, PDB 5VEW) to the active state (right, PDB 5VAI). In particular, the pronounced outward movement of the intracellular tip of helix VI disrupts the HETY motif and the cytoplasmic network, thus enabling the opening of the cytosolic cavity and the interaction with the G protein. Source: (b) Based on Wootten et al. [43, 44].

structural flexibility of helix VI and VII is indicated by the high temperature factors observed for these helices in the crystals of the inactive receptors [19, 20] and weak densities in cryo-EM maps in these regions [28]. Together with a shift of helix I and limited changes in the positions of the other helices, the movement of helix VI and VII opens the TM bundle for ligand binding and yields a characteristic V-shape of the binding pocket (Figure 10.1b).

This opening of the pocket upon ligand binding is a distinctive feature of class B receptors and has no counterpart in the activation of class A GPCRs. On the contrary, the binding pocket of class A ligands in general contracts upon ligand binding, which can happen by inward movements of helices VI and VII [45–47].

In some cases, class A pockets can even close on top of the bound ligand [48, 49]. These marked conformational differences reflect first the difference in the nature of the agonists, with the V-shaped open binding pocket tailored to accommodate large peptides for class B, whereas class A pockets contract upon binding of small-molecule agonists. Second, the conformational rearrangement observed around the ligand binding site of class A receptors is in general smaller compared to the large rearrangement, including important conformational changes in the hinge region, which are observed for class B receptors.

At the intracellular side, the most prominent structural change observed in the active receptors compared to the inactive structures is a large 15–18 Å outward movement of the intracellular half of helix VI with respect to the central axis of the TMD. This is accompanied by a relatively small movement of the cytoplasmic end of helix V. The inactive state of class B receptors is stabilized by conserved polar networks, two of which are located near and at the cytoplasmic face of the TMD, respectively (Figure 10.4b) [43, 44]. The first network, the so-called HETY motif, is established by a group of four polar amino acids distributed between helices II, III, VI, and VII and comprises: $His^{2.50}$, $Glu^{3.50}$, $Thr^{6.42}$, and $Tyr^{7.57}$. This HETY motif is considered the counterpart to the E/DRY motif, which is found in class A GPCRs and likewise locks the base of the receptor in an inactive conformation [50–52]. Indeed, mutations of the HETY motif yield constitutively active receptors [51, 53, 54]. The second network, also indicated as cytoplasmic network, is formed by four amino acids located between helices II, VI, VII further down toward the cytoplasm and helix VIII: $Arg^{2.46}$, $Arg/Lys^{6.37}$, $Asn^{7.61}$, and $Glu^{8.41}$. During receptor activation, the large movement of helix VI induced by agonist binding disrupts these polar networks and opens up the cytoplasmic side of the TMD to enable G protein docking.

The outward movement of both ends of helix VI generate a sharp kink (about 60° in the CTR, near-80° in the PTHR [31]) in the middle of the helix. This kink is enabled by the presence of a highly conserved PXXG ($Pro^{6.47}$-X-X-$Gly^{6.50}$) motif, where X denotes a bulky hydrophobic amino acid, conserved as $Leu/Val^{6.48}$ and $Leu/Phe^{6.49}$. Kinks in helices are often observed at Pro residues, because Pro forms a tertiary amide bond that is not optimally accommodated in a α-helical structure and also lacks a H-bond donor for the stabilization of the helical backbone [55]. The combination of a Pro with a flexible Gly in its proximity confers to the mid portion of helix VI an exceptional flexibility to enable the transition from the inactive to the active state. Indeed, whereas modest kinks have been observed also in helix VI and VII of class A GPCRs, such pronounced kinks have not been observed in any GPCR helix of other classes. In the inactive state, the PXXG motif is stabilized by hydrophobic interactions with a cluster of hydrophobic residues, that are located approximately at the same level in the TMD. Activation of the receptors leads to a major rearrangement of the amino acids of this hydrophobic network, which reorganizes to stabilize the active state [27]. In addition, as highlighted in the work that describes the active structure of the LA-PTH-bound PTHR [31], the backbone of the kink in helix VI is stabilized by interactions of the PXXG motif with $His^{6.52}$, $Gln^{7.49}$, and $Asn^{5.50}$. These residues, which belong to a central polar network (Figure 10.4b, further described in the next paragraph), are conserved among class B receptors, suggesting a common mechanism for the stabilization of the kink in helix VI.

In the TMD core, class B GPCRs feature a central network of four conserved polar residues located between helices II, III, VI, and VII: $Arg^{2.60}$, $Asn^{3.43}$, $His^{6.52}$, and $Gln^{7.49}$. This network plays a crucial role in receptor expression and integrity, as well as in the regulation of receptor activation and downstream signaling [20, 21, 43, 44, 51, 56, 57]. Mutation of any residue in the central polar network results in impaired signaling at the GLP-1R, in a pathway- and ligand-dependent manner. Structural data show that the polar interactions stabilize the receptor conformation in the inactive state. The network is preserved upon agonist binding, but it undergoes major rearrangements, which destabilize the inactive conformation of helix VI and enable the transition to the active state.

All class B peptides penetrate the TMD to a similar depth (Figure 10.3c–e). This is comparable to that of binding sites of orthosteric small-molecule agonists of class A GPCRs, such as the high affinity agonist BI-167107 in the structure of the β_2-adrenergic receptor (AR)-G_s protein complex [58] (Figure 10.3f). Given the importance of the movement of helix VI for activation, it is not surprising that all peptide agonists, independent from the conformation that they assume, reach a similar position near helix VI in the TMD pocket. The peptide binding pocket is located just above the central polar network and is delimited by all TM helices except helix IV. Helix IV is located at the periphery of the TM bundle as observed in class A GPCRs. The ligands engage in important interactions with all helices in the pocket. In the case of helix VI, peptide binding may induce a partial unwinding of the helical tip, which is clearly observed in the GLP-1R and the CTR structures compared to the structure of the antagonist-bound GCGR TMD [21]. Cross-linking experiments at the CRF_1R reveal a stretch of contiguous amino acids interacting with the peptide agonist at the tip helix VI [36], which also supports a partially unwound conformation of helix VI compared to the conformation observed in the antagonist-bound crystal [19].

Overall, the major conformational rearrangements observed between inactive and active structures of class B receptors are very similar to those observed in class A GPCRs. In particular, the intracellular tips of all transmembrane helices are located at the same relative positions, which results in a common structural motif for the interaction of the G protein. Overall, the structural data suggest that the global cytoplasmic changes that accompany receptor activation, especially the large outward moment of helix VI, are universally conserved across GPCR classes.

10.5 Ligand Interactions with the Extracellular Loops

Interactions of class B peptides with the ECLs of their receptors are crucial for correct expression and functioning of their receptors, as supported by a large number of mutagenesis studies, studies with receptor chimeras and cross-linking studies [59].

The structural data show a prominent role for ECL2. In all GPCRs, the conformation of ECL2 is partially constrained by a universally conserved disulfide bridge between a cysteine residue in the loop and a cysteine at the top of helix III, so that the loop is positioned right above the binding pocket. The fold of ECL2 is unstructured in all class B structures. This is a major difference compared to many class A GPCRs, where ECL2 forms a β-hairpin [60]. Key electrostatic interactions between the bound ligands and ECL2 are observed in all structures of peptide-bound GLP-1 and are predicted in the CTR structure [28]. Extensive interactions between the ligand and ECL2 are described also in the structures of the PTHR and in the CGRPR structure. A critical role for the sequence and conformation of ECL2 in the activation of class B receptors is supported by a number of mutagenesis, cross-linking, and *in silico* studies on different receptors, including the GLP-1R [56, 61–63], GCGR [20, 64], CTR [65], CGRPR [66–69], secretin receptor [70], CRF receptors [71–74], and other class B GPCRs [59]. Interestingly, interactions with the ECL2 have been shown to be very important also for the interaction between the CLR and the RAMP1 in the structure of CGRPR [29]. The importance of ECL2 for ligand binding and receptor function has been well described also for class A GPCRs [60, 75].

While the crucial role of ECL2 for ligand binding and GPCR activation is universally recognized, interaction patterns of ligands with ECL1 and ECL3 vary substantially between different receptors. The ECL1 of class B GPCRs is in general longer (>10 residues) compared to class A GPCRs (4–5 residues) [20]. Prominent interactions of ECL1 with the bound ligand are observed in the structures of GLP-1R and GCGR. In these receptors, ECL1 is quite long, compared to the CTR and CLR. It forms a helical segment that is elevated respect to the TMD by an extension of helix II and packs against the C-terminal segment of the ligand (Figure 10.3b). HDX experiments and MD simulations at the GCGR [7] suggest that ligand binding stabilizes the helical conformation of ECL1. An important role of ECL1 in ligand binding is supported by mutagenesis experiments at both the GLP-1R [56, 62, 76] and the GCGR [20, 64, 77]. Only limited contacts are observed between the bound CGRP and ECL1 in the structure of the CGRPR. This is in line with mutagenesis studies, which have shown only limited effects of mutations in the region of ECL1 itself on ligand binding, whereas larger effects were observed for positions at the top of helices II and III [42]. In the structures of the PTHR and CTR, ECL1 is not resolved, which suggests a high conformational flexibility of the loop and weak interactions with the ligands. In the structure of the inactive CRF_1R TMD bound to a small molecule antagonist, the ECL1 of the CRF_1R features a well-structured helical segment [19], although there is no bound peptide to stabilize it. This result suggests that the helical folding of the ECL1 is only partially influenced by the bound ligand and depends in any case on the intrinsic conformational propensity of each receptor. In the extensive photo-cross-linking mapping of the CRF_1R binding pocket, no hits in ECL1 indicating ligand-receptor interactions were found either for peptide agonists or antagonists [37].

ECL3, which is in general shorter than ECL2, is probably the most flexible in class B GPCRs, as it follows the movements of helix VI and VII during receptor activation. High structural flexibility of the top of TM6/ECL3 is supported by high temperature factors in crystal structures and weak densities in many cryo-EM maps. The loop is not resolved in the structures of CTR and CGRPR. It is also not resolved in the structure of the isolated inactive GCGR TMD, which probably indicates only limited interactions between this loop and the bound ligand. Nonetheless, mutations in ECL3 have been shown to both influence the cell surface expression of the CGRPR, probably by hampering the correct targeting to the membrane, and to affect CGRP potency [42].

In the structures of peptide-bound GCGR, GLP-1R, and PTHR, ECL3 appears unstructured and shows notable interactions with the N-terminal segments of the ligands. In the GLP-1R structures, the loop remains very flexible in the ligand-bound state, as indicated by a poor quality of the cryo-EM map for part of the loop, which hampered the accurate modeling of amino acid side chains [26, 27]. In the structure of peptide-bound GCGR, ECL3 engages in a hydrogen bond with the N-terminal region of the glucagon analog [24]. In the structures of the PTHR, ECL3 limits together with ECL2 the binding pocket for PTH in the TMD and interacts with the N-terminal segment of the ligand [30, 31]. MD simulations run on the model of the ligand-bound CRF_1R showed a very high flexibility for ECL3, with the loop swaying a variable distance of ~6–14 Å from the ligand [37]. Indeed, ECL3 is resolved in only one of the three molecules composing the crystallographic unit of the structure of CRF_1R TMD. Nonetheless, the presence of cross-linking hits in ECL3 indicates the loop must come in proximity to the ligand in the associated complex [37]. This is consistent with mutagenesis experiments and studies with truncated receptors showing that ECL3 plays a major role in CRF binding to CRF_1R [78, 79]. The occurrence of cross-linking between the bound peptide and ECL3 has also been shown in disulfide trapping experiments with the secretin receptor [70, 80].

Numerous studies have shown that ECL3, together with the extracellular boundaries of helices VI and VII and ECL2, is a major player in biased signaling at the GLP-1R [27, 56]. Indeed, the comparison of the two structures of GLP-1R bound either to the natural agonist GLP-1 or the biased agonist ExP5 show that the conformation of ECL3 depends on the bound ligand. In particular, ExP5 induces a larger outward shift of the extracellular portions of helix VI compared to GLP-1, which is accompanied by a distinct positioning of the upper portion of helix VII and a drastically different conformation of ECL3 (Figure 10.5a). These findings explain why mutations of specific residues in this region have different effects on binding and signaling of the two ligands. Overall, cross-linking and mutagenesis experiments at several class B receptors speak for the functional importance of ECL3 across the family [59].

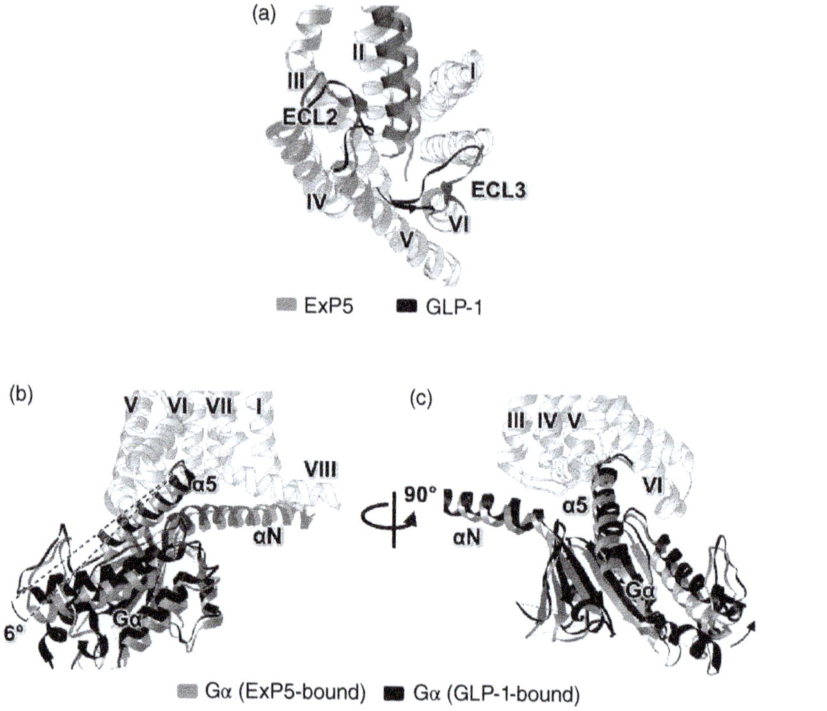

Figure 10.5 Biased agonism at the GLP-1R. Superimposition of GLP-1R bound to either GLP-1 (PDB 5VAI) or the biased agonist ExP5 (PDB 6B3J). (a) At the extracellular side, ExP5 induces a strikingly different conformation of ECL3 compared to that induced by GLP-1. (b and c) At the intracellular side, the two ligands promote a different engagement of the G protein, with the C-terminal α5-helix of the Ras-like domain of Gα docking to the receptor at a different angle and the whole G protein showing distinct orientations.

10.6 Interactions Between Class B GPCRs and G_s Protein

Class B GPCRs signal primarily via G_s (Figure 10.6a). While no class B crystal structure includes the G protein, the five cryo-EM structures of the CT, GLP-1, PTH, and CGRP receptors depict the complex of the ligand-bound GPCR with the G_s protein complete of its subunits α, β, and γ. The Gα subunit of G proteins comprises an alpha helical domain (AHD) and a Ras-like domain, which enclose the nucleotide-binding site at their interface. Gα is very dynamic [81, 82]. When a G protein is activated, the AHD separates from the Ras-like domain to expose the nucleotide-binding site, thus enabling the GDP-to-GTP exchange that initiates the G protein-mediated signaling. In the EM structures, the Ras-like domain of the G protein is, in general, well resolved, whereas the AHD is visible only at lower resolution due to a large variability in its orientation. While in the structure of the β_2-AR-G_s protein complex the AHD was stabilized by crystal contacts in the open conformation [58], the EM maps of the class B receptors reflect the delocalization of the AHD with respect to the Ras-like domain in the absence of bound nucleotide [83, 84].

The overall conformation of the G protein is well conserved across the structures of complexes with class B receptors and is almost identical to the structure observed in the β_2-AR-G_s complex (Figure 10.6b). Likewise, the interactions between the G protein and the class B receptors are very similar as previously observed in the β_2-AR-G_s structure. This is in line with the substantial correspondence of the topology of the TM helices at the intracellular side between class A and class B receptors. Upon agonist binding, the wide outward shift of helix VI and the resulting break of the polar networks that stabilize the inactive state leads to the opening of the cytoplasmic face of the helical bundle. The G protein docks to this cavity mainly by positioning the C-terminal helix (α5-helix) of the Ras-like domain in the place occupied in the inactive state by the tip of helix VI.

The complex is stabilized by both polar and hydrophobic Van der Waals interactions. With the sole exception of helix I, all TM helices of the receptor together with ICL2, ICL3, and helix VIII engage in extensive interactions with the Ras-like domain of the G protein. The most extensive contacts take place with the α5-helix, but also involve the α4-helix, the β6-strand, and the N-terminal αN-helix (Figure 10.6c). Electrostatic interactions or hydrogen bonds between the G protein and amino acids belonging to the cytoplasmic polar network of the receptors ($Arg^{2.46}$, $Arg/Lys^{6.37}$, $Asn^{7.61}$, and $Glu^{8.41}$, see above) stabilize the open conformation of the GPCR base [26]. This is similar to what observed for class A GPCRs for the E/DRY network, which stabilizes the base of the receptor in the inactive state and is involved in the interaction with the G protein upon activation [52, 58]. In general, several of the receptor residues that form interactions with

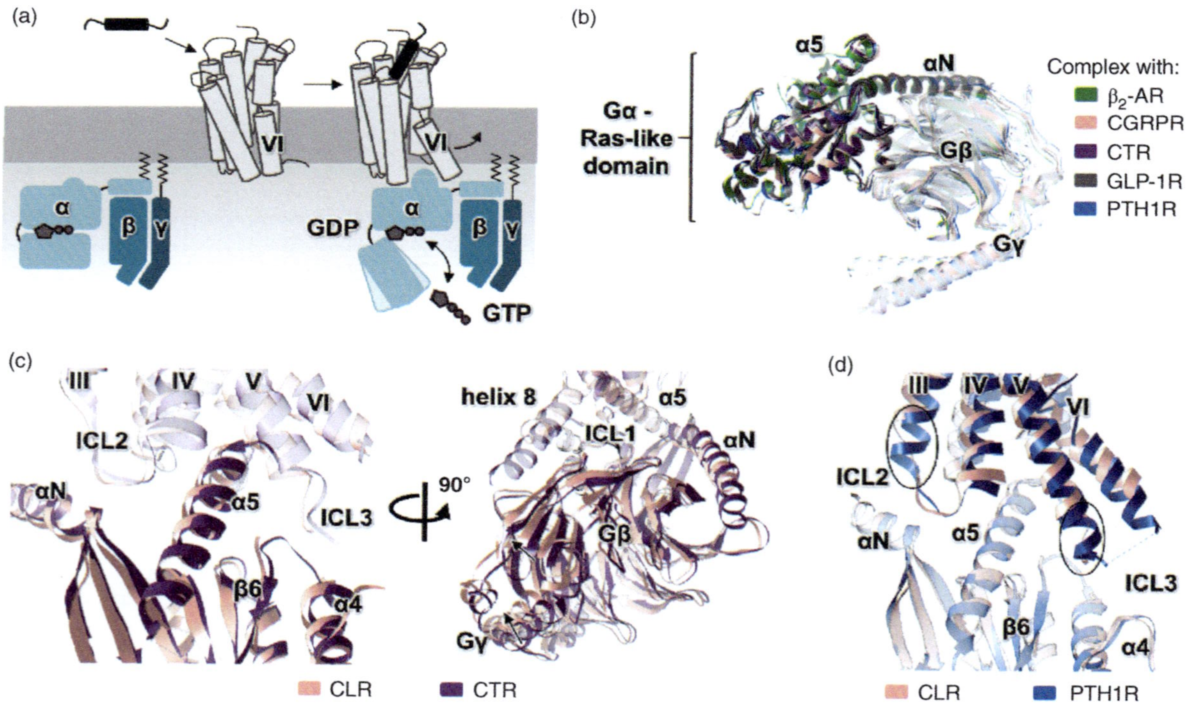

Figure 10.6 High structural homology of G_s proteins in complex with class B GPCRs. (a) Schematic representation of G protein activation. Ligand-binding to a GPCR triggers an outward movement of the intracellular tip of helix VI, which enables the binding of the heterotrimeric G protein to the cytosolic face of the receptor. The G protein-GPCR interaction increases the flexibility of the α-helical domain (AHD) of Gα, which facilitates the nucleotide-exchange. (b) Superimposition of the Ras-like domains of the G_s proteins from the structures of the complexes with the different class B GPCRs. (c) Comparison of G_s proteins in complex with CLR (PDB 6E3Y) and CTR (PDB 5UZ7) reveals the most extensive contact sites of the G protein (α5-helix, α4-helix, β6-strand), which are conserved in the complexes with all GPCRs. Variations in the conformation of the αN-helix is accompanied by a shift in the position of the βγ complex. D The α5-helix of $G\alpha_s$ in complex with CLR (PDB 6E3Y) reveals a slightly altered orientation in comparison to the PTH1R complex (PDB 6NBF) arising from receptor-specific structure and orientation of ICL2 and TM6. Source: (c) Based on Liang et al. [29].

the G_s protein are highly conserved across class B GPCRs [28, 31]. As there is no universally conserved motif across different GPCR classes at the interface with the G protein, the G protein employs the same set of C-terminal amino acids to engage in diverse interactions with different receptors, which is enabled by a certain degree of conformational flexibility [26].

Furthermore, class B GPCRs interact with the subunit β of the G_s protein [24, 26, 28, 31]. This is a major difference compared to the structure of the β_2-AR-G_s complex, where no receptor-Gβ interactions were observed. Compared to the β_2-AR-G_s structure, the direct interaction between the GPCR and the Gβ subunit takes the Gβγ complex slightly closer to the receptor [26]. Contacts with Gβ are mediated by ICL1 and the polar face of helix VIII [26, 28]. The interaction between helix VIII and the β subunit of G_s likely contributes to the efficacy of the receptor-G protein interaction, as truncation of helix VII at the CTR leads to a receptor that elicits only a minor cAMP response upon agonist stimulus [28].

Although the backbones of the Ras-like domains of the G protein of the five available class B structures overlap well with each other (Figure 10.6b), subtle differences can be observed both in the G protein conformation and in the interaction with the receptors. For instance, in the G_s-CLR complex in the CGRPR structure, the αN-helix of the G-protein is oriented differently compared to the G_s complex with the highly homologous CTR receptor [29] (Figure 10.6c). This difference in the α subunit propagates to the βγ complex, which results in a different position. Another difference between these two receptors is the length of the structured segment of the C-tail. CLR features a shorter helix VIII compared to CTR and along with this a less pronounced interaction with the Gβ-subunit [29]. Other differences can be observed in the relative positioning of Gα in respect to the receptors, which arise from a different positioning of α5 in the cytosolic pocket. Different engagement of α5 by the various GPCRs may be due to the peculiar features of ICL2 of the receptors and the extent of the outward shift of helix VI. CTR and CLR feature a longer ICL2 and a wider outward shift of helix VI compared to GLP-1R and PTHR (Figure 10.6d), which leads to a closer contact with the αN- and α5-helices [29, 31].

Comparing the GPCR-G_s interaction in the two structures of GLP-1R bound to either GLP-1 or to the G_s biased agonist ExP5 provides a unique insight into the molecular basis of signaling bias. The overall conformation of the G protein in the two complexes is very similar, with the most prominent difference being a modest shift of the very N-terminal segment of the αN-helix (Figure 10.5b). Instead, there is a notable difference in the binding mode of the Ras-like domain to the receptor. In the ExP5-bound structure, the α5-helix docks to the cytoplasmic cavity with a different angle compared to the GLP-1-bound structure (6° variance), which is accompanied by an overall rotation of the G protein (Figure 10.5c). This leads to the loss of contact with the α4-helix and to weaker interactions between the G protein and ICL2. Moreover, the intracellular tip of helix V and ICL3, which were clearly defined in the GLP-1-bound structure, are not resolved at high resolution in the ExP5-bound structure, which suggest a higher flexibility of this region when the receptor is bound to the biased agonist. Compared to GLP-1, ExP5 shows an enhanced G protein efficacy and limited β-arrestin recruitment [27, 85]. We have described above that there are profound differences in the conformation of the TM bundle of the GLP-1R when bound to GLP-1 rather than ExP5. However, while differences at the extracellular side are striking (especially in the helix VI-ECL3-helix VII region), the differences in the cytosolic side are subtler. Overall, the structural data support the notion that differences in the receptor conformation induced by distinct ligands influence the conformation of the G protein [86, 87], although the structural details that lead to specific effects still remain to be elucidated.

10.7 Conclusion

In summary, the recent structures of class B GPCR represent a real breakthrough in our understanding of the activation mechanism of these receptors. Taken together, the structural data point at a common activation mechanism for class A and class B GPCRs, which includes an outward shift of the intracellular tip of helix VI as the major structural hallmark. The structures provide a solid base for rationalizing a huge body of mutagenesis data and biochemical data generated in the past for class B receptors and will facilitate the rational design of novel therapeutics for these important therapeutic targets.

References

1 Fredriksson, R., Lagerstrom, M.C., Lundin, L.G., and Schioth, H.B. (2003). The G-protein-coupled receptors in the human genome form five main families. Phylogenetic analysis, paralogon groups, and fingerprints. *Mol. Pharmacol.* 63 (6): 1256–1272.

2 Cho, Y.M., Merchant, C.E., and Kieffer, T.J. (2012). Targeting the glucagon receptor family for diabetes and obesity therapy. *Pharmacol. Ther.* 135 (3): 247–278.

3 Karsan, N. and Goadsby, P.J. (2015). Calcitonin gene-related peptide and migraine. *Curr. Opin. Neurol.* 28 (3): 250–254.

4 Kendler, D.L., Marin, F., Zerbini, C.A.F. et al. (2018). Effects of teriparatide and risedronate on new fractures in post-menopausal women with severe osteoporosis (VERO): a multicentre, double-blind, double-dummy, randomised controlled trial. *Lancet* 391 (10117): 230–240.

5 Poyner, D.R., Sexton, P.M., Marshall, I. et al. (2002). International Union of Pharmacology. XXXII. The mammalian calcitonin gene-related peptides,

adrenomedullin, amylin, and calcitonin receptors. *Pharmacol. Rev.* 54 (2): 233–246.

6 Pal, K., Melcher, K., and Xu, H.E. (2012). Structure and mechanism for recognition of peptide hormones by class B G-protein-coupled receptors. *Acta Pharmacol. Sin.* 33 (3): 300–311.

7 Yang, L., Yang, D., de Graaf, C. et al. (2015). Conformational states of the full-length glucagon receptor. *Nat. Commun.* 6: 7859.

8 Runge, S., Gram, C., Brauner-Osborne, H. et al. (2003). Three distinct epitopes on the extracellular face of the glucagon receptor determine specificity for the glucagon amino terminus. *J. Biol. Chem.* 278 (30): 28005–28010.

9 Vaudry, D., Gonzalez, B.J., Basille, M. et al. (2000). Pituitary adenylate cyclase-activating polypeptide and its receptors: from structure to functions. *Pharmacol. Rev.* 52 (2): 269–324.

10 Runge, S., Wulff, B.S., Madsen, K. et al. (2003). Different domains of the glucagon and glucagon-like peptide-1 receptors provide the critical determinants of ligand selectivity. *Br. J. Pharmacol.* 138 (5): 787–794.

11 Gulyas, J., Rivier, C., Perrin, M. et al. (1995). Potent, structurally constrained agonists and competitive antagonists of corticotropin-releasing factor. *Proc. Natl. Acad. Sci. U.S.A.* 92 (23): 10575–10579.

12 Stroop, S.D., Kuestner, R.E., Serwold, T.F. et al. (1995). Chimeric human calcitonin and glucagon receptors reveal two dissociable calcitonin interaction sites. *Biochemistry* 34 (3): 1050–1057.

13 Holtmann, M.H., Hadac, E.M., and Miller, L.J. (1995). Critical contributions of amino-terminal extracellular domains in agonist binding and activation of secretin and vasoactive intestinal polypeptide receptors. Studies of chimeric receptors. *J. Biol. Chem.* 270 (24): 14394–14398.

14 Bergwitz, C., Gardella, T.J., Flannery, M.R. et al. (1996). Full activation of chimeric receptors by hybrids between parathyroid hormone and calcitonin. Evidence for a common pattern of ligand-receptor interaction. *J. Biol. Chem.* 271 (43): 26469–26472.

15 Nielsen, S.M., Nielsen, L.Z., Hjorth, S.A. et al. (2000). Constitutive activation of tethered-peptide/corticotropin-releasing factor receptor chimeras. *Proc. Natl. Acad. Sci. U.S.A.* 97 (18): 10277–10281.

16 Hollenstein, K., de Graaf, C., Bortolato, A. et al. (2014). Insights into the structure of class B GPCRs. *Trends Pharmacol. Sci.* 35 (1): 12–22.

17 Perrin, M.H., Grace, C.R., Riek, R., and Vale, W.W. (2006). The three-dimensional structure of the N-terminal domain of corticotropin-releasing factor receptors: sushi domains and the B1 family of G protein-coupled receptors. *Ann. N.Y. Acad. Sci.* 1070: 105–119.

18 Pioszak, A.A. and Xu, H.E. (2008). Molecular recognition of parathyroid hormone by its G protein-coupled receptor. *Proc. Natl. Acad. Sci. U.S.A.* 105 (13): 5034–5039.

19 Hollenstein, K., Kean, J., Bortolato, A. et al. (2013). Structure of class B GPCR corticotropin-releasing factor receptor 1. *Nature* 499 (7459): 438–443.

20 Siu, F.Y., He, M., de Graaf, C. et al. (2013). Structure of the human glucagon class B G-protein-coupled receptor. *Nature* 499 (7459): 444–449.

21 Jazayeri, A., Dore, A.S., Lamb, D. et al. (2016). Extra-helical binding site of a glucagon receptor antagonist. *Nature* 533 (7602): 274–277.

22 Song, G., Yang, D., Wang, Y. et al. (2017). Human GLP-1 receptor transmembrane domain structure in complex with allosteric modulators. *Nature* 546 (7657): 312–315.

23 Zhang, H., Qiao, A., Yang, D. et al. (2017). Structure of the full-length glucagon class B G-protein-coupled receptor. *Nature* 546 (7657): 259–264.

24 Zhang, H., Qiao, A., Yang, L. et al. (2018). Structure of the glucagon receptor in complex with a glucagon analogue. *Nature* 553 (7686): 106–110.

25 Jazayeri, A., Rappas, M., Brown, A.J.H. et al. (2017). Crystal structure of the GLP-1 receptor bound to a peptide agonist. *Nature* 546 (7657): 254–258.

26 Zhang, Y., Sun, B., Feng, D. et al. (2017). Cryo-EM structure of the activated GLP-1 receptor in complex with a G protein. *Nature* 546 (7657): 248–253.

27 Liang, Y.L., Khoshouei, M., Glukhova, A. et al. (2018). Phase-plate cryo-EM structure of a biased agonist-bound human GLP-1 receptor-G_s complex. *Nature* 555 (7694): 121–125.

28 Liang, Y.L., Khoshouei, M., Radjainia, M. et al. (2017). Phase-plate cryo-EM structure of a class B GPCR-G-protein complex. *Nature* 546 (7656): 118–123.

29 Liang, Y.L., Khoshouei, M., Deganutti, G. et al. (2018). Cryo-EM structure of the active, G_s-protein complexed, human CGRP receptor. *Nature* 561 (7724): 492–497.

30 Ehrenmann, J., Schoppe, J., Klenk, C. et al. (2018). High-resolution crystal structure of parathyroid hormone 1 receptor in complex with a peptide agonist. *Nat. Struct. Mol. Biol.* 25 (12): 1086–1092.

31 Zhao, L.H., Ma, S., Sutkeviciute, I. et al. (2019). Structure and dynamics of the active human parathyroid hormone receptor-1. *Science* 364 (6436): 148–153.

32 Brown, A. and Shao, S. (2018). Ribosomes and cryo-EM: a duet. *Curr. Opin. Struct. Biol.* 52: 1–7.

33 Kuhlbrandt, W. (2014). Biochemistry. The resolution revolution. *Science* 343 (6178): 1443–1444.

34 Danev, R. and Baumeister, W. (2017). Expanding the boundaries of cryo-EM with phase plates. *Curr. Opin. Struct. Biol.* 46: 87–94.

35 Cheng, Y. (2018). Membrane protein structural biology in the era of single particle cryo-EM. *Curr. Opin. Struct. Biol.* 52: 58–63.

36 Coin, I., Katritch, V., Sun, T. et al. (2013). Genetically encoded chemical probes in cells reveal the binding path of urocortin-I to CRF class B GPCR. *Cell* 155 (6): 1258–1269.

37 Seidel, L., Zarzycka, B., Zaidi, S.A. et al. (2017). Structural insight into the activation of a class B G-protein-coupled receptor by peptide hormones in live human cells. *Elife* 6.

38 Coin, I., Perrin, M.H., Vale, W.W., and Wang, L. (2011). Photo-cross-linkers incorporated into G-protein-coupled receptors in mammalian cells: a ligand comparison. *Angew. Chem. Int. Ed.* 50: 8077–8081.

39 Coin, I. (2018). Application of non-canonical crosslinking amino acids to study protein-protein interactions in live cells. *Curr. Opin. Chem. Biol.* 46: 156–163.

40 Pal, K., Melcher, K., and Xu, H.E. (2013). Structure modeling using genetically engineered crosslinking. *Cell* 155 (6): 1207–1208.

41 Koth, C.M., Murray, J.M., Mukund, S. et al. (2012). Molecular basis for negative regulation of the glucagon receptor. *Proc. Natl. Acad. Sci. U.S.A.* 109 (36): 14393–14398.

42 Barwell, J., Conner, A., and Poyner, D.R. (2011). Extracellular loops 1 and 3 and their associated transmembrane regions of the calcitonin receptor-like receptor are needed for CGRP receptor function. *Biochim. Biophys. Acta* 1813 (10): 1906–1916.

43 Wootten, D., Simms, J., Miller, L.J. et al. (2013). Polar transmembrane interactions drive formation of ligand-specific and signal pathway-biased family B G protein-coupled receptor conformations. *Proc. Natl. Acad. Sci. U.S.A.* 110 (13): 5211–5216.

44 Wootten, D., Reynolds, C.A., Smith, K.J. et al. (2016). Key interactions by conserved polar amino acids located at the transmembrane helical boundaries in class B GPCRs modulate activation, effector specificity and biased signalling in the glucagon-like peptide-1 receptor. *Biochem. Pharmacol.* 118: 68–87.

45 Manglik, A. and Kruse, A.C. (2017). Structural basis for G protein-coupled receptor activation. *Biochemistry* 56 (42): 5628–5634.

46 Cao, C., Zhang, H., Yang, Z., and Wu, B. (2018). Peptide recognition, signaling and modulation of class B G protein-coupled receptors. *Curr. Opin. Struct. Biol.* 51: 53–60.

47 Warne, T., Edwards, P.C., Dore, A.S. et al. (2019). Molecular basis for high-affinity agonist binding in GPCRs. *Science* 364 (6442): 775–778.

48 Kruse, A.C., Ring, A.M., Manglik, A. et al. (2013). Activation and allosteric modulation of a muscarinic acetylcholine receptor. *Nature* 504 (7478): 101–106.

49 DeVree, B.T., Mahoney, J.P., Velez-Ruiz, G.A. et al. (2016). Allosteric coupling from G protein to the agonist-binding pocket in GPCRs. *Nature* 535 (7610): 182–186.

50 Rasmussen, S.G., Choi, H.J., Rosenbaum, D.M. et al. (2007). Crystal structure of the human β_2 adrenergic G-protein-coupled receptor. *Nature* 450 (7168): 383–387.

51 Vohra, S., Taddese, B., Conner, A.C. et al. (2013). Similarity between class A and class B G-protein-coupled receptors exemplified through calcitonin gene-related peptide receptor modelling and mutagenesis studies. *J. R. Soc. Interface* 10 (79): 20120846.

52 Vogel, R., Mahalingam, M., Ludeke, S. et al. (2008). Functional role of the "ionic lock" – an interhelical hydrogen-bond network in family A heptahelical receptors. *J. Mol. Biol.* 380 (4): 648–655.

53 Hjorth, S.A., Orskov, C., and Schwartz, T.W. (1998). Constitutive activity of glucagon receptor mutants. *Mol. Endocrinol.* 12 (1): 78–86.

54 Schipani, E., Kruse, K., and Juppner, H. (1995). A constitutively active mutant PTH-PTHrP receptor in Jansen-type metaphyseal chondrodysplasia. *Science* 268 (5207): 98–100.

55 Williams, K.A. and Deber, C.M. (1991). Proline residues in transmembrane helices: structural or dynamic role? *Biochemistry* 30 (37): 8919–8923.

56 Wootten, D., Reynolds, C.A., Smith, K.J. et al. (2016). The extracellular surface of the GLP-1 receptor is a molecular trigger for biased agonism. *Cell* 165 (7): 1632–1643.

57 Wootten, D., Reynolds, C.A., Koole, C. et al. (2016). A hydrogen-bonded polar network in the core of the glucagon-like peptide-1 receptor is a fulcrum for biased agonism: lessons from class B crystal structures. *Mol. Pharmacol.* 89 (3): 335–347.

58 Rasmussen, S.G., DeVree, B.T., Zou, Y. et al. (2011). Crystal structure of the β_2 adrenergic receptor-G_s protein complex. *Nature* 477 (7366): 549–555.

59 Dong, M., Koole, C., Wootten, D. et al. (2014). Structural and functional insights into the juxtamembranous amino-terminal tail and extracellular loop regions of class B GPCRs. *Br. J. Pharmacol.* 171 (5): 1085–1101.

60 Woolley, M.J. and Conner, A.C. (2017). Understanding the common themes and diverse roles of the second extracellular loop (ECL2) of the GPCR super-family. *Mol. Cell. Endocrinol.* 449: 3–11.

61 Koole, C., Wootten, D., Simms, J. et al. (2012). Second extracellular loop of human glucagon-like peptide-1 receptor (GLP-1R) differentially regulates orthosteric but not allosteric agonist binding and function. *J. Biol. Chem.* 287 (6): 3659–3673.

62 Yang, D., de Graaf, C., Yang, L. et al. (2016). Structural determinants of binding the seven-transmembrane domain of the glucagon-like peptide-1 receptor (GLP-1R). *J. Biol. Chem.* 291 (25): 12991–13004.

63 Koole, C., Wootten, D., Simms, J. et al. (2012). Second extracellular loop of human glucagon-like peptide-1 receptor (GLP-1R) has a critical role in GLP-1 peptide binding and receptor activation. *J. Biol. Chem.* 287 (6): 3642–3658.

64 Unson, C.G., Wu, C.R., Jiang, Y. et al. (2002). Roles of specific extracellular domains of the glucagon receptor in ligand binding and signaling. *Biochemistry* 41 (39): 11795–11803.

65 Dal Maso, E., Zhu, Y., Pham, V. et al. (2018). Extracellular loops 2 and 3 of the calcitonin receptor selectively modify agonist binding and efficacy. *Biochem. Pharmacol.* 150: 214–244.

66 Woolley, M.J., Simms, J., Mobarec, J.C. et al. (2017). Understanding the molecular functions of the second extracellular loop (ECL2) of the calcitonin gene-related peptide (CGRP) receptor using a comprehensive mutagenesis approach. *Mol. Cell. Endocrinol.* 454: 39–49.

67 Woolley, M.J., Simms, J., Uddin, S. et al. (2017). Relative antagonism of mutants of the CGRP receptor extracellular loop 2 domain (ECL2) using a truncated competitive antagonist (CGRP8-37): evidence for the dual involvement of ECL2 in the two-domain binding model. *Biochemistry* 56 (30): 3877–3880.

68 Woolley, M.J., Watkins, H.A., Taddese, B. et al. (2013). The role of ECL2 in CGRP receptor activation: a combined modelling and experimental approach. *J. R. Soc. Interface* 10 (88): 20130589.

69 Simms, J., Uddin, R., Sakmar, T.P. et al. (2018). Photoaffinity cross-linking and unnatural amino acid mutagenesis reveal insights into calcitonin gene-related peptide binding to the calcitonin receptor-like receptor/receptor activity-modifying protein 1 (CLR/RAMP1) complex. *Biochemistry* 57 (32): 4915–4922.

70 Dong, M., Lam, P.C., Orry, A. et al. (2016). Use of cysteine trapping to map spatial approximations between residues contributing to the Helix N-capping motif of secretin and distinct residues within each of the extracellular loops of its receptor. *J. Biol. Chem.* 291 (10): 5172–5184.

71 Gkountelias, K., Tselios, T., Venihaki, M. et al. (2009). Alanine scanning mutagenesis of the second extracellular loop of type 1 corticotropin-releasing factor receptor revealed residues critical for peptide binding. *Mol. Pharmacol.* 75 (4): 793–800.

72 Liaw, C.W., Grigoriadis, D.E., Lovenberg, T.W. et al. (1997). Localization of ligand-binding domains of human corticotropin-releasing factor receptor: a chimeric receptor approach. *Mol. Endocrinol.* 11 (7): 980–985.

73 Kraetke, O., Holeran, B., Berger, H. et al. (2005). Photoaffinity cross-linking of the corticotropin-releasing factor receptor type 1 with photoreactive urocortin analogues. *Biochemistry* 44 (47): 15569–15577.

74 Assil-Kishawi, I. and Abou-Samra, A.B. (2002). Sauvagine cross-links to the second extracellular loop of the corticotropin-releasing factor type 1 receptor. *J. Biol. Chem.* 277 (36): 32558–32561.

75 Wheatley, M., Wootten, D., Conner, M.T. et al. (2012). Lifting the lid on GPCRs: the role of extracellular loops. *Br. J. Pharmacol.* 165 (6): 1688–1703.

76 Xiao, Q., Jeng, W., and Wheeler, M.B. (2000). Characterization of glucagon-like peptide-1 receptor-binding determinants. *J. Mol. Endocrinol.* 25 (3): 321–335.

77 Roberts, D.J., Vertongen, P., and Waelbroeck, M. (2011). Analysis of the glucagon receptor first extracellular loop by the substituted cysteine accessibility method. *Peptides* 32 (8): 1593–1599.

78 Sydow, S., Flaccus, A., Fischer, A., and Spiess, J. (1999). The role of the fourth extracellular domain of the rat corticotropin-releasing factor receptor type 1 in ligand binding. *Eur. J. Biochem.* 259 (1, 2): 55–62.

79 Sydow, S., Radulovic, J., Dautzenberg, F.M., and Spiess, J. (1997). Structure-function relationship of different domains of the rat corticotropin-releasing factor receptor. *Brain Res. Mol. Brain Res.* 52 (2): 182–193.

80 Dong, M.Q., Xu, X.Q., Ball, A.M. et al. (2012). Mapping spatial approximations between the amino terminus of secretin and each of the extracellular loops of its receptor using cysteine trapping. *FASEB J.* 26 (12): 5092–5105.

81 Goricanec, D., Stehle, R., Egloff, P. et al. (2016). Conformational dynamics of a G-protein alpha subunit is tightly regulated by nucleotide binding. *Proc. Natl. Acad. Sci. U.S.A.* 113 (26): E3629–E3638.

82 Dror, R.O., Mildorf, T.J., Hilger, D. et al. (2015). Signal transduction. Structural basis for nucleotide exchange in heterotrimeric G proteins. *Science* 348 (6241): 1361–1365.

83 Westfield, G.H., Rasmussen, S.G., Su, M. et al. (2011). Structural flexibility of the Gαs α-helical domain in the β_2-adrenoceptor G_s complex. *Proc. Natl. Acad. Sci. U.S.A.* 108 (38): 16086–16091.

84 Van Eps, N., Preininger, A.M., Alexander, N. et al. (2011). Interaction of a G protein with an activated receptor opens the interdomain interface in the α subunit. *Proc. Natl. Acad. Sci. U.S.A.* 108 (23): 9420–9424.

85 Zhang, H., Sturchler, E., Zhu, J. et al. (2015). Autocrine selection of a GLP-1R G-protein biased agonist with potent antidiabetic effects. *Nat. Commun.* 6: 8918.

86 Furness, S.G.B., Liang, Y.L., Nowell, C.J. et al. (2016). Ligand-dependent modulation of G protein conformation alters drug efficacy. *Cell* 167 (3): 739–49.e11.

87 Gregorio, G.G., Masureel, M., Hilger, D. et al. (2017). Single-molecule analysis of ligand efficacy in β_2AR-G-protein activation. *Nature* 547 (7661): 68–73.

11

Dynamical Basis of GPCR–G Protein Coupling Selectivity and Promiscuity

Nagarajan Vaidehi[1], Fredrik Sadler[2], and Sivaraj Sivaramakrishnan[2,3]

[1] *Department of Computational and Quantitative Medicine, City of Hope Cancer Center, Beckman Research Institute of the City of Hope, Duarte, CA, USA*
[2] *Biochemistry, Molecular Biology and Biophysics Graduate Program, University of Minnesota, Minneapolis, MN, USA*
[3] *Department of Genetics, Cell Biology, and Development, University of Minnesota, Minneapolis, MN, USA*

11.1 Introduction and Scope

G protein coupled receptors (GPCRs) are seven helical transmembrane proteins expressed ubiquitously in many mammalian cell types that play a critical role in cellular signal transduction. By the same reason they are also the largest class of drug targets for blockbuster drugs. Following agonist binding, GPCRs couple and activate trimeric G proteins. Trimeric G proteins consist of three subunits namely: G_α, G_β, and G_γ. Upon activation of a trimeric G protein, the G_α subunit loses affinity for guanidine diphosphate (GDP) and binds to guanidine triphosphate (GTP), and the resulting conformational change leads to G_α subunit dissociation from $G_{\beta\gamma}$. The GTP bound G_α subunit subsequently activates effector proteins that trigger signaling cascades. The G_α subunit of the trimeric G protein is encoded by 18 different genes in mammals that fall into four functional classes, namely $G_{\alpha s,}$ $G_{\alpha i/o}$, $G_{\alpha q/11}$, and $G_{\alpha 12/13}$.

An activated agonist-GPCR pair either specifically couple to one subtype of G_α subunit or promiscuously couple to more than one G protein [1–8]. In addition to activating G protein mediated signaling pathways, GPCRs also activate extracellular signal related kinase (ERK) signaling pathways through coupling to the β-arrestin family of proteins [9]. Certain agonists couple and activate both G protein and β-arrestin signaling pathways with similar efficacy while other agonists show significant differences in their efficacy towards these pathways. Such agonists are known as "functionally selective" or "biased." The ability of functionally selective GPCR agonists to cause differential effects among G protein

GPCRs as Therapeutic Targets, Volume 1, First Edition. Edited by Annette Gilchrist.

and β-arrestin signaling pathways has led to their markedly improved therapeutic index as drugs [10]. Although the discovery of β-arrestin signaling pathway selective agonists has gained traction with a growing body of literature, agonist selectivity for specific G protein subtypes is relatively obscure. Nonetheless, G protein specific agonists have therapeutic advantages. For instance, a switch in the dopamine 1 receptor (D1DR) from $G_{\alpha s/olf}$ to $G_{\alpha q}$ is being investigated as a desirable outcome in the treatment of neurodegenerative diseases [11]. Contrary to their initial classification as antagonists, commonly used muscarinic receptor ligands have been shown to display functional selectivity towards non-preferential G proteins, with implications for the clinical selection of existing compounds targeting this receptor family [8].

This review chapter focuses on the mechanistic basis of GPCR–G protein selectivity and highlights the need to investigate the structural dynamics of receptor–G protein interactions using integrated methodologies (Figure 11.1). G protein

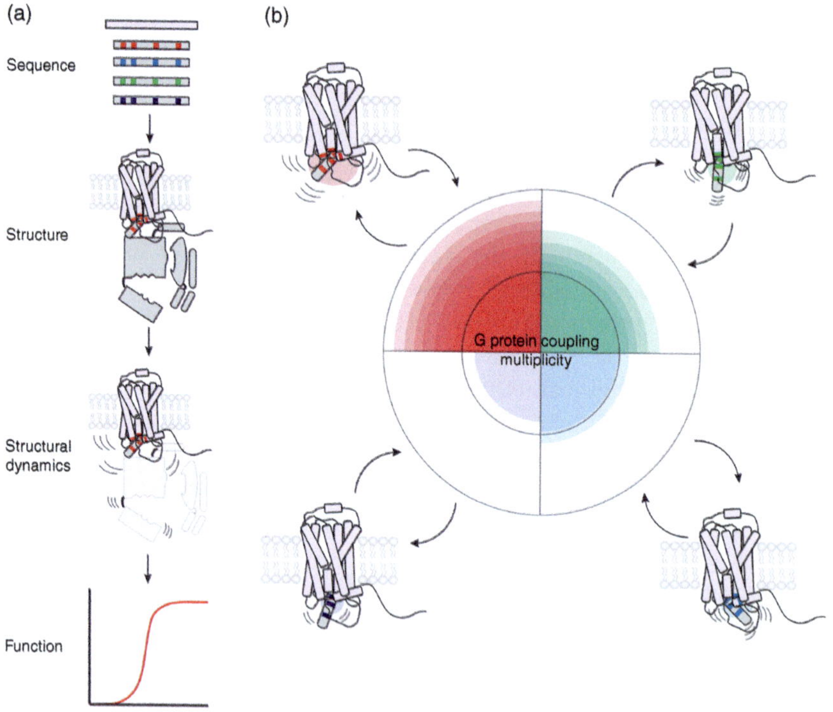

Figure 11.1 (a) Trajectory connecting GPCR and G protein sequence to GPCR function. The structural dynamics of the GPCR-G protein interaction are a key intermediary to mechanistically understanding GPCR function. (b) Structural dynamics of GPCR-G protein interactions are understood through integrated approaches, combining analysis of the dynamics of the GPCR G protein interaction (receptor cartoons, outside) with systematic probing of the G protein coupling multiplicity of a receptor (radial diagram, inside).

selectivity of an agonist-GPCR pair emerges from: (i) cell specific factors such as relative expression levels of GPCRs and G proteins, spatiotemporal localization, and post translational modifications [12], (ii) amino acid sequence composition of the GPCR–G proteins [13, 14], (iii) three-dimensional structural features of the ligand–GPCR–G protein complex [15], and (iv) dynamical features of the GPCR–G protein interactions [7, 16]. We briefly highlight the role of sequence and structure before elaborating on different techniques to probe receptor–G protein interaction dynamics, followed by insights emerging from integrated approaches and conclude with opportunities to leverage insights to understand agonist efficacy and allosteric modulation.

11.2 Delineating G Protein Selectivity Determinants by Combining Sequence Analysis with Structural Information

The human genome encodes 800 distinct GPCRs and 18 different G proteins. Given this vast repertoire, an outstanding topic in GPCR biology are the determinants of selectivity by a GPCR for specific subtypes of G proteins. Pioneering studies by Hamm and coworkers used chimeric constructs of different G_α subunits to show that GPCRs differentially leverage distinct domains on the G_α subunits to mediate selective G protein interactions [17–21]. Other early studies swapping single amino acid residues in the C-terminus of the G_α subunit showed that particular residues are sufficient for switching signaling downstream of a particular GPCR through different G protein-mediated pathways [22, 23]. More recently, Flock et al. performed differential evolution analysis of paralogous and orthologous G protein sequences and identified a G protein based selectivity bar code, relaying the contribution of the C-terminus of the G_α subunit for selectivity [14]. However, the greater sequence diversity across GPCRs does not reveal complementary sequence motifs in the receptors themselves that correspond with their G protein coupling patterns.

The development of novel GPCR purification strategies combined with advances in cryo-electron microscopy over the last decade have yielded atomistic insights into the GPCR–G protein interaction interface. However, the dynamic nature of these proteins and their interactions has necessitated the use of either engineered thermostable mutants of the G_α subunit or the inclusion of exogenous proteins that stabilize both the GPCR and the heterotrimeric G protein, including single fragment nanobodies and chimeric T4 lysozyme-GPCR fusions. Consequently, only a limited number of GPCR–G protein complex structures are available to date, with the vast majority being $G_{\alpha s}$- or $G_{\alpha i}$-coupled receptor complexes. There structures of four class A $G_{\alpha q}$-coupled receptor complexes, and no available structures of GPCR–$G_{\alpha 12/13}$ complexes. Nevertheless, analysis of available GPCR–G protein complex structures have shown that the last 27 amino acids in the C-terminus of the G_α subunit constitute more than 75% of the intramolecular contacts the G protein makes with the GPCR. This C-terminal segment of the G_α subunit adopts an α-helical conformation and is known as the α5 helix in the G protein residue numbering nomenclature. The α5 helix of the G_α subunit inserts into the intracellular regions of transmembrane helices TM3, TM5, TM6, and TM7 of the GPCR.

High-resolution structures reveal that the overall architecture of GPCR–G protein complexes is conserved, providing a structural basis for the significance of the α5 helix for GPCR–G protein selectivity. Analysis of the three-dimensional structures of the two $G_{\alpha s}$-coupled GPCR structures shows that the orientation of the α5 helices is slightly different in different receptors (Figure 11.2, left). In general, $G_{\alpha s}$-coupled receptor structures indicate that TM5, TM6, and intracellular loop 2 (ICL2) are involved in binding the α5 helix of $G_{\alpha s}$ [15, 24]. Additionally, residues in the loop connecting the β4 and β5 strands as well as in the αN helix of the $G_{\alpha s}$ subunit make contact with the intracellular loop 2 (ICL2) in both the β2-adrenergic receptor (β_2AR) and the adenosine 2A receptor (A_{2A}R). In contrast, GPCR–$G_{\alpha i}$ complex structures show that the α5 helix of $G_{\alpha i}$ interacts heavily with ICL1, TM6, and helix8. Additionally, the peripheral interactions between $G_{\alpha i}$ and ICL2 residues are more distant. Comparison of the α5 helix orientations in the three-dimensional structures of 5 GPCRs coupled to $G_{\alpha i}$ also shows subtle variation among these receptors (Figure 11.2, middle).

Despite cohesion between sequence-based and structure-based analyses, there are significant differences in the interaction interfaces between specific receptor sub-types and G_α subunits that muddy a systematic understanding of GPCR–G protein selectivity. Sequence-structure analysis of the interfacial residues from three-dimensional structures of GPCR–G protein complexes failed to reveal a

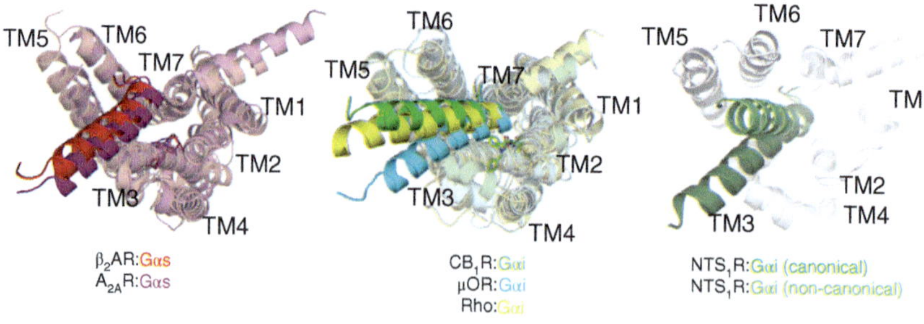

Figure 11.2 GPCR-G protein complex structures show a range of conformational heterogeneity. Receptors bound to the same G protein are shown to display the range in the conformational ensemble for each interaction. The two $G_{\alpha s}$-coupled GPCRs, β_2AR and A_{2A}R are shown on the left. $G_{\alpha i}$-coupled GPCRs are shown at center and right of the figure. In center are the complexes of CB_1R, μOR, and Rhodopsin bound to $G_{\alpha i1}$. The figure on the right shows the NTS_1R bound to the $G_{\alpha i1}$ protein in two different conformations.

"selectivity" bar code [14]. Furthermore, the α5 helices in two different β_2AR structures show distinct orientations that suggest a heterogeneity in interaction states. A similar phenomenon was observed in Cryo-Electron Micrographs of the neurotensin 1 receptor (NTS_1R)-G_i complex (Figure 11.2, right). The diversity of conformational states trapped by high-resolution structures highlights the dynamic nature of the GPCR–G protein interface that is not captured by static "snap shots." Hence, while GPCR sequence and structure provide an essential foundation for G protein selectivity, they do not capture the temporal dimension, namely the sustenance or the persistence time of the GPCR–G protein residue contacts that might play a critical role in determining either G protein selective or promiscuous coupling by GPCRs [16].

11.3 Dynamical Basis of GPCR Signaling

GPCRs are intrinsically dynamic proteins. They exist in multiple conformational states ranging from an inactive resting state to active-intermediate state when bound to an agonist, to a fully active state when both agonist and G protein are bound. There is an equilibrium among these states, even in the absence of any agonist binding to GPCRs. This idea is evident in that many GPCRs exhibit basal activity even when not bound to any agonist. Under this equilibrium model, binding of agonists to β_2AR decreases the likelihood that a receptor will be in the inactive state and increases the likelihood that a receptor will be in the active intermediate state. This phenomenon can be seen as shifts in the relative populations of each of these conformational states when observed using NMR and molecular dynamics simulations [25, 26]. Agonists with different efficacies, such as full and partial agonists, can shift the relative population of these states to different extents, as shown by single molecule studies [27, 28]. GPCRs in the active intermediate state exhibit increased affinity for the G protein, leading to coupling and signaling. G protein-coupled GPCRs also exhibit increased affinity to agonists over antagonists [29–31], highlighting the allosteric communication between the extracellular site and the G protein coupling site(s). While the conformational changes in the transmembrane helices and ICL2 that occur during GPCR activation is relatively well established for class A GPCRs, the role of intracellular loop 3 (ICL3) and C-terminus on receptor activation and coupling to G proteins is relatively unknown. This is because the crystal structures of most class A GPCRs have been solved by eliminating most of the residues in ICL3 and C-terminus. These regions are intrinsically disordered and hence not readily amenable to conventional biophysical structural studies. In the following sections, we first highlight the role of spectroscopic techniques, molecular dynamics simulations, and cell biology approaches in framing the conformational landscape of GPCR signaling.

11.4 Cellular Resonance Energy Transfer Studies Reveal a Spectrum of G Protein Activities

Fluorescence Resonance Energy Transfer (FRET) and Bioluminescence Resonance Energy Transfer (BRET) techniques have been very widely used to probe GPCR–G protein interactions [32]. These two techniques have already provided a wealth of information on coupling propensities of different G proteins to a given GPCR. The advantages of these techniques include that they are: (i) scalable to multiple GPCRs, ligand, and G protein pairings, (ii) the measurements are performed in living cells that capture both the structural and cellular effects on G protein coupling propensity. Further, BRET has been used extensively to study the differential signaling bias between G proteins and β-arrestin [33, 34]. To improve the strength of the BRET signal, the luciferase was engineered to a brighter blue-shifted luciferase NLuc and its use is referred to as NanoBRET [35]. Martemyanov and coworkers used the nanoBRET technique, labeling free dissociated $G_{\beta\gamma}$ and G protein receptor kinase to study the signaling profile of four different GPCRs to 14 different G_α subunit containing trimeric G proteins. They showed that each of the four GPCRs studied have different coupling strengths to each of the 14 G proteins [4]. Additionally, different ligands activating the same GPCR also showed differences in their subsequent G protein coupling profile. While FRET and BRET studies have determined regions of the G protein that mediate selective GPCR interactions, on their own they do not provide any atomic level structural insights concerning the amino acid residues that are important in G protein selectivity of GPCRs.

11.5 Spectroscopic Approaches to Probe the Dynamics of GPCR–G Protein Complexes

Nuclear magnetic resonance (NMR) and Double Electron Energy Resonance (DEER) techniques have proven useful to probe the dynamics of GPCRs. They have been used to study the effect of agonist and antagonist binding on GPCR conformation dynamics [26, 28, 29, 31, 36, 37]. They revealed that agonist binding makes the receptor more flexible and dynamic, whereas G protein association stabilizes receptor conformation and increases the affinity of a GPCR for a full agonist. With regards to G protein selectivity, Hubbell and coworkers derived a structural model of the conformational dynamics of rhodopsin using a combination of inter-residue distances measured from DEER spectroscopy in lipid nanodiscs and MD simulations. Through this approach, they discovered that the receptor adopts a more heterogeneous conformational ensemble when bound to $G_{\alpha i}$ than when bound to G_t [38]. Despite such advances, NMR and

DEER techniques yield sparse information on the inter-residue distances that are specifically labeled for probing. This stems from the limitations on the need to label residues that are directly solvent accessible. Therefore, it is difficult to use these techniques to obtain any information on how the core transmembrane regions move when agonists or antagonists bind. Further, most of the NMR and DEER studies were performed in detergent solubilized receptors, which is a non-native environment for GPCRs. It has been recently shown that the transition rates between different inactive and active states could be significantly impacted by the lipid environment [39]. More recent experiments in nanodiscs encapsulating lipid bilayers better mimic the cell membrane environment [40]. However, the widespread use of lipid nanodiscs is not yet feasible for studying the dynamics of the GPCR–G protein complexes due to the challenges posed in purifying and labeling these complexes. For the same reason, NMR and DEER techniques are not scalable to study multiple ligand–GPCR–G protein pairings.

Hydroxyl-radical mediated protein fingerprinting mass spectrometry (HRF-MS) and hydrogen-deuterium exchange mass spectrometry (HDX-MS) provide complementary approaches to map the temporal dynamics of GPCR–G protein interactions. Leveraging these techniques, Du et al. studied the dynamics and early events in formation of β_2AR–G_s complex [41]. They showed that prior to complex assembly, ICL2 is unstructured, and G_α residues M221 and F376 interact in a hydrophobic pocket within the G_α subunit. During the early events of G protein binding, the receptor initially engages the 5 C-terminal residues of G_α and ICL2 becomes more helical. As technical implementation of these methodologies evolves, HRF-MS and HDX-MS may become more broadly applicable to dissect the activation mechanism of other GPCR–G protein pairs.

Collectively, these spectroscopic techniques highlight the dynamic nature of GPCR–G protein ensembles, while providing little information on changes in the pairwise inter-residue distances that are labeled, changes in protein secondary structure, or changes in solvent accessibility. The sparse information does not provide a holistic view of the possible conformational changes that underlie G protein selectivity. Further, spectroscopic approaches that provide atomistic insights rely on extensive receptor-specific purification strategies that limit comparative assessment of multiple GPCR–G protein interaction pairs.

11.6 Molecular Dynamics Simulations Reveal Allosteric Communication from Ligand to Effector Binding Interface

Molecular dynamics (MD) simulation of GPCRs bound to various types of ligands, G proteins, and peptides have played a vital role in providing atomistic level insight that builds on sparse experimental structural data. Although it has been clear that ligand binding in the extracellular region of a GPCR leads to changes in the G protein coupling region about 30 Å away, until recently the mechanism by which this information transmission happened was relatively unknown. In 2014, Bhattacharya and Vaidehi developed the computational method "*Allosteer*" that uses MD simulation trajectories to calculate the strength of allosteric communication from the ligand binding site to the G protein coupling site in class A GPCRs [42, 43]. This study showed that antagonist bound inactive state of a GPCR showed a strong allosteric communication while an agonist bound active intermediate state showed weaker allosteric communication. Subsequently, other computational studies [44–46] as well as experimental studies [29, 47] showed the existence of the allosteric influence of agonist binding on G protein coupling and inversely the effect of G protein coupling on the agonist binding. While traditionally limited by sampling time (μs) and the accuracy of physical models of atomistic interactions, the geometric progression of computing resources, especially the widespread use of cloud computing, combined with development of more realistic models that capture chemical interactions have made MD simulations a complementary, scalable approach to understand GPCR conformational dynamics and the kinetics of ligand binding. Further, MD simulations of GPCRs can be performed in both explicit lipid bilayer and detergents [48, 49] that allows for mimicking the cellular environment or detergents used for GPCR purification, in order to recapitulate the experimental solvent environment where biophysical and structural studies were done. Lastly, MD simulations continue to play a vital role in drug design in GPCRs, especially with the advent of mega-scale virtual ligand libraries [50]. Despite these advantages, the broad conformational landscape of GPCR–G protein interactions requires experimental techniques that can test alternate models proposed by MD simulations.

11.7 An Integrated Approach Using MD and Experiment Measurements

While MD simulations can be paired with an experimental technique, there is a need to iteratively test the significance of diverse conformational states revealed by MD and also to do so for a range of receptor-effector combinations. As outlined earlier, the need for receptor-specific purification strategies and accessibility to resources and technical know-how limits the use of NMR/DEER/HDX-MS/HRF-MS technologies to gain generalized dynamic insights into receptor-effector interactions. On the other hand, BRET/FRET measurements have demonstrated the ability to probe numerous receptor-effector pairs. However, they are impacted by the relative expression level of labeled versus unlabeled G proteins that could weaken the signal and hence reduce sensitivity. Sivaramakrishnan and coworkers and have developed a genetically encoded FRET biosensor technique called systematic protein affinity strength modulation (SPASM) to study protein-protein interactions. The sensor involves tethering two proteins or protein domains tethered with an ER/K linker that is flanked by a FRET pair (mCerulean, FRET donor, and mCitrine, FRET acceptor) [51]. The FRET intensity measured using the sensor correlates linearly with the fraction of the sensors in the bound state. This technique was applied to measure GPCR–G protein interactions in live cells [6, 52]. The linker length was optimized to mimic the effective concentration of the G protein in the cell in comparison to the GPCR. A peptide making up the last 27 amino acid residues of the C-terminus of $G_{\alpha s}$, $G_{\alpha i}$, or $G_{\alpha q}$ was tethered to 9 different GPCRs to measure the change in FRET intensity in live cells. These measurements showed that predominantly $G_{\alpha s}$-coupled receptors showed higher FRET intensity changes upon agonist treatment when tethered to $G_{\alpha s}$ peptide compared to the sensors tethered to $G_{\alpha i}$ and $G_{\alpha q}$ peptides [7].

While the role of the C-terminal α5 helix of the G_{α} subunit as a determinant of GPCR–G protein pairing has long been established in rhodopsin and other systems, it was not clear until recently which residues were involved as hotspots in G protein selectivity. Specifically, the relative role of sequence conserved and non-conserved residues remained unclear. While mutagenesis studies have established a role for non-conserved residues, potential differences in the temporal sampling of conserved residues emphasize synergies between multiple interactions in G protein selection. To address this knowledge gap, live cell FRET-sensors were integrated with MD simulations for 27 different GPCRs along with the last 27 amino acid residues of the C-terminal α5 helix of these different G_{α} subunits ($G_{\alpha s}$, $G_{\alpha i}$, and $G_{\alpha q}$). Across receptors, the cognate Gα peptides bound in distinct orientations and engaged distinct residues within the GPCR. The binding orientations of $G_{\alpha i}$ and $G_{\alpha q}$ peptides were distinct from the $G_{\alpha s}$ peptide orientation [7, 16]. This work was published prior to the structures of GPCR–$G_{\alpha i}$ and GPCR–$G_{\alpha q}$ complexes. Further, all three cognate G_{α} ($G_{\alpha s}$, $G_{\alpha i}$, and $G_{\alpha q}$) peptides showed favorable binding energies and lower structural flexibility when bound to their cognate GPCRs. The non-cognate G_{α} peptides interact weakly with the GPCRs and eventually dissociate from the GPCR without effective coupling. The $G_{\alpha s}$ peptide interactions with $G_{\alpha s}$-coupled receptors are dominated by electrostatic interactions, while the $G_{\alpha q}$ peptide interactions with $G_{\alpha q}$-coupled GPCRs are predominately hydrophobic. With regards to GPCR conformational dynamics, agonist activation caused the cytosolic surface of the GPCR to exhibit a high degree of structural plasticity characterized by latent G protein binding cavities, each of which is specific for a distinct G protein. While, cognate and non-cognate G proteins interface primarily with their preferred cavity, hot spot residues in the GPCR unique to the cognate G protein lead to strong intermolecular contacts, full complexation, and productive signaling. These studies demonstrated that G protein coupling stems from the dynamic temporal sampling of multiple GPCR–G protein interactions and not from a binary lock-and-key connection between receptor and cognate G protein. A dynamic temporal framework of G protein selection opens the door to address the widespread promiscuity in GPCR signaling as elaborated in the next section.

11.8 Towards Addressing GPCR–G Protein Coupling Promiscuity

GPCRs exhibit a spectrum of coupling strengths towards multiple G proteins ranging from highly selective coupling to promiscuous coupling [53, 54]. Under cellular conditions, the agonist-GPCR pair is presented with both cognate and non-cognate G proteins. The influence of the non-cognate G proteins on the structural ensemble of the agonist-GPCR pairing and its effect on signaling has been highlighted by studies on GPCR priming [55, 56]. GPCR priming refers to the phenomenon of enhanced cognate signaling driven by interactions with non-cognate G proteins. Studies on GPCR priming highlight the interconnected nature of G protein signaling pathways [55, 56]. Nonetheless, the structural basis of non-cognate G protein effects on signaling or G protein coupling promiscuity remains obscure. We propose that the missing link is translating the dynamics of the large intracellular interface of the agonist–GPCR–G protein complex to differential G protein coupling. For instance, when comparing MD simulations on β_2AR coupled to either $G_{\alpha s}$ or the non-cognate $G_{\alpha q}$, we identified weak β_2AR–$G_{\alpha q}$ interactions that can be substituted to strengthen G protein coupling such that an engineered a triple mutant β_2AR-Q142K-R228I-Q229W showed similar signaling strengths through both $G_{\alpha s}$ and $G_{\alpha q}$ signaling pathways [16]. Through generating this promiscuous mutant β_2AR that signals through $G_{\alpha s}$ and $G_{\alpha q}$, this study demonstrated that the "dynamic structural plasticity" of the GPCR cytosolic pocket underlies G protein promiscuity. It also showed that the GPCR intracellular surface is dynamic and adaptive to binding multiple G proteins.

11.9 Towards the Structural Basis of Agonist Efficacy and Allosteric Modulation of GPCR Signaling

A key emphasis of GPCR drug discovery is the tuning of agonist efficacy to match the desired physiological outcomes. Agonist efficacy stems from a range of factors including the abundance of receptor and components of the effector signaling cascade. Nonetheless, with the scaling up of structure-based drug discovery efforts, understanding the structural basis of agonist efficacy at the level of the receptor has gained importance. The weaker coupling of G proteins and non-G protein effectors to the receptor from lower efficacy agonists (partial agonists) has limited structural insights into the influence of agonist efficacy on the GPCR–G protein interface. Similarly, ligands that decrease the constitutive activity of a GPCR (inverse agonists) can bias signaling through secondary signaling pathways, but have the same structural knowledge gap as partial agonists [57]. Single molecule studies demonstrate that agonist efficacy influences both receptor conformational dynamics and G protein activation kinetics. The atomistic effects of partial agonists on receptor conformation remain unclear and represent the next frontier of GPCR structural biology.

Alternatively, allosteric modulators of GPCRs have the potential to modulate context-dependent signaling while increasing receptor specificity. However, the structural similarities of class A GPCRs preclude straightforward identification of unique allosteric binding sites. Moreover, the influence of known allosteric modulators on GPCR conformational dynamics and G protein selection and coupling need to be examined. With continued identification of novel ligands from high-throughput screens of new drug candidate receptors and structure-based drug discovery efforts of established targets, there is an urgent need to develop dynamic structural frameworks of agonist efficacy and receptor allostery.

Acknowledgments

Research in N.V. Lab was funded by NIH R01-GM117923 and R01-GM 097261. Research in S.S. lab was funded by NIH R01-GM117923 and R35-GM126940.

References

1 Inoue, A., Ishiguro, J., Kitamura, H. et al. (2012). TGFα shedding assay: an accurate and versatile method for detecting GPCR activation. *Nat. Methods* 9 (10): 1021–1029.

2 Kandola, M.K., Sykes, L., Lee, Y.S. et al. (2014). EP2 receptor activates dual G protein signaling pathways that mediate contrasting proinflammatory and relaxatory responses in term pregnant human myometrium. *Endocrinology* 155 (2): 605–617.

3 Saulière, A., Bellot, M., Paris, H. et al. (2012). Deciphering biased-agonism complexity reveals a new active AT1 receptor entity. *Nat. Chem. Biol.* 8 (7): 622–630.

4 Masuho, I., Ostrovskaya, O., Kramer, G.M. et al. (2015). Distinct profiles of functional discrimination among G proteins determine the actions of G protein-coupled receptors. *Sci. Signal.* 8 (405): ra123.
5 Okashah, N., Wan, Q., Ghosh, S. et al. (2019). Variable G protein determinants of GPCR coupling selectivity. *Proc. Natl. Acad. Sci. U.S.A.* 116 (24): 12054–12059.
6 Malik, R.U., Ritt, M., DeVree, B.T. et al. (2013). Detection of G protein-selective G protein-coupled receptor (GPCR) conformations in live cells. *J. Biol. Chem.* 288 (24): 17167–17178.
7 Semack, A., Sandhu, M., Malik, R.U. et al. (2016). Structural elements in the $G_{\alpha s}$ and $G_{\alpha q}$ C termini that mediate selective G Protein-coupled Receptor (GPCR) signaling. *J. Biol. Chem.* 291 (34): 17929–17940.
8 Michal, P., El-Fakahany, E.E., and Doležal, V. (2007). Muscarinic M2 receptors directly activate Gq/11 and Gs G-proteins. *J. Pharmacol. Exp. Ther.* 320 (2): 607–614.
9 Tohgo, A., Pierce, K.L., Choy, E.W. et al. (2002). β-arrestin scaffolding of the ERK cascade enhances cytosolic ERK activity but inhibits ERK-mediated transcription following angiotensin AT1a receptor stimulation. *J. Biol. Chem.* 277 (11): 9429–9436.
10 Azzi, M., Charest, P.G., Angers, S. et al. (2003). β-arrestin-mediated activation of MAPK by inverse agonists reveals distinct active conformations for G protein-coupled receptors. *Proc. Natl. Acad. Sci. U.S.A.* 100 (20): 11406–11411.
11 Rashid, A.J., So, C.H., Kong, M.M.C. et al. (2007). D1-D2 dopamine receptor heterooligomers with unique pharmacology are coupled to rapid activation of Gq/11 in the striatum. *Proc. Natl. Acad. Sci. U.S.A.* 104 (2): 654–659.
12 Luttrell, L.M., Maudsley, S., and Gesty-Palmer, D. (2018). Translating in vitro ligand bias into in vivo efficacy. *Cell. Signalling* 41: 46–55.
13 Flock, T., Ravarani, C.N.J.J., Sun, D. et al. (2015). Universal allosteric mechanism for Gα activation by GPCRs. *Nature* 524 (7564): 173–182.
14 Flock, T., Hauser, A.S., Lund, N. et al. (2017). Selectivity determinants of GPCR-G-protein binding. *Nature* 545 (7654): 317–322.
15 García-Nafría, J. and Tate, C.G. (2019). Cryo-EM structures of GPCRs coupled to Gs, Gi and Go. *Mol. Cell. Endocrinol.* 488: 1–13.
16 Sandhu, M., Touma, A.M., Dysthe, M. et al. (2019). Conformational plasticity of the intracellular cavity of GPCR – G-protein complexes leads to G-protein promiscuity and selectivity. *Proc. Natl. Acad. Sci. U.S.A.* 116 (24): 11956–11965.
17 Hamm, H.E., Deretic, D., Arendt, A. et al. (1988). Site of G protein binding to rhodopsin mapped with synthetic peptides from the α subunit. *Science* 241 (4867): 832–835.
18 Lambright, D.G., Noel, J.P., Hamm, H.E., and Sigler, P.B. (1994). Structural determinants for activation of the α-subunit of a heterotrimeric G protein. *Nature* 369 (6482): 621–628.

19 Skiba, N.P., Yang, C.S., Huang, T. et al. (1999). The α-helical domain of Gα(t) determines specific interaction with regulator of G protein signaling 9. *J. Biol. Chem.* 274 (13): 8770–8778.

20 Slessareva, J.E., Ma, H., Depree, K.M. et al. (2003). Closely related G-protein-coupled receptors use multiple and distinct domains on G-protein α-subunits for selective coupling. *J. Biol. Chem.* 278 (50): 50530–50536.

21 Van Eps, N., Oldham, W.M., Hamm, H.E., and Hubbell, W.L. (2006). Structural and dynamical changes in an α-subunit of a heterotrimeric G protein along the activation pathway. *Proc. Natl. Acad. Sci. U.S.A.* 103 (44): 16194–16199.

22 Conklin, B.R. and Bourne, H.R. (1993). Structural elements of Gα subunits that interact with Gβγ, receptors, and effectors. *Cell* 73 (4): 631–641.

23 Conklin, B.R., Herzmark, P., Ishida, S. et al. (1996). Carboxyl-terminal mutations of G(qα) and G(sα) that alter the fidelity of receptor activation. *Mol. Pharmacol.* 50 (4): 885–890.

24 Glukhova, A., Draper-Joyce, C.J., Sunahara, R.K. et al. (2018). Rules of engagement: GPCRs and G proteins. *ACS Pharmacol. Transl. Sci.* 1 (2): 73–83.

25 Niesen, M.J.M., Bhattacharya, S., and Vaidehi, N. (2011). The role of conformational ensembles in ligand recognition in G-protein coupled receptors. *J. Am. Chem. Soc.* 133 (33): 13197–13204.

26 Bokoch, M.P., Kobilka, B.K., Prosser, R.S. et al. (2013). The dynamic process of β2-adrenergic receptor activation. *Cell* 152 (3): 532–542.

27 Ye, L., Van Eps, N., Zimmer, M. et al. (2016). Activation of the A 2A adenosine G-protein-coupled receptor by conformational selection. *Nature* 533: 265–268.

28 Gregorio, G.G., Masureel, M., Hilger, D. et al. (2017). Single-molecule analysis of ligand efficacy in β2AR-G-protein activation. *Nature* 547 (7661): 68–73.

29 Devree, B.T., Mahoney, J.P., Vélez-Ruiz, G.A. et al. (2016). Allosteric coupling from G protein to the agonist-binding pocket in GPCRs. *Nature* 535 (7610): 182–186.

30 Mahoney, J.P. and Sunahara, R.K. (2016). Mechanistic insights into GPCR–G protein interactions. *Curr. Opin. Struct. Biol.* 41: 247–254.

31 Lee, S., Nivedha, A.K., Tate, C.G., and Vaidehi, N. (2019). The dynamic role of G-protein in stabilizing the active state of the adenosine A2A receptor. *Cell Struct.* 24 (4): 703–712.e3.

32 Kobayashi, H., Picard, L.P., Schönegge, A.M., and Bouvier, M. (2019). Bioluminescence resonance energy transfer–based imaging of protein–protein interactions in living cells. *Nat. Protoc.* 14 (4): 1084–1107.

33 Namkung, Y., Le Gouill, C., Lukashova, V. et al. (2016). Monitoring G protein-coupled receptor and β-arrestin trafficking in live cells using enhanced bystander BRET. *Nat. Commun.* 7: 12178.

34 Namkung, Y., LeGouill, C., Kumar, S. et al. (2018). Functional selectivity profiling of the angiotensin II type 1 receptor using pathway-wide BRET signaling sensors. *Sci. Signal.* 11 (559): eaat1631.

35 Hall, M.P., Unch, J., Binkowski, B.F. et al. (2012). Engineered luciferase reporter from a deep sea shrimp utilizing a novel imidazopyrazinone substrate. *ACS Chem. Biol.* 7 (11): 1848–1857.

36 Maguire, M.E., Van Arsdale, P.M., and Gilman, A.G. (1976). An agonist-specific effect of guanine nucleotides on binding to the β2 adrenergic receptor. *Mol. Pharmacol.* 12 (2): 335–339.

37 Manglik, A., Kim, T.H.H., Masureel, M. et al. (2015). Structural insights into the dynamic process of β2-adrenergic receptor signaling. *Cell* 161 (5): 1101–1111.

38 Van Eps, N., Caro, L.N., Morizumi, T. et al. (2017). Conformational equilibria of light-activated rhodopsin in nanodiscs. *Proc. Natl. Acad. Sci. U.S.A.* 114 (16): E3268–E3275.

39 Staus, D.P., Wingler, L.M., Pichugin, D. et al. (2019). Detergent- and phospholipid-based reconstitution systems have differential effects on constitutive activity of G-protein– coupled receptors. *J. Biol. Chem.* 294 (36): 13218–13223.

40 Banerjee, S., Huber, T., and Sakmar, T.P. (2008). Rapid incorporation of functional rhodopsin into nanoscale apolipoprotein bound bilayer (NABB) particles. *J. Mol. Biol.* 377 (4): 1067–1081.

41 Du, Y., Duc, N.M., Lodowski, D.T., and Kobilka, B.K. (2019). Assembly of a GPCR-G Protein Complex. *Cell* 177 (5): 1232–1242.e11.

42 Bhattacharya, S. and Vaidehi, N. (2014). Differences in allosteric communication pipelines in the inactive and active states of a GPCR. *Biophys. J.* 107 (2): 422–434.

43 Vaidehi, N. and Bhattacharya, S. (2016). Allosteric communication pipelines in G-protein-coupled receptors. *Curr. Opin. Pharmacol.* 30: 76–83.

44 Miao, Y., Nichols, S.E., Gasper, P.M. et al. (2013). Activation and dynamic network of the M2 muscarinic receptor. *Proc. Natl. Acad. Sci. U.S.A.* 110 (27): 10982–10987.

45 Marino, K.A. and Filizola, M. (2018). Investigating small-molecule ligand binding to G protein-coupled receptors with biased or unbiased molecular dynamics simulations. In: *Computational Methods for GPCR Drug Discovery* (ed. A. Heifetz), 351–364. New York, NY: Springer New York.

46 Dror, R.O., Arlow, D.H., Maragakis, P. et al. (2011). Activation mechanism of the β2-adrenergic receptor. *Proc. Natl. Acad. Sci. U.S.A.* 108 (46): 18684 LP–18689.

47 Chen, K.-Y.M., Keri, D., and Barth, P. (2020). Computational design of G protein-coupled receptor allosteric signal transductions. *Nat. Chem. Biol.* 16 (1): 77–86.

48 Lee, S., Ghosh, S., Jana, S. et al. (2020). How do branched detergents stabilize GPCRs in micelles? *Biochemistry* 59 (23): 2125–2134.

49 Lee, S., Mao, A., Bhattacharya, S. et al. (2016). How do short chain nonionic detergents destabilize G-protein-coupled receptors? *J. Am. Chem. Soc.* 138 (47): 15425–15433.

50 Shoichet, B.K. and Kobilka, B.K. (2012). Structure-based drug screening for G-protein-coupled receptors. *Trends Pharmacol. Sci.* 33 (5): 268–272.

51 Sivaramakrishnan, S. and Spudich, J.A. (2011). Systematic control of protein interaction using a modular ER/K α-helix linker. *Proc. Natl. Acad. Sci. U.S.A.* 108 (51): 20467–20472.

52 Malik, R.U., Dysthe, M., Ritt, M. et al. (2017). ER/K linked GPCR-G protein fusions systematically modulate second messenger response in cells. *Sci. Rep.* 7 (1): 1–13.

53 Wenzel-Seifert, K. and Seifert, R. (2000). Molecular analysis of β_2 -adrenoceptor coupling to G_s -, G_i -, and G_q -proteins. *Mol. Pharmacol.* 58 (5): 954–966.

54 Stallaert, W., Dorn, J.F., van der Westhuizen, E. et al. (2012). Impedance responses reveal β 2-adrenergic receptor signaling pluridimensionality and allow classification of ligands with distinct signaling profiles. *PLoS One* 7 (1): e29420.

55 Gupte, T.M., Malik, R.U., Sommese, R.F. et al. (2017). Priming GPCR signaling through the synergistic effect of two G proteins. *Proc. Natl. Acad. Sci. U.S.A.* 114 (14): 3756–3761.

56 Gupte, T.M., Ritt, M., Dysthe, M. et al. (2019). Minute-scale persistence of a GPCR conformation state triggered by non-cognate G protein interactions primes signaling. *Nat. Commun.* 10 (1): 4836.

57 Mo, X.L. and Tao, Y.X. (2013). Activation of MAPK by inverse agonists in six naturally occurring constitutively active mutant human melanocortin-4 receptors. *Biochim. Biophys. Acta, Mol. Basis Dis.* 1832 (12): 1939–1948.

12

Virtual Screening and Bioactivity Modeling for G Protein-Coupled Receptors

Wallace Chan[1,2,3], Jiansheng Wu[1,4], Eric Bell[1], and Yang Zhang[1,2]

[1] *Department of Computational Medicine and Bioinformatics, University of Michigan, Ann Arbor, MI, USA*
[2] *Department of Biological Chemistry, University of Michigan, Ann Arbor, MI, USA*
[3] *Department of Pharmacology, University of Michigan, Ann Arbor, MI, USA*
[4] *School of Geographic and Biological Information, Nanjing University of Posts and Telecommunications, Nanjing, China*

12.1 Introduction

G protein-coupled receptors (GPCRs) are a superfamily of integral membrane proteins and consist of over 800 established members, establishing them as the third-largest family of proteins in humans [1, 2]. Marked by their distinctive seven-pass transmembrane domain, they account for almost 5% of the human proteome. Consequently, they have been implicated in a multitude of diseases, including cancer and diabetes [3, 4]. Moreover, almost a third of all drugs in use today target these receptors, accentuating their importance in drug discovery [5]. Given the interest the pharmaceutical industry has in GPCRs, extensive scientific efforts have been made to develop novel drugs for a variety of medical conditions.

Drug discovery has traditionally used high throughput screens (HTSs) as a means to discover hit compounds from enormous chemical libraries. Unfortunately, these large-scale assays are typically very expensive, time-consuming, and laborious. In the years following its explosive beginnings in the pharmaceutical industry in the early 1980s [6], computer-aided drug design (CADD) methods were developed to computationally predict how well a potential drug would bind to a receptor or to model how it binds; predictions effectively compensate for the brute force approach of HTS and help inform further biochemical experiments. In particular, virtual (or *in silico*) screening complements HTS by reducing the chemical space to be explored. Using various CADD approaches, computational chemists could then assign scores to chemical compounds and rank them accordingly, helping prioritize which compounds to experimentally assay.

GPCRs as Therapeutic Targets, Volume 1, First Edition. Edited by Annette Gilchrist.

In the current chapter, we aim to provide the reader with an introduction to virtual screening and its application to GPCRs. In addition to providing an overview of virtual screening and its required components, we will delve into what will be referred to as classical virtual screening; this includes many well-established approaches with which many medicinal chemists will be familiar, such as chemical similarity comparisons and molecular docking. Subsequently, we will survey the use of chemogenomics and machine learning in virtual screening, including bioactivity prediction. Lastly, various topics on inverse virtual screening will be presented to give the audience a sense of how off-target effects of drugs can be computationally examined or addressed.

12.2 Overview of Virtual Screening

12.2.1 Principle of Virtual Screening

In virtual screening, the overall aim is to computationally screen through a database of chemical compounds that would be tested in the wet lab. A typical workflow can be represented as follows:

1) Prepare inputs (receptor, pharmacophore, etc.) for CADD method
2) Format chemical compound database
3) Screen through database with CADD method
4) Rank compounds in database by prediction scores
5) Select top *n* compounds for experimental validation

Compound databases can vary greatly in size depending on the target of interest, ranging from tens of thousands to millions of compounds. There is a general misconception from the scientific community that virtual screening is a complicated process; to an extent, it is beautifully simple. One way to envision screening is the large-scale repetition of a CADD methodology against each compound in the database, analogous to a loop in a computer program. However, the intricacies and challenges of virtual screening are found primarily in the parameterization of the CADD methodology, as well as the way one processes the resulting predictions. After a virtual screen, the CADD methodology will have assigned a metric or score to each compound. Subsequently, the compounds will be ranked from most likely to least likely to bind or interact with the receptor of interest. The top-ranked compounds are then typically chosen for experimental validation, either by selection after clustering or visual inspection of docked poses.

Before going further, it should be noted to the reader that a proper understanding of various computer representations of chemical compounds is useful for troubleshooting errors related to file formats while using a CADD methodology. Moreover, the proper design of a chemical database is paramount to the success of a virtual screening campaign. Finally, a stringent implementation of a retrospective or prospective virtual screening for a drug target of interest is a necessity for validation. These will all be detailed in the subsequent Sections 12.2.2–12.2.4.

12.2.2 Computer Representation of Chemical Compounds

12.2.2.1 Line Notation

Line notation allows for the representation of a chemical compound using a string of ASCII characters. Despite looking rather odd to the untrained eye, they are completely readable, and those familiar with the format would be able to convert between it and the corresponding 2D chemical structure. Nowadays, this representation is primarily used for chemical database searching. The *S*implified *M*olecular-*I*nput *L*ine-*E*ntry *S*ystem (SMILES) and *In*ternational *Ch*emical *I*dentifier (InChI) formats are currently the most widely used.

SMILES strings were initially conceived in the 1980s as a means to make chemical compounds machine readable. Each letter represents an atom (B, C, N, O, P, S, F, Cl, Br, or I), single bonds are usually implicit, and double and triple bonds are represented as "=" and "#", respectively. Aromaticity is denoted with alternating equal signs (e.g. pyridine moiety: C4=CC=CC=N4). Additionally, rings are classified by including an opening and closing number (e.g. thiophene moiety: C1=C(SC=C1)). The use of parentheses indicates branching, and stereochemistry is specified at chiral centers with "@". *SM*iles *AR*bitrary *T*arget *S*pecification (SMARTS) strings were developed by the Daylight Chemical Information Systems as a robust extension of the SMILES string that provided expanded functionality, such as the ability to filter a compound database by substructure. However, one of the biggest drawbacks of this format is that there is no standard way to generate the SMILES string [7]. Thus, the heterogeneity of SMILES strings possible for a single compound can complicate chemical database searching, especially when a compound of interest cannot be found due to this problem.

InChI strings were developed in 2005 by the *I*nternational *U*nion of *P*ure and *A*pplied *C*hemistry (IUPAC) in response to the inconsistencies produced by SMILES strings [8]. Additionally, they were able to express more information than SMILES strings. All InChI strings start with "InChI=", followed by the version number and an "S", which corresponds to its standardization. Subsequently, there are six layers of information; the first layer is the most important and gives the chemical formula, atomic connections, and hydrogen atoms, while the others focus on other chemical aspects such as charge, stereochemistry, and isotopes. Also, it should be noted that the InChI format is conspicuously more difficult to read than SMILES. InChI keys, 27-character hashed versions of InChI strings, allow for extremely fast chemical database searches due to their reduced length. A previous study has demonstrated that a single duplicate for the first 14 characters could theoretically occur 0.014% of the time in a database of 100 million compounds [9]. Given that most chemical databases have well below this number of chemical compounds, it can be assumed that a duplication will likely not occur. A drawback of using the InChI key is that it cannot be converted back to its respective InChI string, thus these two descriptors always need to be paired.

12.2.2.2 Molecular Fingerprints

Molecular fingerprints provide an abstraction of the chemical features of compounds into binary vectors. All have a fixed length for purposes of comparison and can be used to calculate chemical similarity mind-bogglingly fast. Though efficient, they likely have the least specific information packed into their form. Over the years, various developments have aimed to squeeze as much information as possible into small vector lengths.

Substructure key-based fingerprints consist of a predefined set of substructures, and the number of possible bits is defined by the number of substructures. One of the most commonly used fingerprints of this type is *M*olecular *ACC*ess *S*ystem (MACCS), first developed by MDL Information Systems (formerly Molecular Design Limited) in 1979. Interestingly, they were initially intended for use in database searching as opposed to virtual screening [10], which is the common method it is used for today. They assume two different variants: one with 960 substructures, and the other with 166 of the most interesting substructures for drug discovery, paired with corresponding SMARTS strings [10]. Not surprisingly, the latter is far more popular. The principle of how this type of fingerprint works is that each position in the fingerprint corresponds to a substructure. If the compound has the substructure in its chemical structure, then the bit will be set to "1". Otherwise, it would be set to "0". A drawback to using these types of fingerprints is that they are usually relatively sparse in content, such that they will have mostly zeros, as typical molecules will have very few of the substructures.

Path-based fingerprints are constructed by analyzing every possible fragment in a molecule of a given linear path length, then cryptographically mapping them onto a fixed-length fingerprint. An example is given in Figure 12.1a for oliceridine, using a path length of 3. Occasionally, bit collisions occur when the same bit is assigned to two different fragments. However, this is not a common occurrence and can be reduced by increasing the fingerprint length. The Daylight fingerprint, developed by Daylight Chemical Information Systems (hence the namesake), is the most used out of all of the fingerprints of this type and typically consists of 1028 bits.

Circular fingerprints are very similar to path-based fingerprints in that they are mapped from a collection of molecular fragments onto a fixed-length fingerprint. However, their method of fragment analysis is not based on fragments generated

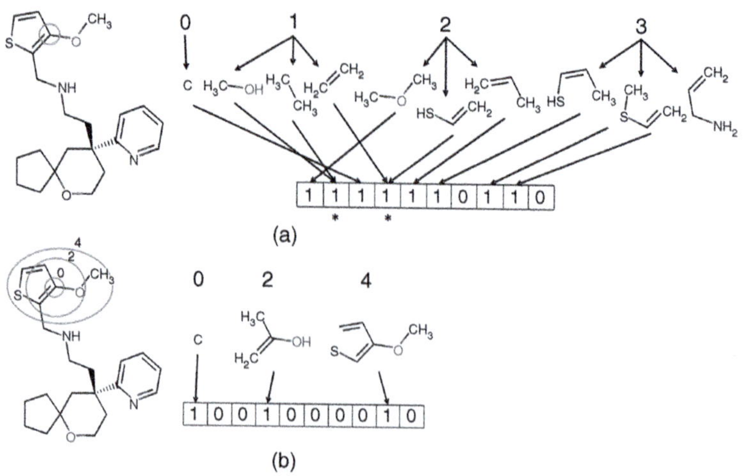

Figure 12.1 Hypothetical 10-bit fingerprints for Oliceridine. (a) A path-based fingerprint with a path length of 3 is used for this example. Only fragments found from a single starting atom (green circle) are shown. The path lengths of the fragments (0, 1, 2, 3) are numbered in bold. The asterisks (*) denote where there are bit collisions. (b) A circular fingerprint with a radius of 2 is used for this example. Only fragments found from a single starting atom (innermost green circle) and onwards are shown. The diameters of the fragments (0, 2, 4) are numbered in bold. For both fingerprints, the fragment-generating process occurs for all atoms on the molecule. MarvinSketch was used for drawing and displaying the chemical structures, MarvinSketch 18.10.0, 2018, ChemAxon (http://www.chemaxon.com).

in a linear path, but rather, the chemical environment centered around each atom within a certain radius. An example for oliceridine is given in Figure 12.1b, where a radius of 2 was used. Here, fragments are generated by moving a certain radius away from a starting atom up until a diameter of 4, resulting in 3 fragments for the specified starting atom. The ECFP4 fingerprint is the industry standard of this type, and not surprisingly, it has been shown to be among the best performing fingerprints in a recent benchmark that ranked diverse structures by similarity [11].

12.2.2.3 Chemical Table Files

Another strategy for storing chemical information in a text file is chemical table file family of file formats. Originally developed by MDL Information Systems starting in the late 1970s [12], they have become one of the most widely used file formats, having been adopted by a vast majority of computational chemistry software. Of those in this family, focus will be upon the *S*tructure-*D*ata *F*ile (SDF) format due to its widespread use. In brief, the file starts with a three-line header block, which is mandatory but can be left empty if desired. This is followed by a "counts" line, which consists of specifications such as number of atoms, number of bonds, and so forth. The "atoms" block provides information about the coordinates and identity of the atoms, while the "bond" block describes the connectivity between atoms. The properties block denotes any existing charges or isotopes, as well as the end of the molecular description. SDF is unique in that the subsequent associated data allow the inclusion of miscellaneous information not allowed in the main form, such as the IUPAC name and database identifiers. The tag for each data type is included inside angle brackets ("<", ">"), and the relevant data is placed on the line immediately following it.

Originating from the now-defunct Tripos, the Mol2 format has achieved a similar level of popularity and usage as the SDF format. Many aspects are almost identical to the SDF format, where various blocks are designated for counts, atom, and bond information, though with different column formatting. Moreover, each block is recognized starting with a record type indicator (e.g. @<TRIPOS>ATOM), followed by the corresponding data. Apart from the main record type indicators, there exists many others not available in SDF format, such as substructures and rotatable bonds.

12.2.2.4 PDB File Format

Most researchers in biochemistry will be fairly acquainted with the Protein Data Bank (PDB) file format, since it has been primarily used to describe the three-dimensional structure of proteins, DNA, and RNA. This file format was first conceived in 1976 as a means to help researchers exchange protein coordinates through a database [13]. Not surprisingly, its format has been revised and updated numerous times over the years. Essentially, a PDB file is a text file that contains various information about the structure provided in specified ranges of columns. The file contains a variety of data, ranging from resolution and method used to solve the structure to atomic coordinate specifications.

One of the most important pieces of information within the PDB file is the "ATOM" record name. An example is shown in Figure 12.2 that depicts the coordinates for two representative amino acids from a PDB structure. Each line depicts a single atom in the structure. For example, the first line corresponds to the backbone nitrogen of Gly-85. Furthermore, the atomic coordinates of this atom (−2.211, 29.344, −42.463) are given so that whichever algorithm or molecular visualization software is used can correctly process this representation.

12.2.3 Chemical and Biological Databases

As the amount of data available to the scientific community has increased over time, there has become a distinct need to catalogue and organize it so that it can be easily accessible. Truly, gone are the days of hours-long expeditions to the library

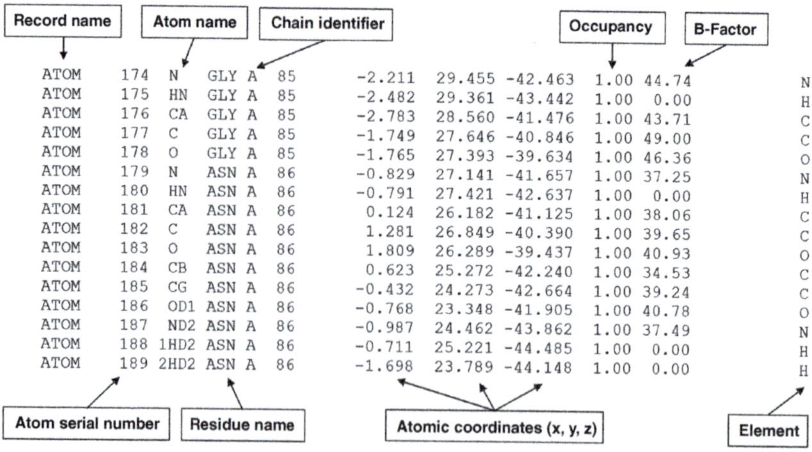

Figure 12.2 Representative portion of PDB file. The portions with the "Record Name" of ATOM helps software understand the identity and location of atoms and therefore help correctly process relevant information from the file. The amino acids, Glycine (GLY) in position 85 and asparagine (ASN) in position 86, from this structure are shown.

in search of publications that may or may not have been helpful to the question at hand. Astoundingly, there now exist public databases that index data anywhere from the primary structures of proteins to various experimental values of ligands for a given receptor.

12.2.3.1 Biological Databases

UniProt is the *de facto* standard source of information for proteins [14]. This database originated from the merging of data from European Bioinformatics Institute (EBI), Swiss Institute of Bioinformatics (SIB), and Protein Information Resource (PIR) into an entity known as the UniProt consortium. The most commonly used portion of the database is referred to as UniProt Knowledgebase (UniProtKB), which is subdivided into Swiss-Prot and TrEMBL. The former collection of data is manually annotated and reviewed by experts of each respective protein, while the latter refers to those that are computationally annotated from genomic data. Not surprisingly, TrEMBL contains a far larger quantity of data than Swiss-Prot. Within Swiss-Prot, a multitude of information about a protein of interest is available, such as primary structure, post-translational modifications, function, subcellular localization, and known protein–protein interactions.

The *P*rotein *D*ata *B*ank (PDB) is the single largest repository for protein, DNA, and RNA structures solved by structural biologists [15]. It began as a united effort in the 1970s to provide the scientific community with protein structures coded into punch cards [13]. As the Internet came into fruition, it became possible to move the data onto an online platform for its higher throughput distribution.

Thus, the first web-server for browsing the PDB was developed at Brookhaven National Laboratory in 1996 [16]. With the explosion of solved structures starting in the 1980s, this resource became increasingly invaluable to life science researchers around the world. In its current state, it serves as the primary resource that provides protein structures for structure-based drug discovery efforts. The G Protein-Coupled Receptors database (GPCRdb) [17] was developed in 1993 as repository for GPCR-related data, and after the GPCR structure boom, it has continually participated in the manual curation of GPCR structures. A more recent effort to catalogue experimental GPCR structures from our group in a user-friendly fashion is GPCR-EXP, which is semi-manually curated and updated weekly (https://zhanglab.ccmb.med.umich.edu/GPCR-EXP/).

12.2.3.2 Chemical Databases

First released in 2009, ChEMBL is arguably the most massive database for molecules with drug-like properties and biological activity [18]. As of its latest release (ChEMBL 25), the database contains 1 879 206 unique compounds corresponding to 12 482 targets and 15 504 603 activities from 72 271 publications, all derived from manual annotation. A similar database founded over a decade earlier at University of California at San Diego is BindingDB [19], which also contains a large amount of manually curated affinity data. However, it has less of a focus on membrane receptors than ChEMBL and more strongly emphasizes enzyme targets [20]. DrugBank is a chemical database whose topic of interest is information on drug molecules and their corresponding targets [21]. Another interesting database of note is Psychoactive Drug Screening Program's (PDSP) K_i database [22], which houses a sizeable number of experimental affinities. A large portion of their data is dedicated to GPCRs. Also, the *I*nternational *U*nion of Basic and Clinical *PHAR*macology's (IUPHAR) Guide to Pharmacology is a chemical database that deals primarily with popular pharmacological targets, such as GPCRs and ion channels [23]. It is manually curated by experts, and only ligands that have been well characterized are included. In contrast, ChEMBL, BindingDB, and PDSP K_i are looser in their criteria for inclusion, where the binding mode or mechanism are largely unknown for most ligands. PubChem is a pure chemical database maintained by the National Center for Biotechnology Information (NCBI) [24], containing approximately 93.9 million chemical compounds. Additionally, they have a gargantuan collection of bioactivity data from about 1.25 million high-throughput screening campaigns, each with several million values.

All of the aforementioned chemical databases contain GPCR-related experimental data. One of the earliest efforts in organizing such data was with G protein-coupled receptor-LIgand DAtabase (GLIDA) [25, 26]. Moreover, our group developed a database called G protein-coupled receptor-Ligand ASSociation (GLASS) database [27], which processes and unifies GPCR experimental data across ChEMBL, BindingDB, IUPHAR, PDSP K_i, and DrugBank, and remains the most comprehensive database of its type. It has been used in other algorithms as input, such as SwissSimilarity [28] and weighted deep learning and random forest (WDL-RF) [29].

12.2.4 Retrospective and Prospective Virtual Screening

Retrospective virtual screening is performed to computationally validate a method's predictive performance based on a set of known active compounds and their associated decoys. Moreover, it is employed as the primary method of theoretical studies in benchmarking virtual screening methods, as well as serves as a calibration of predictive conditions so that prospective virtual screens are optimally successful. In brief, decoys are compounds that likely do not interact with the receptor of interest but are similar in some way with the active compound. A common way to produce decoys is to generate compounds that are similar in one aspect but different in another. For example, directory of useful decoys (DUD) [30] and directory of useful decoys enhanced (DUD-E) [31] are two such datasets that provide 33 and 50 decoys, respectively, per active compound that are chemically similar yet topologically different. Additionally, GPCR-Bench [32] and GPCR ligand library/GPCR decoy database (GLL/GDD) [33] are GPCR-specific datasets created in a similar fashion. It is important to note that the core assumption of using an abundance of decoys over actives is that most compounds will not bind to a given target by sheer chance, in principle making them hypothetical inactive compounds. Furthermore, it is also recommended to add experimentally determined inactive compounds to the decoy set whenever available, as is done by DUD-E.

After all active compounds and decoys are all scored, they will be ranked accordingly. The goal is to try and get as many active compounds into the top-ranking portion as possible. A typical metric for evaluation is the enrichment factor of the top 1% ($EF_{1\%}$), given as follows:

$$EF_{1\%} = \frac{N_{act}^{1\%} / N_{select}^{1\%}}{N_{act} / N_{tot}} \tag{12.1}$$

where N_{act} and N_{tot} are the total numbers of the active and all compounds, respectively. $N_{act}^{1\%}$ and $N_{select}^{1\%}$ are, respectively, the number of active ligands and the number of all candidates in the top 1% of the ranked database. The numerator essentially accounts for the proportion of active compounds found in the top 1%, while the denominator represents the probability of selecting an active compound randomly from the entire database. An $EF_{1\%}$ greater than 1 would mean the screening method performed better than randomness, while if it were less than 1, it would mean it performed worse than randomness.

An important metric typically paired with the enrichment factor is the area under the curve (AUC) of the receiver operator characteristic (ROC) curve, where the true and false positive rates are calculated for both the active and decoys, respectively. A value greater than 0.5 would suggest better performance than randomness, while the inverse would suggest worse than randomness. A drawback to this metric is that the entire list of ranked compounds is taken into account, thereby putting great emphasis on the portions of the database that are unlikely to ever have a compound chosen for experimentation. To account for this, some groups have developed ways to give a higher weight to early enrichment. The Shoichet group utilized the logAUC metric in DUD-E [31], while Schrödinger seems to favor the Boltzmann-enhanced discrimination of the receiver operator characteristic (BEDROC) [34, 35]. Either of these metrics will allow a better examination of how well the CADD methodology is able to distinguish active compounds from decoys.

As opposed to its counterpart, prospective virtual screening is far more straightforward: in a prospective screen, a computational chemist chooses several high-scoring compounds for experimental validation. One way of choosing compounds would be to cluster a subset by molecular frameworks (i.e. Bemis–Murcko scaffolds [36]) or chemical similarity, wherein the top-ranking compound in each cluster would be chosen. Another way would be to visually examine the docking poses of the compounds for important interactions with the receptor. Unfortunately, there is no replacement for a prospective virtual screen, as retrospective virtual screens remain theoretical in nature and serve only as a measure of how believable virtual screening can be.

12.3 Conventional Virtual Screening

In the traditional sense, virtual screens are categorized as ligand based or structure based, depending on which CADD methodology is used; the former utilizes pure chemical information in its search process, whereas the latter uses protein structural information to determine how a compound binds. Most of the major modeling suites from various companies (i.e. molecular operating environment [MOE], Schrodinger, Cambridge Crystallographic Data Centre (CCDC), BIOVIA, etc.) have the capability for all or most of the following methods.

12.3.1 Ligand-Based Approaches

12.3.1.1 Chemical Similarity

Molecular fingerprints are most often used in the calculation of chemical similarity. As a brief reminder, they are composed of a fixed-length string of bits; the presence of "1" in a position denotes the presence of a chemical feature, while "0" denotes its absence. The simplicity of this form allows for the possibility of blazing fast calculations. A multitude of similarity metrics are available, but the Tanimoto coefficient has proven to perform the best and therefore has been most popular [37]. Its calculation is shown as follows:

$$\text{Tanimoto Coefficient} = \frac{c}{a + b - c} \tag{12.2}$$

where a is the number of bits in the first molecule, b is the number of bits in the second molecule, and c is the number of shared bits between the two molecules. Only the same type of molecular fingerprint can be compared between molecules; mixing different types will lead to erroneous results. Free software for chemical similarity screening includes OpenBabel [38] and RDKit [39], both of which are user-friendly standard toolkits in cheminformatics.

12.3.1.2 Ligand-Based Pharmacophores

A pharmacophore is a collection of chemical features (H-bond donor, H-bond acceptor, aromatic, etc.) represented spatially, developed from a set of known bioactive compounds. They are typically used when the protein structure of the target is not known, which was historically the case. In brief, the pharmacophore is constructed by structurally superposing low-energy conformers of the bioactive compounds, whereupon chemical features from superposed moieties among the compounds are assigned to the model. It should be noted that it is assumed that the shared chemical features contribute to the bioactivity. When used in a virtual screen, the target compounds will be spatially matched onto the pharmacophore model and scored.

12.3.1.3 Shape-Based Comparison

Molecular shape has long been established as being an important contribution to bioactivity [40]. Thus, another common method in ligand-based virtual screening has involved the use of shape-based matching. Query conformers are geometrically matched to other target conformers, trying to achieve the best 3D electron density overlap. One of the earliest algorithms to implement this approach was ROCS [41] from OpenEye Scientific Software, while other freely available methods include Ultrafast Shape Recognition (USR) [42], LigSift [43], and LS-align [44].

12.3.2 Structure-Based Approaches

12.3.2.1 Structure-Based Pharmacophores

When the drug target of interest has a protein structure available, then a structure-based pharmacophore can be employed in virtual screening. This type operates similarly to the ligand-based pharmacophore, except the spatial distribution of chemical features are informed by the ligand-binding pocket as opposed to a ligand structural superposition. Moreover, these can be used when there is little to no knowledge of how a ligand binds to a receptor, especially in the case of orphan receptors. However, the selection of chemical features to be used in the model is nontrivial, given the large amount of uncertainty of residue importance in the binding site, and warrants careful decision-making.

12.3.2.2 Molecular Docking

Molecular docking is a method used to predict how a ligand binds with a receptor through conformational search and scoring functions. Prior knowledge of the ligand binding site is typically required in order to optimize the area to be examined. Protocols for most docking programs start with adding non-polar hydrogens and partial charges to both the receptor and the compounds. This is then followed by the docking algorithm performing a conformational search for the most favorable ligand pose, which is evaluated with a scoring function at each step. Subsequently, the top poses are generated for the user, who can then visualize them with a molecular viewer. Additionally, the final scores for each predicted pose are also given.

There have been dozens of docking software programs developed over the years, and each has approached docking in a different way. Some of the major differences between these are: (i) the search algorithm, (ii) scoring function, and (iii) conformational flexibility of the ligand and the receptor. Among the top methods employed for conformational searching are the Lamarckian genetic algorithm (AutoDock [45]), genetic algorithm (GOLD [46]), local search global optimizer (AutoDock Vina [47]), ant colony optimization (PLANTS [48]), anchor-and-grow (DOCK 6 [49]), and exhaustive search (Glide [50, 51]). Though these strategies differ greatly in their search algorithms, their basic premise remains the same: they aim to achieve the most favorable ligand pose. To do so, a scoring function must be calculated at each step of conformational sampling to evaluate the pose. Many of the functions used currently are physics-based force fields that approximate the binding energy of the ligand pose in the binding site. For example, the scoring function from DOCK 6 simply uses van der Waals and electrostatic terms for computational efficiency [52]. Various others take other physical terms into account, such as hydrogen bonding, ligand desolvation, and hydrophobic contributions [46]. Additionally, there exist empirical scoring functions, which estimate the binding energy using a set of weighted energy terms, and knowledge-based scoring functions, which utilize statistical energy potentials derived from experimentally solved structures [53, 54]. Finally, there is the option of how to treat the receptor and ligand during docking with respect to conformational flexibility. Most software packages make the receptor rigid because of the computational rigor involved in sampling all receptor conformations. However, some programs provide an option to make certain side chains of the receptor flexible, such as AutoDock Vina [47] and GOLD [46]. Schrödinger has an induced-fit docking protocol that allows for both ligand flexibility and conformational changes in the binding site, though its application to virtual screening of a large number of compounds is limited due to its computational intensity. Most of the earliest docking methods, such as the original DOCK [55], treated the ligand as rigid in order to find molecules with shape complementarity to the binding site. Nevertheless, this methodology's success is dependent on the conformation of the molecule being docked, which may be a vastly different conformation from what is observed in reality.

12.3.3 Application to GPCRs

To date, there exists a plethora of prospective virtual screening campaigns applied to GPCRs. To list them all would be outside the scope of this chapter, but the reader can find further information from reviews [56–58]. Some particularly interesting studies have included the discovery of: (i) a biased agonist for the μ opioid receptor [59], (ii) selective agonists for the serotonin 1B receptor over the serotonin 2B receptor [60], and (iii) antagonists for the C–X–C chemokine receptor 4 [61].

Several compounds found initially from virtual screen campaigns have actually entered clinical trials. In 2006, a group from Predix Pharmaceuticals (known as Epix Pharmaceuticals before collapsing) performed a docking-based virtual screen on a homology model of the serotonin 1A receptor that resulted in a potent, selective agonist [62]. The reported molecule became the drug candidate, Naluzotan, and proceeded into a phase III clinical trial; ultimately, it failed to perform better than the placebo and was discontinued [63]. From the same company, a separate virtual screening campaign with the serotonin 4 receptor using a similar methodology produced a selective, partial agonist that also made it to clinical trials, though it too failed [64]. In another study from Heptares Therapeutics, a novel adenosine A_{2A} receptor antagonist was discovered through a docking-based virtual screen on homology models, called AZD4635, and is currently in phase II clinical trials for lung cancer [65]. Additionally, they also have a muscarinic M4 receptor agonist, HTL0016878, in phase I clinical trials for Alzheimer's disease, found through similar methods [66]. To the best of the authors' knowledge, there has yet to be an approved drug found from virtual screening targeting GPCRs, though many examples exist for various other targets, such as growth factor-β1 receptor kinase [67]. Despite this, numerous GPCR-targeting drugs resulting from virtual screening await their verdict in clinical trials, and it is only a matter of time before one hits the market, which would undoubtedly validate the current computational methods and increase confidence in virtual screening for GPCRs.

12.3.4 Challenges

Despite the relative successes each type of virtual screening approach has had, there are distinct advantages and disadvantages to each. Ligand-based methods, such as chemical similarity, are computationally inexpensive and can screen millions of compounds within a short time but have the drawback of being biased towards the known ligands used to build the model. Conversely, structure-based methods, such as molecular docking, inherently have no such bias, but they are extremely computationally expensive relative to ligand-based methods. A trend in recent years has culminated in the combination of these methods to address their respective shortcomings [68]. Given the speed of ligand-based methods, several studies experimented with using it to first produce an "enriched" database of top-ranking compounds, followed with molecular docking on this reduced subset [69–71]. This can greatly reduce the computational cost and enable virtual screening for groups without access to high performance computing clusters. Other groups have exploited the complementarity between ligand- and structure-based methods by running them in parallel and employing a consensus scoring system for ranking [72–74]. Regardless of the strategy used, the manual selection of bioactive compounds remains a great challenge.

12.4 Chemogenomics-Based Virtual Screening

Oftentimes, a drug target will have neither structural information nor known active compounds. In cases like this, related proteins can be used to infer what compounds the drug target can bind, based on the assumption that similar receptors bind similar ligands [75]. The sequence of the protein can be used in a sequence-based alignment search to find homologous proteins with sets of bioactive compounds. Alternatively, a structural homology model of the protein can be generated based on the sequence and structurally compared with all known protein structures to find related proteins.

FINDSITE was an early implementation of a chemogenomics-based virtual screening algorithm, which utilized ligand information from structurally homologous receptors found through fold-recognition in a ligand-based virtual screening (VS) [76]. This algorithm later evolved into FINDSITEX, which used modelled structures as structural templates instead [77]. FINDSITE^{comb} and its successor, FINDSITE$^{comb2.0}$, incorporated FINDSITEX along with an improved version of FINDSITE that filters out false-positive ligands (FINDSITE^{filt}), vastly improving its performance in benchmarks [78, 79]. Additionally, another recent algorithm is PoLi, developed from the same lab; it looks for similar receptors by performing binding pocket structure comparison between the target and templates, followed by a ligand-based screening search [80]. SPOT-Ligand [81] is an algorithm that employs global structure alignment for the acquisition of protein structures that are structurally similar and have sets of bioactive ligands, which are then used in a ligand-based virtual screen. An updated version of the method, SPOT-Ligand 2 [82], included a more comprehensive protein–ligand database, and consequently, it achieved a better performance than its predecessor. Recently, our group has developed a pipeline, *M*ichig*a*n *G* prot*e*in-coup*l*ed receptor *l*ig*a*n*d*-based virtual screen (MAGELLAN), that utilizes structure- and sequence-based similarity to find homologous GPCRs, whereupon their ligands are clustered and then used to construct ligand profiles; using a consensus scoring function, a ligand profile-based virtual screen can then be performed against a database of choice [83].

12.5 Bioactivity Modeling with Machine Learning

Machine learning is the study of algorithms and statistical models where computer systems are used to effectively implement a specific task without explicit instructions, instead drawing information from inference and patterns. Machine learning algorithms construct a mathematical model using sample data, called "training data," to improve the performance P at a task T based on experience E [84]. Machine learning has been applied in a wide variety of applications, such as computer vision and e-mail filtering.

12.5.1 Pipeline

The aim of machine learning in virtual screening is to incorporate data from multiple sources into sensible models for describing and screening compounds with the goal of identifying active drug targets. The typical machine learning pipeline begins from data acquisition, proceeds to feature engineering, then to algorithm selection and model construction, and finally to model evaluation and application. Figure 12.3 shows the overview of a typical machine learning-based virtual screening workflow.

12.5.2 Data Preparation

Data preparation is the step of transforming and cleaning raw data prior to processing and further analysis. *Steve Lohr of The New York Times said: "Data scientists,*

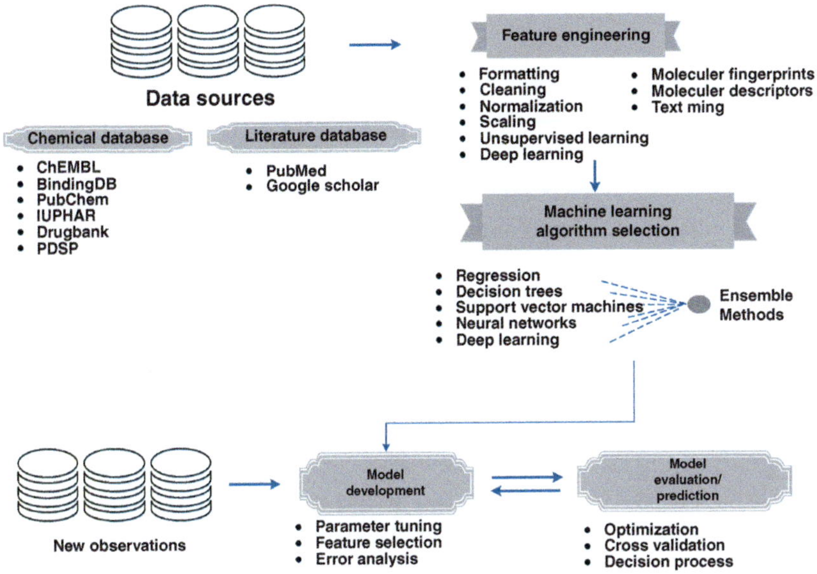

Figure 12.3 Typical machine learning workflow.

according to interviews and expert estimates, spend 50 percent to 80 percent of their time mired in the mundane labor of collecting and preparing unruly digital data, before it can be explored for useful nuggets." Thus, it is important and often involves reformatting data, imputation of missing values, and the combination of datasets to increase sample size. Data preparation usually is a lengthy undertaking for scientists, but it is essential as a prerequisite to put data in its proper context in order to gain meaningful insights and eliminate bias resulting from poor data quality.

12.5.3 Feature Selection

Feature selection aims to reduce the dimensionality of data by using only a subset of the most important features to build a model. Selection criteria usually consist of the minimization of predictive errors for models given diverse feature subsets. Algorithms search for a subset of features that can optimally model measured responses, subject to specific constraints, such as ℓ_1-norm and ℓ_2-norm regularization. As shown in Figure 12.4, there are three general classes of feature selection algorithms: filter methods, wrapper methods and embedded methods.

Filter methods usually adopt a statistical measure to determine a score for each feature. The features can be ranked by the scores and are either selected to be saved or removed from the model. The methods usually are univariate and consider each feature independently or with regards to the dependent variable. Some examples

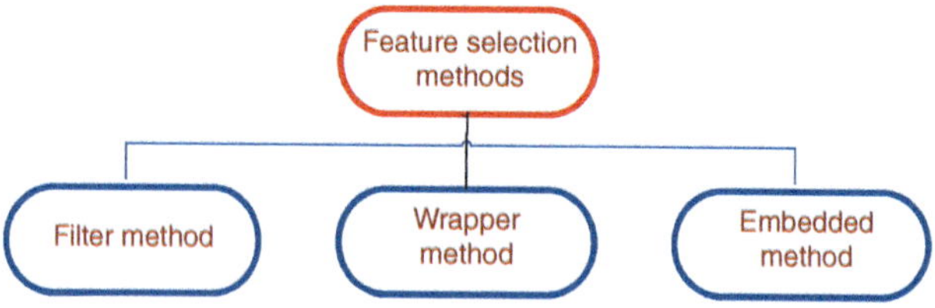

Figure 12.4 Classes of feature selection algorithms.

of filter methods involve the Chi squared test, information gain, and correlation coefficient scores.

Wrapper methods regard the selection of a subset of features as a search problem, in which different combinations are generated, evaluated and compared to each other. A predictive model is built to evaluate the combinations of features and to assign a ranking according to the model accuracy. The search process can be methodical such as a best-first search, or it may be stochastic such as a random hill-climbing algorithm, or it may use heuristics, like forward and backward ways to add and remove features, such as the recursive feature elimination algorithm.

Embedded methods aim to learn which features have the best contribution to the model performance while the model is being built. The most common class of embedded feature selection is regularization-based methods. Regularization methods are also called as penalization methods that introduce additional constraint terms into the objective function of a predictive algorithm that push the model toward lower complexity. Examples of regularization algorithms involve the LASSO, Elastic Net, and Ridge Regression.

12.5.3.1 A Case

We proposed a new method SED to predict ligand bioactivities and to recognize key substructures associated with GPCRs through the coupling of screening for LASSO of long extended-connectivity fingerprints (ECFPs) with deep neural network training [85]. Shown in Figure 12.5, the SED pipeline contains three successive steps: (i) representation of long ECFPs for ligand molecules, (ii) feature selection by screening for LASSO of ECFPs, and (iii) bioactivity prediction through a deep neural network regression model.

12.5.4 Algorithms

12.5.4.1 Traditional Algorithms

Traditional applications of machine learning in virtual screening focus on the use of supervised techniques to train statistical learning algorithms to classify databases of molecules by their activity against a particular drug target.

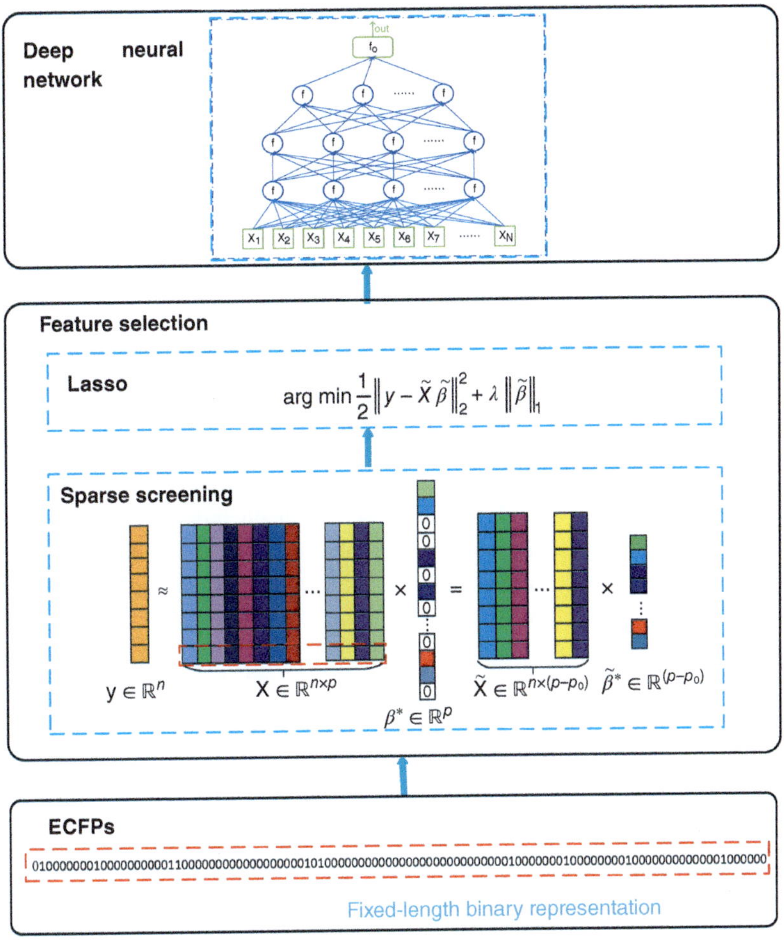

Figure 12.5 Schematic of SED. The approach is composed of three stages: long extended-connectivity fingerprint (ECFP) representation for ligand molecules, feature selection by screening for LASSO, and construction of deep neural network regression prediction models. Source: Wu et al. [85]/with permission of Oxford University Press.

Both ligand-based virtual screening and structure-based docking benefit from machine learning algorithms, including naïve Bayesian classifiers, neural networks, support vector machines, and decision trees, as well as more regression techniques.

The most simple machine learning method in virtual screening is multiple linear regression (MLR), which has been widely used in quantitative structure–activity relationship (QSAR) software [86–89], such as CoMSIA [90]. For instance, Evers et al. proposed a method of integrating linear regression and classification methods with discriminative analysis to find potential ligands targeting GPCRs from the Molecular Design Limited drug data report (MDDR) [89]. For four GPCR drug targets, Feher et al. employed consensus scoring to combine multiple different ligand-based methods using 2D descriptors, and 3D pharmacophore models [91]. Another usage of linear regression in virtual screening is in structure-based docking approaches [92, 93]. For instance, Jacobsson and Karlén employed partial least squares (PLSs) to correct for the size bias of many docking scoring functions [93].

Conceptually, one of the popular methods to model the activity of a molecule is to search a database for the molecules that are the most similar to it. The predicted score values are the activity of these "nearest neighbors," a method referred to as "k nearest neighbors" (kNN). A few attempts have been made to improve the model performance of virtual screening based on the kNN methods, such as CoLiBRI [94], ENTess [95], MFMNN [96] methods and several lazy methods [97, 98].

Naïve Bayesian is among the simplest classifiers. The probability of activity is determined by the ratio of actives to inactives that share the descriptor value. This approach supposes that each descriptor is statistically independent. The earliest usage of a naïve Bayesian method in virtual screening was for Binary QSAR, introduced by Labute [99]. Following this work, several excellent applications of the naïve Bayesian model have been put forward [100–102]. For instance, Glick et al. adopted a naïve Bayesian model with ECFP descriptors as a post-processing method in order to prioritize hits from experimental HTS data [102]. Then, they extended the usage of the naïve Bayesian model in docking to take the place of HTS as a source of potential active compounds [101].

Support vector machines (SVMs) were proposed by Vapnik [103], where a separation hyperplane boundary was defined, and examples were classified depending on which side of the boundary they are located. The initial work on the usage of SVM in virtual screening was carried out by the Willett group, where 35 991 molecules in the National Cancer Institute (NCI) AIDS data set were tested, using UNITY fingerprints as attributes [104]. Following this work, multiple excellent virtual screening methods based on SVM have been proposed [105–108]. For example, a prospective usage of SVM in virtual screening was developed by Schneider et al., who adopted a model which was built from 331 dopamine receptors inhibitors for screening the SPECS and Interbioscreen databases (over 255 000 molecules combined) which resulted in the identification of 11 compounds with a high selectivity for the D3 receptor [108].

Artificial neural networks (ANNs) have played a long-established role in cheminformatics. ANNs are composed of a set of connected artificial "neurons." A single neuron takes multiple numerical inputs, and outputs a transformed and weighted sum of the inputs; through layers of many parallel neurons, complex classification functions can be defined through training. Multiple uses of ANNs to produce consensus scoring functions for docking have been proposed [109, 110].

For instance, Sem et al. adopted ANNs to generate a consensus score for predicting CYP2D6 binding affinity through combination of the AutoDock and XScore tools [110]. ANNs have also been applied for VS beyond the pharmaceutical area, for example in heterogeneous catalysis [111].

A decision tree denotes the conjunction of a series of "rules," each of which is a predicate concerning a subset of descriptors. The process of training a decision tree model determines which rules are involved in the tree, and what classification is made at each leaf. The process is usually carried out by choosing a valuable rule that can divide the training data into two or more groups. The process is then repeated for each of the subsets, until a termination criterion is reached. A few applications of decision trees in VS propose novel classification tools in QSAR [112, 113]. For instance, Jones-Hertzog et al. proposed a decision tree method for screening 23 000 compounds against 14 GPCR drug targets, which outperforms random selection and similarity searching in the majority of cases [112]. Yamakazi developed a single decision tree, which was trained on 130 PDE-5 inhibitors and 10 000 inactives, to screen 50 520 molecules in the SPECS database [113].

Ensemble methods denote a series of classifiers that combine the output of base classifiers to arrive at the final decision. Classical ensemble methods include bootstrapping, boosting, etc. Simple examples of the ensemble method in VS include the study of van Rhee et al., who adopted pairs of decision trees to screen 3000 molecules, which demonstrated a 13-fold enrichment over the hit rate of a 14 000 member HTS when screening against an ion channel target [114]. Random forest (RF) represents one of the most popular ensemble methods; the model consists of an ensemble of many randomly generated decision trees. Due to its ease of use, high accuracy, and robustness to adjustable parameters, RF has become a "gold standard" for QSAR method comparison [115]. Svetnik et al. have examined the usage of RF against several datasets, including the same HTS data which was used to evaluate boosting, and it was shown to have competitive performance [116]. RF has also been applied in conjunction with docking. For example, Teramoto and Fukunishi adopted a RF model to predict the root mean square deviation (RMSD) of a docked conformation relative to the bioactive conformation [117].

12.5.4.2 New Algorithms

Deep learning is a branch of machine learning consisting of a set of algorithms that attempt to model high-level abstractions in data by using multiple processing layers, with complex structures or otherwise, composed of multiple non-linear transformations. Recently, deep learning-based methods have witnessed impressive success in ligand-based virtual screening [118–122]. For instance, in 2012, Merck organized a challenge for the design of machine learning methods to model the bioactivities of ligands acting with target proteins, and methods using deep learning achieved the best performance. Later, Ma et al. (2015) proposed a deep neural net model for determining QSARs, which demonstrated better performance than random forest models for most of the data they studied [115]. Most recently, we proposed a weighted deep learning algorithm that takes arbitrarily sized inputs and generates bioactivity predictions which are significantly more accurate than the control predictors with different molecular fingerprints and descriptors [123].

Applications of deep learning models in *de novo* ligand design have also been developed. There are multiple examples of models which adopt autoencoders and/or recurrent neural networks that can produce new molecules with ideal properties [124–127]. The usage of autoencoders also permit the representation of molecules as short, real-valued vectors, which are extracted from the bottleneck layers, in order to facilitate exploration of the chemical space [125].

Although deep learning is more readily applied in ligand-based VS, there are currently a few interesting examples for structure-based VS applications [128–130]. For instance, in AtomNet, the input of molecular complex is discretized to a 3D grid framework and directly fed into a convolutional neural network model [128]. Another similar work was created by Ragoza et al. (2016) where two independent classification tasks, i.e. activity and pose prediction, were trained and performed [130].

Multi-task learning is a class of machine learning approaches that learns a task together with other related tasks at the same time, with a shared representation. This can usually achieve a better model for the main task, because it allows the models to capture commonality among the tasks. Neural network models allow for the easy construction of multi-task classifiers and regression models, such as those for predicting binding activities against multiple targets at once. It has been shown that such QSAR models can perform better than single-task models [115, 121, 122, 131–133], because they can benefit from more training data, and share internal representations between tasks. In 2012, Merck & Co. hosted a Kaggle challenge where the ability of data science to improve predictive performance of QSAR methods was benchmarked. The winning team used multi-task deep networks, ensembled with other machine-learning techniques, to achieve a 15% relative improvement over the baseline method. The multi-task DNNs, which were called as "joint DNNs" in Ma et al. [115], can simultaneously model more than one molecular activity task. All tasks share the same input and hidden layers, but each task has its own output values [122, 132]. In 2017, the Pande group from Stanford University constructed a ligand-based virtual screening model through multi-task deep learning and established an excellent open source platform DeepChem [121]. In 2017, Xu et al. from Merck Pharmaceutical Company adopted multi-task deep learning to build ligand-based virtual screening models, and tried to analyze how it can improve the model performance [122].

12.5.5 Evaluation

12.5.5.1 Regression Models

In the Kaggle challenge organized by Merck in 2012, the correlation coefficient (r^2) was used to assess the performance of drug activity predictions. This metric is calculated as

$$r^2 = \frac{\left[\sum_{i=1}^{n} (y_i - \bar{y})(\hat{y}_i - \bar{\hat{y}})\right]^2}{\sum_{i=1}^{n} (y_i - \bar{y})^2 \sum_{i=1}^{n} (\hat{y}_i - \bar{\hat{y}})^2} \tag{12.3}$$

where y_i is the true activity, $\bar{y}$ is the mean of the true activity, $\hat{y}_i$ is the predicted activity, $\bar{\hat{y}}$ is the mean of the predicted activity, and n is the number of ligand molecules in the dataset. The larger the value of r^2, the better the prediction performance.

A common metric for evaluating regression models is the root mean square error (RMSE), given by

$$\text{RMSE} = \sqrt{\frac{1}{n}\sum_{i=1}^{n} (y_i - \hat{y}_i)^2} \tag{12.4}$$

where y_i and $\hat{y}_i$ are the true and predicted activity values, respectively, and n is the number of ligand molecules. The smaller the RMSE value, the better the prediction performance.

12.5.5.2 Classification Models

The overall prediction accuracy (Q), sensitivity (Sn), precision (P), specificity (Sp), and Matthew's correlation coefficient (CC) are commonly used for assessment of the classification system.

$$Q = \frac{\text{TP} + \text{TN}}{\text{TP} + \text{TN} + \text{FP} + \text{FN}} \tag{12.5}$$

$$\text{Sn} = \frac{\text{TP}}{\text{TP} + \text{FN}} \tag{12.6}$$

$$P = \frac{\text{TP}}{\text{TP} + \text{FP}} \tag{12.7}$$

$$\text{Sp} = \frac{\text{TN}}{\text{TN} + \text{FP}} \tag{12.8}$$

$$\text{CC} = \frac{\text{TP} \times \text{TN} - \text{FP} \times \text{FN}}{\sqrt{(\text{TP} + \text{FP}) \times (\text{TP} + \text{FN}) \times (\text{TN} + \text{FP}) \times (\text{TN} + \text{FN})}} \tag{12.9}$$

where TP, TN, FP, and FN are true positive, true negative, false positive, and false negative, respectively.

The ROC curve is probably the most robust technique for evaluating classifiers and visualizing their performance. Classification machine learning models can be validated by accuracy estimation techniques like the K-fold cross validation method, where the dataset is randomly partitioned into K subsets, and then K experiments are performed each respectively considering 1 subset for evaluation and the remaining $K - 1$ subsets for training the model.

12.6 Inverse Virtual Screening

The previously described procedure of virtual screening, in which a set of ligands is screened against a single protein target, can alternatively be described as "classical" or "forward" virtual screening. In "inverse" virtual screening (IVS), also known as "virtual target screening" or "target fishing," the roles assigned to the ligand and the protein are reversed: a single ligand of interest is screened against a set of proteins. While the underlying principles and methodologies of IVS are similar to typical virtual screening, a few challenges unique to IVS arise, which have been addressed in some unique ways.

12.6.1 Knowledge-Based Approaches

In order to perform an IVS, the most logically simple approach is to use methods that perform forward virtual screening but modify them so that the receptor is changed instead of the ligand. This means that most categories of forward virtual screening approaches (ligand-based, structure-based, etc.) also apply to IVS. Since many of these approaches determine interaction likelihood through searching of a database of experimentally observed binding events, such "knowledge-based" approaches have also populated the field of IVS.

One simple knowledge-based approach is the Similarity Ensemble Approach (SEA), which evaluates the biological similarity of targets by the chemical similarity of their respective ligands [134]. The chemical similarity is defined as the sum of Tanimoto Coefficients above a tuned threshold between all pairs of ligands, where each ligand is encoded by the Daylight fingerprint. Through this method, one can perform IVS by evaluating the chemical similarity of a query ligand against each target's respective ligand set. Accordingly, SEA has been applied in both retrospective and prospective drug-target prediction [135].

FINDSITE$^{comb2.0}$ is a unique approach that combines both protein structure and chemical similarity in order to predict protein–ligand interactions [79]. The method uses a combination of threading and structure comparison to identify ligand-bound protein structures (both experimental and modeled) from which a set of template ligands can be extracted. This set of template ligands is used to screen for active compounds in a compound library using chemical similarity. While the method is primarily developed and benchmarked as a forward screening approach, it also can be used for IVS through evaluating the chemical similarity of the input ligand to the predicted active ligands for a given target.

Despite the relative speed and simplicity of these approaches, they are ultimately limited by the depth of the knowledgebase on which they are built. For example, a ligand-based approach cannot effectively screen a molecule that is too unlike any ligand in its database; such results would be based on spurious similarity and ultimately would not be trustworthy. Therefore, these methods are only as powerful as the databases on which they are built. This implies that they will only become more accurate over time as more data becomes available. However, there will always exist edge cases that are not properly addressed and documented by the databases of knowledge-based approaches, and so, methods that can make protein–ligand interaction predictions without explicit dependency on available data, such as docking, can potentially provide insight where none currently exists.

12.6.2 Docking Approaches

Given the popularity of protein–ligand docking approaches in forward virtual screening, it is no surprise that many methods for inverse virtual screening are also based on these methods. TarFisDock is one of the most simply constructed of these approaches [136]. It performs screening of a molecule of interest against a potential drug target database (PDTD) through a method based on DOCK 4.0 and returns a set of protein targets ranked by their predicted interaction energy. However, as noted by several other studies, while docking energy functions are reasonable to rank ligands given a single target, they are not necessarily comparable across targets. One way to address this problem is implemented in the first inverse docking program, INVDOCK, which requires that "hits" in the screen meet not only some minimum binding energy threshold, but also meet a threshold based on the target's binding affinity for its native ligand, ensuring that any predicted interactions would be competitive relative to the native interaction [137]. Another method of scoring targets is to implement an interaction fingerprinting technique, in which predicted interactions are evaluated based on the presence or absence of interactions between the ligand and each residue of the protein. Such a technique is implemented in IFPTarget [138], an approach that compares an interaction fingerprint generated from an AutoDock Vina docking result to a database of native interaction fingerprints and combines this comparison with a few more traditional scoring functions to identify likely ligand targets. This method can be seen as a hybrid between dependence on previously observed data and *a priori* docking predictions, where the docking scores can provide a prediction where no interaction is found and the fingerprint score can find near native binding modes that were predicted to be unfavorable by docking.

Since these approaches depend on protein–ligand docking, they inherit all of the shortcomings therein, such as the assumption of receptor rigidity, the lack of a scoring function that perfectly ranks docking results, the relative computational expense of docking, and the need for a high-resolution receptor structure with a well-defined binding pocket. This last point is particularly restrictive in IVS, as the majority of targets that one could logically screen against do not have solved structures, and even less have clearly defined binding pockets. In consideration of this problem, many docking-based IVS approaches only offer screening against a small database of proteins. If one wishes to use IVS to determine how a small molecule will impact a cell at a systems level, screening against such a limited database will not provide a complete view.

12.6.2.1 Applications of IVS

One of the most common applications of IVS is to increase the efficiency of computational drug discovery by assessing a molecule's ability to bind to proteins other than the intended therapeutic target. In these studies, the molecule of interest is typically identified through a previously performed forward virtual screen. When both forward and inverse virtual screens are used in combination, the forward screen can be viewed as a "sensitivity" screen in which a ligand that can tightly bind a given protein target is found, and the inverse virtual screen serves as a "specificity" screen which ensures that the ligand binds tightly only to the target protein and not to any other proteins within the biological context. Such off-target interactions can give rise to consequences other than the intended effect, such as side effects or compound toxicity, and therefore, prediction of these interactions can aid the efficiency of drug design studies by computationally identifying these problems before they are discovered *in vivo*. However, not all interactions other than the intended one are deleterious. In fact, the efficacy of a ligand may be enhanced through the interaction of several proteins at once, giving rise to the principle of "multi-target" design (a.k.a polypharmacology). Understanding the extent to which a compound will impact a biological context requires the application of systems biology models that take into account how the set of predicted binding events from an IVS result will perturb proteomic and metabolomic networks.

Another application of IVS is the discovery of therapeutic targets for a given active ligand. For example, there exist many drugs that are known to be clinically effective, yet their mechanism of action remains nebulous. Through IVS, one can identify a set of candidate proteins which might be the therapeutic target(s) of the drug molecule of interest. However, the mechanism of the drug need not be completely unknown for IVS to be helpful. In fact, IVS has also been shown to be effective for "drug repurposing," in which a clinically approved drug is used to treat some disease other than the one for which it was developed. Through screening the drug to be repurposed against a library of potential therapeutic targets, high-affinity drug–target interactions can be discovered, which can lead to novel therapeutic action.

12.6.3 Challenges

While inverse virtual screening demonstrates promise for improving the drug design process through systems biology, a few challenges need to be overcome before the technique is widely accepted. One of the most pressing issues facing the field is the difficulty of constructing benchmark datasets due to publication bias. Since non-interacting protein–ligand pairs are frequently not reported, benchmarks lack confirmed non-interactions and instead rely on the assumption that if no interaction has been reported, it does not exist. Another challenge is the relative complexity of understanding of protein biochemistry relative to ligand chemistry. When constructing a forward virtual screen, one can choose a protein target that is sufficiently understood (i.e. solved protein structure, clearly defined binding pocket, a breadth of known binding partners, etc.), but in an IVS, no such luxury exists. If one is to gain a fully comprehensive view of how a molecule will impact all proteins within some biological context, all proteins therein must be addressed.

12.7 Conclusion

Hearkening back to its origins, virtual screening has come a long way and has evolved into a mélange of algorithms. Apart from the conventional ligand- and structure-based methods, the field has been burgeoning in recent years with machine learning-based approaches for the modeling of ligand bioactivity. Moreover, chemogenomics has been utilized in virtual screening to attempt to de-orphanize receptors, while inverse virtual screening is opening up new avenues for the prediction of off-target effects. In many ways, it feels as if the field has only just been born. Future developments are eagerly anticipated in the hopes that novel therapeutic compounds can be discovered for one of the largest protein families in human.

Acknowledgments

The study is supported in part by the National Institute of General Medical Sciences (GM070449, GM083107, GM116960), National Institute of Allergy and Infectious Diseases (AI134678), and the National Science Foundation (DBI1564756).

References

1 DeVree, B.T., Mahoney, J.P., Vélez-Ruiz, G.A. et al. (2016). Allosteric coupling from G protein to the agonist-binding pocket in GPCRs. *Nature* 535 (7610): 182.

2 Venter, J.C., Adams, M.D., Myers, E.W. et al. (2001). The sequence of the human genome. *Science* 291 (5507): 1304–1351.

3 O'Hayre, M., Vazquez-Prado, J., Kufareva, I. et al. (2013). The emerging mutational landscape of G proteins and G-protein-coupled receptors in cancer. *Nat. Rev. Cancer* 13 (6): 412–424.

4 Rompler, H., Staubert, C., Thor, D. et al. (2007). G Protein-coupled time travel: evolutionary aspects of GPCR research. *Mol. Interv.* 7 (1): 17–25.

5 Garland, S.L. (2013). Are GPCRs still a source of new targets? *J. Biomol. Screen.* 18 (9): 947–966.

6 Van Drie, J.H. (2007). Computer-aided drug design: the next 20 years. *J. Comput. Aided Mol. Des.* 21 (10, 11): 591–601.

7 O'Boyle, N.M. (2012). Towards a Universal SMILES representation – a standard method to generate canonical SMILES based on the InChI. *J. Cheminf.* 4 (1): 22.

8 Heller, S., McNaught, A., Stein, S. et al. (2013). InChI – the worldwide chemical structure identifier standard. *J. Cheminf.* 5 (1): 7.

9 Pletnev, I., Erin, A., McNaught, A. et al. (2012). InChIKey collision resistance: an experimental testing. *J. Cheminf.* 4 (1): 39.

10 Durant, J.L., Leland, B.A., Henry, D.R. et al. (2002). Reoptimization of MDL keys for use in drug discovery. *J. Chem. Inf. Comput. Sci.* 42 (6): 1273–1280.

11 O'Boyle, N.M. and Sayle, R.A. (2016). Comparing structural fingerprints using a literature-based similarity benchmark. *J. Cheminf.* 8 (1): 36.

12 Dalby, A., Nourse, J.G., Hounshell, W.D. et al. (1992). Description of several chemical structure file formats used by computer programs developed at Molecular Design Limited. *J. Chem. Inf. Comput. Sci.* 32 (3): 244–255.

13 Berman, H.M. (2008). The protein data bank: a historical perspective. *Acta Crystallogr. A* 64 (1): 88–95.

14 Consortium U (2016). UniProt: the universal protein knowledgebase. *Nucleic Acids Res.* 45 (D1): D158–D169.

15 Rose, P.W., Prlic, A., Altunkaya, A. et al. (2017). The RCSB protein data bank: integrative view of protein, gene and 3D structural information. *Nucleic Acids Res.* 45 (D1): D271–D281.

16 Prilusky, J. (1996). OCA, a browser-database for protein structure/function. http://oca.weizmann.ac.il/oca-bin/ocamain.

17 Pándy-Szekeres, G., Munk, C., Tsonkov, T.M. et al. (2017). GPCRdb in 2018: adding GPCR structure models and ligands. *Nucleic Acids Res.* 46 (D1): D440–D446.

18 Gaulton, A., Bellis, L.J., Bento, A.P. et al. (2012). ChEMBL: a large-scale bioactivity database for drug discovery. *Nucleic Acids Res.* 40 (Database issue): D1100–D1107.

19 Gilson, M.K., Liu, T., Baitaluk, M. et al. (2015). BindingDB in 2015: a public database for medicinal chemistry, computational chemistry and systems pharmacology. *Nucleic Acids Res.* 44 (D1): D1045–D1053.

20 Wassermann, A.M. and Bajorath, J. (2011). BindingDB and ChEMBL: online compound databases for drug discovery. *Expert Opin. Drug Discov.* 6 (7): 683–687.

21 Law, V., Knox, C., Djoumbou, Y. et al. (2014). DrugBank 4.0: shedding new light on drug metabolism. *Nucleic Acids Res.* 42 (Database issue): D1091–D1097.

22 Roth, B.L., Lopez, E., Patel, S. et al. (2000). The multiplicity of serotonin receptors: uselessly diverse molecules or an embarrassment of riches? *Neuroscientist* 6 (4): 252–262.

23 Alexander, S.P., Davenport, A.P., Kelly, E. et al. (2015). The Concise Guide to PHARMACOLOGY 2015/16: G protein-coupled receptors. *Br. J. Pharmacol.* 172 (24): 5744–5869.

24 Wang, Y., Xiao, J., Suzek, T.O. et al. (2009). PubChem: a public information system for analyzing bioactivities of small molecules. *Nucleic Acids Res.* 37 (Web Server issue): W623–W633.

25 Okuno, Y., Yang, J., Taneishi, K. et al. (2006). GLIDA: GPCR-ligand database for chemical genomic drug discovery. *Nucleic Acids Res.* 34 (Suppl 1): D673–D677.

26 Okuno, Y., Tamon, A., Yabuuchi, H. et al. (2008). GLIDA: GPCR–ligand database for chemical genomics drug discovery – database and tools update. *Nucleic Acids Res.* 36 (Database issue): D907–D912.

27 Chan, W.K., Zhang, H., Yang, J. et al. (2015). GLASS: a comprehensive database for experimentally validated GPCR-ligand associations. *Bioinformatics* 31 (18): 3035–3042.

28 Zoete, V., Daina, A., Bovigny, C. et al. (2016). SwissSimilarity: a web tool for low to ultra high throughput ligand-based virtual screening. *J. Chem. Inf. Model.* 56 (8): 1399–1404.

29 Wu, J., Zhang, Q., Wu, W. et al. (2018). WDL-RF: predicting bioactivities of ligand molecules acting with G protein-coupled receptors by combining weighted deep learning and random forest. *Bioinformatics* 1: 12.

30 Huang, N., Shoichet, B.K., and Irwin, J.J. (2006). Benchmarking sets for molecular docking. *J. Med. Chem.* 49 (23): 6789–6801.

31 Mysinger, M.M., Carchia, M., Irwin, J.J. et al. (2012). Directory of useful decoys, enhanced (DUD-E): better ligands and decoys for better benchmarking. *J. Med. Chem.* 55 (14): 6582–6594.

32 Weiss, D.R., Bortolato, A., Tehan, B. et al. (2016). GPCR-bench: a benchmarking set and practitioners' guide for G protein-coupled receptor docking. *J. Chem. Inf. Model.* 56 (4): 642–651.

33 Gatica, E.A. and Cavasotto, C.N. (2011). Ligand and decoy sets for docking to G protein-coupled receptors. *J. Chem. Inf. Model.* 52 (1): 1–6.

34 Sastry, G.M., Inakollu, V.S., and Sherman, W. (2013). Boosting virtual screening enrichments with data fusion: coalescing hits from two-dimensional fingerprints, shape, and docking. *J. Chem. Inf. Model.* 53 (7): 1531–1542.

35 Truchon, J.-F. and Bayly, C.I. (2007). Evaluating virtual screening methods: good and bad metrics for the "early recognition" problem. *J. Chem. Inf. Model.* 47 (2): 488–508.

36 Bemis, G.W. and Murcko, M.A. (1996). The properties of known drugs. 1. Molecular frameworks. *J. Med. Chem.* 39 (15): 2887–2893.

37 Bajusz, D., Rácz, A., and Héberger, K. (2015). Why is Tanimoto index an appropriate choice for fingerprint-based similarity calculations? *J. Cheminf.* 7 (1): 20.

38 O'Boyle, N.M., Banck, M., James, C.A. et al. (2011). Open Babel: an open chemical toolbox. *J. Cheminf.* 3 (1): 33.

39 Landrum, G. (2012). RDKit: open-source cheminformatics. http://www rdkit.org (accessed 07 March 2022).

40 Grant, J.A. and Pickup, B. (1995). A Gaussian description of molecular shape. *J. Phys. Chem.* 99 (11): 3503–3510.

41 Hawkins, P.C.D., Skillman, A.G., and Nicholls, A. (2007). Comparison of shape-matching and docking as virtual screening tools. *J. Med. Chem.* 50 (1): 74–82.

42 Ballester, P.J. (2011). Ultrafast shape recognition: method and applications. *Future Med. Chem.* 3 (1): 65–78.

43 Roy, A. and Skolnick, J. (2014). LIGSIFT: an open-source tool for ligand structural alignment and virtual screening. *Bioinformatics* 31 (4): 539–544.

44 Hu, J., Liu, Z., Yu, D.-J. et al. (2018). LS-align: an atom-level, flexible ligand structural alignment algorithm for high-throughput virtual screening. *Bioinformatics* 34 (13): 2209–2218.

45 Morris, G.M., Huey, R., Lindstrom, W. et al. (2009). AutoDock4 and AutoDockTools4: automated docking with selective receptor flexibility. *J. Comput. Chem.* 30 (16): 2785–2791.

46 Jones, G., Willett, P., Glen, R.C. et al. (1997). Development and validation of a genetic algorithm for flexible docking. *J. Mol. Biol.* 267 (3): 727–748.

47 Trott, O. and Olson, A.J. (2010). AutoDock Vina: improving the speed and accuracy of docking with a new scoring function, efficient optimization, and multithreading. *J. Comput. Chem.* 31 (2): 455–461.

48 Korb, O., Stutzle, T., and Exner, T.E. (2009). Empirical scoring functions for advanced protein–ligand docking with PLANTS. *J. Chem. Inf. Model.* 49 (1): 84–96.

49 Allen, W.J., Balius, T.E., Mukherjee, S. et al. (2015). DOCK 6: impact of new features and current docking performance. *J. Comput. Chem.* 36 (15): 1132–1156.

50 Friesner, R.A., Banks, J.L., Murphy, R.B. et al. (2004). Glide: a new approach for rapid, accurate docking and scoring. 1. Method and assessment of docking accuracy. *J. Med. Chem.* 47 (7): 1739–1749.

51 Halgren, T.A., Murphy, R.B., Friesner, R.A. et al. (2004). Glide: a new approach for rapid, accurate docking and scoring. 2. Enrichment factors in database screening. *J. Med. Chem.* 47 (7): 1750–1759.

52 Meng, E.C., Shoichet, B.K., and Kuntz, I.D. (1992). Automated docking with grid-based energy evaluation. *J. Comput. Chem.* 13 (4): 505–524.

53 Huang, S.-Y., Grinter, S.Z., and Zou, X. (2010). Scoring functions and their evaluation methods for protein–ligand docking: recent advances and future directions. *Phys. Chem. Chem. Phys.* 12 (40): 12899–12908.

54 Liu, J. and Wang, R. (2015). Classification of current scoring functions. *J. Chem. Inf. Model.* 55 (3): 475–482.

55 Kuntz, I.D., Blaney, J.M., Oatley, S.J. et al. (1982). A geometric approach to macromolecule–ligand interactions. *J. Mol. Biol.* 161 (2): 269–288.

56 Basith, S., Cui, M., Macalino, S.J. et al. (2018). Exploring G protein-coupled receptors (GPCRs) ligand space via cheminformatics approaches: impact on rational drug design. *Front. Pharmacol.* 9: 128.

57 Yuan, X. and Xu, Y. (2018). Recent trends and applications of molecular modeling in GPCR–ligand recognition and structure-based drug design. *Int. J. Mol. Sci.* 19 (7): 2105.

58 Roth, B.L., Irwin, J.J., and Shoichet, B.K. (2017). Discovery of new GPCR ligands to illuminate new biology. *Nat. Chem. Biol.* 13 (11): 1143.

59 Manglik, A., Lin, H., Aryal, D.K. et al. (2016). Structure-based discovery of opioid analgesics with reduced side effects. *Nature* 537 (7619): 185.

60 Rodriguez, D., Brea, J., Loza, M.I. et al. (2014). Structure-based discovery of selective serotonin 5-HT(1B) receptor ligands. *Structure* 22 (8): 1140–1151.

61 Mysinger, M.M., Weiss, D.R., Ziarek, J.J. et al. (2012). Structure-based ligand discovery for the protein-protein interface of chemokine receptor CXCR4. *Proc. Natl. Acad. Sci. U.S.A.* 109 (14): 5517–5522.

62 Becker, O.M., Dhanoa, D.S., Marantz, Y. et al. (2006). An integrated in silico 3D model-driven discovery of a novel, potent, and selective amidosulfonamide 5-HT1A agonist (PRX-00023) for the treatment of anxiety and depression. *J. Med. Chem.* 49 (11): 3116–3135.

63 Kirchhoff, V.D., Nguyen, H.T., Soczynska, J.K. et al. (2009). Discontinued psychiatric drugs in 2008. *Exp. Opin. Investig. Drugs* 18 (10): 1431–1443.

64 Saha, A.K., Becker, O.M., Noiman, S., et al. (eds.) (2006). PRX-03140: the discovery and development of a novel 5HT4 partial agonist for the treatment of Alzheimer's disease. Abstracts of Papers of the American Chemical Society.

65 Langmead, C.J., Andrews, S.P., Congreve, M. et al. (2012). Identification of novel adenosine A2A receptor antagonists by virtual screening. *J. Med. Chem.* 55 (5): 1904–1909.

66 Verma, S., Kumar, A., Tripathi, T. et al. (2018). Muscarinic and nicotinic acetylcholine receptor agonists: current scenario in Alzheimer's disease therapy. *J. Pharm. Pharmacol.* 70 (8): 985–993.

67 Sliwoski, G., Kothiwale, S., Meiler, J. et al. (2014). Computational methods in drug discovery. *Pharmacol. Rev.* 66 (1): 334–395.

68 Drwal, M.N. and Griffith, R. (2013). Combination of ligand- and structure-based methods in virtual screening. *Drug Discov. Today Technol.* 10 (3): e395–e401.

69 Khan, K.M., Wadood, A., Ali, M. et al. (2010). Identification of potent urease inhibitors via ligand-and structure-based virtual screening and in vitro assays. *J. Mol. Graphics Modell.* 28 (8): 792–798.

70 Weidlich, I.E., Dexheimer, T., Marchand, C. et al. (2010). Inhibitors of human tyrosyl-DNA phospodiesterase (hTdp1) developed by virtual screening using ligand-based pharmacophores. *Bioorg. Med. Chem.* 18 (1): 182–189.

71 Banoglu, E., Çalışkan, B., Luderer, S. et al. (2012). Identification of novel benzimidazole derivatives as inhibitors of leukotriene biosynthesis by virtual screening targeting 5-lipoxygenase-activating protein (FLAP). *Bioorg. Med. Chem.* 20 (12): 3728–3741.

72 Svensson, F., Karlén, A., and Sköld, C. (2011). Virtual screening data fusion using both structure- and ligand-based methods. *J. Chem. Inf. Model.* 52 (1): 225–232.

73 Swann, S.L., Brown, S.P., Muchmore, S.W. et al. (2011). A unified, probabilistic framework for structure- and ligand-based virtual screening. *J. Med. Chem.* 54 (5): 1223–1232.

74 Tan, L., Geppert, H., Sisay, M.T. et al. (2008). Integrating structure- and ligand-based virtual screening: comparison of individual, parallel, and fused molecular docking and similarity search calculations on multiple targets. *ChemMedChem* 3 (10): 1566–1571.

75 Klabunde, T. (2007). Chemogenomic approaches to drug discovery: similar receptors bind similar ligands. *Br. J. Pharmacol.* 152 (1): 5–7.

76 Brylinski, M. and Skolnick, J. (2008). A threading-based method (FINDSITE) for ligand-binding site prediction and functional annotation. *Proc. Natl. Acad. Sci. U.S.A.* 105 (1): 129–134.

77 Zhou, H.Y. and Skolnick, J. (2012). FINDSITEx: a structure-based, small molecule virtual screening approach with application to all identified human GPCRs. *Mol. Pharm.* 9 (6): 1775–1784.

78 Zhou, H. and Skolnick, J. (2012). FINDSITEcomb: a threading/structure-based, proteomic-scale virtual ligand screening approach. *J. Chem. Inf. Model.* 53 (1): 230–240.

79 Zhou, H., Cao, H., and Skolnick, J. (2018). FINDSITE$^{comb2.0}$: a new approach for virtual ligand screening of proteins and virtual target screening of biomolecules. *J. Chem. Inf. Model.* 58 (11): 2343–2354.

80 Roy, A., Srinivasan, B., and Skolnick, J. (2015). PoLi: a virtual screening pipeline based on template pocket and ligand similarity. *J. Chem. Inf. Model.* 55 (8): 1757–1770.

81 Yang, Y., Zhan, J., and Zhou, Y. (2016). SPOT-ligand: fast and effective structure-based virtual screening by binding homology search according to ligand and receptor similarity. *J. Comput. Chem.* 37 (18): 1734–1739.

82 Litfin, T., Zhou, Y., and Yang, Y. (2017). SPOT-ligand 2: improving structure-based virtual screening by binding-homology search on an expanded structural template library. *Bioinformatics* 33 (8): 1238–1240.

83 Chan, W.K. and Zhang, Y. (2020). Virtual screening of human class – a GPCRs using ligand profiles built on multiple ligand–receptor interactions. *J. Mol. Biol.* 432 (17): 4872–4890.

84 Mitchell, T., Buchanan, B., DeJong, G. et al. (1990). Machine learning. *Ann. Rev. Comput. Sci.* 4 (1): 417–433.

85 Wu, J.-S., Liu, B., Chan, W.K.B. et al. (2019). Precise modelling and interpretation of bioactivities of ligands targeting G protein-coupled receptors. *Bioinformatics* 35: i324–i332.

86 Hansch, C. and Fujita, T. (1964). *p*-σ-π Analysis. A method for the correlation of biological activity and chemical structure. *J. Am. Chem. Soc.* 86 (8): 1616–1626.

87 Cramer, R.D., Patterson, D.E., and Bunce, J.D. (1988). Comparative molecular field analysis (CoMFA). 1. Effect of shape on binding of steroids to carrier proteins. *J. Am. Chem. Soc.* 110 (18): 5959–5967.

88 Gedeck, P., Rohde, B., and Bartels, C. (2006). QSAR – how good is it in practice? Comparison of descriptor sets on an unbiased cross section of corporate data sets. *J. Chem. Inf. Model.* 46 (5): 1924–1936.

89 Evers, A., Hessler, G., Matter, H. et al. (2005). Virtual screening of biogenic amine-binding G-protein coupled receptors: comparative evaluation of protein- and ligand-based virtual screening protocols. *J. Med. Chem.* 48 (17): 5448–5465.

90 Klebe, G., Abraham, U., and Mietzner, T. (1994). Molecular similarity indices in a comparative analysis (CoMSIA) of drug molecules to correlate and predict their biological activity. *J. Med. Chem.* 37 (24): 4130–4146.

91 Baber, J.C., Shirley, W.A., Gao, Y. et al. (2006). The use of consensus scoring in ligand-based virtual screening. *J. Chem. Inf. Model.* 46 (1): 277–288.

92 Cherkasov, A., Ban, F., Li, Y. et al. (2006). Progressive docking: a hybrid QSAR/docking approach for accelerating in silico high throughput screening. *J. Med. Chem.* 49 (25): 7466–7478.

93 Jacobsson, M. and Karlén, A. (2006). Ligand bias of scoring functions in structure-based virtual screening. *J. Chem. Inf. Model.* 46 (3): 1334–1343.

94 Oloff, S., Zhang, S., Sukumar, N. et al. (2006). Chemometric analysis of ligand receptor complementarity: identifying Complementary Ligands Based on Receptor Information (CoLiBRI). *J. Chem. Inf. Model.* 46 (2): 844–851.

95 Zhang, S., Golbraikh, A., and Tropsha, A. (2006). Development of quantitative structure-binding affinity relationship models based on novel geometrical chemical descriptors of the protein–ligand interfaces. *J. Med. Chem.* 49 (9): 2713–2724.

96 Miller, D.W. (2001). Results of a new classification algorithm combining K nearest neighbors and recursive partitioning. *J. Chem. Inf. Comput. Sci.* 41 (1): 168–175.

97 Guha, R., Dutta, D., Jurs, P.C. et al. (2006). Local lazy regression: making use of the neighborhood to improve QSAR predictions. *J. Chem. Inf. Model.* 46 (4): 1836–1847.

98 Zhang, S., Golbraikh, A., Oloff, S. et al. (2006). A novel automated lazy learning QSAR (ALL-QSAR) approach: method development, applications, and virtual screening of chemical databases using validated ALL-QSAR models. *J. Chem. Inf. Model.* 46 (5): 1984–1995.

99 Labute, P. (1999). Binary QSAR: a new method for the determination of quantitative structure activity relationships. *Pac. Symp. Biocomput.* 444–455.

100 Klon, A.E., Glick, M., Thoma, M. et al. (2004). Finding more needles in the haystack: a simple and efficient method for improving high-throughput docking results. *J. Med. Chem.* 47 (11): 2743–2749.

101 Klon, A.E., Glick, M., and Davies, J.W. (2004). Application of machine learning to improve the results of high-throughput docking against the HIV-1 protease. *J. Chem. Inf. Comput. Sci.* 44 (6): 2216–2224.

102 Glick, M., Klon, A.E., Acklin, P. et al. (2004). Enrichment of extremely noisy high-throughput screening data using a naive Bayes classifier. *J. Biomol. Screen.* 9 (1): 32–36.

103 Cortes, C. and Vapnik, V. (1995). Support-vector networks. *Mach. Learn.* 20 (3): 273–297.

104 Wilton, D., Willett, P., Lawson, K. et al. (2003). Comparison of ranking methods for virtual screening in lead-discovery programs. *J. Chem. Inf. Comput. Sci.* 43 (2): 469–474.

105 Wilton, D.J., Harrison, R.F., Willett, P. et al. (2006). Virtual screening using binary kernel discrimination: analysis of pesticide data. *J. Chem. Inf. Model.* 46 (2): 471–477.

106 Franke, L., Byvatov, E., Werz, O. et al. (2005). Extraction and visualization of potential pharmacophore points using support vector machines: application to ligand-based virtual screening for COX-2 inhibitors. *J. Med. Chem.* 48 (22): 6997–7004.

107 Lepp, Z., Kinoshita, T., and Chuman, H. (2006). Screening for new antidepressant leads of multiple activities by support vector machines. *J. Chem. Inf. Model.* 46 (1): 158–167.

108 Byvatov, E., Sasse, B.C., Stark, H. et al. (2005). From virtual to real screening for D3 dopamine receptor ligands. *ChemBioChem* 6 (6): 997–999.

109 Betzi, S., Suhre, K., Chétrit, B. et al. (2006). GFscore: a general nonlinear consensus scoring function for high-throughput docking. *J. Chem. Inf. Model.* 46 (4): 1704–1712.

110 Bazeley, P.S., Prithivi, S., Struble, C.A. et al. (2006). Synergistic use of compound properties and docking scores in neural network modeling of CYP2D6 binding: predicting affinity and conformational sampling. *J. Chem. Inf. Model.* 46 (6): 2698–2708.

111 Omata, K., Kobayashi, Y., and Yamada, M. (2007). Artificial neural network aided virtual screening of additives to a Co/$SrCO_3$ catalyst for preferential oxidation of CO in excess hydrogen. *Catal. Commun.* 8 (1): 1–5.

112 Jones-Hertzog, D.K., Mukhopadhyay, P., Keefer, C.E. et al. (1999). Use of recursive partitioning in the sequential screening of G-protein–coupled receptors. *J. Pharmacol. Toxicol. Methods* 42 (4): 207–215.

113 Yamazaki, K., Kusunose, N., Fujita, K. et al. (2006). Identification of phosphodiesterase-1 and 5 dual inhibitors by a ligand-based virtual screening optimized for lead evolution. *Bioorg. Med. Chem. Lett.* 16 (5): 1371–1379.

114 van Rhee, A.M. (2003). Use of recursion forests in the sequential screening process: consensus selection by multiple recursion trees. *J. Chem. Inf. Comput. Sci.* 43 (3): 941–948.

115 Ma, J., Sheridan, R.P., Liaw, A. et al. (2015). Deep neural nets as a method for quantitative structure–activity relationships. *J. Chem. Inf. Model.* 55 (2): 263–274.

116 Svetnik, V., Liaw, A., Tong, C. et al. (2003). Random forest: a classification and regression tool for compound classification and QSAR modeling. *J. Chem. Inf. Comput. Sci.* 43 (6): 1947–1958.

117 Teramoto, R. and Fukunishi, H. (2007). Supervised consensus scoring for docking and virtual screening. *J. Chem. Inf. Model.* 47 (2): 526–534.

118 Wallach, I., Dzamba, M., and Heifets, A. (2015). AtomNet: a deep convolutional neural network for bioactivity prediction in structure-based drug discovery. *Math. Z.* 47 (1): 34–46.

119 Winkler, D.A. and Le, T.C. (2017). Performance of deep and shallow neural networks, the universal approximation theorem, activity cliffs, and QSAR. *Mol. Inform.* 36 (1, 2).

120 Unterthiner, T., Mayr, A., Klambauer, G., et al. (eds.) (2014). Deep learning as an opportunity in virtual screening. *Proceedings of the Deep Learning and Representation Learning Workshop (NIPS 2014)*, Los Angeles, USA.

121 Ramsundar, B., Liu, B., Wu, Z. et al. (2017). Is multitask deep learning practical for pharma? *J. Chem. Inf. Model.* 57 (8): 2068–2076.

122 Xu, Y., Ma, J., Liaw, A. et al. (2017). Demystifying multitask deep neural networks for quantitative structure–activity relationships. *J. Chem. Inf. Model.* 57 (10): 2490–2504.

123 Wu, J., Zhang, Q., Wu, W. et al. (2018). WDL-RF: predicting bioactivities of ligand molecules acting with G protein-coupled receptors by combining weighted deep learning and random forest. *Bioinformatics* 34 (13): 2271–2282.

124 Ertl, P., Lewis, R., Martin, E. et al. (2017). In silico generation of novel, drug-like chemical matter using the LSTM neural network. *arXiv* preprint arXiv:171207449 .

125 Gómez-Bombarelli, R., Wei, J.N., Duvenaud, D. et al. (2018). Automatic chemical design using a data-driven continuous representation of molecules. *ACS Cent. Sci.* 4 (2): 268–276.

126 Olivecrona, M., Blaschke, T., Engkvist, O. et al. (2017). Molecular de-novo design through deep reinforcement learning. *J. Cheminf.* 9 (1): 48.

127 Segler, M.H., Kogej, T., Tyrchan, C. et al. (2017). Generating focused molecule libraries for drug discovery with recurrent neural networks. *ACS Cent. Sci.* 4 (1): 120–131.

128 Wallach, I., Dzamba, M., and Heifets, A. (2015). AtomNet: a deep convolutional neural network for bioactivity prediction in structure-based drug discovery. *arXiv* preprint arXiv:151002855 .

129 Stepniewska-Dziubinska, M.M., Zielenkiewicz, P., and Siedlecki, P. (2018). Development and evaluation of a deep learning model for protein–ligand binding affinity prediction. *Bioinformatics* 34 (21): 3666–3674.

130 Ragoza, M., Hochuli, J., Idrobo, E. et al. (2017). Protein–ligand scoring with convolutional neural networks. *J. Chem. Inf. Model.* 57 (4): 942–957.

131 Rosenbaum, L., Dörr, A., Bauer, M.R. et al. (2013). Inferring multi-target QSAR models with taxonomy-based multi-task learning. *J. Cheminf.* 5 (1): 33.

132 Dahl, G.E., Jaitly, N., and Salakhutdinov, R. (2014). Multi-task neural networks for QSAR predictions. *arXiv* preprint arXiv:14061231 .

133 Ramsundar, B., Kearnes, S., Riley, P. et al. (2015). Massively multitask networks for drug discovery. *arXiv* preprint arXiv:150202072 .

134 Keiser, M.J., Roth, B.L., Armbruster, B.N. et al. (2007). Relating protein pharmacology by ligand chemistry. *Nat. Biotechnol.* 25 (2): 197–206.

135 Keiser, M.J., Setola, V., Irwin, J.J. et al. (2009). Predicting new molecular targets for known drugs. *Nature* 462 (7270): 175–181.

136 Li, H., Gao, Z., Kang, L. et al. (2006). TarFisDock: a web server for identifying drug targets with docking approach. *Nucleic Acids Res.* 34 (Web Server issue): W219–W224.

137 Chen, Y.Z. and Zhi, D.G. (2001). Ligand-protein inverse docking and its potential use in the computer search of protein targets of a small molecule. *Proteins* 43 (2): 217–226.

138 Li, G.-B., Yu, Z.-J., Liu, S. et al. (2017). IFPTarget: a customized virtual target identification method based on protein–ligand interaction fingerprinting analyses. *J. Chem. Inf. Model.* 57 (7): 1640–1651.

13

Importance of Structure and Dynamics in the Rational Drug Design of G Protein-Coupled Receptor (GPCR) Modulators

Raudah Lazim[1], Yoonji Lee[2], Pratanphorn Nakliang[1], and Sun Choi[1]

[1] *College of Pharmacy and Graduate School of Pharmaceutical Sciences, Ewha Womans University, Seoul, Republic of Korea*
[2] *College of Pharmacy, Chung-Ang University, Seoul, Republic of Korea*

13.1 Introduction

G protein-coupled receptors (GPCRs) are a vital family of seven-transmembrane (7TM) proteins that are remarkably versatile as demonstrated by their wide spectrum of stimuli such as photons, odorants, ions, neurotransmitters, and hormones. With more than 800 diverse members, GPCRs are responsible for the modulation of many intracellular signaling cascades concerned with essential human physiology (e.g. taste, vision, and olfactory) as well as cellular activities [1–4]. Binding of an agonist at the extracellular surface prompts a concomitant conformational change, which thereafter dictates proteins (G proteins, G protein-coupled receptor kinases (GRKs) and β-arrestins) binding at the cytoplasmic surface of the activated GPCRs [3, 5]. Owing to advances in structural biology, pharmacology, and biotechnology, the research scope of GPCRs have expanded beyond the early assumption of GPCRs being only "on-off switches" to include receptor activation, receptor sensitization and desensitization, allosteric modulation, biased agonism, GPCR oligomerization and its association to pathologic aggregation, and polypharmacology [4, 6–12]. Currently, around 35% of drugs approved by the FDA are targeting GPCRs, and these not only include the traditional agonists and antagonists but also cover allosteric modulators and inverse agonists [13, 14]. Additionally, the ideation of biased ligands adds another dimension to strategies used to design new GPCR ligands. The discovery of biased ligands affords an attractive mechanism, making it possible for medicinal chemists to manipulate the activation of desired signaling pathway(s) while simultaneously maximizing ligand efficacies and reducing adverse side effects [9, 15–17].

GPCRs as Therapeutic Targets, Volume 1, First Edition. Edited by Annette Gilchrist.

Breakthroughs in X-ray crystallography and other biophysical methods, such as X-ray fast electron laser crystallography, cryo-electron microscopy (cryo-EM), and nuclear magnetic resonance (NMR) spectroscopy, have resulted in the elucidation of many three-dimensional (3D) structures of GPCRs, hence permitting structural-based studies of interactions and communications vital for signal transduction across the cell membrane at the atomic level [18–22]. This pushes the agenda of structure-based drug design (SBDD) that has gained much popularity with the introduction of reverse pharmacology [20]. Even though amassing 3D structures is a vital component of SBDD, protein dynamics is also an important factor in drug design as it is the window towards understanding protein function. Experimental and computational studies exploring the conformational dynamics of GPCRs have propagated the idea of structural heterogeneity governing the dexterity of GPCRs to mediate multiple signaling pathways [3, 19, 21, 23–25]. Henceforth, research focusing on GPCRs gravitated toward learning about the connection between conformational plasticity and function [19]. This subsequently may guide the direction in which rational drug design proceeds, increasing the likelihood of developing new therapeutics for numerous GPCR-related diseases [26–31]. The accessibility to the 3D structures of GPCRs and computational methods such as molecular docking, virtual screening, molecular dynamics (MD), and Monte Carlo (MC) simulations have contributed significantly towards rationally designing GPCR ligands [19, 32–35]. In this chapter, we will briefly highlight recent advances in computational approaches that have facilitated GPCR drug discovery efforts, portraying the significance of protein dynamics, water dynamics, and discovery of diverse binding sites in developing new ligands (Figure 13.1).

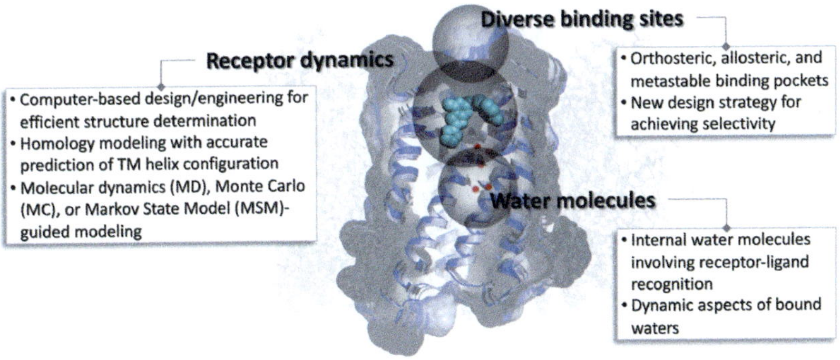

Figure 13.1 Summary of recent topics in computational approaches and ligand discovery related to GPCRs.

13.2 Structure Determination of GPCR

13.2.1 GPCR Engineering: Improving Stability and Reprogramming Function of GPCRs

Recent decades saw an upward trend in the number of GPCR structures elucidated by virtue of advances in protein engineering and crystallographic techniques. However, the number of unique experimentally determined GPCR structures (tallied at 64 to date, https://gpcrdb.org/structure/statistics) pales in comparison to the legions of GPCRs already identified, and there is still an inadequate representation of the active and intermediate states [21, 36, 37]. Protein engineering has facilitated the determination of high-resolution GPCR structures and this often involved perturbations such as mutations, truncations of the N and C termini, binding of high-affinity ligands, use of fusion proteins (e.g. T4 lysozyme and thermostabilized apocytochrome), and the utilization of antibodies/nanobodies as chaperones to stimulate crystallization [38–40]. These perturbations enhance the thermostability of GPCRs, thereby increasing the number of resolved GPCR structures available. However, some of the structural manipulations highlighted could change the intrinsic properties of the proteins leading to structures with altered configurations and dynamics, as well as erroneous ligand binding affinities [41]. Therefore, more studies, both experimental and theoretical, are necessary to overcome these issues for continuous solving of GPCR structures.

White et al. by combining structural, biophysical, and pharmacological information, solved two active-like conformations of human A2A adenosine receptors (A2AAR) with D52$^{2.50}$N and S91$^{3.39}$ A mutations [42]. The D$^{2.50}$ mutation resulted in several anomalies in the regular function of A2AAR that is unlike the S91$^{3.39}$A mutation, namely sodium desensitization and loss of G-protein signaling. Based on the structures solved, it was inferred that the mutation of D$^{2.50}$ to asparagine might affect G-protein signaling by disrupting native interactions between the residue and the highly conserved NPxxY motif. The structural integrity of this motif was suggested to be of importance to signal transduction that proceeds via the orthosteric binding site to the G-protein binding interface. Despite the loss of native receptor function, the D$^{2.50}$ mutation proved a useful scheme for the design of GPCR constructs for crystallization due to better thermostability achieved compared to wild type and the high conservation of this residue in class A GPCRs [42, 43]. The combination of higher melting temperature and preservation of ligand binding configuration in the D$^{2.50}$N mutation is also useful for ligand/fragment screening bioassay experiments using either NMR or surface plasmon resonance (SPR).

Besides greater thermostability, mutation of $D^{2.50}$ to uncharged amino acids such as alanine and valine has shown to affect signaling profiles of GPCRs and alter the pharmacological profile of GPCR ligands [42–45]. An interesting example is a study conducted by Fenalti et al. who observed the change in the pharmacology of cyclopentene-containing δ-opioid receptor ligands (e.g. naltrindole) from an antagonist to a biased agonist that preferred the β-arrestin-mediated signaling pathway when $D^{2.50}$ was mutated to Ala [45]. Therefore, the identification and point mutation of allosteric residues such as $D^{2.50}$ may serve as a standard protocol for the design of thermostable GPCRs with preserved orthosteric binding sites and reprogrammed function. The strategy of mutating $D^{2.50}$ to asparagine was proposed to be applicable across a broad range of class A GPCRs by White et al. hence potentially contributing to the increase in the number of GPCR structures solved and encouraging rational design of new GPCR modulators.

Even though multiple proteins with enhanced stability and novel functions have been reported via protein engineering, most experiments are time-consuming and cost-intensive. Recently, computational methods have found their way into protein engineering in a bid to boost the efficiency of manipulating the stability and function of GPCRs to generate more 3D structures. Chen et al. developed a new method which combines bioinformatics, *in silico* design, and experimental approaches to identify and design stabilizing mutations of large membrane receptors [46]. With the knowledge that the sequences of GPCRs evolved to assume metastable conformations, they have exploited sequence information to identify mutations with substantial stabilizing effect without affecting the overall structure and function of the membrane protein. To do so, they have identified polar residues with low sequence conservation within the TM region via multiple sequence alignment of 270 representative structures of class A GPCRs. Among these polar residues, those that formed suboptimal hydrogen bonds (hydrogen bond energy <0.5 Rosetta units) were selected as potential mutation sites, except for residues that possibly interact with bound ligands. Additionally, appropriate residues for mutation were selected by identifying amino acids positioned at partially and fully buried regions with mediocre packing scores, as determined using RosettaHoles, and exhibited packing defects, as shown via directional atomic contact analysis method [46–48]. Additional residues within 8 Å radius with side chains pointing toward the selected residues were also subjected to *in silico* mutagenesis to perturb the local environment of the selected residues. Hydrogen bonds and van der Waals interactions were targeted in this study since these interactions are typically involved in stabilizing the tertiary structures of membrane proteins [46, 49]. Using RosettaMembrane – an all-atom model describing physical interactions within protein and between protein and solvent in a membrane system – a series of torsion sampling, global and local minimization, as well as refinement of the redesigned metastable region of the mutated receptor were conducted to eliminate steric clashes and unconventional intra-protein and protein-membrane interactions [46, 49]. The redesigned metastable receptors were compared to their native structure, and only those with a stability boost of more than 0.1 Rosetta units and comparable local packing scores were chosen for experimental validation [46]. The application of this method to the β_1-adrenergic receptor (β_1AR) resulted in the design of 13 novel variants that are stabilized in their predetermined, inactive states [46].

Other studies applying similar methodologies (*vide supra*) have also emerged to successfully model GPCRs in active and inactive states [40, 46, 50–52]. Popov et al. developed a comprehensive prediction tool named CompoMug that uses both sequence and structural information to effectively predict a single or a cluster of residues for mutation that would confer thermal stability to GPCRs [52]. CompoMug relies on a combination of knowledge-based, sequence-based, structure-based, and machine-learning-based components to predict desirable mutations. In the knowledge-based module, the tool utilizes known point mutations of highly conserved residues such as $D^{2.50}$ and $S^{3.39}$ (*vide supra*) that were successfully used to solve 3D structures of numerous receptors with improved thermostability. For the sequence-based component of the prediction tool, the developers used multiple sequence alignment and evolutionary tracing to identify amino acids that deviate from known sequence conservation patterns of related GPCRs. The aberrant residues identified are targeted for mutation to their respective conserved amino acids to fulfill the assumption that restoring the conservation pattern could confer stability to the mutated receptors. In terms of structure, the developers aimed to achieve stability by minimizing conformational flexibility through the formation of salt bridges or disulfide bonds. For that, accurate receptor models were acquired through homology modeling using Molsoft ICM-Pro. The refined model was scanned for potential salt bridges using a set of criteria such as residue position, the orientation of side chains, and inter-residue distance. The possible locations of disulfide bonds were predicted using Disulfide by Design (DbD) software [52–54]. Stabilizing mutations were shortlisted from the list of residue pairs identified through free energy calculation available in ICM-Pro.

Machine learning was incorporated in the development of CompoMug to train a classifier using extensive experimental data accrued over the years. Alanine scanning mutagenesis data available for three GPCRs, namely neurotensin receptor, A_{2A} adenosine receptor, and β_1 adrenergic receptor, was used as training benchmarks and translated into feature vectors that store information related to stability. These features were used to train a support vector machine (SVM) classifier to predict potential point mutations that exert stability. The features used to train the SVM classifier include sequence properties (e.g. hydrophobicity, polarity, charge, and size of side chains), structural characteristics (e.g. type of interactions, native contacts, and solvent accessible surface area), as well as energies based on force field descriptors (e.g. hydrogen bonding, van der Waals, electrostatic interactions, solvation, potential energy, and total energies) [51, 52]. Through this method, Peng et al. have solved two crystal structures of the serotonin 2c (5-HT_{2C}) receptor in complex with ergotamine (3.0 Å resolution) and ritanserin (2.7 Å resolution) [51]. Similarly, Popov et al. were able to predict ten new mutations with stabilizing properties that go on to assist the crystallization as well as the solving of the 5-HT_{2C} receptor in inactive and active-like conformations [52].

Based on the case studies discussed above, integrating computational approaches into protein engineering could aid in the determination of inactive, as well as the under-represented active states of GPCRs. GPCR engineering could advance much faster than before, improving rational design of GPCR modulators by providing structural information that previously was not easily accessible due to limitations in experimental procedures. This permits researchers to explore previously elusive GPCRs, thus accelerating drug discovery efforts [38, 40, 55, 56].

13.2.2 Structure Prediction of GPCRs

Crystallization of membrane proteins is a laborious task and often involves intensive protein engineering as well as customized experimental techniques that require time and effort to devise. Predicting the three-dimensional structures of unresolved GPCRs through homology modeling offers a fairly quick and inexpensive alternative to gaining structural knowledge that could be vital for the design of new therapeutics. The reliability of constructed homology models depends heavily on the sequence identity of the template used, and a sequence identity of at least 20% has been shown to generate relatively accurate models [33, 40, 57]. With the remarkable accumulation of GPCR structures over the years, a common strategy employed during homology modeling is to use as much experimental data as possible to prevent the incorporation of undesirable artifacts that could interfere with the reliability of *in silico* drug design and discovery [21, 37]. For instance, the GPCRdb homology model pipeline utilized local templates to model missing loops or malformed sections (e.g. helix distortions) in the main template instead of *ab initio* modeling [37, 58]. The homology modeling of newly released structures of adenosine A_1 receptor (A_1AR), CC chemokine receptor 2, protease-activated receptor 2, and apelin receptor using GPCRdb homology pipeline resulted in structures with better average root mean square deviations (RMSDs) compared to other existing prediction tools [37].

Homology modeling being at the forefront of most structure-based studies of GPCR has also prompted substantial improvements to its general protocol.

Structure prediction servers catering to GPCRs or membrane proteins in general have undergone continuous developments to accommodate the need for accurate GPCR models, thus making high-quality models accessible to researchers with little to no computational background [37, 59–64]. Notwithstanding the convenience endowed by the availability of these structure prediction tools, it is imperative for researchers to select a homology modeling strategy that is best suited for the protein of interest. Nikolaev et al. have evaluated two stages of homology modeling known to greatly affect model quality, namely sequence alignment and structure building [33]. In this study, they have assessed three pairwise sequence alignment algorithms (MP-T, AlignMe, and MUSTER), two multiple sequence alignment algorithms (Clustal Omega and PralineTM), as well as three structure modeling programs, namely Medeller, I-TASSER, and RosettaCM, and their effects on the quality of rhodopsin homology models constructed by using combinations of these tools. The rhodopsin models built in this study are those with experimentally determined structures to allow model quality assessments using RMSD and global distance test-high accuracy (GDT-HA) scores. Twenty-four unique rhodopsin structures were clustered based on sequence identity to set up 18 target-template pairs with sequence identities ranging between 15% and 84%. All three structure modeling programs fared well and provided high-quality predictions of the TM region in spite of low sequence identity, with Medeller generating the best models with the lowest C_α-RMSD values at the TM region [33]. However, Medeller does not perform as well when flexible regions of the rhodopsin model were taken into consideration. In cases where the accuracy of the flexible region is of importance, a combination of I-TASSER and AlignMe offers the best results [33].

Medeller also performed better than its counterparts in terms of reproducing the "protein function" of rhodopsin, which was analyzed through visual observations of the side chains of active site residues [33]. This may be due to the strong geometric restraints imposed on the "core structure" (region of high similarity between target and template, which is usually the TM core in membrane proteins) during model building in Medeller, as opposed to the unconstrained energy minimization of side chains permitted in I-TASSER and RosettaCM [33, 59, 61]. This study revealed the importance of not only selecting a good template but also making the right choices on the methods used for homology modeling. Accurate alignment of the target sequence and template is also crucial to ensure the construction of accurate models, especially when geometric constraints are involved. Therefore, it is crucial for modelers to select the best combination of sequence alignment and structure building algorithms to construct a reliable homology model. Tables 13.1 and 13.2 consist of brief descriptions of structure prediction and sequence alignment programs, respectively, commonly used in performing homology modeling of GPCRs and membrane proteins [37, 59–72].

Table 13.1 Brief descriptions of several structure prediction tools that cater to GPCRs/membrane proteins.

Method	Description	Website
GPCRM	Homology modeling using Modeller.[a] Multiple templates and profile-profile alignment. Implement two loop modeling methods – Modeller and Rosetta. Three different scoring function used – Rosetta, Rosetta-MP, and BCL::Score.	http://gpcrm.biomodellab.eu
GPCR-I-TASSER	Uses I-TASSER iterative threading assembly method to build models. Assimilates experimental mutagenesis data (GPCR-RD)[b] and ab initio TM helix assembly simulation for model construction. Constructs models with or without homologous templates.	https://zhanglab.ccmb.med.umich.edu/GPCR-I-TASSER/
GPCR-ModSim	Single/multiple template-based homology modeling (Modeller). Loop refinement using LoopModel module in Modeller. Equilibration of constructed models in explicit membrane using MD simulation.	http://open.gpcr-modsim.org
GPCR-SSFE 2.0	Fragment-based homology modeling (Modeller) designed for class A GPCRs. Fingerprint correlation scoring to identify the best template. Enable interactive loop modeling using SL2 web service[c] (omicX).	http://www.ssfa-7tmr.de/ssfe2/
GOMoDo	Single template homology modeling using Modeller. User-defined preference for automatic loop refinement using LoopModel. Blind docking of ligands using Autodock Vina[d] or information-based docking using HADDOCK.[e]	http://molsim.sci.univr.it/cgi-bin/cona/begin.php

(Continued)

Table 13.1 (Continued)

Method	Description	Website
Medeller	Single/multiple template homology modeling designed for membrane proteins. Identify common TM core between target and template during alignment. Template's membrane insertion obtained using iMembrane.[f)] Modeling guided by geometric restraints imposed on identified core structure.	http://opig.stats.ox.ac.uk/webapps/medeller/

a) Modeller: a tool used to perform homology modeling of proteins (https://salilab.org/modeller/).
b) GPCR-RD: a database of experimental restraints of GPCRs (https://zhanglab.ccmb.med.umich.edu/GPCR-RD/).
c) SL2: an interactive web service for adding missing parts in a protein such as loops (http://proteinformatics.charite.de/ngl-tools/sl2/start.php).
d) Autodock Vina: an open source molecular docking software (http://proteinformatics.charite.de/ngl-tools/sl2/start.php).
e) HADDOCK: an information-based flexible docking tool for modeling protein-protein, protein-ligand and protein-nucleic acid complexes (http://haddock.science.uu.nl/services/HADDOCK/haddock.php).
f) iMembrane: a homology-based approach to predict part of a protein that is in contact with membrane layer (http://opig.stats.ox.ac.uk/webapps/medeller/).

The binding of different ligands elicits different binding site configurations, which if not modeled precisely could lead to artifacts being inherited into studies associated with SBDD such as receptor-ligand interaction, lead optimization, and rationalization of on-target and off-target effects. Therefore, ensuring the reliability of constructed GPCR models, especially the correctness of binding site geometry, is an essential step toward *in silico* design of novel GPCR drugs. Castleman et al. hypothesized that the use of templates with high sequence identity at binding pockets, i.e. high local sequence similarity, could lead to the development of models with more precise binding site conformation compared to templates with high global sequence similarity [35]. The premise was based on the knowledge that backbone structures of the TM domains of GPCRs are highly preserved across the GPCR superfamily while the orthosteric binding sites underwent substantial evolutionary changes to cater to the diverse functions of the different types of GPCRs. Therefore, for comparison, templates with the highest local and global similarity scores as provided by "CoINPocket" and GPCRdb receptor similarity tool, respectively, were acquired for each of the six

Table 13.2 Brief descriptions of several sequence alignment programs.

Method	Description	Website
MP-T[a)]	Sequence-structure alignment web server for membrane proteins based on multiple sequence alignment.	http://opig.stats.ox.ac.uk/webapps/MPT/php/
AlignMe[a)]	Pairwise alignment – alignment of two sequences using various similarity measures.[b)]	http://www.bioinfo.mpg.de/AlignMe/
	Profile-profile alignment – Generates hydropathy profiles from two multiple sequence alignments	
MUSTER[a)]	Sequence-template alignment generated by combining sequence profile-profile alignment and structural information.[c)]	https://zhanglab.ccmb.med.umich.edu/MUSTER/
Clustal Omega[a)]	Multiple sequence alignment that utilizes guide trees and Hidden Markov Model (HMM) profile-profile method.	https://www.ebi.ac.uk/Tools/msa/clustalo/ http://www.clustal.org/omega/
Praline[TM a)]	Multiple sequence alignment algorithm developed for transmembrane proteins through the integration of membrane-specific scoring matrix.	http://www.ibi.vu.nl/programs/pralinewww/
TM-Aligner	Multiple sequence alignment tool based on "Wu-Manber string matching algorithm" that caters to membrane proteins.	http://lms.snu.edu.in/TM-Aligner/
T-Coffee	Consistency-based multiple sequence alignment of proteins and nucleic acids.	http://www.tcoffee.org/ http://tcoffee.crg.cat https://www.ebi.ac.uk/Tools/msa/tcoffee/
MUSCLE	Multiple sequence alignment algorithm that integrates distance estimation, profile alignment, and refinement.	http://www.drive5.com/muscle/ https://www.ebi.ac.uk/Tools/msa/muscle/

a) Algorithm used in the comparative study performed by Nikolaev et al.
b) Similarity measures include hydrophobicity scale, substitution matrix, secondary structures, etc.
c) Structural information used include secondary structures, solvent accessibility, torsion angles, hydrophobic scoring matrix etc.

class A GPCR targets, namely chemokine receptor 4 (CXCR4), free fatty acid receptor 1 (FFAR1), nociceptin opioid receptor (NOP), P2Y purinoceptor 12 (P2Y12), kappa opioid receptor (OPRK), and muscarinic receptor 1 (M1), and utilized for homology modeling [35, 37, 73]. Docking of ligands obtained from reference structures was performed on the homology models as well as their respective crystal structures for comparison. Analyses of the homology models revealed that majority of the models generated fall within acceptable RMSD threshold (RMSD < 5.3 Å) regardless of templates used. The exceptions were FFAR1, where both local and global templates failed to generate acceptable models, and P2Y12, where only the use of local templates provided models that met the designated RMSD threshold [35]. The docking results, however, provided an interesting look on the importance of template selection for ligand docking and protein-ligand interaction studies. Docking poses of partnering ligands in models which were generated using templates that were selected based on local similarity at putative binding pockets showed better agreement with crystallographic ligand coordinates as reflected by average Tanimoto coefficient and ligand pose RMSD [35].

Based on the studies highlighted (*vide supra*), it is important for modelers to meticulously plan their homology modeling protocol to ensure that reliable information will be acquired when generated models are utilized to investigate mechanisms associated to ligand binding, signal propagation, and receptor activation/deactivation, all of which are crucial for drug discovery. Besides template selection, sequence alignment, and model construction, structure refinement is also an important component of homology modeling. Castleman et al. added a caveat to their study that structure refinement prior to docking simulation is necessary to achieve better ligand RMSDs and to be able to observe protein-ligand interactions that closely resemble native binding mode as reflected by crystallographic structures [35]. Structure refinement after homology modeling was excluded in their study as the focus was on the effect of local and global templates on docking performance. The importance of model refinement in GPCR studies will be discussed in Section 13.3 of this chapter.

13.3 Importance of Dynamics in Characterizing GPCR Structure and Function for the Design of New GPCR Compounds

13.3.1 Significance of Conformational Flexibility in SBDD

Conformational flexibility plays a substantial role in directing the biological activity of GPCRs thus allowing for ligand recognition, receptor activation, and signal transduction across membranes [4, 19, 74]. Protein fluctuations include but are not limited to minute variations in the torsions of amino acid side chains, backbone flexibility, and large movements of protein domains leading to changes in conformational states [19, 74]. Therefore, depending only on static structures captured through X-ray crystallography for structure-based studies of GPCRs is not advisable given the high importance placed on conformational heterogeneity on GPCR function. Incorporating the plasticity of GPCRs into virtual screening (VS) campaigns and docking simulations is essential in SBDD. Protein flexibility has been included in SBDD through several methods such as ensemble docking (using multiple crystal structures or multiple conformations from MD or MC simulation trajectories), rotamer sampling, induced-fit docking (sampling of both backbone and side chain fluctuations), and implementation of soft-core potentials in free energy calculations (truncation of non-bonded interactions to prevent steric clashes) [74–77].

Model refinement after homology modeling is a crucial factor in ensuring the accuracy of the homology models constructed. As mentioned in an earlier section, Castleman et al. provided a caveat that emphasized the importance of model refinement prior to docking experiments [35]. Chen et al. provided a similar conclusion in their study where the effects of structure refinement on the reliability of docking simulations were scrutinized for crystal structures, as well as constructed homology models of angiotensin II type I receptor (AT1R) [78]. To verify the impact of crystal structure refinement on the credibility of docking experiments, they refined two crystal structures of AT1R (PDB id: 4YAY and 4ZUD [79, 80]) using MD simulations with the receptor embedded in an explicit lipid bilayer. Both crystal structures underwent considerable conformational changes during the simulations, and following the stabilization of the receptors after 50 ns, the final snapshots were acquired for docking. The screening power of the original and MD-refined crystal structures was examined by docking a validation set of known antagonists and decoys. The refined crystal structures showed a better enrichment ratio compared to the original crystal structures. Docking of ZD7155 and olmesartan (ligands originally crystallized with AT1Rs) to both MD-refined structures showed consistent agreement with experimental poses across two docking platforms, namely Glide (SP and XP) and Autodock Vina. This contrasted with the substandard performance of Autodock Vina observed for 4ZUD when docking was performed without MD refinement. Besides structure refinement, the study also showed the importance of considering the flexibility of the receptor globally, via conformational sampling using MD simulation, and locally, through optimizing ligand binding site using induced fit docking (IFD). Consideration of an ensemble of structures via conformational clustering of MD trajectories and conducting IFD increased the likelihood of acquiring docking poses that are close to experiment [78]. This is an expected observation given that ligand binding typically involves synchronous changes in configurations of the protein and ligand to enhance affinity at the binding site [19, 74, 81].

More comprehensive conformational sampling methods and statistical analyses are required to increase the credibility of binding pose prediction. Rather than long single-trajectory simulations, running multiple short simulations has shown better quantitative sampling of protein conformational space. Prediction of binding poses by means of scoring functions implemented in molecular docking platforms while practical does not serve as a definitive solution to acquiring binding prediction of reasonable precision. Pairing short MD simulations and typical binding free energy (ΔG_{bind}) estimation methods such as molecular mechanics/Poisson-Boltzmann surface area (MM/PBSA), molecular mechanisms/generalized born surface area (MM/GBSA), and linear interaction energy (LIE) have demonstrated reasonable prediction of binding affinity with acceptable standard error of estimation. However, multiple MD simulations and free energy calculations need to be conducted using a different initial velocity for each binding pose, and the results averaged to acquire a decent ΔG_{bind} prediction. These entailed a huge computational expenditure, which may not be practical for drug discovery where fast and precise predictions are desired. Terayama et al. proposed a solution to this problem which involved the use of best arm identification (BAI) algorithm [82], a machine learning method, to curb the number of MD+MM/PBSA routines used to within the limit of computational resources assigned by users, while keeping the accuracy of ΔG_{bind} predictions [83]. Implementation of the pose prediction using BAI algorithms has reduced the average number of MD+MM/PBSA rounds required from 153 (uniform sampling, 20 poses*n runs of MD+MM/PBSA for every binding pose) to around 40–60 rounds (based on three BAI algorithms), without jeopardizing the accuracy of binding pose prediction. This study also compared the docking poses predicted using the routine MD+MM/PBSA runs and molecular docking. The former correctly determined binding poses for all the complexes, whereas the latter was only able to correctly predict on two occasions. While the subjects tested in this study were not GPCRs, the implementation of this method could ameliorate rational design of GPCR modulators as conformational plasticity is continuously being advocated to be of importance in drug design.

Monitoring the dynamics of GPCRs could also provide insights on the different ways of stabilizing GPCRs in various conformational states hence equipping researchers with useful structural information for drug development that could not be observed through static crystal structures. Lückmann et al. with the aid of MD simulation and structure-based VS, have discovered an ago-allosteric modulator for FFAR1 [84]. Ago-allosteric modulators are compounds that on their own act as agonists, whereas in the presence of endogenous agonists behave as positive allosteric modulators (PAMs) [85]. Unbiased MD simulations of FFAR1 in the absence of an agonist, TAK-875, revealed a significant rearrangement of polar and charged residues at the inter-helical binding pocket of FFAR1 between the extracellular side of TM3 and TM4 [84]. This event leads to the closing of an extracellular pocket that is surrounded by TM1, TM2, and TM7, which was initially exposed to solvent. Subsequent metadynamics simulation of free FFAR1 and re-docking of TAK-875 to FFAR1 after the unbiased MD simulation affirmed the dynamic connectivity between the two pockets, hence raising the speculation that the stabilization of FFAR1 in the active state could be achieved by designing ligands that bind to the neighboring extracellular pocket, preventing its closure. SBDD of the ago-allosteric modulator was conducted by exploring the extracellular pocket via virtual screening of commercially available synthetic molecules onto the structure ensembles of TAK-875-bound FFAR1. A potential compound showing promising affinity to the extracellular pocket was identified, and optimization of this ligand could lead to the design of a therapeutic agent that targets alternative binding sites on GPCRs while exerting agonist-like behavior. This study is a testament to the importance of examining GPCR dynamics to further comprehend its mode of action for more efficient drug design efforts.

Recently, the Markov state model (MSM) has been well-received as a computational tool for exploring protein dynamics by filtering noise in simulation trajectories to acquire kinetically relevant states [86]. MSM has far-reaching applications in the SBDD of GPCR ligands, given that these receptors are highly dynamic. MSM could improve the accuracy of VS campaigns and docking simulations by providing a starting structure that is functionally relevant since conformational sampling was performed in timescales consistent with experimental data, which might not be possible through just conventional MD simulations. Combination of MD simulation and/or enhanced sampling methods with MSM enabled the study of slow dynamical processes such as ligand binding and unbinding thus allowing the inspection of kinetically relevant states, i.e. bulk, intermediate, bound, metastable and off-pathway states, that could provide critical information for drug design [87, 88]. Using this method via a high-throughput molecular dynamics (HTMD) platform designed by Doerr et al. which involves iterative adaptive sampling and MSM analysis, Ferruz et al. observed the presence of a cryptic pocket in dopamine D3 receptor (D3R), a member of the aminergic family of GPCRs, upon binding of an antagonist [87, 88]. Cryptic pockets are transient sites that are discernable in ligand-bound protein structures but are obstructed in ligand-free proteins by loops, side chains, or in some cases, helical domains [89]. With the aid of a computational tool called FTMap (predict binding hotspot residues), Beglov et al. inspected multiple crystal structures with validated cryptic pockets. They observed that these structures predominantly possessed strong binding hotspots and significant structural flexibility near the cryptic pocket [89, 90]. The role of protein flexibility in uncovering cryptic pockets was substantiated by Ferruz et al. who revealed the repositioning of the side chain of a highly conserved $F346^{6.52}$ residue to form a cryptic pocket between TM5 and TM6, a structural feature that was previously not observed in published crystal structures of aminergic GPCRs [87]. This event facilitated the binding of D3R antagonists through a new binding mechanism that agreed well with site-directed mutational data. The importance of GPCR dynamics was also emphasized in this study by a series of rigid docking simulations performed on the crystal structure of D3R. Without considering protein flexibility, the docking poses of D3R compounds on the receptor diverged from results portrayed by the mutagenesis experiments performed.

13.3.2 Dynamics of Buried Waters in GPCR

The functional role of water in receptor activation was a matter up for debate since the discovery of ordered waters buried inside reported crystal structures of GPCRs [34, 91–93]. Water molecules have been proposed to play an active role in receptor activation by allowing communication among TM helices and establishing a hydrogen bond network between the receptor and ligand to stabilize binding [91]. However, suggesting the influence of water in GPCR activation through a static picture is not ideal, and this query could be conclusively answered by considering the dynamics of GPCRs in water. Yuan et al. performed long scale MD simulations and well-tempered metadynamics simulations to understand the mutual correlation between the conformational transition of GPCRs and the entry of water into the internal space of the receptor [92]. The dynamics of three class A GPCRs, namely A2AAR, β_2-adrenergic receptor (β_2AR), and rhodopsin (Rho) were explored in three distinct conformational states *viz.* active, inactive, and metastates. All the receptors in the active state presented a continuous intramolecular water channel driven by the conformational shift of $Tyr^{7.53}$ of the highly conserved NPxxY motif. The water channel was regulated by the formation of stable interactions between $Tyr^{7.53}$ and G protein residues, and these observations made through MD simulations were consistent with crystal structures. Interestingly, for activated Rho, the side chain of $Tyr^{7.53}$ was maintained at torsions necessary for open channel configuration in the presence and absence of G protein. In the absence of G protein, $Tyr^{7.53}$ established hydrophobic interactions with three methionine residues ($M^{6.40, 6.36, 7.56}$), which were pointed out through mutagenesis to be of importance in Rho activation [92, 94, 95].

Using microsecond long MD simulations, the group also studied the water penetration pathways in different types of class A GPCRs. Two pathways were proposed, namely the "intracellular water penetration" seen in Rho and human sphingosine-1-phosphate receptor 1 ($S1PR_1$) and the "extracellular water penetration" observed for opioid receptors and A2AAR [92]. The large N-termini and ECL2 loops hovering over the allosteric binding pockets and the hydrophobic orthosteric sites of $S1PR_1$ and Rho were determined to be the causal effect of water influx into the channel via the intracellular side of the membrane when the receptors are in the active state. Additionally, two hydrophobic layers were observed in the inactive state of A2AAR, which are located (i) between the allosteric and orthosteric binding sites and (ii) near the G protein binding site, close to the NPxxY motif. Only the second hydrophobic layer was observed in meta states. Sequence alignment performed across 20 diverse representatives of solved class A GPCRs indicated that hydrophobic amino acids are majorly found in the second hydrophobic layer across all receptors, possibly suggesting a common architecture among class A GPCRs when in the inactivated state [92]. Consolidating data acquired through experiments and MD simulations broadens the structural information available, permitting the correlation between structure and function of GPCRs. The study by Yuan et al. harvested several crucial structural features that assisted the important role played by internal waters in GPCR activation, which subsequently could serve as a catalyst for the design of new GPCR drugs that target the water channel to prevent receptor activation.

A recent study done by Venkatakrishnan et al. also evinced the importance of water found within the internal space of GPCRs for receptor activation [93]. Comprehensive studies covering four different types of GPCRs (A2AAR, β_2AR, μ-opioid receptor [MOR], and M2 muscarinic receptor [M2R]) in both active and inactive states concurred with the study done by Yuan et al. alluding to the importance of $Tyr^{7.53}$ in GPCR activation [92, 93]. A stable water network that potentially facilitates GPCR activation was determined through the scrutiny of MD trajectories for stable waters that continuously occupy specific coordinates within the putative GPCR water channel, and inspecting existing crystal structures for conserved crystallized waters [93]. They discovered a stable water network near the G protein binding site, and this structural feature is highly conserved across a diverse range of class A GPCRs. The similarity in binding partners at the G protein binding interface among class A GPCRs may explain the conserved water network at the intracellular half of the receptor. The group also observed similar rearrangements of the water network near the G protein binding site during activation resulting in the identification of categorical differences in stable, internal waters found in active and inactive states [93].

Interactions established between protein and water molecules have been proposed as one of the factors contributing to receptor activation and ligand binding. The studies highlighted above, as well as many others, provided insights into the mechanism driving receptor activation by looking at the atomistic details of both GPCR and internal waters, thus supplementing dynamic information that could not be attained through X-ray crystal structures [19, 92, 93, 96, 97]. Knowledge of the water penetration pathways, as well as the participation of water molecules as mediators in G protein and ligand binding, could lead researchers in rationally designing GPCR ligands. Information acquired pertaining to stable waters within the putative GPCR water channel and G protein binding site could also be utilized in homology modeling to acquire a more reliable model for SBDD, especially with the identification of distinct water networks in inactive and active states of class A GPCRs by Venkatakrishnan and coworkers [93]. The difference in the mobility of crystallized internal waters during MD simulations also forewarned the importance of treating crystallized water with caution as the deletion of crucial internal waters might interfere with the intramolecular communications, affecting the accuracy of *in silico* SBDD.

13.4 Strategies for the Design of New GPCR Modulators

13.4.1 Discovery of New Binding Sites by Considering GPCR Plasticity

Novel modulators of GPCRs are increasingly in demand as the link between GPCRs, and a broad spectrum of diseases such as cancers, neurodegenerative diseases, metabolic diseases, psychological disorders, and cardiac disorders continues to be established [26–31]. Most approved drugs available mainly target aminergic receptors from the class A rhodopsin family [37, 98]. However, due to multiple binding partners, most of these drugs exhibit adverse side effects [98, 99]. This situation arises due to the highly conserved sequence of the traditional orthosteric binding sites across a GPCR subtype. Recently, medicinal chemists have rerouted their drug design strategy in the direction of allosteric modulation. This change in perspective is driven by the fact that allosteric binding sites, compared to orthosteric binding sites, are more structurally diverse, hence enabling the design of drugs with improved subtype selectivity, leading to lesser side effects [36, 100–102]. With the paradigm shift in the design plan, recent years saw an increase in the number of approved allosteric modulators of GPCRs (currently making up 3% of approved drugs) [21, 37]. Several types of allosteric ligands have been introduced, namely PAMs, negative allosteric modulators (NAMs), and neutral allosteric ligands (NALs), which increase, decrease, or have no effect respectively, on the affinity and/or efficacy of orthosteric ligands on the activity of the targeted receptors [36, 103].

Technological advances in biochemistry and biophysical methods have led to the discovery of several allosteric sites, as well as critical allosteric residues that we could target to modulate or stabilize GPCRs [36, 42, 87]. The allosteric sites could be categorized to four regions namely, (i) preceding orthosteric sites that were left vacant as endogenous ligands evolved to bind at other sites, (ii) metastable

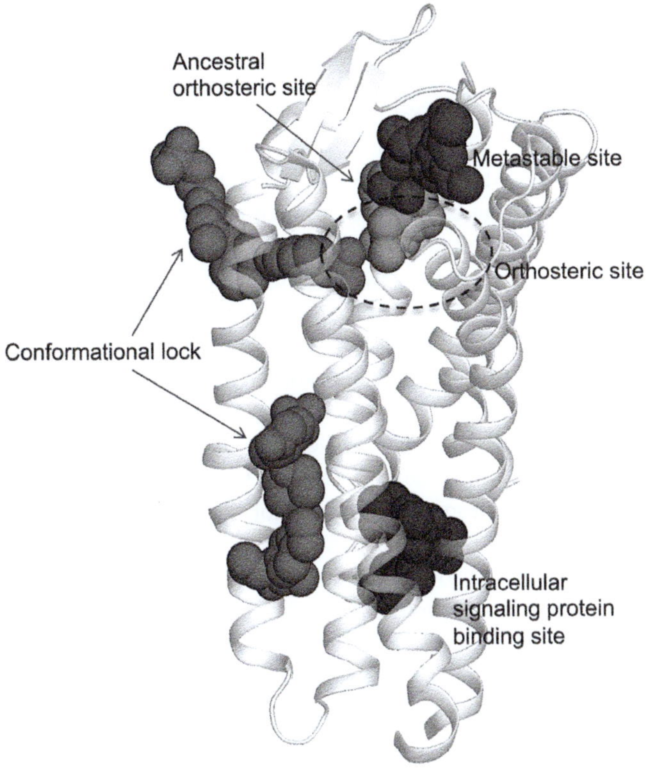

Figure 13.2 Schematic representation of the different allosteric binding sites reported in X-ray crystal structures. Ancestral orthosteric binding sites represent preceding orthosteric binding sites vacated because of evolution (PDB id: 3ODU). Metastable binding sites correspond to transient binding sites, usually located at the extracellular vestibule, on path to orthosteric binding sites (PDB id: 4XNW). Conformational locks correspond to binding sites that when bound by a ligand fixed the receptor in a specific state i.e. active, inactive or metastable. Ligands possessing conformational lock property bind at diverse locations but are commonly located between the protein-lipid bilayer interface (PDB id: 4PHU and 5TZY). Ligands also bind from the intracellular side of the lipid bilayer to enhance or interfere with the binding of partnering intracellular signal proteins (PDB id: 5LWE). Source: Based on Congreve et al. [36] and Wakefield et al. [104].

binding sites at which ligands bind transiently on path to the orthosteric binding site, (iii) sites that serve as conformational locks, which when bound could regulate the conformational states and signaling activities of GPCRs, and (iv) binding interface of intracellular signaling proteins (Figure 13.2) [36, 104]. The first two allosteric sites are usually located close to the native orthosteric binding sites, facing the extracellular environment and within the TM domain. The conformational locks, on the other hand, could be found either within the hydrophobic core of the TM domain or on the exterior of the receptor that is in contact with the lipid bilayer. The last allosteric site faces the intracellular side of the membrane, and allosteric ligands binding to this site could be designed to either promote or interfere with the binding of partnering signaling proteins such as G protein and β-arrestin [36, 103, 104].

In addition to the allosteric binding sites highlighted above, the introduction of MD simulations to account for the correlation between conformational heterogeneity and activity of GPCRs has led to the discovery of binding sites that eluded detection via X-ray crystallography. Access to powerful computational resources has enabled researchers to conduct large-scale MD simulations exploring the binding mechanisms of orthosteric ligands [105]. Through unbiased simulations and enhanced sampling methods, numerous studies have uncovered hidden allosteric pockets and new binding mechanisms of ligands that we could utilize to design new therapeutic ligands or drugs. Dror et al. conducted unbiased MD simulations in the microsecond timescale to investigate the binding mechanism and kinetics of three antagonists (propranolol, alprenolol, and dihydroalprenolol) and an agonist (isoproterenol) of β_1AR and β_2AR, allowing the observation of spontaneous binding of the ligands to the receptors without external harmonic restraints [106]. Binding kinetics and protein-ligand interactions from MD simulations are in agreement with that observed experimentally. Further inspection of the MD trajectories revealed the dominant binding pathway of the drugs to the receptors, which coincides with the metastable binding of the ligand at the "extracellular vestibule" prior to assuming their respective crystallographic binding pose at the orthosteric site. The ligands surpassed two energy barriers to undertake their respective crystallographic binding pose. The first energy barrier involved the water sequestration of the ligand and the extracellular vestibule pocket to initiate ligand entry into the receptor followed by a second energy barrier, which calls for structural changes and dehydration to accommodate ligand diffusion from the extracellular vestibule site to the orthosteric pocket via a tapered passage [106]. Metastable binding sites such as the extracellular vestibule pockets of β_1AR and β_2AR revealed through unbiased MD simulations, have been increasingly investigated as a potential target for the design of allosteric modulators, as well as bitopic ligands, possibly supplying a new pool of GPCR ligands [105, 107, 108].

Besides the "extracellular vestibule" metastable pocket (*vide supra*), using MD simulations and mutagenesis experiments, Chan et al. have also revealed the presence of another binding pocket at the intracellular half of M3 and M4 muscarinic acetylcholine receptors (mAChRs) that are situated near the middle of the TM core and beside the highly conserved $D^{2.50}$ residue [109]. Microsecond timescale MD simulations of ligand-bound M3 and M4 mAChRs, with the ligands, tiotropium antagonist (TTP) and acetylcholine(Ach), positioned at the extracellular vestibule of the receptors, uncovered the presence of the new binding pocket in the presence of ACh molecule. Comparison between trajectories acquired from the two systems revealed that both TTP and ACh molecules bind to the orthosteric site. However, as the simulations progressed, TTP remained stably bound at the orthosteric site while ACh moved on to explore an adjacent pocket that opened up during the simulation by the flipping of a highly conserved microswitch, $Trp^{6.48}$, in adenosine receptors. This observation was corroborated using the well-tempered metadynamics simulations of ACh-bound receptors, which sampled three distinct energy states, namely the transient binding at the extracellular vestibule, binding at the orthosteric site of mAChR, and binding at the newly observed binding pocket. Subsequent inspections of 39 unique crystal structures of class A GPCRs alluded to the presence of this new binding pocket in most of the crystal structures, albeit with smaller volume compared to the ones observed in MD-derived structures of agonist-bound M3 and M4 mAChRs [109]. This observation could be due to the induced-fit mechanisms triggered by the binding of orthosteric ligands, which cannot be observed sans conformational sampling via either MD simulations or other enhanced sampling methods.

Exploring GPCR dynamics through MD simulations has also enabled researchers to uncover cryptic binding pockets (*vide supra*) that may come in handy when designing subtype-selective drugs. Hollingsworth et al. with the aid of MD simulations, investigated the preference of a PAM, known as benzoquinaxoline-12 (BQZ12), for M1 mAChR over other mAChRs [110]. MD simulation initiated from the crystal structure of M1 mAChR revealed the presence of a cryptic pocket adjacent to the putative allosteric binding site between TM2 and extracellular loop 2 (ECL2), which was not seen in the crystallized structure of mAChRs. The docking of BQZ12 using the new, MD-derived conformational state of the receptor enabled the determination of a binding pose that corresponds to the binding of the planar aromatic core of BQZ12 at the allosteric site and the non-planar component at the cryptic pocket. The movement of $Tyr^{2.64}$ and $Glu^{7.36}$ to establish a hydrogen bond was proposed to play a vital role in keeping the cryptic pocket open, which in turn contributes to the selectivity of BQZ12 for M1 mAChR. This correlation was inferred due to the more frequent observation of the cryptic pocket in M1 rather than M2-M4 mAChRs during the simulations performed. Mutational studies targeting the hydrogen bond pair and ligand modification experiments attributed the subtype-selectivity of BQZ12 to the non-planar component of the modulator that preferentially binds to the cryptic pocket, concurring with the significant potency seen for non-planar M1 mAChR PAMs. Besides mechanistic elucidation of the subtype-selectivity of BQZ12, this study also reconciled the difference in observation between crystallized structure and mutagenesis data, wherein the latter supports the binding of BQZ12 to M1 mAChR but was portrayed otherwise three-dimensionally [110]. Based on the case studies highlighted above, it is imperative for us to consider protein dynamics to obtain a comprehensive understanding of GPCR activity. Structural features that are obscured in static structures could be uncovered through conformational sampling, potentially securing a new binding pocket that could be targeted for the design of new drugs with specific therapeutic profile.

13.4.2 Biased Ligands: Adding a New Dimension to the Design and Discovery of GPCR Ligands

Progress in X-ray crystallography and other biophysical techniques have revealed the intricacies of GPCR functions (*vide supra*), and this includes the ability of GPCR to participate in biased signaling [16]. Biased signaling or functional selectivity of GPCR is prompted through the binding of a ligand to the receptor, thereby preferentially regulating desired downstream signaling events by G-protein and/or GRKs/arrestins [3, 9, 11, 111]. Due to the functional complexity of GPCRs, the incorporation of biased signaling into drug design efforts has been plagued by numerous challenges [9, 15, 16, 111]. However, this does not hamper the progress of developing "biased GPCR ligands" due to their appealing therapeutic benefits such as better ligand efficiency and lesser adverse side effects [15, 111]. The design of biased ligands requires comprehensive scrutinization of the binding site activities of GPCRs. With the increasing number of co-crystal structures of GPCR-ligand complexes solved (220 unique complexes reported, https://gpcrdb.org/structure/statistics), more information pertaining to hotspot residues at binding pockets could be harnessed.

By analyzing crystal structures and experimental information gathered through mutagenesis and NMR, McCorvy et al. hypothesized that interactions established by orthosteric ligands at two specific regions in aminergic GPCRs, namely TM5 and ECL2, could selectively control the preferred downstream signals to either G-protein activation or β-arrestin recruitment [111]. To prove their hypothesis, the group used the D2 dopamine receptor (D2R), as a model system, and analogues of indole-aripiprazole hybrid to examine the functional selectivity rendered by ligands based on interactions established with the receptor. The designed indole-aripiprazole hybrid was inspired by co-crystal structures of indole-piperazine and turkey β1AR, in which the NH group of the indole moiety interacts with a highly conserved $S^{5.42}$ via hydrogen bonding, an interaction that preferentially activates the G-protein signaling pathway. As such, the dichlorophenyl-piperazine fragment of aripiprazole was replaced with an indole-piperazine fragment to afford the hybrid ligand. To examine whether disrupting ligand interactions with TM5 will shift the equilibrium toward the β-arrestin pathway, ligand modifications through N-substitutions of the indole moiety of the hybrid ligand were performed to induce steric hindrance, prohibiting hydrogen bond formation between the ligand and TM5. Binding assays revealed that N-methylation/propylation of the indole-aripiprazole hybrid suppressed the G protein-signaling pathway and preserved the β-arrestin recruitment activity of D2R. MD simulations and mutagenesis experiments further affirmed the hypothesis that inhibiting hydrogen bond formation between ligand and conserved serine residues in TM5 and ameliorating interactions between ligand and ECL2 ($Ile184^{EL2}$ in D2R) enhanced the bias profile of ligands toward β-arrestin recruitment. Through a comprehensive study of the binding mechanism of ligands to D2R using both experimental and computational approaches, the study provided an exemplary protocol that could be adapted to guide the design of biased ligands for other targets.

The heightened interest in allosteric modulation (*vide supra*) has also led to the development of bitopic ligands, a subclass of bivalent compounds that simultaneously bind to both orthosteric and allosteric sites [36, 105, 109, 112]. Having fragments that bind preferably to both orthosteric and allosteric pockets, bitopic ligands are expected to display desirable properties such as high affinity, subtype selectivity, and/or functional selectivity (e.g. biased agonism) [105]. Valant et al. rationally designed a novel bitopic ligand, VCP746, that demonstrated biased agonism at A_1AR [108]. A_1AR is a significant therapeutic target for cardiac-related diseases, such as angina and chronic heart failure, among others. However, adverse dose-related side effects such as bradycardia, atrial fibrillation, atrioventricular blocks, and hypotension delayed the progress of A_1AR agonists to clinic [107, 108, 113]. VCP746, an adenosine moiety linked to a PAM, was designed to block such on-target side effects mediated by A_1AR while exerting desired therapeutic effects such as anti-ischemic effect and cytoprotection. The biased signaling of A_1AR by the bitopic ligand was proven through its preference to activate the inhibition of forskolin-stimulated cAMP accumulation rather than the noncanonical pathway of calcium mobilization [107]. An extended study conducted by Aurelio et al. also revealed the structure-activity relationship of the bitopic ligand that governed its ligand bias property [107]. Through structural modifications of the three components of the ligand namely, adenosine moiety, linker, and allosteric pharmacophore, it was concluded that the presence and orientation of the allosteric molecule in the bitopic compound played a vital role toward gaining biased agonism at the receptor.

13.4.3 Quantum Mechanics/Molecular Mechanics (QM/MM) Approach Enhances Ligand Design

Comprehending biochemical reactions participating in drug metabolism is a vital process for elucidating the mechanisms involved in the biotransformation of drugs as well as the pharmacological effect of drugs on protein function. Therefore, performing simulations that probe biochemical reactions of drugs/ligands may provide necessary insights that could boost drug design and development. The computational method that is capable of modeling chemical reactions is the quantum mechanics (QM) approach, in which the electronic properties of the molecule were taken into account. Due to the relatively large size of biomolecules that consist of more than hundreds of atoms (e.g. an enzyme, GPCR), QM calculations could not handle the entire system. To overcome this limitation, the hybrid QM/MM approach, whereby the region of interest will be determined using an accurate QM method, and the rest of the system will be treated by the classical MM force field, could be applied for drug discovery [114–117].

Several studies have demonstrated the potential of the QM/MM strategy in elevating the quality of molecular docking simulations and experiments, thereby being a crucial contributor to the development and design of drugs with high potency and selectivity. Zanatta et al. utilized the combination of X-ray co-crystal structure of an antagonist, eticlopride, bound to D3R and QM/MM simulation to enhance the accuracy of classical docking of the antipsychotic drug called haloperidol, which is used in schizophrenia treatment, to the orthosteric binding site of D3R [118]. The QM/MM enhanced docking refinement disclosed several important insights into the ligand-binding conformation and the roles of amino acid residues around the binding site. The interaction analysis suggested that the attractive interaction between haloperidol and D3R helices followed the sequence of TM3 > TM7 > TM6 > ECL2 > TM5 ≫ TM2, where the interaction between haloperidol and TM2 of D3R is repulsive. This approach was also employed by Sekharam et al. to model the mouse olfactory receptor, MOR244-3, which is responsible for organosulfur odorants. The QM/MM optimization provided decisive Cu-ligand interactions that are consistent to site-directed mutagenesis experiment of the receptor activation [119–121]. Furthermore, the results would shed some light on mammalian olfaction studies at the atomistic level. Recently, the activation of human odorant receptors OR5AN1 and OR1A1 was studied [122]. The comparison of calculated binding energies of (*R*)-muscone and related compounds is in good agreement with experimental results that preferred (*R*)- over (*S*)-enantiomer. The structural observation revealed that the ligand was stabilized by forming a hydrogen bond with Tyr260 and other surrounding aromatic residues. This valuable finding may lead to the instructive development of the quantitative structure-activity relationship (QSAR) model.

Besides docking structure refinement, the QM/MM approach is commonly applied for reaction mechanism exploration. Singh et al. studied the mechanism of phosphatidyglycerol activation catalyzed by prolipoprotein diacylglyceryl transferase (Lgt) [123]. They suggested that Arg143, Arg239, and Glu202 are important in substrate accommodation and C—O ester bond activation. His103 aids in stabilizing the thiolate anion formation, triggering the desired chemical reaction. In addition, the calculated activation energy was 18.6 kcal/mol, which is comparable to the experimental data [124]. Therefore, the advantage of enzymatic reaction validation would serve as important information for novel antimicrobial drug design. This *in silico* transmembrane simulation could be considered as an example model for future application on GPCR mechanistic studies.

The largest class of GPCRs is class A rhodopsin-like receptor that is an important membrane photoresponsive protein. Therefore, a suitable tool for investigating the optical properties of this class is the QM method. However, pure QM calculation is not applicable due to the enormous size of GPCR. Thus, QM/MM simulation has become a practical tool that reduces the computational cost while preserving to a certain extent the accuracy of a full QM calculation. Pedraza-González et al. proposed a computational protocol called automatic rhodopsin modeling (ARM) which (semi) automatically prepares and performs QM/MM simulations to study and predict the optical properties of class A rhodopsin-like system [125]. Based on their benchmark test set, the computed maximum absorption wavelength (λ^a_{max}) showed an excellent agreement with observed experimental data. Moreover, the ARM also provided a practical consideration for class A and other classes of GPCR.

13.5 Concluding Remarks

Advances in crystallography and other biophysical methods have led to a pronounced increase in the number of resolved GPCR structures. Despite the increasing number of elucidated structures, the number of unique GPCR structures resolved are still lacking, especially for the non-olfactory receptors [37]. Homology modeling closes the gap in information by using data available from resolved structures to model a receptor of interest in the desired conformational state. Continuous improvements in homology modeling protocols and accessibility to more templates have resulted in better quality models hence enhancing computationally driven SBDD efforts such as virtual screening, molecular docking, and lead optimization. Multitudes of computational studies have also been conducted to better comprehend the dynamics of GPCRs, which importance in governing GPCR functionality have been delineated in both experimental and computational studies. The implementation of MD simulations and enhanced sampling methods have permitted the sampling of secluded conformations, which are not visible through X-ray crystallography. The continuous influence of protein engineering in GPCR research may prophesize the ushering of a variety of engineered GPCRs with enhanced stability and reprogrammed functions, hence allowing researchers to explore previously elusive GPCRs and accelerate drug discovery [38, 40, 55, 56].

Acknowledgment

This work was supported by the Mid-career Researcher Program (NRF-2020R1A2C2101636), Brain Pool Program (NRF-2020H1D3A1A02080803), and Bio & Medical Technology Development Program (NRF-2019M3E5D4065251) funded by the Ministry of Science and ICT (MSIT) and the Ministry of Health and Welfare (MOHW) through the National Research Foundation of Korea (NRF). It was also supported by RP-Grant 2020 of Ewha Womans University and the Ewha Womans University Research Grant of 2021.

Glossary

Orthosteric binding site An evolutionarily conserved binding pocket which houses endogenous ligands.

Allosteric binding site A binding pocket other than the orthosteric binding site that can modulate ligand binding, as well as receptor activity.

Metastable binding site A transient binding site that is often found near the opening of the orthosteric binding site. This binding site is a target for the design of bitopic (bivalent) ligands and is often involved in weak binding interactions.

Homology modeling Construction of a model based on information harnessed from known experimental structures that have reasonable sequence similarity to the protein of interest. Advances in homology modeling techniques (i.e. sequence alignment and model building) have enabled the modeling of proteins using templates with sequence identity of as low as 20%.

Molecular dynamics (MD) simulation Computer simulation utilized to monitor the physical movement of atoms interacting in a system through the numerical solution of Newton's classical equation of motion. Force field (i.e. interatomic potential), which classically are based on empirical approximations, provides the means to evaluate the potential as well as interatomic (bonded and nonbonded) interactions between atoms in a multi-body system.

Metadynamics simulation A technique used to enhance sampling in MD simulation. This method is used to overcome large energy barriers and estimates the free energy surface of a system as a function of collective variables (CVs). CVs are carefully selected degrees of freedom such as distance, torsion angles, etc. that could effectively describe a slow occurring dynamic process (rare events) of a system without bias.

Markov state model (MSM) A mathematical technique used to model long-timescale processes of a system. Used as an extension to MD simulation from which kinetic models were constructed to map the conformational space of the system. This method follows a comprehensive statistical approach to obtain a coarse-grained representation of noisy simulation data for easy interpretation of the system's dynamics. Instead of large-scale single-trajectory simulation, MSM construction usually involves data derived from a series of short MD simulations.

References

1 Rosenbaum, D.M., Rasmussen, S.G.F., and Kobilka, B.K. (2009). The structure and function of G-protein-coupled receptors. *Nature* 459 (7245): 356–363.

2 Weis, W.I. and Kobilka, B.K. (2014). The molecular basis of G protein–coupled receptor activation. *Annu. Rev. Biochem.* 87 (1): 897–919.

3 Gurevich, V.V. and Gurevich, E.V. (2017). Molecular mechanisms of GPCR signaling: a structural perspective. *Int. J. Mol. Sci.* 18 (12): 2519.

4 Hilger, D., Masureel, M., and Kobilka, B.K. (2018). Structure and dynamics of GPCR signaling complexes. *Nat. Struct. Mol. Biol.* 25 (1): 4–12.

5 Shukla, A.K., Singh, G., and Ghosh, E. (2014). Emerging structural insights into biased GPCR signaling. *Trends Biochem. Sci.* 39 (12): 594–602.

6 Ferré, S., Casadó, V., Devi, L.A. et al. (2014). G Protein–coupled receptor oligomerization revisited: functional and pharmacological perspectives. *Pharmacol. Rev.* 66 (2): 413.

7 Wacker, D., Stevens, R.C., and Roth, B.L. (2017). How ligands illuminate GPCR molecular pharmacology. *Cell* 170 (3): 414–427.

8 Dijkman, P.M., Castell, O.K., Goddard, A.D. et al. (2018). Dynamic tuneable G protein-coupled receptor monomer-dimer populations. *Nat. Commun.* 9 (1): 1710.

9 Bermudez, M., Nguyen, T.N., Omieczynski, C. et al. (2019). Strategies for the discovery of biased GPCR ligands. *Drug Discovery Today* 24 (4): 1031–1037.

10 Felsing, D.E., Canal, C.E., and Booth, R.G. (2019). Ligand-directed serotonin 5-HT2C receptor desensitization and sensitization. *Eur. J. Pharmacol.* 848: 131–139.

11 Gurevich, V.V. and Gurevich, E.V. (2019). The structural basis of the arrestin binding to GPCRs. *Mol. Cell. Endocrinol.* 484: 34–41.

12 Quitterer, U. and AbdAlla, S. (2019). Discovery of pathologic GPCR aggregation. *Front. Med.* 6: 9.

13 Santos, R., Ursu, O., Gaulton, A. et al. (2016). A comprehensive map of molecular drug targets. *Nat. Rev. Drug Discovery* 16: 19.

14 Hauser, A.S., Attwood, M.M., Rask-Andersen, M. et al. (2017). Trends in GPCR drug discovery: new agents, targets and indications. *Nat. Rev. Drug Discovery* 16: 829.

15 Tan, L., Yan, W., McCorvy, J.D. et al. (2018). Biased ligands of G protein-coupled receptors (GPCRs): structure–functional selectivity relationships (SFSRs) and therapeutic potential. *J. Med. Chem.* 61 (22): 9841–9878.

16 Wootten, D., Christopoulos, A., Marti-Solano, M. et al. (2018). Mechanisms of signalling and biased agonism in G protein-coupled receptors. *Nat. Rev. Mol. Cell Biol.* 19 (10): 638–653.

17 Kenakin, T. (2011). Functional selectivity and biased receptor signaling. *J. Pharmacol. Exp. Ther.* 336 (2): 296.

18 Gao, Y., Westfield, G., Erickson, J.W. et al. (2017). Isolation and structure-function characterization of a signaling-active rhodopsin-G protein complex. *J. Biol. Chem.* 292 (34): 14280–14289.

19 Latorraca, N.R., Venkatakrishnan, A.J., and Dror, R.O. (2017). GPCR dynamics: structures in motion. *Chem. Rev.* 117 (1): 139–155.

20 van Montfort Rob, L.M. and Workman, P. (2017). Structure-based drug design: aiming for a perfect fit. *Essays Biochem.* 61 (5): 431.

21 Munk, C., Mutt, E., Isberg, V. et al. (2019). An online resource for GPCR structure determination and analysis. *Nat. Methods* 16 (2): 151–162.

22 Shimada, I., Ueda, T., Kofuku, Y. et al. (2019). GPCR drug discovery: integrating solution NMR data with crystal and cryo-EM structures. *Nat. Rev. Drug Discovery* 18: 59–82.

23 Yao, X.J., Vélez Ruiz, G., Whorton, M.R. et al. (2009). The effect of ligand efficacy on the formation and stability of a GPCR-G protein complex. *Proc. Natl. Acad. Sci. U.S.A.* 106 (23): 9501.

24 Manglik, A., Kim Tae, H., Masureel, M. et al. (2015). Structural insights into the dynamic process of β2-adrenergic receptor signaling. *Cell* 161 (5): 1101–1111.

25 Sena, D.M. Jr., Cong, X., Giorgetti, A. et al. (2017). Structural heterogeneity of the μ-opioid receptor's conformational ensemble in the apo state. *Sci. Rep.* 7: 45761.

26 Szewczyk, J.W., Acton, J., Adams, A.D. et al. (2011). Design of potent and selective GPR119 agonists for type II diabetes. *Bioorg. Med. Chem. Lett.* 21 (9): 2665–2669.

27 Capote, L.A., Mendez Perez, R., and Lymperopoulos, A. (2015). GPCR signaling and cardiac function. *Eur. J. Pharmacol.* 763: 143–148.

28 Franco, R., Martínez-Pinilla, E., Navarro, G. et al. (2017). Potential of GPCRs to modulate MAPK and mTOR pathways in Alzheimer's disease. *Prog. Neurobiol.* 149-150: 21–38.

29 Insel, P.A., Sriram, K., Wiley, S.Z. et al. (2018). GPCRomics: GPCR expression in cancer cells and tumors identifies new, potential biomarkers and therapeutic targets. *Front. Pharmacol.* 9: 431.
30 Nebigil, C.G. and Désaubry, L. (2019). The role of GPCR signaling in cardiac epithelial to mesenchymal transformation (EMT). *Trends Cardiovasc. Med.* 29 (4): 200–204.
31 Sebastiani, G., Ceccarelli, E., Castagna, M.G. et al. (2018). G-protein-coupled receptors (GPCRs) in the treatment of diabetes: current view and future perspectives. *Best Pract. Res. Clin. Endocrinol. Metab.* 32 (2): 201–213.
32 Kaczor, A.A., Rutkowska, E., Bartuzi, D. et al. (2016). Computational methods for studying G protein-coupled receptors (GPCRs). *Method Cell Biol.* 132: 359–399.
33 Nikolaev, D.M., Shtyrov, A.A., Panov, M.S. et al. (2018). A comparative study of modern homology modeling algorithms for rhodopsin structure prediction. *ACS Omega* 3 (7): 7555–7566.
34 Yuan, X.J. and Xu, Y.C. (2018). Recent trends and applications of molecular modeling in GPCR-ligand recognition and structure-based drug design. *Int. J. Mol. Sci.* 19 (7): 2105.
35 Castleman, P.N., Sears, C.K., Cole, J.A. et al. (2019). GPCR homology model template selection benchmarking: Global versus local similarity measures. *J. Mol. Graphics Modell.* 86: 235–246.
36 Congreve, M., Oswald, C., and Marshall, F.H. (2017). Applying structure-based drug design approaches to allosteric modulators of GPCRs. *Trends Pharmacol. Sci.* 38 (9): 837–847.
37 Pandy-Szekeres, G., Munk, C., Tsonkov, T.M. et al. (2018). GPCRdb in 2018: adding GPCR structure models and ligands. *Nucleic Acids Res.* 46 (D1): D440–D446.
38 Rosenbaum, D.M., Cherezov, V., Hanson, M.A. et al. (2007). GPCR engineering yields high-resolution structural insights into β2-adrenergic receptor function. *Science* 318 (5854): 1266.
39 Xiang, J., Chun, E., Liu, C. et al. (2016). Successful strategies to determine high-resolution structures of GPCRs. *Trends Pharmacol. Sci.* 37 (12): 1055–1069.
40 Keri, D. and Barth, P. (2018). Reprogramming G protein coupled receptor structure and function. *Curr. Opin. Struct. Biol.* 51: 187–194.
41 Bill, R.M., Henderson, P.J.F., Iwata, S. et al. (2011). Overcoming barriers to membrane protein structure determination. *Nat. Biotechnol.* 29 (4): 335–340.
42 White, K.L., Eddy, M.T., Gao, Z.G. et al. (2018). Structural connection between activation microswitch and allosteric sodium site in GPCR signaling. *Structure* 26 (2): 259–269.

43 Katritch, V., Fenalti, G., Abola, E.E. et al. (2014). Allosteric sodium in class A GPCR signaling. *Trends Biochem. Sci.* 39 (5): 233–244.

44 Massink, A., Gutiérrez-de-Terán, H., Lenselink, E.B. et al. (2015). Sodium ion binding pocket mutations and adenosine A2A receptor function. *Mol. Pharmacol.* 87 (2): 305–313.

45 Fenalti, G., Giguere, P.M., Katritch, V. et al. (2014). Molecular control of δ-opioid receptor signalling. *Nature* 506: 191–196.

46 Chen, K.-Y.M., Zhou, F., Fryszczyn, B.G. et al. (2012). Naturally evolved G protein-coupled receptors adopt metastable conformations. *Proc. Natl. Acad. Sci. U.S.A.* 109 (33): 13284.

47 Vriend, G. and Sander, C. (1993). Quality control of protein models: directional atomic contact analysis. *J. Appl. Crystallogr.* 26 (1): 47–60.

48 Sheffler, W. and Baker, D. (2009). RosettaHoles: Rapid assessment of protein core packing for structure prediction, refinement, design, and validation. *Protein Sci.* 18 (1): 229–239.

49 Barth, P., Schonbrun, J., and Baker, D. (2007). Toward high-resolution prediction and design of transmembrane helical protein structures. *Proc. Natl. Acad. Sci. U.S.A.* 104 (40): 15682.

50 Chen, K.-Y.M., Sun, J., Salvo, J.S. et al. (2014). High-resolution modeling of transmembrane helical protein structures from distant homologues. *PLoS Comput. Biol.* 10 (5): e1003636.

51 Peng, Y., McCorvy, J.D., Harpsøe, K. et al. (2018). 5-HT2C receptor structures reveal the structural basis of GPCR polypharmacology. *Cell* 172 (4): 719–30.e14.

52 Popov, P., Peng, Y., Shen, L. et al. (2018). Computational design of thermostabilizing point mutations for G protein-coupled receptors. *eLife* 7: e34729.

53 Cardozo, T., Totrov, M., and Abagyan, R. (1995). Homology modeling by the ICM method. *Proteins Struct. Funct. Bioinf.* 23 (3): 403–414.

54 Craig, D.B. and Dombkowski, A.A. (2013). Disulfide by Design 2.0: a web-based tool for disulfide engineering in proteins. *BMC Bioinf.* 14 (1).

55 Arber, C., Young, M., and Barth, P. (2017). Reprogramming cellular functions with engineered membrane proteins. *Curr. Opin. Biotechnol.* 47: 92–101.

56 Feng, X., Ambia, J., Chen, K.-Y.M. et al. (2017). Computational design of ligand-binding membrane receptors with high selectivity. *Nat. Chem. Biol.* 13: 715.

57 Schmidt, T., Bergner, A., and Schwede, T. (2014). Modelling three-dimensional protein structures for applications in drug design. *Drug Discovery Today* 19 (7): 890–897.

58 Munk, C., Isberg, V., Mordalski, S. et al. (2016). GPCRdb: the G protein-coupled receptor database – an introduction. *Br. J. Pharmacol.* 173 (14): 2195–2207.

59 Kelm, S., Shi, J., and Deane, C.M. (2010). MEDELLER: homology-based coordinate generation for membrane proteins. *Bioinformatics* 26 (22): 2833–2840.

60 Sandal, M., Duy, T.P., Cona, M. et al. (2013). GOMoDo: a GPCRs online modeling and docking webserver. *PLoS One* 8 (9): e74092.

61 Zhang, J., Yang, J.Y., Jang, R. et al. (2015). GPCR-I-TASSER: a hybrid approach to G protein-coupled receptor structure modeling and the application to the human genome. *Structure* 23 (8): 1538–1549.

62 Esguerra, M., Siretskiy, A., Bello, X. et al. (2016). GPCR-ModSim: a comprehensive web based solution for modeling G-protein coupled receptors. *Nucleic Acids Res.* 44 (W1): W455–W462.

63 Worth, C.L., Kreuchwig, F., Tiemann, J.K.S. et al. (2017). GPCR-SSFE 2.0-a fragment-based molecular modeling web tool for Class A G-protein coupled receptors. *Nucleic Acids Res.* 45 (W1): W408–W415.

64 Sztyler, A., Latek, D., Jakowiecki, J. et al. (2018). GPCRM: a homology modeling web service with triple membrane-fitted quality assessment of GPCR models. *Nucleic Acids Res.* 46 (W1): W387–W395.

65 Edgar, R.C. (2004). MUSCLE: multiple sequence alignment with high accuracy and high throughput. *Nucleic Acids Res.* 32 (5): 1792–1797.

66 Heringa, J., Feenstra, K.A., and Pirovano, W. (2008). PRALINE™: a strategy for improved multiple alignment of transmembrane proteins. *Bioinformatics* 24 (4): 492–497.

67 Wu, S. and Zhang, Y. (2008). MUSTER: Improving protein sequence profile-profile alignments by using multiple sources of structure information. *Proteins* 72 (2): 547–556.

68 Montanyola, A., Xenarios, I., Taly, J.-F. et al. (2011). T-Coffee: a web server for the multiple sequence alignment of protein and RNA sequences using structural information and homology extension. *Nucleic Acids Res.* 39 (suppl_2): W13–W17.

69 Deane, C.M. and Hill, J.R. (2012). MP-T: improving membrane protein alignment for structure prediction. *Bioinformatics* 29 (1): 54–61.

70 Stamm, M., Staritzbichler, R., Khafizov, K. et al. (2014). AlignMe–a membrane protein sequence alignment web server. *Nucleic Acids Res.* 42 (Web Server issue): W246–W251.

71 Bhat, B., Ganai, N.A., Andrabi, S.M. et al. (2017). TM-Aligner: multiple sequence alignment tool for transmembrane proteins with reduced time and improved accuracy. *Sci. Rep.* 7 (1): 12543.

72 Sievers, F. and Higgins, D.G. (2018). Clustal Omega for making accurate alignments of many protein sequences. *Protein Sci.* 27 (1): 135–145.

73 Ngo, T., Ilatovskiy, A.V., Stewart, A.G. et al. (2017). Orphan receptor ligand discovery by pickpocketing pharmacological neighbors. *Nat. Chem. Biol.* 13 (2): 235–242.

74 Tarcsay, Á., Paragi, G., Vass, M. et al. (2013). The impact of molecular dynamics sampling on the performance of virtual screening against GPCRs. *J. Chem. Inf. Modell.* 53 (11): 2990–2999.

75 Verdonk, M.L., Cole, J.C., Hartshorn, M.J. et al. (2003). Improved protein–ligand docking using GOLD. *Proteins Struct. Funct. Bioinf.* 52 (4): 609–623.

76 Hornak, V. and Simmerling, C. (2004). Development of softcore potential functions for overcoming steric barriers in molecular dynamics simulations. *J. Mol. Graphics Modell.* 22 (5): 405–413.

77 Sherman, W., Day, T., Jacobson, M.P. et al. (2006). Novel procedure for modeling ligand/receptor induced fit effects. *J. Med. Chem.* 49 (2): 534–553.

78 Chen, H., Fu, W., Wang, Z. et al. (2019). Reliability of docking-based virtual screening for GPCR ligands with homology modeled structures: a case study of the angiotensin II type I receptor. *ACS Chem. Neurosci.* 10 (1): 677–689.

79 Zhang, H., Unal, H., Gati, C. et al. (2015). Structure of the angiotensin receptor revealed by serial femtosecond crystallography. *Cell* 161 (4): 833–844.

80 Zhang, H., Unal, H., Desnoyer, R. et al. (2015). Structural basis for ligand recognition and functional selectivity at angiotensin receptor. *J. Biol. Chem.* 290 (49): 29127–29139.

81 Śledź, P. and Caflisch, A. (2018). Protein structure-based drug design: from docking to molecular dynamics. *Curr. Opin. Struct. Biol.* 48: 93–102.

82 Bubeck, S., Munos, R., and Stoltz, G. (ed.) (2009). *Pure Exploration in Multi-armed Bandits Problems.* Berlin, Heidelberg: Springer Berlin Heidelberg.

83 Terayama, K., Iwata, H., Araki, M. et al. (2017). Machine learning accelerates MD-based binding pose prediction between ligands and proteins. *Bioinformatics* 34 (5): 770–778.

84 Lückmann, M., Trauelsen, M., Bentsen, M.A. et al. (2019). Molecular dynamics-guided discovery of an ago-allosteric modulator for GPR40/FFAR1. *Proc. Natl. Acad. Sci. U.S.A.* 116 (14): 7123–7128.

85 Schwartz, T.W. and Holst, B. (2006). Ago-allosteric modulation and other types of allostery in dimeric 7TM receptors. *J. Recept. Signal Transduction* 26 (1-2): 107–128.

86 Shukla, D., Hernández, C.X., Weber, J.K. et al. (2015). Markov state models provide insights into dynamic modulation of protein function. *Acc. Chem. Res.* 48 (2): 414–422.

87 Ferruz, N., Doerr, S., Vanase-Frawley, M.A. et al. (2018). Dopamine D3 receptor antagonist reveals a cryptic pocket in aminergic GPCRs. *Sci. Rep.* 8 (1): 897.

88 Doerr, S., Harvey, M.J., Noé, F. et al. (2016). HTMD: high-throughput molecular dynamics for molecular discovery. *J. Chem. Theory Comput.* 12 (4): 1845–1852.

89 Beglov, D., Hall, D.R., Wakefield, A.E. et al. (2018). Exploring the structural origins of cryptic sites on proteins. *Proc. Natl. Acad. Sci. U.S.A.* 115 (15): E3416.

90 Ngan, C.H., Bohnuud, T., Mottarella, S.E. et al. (2012). FTMAP: extended protein mapping with user-selected probe molecules. *Nucleic Acids Res.* 40 (Web Server issue): W271–W275.

91 Palczewski, K., Kumasaka, T., Hori, T. et al. (2000). Crystal structure of rhodopsin: a G protein-coupled receptor. *Science* 289 (5480): 739.

92 Yuan, S., Filipek, S., Palczewski, K. et al. (2014). Activation of G-protein-coupled receptors correlates with the formation of a continuous internal water pathway. *Nat. Commun.* 5: 4733.

93 Venkatakrishnan, A.J., Ma, A.K., Fonseca, R. et al. (2019). Diverse GPCRs exhibit conserved water networks for stabilization and activation. *Proc. Natl. Acad. Sci. U.S.A.* 116 (8): 3288.

94 Han, M., Lin, S.W., Minkova, M. et al. (1996). Functional interaction of transmembrane Helices 3 and 6 in rhodopsin: replacement of phenylalanine 261 by alanine causes reversion of phenotype of a glycine 121 replacement mutant. *Journal of Biological Chemistry.* 271 (50): 32337–32342.

95 Altenbach, C., Yang, K., Farrens, D.L. et al. (1996). Structural features and light-dependent changes in the cytoplasmic interhelical E–F loop region of rhodopsin: a site-directed spin-labeling study. *Biochemistry* 35 (38): 12470–12478.

96 Lee, Y., Kim, S., Choi, S. et al. (2016). Ultraslow water-mediated transmembrane interactions regulate the activation of A2A adenosine receptor. *Biophys. J.* 111 (6): 1180–1191.

97 Tomobe, K., Yamamoto, E., Kholmurodov, K. et al. (2017). Water permeation through the internal water pathway in activated GPCR rhodopsin. *PLoS One.* 12 (5): e0176876.

98 Michino, M., Beuming, T., Donthamsetti, P. et al. (2015). What can crystal structures of aminergic receptors tell us about designing subtype-selective ligands? *Pharmacol. Rev.* 67 (1): 198.

99 Korczynska, M., Clark, M.J., Valant, C. et al. (2018). Structure-based discovery of selective positive allosteric modulators of antagonists for the M2 muscarinic acetylcholine receptor. *Proc. Natl. Acad. Sci. U.S.A.* 115 (10): E2419–E2428.

100 Miao, Y.L., Goldfeld, D.A., Von Moo, E. et al. (2016). Accelerated structure-based design of chemically diverse allosteric modulators of a muscarinic G protein-coupled receptor. *Proc. Natl. Acad. Sci. U.S.A.* 113 (38): E5675–E5684.

101 Bock, A., Schrage, R., and Mohr, K. (2018). Allosteric modulators targeting CNS muscarinic receptors. *Neuropharmacology* 136: 427–437.

102 Wold, E.A., Chen, J., Cunningham, K.A. et al. (2019). Allosteric modulation of class A GPCRs: targets, agents, and emerging concepts. *J. Med. Chem.* 62 (1): 88–127.

103 Wootten, D., Christopoulos, A., and Sexton, P.M. (2013). Emerging paradigms in GPCR allostery: implications for drug discovery. *Nat. Rev. Drug Discovery.* 12 (8): 630–644.

104 Wakefield, A.E., Mason, J.S., Vajda, S. et al. (2019). Analysis of tractable allosteric sites in G protein-coupled receptors. *Sci. Rep.* 9 (1): 6180.

105 Fronik, P., Gaiser, B.I., and Sejer, P.D. (2017). Bitopic ligands and metastable binding sites: opportunities for G protein-coupled receptor (GPCR) medicinal chemistry. *J. Med. Chem.* 60 (10): 4126–4134.

106 Dror, R.O., Pan, A.C., Arlow, D.H. et al. (2011). Pathway and mechanism of drug binding to G-protein-coupled receptors. *Proc. Natl. Acad. Sci. U.S.A.* 108 (32): 13118–13123.

107 Aurelio, L., Baltos, J.-A., Ford, L. et al. (2018). A structure–activity relationship study of bitopic N6-substituted adenosine derivatives as biased adenosine A1 receptor agonists. *J. Med. Chem.* 61 (5): 2087–2103.

108 Valant, C., May, L.T., Aurelio, L. et al. (2014). Separation of on-target efficacy from adverse effects through rational design of a bitopic adenosine receptor agonist. *Proc. Natl. Acad. Sci.* 111 (12): 4614–4619.

109 Chan, H.C.S., Wang, J., Palczewski, K. et al. (2018). Exploring a new ligand binding site of G protein-coupled receptors. *Chem. Sci.* 9 (31): 6480–6489.

110 Hollingsworth, S.A., Kelly, B., Valant, C. et al. (2019). Cryptic pocket formation underlies allosteric modulator selectivity at muscarinic GPCRs. *Nat. Commun.* 10 (1): 3289.

111 McCorvy, J.D., Butler, K.V., Kelly, B. et al. (2018). Structure-inspired design of β-arrestin-biased ligands for aminergic GPCRs. *Nat. Chem. Biol.* 14 (2): 126–134.

112 Deganutti, G., Welihinda, A., and Moro, S. (2017). Comparison of the human A2A adenosine receptor recognition by adenosine and inosine: new insight from supervised molecular dynamics simulations. *ChemMedChem* 12 (16): 1319–1326.

113 Jacobson, K.A., Tosh, D.K., Jain, S. et al. (2019). Historical and current adenosine receptor agonists in preclinical and clinical development. *Front. Cell. Neurosci.* 13: 124.

114 Vreven, T., Byun, K.S., Komaromi, I. et al. (2006). Combining quantum mechanics methods with molecular mechanics methods in ONIOM. *J. Chem. Theory Comput.* 2 (3): 815–826.

115 Amaro, R.E. and Mulholland, A.J. (2018). Multiscale methods in drug design bridge chemical and biological complexity in the search for cures. *Nat. Rev. Chem.* 2 (4).

116 Ryazantsev, M.N., Nikolaev, D.M., Struts, A.V. et al. (2019). Quantum mechanical and molecular mechanics modeling of membrane-embedded rhodopsins. *J. Membr. Biol.* 252 (4-5, 425): –49.

117 Senn, H.M. and Thiel, W. (2009). QM/MM methods for biomolecular systems. *Angew. Chem. Int. Ed.* 48 (7): 1198–1229.

118 Zanatta, G., Nunes, G., Bezerra, E.M. et al. (2014). Antipsychotic haloperidol binding to the human dopamine D3 receptor: beyond docking through QM/MM refinement toward the design of improved schizophrenia medicines. *ACS Chem. Neurosci.* 5 (10): 1041–1054.

119 Sekharan, S., Ertem, M.Z., Zhuang, H.Y. et al. (2014). QM/MM model of the mouse olfactory receptor MOR244-3 validated by site-directed mutagenesis experiments. *Biophys. J.* 107 (5): L05–L08.

120 Gelis, L., Wolf, S., Hatt, H. et al. (2012). Prediction of a ligand-binding niche within a human olfactory receptor by combining site-directed mutagenesis with dynamic homology modeling. *Angew. Chem. Int. Ed.* 51 (5): 1274–1278.

121 Duan, X.F., Block, E., Li, Z. et al. (2012). Crucial role of copper in detection of metal-coordinating odorants. *Proc. Natl. Acad. Sci. U.S.A.* 109 (9): 3492–3497.

122 Ahmed, L., Zhang, Y.T., Block, E. et al. (2018). Molecular mechanism of activation of human musk receptors OR5AN1 and OR1A1 by (R)-muscone and diverse other musk-smelling compounds. *Proc. Natl. Acad. Sci. U.S.A.* 115 (17): E3950–E3958.

123 Singh, W., Bilal, M., McClory, J. et al. (2019). Mechanism of phosphatidylglycerol activation catalyzed by prolipoprotein diacylglyceryl transferase. *J. Phys. Chem. B* 123 (33): 7092–7102.

124 Sankaran, K., Gan, K., Rash, B. et al. (1997). Roles of histidine-103 and tyrosine-235 in the function of the prolipoprotein diacylglyceryl transferase of *Escherichia coli*. *J. Bacteriol.* 179: 2944–2948.

125 Pedraza-Gonzalez, L., De Vico, L., Marin, M.D. et al. (2019). a-ARM: automatic rhodopsin modeling with chromophore cavity generation, ionization state selection, and external counterion placement. *J. Chem. Theory Comput.* 15 (5): 3134–3152.

14

Signaling, Physiology, and Targeting of GPCR-Regulated Phospholipase C Enzymes

Naincy R. Chandan, Alan V. Smrcka, and Hoa T.N. Phan

Department of Pharmacology, University of Michigan Medical School, Ann Arbor, MI, USA

14.1 Background

14.1.1 A Brief History of the Discovery of Phospholipase Cs (PLCs)

The seminal discoveries of the second messengers inositol 1,4,5-trisphosphate (IP3) [1, 2], and diacylglycerol (DAG) [3] in the late 1970s and early 1980s led to efforts to isolate and identify inositol-phospholipid-specific phospholipase C (PLC) enzymes. Multiple PLC enzymes of varying molecular weights were initially purified in the late 1980s [4–7]. Based on their molecular weights, various PLC isozymes were classified as β for the 150–154 kDa enzyme, γ for the 130–145 kDa enzyme, and δ for the 85–88 kDa enzyme [8]. Subsequent molecular cloning [9, 10] and sequencing [11, 12] efforts identified conserved regions of sequence homology termed X and Y domains that, together, constitute the catalytic domain of PLCs [8]. Later, three additional classes of PLCs were discovered: PLCε [13–16], ζ [17], and η [18–20].

All PLCs have a core domain structure containing a pleckstrin homology (PH) domain, followed by four tandem EF-hand repeats (EF1-4), a triose phosphate isomerase (TIM) barrel that houses the active site composed of split X and Y domains connected by an X–Y linker, and a C2 domain [21]. Beyond the conserved core, each isoform has distinct regulatory elements that are involved with their distinct modes of activation and biological functions (Figure 14.1).

PLC enzymes hydrolyze the membrane phospholipid, phosphatidylinositol 4,5-bisphosphate (PIP_2) and/or phosphatidylinositol 4-phosphate (PI4P) into second messengers IP3 (or inert IP2 in the case of PI4P hydrolysis) and DAG

☆ Naincy R. Chandan and Hoa T.N. Phan contributed equally to review.

GPCRs as Therapeutic Targets, Volume 1, First Edition. Edited by Annette Gilchrist.

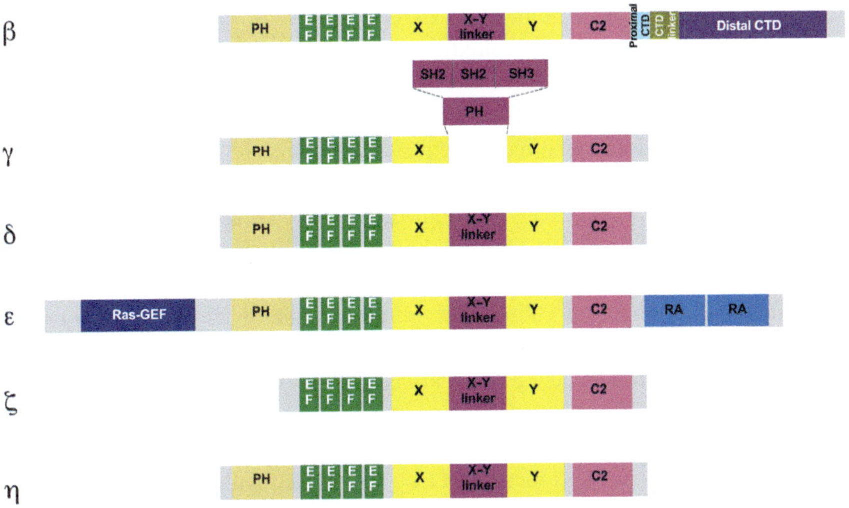

Figure 14.1 Domain structures of PLC subtypes. All PLC subtypes have 4 EF hands, the X and Y domains split by an XY linker to form the catalytic core and the C2 domain. Besides those conserved domains, each isoform has other distinct regulatory domains. PH: pleckstrin homology domain, CTD: C terminal domain, GEF: guanine exchange factor.

[4–7, 22, 23]. The regulatory mechanisms for PLCγ isoforms were the first to be identified and are activated by receptor or nonreceptor protein-tyrosine kinases such as nerve growth factor (NGF) and platelet-derived growth factor (PDGF) receptors [24–27]. However, the direct connection between PLC and G protein signaling was only established when the $G\alpha_q$ class of pertussis toxin (PTX) insensitive G protein α subunits was discovered [28, 29]. Multiple studies demonstrated that PLCβ was a direct effector of $G\alpha_q$ [30–32]. When PLCβ subtypes β2, β3, and β4 were later identified, they were also shown to be activated by $G\alpha_q$ [33–37]. The Gβγ dimer, at a higher concentration than that of $G\alpha_q$ (20–200 nM range as compared with 1–5 nM of $G\alpha_q$), could also activate PLCβ1, 2, and 3 to various degrees, but not PLCβ4 [36–39]. Interestingly, PLCβ3 is the only isoform that can be synergistically activated by $G\alpha_q$ and Gβγ. The source of Gβγ in cells is from Gi/o-coupled receptor stimulation since Gβγ-mediated PLCβ activation is PTX sensitive [40, 41] and Gi is more abundant than other G proteins in cells, providing sufficient concentrations of Gβγ to activate PLC [42]. PLCβ2 is also activated by the small GTPase protein Rac [43, 44].

In 1996, an *in silico* search for novel Ras association (RA) domains in available databases identified an open reading frame containing two RA domains, a TIM-barrel fold of the catalytic type associated with PLC isoforms, and a putative CDC25 domain [45]. PLC210, the first PLCε ortholog, was discovered in 1998 as a novel Ras-regulated PLC in *Caenorhabditis elegans*. It was identified as a novel binder of LET-60 (Ras ortholog in *C. elegans*) in a yeast two-hybrid screen of a *C. elegans* cDNA library [46]. Soon after its discovery, in 2001, multiple laboratories [13–15] independently used expressed sequence tag databases to clone mammalian ortholog of PLC210, designated as PLCε.

PLCε was found to be directly activated by multiple monomeric GTPases and was later shown to directly activated by Gβγ [47]. It has also been shown to be regulated by $G_{12/13}$ likely through their ability to activate Rho which, in turn, directly activates PLCε. This will be discussed in more detail later. Besides being an effector of Ras, the CDC25 domain of PLCε is a GEF for Rap, leading to activation of Rap and subsequent reinforcement of its own activity [48]. Alternatively, Rap activation has been shown to mediate downstream activities such as the mitogen-activated protein kinase (MAPK) activation, independent of its lipase activity [13].

Given the extensive literature on PLC regulation and physiological function, within this chapter, we will focus on PLCβ and PLCε regulation and function as they relate to GPCR signaling, while bearing in mind that there may be other PLC types, such as PLCδ or PLCη, that may also be involved in GPCR signaling.

14.2 Regulation of PLCβ and PLCε by G Proteins and GPCRs

While it is well known that PLCβ isoforms are regulated by GPCRs, less well appreciated is the extent to which PLCε contributes to GPCR stimulated phosphoinositide (PI) hydrolysis, particularly during sustained GPCR stimulation. In this section, we discuss what is known about the regulatory features of these enzymes in terms of biochemical and cell biological roles of individual PLC isoforms.

14.2.1 Structural Basis for Regulation of PLC

14.2.1.1 Insights from $G\alpha_q$-PLCβ Co-Crystal Structures

Several X-ray crystal structures of PLCβ have been solved, consistently presenting a compact, globular structure of the core of the enzyme consisting of PH, EF-hand, X–Y and C2 domains, as discussed earlier, and as shown in Figure 14.2a [43, 49–51]. Initial structures lacked the critical 400 amino acids of the C-terminus. The structure of the C-terminal domain (CTD) was solved in isolation revealing an extended coiled-coil bin/amphiphysin/rv (BAR) like domain that crystallized as a dimer. Initial biochemical work suggested that the primary binding site for $G\alpha_q$ was the C-terminus, largely because deletion of this domain eliminated $G\alpha_q$-dependent regulation. The first structure of PLCβ complexed to $G\alpha_q$, solved by the Harden and Sondek groups, surprisingly revealed

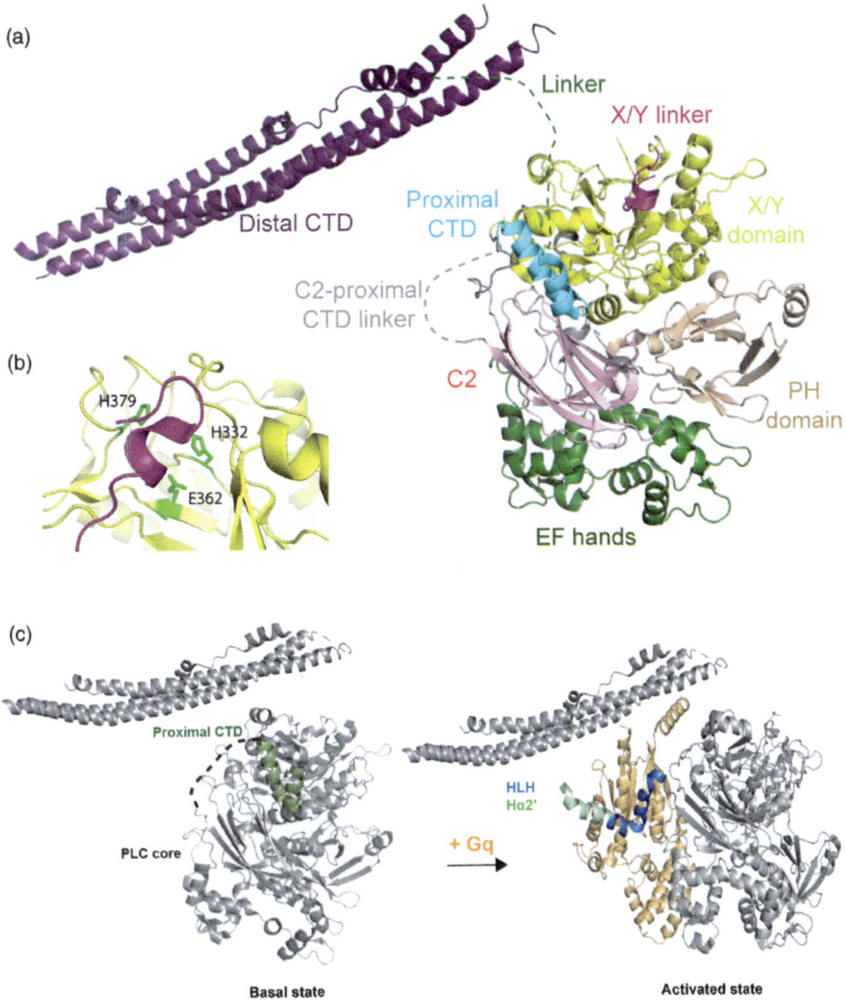

Figure 14.2 Structure of PLCβ. (a) Compact globular packaging of PH, EF-hand, X–Y, and C2 domains forms the core enzyme. (b) PLC autoinhibition by the PLCβ X–Y linker (top view). The ordered region of the X–Y linker (pink) lies on top of the active site (yellow) containing the catalytic residues H332, H379, and E362 (green), preventing the access to PI substrate in the membrane. (c) A model for PLC activation by $G\alpha_q$. (left) The HLH (green) region folds against the catalytic domain, inhibiting enzymatic activity. (right) Upon binding $G\alpha_q$ (orange) the HLH domain (blue and green) swings away from the catalytic domain to bind to $G\alpha_q$, releasing autoinhibition of catalytic activity by the $H\alpha2'$ region (green).

binding of Gα_q-GTP to a truncated variant of PLCβ3, lacking the C-terminus. Rather than binding to the C-terminus, Gα_q was bound to a Helix–Loop–Helix (HLH) structure immediately following the C2 domain. This construct was still not activated by Gα_q showing that, despite clear high-affinity binding, the CTD is absolutely required for enzyme activation. Later, full-length PLCβ3 was crystalized in a complex with Gα_q by the Tesmer laboratory, showing the same binding to the HLH, but the coiled-coil C terminus appears as a somewhat independent domain, making limited contacts with the core, and is not a dimer (Figure 14.2d). Interestingly, limited contacts between the Gα_q amino terminus and the CTD were identified in this structure.

In the structures of Gα_q-bound PLCβs, and in Rac-bound PLCβ2, no significant conformational changes in the catalytic site were observed that could explain the increased activity observed upon G protein binding. This has led to two general models for PLC activation by G proteins. In the basal state, the PLC structure is packed in a way that would preclude substrate binding. The catalytic site, within the TIM barrel split into X and Y domains, is capped by the linker between the X and Y domains, blocking the active site from accessing its PI substrate in the membrane [49] (Figure 14.2b). In one model for Gα_q-dependent activation, and proposed as a general model for PLC activation, G protein binding promotes PLCβ binding to the negatively charged inner surface of the plasma membrane (PM) [52, 53]. In this model, the core is optimally positioned by the G protein at the membrane such that the X–Y linker is displaced by electrostatic repulsion between an acidic stretch in the X–Y linker, and the negatively charged surface of the membrane, removing the lid leading to enzyme activation [51]. In support of this model, activators such as Gα_q, Gβγ, or small G protein (Rac, CDC42) bind to different sites to activate PLCβ2 or β3 [54]. The membrane plays a central role because, without it, activators such as Gα_q or Gβγ cannot activate PLCβ isoforms. The electrostatic potential, stored curvature elastic stress, and lipid composition of the membrane have been shown to be important in Rac-mediated PLCβ2 activation [55] and membrane adsorption as well [53]. On the other hand, PLCβ2 already bound to vesicle membrane surfaces can be activated by Gβγ [56], suggesting either an allosteric mechanism or a mechanism for activation that involves reorientation on the membrane surface.

A second model is based on several observations from the full-length structures of PLCβ that were solved by the Tesmer laboratory [57]. In the full-length structure of PLCβ in the absence of Gα_q, a helix (Hα2′) was observed, immediately C-terminal to the portion of the Gα_q binding HLH. Hα2′ packs directly against the PLC catalytic domain, docking in a conserved cleft formed between the TIM barrel and C2 domains, in close proximity to the active site and the X–Y linker [58] (Figure 14.2c). In the full-length Gα_q-bound PLCβ3 complex, Gα_q binding to

HLH displaces the Hα2′ element from the catalytic core [52] (Figure 14.2c). Mutagenic studies indicate that Hα2′ autoinhibits PLC activity. Thus, displacement of this domain away from the catalytic core has been suggested as a mechanism for activation, although how this relieves inhibition of catalytic activity is not clear since no fundamental rearrangement of catalytic amino acids is observed.

An additional level of regulation may involve the CTD. In the full-length PLCβ3-Gα_q complex, the Dα3 helix of the distal CTD forms a transient interaction with the top of the catalytic core to help partition the enzyme between membrane-associated and cytosolic populations and may add another layer of inhibition [57]. The palmitoylated N-terminal helix of Gα_q has additional interactions with a conserved hydrophobic patch on the distal CTD [57]. Recent deuterium exchange mass spectrometry (DXMS) analysis of PLC on a membrane surface confirmed an interaction between Gα_q and the CTD. Biochemical analysis demonstrated that the distal CTD autoinhibits PLC activation, and DXMS analysis shows striking rearrangement of the CTD upon binding of Gα_q or Gβγ [59].

Finally, studies of a PLCβ4 point mutation found in uveal melanoma patients that results in conversion of Asp 630 to Tyr (PLCβ4(D630Y)) reveals the possibility of another mechanism for PLC activation [60]. PLCβ4(D630Y) is constitutively active to the extent that it cannot be further activated by Gα_q in cells. This amino acid is highly conserved in all PLC isoforms and mutation of this amino acid to Tyr in PLCβ2, PLCβ3, and PLCε results in constitutive activation suggesting a common conserved mechanism. Mutation to any amino acid that is not negatively charged results in activation indicating a requirement for a negative charge at this position. Purified PLCβ4(D630Y) also shows high activity *in vitro* analysis using a water-soluble PLC substrate indicates that this activation does not require a lipid bilayer surface. This observation seems to rule out the membrane dependent activation dependent mechanism described earlier in this section, and further mutagenesis suggest that X–Y linker displacement is not involved. Modeling PLCβ4 based on the high-resolution structure of PLCβ3 shows that D630 on the surface of the Y domain and does not interact with other domains within PLCβ including the X–Y linker and the Hα2′ helix discussed earlier. Thus, this elevated activity of PLC cannot be explained by current models for PLC activation suggesting that there is an entirely different mechanism for PLC activation that could be utilized by upstream regulators. The atomic mechanistic explanation of this activation, and its importance in regulation by G proteins subunit will require further investigation.

14.2.2 Activation of PLCβ by Gβγ

Although individual X-ray crystal structures of Gβγ and PLCβs are available (reviewed in [61, 62]), a structure of the Gβγ-PLCβ complex has not been determined. Instead, various biochemical approaches have investigated Gβγ-PLC binding interactions. Several studies indicate that Gβγ binds to the PH domain of PLCβ. One study proposes the intriguing idea that Gβγ engages PLCβ by binding to the PH domain at a surface that is buried in an intramolecular interaction with the EF-hand domains. It was proposed that this interface is dynamically exposed in solution, to reveal a binding site for Gβγ to bind, recruit PLC to the membrane, and re-position the PH domain relative to the catalytic core [63–65]. Another proposed binding site is the α5 helix in the Y box of the TIM barrel of PLCβ with either the N terminus or the propeller of Gβγ [66], allowing enzyme activation through repositioning of the core or an allosteric mechanism [62]. A recent DXMS study has shown that upon Gβγ binding, there is a dramatic conformational change in the distal CTD of PLCβ2, and that the CTD autoinhibits the catalytic activity of the PLCβ2 enzyme [59]. This leads to a model where Gβγ engagement with PLC relieves CTD autoinhibitory activity as part of the mechanism for activation. Unfortunately, these studies did not reveal a Gβγ binding site and were unable to provide support for either the PH domain or the α5-helix as binding sites. Altogether, however, it is likely that Gβγ engages multiple domains in PLCβ to not only recruit the enzyme to the membrane but also induce an allosteric rearrangement of at least the distal CTD to activate the core enzyme.

PLCβ3 activity has been demonstrated both biochemically [67, 68] and in physiological systems [69] to be regulated in a synergistic manner by $G\alpha_q$ and Gβγ. This supports the idea that Gβγ and $G\alpha_q$ bind to different sites on the protein. Biochemical studies indicate that the synergy can be explained by a two-state mechanism but the structural basis for the synergy is unknown. Recent studies have built on this observation to suggest that in many cell types Gβγ-regulation of PLC types requires a coincident $G\alpha_q$ stimulus [70]. While this may be true in many cell types, in cells that express predominantly PLCβ2, coincident $G\alpha_q$ stimulation is likely not required to observe Gβγ-regulation, since PLCβ2 is primarily regulated by Gβγ and does not display synergy with $G\alpha_q$ [67, 71]. As discussed in the following text, PLCβ2 is expressed primarily in myeloid cells where G_i-coupled receptor-dependent PLC activation is prominently observed.

14.2.2.1 Activation of PLCβ by the Small GTPase, Rac1

The X-ray crystal structure of a Rac1-PLCβ2 complex showed a binding interface between the Switch I and II region of Rac with a hydrophobic ridge on the surface of the PH domain of PLCβ2 without changing the conformational state of PLCβ2 [43]. Activated Rac1, anchored to the membrane via its prenylated C-terminus, binds PLCβ2 leading to its redistribution from the cytosol to the PM and optimizes the orientation of the Rac1 catalytic core complex at the membrane, possibly promoting interfacial activation via the X–Y linker as discussed earlier [43, 44]. Rac1 and Gβγ bind to distinct regions in the PH domain of PLCβ2, with the binding site for Rac on the surface of the PH domain, while the binding site for Gβγ is more buried [65], which could explain the additive increase in activation of PLCβ2 by both Rac2 and Gβγ [72].

14.2.2.2 Regulation of PLCε

PLCε has a similar core domain structure to PLCβ but is substantially larger. At its N-terminus is a CDC25 homology domain that has specific GEF activity toward Rap1 [73]. This is followed by the PH-EF-XY-C2 common PLC domain structure. At the C terminus, instead of the coiled-coil domain found in PLCβ, there are tandem Ras association domains, RA1 and RA2 that bind to small Ras-related GTPases. There are two splice variants of PLCε differing at the amino terminus [74]. These splice variants have unique tissue distributions implying unique functions, but specific biochemical or cellular functional differences have not been identified as will be discussed in the following text.

Multiple signaling pathways downstream of GPCRs and receptor tyrosine kinases (RTKs) converge on PLCε. PLCε was initially identified as a novel Ras effector and is significantly activated when co-expressed with activated H-Ras in cells [14, 15]. Later it was shown that multiple other Ras-related proteins activate PLCε downstream of GPCRs or RTKs. Activation occurs through two distinct mechanisms that are either dependent on direct interaction with the RA2 domain, or through other not clearly defined regions of the protein. In transfected COS7 cells, Ras, Rap1A, Rap2A, Rap2B, and TC21 (R-Ras) mediated PLCε activation depends on direct binding to the RA2 domain. In contrast, Rho, Ral, and Rac stimulate PLCε activation in an RA independent manner (discussed further in the following text) [75, 76]. PLCε was also found to be activated by Gβγ in an RA domain-independent manner [77]. The binding site for Gβγ is not as clearly defined as for small GTPases and appears to involve multiple regions of the protein [47]. $G\alpha_q$ does not directly activate PLCε. Later work demonstrated the involvement of many of these biochemically characterized regulatory mechanisms as critical components of physiological receptor signaling as described in later sections of this review.

Since PLCε contains a RapGEF (CDC25) domain at its N-terminus, it can be both upstream and downstream of Rap and other small GTPases (Figure 14.3a). The mechanisms for how PLCε-RapGEF activity is regulated is unknown.

Comparatively little is known about the mechanism for activation of PLCε, although much of the core structure is the same as for PLCβ discussed earlier [78]. Thus, it is likely that lessons learned for PLCβ activation by G proteins can be applied to PLCε. However, unlike PLCβ isoforms, biochemical regulation *in vitro* by upstream regulators has been difficult to demonstrate and is of much lower magnitude (four- to fivefold for PLCε compared with up to 600-fold for PLCβ) indicating that PLCε activation may also involve mechanisms that are unique to this isoform. Ras-dependent PLCε translocation to the PM and Rap-dependent translocation to the Golgi apparatus have been shown to activate PLCε in cells [14, 79]. Such translocation-dependent mechanisms may not have been accurately mimicked with *in vitro* assays. However, the activation process is likely more complex than simple translocation. It has been postulated that PLCε is inactive in the cytosol and Ras binding results in translocation to the PM and changes in the conformation of PLCε leading to activation [14, 79]. The autoinhibitory X–Y linker in PLCε may also be linked to the mechanism for activation. As discussed earlier, the RA2 domain is required for Ras-mediated activation, and a

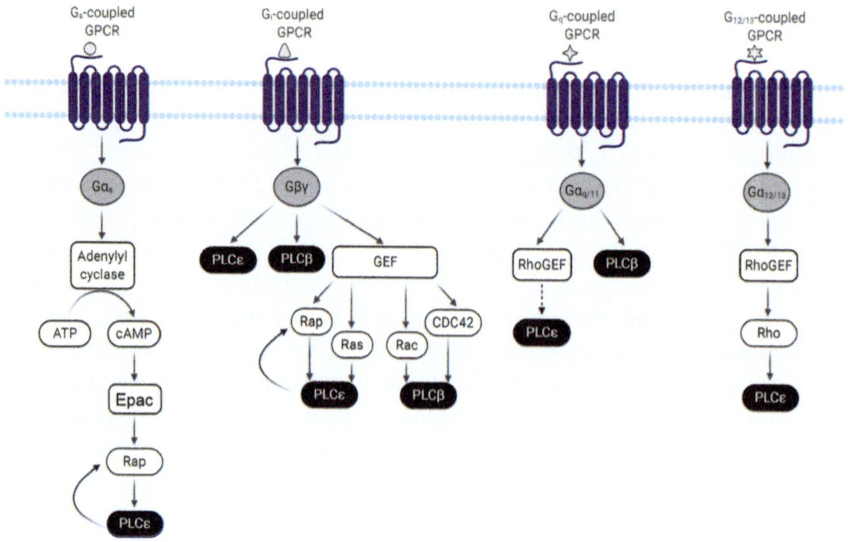

Figure 14.3 Schematic diagram of signaling pathways downstream of four GPCR classes through PLCβ and PLCε. (a) Multiple extracellular stimuli activate PLC isoforms directly or indirectly via G proteins. PLCβ isoforms are well known to be directly downstream of most GPCRs, while PLCε is downstream of all GPCR families via small GTPases or Gβγ. Activation of PLCε downstream of $G\alpha_s$, adenylate cyclase, cyclic AMP and Epac as well as $G\alpha_{12/13}$, RhoGEF, and Rho are recurring signaling pathways reported in many systems. Note: Possible signaling pathway is depicted as a dotted arrow, but has not been demonstrated. (b) Activated PLCs catalyze hydrolysis of PIP_2, at the PM, and/or PI4P present both at the PM as well as internal membranes (Golgi or sarcoplasmic reticulum). Hydrolysis of PIP_2 generates two second messengers, DAG and IP3 whereas PI4P hydrolysis generates local DAG pools. These second messengers can activate a myriad of signaling molecules, including protein kinase C (PKC) and protein kinase D (PKD), which regulate expression of multiple genes involved in (patho)physiological processes. (HDAC): histone deacetylase, (CaMKII): calcium-calmodulin dependent protein kinase II, (RyR2) ryanodine receptor type 2, (NFκB): nuclear factor kappa-light-chain-enhancer of activated B cells. Source: Created with BioRender.com.

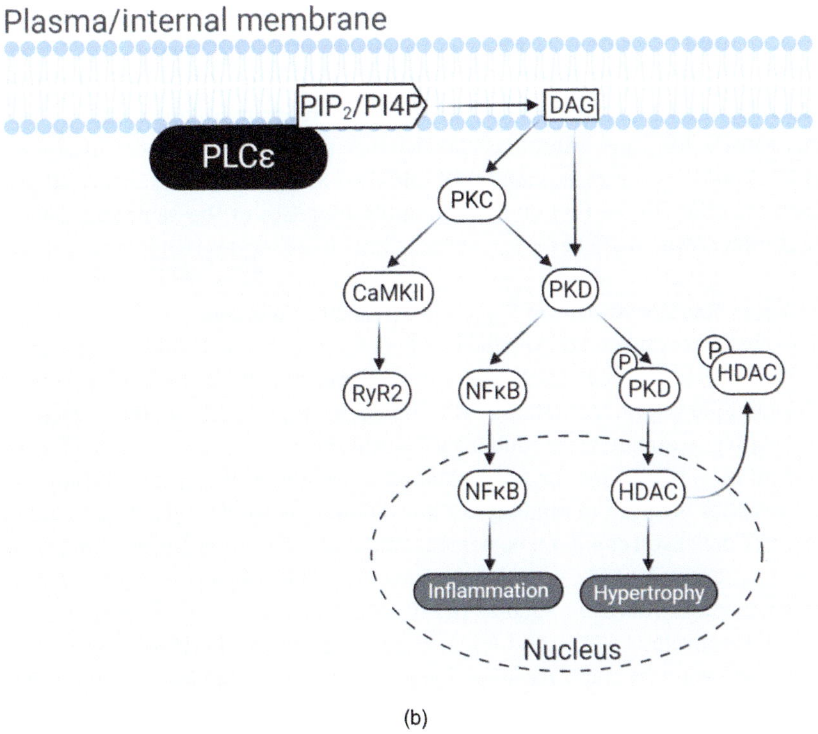

Figure 14.3 (*Continued*)

single point mutation at a conserved lysine residue (K2150E) in the RA2 domain inhibits Ras binding and activation [15, 79]. The structure of the RA2 domain and Ras complex has been obtained by X-ray crystallography, which showed that RA2 makes contacts with both switch I and switch II regions of Ras, a common feature of Ras effectors [79].

14.2.3 Evidence for PLCε Involvement in GPCR Signaling

The evidence for PLCβ regulation downstream of GPCRs is overwhelming and involves direct interactions of heterotrimeric G proteins with PLCβ enzymes. PLCε activation is mediated largely by small GTPases and coupling to GPCRs is indirect in most cases. Evidence for this comes primarily from knockout (KO) and overexpression studies. Some examples of these studies are described later in the context of particular groups of GPCRs.

14.2.3.1 Ras and Gβγ Downstream of G_i-Coupled Receptors

Ras can be activated by GPCRs coupled to G_i. Gβγ subunits involved in effector regulation are primarily derived from activation of G_i protein heterotrimers.

Thus, there are mechanistic routes for activation of PLCε signaling downstream of G_i-coupled receptors. In astrocytes isolated from PLCε$^{+/+}$ and PLCε$^{-/-}$ mice, lysophosphatidic acid (LPA), sphingosine-1-phosphate (S1P), and thrombin dependent inositol phosphate (IP) production were strongly dependent on PLCε. In contrast, M1 muscarinic receptor-dependent IP responses were insensitive to PLCε deletion. In these cases PTX inhibited S1P and LPA-dependent IP production, while Rho inhibition did not, implicating Gβγ or Ras signaling in PLCε activation downstream of these GPCRs, at least in astrocytes [80] (Figure 14.3a).

14.2.3.2 Rho Downstream of $G_{12/13}$ and G_q-Coupled Receptors

Activated RhoA when co-expressed with PLCε in COS7 cells leads to activation of PLCε [81]. Purified RhoA also activated purified PLCε reconstituted with phospholipid vesicles containing PIP_2, confirming that PLCε is a direct effector of RhoA [82]. A unique 65 residue insert within the Y box of the catalytic domain (Figure 14.1) is required for Rho mediated activation of PLCε [81, 82]. However, the residues involved in binding to Rho were not identified since direct binding to this Y box insert could not be demonstrated [81]. This same insert is important for $G_{12/13}$-dependent PLCε activation. Thus, $G_{12/13}$ likely activates PLCε via RhoA downstream of RhoGEFs such as p115RhoGEF or LARG [13]. Thrombin, LPA, and S1P activate G_q, $G_{i/o}$, and $G_{12/13}$-coupled receptors; therefore, downstream PLCε activation by these receptors could potentially be mediated through Ras, Rho, or Gβγ [75, 76, 80, 83]. In the example of astrocytes discussed in the previous paragraph, thrombin stimulated IP production was not inhibited by PTX but instead was inhibited by C3 toxin implicating $G_{12/13}$-dependent regulation by thrombin in astrocytes. In COS7 cells, LPA and thrombin-dependent IP responses were strongly enhanced by overexpression of PLCε and was shown to be Rho dependent [75]. In another example, in HEK293 cells, activation of the $G\alpha_q$-coupled M3 muscarinic acetylcholine receptor activates PLCε via Ca^{2+}-dependent regulation of adenylate cyclase leading to exchange protein activated by cAMP (Epac) and Rap dependent PLCε activation as discussed in the next section [84] (Fig. 14.3a).

14.2.3.3 Rap Downstream of G_s-Coupled Receptors

As discussed earlier, PLCε can be activated by Rap (Rap1A, Rap2A, and Rap2B). Multiple reports indicate that G_s-coupled receptors activate PLCε via cAMP-dependent activation of Epac, an exchange factor for Rap, yielding activated Rap which then binds to and directly activates PLCε [16, 84–86]. This seems to be a major mechanism for cAMP-Epac regulation of cellular functions in many cell types (Figure 14.3b) [87–89].

As described earlier, different GPCRs appear to use distinct signaling pathways to signal to small G proteins to generate downstream signals through PLCε. These well-documented pathways are critical for understanding not only the role of PLCε in the distinct pathophysiological processes but also the contribution of PLCε to overall PLC activity downstream of GPCRs [90, 91] (Figure 14.3a).

14.2.4 Temporal Control of PLC Signaling

Activation of PLC isoforms displays agonist-specific temporal profiles. One study showed that PLCβ3 is predominantly involved in acute (1–3 minutes) and PLCε in sustained (10–60 minutes) IP accumulation [83]. In astrocytes, for example, $G\alpha_{12/13}$-coupled receptor-dependent activation of PLCε mediates sustained signaling through the generation of DAG whereas, G_q-coupled receptors initiate more transient responses [92]. PLCβ is a strong activator of $G\alpha_q$'s GTPase activity (GAP), which could limit the duration PLC activation, especially downstream of GPCRs that desensitize rapidly. PLCε is not a GAP for its activators and in fact, the GEF activity of the CDC25-homology domain is possibly involved in the extended time course of PLCε signaling [48, 80]. Upon activation of PLC, the GEF domain also becomes activated by an unknown mechanism resulting in activation of Rap. Two general pieces of evidence support this notion. In the heart, deletion of PLCε dramatically reduces Epac-dependent Rap activation, suggesting that PLCε amplifies Rap activation downstream of cAMP and Epac (Figure 14.3b) [93]. Another study showed that CDC25 is required for thrombin-dependent sustained ERK1/2 activation and thymidine incorporation in astrocytes [80]. Thus, a clear mechanism exists for PLCε participation in long term PLC signaling downstream of GPCRs.

14.2.5 Tissue and Cellular Expression of PLC Isoforms

14.2.5.1 Phospholipase Cβ

PLCβ isoforms have significantly different tissue distributions, along with different physiological functions. PLCβ1 is expressed in the cerebral cortex and hippocampus [94], as well as in the heart [95, 96]. On the other hand, PLCβ2 is enriched in cells from hematopoietic origin and platelets, where they regulate chemotaxis and other innate immune cell functions [62, 97]. PLCβ3 is more ubiquitously expressed and assumes different roles depending on the tissue [62]. Finally, PLCβ4 is expressed in the Purkinje cells of the cerebellum and in the retina for visual processing events after phototransduction [98]. PLCβ4 is even more highly expressed than PLCβ1 and β3 in rat, mouse, and human hearts, and less than other PLCβ isoforms in platelets [99–101]. However, its significance in cardiac and platelet signaling remains to be determined.

PLCβ1, 2, and 4 isoforms are also expressed as splice variants. These variants differ in their C- terminal domain length and sequences, which impact $G\alpha_q$ binding, interactions with scaffolding proteins, and cellular localization, and

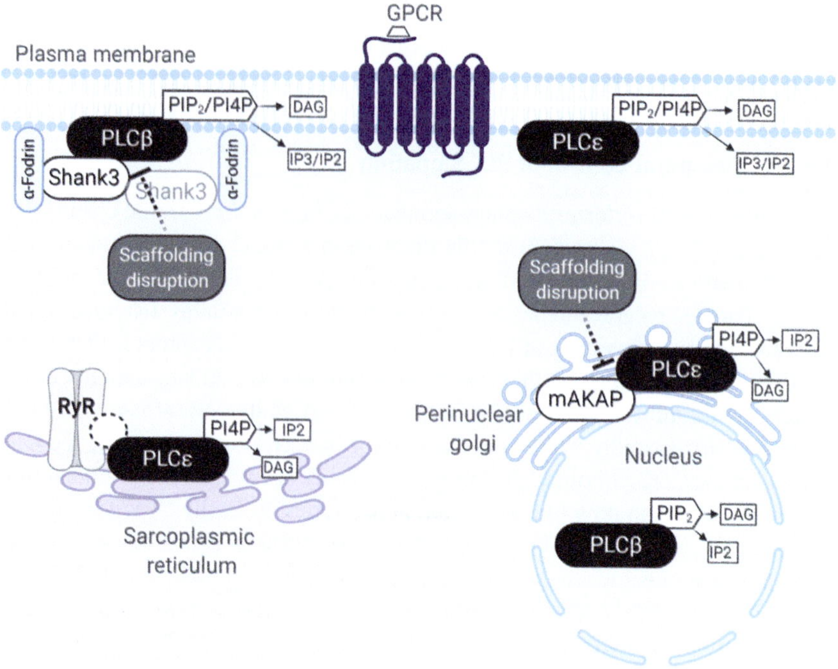

Figure 14.4 Subcellular localization of PLCβ and PLCε. Subcellular compartmentalization of PLCs leads to the localized production of the second messengers to facilitate modulation of specific cellular functions. PIP_2 is present both at the PM and in the nuclear matrix but is either not present, or is of very low abundance on internal membranes. PI4P is present on the perinuclear Golgi in myocytes and internal membranes such as sarcoplasmic reticulum (SR). PIP_2 and PI4P are depicted as substrates of PLC at the respective subcellular membranes where they are present. Source: Created with BioRender.com.

as a result, influence their cellular roles and functions (discussed in detail in Section 14.2.6.1). PLCβ1 has two main splice variants. PLCβ1a has a Post synaptic density protein PSD95/Drosophila disc large tumor suppressor (Dlg1)/and Zonula occludens-1 protein (zo-1) (PDZ) binding motif while the PLCβ1b variant has a proline-rich sequence at its C-terminus [102, 103]. Localization of PLCβ1a varies depending on cell type, while PLCβ1b mostly localizes to the nucleus (Figure 14.4) [103–105]. Detailed analysis of human PLCβ1 has identified more PLCβ1 variants, among which is a variant that lacks part of the PH domain [106]. PLCβ2a and PLCβ2b, which differ by deletion of 15 amino acids 864–878 in the CTD, were identified in hematopoietic cells and platelets [107, 108]. Their activities were similar when co-transfected with active $G\alpha_q$, while the PLCβ2a variant accumulates more in the nucleus than PLCβ2b [108]. Three splice variants of PLCβ4 have been identified [109, 110]. In the PLCβ4b variant, 162 amino acids at the C terminus of the full-length PLCβ4a variant is replaced by a unique 10 amino acid sequence, which makes PLCβ4b unable to localize to the membrane and insensitive to activation by $G\alpha_q$ [110]. The PLCβ4c variant, enriched in the cerebellum and eye, has 41 unique amino acid residues at its C terminus replacing the last 22 residues in PLCβ4a containing a triple PDZ domain interacting motif in PLCβ4a [109]. Variants of PLCβs have been reviewed in detail previously [97]. Studies of VFP-fused PLCβ isoform variants in HeLa cells have shown a PM localization of PLCβ1a, PLCβ1b, and PLCβ4a and a cytosolic localization of PLCβ2 and PLCβ3 [105].

At the membrane, PLCβs rely on their canonical lipase function to mediate signals from GPCRs to catalyze PIP_2 or PI4P hydrolysis to IP_3 or IP_2 respectively, and DAG [23]. PLCβ isoforms (mostly PLCβ1a) have also been shown to function in the cytosol independently of GPCR stimulation by binding with α- and γ-synuclein, component 3 of RISC, cyclin-dependent kinase 16 and stress granule proteins through the CTD, regulating siRNA processing or cell differentiation (reviewed by Scarlata [111]). PLCβ isoforms associated with the nuclear matrix such as PLCβ1b, primarily regulate nuclear inositide-dependent signaling involved the cell cycle and cell differentiation [112, 113].

14.2.5.2 Phospholipase Cε

PLCε mRNA distribution has been assessed by northern blotting [13, 15] and quantitative real-time PCR [74], revealing a relatively ubiquitous expression pattern. PLCε mRNA is abundant in the liver, lung, and heart compared with other tissues. Two splice variants of PLCε have been identified, a short PLCε1a (1994 amino acids) and a long PLCε1b (2303 amino acids) that differ at their N-termini [74]. These splice variants are differentially distributed in various tissues; PLCε1a mRNA is found in all major organs including the heart, brain, skeletal muscle, spleen, lungs, kidneys, thymus, intestine, pancreas, testis, and uterus except in peripheral leukocytes. PLCε1b is highly expressed in the placenta, lung, and spleen but absent in the peripheral blood leukocytes, skeletal muscle, thymus, and testis [74]. Moreover, PLCε1a is expressed in both HEK293T and HEK293A cells, whereas only HEK293A cells express PLCε1b. Neither variant is detected in Jurkat, HeLa, or MCF7 cell lines [74]. Regulatory and functional differences between the variants have not been identified [74]. PLCε orthologs have been identified in rat (PLCε1c, 2281 amino acids.) [97] as well as zebrafish (2248 amino acids) [114].

PLCε is often present in relatively low abundance in some tissues compared with other PLC isoforms when assessed by western blotting. A case in point is in cardiac myocytes, where immunoprecipitation and western blotting is required for detection [74, 115]. Nevertheless, PLCε plays multiple critical roles in cardiac myocyte physiology, in part, through its scaffolding to subcellular compartments as discussed later. Thus, abundance alone does not necessarily predict the importance of a given PLC isoform in a particular tissue.

While phospholipase C isoforms are widely expressed, their cell and tissue-specific expression profiles determine the profile for GPCR dependent PLC responses. For example, PLCβ2 primarily mediates $G\alpha_i/G\beta\gamma$ dependent responses downstream of G_i-coupled receptors and is found primarily in cells of hematopoietic lineage where G_i-coupled chemokine receptors play prominent roles. PLCβ1 and PLCβ4 are primarily $G\alpha_q$ responsive and more widely distributed. These two enzymes have distinct biochemical properties, but how they tailor cell-specific responses have not been clarified. Finally, PLCβ3 is uniquely activated in a synergistically by $G\alpha_q$ and $G\beta\gamma$ positioning it to respond as a coincidence detector for simultaneous stimulus by G_q and Gi-coupled receptors. PLCβ3 is widely expressed and may act to provide coincidence detection in many cell types.

14.2.6 Scaffolded PLC Signaling Complexes

14.2.6.1 Phospholipase Cβ

Differences in the CTD domain of PLCβ isoforms dictate their specific interaction with scaffolding proteins. PLCβ1a has a PDZ binding sequence that can bind the first PDZ domain of Na^+–H^+ exchange regulatory factor 1 (NHERF1) [116, 117], while PLCβ1b directly binds to Shank3 through a proline-rich motif at its C terminus, specifically targeting it to the sarcolemma in cardiac myocytes [118, 119] (Figure 14.4). NHERF1, which contains two PDZ domains, mediates the formation of a ternary complex with PLCβ1 and the parathyroid hormone receptor PTH1R, in an *in vitro* system using purified proteins [120]. In neutrophils, CXCR2 and PLCβ2 preferably interact with the PDZ2 domain of NHERF1 to form a signaling complex in which PLCβ2 specifically and efficiently mediates signaling by CXCR2 [121, 122]. In pancreatic cancer cells, PLCβ3 forms a complex with CXCR2 and the PDZ1 or PDZ2 domain of NHERF1 [123]. PLCβ3 binds to the second PDZ domain of E3KARP in the brush border membranes of the intestine, colon, and kidney, and to Shank2 to mediate metabotropic Glutamate receptor (m-GluR) signaling in the postsynaptic density region of neuronal synapses [116, 117, 124]. PLCβ3 forms a ternary complex with somatostatin receptor (SSTR) and PDZ domain-containing protein 1 (PDZK1) to specifically mediate SST responses in HEK293 cells [125]. PLCβ3 has been reported to bind the second PDZ domain of NHERF2 to potentiate muscarinic receptor signaling [126]. Interestingly, PLCβ3 can also directly bind to the M3 muscarinic receptor to mediate its signaling without any scaffolding protein [127]. NHERF2 and membrane-associated guanylate kinase, WW and PDZ domain-containing protein (MAGI3) can competitively bind to PLCβ3 to enhance, or inhibit, LPA-mediated signaling through $G\alpha_q$ or $G\alpha_{12}$ signaling in colon cancer cells [128].

PLCβ4 is highly homologous to the NorpA PLC protein that mediates the phototransduction cascade in Drosophila [109, 129]. Here NorpA binds to a multi-PDZ domain-containing protein inactivation no afterpotential D (INAD), which also binds to G_q, PKC, and the downstream effector TrpC. In a recent study reporting the crystal structure of NorpA bound to the INAD PDZ45 domain, it was found that PLCβ4 also binds to a mammalian INAD-like protein (INAD PDZ89) suggesting a similar role for this complex in mammalian phototransduction [130]. PLCβ4 also forms a signaling complex with the metabotropic glutamate receptor 1α (mGluR1α), the inositol 1,4,5 trisphosphate receptor 1 (IP3R1), and Homer 1 around the perisynapse and smooth endoplasmic reticulum of Purkinje cells and thalamocortical neurons [131, 132]. Overall, the specific interaction between PLC isoforms with PDZ-containing proteins, signaling complex formation between PDZ binding proteins and certain receptors, and the differential expression of PDZ-containing proteins in multiple tissues, allow tissue, subcellular-specific, and receptor-specific PLCβ signaling. Interrogating these interactions may be a viable strategy for developing therapies for different diseases, which will be discussed later in this chapter.

14.2.6.2 Phospholipase Cε

PLCε localizes to multiple subcellular compartments, including the Golgi apparatus, nuclear membrane, cytoplasm, and PM, depending on its activation status, activators, and scaffolding interactions (Figure 14.4). In COS7 cells following Ras stimulation, it localizes to the PM, while Rap1 mediated activation brings it to the perinuclear region [14]. Our laboratory has shown that there are multiple pools of PLCε, localized to different compartments in cardiomyocytes. Apart from its localization at the PM, localization to the perinuclear region, and sarcoplasmic reticulum (SR) are important for regulating cardiac hypertrophy and contractility, respectively. PLCε interacts with muscle-specific A-kinase anchoring protein β (mAKAP) which also scaffolds Epac, PKA, and adenylate cyclase as a part of a multiprotein complex at the nuclear envelope in myocytes [133, 134]. The interaction between PLCε-mAKAP is mediated by the RA1 domain, and to a lesser extent the RA2 domain of PLCε, binding to the Spectrin Repeat-1 (SR1) domain of mAKAP. Another pool of PLCε is localized at the SR, bound to the type 2 ryanodine receptor (RyR2) [22, 85, 93]. The interaction between PLCε and RyR2 is less well-defined molecularly and seems to be indirect.

14.3 Diseases and Phenotypes Associated with PLCβ and PLCε

When considering whether PLC inhibition might be a therapeutic target it is instructive to understand the physiological functions and potential PLC pathophysiology. The examples in the following text are from KO animals, cell models, or human disease association studies. We focus on roles that have relatively clear possible connections to GPCR-dependent regulation.

14.3.1 Phospholipase Cβ1

14.3.1.1 Central Nervous System

In the central nervous system, PLCβ1 may contribute to genetic predisposition for schizophrenia. A fraction of schizophrenia patients has mono- or bi-allelic deletion of PLCB1 gene [135, 136]. Studies in mice showed that PLCβ1 signaling in the medial prefrontal cortex was necessary for working memory [137] and global PLCβ1 KO mice had schizophrenia-like behavioral endophenotypes [138, 139]. It was further identified that the reduction of the PLCβ1a variant is involved in schizophrenia while the PLCβ1b expression level was reduced in suicide completers [140].

Loss-of-function mutations in PLCB1 are associated with infantile epileptic encephalopathy [141, 142]. In agreement with human disease characteristics, PLCβ1 deficient mice developed epilepsy [143] and aberrant mossy fiber projections [94]. This loss of PLCβ1 in the hippocampus might interfere with the function of inhibitory neuronal circuitry in the hippocampus by multiple $G\alpha_q$-coupled receptor neurotransmitters [143].

14.3.1.2 Cardiac

Higher expression of PLCβ1b, but not PLCβ1a or other isoforms, in left atrial tissue was found in mitral valve disease patients [144]. While in other cell types the PLCβ1b variant mostly localizes to the nucleus (Figure 14.4), it specifically targets the sarcolemma in cardiomyocytes via its direct binding with Shank3 through the C-terminal proline-rich motif on PLCβ1b [119] (Figure 14.4).This targeting is required for $G\alpha_q$-mediated hypertrophy and apoptosis in neonatal rat cardiomyocytes (NRVMs) [145]. PLCβ1a, PLCβ3, PLCδ1, or PLCγ1 are also expressed in cardiomyocytes but do not mediate this phenotype. Overexpression of PLCβ1b in mice also induced cardiac contractile dysfunction via activation of PKCα [146]. There are various hormones that signal through $G\alpha_q$ in the heart to promote hypertrophy including norepinephrine (α1AR), endothelin (ET), and Angiotensin II (AT2R). Therapeutic strategies to block these receptors or $G\alpha_q$ have their limits [147], mostly due to the selectivity of antagonists targeting these receptors and a wide expression and opposing effects of $G\alpha_q$ at the sarcolemma versus at the nuclear envelope [148, 149]. The unique localization PLCβ1b at the sarcolemma via binding with Shank3, and its specificity over other PLC isozymes in mediating cardiac hypertrophy render it a potential unique target for the treatment of heart failure as will be discussed later (Figure 14.4).

There is also evidence that PLCβ1, along with $G\alpha_q$ and PIP_2, is associated with the nuclear envelope in cardiac myocytes and responds to α1-adrenergic receptors (α1ARs) at the nuclear envelope [149]. In this study, activation of α1ARs induced histone deacetylase 5 (HDAC5) export that was inhibited by the IP3 receptor antagonist, 2-Aminoethoxydiphenylborane (2-APB), without a significant activation of PI hydrolysis at the sarcolemma. Activation of α1-ARs also induced a robust transcriptional response associated with cardioprotective hypertrophic, survival, inotropic, and antifibrotic gene programs that could not be attributed to PM PIP_2 hydrolysis [149]. Conventionally, antagonists of the αAR have been used to manage maladaptive response in heart failure; however, α1AR agonists that target nuclear membrane localized α1ARs could be beneficial to treat cardiac injury [150].

14.3.1.3 Cancer

PLCβ1 can regulate cell cycle and differentiation [151] and therefore could be involved in cancer development. An intronic insertion/deletion mutation has been found predominantly in patients with euthyroid multinodular goiter. This mutation results in a higher expression of both PLCβ1a and PLCβ1b splice variants [152]. PLCβ1 amplification, which leads to higher PLCβ1 protein expression, also correlates with high proliferative activity of breast cancer [153]. PLCβ1 is highly expressed in human glioma tissue. It mediates growth, invasion, and migration of glioma cell lines, in part, through the ERK pathway, and can be suppressed by the expression of microRNA miR-423-5p [154].

14.3.2 Phospholipase Cβ2

14.3.2.1 Platelets

PLCβ2 expression is largely restricted to cells of hematopoietic origin. Platelets isolated from patients with a mild bleeding disorder had decreased expression of PLCβ2 and increased expression of PLCβ4. These platelets also showed impaired aggregation and secretion upon stimulation by multiple agonists including both G_q and G_i agonists [99, 155]. This presents a non-interchangeable function of PLC isoforms (PLCβ2 and PLCβ4), which makes sense given their different mechanisms for activation and regulation. Knockout mice lacking PLCβ2/β3 could not form stable thrombi upon chemical-induced carotid injury. Platelets isolated from these mice failed to raise cytosolic calcium levels following agonist stimulation [101].

14.3.2.2 Chemokine Directed Cell Migration

In PLCβ2 KO mice, IP_3 production and superoxide production in response to fMLP were strongly suppressed. Double KO of PLCβ2 and PLCβ3 completely blocked fMLP-dependent IP_3 production and superoxide release but chemotaxis in response to fMLP was unaffected or surprisingly increased in response to CCL3 [156]. This contrasts with a later study showing that CXCL12α-dependent T-cell migration is suppressed but fMLP-dependent changes are unaffected [157]. It surprising and interesting that certain cell types have distinct signaling pathways downstream of Gi-coupled receptors that are differentially required for cell migration.

14.3.3 Phospholipase Cβ3

14.3.3.1 Opioid Analgesia

PLCβ3 mediates nociception in dorsal root ganglion (DRG) neurons. Pharmacological inhibition of PLC activity by the nonspecific PLC inhibitor U73122 produces antinociception in mice [158]. Chronic μ-opioid receptor (MOR) activation and very-low-dose morphine have been shown to promote hyperalgesia. MOR, in turn, can activate PLC through the release of Gβγ from $G\alpha_{i/o}$ [159]. Of the four PLCβ isoforms, PLCβ1, PLCβ3, and PLCβ4 are found in the CNS but only PLCβ3 is responsive to Gβγ and thus would be the only candidate PLCβ isoform downstream of opioid receptors [71]. Genetic deletion of PLCβ3 in mice potentiates opioid analgesia [160]. Hyperalgesia upon chronic MOR stimulation was suggested to involve activation of PLCβ3 and subsequent downstream PKC activation [161].

One study has shown that PLC can be activated by MOR [162], while another has reported MOR does not stimulate DAG generation in HEK293 cells [163]. The differences in assay conditions and cell types aside, one explanation of these contradictory findings is that a low level of basal $G\alpha_q$ signaling is required to observe significant $G\alpha_i$/Gβγ dependent stimulation of PLCβ3 which is the predominant isoform in many cell types. Recent studies from the Martemyanov group showed that GPR139 – an orphan GPCR, counteracts the MOR effect on firing of medial habenular neurons, primarily by signaling through $G\alpha_q$. This study suggests a role for PLC, possibly PLCβ3, downstream of $G\alpha_q$ in inhibiting opioid analgesia [164].

14.3.3.2 Atherosclerosis

PLCβ3 deficiency has been shown to be associated with reduced murine macrophage apoptosis, in part, by regulating Bcl-XL expression via PKC, without affecting macrophage functions such as migration, adhesion, or phagocytosis. The significance of PLCβ3/Bcl-XL signaling was confirmed in an ApoE-deficient atherosclerosis mouse model, where PLCβ3/ApoE KO animals showed decreased atherogenesis, increased macrophage apoptosis in the atheromas, and lower Bcl-XL expression in the lesions [165].

14.3.3.3 Spondylometaphyseal Dysplasia

In Emirati patients with spondylometaphyseal dysplasia (a form of bone dysplasia) and corneal dystrophy, a homozygous mutation in the Hα2′ helix of PLCβ3 has been identified [166]. The Hα2′ helix is an autoinhibitory element of PLC that is involved in regulation of PLCβ3 by $G\alpha_q$, and controls PLC basal activity. The mutation results in significantly reduced protein expression compared with the wild-type PLCβ3, thus it was characterized as LOF. Phenotypically, the authors interpreted the LOF mutation to cause an accumulation of PIP_2 and dysregulation of the actin cytoskeleton and aberrant fibroblast morphology.

14.3.4 Phospholipase Cβ4

14.3.4.1 Craniofacial Malformation

A coding mutation in PLCβ4, the R621H mutation, has been identified as one of the genetic determinants of auriculocondylar syndrome 2- a rare craniofacial malformation of first and second pharyngeal arches [167]. This mutated arginine, in the Y box of the TIM barrel, is conserved among all types of PLCβ. In the structure of full-length PLCβ3 in complex with $G\alpha_q$, the corresponding R646 residue (R621 in PLCβ4) lies in a ß-sheet of the core enzyme [57] and is positioned very close to the catalytic E362 residue [62] (E358 in PLCβ4). The R621 may also contact with Ca^{2+} or substrate in the catalytic site [167]. The R621H mutation likely destabilizes the catalytic site. It was unclear from this report whether the subjects showed any other abnormal traits.

14.3.4.2 Neuronal Phenotypes

PLCβ4 KO mice developed locomotor ataxia [143], impaired long term depression in the rostral cerebellum [168], and impaired visual processing [98]. PLCβ4-deficient mice developed irregular sleep sequences and ultradian body temperature rhythms due to abnormality in the thalamic mGluR1-PLCβ4 pathway [132, 169].

14.3.4.3 Thrombosis

PLCβ4 is expressed in mouse platelets and at a very low level in human platelets [99, 101]. A *de novo* mutation in PLCβ4 (R335Q) has been identified as a novel thrombosuppressor [170] suggesting that PLCβ4 may be in the thrombin-protease activated receptor (PAR) stimulated aggregation cascade.

14.3.4.4 Uveal Melanoma

A recurrent mutation at Asp 630 in the TIM barrel of PLCβ4 has been identified in human uveal melanoma (UM) tumor samples [171]. The genetic basis of uveal melanoma is relatively well-characterized [172–174]. Mutations in genes encoding proteins in the same pathway, the $G\alpha_q$-coupled receptor $CysLT_2R$, $G\alpha_{q/11}$, and PLCβ4 have been identified in uveal melanoma samples. The activating mutation in $CysLT_2R$ has been shown to bias toward $G\alpha_q$ signaling [175]. Constitutively active mutations in $G\alpha_q$ and $G\alpha_{11}$ transduce mitogenic signals through a Trio/Rho pathway that was reported to be independent of PLCβ [176–178]. $CysLT_2R$, $G\alpha_{q/11}$ and PLCβ4 are in the same pathway, and mutations of these proteins are expressed in a mutually exclusive manner in uveal melanoma samples [179]. This suggests that PLCβ4 downstream of $G\alpha_q$ could contribute to UM development. A recent study from our laboratory demonstrated that Asp 630 to Tyr (D630Y) mutation results in strong activation of PLCβ4 as well as other PLC isoforms (see discussion in Section 14.2.1 of this chapter). Expression of PLCβ4 D630Y in a cutaneous melanoma cell line promoted cell proliferation *in vitro* and tumor formation in mice. Additionally, inhibition of PLC reduced proliferation of uveal melanoma cell lines with constitutively activate $G\alpha_q$ as a driver but not in cells with a B-Raf mutation as a driver. These data strongly support the idea that PLC activation downstream of $G\alpha_q$ is a contributing pathway to UM development and that constitutive activation of PLC is sufficient to drive to tumor formation. Another recent study demonstrated that PLCβ4(D630Y) leads to activation of ERK signaling in UM cell lines and that this pathway is important for UM cell proliferation. A better understanding of the pathways driven by these mutations will offer a more thorough personalized medicine strategy for better disease management [180–182].

14.3.5 Phospholipase Cε

Multiple studies using PLCε knockdown and KO in cell and in animals, respectively, have provided insight into the role of PLCε in physiology and disease. In addition, GWAS and other genetic studies have implicated PLCε in human disease.

14.3.5.1 Cardiovascular Functions

$G_{12/13}$ and G_s-Coupled Receptors and PLCε in Cardiac Hypertrophy PLCε KO mice (PLCε$^{-/-}$) do not have a clearly observable baseline phenotype [115]. However, when subjected to chronic adrenergic stimulation, PLCε$^{-/-}$ mice developed more severe hypertrophy than PLCε$^{+/+}$ mice. Later studies with isolated cardiac myocytes directly contrasted with this observation, where siRNA mediated knockdown of PLCε in neonatal rat ventricular myocytes (NRVMs) inhibited hypertrophic growth in response to stimulation of various GPCRs and RTKs [134]. To address this conundrum, PLCε$^{flox/flox}$ mice were crossed with αMHC-Mer-Cre-Mer mice to KO PLCε specifically and temporally in cardiac myocytes. These mice were protected from pressure-overload-induced hypertrophy indicating that PLCε mediates hypertrophy [22]. The incongruity in the role of PLCε observed between global and tissue-specific PLCε KO can potentially be attributed to the upregulation of compensatory signaling molecules, such as CAMKII, in PLCε global KO mouse during development [22, 134].

In this study, we found that PLCε localized to the nuclear envelope via scaffolding with muscle-specific AKAP (mAKAP) is required for its hypertrophic function [134]. A potential problem with this model was that little or no PIP_2 was detectable surrounding the nucleus. On the other hand, PI4P is highly enriched in the Golgi apparatus surrounding the nucleus and PLCε can utilize this alternative substrate, PI4P, at the Golgi apparatus to generate DAG locally [22]. This pool of DAG is required for nuclear PKD activation and subsequent induction of hypertrophic genes. Thus, this study identified PLCε as a novel target for treatment of the heart failure [22]. Upstream of PLCε, multiple mechanisms for activation were found. District pools of cyclic nucleotides downstream of the α-adrenergic receptor (αAR) were found to differently regulate PLCε through Epac and Rap-dependent activation, or PKA-dependent inhibition [183]. Interestingly, cell surface α1AR activation was unable to activate PLCε at the Golgi. On the other hand, β1-adrenergic receptors (β1ARs) residing in the Golgi apparatus stimulate PLCε through a local pool of cAMP that has privileged access to the Epac/mAKAP/PLCε complex and is required for catecholamine dependent cardiomyocyte hypertrophy [184].

β1AR-PLCε and Cardiac Contractility We also investigated the mechanism for decreased cardiac contractile function in response to adrenergic stimulation in the global PLCε$^{-/-}$ mouse. A significant decrease in adrenergic stimulation of Ca^{+2}-induced Ca^{+2} release (CICR) in isolated myocytes, and contraction in intact heart, was observed in PLCε$^{-/-}$ mice [85]. However, no difference in contractile function between PLCε$^{-/-}$ and wild type (WT) mice was seen at baseline. A mechanistic pathway was identified where the β-adrenergic receptor regulates CICR via a cAMP/Epac/Rap/PLCε/PKCε/CamKII signaling axis that results in phosphorylation of RyR2 to regulate CICR [85, 93]. PLCε binds to the RYR2 complex at the sarcoplasmic reticulum, and this likely directs local PLCε activation to direct activity of the PLC signaling pathway to RyR2. In PLCε KO mouse, this signaling circuit is interrupted leading to a decrease in cardiac contractility in response to AR stimulation [91, 93].

A common theme emerging from these studies is that in the heart cAMP-dependent Epac activation is a major mechanism for engaging PLCε and that the subcellular location where this engagement occurs in the myocyte determines the physiological output of the pathway.

$G_{12/13}$-Coupled Receptors and PLCε in Cardio Protection PLCε has been shown to play a role in cardio-protection during ischemia/reperfusion (I/R) injury [185]. Deletion of PLCε prevents S1P-dependent protection in a mouse I/R injury model. Circulating S1P activates PLCε via RhoA. PLCε is responsible for PKD activation which in turn which prevents translocation of cofilin-2 and Bax to the mitochondria resulting in protection of the heart from oxidation-induced cell-death [185]. In contrast, inflammation is a major factor regulating myocardial I/R injury, and one report suggests that PLCε increases expression of pro-inflammatory cytokines that promote I/R injury [186]. It is possible that both of these mechanisms are operating and oppose each other. Depending on the physiological model system one process may dominate the overall phenotype.

14.3.5.2 Insulin Secretion

GLP1 Receptors and PLCε in Pancreatic β Cells In pancreatic β-cells, glucagon-like peptide-1 (GLP-1), binds to GLP-1 receptor (GLP-1R) and the signaling pathway downstream of the receptor ultimately potentiates the release of insulin from islet β cells in response to glucose. Activation of GLP-1R, a G_s-coupled GPCR leads to increases cAMP and activation of both PKA and Epac, which independently regulate CICR. Additionally, Ca^{2+}-dependent exocytosis of insulin can be stimulated by Ca^{2+} release from intracellular Ca^{2+} stores. Exendin-4, a GLP-1 agonist, failed to facilitate CICR in PLCε KO mice and recombinant PLC-ε expression rescued CICR [87]. In addition, KO of PLC-ε gene expression prevented Epac activator (8-pCPT-2′-O-Me-cAMP-AM)-dependent potentiation of glucose-stimulated insulin secretion (GSIS) from mouse islets [187]. Therefore, increasing the function of PLCε can be a potential approach for the treatment of Type 2 diabetes leading to increased insulin secretion.

14.3.5.3 Inflammation

Multiple lines of evidence support a role for PLCε in inflammatory responses, some examples of which are discussed here, and some common themes emerge.

Skin In PLCε$^{-/-}$ mice, proinflammatory cytokine and the local inflammatory reactions in response to hapten sensitization were suppressed, proinflammatory cytokine production was reduced and allergic sensitivity was decreased [188]. Overexpression of PLCε resulted in spontaneous skin inflammation and increased cytokine production and T-cell infiltration [189]. The proinflammatory role of PLCε was observed in the 12-*O*-tetradecanoylphorbol-13-acetate (TPA)-induced skin inflammation model as well [190]. A subsequent study in cultured human keratinocytes indicated that the PLCε downstream signaling pathway may cooperate with downstream NF-κB to augment tumor necrosis factor (TNFα)-induced chemokine (C—C motif) ligand 2 (CCL2) expression [191]. Another study of PLCε$^{-/-}$ mice and cultured epidermal keratinocytes and dermal fibroblasts from these mice showed that PLCε is crucial for ultraviolet B (UVB) light-induced neutrophil infiltration in the skin by regulating the expression of CXCL1/keratinocyte-derived chemokine (KC). The PLC inhibitor, U73122, significantly attenuated the UVB-induced upregulation of CXCL1/KC in cultured cells [192]. PLCε deficient mice had reduced cutaneous wound healing and reduced proinflammatory cytokine production, indicating its role in the early stage of cutaneous wound healing [193].

Astrocytes PLCε is a crucial regulator of neuroinflammation downstream of GPCR activation by Thrombin, LPA, or S1P in primary astrocytes. Activated PLCε mediated sustained activation PKD, NF-κB, and production of cyclooxygenase (COX-2). An inhibitor of the upstream IκB kinase (IKK) inhibited PLCε-mediated inflammatory responses [92].

Lung An autocrine TNFα-LPA pathway involves PLCε in the lung. PLCε is highly expressed in lung tissue based on western blotting [194]. To understand roles for PLCε in lung function a number of studies were done. In an ovalbumin-induced allergic bronchial asthma mouse model, PLCε KO mice had reduced airway hyperresponsiveness and bronchial inflammation. Primary bronchial alveolar cells from these mice had a large reduction in TNFα-dependent cytokine production [195]. A later study in human colon epithelial Caco2 cells suggested a mechanism that could explain PLCε involvement in the TNFα pathway in the lung. It was shown that TNFα induced production of LPA, which in turn activated PKD/NF-κB via PLCε, to both further increase LPA levels, and promote inflammatory cytokine production. This led to the recruitment and activation of immune cells to initiate and sustain inflammation [196]. In two separate studies in bacterial LPS models of acute lung injury, PLCε$^{-/-}$ mice had reduced inflammatory cytokine production and vascular leakage [194, 197]. In PLCε KO mice in an LPS-induced allergic bronchial asthma mouse model, it was shown that loss of PLCε activity substantially alleviated airway hyperresponsiveness, bronchial inflammation, and reduced inflammatory cytokine production [195]. PLCε, downstream of Epac1, but not Epac2 enhanced cigarette smoke-induced inflammatory responses and an increase in neutrophil numbers *in vivo* [198].

A common theme in these studies is that PLCε is involved in inflammatory processes downstream of multiple stimuli either directly, or indirectly through autocrine secretory mechanisms. In all cases, this involves PKD regulation of NFκB (Figure 14.3b). PKD is directly activated by DAG a product of PLC activity. Of the PLC isoforms, PLCε is uniquely positioned to mediate these responses due to its responsiveness to multiple upstream stimuli, and the sustained nature of PLCε-mediated DAG production.

Inflammation and Cancer Phospholipase C E1 (*PLCE1*) gene polymorphisms and PLCε expression alterations are involved in many cancers (see Table 14.1 and Section 14.3.5.4). The inflammatory response is often associated with cancer development and progression. The association of *PLCE1* with various cancers is a result of its function in regulating inflammatory responses through mechanisms related to those described earlier.

Kidney Physiology and Disease The involvement of PLCε in kidney function was revealed in a seminal study identifying loss of function mutations of PLCε

Table 14.1 Summary of physiological roles of PLCε in cancer.

Tumor type	Proposed role in tumorigenesis	*In vitro* and *in vivo* models		PLCε expression and genetic manipulation or treatment phenotype	Proposed signaling pathways	References
Skin cancer	Promoter	$PLC\varepsilon^{\Delta X/\Delta X}$ mice (Catalytically inactive)	Two-stage skin chemical carcinogenesis model (2SCC)	• Delayed onset and reduced incidence of skin squamous tumors	G_i-coupled-GPCR→Ras→PLCε	[199]
	Promoter	$PLC\varepsilon^{\Delta X/\Delta X}$ mice	2SCC	• Resistance to TPA-induced skin inflammation	TPA→PKC and RasGRP3→Rap1A→PLCε→IL-1α and CXCL-2	[190]
	Suppressor	$PLC\varepsilon^{-/-}$ and $PLC\varepsilon^{RAm/Ram}$ mice	2SCC	• Increased susceptibility to tumor formation	PKC	[200]
	Suppressor	$PLC\varepsilon^{-/-}$ mice	Ultraviolet B (UVB)-induced skin carcinogenesis	• Increased neoplasms development including malignant tumors • Resistance to cell death and attenuation of inflammation induced by UVB	—	[201]
Lung cancer	Suppressor	Lung adenocarcinoma tissues samples	—	• Reduced PLCε expression in tissues	—	[202]
	Oncogene	NSCLC tissue samples	—	• Increased PLCε expression in tissues • Reduced p53 expression, inhibition of apoptosis	—	[203]

	Oncogene	NSCLC tissue samples and cell lines	—	• Increased PLCε expression in tissues and cell lines • Increased PTEN promoter methylation, reduced PTEN expression	—	[204]
	Tumor promoter	Lung epithelial cells and A/J mice	4-(Methylnitrosamino)-1-(3- pyridyl)-1-butanone (NNK) mediated carcinogenesis	• Increased colony and foci formation in cell lines • Increased tumor burden in mice	β1AR→Gβγ→PLCε→IP3→Ca^{+2}→SMAD, NF-kB→IGF2→IGFR1→IGF-R1-P (activation)	[205]
Colorectal cancer	Support tumorigenesis	$PLC\varepsilon^{-/-}$ in $Apc^{Min/+}$ mice	Spontaneously develops multiple intestinal neoplasms in $Apc^{Min/+}$	• Augmentation of inflammation and angiogenesis	—	[206]
	Suppressor	Colorectal cell lines	Expression of exogenous PLCε in the cell line	• Reduced PLCε expression • Expression of exogenous PLCε decreased cell viability and proliferation	Epigenetic silencing of PLCε expression through hypermethylation	[74]
	Suppressor	Sporadic colorectal cancer tissues and nude mice tumor xenograft model	PLCε overexpression in cell line	• Reduced PLCε expression in tissues • Small tumor size in mice inoculated with PLCε overexpressing cells	—	[207]

(Continued)

Table 14.1 (Continued)

Tumor type	Proposed role in tumorigenesis	*In vitro* and *in vivo* models		PLCε expression and genetic manipulation or treatment phenotype	Proposed signaling pathways	References
Prostate Cancer (PCa)	Putative oncogene	PCa tissues and cell lines	PLCε knockdown in cell line	• Increased PLCε, AR and Notch1 expression in tissues • PLCε knockdown decreased cell growth and colony formation	Increase in Notch signaling and nuclear translocation of the androgen receptor (AR)	[208]
	Promoter	PCa tissue, cell lines and nude mice tumor xenograft model	PLCε knockdown in cell line	• Increased PLCε and Twist1 expression • Increase mitochondrial oxidative metabolism and migration in PLCε knockdown cell lines • Small tumor volume and weight in mice inoculated with PLCε knockdown cells	PLCε→PPARβ→ Transcription of Twsit1 PLCε→MAPK→ decrease in ubiquitination of Twist1	[209]
	Oncogene	PCa tissues and cell lines	PLCε knockdown in cell line	• Reduced PLCε, increased PTEN expression in tissues • PLCε knockdown inhibited cell proliferation	PTEN/AKT signaling	[210]

Oncogene	PCa cell line	Metformin treatment	• Inhibit proliferation, invasion, migration, and apoptosis	Notch and AR signaling	[211]
Oncogene	PCa cell line	PLCε knockdown in cell line	• PLCε knockdown suppressed the proliferation and invasion • Reduced resistance to AR antagonist, enzalutamide, in castration-resistant prostate cancer (CRPC) cells	Bone morphogenetic protein (BMP)-6/SMAD signaling	[212]
Oncogene	PCa tissues, cell lines and nude mice tumor xenograft model	PLCε knockdown in cell line	• Increased PLCε and Gli-1/Gli-2 expression in tissues • PLCε knockdown suppressed CRPC cell proliferation, invasion. • Increased the sensitivity to enzalutamide both *in vitro* and *in vivo*	AR signaling	[213]
Oncogene	PCa tissues, cell lines and nude mice tumor xenograft model	PLCε knockdown in cell line	• Increased expression of PLCε, wnt3a, and AR in tissues • Expression of exogenous PLCε decreased cell proliferation, tumor growth, and bone metastasis. Increased the sensitivity to enzalutamide both *in vitro* and *in vivo*	wnt3a/β-catenin/AR signaling	[214]

(Continued)

Table 14.1 (Continued)

Tumor type	Proposed role in tumorigenesis	*In vitro* and *in vivo* models		PLCε expression and genetic manipulation or treatment phenotype	Proposed signaling pathways	References
	Oncogene	PCa cell lines and nude mice tumor xenograft model	Inhibitor 9 combined with androgen deprivation therapy or chemotherapy	• Increased KRas, PLCε and PKCε expression in tissues • Decreased invasion, migration, and drug resistance of CRPC cell line and invasion and metastasis *in vivo*	KRas→ PLCε→PKCε	[215]
	Oncogene	PCa tissues, cell lines and nude mice tumor xenograft model	PLCε knockdown in cell line	• Increased PLCε and YAP expression in tissues • I PLCε knockdown inhibited serine/glycine production and proliferation, tumor growth *in vitro* and *in vivo*	Yes- associated protein (YAP) nuclear translocation	[216]
	Oncogene	PCa tissues, cell lines and nude mice tumor xenograft model	PLCε knockdown in cell line	• Increased PLCε expression in tissues • PLCε knockdown reduced cell metastasis, glucose consumption and lactate production *in vitro* and tumor growth *in vivo*	HIF-1α/MEK/ERK	[217]

	Oncogene	PCa tissues, cell lines and mice tumor xenograft model	PLCε knockdown in cell line	• Increased PLCε, AR and DNA-PKcs expression in tissues • PLCε knockdown increased radio sensitivity *in vitro* and decreased CRPC growth *in vivo*	AR/PARP1/DNA-PKCs	[218]
Bladder cancer	Oncogene	Bladder cancer cell line	PLCε knockdown in cell line	• PLCε knockdown decreases cell invasion	—	[219]
	Oncogene	Bladder cancer cell line and mice tumor xenograft model	PLCε knockdown in cell line	• PLCε knockdown inhibited proliferation and caused significant cell cycle arrest • Reduced tumor xenograft growth	—	[220]
	Oncogene	Bladder cancer cell line	PLCε knockdown in cell line	• PLCε knockdown reduced cell proliferation and arrested the cell cycle at G0/G1-phase • Down-regulated oncogenes c-fos and c-jun	PKCα	[221]
	Oncogene	Bladder cancer cell line and mice tumor xenograft model	PLCε knockdown in cell line	• PLCε knockdown decreased cell invasion and migration *in vitro* • Suppressed tumor growth *in vivo*	PKCα/β/TBX3/E-cadherin	[222]

(Continued)

Table 14.1 (Continued)

Tumor type	Proposed role in tumorigenesis	*In vitro* and *in vivo* models		PLCε expression and genetic manipulation or treatment phenotype	Proposed signaling pathways	References
	Oncogene	PLCε$^{-/-}$ mice	*N*-butyl-*N*-(4-hydroxybutyl) nitrosamine-induced bladder tumorigenesis model	• Resistant to BBN-induced bladder carcinogenesis and decrease in inflammatory responses and angiogenesis-associated molecules COX2 and VEGF-A	—	[223]
	Oncogene	Bladder cancer tissues and cell line	PLCε knockdown in cell line	• Increased PLCε and p-STAT-3 expression in tissues • PLCε knockdown decreased inflammatory cytokine production and inflammation-associated genes	STAT3 phosphorylation leading to inflammatory cytokine release	[224]
	Oncogene	Bladder cancer tissues and mice tumor xenograft model	PLCε knockdown in cell line	• Increased PLCε and lactate dehydrogenase (LDHA) expression in tissues • PLCε knockdown inhibited cell proliferation, glucose consumption and lactate production • Inhibition of bladder cancer cell growth *in vivo*	STAT3/LDHA	[225]

Esophageal cancer	Susceptibility loci	Tumor tissues-GWAS study	—	—	—	[226–228]
	Oncogene	Esophageal Squamous Cell Carcinoma (ESCC) cell lines and mice tumor xenograft model	PLCε knock out in cell line	• PLCε knockdown decreased invasion, proliferation capacity of ESCC *in vitro*, and decreased tumor size *in vivo*	PLCε →Snail expression	[229]
	Promoter	Esophageal carcinoma cell line	PLCε knockdown in cell line	• Increased PLCε, PKCα expression • PLCε knockdown decreased cell invasion and migration	PKCα/NF-κB	[230]
	Proto-oncogene	Tumor tissues, cell line and nude mice tumor xenograft model	PLCε knockdown in cell line	• Increased PLCε, p65, IKK, and IκBα expression in tissues • PLCε knockdown reduced cell proliferation, angiogenesis and increased apoptosis *in vitro* and *in vivo*	PLCε/NF-κB and VEGF-C/Bcl-2	[231]
	Tumor promoter	ESCC cell line	PLCε knockdown in cell line	• Increased PLCε expression • PLCε knockdown increased apoptosis	p53 promoter methylation, decreased expression	[232]

(Continued)

Table 14.1 (Continued)

Tumor type	Proposed role in tumorigenesis	*In vitro* and *in vivo* models		PLCε expression and genetic manipulation or treatment phenotype	Proposed signaling pathways	References
	Potential oncogene	ESCC cell tissues, line and mice tumor xenograft model	PLCε knockdown in cell line	• Increased PLCε expression in tissues • PLCε knockdown increased autophagy and apoptosis in *vitro* and *in vivo*	MDM2-dependent ubiquitination → decreased p53 expression	[233]
	Biomarker	ESCC tissues	—	• Increased PLCε expression in tissues	—	[234]
	Positively correlated with cancer	Esophageal cancer cell line, PLCε$^{-/-}$ mice, NMBA-treated rat model	PLCε knockdown in cell line	• Increased PLCε and PRKCA expression in rat esophageal epithelial cells and ESCC • PLCε knockdown reduced cytokines and PRKCA expression in cell lines and KO mice	PLCε→ PRKCA	[235]
	Oncogene	ESCC tissues and cell lines	PLCε knockdown in cell line	• Increased PLCε expression in tissues • PLCε knockdown increased apoptosis, sensitivity of cancer cells to chemotherapeutic drugs and decreased proliferation, invasion, migration	miR-145→inhibit PLCε expressions	[236]

	Oncogene	ESCC cells lines	PLCε knockdown in cell line	• PLCε knockdown inhibited proliferation, migration, and promote apoptosis	MiR-34a and MiR-328→ inhibit PLCε expression	[237, 238]
Gastric cancer	Susceptibility loci	Tumor tissues-GWAS study	—	—	—	[226, 228]
	Positive association	Tumor tissues and cell lines	—	• Increased PLCε expression	—	[239–241]
Head and neck squamous cell carcinoma (HNSCC)	Tumor promoter	Human squamous carcinoma cell line	—	• Tumor cell growth and migration	CD44-LARG-EGFR/RhoA→PLCε→CaMKII/Filamin→Cytoskeleton reorganization	[242]
Renal cell carcinoma	Tumor promoter	Renal cell carcinoma tissues and cell line	—	• Increased PLCε expression • Repressed growth and induced apoptosis	PLCε→NFκB→VEGF	[243]

PPARβ, peroxisome proliferator-activated receptor β; IL-1α, interleukin-1α; SMAD, SMAD transcription factor; TPA, 12-*O*-tetradecanoylphorbol-13-acetate; RasGRP3, RAS Guanyl Releasing Protein 3; PKC, protein kinase C; IGF2, insulin like growth factor 2; IGF-RI, Insulin like growth factor receptor I; NF-kB, Nuclear Factor Of Kappa Light Polypeptide Gene Enhancer In B-Cells; IGR-RI-P, phosphorylated insulin like growth factor receptor I; PTEN, phosphatase and tensin homology; Akt, Akt serine/threonine kinase; HIF-1α, hypoxia inducible factor-1α; AR, androgen receptor; COX2, cyclooxygenase-2; VEGF-A, vascular endothelial growth factor-A; STAT, signal transducer and activator of transcription; GWAS, genome-wide association study; LARG, Leukemia-associated Rho guanine nucleotide exchange factor; PRKCA, protein kinase Ca; EGFR, epidermal growth factor receptor; CamKII, calcium calmodulin-dependent protein kinase II as causative for childhood nephrotic syndrome [244]. This finding was then supported by multiple other studies [244–247]. These mutations include truncating and missense mutations, which are associated with diffuse mesangial sclerosis, and focal segmental glomerulosclerosis, respectively. One study has identified an insertion in the *PLCE1* gene in steroid-resistant nephrotic syndrome [248]. PLCε is highly enriched in podocytes that are the major cellular components of the glomerular filtration barrier. The exact role of PLCε in these cells is not clear but several observations have been made. PLCε has been shown to directly bind to IQ motif-containing GTPase-activating protein 1 (IQGAP1), which interacts with nephrin a key structural component of the glomerular slit diaphragm involved in urine filtration [244]. This protein complex has been postulated to regulate morphogenetic processes during glomerular development. PLCε also directly interacts with BRAF (v-raf murine sarcoma viral oncogene homolog B1) [249], a DAG sensitive Trp channel (TrpC6) involved in Ca^{+2} signaling [250], advillin, which is involved in actin binding [251] and non-catalytic region of tyrosine kinase adaptor protein (NCK), ubiquitously expressed adaptor proteins [252] in podocytes. Exactly how all these interactions are coordinated to regulate podocyte functions is not completely clear.

However, some individuals with loss-of-function mutations in *PLCE1* showed no detectable kidney disease [253]. Additionally, global PLCε$^{-/-}$ mice exhibit normal kidney structure and function. Thus, lack of PLCε is insufficient to produce nephrotic syndrome. This indicates that other genetic or environmental factors influence the penetrance of *PLCE1* mutations in this disease the phenotype [244]. This scenario fits the "two-hit" hypothesis in which loss-of-function mutations in *PLCE1* predisposes individuals to development of nephrosis but a second pathophysiological stressor is needed for the disease to emerge. In support of this hypothesis, a recent report showed that under conditions of hypertensive stress induced by either Angiotensin II or a high salt diet, PLCε KO mice develop albuminuria and glomerular damage that is not seen in WT mice [254]. Overall, these studies confirm a central role of PLCε development of kidney disease in humans.

14.3.5.4 Cancer

PLCE1 is frequently mutated in the different types of tumors, as observed in The Cancer Genome Atlas (TCGA) database. Polymorphisms in the *PLCE1* gene play a crucial role in the development and progression of several types of cancer. As the small GTPase Rho is important for cell migration, Rap regulates cell junctions and cell adhesion, and Ras is linked to cell proliferation and survival. The ability of PLCε to be regulated by Rho, Ras, and Rho suggest that it can contribute to both proliferation and migration (Table 14.1).

From these numerous studies, it appears that PLCε plays a critical role in development of many cancer types. PLCε can be either an oncogene or tumor suppressor depending on the type of cancer. However, detailed mechanisms underlying the function of PLCε in carcinogenesis remain to be fully elucidated. Many of the current studies don't clarify whether PLCε is required in the tumor cells themselves or in cells that constitute the tumor microenvironment. In addition, SNP polymorphisms in the *PLCE1* gene alone might have a little impact on cancer development, and therefore, studies on *PLCE1* cooperation with other genes are warranted to consolidate the currently incomplete picture.

14.3.5.5 Common Themes for PLCε Regulation and Function

Some common themes emerge from these studies (see Figure 14.3): (i) cAMP dependent EPAC and Rap pathways are commonly found upstream of PLCε confirming physiological roles downstream of G_s-coupled receptors. (ii) Rho activation is also a commonly observed upstream pathway placing PLCε downstream of $G_{12/13}$-coupled GPCRs. (iii) PLCε pathways often converge on sustained PKD activation, likely by virtue of its ability to generate DAG over longer time frames. In the context of inflammation PLCε-dependent PKD activation results in activation of NF-kB and inflammatory cytokine production in multiple cellular and tissue contexts. In the heart, PKD activation regulates HDAC dependent de-repression of hypertrophic gene expression. (iv) In at least two systems, cardiac myocytes and pancreatic β cells, G_s-coupled receptors regulate CICR via EPAC and PLCε.

14.4 Opportunities to Therapeutically Target GPCR–G Protein–PLC Signaling

PLC enzymes mediate a pleiotropic range of effects downstream of GPCRs and G proteins. GPCR targeting drugs account for 30% of the commercial drug market and will continue to grow when more is known about each receptor's biology. Potential approaches include but are not limited to biased, allosteric, and compartmentalized signaling. Efforts to develop small molecules, protein-based, and viral-based therapies targeting G proteins are also underway [161]. Given the multitude of physiologies and pathophysiologies where PLC function is involved, successful specific targeting of PLC enzymes would likely be beneficial in many cases. Unfortunately, only tool compounds are currently available to modulate PLC, and these are poorly selective [62]. It is somewhat surprising that specific PLC inhibitors have yet to be developed. Some of the roadblocks and alternative strategies are discussed in the following text. Because PLC activity is important in many cell types, strategies for selective targeting are imperative to avoid side effects.

Some possible strategies to target PLC-mediated signaling include targeting (i) their upstream activators (GPCRs, G proteins); (ii) PLC themselves; or (iii) their downstream effectors (PKC, MEK). Moreover, localization of certain PLC enzymes in a cell (PLCβ1b in cardiomyocyte sarcolemma, PLCε in the Golgi), or their interaction with PDZ binding proteins also provide a means to more specifically target PLC and will be discussed in more detail in the following text.

14.4.1 Selective Competitive Inhibitors. Why or Why not?

As discussed earlier PLCs regulate a gamut of cellular signaling pathways and are correlated both positively and negatively with disease pathophysiology. There are currently no specific small molecule inhibitors available. A widely used thiol-reactive alkylating compound U73122 has been used as an inhibitor of PLC and is useful in some contexts [255], but it has been shown to have multiple off-target effects [256] and can even stimulate some isoforms of PLC [257]. For this reason, this compound must be used with great care and appropriate controls. Efforts towards development of more specific PLC inhibitors have used a robust fluorogenic high throughput screening method using a water-soluble and fluorogenic analog of PIP_2, WH-15, as a reporter [256, 258]. With a small compound library (6280 compounds), only moderate potency compounds with poor cell permeability were identified. It is possible that with a larger screen more suitable compounds could be identified. As an enzyme, it might be expected that identification of catalytic site-directed competitive inhibitors of PLC would be relatively straightforward, so it is somewhat surprising that high-quality PLC inhibitors have yet to be developed. Although, targeting of the catalytic site has the drawback that catalytic sites of PLC isoforms are highly conserved, and it might be difficult to identify isozyme-specific inhibitors.

WH-15 is a useful reporter of PLC catalytic activity but various data show that modulation of PLCs by allosteric modulators such as $G\alpha_q$ requires a membrane interface [259]. Thus, as a soluble substrate WH-15 does not allow monitoring of G protein-dependent activation of PLCs. More recently, a lipid bilayer associated fluorogenic substrate XY-69 was developed and used to measure $G\alpha_q$-dependent activation of PLCβ3. Using this platform, it would be possible to identify inhibitors that act at allosteric sites required for G protein-dependent activation. These sites might be less conserved than the orthosteric site, providing room for the selectivity towards various isoforms.

14.4.2 Targeting Upstream Activators

14.4.2.1 Phospholipase Cβ

One way to target PLC activity is to manipulate the activity of PLC activators such as $G\alpha_q$ or Gβγ for PLCβ, or Epac, for example, for PLCε. One area of research in our laboratory has been to explore inhibiting PLCβ3-dependent hyperalgesia by inhibiting Gβγ subunits to prevent PLC activation downstream of MOR. This approach aims to improve opioid analgesia by altering or "biasing" the signaling processes after MOR engagement by an agonist, by targeting the G proteins that are coupled to opioid receptors. The role of PLCβ3 in opioid-induced analgesia was discussed earlier in this chapter. PLCβ3 is unique among phospholipase C enzymes in that it directly activated by Gβγ and $G\alpha_q$ subunits in a synergistic fashion [67, 68, 71]. Administration of M119, a Gβγ inhibitor, potentiated opioid analgesia in wild type mice but not in PLCβ3 deficient mice [260], suggesting that the potentiating action of M119 is mediated, at least in part, through inhibition of Gβγ activation of PLCβ3. PLCβ3 was shown to be a key mediator of morphine-induced hyperalgesia in mice, and gallein, another Gβγ inhibitor, inhibits morphine-induced hyperalgesia [261]. Identification of more effective "drug-like" Gβγ inhibitors could lead to the development of a strong lead candidate for therapeutic intervention.

PLCβ activation downstream of GPCRs has also been inhibited by the use of natural toxins that have been characterized as specific inhibitors of $G\alpha_{q/11}$. These have been explored for potential utility in inhibition of thrombosis [262, 263], airway inflammation [264, 265], and uveal melanoma [266, 267]. This approach has the drawback that broad inhibition of $G\alpha_q$ could have significant side effects such as hypotension that has been observed in animal models. An approach that overcomes these concerns, that could also be applied to PLC inhibitors, is to restrict the route of delivery to target the specific tissues of interest, i.e. inhalation for airway inflammation, or topical application for ocular melanoma.

14.4.2.2 Phospholipase Cε

As has been discussed, PLCε can be activated by multiple upstream stimuli. Here we will discuss some examples.

Gβγ Inhibition As discussed earlier, Gβγ inhibition can block the regulation of certain PLCε isoforms. We have also shown that Gβγ inhibition can inhibit PLCε activation in the heart. In this study, we examined the regulation of the Golgi-PLCε cardiac hypertrophic pathway driven by endothelin 1 (ET-1). We found that ET-1 stimulates PLCε-dependent hydrolysis of PI4P in the Golgi, and this could be blocked by treatment of the myocytes with the Gβγ inhibitors gallein, or the c-terminal PH domain of G protein-coupled receptor kinase 2 (GRK2ct) [268]. These treatments also blocked cardiac hypertrophy driven by ET-1 leading us to propose that Gβγ-dependent regulation of PLCε at the Golgi is critical for ET-1 driven cardiac hypertrophy.

Epac and Rap1 As discussed earlier, one canonical mechanism for PLCε activation is via the Epac dependent stimulation of Rap. Multiple small-molecule Epac modulators; agonists and antagonists have been discovered. There are two isoforms of Epac, Epac1, and Epac2 that have different cellular functions. Agonists and antagonists have been developed that are selective these for Epac isoforms (reviewed in [269, 270]). As an example, a recent study showed that Epac1 inhibition protects the heart from acute and chronic stress in mice [271]. PLCε was not examined in this study but is likely involved at some level. Another recent study showed that the deletion of Epac1, or treatment with an Epac inhibitor, inhibits retinal degeneration in a high intraocular pressure model of glaucoma [272]. Again, PLCε was not directly implicated in this study, but its involvement was proposed.

14.4.3 Scaffolding Interactions

One way to specifically target PLCs downstream of GPCRs would be to disrupt scaffolding interactions that are specific to certain tissues and processes. Two of many examples of scaffolding interactions that could be disrupted are described here as examples.

As discussed earlier, the interaction between PLCβ1b and Shank3 in cardiomyocytes is important for PLCβ1b sarcolemmal localization and coupling to $G\alpha_q$ coupled receptors. Disrupting the binding of PLCβ1b to Shank3 with a peptide corresponding to the PDZ binding sequence of PLCβ1b, prevented contractile dysfunction following pressure overload *in vivo* [273] (Figure 14.4). This specific inhibition of PLCβ1b in the heart to attenuate cardiac hypertrophy is not achievable by other methods such as with the use of $G\alpha_q$ or PKC inhibitors. Another receptor-PLCβ-PDZ binding protein complex that presents therapeutic opportunities is the PTH1R-PLCβ-NHERF1 complex [120]. Interrogating the interaction between PTH1R and PDZ binding protein NHERF1 by small molecules fine-tunes activation of PLCβ for the treatment of osteoporosis [274].

There are at least two different roles for PLCε in cardiac myocytes; regulation of cardiac contractility through regulation of RyR2 at the SR, and hypertrophy through scaffolding to mAKAP at the nuclear envelope. We have shown that disruption of PLCε binding to mAKAP at the nuclear envelope inhibits development of cardiomyocyte hypertrophy in response to multiple GPCR-dependent hypertrophic stimuli [134] (Figure 14.4). The development of small-molecule inhibitors of this interaction could be a possible approach to selective blockade of this pool of PLCε for the treatment of cardiac hypertrophy.

14.4.4 Substrate Availability

One novel strategy to target the PLC function is to control its substrate availability. Oxysterol-binding protein (OSBP)-related protein 4 (ORP4L) is selectively expressed in acute T-lymphoblastic leukemia (T-ALL) cells but not normal T-cells [275]. ORP4L acts as an adaptor/scaffold assembling CD3ε, $G\alpha_{q/11}$, and PLCβ3 into a complex that activates PLCβ3, which, in turn, induces IP3-mediated endoplasmic reticulum Ca^{2+} release, and oxidative phosphorylation, resulting in an increase of mitochondrial respiration for cell survival [275]. ORP4L is required for activation of PLCβ3 in T-ALL cells and this signaling is GPCR-independent. In leukemia stem cell (LSC), ORP4L was proposed to extract PIP_2 from the PM and present this substrate to PLCβ3. Blockage of the substrate (PIP_2) presenting protein ORP4L by LYZ-81 blocks PLCβ3's role in LSC bioenergetics and survival, which could be useful for LSC elimination in leukemia therapy [276]. While interesting, this model has not yet been directly tested in a reconstitution system.

14.4.5 Expression Level

Several drugs have been shown to modulate PLC expression levels. Induction of nuclear PLCβ1b expression by azacitidine, a hypomethylating agent, or upregulation of cytoplasmic PLCβ1a by lenalidomide [277] have been shown to be beneficial in patients with myelodysplastic syndrome (MDS). In glioma, the high expression of PLCβ1 is targeted by the expression of a micro RNA-miR-423-5p. miR-423-5p binds to the 3′UTR of PLCβ1 to suppress PLCβ1 expression and PLCβ1-mediated tumor growth and invasion [154]. Given the clinical advances in miRNA therapeutics [278], this could be a promising strategy to modulate PLC expression in diseases.

14.5 Conclusion and Perspectives

In this review, we present roles for the well-studied PLCβ isoforms in GPCR-dependent regulation of phosphoinositide hydrolysis. We also discuss the perhaps underappreciated role for PLCε in regulating longer-term PI hydrolysis in response to GPCR stimulation. Mechanisms for regulation of these enzyme families by GPCRs are emerging but the exact mechanism for G protein-dependent regulation of these isoforms remains to be clarified. New structural data may shed light on this and in particular, cryo-EM studies may reveal PLC conformations that have not been observable by X-ray crystallography. We discuss the plethora of physiological and pathophysiological roles for these PLC isoforms, and likely more will emerge. Thus, selective targeting of these enzymes could be efficacious in a number of therapeutic conditions and is an important goal. It is surprising that no inhibitors specific for PLC, in general, have been identified given the number of years of study, and our in-depth knowledge of the enzyme's structure. It would be more important, however, to develop strategies to target specific PLC isoforms to avoid potential side effects of inhibiting these ubiquitously expressed enzymes. Several strategies are proposed that avoid targeting the catalytic site to selectively inhibit PLCs, and new assays have emerged that have the potential for high throughput screening for PLC inhibitors that are isotype selective. Despite the years of study of these enzymes, there is much that remains to be done to understand their functions, regulation, and therapeutic potential.

Abbreviations

α1AR	α1 adrenergic receptor
cAMP	cyclic adenosine 3′5′ monophosphate
DAG	diacylglycerol
Epac	exchange factor activated by cAMP
GPCR	G protein-coupled receptor
HLH	Helix–Loop–Helix
IP3	inositol 1,4,5-trisphosphate
mAKAPβ	muscle-specific A-kinase anchoring protein β
NHERF	Na^+–H^+ exchange regulatory factor
ORP4L	oxysterol-binding protein (OSBP)-related protein 4
PDZ	post synaptic density protein (PSD95)/Drosophila disc large tumor suppressor (Dlg1)/and Zonula occludens-1 protein (zo-1)
PH	pleckstrin homology
PI4P	phosphatidyl inositol 4-phosphate
PIP_2	phosphatidylinositol 4,5 bisphosphate
PKA	protein kinase A
PKC	protein kinase C
PKD	protein kinase D
PLC	phospholipase C
PM	plasma membrane
PTX	pertussis toxin
RA	Ras association domain
RTK	receptor tyrosine kinase
RyR2	type 2 ryanodine receptor
SR	sarcoplasmic reticulum
TIM	triose phosphate isomerase
TrpC	transient receptor potential cation channel
UM	uveal melanoma

References

1 Fain, J.N. and Berridge, M.J. (1979). Relationship between hormonal activation of phosphatidylinositol hydrolysis, fluid secretion and calcium flux in the blowfly salivary gland. *Biochem. J* 178: 45–58.

2 Streb, H., Irvine, R.F., Berridge, M.J., and Schulz, I. (1983). Release of Ca2+ from a nonmitochondrial intracellular store in pancreatic acinar cells by inositol-1,4,5-trisphosphate. *Nature* 306: 67–69.

3 Takai, Y., Kishimoto, A., Kikkawa, U. et al. (1979). Unsaturated diacylglycerol as a possible messenger for the activation of calcium-activated, phospholipid-dependent protein kinase system. *Biochem. Biophys. Res. Commun.* 91: 1218–1224.

4 Ryu, S.H., Cho, K.S., Lee, K.Y. et al. (1986). Two forms of phosphatidylinositol-specific phospholipase C from bovine brain. *Biochem. Biophys. Res. Commun.* 141: 137–144.

5 Ryu, S.H., Suh, P.G., Cho, K.S. et al. (1987). Bovine brain cytosol contains three immunologically distinct forms of inositolphospholipid-specific phospholipase C. *Proc. Natl. Acad. Sci. U.S.A.* 84: 6649–6653.

6 Katan, M. and Parker, P.J. (1987). Purification of phosphoinositide-specific phospholipase C from a particulate fraction of bovine brain. *Eur. J. Biochem.* 168: 413–418.

7 Fukui, T., Lutz, R.J., and Lowenstein, J.M. (1988). Purification of a phospholipase C from rat liver cytosol that acts on phosphatidylinositol 4,5-bisphosphate and phosphatidylinositol 4-phosphate. *J. Biol. Chem.* 263: 17730–17737.

8 Rhee, S.G., Suh, P.G., Ryu, S.H., and Lee, S.Y. (1989). Studies of inositol phospholipid-specific phospholipase C. *Science* 244: 546–550.

9 Suh, P.G., Ryu, S.H., Moon, K.H. et al. (1988). Cloning and sequence of multiple forms of phospholipase C. *Cell* 54: 161–169.

10 Suh, P.G., Ryu, S.H., Moon, K.H. et al. (1988). Inositol phospholipid-specific phospholipase C: complete cDNA and protein sequences and sequence homology to tyrosine kinase-related oncogene products. *Proc. Natl. Acad. Sci. U.S.A.* 85: 5419–5423.

11 Katan, M., Kriz, R.W., Totty, N. et al. (1988). Determination of the primary structure of PLC-154 demonstrates diversity of phosphoinositide-specific phospholipase C activities. *Cell* 54: 171–177.

12 Stahl, M.L., Ferenz, C.R., Kelleher, K.L. et al. (1988). Sequence similarity of phospholipase C with the non-catalytic region of src. *Nature* 332: 269–272.

13 Lopez, I., Mak, E.C., Ding, J. et al. (2001). A novel bifunctional phospholipase c that is regulated by Galpha 12 and stimulates the Ras/mitogen-activated protein kinase pathway. *J. Biol. Chem.* 276: 2758–2765.

14 Song, C., Hu, C.D., Masago, M. et al. (2001). Regulation of a novel human phospholipase C, PLCepsilon, through membrane targeting by Ras. *J. Biol. Chem.* 276: 2752–2757.

15 Kelley, G.G., Reks, S.E., Ondrako, J.M., and Smrcka, A.V. (2001). Phospholipase C(epsilon): a novel Ras effector. *EMBO J.* 20: 743–754.

16 Schmidt, M., Evellin, S., Weernink, P.A. et al. (2001). A new phospholipase-C-calcium signalling pathway mediated by cyclic AMP and a Rap GTPase. *Nat. Cell Biol.* 3: 1020–1024.

17 Parrington, J., Jones, M.L., Tunwell, R. et al. (2002). Phospholipase C isoforms in mammalian spermatozoa: potential components of the sperm factor that causes Ca2+ release in eggs. *Reproduction* 123: 31–39.

18 Hwang, J.I., Oh, Y.S., Shin, K.J. et al. (2005). Molecular cloning and characterization of a novel phospholipase C. *PLC-eta. Biochem. J.* 389: 181–186.

19 Nakahara, M., Shimozawa, M., Nakamura, Y. et al. (2005). A novel phospholipase C, PLC(eta)2, is a neuron-specific isozyme. *J. Biol. Chem.* 280: 29128–29134.

20 Stewart, A.J., Mukherjee, J., Roberts, S.J. et al. (2005). Identification of a novel class of mammalian phosphoinositol-specific phospholipase C enzymes. *Int. J. Mol. Med.* 15: 117–121.

21 Garland-Kuntz, E.E., Vago, F.S., Sieng, M. et al. (2018). Direct observation of conformational dynamics of the PH domain in phospholipases C and beta may contribute to subfamily-specific roles in regulation. *J. Biol. Chem.* 293: 17477–17490.

22 Zhang, L., Malik, S., Pang, J. et al. (2013). Phospholipase Cepsilon hydrolyzes perinuclear phosphatidylinositol 4-phosphate to regulate cardiac hypertrophy. *Cell* 153: 216–227.

23 de Rubio, R.G., Ransom, R.F., Malik, S. et al. (2018). Phosphatidylinositol 4-phosphate is a major source of GPCR-stimulated phosphoinositide production. *Sci. Signal.* 11.

24 Kim, H.K., Kim, J.W., Zilberstein, A. et al. (1991). PDGF stimulation of inositol phospholipid hydrolysis requires PLC-gamma 1 phosphorylation on tyrosine residues 783 and 1254. *Cell* 65: 435–441.

25 Kim, U.H., Fink, D. Jr., Kim, H.S. et al. (1991). Nerve growth factor stimulates phosphorylation of phospholipase C-gamma in PC12 cells. *J. Biol. Chem.* 266: 1359–1362.

26 Park, D.J., Rho, H.W., and Rhee, S.G. (1991). CD3 stimulation causes phosphorylation of phospholipase C-gamma 1 on serine and tyrosine residues in a human T-cell line. *Proc. Natl. Acad. Sci. U.S.A.* 88: 5453–5456.

27 Carter, R.H., Park, D.J., Rhee, S.G., and Fearon, D.T. (1991). Tyrosine phosphorylation of phospholipase C induced by membrane immunoglobulin in B lymphocytes. *Proc. Natl. Acad. Sci. U.S.A.* 88: 2745–2749.

28 Pang, I.H. and Sternweis, P.C. (1990). Purification of unique alpha subunits of GTP-binding regulatory proteins (G proteins) by affinity chromatography with immobilized beta gamma subunits. *J. Biol. Chem.* 265: 18707–18712.

29 Strathmann, M. and Simon, M.I. (1990). G protein diversity: a distinct class of alpha subunits is present in vertebrates and invertebrates. *Proc. Natl. Acad. Sci. U.S.A.* 87: 9113–9117.

30 Taylor, S.J., Chae, H.Z., Rhee, S.G., and Exton, J.H. (1991). Activation of the beta 1 isozyme of phospholipase C by alpha subunits of the Gq class of G proteins. *Nature* 350: 516–518.

31 Blank, J.L., Ross, A.H., and Exton, J.H. (1991). Purification and characterization of two G-proteins that activate the beta 1 isozyme of phosphoinositide-specific phospholipase C. Identification as members of the Gq class. *J. Biol. Chem.* 266: 18206–18216.

32 Smrcka, A.V., Hepler, J.R., Brown, K.O., and Sternweis, P.C. (1991). Regulation of polyphosphoinositide-specific phospholipase C activity by purified Gq. *Science* 251: 804–807.

33 Park, D., Jhon, D.Y., Kriz, R. et al. (1992). Cloning, sequencing, expression, and Gq-independent activation of phospholipase C-beta 2. *J. Biol. Chem.* 267: 16048–16055.

34 Jhon, D.Y., Lee, H.H., Park, D. et al. (1993). Cloning, sequencing, purification, and Gq-dependent activation of phospholipase C-beta 3. *J. Biol. Chem.* 268: 6654–6661.

35 Kim, M.J., Bahk, Y.Y., Min, D.S. et al. (1993). Cloning of cDNA encoding rat phospholipase C-beta 4, a new member of the phospholipase C. *Biochem. Biophys. Res. Commun.* 194: 706–712.

36 Lee, C.W., Lee, K.H., Lee, S.B. et al. (1994). Regulation of phospholipase C-beta 4 by ribonucleotides and the alpha subunit of Gq. *J. Biol. Chem.* 269: 25335–25338.

37 Jiang, H., Wu, D., and Simon, M.I. (1994). Activation of phospholipase C beta 4 by heterotrimeric GTP-binding proteins. *J. Biol. Chem.* 269: 7593–7596.

38 Park, D., Jhon, D.Y., Lee, C.W. et al. (1993). Activation of phospholipase C isozymes by G protein beta gamma subunits. *J. Biol. Chem.* 268: 4573–4576.

39 Blank, J.L., Brattain, K.A., and Exton, J.H. (1992). Activation of cytosolic phosphoinositide phospholipase C by G-protein beta gamma subunits. *J. Biol. Chem.* 267: 23069–23075.

40 Camps, M., Carozzi, A., Schnabel, P. et al. (1992). Isozyme-selective stimulation of phospholipase C-beta 2 by G protein beta gamma-subunits. *Nature* 360: 684–686.

41 Katz, A., Wu, D., and Simon, M.I. (1992). Subunits beta gamma of heterotrimeric G protein activate beta 2 isoform of phospholipase C. *Nature* 360: 686–689.

42 Kadamur, G. and Ross, E.M. (2013). Mammalian phospholipase C. *Annu. Rev. Physiol.* 75: 127–154.

43 Jezyk, M.R., Snyder, J.T., Gershberg, S. et al. (2006). Crystal structure of Rac1 bound to its effector phospholipase C-beta2. *Nat. Struct. Mol. Biol.* 13: 1135–1140.

44 Snyder, J.T., Singer, A.U., Wing, M.R. et al. (2003). The pleckstrin homology domain of phospholipase C-beta2 as an effector site for Rac. *J. Biol. Chem.* 278: 21099–21104.

45 Ponting, C.P. and Benjamin, D.R. (1996). A novel family of Ras-binding domains. *Trends Biochem. Sci* 21: 422–425.

46 Shibatohge, M., Kariya, K., Liao, Y. et al. (1998). Identification of PLC210, a *Caenorhabditis elegans* phospholipase C, as a putative effector of Ras. *J. Biol. Chem.* 273: 6218–6222.

47 Madukwe, J.C., Garland-Kuntz, E.E., Lyon, A.M., and Smrcka, A.V. (2018). G protein betagamma subunits directly interact with and activate phospholipase C. *J. Biol. Chem.* 293: 6387–6397.

48 Jin, T.G., Satoh, T., Liao, Y. et al. (2001). Role of the CDC25 homology domain of phospholipase Cepsilon in amplification of Rap1-dependent signaling. *J. Biol. Chem.* 276: 30301–30307.

49 Hicks, S.N., Jezyk, M.R., Gershburg, S. et al. (2008). General and versatile autoinhibition of PLC isozymes. *Mol. Cell* 31: 383–394.

50 Waldo, G.L., Ricks, T.K., Hicks, S.N. et al. (2010). Kinetic scaffolding mediated by a phospholipase C-beta and Gq signaling complex. *Science* 330: 974–980.

51 Lyon, A.M., Begley, J.A., Manett, T.D., and Tesmer, J.J.G. (2014). Molecular mechanisms of phospholipase C beta3 autoinhibition. *Structure* 22: 1844–1854.

52 Hudson, B.N., Jessup, R.E., Prahalad, K.K., and Lyon, A.M. (2019). Galphaq and the phospholipase Cbeta3 X-Y linker regulate adsorption and activity on compressed lipid monolayers. *Biochemistry* 58: 3454–3467.

53 Hudson, B.N., Hyun, S.H., Thompson, D.H., and Lyon, A.M. (2017). Phospholipase Cbeta3 membrane adsorption and activation are regulated by its

C-terminal domains and phosphatidylinositol 4,5-bisphosphate. *Biochemistry* 56: 5604–5614.

54 Gutman, O., Walliser, C., Piechulek, T. et al. (2010). Differential regulation of phospholipase C-beta2 activity and membrane interaction by Galphaq, Gbeta1gamma2, and Rac2. *J. Biol. Chem.* 285: 3905–3915.

55 Arduin, A., Gaffney, P.R., and Ces, O. (2015). Regulation of PLCbeta2 by the electrostatic and mechanical properties of lipid bilayers. *Sci. Rep.* 5: 12628.

56 Romoser, V., Ball, R., and Smrcka, A.V. (1996). Phospholipase C beta2 association with phospholipid interfaces assessed by fluorescence resonance energy transfer. G protein betagamma subunit-mediated translocation is not required for enzyme activation. *J. Biol. Chem.* 271: 25071–25078.

57 Lyon, A.M., Dutta, S., Boguth, C.A. et al. (2013). Full-length Galpha(q)-phospholipase C-beta3 structure reveals interfaces of the C-terminal coiled-coil domain. *Nat. Struct. Mol. Biol.* 20: 355–362.

58 Lyon, A.M., Tesmer, V.M., Dhamsania, V.D. et al. (2011). An autoinhibitory helix in the C-terminal region of phospholipase C-beta mediates Galphaq activation. *Nat. Struct. Mol. Biol.* 18: 999–1005.

59 Fisher, I.J., Jenkins, M.L., Tall, G.G. et al. (2020). Activation of phospholipase C beta by Gbetagamma and Galphaq involves C-terminal rearrangement to release autoinhibition. *Structure* 28 (7): 810–819.

60 Phan, H.T.N., Kim, N.H., Wei, W. et al. (2021). Uveal melanoma-associated mutations in PLCβ4 are constitutively activating and promote melanocyte proliferation and tumorigenesis. *Sci. Signal.* 14: eabj4243.

61 Smrcka, A.V. and Fisher, I. (2019). G-protein betagamma subunits as multi-functional scaffolds and transducers in G-protein-coupled receptor signaling. *Cell. Mol. Life Sci.* 76: 4447–4459.

62 Lyon, A.M. and Tesmer, J.J. (2013). Structural insights into phospholipase C-beta function. *Mol. Pharmacol.* 84: 488–500.

63 Wang, T., Dowal, L., El-Maghrabi, M.R. et al. (2000). The pleckstrin homology domain of phospholipase C-beta(2) links the binding of gbetagamma to activation of the catalytic core. *J. Biol. Chem.* 275: 7466–7469.

64 Barr, A.J., Ali, H., Haribabu, B. et al. (2000). Identification of a region at the N-terminus of phospholipase C-beta 3 that interacts with G protein beta gamma subunits. *Biochemistry* 39: 1800–1806.

65 Kadamur, G. and Ross, E.M. (2016). Intrinsic pleckstrin homology (PH) domain motion in phospholipase C-beta exposes a Gbetagamma protein binding site. *J. Biol. Chem.* 291: 11394–11406.

66 Bonacci, T.M., Ghosh, M., Malik, S., and Smrcka, A.V. (2005). Regulatory interactions between the amino terminus of G-protein betagamma subunits and the catalytic domain of phospholipase Cbeta2. *J. Biol. Chem.* 280: 10174–10181.

67 Philip, F., Kadamur, G., RGl, S. et al. (2010). Synergistic activation of phospholipase C-β3 by Gαq and Gβγ describes a simple two-state coincidence detector. *Curr. Biol.* 20: 1327–1335.

68 Rebres, R.A., Roach, T.I.A., Fraser, I.D.C. et al. (2011). Synergistic Ca2+ responses by Gβγ and Gαq-coupled G-protein-coupled receptors require a single PLCβ isoform that is sensitive to both Gβγ and Gαq. *J. Biol. Chem.* 286: 942–951.

69 Roach, T.I., Rebres, R.A., Fraser, I.D. et al. (2008). Signaling and cross-talk by C5a and UDP in macrophages selectively use PLCbeta3 to regulate intracellular free calcium. *J. Biol. Chem.* 283: 17351–17361.

70 Pfeil, E.M., Brands, J., Merten, N. et al. (2020). Heterotrimeric G Protein Subunit Gαq Is a Master Switch for Gβγ-Mediated Calcium Mobilization by Gi-Coupled GPCRs. *Mol. Cell* 80: 940–54.e6.

71 Smrcka, A.V. and Sternweis, P.C. (1993). Regulation of purified subtypes of phosphatidylinositol specific phospholipase Cβ by G protein α and βγ subunits. *J. Biol. Chem.* 268: 9667–9674.

72 Illenberger, D., Walliser, C., Nurnberg, B. et al. (2003). Specificity and structural requirements of phospholipase C-beta stimulation by Rho GTPases versus G protein beta gamma dimers. *J. Biol. Chem.* 278: 3006–3014.

73 Song, C., Satoh, T., Edamatsu, H. et al. (2002). Differential roles of Ras and Rap1 in growth factor-dependent activation of phospholipase C epsilon. *Oncogene* 21: 8105–8113.

74 Sorli, S.C., Bunney, T.D., Sugden, P.H. et al. (2005). Signaling properties and expression in normal and tumor tissues of two phospholipase C epsilon splice variants. *Oncogene* 24: 90–100.

75 Hains, M.D., Wing, M.R., Maddileti, S. et al. (2006). Galpha12/13- and rho-dependent activation of phospholipase C-epsilon by lysophosphatidic acid and thrombin receptors. *Mol. Pharmacol.* 69: 2068–2075.

76 Kelley, G.G., Reks, S.E., and Smrcka, A.V. (2004). Hormonal regulation of phospholipase Cepsilon through distinct and overlapping pathways involving G12 and Ras family G-proteins. *Biochem. J* 378: 129–139.

77 Wing, M.R., Houston, D., Kelley, G.G. et al. (2001). Activation of phospholipase C-epsilon by heterotrimeric G protein betagamma-subunits. *J. Biol. Chem.* 276: 48257–48261.

78 Rugema, N.Y., Garland-Kuntz, E.E., Sieng, M. et al. (2020). Structure of phospholipase Cε reveals an integrated RA1 domain and previously unidentified regulatory elements. *Commun. Biol.* 3: 445.

79 Bunney, T.D., Harris, R., Gandarillas, N.L. et al. (2006). Structural and mechanistic insights into ras association domains of phospholipase C epsilon. *Mol. Cell* 21: 495–507.

80 Citro, S., Malik, S., Oestreich, E.A. et al. (2007). Phospholipase Cepsilon is a nexus for Rho and Rap-mediated G protein-coupled receptor-induced astrocyte proliferation. *Proc. Natl. Acad. Sci. U.S.A.* 104: 15543–15548.

81 Wing, M.R., Snyder, J.T., Sondek, J., and Harden, T.K. (2003). Direct activation of phospholipase C-epsilon by Rho. *J. Biol. Chem.* 278: 41253–41258.

82 Seifert, J.P., Wing, M.R., Snyder, J.T. et al. (2004). RhoA activates purified phospholipase C-epsilon by a guanine nucleotide-dependent mechanism. *J. Biol. Chem.* 279: 47992–47997.

83 Kelley, G.G., Kaproth-Joslin, K.A., Reks, S.E. et al. (2006). G-protein-coupled receptor agonists activate endogenous phospholipase Cepsilon and phospholipase Cbeta3 in a temporally distinct manner. *J. Biol. Chem.* 281: 2639–2648.

84 Evellin, S., Nolte, J., Tysack, K. et al. (2002). Stimulation of phospholipase C-epsilon by the M3 muscarinic acetylcholine receptor mediated by cyclic AMP and the GTPase Rap2B. *J. Biol. Chem.* 277: 16805–16813.

85 Oestreich, E.A., Wang, H., Malik, S. et al. (2007). Epac-mediated activation of phospholipase C(epsilon) plays a critical role in beta-adrenergic receptor-dependent enhancement of Ca2+ mobilization in cardiac myocytes. *J. Biol. Chem.* 282: 5488–5495.

86 Keiper, M., Stope, M.B., Szatkowski, D. et al. (2004). Epac- and Ca2+ -controlled activation of Ras and extracellular signal-regulated kinases by Gs-coupled receptors. *J. Biol. Chem.* 279: 46497–46508.

87 Dzhura, I., Chepurny, O.G., Kelley, G.G. et al. (2010). Epac2-dependent mobilization of intracellular Ca(2)+ by glucagon-like peptide-1 receptor agonist exendin-4 is disrupted in beta-cells of phospholipase C-epsilon knockout mice. *J. Phys.* 588: 4871–4889.

88 Tong, J., Liu, X., Vickstrom, C. et al. (2017). The Epac-phospholipase Cepsilon pathway regulates endocannabinoid signaling and cocaine-induced disinhibition of ventral tegmental area dopamine neurons. *J. Neurosci.* 37: 3030–3044.

89 Ma, W. and St-Jacques, B. (2018). Signalling transduction events involved in agonist-induced PGE2/EP4 receptor externalization in cultured rat dorsal root ganglion neurons. *Eur. J. Pain* 22: 845–861.

90 Balla, T. (2013). Phosphoinositides: tiny lipids with giant impact on cell regulation. *Physiol. Rev.* 93: 1019–1137.

91 Smrcka, A.V., Brown, J.H., and Holz, G.G. (2012). Role of phospholipase Cepsilon in physiological phosphoinositide signaling networks. *Cell. Signalling* 24: 1333–1343.

92 Dusaban, S.S., Purcell, N.H., Rockenstein, E. et al. (2013). Phospholipase C epsilon links G protein-coupled receptor activation to inflammatory astrocytic responses. *Proc. Natl. Acad. Sci. U.S.A.* 110: 3609–3614.

93 Oestreich, E.A., Malik, S., Goonasekera, S.A. et al. (2009). Epac and phospholipase Cepsilon regulate Ca2+ release in the heart by activation of protein kinase Cepsilon and calcium-calmodulin kinase II. *J. Biol. Chem.* 284: 1514–1522.

94 Bohm, D., Schwegler, H., Kotthaus, L. et al. (2002). Disruption of PLC-beta 1-mediated signal transduction in mutant mice causes age-dependent hippocampal mossy fiber sprouting and neurodegeneration. *Mol. Cell Neurosci.* 21: 584–601.

95 Mende, U., Kagen, A., Meister, M., and Neer, E.J. (1999). Signal transduction in atria and ventricles of mice with transient cardiac expression of activated G protein alpha(q). *Circ. Res.* 85: 1085–1091.

96 Arthur, J.F., Matkovich, S.J., Mitchell, C.J. et al. (2001). Evidence for selective coupling of alpha 1-adrenergic receptors to phospholipase C-beta 1 in rat neonatal cardiomyocytes. *J. Biol. Chem.* 276: 37341–37346.

97 Suh, P.G., Park, J.I., Manzoli, L. et al. (2008). Multiple roles of phosphoinositide-specific phospholipase C isozymes. *BMB Rep.* 41: 415–434.

98 Jiang, H., Lyubarsky, A., Dodd, R. et al. (1996). Phospholipase C beta 4 is involved in modulating the visual response in mice. *Proc. Natl. Acad. Sci. U.S.A.* 93: 14598–14601.

99 Lee, S.B., Rao, A.K., Lee, K.H. et al. (1996). Decreased expression of phospholipase C-beta 2 isozyme in human platelets with impaired function. *Blood* 88: 1684–1691.

100 Otaegui, D., Querejeta, R., Arrieta, A. et al. (2010). Phospholipase Cbeta4 isozyme is expressed in human, rat, and murine heart left ventricles and in HL-1 cardiomyocytes. *Mol. Cell. Biochem.* 337: 167–173.

101 Lian, L., Wang, Y., Draznin, J. et al. (2005). The relative role of PLCbeta and PI3Kgamma in platelet activation. *Blood* 106: 110–117.

102 Bahk, Y.Y., Lee, Y.H., Lee, T.G. et al. (1994). Two forms of phospholipase C-beta 1 generated by alternative splicing. *J. Biol. Chem.* 269: 8240–8245.

103 Bahk, Y.Y., Song, H., Baek, S.H. et al. (1998). Localization of two forms of phospholipase C-beta1, a and b, in C6Bu-1 cells. *Biochim. Biophys. Acta* 1389: 76–80.

104 Faenza, I., Matteucci, A., Manzoli, L. et al. (2000). A role for nuclear phospholipase Cbeta 1 in cell cycle control. *J. Biol. Chem.* 275: 30520–30524.

105 Adjobo-Hermans, M.J., Crosby, K.C., Putyrski, M. et al. (2013). PLCbeta isoforms differ in their subcellular location and their CT-domain dependent interaction with Galphaq. *Cell. Signalling* 25: 255–263.

106 Peruzzi, D., Aluigi, M., Manzoli, L. et al. (2002). Molecular characterization of the human PLC beta1 gene. *Biochim. Biophys. Acta* 1584: 46–54.

107 Mao, G.F., Kunapuli, S.P., and Koneti, R.A. (2000). Evidence for two alternatively spliced forms of phospholipase C-beta2 in haematopoietic cells. *Br. J. Haematol.* 110: 402–408.

108 Sun, L., Mao, G., Kunapuli, S.P. et al. (2007). Alternative splice variants of phospholipase C-beta2 are expressed in platelets: effect on Galphaq-dependent activation and localization. *Platelets* 18: 217–223.

109 Adamski, F.M., Timms, K.M., and Shieh, B.H. (1999). A unique isoform of phospholipase Cbeta4 highly expressed in the cerebellum and eye. *Biochim. Biophys. Acta* 1444: 55–60.

110 Kim, M.J., Min, D.S., Ryu, S.H., and Suh, P.G. (1998). A cytosolic, galphaq- and betagamma-insensitive splice variant of phospholipase C-beta4. *J. Biol. Chem.* 273: 3618–3624.

111 Scarlata, S. (2019). The role of phospholipase Cbeta on the plasma membrane and in the cytosol: How modular domains enable novel functions. *Adv. Biol. Regul.* 73: 100636.

112 Cocco, L., Martelli, A.M., Mazzotti, G. et al. (2000). Inositides and the nucleus: phospholipase Cbeta family localization and signaling activity. *Adv. Enzyme Regul.* 40: 83–95.

113 Follo, M.Y., Ratti, S., Manzoli, L. et al. (2020). Inositide-Dependent Nuclear Signalling in Health and Disease. *Handb. Exp. Pharmacol.* 259: 291–308.

114 Zhou, W. and Hildebrandt, F. (2009). Molecular cloning and expression of phospholipase C epsilon 1 in zebrafish. *Gene Expr. Patterns* 9: 282–288.

115 Wang, H., Oestreich, E.A., Maekawa, N. et al. (2005). Phospholipase C epsilon modulates beta-adrenergic receptor-dependent cardiac contraction and inhibits cardiac hypertrophy. *Circ. Res.* 97: 1305–1313.

116 Suh, P.G., Hwang, J.I., Ryu, S.H. et al. (2001). The roles of PDZ-containing proteins in PLC-beta-mediated signaling. *Biochem. Biophys. Res. Commun.* 288: 1–7.

117 Kim, J.K., Lim, S., Kim, J. et al. (2011). Subtype-specific roles of phospholipase C-beta via differential interactions with PDZ domain proteins. *Adv. Enzyme Regul.* 51: 138–151.

118 Grubb, D.R., Iliades, P., Cooley, N. et al. (2011). Phospholipase Cbeta1b associates with a Shank3 complex at the cardiac sarcolemma. *FASEB J.* 25: 1040–1047.

119 Grubb, D.R., Luo, J., and Woodcock, E.A. (2015). Phospholipase Cbeta1b directly binds the SH3 domain of Shank3 for targeting and activation in cardiomyocytes. *Biochem. Biophys. Res. Commun.* 461: 519–524.

120 Sun, C. and Mierke, D.F. (2005). Characterization of interactions of Na+/H+ exchanger regulatory factor-1 with the parathyroid hormone receptor and phospholipase C. *J. Pept. Res.* 65: 411–417.

121 Wu, Y., Wang, S., Farooq, S.M. et al. (2012). A chemokine receptor CXCR2 macromolecular complex regulates neutrophil functions in inflammatory diseases. *J. Biol. Chem.* 287: 5744–5755.

122 Holcomb, J., Jiang, Y., Guan, X. et al. (2014). Crystal structure of the NHERF1 PDZ2 domain in complex with the chemokine receptor CXCR2 reveals probable modes of PDZ2 dimerization. *Biochem. Biophys. Res. Commun.* 448: 169–174.

123 Jiang, Y., Wang, S., Holcomb, J. et al. (2014). Crystallographic analysis of NHERF1-PLCbeta3 interaction provides structural basis for CXCR2 signaling in pancreatic cancer. *Biochem. Biophys. Res. Commun.* 446: 638–643.

124 Hwang, J.I., Kim, H.S., Lee, J.R. et al. (2005). The interaction of phospholipase C-beta3 with Shank2 regulates mGluR-mediated calcium signal. *J. Biol. Chem.* 280: 12467–12473.

125 Kim, J.K., Kwon, O., Kim, J. et al. (2012). PDZ domain-containing 1 (PDZK1) protein regulates phospholipase C-beta3 (PLC-beta3)-specific activation of somatostatin by forming a ternary complex with PLC-beta3 and somatostatin receptors. *J. Biol. Chem.* 287: 21012–21024.

126 Hwang, J.I., Heo, K., Shin, K.J. et al. (2000). Regulation of phospholipase C-beta 3 activity by Na+/H+ exchanger regulatory factor 2. *J. Biol. Chem.* 275: 16632–16637.

127 Kan, W., Adjobo-Hermans, M., Burroughs, M. et al. (2014). M3 muscarinic receptor interaction with phospholipase C beta3 determines its signaling efficiency. *J. Biol. Chem.* 289: 11206–11218.

128 Lee, S.J., Ritter, S.L., Zhang, H. et al. (2011). MAGI-3 competes with NHERF-2 to negatively regulate LPA2 receptor signaling in colon cancer cells. *Gastroenterology* 140: 924–934.

129 Bloomquist, B.T., Shortridge, R.D., Schneuwly, S. et al. (1988). Isolation of a putative phospholipase C gene of Drosophila, norpA, and its role in phototransduction. *Cell* 54: 723–733.

130 Ye, F., Huang, Y., Li, J. et al. (2018). An unexpected INAD PDZ tandem-mediated plcbeta binding in Drosophila photo receptors. *eLife* 7.

131 Nakamura, M., Sato, K., Fukaya, M. et al. (2004). Signaling complex formation of phospholipase Cbeta4 with metabotropic glutamate receptor type 1alpha and 1,4,5-trisphosphate receptor at the perisynapse and endoplasmic reticulum in the mouse brain. *Eur. J. Neurosci.* 20: 2929–2944.

132 Hong, J., Lee, J., Song, K. et al. (2016). The thalamic mGluR1-PLCbeta4 pathway is critical in sleep architecture. *Mol. Brain* 9: 100.

133 Dodge-Kafka, K.L., Soughayer, J., Pare, G.C. et al. (2005). The protein kinase A anchoring protein mAKAP coordinates two integrated cAMP effector pathways. *Nature* 437: 574–578.

134 Zhang, L., Malik, S., Kelley, G.G. et al. (2011). Phospholipase C epsilon scaffolds to muscle-specific A kinase anchoring protein (mAKAPbeta) and integrates multiple hypertrophic stimuli in cardiac myocytes. *J. Biol. Chem.* 286: 23012–23021.

135 Lo Vasco, V.R., Cardinale, G., and Polonia, P. (2012). Deletion of PLCB1 gene in schizophrenia-affected patients. *J. Cell. Mol. Med.* 16: 844–851.

136 Arinami, T., Ohtsuki, T., Ishiguro, H. et al. (2005). Genomewide high-density SNP linkage analysis of 236 Japanese families supports the existence of schizophrenia susceptibility loci on chromosomes 1p, 14q, and 20p. *Am. J. Hum. Genet.* 77: 937–944.

137 Kim, S.W., Seo, M., Kim, D.S. et al. (2015). Knockdown of phospholipase C-beta1 in the medial prefrontal cortex of male mice impairs working memory among multiple schizophrenia endophenotypes. *J. Psychiatry Neurosci.* 40: 78–88.

138 Manning, E.E., Ransome, M.I., Burrows, E.L., and Hannan, A.J. (2012). Increased adult hippocampal neurogenesis and abnormal migration of adult-born granule neurons is associated with hippocampal-specific cognitive deficits in phospholipase C-beta1 knockout mice. *Hippocampus* 22: 309–319.

139 Kim, S.W., Cho, T., and Lee, S. (2015). Phospholipase C-beta1 Hypofunction in the Pathogenesis of Schizophrenia. *Front. Psychiatry* 6: 159.

140 Udawela, M., Scarr, E., Boer, S. et al. (2017). Isoform specific differences in phospholipase C beta 1 expression in the prefrontal cortex in schizophrenia and suicide. *NPJ Schizophr.* 3: 19.

141 Schoonjans, A.S., Meuwissen, M., Reyniers, E. et al. (2016). PLCB1 epileptic encephalopathies; Review and expansion of the phenotypic spectrum. *Eur. J. Paediatr. Neurol.* 20: 474–479.

142 Desprairies, C., Valence, S., Maurey, H. et al. (2020). Three novel patients with epileptic encephalopathy due to biallelic mutations in the PLCB1 gene. *Clin. Genet.* 97: 477–482.

143 Kim, D., Jun, K.S., Lee, S.B. et al. (1997). Phospholipase C isozymes selectively couple to specific neurotransmitter receptors. *Nature* 389: 290–293.

144 Woodcock, E.A., Grubb, D.R., Filtz, T.M. et al. (2009). Selective activation of the "b" splice variant of phospholipase Cbeta1 in chronically dilated human and mouse atria. *J. Mol. Cell Cardiol.* 47: 676–683.

145 Filtz, T.M., Grubb, D.R., McLeod-Dryden, T.J. et al. (2009). Gq-initiated cardiomyocyte hypertrophy is mediated by phospholipase Cbeta1b. *FASEB J.* 23: 3564–3570.

146 Grubb, D.R., Crook, B., Ma, Y. et al. (2015). The atypical 'b' splice variant of phospholipase Cbeta1 promotes cardiac contractile dysfunction. *J. Mol. Cell Cardiol.* 84: 95–103.

147 Woodcock, E.A., Grubb, D.R., and Iliades, P. (2010). Potential treatment of cardiac hypertrophy and heart failure by inhibiting the sarcolemmal binding of phospholipase Cbeta1b. *Curr. Drug Targets* 11: 1032–1040.

148 Myagmar, B.E., Ismaili, T., Swigart, P.M. et al. (2019). Coupling to Gq signaling is required for cardioprotection by an alpha-1A-adrenergic receptor agonist. *Circ. Res.* 125: 699–706.

149 Dahl, E.F., Wu, S.C., Healy, C.L. et al. (2018). Subcellular compartmentalization of proximal Galphaq-receptor signaling produces unique hypertrophic phenotypes in adult cardiac myocytes. *J. Biol. Chem.* 293: 8734–8749.

150 Akinaga, J., Garcia-Sainz, J.A., and A SP. (2019). Updates in the function and regulation of alpha1 -adrenoceptors. *Br. J. Pharmacol.* 176: 2343–2357.

151 Ratti, S., Mongiorgi, S., Ramazzotti, G. et al. (2017). Nuclear inositide signaling via phospholipase C. *J. Cell. Biochem.* 118: 1969–1978.

152 Bakhsh, A.D., Ladas, I., Hamshere, M.L. et al. (2018). An InDel in phospholipase-C-B-1 is linked with euthyroid multinodular goiter. *Thyroid* 28: 891–901.

153 Molinari, C., Medri, L., Follo, M.Y. et al. (2012). PI-PLCbeta1 gene copy number alterations in breast cancer. *Oncol. Rep.* 27: 403–408.

154 Zhao, P., Sun, S., Zhai, Y. et al. (2019). miR-423-5p inhibits the proliferation and metastasis of glioblastoma cells by targeting phospholipase C beta 1. *Int. J. Clin. Exp. Pathol.* 12: 2941–2950.

155 Yang, X., Sun, L., Ghosh, S., and Rao, A.K. (1996). Human platelet signaling defect characterized by impaired production of inositol-1,4,5-triphosphate and phosphatidic acid and diminished Pleckstrin phosphorylation: evidence for defective phospholipase C activation. *Blood* 88: 1676–1683.

156 Li, Z., Jiang, H., Xie, W. et al. (2000). Roles of PLC-beta2 and -beta3 and PI3Kgamma in chemoattractant-mediated signal transduction. *Science* 287: 1046–1049.

157 Bach, T.L., Chen, Q.M., Kerr, W.T. et al. (2007). Phospholipase cbeta is critical for T cell chemotaxis. *J. Immunol.* 179: 2223–2227.

158 Shi, T.J., Liu, S.X., Hammarberg, H. et al. (2008). Phospholipase C{beta}3 in mouse and human dorsal root ganglia and spinal cord is a possible target for treatment of neuropathic pain. *Proc. Natl. Acad. Sci. U.S.A.* 105: 20004–20008.

159 Mathews, J.L., Smrcka, A.V., and Bidlack, J.M. (2008). A novel Gβγ subunit inhibitor selectively modulates μ-opioid-dependent antinociception and attenuates acute morphine-induced antinociceptive tolerance and dependence. *J. Neurosci.* 28: 12183–12189.

160 Xie, W., Samoriski, G.M., McLaughlin, J.P. et al. (1999). Genetic alteration of phospholipase C beta3 expression modulates behavioral and cellular responses to mu opioids. *Proc. Natl. Acad. Sci. U.S.A.* 96: 10385–10390.

161 Campbell, A.P. and Smrcka, A.V. (2018). Targeting G protein-coupled receptor signalling by blocking G proteins. *Nat. Rev. Drug Discovery* 17: 789–803.

162 Mathews, J.L., Smrcka, A.V., and Bidlack, J.M. (2008). A novel Gbetagamma-subunit inhibitor selectively modulates mu-opioid-dependent antinociception and attenuates acute morphine-induced antinociceptive tolerance and dependence. *J. Neurosci.* 28: 12183–12189.

163 Arttamangkul, S., Birdsong, W., and Williams, J.T. (2015). Does PKC activation increase the homologous desensitization of mu opioid receptors? *Br. J. Pharmacol.* 172: 583–592.

164 Stoveken, H.M., Zucca, S., Masuho, I. et al. (2020). The orphan receptor GPR139 signals via Gq/11 to oppose opioid effects. *J. Biol. Chem.* 295 (31): 10822–10830.

165 Wang, Z., Liu, B., Wang, P. et al. (2008). Phospholipase C beta3 deficiency leads to macrophage hypersensitivity to apoptotic induction and reduction of atherosclerosis in mice. *J. Clin. Invest.* 118: 195–204.

166 Ben-Salem, S., Robbins, S.M., Lm Sobreira, N. et al. (2018). Defect in phosphoinositide signalling through a homozygous variant in PLCB3 causes a new form of spondylometaphyseal dysplasia with corneal dystrophy. *J. Med. Genet.* 55: 122–130.

167 Nabil, A., El Shafei, S., El Shakankiri, N.M. et al. (2020). A familial PLCB4 mutation causing auriculocondylar syndrome 2 with variable severity. *Eur. J. Med. Genet.* 103917.

168 Miyata, M., Kim, H.T., Hashimoto, K. et al. (2001). Deficient long-term synaptic depression in the rostral cerebellum correlated with impaired motor learning in phospholipase C beta4 mutant mice. *Eur. J. Neurosci.* 13: 1945–1954.

169 Ikeda, M., Hirono, M., Sugiyama, T. et al. (2009). Phospholipase C-beta4 is essential for the progression of the normal sleep sequence and ultradian body temperature rhythms in mice. *PLoS One* 4: e7737.

170 Tomberg, K., Westrick, R.J., Kotnik, E.N. et al. (2018). Whole exome sequencing of ENU-induced thrombosis modifier mutations in the mouse. *PLos Genet.* 14: e1007658.

171 Johansson, P., Aoude, L.G., Wadt, K. et al. (2016). Deep sequencing of uveal melanoma identifies a recurrent mutation in PLCB4. *Oncotarget* 7: 4624–4631.

172 Bakhoum, M.F. and Esmaeli, B. (2019). Molecular characteristics of uveal melanoma: insights from the cancer genome atlas (TCGA) project. *Cancers (Basel)* 11.

173 Shain, A.H., Bagger, M.M., Yu, R. et al. (2019). The genetic evolution of metastatic uveal melanoma. *Nat. Genet.* 51: 1123–1130.

174 Karlsson, J., Nilsson, L.M., Mitra, S. et al. (2020). Molecular profiling of driver events in metastatic uveal melanoma. *Nat. Commun.* 11: 1894.

175 Ceraudo, E., Horioka, M., Mattheisen, J.M. et al. (2021). Direct evidence that the GPCR CysLTR2 mutant causative of uveal melanoma is constitutively active with highly biased signaling. *J. Biol. Chem.* 296: 100163.

176 Vaque, J.P., Dorsam, R.T., Feng, X. et al. (2013). A genome-wide RNAi screen reveals a Trio-regulated Rho GTPase circuitry transducing mitogenic signals initiated by G protein-coupled receptors. *Mol. Cell* 49: 94–108.

177 Feng, X., Degese, M.S., Iglesias-Bartolome, R. et al. (2014). Hippo-independent activation of YAP by the GNAQ uveal melanoma oncogene through a trio-regulated rho GTPase signaling circuitry. *Cancer Cell* 25: 831–845.

178 Feng, X., Arang, N., Rigiracciolo, D.C. et al. (2019). A platform of synthetic lethal gene interaction networks reveals that the GNAQ uveal melanoma oncogene controls the hippo pathway through FAK. *Cancer Cell* 35 (457-72): e5.

179 Moore, A.R., Ceraudo, E., Sher, J.J. et al. (2016). Recurrent activating mutations of G-protein-coupled receptor CYSLTR2 in uveal melanoma. *Nat. Genet.* 48: 675–680.

180 Luke, J.J., Triozzi, P.L., McKenna, K.C. et al. (2015). Biology of advanced uveal melanoma and next steps for clinical therapeutics. *Pigment Cell Melanoma Res.* 28: 135–147.

181 Carvajal, R.D., Schwartz, G.K., Tezel, T. et al. (2017). Metastatic disease from uveal melanoma: treatment options and future prospects. *Br. J. Ophthalmol.* 101: 38–44.

182 Luke, J.J. (2019). The newest treatments for uveal melanoma. *Clin. Adv. Hematol. Oncol.* 17: 490–493.

183 Nash, C.A., Brown, L.M., Malik, S. et al. (2018). Compartmentalized cyclic nucleotides have opposing effects on regulation of hypertrophic phospholipase Cepsilon signaling in cardiac myocytes. *J. Mol. Cell Cardiol.* 121: 51–59.

184 Nash, C.A., Wei, W., Irannejad, R., and Smrcka, A.V. (2019). Golgi localized beta1-adrenergic receptors stimulate Golgi PI4P hydrolysis by PLCepsilon to regulate cardiac hypertrophy. *eLife* 8.

185 Xiang, S.Y., Ouyang, K., Yung, B.S. et al. (2013). PLCepsilon, PKD1, and SSH1L transduce RhoA signaling to protect mitochondria from oxidative stress in the heart. *Sci. Signal.* 6: ra108.

186 Li, W., Li, Y., Chu, Y. et al. (2019). PLCE1 promotes myocardial ischemia-reperfusion injury in H/R H9c2 cells and I/R rats by promoting inflammation. *Biosci. Rep.* 39.

187 Dzhura, I., Chepurny, O.G., Leech, C.A. et al. (2011). Phospholipase C-epsilon links Epac2 activation to the potentiation of glucose-stimulated insulin secretion from mouse islets of Langerhans. *Islets* 3: 121–128.

188 Hu, L., Edamatsu, H., Takenaka, N. et al. (2010). Crucial role of phospholipase Cepsilon in induction of local skin inflammatory reactions in the elicitation stage of allergic contact hypersensitivity. *J. Immunol.* 184: 993–1002.

189 Takenaka, N., Edamatsu, H., Suzuki, N. et al. (2011). Overexpression of phospholipase Cepsilon in keratinocytes upregulates cytokine expression and causes dermatitis with acanthosis and T-cell infiltration. *Eur. J. Immunol.* 41: 202–213.

190 Ikuta, S., Edamatsu, H., Li, M. et al. (2008). Crucial role of phospholipase C epsilon in skin inflammation induced by tumor-promoting phorbol ester. *Cancer Res.* 68: 64–72.

191 Harada, Y., Edamatsu, H., and Kataoka, T. (2011). PLCepsilon cooperates with the NF-kappaB pathway to augment TNFalpha-stimulated CCL2/MCP1 expression in human keratinocyte. *Biochem. Biophys. Res. Commun.* 414: 106–111.

192 Oka, M., Edamatsu, H., Kunisada, M. et al. (2011). Phospholipase Cvarepsilon has a crucial role in ultraviolet B-induced neutrophil-associated skin inflammation by regulating the expression of CXCL1/KC. *Lab. Invest.* 91: 711–718.

193 Zhu, X., Sun, Y., Mu, X. et al. (2017). Phospholipase Cepsilon deficiency delays the early stage of cutaneous wound healing and attenuates scar formation in mice. *Biochem. Biophys. Res. Commun.* 484: 144–151.

194 Bijli, K.M., Fazal, F., Slavin, S.A. et al. (2016). Phospholipase C-epsilon signaling mediates endothelial cell inflammation and barrier disruption in acute lung injury. *Am. J. Physiol. Lung Cell Mol. Physiol.* 311: L517–L524.

195 Nagano, T., Edamatsu, H., Kobayashi, K. et al. (2014). Phospholipase cepsilon, an effector of ras and rap small GTPases, is required for airway inflammatory response in a mouse model of bronchial asthma. *PLoS One* 9: e108373.

196 Wakita, M., Edamatsu, H., Li, M. et al. (2016). Phospholipase C activates nuclear factor-kappaB signaling by causing cytoplasmic localization of ribosomal S6 kinase and facilitating its phosphorylation of inhibitor kappaB in colon epithelial cells. *J. Biol. Chem.* 291: 12586–12600.

197 Umezawa, K., Nagano, T., Kobayashi, K. et al. (2019). Phospholipase Cepsilon plays a crucial role in neutrophilic inflammation accompanying acute lung injury through augmentation of CXC chemokine production from alveolar epithelial cells. *Respir. Res.* 20: 9.

198 Oldenburger, A., Timens, W., Bos, S. et al. (2014). Epac1 and Epac2 are differentially involved in inflammatory and remodeling processes induced by cigarette smoke. *FASEB J.* 28: 4617–4628.

199 Bai, Y., Edamatsu, H., Maeda, S. et al. (2004). Crucial role of phospholipase Cepsilon in chemical carcinogen-induced skin tumor development. *Cancer Res.* 64: 8808–8810.

200 Martins, M., McCarthy, A., Baxendale, R. et al. (2014). Tumor suppressor role of phospholipase C epsilon in Ras-triggered cancers. *Proc. Natl. Acad. Sci. U.S.A.* 111: 4239–4244.

201 Oka, M., Edamatsu, H., Kunisada, M. et al. (2010). Enhancement of ultraviolet B-induced skin tumor development in phospholipase Cepsilon-knockout mice is associated with decreased cell death. *Carcinogenesis* 31: 1897–1902.

202 Rhodes, D.R., Yu, J., Shanker, K. et al. (2004). ONCOMINE: a cancer microarray database and integrated data-mining platform. *Neoplasia* 6: 1–6.

203 Luo, X.P. (2014). Phospholipase C epsilon-1 inhibits p53 expression in lung cancer. *Cell Biochem. Funct.* 32: 294–298.

204 Yue, Q.Y., Zhao, W., Tan, Y. et al. (2019). PLCE1 inhibits apoptosis of non-small cell lung cancer via promoting PTEN methylation. *Eur. Rev. Med. Pharmacol. Sci.* 23: 6211–6216.

205 Min, H.Y., Boo, H.J., Lee, H.J. et al. (2016). Smoking-associated lung cancer prevention by blockade of the beta-adrenergic receptor-mediated insulin-like growth factor receptor activation. *Oncotarget* 7: 70936–70947.

206 Li, M., Edamatsu, H., Kitazawa, R. et al. (2009). Phospholipase Cepsilon promotes intestinal tumorigenesis of Apc(Min/+) mice through augmentation of inflammation and angiogenesis. *Carcinogenesis* 30: 1424–1432.

207 Wang, X., Zhou, C., Qiu, G. et al. (2012). Phospholipase C epsilon plays a suppressive role in incidence of colorectal cancer. *Med. Oncol.* 29: 1051–1058.

208 Wang, Y., Wu, X., Ou, L. et al. (2015). PLCepsilon knockdown inhibits prostate cancer cell proliferation via suppression of Notch signalling and nuclear translocation of the androgen receptor. *Cancer Lett.* 362: 61–69.

209 Fan, J., Fan, Y., Wang, X. et al. (2019). PLCε regulates prostate cancer mitochondrial oxidative metabolism and migration via upregulation of Twist1. *J. Exp. Clin. Cancer Res.* 38: 337.

210 Wang, X., Fan, Y., Du, Z. et al. (2018). Knockdown of phospholipase Cepsilon (PLCepsilon) inhibits cell proliferation via phosphatase and tensin homolog deleted on chromosome 10 (PTEN)/AKT signaling pathway in human prostate cancer. *Med. Sci. Monit.* 24: 254–263.

211 Yang, Y. and Wu, X.H. (2017). Study on the influence of metformin on castration-resistant prostate cancer PC-3 cell line biological behavior by its inhibition on PLCepsilon gene-mediated Notchl/Hes and androgen receptor signaling pathway. *Eur. Rev. Med. Pharmacol. Sci.* 21: 1918–1923.

212 Yuan, M., Gao, Y., Li, L. et al. (2019). Phospholipase C (PLC)epsilon promotes androgen receptor antagonist resistance via the bone morphogenetic protein (BMP)-6/SMAD axis in a castration-resistant prostate cancer cell line. *Med. Sci. Monit.* 25: 4438–4449.

213 Sun, W., Li, L., Du, Z. et al. (2019). Combination of phospholipase Cepsilon knockdown with GANT61 sensitizes castrationresistant prostate cancer cells to enzalutamide by suppressing the androgen receptor signaling pathway. *Oncol. Rep.* 41: 2689–2702.

214 Li, L., Du, Z., Gao, Y. et al. (2019). PLCepsilon knockdown overcomes drug resistance to androgen receptor antagonist in castration-resistant prostate cancer by suppressing the wnt3a/beta-catenin pathway. *J. Cell. Physiol.*

215 Liu, J., Zheng, Y., Gao, Y. et al. (2020). Inhibitor 9 combined with androgen deprivation therapy or chemotherapy delays the malignant behavior of castration-resistant prostate cancer through K-Ras/PLCepsilon/PKCepsilon signaling pathway. *Front. Oncol.* 10: 75.

216 Duan, L.M., Liu, J.Y., Yu, C.W. et al. (2020). PLCepsilon knockdown prevents serine/glycine metabolism and proliferation of prostate cancer by suppressing YAP. *Am. J. Cancer Res.* 10: 196–210.

217 Fan, Y., Ou, L., Fan, J. et al. (2020). PLCepsilon regulates metabolism and metastasis signaling via HIF-1alpha/MEK/ERK pathway in prostate cancer. *J. Cell. Physiol.*

218 Pu, J., Li, T., Liu, N. et al. (2020). PLCepsilon knockdown enhances the radiosensitivity of castrationresistant prostate cancer via the AR/PARP1/DNAPKcs axis. *Oncol. Rep.* 43: 1397–1412.

219 Ou, L., Guo, Y., Luo, C. et al. (2010). RNA interference suppressing PLCE1 gene expression decreases invasive power of human bladder cancer T24 cell line. *Cancer Genet. Cytogenet.* 200: 110–119.

220 Cheng, H., Luo, C., Wu, X. et al. (2011). shRNA targeting PLCepsilon inhibits bladder cancer cell growth in vitro and in vivo. *Urology* 78 (474): e7–e11.

221 Ling, Y., Chunli, L., Xiaohou, W., and Qiaoling, Z. (2011). Involvement of the PLCepsilon/PKCalpha pathway in human BIU-87 bladder cancer cell proliferation. *Cell Biol. Int.* 35: 1031–1036.

222 Du, H.F., Ou, L.P., Yang, X. et al. (2014). A new PKCalpha/beta/TBX3/E-cadherin pathway is involved in PLCepsilon-regulated invasion and migration in human bladder cancer cells. *Cell. Signalling* 26: 580–593.

223 Jiang, T., Liu, T., Li, L. et al. (2016). Knockout of phospholipase Cepsilon attenuates *N*-butyl-*N*-(4-hydroxybutyl) nitrosamine-induced bladder tumorigenesis. *Mol. Med. Rep.* 13: 2039–2045.

224 Yang, X., Ou, L., Tang, M. et al. (2015). Knockdown of PLCepsilon inhibits inflammatory cytokine release via STAT3 phosphorylation in human bladder cancer cells. *Tumour Biol.* 36: 9723–9732.

225 Cheng, H., Hao, Y., Gao, Y. et al. (2019). PLCepsilon promotes urinary bladder cancer cells proliferation through STAT3/LDHA pathwaymediated glycolysis. *Oncol. Rep.* 41: 2844–2854.

226 Wang, L.D., Zhou, F.Y., Li, X.M. et al. (2010). Genome-wide association study of esophageal squamous cell carcinoma in Chinese subjects identifies susceptibility loci at PLCE1 and C20orf54. *Nat. Genet.* 42: 759–763.

227 Wu, C., Hu, Z., He, Z. et al. (2011). Genome-wide association study identifies three new susceptibility loci for esophageal squamous-cell carcinoma in Chinese populations. *Nat. Genet.* 43: 679–684.

228 Abnet, C.C., Freedman, N.D., Hu, N. et al. (2010). A shared susceptibility locus in PLCE1 at 10q23 for gastric adenocarcinoma and esophageal squamous cell carcinoma. *Nat. Genet.* 42: 764–767.

229 Zhai, S., Liu, C., Zhang, L. et al. (2017). PLCE1 Promotes Esophageal Cancer Cell Progression by Maintaining the Transcriptional Activity of Snail. *Neoplasia* 19: 154–164.

230 Li, Y. and Luan, C. (2018). PLCE1 Promotes the Invasion and Migration of Esophageal Cancer Cells by Up-Regulating the PKCalpha/NF-kappaB Pathway. *Yonsei Med. J.* 59: 1159–1165.

231 Chen, Y., Wang, D., Peng, H. et al. (2019). Epigenetically upregulated oncoprotein PLCE1 drives esophageal carcinoma angiogenesis and proliferation via activating the PI-PLCepsilon-NF-kappaB signaling pathway and VEGF-C/Bcl-2 expression. *Mol. Cancer* 18: 1.

232 Li, Y., An, J., Huang, S. et al. (2014). PLCE1 suppresses p53 expression in esophageal cancer cells. *Cancer Invest.* 32: 236–240.

233 Chen, Y., Xin, H., Peng, H. et al. (2020). Hypomethylation-linked activation of PLCE1 impedes autophagy and promotes tumorigenesis through MDM2-mediated ubiquitination and destabilization of p53. *Cancer Res.*

234 Yu, J., Zheng, Y., Han, X.P. et al. (2019). Three-gene immunohistochemical panel predicts progression and unfavorable prognosis in esophageal squamous cell carcinoma. *Hum. Pathol.* 88: 7–17.

235 Guo, Y., Bao, Y., Ma, M. et al. (2017). Clinical significance of the correlation between PLCE 1 and PRKCA in esophageal inflammation and esophageal carcinoma. *Oncotarget* 8: 33285–33299.

236 Cui, X.B., Li, S., Li, T.T. et al. (2016). Targeting oncogenic PLCE1 by miR-145 impairs tumor proliferation and metastasis of esophageal squamous cell carcinoma. *Oncotarget* 7: 1777–1795.

237 Cui, X.B., Peng, H., Li, R.R. et al. (2017). MicroRNA-34a functions as a tumor suppressor by directly targeting oncogenic PLCE1 in Kazakh esophageal squamous cell carcinoma. *Oncotarget* 8: 92454–92469.

238 Han, N., Zhao, W., Zhang, Z., and Zheng, P. (2016). MiR-328 suppresses the survival of esophageal cancer cells by targeting PLCE1. *Biochem. Biophys. Res. Commun.* 470: 175–180.

239 Wang, M., Zhang, R., He, J. et al. (2012). Potentially functional variants of PLCE1 identified by GWASs contribute to gastric adenocarcinoma susceptibility in an eastern Chinese population. *PLoS One* 7: e31932.

240 Luo, D., Gao, Y., Wang, S. et al. (2011). Genetic variation in PLCE1 is associated with gastric cancer survival in a Chinese population. *J. Gastroenterol.* 46: 1260–1266.

241 Chen, J., Wang, W., Zhang, T. et al. (2012). Differential expression of phospholipase C epsilon 1 is associated with chronic atrophic gastritis and gastric cancer. *PLoS One* 7: e47563.

242 Bourguignon, L.Y., Gilad, E., Brightman, A. et al. (2006). Hyaluronan-CD44 interaction with leukemia-associated RhoGEF and epidermal growth factor receptor promotes Rho/Ras co-activation, phospholipase C epsilon-Ca2+ signaling, and cytoskeleton modification in head and neck squamous cell carcinoma cells. *J. Biol. Chem.* 281: 14026–14040.

243 Du, H.F., Ou, L.P., Song, X.D. et al. (2014). Nuclear factor-kappaB signaling pathway is involved in phospholipase Cepsilon-regulated proliferation in human renal cell carcinoma cells. *Mol. Cell. Biochem.* 389: 265–275.

244 Hinkes, B., Wiggins, R.C., Gbadegesin, R. et al. (2006). Positional cloning uncovers mutations in PLCE1 responsible for a nephrotic syndrome variant that may be reversible. *Nat. Genet.* 38: 1397–1405.

245 Gbadegesin, R., Hinkes, B.G., Hoskins, B.E. et al. (2008). Mutations in PLCE1 are a major cause of isolated diffuse mesangial sclerosis (IDMS). *Nephrol. Dial Transplant* 23: 1291–1297.

246 Boyer, O., Benoit, G., Gribouval, O. et al. (2010). Mutational analysis of the PLCE1 gene in steroid resistant nephrotic syndrome. *J. Med. Genet.* 47: 445–452.

247 Warejko, J.K., Tan, W., Daga, A. et al. (2018). Whole Exome Sequencing of Patients with Steroid-Resistant Nephrotic Syndrome. *Clin. J. Am. Soc. Nephrol.* 13: 53–62.

248 Hashmi, J.A., Safar, R.A., Afzal, S. et al. (2018). Whole exome sequencing identification of a novel insertion mutation in the phospholipase C epsilon1 gene in a family with steroid resistant inherited nephrotic syndrome. *Mol. Med. Rep.* 18: 5095–5100.

249 Chaib, H., Hoskins, B.E., Ashraf, S. et al. (2008). Identification of BRAF as a new interactor of PLCepsilon1, the protein mutated in nephrotic syndrome type 3. *Am. J. Physiol. Renal. Physiol.* 294: F93–F99.

250 Kalwa, H., Storch, U., Demleitner, J. et al. (2015). Phospholipase C epsilon (PLCepsilon) induced TRPC6 activation: a common but redundant mechanism in primary podocytes. *J. Cell. Physiol.* 230: 1389–1399.

251 Rao, J., Ashraf, S., Tan, W. et al. (2017). Advillin acts upstream of phospholipase C 1 in steroid-resistant nephrotic syndrome. *J. Clin. Invest.* 127: 4257–4269.

252 Yu, S., Choi, W.I., Choi, Y.J. et al. (2020). PLCE1 regulates the migration, proliferation, and differentiation of podocytes. *Exp. Mol. Med.* 52: 594–603.

253 Gilbert, R.D., Turner, C.L., Gibson, J. et al. (2009). Mutations in phospholipase C epsilon 1 are not sufficient to cause diffuse mesangial sclerosis. *Kidney Int.* 75: 415–419.

254 Atchison, D.K., O'Connor, C.L., Menon, R. et al. (2020). Hypertension induces glomerulosclerosis in phospholipase C-epsilon1 deficiency. *Am. J. Physiol. Renal. Physiol.* 318: F1177–F1187.

255 Bleasdale, J.E., Bundy, G.L., Bunting, S. et al. (1989). Inhibition of phospholipase C dependent processes by U-73, 122. *Adv. Prostaglandin Thromboxane Leukot. Res.* 19: 590–593.

256 Huang, W., Barrett, M., Hajicek, N. et al. (2013). Small molecule inhibitors of phospholipase C from a novel high-throughput screen. *J. Biol. Chem.* 288: 5840–5848.

257 Klein, R.R., Bourdon, D.M., Costales, C.L. et al. (2011). Direct activation of human phospholipase C by its well known inhibitor u73122. *J. Biol. Chem.* 286: 12407–12416.

258 Huang, W., Hicks, S.N., Sondek, J., and Zhang, Q. (2011). A fluorogenic, small molecule reporter for mammalian phospholipase C isozymes. *ACS Chem. Biol.* 6: 223–228.

259 Charpentier, T.H., Waldo, G.L., Barrett, M.O. et al. (2014). Membrane-induced allosteric control of phospholipase C-beta isozymes. *J. Biol. Chem.* 289: 29545–29557.

260 Bonacci, T.M., Mathews, J.L., Yuan, C. et al. (2006). Differential targeting of Gbetagamma-subunit signaling with small molecules. *Science* 312: 443–446.

261 Bianchi, E., Norcini, M., Smrcka, A., and Ghelardini, C. (2009). Supraspinal Gbetagamma-dependent stimulation of PLCbeta originating from G inhibitory protein-mu opioid receptor-coupling is necessary for morphine induced acute hyperalgesia. *J. Neurochem.* 111: 171–180.

262 Kawasaki, T., Taniguchi, M., Moritani, Y. et al. (2003). Antithrombotic and thrombolytic efficacy of YM-254890, a G q/11 inhibitor, in a rat model of arterial thrombosis. *Thromb. Haemost.* 90: 406–413.

263 Kawasaki, T., Taniguchi, M., Moritani, Y. et al. (2005). Pharmacological properties of YM-254890, a specific G(alpha)q/11 inhibitor, on thrombosis and neointima formation in mice. *Thromb. Haemost.* 94: 184–192.

264 Carr, R. 3rd, Koziol-White, C., Zhang, J. et al. (2016). Interdicting Gq Activation in Airway Disease by Receptor-Dependent and Receptor-Independent Mechanisms. *Mol. Pharmacol.* 89: 94–104.

265 Matthey, M., Roberts, R., Seidinger, A. et al. (2017). Targeted inhibition of Gq signaling induces airway relaxation in mouse models of asthma. *Sci. Transl. Med.* 9.

266 Onken, M.D., Makepeace, C.M., Kaltenbronn, K.M. et al. (2018). Targeting nucleotide exchange to inhibit constitutively active G protein alpha subunits in cancer cells. *Sci. Signal.* 11.

267 Lapadula, D., Farias, E., Randolph, C.E. et al. (2019). Effects of Oncogenic Galphaq and Galpha11 Inhibition by FR900359 in Uveal Melanoma. *Mol. Cancer Res.* 17: 963–973.

268 Malik, S., deRubio, R.G., Trembley, M. et al. (2015). G protein betagamma subunits regulate cardiomyocyte hypertrophy through a perinuclear Golgi phosphatidylinositol 4-phosphate hydrolysis pathway. *Mol. Biol. Cell* 26: 1188–1198.

269 Bouvet, M., Blondeau, J.P., and Lezoualc'h, F. (2019). The Epac1 Protein: Pharmacological Modulators, Cardiac Signalosome and Pathophysiology. *Cells* 8.

270 Robichaux, W.G. 3rd, and Cheng, X. (2018). Intracellular cAMP Sensor EPAC: Physiology, Pathophysiology, and Therapeutics Development. *Physiol. Rev.* 98: 919–1053.

271 Laudette, M., Coluccia, A., Sainte-Marie, Y. et al. (2019). Identification of a pharmacological inhibitor of Epac1 that protects the heart against acute and chronic models of cardiac stress. *Cardiovasc. Res.* 115: 1766–1777.

272 Liu, W., Ha, Y., Xia, F. et al. (2020). Neuronal Epac1 mediates retinal neurodegeneration in mouse models of ocular hypertension. *J. Exp. Med.* 217.

273 Grubb, D.R., Gao, X.M., Kiriazis, H. et al. (2016). Expressing an inhibitor of PLCbeta1b sustains contractile function following pressure overload. *J. Mol. Cell Cardiol.* 93: 12–17.

274 Fitzpatrick, J.M., Pellegrini, M., Cushing, P.R., and Mierke, D.F. (2014). Small molecule inhibition of the Na(+)/H(+) exchange regulatory factor 1 and parathyroid hormone 1 receptor interaction. *Biochemistry* 53: 5916–5922.

275 Zhong, W., Yi, Q., Xu, B. et al. (2016). ORP4L is essential for T-cell acute lymphoblastic leukemia cell survival. *Nat. Commun.* 7: 12702.

276 Zhong, W., Xu, M., Li, C. et al. (2019). ORP4L extracts and presents PIP2 from plasma membrane for PLCbeta3 catalysis: targeting it eradicates leukemia stem cells. *Cell Rep.* 26 (2166-77): e9.

277 Ratti, S., Follo, M.Y., Ramazzotti, G. et al. (2019). Nuclear phospholipase C isoenzyme imbalance leads to pathologies in brain, hematologic, neuromuscular, and fertility disorders. *J. Lipid Res.* 60: 312–317.

278 Hanna, J., Hossain, G.S., and Kocerha, J. (2019). The Potential for microRNA Therapeutics and Clinical Research. *Front. Genet.* 10: 478.

GPCRs as Therapeutic Targets

GPCRs as Therapeutic Targets

Volume 2

Edited by

Annette Gilchrist
College of Pharmacy-Downers Grove, Midwestern University,
Downers Grove, IL, USA

WILEY

This edition first published 2023

Registered Office
John Wiley & Sons, Inc., 111 River Street, Hoboken, NJ 07030, USA

Editorial Office
John Wiley & Sons, Inc., 111 River Street, Hoboken, NJ 07030, USA

For details of our global editorial offices, customer services, and more information about Wiley products visit us at www.wiley.com.

Wiley also publishes its books in a variety of electronic formats and by print-on-demand. Some content that appears in standard print versions of this book may not be available in other formats.

Library of Congress Cataloging-in-Publication Data

Names: Gilchrist, Annette, editor.
Title: GPCRs as therapeutic targets / edited by Annette Gilchrist.
Other titles: G-protein-coupled receptors as therapeutic targets
Description: Hoboken, NJ : Wiley, 2023. | Includes bibliographical references and index.
Identifiers: LCCN 2022016266 (print) | LCCN 2022016267 (ebook) | ISBN 9781119564744 (cloth) | ISBN 9781119564799 (adobe pdf) | ISBN 9781119564720 (epub)
Subjects: MESH: Receptors, G-Protein-Coupled | Drug Delivery Systems
Classification: LCC RS199.5 (print) | LCC RS199.5 (ebook) | NLM QU 55.7 | DDC 615/.6–dc23/eng/20220608
LC record available at https://lccn.loc.gov/2022016266
LC ebook record available at https://lccn.loc.gov/2022016267

Cover image: © Jon Burkhart; acrylic (Media)
Cover design by Wiley

Set in 9.5/12.5pt STIXTwoText by Straive, Chennai, India

Contents

Preface

G protein-coupled receptors (GPCRs) are the largest group of cell surface receptors. They regulate nearly all known human physiological processes from the sensory modalities of vision, taste, and smell to hormones that control our growth and development to neurotransmitters that govern behavior. Given their role in normal homeostasis and a broad array of pathological conditions including cancer, diabetes, cardiovascular disease, and asthma to name just a few, they serve as the targets for hundreds of drugs and more recently biologics such as monoclonal antibodies. Yet our understanding of GPCRs continues to evolve and texts that discuss these receptors must constantly be revisited. For example, we are just beginning to appreciate the importance of genetic variation in targeted GPCRs.

In the 30 years I have studied GPCRs many novel pharmacological concepts have been advanced. We have seen the emergence of constitutively active GPCRs and inverse agonism, arrestin signaling and functional selectivity, receptor dimerization, and signaling through subcellular receptors. We have seen many GPCRs without known endogenous ligands (orphans) undergo deorphanization, and accepted that some orphan receptors may only function constitutively in a ligand-independent manner. Advances in our understanding of their basal activity, their ability to bind a diverse array of ligands, how they communicate a signal across the cell membrane or within the cells, when they dimerize, their crosstalk with other receptors, and that their genetic variation can lead to disease or differences in drug response has expanded our appreciation for GPCRs. In addition, the depth of our understanding of GPCR pharmacology has in turn altered the drug discovery process itself, expanding the ways in which they are screened for compounds that modulate their signaling.

This two volume book set is organized into 26 chapters and will serve as a resource for any scientists investigating GPCRs, be it in academia or industry. The first volume provides in-depth information about the molecular pharmacology of this important target class and presents up-to-date material on GPCR structures and structure based drug design. There are eight chapters on the evolving pharmacology for GPCRs, including chapters discussing allosteric modulation, receptor dimerization, deorphanization, ubiquitination, intracellular trafficking, and subcellular GPCR signaling. The next six chapters discuss the rapidly growing field of GPCR structures and structure based drug design. Included in this section are chapters on the structural basis of G protein selectivity, as well as rational drug design for not only GPCRs but downstream signaling molecules such as phospholipase C. The second volume includes information on the role of GPCRs in disease and novel approaches for studying this receptor family. There are seven chapters addressing how GPCRs play a role in a wide range of pathological states including cancer, substance use disorders, cerebrovascular disease, and metabolic disease. The final five chapters present recent approaches employed to study GPCRs including mass spectrometry, bioluminescence, single molecule microscopy, and optogenetics. Together, the two volume book set provides a thorough overview of GPCRs in terms of their structure, pharmacology, function, and role in disease states, and provides information on novel approaches to measure GPCR activity.

Annette Gilchrist
Midwestern University

List of Contributors

John A. Allen
Department of Pharmacology and Toxicology
University of Texas Medical Branch
Galveston, TX
USA

and

Department of Neuroscience and Cell Biology
University of Texas Medical Branch
Galveston, TX
USA

and

Center for Addiction Research
University of Texas Medical Branch
Galveston, TX
USA

Salete J. Baptista
Data-Driven Molecular Design Group
CNC – Center for Neuroscience and Cell Biology
University of Coimbra
Coimbra
Portugal

and

Centro de Ciências e Tecnologias Nucleares, Instituto Superior Técnico
Universidade de Lisboa
Bobadela LRS
Portugal

Carlos A.V. Barreto
Data-Driven Molecular Design Group
CNC – Center for Neuroscience and Cell Biology
University of Coimbra
Coimbra
Portugal

and

Institute for Interdisciplinary Research, University of Coimbra
PhD Programme in Experimental Biology and Biomedicine
Coimbra
Portugal

Eric Bell
Department of Computational Medicine and Bioinformatics
University of Michigan
Ann Arbor, MI
USA

Kyla Bourque
Department of Pharmacology and Therapeutics
McGill University
Montréal, Québec
Canada

Mariaelana Brenner
Department of Medicine, Division of Endocrinology, Diabetes and Metabolism
University of Illinois College of Medicine
Chicago, IL
USA

Seema Briyal
Pharmaceutical Sciences
Midwestern University
Downers Grove, IL
USA

Michael R. Bruchas
Center of Excellence in the Neurobiology of Addiction, Pain, and Emotion, Department of Anesthesiology
University of Washington
Seattle, WA
USA

and

Department of Pharmacology
University of Washington
Seattle, WA
USA

Beatriz Bueschbell
Department of Pharmaceutical and Medicinal Chemistry
Pharmaceutical Institute, University of Bonn
Bonn
Germany

Davide Calebiro
College of Medical and Dental Sciences
Institute of Metabolism and Systems Research, University of Birmingham
Birmingham
UK

and

Centre of Membrane Proteins and Receptors (COMPARE)
Universities of Nottingham and Birmingham
UK

Wallace Chan
Department of Computational Medicine and Bioinformatics
University of Michigan
Ann Arbor, MI
USA

and

Department of Biological Chemistry
University of Michigan
Ann Arbor, MI
USA

and

Department of Pharmacology
University of Michigan
Ann Arbor, MI
USA

Naincy R. Chandan
Department of Pharmacology
University of Michigan Medical School
Ann Arbor, MI
USA

Sun Choi
College of Pharmacy and Graduate School of Pharmaceutical Sciences
Ewha Womans University
Seoul
Republic of Korea

Irene Coin
Faculty of Life Sciences
Institute of Biochemistry, Leipzig University
Leipzig
Germany

Bryan A. Copits
Pain Center, Department of Anesthesiology
Washington University School of Medicine
St. Louis, MO
USA

Natasha C. Dale
Molecular Endocrinology and Pharmacology
Harry Perkins Institute of Medical Research, QEII Medical Centre
Nedlands, Western Australia
Australia

and

Centre for Medical Research
The University of Western Australia
Crawley, Western Australia
Australia

and

National Centre
Australian Research Council Centre for Personalised Therapeutics Technologies
Australia

Michael R. Dores
Department of Biology
Hofstra University
Hempstead, NY
USA

Stefan Ernicke
Faculty of Life Sciences, Institute of Biochemistry
Leipzig University
Leipzig
Germany

Daniel E. Felsing
Department of Pharmacology and Toxicology
University of Texas Medical Branch
Galveston, TX
USA

and

Center for Addiction Research
University of Texas Medical Branch
Galveston, TX
USA

Geordi Frere
Department of Chemistry, Chemical and Physical Sciences
University of Toronto
Mississauga, ON
Canada

Elyssa Frohlich
Department of Pharmacology and Therapeutics
McGill University
Montréal, Québec
Canada

Annette Gilchrist
Department of Pharmaceutical Sciences, College of Pharmacy-Downers Grove
Midwestern University
Downers Grove, IL
USA

Sophie Gough
Department of Medicine, Division of Endocrinology, Diabetes and Metabolism
University of Illinois College of Medicine
Chicago, IL
USA

Jerome Gould
Department of Chemistry, Chemical and Physical Sciences
University of Toronto
Mississauga, ON
Canada

Karen J. Gregory
Drug Discovery Biology
Monash Institute of Pharmaceutical Science, Monash University
Parkville
Australia

Jak Grimes
College of Medical and Dental Sciences
Institute of Metabolism and Systems Research, University of Birmingham
Birmingham
UK

and

Centre of Membrane Proteins and Receptors (COMPARE)
Universities of Nottingham and Birmingham
UK

Anil Gulati
Pharmaceutical Sciences
Midwestern University
Downers Grove, IL
USA

and

Pharmazz, Inc.
Willowbrook, IL
USA

Patrick T. Gunning
Department of Chemistry, Chemical and Physical Sciences
University of Toronto
Mississauga, ON
Canada

Michael A. Hanson
SB SciTech
San Marcos, CA
USA

Steven K. Harmon
Department of Neuroscience
Washington University School of Medicine
Saint Louis, MO
USA

Advait Hasabnis
Department of Chemistry, Chemical and Physical Sciences
University of Toronto
Mississauga, ON
Canada

Terence E. Hébert
Department of Pharmacology and Therapeutics
McGill University
Montréal, Québec
Canada

Laura H. Heitman
Division of Drug Discovery and Safety
LACDR
Leiden University
The Netherlands

and

Oncode Institute
Leiden
The Netherlands

Carole Le Henaff
Department of Molecular Pathobiology
New York University College of Dentistry
New York, NY
USA

Iara C. Ibay
Department of Pharmaceutical Sciences, College of Pharmacy-Downers Grove
Midwestern University
Downers Grove, IL
USA

Adriaan P. IJzerman
Division of Drug Discovery and Safety, LACDR
Leiden University
The Netherlands

Roshanak Irannejad
Department of Biochemistry and Biophysics
Cardiovascular Research Institute
University of California
San Francisco, CA
USA

Ralf Jockers
Institut Cochin, CNRS, INSERM
Université de Paris
Paris
France

Elizabeth K.M. Johnstone
Molecular Endocrinology and Pharmacology
Harry Perkins Institute of Medical Research, QEII Medical Centre
Nedlands, Western Australia
Australia

and

Centre for Medical Research
The University of Western Australia
Crawley, Western Australia
Australia

and

National Centre
Australian Research Council Centre
for Personalised Therapeutics
Technologies
Australia

Yuh-Jiin I. Jong
Department of Neuroscience
Washington University School of
Medicine
Saint Louis, MO
USA

Tracy M. Josephs
Drug Discovery Biology
Monash Institute of Pharmaceutical
Science, Monash University
Parkville
Australia

Andrew N. Keller
Drug Discovery Biology
Monash Institute of Pharmaceutical
Science, Monash University
Parkville
Australia

Terry Kenakin
Department of Pharmacology
University of North Carolina School of
Medicine
Chapel Hill, NC
USA

Brian T. Layden
Department of Medicine, Division of
Endocrinology, Diabetes and
Metabolism
University of Illinois College of
Medicine
Chicago, IL
USA

and

Jesse Brown Veterans Affairs Medical
Center
Department of Medicine, Section of
Endocrinology
Chicago, IL
USA

Raudah Lazim
College of Pharmacy and Graduate
School of Pharmaceutical Sciences
Ewha Womans University
Seoul
Republic of Korea

Kristen R. Lednovich
Department of Medicine, Division of
Endocrinology, Diabetes and
Metabolism
University of Illinois College of
Medicine
Chicago, IL
USA

Yoonji Lee
College of Pharmacy
Chung-Ang University
Seoul
Republic of Korea

Katie Leach
Drug Discovery Biology
Monash Institute of Pharmaceutical Science, Monash University
Parkville
Australia

Agostinho Lemos
Data-Driven Molecular Design Group
CNC – Center for Neuroscience and Cell Biology
University of Coimbra
Coimbra
Portugal

Braden T. Lobingier
Department of Chemical Physiology and Biochemistry
Oregon Health and Sciences University
Portland, OR
USA

Yan Lu
iHuman Institute, ShanghaiTech University
Shanghai
China

and

School of Life Science and Technology
ShanghaiTech University
Shanghai
China

Miguel Machuqueiro
Departmento de Química e Bioquímica, Faculdade de Ciências
Universidade de Lisboa
BioISI-Biosystems and Integrative Sciences Institute
Lisboa
Portugal

Pedro R. Magalhães
Departmento de Química e Bioquímica, Faculdade de Ciências
Universidade de Lisboa
BioISI-Biosystems and Integrative Sciences Institute
Lisboa
Portugal

Rita Melo
Data-Driven Molecular Design Group
CNC – Center for Neuroscience and Cell Biology
University of Coimbra
Coimbra
Portugal

and

Centro de Ciências e Tecnologias Nucleares, Instituto Superior Técnico
Universidade de Lisboa
Bobadela LRS
Portugal

Ravi Mistry
College of Medical and Dental Sciences
Institute of Metabolism and Systems Research, University of Birmingham
Birmingham
UK

and

Centre of Membrane Proteins and Receptors (COMPARE)
Universities of Nottingham and Birmingham
UK

Irina S. Moreira
Data-Driven Molecular Design Group
CNC – Center for Neuroscience and Cell Biology
University of Coimbra
Coimbra
Portugal

and

Department of Life Sciences, Faculty of Science and Technology
University of Coimbra
Coimbra
Portugal

and

CIBB – Center for Innovative Biomedicine and Biotechnology
University of Coimbra
Coimbra
Portugal

Pratanphorn Nakliang
College of Pharmacy and Graduate School of Pharmaceutical Sciences
Ewha Womans University
Seoul
Republic of Korea

Ashley N. Nilson
Department of Neuroscience and Cell Biology
University of Texas Medical Branch
Galveston, TX
USA

and

Center for Addiction Research
University of Texas Medical Branch
Galveston, TX
USA

Atsuro Oishi
Institut Cochin, CNRS, INSERM
Université de Paris
Paris
France

and

Department of Anatomy
Kyorin University Faculty of Medicine
Tokyo
Japan

and

Cancer RNA Research Unit
National Cancer Center Research Institute
Tokyo
Japan

Karen L. O'Malley
Department of Neuroscience
Washington University School of Medicine
Saint Louis, MO
USA

Patrick R. O'Neill
Hatos Center for Neuropharmacology
Department of Psychiatry and
Biobehavioral Sciences
University of California Los Angeles
Los Angeles, CA
USA

Maria C. Orencia
GPCR Consortium
San Marcos, CA
USA

Aditya Pandey
Department of Chemistry, Chemical
and Physical Sciences
University of Toronto
Mississauga, ON
Canada

and

Department of Biochemistry
University of Toronto
Toronto, ON
Canada

Nicola C. Partridge
Department of Molecular Pathobiology
New York University College of
Dentistry
New York, NY
USA

Frank Park
Department of Pharmaceutical
Sciences
University of Tennessee Health
Science Center
Memphis, TN
USA

Kevin D.G. Pfleger
Molecular Endocrinology and
Pharmacology
Harry Perkins Institute of Medical
Research
QEII Medical Centre
Nedlands, Western Australia
Australia

and

Centre for Medical Research
The University of Western Australia
Crawley, Western Australia
Australia

and

National Centre
Australian Research Council Centre
for Personalised Therapeutics
Technologies
Australia

and

Dimerix Limited
Nedlands, Western Australia
Australia

Hoa T.N. Phan
Department of Pharmacology
University of Michigan Medical School
Ann Arbor, MI
USA

António J. Preto
Data-Driven Molecular Design Group
CNC – Center for Neuroscience and
Cell Biology
University of Coimbra
Coimbra
Portugal

and

Institute for Interdisciplinary Research, University of Coimbra
PhD Programme in Experimental Biology and Biomedicine
Coimbra
Portugal

Robert S. Prosser
Department of Chemistry, Chemical and Physical Sciences
University of Toronto
Mississauga, ON
Canada

and

Department of Biochemistry
University of Toronto
Toronto, ON
Canada

Talha Qadri
Department of Medicine, Division of Endocrinology, Diabetes and Metabolism
University of Illinois College of Medicine
Chicago, IL
USA

Shanshan Qin
iHuman Institute, ShanghaiTech University
Shanghai
China

Amaresh Ranjan
Pharmaceutical Sciences
Midwestern University
Downers Grove, IL
USA

Nícia Rosário-Ferreira
Data-Driven Molecular Design Group
CNC – Center for Neuroscience and Cell Biology
University of Coimbra
Coimbra
Portugal

and

Chemistry Department, Faculty of Science and Technology, Coimbra Chemistry Center
University of Coimbra
Coimbra
Portugal

Fredrik Sadler
Biochemistry, Molecular Biology and Biophysics Graduate Program
University of Minnesota
Minneapolis, MN
USA

Anke C. Schiedel
Department of Pharmaceutical and Medicinal Chemistry, Pharmaceutical Institute
University of Bonn
Bonn
Germany

Wenqing Shui
iHuman Institute
ShanghaiTech University
Shanghai
China

and

School of Life Science and Technology
ShanghaiTech University
Shanghai
China

Sivaraj Sivaramakrishnan
Biochemistry, Molecular Biology and
Biophysics Graduate Program
University of Minnesota
Minneapolis, MN
USA

and

Department of Genetics
Cell Biology, and Development
University of Minnesota
Minneapolis, MN
USA

Alan V. Smrcka
Department of Pharmacology
University of Michigan Medical School
Ann Arbor, MI
USA

Emma Tripp
College of Medical and Dental
Sciences
Institute of Metabolism and Systems
Research, University of Birmingham
Birmingham
UK

and

Centre of Membrane Proteins and
Receptors (COMPARE)
Universities of Nottingham and
Birmingham
UK

Nagarajan Vaidehi
Department of Computational and
Quantitative Medicine, City of Hope
Cancer Center
Beckman Research Institute of the
City of Hope
Duarte, CA
USA

Xuesong Wang
Division of Drug Discovery and Safety
LACDR
Leiden University
The Netherlands

Gerard J. P. van Westen
Division of Drug Discovery and Safety
ALCDR
Leiden University
The Netherlands

Carl W. White
Molecular Endocrinology and
Pharmacology
Harry Perkins Institute of Medical
Research, QEII Medical Centre
Nedlands, Western Australia
Australia

and

Centre for Medical Research
The University of Western Australia
Crawley, Western Australia
Australia

and

National Centre
Australian Research Council Centre
for Personalised Therapeutics
Technologies
Australia

Jiansheng Wu
Department of Computational Medicine and Bioinformatics
University of Michigan
Ann Arbor, MI
USA

and

School of Geographic and Biological Information
Nanjing University of Posts and Telecommunications
Nanjing
China

Yang Zhang
Department of Computational Medicine and Bioinformatics
University of Michigan
Ann Arbor, MI
USA

and

Department of Biological Chemistry
University of Michigan
Ann Arbor, MI
USA

Part III

GPCRs and Disease

15

G Protein-Coupled Receptors in Metabolic Disease

Kristen R. Lednovich[1], Sophie Gough[1], Mariaelana Brenner[1], Talha Qadri[1], and Brian T. Layden[1,2]

[1] *Department of Medicine, Division of Endocrinology, Diabetes and Metabolism, University of Illinois College of Medicine, Chicago, IL, USA*
[2] *Jesse Brown Veterans Affairs Medical Center, Department of Medicine, Section of Endocrinology, Chicago, IL, USA*

15.1 Introduction

Since the Nobel-prize winning discovery of G protein-coupled receptors (GPCRs) in 1994, our understanding of their critical contributions to physiology has expanded substantially. With over 800 unique receptors in the human genome and 200 known ligands, GPCRs have been found to play essential roles in almost all known physiological processes, including metabolic homeostasis [1]. Great strides in biomedical discovery have allowed for extensive characterization of the metabolic functions of GPCRs including energy homeostasis, glucose and fat metabolism, and more.

Due to their critical involvement in a diverse range of physiological processes, GPCRs represent the largest and most comprehensively studied family of therapeutic drug targets, compromising approximately 35% of all FDA-approved drugs [2]. The number of GPCR-based therapies in development has rapidly expanded in the recent years due to advances in structural biology and receptor pharmacology. With the emergence of metabolic diseases such as obesity and diabetes, many of these new therapies target metabolic functions. Of the current GPCR-based therapies in clinical trials, the highest number of pharmaceutical agents are for use in diabetes patients, underscoring the importance of this family of receptors in metabolic function [3].

GPCRs as Therapeutic Targets, Volume 2, First Edition. Edited by Annette Gilchrist.

In this chapter, we begin with an overview of major metabolic diseases with an emphasis on highly prevalent disorders including obesity and its associated conditions, which has rapidly burgeoned into a global epidemic affecting up to nearly 40% of the world's adult population [4]. We then describe the basic structural and biochemical properties of GPCRs, their associated heterotrimeric proteins, and signaling pathways. Finally, we conclude with an in-depth discussion of the major GPCRs expressed in each major metabolic organic, including the pancreas, intestines, liver, kidneys, and adipose tissue. We describe the physiology and predominant metabolic roles of each organ, emphasizing the contributions of key GPCRs and signaling pathways involved in tissue function. Additionally, we discuss both emerging therapeutic targets as well as therapies currently on the market, thereby providing a comprehensive review of GPCRs and their role in metabolic disease.

15.2 Metabolic Disease

15.2.1 Overview

Metabolic diseases comprise some of the most common human diseases of the modern era, most notably in the United States and other Westernized countries – though prevalence and incidence are increasing in many parts of the world [5]. Of note, the global rise of obesity and metabolic syndrome is one of the most concerning current public health crises. This epidemic can be – in part – attributed to Westernized high-fat diets and sedentary lifestyles.

Here, we will focus on adult endocrine disorders, as childhood endocrine and metabolic disorders encompass a broader scope, including inborn errors of metabolism. There are several different categories of adult endocrine and metabolic disorders: hypothalamic-pituitary disorders, thyroid disorders, female and male endocrine disorders, calcium and metabolic bone disorders, diabetes (type 1 diabetes (T1D) and type 2 diabetes (T2D) and associated disorders), dyslipidemias, and obesity. The most prevalent endocrine disorders in the US – with prevalence estimates over 5% in adults – include diabetes mellitus (DM), impaired fasting glucose, impaired glucose tolerance, obesity, metabolic syndrome, osteoporosis, osteopenia, mild-moderate hypovitaminosis D, erectile dysfunction (ED), dyslipidemia, and thyroiditis. The prevalence of DM (defined as patients who report of ever being told of having diabetes and/or use of oral hypoglycemic medications or insulin or 1997 American Diabetes Association criteria) in the US is 24.2%. According to the National Cholesterol Education Program (NCEP) Adult Treatment Panel III (ATP III) criteria, the prevalence of Metabolic Syndrome is 34%, and according to the International Diabetes Foundation (IDF) criteria, the prevalence is 39% [6].

Metabolic syndrome is particularly threatening as both glucose-modulating and cardiovascular states are compromised, thereby leading to microvascular and macrovascular consequences. Thus, metabolic risk factors promote the development of both T2D and atherosclerotic cardiovascular disease (CVD). These conditions are of particular interest for scientific study as their prevalence, mortality, and morbidity is ever-increasing in our populations. Establishing the pathophysiological mechanisms can elucidate better treatment options and improve health-outcomes for patients.

15.2.2 Metabolic Syndrome

The pathogenesis of metabolic syndrome (MetS) is primarily driven by energy excess, physical inactivity, and obesity; though there are additional risk-attributable genetic and metabolic susceptibilities. The metabolic syndrome is a constellation of symptoms and pathologies which confer an increased risk for atherosclerotic cardiovascular disease (ASCVD) and type 2 diabetes. Of note, the usefulness of identifying this constellation of symptoms beyond their individual effects in identifying patients at risk for ASCVD and T2D has recently been put up for debate. There are several definitions for metabolic syndrome, but the most commonly used are the NCEP, ATP III, and IDF criteria. Most criteria include abnormalities of glucose metabolism, dyslipidemia, obesity (particularly upper-body and central adiposity), and hypertension (HTN).

MetS has several important risk factors for ASCVD: atherogenic dyslipidemia, HTN, dysglycemia, a pro-thrombotic state, and a pro-inflammatory state. Dyslipidemia is thought to be the primarily feature putting patients at risk for ASCVD – etiology explained below. Obesity and calorie-excess increases renal absorption of sodium (to which insulin resistance also contributes), expands intravascular volume, and activates the renin-angiotensin-aldosterone system (RAAS). Hyperglycemia resulting from insulin-resistance puts most patients with MetS in the prediabetic or diabetic range. Both prediabetes and diabetes put patients at risk for microvascular disease – namely, chronic kidney disease and diabetic retinopathy (the leading cause of blindness in the US) [7]. The obesity-induced pro-inflammatory state is associated with an increase in acute-phase reactants and persistent low-grade inflammation. Additionally, cytokine release from excess adipose tissue contributes to insulin resistance in skeletal muscle, pancreatic beta cell dysfunction, and pituitary-adrenal axis dysfunction. Patients with MetS are at high risk for ASCVD and venous thrombosis as a result of endothelial dysfunction, coagulative state, impaired fibrinolysis, and platelet dysfunction.

As previously stated, metabolic syndrome results primarily from a state of calorie excess and physical inactivity. The Western Diet – primarily composed of high amounts of fat and sugars – has led the metabolic syndrome pandemic particularly in the US, but it is spreading globally. Calorie restriction has traditionally been the primary intervention for patients. However, traditional low-calorie diets are very difficult for patients to adhere to long term – resulting in weight re-gain in most patients. Long term changes require behavioral and lifestyle modifications. The best method for patient intervention is a team approach involving the physician team and a medical nutrition therapist. Dietary changes should be centered around reducing – or better, eliminating – saturated fat and cholesterol-laden foods. Randomized trials have shown that primarily plant-based diets result in sustained weight loss, improved insulin sensitivity, and decreased cardiovascular events when compared to a traditional method of weight loss (i.e. calorie restriction) [8, 9]. The DASH diet and Mediterranean diet have also been described as acceptable methods of sustained dietary modification [10, 11]. Though regular physical activity and dietary modifications are the mainstay of treatment, bariatric surgery may also be considered for means of weight loss. Medical management of related conditions including hypertension, dyslipidemia, and impaired glucose tolerance may require ACE inhibitors, statins, and metformin (respectively) treatment, though lifestyle intervention is the primary step before medications [12].

15.2.3 Type 2 Diabetes

The pathogenesis of Type 2 Diabetes (T2D) is complicated as patients present with varying degrees of insulin resistance and insulin deficiency. Additionally, pathogenesis and presentation are multifactorial – often due to lifestyle, genetic predisposition, and environmental influences. T2D is generally defined by a hemoglobin A1C (glycated hemoglobin which measures average glycemia over three months) greater than 6.5%. Skeletal muscle and liver tissue become resistant to the glycemic modulating activities of insulin, causing the pancreatic beta cells pump out more insulin to lower blood sugar. Hyperactivity of beta cells is thought to progressively impair pancreatic function. After prolonged insulin hyperproduction and insulin resistant tissues, pancreatic beta cell function declines, resulting in a marked decrease in insulin production [13, 14]. Thus, T2D pathogenesis is a vicious cycle of worsening metabolic function and hyperglycemia – making it difficult to isolate the precise pathophysiological mechanisms of disease. Metabolic syndrome puts patients at very high risk for progression to T2D; treatment goals of MetS may be focused around prevention of this progression. The features of MetS – hyperinsulinemia secondary to insulin resistance, dyslipidemia, inflammatory cytokines from adipose tissue, and oxidative factors have all been implicated in the pathogenesis of T2D and cardiovascular complications.

The post-prandial state increases blood-glucose levels, and glucose is transported into pancreatic beta cells via the glucose transporter 2 (GLUT-2). As a result, insulin is secreted from beta cells, allowing for glucose-uptake by hepatic and skeletal muscle tissue. Mouse models of reduced GLUT-2 expression result in mice with glucose intolerance, and notably, similar reductions in GLUT-2 expression are induced when mice are fed a high-fat diet [15, 16].

Insulin resistance is thought to be the most important predictor of T2D. Development of insulin resistance is multifactorial process including factors such as increased myocellular lipid content, hyperglycemia itself, altered levels of adipokines (such as leptin, adiponectin, and tumor necrosis factor alpha [TNFa]), genetic factors, and further increasing with advancing age and increased weight. Moreover, the relationship between genetic and environmental factors in the role of diabetic pathogenesis is complex and the focus of many scientific studies. T2D is two to six times more prevalent in African Americans, Native Americans, Pima Indians, and Hispanic Americans in the US than in caucasains [17, 18]. Thirty-nine percent of patients with T2D have at least one parent with the same disease; the lifetime risk for a first-degree relative of a patient with T2D is 5–10 times higher than that of an age- and weight-matched subject without a family history of diabetes [19].

15.2.4 Atherosclerotic Cardiovascular Disease

Three meta-analyses found that metabolic syndrome increases the risk for CVD incidents and all-cause mortality [20–22]. The increased risk of CVD events has shown to be associated with the constellation of metabolic syndrome features rather than simply to obesity. One study found that when compared to obese people without MetS, obese people with MetS had a 10-fold increased risk for diabetes and a 2-fold increased risk for CVD compared to normal weight people without MetS [23].

Dyslipidemia – primarily elevation of apo-B containing lipoproteins LDL and VLDL, elevated triglycerides, and reduced HDL – results in the particularly threating atherosclerotic state. Circulating lipids are trapped in the arterial wall in foamy lipid-laden macrophages which form the first stage of atherosclerosis, the fatty streak. As connective tissue forms around the streak, a fibrous plaque then forms around the cholesterol-rich core in the arterial wall. These plaques are at risk for rupture based on how thin the layer of connective tissue is. Additionally, the plaques can occlude the arterial circumference, eventually resulting in decreased blood flow to the organs supplied by said artery. When atherosclerotic plaques rupture or completely occlude vessels, the patient suffers acute carotid and cerebral vascular events – namely, myocardial infarctions (MI) or strokes. Thus, patients with metabolic syndrome are at significantly increased risk for atherosclerotic cardiovascular events and all-cause mortality.

15.2.5 Metabolic Syndrome Associated Diseases

Metabolic syndrome is also associated with other obesity-related disorders. MetS confers an increased risk for developing non-alcoholic fatty liver disease (NAFLD) with steatosis, fibrosis, and cirrhosis. In a study of 304 patients with NAFLD but without overt diabetes, after correcting for age, sex, and BMI, MetS was associated with an increased risk for severe fibrosis – (odds ratio [OR] 3.5, 95% confidence interval [CI] 1.1–11.2) [24]. MetS also increases the risk for hepatocellular carcinoma and intrahepatic cholangiocarcinoma. Risk of developing chronic kidney disease (CKD) is also increased (OR 2.6) in MetS, and risk increased as the number of features of metabolic syndrome increased respectively [25]. CKD is a major complication of T2D – in fact, hypertension and diabetes (two features/consequences of MetS) are the most common causes of CKD [26]. Finally, metabolic syndrome is also associated with increased risk of polycystic ovary syndrome, obstructive sleep apnea, and hyperuricemia and gout [27–30]. Metabolic syndrome increases the risk for developing other endocrine, metabolic, and CVDs. Thus, it is of increasing urgency to elucidate the molecular etiology of metabolic syndrome to improve therapeutic options.

15.3 G Protein-Coupled Receptors

15.3.1 Overview

Transmembrane proteins are integral membrane proteins that stretch across the entirety of the cell. GPCRs are considered to be the largest group of transmembrane proteins found in human genes. They are cell surface receptors, activated by either a ligand or an extracellular stimulus. This activation produces a conformational change in the protein, which, in turn, leads to the transduction of the signal into the cell through signaling cascades. The term "7TM receptor" (7 transmembrane domains) is also used to define GPCRs, although there are some 7TMs that are not part of the GPCR family. There are about eight hundred different GPCRs identified within humans. The majority of these preserve a role in sensory functions, including smell (400), taste (33), light perception (10) and pheromone signaling (5). The rest (~350) are non-sensory GPCRs, which mediate signaling through ligands ranging in size from small molecules to large proteins.

15.3.2 Structure

GPCRs are composed of seven transmembrane α-helices, which are embedded within the plasma membrane. They are considered to be amphipathic given their transmembrane structure [31]. Interconnecting each of the 7TMs are loops that connect them from beyond the membrane. The three loops found on the intracellular side are called cytosolic loops and the three loops found on the extracellular side are called extracellular loops. While all α-helices are relatively close to each other, there is a sizeable gap between helix five and helix six. This gap is reserved for a G-protein which interacts with a segment of the cytosolic loop found there. Similarly, the extracellular loop found between helix six and seven is accounted as the messenger binding site where a signal is received. The amino terminal segment (N-terminus) is found on the extracellular side of the membrane, and the carboxy terminal tail (C-terminus) is found on the cytosolic side.

15.3.3 Receptor Families

There are six major classifications of GPCRs: Class A (Rhodopsin-like), Class B (Secretin family), Class C (glutamate family), Class D (fungal mating receptors), Class E (cAMP receptors) and Class F(frizzled or smooth receptors) [32]. The largest is Class A (rhodopsin analogous), which accounts for eighty percent of all GPCRs. This class includes a variety of small molecules such as hormones, neuropeptides, light receptors, olfactory receptors, and visual pigments. Class B is known as the "secretin receptor family" and contains about 70 receptors [33]. They have seven TM helices and an extended N-terminal domain containing around 120 residues, which are stabilized by disulfide bonds. Secretin receptors are regulated by peptide hormones by the glucagon hormone family. They also encode for 15 genes in humans. Class C consists of the glutamate family and includes metabotropic glutamate receptors, GABA (gamma-aminobutyric acid), calcium-sensing receptors and taste receptors. These receptors are also responsible for prompting the inositol phosphate/Ca2+ intracellular pathway. They can be distinguished by seven TM helices on an extracellular N-terminus. Class D consists of fungal mating receptors, class E contains cAMP receptors, and lastly, class F consists of frizzled or smooth receptors. The last three classes mentioned (D-F) are structurally similar in that they have 7 hydrophobic domains that are regarded as 7-TM helices.

15.3.4 Signaling Pathways

When a ligand or a signaling mediator attaches to the extracellular loop between helix six and seven, a conformational change is produced in the receptor and an interaction is triggered between the GPCR and a nearby G-protein, resulting in its activation. This reaction is facilitated by a guanine nucleotide-binding complex consisting of three subunits: Gα, Gβ, and Gγ [34]. These three remain firmly bound to each other and are attached by the Gα and Gγ subunits to the plasma membrane via lipid anchor proteins. GDP (guanosine-diphosphate) is found attached to the alpha subunit and the G-protein complex remains dormant while GDP is present. However, when the G-protein is activated, the Gα subunit is activated through the exchanging of GDP for GTP. The Gα subunit and the Gβ/Gγ dimer then break off from each other and detach from the plasma membrane to transduce the signal within the cell. EGFRs (epidermal growth factor receptors) are a type of tyrosine-kinase, which can co-assist GPCRs in the transduction of this signal and act as a downstream signaling partner in the mediation of signals. The phosphorylation of GPCRs, carried out by GPCR kinases (GRKs), protein kinase A (PKA), and protein kinase C (PKC), promotes the termination of this signal by replacing the present GTP with GDP to return to its dormant state. A few notable "downstream" transduction pathways are: cAMP pathway, PLC-γ, GEFs, and ASK1. cAMP is a secondary messenger which has an integral role for cellular responses to many hormones and neurotransmitters [35–37]. Activation occurs when the Gα subunit binds to cAMP after the appropriate ligand binds to the GPCR. Protein kinase (PKA) is cAMP's most significant target. Inactive PKA consists of two C subunits and two R subunits. Activated cAMP binds to the R subunit causing it to dissociate from the C subunit resulting in the activation of PKA. Activated PKA has a multitude of purposes, such as playing a role in the breakdown of glycogen pathway and the activation of CREB, which assists in the transcription of genes [38].

15.4 Discussion by Organ

Metabolic homeostasis is a whole-body process and requires the coordination of many different organs working in synchrony to carry out essential tasks such as maintaining glucose homeostasis and insulin secretion, heart rate and blood pressure, energy expenditure, lipid metabolism, and more. This involves the sensing of both internal and external cues by GPCRs, which respond to a diverse array of ligands to trigger intracellular responses, therefore allowing the body to maintain homeostasis while adapting to different environmental conditions. Here, we summarize the major contributions of key GPCRs to metabolic regulation within each body organ. Additionally, we discuss their roles in metabolic diseases and describe both existing and emerging GPCR-based therapies. Of note, because almost all body organs exhibit some metabolic function to an extent, we have limited our discussion to the major organs implicated in the diseases highlighted in the first part of this chapter.

15.4.1 Pancreas

Located in the upper intestinal tract between the duodenum and stomach, the pancreas is a digestive organ that has two major physiological functions: to secrete enzymes to aid in the digestion of food and to maintain glucose homeostasis throughout the body. The pancreas works in tandem with the duodenum to break down carbohydrates, proteins, and lipids after they exit the stomach. This is achieved through the secretion of digestive enzymes, such as trypsinogen, chymotrypsinogen, lipase, and amylase, which are produced by specialized acinar cells within the pancreas as pro-peptides and are released into the duodenum where they are then activated [39]. Secreted in an acid-neutralizing bicarbonate fluid, approximately 1.5–3 L of enzyme-rich pancreatic juice is drained directly into the duodenum per day and is largely regulated by a number of peptide hormones, such as secretin, cholecystokinin (CCK), and vasoactive intestinal peptide (VIP), which are produced in the upper intestinal tract in response to food. This bidirectional communicative structure allows for the modulation of pancreatic digestive functions in response to physiological changes such as stress, quantity of food, and presence of chemicals such as alcohol.

In addition to its role in the process of digestion, the pancreas is also classified as a major endocrine organ and carries out the complex task of regulating sugar metabolism while maintaining glucose homeostasis throughout the body [40]. Embedded within the pancreas are highly specialized clusters of cells called islets of Langerhans, which consist of five different cell types – alpha, beta, gamma, delta, and epsilon cells – that secrete an important group of hormones that are released into the blood stream in response to different environmental conditions. Alpha cells are responsible for the secretion of glucagon, a hormone which is produced during a fasting state in which circulating blood glucose levels are low. Glucagon is transported directly to the liver, where it increases the breakdown of glycogen (glycogenolysis) and production of glucose (gluconeogenesis), allowing for the generation of endogenous glucose when blood levels of glucose are low [41]. Pancreatic beta cells make up approximately 70% of islets, and secrete a hormone called insulin, which acts reciprocally to glucagon. Insulin is produced in a dose-dependent manner under fed conditions, in which high levels of glucose levels are present in the blood, and facilitates the uptake of glucose into hepatic, muscle, and adipocyte cells for energy storage. Together, glucagon and insulin act as the master regulators for glucose homeostasis by modulating glucose production and storage in response to changing blood glucose levels. This process is further aided by additional hormones produced in the pancreas, such as somatostatin, which regulates alpha and beta cell activity, and ghrelin and amylin, which control satiety.

GPCRs expressed in the pancreas play pivotal roles in carrying out the organ's major functions in both digestion and glucose homeostasis. Communication of gut-produced peptide hormones and their effects on the pancreas is largely mediated by this family of receptors. The secretin receptor is a class B GPCR that is present on pancreatic ductular cells, acinar cells, and even some cells within the islets and becomes activated in the presence of secretin, which is produced by the duodenum in response to high levels of gastric acid [42]. Upon activation, the secretin receptor may couple to either a cAMP-mediated Gαs pathway, or a Ca^{2+}-mediated Gαq pathway, which results in the release of pancreatic juice into the duodenum to neutralize acidic conditions and provide essential digestive enzymes to the intestinal tract. Similarly, pancreatic CCK receptors are responsible for mediating the effects of CCK and gastrin produced in the duodenum. There are two subtypes of CCK receptors which share 50% homology: CCK_A, which is predominant found in the intestinal tract and stimulates bicarbonate secretion, gall bladder emptying, and inhibits gut motility; and CCK_B, which regulates gut-brain functions and is found primarily in the central nervous system. Pancreatic CCK_A generally acts via a Gαq pathway and may act via Gαs pathway when ligand concentrations are high [43].

GPCRs also play an essential role in mediating the function of pancreatic islets, primarily by sensing many of the circulating factors that control secretion of insulin, such as catecholamines, free fatty acids, and glucagon-like peptide-1 (GLP-1), which are released into the blood by other body organs in response to changing glucose levels [44]. There are over 30 characterized GPCRs expressed on pancreatic islets in addition to a myriad of orphan receptors, whose roles are just recently emerging. Based on their downstream signaling pathways, these receptors may either stimulate or inhibit insulin secretion. Typically, the stimulatory Gαs and Gαq-signaling pathways result in an increased secretion of insulin, while the inhibitory Gαi-signaling pathway results in a decreased secretion of insulin. One important group of GPCRs present in pancreatic islets is the adrenergic family of receptors, specifically the α2 and β2 receptors, which have also been shown to regulate islet function [45]. These receptors belong to the rhodopsin-like class A family, and their ligands are catecholamines, including epinephrine and norepinephrine. Activation of the α2 receptor triggers a Gai-mediated inhibition of insulin secretion, and activation of glucagon secretion, while activation of the β2 receptor results in Gαs-signaling and stimulates insulin and glucagon secretion. Additionally, the α2 receptor has been implicated in the pathogenesis of type 2 diabetes (T2D) [46]. Individuals with increased α2 receptor expression have been shown to have impaired glucose-stimulated insulin secretion, elevated fasting blood glucose levels, and an increased risk of T2D. Therefore, it is thought that an α2 receptor antagonist could improve insulin secretion, however, no current therapies for metabolic disorders exist in humans, and usefulness may be limited by both genetic polymorphisms and physiological variations in metabolism between person to person. Nonetheless, α2 receptor antagonists have been developed for use in research.

Interestingly, several types of melatonin receptors are expressed within pancreatic islets and have been shown to influence insulin secretion [47]. Melatonin is a circulating hormone that is produced by the pineal gland and is well known for its ability to regulate circadian rhythm and sleep patterns. Its levels are highest during the night. Recent research has shown that melatonin receptors MTNR1A and MTNR1B bind to nocturnally circulating melatonin within the pancreatic islet, which results in a sharp decline in insulin production thought to be mediated by a Gαi signaling pathway [48]. This observation became the basis for the physiological link between sleep disruption and development of T2D, as sleep disturbances are considered a risk factor for T2D and obesity – a finding also supported by the results of a GWAS study which identified a mutation in the MTNR1B receptor in T2D patients [49]. This exciting new field of research has now implicated pancreatic melatonin receptors as novel drug targets for T2D, and antagonists for MTNR1B are currently under development.

While free fatty acids (FFAs) are mostly known for their role as metabolic substrates, they have also been identified as signaling molecules with the ability to influence insulin and glucagon secretion, and several of their receptors are expressed within the pancreatic islet [50]. Moreover, due to their role as nutrient sensors and mediators of glucose homeostasis, FFA receptors have been identified as promising drug targets for metabolic diseases such as T2D and obesity. Free fatty acids receptors 1(FFA1) and FFA4 respond to long-chain fatty acids, which have a carbon length of >12. FFA1, previously known as GPR40, has been shown act via Gαq-mediated signaling in pancreatic islets to stimulate insulin production within the beta cell. Based on this finding, efforts have been made to develop a small-molecule agonist for FFA1 for use in patients with T2D, however, it has been debated whether an antagonist may be more effective, as T2D patients typically have higher levels of circulating FFAs [51]. FFA4, previously known as GPR120, is also expressed in islets, where it is thought to inhibit the secretion of somatostatin (SST) by delta cells. Though its role in the pancreas is still emerging, studies in mice have revealed activation of the receptor potentiates insulin and glucagon secretion and improves glycemic control via Gαs-mediated signaling. FFA2 and FFA3 (formerly GPR43 and GPR41, respectively) are short-chain fatty acid receptors, whose ligands have a carbon chain length of <4 and consist predominantly of acetate, propionate, and butyrate. FFA2 and FFA3 are expressed in multiple tissues throughout the body, including pancreatic islets, where they have been recently characterized. Studies have shown that FFA2 and FFA3 work reciprocally to modulate insulin secretion, with FFA2 acting via a Gαq-mediated signaling pathway to increase insulin secretion, while FFA3 acts via a Gαi-mediated signaling pathway to inhibit insulin secretion [52]. Additionally, FFA2 and FFA3 are thought to influence β-cell survival and proliferation. While these receptors have a considerable potential as targets for therapeutic drugs, many other factors influencing their function, such as the contributions of diet and gut microbiome, remain unclear. Furthermore, pharmacological differences exist between human and mouse isoforms of these receptors but are yet to be fully clarified. Despite these challenges, targeting the FFA family of receptors, both within the pancreatic beta cell and in other cell types, represents a promising therapeutic strategy for treatment of obesity and T2D.

The GLP-1 receptor (GLP-1R) is perhaps the most well-studied pancreatic GPCR and acts as a major regulator of glucose homeostasis via its ability to sense circulating blood levels of GLP-1, a key insulinotrophic hormone produced in enteroendocrine cells within the intestine, as well as a small subset of neurons upon the consumption of food [53]. GLP1R belongs to the class B family of GPCRs and is coupled to Gαs, acting via an influx of intracellular cAMP to result in insulin secretion from the beta cell, as well as beta cell proliferation and neogenesis. While GLP-1R is able to bind to glucagon, its affinity is approximately 100-1000-fold lower, rendering it physiologically irrelevant. The gastric inhibitory protein receptor (GIP-R) is another important receptor that acts similarly to GLP-1R, however its ligand is GIP, which is also produced by enteroendocrine cells in the intestine and exerts an insulinotropic effect on pancreatic islet beta-cells. GIP produces a number of other diverse physiological response that occur following a meal, including inhibiting gastrointestinal motility and secretion of acid. Its receptors are also expressed in other tissues throughout the body. Like GLP-1R, GIP-R couples to Gαs, and in addition to stimulating the release of insulin, has an anti-apoptotic effect on pancreatic beta cells [54].

GLP-1 and GIP are classified as incretin hormones, meaning that they are able to augment insulin secretion when released from the intestine following the presence of glucose. Their contribution to increased stimulation of insulin secretion elicited by oral glucose as compared with intravenous administration of glucose under similar plasma glucose level is defined as "the incretin effect," and accounts for up to 60% of total insulin secretion following oral glucose administration. Importantly, this is due to the release of GLP-1 and GIP from enteroendocrine cells into the bloodstream, where they go on to stimulate receptors on the pancreatic beta cell, resulting in an increase in insulin secretion [55]. It is now well-understood that the incretin effect is drastically reduced in patients with T2D, and recent studies have uncovered that this is primarily due to defective GLP-1 and GIP production as a consequence of the diabetic state [56]. As such, GLP-1R agonists (GLP1RA) were identified as a novel strategy to treat symptoms of T2D, thereby restoring the metabolic benefits of GLP-1. This strategy has seen remarkable clinical success, starting with the approval of the first GLP1RA, Exenatide, by the FDA in 2005 [57]. Patients administered a twice-daily subcutaneous injection of this GLP-1 analog saw increased glucose-dependent insulin secretion and decreased glucagon secretion as well as delayed gastric emptying and increased satiety, resulting in reduced glucose levels, A1C levels, and body weight. Following the success of Exenatide, a number of other GLP1RA have been developed, each with differences in terms of molecular structure, pharmacokinetics, dose, and administration. While a similar effort has been made to develop agonists for GIP, its expression across multiple tissues, particularly within adipose tissue, have complicated drug development, and unlike GLP-1, pharmacological doses of exogenous GIP were not found to be insulinotropic in patients with T2D [58]. Though targeting the GIP receptor directly remains controversial, recent findings indicate that a dual GLP-1R/GIPR agonist strategy may still have utility in the treatment of T2D.

15.4.2 Liver

The liver is a vital organ that carries out over 500 essential bodily functions, ranging from detoxification, protein synthesis, vitamin and mineral storage, to digestion and bile production. Additionally, the liver plays a central role in maintaining metabolic homeostasis, acting as a hub to metabolically connect various tissues throughout the body, and is the major site for glucose, in the form of glycogen [59]. Following the digestion of food, nutrient-rich blood containing glucose, amino acids, and lipids enters the liver through the portal vein, along with chemical messengers in the form of hormones and fatty acids. Specialized cells within the liver called hepatocytes sense the presence of nutrients, metabolites, and chemical messengers, and initiate a series of metabolic changes in response to environmental conditions. Under a fasting state, the liver releases energy substrates in the form of glucose, triglycerides, and ketone bodies, thereby providing essential metabolic fuel to other tissues throughout the body [60]. Conversely, during a postprandial state, the liver converts glucose to glycogen for storage, and free fatty acids are esterified to form triglycerides for lipid storage. Through this process, the liver is able to tightly control energy and metabolic homeostasis.

One major function of hepatic GPCRs is to mediate the effects of the variety of metabolites, nutrients, and chemical signals present in circulating blood from the portal vein. The liver responds to peptide hormones produced in other tissues, including insulin and glucagon. The glucagon receptor (GCGR) is a class B GPCR and is abundantly expressed within hepatic tissue and is essential for the hepatic glucose production (HPG), which occurs during a fasting or hypoglycemic state [61]. Circulating glucagon binds to GCGR, which is primarily coupled to Gas, and stimulates a cAMP-mediated cascade to ultimately initiate glycogenolysis and gluconeogenesis. The breakdown of glycogen into glucose supplies the body with energy and maintains normal glucose levels during a fasting or low energy state. Of note, hepatic GCGR has also been described to act via Gαq-mediated signaling. In addition to regulating blood glucose levels, hepatic GCGR has also been shown to control hepatic amino acid catabolism and serum amino acid levels [62].

Finally, hepatic GCGR plays a key role in lipid metabolism through its ability to initiate the breakdown of lipids and subsequent mobilization of free fatty acids to other tissues during a fasting state [63]. Multiple studies have shown that elevated glucagon levels are present in individuals with T2D, suggesting that increased signaling via GCGR in the liver plays a role in the development of T2D. Indeed, GCGR polymorphisms have also been linked to diabetic populations [64]. Based on these observations, considerable effort has been made to produce a small-molecule antagonist for GCGR (GRA) for the treatment of T2D. Several decades of development have revealed obstacles in this strategy, most notably the adverse metabolic effects observed in response to GRAs, which include the potential development of glycogen storage disease, increase in serum lipids, and risk of liver toxicity due to overproduction of liver enzymes, all of which have been reported in various developmental experiments involving GRA administration [65]. Another concern with GRA therapy is the potential for malignancies to develop within pancreatic alpha cells, which undergo hyperplasia in response to the blockage of their secreted product. Despite these challenges, the effort to develop GRAs for clinical use has not yet been abandoned.

15.4.3 Intestine

In addition to its central role in digestion, the intestine is a vital metabolic organ that has recently garnered attention for its newly identified functions in regulating metabolic homeostasis, as well as its role in the pathophysiology of metabolic diseases. The intestine is part of the gastrointestinal tract, an organ system also consisting of the mouth, esophagus, and stomach, which ingests food, breaks down its components into useable energy and nutrients for the body, and then expels remaining waste as feces [66]. The lower gastrointestinal tract, collectively referred to as the intestine, is composed of the small intestine and large intestine and is the major site of breakdown and absorption of nutrients. In humans, the small intestine is divided into three portions: the duodenum which receives the contents of the stomach, the jejunum and the ileum, which passes contents into the large intestine, also known as the colon. The epithelial lining of the intestine is composed of a thick, sticky surface of mucus membrane called mucosa. This surface contains a multitude of highly specialized cell types, which varies widely depending on the portion of the intestine. In addition to mucus-secreting goblet cells and immunogenic Paneth cells, the mucosa is home to hormone-secreting enteroendocrine cells [67]. Although enteroendocrine cells only represent approximately 1% of the mucosal cells, their collective control over metabolism has coined the intestine as "the largest endocrine organ in the body." Indeed, the major function of enteroendocrine cells is to produce a large variety of highly specialized peptide hormones which interact with other organs to carry out pivotal roles including insulin secretion, appetite regulation, and digestion and absorption of nutrients. Enteroendocrine cells are stimulated by products of food digestion passing through the intestinal lumen, including glucose, fatty acids and amino acids, as well as metabolites produced by the gut microbiota, thereby acting as nutrient sensors. As such, enteroendocrine cells are lined with receptors involved in a range of signaling pathways that mediate secretion of hormones. In addition to ion channels and transporters, GPCRs play an essential role in enteroendocrine cell activity through their ability to sense nutrients and mediate their responses. In the small intestine, GPCRs sense and respond predominately to nutrients as they are broken down and absorbed, while in the colon, GPCRs are activated mostly by metabolites generated by the microbiome from remaining intestinal content [68]. In this section, we will focus on nutrient and metabolite sensing GPCRs within enteroendocrine cells.

Within the small intestine, GPCRs respond to a broad range of nutrients that are absorbed by the epithelium from food components as they move through the gastrointestinal tract, producing hormonal signals that enter the bloodstream and circulate to peripheral tissues to carry out postprandial functions, including appetite regulation and glucose homeostasis [69]. The ingestion of fat and subsequent breakdown of lipid products in the small intestine has been shown to result in the secretion of hormones, including CCK, GLP-1, and GIP from enteroendocrine cells, an effect largely mediated by GPCRs [70]. Among these are long-chain fatty acid receptors FFA1 and FFA4 – also expressed in the pancreas and adipose tissue – which couple to Gαq signaling pathways and trigger the release of CCK, GLP-1, and GIP to coordinate postprandial responses. GPR119 is another fat sensing GPCR that is highly expressed on enteroendocrine cells in the small intestine which responds to lipid derived Oleoylethanolamine (OEA) and couples to a Gαs coupled signaling pathway. While its precise effects on hormonal secretion are unclear, GLP-1 and GIP1 are thought to be the primary hormones released based on a number of *in vitro* and *in vivo* experiments [71]. Interestingly, GPR119 is also expressed on the basolateral surface of enteroendocrine cells, and it is hypothesized that some of its sensing occurs only after lipids are absorbed through the endocrine cell and hydrolyzed in the basolateral space. This may also be true for FFA1 and FFA4, although apical vs basolateral expression of these receptors is still unclear [72]. FFA1, FFA4, and GPR119 are thought to contribute substantially to the incretin effect through their ability to mediate GLP-1 and GIP release, and similarly to other free fatty acid receptors, have been identified as potential drug targets for the treatment of T2D. Multiple synthetic agonists and antagonists have been developed for these receptors, though due to challenges including binding specificity and receptor affinity, they have yet to be tested in humans [73].

Amino acid receptors including CaSR and GPR142 are highly expressed in the small intestine and have been strongly linked to postprandial gut hormone release.

The canonical role of CaSR is as a calcium sensing receptor, but it has also been shown to respond to amino acids including glutamine (Gln), phenylalanine (Phe), and tryptophan (Trp). Though literature on this receptor's role in gut peptide hormone secretion is scarce, it has been found to signal via a Gαq coupled pathway to result in secretion of GLP-1 and GIP in addition to PYY, a powerful anorexigenic hormone involved in regulating satiety, food intake, and intestinal transit following a meal [74]. Like CaSR, GPR142 also responds to l-amino acids, and through a Gαq coupled signaling response, is thought to trigger release of GLP-1 and GIP from enteroendocrine cells. These receptors are not fully characterized, and ongoing research focuses on elucidating their tissue-specific effects, as well as their contribution to insulin secretion when challenged with l-amino acids [75]. Likewise, there are a number of orphaned nutrient sensing GPCRs that are thought to influence gut hormone secretion in the small intestine, including taste receptors T1R2 and T1R3 and amino acid sensing GPRC6A. It is clear, however, that enteroendocrine cell derived hormones are produced via a cumulative effect of many different nutrient sensing receptors, each responding to their unique ligands.

The gut microbiome has received considerable attention in the recent years for its newly identified ability to influence host metabolic homeostasis, contributing to pivotal processes, such as glucose and fat metabolism as well as energy homeostasis. Furthermore, the gut microbiome has emerged as a key factor in the development of major metabolic diseases including obesity, T2D, and CVD. One of the many mechanisms through which this occurs is through the production of microbial metabolites via bacteria in the distal intestine, which complements mammalian enzymatic activities to generate a broad metabolic repertoire as well [76]. By providing enzymes otherwise absent from the human genome, the gut microbiome contributes substantially to human metabolism, and produces key metabolites including short-chain fatty acids, indole, secondary bile acids, and lipopolysaccharide – all of which directly influence hormone release from enteroendocrine cells.

Short-chain fatty acids (SCFAs) are a group of key metabolites broken down from dietary fibers in the distal colon via the gut microbiome and consist primarily of acetate (C_2), propionate (C_3), and butyrate (C_4). SCFAs are considered to be metabolically beneficial, as well as protective of gut dysbiosis, and their effects on human physiology and metabolism are vast. In addition to serving as a vital source of energy for colonocytes, SCFAs contribute to a variety of physiological processes including inhibition of histone deacetylases, mediation of immune responses, regulation of gut barrier function, glycemic control, fat accumulation, and more [77]. In the distal colon, SCFAs are sensed, in part, by several different GPCRs, including FFA2 and FFA3 as well as orphan receptors GPR109a and OlfR78. FFA2 and FFA3 are expressed in a variety of tissues throughout the body, including the pancreatic beta cell, immune cells, adipocytes, and muscle. Additionally, they are expressed along the intestinal tract, with their highest expression in the distal colon where the levels of their endogenous ligands are over 100-fold higher than circulating blood levels [78]. FFA2 and FFA3 have a 43% sequence homology and share overlapping ligands: acetate, propionate, and butyrate, though their affinities for each ligand differ slightly. FFA2 has the highest affinity for acetate and propionate followed by butyrate, while FFA3 has the highest affinity for butyrate and propionate followed by acetate. Of note, FFA2 and FFA3 can also respond to other SCFAs: formate (C_1), pentanoate (C_5), and hexanoate (C_6), however levels of these metabolites in the colon are typically much lower. The signaling capabilities of each receptor is thought to vary depending on their physiological location, and while we have previously described FFA2 and FFA3 within the pancreas, it is important to indicate differences within each organ. FFA2 has been reported to participate in both Gαq and Gαi coupled signaling pathways, however, signaling in the intestine appears to be limited to a Gαq pathway. The signaling of FFA3 in the intestine is still unclear, as a paucity of data exists for this receptor. In extra-intestinal tissues, FFA3 has been observed to signal via an inhibitory Gαi-coupled pathway. The precise effects of FFA2 and FFA3 on hormonal secretion within enteroendocrine cells is still poorly understood, and there is a growing body of conflicting data obfuscating the actual roles of these receptors. FFA2 and FFA3 were initially reported to mediate GLP-1 release and have been suggested to contribute to glucose homeostasis by modulating insulin secretion and limiting insulin resistance [79]. Other groups have observed FFA2 and FFA3-mediated release of PYY and an absence of GLP-1 release from enteroendocrine cells, indicating that these receptors influence metabolic homeostasis primarily through modulating intestinal function and coordinating postprandial gut-brain roles, such as satiety and food intake. While FFA2's ability to mediate PYY release is generally accepted, the role of FFA3 is obscured by its dual expression in enteric nerves underlying the basolateral surface of enteroendocrine cells. Therefore, it is possible that FFA3 acts primarily through gut-brain circuitry as opposed to directly sensing SCFAs within the lumen of the intestine [80]. Nonetheless, FFA2 and FFA3 have been clearly implicated in mediating the effects of SCFAs and represent a novel link between the gut microbiome and metabolic homeostasis. Elucidating their precise tissue-specific effects is essential for the development of novel drug targets for these receptors.

15.4.4 Kidney and Adrenal Glands

The kidneys are important for blood filtration and body homeostasis. They actively work to filter waste from the blood while resorbing nutrients. They are also key regulators of the body's salt and water balance. Through various signaling mechanisms and transport channels, the kidneys regulate their resorption of water to maintain homeostasis. Consequently, the kidneys are intimately associated with blood pressure because of their ability to regulate fluid volume. This is of importance when discussing metabolic syndrome which includes hypertension as one of its components. The kidney contains several GPCRs that sense the body's overall physiological state and help to maintain balance.

One GPCR that has been well-documented in renin-induced hypertension is Succinate Receptor 1 (SUCNR1 or GPR91). As its name implies, the central ligand for this receptor is succinate, a key intermediate in the citric acid cycle. GPR91 is found throughout the tubules of the kidney, but most notably in the macula densa cells and the glomerular vasculature. Through activation of this receptor, signaling cascades eventually lead to the production of prostaglandins E2 and I2 and nitric oxide, which together act via paracrine signaling to stimulate the juxtaglomerular apparatus to release renin [81, 82]. Renin is part of the RAAS which is ultimately responsible for increasing blood pressure by increasing the level of water resorbed from the kidney. Succinate levels are increased in conditions of local stress such as ischemia or energy imbalances. In particular, succinate is increased in chronic hyperglycemic states such as those found in diabetes mellitus or metabolic disorder [83]. Currently, some studies are also showing increased circulating succinate levels in specific gut microbiome profiles associated with human obesity [84]. Increased levels of succinate seen in conditions such as diabetes, metabolic disorder, and obesity lead to increased activation of this receptor which in turn, may cause hypertension and glomerular hyperfiltration.

Another GPCR located in the kidney is the type 2 vasopressin receptor (V2R). This receptor is located in the principal cells of the collecting duct and is responsible for increasing water resorption from the tubules of the kidney. Vasopressin, or antidiuretic hormone, is released from the posterior pituitary gland when there is an increased plasma osmolality or a decreased intracellular fluid volume. Upon binding of vasopressin to the V2R, aquaporin 2 channels are localized to the plasma membrane from intracellular vesicles. This increases the membrane's permeability to water and allows for increased water uptake [85]. Although not fully understood, it is well-documented that individuals with type 1 or type 2 diabetes have elevated levels of circulating vasopressin. Current studies using animal models suggest causal links between the increased antidiuretic action of vasopressin and early stage diabetic nephropathy. Although further studies are needed, V2R may be a promising target for future therapeutics in preventing early stage diabetic kidney disease [86].

The adrenal glands are located directly superior to the kidneys and are an integral part of the hypothalamic–pituitary–adrenal (HPA) axis. The HPA axis regulates the body's stress response. When the body experiences stress, the hypothalamus releases corticotropin-releasing hormone which then stimulates the anterior pituitary gland to release adrenocorticotropic hormone (ACTH) into the bloodstream. ACTH travels through the bloodstream to the adrenal glands where it activates the adrenocorticotropic hormone receptor, also known as Melanocortin receptor 2. The ACTH receptor is a Gs-coupled protein receptor located in the adrenocortical cells of the adrenal glands. Upon activation, levels of cAMP increase which ultimately stimulates steroidogenesis and the release of glucocorticoids, namely cortisol [87]. Cortisol is released in response to stress and drives the increased production of glucose to provide energy for the body. It does this by stimulating lipolysis and proteolysis in various tissues to provide for increased gluconeogenesis in the liver, thereby increasing glucose levels in the blood. Cortisol also induces insulin-resistance. Increased activation of this receptor through chronic stress can ultimately lead to insulin resistance and chronically elevated blood glucose levels.

The adrenal medulla also synthesizes catecholamines such as epinephrine and norepinephrine. These catecholamines are important in triggering the body's "fight or flight" response when the sympathetic nervous system is activated. When these catecholamines are released from post-ganglionic neurons, they stimulate a variety of adrenergic receptors found on various organs that trigger the body's sympathetic response. Adrenergic receptors, also called adrenoreceptors, are a class of GPCRs that include α1, α2, β1, β2, and β3 receptors. Each of these receptors signal through a specific G protein class and therefore act through various signal transduction pathways to produce a response. The specific response is dependent on the location of the receptors throughout the body. The α1 receptors function through the Gαq pathway and stimulate vascular smooth muscle contraction and pupillary dilation. αAll the β receptors are coupled to the Gαs pathway and activate adenylyl cyclase to increase cAMP. β1 receptors are primarily found in the heart and increase heart rate, contractility, and renin release from the kidneys. β2 receptors are primarily found in the lungs and cause bronchodilation. Finally, β3 receptors are mainly found in adipose tissue and increase lipolysis. Although not a comprehensive list, this highlights the variety of roles that the adrenoreceptors play in activating the sympathetic response. Specifically, in terms of metabolism, the adrenoreceptors raise blood glucose levels by increasing lipolysis and glycogenolysis to provide the body with the needed energy during the "fight or flight" response [88]. The ability of the adrenoreceptors to regulate blood glucose levels suggest some potential as therapeutics in metabolic disorders; however, their other biological roles may limit their utility in metabolic disorders.

15.4.5 Adipose Tissue

Adipose tissue, colloquially known as body fat, is a loose, fibrous type of connective tissue packed with cells called adipocytes in addition to specialized macrophages, fibroblasts, and endothelial cells. Collectively, adipose tissue is considered to be a major endocrine organ and plays a central role in regulating processes such as energy, lipid, and glucose homeostasis [89]. In humans, there are three major types of adipose tissue that differ in morphology and gene expression, each with specialized metabolic functions. White adipose tissue (WAT) is composed of single large lipid droplets and makes up approximately 20–25% of an individual's total body weight. The primary functions in WAT are to store energy in the form of triglycerides and insulin in the body's other tissues. WAT is found in two distinct types of deposits throughout the body: subcutaneous adipose tissue is found beneath the skin, while visceral adipose tissue is found lining internal organs such as the heart, abdomen, and urogenital region [90]. While the precise differences between these two types of WAT are still unclear, it is generally accepted that fat distribution, rather than mass, plays an important role in pathophysiology of obesity and associated diseases. Deposits of visceral fat, particularly surrounding the abdomen (mesenteric and ornamental), are associated with the development of obesity-related diseases, such as NAFLD and CVD. Brown adipose tissue (BAT) is composed of multilocular cells rich in mitochondria that release heat and play an important role in thermal regulation. In humans, BAT is most abundant in newborn babies and decreases with age. In adults, deposits of BAT can be found in the lower neck and supraclavicular areas. Due to its ability to rapidly convert glucose and fat into energy, BAT is well known to exert a myriad of benefits on human metabolism, including improved glucose and insulin sensitivity, and has subsequently emerged as a major target for obesity, T2D, and other obesity-related diseases [91]. Beige adipose tissue, with an intermediate morphology to WAT and BAT, has recently been identified as a third type of fat and can display the phenotype of either type of tissue depending on environmental conditions. Beige adipocytes are thought to originate from a transition from WAT to BAT in a process known as "browning," driven by cold exposure as well as other environmental signals [92]. A multitude of GPCRs are found in all types of adipose tissue and critically contribute to their specialized functions.

Adipose tissue is considered an endocrine organ and secretes two key peptide hormones: leptin and adiponectin, which act as essential regulators of energy homeostasis [93]. Dysfunction in leptin and adiponectin signaling has been linked to major metabolic diseases such as obesity, T2D, and CVD. Adiponectin is found circulating in whole blood and stimulates fatty acid oxidation in skeletal muscle while inhibiting glucose production in the liver to coordinate whole-body energy homeostasis [94]. Additionally, adiponectin is a powerful anti-inflammatory agent. Release of adiponectin from WAT is stimulated by the binding of insulin to insulin receptors on the surface of adipocytes. While the insulin receptor itself is a tyrosine kinase receptor, a number of GPCRs also contribute to adiponectin release in response to insulin, including ADORA1, APLNR, EDNRA, GPR116, and TSHR, all of which act via a Gαs coupled signaling pathway to result in an intracellular release of cAMP. This results in secretion of adiponectin from adipose tissue, which circulates in high abundance to skeletal muscle, the liver, and the heart. Leptin is the other major peptide hormone secreted by adipose tissue in levels that are correlated with body fat [95]. It plays a central role in major metabolic processes including regulation of energy homeostasis and function of neuroendocrine hormones by binding to its receptors on the surface of the hypothalamus.

The release of leptin from WAT is also controlled by insulin, in addition to a number of other circulating hormones and signaling molecules. Unlike adiponectin secretion, GPCRs that signal via an inhibitory Gαi coupled pathway contribute to an increase in secretion, while GPCRs that signal via the stimulatory Gαs pathway inhibit secretion. There are 20 identified GPCRs that influence leptin secretion, with an additional 20 GPCRs that have implicated involvement. Additionally, exercise is known to be a major inhibitor of leptin secretion, acting through beta adrenergic receptors (betaARs) expressed in WAT [96]. There are three betaAR subtypes: beta1AR, beta2AR, and beta3AR, all of which couple to Gαs. While beta1AR and beta2AR are broadly expressed throughout the body, beta3AR is found primarily in adipose tissue, including both WAT and BAT. In addition to its effects on leptin release, beta3AR has also been shown to contribute to the process of lipolysis in WAT and thermogenic regulation in BAT. Due to its major effects on metabolic regulation, beta3AR has been identified as a potential drug target for obesity and obesity-related diseases. There has been an ongoing effort to develop an agonist for beta3AR for over three decades with no current therapies currently on the market for treatment of obesity. This is due to a number of complications that have emerged during drug development, notably, the realization of major pharmacological differences between the human and murine isoforms of the receptor [97]. While clinical trials for this receptor have been largely unsuccessful in the treatment of obesity, a successful beta3AR agonist was developed for overactive bladder (OAB) and approved by the FDA in 2012 under the name Mirabegron. For several decades, interest in developing a beta3AR agonist for weight loss has waned, though the recent discoveries involving the importance of BAT and thermogenesis in obesity has re-sparked interested in targeting this receptor.

The regulation of lipolysis in WAT is also mediated, in part, by GPCRs. Lipolysis consists of the hydrolysis of stored triacytlglycerol (TAGs) into free fatty acids and glycerol, which is then released into the bloodstream to supply energy for the body. This action is dependent on three major enzymes: adipose triglyceride lipase (ATGL), hormone-sensitive lipase (HSL), and monoacylglycerol lipase and is regulated by GPCR-mediated levels intracellular cAMP levels. There are a multitude of GPCRs involved in this process, most notably adrenergic receptors α1A

(ADRA1A), β1 (ADRB1), and β2 (ADRB2) [98]. Numerous other GPCRs, mostly partially characterized, are also implicated to influence lipolysis, and therefore represent additional therapeutic targets for obesity. More research is needed in order to understand both the individual and collective impacts of these receptors on adipose tissue function.

15.5 Summary

GPCRs critically contribute to a large variety of vital processes in human physiology, including metabolism and energy homeostasis. Considered the largest and well-studied family of receptors, GPCRs play pivotal roles in glucose metabolism, fat metabolism, digestion and utilization of nutrients, and more. These functions are carried out in a coordinated manner by receptors on many different organs throughout the body, including the intestines, pancreas, liver, kidney, and adipose tissue. One major role of GPCRs within these organs is to sense the presence of nutrients and then modulate metabolic processes based on their availability. Dysfunction in GPCR signaling is a known contributor to major metabolic diseases such as diabetes, obesity, CVDs, and other disorders related to metabolic syndrome. The global prevalence of metabolic diseases has increased dramatically within the past few decades, and a major effort has been made to develop novel therapies. Because GPCRs are highly amenable pharmaceutical targets, recent research has focused on identifying receptors that may improve the prognosis for individuals with metabolic diseases. One such strategy that has seen clinical success involves targeting the receptors of an important incretin hormone, GLP-1, which stimulates the release of insulin to regulate glucose metabolism. GLP-1 receptor agonists are currently approved for the treatment of type 2 diabetes, and due to the success of GLP-1 receptor agonists, a considerable effort has been made to target receptors of other incretin hormones. Pharmaceutical agents targeting the receptors for glucagon and GIP are under development, though no FDA-approved therapies currently exist. Other strategies involve targeting nutrient-sensing families of receptors, such as receptors for free fatty acids and amino acids. Due to their abundant and diverse nature, targeting GPCRs represents a promising strategy for the treatment of metabolic diseases, and advances in research will lead to the development of novel therapeutic strategies.

Acknowledgments

BTL is supported by the National Institutes of Health under award number R01DK104927-01A1 and Department of Veterans Affairs, Veterans Health

Administration, Office of Research and Development, VA merit (Grant no. 1I01BX003382).

References

1 Hill, S. (2006). G-protein-coupled receptors: past, present and future. *Br. J. Pharmacol.* 147: S27–S37. https://doi.org/10.1038/sj.bjp.0706455.

2 Sriram, K. and Insel, P.A. (2018). G protein-coupled receptors as targets for approved drugs: how many targets and how many drugs? *Mol. Pharmacol.* 93 (4): 251–258. https://doi.org/. doi: 10.1124/mol.117.111062.

3 Hauser, A., Attwood, M.M., Rask-Andersen, M. et al. (2017). Trends in GPCR drug discovery: new agents, targets and indications. *Nat. Rev. Drug Discovery* 16 (12): 829–842. https://doi.org/10.1038/nrd.2017.178.

4 Center for Disease Control (2018). Adult obesity facts. https://www.cdc.gov/obesity/data/adult.html (accessed 30 November 2019).

5 Afshin, A., Forouzanfar, M.H., Reitsma, M.B. et al. (2017). Health effects of overweight and obesity in 195 countries over 25 years*. *N. Engl. J. Med.* 377 (1): 13–27. https://doi.org/10.1056/NEJMoa1614362.

6 Golden, S.H., Robinson, K.A., Saldanha, I. et al. (2009). Prevalence and incidence of endocrine and metabolic disorders in the United States: a comprehensive review. *J. Clin. Endocrinol. Metab.* 94 (6): 1853–1878. https://doi.org/10.1210/jc.2008-2291.

7 Alexander, S.P., Christopoulos, A., Davenport, A.P. et al. (2017). THE CONCISE GUIDE TO PHARMACOLOGY 2017/18: G protein-coupled receptors. *Br. J. Pharmacol.* 174 (Suppl 1): S17–S129. https://doi.org/10.1111/bph.13878.

8 Barnard, N.D., Cohen, J., Jenkins, D.J. et al. (2006). A low-fat vegan diet improves glycemic control and cardiovascular risk factors in a randomized clinical trial in individuals with type 2 diabetes. *Diabetes Care* 29 (8): 1777–1783. https://doi.org/10.2337/dc06-0606.

9 Kahleova, H., Matoulek, M., Malinska, H. et al. (2011). Vegetarian diet improves insulin resistance and oxidative stress markers more than conventional diet in subjects with Type 2 diabetes. *Diabetes Med.* 28 (5): 549–559. https://doi.org/10.1111/j.1464-5491.2010.03209.x.

10 Hikmat, F. and Appel, L.J. (2014). Effects of the DASH diet on blood pressure in patients with and without metabolic syndrome: results from the DASH trial. *J. Hum. Hypertens.* 28 (3): 170–175. https://doi.org/10.1038/jhh.2013.52.

11 Carey, V.J., Bishop, L., Charleston, J. et al. (2005). Rationale and design of the optimal macro-nutrient intake heart trial to prevent heart disease (OMNI-Heart). *Clin. Trials* 2 (6): 529–537. https://doi.org/10.1191/1740774505cn123oa.

12 Knowler, W.C., Fowler, S.E., Hamman, R.F. et al. (2009). 10-year follow-up of diabetes incidence and weight loss in the Diabetes Prevention Program Outcomes Study. *Lancet* 374 (9702): 1677–1686. https://doi.org/10.1016/S0140-6736(09)61457-4.

13 Beck-Nielsen, H. and Groop, L.C. (1994). Metabolic and genetic characterization of prediabetic states: sequence of events leading to non-insulin-dependent diabetes mellitus. *J. Clin. Invest.* 94 (5): 1714–1721. https://doi.org/10.1172/JCI117518.

14 Robertson, R.P. (1995). Antagonist: diabetes and insulin resistance - philosophy, science, and the multiplier hypothesis. *J. Lab. Clin. Med.* 125 (5): 560–565.

15 Ohtsubo, K., Takamatsu, S., Minowa, M.T. et al. (2005). Dietary and genetic control of glucose transporter 2 glycosylation promotes insulin secretion in suppressing diabetes. *Cell* 123 (7): 1307–1321. https://doi.org/10.1016/j.cell.2005.09.041.

16 Thorens, B. (2006). A toggle for type 2 diabetes? *N. Engl. J. Med.* 354 (15): 1636–1638. https://doi.org/10.1056/NEJMcibr060422. PubMed PMID: 16611957.

17 Carter, J.S., Pugh, J.A., and Monterrosa, A. (1996). Non-insulin-dependent diabetes mellitus in minorities in the United States. *Ann. Intern. Med.* 125 (3): 221–232. https://doi.org/10.7326/0003-4819-125-3-199608010-00011.

18 Harris, M.I., Flegal, K.M., Cowie, C.C. et al. (1998). Prevalence of diabetes, impaired fasting glucose, and impaired glucose tolerance in U.S. adults: The Third National Health and Nutrition Examination Survey, 1988-1994. *Diabetes Care* 21 (4): 518–524. https://doi.org/10.2337/diacare.21.4.518.

19 Bennett, P.H. (1990). *Ellenberg and Rifkin's Diabetes Mellitus*. New York: Elsevier.

20 Ford, E.S. (2005). Risks for all-cause mortality, cardiovascular disease, and diabetes associated with the metabolic syndrome: a summary of the evidence. *Diabetes Care* 28 (7): 1769–1778. https://doi.org/10.2337/diacare.28.7.1769. Review. PubMed PMID: 15983333.

21 Meigs, J.B., Wilson, P.W., Fox, C.S. et al. (2006). Body mass index, metabolic syndrome, and risk of type 2 diabetes or cardiovascular disease. *J. Clin. Endocrinol. Metab.* 91 (8): 2906–2912. https://doi.org/10.1210/jc.2006-0594.

22 Marchesini, G., Bugianesi, E., Forlani, G. et al. (2003). Nonalcoholic fatty liver, steatohepatitis, and the metabolic syndrome. *Hepatology* 37 (4): 917–923. doi: 10.1053/jhep.2003.50161.

23 Chen, J., Muntner, P., Hamm, L.L. et al. (2004). The metabolic syndrome and chronic kidney disease in U.S. adults. *Ann. Intern. Med.* 140 (3): 167–174. https://doi.org/10.7326/0003-4819-140-3-200402030-00007.

24 The National Institute of Diabetes and Digestive and Kidney Diseases (2016). Causes of chronic kidney disease. https://www.niddk.nih.gov/health-information/kidney-disease/chronic-kidney-disease-ckd/causes. (accessed 1 Sept 2019).

25 Pasquali, R., Gambineri, A., Anconetani, B. et al. (1999). The natural history of the metabolic syndrome in young women with the polycystic ovary syndrome and the effect of long-term oestrogen-progestagen treatment. *Clin. Endocrinol. (Oxford)* 50 (4): 517–527. https//doi.org/10.1046/j.1365-2265.1999.00701.x.

26 Vgontzas, A.N., Papanicolaou, D.A., Bixler, E.O. et al. (2000). Sleep apnea and daytime sleepiness and fatigue: relation to visceral obesity, insulin resistance, and hypercytokinemia. *J. Clin. Endocrinol.* 85 (3): 1151–1158. https://doi.org/10.1210/jcem.85.3.6484.

27 Choi, H.K., Ford, E.S., Li, C., and Curhan, G. (2007). Prevalence of the metabolic syndrome in patients with gout: the Third National Health and Nutrition Examination Survey. *Arthritis Rheumatol.* 57 (1): 109–115. https://doi.org/10.1002/art.22466.

28 Choi, H.K. and Ford, E.S. (2007). Prevalence of the metabolic syndrome in individuals with hyperuricemia. *Am. J. Med.* 120 (5): 442–447. https://doi.org/10.1016/j.amjmed.2006.06.040.

29 American Diabetes Association (2015). Standards of medical care in diabetes. *Diabetes Care* 38 Suppl:S4. https://doi.org/10.2337/dc15-S003. [GPCR Background Talha].

30 Naga Prasad, S.V. (2017). Changing paradigms for G-protein-coupled receptor signaling. *J. Cardiovasc. Pharmacol.* 70 (1): 1–2. https://doi.org/10.1097/FJC.0000000000000498.

31 Hu, G.M., Mai, T.L., and Chen, C.M. (2017). Visualizing the GPCR network: classification and evolution. *Sci. Rep.* 7 (1): 15495. doi: 10.1038/s41598-017-15707-9.

32 Creative Diagnostics (2019). GPCR pathway. https://www.creative-diagnostics.com/gpcr-pathway.htm (accessed 30 August 2019).

33 Sassone-Corsi, P. (2012). The cyclic AMP pathway. *Cold Spring Harbor Perspect. Biol.* 4 (12): https://doi.org/10.1101/cshperspect.a011148.

34 Lan, M. and Pei, G. (2007). B-arrestin signaling and regulation of transcription. *J. Cell Sci.* 120: 213–218. https://doi.org/10.1242/jcs.03338.

35 Smith, F.D. and Scott, J.D. (2006). Anchored cAMP signaling: onward and upward – a short history of compartmentalized cAMP signal transduction. *Eur. J. Cell Biol.* 85 (7): 585–592. https://doi.org/10.1016/j.ejcb.2006.01.011.

36 Lounsbury, K. (2009). Signal transduction and second messengers. Chapter 6. In: *Pharmacology: Principles and Practice* (ed. M. Hacker, W. Messer and K. Bachmann), 103–112. Academic Press https://doi.org/10.1016/B978-0-12-369521-5.00006-3.

37 Hattori, K., Naguro, I., Okabe, K. et al. (2016). ASK1 signalling regulates brown and beige adipocyte function. *Nat. Commun.* 7: 11158. https://doi.org/10.1038/ncomms11158.

38 Wu, H.Y., Tang, X.Q., Liu, H. et al. (2018). Both classic Gs-cAMP/PKA/CREB and alternative Gs-cAMP/PKA/p38β/CREB signal pathways mediate exenatide-stimulated expression of M2 microglial markers. *J. Neuroimmunol.* 316: 17–22. https://doi.org/10.1016/j.jneuroim.2017.12.005.

39 Morisset, J. (2014). Seventy years of pancreatic physiology: take a look back. *Pancreas* 43 (8): 1172–1184. https://doi.org/10.1097/MPA.0000000000000226.

40 El Sayed, S.A. and Mukherjee, S. (2019). *Physiology, Pancreas*. Florida: StatPearls Publishing.

41 Kulina, G.R. and Rayfield, E.J. (2016). The role of glucagon in the pathophysiology and management of diabetes. *Endocr. Pract.* 22 (5): 612–621. https://doi.org/10.4158/EP15984.RA.

42 Afroze, S., Meng, F., Jensen, K. et al. (2013). The physiological roles of secretin and its receptor. *Ann. Transl. Med.* 1 (3): 29. https://doi.org/10.3978/j.issn.2305-5839.2012.12.01.

43 Pancreapedia (2013). Cholecystokinin type 1 receptor. https://www.pancreapedia.org/molecules/cholecystokinin-type-1-receptor (accessed 30 November 2019).

44 Layden, B.T., Durai, V., and Lowe, W.L. Jr., (2010). G-protein-coupled receptors, pancreatic islets, and diabetes. *Nat. Educ.* 3 (9): 13.

45 Winzell, M.S. and Ahren, B. (2007). G-protein-coupled receptors and islet function – implication for treatment of type 2 diabetes. *Pharmacol. Ther.* 116 (3): 437–448. https://doi.org/10.1016/j.pharmthera.2007.08.002.

46 Rosengren, A.H., Jokubka, R. et al. (2010). Overexpression of alpha2A-adrenergic receptors contributes to type 2 diabetes. *Science* 327: 217–220. https://doi.org/10.1126/science.1176827.

47 Mulder, H., Nagorny, C.L., Lyssenko, V., and Groop, L. (2009). Melatonin receptors in pancreatic islets: good morning to a novel type 2 diabetes gene. *Diabetologia* 52 (7): 1240–1249. https://doi.org/10.1007/s00125-009-1359-y.

48 von Gall, C., Stehle, J.H., and Weaver, D.R. (2002). Mammalian melatonin receptors: molecular biology and signal transduction. *Cell Tissue Res.* 309: 151–162.

49 Prokopenko, I., Langenberg, C., Florez, J.C. et al. (2009). Variants in *MTNR1B* influence fasting glucose levels. *Nat. Genet.* 41: 77–81.

50 Lee Kennedy, R. et al. (2010). Review: free fatty acid receptors: emerging targets for treatment of diabetes and its complications. *Therap. Adv. Endocrinol. Metab.* 165–175: https://doi.org/10.1177/2042018810381066.

51 Sorhede Winzell, M. and Bo, A. (2007). G-protein-coupled receptors and islet function, Implications for treatment of type 2 diabetes. *Pharmacol. Ther.* https://doi.org/10.1016/j.pharmthera.2007.08.002.

52 Tang, C., Ahmed, K., Gille, A. et al. (2015). Loss of FFA2 and FFA3 increases insulin secretion and improves glucose tolerance in type 2 diabetes. *Nat. Med.* 21 (2): 173–177. https://doi.org/10.1038/nm.3779.

53 Doyle, M. and Egan, J. (2007). Mechanisms of action of GLP-1 in the pancreas. *Pharmacol. Ther.* 113 (3): 546–593. https://doi.org/10.1016/j.pharmthera.2006.11.007.

54 Harada, N. and Inagaki, N. (2017). Role of GIP receptor signaling in β-cell survival. *Diabetol Int.* 8 (2): 137–138. https://doi.org/10.1007/s13340-017-0317-z.

55 Nauck, M.A. and Meier, J.J. (2018). Incretin hormones: their role in health and disease. *Diabetes Obes. Metab.* 20 (Suppl 1): 5–21. https://doi.org/10.1111/dom.13129.

56 Knop, F.K., Vilsbøll, T., Højberg, P.V. et al. (2007). Reduced incretin effect in type 2 diabetes: cause or consequence of the diabetic state? *Diabetes* 56 (8): 1951–1959. https://doi.org/10.2337/db07-0100.

57 Trujillo, J.M., Nuffer, W., and Ellis, S.L. (2015). GLP-1 receptor agonists: a review of head-to-head clinical studies. *Ther. Adv. Endocrinol. Metab.* 6 (1): 19–28. https://doi.org/10.1177/2042018814559725.

58 Al-Zamel, N., Al-Sabah, S., Luqmani, Y. et al. (2019). A dual GLP-1/GIP receptor agonist does not antagonize glucagon at its receptor but may act as a biased agonist at the GLP-1 receptor. *Int. J. Mol. Sci.* 20 (14): https://doi.org/10.3390/ijms20143532.

59 Rui, L. (2014). Energy metabolism in the liver. *Compr. Physiol.* 4 (1): 177–197. https://doi.org/10.1002/cphy.c130024.

60 Fu, Z.D. and Cui, J.Y. (2017). Remote sensing between liver and intestine: importance of microbial metabolites. *Curr. Pharmacol. Rep.* 3 (3): 101–113. https://doi.org/10.1007/s40495-017-0087-0.

61 Kim, T. et al. (2018). Glucagon receptor signaling regulates energy metabolism via hepatic farnesoid X receptor and fibroblast growth factor 21. *Diabetes* 67 (9): 1773–1782. https://doi.org/10.2337/db17-1502.

62 Solloway, M.J., Madjidi, A., Gu, C. et al. (2015). Glucagon couples hepatic amino acid catabolism to mTOR-dependent regulation of α-cell mass. *Cell Rep.* 12 (3): 495–510. https://doi.org/10.1016/j.celrep.2015.06.034.

63 Galsgaard, K.D., Pedersen, J., Knop, F.K. et al. (2019). Glucagon receptor signaling and lipid metabolism. *Front. Physiol.* 10: 413. https://doi.org/10.3389/fphys.2019.00413.

64 Glucagon.com (2019). Glucagon Receptor Antagonists. http://glucagon.com/glucagonreceptorantag.html (accessed 30 November 2019).

65 Pearson, M.J., Unger, R.H., and Holland, W.L. (2016). Clinical trials, triumphs, and tribulations of glucagon receptor antagonists. *Diabetes Care* 39 (7): 1075–1077. https://doi.org/10.2337/dci15-0033.

66 Campbell, J., Berry, J., and Liang, Y. (2019). Anatomy and physiology of the small intestine. Chapter 71. In: *Shackelford's Surgery of the Alimentary Tract* (ed. C.J. Yeo), 2 Volume Set (8) 1, 817–841. Elsevier. Volume 1, 2019 https://doi.org/10.1016/B978-0-323-40232-3.00071-6.

67 Gribble, F.M. and Reimann, F. (2019). Function and mechanisms of enteroendocrine cells and gut hormones in metabolism. *Nat. Rev. Endocrinol.* 15 (4): 226–237. https://doi.org/10.1038/s41574-019-0168-8.

68 Maillet, E.L. (2011). Modulation of T1R chemosensory receptors for sweet nutrients – new paradigms in metabolic regulation. *Med. Sci. (Paris)* 27 (2): 177–182. https://doi.org/10.1051/medsci/2011272177.

69 Elliott, R.M., Morgan, L.M., Tredger, J.A. et al. (1993). Glucagon-like peptide-1 (7-36)amide and glucose-dependent insulinotropic polypeptide secretion in response to nutrient ingestion in man: acute post-prandial and 24-h secretion patterns. *J. Endocrinol.* 138: 159–166. https://doi.org/10.1677/joe.0.1380159.

70 Gribble, F.M., Diakogiannaki, E., and Reimann, F. (2016). Gut hormone regulation and secretion via FFA1 and FFA4. In: *Free Fatty Acid Receptors. Handbook of Experimental Pharmacology*, vol. 236 (ed. G. Milligan and I. Kimura). Cham: Springer.

71 Hansen, H.S., Rosenkilde, M.M., Holst, J.J., and Schwartz, T.W. (2012). GPR119 as a fat sensor. *Trends Pharmacol. Sci.* 33 (7): 374–381. https://doi.org/10.1016/j.tips.2012.03.014.

72 Holliday, N.D., Watson, S.J., and Brown, A.J. (2012). Drug discovery opportunities and challenges at g protein coupled receptors for long chain free Fatty acids. *Front. Endocrinol. (Lausanne)* 2: 112. https://doi.org/10.3389/fendo.2011.00112.

73 Psichas, A., Larraufie, P.F., Goldspink, D.A. et al. (2017). Chylomicrons stimulate incretin secretion in mouse and human cells. *Diabetologia* 60: 2475–2485. https://doi.org/10.1007/s00125-017-4420-2.

74 Mace, O.J., Schindler, M., and Patel, S. (2012). The regulation of K- and L-cell activity by GLUT2 and the calcium-sensing receptor CasR in rat small intestine. *J. Physiol.* 590 (12): 2917–2936. https://doi.org/10.1113/jphysiol.2011.223800.

75 Ueda, Y., Iwakura, H., Bando, M. et al. (2018). Differential role of GPR142 in tryptophan-mediated enhancement of insulin secretion in obese and lean mice. *PLoS One* 13 (6): e0198762. https://doi.org/10.1371/journal.pone.0198762.

76 Rowland, I., Gibson, G., Heinken, A. et al. (2018). Gut microbiota functions: metabolism of nutrients and other food components. *Eur. J. Nutr.* 57: 1–24. https://doi.org/10.1007/s00394-017-1445-8.

77 Tan, J., McKenzie, C., Potamitis, M. et al. (2014). The role of short-chain fatty acids in health and disease. *Adv. Immunol.* 121: 91–119. https://doi.org/10.1016/B978-0-12-800100-4.00003-9.

78 Priyadarshini, M. and Layden, B.T. (2015). FFAR3 modulates insulin secretion and global gene expression in mouse islets. *Islets* 7: e1045182. https://doi.org/10.1080/19382014.2015.1045182.

79 Tolhurst, G., Heffron, H., Lam, Y.S. et al. (2012). Short-chain fatty acids stimulate glucagon-like peptide-1 secretion via the G-protein-coupled receptor FFAR2. *Diabetes* 61 (2): 364–371. https://doi.org/10.2337/db11-1019.

80 Haghikia, A. et al. (2015). Dietary fatty acids directly impact central nervous system autoimmunity via the small intestine. *Immunity* 43 (4): 817–829. https://doi.org/10.1016/j.immuni.2015.09.007.

81 de Castro Fonseca, M., Aguiar, C.J., da Rocha Franco, J.A. et al. (2016). GPR91: expanding the frontiers of Krebs cycle intermediates. *Cell Commun. Signal* 14: 3. https://doi.org/10.1186/s12964-016-0126-1.

82 Gilissen, J., Jouret, F., Pirotte, B., and Hanson, J. (2016). Insight into SUCNR1 (GPR91) structure and function. *Pharmacol. Therap.* 159: 56–65. https://doi.org/10.1016/j.pharmthera.2016.01.008.

83 Deen, P.M. and Robben, J.H. (2011). Succinate receptors in the kidney. *J. Am. Soc. Nephrol.* 22 (8): 1416–1422. https://doi.org/. doi: 10.1681/ASN.2010050481.

84 Serena, C. et al. (2018). Elevated circulating levels of succinate in human obesity are linked to specific gut microbiota. *ISME J.* 12 (7): 1642–1657. https://doi.org/10.1038/s41396-018-0068-2.

85 Brown, D., Breton, S., Ausiello, D.A., and Marshansky, V. (2009). Sensing, signaling and sorting events in kidney epithelial cell physiology. *Traffic* 10 (3): 275–284. https://doi.org/10.1111/j.1600-0854.2008.00867.x.

86 El Boustany, R. (2018). Vasopressin and diabetic kidney disease. *Ann. Nutr. Metab.* 72 (suppl 2): 17–20. https://doi.org/10.1159/000488124.

87 Limbird, L.E. (2011). Historical perspective for understanding of adrenergic receptors. *Curr. Top. Membr.* 67: 1–17. https://doi.org/10.1016/B978-0-12-384921-2.00001-X.

88 Riddy, D.M., Delerive, P., Summers, R.J. et al. (2018). G protein-coupled receptors targeting insulin resistance, obesity, and type 2 diabetes mellitus. *Pharmacol. Rev.* 70 (1): 39–67. https://doi.org/10.1124/pr.117.014373.

89 Luo, L. and Liu, M. (2016). Adipose tissue in control of metabolism. *J. Endocrinol.* 231 (3): R77–R99. https://doi.org/10.1530/JOE-16-0211.

90 Bjørndal, B., Burri, L., Staalesen, V. et al. (2011). Different adipose depots: their role in the development of metabolic syndrome and mitochondrial response to hypolipidemic agents. *J. Obes.* 2011: 490650. https://doi.org/10.1155/2011/490650.

91 National Institute of Heath (2019). How brown fat improves metabolism. https://www.nih.gov/news-events/nih-research-matters/how-brown-fat-improves-metabolism (accessed 30 November 2019).

92 Zoico, E., Rubele, S., De Caro, A. et al. (2019). Brown and beige adipose tissue and aging. *Front. Endocrinol. (Lausanne)* 10: 368. https://doi.org/10.3389/fendo.2019.00368.

93 Amisten, S., Neville, M., Hawkes, R. et al. (2015). An atlas of G-protein coupled receptor expression and function in human subcutaneous adipose tissue. *Pharmacol. Ther.* 146: 61–93. https://doi.org/10.1016/j.pharmthera.2014.09.007.

94 Fang, H. and Judd, R.L. (2018). Adiponectin regulation and function. *Compr. Physiol.* 8 (3): 1031–1063. https://doi.org/10.1002/cphy.c170046.

95 Kelesidis, T., Kelesidis, I., Chou, S., and Mantzoros, C.S. (2010). Narrative review: the role of leptin in human physiology: emerging clinical applications. *Ann. Intern. Med.* 152 (2): 93–100. https://doi.org/10.7326/0003-4819-152-2-201001190-00008.

96 Collins, S. and Surwit, R.S. (2001). The beta-adrenergic receptors and the control of adipose tissue metabolism and thermogenesis. *Recent Prog. Horm. Res.* 56: 309–328. https://doi.org/10.1210/rp.56.1.309.

97 Peng, X.R., Gennemark, P., O'Mahony, G., and Bartesaghi, S. (2015). Unlock the thermogenic potential of adipose tissue: pharmacological modulation and implications for treatment of diabetes and obesity. *Front. Endocrinol. (Lausanne)* 174: https://doi.org/10.3389/fendo.2015.00174.

98 Flechtner-Mors, M., Jenkinson, C.P., Alt, A. et al. (2002). In vivo alpha(1)-adrenergic lipolytic activity in subcutaneous adipose tissue of obese subjects. *J. Pharmacol. Exp. Ther.* 301: 229–233.

16

Endothelin Receptors in Cerebrovascular Diseases

Seema Briyal[1], *Amaresh Ranjan*[1], *and Anil Gulati*[1,2]

[1] *Pharmaceutical Sciences, Midwestern University, Downers Grove, IL, USA*
[2] *Pharmazz, Inc., Willowbrook, IL, USA*

16.1 Introduction of Endothelin and Its Receptors

A potent vasoconstrictor peptide, endothelin (ET), was discovered from the culture supernatant of porcine endothelial cells in 1988 [1]. Soon after its discovery, further studies revealed the existence of two other genes encoding ET-like peptides. Each of the peptides had 21 amino acid residues and their amino acid sequences were very similar to that of the first reported ET. Therefore, these three endogenous isoforms were designated as ET-1, ET-2, and ET-3 [2]. Among all the three isoforms, ET-1 is the most well-studied and is known to primarily maintain the firmness of vasculatures in normal condition but during pathological cardiovascular conditions, its expression is upregulated and correlates with worsening patients' condition [3, 4]. The biosynthetic route for ET-1 begins with an inactive precursor peptide known as prepro-ET-1. It is synthesized and released continuously from endothelial cells, and its level is modulated in response to major cardiovascular risk factors [5]. The 212 amino acid prepro-ET-1 is cleaved proteolytically to an intermediate "Big ET-1" (pro-ET-1) of 39-amino acid residues, which is subsequently converted to a mature 21-amino acid peptide, ET-1 by endothelin converting enzyme (ECE) [6] (Figure 16.1). ECE belongs to membrane bound zinc metalloproteases from the neprilysin superfamily [5]. On release from endothelial cells, about one in five molecules of Big ET-1 escapes conversion but further processing to ET-1 may occur by smooth muscle ECE or via alternative pathways catalyzed by chymase [9]. Chymases are known to contribute to the final processing step of ET synthesis. Deletion of chymase in mast cells led to reduced pulmonary ET-1 level by almost half [10, 11], while its overexpression in vascular smooth muscle cells increased the level of ET-1 peptide. Moreover, chymase

GPCRs as Therapeutic Targets, Volume 2, First Edition. Edited by Annette Gilchrist.

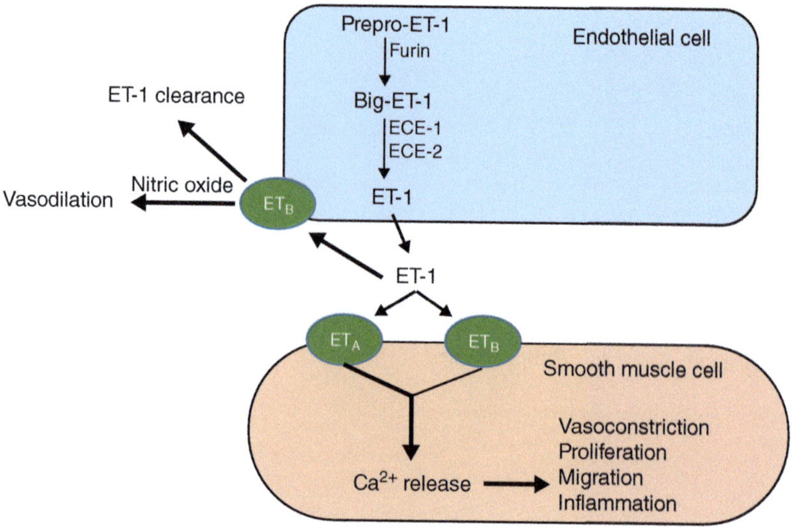

Figure 16.1 Diagram illustrating synthesis of ET peptide and interaction with its receptors. Source: Based on Refs [6–8].

knock out mice in the cardiovascular pathological condition has been observed with attenuated progression of ET-1 dependent diseases [12].

The physiological effects of ETs are known to be activated through two different types of receptors, ET_A and ET_B, both belonging to the superfamily of G-protein coupled receptors (GPCRs). Gene expression analysis studies of these two receptors reflect a ubiquitous expression of ET_A on vascular smooth muscle cells and ET_B on endothelial cells [13–15]. Their relative expression in different organs varies, with higher expression of ET_A mRNA in the heart and lungs and relatively lower expression in the central nervous system (CNS) [16]. On the other hand, high expression of ET_B receptors in the brain, kidney, lungs, and liver has been reported [17].

The binding affinity of ET peptide subtypes to ET receptors are different, specifically, the binding affinity of ET_A: ET-1 ≥ ET-2 ≫ ET-3, whereas ET_B has an equal affinity to all subtypes (ET-1 = ET-2 = ET-3) [13, 14]. The ET receptors produce a variety of effects depending on their interaction with the different ET subtypes. All the three ETs play important roles at cellular and molecular levels; however, ET-1 is the most studied among them. For example, the binding of ET-1 to ET_A receptors activates phospholipase C resulting in accumulation of inositol triphosphate as well as intracellular calcium and leading to long-lasting vasoconstriction [18]. The activation of ET_A receptors is also known to induce cell proliferation in different cell types in tumors such as smooth muscle cells, epithelial cells in colon and ovarian cancers [19, 20]. On the other hand, stimulation of the endothelial ET_B receptors mediates the production of nitric oxide (NO) and prostacyclin and induces vasodilatation [21]. ET_B receptors are also known to mediate pulmonary clearance of circulating ET-1 as well as reuptake of ET-1 by endothelial cells [7, 8] (Figure 16.1). ET-1 binding to ET_B receptor leads to phosphoinositide 3-kinase (PI3K) activation and subsequent production of phosphatidylinositol-3,4,5-trisphosphate, which results in recruitment and activation of protein kinase B/Akt [22]. The PI3K/Akt pathway mediates activation of endothelial nitric oxide synthase (eNOS), NO production, and prevents apoptosis [23].

Apart from the well-known vasoconstrictive action, the role of ET-1 and its receptors have also been shown in the development and pathophysiology of the cerebrovascular system [24, 25]. They are expressed in various types of cells e.g. neurons, astrocytes, and glial cells in the CNS [26]. A wide-spread distribution of ET-1 and its receptors (ET_A and ET_B) in the CNS is associated with various functions including: regulation of sympathetic nervous system, blood flow and blood pressure, apoptosis, cellular proliferation and migration in the CNS [27–30]. Development and use of selective and non-selective agonists and antagonists for the ET_A and/or ET_B receptors has allowed researchers to delineate the action of these receptors with regards to CNS development, pathogenesis, and repair. Translational studies have identified important roles for the ET isoforms as new therapeutic targets in cerebrovascular diseases and clinical trials continue to explore new applications of ET receptor agonists and antagonists. This chapter highlights the potential for utilizing ET receptors as a target for the treatment of cerebrovascular diseases.

16.2 Pathophysiological Role of ET Receptors in Cerebrovascular Diseases: Distribution and Their Importance in Development of the CNS

Cerebrovascular disease is a leading cause of serious long-term disabilities, and the second leading cause of death worldwide [31]. It is a common endpoint of many acute and chronic diseases [32]. ET-1 has been implicated in the pathological conditions associated with cerebrovascular dysfunction and damage, including ischemic stroke, brain trauma, and Alzheimer's disease (AD) [33, 34]. Studies have shown a correlation between increased plasma level of ET-1 and acute ischemic stroke (AIS) and demonstrated an involvement of ET-1 and its receptors with other cerebrovascular diseases such as subarachnoid hemorrhage, and brain trauma [35–37]. As a result, researchers sought to develop ET receptor antagonists/agonists as new therapeutic tools for cerebrovascular diseases. Pathophysiological responses to ET-1 in various organs are mediated by interactions with ET_A and ET_B receptors subtypes. The human brain has a high density of ET_B receptors accounting for 90% of total ET receptors in the brain where they are mainly concentrated in the cerebral cortex [38]. ET_A receptors in the brain constitute a smaller portion of the ET receptors and are largely restricted to the smooth muscle cells of the cerebral vasculature [3, 4, 16].

The physiological action of ET-1 is mediated mainly via ET_A receptors [39]. ET-1 potently constricts brain vessels and is thought to be a mediator of cerebrovascular disorders including delayed vasospasm associated with subarachnoid hemorrhage and ischemic stroke [40]. Therefore, development of ET receptors antagonists has facilitated investigations related to the role of ET-1 in cerebrovascular diseases. Although, some ET_A specific and $ET_{A/B}$ non-specific antagonists showed promising effects in an animal model of stroke and AD, others have failed [41–43]. Additionally, ET_A receptor antagonists ameliorated neuronal damage following experimental focal and global cerebral ischemia [44–46]. These actions have highlighted the therapeutic potential of ET_A receptor antagonists in cerebrovascular diseases. Historically, most research focused on selectively antagonizing ET_A receptors to prevent vasoconstriction, however, this has now shifted towards a therapeutic approach that targets ET_B receptors, and their role in the development and regeneration of CNS [47, 48]. The role of the two types of ET receptors has been delineated in preclinical and acute experimental studies using highly selective ET_A antagonists (e.g. BQ123 and TAK-044) and ET_B antagonist (e.g. BQ788). Three nonpeptide antagonists, bosentan, macitentan, and ambrisentan, that are either mixed ET_A/ET_B antagonists or display greater ET_A selectivity, have been approved by the US FDA for clinical use in the treatment of pulmonary arterial hypertension [3, 4]. On the other hand, studies have shown that ET_B receptors play more important roles in cerebrovascular system than ET_A receptors. They regulate the differentiation, proliferation and migration of neurons, melanocytes, and glia of both the enteric and central nervous systems during pre- as well as post-natal development [24, 25, 49, 50].

The ET_B receptor knock out animals are observed with craniofacial malformation [5, 51] and mortality in ET_B receptor knock out rodents within four weeks of birth is reported [52, 53]. Moreover, they had high levels of ET-1 in the brain and increased apoptosis with a significantly decreased number of neural progenitor cells in the CNS compared to control wild rats. The expression pattern of ET_B receptors indicated their role in early stages of brain development, which mainly involves neural cell proliferation and maturation. To further explore the role of ET_B receptors in developed brain we administered an ET_B receptor specific agonist, IRL-1620 (INN: Sovateltide) in 21 days old rats and observed a significant increase in the levels of ET_B receptors, nerve growth factor (NGF) and vascular endothelial growth factor (VEGF) in the cerebrovascular system [54, 55]. Thus, ET_B receptors play an important role in the pre- and post-natal brain development and their stimulation in post-natal brains may help in neural cell proliferation and maturation, which indicates their potential to be a target for developing therapeutic agents for neurovascular remodeling and regeneration of the adult CNS after damage.

16.3 Role of ET-Receptor Agonists and Antagonists in the Management of Cerebrovascular Diseases

Cerebrovascular disease refers to a group of conditions, diseases, and disorders that affect the blood vessels and blood supply to and in the brain [32]. Its complex pathophysiology includes hypoxia, excitotoxicity, oxidative stress, vascular damage, inflammation, apoptosis, and other events, which causes neural damage and functional impairment of the CNS (Figure 16.2). The list of important cerebrovascular diseases includes ischemic and hemorrhagic stroke, traumatic brain injuries, multiple sclerosis, neonatal hypoxic–ischemic encephalopathy, amyotrophic lateral sclerosis, Parkinson's disease, spinal cord injury (SCI), and AD. ET-1 and its receptors are known to be highly expressed in the brain and play an important role in cerebrovascular system [56]. Therefore, physiological and pathophysiological roles for ET and their receptors in the CNS have been postulated. Accumulating evidence indicates that the level of ET-1 in plasma and brain tissue is increased

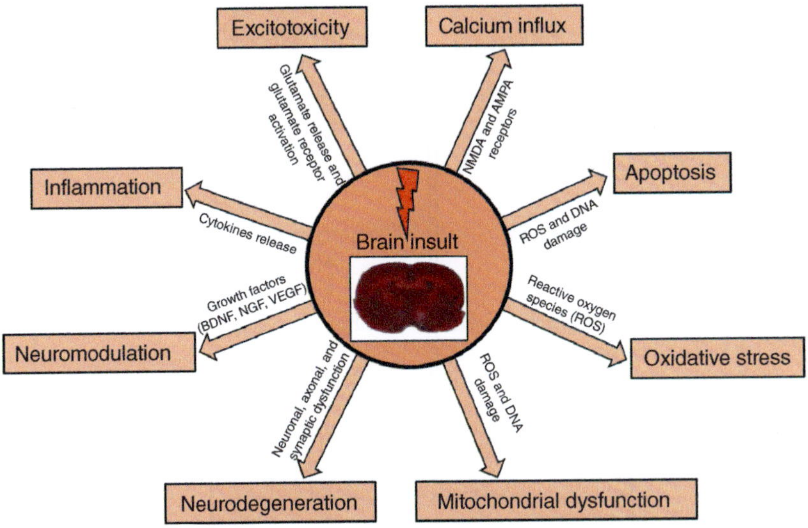

Figure 16.2 Schematic diagram illustrating the main contributory factors involved in the pathophysiology of cerebrovascular diseases.

in patients as well as in animal models in various cerebrovascular diseases such as stroke and AD [35, 57–59]. ET-1 is known to increase the blood–brain barrier permeability and induce neuronal damage [59]. The pathophysiological responses of ET-1 are mediated by their interaction with ET_A and ET_B receptors. ET_B receptors located on vascular endothelial cells contribute to vasodilatation and ET-1 clearance, while both ET_A and ET_B receptors located on vascular smooth muscle cells mediate vasoconstriction [3, 60]. Increasing evidence suggests that selective or nonselective ET_A and ET_B receptor agonists and antagonists may have potential in the treatment of several cerebrovascular diseases based on clinical and animal experiments. In the Sections 16.3.1–16.3.3, we have summarized the current understanding regarding the role of ET receptors, especially ET_B receptors, in cerebrovascular diseases (Figure 16.3).

16.3.1 Acute Ischemic Stroke

AIS is a major cause of disabilities and mortality and remains a serious and significant global health problem [61]. According to recent statistics, the incidence of stroke is around 800 000 new or recurrent events per year and kills an average of 140 000 Americans annually. Approximately 85% strokes are ischemic, and

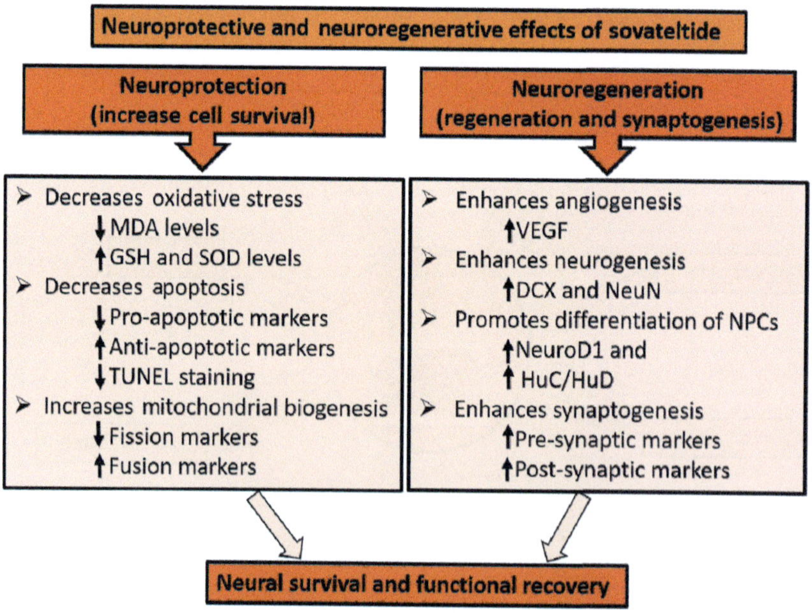

Figure 16.3 Overview of the effect of ET_B receptor agonist sovateltide on neuroprotection and neuro-regeneration in cerebrovascular diseases.

rest are hemorrhagic [62]. The currently available treatment includes tissue plasminogen activator (tPA) and Stentriever to remove thrombi. However, because use of tPA is limited to short time window of <4.5 hours of onset of symptoms and a risk of hemorrhagic transformation, efforts to develop better therapies are needed. Unfortunately, despite many encouraging preclinical results showing numerous drugs as potential neurovascular protectants have failed to demonstrate their beneficial effects in clinical trials. One of the reasons of these failure could be the complex pathophysiology of ischemic stroke, which involves hypoxia, vascular damage, inflammation, apoptosis and other events leading to neural cell damage and functional impairment of the brain. Since, ETs and their receptors are known to regulate multiple cellular functions in neural cells, they could be useful targets to develop effective drugs to treat ischemic stroke.

The ET-1 level in blood and brain tissues are found to be elevated following cerebral ischemia [35, 63]. Higher levels of ET-1 in the damaged brain have been correlated with hypoperfusion, excitotoxicity, blood brain barrier (BBB) disruption, inflammation, and edema. Due to the highly potent vasoconstriction caused by ET-1 via the ET_A receptors, it was postulated that administration of antagonists of ET_A receptors would decrease the damage associated with ischemic stroke. Initial studies targeting this elevation in ET following stroke focused on ET_A receptor antagonist. ET_A receptor antagonists, BQ123, SB234551, A-127722, and S-1039, demonstrated a reduction in infarct area, edema, and neurological deficit following experimental cerebral ischemia [41, 42, 44, 45, 64]. However, mixed $ET_{A/B}$ receptor antagonists have not been consistently effective. While TAK-044 decreased oxidative stress and reduced ischemia, bosentan and SB209670 had no effect [65, 66]. Conversely, deficiency or antagonism of ET_B receptors exacerbates ischemic injury, leading to poorer outcomes [49, 67], indicating that functional ET_B receptors may play a critical role in recovery from cerebral ischemia.

To explore its role in adult cerebral regeneration and repair, we stimulated ET_B receptors in the ischemic brains of middle cerebral artery occluded (MCAO) rats with intravenous administration of sovateltide, a highly selective ET_B receptor agonist. Sovateltide significantly improved neurological and motor functions, with concurrent decrease in infarct volume, oxidative stress, and apoptotic damage. An increase in cell proliferation and angiogenesis was also observed in the sovateltide treated MCAO rat brains [68–70]. These improvements were blocked with ET_B receptor antagonist, BQ788, thus confirming that the effects were due to selective stimulation of the ET_B receptors by sovateltide. Sovateltide appeared to provide both protection of neural cells as well as functional improvement.

The neurological and motor functions in the brain are known to be associated with mitochondrial activity and fate of mitochondria in neurons. Treatment with sovateltide in ischemic stroke brains showed improved mitochondrial fusion, decreased fission, increased size, and biogenesis [71]. The role of sovateltide in neuronal progenitor cells (NPCs) mediated regeneration and repair was also evaluated. Sovateltide upregulated the expression of neuronal differentiation markers, NeuroD1, DCX and HuC/HuD and mature neuronal marker, NeuN in the ischemic cerebral hemisphere. An increase in the differentiation of NPCs and mature neuronal marker suggests a role of ET_B receptors in inducing NPCs to generate new mature neuronal cells in the damaged adult brains. Furthermore, no change in the expression of the pluripotent marker, Oct4 and multipotent neural stem cell marker, Sox2 was observed. These observations suggest a role of ET_B receptors in protection of neural cells against ischemic apoptotic damage and neural function improvement, probably by enhancing mitochondrial activity and their biogenesis. It also helps in progenitor cell mediated brain regeneration and maintenance of pluripotent and multipotent stem cells in an equilibrium [72].

Thus, with an advancement of research in the ET and neurovascular system, it appears that ET and its receptors have a role not only in vascular tone and blood pressure regulation but also in neural cell protection and neuro-regeneration. Studies from our laboratory have specifically demonstrated the importance of ET_B receptors in improving the outcome after damage to the brain following ischemic stroke. ET_B receptors are essential for maturation of progenitor cells, generation of new mature neuronal cells, improved mitochondrial functions, and biogenesis in ischemic brain to restore the neurological and motor functions. Hopefully, further research to explore the signaling pathways related to ET_B receptors and NPC differentiation as well as mitochondrial function and biogenesis will lead us to discover better therapeutic agents to treat ischemic stroke.

16.3.2 Alzheimer's Disease

Alzheimer's disease (AD), an age-related neurodegenerative disease, is characterized by progressive neuronal loss and cognitive decline [73]. In the year 2019, an estimated 5.8 million Americans of all ages are suffering from AD. The numbers are projected to rise nearly 14 million by 2050. The cost in 2019 for all individuals with AD and other dementias is estimated at $277 billion. Medicare and Medicaid are expected to cover $175 billion, or 67%, of the total health care and long-term care payments for people with AD and other dementias [74]. This neurodegenerative disease process is typically characterized by two hallmark pathologies: β-amyloid plaque deposition and neurofibrillary tangles of hyperphosphorylated tau [71, 75, 76]. Additionally, there are other hypothesis that also contributes to the pathophysiology of Alzheimer's disease. Such as cholinergic and glutamate hypothesis. Cholinergic hypothesis states that the possible cause of AD is the loss of central cholinergic neurons and showing the most striking neurochemical disturbance in AD referred as "a deficiency of acetylcholine" (ACh), an important neurotransmitter, involved in the learning and memory formation. This deficiency is caused due to atrophy and degeneration of subcortical cholinergic neurons. Whereas, glutamate is the major excitatory neurotransmitter in the cortex and hippocampus. It is a neurotransmitter (NT) of many neuronal pathways essential to learning and memory. Excessive glutamate activity via *N*-methyl-D-aspartate (NMDA) receptors is toxic and referred as excitotoxicity. Where the cells get too stimulated and may die as a result. Diagnosis is based upon clinical presentation, fulfilling several criteria as well as chemical and imaging biomarkers. Although the overall death rate in the United States from stroke and cardiovascular disease is decreasing, the proportion of deaths related to AD is going up, increasing by 89% between 2000 and 2014 [77, 78]. At present, the only two classes of pharmacological approved therapies are available; cholinesterase inhibitors and NMDA receptor antagonists. The cholinesterase inhibitors donepezil, rivastigmine, and galantamine are recommended therapy for patients with mild, moderate, or severe AD dementia as well as Parkinson's disease dementia [78]. Memantine, which has activity as both a non-competitive NMDA receptor antagonist and a dopamine agonist, is approved for use in patients with moderate-to-severe AD. Existing drugs help mask the symptoms of AD, but fail to cure or delay its progression. The complex neuropathology involved in AD is one of the hurdles in successful drug development and most of the AD drug trials in past have failed. Of the 413 AD trials that were performed during 2002–2012, 99.6% failed to meet their set endpoints [79]. Most of the drugs in these AD trials were developed to target beta amyloid (Aβ) and/or Tau related AD pathological pathways, thus, future research is needed to examine the neural repair process involved in cell survival, regeneration, re-networking of neuronal circuitry and functional recovery in an effort to develop more effective drugs for AD treatment. Studies have shown that ET and its receptors are involved in several pathways related to neural cell survival, regeneration and repair of brains after damage, therefore testing its potential as a drug target for AD needs to be evaluated.

Aggregation of amyloid β (Aβ) plays a key role in the pathogenesis of Alzheimer's disease (AD). There are two major isoforms of Aβ: the 42-residue Aβ42 and the 40-residue Aβ40. The interplay between Aβ42 and Aβ40 has been generally considered to play a critical role in AD [80]. Endothelin-converting enzyme-1 and 2 (ECE-1 and ECE-2) are expressed in endothelial cells and neurons, respectively. They cleave "big ET" to produce the ET-1 [81]. Studies have shown elevated level of ECE-2 and ET-1 in postmortem temporal cortex from AD patients, and ECE-2 expression and ET-1 release to be upregulated by Aβ42 *in vitro*. Both ECE-1 and ECE-2 are capable of cleaving amyloid-β (Aβ) into non-toxic product and enhance Aβ clearance. Additionally, abnormal production of ET-1 has been reported in mice overexpressing amyloid precursor protein. Deletion mutations of either ECE-1 or ECE-2 increase Aβ40 and Aβ42 proteins in mouse brain [81, 82]. ECE-2 knockout mice show impaired learning and memory [83].

Based on these findings, brain ECEs are thought to regulate Aβ protein turnover and highlight a role for ET-1 in the pathogenesis of AD [84, 85]. AD patients have regionally reduced cerebral blood flow and vascular dysfunction that contribute to the development and progression of AD [86]. ET-1 plays a central role in the regulation of cardiovascular functions and regional blood flow [87–89] and may be part of the mechanism by which Aβ interferes with vascular function, because ET-1-induced vasoconstriction in middle cerebral and basilar arteries has been found to be enhanced following exposure to Aβ [90]. Moreover, elevated levels of ET-1 have been detected in the post-mortem brain of individuals with AD when compared to the age-match controls, with a more significant impact noted in the neurons, reactive astrocytes, and cerebral vessel walls. This suggests that high level of ET-1 in the brain might induce neuronal dysfunction in the early stage of AD and indicates the potential benefit of ET receptor antagonists for treatment. Our group examined the effect of ET_A receptor antagonists on Aβ-induced oxidative stress and cognitive dysfunction in an adult rat model of AD. Specific ET_A receptor antagonists, BQ123 and BMS182874, significantly improved the spatial memory deficit and reduced oxidative stress caused by Aβ. Significantly reduced escape latencies and increased preference for the target quadrant were also observed in treated AD rats. On the other hand, a nonspecific ET_A/ET_B receptor antagonist, TAK-044, did not show any improvement in spatial memory deficits caused by Aβ [43]. Lack of improvement with nonspecific ET_A/ET_B antagonist indicates selective involvement of ET_B receptors and suggest that ET_B receptors may be an important target to improve functional recovery in AD.

Stimulation of ET_B receptors have shown anti-apoptotic activity in cultured neurons and against the neurotoxicity of Aβ [24, 25, 28, 47]. Cortical neural progenitor cells show high level of ET_B receptors and their stimulation leads to proliferation and migration, indicating that stimulation of ET_B receptors enhances neuro-regeneration by directly acting on neural progenitors [27, 91, 92]. Additionally, stimulation of ET_B receptors is known to elicit vasodilatation, and previous studies in our laboratory found that intravenous administration of an ET_B receptor agonist, IRL-1620, increased cerebral blood flow (CBF) in normal rats [93] and that the expression of anti-apoptotic marker, Bcl-2 was found to be increased, and pro-apoptotic marker, Bax was found to be decreased with liposomal IRL-1620 in neuronal PC-12 cells [94], indicating a possible role of ET_B receptor agonists in the treatment of AD.

Our group has studied the effect of sovateltide mediated ET_B receptors agonism in a rat model of AD created with intracerebroventricular administration of Aβ40. We administered sovateltide (5 μg/kg, i.v.) and assessed their neurological functions. In behavioral studies using the Morris Water Maze, Aβ produced a significant impairment in spatial memory as evidenced by significantly longer escape latencies and no preference for the quadrant, which previously contained the platform in the probe trial [95]. The observed functional changes after sovateltide treatment concurred with the reduced oxidative stress as measured by decreased levels of malondialdehyde (MDA) and increased levels of reduced glutathione (GSH) and superoxide dismutase (SOD) compared to the vehicle group. Increased level of MDA along with decreased levels of antioxidants GSH and SOD are all hallmarks of oxygen free-radicals generation occurring as an early event in AD pathology [96–99]. Administration of sovateltide produced a significant preventative effect in Aβ induced cognitive impairment in rats, decreased oxidative stress, and improved the spatial memory deficit. These improvements were blocked when animals were administered ET_B receptor antagonist, BQ788, thus confirming that the effects were due to selective stimulation of ET_B receptors by sovateltide [95].

Neuro-regeneration after CNS injury has recently become a focal point for various interventions. Our studies related to cerebral stroke indicated that sovateltide treatment could increase in NGF and VEGF expression in rat brains [100]. These factors are well known for their roles in neurogenesis, therefore an increased expression of these markers suggests that ET_B receptors could be augmenting neurogenesis in the damaged brains. We assessed neurogenesis in an amyloid precursor protein (APP)/PS1 mouse model of AD after treatment with sovateltide. Significantly increased expression of neural progenitor markers NeuroD1 and Doublecortin along with elevated expression of nuclear NeuN (a marker for mature neurons) in the mouse brain was observed. Also, increased expression of pre-synaptic (synapsin1 and synaptophysin) and post-synaptic markers (PSD95) was seen. The effect of neuro-regeneration and repair was tested with learning and memory assessment. APP/PS1 transgenic mice presented with significant impairment in spatial memory as evidenced by significantly longer escape latencies and no preference for the quadrant which previously contained the platform in the probe trial. Treatment with sovateltide significantly reduced learning and memory deficit in 6, 9, and 12 months aged transgenic mice.

In summary, the ET_B receptors agonist sovateltide significantly reduces the AD associated progressive neurodegeneration and helps in neural functional recovery via promoting neuro- and synapto-genesis, ultimately leading to improved functional recovery. Therefore, ET receptors may be considered as a potential pharmacological target for treatment and management of AD.

16.3.3 Spinal Cord Injury

SCI is a debilitating neurological condition with tremendous socioeconomic impact on affected individuals and the health care system. According to the National Spinal Cord Injury Statistical Center, there are 12 500 new cases of SCI each year in North America [101]. Etiologically, more than 90% of SCI cases are traumatic and caused by tragic incidences such as traffic accidents, violence, sports or falls [102]. Clinical outcomes of SCI depend on the severity and location of lesions and may include partial or complete loss of sensory and/or motor function. The estimated life-time cost of a SCI patient is approximately \$2.35 million per patient. Recent advances in medical management of SCI has significantly improved diagnosis, stabilization, survival rate, and well-being of SCI patients [103]. However, there has been small progress on treatment regimens for improving the neurological outcomes of SCI patients. Therefore, it is critical to unravel the cellular and molecular mechanisms of SCI and develop new better and effective treatment for SCI.

Endothelin peptides, expressed in the peripheral and the central nervous systems, can contribute to pathophysiological changes following SCI. After SCI, the disruption of blood–spinal cord barrier by activation of ET and its receptors is a critical event that leads to inflammatory response and oxidative stress, which exacerbates damage and impairs neurological recovery [104, 105]. Receptor binding studies in mammals, including humans demonstrated that ET_A and ET_B receptors are distributed throughout the spinal cord [106]. Guo et al. demonstrated that the blockade of ET_A receptors significantly reduced the expression levels of TNF-α, IL-1β, and IL-6, which could possibly influence functional recovery exerted by the ET system [107]. Oral treatment with bosentan, which blocks ET_A/ET_B receptors, inhibited SCI-induced pain responses, suggesting the participation of ET receptors [108]. Application to the lesion sites of SB209670, an ET_A/ET_B receptor antagonist, markedly inhibited axonal damage after injury, suggested that endogenous ET plays a role in axonal degeneration after SCI [109]. These reports confirm additive pathogenic roles for ET_A and ET_B receptors in the injured spinal cord and may aid in the identification of a set of potential therapeutic targets for neural tissue damage after SCI. Thus, ET receptors might constitute an attractive pharmacological tool for the treatment following SCI and may increase neuronal survival, regeneration, and function after injury.

While numerous studies have been performed to determine the role of central ET_A receptors by pharmacologically stimulating and blocking these receptors to promote functional recovery, none examined the effect of selectively stimulating ET_B receptors to influence functional recovery after SCI. Functional ET_B receptors have been shown to reduce oxidative stress, enhance proliferation of neuronal progenitors and protect against apoptosis in cortical neurons [27, 110, 111]. Oxidative stress plays a critical role in secondary pathogenesis following SCI as it can lead to inflammation and apoptotic cell death. Moreover, our findings showing neuro-regeneration and functional recovery following selective ET_B receptor stimulation in animal models of cerebral ischemia and AD, encouraged us to explore the efficacy of sovateltide in an experimental model of SCI. Adult male Sprague–Dawley rats were subjected to a moderate spinal contusion of

150 kdyn at the thoracic level (T10) and treated with either saline or sovateltide (1, 3, or 5 μg/kg) intravenously at two hours intervals on days 1, 3, and 6 post injury. Motor functions, as determined by Basso, Beattie, Bresnahan scale, significantly improved in the hind limb of rats treated with sovateltide as compared to vehicle-treated animals (unpublished observation). Sovateltide increased the expression of ET_B receptors and synapsin in the spinal cords of rats following SCI. Morphological investigation at 60 days post trauma revealed a significantly higher number of NeuN-positive neurons and BMP-positive myelinated fibers in the proximity of the site of lesion in animal treated with sovateltide as compared to control, suggesting a neuroprotective role of sovateltide and its ability to restrict the degenerative process caused by SCI. Mechanistically, we found that sovateltide enhances stem cell self-renewal via increased expression of SOX-2, Nanog, OCT-4 and activation of the PI3K-AKT pathway in murine spinal cord explants grown for one to three days *ex vivo*. Overall, these observations indicated that sovateltide could attribute to neuro-regeneration, synaptic remodeling and improve motor functions. These findings may serve as a foundation for developing ET-based therapy against multiple targets that alleviate early inflammatory response and oxidative stress and promote long-term functional recovery following SCI.

16.4 Clinical Development of Sovateltide, ET_B Receptor Agonist, as a Drug for Cerebral Ischemia

Cerebral ischemia remains a major cause of morbidity and mortality with little advancement in subacute treatment options. Because of the severity and complexity of acute cerebral ischemia, many clinical trials are currently underway for the treatment of AIS focusing on a range of mechanisms from restoration of blood flow to neuroprotection and to neuro-regeneration (Table 16.1). Therefore, a deeper understanding of neuro-regeneration or brain repair mechanisms in brain pathophysiology will allows physicians to attain more favorable treatment outcomes.

Preclinical studies using rodent model for cerebral ischemia indicate that selective stimulation of ET_B receptors via its agonist, sovateltide (PMZ-1620, IRL-1620), is highly beneficial. Therefore, in view of our preclinical findings, we produced sovateltide suitable for human use and carried out its toxicological testing in mice, rats, and dogs. Phase I (CTRI/2016/11/007509) study to determine the safety, tolerability, and pharmacodynamics of multiple ascending doses of sovateltide in healthy male human volunteers was performed. Each subject received three doses of either 0.3, 0.6, or 0.9 μg/kg administered at an interval of four hours as an intravenous bolus over one minute. They had no adverse effects with 0.3 and 0.6 μg/kg doses; however, nausea and vomiting occurred with the highest dose of 0.9 μg/kg, which resolved within minutes without any intervention. The vital signs, ECGs

Table 16.1 This table summarizes the active and completed clinical trials for the treatment of acute cerebral ischemia (as of 2/28/2020 according to clinicaltrials.gov).

Agent	Mechanism of action	Phase	Clinical trial identifier	Sponsor
Alteplase: Recombinant tissue plasminogen activator (tPA)	Thrombolysis	3	NCT00147316	Mitsubishi Tanabe Pharma Corporation
DLBS1033 (lumbrokinase isolated from *Lumbricus rubellus*)	Anticoagulant	3	NCT01790997	Dexa Medica Group
Edaravone	Antioxidant and anti-edema	3	NCT02430350 NCT01929096	Jiangsu Simcere Pharmaceutical Co., Ltd.
NeuroThera® Laser System (NTS)	Infrared laser technology	3	NCT00419705	PhotoThera, Inc.
Stent-retriever and/or thromboaspiration	Mechanical thrombectomy	3	NCT02216643	Hospital de Clinicas de Porto Alegre
Transcranial laser therapy	Infrared laser technology	3	NCT01120301	PhotoThera, Inc.
Ticagrelor (P2Y12) antagonist	Antiplatelet activity	3	NCT03354429 NCT01994720	AstraZeneca
NXY-059 (Disufenton sodium)	Antioxidant	3	NCT00119626 NCT00061022	AstraZeneca
Cerebrolysin®	Neuroprotection	3	NCT00840671	Ever Neuro Pharma GmbH
Tanakan (Ginkgo biloba)	Neuroprotection and antioxidant	3	NCT00276380	Ipsen
MEXIDOL® (ethylmethylhydroxypyridine succinate)	Antioxidant	3	NCT02793687	Pharmasoft
PMZ-1620 (INN: Sovateltide)	Neural regeneration anti-apoptotic	2/3	NCT04046484 NCT04047563	Pharmazz, Inc.

Erythropoietin (EPO: recombinant human protein)	Neuroprotection	2/3	NCT00604630	Max-Planck-Institute of Experimental Medicine
ONO-2506 (Arundic acid)	Neuroprotection	2/3	NCT00229177	Ono Pharmaceutical Co. Ltd
Desmoteplase (Analogue of tPA)	Thrombolysis	2/3	NCT01104467 NCT00111852	Lundbeck Japan Forest Laboratories
DM199 (Recombinant human tissue kallikrein)	Regeneration	2	NCT03290560	DiaMedica Therapeutics Inc.
Natalizumab (monoclonal antibody)	Anti-inflammatory	2	NCT02730455 NCT01955707	Biogen
Repinotan (5-HT1A receptor agonist)	Neuroprotection	2	NCT00044915	Bayer
MCI-186 (Edaravone)	Antioxidant	2	NCT00821821	Mitsubishi Tanabe Pharma Corporation
AXIS 2: AX200 (filgrastim-recombinant-DNA)	Neuro-regeneration	2	NCT00927836	Sygnis Bioscience GmbH & Co KG
AX200 (Filgrastim)	Regeneration	2	NCT00132470	Axaron Bioscience AG
DP-b99 (zinc and calcium ions Chelator)	Neuroprotection	2	NCT00190047	D-Pharm Ltd.
THR-18's (plasminogen activator inhibitor)	Thrombolysis	2	NCT01957774	D-Pharm Ltd.
3K3A-APC (RHAPSODY) recombinant variant of human activated protein C	Anticoagulant	2	NCT02222714	ZZ Biotech, LLC
TripleAXEL (Rivaroxaban and Warfarin)	Anticoagulant	2	NCT02042534	Asan Medical Center
SA4503 (Sigma-1 receptor agonist cutamesine)	Antioxidant	2	NCT00639249	M's Science Corporation

(continued)

Table 16.1 (Continued)

Agent	Mechanism of action	Phase	Clinical trial identifier	Sponsor
GM602 (GMAIS)	Neuroprotection	2	NCT01221246	Genervon Biopharmaceuticals, LLC
YM872 (zonampanel)	Neuroprotection	2	NCT00044070 NCT00044057	Astellas Pharma Inc.
Microplasmin	Anticoagulant	2	NCT00123305	ThromboGenics
V10153 (INN: Troplasminogen alfa)	Thrombolysis	2	NCT00144014	Vernalis (R&D) Ltd
BIII 890 CL (sodium channel blocker)	Neuroprotection	2	NCT02251197	Boehringer Ingelheim
3K3A-APC (Recombinant variant of human activated protein C)	Neuroprotection	2	NCT02222714	ZZ Biotech, LLC
BMS-986141 (protease activated receptor-4 antagonist)	Thrombolysis	2	NCT02671461	Bristol-Myers Squibb
Desmoteplase (Analogue of tPA)	Thrombolysis	1/2	NCT00638248 NCT00638781	PAION Deutschland GmbH
DS-1040b (inhibitor of thrombin-activatable fibrinolysis inhibitor)	Anticoagulant	1/2	NCT02586233 NCT02071004	Daiichi Sankyo, Inc.
Plasmin (Human)	Anticoagulant	1/2	NCT01014975	Grifols Therapeutics LLC
Lu AA24493 (carbamylated erythropoietin)	Neuroprotection	1	NCT00756249 NCT00870844	H. Lundbeck A/S
ILS-920 (rapamycin analog)	Neuroprotection and regeneration	1	NCT00827190	Wyeth is now a wholly owned subsidiary of Pfizer

or laboratory parameters of all the volunteers were normal. The results of the phase I, the Minimum Intolerable Dose (MID) and Maximum Tolerated Dose (MTD) were established as 0.9 and 0.6 μg/kg of body weight, respectively [25, 112]. We performed phase II trial (CTRI/2017/11/010654; NCT04046484) and investigated the safety, tolerability, and efficacy of sovateltide as an addition to standard of care (SOC) in patients with acute cerebral ischemic stroke. The used therapeutic dose of sovateltide in the phase II trial was 0.3 μg/kg body weight (lower than the established MTD). It was a prospective, multicentric, randomized, double-blind, placebo-controlled study. Adult males or females aged 18–70 years who had radiologically confirmed ischemic stroke within 24 hours were included in the study. Subjects were excluded if they had intracranial hemorrhage and receiving the endovascular therapy. Sovateltide group received three doses of sovateltide administered as an intravenous bolus over one minute at an interval of 3 ± 1 hours on day 1, day 3, and day 6 (total dose of 0.9 μg/kg/day), while the placebo group received an equal volume of saline. A total of 40 patients were enrolled in the study, of whom 36 completed the 90-day follow-up, saline group had $n = 18$ (11 male and 7 female) and sovateltide group had $n = 18$ (15 male and 3 female). All the patients of either group received SOC for ischemic stroke. Efficacy of sovateltide treatment was evaluated using neurological outcomes based on National Institute of Health Stroke Scale (NIHSS), modified Rankin Scale (mRS), and Barthel Index (BI) scores from day 1 to day 90. The EuroQoL-5 Dimensions (EQ-5D) and Stroke-Specific Quality of Life (SSQoL) were used to measure the quality of life of every patient at 60 and 90 days of follow-up. The results of the trial showed improved mRS and BI scores on day 6 compared with day 1 ($p < 0.0001$) in sovateltide treatment group, an effect not seen in the saline group. Sovateltide also increased the frequency of favorable outcomes at three months: 60% of sovateltide, while 40% of saline group patients showed improvement ($p = 0.0519$; odds ratio [OR] 5.25) on the mRS (≥2 points) score, 64% of sovateltide while 36% of saline group patients showed improvement ($p = 0.0112$; OR 12.44) on BI (≥40 points) score, 56% of sovateltide while 43% of saline group patients showed improvement ($p = 0.2714$; OR 2.275) on NIHSS (≥6 points) score. At the same time, the number of patients with complete recovery (defined as an NIHSS score of 0 and a BI of 100) was significantly greater ($p < 0.05$) in the sovateltide than saline group, while an assessment of complete recovery using an mRS score of 0 did not show a statistically significant difference between the treatment groups. Moreover, sovateltide treatment resulted in improved quality of life of the ischemic stroke patients as measured by the EQ-5D and SSQoL on day 90. Thus, the phase II trial of sovateltide for ischemic stroke was successful and the results showed that sovateltide was safe and well-tolerated and resulted in improved neurological outcomes in patients at 90 days post-treatment. The promising results of both phase I human safety and tolerability and phase II studies have

Table 16.2 Summary of clinical trial study design and results of sovateltide.

Stage	ID	Status	*N* and dosages	Objectives	Results
Clinical phase I	CTRI/2016/ 11/007509	Completed	$N = 7$ No. of dosages – 3 Intervals – 4 h Cohort 1 (0.3 μg/kg), $n = 3$. Cohort 2 (0.6 μg/kg), $n = 3$ Cohort 3 (0.9 μg/kg), $n = 1$	Dosages, safety and tolerability	Minimum intolerable dose (MID) – 0.9 μg/kg. Maximum tolerated dose (MTD) – 0.6 μg/kg. Optimal proposed dose for clinical trials – 0.3 μg/kg
Clinical phase II	NCT04046484	Completed	$N = 36$ Dosages – 3 dosages of 0.3 μg/kg at intervals of 3 ± 1 h $n = 18$ (Test) $n = 18$ (Control)	Efficacy at day 6 and 90	Significant improvements in neurological and clinical outcomes in acute ischemic stroke patients at day 6 and 90
Clinical phase III	NCT04047563	Recruitment	$N = 110$ Dosages – 3 dosages of 0.3 μg/kg at intervals of 3 ± 1 h	Efficacy time frame 90 days	N/A

encouraged us to further investigate the efficacy of sovateltide in human phase III studies (NCT04047563) in patients with cerebral ischemia (Table 16.2).

16.5 Conclusions and Perspectives

In the 30 years since its initial discovery, it has become clear that ET is not merely a vasoconstrictor but that it is a multifunctional peptide which affects almost all aspects of cell function. The preclinical and clinical research of more than two decades, has demonstrated an essential role of the ET system in pathophysiology of various diseases including several cerebrovascular disorders. ET receptors antagonists have been effective in improving vascular function in some of the diseases, however their clinical use remains limited with current approval for treatment of pulmonary arterial hypertension only. The ET antagonistic approach was tested in some cerebrovascular diseases such as stroke and AD with the use of ET_A specific or $ET_{A/B}$ non-specific antagonists. They showed promising results in experimental animal models, but none were effective in clinical settings. Nonetheless, the importance of ET in cerebrovascular system could not be ignored, and led to further research on the complex relationship between ET_A and ET_B receptor mediated signaling. Our group as well as others have shown an important role for ET_B receptors in the development of the brain. We have also demonstrated that activation of ET_B receptors by sovateltide (an ET_B receptor specific agonist) effectively protected the brain from ischemic damage, promoted healing, and helped in restoring neurological and motor functions in an animal model as well as in humans. Moreover, we have shown the beneficial effects of ET_B activation on reducing Aβ mediated oxidative stress as well as improvement in learning and memory retention in Alzheimer's rat and mouse models. Thus, these encouraging results indicate that the therapeutic approach targeting ET_B receptors in cerebrovascular system could open new avenues for developing effective drugs to treat various cerebrovascular diseases.

References

1. Yanagisawa, M., Kurihara, H., Kimura, S. et al. (1988). A novel potent vasoconstrictor peptide produced by vascular endothelial cells. *Nature* 332 (6163): 411–415.
2. Inoue, A., Yanagisawa, M., Kimura, S. et al. (1989). The human endothelin family: three structurally and pharmacologically distinct isopeptides predicted by three separate genes. *Proc. Natl. Acad. Sci. U.S.A.* 86 (8): 2863–2867.
3. Barton, M. and Yanagisawa, M. (2019). Endothelin: 30 years from discovery to therapy. *Hypertension* 74 (6): 1232–1265.
4. Davenport, A.P., Hyndman, K.A., Dhaun, N. et al. (2016). Endothelin. *Pharmacol. Rev.* 68 (2): 357–418.
5. Kedzierski, R.M. and Yanagisawa, M. (2001). Endothelin system: the double-edged sword in health and disease. *Annu. Rev. Pharmacol. Toxicol.* 41: 851–876.
6. Xu, D., Emoto, N., Giaid, A. et al. (1994). ECE-1: a membrane-bound metalloprotease that catalyzes the proteolytic activation of big endothelin-1. *Cell* 78 (3): 473–485.
7. Fukuroda, T., Fujikawa, T., Ozaki, S. et al. (1994). Clearance of circulating endothelin-1 by ET_B receptors in rats. *Biochem. Biophys. Res. Commun.* 199 (3): 1461–1465.
8. Luscher, T.F., Yang, Z., Tschudi, M. et al. (1990). Interaction between endothelin-1 and endothelium-derived relaxing factor in human arteries and veins. *Circ. Res.* 66 (4): 1088–1094.

9 Wypij, D.M., Nichols, J.S., Novak, P.J. et al. (1992). Role of mast cell chymase in the extracellular processing of big-endothelin-1 to endothelin-1 in the perfused rat lung. *Biochem. Pharmacol.* 43 (4): 845–853.

10 Guo, C., Ju, H., Leung, D. et al. (2001). A novel vascular smooth muscle chymase is upregulated in hypertensive rats. *J. Clin. Invest.* 107 (6): 703–715.

11 Ju, H., Gros, R., You, X. et al. (2001). Conditional and targeted overexpression of vascular chymase causes hypertension in transgenic mice. *Proc. Natl. Acad. Sci. U.S.A.* 98 (13): 7469–7474.

12 D'Orleans-Juste, P., Houde, M., Rae, G.A. et al. (2008). Endothelin-1 (1–31): from chymase-dependent synthesis to cardiovascular pathologies. *Vascul. Pharmacol.* 49 (2, 3): 51–62.

13 Arai, H., Hori, S., Aramori, I. et al. (1990). Cloning and expression of a cDNA encoding an endothelin receptor. *Nature* 348 (6303): 730–732.

14 Sakurai, T., Yanagisawa, M., Takuwa, Y. et al. (1990). Cloning of a cDNA encoding a non-isopeptide-selective subtype of the endothelin receptor. *Nature* 348 (6303): 732–735.

15 Sakamoto, A., Yanagisawa, M., Sakurai, T. et al. (1991). Cloning and functional expression of human cDNA for the ET_B endothelin receptor. *Biochem. Biophys. Res. Commun.* 178 (2): 656–663.

16 Davenport, A.P., Kuc, R.E., Maguire, J.J., and Harland, S.P. (1995). ET_A receptors predominate in the human vasculature and mediate constriction. *J. Cardiovasc. Pharmacol.* 26 (Suppl 3): S265–S267.

17 Regard, J.B., Sato, I.T., and Coughlin, S.R. (2008). Anatomical profiling of G protein-coupled receptor expression. *Cell* 135 (3): 561–571.

18 Miwa, S., Iwamuro, Y., Zhang, X.F. et al. (1999). Ca^{2+} entry channels in rat thoracic aortic smooth muscle cells activated by endothelin-1. *Jpn. J. Pharmacol.* 80 (4): 281–288.

19 Bagnato, A., Salani, D., Di Castro, V. et al. (1999). Expression of endothelin 1 and endothelin A receptor in ovarian carcinoma: evidence for an autocrine role in tumor growth. *Cancer Res.* 59 (3): 720–727.

20 Maffei, R., Bulgarelli, J., Fiorcari, S. et al. (2014). Endothelin-1 promotes survival and chemoresistance in chronic lymphocytic leukemia B cells through ET_A receptor. *PLoS One* 9 (6): e98818.

21 Barton, M. and Yanagisawa, M. (2008). Endothelin: 20 years from discovery to therapy. *Can. J. Physiol. Pharmacol.* 86 (8): 485–498.

22 Liu, S., Premont, R.T., Kontos, C.D. et al. (2003). Endothelin-1 activates endothelial cell nitric-oxide synthase via heterotrimeric G-protein betagamma subunit signaling to protein jinase B/Akt. *J. Biol. Chem.* 278 (50): 49929–49935.

23 Wedgwood, S. and Black, S.M. (2005). Endothelin-1 decreases endothelial NOS expression and activity through ET_A receptor-mediated generation

of hydrogen peroxide. *Am. J. Physiol. Lung Cell. Mol. Physiol.* 288 (3): L480–L487.

24 Gulati, A. (2016). Endothelin receptors, mitochondria and neurogenesis in cerebral ischemia. *Curr. Neuropharmacol.* 14 (6): 619–626.

25 Gulati, A., Hornick, M.G., Briyal, S., and Lavhale, M.S. (2018). A novel neuroregenerative approach using ET_B receptor agonist, IRL-1620, to treat CNS disorders. *Physiol. Res.* 67 (Suppl 1): S95–S113.

26 MacCumber, M.W., Ross, C.A., and Snyder, S.H. (1990). Endothelin in brain: receptors, mitogenesis, and biosynthesis in glial cells. *Proc. Natl. Acad. Sci. U.S.A.* 87 (6): 2359–2363.

27 Ehrenreich, H., Nau, T.R., Dembowski, C. et al. (2000). Endothelin B receptor deficiency is associated with an increased rate of neuronal apoptosis in the dentate gyrus. *Neuroscience* 95 (4): 993–1001.

28 Vidovic, M., Chen, M.M., Lu, Q.Y. et al. (2008). Deficiency in endothelin receptor B reduces proliferation of neuronal progenitors and increases apoptosis in postnatal rat cerebellum. *Cell. Mol. Neurobiol.* 28 (8): 1129–1138.

29 Gulati, A. and Srimal, R.C. (1993). Endothelin antagonizes the hypotension and potentiates the hypertension induced by clonidine. *Eur. J. Pharmacol.* 230 (3): 293–300.

30 Gulati, A., Rebello, S., Chari, G., and Bhat, R. (1992). Ontogeny of endothelin and its receptors in rat brain. *Life Sci.* 51 (22): 1715–1724.

31 Fisher, M. (2017). Introducing focused updates in cerebrovascular disease. *Stroke* 48 (10): 2653.

32 Chandra, A., Stone, C.R., Li, W.A. et al. (2017). The cerebral circulation and cerebrovascular disease II: pathogenesis of cerebrovascular disease. *Brain Circ.* 3 (2): 57–65.

33 Minami, M., Kimura, M., Iwamoto, N., and Arai, H. (1995). Endothelin-1-like immunoreactivity in cerebral cortex of Alzheimer-type dementia. *Prog. Neuro-Psychopharmacol. Biol. Psychiatry* 19 (3): 509–513.

34 Reid, J.L., Dawson, D., and Macrae, I.M. (1995). Endothelin, cerebral ischaemia and infarction. *Clin. Exp. Hypertens.* 17 (1, 2): 399–407.

35 Ziv, I., Fleminger, G., Djaldetti, R. et al. (1992). Increased plasma endothelin-1 in acute ischemic stroke. *Stroke* 23 (7): 1014–1016.

36 Fujimori, A., Yanagisawa, M., Saito, A. et al. (1990). Endothelin in plasma and cerebrospinal fluid of patients with subarachnoid haemorrhage. *Lancet* 336 (8715): 633.

37 Faraco, G., Moraga, A., Moore, J. et al. (2013). Circulating endothelin-1 alters critical mechanisms regulating cerebral microcirculation. *Hypertension* 62 (4): 759–766.

38 Harland, S.P., Kuc, R.E., Pickard, J.D., and Davenport, A.P. (1995). Characterization of endothelin receptors in human brain cortex, gliomas, and meningiomas. *J. Cardiovasc. Pharmacol.* 26 (Suppl 3): S408–S411.

39 Maguire, J.J. and Davenport, A.P. (2015). Endothelin receptors and their antagonists. *Semin. Nephrol.* 35 (2): 125–136.

40 Mascia, L., Fedorko, L., Stewart, D.J. et al. (2001). Temporal relationship between endothelin-1 concentrations and cerebral vasospasm in patients with aneurysmal subarachnoid hemorrhage. *Stroke* 32 (5): 1185–1190.

41 Barone, F.C., Ohlstein, E.H., Hunter, A.J. et al. (2000). Selective antagonism of endothelin-A-receptors improves outcome in both head trauma and focal stroke in rat. *J. Cardiovasc. Pharmacol.* 36 (5 Suppl 1): S357–S361.

42 Briyal, S. and Gulati, A. (2010). Endothelin-A receptor antagonist BQ123 potentiates acetaminophen induced hypothermia and reduces infarction following focal cerebral ischemia in rats. *Eur. J. Pharmacol.* 644 (1–3): 73–79.

43 Briyal, S., Philip, T., and Gulati, A. (2011). Endothelin-A receptor antagonists prevent amyloid-β-induced increase in ET_A receptor expression, oxidative stress, and cognitive impairment. *J. Alzheimers Dis.* 23 (3): 491–503.

44 Tatlisumak, T., Carano, R.A., Takano, K. et al. (1998). A novel endothelin antagonist, A-127722, attenuates ischemic lesion size in rats with temporary middle cerebral artery occlusion: a diffusion and perfusion MRI study. *Stroke* 29 (4): 850–857; discussion 857–858.

45 Legos, J.J., Lenhard, S.C., Haimbach, R.E. et al. (2008). SB 234551 selective ET_A receptor antagonism: perfusion/diffusion MRI used to define treatable stroke model, time to treatment and mechanism of protection. *Exp. Neurol.* 212 (1): 53–62.

46 Feuerstein, G., Gu, J.L., Ohlstein, E.H. et al. (1994). Peptidic endothelin-1 receptor antagonist, BQ-123, and neuroprotection. *Peptides* 15 (3): 467–469.

47 Druckenbrod, N.R., Powers, P.A., Bartley, C.R. et al. (2008). Targeting of endothelin receptor-B to the neural crest. *Genesis* 46 (8): 396–400.

48 Dembowski, C., Hofmann, P., Koch, T. et al. (2000). Phenotype, intestinal morphology, and survival of homozygous and heterozygous endothelin B receptor–deficient (spotting lethal) rats. *J. Pediatr. Surg.* 35 (3): 480–488.

49 Ehrenreich, H., Oldenburg, J., Hasselblatt, M. et al. (1999). Endothelin B receptor-deficient rats as a subtraction model to study the cerebral endothelin system. *Neuroscience* 91 (3): 1067–1075.

50 Baynash, A.G., Hosoda, K., Giaid, A. et al. (1994). Interaction of endothelin-3 with endothelin-B receptor is essential for development of epidermal melanocytes and enteric neurons. *Cell* 79 (7): 1277–1285.

51 Shin, M.K., Levorse, J.M., Ingram, R.S., and Tilghman, S.M. (1999). The temporal requirement for endothelin receptor-B signalling during neural crest development. *Nature* 402 (6761): 496–501.

52 Riechers, C.C., Knabe, W., Siren, A.L. et al. (2004). Endothelin B receptor deficient transgenic rescue rats: a rescue phenomenon in the brain. *Neuroscience* 124 (4): 719–723.

53 Brand, M., Le Moullec, J.M., Corvol, P., and Gasc, J.M. (1998). Ontogeny of endothelins-1 and -3, their receptors, and endothelin converting enzyme-1 in the early human embryo. *J. Clin. Invest.* 101 (3): 549–559.

54 Puppala, B., Awan, I., Briyal, S. et al. (2015). Ontogeny of endothelin receptors in the brain, heart, and kidneys of neonatal rats. *Brain Dev.* 37 (2): 206–215.

55 Leonard, M.G., Prazad, P., Puppala, B., and Gulati, A. (2015). Selective endothelin-B receptor stimulation increases vascular endothelial growth factor in the rat brain during postnatal development. *Drug Res.* 65 (11): 607–613.

56 Nie, X.J. and Olsson, Y. (1996). Endothelin peptides in brain diseases. *Rev. Neurosci.* 7 (3): 177–186.

57 Benigni, A. and Remuzzi, G. (1999). Endothelin antagonists. *Lancet* 353 (9147): 133–138.

58 Kessler, I.M., Pacheco, Y.G., Lozzi, S.P. et al. (2005). Endothelin-1 levels in plasma and cerebrospinal fluid of patients with cerebral vasospasm after aneurysmal subarachnoid hemorrhage. *Surg. Neurol.* 64 (Suppl 1): S1:2–S1:5; discussion S1:5.

59 Macrae, I.M., Robinson, M.J., Graham, D.I. et al. (1993). Endothelin-1-induced reductions in cerebral blood flow: dose dependency, time course, and neuropathological consequences. *J. Cereb. Blood Flow Metab.* 13 (2): 276–284.

60 Yanagisawa, M. and Masaki, T. (1989). Molecular biology and biochemistry of the endothelins. *Trends Pharmacol. Sci.* 10 (9): 374–378.

61 Benjamin, E.J., Blaha, M.J., Chiuve, S.E. et al. (2017). Heart disease and stroke statistics-2017 update: a report from the American Heart Association. *Circulation* 135 (10): e146–e603.

62 Mozaffarian, D. (2016). Dietary and policy priorities for cardiovascular disease, diabetes, and obesity: a comprehensive review. *Circulation* 133 (2): 187–225.

63 Lampl, Y., Fleminger, G., Gilad, R. et al. (1997). Endothelin in cerebrospinal fluid and plasma of patients in the early stage of ischemic stroke. *Stroke* 28 (10): 1951–1955.

64 Zhang, R.L., Zhang, C., Zhang, L. et al. (2008). Synergistic effect of an endothelin type A receptor antagonist, S-0139, with rtPA on the neuroprotection after embolic stroke. *Stroke* 39 (10): 2830–2836.

65 Briyal, S., Gulati, A., and Gupta, Y.K. (2007). Effect of combination of endothelin receptor antagonist (TAK-044) and aspirin in middle cerebral artery occlusion model of acute ischemic stroke in rats. *Methods Find. Exp. Clin. Pharmacol.* 29 (4): 257–263.

66 Briyal, S., Pant, A.B., and Gupta, Y.K. (2006). Protective effect of endothelin antagonist (TAK-044) on neuronal cell viability in in vitro oxygen-glucose deprivation model of stroke. *Indian J. Physiol. Pharmacol.* 50 (2): 157–162.

67 Chuquet, J., Benchenane, K., Toutain, J. et al. (2002). Selective blockade of endothelin-B receptors exacerbates ischemic brain damage in the rat. *Stroke* 33 (12): 3019–3025.

68 Leonard, M.G., Briyal, S., and Gulati, A. (2011). Endothelin B receptor agonist, IRL-1620, reduces neurological damage following permanent middle cerebral artery occlusion in rats. *Brain Res.* 1420: 48–58.

69 Leonard, M.G., Briyal, S., and Gulati, A. (2012). Endothelin B receptor agonist, IRL-1620, provides long-term neuroprotection in cerebral ischemia in rats. *Brain Res.* 1464: 14–23.

70 Leonard, M.G. and Gulati, A. (2013). Endothelin B receptor agonist, IRL-1620, enhances angiogenesis and neurogenesis following cerebral ischemia in rats. *Brain Res.* 1528: 28–41.

71 Ranjan, A.K., Briyal, S., and Gulati, A. (2020). Sovateltide (IRL-1620) activates neuronal differentiation and prevents mitochondrial dysfunction in adult mammalian brains following stroke. *Sci. Rep.* 10 (1): 12737.

72 Ranjan, A.K., Briyal, S., Khandekar, D., and Gulati, A. (2020). Sovateltide (IRL-1620) affects neuronal progenitors and prevents cerebral tissue damage after ischemic stroke. *Can. J. Physiol. Pharmacol.* 98 (9): 659–666.

73 DeTure, M.A. and Dickson, D.W. (2019). The neuropathological diagnosis of Alzheimer's disease. *Mol. Neurodegener.* 14 (1): 32.

74 Alzheimer's Association Report (2020). 2020 Alzheimer's disease facts and figures. *Alzheimers Dement.* https://doi.org/10.1002/alz.12068.

75 Ballard, C., Gauthier, S., Corbett, A. et al. (2011). Alzheimer's disease. *Lancet* 377 (9770): 1019–1031.

76 Nisbet, R.M., Polanco, J.C., Ittner, L.M., and Gotz, J. (2015). Tau aggregation and its interplay with amyloid-β. *Acta Neuropathol.* 129 (2): 207–220.

77 Crous-Bou, M., Minguillon, C., Gramunt, N., and Molinuevo, J.L. (2017). Alzheimer's disease prevention: from risk factors to early intervention. *Alzheimers Res. Ther.* 9 (1): 71.

78 Weller, J. and Budson, A. (2018). Current understanding of Alzheimer's disease diagnosis and treatment. *F1000Res.* 7: https://doi.org/10.12688/f1000research.14506.1.

79 Cummings, J.L., Ringman, J., and Vinters, H.V. (2014). Neuropathologic correlates of trial-related instruments for Alzheimer's disease. *Am. J. Neurodegener. Dis.* 3 (1): 45–49.

80 Gu, L. and Guo, Z. (2013). Alzheimer's Aβ42 and Aβ40 peptides form interlaced amyloid fibrils. *J. Neurochem.* 126 (3): 305–311.

81 Turner, A.J. and Murphy, L.J. (1996). Molecular pharmacology of endothelin converting enzymes. *Biochem. Pharmacol.* 51 (2): 91–102.

82 Koyama, Y. (2013). Endothelin systems in the brain: involvement in pathophysiological responses of damaged nerve tissues. *Biomol. Concepts* 4 (4): 335–347.

83 Rodriguiz, R.M., Gadnidze, K., Ragnauth, A. et al. (2008). Animals lacking endothelin-converting enzyme-2 are deficient in learning and memory. *Genes Brain Behav.* 7 (4): 418–426.

84 Eckman, E.A., Watson, M., Marlow, L. et al. (2003). Alzheimer's disease β-amyloid peptide is increased in mice deficient in endothelin-converting enzyme. *J. Biol. Chem.* 278 (4): 2081–2084.

85 Palmer, J.C., Baig, S., Kehoe, P.G., and Love, S. (2009). Endothelin-converting enzyme-2 is increased in Alzheimer's disease and up-regulated by Aβ. *Am. J. Pathol.* 175 (1): 262–270.

86 Yamashita, K.I., Taniwaki, Y., Utsunomiya, H., and Taniwaki, T. (2014). Cerebral blood flow reduction associated with orientation for time in amnesic mild cognitive impairment and Alzheimer disease patients. *J. Neuroimaging* 24 (6): 590–594.

87 Gulati, A., Kumar, A., Morrison, S., and Shahani, B.T. (1997). Effect of centrally administered endothelin agonists on systemic and regional blood circulation in the rat: role of sympathetic nervous system. *Neuropeptides* 31 (4): 301–309.

88 Gulati, A., Kumar, A., and Shahani, B.T. (1996). Cardiovascular effects of centrally administered endothelin-1 and its relationship to changes in cerebral blood flow. *Life Sci.* 58 (5): 437–445.

89 Gulati, A., Rebello, S., Roy, S., and Saxena, P.R. (1995). Cardiovascular effects of centrally administered endothelin-1 in rats. *J. Cardiovasc. Pharmacol.* 26 (Suppl 3): S244–S246.

90 Paris, D., Humphrey, J., Quadros, A. et al. (2003). Vasoactive effects of Aβ in isolated human cerebrovessels and in a transgenic mouse model of Alzheimer's disease: role of inflammation. *Neurol. Res.* 25 (6): 642–651.

91 Nishikawa, K., Ayukawa, K., Hara, Y. et al. (2011). Endothelin/endothelin-B receptor signals regulate ventricle-directed interkinetic nuclear migration of cerebral cortical neural progenitors. *Neurochem. Int.* 58 (3): 261–272.

92 Yagami, T., Ueda, K., Asakura, K. et al. (2002). Effects of endothelin B receptor agonists on amyloid β protein (25–35)-induced neuronal cell death. *Brain Res.* 948 (1, 2): 72–81.

93 Leonard, M.G. and Gulati, A. (2009). Repeated administration of ET_B receptor agonist, IRL-1620, produces tachyphylaxis only to its hypotensive effect. *Pharmacol. Res.* 60 (5): 402–410.

94 Joshi, M.D., Oesterling, B.M., Wu, C. et al. (2016). Evaluation of liposomal nanocarriers loaded with ET_B receptor agonist, IRL-1620, using cell-based assays. *Neuroscience* 312: 141–152.

95 Briyal, S., Shepard, C., and Gulati, A. (2014). Endothelin receptor type B agonist, IRL-1620, prevents β amyloid (Aβ) induced oxidative stress and cognitive impairment in normal and diabetic rats. *Pharmacol. Biochem. Behav.* 120: 65–72.

96 Cutler, R.G., Kelly, J., Storie, K. et al. (2004). Involvement of oxidative stress-induced abnormalities in ceramide and cholesterol metabolism in brain aging and Alzheimer's disease. *Proc. Natl. Acad. Sci. U.S.A.* 101 (7): 2070–2075.

97 Mark, R.J., Lovell, M.A., Markesbery, W.R. et al. (1997). A role for 4-hydroxynonenal, an aldehydic product of lipid peroxidation, in disruption of ion homeostasis and neuronal death induced by amyloid β-peptide. *J. Neurochem.* 68 (1): 255–264.

98 Hensley, K., Carney, J.M., Mattson, M.P. et al. (1994). A model for β-amyloid aggregation and neurotoxicity based on free radical generation by the peptide: relevance to Alzheimer disease. *Proc. Natl. Acad. Sci. U.S.A.* 91 (8): 3270–3274.

99 Murray, I.V., Liu, L., Komatsu, H. et al. (2007). Membrane-mediated amyloidogenesis and the promotion of oxidative lipid damage by amyloid β proteins. *J. Biol. Chem.* 282 (13): 9335–9345.

100 Briyal, S., Nguyen, C., Leonard, M., and Gulati, A. (2015). Stimulation of endothelin B receptors by IRL-1620 decreases the progression of Alzheimer's disease. *Neuroscience* 301: 1–11.

101 Hachem, L.D., Ahuja, C.S., and Fehlings, M.G. (2017). Assessment and management of acute spinal cord injury: from point of injury to rehabilitation. *J. Spinal Cord Med.* 40 (6): 665–675.

102 Alizadeh, A., Dyck, S.M., and Karimi-Abdolrezaee, S. (2019). Traumatic spinal cord injury: an overview of pathophysiology, models and acute injury mechanisms. *Front. Neurol.* 10: 282.

103 Stein, D.M., Pineda, J.A., and Roddy, V. (2015). Knight WAt: Emergency Neurological Life Support: traumatic spine injury. *Neurocrit. Care* 23 (Suppl 2): S155–S164.

104 Kallakuri, S., Kreipke, C.W., Schafer, P.C. et al. (2010). Brain cellular localization of endothelin receptors A and B in a rodent model of diffuse traumatic brain injury. *Neuroscience* 168 (3): 820–830.

105 McKenzie, A.L., Hall, J.J., Aihara, N. et al. (1995). Immunolocalization of endothelin in the traumatized spinal cord: relationship to blood-spinal cord barrier breakdown. *J. Neurotrauma* 12 (3): 257–268.

106 Peters, C.M., Rogers, S.D., Pomonis, J.D. et al. (2003). Endothelin receptor expression in the normal and injured spinal cord: potential involvement in injury-induced ischemia and gliosis. *Exp. Neurol.* 180 (1): 1–13.

107 Guo, J., Li, Y., He, Z. et al. (2014). Targeting endothelin receptors A and B attenuates the inflammatory response and improves locomotor function following spinal cord injury in mice. *Int. J. Mol. Med.* 34 (1): 74–82.

108 Forner, S., Martini, A.C., Andrade, E.L., and Rae, G.A. (2016). Neuropathic pain induced by spinal cord injury: role of endothelin ET_A and ET_B receptors. *Neurosci. Lett.* 617: 14–21.

109 Uesugi, M., Kasuya, Y., Hayashi, K., and Goto, K. (1998). SB209670, a potent endothelin receptor antagonist, prevents or delays axonal degeneration after spinal cord injury. *Brain Res.* 786 (1, 2): 235–239.

110 Laziz, I., Larbi, A., Grebert, D. et al. (2011). Endothelin as a neuroprotective factor in the olfactory epithelium. *Neuroscience* 172: 20–29.

111 Yagami, T., Ueda, K., Sakaeda, T. et al. (2005). Effects of an endothelin B receptor agonist on secretory phospholipase A_2-IIA-induced apoptosis in cortical neurons. *Neuropharmacology* 48 (2): 291–300.

112 Gulati, A., Agrawal, N., Vibha, D. et al. (2021). Safety and efficacy of sovateltide (IRL-1620) in a multicenter randomized controlled clinical trial in patients with acute cerebral ischemic stroke. *CNS Drugs* 35 (1): 85–104.

17

The Calcium-Sensing Receptor (CaSR) in Disease

Andrew N. Keller, Tracy M. Josephs, Karen J. Gregory, and Katie Leach

Drug Discovery Biology, Monash Institute of Pharmaceutical Science, Monash University, Parkville, Australia

17.1 Biochemical Features of the CaSR

17.1.1 Endogenous Agonists and Allosteric Modulators

Physiologically, the CaSR's best-characterized endogenous agonist is calcium ($Ca^{2+}{}_{o}$), which is considered the orthosteric agonist. However, $Ca^{2+}{}_{o}$ binds to at least five distinct sites (See Section 17.2), allosterically linked to one another, such that $Ca^{2+}{}_{o}$ binding to one site enhances the binding or efficacy of $Ca^{2+}{}_{o}$ at another [2]. Thus, while $Ca^{2+}{}_{o}$ is an agonist, it is also an allosteric modulator. By definition, allosteric modulators stabilize receptor conformations that have altered orthosteric agonist affinity or efficacy via allosteric cooperativity (the magnitude and direction of allosteric effect on agonist affinity or efficacy). An allosteric modulator that has positive cooperativity (i.e. increases the binding or efficacy of the orthosteric agonist) is called a positive allosteric modulator (PAM), while an allosteric modulator that has negative cooperativity is called a negative allosteric modulator (NAM). Additionally, an allosteric modulator can independently activate or inactivate the receptor. If such allosteric agonists are also PAMs, they are referred to as PAM-agonists. Therefore, by definition, $Ca^{2+}{}_{o}$ is a CaSR PAM-agonist. The cooperative nature of $Ca^{2+}{}_{o}$ binding to the CaSR facilitates its exquisite sensitivity to detect very small fluctuations in $Ca^{2+}{}_{o}$ levels.

In addition to $Ca^{2+}{}_{o}$, a variety of other di- and trivalent cations activate the CaSR, including magnesium (Mg^{2+}), strontium (Sr^{2+}), manganese, ferrous iron, cobalt, nickel, zinc, yttrium, cadmium, barium, europium, gadolinium, terbium, and lead. Despite $Ca^{2+}{}_{o}$ being the most physiologically relevant divalent

Andrew N. Keller and Tracy M. Josephs were contributed equally.

GPCRs as Therapeutic Targets, Volume 2, First Edition. Edited by Annette Gilchrist.

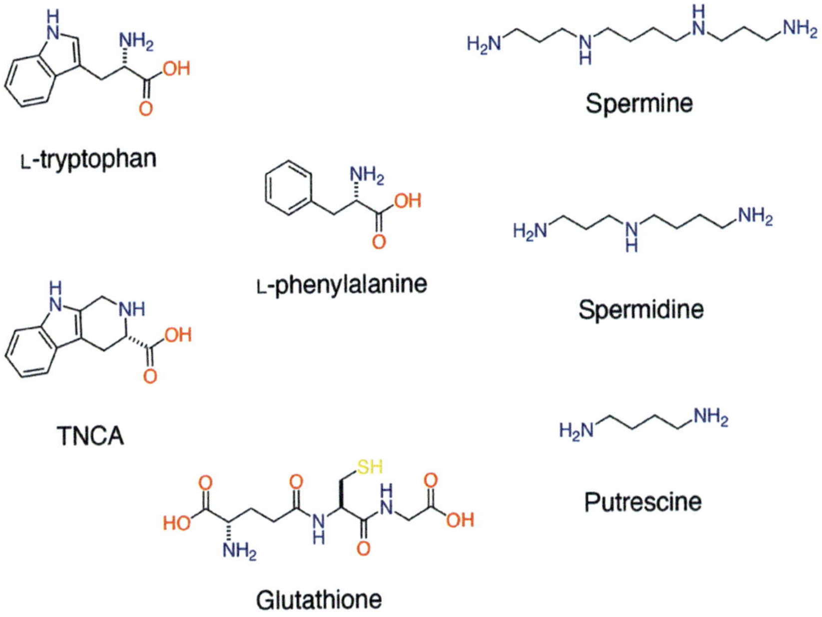

Figure 17.1 Endogenous CaSR agonists and allosteric modulators.

cation agonist, trivalent cations are more potent CaSR agonists than divalent cations [3, 4].

More chemically complex positively charged CaSR agonists also activate the CaSR (Figure 17.1). These include the polyamines spermine, spermidine, and putrescine [5, 6], which are naturally abundant in eukaryotes at concentrations that activate the CaSR physiologically [5, 6]. Polyamines serve as important CaSR agonists in the lungs (See Section 17.3.2.3). In addition to CaSR agonists, L-amino acids and γ-glutamyl peptides are endogenous CaSR PAMs [7, 8], while phosphate ions, sodium chloride, and protons are endogenous CaSR NAMs [9–11]. The ability of the CaSR to respond to multiple diverse endogenous ligands, alone or in combination, drives tissue specific CaSR responses.

17.1.2 Synthetic Agonists and Allosteric Modulators

Mirroring the multitude of endogenous CaSR ligands, an array of structurally diverse exogenous or synthetic CaSR ligands have been identified. Exogenous CaSR agonists include aminoglycoside antibiotics such as neomycin [12] and the mimetic of major basic protein, poly-L-arginine [5, 6]. The importance of both these exogenous ligands has only tentatively been described.

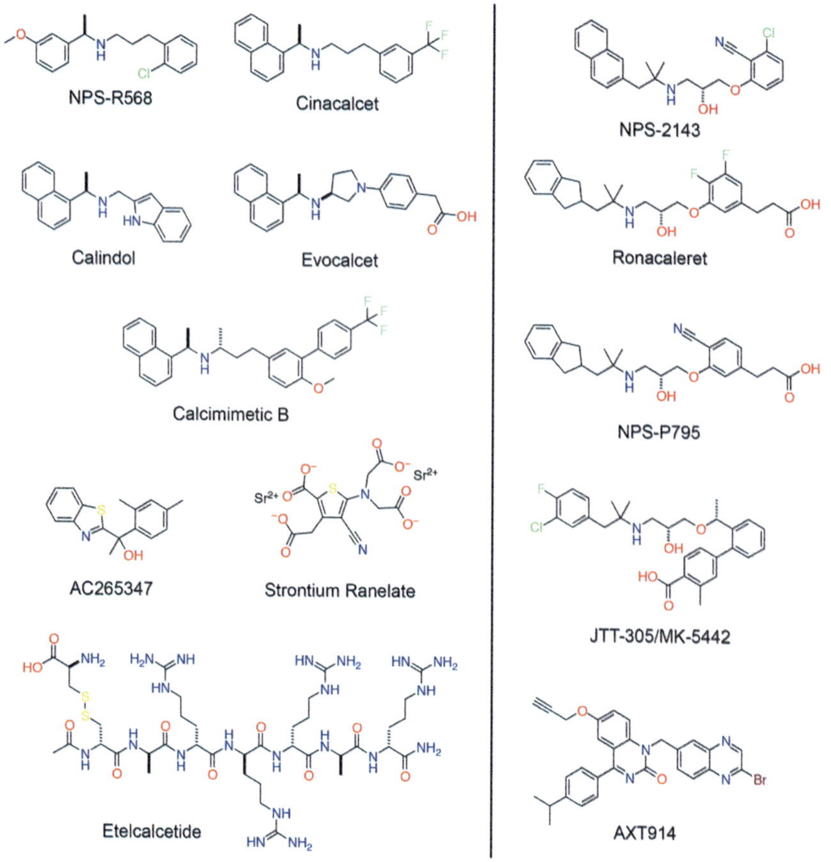

Figure 17.2 Synthetic CaSR allosteric modulators.

In contrast, both small molecule PAMs and NAMs have been developed and characterized for the CaSR. The CaSR PAM cinacalcet (Figure 17.2) was the first allosteric modulator to achieve therapeutic approval for a G protein-coupled receptor (GPCR). The PAMs evocalcet and etelcalcetide are now also approved clinically (See Section 17.4). Cinacalcet and evocalcet possess an aryl-alkylamine core (Figure 17.2; for review see Nemeth et al. [13]) common to many CaSR PAMs, including calindol and NPS R568 (Figure 17.2). The activity of all aryl-alkylamine PAMs is dependent on the stereochemistry of the methyl between the aromatic and secondary nitrogen being in the *R*-configuration [13], demonstrating how subtle changes in these PAMs are detrimental to PAM actions.

In the absence of cations, the aryl-alkylamine PAMs exhibit agonism, and are therefore most accurately described as PAM-agonists. However, aryl-alkylamine agonist activity is mediated at lower concentrations than those required to potentiate cation activity [14, 15]. These observations are most likely explained by reduced PAM affinity in the absence versus the presence of cations, consistent with a positive binding interaction between aryl-alkylamine PAMs and cations.

In addition to aryl-alkylamine PAMs, several structurally distinct CaSR PAMs have been identified, most notably calcimimetic B and AC265347 (Figure 17.2). Similar to the aryl-alkylamines, calcimimetic B and AC265347 are PAM-agonists; however their agonism is mediated at PAM concentrations that are closer to those required for potentiating CaSR activity [16]. Therefore, for structurally diverse PAMs, such as calcimimetic B and AC265347, binding to the CaSR is less dependent on the presence of cations, indicating negligible cooperativity at the level of binding.

Finally, a unique peptide PAM named etelcalcetide was identified several years ago [17]. Etelcalcetide consists of seven D-amino acids, with the N-terminal D-cysteine linked to an L-cysteine via a disulfide bond [17]. Etelcalcetide is unique among CaSR ligands, being the first described synthetic peptide PAM.

Despite being chemically and somewhat structurally similar to aryl-alkylamine PAMs, aryl-alkylamine NAMs have also been developed. The first CaSR NAM discovered was the aryl-alkylamine NAM, NPS2143; identified from the same drug discovery program as cinacalcet [13]. Ronacaleret, JTT-305/MKK-5442, and NPSP795 (Figure 17.2) were later developed as NAMs with more favorable pharmacokinetic profiles (See Section 17.3.2). Several additional CaSR NAMs based on alternative pharmacophores also exist, including the quinazolinones (AXT914 and ATF936) [18, 19], pyridines (BMS compound 1) [20], and 3*H*-quinazoline-4-ones and 3*H*-pyrimidine-4-ones (Pfizer compound 1) [21, 22] (Figure 17.2) (for detailed review see Nemeth et al. [13]). These NAM's were originally developed for the treatment of osteoporosis, however more recently the repurposing of these NAMs for the treatment of calcium handling disorders has gained traction (see Section 17.4).

Unfortunately, no commercially available CaSR radioligands are available, which has hindered CaSR drug discovery by prohibiting quantification of drug binding to the CaSR. Thus, most CaSR discovery efforts have relied upon measures of PAM or NAM potency for potentiating or inhibiting a single $Ca^{2+}{}_{o}$ concentration in a functional assay to guide drug development. However, in more recent years, the use of an allosteric ternary complex model as well as operational models of agonism and allosterism has enabled quantification of PAM and NAM affinity, cooperativity, and efficacy at the CaSR [2, 14–16, 23–26]. These models have revealed key differences in affinity, cooperativity, and agonism between different PAMs and NAMs that were not apparent from studies that quantified only PAM and NAM potency [27]. This is because PAM and NAM potency is a composite of affinity, cooperativity, and efficacy, and it is therefore difficult to improve each of these parameters individually using structure activity relationship (SAR)-based drug optimization. Thus, such mathematical models should inform future CaSR discovery efforts by providing a more accurate understanding of the structural and chemical attributes that govern CaSR PAM and NAM actions.

17.1.3 CaSR-Mediated Signaling Pathways

The CaSR is promiscuously coupled to multiple G proteins, primarily from these the $G\alpha_{q/11}$ and $G\alpha_{i/o}$ family. CaSR coupling to $G\alpha_{q/11}$ activates phospholipase C (PLC), and subsequent phosphatidylinositol 4,5-bisphosphate hydrolysis to the second messengers, inositol trisphosphate (IP_3), and diacylglycerol (DAG) [28, 29]. In turn, IP_3 triggers release of intracellular calcium ($Ca^{2+}{}_i$) from stores in the endoplasmic reticulum [28, 29], and both $Ca^{2+}{}_i$ and DAG stimulate protein kinase C (PKC) downstream of CaSR activation [28, 29]. Furthermore, post CaSR coupling and activation, $G\alpha_{11}$ interacts directly with the transient receptor potential channel 1 (TRPC1) ion channel [30]. In contrast, CaSR coupling to $G\alpha_{i/o}$ inhibits adenylate cyclase, decreasing cyclic adenosine monophosphate (cAMP) levels [29, 31]. Further, both $G\alpha_{i/o}$- and $G\alpha_{q/11}$-mediated signaling pathways lead to phosphorylation of the mitogen-activated protein kinase (MAPK), extracellular signal-regulated kinase 1/2 (pERK1/2) [31], although CaSR-mediated pERK1/2 also occurs in part by a β-arrestin-dependent pathway [32, 33]. Promiscuous CaSR coupling offers an additional means to engender tissue specific signaling by the CaSR, which may be perturbed in different disease contexts.

In human breast cancer cells and in murine pituitary corticotrope-derived AtT-20 cells, the CaSR couples to $G\alpha_s$ [34, 35]. However, CaSR-mediated stimulation of cAMP accumulation is not observed in (human embryonic kidney) HEK293 cells [36], and whether the CaSR couples to $G\alpha_s$ under normal physiological conditions is unknown. Further, the CaSR activates phospholipase D and Rho kinase signaling independently of $G\alpha_{q/11}$ or $G\alpha_{i/o}$, suggesting the CaSR may also couple to $G\alpha_{12/13}$ in certain cells [37]. Given the wide tissue distribution and the many physiological processes regulated by the CaSR, it is likely that many of the CaSR signaling pathways remain undefined. As such, the CaSR may interact with multiple signaling proteins, including those that are independent of heterotrimeric G proteins.

CaSR signaling is attenuated by phosphorylation, principally at S875 and T888 in the CaSR C-terminal tail [38–40]. T888 phosphorylation is mediated by PKC [38], while dephosphorylation is driven by a calyculin-A sensitive protein phosphatase [41]. The phosphorylation-deficient clinical mutant, T888M, causes a gain-of-function *in vitro*, while in humans it leads to reduced parathyroid hormone (PTH) levels [42].

CaSR signaling is also regulated by an intricate relationship between internalization via β-arrestin 1 and the clathrin-binding adapter protein 2 (AP2) heterotetramer [32, 33], and via a phenomenon termed agonist-driven insertional signaling (ADIS). Cell surface CaSRs are constitutively internalized and replaced, but agonists and PAMs increase internalization [33]. CaSRs internalized in response to agonist exposure continue to signal from endosomes [43]. Intriguingly, when cell membrane-localized CaSRs are exposed to agonists or PAMs, ADIS drives mobilization of an intracellular pool of receptors to the cell surface and new receptors are synthesized to replenish the pool [44]. In this way, the CaSR maintains a constant capacity to sense $Ca^{2+}{}_o$ in the extracellular environment.

17.1.4 CaSR Biased Agonism

Due to the CaSR's varied natural ligands and promiscuous coupling to multiple transducers, it is no surprise that the CaSR regulates signaling in a highly specific manner via biased agonism. Biased agonism is the phenomenon where different agonists differentially stabilize distinct receptor conformations that preferentially activate a subset of a receptor's possible signaling outputs to the exclusion of others [45]. For instance, in CaSR-HEK293 cells $Ca^{2+}{}_o$ preferentially stimulates $Ca^{2+}{}_i$ mobilization over pERK1/2, while spermine preferentially activates pERK1/2 [36]. Similarly, L-amino acids activate $Ca^{2+}{}_i$ mobilization and pERK1/2 [46], and also inhibit cAMP synthesis. However, L-amino acids do not stimulate certain signaling events activated in response to $Ca^{2+}{}_o$, including Rho-dependent actin stress fiber formation [47] and cAMP response element-binding protein (CREB) phosphorylation [48]. Evidence in HEK293 cells suggests that the subcellular location of CaSR expression may enable selective activation of intracellular signaling pathways via "location bias." Thus, while CaSRs localized to the cell membrane signal via $G\alpha_{q/11}$ and $G\alpha_{i/o}$, CaSRs that are internalized into endosomes predominantly signal via $G\alpha_{q/11}$ [43]. It must be noted that many studies reporting CaSR-mediated biased agonism have not been performed in a systematic manner using identical conditions across assays (e.g. buffers, duration of agonist stimulation) or the same cellular background. Thus, there remains a need to systematically quantify CaSR preferences for different intracellular signaling pathways. Nonetheless, if real, biased agonism, due to either differential coupling or location, offers a means to fine-tune cellular responses activated by the CaSR and presents an attractive modality for targeted therapeutic intervention.

Functional studies in CaSR-HEK293 cells have demonstrated that loss- and gain-of-function mutations (See Section 17.3.1) also alter CaSR signaling in a biased manner [23, 49]. For instance, while the wild type CaSR couples preferentially to $Ca^{2+}{}_i$ versus pERK1/2, some loss-of-function mutants signal equally via the $Ca^{2+}{}_i$ and pERK1/2 pathways, or predominantly via pERK1/2, while several gain-of-function mutants couple more strongly to $Ca^{2+}{}_i$ than the wild type CaSR [50–52]. Although the pathophysiological relevance of mutation-induced bias is at present unknown, it likely plays an important role in perturbing CaSR function in endocrine disorders (See Section 17.3.1).

Importantly, CaSR PAMs and NAMs engender "biased modulation" either via differences in their magnitude of cooperativity between pathways or via distinct PAM and NAM affinities for receptor states that couple to diverse signal transducers [14, 16, 23, 25, 51]. For instance, cinacalcet preferentially potentiates $Ca^{2+}{}_o$-mediated $Ca^{2+}{}_i$ mobilization over pERK1/2, in contrast to AC265347, which preferentially potentiates pERK1/2 over $Ca^{2+}{}_i$-mobilization [14, 16]. Similarly the NAM NPS2143 preferentially inhibits $Ca^{2+}{}_o$-mediated $Ca^{2+}{}_i$ mobilization over pERK1/2 [14, 16]. Intriguingly, biased modulation extends beyond alterations in receptor signaling. The naturally occurring loss-of-function mutation, G670E, traps CaSR in the endoplasmic reticulum. Cinacalcet rescues the surface expression of this mutant, while AC265347 does not [51]. Interestingly, the NAM NPS2143 also rescues the expression of the G670E mutant, despite being a NAM for receptor signaling [51].

17.2 CaSR Structure

The CaSR possesses the prototypical GPCR 7-transmembrane (7TM) spanning alpha helices (TM1–7), linked by three intracellular and three extracellular loops (intracellular loops [ICLs] and ECLs, respectively) (Figure 17.3a). However, the CaSR and related Class C GPCRs such as the metabotropic glutamate receptors ($mGlu_{1-5}$), γ-aminobutyric acid subtype B ($GABA_B$) receptor, and type 1 taste ($T1R_{1-3}$) receptors, have additional distinguishing structural features from other GPCRs [53, 54]. Principally, Class C GPCRs have a large extracellular domain, which contains the agonist binding sites, held distal from the 7TM bundle by a rigid cysteine rich sub-domain (excepting $GABA_B$ receptors) (Figure 17.3a; [54, 55]). Further, the CaSR is an obligate homodimer, covalently linked by two disulfide bonds between the extracellular domains [56–61].

17.2.1 The CaSR Extracellular Domain

The CaSR extracellular domain contains binding sites for the orthosteric agonist, $Ca^{2+}{}_o$, as well as for amino acids, phosphates, and sulfates. Two independent groups determined the structure of the CaSR extracellular domain by X-ray crystallography [60, 61]. Both structures showed that the extracellular domain consists of two "lobes" that constitute a so-called venus-flytrap (VFT) domain, consistent with the extracellular domain structures of $mGlu_{1-5}$, $GABA_B$, and T1R1 [55, 61–68]. The VFT domain is named based on its three-dimensional

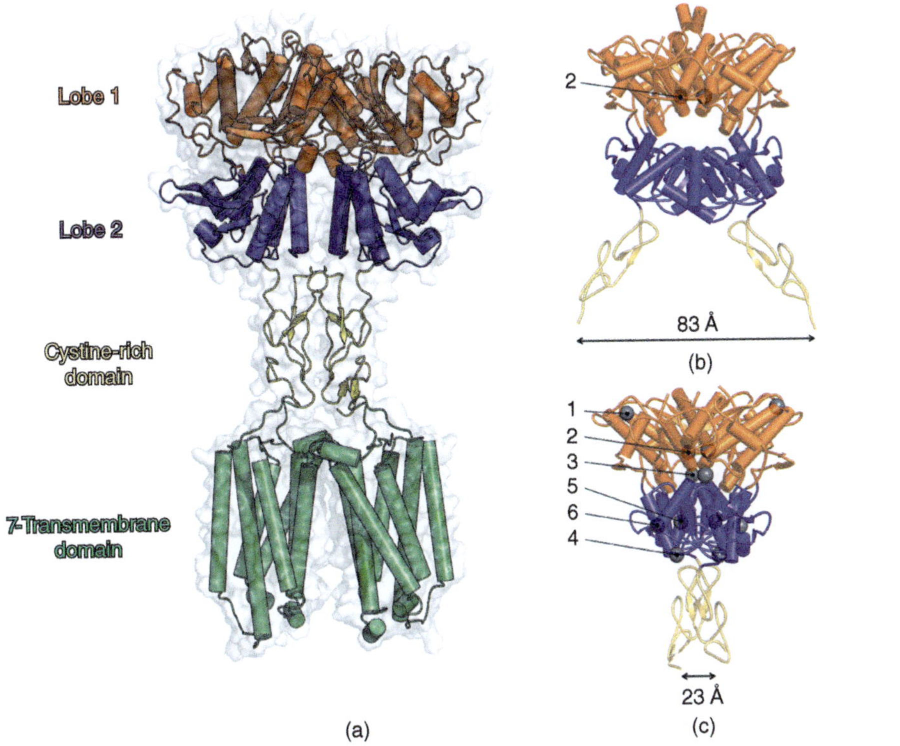

Figure 17.3 Structural features of the CaSR. (a) The CaSR is an obligate homodimer with each protomer comprising a large extracellular domain tethered to the prototypical 7TM bundle by a cysteine rich subdomain. (b) The two lobes of the VFT domains are joined by a flexible "hinge" region facilitating opening and closing of the VFT domain within each protomer. In the inactive conformation (PDB: 5k5t) the VFT lobes and cysteine rich domains adopt an open conformation where $Ca^{2+}{}_{o}$ is only present at cation site 2 (labeled and represented as a gray sphere). (c) In the active conformation, the CaSR adopts a closed conformation whereby the VFT domain closes and the cysteine rich domains move toward their dimeric counterpart. The cation binding sites (labeled and represented as gray spheres) differ depending on the crystallization conditions, where cation site 1 contains $Ca^{2+}{}_{o}$ (PDB: 5k5s) or Mg^{2+} (PDB: 5fbk); cation site 2 contains $Ca^{2+}{}_{o}$ (PDB: 5k5s and 5k5t); cation site 3 and 4 contains $Ca^{2+}{}_{o}$ (PDB: 5k5s); cation site 5 contains Mg^{2+} (PDB: 5fbk and 5fbh); or cation site 6 contains Gd^{2+} (PDB: 5fbh).

structure being reminiscent of the two leaves of a VFT plant. In class C GPCR VFT domains, the lobes are joined by a flexible "hinge" region that allows the lobes to open and close around ligands that bind in a cleft formed between the lobe interfaces (Figure 17.3b,c) [60, 61]. In the absence of agonists or PAMs, the CaSR VFT is open. CaSR activation stabilizes interactions between the VFT lobes, and the structure closes (Figure 17.3b,c) [60, 61]. In other class C GPCR extracellular domains, stabilization of the closed VFT domain is mediated by the orthosteric agonist binding in the cleft. However, $Ca^{2+}{}_{o}$ binding sites are not located in this cleft. Rather $Ca^{2+}{}_{o}$ decorates the extracellular domain, with up to five binding sites suggested from the crystal structures [60, 61]. It is unclear how $Ca^{2+}{}_{o}$ binding favors the closed VFT conformation, or whether it can do so in the absence of a ligand binding in the cleft [60, 61]. The CaSR VFT cleft, which is analogous to the amino acid binding site in $mGlu_{1-5}$ and T1R1, constitutes a binding site for L-amino acids [55, 61–66]. L-amino acids do not activate the CaSR in the absence of $Ca^{2+}{}_{o}$ or other cations, but allosterically potentiate cation activity [7, 46, 62]. However, given that a closed CaSR VFT conformation was only observed when the cleft was occupied, L-amino acids have been termed CaSR co-agonists [61].

Interestingly, one X-ray crystallography structure was solved with a naturally occurring tryptophan derivative, L-1,2,3,4-tetrahydronorharman-3-carboxylic acid (TNCA), bound within the amino acid binding site despite TNCA not being added during protein purification or crystallography (Figure 17.1) [60, 61]. Similarly, γ-glutamyl peptides (Figure 17.1) also bind to the VFT cleft and act as CaSR PAMs, consistent with their interaction with other class C chemosensory receptors [69, 70]. Thus, while the canonical VFT binding site for class C GPCRs is not the binding site for $Ca^{2+}{}_{o}$ in the CaSR, it remains an important binding site for other endogenous allosteric molecules [8, 69].

17.2.2 CaSR 7TM

At the time of writing, there are no published structures of the CaSR 7TM, ICLs, or ECLs, and information about these regions have been inferred from closely related class C GPCR structures [71, 72]. For GPCR classes A, B, and F, agonists stabilize active 7TM domain conformations via direct interactions with the 7TM helices and ECLs [73]. However, for class C GPCRs, active 7TM domain states are stabilized by large conformational rearrangements in the VFT domain [55, 60, 61]. Nevertheless, the CaSR does retain $Ca^{2+}{}_{o}$ activity in the absence of the extracellular domain, suggesting that there is at least one $Ca^{2+}{}_{o}$ binding site in the 7TM domain, in contrast to other members of the class C GPCR family [14]. While it is yet to be directly established for the CaSR, activation of the 7TM domain is likely driven by a similar mechanism elucidated for $mGlu_5R$, which was determined using single particle reconstructions from cryogenic electron-microscopy imaging [55]. The proposed mechanism of 7TM activation involves a combination of interactions across the dimer interface of both the VFT and 7TM and stabilization of the ECLs with the cysteine rich domain [55]. Crystal structures of the CaSR extracellular domain reveal that the cysteine rich domains move toward one another by approximately 60 Å upon $Ca^{2+}{}_{o}$ and amino acid binding (Figure 17.3b,c; [61]). The large change in distance between the cysteine rich domains is thought to either bring the 7TM domains together or change an existing 7TM dimer interface.

17.2.3 Small Molecule Allosteric Binding Sites

In addition to multiple $Ca^{2+}{}_{0}$ binding sites, the CaSR possesses a number of distinct binding sites for synthetic allosteric modulators. Homology modeling of the CaSR 7TM domain based on the $mGlu_1R$ and $mGlu_5R$ X-ray crystal structures reveals an extended cavity within the CaSR 7TM bundle [14, 15, 24, 26]. This is the binding site for some synthetic small molecule allosteric modulators of the CaSR, including cinacalcet, the therapeutically approved PAM.

By combining the quantification of allosteric modulator affinity and cooperativity using pharmacological models with comprehensive mutagenesis studies and molecular modeling, it has been possible to elucidate how allosteric modulators interact with the CaSR [14–16, 23–26]. The activity of cinacalcet and other aryl-alkylamine PAMs depends on a key salt-bridge interaction between the ionizable nitrogen in the aryl-alkylamine core structure and E837 within the CaSR 7TM domain allosteric site [14, 15]. However, there are alternative chemical scaffolds that act as CaSR PAMs, whose binding is not dependent on the salt-bridge interaction with E837 (for review see Nemeth et al. [13]). Notably, the benzothiazole CaSR PAM, AC265347 (Figure 17.2), interacts with the CaSR independently of E837 by binding lower in the 7TM domain allosteric site [14].

The allosteric site in the CaSR 7TM domain is also the binding site of small molecule CaSR NAMs. Structure–function mutagenesis, analytical pharmacology, and molecular modeling studies have shown that aryl-alkylamine NAMs bind the same allosteric site as aryl-alkylamine PAMs and are also reliant on the salt–bridge interaction with E837, despite their different direction in cooperativity [14, 26]. Similarly, alternative NAM chemical classes utilize allosteric binding sites in the CaSR 7TM, but their binding is not dependent on E837 [26]. Furthermore, the dimeric structure of the CaSR has an important influence on the ability of NAMs to reduce CaSR activity through 7TM domain interactions. Thus, a NAM must inhibit activation of both protomers simultaneously [74]. Indeed, the inhibitory activity of the NAM NPS2143 is reduced when it can only occupy one protomer in the CaSR dimer. In contrast, the CaSR PAM NPS R568 potentiates the CaSR when bound to a single protomer in the dimer [74]. Interestingly, the CaSR allosteric modulator calhex231 inhibits CaSR activity when it binds both protomers in a dimer, but potentiates the CaSR when only one protomer is occupied [24].

Beyond the 7TM domain, the covalent PAM etelcalcetide interacts with the CaSR extracellular domain. Etelcalcetide binding is driven by exchange of a disulfide bond between the L-cysteine and N-terminal D-cysteine in etelcalcetide for a reversible covalent attachment of the D-cysteine to C482 in the CaSR extracellular domain [17]. C482 is located in the VFT hinge region, demonstrating the importance of the hinge in regulating CaSR activity.

There is scope for discovering yet unidentified allosteric binding sites in the full-length dimeric CaSR structure. To date, no NAM has been described that targets the extracellular domain, while evidence suggests that BMS compound 1 does not act via the "common" 7TM allosteric binding cavity used by other 7TM-binding small molecule NAMs [26]. Interaction studies *in vitro* between BMS compound 1 and an aryl-alkylamine NAM demonstrated a neutral allosteric interaction [75]. Allosteric interactions via distinct sites are anticipated to stabilize different CaSR conformations and may therefore offer unique pharmacological profiles with respect to biased agonism and allosterism.

17.3 CaSR (Patho)physiology

17.3.1 Bone and Serum Mineral Homeostasis

17.3.1.1 CaSR in the Parathyroid, Kidney, and Bone

The CaSR's principal function in humans is to keep $Ca^{2+}{}_{o}$ levels within a narrow range of 1.1–1.3 mM [76] (Figure 17.4). To maintain this narrow range, the CaSR responds to $Ca^{2+}{}_{o}$ fluctuations of less than 100 μM [76–78]. The detection of such small changes in $Ca^{2+}{}_{o}$ is achieved through multiple $Ca^{2+}{}_{o}$ binding sites and their positive cooperativity with one another, which results in a $Ca^{2+}{}_{o}$ concentration response relationship characterized by a Hill coefficient between 2 and 4 [2, 23, 76].

CaSRs expressed in the parathyroid glands, kidney, and bone are central to maintaining $Ca^{2+}{}_{o}$ homeostasis. Global heterozygous *CASR* knockout mice exhibit mild hypercalcemia and hypercalciuria; whereas homozygous knockouts display parathyroid hyperplasia, severe hypercalcemia, bone abnormalities, retarded growth, and early lethality [79]. Similarly, mice with parathyroid-specific *CASR* ablation develop severe hypercalcemia and hyperparathyroidism [80, 81], while mice with kidney-specific *CASR* ablation are hypocalciuric [82], supporting the essential role of the kidney CaSR in the regulation of urinary calcium (calcium denotes total as opposed to free ions) excretion.

In the parathyroid glands, the CaSR is abundant in chief cells, one of two parathyroid gland cell types with the other being oxyphil cells [83]. Under hypocalcemic conditions and consequent reduced CaSR activation, PTH secretion from the parathyroid glands is elevated. PTH acts on renal tubules in the kidney to drive calcium reabsorption from the urine and stimulates bone forming osteoblasts. If PTH release and osteoblast activation is sustained, osteoblasts stimulate osteoclasts to release calcium via bone resorption. As serum $Ca^{2+}{}_{o}$ concentrations rise, CaSRs in the parathyroid gland inhibit PTH transcript levels [84],

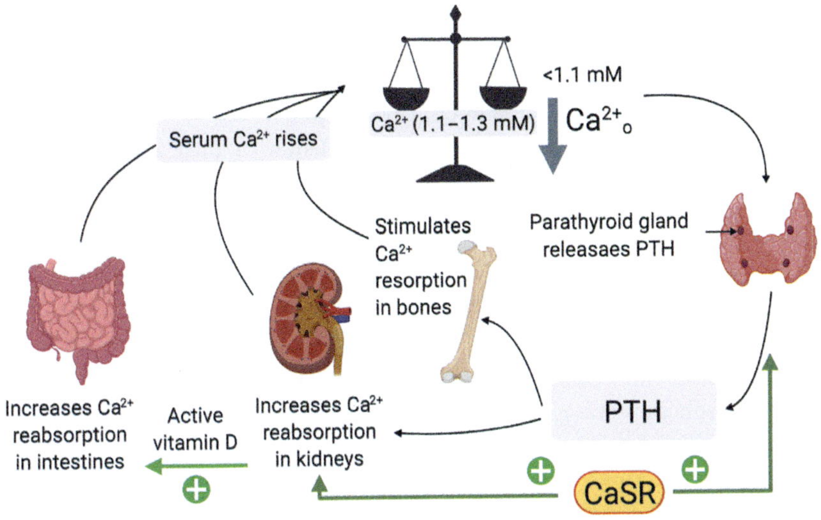

Figure 17.4 Extracellular calcium ($Ca^{2+}{}_o$) is tightly regulated. Reduced serum $Ca^{2+}{}_o$ results in increased parathyroid hormone (PTH) secretion from the parathyroid glands. PTH acts on the PTH1 receptor (PTH1R) in the bone to activate osteoblasts and osteocytes, which release cytokines that stimulate osteoclast activity and enhance bone resorption and $Ca^{2+}{}_o$ release. In the kidney, PTH increases tubular $Ca^{2+}{}_o$ reabsorption, phosphate excretion and stimulates 25-hydroxyvitamin D 1-α-hydroxylase (CYP27B1) enzyme, which promotes the conversion to active 1,25-hydroxyvitamin D (vitamin D). Vitamin D acts on the intestine to increase absorption of dietary $Ca^{2+}{}_o$ via the vitamin D receptor. Feedback is completed upon the increase in $Ca^{2+}{}_o$ and vitamin D reducing PTH secretion from the parathyroid gland. Source: Figure created with http://BioRender.com.

leading to a reduction in PTH synthesis and reduced PTH secretion. Further, under conditions of prolonged exposure to high serum calcium concentrations, CaSR activation drives a reduction in overall parathyroid gland size, diminishing the capacity of the parathyroid to produce PTH [81] and thus helping to ensure the integrity of the negative feedback loop that maintains calcium homeostasis.

While PTH protects against hypocalcemia, protection against hypercalcemia is less dependent on PTH and instead corrected by CaSRs widely expressed in the kidney [82, 85–87]. Renal CaSRs monitor urine and plasma $Ca^{2+}{}_o$ concentrations [82, 88–90]. Under hypercalcemic conditions, renal CaSR activation inhibits the reabsorption of calcium from the urine, thus increasing the rates of renal excretion (for review see Riccardi and Valenti [91]). This occurs in concert with reduced PTH-mediated bone demineralization and calcium absorption in the gut.

In addition to parathyroid and kidney CaSRs, CaSR is expressed in bone cells, where it provides a direct mechanism for driving bone remineralization to protect against hypercalcemia [80, 92–95]. CaSR is expressed on osteoblasts and osteoclasts, as well as other cells that contribute to bone remodeling such as osteocytes and chondrocytes [92]. In these cells, CaSR contributes directly to skeletal development and bone remodeling [80, 93, 94], in turn providing another mechanism for maintaining calcium homeostasis [95].

Mice with osteoblast-specific homozygous deletion of *Casr* exon 7, which encodes the 7TM and C-terminal tail, suffer impaired osteoblast differentiation and increased osteoblast apoptosis, resulting in significantly impaired skeletal growth, severely under mineralized skeletons, and reduced trabecular and cortical bone volume [80, 96]. Most mice die with multiple fractures by three to four weeks after birth [80, 96]. Further, in bone from *Casr* exon 7 knockout mice, mRNA was increased for genes that regulate bone mineralization, including osteopontin (OPN), ankylosis protein (ANK), and nucleotide pyrophosphatase/phosphodiesterase 1 (NPP1), as well as for the gene encoding receptor activator of NFκB ligand (*RANKL*), which promotes bone resorption [80, 96]. Accordingly, these knockout mice showed a doubling of osteoclast numbers and activity with associated trabecular and cortical bone loss [80, 96].

Similarly, the CaSR is expressed in chondrocytes at the end stage of differentiation called hypertrophic chondrocytes, which are located in the growth plate of long bones and are essential for bone growth [92]. Mice with chondrocyte-specific *Casr* exon 7 deletion die in utero [97]. When *Casr* exon 7 chondrocyte deletion was induced before birth (embryonic day 18–19), offspring survived but were small and had under mineralized skeletons, particularly at the hypertrophic zone of the growth plate [97]. Thus, CaSRs in chondrocytes also play a crucial role in skeletal development. Findings from osteoblast- and chondrocyte-specific CaSR knockout mice demonstrate the crucial role of the CaSR in skeletal development.

The CaSR also likely contributes to bone integrity in adults. The bone undergoes constant remodeling, regulated by a variety of hormones (including PTH and calcitonin), cytokines and growth factors that control the number and function of osteoclasts and osteoblasts [92]. The tight coupling between bone resorption by osteoclasts and bone formation by osteoblasts is critical to the maintenance of bone mass and density, and contributes to calcium homeostasis by regulating the discharge of calcium from bone into the circulation. When serum $Ca^{2+}{}_{o}$ levels drop, release of the CaSR's inhibitory control over PTH promotes PTH secretion and consequent osteoblast differentiation [98]. Although osteoblasts initially stimulate bone formation, when PTH concentrations remain elevated for several hours or more and the number of mature osteoblasts increases, osteoblasts produce pro-osteoclast hormones (e.g. RANKL and osteoprotegerin) [99, 100]. These hormones enhance osteoclast numbers and consequent liberation of calcium from the bone matrix via resorption. In osteoblast-specific CaSR knockout mice, the bone anabolic responses to PTH are abolished, demonstrating that the CaSR plays a crucial role in PTH-mediated bone formation [95].

In vitro evidence suggests the CaSR also helps to regulate osteoclast numbers and activity. Elevations in $Ca^{2+}{}_{o}$ concentrations of up to 40 mM are achieved at sites of bone resorption [101]. Such elevated $Ca^{2+}{}_{o}$ concentrations acting via the CaSR inhibit osteoclast maturation and secretion of acid phosphatase (which is critical for bone resorption), increase apoptosis of mature osteoclasts, and promote the secretion of hormones that stimulate osteoblast differentiation and reduce osteoclast differentiation [25, 102–105]. Thus, the CaSR suppresses bone resorption, in turn completing the bone remodeling cycle. Taken together, these findings demonstrate the crucial role of CaSRs expressed in the bone. Indeed, given the critical role of CaSR in the bone, parathyroid glands, and kidneys, it is not surprising that disruption of CaSR function underpins a number of severe human diseases.

17.3.1.2 Primary and Secondary Hyperparathyroidism

The pivotal role of the CaSR in $Ca^{2+}{}_{o}$ homeostasis is demonstrated by numerous endocrine disorders characterized by disruption of the $Ca^{2+}{}_{o}$-PTH relationship. Primary hyperparathyroidism (PHPT) is commonly caused by parathyroid adenoma, hyperplasia or carcinoma, although germline heterozygous or homozygous inactivating *CASR* mutations are occasionally reported in adult-onset PHPT patients and influence the clinical severity of PHPT [106, 107]. Under normal physiological conditions, the $Ca^{2+}{}_{o}$ "set point" (i.e. the serum $Ca^{2+}{}_{o}$ concentration that suppresses 50% of the maximum PTH secretion) is between 1.1–1.3 mM [76]. However, in PHPT the $Ca^{2+}{}_{o}$ set point is increased by 10–30% [108], resulting in hypercalcemia. This hypercalcemia drives a decrease in CaSR expression, making the parathyroid glands less sensitive to $Ca^{2+}{}_{o}$ and therefore unable to suppress PTH synthesis and secretion [109–114], demonstrating the importance of maintaining the intricate relationship between CaSR expression and signaling and PTH secretion.

A similar disruption to the $Ca^{2+}{}_{o}$-PTH relationship is observed in secondary hyperparathyroidism (SHPT). SHPT is a common complication of chronic kidney disease (CKD), and may arise early in the pathogenesis of this disorder [115]. CKD impairs renal $Ca^{2+}{}_{o}$ excretion and absorption, renal phosphate excretion, as well as synthesis of the active vitamin D3 metabolite, 1,25-dihydroxyvitamin D3 (vitamin D3). Vitamin D3 is important for $Ca^{2+}{}_{o}$ homeostasis because it regulates $Ca^{2+}{}_{o}$ absorption in the intestine and kidneys, and also acts to upregulate CaSR expression via vitamin D3 *cis*-response elements in the *CASR* P1 and P2 promoters [116, 117]. The consequences of these effects are fourfold: (i) serum calcium concentrations are reduced, diminishing activation of parathyroid CaSRs [118–120]; (ii) serum phosphate is elevated, further inhibiting parathyroid gland CaSR function [9]; (iii) a decrease in vitamin D3 levels reduces parathyroid CaSR expression [113, 121, 122]; (iv) as the CaSR activates the local synthesis of vitamin D3, impaired CaSR activity and expression further reduces vitamin

D3 levels [123]. Thus, the significant perturbation in CaSR activation in the parathyroid glands results in hyperparathyroidism, which is further exacerbated by parathyroid cell proliferation and parathyroid gland hyperplasia [124, 125] caused by reduced CaSR signaling and expression. Nodular formations develop within enlarged parathyroid glands in chronic dialysis patients in the advanced stages of renal failure with severe SHPT. Reduced CaSR expression is most apparent in nodular parathyroid gland hyperplasia [111–114]. Freshly excised parathyroid glands from hemodialysis patients with SHPT show CaSR expression inversely correlates with parathyroid gland weight and an increase in the set point for $Ca^{2+}{}_{o}$-mediated PTH release *ex vivo* [126]. In a rat model of CKD, renal CaSR expression is also reduced and may contribute to hypercalciuria associated with CKD [127].

17.3.1.3 Familial Hypocalciuric Hypercalcemia (FHH) and Neonatal Severe Hyperparathyroidism (NSHPT)

The essential role of the CaSR is also highlighted by the many naturally occurring mutations in the *CASR* gene, as well as mutations in genes encoding proteins that mediate CaSR signaling and internalization, including G_{11} (encoded by *GNA11*) and AP2 sigma subunit 1 (encoded by *AP2S1*). Loss-of-function *CASR*, *GNA11* or *AP2S1* mutations cause familial hypocalciuric hypercalcemia type 1 (FHH1) and neonatal severe hyperparathyroidism (NSHPT) (*CASR*), FHH2 (*GNA11*), or FHH3 (*AP2S1*). FHH1–3 are characterized by mild–moderate elevations in serum calcium and sometimes magnesium concentrations, increased or normal PTH concentrations, as well as low urinary calcium excretion (hypocalciuria) [128]. While FHH1 and FHH2 are usually relatively benign (reviewed in [128]) FHH3 is most severe and patients present with additional symptoms, including low bone mass and density and cognitive impairments [129–131]. However, given that AP2 is involved in the internalization of many proteins, it is unclear whether the additional symptoms in FHH3 patients are a direct result of disruption to CaSR signaling.

FHH1 is generally caused by heterozygous *CASR* mutations, of which almost 300 have been identified [132]. These mutations have a variety of effects on the CaSR, including loss of CaSR sensitivity to $Ca^{2+}{}_{o}$, as well as a reduction in cell surface expression [52, 133, 134]. Only four FHH2 mutations and three FHH3 mutations have been identified to date [130, 131, 135–142]. While FHH2-causing mutations in *GNA11* disrupt CaSR signaling by impairing G_{11} function [136, 137], FHH3 mutations in *AP2S1* inhibit CaSR internalization and attenuate CaSR signaling from endosomes [32].

Compound heterozygous, dominant negative heterozygous, or homozygous *CASR* mutations cause the more severe form of hypercalcemia, NSHPT. NSHPT is usually present at birth, although it has also been reported in a young child [143].

Infants with NSHPT have severely elevated PTH concentrations, hypercalcemia, and low bone mass and density that often results in fractures at birth, and respiratory distress caused by rib cage deformities [144]. Unlike FHH1–3, NSHPT is fatal if left untreated.

17.3.1.4 Autosomal Dominant Hypocalcemia (ADH) and Bartter Syndrome Type V

Gain-of-function *CASR* or *GNA11* mutations cause ADH1 and Bartter syndrome type V (*CASR*), or ADH2 (*GNA11*). More than 100 *CASR* mutations have been described in ADH1 patients, five of which also cause a Bartter syndrome, and most of these mutations were identified in heterozygous individuals [128]. In contrast, only six different ADH2-causing *GNA11* mutations have been identified, and all of these were heterozygous mutations [128]. ADH1 and 2 are generally characterized by symptomatic hypocalcemia (muscle spasms and cramps, pins and needles, seizures) [145], and sometimes patients present with basal ganglia calcifications and nephrocalcinosis. ADH1 and to a lesser extent ADH2 patients also display hypercalciuria [128, 132]. Bartter syndrome type V that can accompany ADH1 is characterized by hypokalemic alkalosis, renal salt wasting, and hyperreninemic hyperaldosteronism [146–148]. *CASR* mutations that cause ADH1 with Bartter syndrome appear to have a greater impact on CaSR function than those that cause ADH1 alone [52, 148, 149]. At least one such mutation, A843E, results in constitutive CaSR signaling [52].

17.3.1.5 CASR Polymorphisms

In addition to *CaSR* mutations, there are a number of single nucleotide polymorphisms described in the *CASR* gene and its upstream 5′ promoter region, present in both healthy and diseased individuals. Notable polymorphisms in the *CASR* coding region are rs1801725, rs1042636, and rs1801726 located in exon 7, which result in amino acid changes Ala 986 (A986) to Ser (denoted 986S), Arg 990 (R990) to Gly (990G), and Gln 1011 (Q1011) to Glu (1011E) in the CaSR intracellular tail [150–153]. Polymorphisms can cause perturbed serum calcium levels and be pathogenic. For instance, 986S polymorphic individuals have higher serum calcium concentrations than those with the predominant allele encoding A986 [151, 154–156]. In contrast, 990G is associated with decreased serum calcium levels [154]. While 986S is associated with increased risk of advanced SHPT [157], despite the contrasting effects of 990G and 986S on serum calcium levels, both polymorphisms are found more frequently in PHPT patients than in healthy controls [158–161]. Further, in PHPT patients, although 990G increases the risk of kidney stones [162], 986S decreases this risk [159].

Gene association studies have sought to determine whether *CASR* gene variants are linked to low bone mass and density, or whether *CASR* polymorphisms alter the risk of osteoporosis. Rs1801725 (986S) was associated with an increased risk of vertebral fractures in a cohort of PHPT patients [163] and lower bone mass and density in pre-menopausal women and adolescents [155, 164]. In addition the 986S polymorphism was more frequent in 248 aged males with osteoporosis [165]. However, no such associations have been found in older women [166–169]. In contrast the rs1042636 (990G) polymorphism safeguarded against low bone mass and density in one cohort of PHPT patients [170]. Given that gene association studies for effects of CaSR polymorphisms on bone mass and density have not been reproducible across all studies, large hypothesis-free genome wide association studies are needed to confirm the putative contribution of *CASR* genotypes to bone phenotypes. Further, most studies suggest that while CaSR polymorphisms alter serum calcium concentrations, there are no significant changes in circulating PTH levels. Therefore, it remains to be determined whether alterations in CaSR activity in bone cells directly mediates putative polymorphism-induced changes in bone mass and density, or whether perturbed serum calcium concentrations and consequent alterations in biochemical phenotypes drive such changes.

A number of *CASR* polymorphisms are also located in the 5′ untranslated gene region, in the transcription promoters that drive *CASR* expression. Several of these polymorphisms are associated with an increased risk of renal tubule calcification, likely associated with reduced CaSR expression [171, 172], as well as an increased risk of kidney stones [172, 173]. While *in vitro* studies indicate the 990G polymorphism may cause a gain of CaSR function [152, 159, 162, 174], at present, it is unclear to what extent the other aforementioned *CaSR* polymorphisms perturb CaSR expression or signaling. Future studies are warranted that examine the effect of these polymorphisms to establish their molecular effects on CaSR function.

17.3.1.6 Autoimmune Diseases

Inhibitory and stimulatory autoantibodies directed against the CaSR have been identified in patients with autoimmune hypocalciuric hypercalcemia (AHH) and acquired autoimmune hypoparathyroidism, respectively. The clinical presentation of AHH is similar to FHH or NSHPT, with hyperparathyroidism and hypercalciuria [175]. However, the symptom severity varies from mild, similar to those seen in FHH, to severe, akin to NSHPT, depending on the autoantibody titer [176, 177]. CaSR inhibitory autoantibodies bind to the CaSR extracellular domain [178]. Some AHH autoantibodies are biased allosteric modulators, augmenting $G\alpha_q$/inositol phosphate (IP) pathway signaling while attenuating $G\alpha_i$/pERK1/2 [179, 180]. Other auto-antibodies have no effect on IP accumulation, but attenuate pERK1/2 [181].

Stimulatory CaSR autoantibodies have been identified in 86% of autoimmune polyendocrine syndrome (APS) type 1 (APS1) patients, which is caused by mutations in the *AIRE* gene encoding an autoimmune regulator protein [182–185].

A deficiency in this protein results in the generation of T lymphocytes that recognize self-antigens [186]. CaSR peptides are among the self-antigens presented for immune surveillance by major histocompatibility complexes and are recognized by auto-reactive $CD8^+$ T cells, which illicit a proinflammatory response [187]. Under these circumstances, hypoparathyroidism results from aberrant immune responses against the parathyroid glands [182–185]. However, while CaSR autoantibodies in APS1 patients have been identified against four distinct extracellular domain sites, only antibodies directed against two of these sites potentiated CaSR signaling in HEK293 cells and PTH secretion in a rat parathyroid gland derived epithelial cells [188], despite there being no apparent differences in the clinical details of patients with or without functional autoantibodies [188]. Therefore, the true relevance of CaSR autoantibodies in APS1 requires further validation.

17.3.2 (Patho)physiological Roles of the CaSR Unrelated to Calcium Homeostasis

17.3.2.1 CaSRs in the Mammary Gland

In the mouse mammary gland, CaSR mRNA levels peak during lactation [189]. The CaSR's main role is to ensure sufficient transport of calcium into milk, primarily by regulating the secretion of parathyroid hormone related peptide (PTHrP) from mammary epithelial cells [189–193]. Circulating PTHrP activates bone resorption during lactation to liberate maternal skeletal $Ca^{2+}{}_{o}$ stores that are transported into milk in the mammary gland [194]. Activation of the CaSR in mammary epithelial cells suppresses PTHrP production and secretion into the circulation [195], thus diminishing the release of $Ca^{2+}{}_{o}$ from the bone when mammary gland and milk $Ca^{2+}{}_{o}$ concentrations are sufficient [190].

In a mouse model of breast cancer, in which introduction of the PyMY oncogene into mammary epithelial cells drives mammary tumor progression and metastasis, CaSR ablation from PyMY-expressing epithelial cells prolonged mouse survival, slowed rates of tumor growth and reduced tumor burden [34]. Intriguingly, while CaSR-mediated inhibition of PTHrP secretion from normal mammary epithelial cells occurs via a $G\alpha_{i/o}$-mediated pathway during lactation, a switch in G protein coupling to $G\alpha_s$ results in CaSR-mediated stimulation of PTHrP secretion in neoplastic mammary epithelial cells [34]. Thus, in breast cancer, the CaSR stimulates PTHrP secretion in a cAMP dependent manner [195], which in turn suppresses the cell cycle inhibitor $p27^{kip1}$, promoting breast cancer cell proliferation [196]. Further, PTHrP-mediated osteoclast activation and consequent bone resorption releases transforming growth factor β, bone morphogenic proteins, and insulin-like growth factors [197]. In turn, these factors activate circulating and primary tumor cells contributing to further bone metastasis [194]. Thus, *in vitro* findings suggest a possible role for the CaSR in breast cancer via dysregulated PTHrP secretion.

There is further evidence for pro-tumorigenic effect of the CaSR in breast cancer. Several studies have suggested that the CaSR promotes the growth of aggressive breast tumors via cancer-induced hypercalcemia [198, 199]. In a cohort of 199 breast cancer patients, the 986S *CASR* polymorphism was associated with elevated serum calcium and consequently larger or more aggressive breast tumors [200]. Further, breast cancer cells commonly metastasize to the bone by osteolytic lesions [201], and a positive correlation has been found between CaSR expression in primary breast tumors and the development of osteolytic bone metastasis [202]. Similarly, *in vitro* the CaSR promotes the migration of breast cancer cells capable of forming bone metastases (MDA-MB-231; MCF-7), but not those with limited bone-metastatic potential (BT474) [198]. In MCF-7 breast cancer cells, the CaSR stimulated proliferation is via activation of PLC-, PKC-, and ERK-dependent signaling [203, 204]. These findings suggest that elevated CaSR expression and signaling specifically drives breast cancer tumor growth and bone metastasis.

However, converse to the previous findings, a number of studies have found an inverse correlation between CaSR expression or activation and breast cancer progression. Lower CaSR expression in breast cancer tumors has been associated with poor overall survival, cause-specific survival, and distant metastasis-free survival of breast cancer patients [205, 206]. Further, a large cohort of post-menopausal women with dietary calcium intakes of >1250 mg/d had reduced risk of developing breast cancer [207]. This association was stronger for estrogen receptor positive (ER+) tumors, which comprise nearly 70% of all breast cancers [208]. Treating an ER+ breast cancer cell line (MCF-7) with high $Ca^{2+}{}_{o}$ concentrations (>15 mM) decreased ER expression [209], an effect that was mimicked by the CaSR PAM NPS R467 [209]. Similarly, another study suggested high $Ca^{2+}{}_{o}$ inhibits breast cancer proliferation, cell invasion, and anchorage-dependent growth via the CaSR [210]. $Ca^{2+}{}_{o}$ also reduced the malignant behavior of MCF-7, MDA-MB-435, and MDA-MB231 breast cancer cells, and downregulated the expression of anti-apoptotic proteins in these cells [210]. These effects rendered the cells prone to apoptosis and enhanced the cytotoxic effects of chemotherapeutics drugs such as paclitaxel [211]. Interestingly, the surviving paclitaxel-resistant cells expressed no CaSR [210]. Finally, CaSR expression rescued BRCA1 defective breast cancer cells, suggesting that the CaSR could contribute to the tumor suppressor effects of BRCA1 [212]. *BRCA1* loss-of-function mutations greatly increase the risk of developing breast cancer by impairing the transcription of genes that control DNA damage repair.

17.3.2.2 CaSR in the Gastrointestinal Tract

The CaSR is expressed along the entire gastrointestinal tract, where it acts as a nutrient sensor, using its multi-modal chemosensory activity to detect $Ca^{2+}{}_{o}$, Mg^{2+}, L-amino acids, and other polypeptides [213–217]. Global *Casr* knockout mice exhibit reduced numbers of gastric G cells. While the CaSR NAM NPS2143 inhibits gastrin secretion in response to $Ca^{2+}{}_{0}$, L-amino acids or cinacalcet gavage in wild type mice, the effects of NPS2143 on gastrin secretion are lost in *CASR* knockout mice [214]. These findings indicate that the CaSR promotes G cell growth, and mediates gastrin secretion in response to nutrients. Studies in *ex vivo* tissues or in enteroendocrine cells and cell lines, suggest the CaSR also stimulates the release of the satiety hormones, protein YY, glucagon-like peptide 1 and cholecystokinin, and inhibits secretion of the appetite-stimulating hormone ghrelin in response to nutrients such as L-amino acids [215, 216, 218–221].

In addition to nutrient-sensing in the gut, the CaSR inhibits proliferation and induces differentiation of colon epithelial cells in colonic crypts [222–224], specialized structures that increase the surface area of the gut to promote nutrient absorption. Rapidly proliferating cells at the base of the colonic crypt that originate as dividing stem cells show no CaSR expression. However, as these cells migrate toward the crypt apex and differentiate, cell proliferation stops and CaSR expression increases as the cells become increasingly sensitive to apoptotic death, proceeded by cell extrusion into the lumen [225]. Knockout of the *Casr* in mouse colonic epithelial cells leads to abnormal crypt structure, which negatively impacts transepithelial barrier function and overall healthy gut microflora [225]. Similarly, esophageal-specific *Casr* knockout impairs barrier function in the mouse esophagus [226]. Colons from global *CASR/PTH* double knockout mice and intestine-specific *Casr* knockout mice, exhibit increased expression of proliferation markers and decreased differentiation and apoptotic markers [227].

The findings in *Casr* knockout mice described earlier are consistent with observations in human colon tumors and colon carcinoma cell lines, where *CASR* promoter hypermethylation, histone deacetylation and expression of microRNAs downregulate CaSR expression during colorectal tumorigenesis [212, 223, 224, 228–233]. While differentiated tumors express low CaSR levels, undifferentiated colon carcinomas lose all CaSR expression [224, 228]. Similarly, there is a well-established correlation between high dietary $Ca^{2+}{}_{0}$ intake and reduced risk of colorectal cancer [234], and several studies suggest that the protective effects of $Ca^{2+}{}_{0}$ are mediated by the CaSR. Like most adherent cells, normal colon epithelial cells in culture require attachment to a matrix in order to grow. However, human colon carcinoma cell lines can undergo anchorage-independent growth, which correlates with tumorigenic potential. When human colon carcinoma cell lines are exposed to elevated $Ca^{2+}{}_{0}$ concentrations, expression of E-cadherin, which acts as a tumor suppressor in colon cancer, is upregulated via the CaSR and cells exhibit reduced anchorage-independent growth [224, 228, 235]. Further, the PAM NPS R568 enhances $Ca^{2+}{}_{0}$-mediated suppression of cell proliferation, and induces cell differentiation and apoptosis, while the NAM NPS2143 has the opposite effect [236]. Changes in colon carcinoma cell growth are accompanied by activation of MAPK, the cell cycle inhibitors $p21^{Waf11}$ and $p27^{Kip1}$ as well as β-catenin, while β-catenin/Wnt, c-myc, and cyclin D1 pathways, which drive malignancy, are suppressed [224, 228, 231, 235, 236]. The involvement of these pathways in CaSR-mediated colon epithelial cell proliferation have been validated by immunohistochemical analysis of human colorectal tumor sections, where loss of CaSR expression is associated with undifferentiated tumors [224, 230]. Importantly, CaSR activation enhances colon carcinoma cell sensitivity to cytotoxic drugs used in cancer therapies [210]. Therefore, CaSR activation in colon carcinomas may improve therapeutic efficiency.

Like the CaSR, vitamin D protects against colon cancer by inhibiting cell proliferation and promoting differentiation of colon cancer cells [237], and it promotes chemotherapy efficacy in colon carcinoma cells [238]. Also akin to the CaSR, vitamin D upregulates expression of E-cadherin and p21/Waf1 and suppresses the β-catenin/T-cell factor (TCF) associated Wnt signaling pathways [239, 240]. Silencing CaSR expression abolished the growth inhibitory and tumor-suppressive effects of vitamin D [241]. These findings suggest that the effects of vitamin D may be, in part, due to regulation of CaSR expression [242]. Collectively, these studies demonstrate that the CaSR plays a central role in mediating anti-tumorigenic effects of $Ca^{2+}{}_{o}$ and vitamin D in the colon, and indicate that loss of colonic CaSR function is a significant risk factor for colon cancer and reduced responsiveness to chemotherapy.

In addition to a role in colon cancer, animal studies have implicated a loss in intestinal CaSR expression in intestinal inflammation. Mice with intestinal epithelial cell *CASR* deletion were more susceptible to dextran-sulfate sodium (DSS)-induced colitis [225]. The increased susceptibility to DSS was due to impaired epithelial barrier integrity, which allowed infiltration of pathogens and inflammatory cells into the gut. In separate studies, CaSR activators including L-amino acids and γ-glutamyl peptides, alleviated DSS-induced colitis [243, 244] or diminished intestinal inflammation in piglets induced by lipopolysaccharide challenge [245]. However, a subsequent study revealed that while dietary protein exacerbated DSS-induced colitis and the CaSR PAM cinacalcet exhibited pro-inflammatory effects in this colitis model, the CaSR NAM NPS2143 alleviated colitis symptoms [246]. Given that there are currently no apparent implications of the CaSR in intestinal inflammation in humans, the (patho)physiological implications of these animal findings remain to be determined.

17.3.2.3 Emerging (Patho)physiological Roles of the CaSR

Pancreas In a cohort of 284 renal transplant recipients, patients homozygous for the 986S polymorphism had elevated serum glucose concentrations in comparison to patients homozygous for A986 [247]. These findings are consistent with observations in a mouse model of ADH1, in which a L723Q gain-of-function mutation causes hypocalcemia. L723Q mutant mice have impaired glucose tolerance and insulin secretion, as well as a reduction in pancreatic islet mass and pancreatic β cell proliferation [248]. Additionally, L723Q mutant mice show no suppression of glucagon secretion in response to glucose and have an increase in pancreatic α cell proliferation [248]. These findings suggest a possible (patho)physiological role for the CaSR in the regulation of blood glucose concentrations, although a small study of FHH1 patients showed no alterations in insulin secretion or β cell function [249].

Lungs The CaSR is expressed in bronchial epithelial cells and airway smooth muscle (ASM) cells, where it mediates healthy growth and development, epithelium wound repair and airway contraction [6, 250–252]. However, in human ASM cells from asthmatics, the CaSR is upregulated [6]. Further, the CaSR agonists, polyamines, are elevated in human asthmatics and mediate airway inflammation, constriction, and pathological airway remodeling [6, 253]. Selective ASM cell CaSR ablation abolished polyamine-mediated ASM contraction in mice [6]. Studies in mice also indicate that CaSRs in the lung promote allergen-induced cytokine secretion, enhance inflammatory cell influx upon airway injury, and augment muscarinic acetylcholine receptor-mediated airway resistance [6, 254–256]. Based on this emerging evidence, CaSR NAMs offer promise as potential therapeutics for pulmonary disorders such as asthma, pulmonary fibrosis, and chronic obstructive pulmonary disease.

Vasculature The CaSR is widely expressed in the vasculature, including in the endothelium, its surrounding smooth muscle, and perivascular neurons [257–262]. Mice with *CASR* exon 7-targeted deletion in vascular smooth muscle (VSM) cells exhibited hypotension, consistent with the reduced contractility of isolated blood vessels from these mice in response to contractile stimuli. Further, while aortas from wildtype mice contracted in response to elevated $Ca^{2+}{}_o$, aortas from mice with *CASR* exon 7 ablation in VSM cells relaxed [263]. These findings suggest that VSM cell CaSRs sensitize blood vessels to contractile stimuli and mediate vascular contraction in response to $Ca^{2+}{}_o$. In humans, the *CASR* 986S polymorphism is a risk factor for cardiovascular disease due to associated hypercalcemia [264, 265]. Vascular calcification is a common consequence of end-stage CKD, and patients have significantly reduced CaSR expression in blood vessels at sites of calcification [257]. Further, overexpression of a dominant-negative mutant (R185Q) CaSR in VSM cells in culture enhanced mineral deposition, an indicator or vascular calcification, while the CaSR PAM NPS R568 attenuated mineral deposition [257]. Therefore, the CaSR in VSM cells may protect against vascular calcification in CKD, and the 986S polymorphism may have a direct effect on CaSR function in VSM cells.

Brain The role of the CaSR in the human brain has not been extensively studied, but work in rats and mice implicates the CaSR in healthy brain development [266]. While global *Casr* deletion is lethal due to hyperparathyroidism, mice with concomitant *Casr* and *Pth* deletion survive to adulthood but display impaired neuron and glial cell differentiation, as well as delayed neural stem cell differentiation [267]. Further, hippocampal-specific CaSR ablation induced three weeks post birth protected mice from ischemia-induced neuronal damage in the hippocampus [268], suggesting CaSR activation is detrimental during brain injury.

Others Consistent with CaSR expression in the skin epidermis and the CaSR's role in driving epidermal cell differentiation [269, 270], recurrent *CASR* mutations have been identified in melanoma [271]. CaSR polymorphisms or aberrant DNA methylation are also associated with neuroblastomas [242] and psoriasis [272]. Importantly, with widespread expression and an ability to respond to diverse stimuli, there are likely many more functions of the CaSR in human physiology that have yet to be elucidated.

17.4 Therapeutic Effects of Drugs Targeting the CaSR

17.4.1 Hyperparathyroidism

The small molecule PAM cinacalcet is Food and Drug Administration (FDA) approved to treat PHPT where parathyroidectomy cannot be performed, to manage hypercalcemia in patients with parathyroid carcinoma, and to normalize SHPT in CKD patients on dialysis [273]. By positively modulating $Ca^{2+}{}_{o}$ at CaSRs expressed in the parathyroid glands, cinacalcet suppresses PTH secretion [16, 23, 274]. Orally administered, cinacalcet reaches maximal concentrations 2–6 hours post administration, with a half-life of approximately 20 hours [273, 275–277]. Although cinacalcet has a good safety profile, it has limitations that constrain its utility. First, cinacalcet causes nausea and vomiting, which reduce patient compliance [278]. The underlying cause of these adverse effects is poorly understood, but CaSR expression in the gastrointestinal tract is likely involved. Additionally, cinacalcet is cleared renally after metabolism in the liver through the activity of the CYP P450 enzymes; CYP3A4, CYP2D6, and CYP1A2 [13, 277, 279]. Consequently, decreased kidney or liver function or consumption of additional drugs that are metabolized by CYP P450 enzymes reduce cinacalcet clearance [13, 279]. Finally, cinacalcet adequately reduces PTH levels in only approximately 60% of SHPT patients [280], and there is considerable patient variation in the cinacalcet dose required to achieve PTH normalization. This variation may be due to *CASR* polymorphisms, which can alter the effectiveness of cinacalcet [281–283].

Cinacalcet has been used off-label to treat FHH and NSHPT caused by naturally occurring loss-of-function mutations in the *CASR*, *GNA11*, or *AP2S1* genes (reviewed by Hannan et al. [284]). Regrettably, there are some loss-of-function mutations where cinacalcet has not proven to be effective [143, 285–288]. Nonetheless, *in vitro* studies have demonstrated that cinacalcet rescues CaSR signaling impairments caused by many mutations, or it recovers loss-of-expression *CASR* mutants by facilitating export of trapped receptors from the Golgi or endoplasmic reticulum [51, 133, 134, 141, 289]. Cinacalcet also successfully corrected the severe hypercalcemia associated with AHH [178].

Evocalcet was recently approved for use in Japan for the treatment of SHPT patients on dialysis [290, 291]. Based on modifications to cinacalcet's arylalkylamine pharmacophore, evocalcet has no interactions with CYP P450 enzymes and improved bioavailability [290]. Consequently, evocalcet retains comparable clinical efficacy to cinacalcet, but its efficacy is achieved at doses that are 10-fold lower than the effective cinacalcet dose [290, 292]. The lower required evocalcet dose may account for the improved gastrointestinal side effect profile of evocalcet versus cinacalcet [290]. With cinacalcet already widely used clinically, it remains to be seen if evocalcet can achieve FDA approval and widespread usage outside of Japan.

Etelcalcetide is approved by the FDA for patients with CKD on dialysis [293]. Further, etelcalcetide does not interact with CYP P450 enzymes in the liver, thus avoiding the same drug interactions observed with cinacalcet. Instead, etelcalcetide is removed by the kidneys (in patients with some remaining kidney function) or by hemodialysis in patients with end-stage kidney disease. This clearance pathway gives etelcalcetide a drug half-life of three to four days [294]. The D-amino acids that comprise etelcalcetide greatly improve its half-life by preventing proteolytic degradation [295]. As Etelcalcetide is administered intravenously immediately post-dialysis, etelcalcetide has a higher rate of patient compliance than cinacalcet and evocalcet [296].

17.4.2 Osteoporosis

CaSR NAM discovery efforts were originally focused on targeting the CaSR in osteoporosis, where CaSR NAMs have the potential to mimic the bone anabolic effects of recombinant PTH1–34 (teriparatide) [297]. Teriparatide replicates a transient increase in PTH, thus stimulating osteoblast activity [298, 299]. Like PTH, teriparatide is rapidly metabolized [299] and thus cleared from the body before osteoclasts are activated. However, teriparatide must be intravenously administered daily, limiting its therapeutic utility as an ongoing treatment option for osteoporosis. To reproduce the clinical benefits of teriparatide, the "ideal" CaSR NAM should inhibit CaSR activity to mimic a drop in $Ca^{2+}{}_o$, consequently stimulating PTH secretion and subsequent osteoblast activity, with rapid NAM clearance to ensure PTH levels decline before osteoclasts are activated.

Intravenous injection of NPS2143 in mice increased PTH secretion and consequent osteoblast activity. Unfortunately, NPS2143 clearance rates were slow, with a drug half-life of two hours in rats, resulting in a sustained increase in PTH secretion that exceeded four hours [300]. While NPS2143 increased bone formation in rats, osteoclast-mediated bone resorption is also activated, offsetting any increase in bone mass and density [301]. To counteract prolonged NAM-mediated elevations in PTH, subsequent NAM discovery efforts focused on increasing drug clearance rates. Using this approach, three CaSR NAMs entered clinical trials for osteoporosis, ronacaleret, AXT914, and JTT-305/MK-5442. However, despite increased clearance compared with NPS2143, all three NAMs failed to improve bone mass and density [19, 302–304]. Further, AXT914 caused hypocalcemia in some patients, resulting in early trial termination [19]. The reasons for NAM failure in osteoporosis are unclear, but they could involve inhibition of CaSRs expressed in osteoblasts. Indeed, strontium ranelate, which received European approval for the treatment of osteoporosis [25] and acts by delivering chelated Sr^{2+} to the bone, inhibits osteoclast activity and promotes osteoblast differentiation in part via interactions with the CaSR [3, 25, 105, 305–310]. However, cardiovascular side effects severely limited clinical usefulness of strontium ranelate leading to its recent withdrawal [311].

17.4.3 ADH and Bartter Syndrome Type V

Although CaSR NAMs did not recapitulate the effect of teriparatide in osteoporosis, other calciotropic diseases may benefit from a clinically available CaSR NAM. As discussed earlier, NPS2143 normalizes *in vitro* signaling responses associated with ADH-causing *CASR* and *GNA11* mutations, increases $Ca^{2+}{}_o$ and PTH concentrations in ADH type 1 and type 2 mouse models, and prevents nephrocalcinosis in ADH type 1 mouse models [51, 312–315]. However, NPS2143 is less effective at gain-of-function CaSR mutations that cause Bartter syndrome type V [51, 314]. Nevertheless, NPSP795, an aryl-alkylamine CaSR NAM, entered phase II clinical trials for the treatment of ADH1. Promisingly, NPSP795 increased PTH in three out of five ADH 1 trial patients and caused a small reduction in renal $Ca^{2+}{}_o$ excretion [316]. However, there was wide variability in the maximal stimulatory effect of NPSP795 on PTH secretion between patients [316]. Nonetheless, the potential for a CaSR NAM to provide relief for sufferers of ADH and Bartter syndrome type V remains a possibility.

17.5 Concluding Remarks

The CaSR is widely distributed throughout the body and is involved in many (patho)physiological processes. The CaSR remains a clinically relevant drug target in endocrine diseases. However, the ever-expanding roles of the CaSR in diverse tissues suggest the CaSR could serve as a putative therapeutic target in numerous diseases. While drugs that target the CaSR alter the secretion of PTH via CaSRs expressed in the parathyroid glands, it has so far not been possible to target the CaSR in tissues outside the parathyroid glands. With major advances in GPCR structural biology, a detailed molecular understanding of the complex CaSR conformational landscape and structural phenomena underpinning biased agonism and allosteric modulation is within reach. As such, the structural elucidation of the CaSR will aid structure-based drug design at the CaSR with the hope to develop novel pharmacophores in the future, with the possibility of tissue-selective CaSR small molecules.

References

1 Riccardi, D. and Kemp, P.J. (2012). The calcium-sensing receptor beyond extracellular calcium homeostasis: conception, development, adult physiology, and disease. *Annu. Rev. Physiol.* 74: 271–297.

2 Gregory, K.J., Giraldo, J., Diao, J. et al. (2020). Evaluation of operational models of agonism and allosterism at receptors with multiple orthosteric binding sites. *Mol. Pharmacol.* 97 (1): 35–45.

3 Brown, E.M., Fuleihan Ge-H, Chen, C.J., and Kifor, O. (1990). A comparison of the effects of divalent and trivalent cations on parathyroid hormone release, 3′,5′-cyclic-adenosine monophosphate accumulation, and the levels of inositol phosphates in bovine parathyroid cells. *Endocrinology* 127 (3): 1064–1071.

4 Handlogten, M.E., Shiraishi, N., Awata, H. et al. (2000). Extracellular Ca^{2+}-sensing receptor is a promiscuous divalent cation sensor that responds to lead. *Am. J. Physiol. Renal Physiol.* 279 (6): F1083–F1091.

5 Quinn, S.J., Ye, C.P., Diaz, R. et al. (1997). The Ca^{2+}-sensing receptor: a target for polyamines. *Am. J. Physiol.* 273 (4 Pt 1): C1315–C1323.

6 Yarova, P.L., Stewart, A.L., Sathish, V. et al. (2015). Calcium-sensing receptor antagonists abrogate airway hyperresponsiveness and inflammation in allergic asthma. *Sci. Transl. Med.* 7 (284): 284ra60.

7 Conigrave, A.D., Quinn, S.J., and Brown, E.M. (2000). L-amino acid sensing by the extracellular Ca^{2+}-sensing receptor. *Proc. Natl. Acad. Sci. U.S.A.* 97 (9): 4814–4819.

8 Broadhead, G.K., Mun, H.C., Avlani, V.A. et al. (2011). Allosteric modulation of the calcium-sensing receptor by γ-glutamyl peptides: inhibition of PTH secretion, suppression of intracellular cAMP levels, and a common mechanism of action with L-amino acids. *J. Biol. Chem.* 286 (11): 8786–8797.

9 Centeno, P.P., Herberger, A., Mun, H.C. et al. (2019). Phosphate acts directly on the calcium-sensing receptor to stimulate parathyroid hormone secretion. *Nat. Commun.* 10 (1): 4693.

10 Quinn, S.J., Kifor, O., Trivedi, S. et al. (1998). Sodium and ionic strength sensing by the calcium receptor. *J. Biol. Chem.* 273 (31): 19579–19586.

11 Campion, K.L., WD, M.C., Warwicker, J. et al. (2015). Pathophysiologic changes in extracellular pH modulate parathyroid calcium-sensing receptor activity and secretion via a histidine-independent mechanism. *J. Am. Soc. Nephrol.* 26 (9): 2163–2171.

12 McLarnon, S.J. and Riccardi, D. (2002). Physiological and pharmacological agonists of the extracellular Ca^{2+}-sensing receptor. *Eur. J. Pharmacol.* 447 (2, 3): 271–278.

13 Nemeth, E.F., Van Wagenen, B.C., and Balandrin, M.F. (2018). Discovery and development of calcimimetic and calcilytic compounds. *Prog. Med. Chem.* 57 (1): 1–86.

14 Leach, K., Gregory, K.J., Kufareva, I. et al. (2016). Towards a structural understanding of allosteric drugs at the human calcium-sensing receptor. *Cell Res.* 26 (5): 574–592.

15 Keller, A.N., Kufareva, I., Josephs, T.M. et al. (2018). Identification of global and ligand-specific calcium sensing receptor activation mechanisms. *Mol. Pharmacol.* 93 (6): 619–630.

16 Cook, A.E., Mistry, S.N., Gregory, K.J. et al. (2015). Biased allosteric modulation at the CaS receptor engendered by structurally diverse calcimimetics. *Br. J. Pharmacol.* 172 (1): 185–200.

17 Alexander, S.T., Hunter, T., Walter, S. et al. (2015). Critical cysteine residues in both the calcium-sensing receptor and the allosteric activator AMG 416 underlie the mechanism of action. *Mol. Pharmacol.* 88 (5): 853–865.

18 John, M.R., Widler, L., Gamse, R. et al. (2011). ATF936, a novel oral calcilytic, increases bone mineral density in rats and transiently releases parathyroid hormone in humans. *Bone* 49 (2): 233–241.

19 John, M.R., Harfst, E., Loeffler, J. et al. (2014). AXT914 a novel, orally-active parathyroid hormone-releasing drug in two early studies of healthy volunteers and postmenopausal women. *Bone* 64: 204–210.

20 Yang, W., Ruan, Z., Wang, Y. et al. (2009). Discovery and structure-activity relationships of trisubstituted pyrimidines/pyridines as novel calcium-sensing receptor antagonists. *J. Med. Chem.* 52 (4): 1204–1208.

21 Didiuk, M.T., Griffith, D.A., Benbow, J.W. et al. (2009). Short-acting 5-(trifluoromethyl)pyrido[4,3-*d*]pyrimidin-4(3*H*)-one derivatives as orally-active calcium-sensing receptor antagonists. *Bioorg. Med. Chem. Lett.* 19 (16): 4555–4559.

22 Kalgutkar, A.S., Griffith, D.A., Ryder, T. et al. (2010). Discovery tactics to mitigate toxicity risks due to reactive metabolite formation with 2-(2-hydroxyaryl)-5-(trifluoromethyl)pyrido[4,3-*d*]pyrimidin-4(3*H*)-one derivatives, potent calcium-sensing receptor antagonists and clinical candidate(s) for the treatment of osteoporosis. *Chem. Res. Toxicol.* 23 (6): 1115–1126.

23 Davey, A.E., Leach, K., Valant, C. et al. (2012). Positive and negative allosteric modulators promote biased signaling at the calcium-sensing receptor. *Endocrinology* 153 (3): 1232–1241.

24 Gregory, K.J., Kufareva, I., Keller, A.N. et al. (2018). Dual action calcium-sensing receptor modulator unmasks novel mode-switching mechanism. *ACS Pharmacol. Transl. Sci.* 1 (2): 96–109.

25 Diepenhorst, N.A., Leach, K., Keller, A.N. et al. (2018). Divergent effects of strontium and calcium-sensing receptor positive allosteric modulators (calcimimetics) on human osteoclast activity. *Br. J. Pharmacol.* 175 (21): 4095–4108.

26 Josephs, T.M., Keller, A.N., Khajehali, E. et al. (2020). Negative allosteric modulators of the human calcium-sensing receptor bind to overlapping and distinct sites within the 7 transmembrane domain. *Br. J. Pharmacol.* 177 (8): 1917–1930.

27 Leach, K., Hannan, F.M., Josephs, T.M. et al. (2020). International Union of Basic and Clinical Pharmacology XXX: calcium sensing receptor nomenclature, pharmacology, and function. *Pharmacol. Rev.* 72: 1–48.

28 Brown, E.M., Gamba, G., Riccardi, D. et al. (1993). Cloning and characterization of an extracellular Ca^{2+}-sensing receptor from bovine parathyroid. *Nature* 366 (6455): 575–580.

29 Chang, W., Pratt, S., Chen, T.H. et al. (1998). Coupling of calcium receptors to inositol phosphate and cyclic AMP generation in mammalian cells and *Xenopus laevis* oocytes and immunodetection of receptor protein by region-specific antipeptide antisera. *J. Bone Miner. Res.* 13 (4): 570–580.

30 Onopiuk, M., Eby, B., Nesin, V. et al. (2020). Control of PTH secretion by the $TRPC_1$ ion channel. *JCI Insight* 5 (8): e132496.

31 Kifor, O., RJ, M.L., Diaz, R. et al. (2001). Regulation of MAP kinase by calcium-sensing receptor in bovine parathyroid and CaR-transfected HEK293 cells. *Am. J. Physiol. Renal Physiol.* 280 (2): F291–F302.

32 Gorvin, C.M., Babinsky, V.N., Malinauskas, T. et al. (2018). A calcium-sensing receptor mutation causing hypocalcemia disrupts a transmembrane salt bridge to activate β-arrestin-biased signaling. *Sci. Signal.* 11 (518): eaan3714.

33 Mos, I., Jacobsen, S.E., Foster, S.R., and Brauner-Osborne, H. (2019). Calcium-sensing receptor internalization is β-arrestin-dependent and modulated by allosteric ligands. *Mol. Pharmacol.* 96 (4): 463–474.

34 Mamillapalli, R., VanHouten, J., Zawalich, W., and Wysolmerski, J. (2008). Switching of G-protein usage by the calcium-sensing receptor reverses its effect on parathyroid hormone-related protein secretion in normal versus malignant breast cells. *J. Biol. Chem.* 283 (36): 24435–24447.

35 Mamillapalli, R. and Wysolmerski, J. (2010). The calcium-sensing receptor couples to $G\alpha_s$ and regulates PTHrP and ACTH secretion in pituitary cells. *J. Endocrinol.* 204 (3): 287–297.

36 Thomsen, A.R., Hvidtfeldt, M., and Brauner-Osborne, H. (2012). Biased agonism of the calcium-sensing receptor. *Cell Calcium* 51 (2): 107–116.

37 Huang, C., Handlogten, M.E., and Miller, R.T. (2002). Parallel activation of phosphatidylinositol 4-kinase and phospholipase C by the extracellular calcium-sensing receptor. *J. Biol. Chem.* 277 (23): 20293–20300.

38 Bai, M., Trivedi, S., Lane, C.R. et al. (1998). Protein kinase C phosphorylation of threonine at position 888 in ${Ca^{2+}}_o$-sensing receptor (CaR) inhibits coupling to Ca^{2+} store release. *J. Biol. Chem.* 273 (33): 21267–21275.

39 Jiang, Y.F., Zhang, Z., Kifor, O. et al. (2002). Protein kinase C (PKC) phosphorylation of the ${Ca^{2+}}_o$-sensing receptor (CaR) modulates functional interaction of G proteins with the CaR cytoplasmic tail. *J. Biol. Chem.* 277 (52): 50543–50549.

40 Binmahfouz, L.S., Centeno, P.P., Conigrave, A.D., and Ward, D.T. (2019). Identification of Serine-875 as an inhibitory phosphorylation site in the calcium-sensing receptor. *Mol. Pharmacol.* 96 (2): 204–211.

41 McCormick, W.D., Atkinson-Dell, R., Campion, K.L. et al. (2010). Increased receptor stimulation elicits differential calcium-sensing receptor (T888) dephosphorylation. *J. Biol. Chem.* 285 (19): 14170–14177.

42 Lazarus, S., Pretorius, C.J., Khafagi, F. et al. (2011). A novel mutation of the primary protein kinase C phosphorylation site in the calcium-sensing receptor causes autosomal dominant hypocalcemia. *Eur. J. Endocrinol.* 164 (3): 429–435.

43 Gorvin, C.M., Rogers, A., Hastoy, B. et al. (2018). AP2σ mutations impair calcium-sensing receptor trafficking and signaling, and show an endosomal pathway to spatially direct G-protein selectivity. *Cell Rep.* 22 (4): 1054–1066.

44 Grant, M.P., Stepanchick, A., Cavanaugh, A., and Breitwieser, G.E. (2011). Agonist-driven maturation and plasma membrane insertion of calcium-sensing receptors dynamically control signal amplitude. *Sci. Signal.* 4 (200): ra78.

45 Kenakin, T. and Christopoulos, A. (2013). Signalling bias in new drug discovery: detection, quantification and therapeutic impact. *Nat. Rev. Drug Discovery* 12 (3): 205–216.

46 Lee, H., Mun, H.-C., Lewis, N.C. et al. (2007). Allosteric activation of the extracellular Ca^{2+}-sensing receptor by L-amino acids enhances ERK1/2 phosphorylation. *Biochem. J.* 404: 141–149.

47 Davies, S.L., Gibbons, C.E., Vizard, T., and Ward, D.T. (2006). Calcium-sensing receptor induces Rho kinase-mediated actin stress fiber assembly and altered cell morphology though not in response to aromatic amino acids. *Am. J. Physiol. Cell Physiol.* 290: 1543–1551.

48 Avlani, V., Ma, W., Mun, H.-C. et al. (2013). Calcium-sensing receptor-dependent activation of CREB phosphorylation in HEK-293 cells and human parathyroid cells. *Am. J. Physiol. Endocrinol. Metab.* 304: E1097–E1104.

49 Hofer, A.M. and Brown, E.M. (2003). Extracellular calcium sensing and signalling. *Nat. Rev. Mol. Cell Biol.* 4 (7): 530–538.

50 Leach, K. and Gregory, K.J. (2017). Molecular insights into allosteric modulation of Class C G protein-coupled receptors. *Pharmacol. Res.* 116: 105–118.

51 Leach, K., Wen, A., Cook, A.E. et al. (2013). Impact of clinically relevant mutations on the pharmacoregulation and signaling bias of the calcium-sensing receptor by positive and negative allosteric modulators. *Endocrinology* 154 (3): 1105–1116.

52 Leach, K., Wen, A., Davey, A.E. et al. (2012). Identification of molecular phenotypes and biased signaling induced by naturally occurring mutations of the human calcium-sensing receptor. *Endocrinology* 153 (9): 4304–4316.

53 Attwood, T.K. and Findlay, J.B. (1994). Fingerprinting G-protein-coupled receptors. *Protein Eng.* 7 (2): 195–203.

54 Kolakowski, L.F. Jr. (1994). GCRDb: a G-protein-coupled receptor database. *Recept. Channels* 2 (1): 1–7.

55 Koehl, A., Hu, H., Feng, D. et al. (2019). Structural insights into the activation of metabotropic glutamate receptors. *Nature* 566 (7742): 79–84.

56 Bai, M., Trivedi, S., and Brown, E.M. (1998). Dimerization of the extracellular calcium-sensing receptor (CaR) on the cell surface of CaR-transfected HEK293 cells. *J. Biol. Chem.* 273 (36): 23605–23610.

57 Pidasheva, S., Grant, M., Canaff, L. et al. (2006). Calcium-sensing receptor dimerizes in the endoplasmic reticulum: biochemical and biophysical characterization of CASR mutants retained intracellularly. *Hum. Mol. Genet.* 15 (14): 2200–2209.

58 Ward, D.T., Brown, E.M., and Harris, H.W. (1998). Disulfide bonds in the extracellular calcium-polyvalent cation-sensing receptor correlate with dimer formation and its response to divalent cations in vitro. *J. Biol. Chem.* 273 (23): 14476–14483.

59 Zhang, Z., Sun, S., Quinn, S.J. et al. (2001). The extracellular calcium-sensing receptor dimerizes through multiple types of intermolecular interactions. *J. Biol. Chem.* 276 (7): 5316–5322.

60 Zhang, C., Zhang, T., Zou, J. et al. (2016). Structural basis for regulation of human calcium-sensing receptor by magnesium ions and an unexpected tryptophan derivative co-agonist. *Sci. Adv.* 2 (5): e1600241.

61 Geng, Y., Mosyak, L., Kurinov, I. et al. (2016). Structural mechanism of ligand activation in human calcium-sensing receptor. *Elife* 5: e13662.

62 Mun, H.C., Franks, A.H., Culverston, E.L. et al. (2004). The Venus Fly Trap domain of the extracellular Ca^{2+}-sensing receptor is required for L-amino acid sensing. *J. Biol. Chem.* 279 (50): 51739–51744.

63 Nuemket, N., Yasui, N., Kusakabe, Y. et al. (2017). Structural basis for perception of diverse chemical substances by T1r taste receptors. *Nat. Commun.* 8: 15530.

64 Monn, J.A., Prieto, L., Taboada, L. et al. (2015). Synthesis and pharmacological characterization of C_4-(thiotriazolyl)-substituted-2-aminobicyclo[3.1.0] hexane-2,6-dicarboxylates. Identification of (1*R*,2*S*,4*R*,5*R*,6*R*)-2-amino-4-(1*H*-1,2,4-triazol-3-ylsulfanyl)bicyclo[3.1.0]hexane-2, 6-dicarboxylic acid (LY2812223), a highly potent, functionally selective $mGlu_2$ receptor agonist. *J. Med. Chem.* 58 (18): 7526–7548.

65 Kunishima, N., Shimada, Y., Tsuji, Y. et al. (2000). Structural basis of glutamate recognition by a dimeric metabotropic glutamate receptor. *Nature* 407 (6807): 971–977.

66 Muto, T., Tsuchiya, D., Morikawa, K., and Jingami, H. (2007). Structures of the extracellular regions of the group II/III metabotropic glutamate receptors. *Proc. Natl. Acad. Sci. U.S.A.* 104 (10): 3759–3764.

67 Geng, Y., Bush, M., Mosyak, L. et al. (2013). Structural mechanism of ligand activation in human $GABA_B$ receptor. *Nature* 504 (7479): 254–259.

68 Geng, Y., Xiong, D., Mosyak, L. et al. (2012). Structure and functional interaction of the extracellular domain of human $GABA_B$ receptor GBR2. *Nat. Neurosci.* 15 (7): 970–978.

69 Wang, M., Yao, Y., Kuang, D., and Hampson, D.R. (2006). Activation of family C G-protein-coupled receptors by the tripeptide glutathione. *J. Biol. Chem.* 281 (13): 8864–8870.

70 Wang, M. and Hampson, D.R. (2006). An evaluation of automated in silico ligand docking of amino acid ligands to family C G-protein coupled receptors. *Bioorg. Med. Chem.* 14 (6): 2032–2039.

71 Dore, A.S., Okrasa, K., Patel, J.C. et al. (2014). Structure of Class C GPCR metabotropic glutamate receptor 5 transmembrane domain. *Nature* 511 (7511): 557–562.

72 Wu, H., Wang, C., Gregory, K.J. et al. (2014). Structure of a Class C GPCR metabotropic glutamate receptor 1 bound to an allosteric modulator. *Science* 344 (6179): 58–64.

73 Weis, W.I. and Kobilka, B.K. (2018). The molecular basis of G protein-coupled receptor activation. *Annu. Rev. Biochem.* 87: 897–919.

74 Jacobsen, S.E., Gether, U., and Brauner-Osborne, H. (2017). Investigating the molecular mechanism of positive and negative allosteric modulators in the calcium-sensing receptor dimer. *Sci. Rep.* 7: 46355.

75 Arey, B.J., Seethala, R., Ma, Z. et al. (2005). A novel calcium-sensing receptor antagonist transiently stimulates parathyroid hormone secretion in vivo. *Endocrinology* 146 (4): 2015–2022.

76 Brown, E.M. (1983). Four-parameter model of the sigmoidal relationship between parathyroid hormone release and extracellular calcium concentration in normal and abnormal parathyroid tissue. *J. Clin. Endocrinol. Metab.* 56 (3): 572–581.

77 Brown, E.M. (1991). Extracellular Ca^{2+} sensing, regulation of parathyroid cell function, and role of Ca^{2+} and other ions as extracellular (first) messengers. *Physiol. Rev.* 71 (2): 371–411.

78 Ramirez, J.A., Goodman, W.G., Gornbein, J. et al. (1993). Direct in vivo comparison of calcium-regulated parathyroid hormone secretion in normal volunteers and patients with secondary hyperparathyroidism. *J. Clin. Endocrinol. Metab.* 76 (6): 1489–1494.

79 Ho, C., Conner, D.A., Pollak, M.R. et al. (1995). A mouse model of human familial hypocalciuric hypercalcemia and neonatal severe hyperparathyroidism. *Nat. Genet.* 11 (4): 389–394.

80 Chang, W., Tu, C., Chen, T.H. et al. (2008). The extracellular calcium-sensing receptor (CaSR) is a critical modulator of skeletal development. *Sci. Signal.* 1 (35): ra1.

81 Fan, Y., Liu, W., Bi, R. et al. (2018). Interrelated role of Klotho and calcium-sensing receptor in parathyroid hormone synthesis and parathyroid hyperplasia. *Proc. Natl. Acad. Sci. U.S.A.* 115 (16): E3749–E3758.

82 Toka, H.R., Al-Romaih, K., Koshy, J.M. et al. (2012). Deficiency of the calcium-sensing receptor in the kidney causes parathyroid hormone-independent hypocalciuria. *J. Am. Soc. Nephrol.* 23 (11): 1879–1890.

83 Ritter, C.S., Haughey, B.H., Miller, B., and Brown, A.J. (2012). Differential gene expression by oxyphil and chief cells of human parathyroid glands. *J. Clin. Endocrinol. Metab.* 97 (8): E1499–E1505.

84 Moallem, E., Kilav, R., Silver, J., and Naveh-Many, T. (1998). RNA-protein binding and post-transcriptional regulation of parathyroid hormone gene expression by calcium and phosphate. *J. Biological Chem.* 273 (9): 5253–5259.

85 Loupy, A., Ramakrishnan, S.K., Wootla, B. et al. (2012). PTH-independent regulation of blood calcium concentration by the calcium-sensing receptor. *J. Clin. Invest.* 122 (9): 3355–3367.

86 Kantham, L., Quinn, S.J., Egbuna, O.I. et al. (2009). The calcium-sensing receptor (CaSR) defends against hypercalcemia independently of its regulation of parathyroid hormone secretion. *Am. J. Physiol. Endocrinol. Metab.* 297 (4): E915–E923.

87 Kos, C.H., Karaplis, A.C., Peng, J.B. et al. (2003). The calcium-sensing receptor is required for normal calcium homeostasis independent of parathyroid hormone. *J. Clin. Invest.* 111 (7): 1021–1028.

88 Riccardi, D., Park, J., Lee, W.S. et al. (1995). Cloning and functional expression of a rat kidney extracellular calcium/polyvalent cation-sensing receptor. *Proc. Natl. Acad. Sci. U.S.A.* 92 (1): 131–135.

89 Gong, Y., Renigunta, V., Himmerkus, N. et al. (2012). Claudin-14 regulates renal Ca^{++} transport in response to CaSR signalling via a novel microRNA pathway. *EMBO J.* 31 (8): 1999–2012.

90 Graca, J.A., Schepelmann, M., Brennan, S.C. et al. (2016). Comparative expression of the extracellular calcium-sensing receptor in the mouse, rat, and human kidney. *Am. J. Physiol. Renal Physiol.* 310 (6): F518–F533.

91 Riccardi, D. and Valenti, G. (2016). Localization and function of the renal calcium-sensing receptor. *Nat. Rev. Nephrol.* 12 (7): 414–425.

92 Santa Maria, C., Cheng, Z., Li, A. et al. (2016). Interplay between CaSR and PTH1R signaling in skeletal development and osteoanabolism. *Semin. Cell Dev. Biol.* 49: 11–23.

93 Goltzman, D. and Hendy, G.N. (2015). The calcium-sensing receptor in bone – mechanistic and therapeutic insights. *Nat. Rev. Endocrinol.* 11 (5): 298–307.

94 Hannan, F.M., Kallay, E., Chang, W. et al. (2018). The calcium-sensing receptor in physiology and in calcitropic and noncalcitropic diseases. *Nat. Rev. Endocrinol.* 15 (1): 33–51.

95 Al-Dujaili, S.A., Koh, A.J., Dang, M. et al. (2016). Calcium sensing receptor function supports osteoblast survival and acts as a co-factor in PTH anabolic actions in bone. *J. Cell. Biochem.* 117 (7): 1556–1567.

96 Dvorak-Ewell, M.M., Chen, T.H., Liang, N. et al. (2011). Osteoblast extracellular Ca^{2+}-sensing receptor regulates bone development, mineralization, and turnover. *J. Bone Miner. Res.* 26 (12): 2935–2947.

97 Cheng, Z., Tu, C., Rodriguez, L. et al. (2007). Type B γ-aminobutyric acid receptors modulate the function of the extracellular Ca^{2+}-sensing receptor and cell differentiation in murine growth plate chondrocytes. *Endocrinology* 148 (10): 4984–4992.

98 Dobnig, H. and Turner, R.T. (1997). The effects of programmed administration of human parathyroid hormone fragment (1–34) on bone histomorphometry and serum chemistry in rats. *Endocrinology* 138 (11): 4607–4612.

99 Weir, E.C., Lowik, C.W., Paliwal, I., and Insogna, K.L. (1996). Colony stimulating factor-1 plays a role in osteoclast formation and function in bone resorption induced by parathyroid hormone and parathyroid hormone-related protein. *J. Bone Miner. Res.* 11 (10): 1474–1481.

100 Ma, Y.L., Cain, R.L., Halladay, D.L. et al. (2001). Catabolic effects of continuous human PTH (1–38) in vivo is associated with sustained stimulation of RANKL and inhibition of osteoprotegerin and gene-associated bone formation. *Endocrinology* 142 (9): 4047–4054.

101 Silver, I.A., Murrills, R.J., and Etherington, D.J. (1988). Microelectrode studies on the acid microenvironment beneath adherent macrophages and osteoclasts. *Exp. Cell. Res.* 175 (2): 266–276.

102 Kameda, T., Mano, H., Yamada, Y. et al. (1998). Calcium-sensing receptor in mature osteoclasts, which are bone resorbing cells. *Biochem. Biophys. Res. Commun.* 245 (2): 419–422.

103 Kanatani, M., Sugimoto, T., Kanzawa, M. et al. (1999). High extracellular calcium inhibits osteoclast-like cell formation by directly acting on the calcium-sensing receptor existing in osteoclast precursor cells. *Biochem. Biophys. Res. Commun.* 261 (1): 144–148.

104 Mentaverri, R., Yano, S., Chattopadhyay, N. et al. (2006). The calcium sensing receptor is directly involved in both osteoclast differentiation and apoptosis. *FASEB J.* 20 (14): 2562–2564.

105 Zaidi, M., Kerby, J., Huang, C.L. et al. (1991). Divalent cations mimic the inhibitory effect of extracellular ionised calcium on bone resorption by isolated rat osteoclasts: further evidence for a "calcium receptor". *J. Cell. Physiol.* 149 (3): 422–427.

106 Corbetta, S., Lania, A., Filopanti, M. et al. (2002). Mitogen-activated protein kinase cascade in human normal and tumoral parathyroid cells. *J. Clin. Endocrinol. Metab.* 87 (5): 2201–2205.

107 Yamauchi, M., Sugimoto, T., Yamaguchi, T. et al. (2001). Association of polymorphic alleles of the calcium-sensing receptor gene with the clinical severity of primary hyperparathyroidism. *Clin. Endocrinol.* 55 (3): 373–379.

108 Bilezikian, J.P., Cusano, N.E., Khan, A.A. et al. (2016). Primary hyperparathyroidism. *Nat. Rev. Dis. Primers* 2: 16033.

109 Cetani, F., Pinchera, A., Pardi, E. et al. (1999). No evidence for mutations in the calcium-sensing receptor gene in sporadic parathyroid adenomas. *J. Bone Miner. Res.* 14 (6): 878–882.

110 Newey, P.J., Nesbit, M.A., Rimmer, A.J. et al. (2012). Whole-exome sequencing studies of nonhereditary (sporadic) parathyroid adenomas. *J. Clin. Endocrinol. Metab.* 97 (10): E1995–E2005.

111 Fukuda, N., Tanaka, H., Tominaga, Y. et al. (1993). Decreased 1,25-dihydroxyvitamin D_3 receptor density is associated with a more severe form of parathyroid hyperplasia in chronic uremic patients. *J. Clin. Invest.* 92 (3): 1436–1443.

112 Gogusev, J., Duchambon, P., Hory, B. et al. (1997). Depressed expression of calcium receptor in parathyroid gland tissue of patients with hyperparathyroidism. *Kidney Int.* 51 (1): 328–336.

113 Kifor, O., Moore, F.D. Jr., Wang, P. et al. (1996). Reduced immunostaining for the extracellular Ca^{2+}-sensing receptor in primary and uremic secondary hyperparathyroidism. *J. Clin. Endocrinol. Metab.* 81 (4): 1598–1606.

114 Tominaga, Y., Kohara, S., Namii, Y. et al. (1996). Clonal analysis of nodular parathyroid hyperplasia in renal hyperparathyroidism. *World J. Surg.* 20 (7): 744–750; discussion 50–52.

115 Wei, Y., Lin, J., Yang, F. et al. (2016). Risk factors associated with secondary hyperparathyroidism in patients with chronic kidney disease. *Exp. Ther. Med.* 12 (2): 1206–1212.

116 Li, Y.C., Pirro, A.E., Amling, M. et al. (1997). Targeted ablation of the vitamin D receptor: an animal model of vitamin D-dependent rickets type II with alopecia. *Proc. Natl. Acad. Sci. U.S.A.* 94 (18): 9831–9835.

117 Yoshizawa, T., Handa, Y., Uematsu, Y. et al. (1997). Mice lacking the vitamin D receptor exhibit impaired bone formation, uterine hypoplasia and growth retardation after weaning. *Nat. Genet.* 16 (4): 391–396.

118 Trechsel, U., Eisman, J.A., Fischer, J.A. et al. (1980). Calcium-dependent, parathyroid hormone-independent regulation of 1,25-dihydroxyvitamin D. *Am. J. Physiol.* 239 (2): E119–E124.

119 Matsumoto, T., Ikeda, K., Morita, K. et al. (1987). Blood Ca^{2+} modulates responsiveness of renal 25(OH)D_3-1 α-hydroxylase to PTH in rats. *Am. J. Physiol.* 253 (5 Pt 1): E503–E507.

120 Weisinger, J.R., Favus, M.J., Langman, C.B., and Bushinsky, D.A. (1989). Regulation of 1,25-dihydroxyvitamin D_3 by calcium in the parathyroidectomized, parathyroid hormone-replete rat. *J. Bone Miner. Res.* 4 (6): 929–935.

121 Cetani, F., Picone, A., Cerrai, P. et al. (2000). Parathyroid expression of calcium-sensing receptor protein and in vivo parathyroid hormone-Ca^{2+} set-point in patients with primary hyperparathyroidism. *J. Clin. Endocrinol. Metab.* 85 (12): 4789–4794.

122 Corbetta, S., Mantovani, G., Lania, A. et al. (2000). Calcium-sensing receptor expression and signalling in human parathyroid adenomas and primary hyperplasia. *Clin. Endocrinol.* 52 (3): 339–348.

123 Maiti, A., Hait, N.C., and Beckman, M.J. (2008). Extracellular calcium-sensing receptor activation induces vitamin D receptor levels in proximal kidney HK-2G cells by a mechanism that requires phosphorylation of p38α MAPK. *J. Biol. Chem.* 283 (1): 175–183.

124 Pollak, M.R., Brown, E.M., Chou, Y.H. et al. (1993). Mutations in the human Ca^{2+}-sensing receptor gene cause familial hypocalciuric hypercalcemia and neonatal severe hyperparathyroidism. *Cell* 75 (7): 1297–1303.

125 Tominaga, Y., Tanaka, Y., Sato, K. et al. (1997). Histopathology, pathophysiology, and indications for surgical treatment of renal hyperparathyroidism. *Semin. Surg. Oncol.* 13 (2): 78–86.

126 Malberti, F., Farina, M., and Imbasciati, E. (1999). The PTH-calcium curve and the set point of calcium in primary and secondary hyperparathyroidism. *Nephrol. Dial. Transplant.* 14 (10): 2398–2406.

127 Mathias, R.S., Nguyen, H.T., Zhang, M.Y., and Portale, A.A. (1998). Reduced expression of the renal calcium-sensing receptor in rats with experimental chronic renal insufficiency. *J. Am. Soc. Nephrol.* 9 (11): 2067–2074.

128 Hannan, F.M., Babinsky, V.N., and Thakker, R.V. (2016). Disorders of the calcium-sensing receptor and partner proteins: insights into the molecular basis of calcium homeostasis. *J. Mol. Endocrinol.* 57 (3): R127–R142.

129 McMurtry, C.T., Schranck, F.W., Walkenhorst, D.A. et al. (1992). Significant developmental elevation in serum parathyroid hormone levels in a large kindred with familial benign (hypocalciuric) hypercalcemia. *Am. J. Med.* 93 (3): 247–258.

130 Hannan, F.M., Howles, S.A., Rogers, A. et al. (2015). Adaptor protein-2 sigma subunit mutations causing familial hypocalciuric hypercalcaemia type 3 (FHH3) demonstrate genotype-phenotype correlations, codon bias and dominant-negative effects. *Hum. Mol. Genet.* 24 (18): 5079–5092.

131 Vargas-Poussou, R., Mansour-Hendili, L., Baron, S. et al. (2016). Familial hypocalciuric hypercalcemia types 1 and 3 and primary hyperparathyroidism: similarities and differences. *J. Clin. Endocrinol. Metab.* 101 (5): 2185–2195.

132 Gorvin, C.M. (2019). Molecular and clinical insights from studies of calcium-sensing receptor mutations. *J. Mol. Endocrinol.* 63 (2): R1–R16.

133 Huang, Y. and Breitwieser, G.E. (2007). Rescue of calcium-sensing receptor mutants by allosteric modulators reveals a conformational checkpoint in receptor biogenesis. *J. Biol. Chem.* 282 (13): 9517–9525.

134 White, E., McKenna, J., Cavanaugh, A., and Breitwieser, G.E. (2009). Pharmacochaperone-mediated rescue of calcium-sensing receptor loss-of-function mutants. *Mol. Endocrinol.* 23 (7): 1115–1123.

135 Gorvin, C.M., Cranston, T., Hannan, F.M. et al. (2016). A G-protein subunit-α_{11} loss-of-function mutation, Thr54Met, causes familial hypocalciuric hypercalcemia type 2 (FHH2). *J. Bone Miner. Res.* 31 (6): 1200–1206.

136 Gorvin, C.M., Hannan, F.M., Howles, S.A. et al. (2017). Gα_{11} mutation in mice causes hypocalcemia rectifiable by calcilytic therapy. *JCI Insight* 2 (3): e91103.

137 Nesbit, M.A., Hannan, F.M., Howles, S.A. et al. (2013). Mutations affecting G-protein subunit α_{11} in hypercalcemia and hypocalcemia. *N. Engl. J. Med.* 368 (26): 2476–2486.

138 Fujisawa, Y., Yamaguchi, R., Satake, E. et al. (2013). Identification of AP2S1 mutation and effects of low calcium formula in an infant with hypercalcemia and hypercalciuria. *J. Clin. Endocrinol. Metab.* 98 (12): E2022–E2027.

139 Hendy, G.N., Canaff, L., Newfield, R.S. et al. (2014). Codon Arg15 mutations of the AP2S1 gene: common occurrence in familial hypocalciuric hypercalcemia cases negative for calcium-sensing receptor (CASR) mutations. *J. Clin. Endocrinol. Metab.* 99 (7): E1311–E1315.

140 Hovden, S., Rejnmark, L., Ladefoged, S.A., and Nissen, P.H. (2017). AP2S1 and GNA11 mutations – not a common cause of familial hypocalciuric hypercalcemia. *Eur. J. Endocrinol.* 176 (2): 177–185.

141 Howles, S.A., Hannan, F.M., Babinsky, V.N. et al. (2016). Cinacalcet for symptomatic hypercalcemia caused by AP2S1 mutations. *N. Engl. J. Med.* 374 (14): 1396–1398.

142 Nesbit, M.A., Hannan, F.M., Howles, S.A. et al. (2013). Mutations in AP2S1 cause familial hypocalciuric hypercalcemia type 3. *Nat. Genet.* 45 (1): 93–97.

143 Schnabel, D., Letz, S., Lankes, E. et al. (2014). Severe but not neonatally lethal. A homozygous inactivating CaSR mutation in a 3 year old child. *Exp. Clin. Endocrinol. Diabetes* 122 (03): P041.

144 Hannan, F.M. and Thakker, R.V. (2013). Calcium-sensing receptor (CaSR) mutations and disorders of calcium, electrolyte and water metabolism. *Best Pract. Res. Clin. Endocrinol. Metab.* 27 (3): 359–371.

145 Raue, F., Pichl, J., Dorr, H.G. et al. (2011). Activating mutations in the calcium-sensing receptor: genetic and clinical spectrum in 25 patients with autosomal dominant hypocalcaemia – a German survey. *Clin. Endocrinol.* 75 (6): 760–765.

146 Vargas-Poussou, R., Huang, C., Hulin, P. et al. (2002). Functional characterization of a calcium-sensing receptor mutation in severe autosomal dominant hypocalcemia with a Bartter-like syndrome. *J. Am. Soc. Nephrol.* 13 (9): 2259–2266.

147 Watanabe, S., Fukumoto, S., Chang, H. et al. (2002). Association between activating mutations of calcium-sensing receptor and Bartter's syndrome. *Lancet* 360 (9334): 692–694.

148 Kinoshita, Y., Hori, M., Taguchi, M. et al. (2014). Functional activities of mutant calcium-sensing receptors determine clinical presentations in patients with autosomal dominant hypocalcemia. *J. Clin. Endocrinol. Metab.* 99 (2): E363–E368.

149 Tan, Y.M., Cardinal, J., Franks, A.H. et al. (2003). Autosomal dominant hypocalcemia: a novel activating mutation (E604K) in the cysteine-rich domain of the calcium-sensing receptor. *J. Clin. Endocrinol. Metab.* 88 (2): 605–610.

150 Yun, F.H., Wong, B.Y., Chase, M. et al. (2007). Genetic variation at the calcium-sensing receptor (CASR) locus: implications for clinical molecular diagnostics. *Clin. Biochem.* 40 (8): 551–561.

151 Cole, D.E., Peltekova, V.D., Rubin, L.A. et al. (1999). A986S polymorphism of the calcium-sensing receptor and circulating calcium concentrations. *Lancet* 353 (9147): 112–115.

152 Vezzoli, G., Terranegra, A., Arcidiacono, T. et al. (2007). R990G polymorphism of calcium-sensing receptor does produce a gain-of-function and predispose to primary hypercalciuria. *Kidney Int.* 71 (11): 1155–1162.

153 Hu, J. and Spiegel, A.M. (2007). Structure and function of the human calcium-sensing receptor: insights from natural and engineered mutations and allosteric modulators. *J. Cell. Mol. Med.* 11 (5): 908–922.

154 Kapur, K., Johnson, T., Beckmann, N.D. et al. (2010). Genome-wide meta-analysis for serum calcium identifies significantly associated SNPs near the calcium-sensing receptor (CASR) gene. *PLos Genet.* 6 (7): e1001035.

155 Lorentzon, M., Lorentzon, R., Lerner, U.H., and Nordstrom, P. (2001). Calcium sensing receptor gene polymorphism, circulating calcium concentrations and bone mineral density in healthy adolescent girls. *Eur. J. Endocrinol.* 144 (3): 257–261.

156 Laaksonen, M.M., Outila, T.A., Karkkainen, M.U. et al. (2009). Associations of vitamin D receptor, calcium-sensing receptor and parathyroid hormone gene polymorphisms with calcium homeostasis and peripheral bone density in adult Finns. *J. Nutrigenet. Nutrigen.* 2 (2): 55–63.

157 Grzegorzewska, A.E., Bednarski, D., Swiderska, M. et al. (2018). The calcium-sensing receptor gene polymorphism rs1801725 and calcium-related phenotypes in hemodialysis patients. *Kidney Blood Press Res.* 43 (3): 719–734.

158 Cetani, F., Borsari, S., Vignali, E. et al. (2002). Calcium-sensing receptor gene polymorphisms in primary hyperparathyroidism. *J. Endocrinol. Invest.* 25 (7): 614–619.

159 Scillitani, A., Guarnieri, V., Battista, C. et al. (2007). Primary hyperparathyroidism and the presence of kidney stones are associated with different haplotypes of the calcium-sensing receptor. *J. Clin. Endocrinol. Metab.* 92 (1): 277–283.

160 Miedlich, S., Lamesch, P., Mueller, A., and Paschke, R. (2001). Frequency of the calcium-sensing receptor variant A986S in patients with primary hyperparathyroidism. *Eur. J. Endocrinol.* 145 (4): 421–427.

161 Wang, X.M., Wu, Y.W., Li, Z.J. et al. (2016). Polymorphisms of CASR gene increase the risk of primary hyperparathyroidism. *J. Endocrinol. Invest.* 39 (6): 617–625.

162 Vezzoli, G., Tanini, A., Ferrucci, L. et al. (2002). Influence of calcium-sensing receptor gene on urinary calcium excretion in stone-forming patients. *J. Am. Soc. Nephrol.* 13 (10): 2517–2523.

163 Eller-Vainicher, C., Battista, C., Guarnieri, V. et al. (2014). Factors associated with vertebral fracture risk in patients with primary hyperparathyroidism. *Eur. J. Endocrinol.* 171 (3): 399–406.

164 Eckstein, M., Vered, I., Ish-Shalom, S. et al. (2002). Vitamin D and calcium-sensing receptor genotypes in men and premenopausal women with low bone mineral density. *Isr. Med. Assoc. J.* 4 (5): 340–344.

165 Di Nisio, A., Rocca, M.S., Ghezzi, M. et al. (2018). Calcium-sensing receptor polymorphisms increase the risk of osteoporosis in ageing males. *Endocrine* 61 (2): 349–352.

166 Bollerslev, J., Wilson, S.G., Dick, I.M. et al. (2004). Calcium-sensing receptor gene polymorphism A986S does not predict serum calcium level, bone mineral density, calcaneal ultrasound indices, or fracture rate in a large cohort of elderly women. *Calcif. Tissue Int.* 74 (1): 12–17.

167 Cetani, F., Pardi, E., Borsari, S. et al. (2003). Calcium-sensing receptor gene polymorphism is not associated with bone mineral density in Italian postmenopausal women. *Eur. J. Endocrinol.* 148 (6): 603–607.

168 Young, R., Wu, F., Van de Water, N. et al. (2003). Calcium sensing receptor gene A986S polymorphism and responsiveness to calcium supplementation in postmenopausal women. *J. Clin. Endocrinol. Metab.* 88 (2): 697–700.

169 Takacs, I., Speer, G., Bajnok, E. et al. (2002). Lack of association between calcium-sensing receptor gene "A986S" polymorphism and bone mineral density in Hungarian postmenopausal women. *Bone* 30 (6): 849–852.

170 Han, G., Wang, O., Nie, M. et al. (2013). Clinical phenotypes of Chinese primary hyperparathyroidism patients are associated with the calcium-sensing receptor gene R990G polymorphism. *Eur. J. Endocrinol.* 169 (5): 629–638.

171 Vezzoli, G., Terranegra, A., Arcidiacono, T. et al. (2010). Calcium kidney stones are associated with a haplotype of the calcium-sensing receptor gene regulatory region. *Nephrol. Dial. Transplant.* 25 (7): 2245–2252.

172 Vezzoli, G., Scillitani, A., Corbetta, S. et al. (2011). Polymorphisms at the regulatory regions of the CASR gene influence stone risk in primary hyperparathyroidism. *Eur. J. Endocrinol.* 164 (3): 421–427.

173 Vezzoli, G., Terranegra, A., Aloia, A. et al. (2013). Decreased transcriptional activity of calcium-sensing receptor gene promoter 1 is associated with calcium nephrolithiasis. *J. Clin. Endocrinol. Metab.* 98 (9): 3839–3847.

174 Corbetta, S., Eller-Vainicher, C., Filopanti, M. et al. (2006). R990G polymorphism of the calcium-sensing receptor and renal calcium excretion in patients with primary hyperparathyroidism. *Eur. J. Endocrinol.* 155 (5): 687–692.

175 Kifor, O., Moore, F.D. Jr., Delaney, M. et al. (2003). A syndrome of hypocalciuric hypercalcemia caused by autoantibodies directed at the calcium-sensing receptor. *J. Clin. Endocrinol. Metab.* 88 (1): 60–72.

176 Kifor, O., McElduff, A., LeBoff, M.S. et al. (2004). Activating antibodies to the calcium-sensing receptor in two patients with autoimmune hypoparathyroidism. *J. Clin. Endocrinol. Metab.* 89 (2): 548–556.

177 Posillico, J.T., Wortsman, J., Srikanta, S. et al. (1986). Parathyroid cell surface autoantibodies that inhibit parathyroid hormone secretion from dispersed human parathyroid cells. *J. Bone Miner. Res.* 1 (5): 475–483.

178 Makita, N., Ando, T., Sato, J. et al. (2019). Cinacalcet corrects biased allosteric modulation of CaSR by AHH autoantibody. *JCI Insight* 4 (8): e126449.

179 Makita, N. and Iiri, T. (2014). Biased agonism: a novel paradigm in G protein-coupled receptor signaling observed in acquired hypocalciuric hypercalcemia. *Endocr. J.* 61 (4): 303–309.

180 Makita, N., Sato, J., Manaka, K. et al. (2007). An acquired hypocalciuric hypercalcemia autoantibody induces allosteric transition among active human Ca-sensing receptor conformations. *Proc. Natl. Acad. Sci. U.S.A.* 104 (13): 5443–5448.

181 Pallais, J.C., Kemp, E.H., Bergwitz, C. et al. (2011). Autoimmune hypocalciuric hypercalcemia unresponsive to glucocorticoid therapy in a patient with blocking autoantibodies against the calcium-sensing receptor. *J. Clin. Endocrinol. Metab.* 96 (3): 672–680.

182 Li, Y., Song, Y.H., Rais, N. et al. (1996). Autoantibodies to the extracellular domain of the calcium sensing receptor in patients with acquired hypoparathyroidism. *J. Clin. Invest.* 97 (4): 910–914.

183 Mayer, A., Ploix, C., Orgiazzi, J. et al. (2004). Calcium-sensing receptor autoantibodies are relevant markers of acquired hypoparathyroidism. *J. Clin. Endocrinol. Metab.* 89 (9): 4484–4488.

184 Gavalas, N.G., Kemp, E.H., Krohn, K.J. et al. (2007). The calcium-sensing receptor is a target of autoantibodies in patients with autoimmune polyendocrine syndrome type 1. *J. Clin. Endocrinol. Metab.* 92 (6): 2107–2114.

185 Tomar, N., Gupta, N., and Goswami, R. (2013). Calcium-sensing receptor autoantibodies and idiopathic hypoparathyroidism. *J. Clin. Endocrinol. Metab.* 98 (9): 3884–3891.

186 Kuroda, N., Mitani, T., Takeda, N. et al. (2005). Development of autoimmunity against transcriptionally unrepressed target antigen in the thymus of Aire-deficient mice. *J. Immunol.* 174 (4): 1862–1870.
187 Mahtab, S., Vaish, U., Saha, S. et al. (2017). Presence of autoreactive, MHC class I-restricted, calcium-sensing receptor (CaSR)-specific $CD8^+$ T cells in idiopathic hypoparathyroidism. *J. Clin. Endocrinol. Metab.* 102 (1): 167–175.
188 Habibullah, M., Porter, J.A., Kluger, N. et al. (2018). Calcium-sensing receptor autoantibodies in patients with autoimmune polyendocrine syndrome type 1: epitopes, specificity, functional affinity, IgG subclass, and effects on receptor activity. *J. Immunol.* 201 (11): 3175–3183.
189 VanHouten, J.N., Dann, P., McGeoch, G. et al. (2004). The calcium-sensing receptor regulates mammary gland parathyroid hormone-related protein production and calcium transport. *J. Clin. Invest.* 113 (4): 598–608.
190 Mamillapalli, R., VanHouten, J.N., Dann, P. et al. (2013). Mammary-specific ablation of the calcium-sensing receptor during lactation alters maternal calcium metabolism, milk calcium transport, and neonatal calcium accrual. *Endocrinology* 154 (9): 3031–3042.
191 Cheng, I., Klingensmith, M.E., Chattopadhyay, N. et al. (1998). Identification and localization of the extracellular calcium-sensing receptor in human breast. *J. Clin. Endocrinol. Metab.* 83 (2): 703–707.
192 Kim, W. and Wysolmerski, J.J. (2016). Calcium-sensing receptor in breast physiology and cancer. *Front. Physiol.* 7: 440.
193 VanHouten, J.N., Neville, M.C., and Wysolmerski, J.J. (2007). The calcium-sensing receptor regulates plasma membrane calcium adenosine triphosphatase isoform 2 activity in mammary epithelial cells: a mechanism for calcium-regulated calcium transport into milk. *Endocrinology* 148 (12): 5943–5954.
194 Wysolmerski, J.J. (2012). Parathyroid hormone-related protein: an update. *J. Clin. Endocrinol. Metab.* 97 (9): 2947–2956.
195 Vanhouten, J.N. and Wysolmerski, J.J. (2013). The calcium-sensing receptor in the breast. *Best Pract. Res. Clin. Endocrinol. Metab.* 27 (3): 403–414.
196 Kim, W., Takyar, F.M., Swan, K. et al. (2016). Calcium-sensing receptor promotes breast cancer by stimulating intracrine actions of parathyroid hormone-related protein. *Cancer Res.* 76 (18): 5348–5360.
197 Sanders, J.L., Chattopadhyay, N., Kifor, O. et al. (2000). Extracellular calcium-sensing receptor expression and its potential role in regulating parathyroid hormone-related peptide secretion in human breast cancer cell lines. *Endocrinology* 141 (12): 4357–4364.
198 Saidak, Z., Boudot, C., Abdoune, R. et al. (2009). Extracellular calcium promotes the migration of breast cancer cells through the activation of the calcium sensing receptor. *Exp. Cell Res.* 315 (12): 2072–2080.

199 Sanders, J.L., Chattopadhyay, N., Kifor, O. et al. (2001). Ca^{2+}-sensing receptor expression and PTHrP secretion in PC-3 human prostate cancer cells. *Am. J. Physiol. Endocrinol. Metab.* 281 (6): E1267–E1274.

200 Wang, L., Widatalla, S.E., Whalen, D.S. et al. (2017). Association of calcium sensing receptor polymorphisms at rs1801725 with circulating calcium in breast cancer patients. *BMC Cancer* 17 (1): 511.

201 Wysolmerski, J.J. (2012). Osteocytic osteolysis: time for a second look? *Bonekey Rep.* 1: 229.

202 Mihai, R., Stevens, J., McKinney, C., and Ibrahim, N.B. (2006). Expression of the calcium receptor in human breast cancer – a potential new marker predicting the risk of bone metastases. *Eur. J. Surg. Oncol.* 32 (5): 511–515.

203 El Hiani, Y., Ahidouch, A., Lehen'kyi, V. et al. (2009). Extracellular signal-regulated kinases 1 and 2 and $TRPC_1$ channels are required for calcium-sensing receptor-stimulated MCF-7 breast cancer cell proliferation. *Cell. Physiol. Biochem.* 23 (4–6): 335–346.

204 El Hiani, Y., Lehen'kyi, V., Ouadid-Ahidouch, H., and Ahidouch, A. (2009). Activation of the calcium-sensing receptor by high calcium induced breast cancer cell proliferation and $TRPC_1$ cation channel over-expression potentially through EGFR pathways. *Arch. Biochem. Biophys.* 486 (1): 58–63.

205 Li, X., Kong, X., Jiang, L. et al. (2014). A genetic polymorphism (rs17251221) in the calcium-sensing receptor is associated with breast cancer susceptibility and prognosis. *Cell. Physiol. Biochem.* 33 (1): 165–172.

206 Li, X., Li, L., Moran, M.S. et al. (2014). Prognostic significance of calcium-sensing receptor in breast cancer. *Tumour Biol.* 35 (6): 5709–5715.

207 McCullough, M.L., Rodriguez, C., Diver, W.R. et al. (2005). Dairy, calcium, and vitamin D intake and postmenopausal breast cancer risk in the Cancer Prevention Study II Nutrition Cohort. *Cancer Epidemiol. Biomarkers Prev.* 14 (12): 2898–2904.

208 Lim, E., Metzger-Filho, O., and Winer, E.P. (2012). The natural history of hormone receptor-positive breast cancer. *Oncology (Williston Park)* 26 (8): 688–694, 696.

209 Journe, F., Dumon, J.C., Kheddoumi, N. et al. (2004). Extracellular calcium downregulates estrogen receptor α and increases its transcriptional activity through calcium-sensing receptor in breast cancer cells. *Bone* 35 (2): 479–488.

210 Liu, G., Hu, X., Varani, J., and Chakrabarty, S. (2009). Calcium and calcium sensing receptor modulates the expression of thymidylate synthase, NAD(P)H:quinone oxidoreductase 1 and survivin in human colon carcinoma cells: promotion of cytotoxic response to mitomycin C and fluorouracil. *Mol. Carcinog.* 48 (3): 202–211.

211 Promkan, M., Liu, G., Patmasiriwat, P., and Chakrabarty, S. (2011). $BRCA_1$ suppresses the expression of survivin and promotes sensitivity to paclitaxel through the calcium sensing receptor (CaSR) in human breast cancer cells. *Cell Calcium* 49 (2): 79–88.

212 Singh, N., Promkan, M., Liu, G. et al. (2013). Role of calcium sensing receptor (CaSR) in tumorigenesis. *Best Pract. Res. Clin. Endocrinol. Metab.* 27 (3): 455–463.

213 Busque, S.M., Kerstetter, J.E., Geibel, J.P., and Insogna, K. (2005). L-type amino acids stimulate gastric acid secretion by activation of the calcium-sensing receptor in parietal cells. *Am. J. Physiol. Gastrointest. Liver Physiol.* 289 (4): G664–G669.

214 Feng, J., Petersen, C.D., Coy, D.H. et al. (2010). Calcium-sensing receptor is a physiologic multimodal chemosensor regulating gastric G-cell growth and gastrin secretion. *Proc. Natl. Acad. Sci. U.S.A.* 107 (41): 17791–17796.

215 Alamshah, A., Spreckley, E., Norton, M. et al. (2017). L-Phenylalanine modulates gut hormone release and glucose tolerance, and suppresses food intake through the calcium-sensing receptor in rodents. *Int. J. Obes.* 41 (11): 1693–1701.

216 Wang, Y., Chandra, R., Samsa, L.A. et al. (2011). Amino acids stimulate cholecystokinin release through the Ca^{2+}–sensing receptor. *Am. J. Physiol. Gastrointest. Liver Physiol.* 300 (4): G528–G537.

217 Geibel, J., Sritharan, K., Geibel, R. et al. (2006). Calcium-sensing receptor abrogates secretagogue- induced increases in intestinal net fluid secretion by enhancing cyclic nucleotide destruction. *Proc. Natl. Acad. Sci. U.S.A.* 103 (25): 9390–9397.

218 Zhao, X., Xian, Y., Wang, C. et al. (2018). Calcium-sensing receptor-mediated L-tryptophan-induced secretion of cholecystokinin and glucose-dependent insulinotropic peptide in swine duodenum. *J. Vet. Sci.* 19 (2): 179–187.

219 Engelstoft, M.S., Park, W.M., Sakata, I. et al. (2013). Seven transmembrane G protein-coupled receptor repertoire of gastric ghrelin cells. *Mol Metab.* 2 (4): 376–392.

220 Liou, A.P., Sei, Y., Zhao, X. et al. (2011). The extracellular calcium-sensing receptor is required for cholecystokinin secretion in response to L-phenylalanine in acutely isolated intestinal I cells. *Am. J. Physiol. Gastrointest. Liver Physiol.* 300 (4): G538–G546.

221 Pais, R., Gribble, F.M., and Reimann, F. (2016). Signalling pathways involved in the detection of peptones by murine small intestinal enteroendocrine L-cells. *Peptides* 77: 9–15.

222 Kallay, E., Kifor, O., Chattopadhyay, N. et al. (1997). Calcium-dependent c-myc proto-oncogene expression and proliferation of Caco-2 cells: a role for a luminal extracellular calcium-sensing receptor. *Biochem. Biophys. Res. Commun.* 232 (1): 80–83.

223 Kallay, E., Bajna, E., Wrba, F. et al. (2000). Dietary calcium and growth modulation of human colon cancer cells: role of the extracellular calcium-sensing receptor. *Cancer Detect. Prev.* 24 (2): 127–136.

224 Chakrabarty, S., Radjendirane, V., Appelman, H., and Varani, J. (2003). Extracellular calcium and calcium sensing receptor function in human colon carcinomas: promotion of E-cadherin expression and suppression of β-catenin/TCF activation. *Cancer Res.* 63 (1): 67–71.

225 Cheng, S.X., Lightfoot, Y.L., Yang, T. et al. (2014). Epithelial CaSR deficiency alters intestinal integrity and promotes proinflammatory immune responses. *FEBS Lett.* 588 (22): 4158–4166.

226 Nakhoul, N.L., Tu, C.L., Brown, K.L. et al. (2020). Calcium-sensing receptor deletion in the mouse esophagus alters barrier function. *Am. J. Physiol. Gastrointest. Liver Physiol.* 318 (1): G144–G161.

227 Aggarwal, A., Prinz-Wohlgenannt, M., Groschel, C. et al. (2015). The calcium-sensing receptor suppresses epithelial-to-mesenchymal transition and stem cell- like phenotype in the colon. *Mol. Cancer* 14: 61.

228 Chakrabarty, S., Wang, H., Canaff, L. et al. (2005). Calcium sensing receptor in human colon carcinoma: interaction with Ca^{2+} and 1,25-dihydroxyvitamin D_3. *Cancer Res.* 65 (2): 493–498.

229 Fetahu, I.S., Hobaus, J., Aggarwal, A. et al. (2014). Calcium-sensing receptor silencing in colorectal cancer is associated with promoter hypermethylation and loss of acetylation on histone 3. *Int. J. Cancer* 135 (9): 2014–2023.

230 Sheinin, Y., Kallay, E., Wrba, F. et al. (2000). Immunocytochemical localization of the extracellular calcium-sensing receptor in normal and malignant human large intestinal mucosa. *J. Histochem. Cytochem.* 48 (5): 595–602.

231 Bhagavathula, N., Hanosh, A.W., Nerusu, K.C. et al. (2007). Regulation of E-cadherin and β-catenin by Ca^{2+} in colon carcinoma is dependent on calcium-sensing receptor expression and function. *Int. J. Cancer.* 121 (7): 1455–1462.

232 Hizaki, K., Yamamoto, H., Taniguchi, H. et al. (2011). Epigenetic inactivation of calcium-sensing receptor in colorectal carcinogenesis. *Mod. Pathol.* 24 (6): 876–884.

233 Singh, N. and Chakrabarty, S. (2013). Induction of CaSR expression circumvents the molecular features of malignant CaSR null colon cancer cells. *Int. J. Cancer.* 133 (10): 2307–2314.

234 Garland, C., Shekelle, R.B., Barrett-Connor, E. et al. (1985). Dietary vitamin D and calcium and risk of colorectal cancer: a 19-year prospective study in men. *Lancet* 1 (8424): 307–309.

235 Bhagavathula, N., Kelley, E.A., Reddy, M. et al. (2005). Upregulation of calcium-sensing receptor and mitogen-activated protein kinase signalling in the regulation of growth and differentiation in colon carcinoma. *Br. J. Cancer* 93 (12): 1364–1371.

236 Aggarwal, A., Prinz-Wohlgenannt, M., Tennakoon, S. et al. (2015). The calcium-sensing receptor: a promising target for prevention of colorectal cancer. *Biochim. Biophys. Acta* 1853 (9): 2158–2167.

237 Thomas, M.G., Tebbutt, S., and Williamson, R.C. (1992). Vitamin D and its metabolites inhibit cell proliferation in human rectal mucosa and a colon cancer cell line. *Gut* 33 (12): 1660–1663.

238 Taghizadeh, F., Tang, M.J., and Tai, I.T. (2007). Synergism between vitamin D and secreted protein acidic and rich in cysteine-induced apoptosis and growth inhibition results in increased susceptibility of therapy-resistant colorectal cancer cells to chemotherapy. *Mol. Cancer Ther.* 6 (1): 309–317.

239 Palmer, H.G., Gonzalez-Sancho, J.M., Espada, J. et al. (2001). Vitamin D_3 promotes the differentiation of colon carcinoma cells by the induction of E-cadherin and the inhibition of β-catenin signaling. *J. Cell Biol.* 154 (2): 369–387.

240 Shah, S., Islam, M.N., Dakshanamurthy, S. et al. (2006). The molecular basis of vitamin D receptor and β-catenin crossregulation. *Mol. Cell* 21 (6): 799–809.

241 Liu, G., Hu, X., and Chakrabarty, S. (2010). Vitamin D mediates its action in human colon carcinoma cells in a calcium-sensing receptor-dependent manner: downregulates malignant cell behavior and the expression of thymidylate synthase and survivin and promotes cellular sensitivity to 5-FU. *Int. J. Cancer.* 126 (3): 631–639.

242 Hendy, G.N. and Canaff, L. (2016). Calcium-sensing receptor gene: regulation of expression. *Front. Physiol.* 7: 394.

243 Zhang, H., Kovacs-Nolan, J., Kodera, T. et al. (2015). γ-Glutamyl cysteine and γ-glutamyl valine inhibit TNF-α signaling in intestinal epithelial cells and reduce inflammation in a mouse model of colitis via allosteric activation of the calcium-sensing receptor. *Biochim. Biophys. Acta* 1852 (5): 792–804.

244 Mine, Y. and Zhang, H. (2015). Calcium-sensing receptor (CaSR)-mediated anti-inflammatory effects of L-amino acids in intestinal epithelial cells. *J. Agric. Food. Chem.* 63 (45): 9987–9995.

245 Liu, H., Tan, B., Huang, B. et al. (2018). Involvement of calcium-sensing receptor activation in the alleviation of intestinal inflammation in a piglet model by dietary aromatic amino acid supplementation. *Br. J. Nutr.* 120 (12): 1321–1331.

246 Elajnaf, T., Iamartino, L., Mesteri, I. et al. (2019). Nutritional and pharmacological targeting of the calcium-sensing receptor influences chemically induced colitis in mice. *Nutrients* 11 (12): 3072.

247 Babinsky, V.N., Hannan, F.M., Youhanna, S.C. et al. (2015). Association studies of calcium-sensing receptor (CaSR) polymorphisms with serum concentrations of glucose and phosphate, and vascular calcification in renal transplant recipients. *PLoS One* 10 (3): e0119459.

248 Babinsky, V.N., Hannan, F.M., Ramracheya, R.D. et al. (2017). Mutant mice with calcium-sensing receptor activation have hyperglycemia that is rectified by calcilytic therapy. *Endocrinology* 158 (8): 2486–2502.

249 Wolf, P., Krssak, M., Winhofer, Y. et al. (2014). Cardiometabolic phenotyping of patients with familial hypocalcuric hypercalcemia. *J. Clin. Endocrinol. Metab.* 99 (9): E1721–E1726.

250 Brennan, S.C., Wilkinson, W.J., Tseng, H.E. et al. (2016). The extracellular calcium-sensing receptor regulates human fetal lung development via CFTR. *Sci. Rep.* 6: 21975.

251 Finney, B.A., del Moral, P.M., Wilkinson, W.J. et al. (2008). Regulation of mouse lung development by the extracellular calcium-sensing receptor. *CaR. J. Physiol.* 586 (24): 6007–6019.

252 Milara, J., Mata, M., Serrano, A. et al. (2010). Extracellular calcium-sensing receptor mediates human bronchial epithelial wound repair. *Biochem. Pharmacol.* 80 (2): 236–246.

253 Kurosawa, M., Shimizu, Y., Tsukagoshi, H., and Ueki, M. (1992). Elevated levels of peripheral-blood, naturally occurring aliphatic polyamines in bronchial asthmatic patients with active symptoms. *Allergy* 47 (6): 638–643.

254 Lee, J.W., Park, J.W., Kwon, O.K. et al. (2017). NPS2143 inhibits MUC5AC and proinflammatory mediators in cigarette smoke extract (CSE)-stimulated human airway epithelial cells. *Inflammation* 40 (1): 184–194.

255 Lee, J.W., Park, H.A., Kwon, O.K. et al. (2017). NPS 2143, a selective calcium-sensing receptor antagonist inhibits lipopolysaccharide-induced pulmonary inflammation. *Mol. Immunol.* 90: 150–157.

256 Cook, D.P., Rector, M.V., Bouzek, D.C. et al. (2016). Cystic fibrosis transmembrane conductance regulator in sarcoplasmic reticulum of airway smooth muscle. Implications for airway contractility. *Am. J. Respir. Crit. Care Med.* 193 (4): 417–426.

257 Alam, M.U., Kirton, J.P., Wilkinson, F.L. et al. (2009). Calcification is associated with loss of functional calcium-sensing receptor in vascular smooth muscle cells. *Cardiovasc. Res.* 81 (2): 260–268.

258 Smajilovic, S., Hansen, J.L., Christoffersen, T.E. et al. (2006). Extracellular calcium sensing in rat aortic vascular smooth muscle cells. *Biochem. Biophys. Res. Commun.* 348 (4): 1215–1223.

259 Wonneberger, K., Scofield, M.A., and Wangemann, P. (2000). Evidence for a calcium-sensing receptor in the vascular smooth muscle cells of the spiral modiolar artery. *J. Membr. Biol.* 175 (3): 203–212.

260 Ohanian, J., Gatfield, K.M., Ward, D.T., and Ohanian, V. (2005). Evidence for a functional calcium-sensing receptor that modulates myogenic tone in rat subcutaneous small arteries. *Am. J. Physiol. Heart Circ. Physiol.* 288 (4): H1756–H1762.

261 Wang, Y. and Bukoski, R.D. (1998). Distribution of the perivascular nerve Ca^{2+} receptor in rat arteries. *Br. J. Pharmacol.* 125 (7): 1397–1404.

262 Bukoski, R.D., Bian, K., Wang, Y., and Mupanomunda, M. (1997). Perivascular sensory nerve Ca^{2+} receptor and Ca^{2+}-induced relaxation of isolated arteries. *Hypertension* 30 (6): 1431–1439.

263 Schepelmann, M., Yarova, P.L., Lopez-Fernandez, I. et al. (2016). The vascular Ca^{2+}-sensing receptor regulates blood vessel tone and blood pressure. *Am. J. Physiol. Cell Physiol.* 310 (3): C193–C204.

264 Marz, W., Seelhorst, U., Wellnitz, B. et al. (2007). Alanine to serine polymorphism at position 986 of the calcium-sensing receptor associated with coronary heart disease, myocardial infarction, all-cause, and cardiovascular mortality. *J. Clin. Endocrinol. Metab.* 92 (6): 2363–2369.

265 Larsson, S.C., Burgess, S., and Michaelsson, K. (2017). Association of genetic variants related to serum calcium levels with coronary artery disease and myocardial infarction. *JAMA* 318 (4): 371–380.

266 Ruat, M., Molliver, M.E., Snowman, A.M., and Snyder, S.H. (1995). Calcium sensing receptor: molecular cloning in rat and localization to nerve terminals. *Proc. Natl. Acad. Sci. U.S.A.* 92 (8): 3161–3165.

267 Liu, X.L., Lu, Y.S., Gao, J.Y. et al. (2013). Calcium sensing receptor absence delays postnatal brain development via direct and indirect mechanisms. *Mol. Neurobiol.* 48 (3): 590–600.

268 Kim, J.Y., Ho, H., Kim, N. et al. (2014). Calcium-sensing receptor (CaSR) as a novel target for ischemic neuroprotection. *Ann. Clin. Transl. Neurol.* 1 (11): 851–866.

269 Tu, C.L., Celli, A., Mauro, T., and Chang, W. (2019). Calcium-sensing receptor regulates epidermal intracellular Ca^{2+} signaling and re-epithelialization after wounding. *J. Invest. Dermatol.* 139 (4): 919–929.

270 Komuves, L., Oda, Y., Tu, C.L. et al. (2002). Epidermal expression of the full-length extracellular calcium-sensing receptor is required for normal keratinocyte differentiation. *J. Cell. Physiol.* 192 (1): 45–54.

271 Zhang, D. and Xia, J. (2020). Somatic synonymous mutations in regulatory elements contribute to the genetic aetiology of melanoma. *BMC Med. Genet.* 13 (Suppl 5): 43.

272 Zuo, X., Sun, L., Yin, X. et al. (2015). Whole-exome SNP array identifies 15 new susceptibility loci for psoriasis. *Nat. Commun.* 6: 6793.

273 Block, G.A., Martin, K.J., de Francisco, A.L. et al. (2004). Cinacalcet for secondary hyperparathyroidism in patients receiving hemodialysis. *N. Engl. J. Med.* 350 (15): 1516–1525.

274 Nagano, N. (2006). Pharmacological and clinical properties of calcimimetics: calcium receptor activators that afford an innovative approach to controlling hyperparathyroidism. *Pharmacol. Ther.* 109 (3): 339–365.

275 Harris, R.Z., Padhi, D., Marbury, T.C. et al. (2004). Pharmacokinetics, pharmacodynamics, and safety of cinacalcet hydrochloride in hemodialysis patients at doses up to 200 mg once daily. *Am. J. Kidney Dis.* 44 (6): 1070–1076.

276 Padhi, D. and Harris, R. (2009). Clinical pharmacokinetic and pharmacodynamic profile of cinacalcet hydrochloride. *Clin. Pharmacokinet.* 48 (5): 303–311.

277 Kumar, G.N., Sproul, C., Poppe, L. et al. (2004). Metabolism and disposition of calcimimetic agent cinacalcet HCl in humans and animal models. *Drug Metab. Dispos.* 32 (12): 1491–1500.

278 Gincherman, Y., Moloney, K., McKee, C., and Coyne, D.W. (2010). Assessment of adherence to cinacalcet by prescription refill rates in hemodialysis patients. *Hemodial. Int.* 14 (1): 68–72.

279 Nakashima, D., Takama, H., Ogasawara, Y. et al. (2007). Effect of cinacalcet hydrochloride, a new calcimimetic agent, on the pharmacokinetics of dextromethorphan: in vitro and clinical studies. *J. Clin. Pharmacol.* 47 (10): 1311–1319.

280 Lindberg, J.S., Culleton, B., Wong, G. et al. (2005). Cinacalcet HCl, an oral calcimimetic agent for the treatment of secondary hyperparathyroidism in hemodialysis and peritoneal dialysis: a randomized, double-blind, multicenter study. *J. Am. Soc. Nephrol.* 16 (3): 800–807.

281 Jeong, S., Kim, I.W., Oh, K.H. et al. (2016). Pharmacogenetic analysis of cinacalcet response in secondary hyperparathyroidism patients. *Drug Des. Devel. Ther.* 10: 2211–2225.

282 Moe, S.M., Wetherill, L., Decker, B.S. et al. (2017). Calcium-sensing receptor genotype and response to cinacalcet in patients undergoing hemodialysis. *Clin. J. Am. Soc. Nephrol.* 12 (7): 1128–1138.

283 Rothe, H.M., Shapiro, W.B., Sun, W.Y., and Chou, S.Y. (2005). Calcium-sensing receptor gene polymorphism Arg990Gly and its possible effect on response to cinacalcet HCl. *Pharmacogenet. Genomics* 15 (1): 29–34.

284 Hannan, F.M., Olesen, M.K., and Thakker, R.V. (2018). Calcimimetic and calcilytic therapies for inherited disorders of the calcium-sensing receptor signalling pathway. *Br. J. Pharmacol.* 175 (21): 4083–4094.

285 Capozza, M., Chinellato, I., Guarnieri, V. et al. (2018). Case report: acute clinical presentation and neonatal management of primary hyperparathyroidism due to a novel CaSR mutation. *BMC Pediatr.* 18 (1): 340.

286 Garcia Soblechero, E., Ferrer Castillo, M.T., Jimenez Crespo, B. et al. (2013). Neonatal hypercalcemia due to a homozygous mutation in the calcium-sensing receptor: failure of cinacalcet. *Neonatology* 104 (2): 104–108.

287 Ahmad, N., Bahasan, M., Al-Ghamdi, B.A.A. et al. (2017). Neonatal severe hyperparathyroidism secondary to a novel homozygous CASR gene mutation. *Clin. Cases Miner. Bone Metab.* 14 (3): 354–358.

288 Murphy, H., Patrick, J., Baez-Irizarry, E. et al. (2016). Neonatal severe hyperparathyroidism caused by homozygous mutation in CASR: a rare cause of life-threatening hypercalcemia. *Eur. J. Med. Genet.* 59 (4): 227–231.

289 Gorvin, C.M., Hannan, F.M., Cranston, T. et al. (2018). Cinacalcet rectifies hypercalcemia in a patient with familial hypocalciuric hypercalcemia type 2 (FHH2) caused by a germline loss-of-function $G\alpha_{11}$ mutation. *J. Bone Miner. Res.* 33 (1): 32–41.

290 Miyazaki, H., Ikeda, Y., Sakurai, O. et al. (2018). Discovery of evocalcet, a next-generation calcium-sensing receptor agonist for the treatment of hyperparathyroidism. *Bioorg. Med. Chem. Lett.* 28 (11): 2055–2060.

291 Tsuruya, K., Shimazaki, R., Fukagawa, M. et al. (2019). Efficacy and safety of evocalcet in Japanese peritoneal dialysis patients. *Clin. Exp. Nephrol.* 23 (6): 739–748.

292 Kawata, T., Tokunaga, S., Murai, M. et al. (2018). A novel calcimimetic agent, evocalcet (MT-4580/KHK7580), suppresses the parathyroid cell function with little effect on the gastrointestinal tract or CYP isozymes in vivo and in vitro. *PLoS One* 13 (4): e0195316.

293 Patel, J. and Bridgeman, M.B. (2018). Etelcalcetide (Parsabiv) for secondary hyperparathyroidism in adults with chronic kidney disease on hemodialysis. *P T* 43 (7): 396–399.

294 Chen, P., Melhem, M., Xiao, J. et al. (2015). Population pharmacokinetics analysis of AMG 416, an allosteric activator of the calcium-sensing receptor, in subjects with secondary hyperparathyroidism receiving hemodialysis. *J. Clin. Pharmacol.* 55 (6): 620–628.

295 Subramanian, R., Zhu, X., Kerr, S.J. et al. (2016). Nonclinical pharmacokinetics, disposition, and drug-drug interaction potential of a novel d-amino acid peptide agonist of the calcium-sensing receptor AMG 416 (Etelcalcetide). *Drug Metab. Dispos.* 44 (8): 1319–1331.

296 Piccoli, G.B., Trabace, T., Chatrenet, A. et al. (2020). New intravenous calcimimetic agents: new options, new problems. An example on how clinical, economical and ethical considerations affect choice of treatment. *Int. J. Environ. Res. Public Health* 17 (4): 1238.

297 Lindsay, R., Nieves, J., Formica, C. et al. (1997). Randomised controlled study of effect of parathyroid hormone on vertebral-bone mass and fracture incidence among postmenopausal women on oestrogen with osteoporosis. *Lancet* 350 (9077): 550–555.

298 Neer, R.M., Arnaud, C.D., Zanchetta, J.R. et al. (2001). Effect of parathyroid hormone (1–34) on fractures and bone mineral density in postmenopausal women with osteoporosis. *N. Engl. J. Med.* 344 (19): 1434–1441.

299 Tam, C.S., Heersche, J.N., Murray, T.M., and Parsons, J.A. (1982). Parathyroid hormone stimulates the bone apposition rate independently of its resorptive action: differential effects of intermittent and continuous administration. *Endocrinology* 110 (2): 506–512.

300 Marquis, R.W., Lago, A.M., Callahan, J.F. et al. (2009). Antagonists of the calcium receptor I. Amino alcohol-based parathyroid hormone secretagogues. *J. Med. Chem.* 52 (13): 3982–3993.

301 Gowen, M., Stroup, G.B., Dodds, R.A. et al. (2000). Antagonizing the parathyroid calcium receptor stimulates parathyroid hormone secretion and bone formation in osteopenic rats. *J. Clin. Invest.* 105 (11): 1595–1604.

302 Halse, J., Greenspan, S., Cosman, F. et al. (2014). A phase 2, randomized, placebo-controlled, dose-ranging study of the calcium-sensing receptor antagonist MK-5442 in the treatment of postmenopausal women with osteoporosis. *J. Clin. Endocrinol. Metab.* 99 (11): E2207–E2215.

303 Cosman, F., Gilchrist, N., McClung, M. et al. (2016). A phase 2 study of MK-5442, a calcium-sensing receptor antagonist, in postmenopausal women with osteoporosis after long-term use of oral bisphosphonates. *Osteoporos. Int.* 27 (1): 377–386.

304 Fitzpatrick, L.A., Dabrowski, C.E., Cicconetti, G. et al. (2011). The effects of ronacaleret, a calcium-sensing receptor antagonist, on bone mineral density and biochemical markers of bone turnover in postmenopausal women with low bone mineral density. *J. Clin. Endocrinol. Metab.* 96 (8): 2441–2449.

305 Bonnelye, E., Chabadel, A., Saltel, F., and Jurdic, P. (2008). Dual effect of strontium ranelate: stimulation of osteoblast differentiation and inhibition of osteoclast formation and resorption in vitro. *Bone* 42 (1): 129–138.

306 Brennan, T.C., Rybchyn, M.S., Green, W. et al. (2009). Osteoblasts play key roles in the mechanisms of action of strontium ranelate. *Br. J. Pharmacol.* 157 (7): 1291–1300.

307 Caudrillier, A., Hurtel-Lemaire, A.S., Wattel, A. et al. (2010). Strontium ranelate decreases receptor activator of nuclear factor-κB ligand-induced osteoclastic differentiation in vitro: involvement of the calcium-sensing receptor. *Mol. Pharmacol.* 78 (4): 569–576.

308 Chattopadhyay, N., Quinn, S.J., Kifor, O. et al. (2007). The calcium-sensing receptor (CaR) is involved in strontium ranelate-induced osteoblast proliferation. *Biochem. Pharmacol.* 74 (3): 438–447.

309 Fromigue, O., Hay, E., Barbara, A. et al. (2009). Calcium sensing receptor-dependent and receptor-independent activation of osteoblast replication and survival by strontium ranelate. *J. Cell. Mol. Med.* 13 (8B): 2189–2199.

310 Hurtel-Lemaire, A.S., Mentaverri, R., Caudrillier, A. et al. (2009). The calcium-sensing receptor is involved in strontium ranelate-induced osteoclast apoptosis. New insights into the associated signaling pathways. *J. Biological Chem.* 284 (1): 575–584.

311 EMA. (2014). European Medicines Agency recommends that Protelos/Osseor remain available but with further restrictions. London, UK.

312 Babinsky, V.N., Hannan, F.M., Gorvin, C.M. et al. (2016). Allosteric modulation of the calcium-sensing receptor rectifies signaling abnormalities associated with G-protein α_{11} mutations causing hypercalcemic and hypocalcemic disorders. *J. Biol. Chem.* 291: 10876–10885.

313 Hannan, F.M., Walls, G.V., Babinsky, V.N. et al. (2015). The calcilytic agent NPS 2143 rectifies hypocalcemia in a mouse model with an activating calcium-sensing receptor (CaSR) mutation: relevance to autosomal dominant hypocalcemia type 1 (ADH1). *Endocrinology* 156 (9): 3114–3121.

314 Letz, S., Rus, R., Haag, C. et al. (2010). Novel activating mutations of the calcium-sensing receptor: the calcilytic NPS-2143 mitigates excessive signal transduction of mutant receptors. *J. Clin. Endocrinol. Metab.* 95 (10): E229–E233.

315 Dong, B., Endo, I., Ohnishi, Y. et al. (2015). Calcilytic ameliorates abnormalities of mutant calcium-sensing receptor (CaSR) Knock-in mice mimicking autosomal dominant hypocalcemia (ADH). *J. Bone Miner. Res.* 30: 1980–1993.

316 Roberts, M.S., Gafni, R.I., Brillante, B. et al. (2019). Treatment of autosomal dominant hypocalcemia type 1 with the calcilytic NPSP795 (SHP635). *J. Bone Miner. Res.* 34 (9): 1609–1618.

18

G Protein-Coupled Receptors and Their Mutations in Cancer – A Focus on Adenosine Receptors

Xuesong Wang[1], Gerard J. P. van Westen[1], Adriaan P. IJzerman[1], and Laura H. Heitman[1,2]

[1] *Division of Drug Discovery and Safety, LACDR, Leiden University, The Netherlands*
[2] *Oncode Institute, Leiden, The Netherlands*

18.1 Introduction

Drug development is mostly geared toward members of one of the following five protein families: kinases, G protein-coupled receptors (GPCRs), nuclear hormone receptors, ion channels, or proteases [1]. GPCRs are the largest family of membrane-bound proteins and include approximately 800 members accounting for around 4% of encoded human genes [2]. They can be subdivided into five main families: glutamate, rhodopsin, adhesion, frizzled/taste, and secretin (GRAFS) [3]. An alternative subdivision is in three main classes (A, B, and C) [4].

GPCRs share a common structure that consists of seven-transmembrane (7TM) helices, connected by three intracellular (IL) and three extracellular (EL) loops, an extracellular amino terminus, and an intracellular carboxyl terminus [2]. GPCRs respond to a wide diversity of physiological endogenous ligands, including neurotransmitters and hormones. GPCRs are coupled to different families of heterotrimeric G proteins, which consist of three subunits, α, β, and γ [2]. Extracellular signaling leads to conformational changes in GPCRs, causing the replacement of GDP for GTP at the G_α subunit. This exchange makes the $G_{\beta\gamma}$ subunit dissociate from G_α, which leads to interaction with effector proteins in the cell (Figure 18.1) [5, 6]. Based on sequence similarity, the G_α-subunit family is divided into four major subfamilies, $G_{\alpha s}$, $G_{\alpha i}$, $G_{\alpha q/11}$, and $G_{\alpha 12/13}$. Downstream signaling pathways can be regulated through both G_α subunits and $G_{\beta\gamma}$-dimers of G proteins by coupling to different effector molecules, such as phospholipase C (activated by $G_{\alpha q}$ or $G_{\alpha 11}$), or adenylyl cyclase (activated or inhibited by $G_{\alpha s}$ and $G_{\alpha i}$). More downstream cellular signaling cascades involve second messengers,

GPCRs as Therapeutic Targets, Volume 2, First Edition. Edited by Annette Gilchrist.

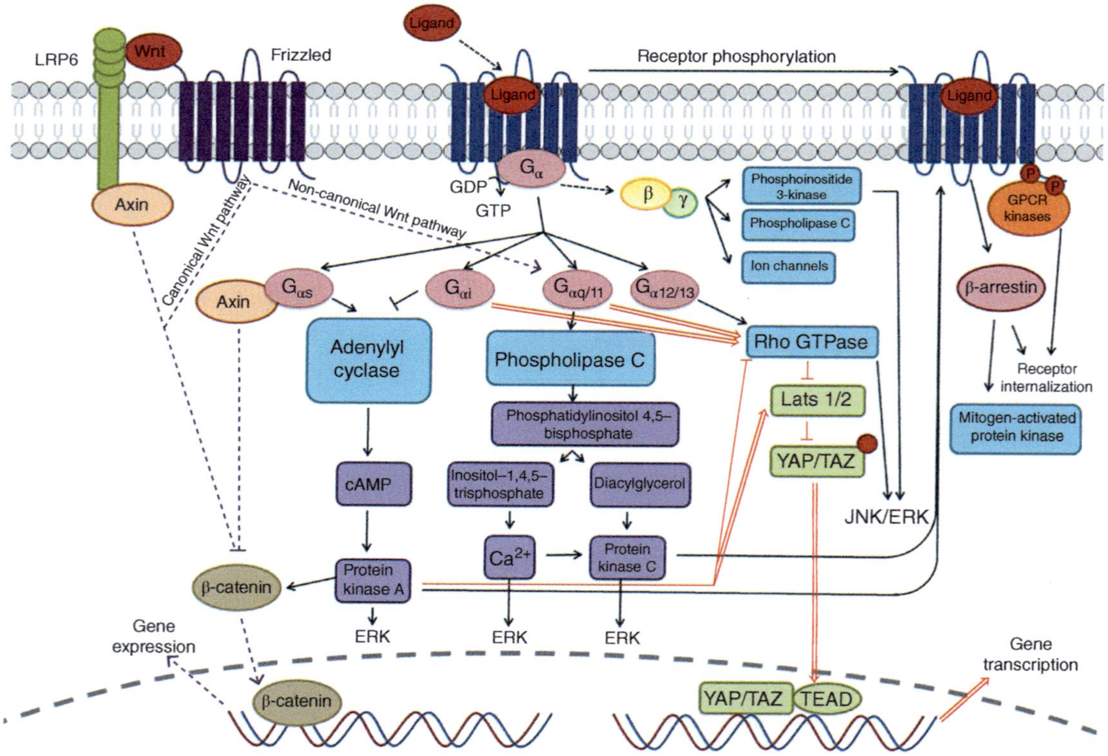

Figure 18.1 GPCR signaling from extracellular to intracellular. Upon receptor activation, GPCRs interact with G proteins, resulting in the dissociation of the α and $\beta\gamma$ subunits and subsequent activation of downstream signaling effectors (black arrows). Subsequently, protein kinase A and protein kinase C regulate receptor phosphorylation and turn off the G protein signaling. GPCR kinases phosphorylation of GPCR leads to β-arrestin recruitment and eventually receptor desensitization and internalization. β-arrestin engagement with the receptor also initiates the activation of β-arrestin–mediated signaling. Hippo signaling pathway can be activated by GPCR activation via several G proteins (red arrows). The Wnt pathway is mainly regulated by Frizzled receptors (purple arrows), where canonical and noncanonical signaling pathways can be activated. Source: Based on Refs [5–12].

namely intracellular Ca^{2+} ($G_{\alpha q}$ or $G_{\alpha 11}$) and cyclic adenosine monophosphate (cAMP) ($G_{\alpha s}$ or $G_{\alpha i}$) [13, 14]. Once the receptor is activated, this is often followed by receptor desensitization and internalization via G protein-coupled receptor kinase (GRK)-mediated phosphorylation of the agonist-occupied receptor. GPCR phosphorylation regulates β-arrestin recruitment from the cytosol to the receptor, resulting in the termination of G protein-mediated signaling [15, 16].

GPCRs are distributed throughout the human body and in combination with these various GPCR-related downstream pathways they have a crucial role in numerous physiological functions. However, GPCRs also make a substantial contribution to human pathophysiology [6]. This protein family has thus been investigated as pharmacological targets for decades, focusing on their ligand-binding site that often can be accessed from the cell surface [17]. The major disease indications for GPCR modulators have shifted over the years from, e.g. high blood pressure to metabolic diseases such as diabetes and obesity, as well as several central nervous system (CNS) disorders [6]. Recently, GPCRs have become the targets for new indications, such as smoking cessation, hypocalcaemia, short bowel syndrome/Crohn's disease, and multiple sclerosis. More recently, they are also seen as regulators of tumor initiation and progression including cell death, cell proliferation, invasion, angiogenesis, metastasis, stress signaling, and immune evasion, making GPCRs attractive cancer drug targets [13, 14, 18–20]. Therefore, intervening with GPCRs and their distant regulatory pathways provides an opportunity for developing approaches for cancer prevention, diagnosis, and treatment [21]. Examples of anti-cancer drugs targeting GPCRs undergoing clinical trials are summarized in Table 18.1, though this table is not exhaustive.

Crystal structures of GPCRs in complex with various ligands and/or G proteins provide numerous templates for structure-based drug design and discovery [6]. Hauser et al. have recently reviewed all GPCR approved agents and drugs in clinical trials, which accounts for 475 chemical entities acting at 108 unique GPCRs [6]. Although around 30% of currently used therapeutic drugs are targeting GPCRs, only around 10% of the superfamily is being addressed and only few GPCRs are being explored as oncology drug targets [17], leaving many opportunities for cancer drug discovery.

18.2 GPCRs and Cancer

Cancer development consists of multiple steps, where the hallmarks of cancer comprise 10 biological aspects, i.e. sustaining proliferative signaling, enabling replicative immortality, resisting cell death, evading growth suppressors, activating invasion and metastasis, avoiding immune destruction, inducing angiogenesis, tumor-promoting inflammation, deregulating cellular energetics,

Table 18.1 Examples of anti-cancer drugs and antibodies currently under clinical trials targeting GPCRs [22].

Receptor	Cancer type	Drug	Type of molecule	Phase	Sponsor
Adenosine receptors	Prostatic cancer	AB928	Small molecule	I (combined with zimberelimab)	Arcus Biosciences
Androgen receptor	Prostatic cancer	Apalutamide	Small molecule	I	Janssen Research & Development
CCR2	Pancreatic cancer	CCX872	small molecule	I	ChemoCentryx
	Melanoma	Plozalizumab	Humanized monoclonal antibody	I	Millennium Pharmaceuticals
CCR4	Adult T-cell leukemia and lymphoma	Mogamulizumab (KW-0761)	Humanized, afucosylated monoclonal antibody	II	Kyowa Hakko Kirin
Cholecystokinin-2 receptor	Pancreatic cancer	G17DT	immunogen	III	Cancer Advances
CXCR4	Multiple myeloma	Ulocuplumab (BMS-936564)	Fully human monoclonal antibody	I	Bristol-Myers Squibb
	Advanced or Metastatic cancer	LY-2624587	Monoclonal antibody	I	Eli Lilly
Dopamine D2 receptor	Endometrial cancer	ONC201	Small molecule	II	Fox Chase Cancer Center with Oncoceutics
Endothelin A receptor	Prostate cancer	Zibotentan (ZD4054)	small molecule	I, II, III	AstraZeneca
		Atrasentan (ABT-627)	small molecule	II	Abbott
				III (combined with docetaxel and prednisone)	Southwest Oncology Group
Endothelin-1 receptor	Ovarian cancer	GDC-0449 (Vismodegib)	Small molecule	II	Genentech

Epidermal growth factor receptor 2	Breast cancer	TALAZOPARIB	Small molecule	II	Pfizer
		Atezolizumab	Monoclonal antibody	II	Renske Altena
	Metastatic colorectal cancer	KL-140	Small molecule	III (combined with antibody)	Sichuan Kelun Pharmaceutical Research Institute
	Nonsmall cell lung cancer	Osimertinib (AZD9291)	Small molecule	III	AstraZeneca
GPR20	Gastrointestinal stromal tumors	DS-6157a	Antibody-drug conjugate	I	Daiichi Sankyo
Frizzled receptor (FZD1, 2, 5, 7, 8)	Metastatic breast cancer	Vantictumab (OMP-18R5)	Human monoclonal antibody	I (combined with paclitaxel)	OncoMed Pharmaceuticals
	Nonsmall cell lung carcinoma	Vantictumab (OMP-18R5)	Human monoclonal antibody	I (combined with Docetaxel)	OncoMed Pharmaceuticals
Frizzled receptor FZD7	Pancreatic cancer	Vantictumab (OMP-18R5)	Human monoclonal antibody	I (combined with Nab-Paclitaxel and Gemcitabine)	OncoMed Pharmaceuticals
Prostaglandin E2 receptor 4	Advanced solid tumors	AAT-007	Small molecule	II	University of Maryland
Smoothened receptor	Head and neck cancer	GDC-0449 (Vismodegib)	Small molecule	II	Private person in collaboration with Genentech
β-adrenergic receptor	Ovarian cancer	Propranolol (β-adrenergic receptor antagonists)	Small molecule	I	Washington University School of Medicine
	Metastatic breast cancer	β-adrenergic receptor antagonists	Small molecule	II	Columbia University

Source: Adapted and updated from Usman et al. [23] with Adapted from Ref. [22] and Usman et al. [23].

and genome instability and mutation [24]. GPCRs are traditionally connected to many physiological functions demonstrated by postmitotic, differentiated cells, but are also present on proliferating cells, and involved in cancer development and cancer metastasis [21, 25]. Current treatments are targeted toward only a small portion of the GPCR family (Table 18.1). Therefore, fundamental research is essential to obtain further insight in the roles of GPCRs in this disease area [25]. In the next paragraphs, we will discuss GPCRs and their signaling pathways for which a role in tumor biology has been firmly established (Figure 18.2).

18.2.1 GPCR Signaling in Cancer

Following GPCR activation, several mechanisms, and modulatory proteins are involved in preventing hyperactivation of GPCR signaling, including GTP hydrolysis, second messenger-related protein kinases, G protein-coupled receptor kinases (GRKs), and arrestins (Figure 18.1) [7]. As a result of phosphorylation of specific serine and threonine residues in the C-terminus of GPCRs, GPCR-mediated activity is abolished by GRKs followed by β-arrestin recruitment, which precedes cytosolic internalization and degradation of GPCRs by lysosomes [27]. Based on sequence similarity, mammalian GRKs have been classified into three subgroups, namely the rhodopsin kinases (GRK1 and GRK7), the β-ARK subgroup (GRK2 and GRK3), as well as the GRK4 subgroup (GRK4, GRK5, and GRK6) [28, 29]. GRKs, acting as negative regulators of GPCR activity, may participate in cancer progression and tumor vascularization in a cell type-dependent manner [30–34]. Specifically, GRK1/7 is involved in cancer-associated retinopathy found in lung cancer patients via the interaction with recoverin, although direct evidence of reduced GRK1/7 activity in cancer progression has not yet been established [35].

In contrast, the involvement of GRK2 in cancer has been well-characterized with opposing effects in different cancer types. Poor survival rates and a high tumor stage of patients with pancreatic cancer are correlated with high GRK2 expression in clinical studies [36]. Overexpressed GRK2 in differentiated thyroid carcinoma reduces cancer cell proliferation through rapid desensitization of the thyroid-stimulating hormone receptor (TSHR) [37]. GRK2 acts as a negative regulator of cell cycle progression in human hepatocellular carcinoma HepG2 cells (HCCs) [31, 38]. Besides, GRK2 participates in inhibiting Kaposi's-associated sarcoma herpes virus (KSHV/HHV-8)-associated tumor progression as well as breast cancer and gastric cancer progression [39–41].

The role for GRK3 in breast cancer progression has been implicated to regulate CXCL12-mediated CXCR4 activation [42]. In prostate cancer, GRK3 is overexpressed in both primary tumor and metastatic cells [43]. Moreover, decreased GRK3 levels may be beneficial for cancer cell survival by increasing stress adaptation in cancer cells [44].

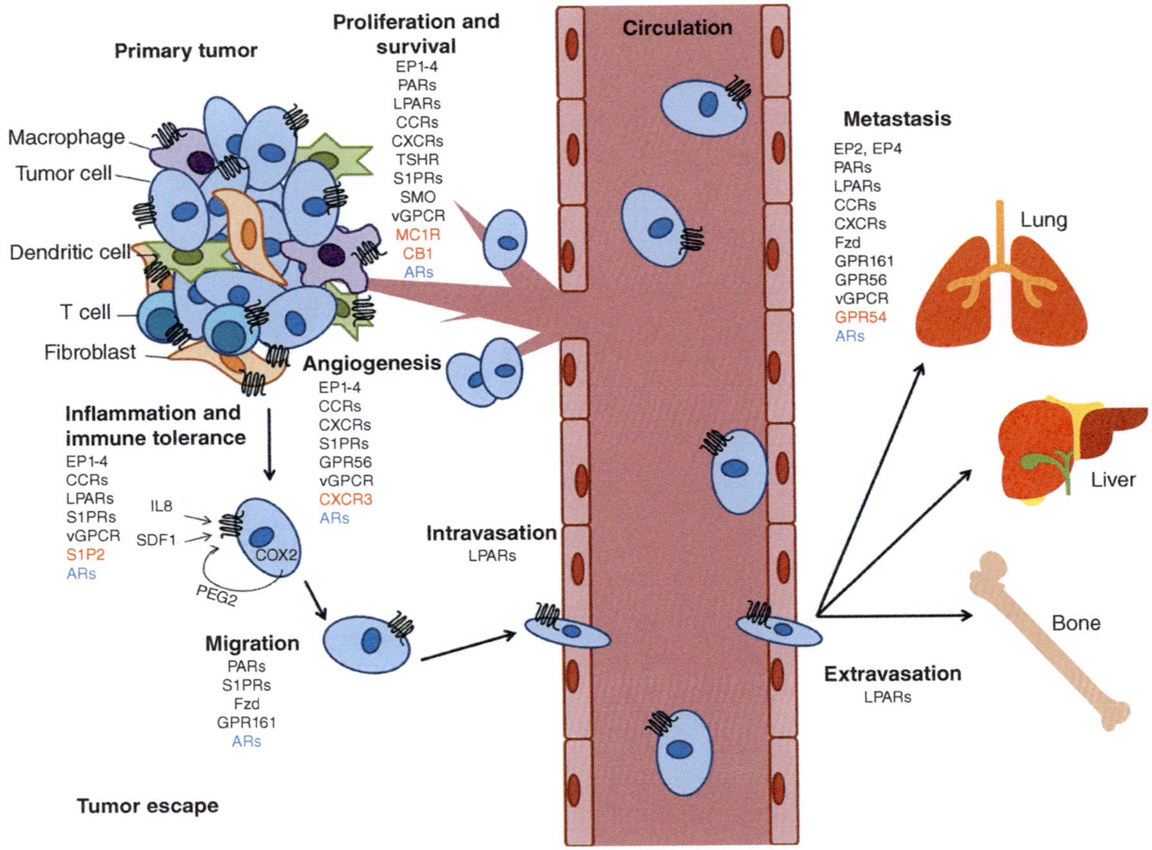

Figure 18.2 GPCRs and their roles in cancer hallmarks, adapted from Nieto Gutierrez et al. (2018) [26] and Dorsam et al. (2007) [21]. GPCRs with stimulating effects are marked in black, while GPCRs with suppressive effects are marked in red. Of note, adenosine receptors (blue) have both pro- and anti-tumoral effects depending on cancer type.

GRK4 is mainly expressed in the testis, kidney, brain, ovaries, and myometrium, although further research is warranted to characterize the role of GRK4 in cancer [45, 46].

GRK5 plays diverse roles during tumorigenesis. In thyroid carcinoma, GRK5 inhibits the desensitization of TSHR [37]. GRK5 expression in glioblastoma cells is associated with a worsened prognosis in patients with stage II–IV glioblastoma [47]. It has been reported that induced expression of GRK5 is linked with oncogenesis and tumorigenesis via the regulation of cell cycle progression in prostate cancer [48]. GRK6 might be a promising biomarker for the early diagnosis of hepatocellular carcinoma [49].

Frizzled (Fzd) receptors, a subgroup of GPCRs, regulate normal development, tissue homeostasis, and pathological processes, such as in cancer, through the interaction with Wnt proteins [50]. It is known that one single Wnt ligand is able to bind multiple different Fzd receptors. Moreover, the Wnt signaling pathway, highly conserved among species, regulates critical aspects of cell proliferation, polarity, cell fate determination and development, and stem cell differentiation [51, 52]. The canonical Wnt signaling pathway involves β-catenin (Figure 18.1), and this pathway has long been implicated in cancer, including colon cancer, ovarian carcinomas, hepatocellular carcinomas, melanomas, and prostate cancers [50]. Wnt signaling via Fzd receptors is known to stabilize β-catenin and the low-density lipoprotein-related protein 6 receptor (LRP6) complex that inhibits β-catenin degradation [53]. Stabilized β-catenin, expressed in a wide range of cancers, is translocated to the cell nucleus and functions as a co-transcription factor [54, 55]. In a diverse set of cancer types, the most commonly upregulated Wnt receptor, Fzd7, is therefore an attractive target for anti-cancer therapeutic strategy [56]. Noncanonical Wnt signaling consists of both the planar cell polarity pathway mediated via JNK signaling and the Ca^{2+}/PKC pathway via $G_{\alpha q/11}$ (Figure 18.1) [8, 9]. The Wnt family also plays a prominent role in cancer stem cell (CSC) function, suggesting that deviant Wnt signaling might lead to tumorigenesis [54]. Taken together, blockage of the Wnt pathway may eventually inhibit tumor growth and tumor initiation [56]. Targeting the Wnt pathway-related Fzd receptor family members is therefore a potential anti-cancer strategy.

GPCRs are known as one of the multiple upstream modulators of the recently discovered Hippo signaling pathway that plays a crucial role in coordinating cell proliferation, autophagy, cell growth, and apoptosis to establish and maintain particular control of organ size [26]. The mammalian Hippo signaling pathway commonly consists of serine/threonine kinase mammalian Ste2-like kinases 1 and 2 (MST1 and 2) that activate large tumor suppressor kinase 1/2 (LATS1 and 2), which phosphorylate the transcriptional coactivators Yes-associated protein (YAP) and transcriptional coactivator with PDZ-binding motif (TAZ) [26]. Phosphorylated YAP and TAZ are kept in the cytosol thereby preventing their access to the nucleus and inhibiting their transcriptional activity [57, 58]. Overgrowth of both tissue and organ always involves unwanted inhibition or loss of function of Hippo signaling, which is associated with a wide variety of cancers [59–61]. YAP and TAZ functionality can either be induced or blocked depending on the different type of G protein that is coupled to a GPCR (Figure 18.1) [10, 11]. In general, $G_{\alpha s}$-coupled GPCRs induce YAP phosphorylation preventing YAP-mediated transcriptional activation and YAP accumulation in nuclei, while $G_{\alpha 12/13}$, $G_{\alpha q/11}$, or $G_{\alpha i/o}$-coupled GPCRs result in dephosphorylation of YAP [62]. Therefore, the regulation of the Hippo signaling pathway via GPCRs may be a useful strategy in the treatment of certain cancer types.

Taken together, research on GPCR signaling networks is warranted to provide a better understanding of the involvement of these networks in cancer progression.

18.2.2 Inflammatory Role of GPCRs in Cancer

GPCRs are known to play an important role in the modulation of key inflammatory mediators [63], thus suggesting a possible link between cancer and chronic inflammation (Figure 18.2). Additionally, GPCRs have a crucial role in tumor-induced angiogenesis and migration of cancer cells to cause tumor metastasis [21]. Angiogenic factors produced from solid tumors are known to promote the proliferation of endothelial cells, therefore, leading to the formation of tumor vascularization to increase oxygen and nutrients supply for the tumor cells and routes for invasion and metastasis [19]. GPCRs and their related ligands are known to enhance angiogenesis by either directly inducing endothelial cell proliferation or indirectly by promoting release of vascular endothelial growth factor (VEGF) and other angiogenic factors from immune, stromal, or cancer cells [21].

Chemokines and chemokine (CC and CXC) receptors play an essential role in regulating the function of the immune system [64]. Notably, most chemokine receptors have been reported to bind several chemokines, while some are more selective, namely CXCR4, which only binds to CXCL12 [65]. In cancer, they participate either in several steps of the antitumor immunity or in the coordination of the release of several mediators to promote angiogenesis, thus facilitating tumor development [66]. For instance, CCL2 recruits CCR2-bearing tumor-associated macrophages, which are known to modulate tumor vascularization and growth [65, 67]. A similar effect has been suggested for CCL5, the chemokine ligand for CCR5, which is also related to macrophage recruitment as well as the recruitment of other leukocytes into inflammatory sites including eosinophils, T cells, and basophils [68]. Furthermore, CCL5/CCR5 interactions may regulate tumor development in several ways, e.g. by stimulating angiogenesis, acting as growth factors, modulating the extracellular matrix, and taking part in immune evasion mechanisms [69]. Moreover, killing of tumor cells can be promoted by some immune cells; therefore, the presenting chemokine in the tumor microenvironment (TME) may help tumor cells escape the immune surveillance system through a less effective humoral response [65, 67]. CXCL12, among other CXC-chemokines, is known to alter the local immune response and is a potent chemoattractant for pre-B lymphocytes, T cells, and dendritic cells. One of the receptors of CXCL12, CXCR4, is expressed on monocytes, T lymphocytes, endothelial cells, and neutrophils. This chemokine produced by various cell types in the TME regulates the activity of immunosuppressive cells and thus contributes to tumor progression [66]. Therefore, the characterization of the different chemokine networks in various cancer types may provide a better understanding of the immune-related mechanisms in cancer development.

The interaction between prostaglandin (PG) production and tumor progression is one of the most intensively investigated among the effectors linking inflammation and cancer. Cyclooxygenases (COX-1 and COX-2) produce PGs and thromboxane A_2 (TXA_2), and the pro-inflammatory effects of COX-1/2 are mediated upon binding of PGs to their respective GPCRs. Therefore, in numerous cancer types, inhibiting COX-1/2 using nonsteroidal anti-inflammatory drugs (NSAIDs) can reduce the incidence and risk of cancer [70, 71]. For instance, NSAIDs inhibiting COX-2 can reduce the overall size and amount of adenomas in colorectal cancer patients. Furthermore, these drugs are used as a chemopreventive approach for colon cancer in healthy individuals [70, 71]. Therefore, for early and advanced cancer, the effect of COX-2 inhibition in cancer prevention is under investigation in many clinical trials [72–74]. However, due to the potential cardiovascular complications of COX-2 inhibitors [75], the (downstream/direct) inhibition of prostanoid receptors, yet another subfamily of GPCRs, might serve as an alternative strategy for cancer prevention and treatment.

Prostaglandin E_2 (PGE_2) and its GPCRs, E prostanoid receptors (EP_1–EP_4), are involved in tumor progression [76–78]. The EP_1 receptor is abundantly expressed, while the EP_3 receptor is only expressed at high level in certain tissues, such as pancreas, adipose tissues, vena cava, and kidney. The EP_4 receptor is mainly expressed in the uterus, gastrointestinal tract, skin, and hematopoietic tissues, while the EP_2 receptor is the least abundantly expressed EP receptor, although widely distributed in many tissues [26]. In colon cancer, using an EP_1 receptor knockout mice model, EP_1 receptors were shown to play an essential role in cancer progression, and similar results were obtained in wildtype mice treated with an EP_1 receptor antagonist [79]. The involvement of EP_4 receptors is widely established in tumorigenesis of multiple malignancies [80]. Together, COX-2 overexpression and PGE_2-mediated activation of EP_2 and EP_4 receptors contribute to the abnormal growth, and metastatic and angiogenesis potential of many highly prevalent cancers other than colon cancer [73, 81–83]. PGE_2 and antigen presenting cell (APC)-regulated mechanisms are also known to be intimately related. APC inactivation is known as an early event in progression of colon cancer, which results in β-catenin stabilization in cytoplasm [78, 84]. Similar to Wnt/Fzd receptors, PGE_2 stimulates the β-catenin pathway through the EP_2 receptor in colon cancer cells [85, 86]. Even the EP_2 receptor is also involved in VEGF expression to promote angiogenesis [87]. The least commonly cancer-linked EP receptor, EP_3 receptor, may indirectly modulate pathways involved in tumor angiogenesis [88–90].

Moreover, PGE_2 initiates crosstalk with other GPCRs. For instance, the endogenous ligand of the endothelin 1 receptor, endothelin 1, is expressed at a high level in over 80% of colon cancers and can rescue colon cells from apoptosis by β-catenin inhibition as it is also a downstream transcriptional target of β-catenin [72].

Investigation of cancer-related inflammation and the role of GPCRs therein may thus provide insights in the immune-related mechanisms of cancer progression and suggest novel strategies in cancer immunotherapy.

18.2.3 Aberrant Expression of GPCRs in Cancer

Overexpression of many GPCRs occurs in a wide variety of cancer types, suggesting their contribution to tumor cell growth upon activation of their ligands whether produced locally or circulating. Among these, chemokine receptors, protease-activated receptors (PARs), as well as receptors for bioactive lipids (sphingosine-1-phosphate receptors [S1PRs] and lysophosphatidic acid receptors [LPARs]), increase cell proliferation in a wide array of cancer cells [21].

In highly invasive breast carcinomas, PAR1 overexpression has been observed, and this increased expression in mammary glands in a mouse model results in premalignant atypical intraductal hyperplasia [91]. The increase of PAR1 expression is also seen in advanced stage prostate cancer [92].

Quite often, the altered expression of chemokine receptors in cancer cells together with the release of chemokines from secondary organs causes organ-specific metastasis [21]. High expression levels of CXCR4 in prostate cancer cells increase blood-vessel density and muscle invasion [93]. CCR10, the receptor for CCL27 and CCL28, has been found expressed in human melanoma cells. In a common site of metastasis for melanoma, i.e. skin, CCL27 and CCL28 are highly expressed [94]. Constitutive production of CXCL1 and CXCL8 is observed in melanoma cells. Furthermore, the same cell types also highly express the receptor of these chemokines, CXCR2, and stimulation of autocrine chemokine increases proliferation, migration, and survival of tumor cells [95]. Many different tumor cells aberrantly express CXCR4, and cancer cells with a high level of CXCR4 expression are more prone to undergo metastases [96]. In some cancer types, CXCR4 is coexpressed with other CC or CXC chemokine receptors on cancer cells, and the combination has been suggested to be related to cancer progression [97].

In ovarian cancer, LPA is one of the most potent mitogens secreted by cancer cells to the ascites fluid. The effect of LPA in promoting growth, survival, and resistance to chemotherapy is via the stimulation of the LPARs that are frequently overexpressed in ovarian cancer cells [98].

S1P regulates central cellular processes and affects all steps of tumor growth and metastasis [99]. S1P has been shown to be important in the metastatic behavior of aggressive thyroid tumors, where the expression of S1P receptors is upregulated [100]. High expression of S1P together with upregulation of S1P receptors is observed in the human prostate cancer cell line PC3, where together they protect cells from apoptosis induced by camptothecin, a topoisomerase inhibitor [101]. The pathological diversities in gastric cancers are largely dependent on the expression profiles of the S1P receptor subtype and thus therapeutic interventions targeting each S1P receptor might be clinically effective in preventing metastasis [102].

Therefore, overexpression of a certain GPCR may serve as a biomarker for diagnostic purposes in some cancer types and a target for cancer treatment.

18.2.4 Role of GPCRs in Cancer Cell Proliferation and Metastasis

Metastasis, as one of the more serious challenges for cancer therapy, significantly reduces life quality and overall long-term survival of patients with malignancies [103]. Cancer cells metastasize to specific organs with a much greater incidence than would be expected from the primary tumor site and the secondary organs (Figure 18.2). GPCRs are considered as attractive targets to prevent and treat metastasis.

PARs are activated by a unique proteolytic cleavage mechanism of their N-terminus, leading to the activation of a diverse network of signals [104]. PARs sense and respond to proteases activated in the TME and are thus pivotally involved in tumor invasion and metastatic efficiency [104]. In an animal melanoma model, PAR1 knock-down led to reduced tumor growth and metastases to the lung [26]. Expression of both PAR1 and PAR2 has been established in different types of cancer cells, and evidence suggests the contribution of PAR2 in tumor development. For example, in a mouse model with PAR2 deletion breast tumor development was delayed with decreased metastasis formation [105]. A similar observation was noted in breast xenograft models in which inhibitory antibodies of PAR2 weakened tumor growth and metastasis [106].

Chemokines and chemokine receptors also direct the traffic of leukocytes and their progenitor cells between the lymphoid organs and the blood and the migration of leukocytes to sites of inflammation [65]. As mentioned before multiple chemokine receptors are expressed in tumor cells and are activated in response to released chemokines in the TME from tumor-infiltrating leukocytes, macrophages, stromal cells, and even cancer cells, thus promoting the survival and motility of cancer cells in both an autocrine and paracrine pattern [65]. Although few studies comprehensively characterized all chemokine receptors present on cancer cells, one single chemokine receptor is still able to influence the spreading direction of a cancer cell. Multiple chemokine receptors possibly participate in metastases of the same cancer to different sites [97]. CXCR4 is highly expressed in many tumor cells, and CXCR4 activation stimulates directed cancer cell migration and induces their filtration through bone marrow stromal cells, endothelial cells, and fibroblasts [97]. Breast tumors with high CXCR4 expression showed more extended metastasis in comparison to tumors with low expression of CXCR4, but no significant correlation was found with blood-born metastasis [107]. Furthermore, CXCR4-mediated metastasis was observed from melanoma cells to the lungs [108]. Pancreatic cancer cells with CCR6 expression show enhanced proliferation in response to CCL20 [109]. CCR7 has been found expressed in some cancer types, which was correlated to metastatic potential and poor prognosis. The ligand of CCR7, CCL12, has been found at high levels in the lymph nodes that connect many cancers [94].

Similar to chemokine receptors LPARs are reported to have proliferative, prosurvival and promigratory effects in ovarian cancer cells, while LPARs also stimulate further LPA release [98]. In this case, an autocrine loop occurs to drive the unrestricted ovarian cancer cell growth [98, 110].

As the GPCRs described above promote cancer cell proliferation and metastasis, inhibitors of these (and potentially several other) GPCRs have the potential of decreasing the survival, uncontrolled proliferation, and organ-specific metastasis of cancer cells.

18.2.5 Viral GPCRs in Cancer

In the examples above, human GPCRs played a role in cancer development. However, many human cancer-associated viruses hijack GPCR signaling to enhance their life cycle [111]. Among them, herpes viruses can set up life-long persistent infection in humans with normal immune response. Particularly, reactivation of herpes viruses in patients with an impaired immune system may cause severe morbidity and mortality [112, 113]. These viral GPCRs, sharing sequence homology with human chemokine receptors, are able to direct immune cells to the inflammation site; they are also actively involved in different pathological processes, including tumor growth, survival, and metastasis (Figure 18.2) [65, 66]. Different from predominantly $G_{\alpha i}$ coupled human chemokine receptors, viral GPCRs may activate different downstream signaling pathways through several G_{α} proteins even in the absence of ligand activation, meaning they are constitutively activated [111, 114]. Moreover, viral GPCRs promiscuously respond to a wide array of chemokines, indicating that they can utilize the immune system of the host to modulate viral dissemination as part of the immunopathology of the viral infection [115]. Early studies of virally encoded oncogenes have provided evidence that at least seven human viruses, hepatitis B virus (HBV), Epstein–Barr virus (EBV/HHV-4), human papilloma virus (HPV), human T-cell lymphotropic virus (HTLV-1), hepatitis C virus (HCV), Merkel cell polyomavirus, and Kaposi's associated sarcoma herpes virus (KSHV/HHV-8), are involved in 10–15% of cancers [116, 117]. Interestingly, open reading frames encoding GPCRs have been observed in the viral genomes of many human viruses, suggesting that replicative success benefits from these signaling pathways [118]. For instance, human cytomegalovirus (HCMV/HHV-5) regulates the expression of at least four GPCRs (US27, US28, UL33, and UL78), and EBV encodes one GPCR (BILF1). The receptor expressed by KSHV, commonly known as KSHV vGPCR (or ORF74), shares similar structure and functionality to CXCR1 and CXCR2 [119]. Constitutive activity of KSHV vGPCR gives it potent transforming and proangiogenic properties, and contributes to Kaposi's sarcoma development [120]. It promotes the function of a complex signaling network to induce sarcomagenesis. KSHV vGPCR-expressing cells activate the dysregulated growth of distant and surrounding endothelial cells, thus representing an example of paracrine neoplasia [120]. Hence, inhibition of vGPCRs and their downstream signaling networks may provide new potential treatment for KSHV-associated cancers.

18.2.6 Role of GPCRs in Tumor Suppression

GPCRs mainly show protumoral effects, but in certain malignancies, some GPCRs and related G proteins may actually show tumor suppressive effects (Figure 18.2). As an example, inactivating mutations of the melanocortin 1 receptor (MC1R) increase the risk of melanoma development [121], suggesting the wildtype receptor to be tumor suppressive. CXCR3 ligands suppress tumor progression by indirectly mediating anti-angiogenic effects, while in gliomas, colorectal, skin, and breast cancer the cannabinoid receptors (CB1 and CB2) play tumor suppressive roles [122]. In addition, SIP_2 receptor signaling via $G_{\alpha 13}$ acts as a tumor suppressor in diffuse large B cell lymphoma (DLBCL) [123]. Lastly, *KiSS1*-derived peptide receptor (GPR54) suppresses metastasis in melanoma and breast cancer cells [124]. These are certainly not the only anti-tumorigenic GPCR/G protein signaling pathways in different cancers [19], and many anti-tumoral GPCRs are likely to be discovered in the future.

18.3 GPCR Mutations in Cancer

Large-scale sequencing efforts of cancer genomes combined with unbiased systematic approaches, have identified a large number of mutations in GPCRs and G proteins. Specifically, it was found that in approximately 20% of all cancers GPCRs are mutated [125, 126], making research into the potential oncogenic effects of GPCR mutations paramount. Cancer-related mutation data have, for example been collected in the Cancer Genome Atlas (TCGA) [127] and the National Cancer Institute Genomic Data Commons (GDC) [128]. Of note, higher mutation rates are often observed for certain conserved residues and given the (evolutionary) importance of these residues the exact impact of these mutations in receptor pharmacology warrants considerable investigation [125, 129]. In tumor samples, analysis of GPCR somatic mutation rates in comparison with the background mutation rates has identified several frequently mutated GPCRs, suggesting their involvement in cancer [125]. In this section, we will first discuss the general consequences of a mutation for receptor behavior (Figure 18.3), be it somatic or experimentally induced. Thereafter, we will discuss in depth some of the naturally occurring mutations in cancer tissue.

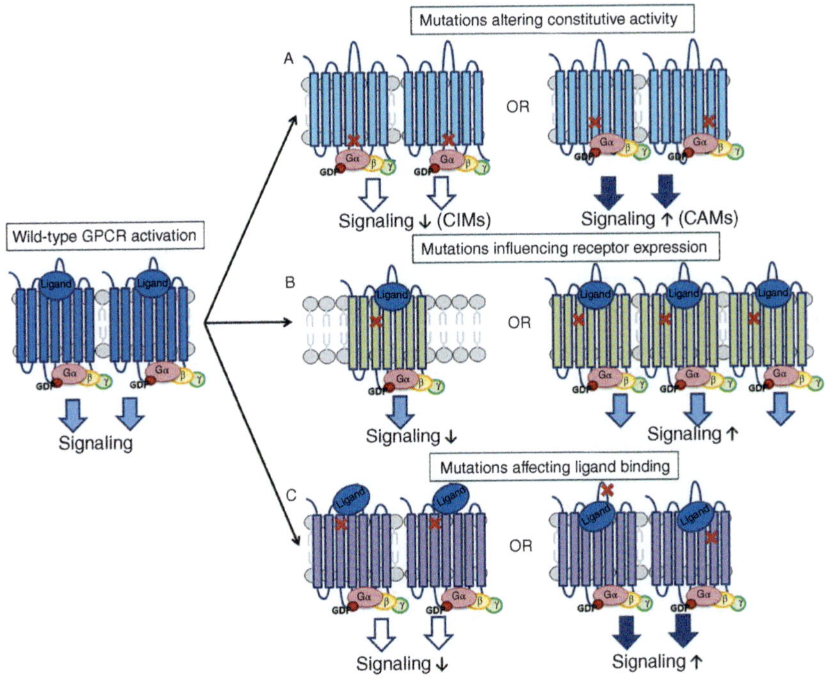

Figure 18.3 Effects of GPCR mutations on receptor pharmacology, such as (A) constitutive activity, (B) receptor expression, and (C) ligand binding and their concomitant effect on receptor signaling. Red crosses indicate potential mutation sites. Source: Based on Peeters et al. [130].

18.3.1 GPCR Mutations Affecting Constitutive Activity

Several detailed three-dimensional structures of various GPCRs in different activation states have been published, providing us with an unprecedented understanding of GPCR structure and function [131]. GPCRs harbor various signaling states and structural conformations [132]. Based on the simplified two-state-receptor model, all states that are unfavorable for G protein binding are referred to as inactive (R), and all states that couple to G proteins as "active" (R*) [133]. Since GPCRs are flexible, the equilibrium between R and R* states provides room for the presence of the R* state even without agonist binding. This brings constitutive or basal activity into play, which varies highly among wildtype GPCRs and has physiological importance in many cases. Both decreased and increased constitutive activity are known to cause disease phenotypes, such as neuropsychiatric and cardiovascular diseases [134]. Mutant receptors that show induced basal activity in the absence of an agonist, are referred as constitutively active mutants (CAMs), while those with decreased basal activity are termed constitutively inactive mutants (CIMs) (Figure 18.3) [130]. In many GPCRs, an aspartate acid residue in TM6 (Asp6.30) and an arginine in the TM3 DRY motif (Arg3.50) form a salt bridge, also called "ionic lock," which has been linked to the modulation of basal activity by limiting GPCR flexibility [135–137]. This salt bridge can be disrupted by mutation of the aspartate acid at residue 6.30 into different amino acids, resulting in increased constitutive activity in several GPCRs [138–141]. This increased flexibility in GPCRs can preclude the necessity for ligand binding to open the cavity of G protein binding.

In general, at the amino acid side chains within the helical bundle a mutation with increased hydrophilicity can destabilize the receptor [142]. Increasing the size of the amino acid side chain may lead to clashes with surroundings, which lead to small conformational changes in the immediate surroundings. This is especially important for the residues located around the kink area in TM6, where a small conformational change can cause the movement of the helix and open up the cavity for G protein coupling [143, 144]. Several CAMs are caused by an altered charge of the affected amino acid, either a change from negative to neutral or from negative to positive. In general, charged side chains in the helical bundle likely participate in electrostatic interactions. Therefore, mutations with charge-altered side chain break these interactions, releasing a constraint that otherwise keeps the receptor in the R state [145]. Mutations leading to decreased basal activity have also been found in many GPCRs by restraining interactions, while most of them also cause other defects, such as impaired agonist binding, impaired G protein coupling, or a generally decreased response to agonist-mediated activation [145].

18.3.2 GPCR Mutations Affect Receptor Expression

GPCR mutations can affect receptor expression (Figure 18.3b). Although increased GPCR expression in cancer is regularly reported (see above), it is yet unclear if mutations play a role in this. In contrast, among all impacts of GPCR mutations, the most common defect is impaired receptor expression [146]. Even though the biosynthesis of mutant receptors does not seem to be much affected, the trafficking of mutant receptors has been altered in that nascent receptors with misfolded structures are not moved to the Golgi apparatus, but instead, transported to lysosomes for degradation [147, 148]. Any mutation-causing disturbance of the disulfide bridge between the second extracellular loop and residue Cys3.25 in TM3 has been suggested to cause receptor malfunction and instability [149]. Other causes of faulty trafficking are the deletion or disruption of signal peptide motifs. For example in the follicle-stimulating hormone receptor (FSHR), a motif within the C-terminal tail has been reported to be important for targeting the plasma membrane, and mutations found in this motif often result in intracellular retention [150]. Thus, mutations-altering receptor expression may also affect the functionality of a GPCR.

18.3.3 GPCR Mutations Affect Agonist-Dependent Activation

Upon ligand binding, a GPCR mutation can modify the response by altering agonist affinity, efficacy, and/or receptor selectivity (Figure 18.3). During receptor activation, the conformation of intracellular parts of TMs 5 and 6 change considerably to generate an interaction with G protein and to process activation [142]. It is possible that a mutation located in or near key positions, including microswitches, GPCR-G protein-interaction interface, and ligand-dependent trigger residues, can partially mimic this process, resulting in altered receptor functionality. Residues directly and indirectly essential for agonist binding are usually found within the ELs and in the (top half of the) 7TM domains, and are expected to modulate affinity [136, 149]. Upon agonist binding, a mutation facilitating the R* state formation of the receptor provides a more preferred interface interaction for G-protein activation and thus increases agonist affinity and efficacy. However, depending on the type of ligand and receptor, GPCRs can engage a G protein and/or β-arrestin, or prefer one over the other. This ability of a ligand is called functional selectivity or biased signaling, which directs a GPCR toward a specific conformation selectively linked to a particular activation pathway [132]. In general, mutations located within the 7TM domain can be expected to disrupt the energy barrier for agonist-mediated activation, thereby modulating functional selectivity of the receptor [145]. Residues along TM6 are especially interesting, as these residues experience the most dramatic structural/spatial change upon receptor activation [151]. Moreover, the residues located around the kink area of TM6, including clusters between TMs 6 and 7 and between TMs 3 and 6, have been suggested in movement regulation [149]. Taken together, agonist-dependent activation of mutant GPCRs and changes in the basal activity of such receptors (as described in the preceding paragraph) are inextricably linked.

18.3.4 Role of GPCR Mutations in Cancer

One of the most frequently mutated GPCRs in tumors is smoothened (SMO), a class F GPCR, for which the 12-transmembrane receptor Patched (PTCH) acts as a negative regulator [152, 153]. This SMO inhibition is relieved upon the binding Hedgehog (Hh) to PTCH, which results in downstream stimulation of the transcription factor glioma-associated oncogene (GLI) [152, 153]. Mutations in PTCH and SMO have been suggested to initiate sporadic basal cell carcinoma [154–156]. An activating mutation of SMO, W535L located at the bottom part of TM7, initially identified from basal-cell carcinoma, has recently also been found in meningiomas [157, 158]. In addition, mutations of SMO have been identified in colon cancer and cancers in the CNS and emerging evidence strongly supports continuous SMO signaling is involved in tumor progression [159].

The second most frequently mutated GPCRs are the metabotropic glutamate receptor (mGlu) family members, mGlu1–8, which have a significant cancer-specific distribution [125]. Mutations in mGlu8, mGlu1, and mGlu3 have been identified in squamous nonsmall cell lung cancer (NSCLC), melanomas, and NSCLC adenocarcinomas, respectively [125]. Mutated mGlu8 was found in 8% of NSCLCs of the squamous subtype, while mutated mGlu1 was found in 7% of NSCLC adenocarcinomas [126]. However, the impact of these mutations at a molecular and cellular level still needs to be fully characterized to understand their subsequent effects on tumor progression [19]. Mutated mGlu3 has been found in 16% of examined melanomas in a study in which endogenously expressed mutant GPCRs were linked to the progression of melanoma by using a systematic exon-capture analysis together with a massively parallel sequencing approach on 734 GPCRs [160]. The mGlu receptor family is of particular interest due to the increased availability of glutamate, its endogenous ligand, in the TME [161]. Together, this suggests that at the surface of tumor cells both wildtype and mutant mGlu receptors may be expressed and activated.

The GPCR adhesion receptor family, consisting of 33 members, also presents frequent mutations. The adhesion receptors, including GPR98, GPR112, and brain-specific angiogenesis inhibitor (BAI) members, are involved in regulating cell–cell and cell–matrix interactions [162]. Among these receptors, the most frequently mutated GPCR in cancer is GPR98 [125], for instance in melanoma progression it was mutated in 28% of the melanomas examined [160].

Syndromes caused by unrestrained hormonal secretion often involve activating mutations in GPCRs, which are also found in endocrine tumors [21]. For instance, TSHR mutations with activating effects are found in around 80% of thyroid adenomas and in some thyroid carcinomas, and TSHR mutations in germ cells result in familial nonautoimmune hyperthyroidism [75]. In thyroid cancer, TSHR is found to be the most frequently mutated GPCR, while it is also mutated in ovarian, lung, and large intestine cancers. These TSHR variants need further investigation to unravel their precise role [125]. Mutations in the TSHR are located throughout the receptor structure. The effects of these variants vary from completely abolished to slightly decreased TSH response, to those with increased constitutive activity [75]. The activating TSHR mutants promote the constant activation of adenylate cyclase via $G_{\alpha s}$, leading to hyperfunctional thyroid adenomas [75]. A close homolog of TSHR, luteinizing hormone receptor (LHCGR), is particularly evident in colon, lung, and breast cancers, whereas another related GPCR, FSHR, is known to be mutated in large intestine cancers [125]. Some subtypes of adult stem cells express other TSHR-related receptors, such as leucine-rich repeat-containing GPCR 4 (LGR4), LGR5, and LGR6 [163]. Mutations of these receptors have also been identified in melanoma and in colon carcinoma. This implicates that these stem cell populations play a potential role in cancer initiation. Taken together, cancer-specific GPCR mutations offer a novel approach for the development of strategies that target both cancer prevention and treatment.

18.4 Adenosine Receptors and Cancer

The adenosine receptors (ARs) belong to Class A, rhodopsin-like GPCRs and there are four subtypes, A_1AR, $A_{2A}AR$, $A_{2B}AR$, and A_3AR. ARs have attracted much attention in the recent years as therapeutic targets [3]. As they play an important role in both physiological and pathophysiological conditions, in-depth investigation of these GPCRs is required. Additionally, all four AR subtypes have been detected in different human tumor tissues [164]. Dependent on the adenosine receptor subtype, extracellular binding of adenosine leads to activation of different downstream signaling cascades (Figure 18.1). Through G_i-coupling, the A_1AR and A_3AR inhibit adenylate cyclase and reduce cAMP levels [165–168]. $A_{2A}AR$ and $A_{2B}AR$ are coupled to G_S proteins and increase the levels of cAMP [169, 170]. In addition, $A_{2B}AR$ can also couple to G_q proteins, which causes the activation of phospholipase C and mobilization of calcium [171, 172]. We will first briefly discuss the general (patho)physiological roles of ARs and their sometimes contradicting roles, i.e. pro- and anti-tumoral effects in cancer cells, as well as their involvement in the TME. Then we will discuss potential cancer treatments targeting ARs. Afterwards, we will take $A_{2B}AR$ as an example to discuss the effect of cancer-related mutations in receptor pharmacology.

18.4.1 Distribution and (Patho)physiological Roles of Adenosine Receptors

The A_1AR is abundant in the CNS, with high expression levels in the hippocampus, cerebral cortex, thalamus, cerebellum, spinal cord, and brain stem. The A_1AR has also been identified in numerous peripheral tissues, including testis, vas deferens, stomach, white adipose tissue, pituitary, spleen, heart, adrenal gland, aorta, bladder, eye, and liver. In tissues, such as kidney, lung, and small intestine, A_1AR is expressed at low levels [173–175]. A_1AR has a pivotal role in neuronal, renal, and cardiac processes [176]. Based on the receptors involved or the site of application, adenosine causes either pro- or anti-nociceptive effects on pain. A_1AR is mostly suggested to be involved in pain pathways and has been reported in animal models of inflammatory and neuropathic pain [177]. Moreover, during periods of decreased sleep duration and sustained wakefulness, accumulated adenosine acts as a natural sleep-promoting agent [178]. Indeed, A_1ARs expressed in the suprachiasmatic nucleus regulate the circadian clock response to light [179]. Endogenous adenosine, via A_1AR activation, regulates long-term synaptic plasticity phenomena, such as depotentiation, long-term depression, and long-term potentiation [180]. Moreover, A_1AR antagonists have been suggested to be beneficial in memory disorders. In the CNS, acute administration of A_1AR agonists is neuroprotective, while A_1AR antagonists promote the death of neuronal cells in ischemic models [181]. In addition, selective agonists of A_1AR may be utilized as pharmacologic preconditioning agents for lung transplantation to prevent ischemia–reperfusion injury, as well as other forms of pulmonary vascular ischemia [182].

In the CNS, the $A_{2A}AR$ is expressed at high levels in the olfactory tubercle and striatum [174]. It is also highly expressed in blood platelets, leucocytes, spleen, and thymus in the periphery. Intermediate expression levels of $A_{2A}AR$ have been found in blood vessels, heart, and lung [172, 173]. The $A_{2A}AR$ is involved in the onset of vasodilation, exploratory activity, aggressiveness, hypoalgesia, and inhibition of platelet aggregation [173]. Additionally, $A_{2A}AR$ is involved in the progression of Parkinson's disease, Alzheimer's disease, Huntington's disease, and in the attenuation of neuroprotection, ischemia, and inflammation, particularly in peripheral tissues [172, 173]. More recently, $A_{2A}AR$ antagonists have been suggested for the management of chronic pain [183]. For example, in two acute thermal pain tests, the $A_{2A}AR$ selective antagonist SCH 58261 produced antinociception [184]. The impact of $A_{2A}ARs$ in controlling neuronal damage was first proposed in a cerebral ischemic injury model [184]. Studies have shown that either the genetic elimination or the pharmacological blockade of $A_{2A}ARs$ in brain ischemia animal models provided robust neuroprotection [184]. Furthermore, the excessive activity of the thalamic cortex can be reduced through blockage of $A_{2A}ARs$ on striatopallidal neurons. This is known to restore balance between striatopallidal and striatonigral neurons and consequently influence the efficacy of $A_{2A}AR$ antagonist in Parkinson's disease [185]. Several lines of evidence support a role for $A_{2A}AR$ in abuse substance pathologies, thereby suggesting this receptor as a possible target for the treatment of drug addiction [186].

$A_{2B}ARs$ receptors are expressed in many organs, including kidney, colon, lung, and spleen, and the vasculature is the primary site of expression in all of these organs. In addition, endothelial cells, macrophages, and smooth muscle cells display a high level of $A_{2B}AR$ expression [187], and colonic epithelial cells also express this receptor [188]. $A_{2B}AR$ expression has been shown in isolated primary cells, as well as cell lines for other cell types, such as dendritic cells, lymphocytes, and mast cells [188–190]. It has long been implicated that extracellular adenosine is related to adaptation to hypoxic conditions and $A_{2B}ARs$ play an essential role in inhibiting hypoxia-induced vascular leak *in vivo* [191]. In addition, extracellular adenosine stimulates cell death of human arterial smooth muscle cells through a cAMP-dependent pathway mediated by $A_{2B}AR$ [192].

The A_3AR is expressed at relatively low levels in the CNS, particularly in the thalamus and the hypothalamus [190]. Liver and lung express A_3AR with the highest reported levels, while the aorta expresses A_3AR at intermediate levels [174]. Additionally, the A_3AR has been found in mast cells, eosinophils, kidney, testis, heart, placenta, spleen, bladder, uterus, aorta, jejunum, eye, and proximal colon, although with differences in expression level, and among species [173–175]. The A_3AR is suggested to mediate apoptotic events in certain cell types, allergic responses, and airway inflammation [174, 175]. Despite the low expression in the brain, the involvement of A_3AR in both normal and pathological conditions in the CNS has been of considerable interest [192]. Furthermore, higher expression levels of A_3ARs have been determined in tumor cells than in healthy cells, demonstrating its potential role as a tumor marker [193].

18.4.2 Role of Adenosine Receptors in Cancer Cells

ARs have been associated with carcinogenesis, where both pro- and anti-tumoral effects have been identified (Figure 18.4). Although multiple studies addressed the role of A_1AR in cancer progression, its precise role has not been fully characterized. Increased expression levels of the A_1AR have been observed in diverse

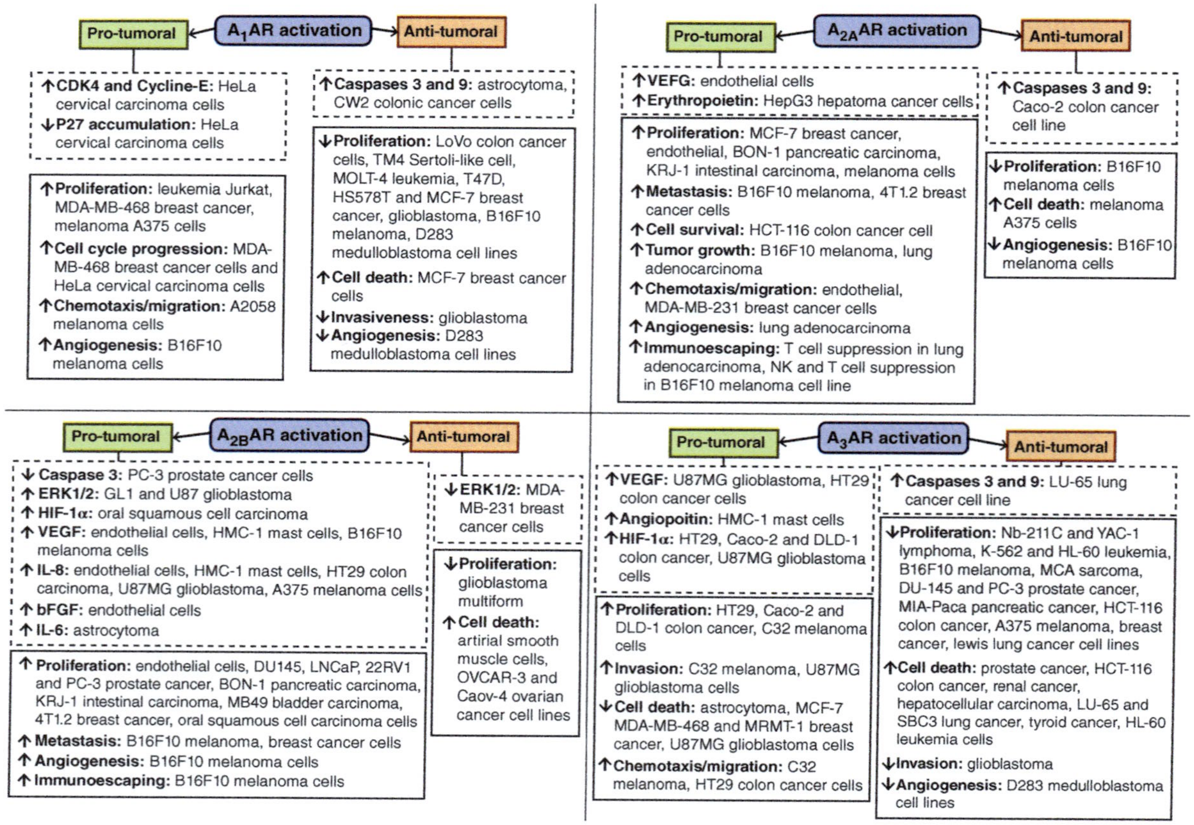

Figure 18.4 Pro- and anti-tumoral effects of adenosine receptors, A_1AR, A_{2A}AR, A_{2B}AR, and A_3AR adapted from Gessi et al. (2011) [164] and Kazemi et al. (2018) [194]. Effects on cellular levels are shown in dashed boxes and effects in cancer hallmarks are shown in solid boxes.

cancer cells, such as colorectal adenocarcinomas, leukemia Jurkat, and melanoma A375 cell lines [12, 195, 196]. Overexpression of A_1AR in MDA-MB-468 breast cancer cells enhances cell growth, survival, and proliferation [195]. In renal cell carcinoma, blockage of A_1AR signaling inhibits cell proliferation and migration by regulating the ERK/JNK signaling pathway (Figure 18.1) [12]. Lastly, it has been shown that adenosine-mediated tumor cell chemotaxis could be inhibited by an A_1AR antagonist [197]. On the other hand, in CW2 colonic cancer cells and rat astrocytoma cells, adenosine induces cell death through activation of caspases 3 and 9 via A_1AR [198, 199]. It has been shown that activation of A_1AR causes increased apoptosis and thus inhibits tumor growth in MCF7 breast cancer cell lines [196]. In breast tumor cell lines T74D and HS578T, leukemia MOLT-4, and human LoVo metastatic cell lines, stimulation of A_1AR inhibits cell proliferation [164, 200]. In addition, activation of A_1AR expressed on microglia dampens the growth of glioblastoma [201].

As $A_{2A}AR$ has been discovered on the cell surface of diverse human tumor cells, this receptor is expected to play a role in cancer [202, 203]. $A_{2A}AR$ is involved in promotion of angiogenesis inducing endothelial cell proliferation and migration, as well as synthesis of important angiogenic growth factors such as VEGF [204, 205]. Adenosine-mediated $A_{2A}AR$ activation produces cAMP and inhibits the destruction of cancer cells initiated by lymphokine-activated killer cells [206]. Activation of $A_{2A}AR$ also stimulates cell proliferation of endothelial and melanoma cells [207]. Under both normal and hypoxic conditions in hepatocellular carcinoma cells (Hep3B) *in vitro* and *in vivo*, $A_{2A}AR$ activation has been shown to increase erythropoietin production, resulting in increased cell survival [208, 209]. Additionally, an $A_{2A}AR$ agonist induces proliferation of MCF-7 breast cancer cells and interferes with the ethanol-induced activation of estrogen receptor signal transduction [202]. In contrast, adenosine promotes cell death via $A_{2A}AR$ by activating caspase 3 and 9 in association with mitochondrial damage in the Caco-2 colon cancer cell line [210]. $A_{2A}AR$ activation in human A375 melanoma cells has been shown to promote cell death [207].

$A_{2B}AR$ has been reported to be expressed in many tissues and cell types and is only activated by high concentrations of adenosine, such as are present in certain pathological conditions [211, 212]. In HT29 human colon cancer cells activation of $A_{2B}AR$ via adenosine increased IL-8 expression, thus promoting angiogenesis [213, 214]. In a triple negative breast cancer model with MDA-MD-231 cells, adenosine increased proliferation and migration via activation of the $A_{2B}AR$ [215]. Besides, it has been shown that proangiogenic dendritic cells can develop an anomalous phenotype via the stimulation of $A_{2B}AR$ [216]. In mouse cancer models, experimental and spontaneous metastasis formation has been induced by activation of $A_{2B}AR$, which is also known to worsen the efficacy of classical chemotherapy drugs [217]. Additionally, $A_{2B}AR$-stimulated metastasis has been observed in melanoma, breast, ovarian, and blood carcinomas [218, 219]. In contrast, $A_{2B}AR$ activation inhibits the ERK1/2 pathway, which results in an anti-proliferative effect in MDA-MB-231 breast cancer cells [220].

A_3AR has been investigated for its role in diverse types of cancer cells. The receptor is involved in both pro- and anti-proliferative effects, as well as in cell migration and cell death [203, 221–223]. It has been shown that A_3AR activation leads to induced expression of angiopoietin 2 in HMC-1 human mastocytoma cancer cells and melanoma cells [224, 225], resulting in a protumor effect. In addition, in glioblastoma cells an increase in expression of multidrug resistance-associated protein 1 via A_3AR activation was inhibited by A_3AR antagonist administration, which also increased the anti-tumoral effect of the chemotherapy drug vincristine [226]. On the other hand, A_3AR is also involved in controlling the cell cycle and inhibiting tumor growth both *in vitro* and *in vivo* [173]. Experimental animal studies have validated the therapeutic efficacy of orally administered A_3AR agonists in multiple pathological models, including xenograft, syngeneic, metastatic, and orthotopic models of melanoma, hepatocellular, prostate, and colon carcinomas. It was shown that these drugs decreased cell proliferation in murine melanoma of both syngeneic and lung metastatic models [227]. In summary, as all four subtypes of ARs regulate cancer progression in diverse type of cancers, the incorporation of adenosine ligands and pharmacological inhibitors of AR signaling into preclinical studies may provide for novel therapeutic approaches for cancer treatment.

18.4.3 Roles of Adenosine Receptors in the Tumor Microenvironment

In addition to their role(s) in cancer cells, all four AR subtypes are involved in the control of inflammatory responses within the cancer cell environment. For example, $A_{2A}AR$ and $A_{2B}AR$ are known to exert immunosuppression during conditions of inflammation, hypoxia, and cellular stress, when endogenous adenosine accumulates extracellularly [228–230]. In conjunction, the TME exhibits high concentrations of adenosine produced by stromal and immune cells, inflammation, and tissue disruption. Due to the lack of perfusion, hypoxia is a predominant driver that can cause cellular stress [231, 232] and secretion of a large amount of ATP [233]. CD39 catalyzes the conversion of ATP and ADP into AMP, whereas CD73 catalyzes the irreversible conversion of AMP into adenosine [234]. CD38 generates cyclic ADP-ribose from NAD^+, which is essential for the regulation of intracellular Ca^{2+} [235]. Tumor cells, NK cells, and T cells all express CD38 which can promote adenosine generation and ultimately suppression of T cell proliferation and function [236]. Adenosine and enzymatically active CD39/CD73 are also carried by tumor-derived exosomes (TEX), which promote cancer progression via angiogenesis stimulation [237]. Thus, the accumulation of adenosine in the TME is predominantly via the catabolism of extracellular ATP to adenosine by CD38, CD39, and CD73, and suppression of anti-tumor immunity is via the activation of ARs [238]. Furthermore, accumulation of adenosine causes a reduced anti-tumoral immune response via adenosine receptors in hypoxic tumoral environments [239]. This promotes hypoxic tumor cell survival and immunoescape [164].

Although the exact inflammatory roles of A_1AR have not been fully characterized in the TME, deletion of A_1AR has recently been demonstrated to upregulate the expression of programmed death-ligand 1 (PD-L1 or CD274) [240]. Therapeutic strategies blocking the interaction between PD-L1 and its receptor, programmed cell death protein 1 (PD-1), have been developed for multiple tumor types [241, 242]. Moreover, the combination of PD-1 monoclonal antibody and selective A_1AR antagonist induces the activity of $CD8^+$ T cells, resulting in increased treatment efficacy for melanoma and NSCLC [240].

It has been proven in a large number of studies that adenosine, through the activation of A_{2A}AR, also participates in T cell anergy induction, Treg response stimulation, and the inhibition of natural killer cell activity. Combined this results in enhanced escape of tumor cells from the immune system with subsequent metastasis formation [243, 244]. In large solid tumors where extracellular adenosine accumulates under hypoxic conditions, A_{2A}AR expressed on the T cell surface inhibits incoming antitumor cytotoxic T lymphocytes from destroying the tumor [245]. Recently, A_{2A}ARs have also been described to protect tumors from anti-tumor T cells, which may form the framework for an A_{2A}AR antagonist-based cancer immunotherapy to prevent the inhibition of anti-tumor T cells in the TME [246].

Similar to A_{2A}AR, A_{2B}AR abrogates immune responses by stimulating the production of cAMP, and thus promoting immunoescape [218]. The pro-tumoral effect of A_{2B}AR is in the activation of myeloid-derived suppressor cells and M2 macrophages, which are essential for proliferation, angiogenesis, and metastasis [218, 247]. Besides, adenosine produced by TEX promotes angiogenesis via A_{2B}AR [237]. Under hypoxic conditions, high levels of adenosine in the TME lead to the release of angiogenic factors via an A_{2B}AR response which further promotes tumor growth [248].

The involvement of the A_3AR in hypoxia, typical of solid tumors [249], has been reported in the TME in an *in vitro* model as well as solid tumors *in vivo* [211]. One of the first lines of evidence showed that activation of A_3AR in mast cells was responsible for the release of allergic mediators and mast cell degranulation [250]. As a result, all these subtypes of ARs provide protection against excessive inflammation and tissue injury. Thus, AR antagonists may be useful for increasing the immune response in the TME and provide a potential approach for cancer treatment, which will be discussed in more details in the next paragraph.

18.4.4 Potential Anti-cancer Therapies Targeting Adenosine Receptors

Multiple antagonistic antibodies and small molecule inhibitors against adenosine receptors have been developed and display therapeutic efficacy in clinical trials against different solid tumors (Table 18.1) [238]. Several $A_{2A}AR$ antagonists have shown promising anti-cancer effects in preclinical models of melanoma, renal cell cancer, TNBC, colorectal, bladder, and prostate cancers [251, 252], which will be discussed below. Interestingly, $A_{2A}AR$ agonists are also being evaluated as a novel pharmacological approach to improve drug delivery to the brain due to their ability to transiently increase BBB permeability, especially for drugs targeting brain tumors [253]. Specifically, using microdialysis in glioblastoma patients regadenoson was tested with increasing doses of the anti-cancer drug temozolomide in the brain interstitium [254]. Therefore, to reduce neurological side effects, anti-cancer molecules targeting $A_{2A}AR$ should be unable to cross the blood–brain barrier [255].

Although the majority of clinical development has targeted $A_{2A}AR$, compounds targeting other subtypes of ARs are also being obtained. Inhibiting the $A_{2B}AR$ receptor with antagonists has been reported to reduce tumor metastasis load and to inhibit the growth of melanoma, breast, and prostate cancer in mouse models [217, 256, 257]. Whereas the importance of $A_{2A}AR$ and $A_{2B}AR$ compounds in chemotherapy has slowly emerged, the potential role of A_1AR or A_3AR agonists/antagonists still remains to be elucidated in preclinical studies, either on their own or in combination treatment.

Cancer immunotherapies are now considered as a pillar of cancer treatment [238]. Blockade of checkpoint receptor can cause durable responses in various cancers; however, this treatment does not work for all patients. Further investigations are needed to unravel the mechanisms of tumor evasion and to identify other potential targets that can overcome these "brakes" on the immune response [238]. Adenosine is known as a negative regulator of NK cell and T cell responses in the TME via $A_{2A}AR$, thus, targeting this $A_{2A}AR$-involved pathway in the clinic may be beneficial for increasing the efficacy of immunotherapies. This hypothesis was substantiated in a study showing that in an $A_{2A}AR$ knock-out mouse model the activation or increase of T cells in the TME was induced, leading to potently induced anti-tumoral effects [258, 259]. The combination of anti-PD-1 with inhibition of CD73, $A_{2A}AR$, or CD38 leads to stimulated anti-tumor T cell responses mediated by induced expression of Granzyme B and IFNγ in CD8+ T cells [218, 260, 261]. To further increase anti-tumoral effects, dual checkpoint blockade was combined with $A_{2A}AR$ knock-out in T cells, which led to better penetration into hypoxic tumors [231]. The majority of combination approaches has long been focused on blocking the function of adenosine in order to increase T cell responses within the TME, while the adenosine axis within NK cells can also be targeted [218]. Hence, future combination approaches may explore compounds that stimulate other subtypes of immune cells modulated by adenosine, potentially via one of the other adenosine receptor subtypes.

18.4.5 Adenosine Receptor Mutations in Cancer

Although abnormal functions of the ARs are observed in cancer tissues, the role of AR mutations present in human cancers is not well characterized yet. Herein, we will take the A_{2B}AR as an example to discuss the effects of cancer-related mutations on receptor pharmacology.

To gather a selection of interesting cancer-related mutations, insight in the ligand-receptor and receptor-G protein interaction of these receptors is essential. Previous studies focusing on site-directed mutagenesis and docking studies of the A_1AR, A_{2A}AR, and A_{2B}AR have identified residues involved in functional activity and ligand recognition [151, 262–269]. Alanine scanning is generally used to examine the contribution of a single amino acid to receptor pharmacology. Substitution by alanine represents a replacement of the side chain at the β-carbon by a methyl group, thus removing the properties and functionality of the original amino acid [270]. Although alanine scanning is a useful tool to identify pharmacologically important positions [271], cancer-related mutations rarely involve a simple alanine substitution. Therefore, the influence of cancer-related somatic mutations on ligand binding and functional activity is likely more intricate.

Recently, cancer-related mutations of A_{2B}AR, derived from TCGA, have been characterized in an engineered yeast system [272]. Mutations with dramatic impact in receptor pharmacology were found in highly conserved residues, as well as residues that were directly involved in receptor-ligand and receptor-G protein interaction. Interestingly, none of these cancer-related mutations were found to match the natural variants of A_{2B}AR existing in the human population, indicating that they could indeed be cancer-specific. Yet, data obtained from yeast assays might not be representative for mammalian species. Therefore, further studies in mammalian cells are needed to provide insight in receptor pharmacology and ligand binding of these cancer-related mutations in physiological and/or pathological conditions. Since adenosine, as pointed out above, is an anti-inflammatory stimulus in the tumor microenvironment [164], both wildtype and mutant adenosine receptors may play an important, yet largely undefined role in cancer progression, which eventually may be modulated with medicinal products.

18.5 Conclusion

Although a large amount of evidence supports the role of GPCRs in all stages of cancer progression, the knowledge of genetically altered variants in G proteins and GPCRs was initially limited to only a few neoplastic lesions, mainly endocrine tumors. Currently, GPCRs together with their downstream signaling pathways have been directly targeted for anti-cancer treatments. However, a pharmacological characterization is still needed to improve anti-cancer drug design. Investigation of cancer-related inflammation and GPCRs involved therein will provide a better understanding of the immune-related mechanisms of cancer development. In numerous different cancer types, GPCRs as well as viral GPCRs, have been shown to be involved in multiple hallmarks of cancer, suggesting their importance in cancer progression. As an example, adenosine and adenosine receptors are known to regulate the immune response in the TME, while they are also directly involved in cancer hallmarks. Additionally, cancer-specific mutations influencing receptor pharmacology may also alter the cancer-related roles of GPCRs. Thus, drug discovery targeting both wildtype and mutant (adenosine) receptors might be a novel cancer-therapeutic approach. Overall, while being targets of many drugs on the market already, GPCRs are promising targets for developing novel strategies for cancer prevention and treatment.

References

1 Rask-Andersen, M., Masuram, S., and Schiöth, H.B. (2014). The druggable genome: evaluation of drug targets in clinical trials suggests major shifts in molecular class and indication. *Annu. Rev. Pharmacol. Toxicol.* 54: 9–26.

2 Vassilatis, D.K., Hohmann, J.G., Zeng, H. et al. (2003). The G protein-coupled receptor repertoires of human and mouse. *Proc. Natl. Acad. Sci. U.S.A.* 100: 4903–4908.

3 Fredriksson, R., Lagerström, M.C., Lundin, L.-G., and Schiöth, H.B. (2003). The G-protein-coupled receptors in the human genome form five main families. Phylogenetic analysis, paralogon groups, and fingerprints. *Mol. Pharmacol.* 63: 1256–1272.

4 Kolakowski, L.F.J. (1994). GCRDb: a G-protein-coupled receptor database. *Recept. Channels* 2: 1–7.

5 Simon, M., Strathmann, M., and Gautam, N. (1991). Diversity of G proteins in signal transduction. *Science* 252: 802–808.

6 Hauser, A.S., Attwood, M.M., Rask-Andersen, M. et al. (2017). Trends in GPCR drug discovery: new agents, targets and indications. *Nat. Rev. Drug Discovery* 16: 829–842.

7 Yu, S., Sun, L., Jiao, Y., and Lee, L.T.O. (2018). The role of G protein-coupled receptor kinases in cancer. *Int. J. Biol. Sci.* 14: 189–203.

8 O'Connell, M.P. and Weeraratna, A.T. (2009). Hear the Wnt Ror: how melanoma cells adjust to changes in Wnt. *Pigm. Cell Melanoma Res.* 22: 724–739.

9 Chien, A.J., Conrad, W.H., and Moon, R.T. (2009). A Wnt survival guide: from flies to human disease. *J. Invest. Dermatol.* 129: 1614–1627.

10 Guo, L. and Teng, L. (2015). YAP/TAZ for cancer therapy: opportunities and challenges (review). *Int. J. Oncol.* 46: 1444–1452.

11 Yu, F.X., Zhao, B., Panupinthu, N. et al. (2012). Regulation of the Hippo-YAP pathway by G-protein-coupled receptor signaling. *Cell* 150: 780–791.

12 Zhou, Y., Tong, L., Chu, X. et al. (2017). The adenosine A_1 receptor antagonist DPCPX inhibits tumor progression via the ERK/JNK pathway in renal cell carcinoma. *Cell. Physiol. Biochem.* 43: 733–742.

13 Hollenberg, M.D., Mihara, K., Polley, D. et al. (2014). Biased signalling and proteinase-activated receptors (PARs): targeting inflammatory disease. *Br. J. Pharmacol.* 171: 1180–1194.

14 Kenakin, T. (2012). The potential for selective pharmacological therapies through biased receptor signaling. *BMC Pharmacol. Toxicol.* 13: 1–8.

15 Cattaneo, F., Guerra, G., Parisi, M. et al. (2014). Cell-surface receptors transactivation mediated by G protein-coupled receptors. *Int. J. Mol. Sci.* 15: 19700–19728.

16 Heitzler, D., Crépieux, P., Poupon, A. et al. (2009). Towards a systems biology approach of G protein-coupled receptor signalling: challenges and expectations. *C.R. Biol.* 332: 947–957.

17 Lagerström, M.C. and Schiöth, H.B. (2008). Structural diversity of G protein-coupled receptors and significance for drug discovery. *Nat. Rev. Drug Discovery* 7: 339–357.

18 Lynch, J.R. and Wang, J.Y. (2016). G protein-coupled receptor signaling in stem cells and cancer. *Int. J. Mol. Sci.* 17.

19 O'Hayre, M., Degese, M.S., and Gutkind, J.S. (2014). Novel insights into G protein and G protein-coupled receptor signaling in cancer. *Curr. Opin. Cell Biol.* 27: 126–135.

20 Sever, R. and Brugge, J.S. (2015). Signal transduction in cancer. *Cold Spring Harb. Perspect. Med.* 5.

21 Dorsam, R.T. and Gutkind, J.S. (2007). G-protein-coupled receptors and cancer. *Nat. Rev. Cancer* 7: 79–94.

22 Home – http://ClinicalTrials.gov, https://clinicaltrials.gov/ct2/home (accessed 20 August 2020).

23 Usman, S., Khawer, M., Rafique, S. et al. (2020). The current status of anti GPCR drugs against different cancers. *J. Pharm. Anal.* 10: 517–521.

24 Hanahan, D. and Weinberg, R.A. (2011). Hallmarks of cancer: the next generation. *Cell* 144: 646–674.

25 Lappano, R. and Maggiolini, M. (2011). G protein-coupled receptors: novel targets for drug discovery in cancer. *Nat. Rev. Drug Discovery* 10: 47–60.

26 Nieto Gutierrez, A. and McDonald, P.H. (2018). GPCRs: emerging anti-cancer drug targets. *Cell. Signalling* 41: 65–74.

27 Audet, M. and Bouvier, M. (2012). Restructuring G-protein-coupled receptor activation. *Cell* 151: 14–23.

28 Premont, R.T., Macrae, A.D., Stoffel, R.H. et al. (1996). Characterization of the G protein-coupled receptor kinase GRK4: identification of four splice variants. *J. Biol. Chem.* 271: 6403–6410.

29 Hall, R.A., Spurney, R.F., Premont, R.T. et al. (1999). G protein-coupled receptor kinase 6A phosphorylates the Na^+/H^+ exchanger regulatory factor via a PDZ domain-mediated interaction. *J. Biol. Chem.* 274: 24328–24334.

30 Penela, P., Murga, C., Ribas, C. et al. (2010). The complex G protein-coupled receptor kinase 2 (GRK2) interactome unveils new physiopathological targets. *Br. J. Pharmacol.* 160: 821–832.

31 Ma, Y., Han, C.C., Huang, Q. et al. (2016). GRK2 overexpression inhibits IGF1-induced proliferation and migration of human hepatocellular carcinoma cells by downregulating EGR1. *Oncol. Rep.* 35: 3068–3074.

32 Miyagawa, Y., Ohguro, H., Odagiri, H. et al. (2003). Aberrantly expressed recoverin is functionally associated with G-protein-coupled receptor kinases in cancer cell lines. *Biochem. Biophys. Res. Commun.* 300: 669–673.

33 Billard, M.J., Fitzhugh, D.J., Parker, J.S. et al. (2016). G protein coupled receptor kinase 3 regulates breast cancer migration, invasion, and metastasis. *PLoS One* 11: 1–23.

34 Gurevich, E.V., Tesmer, J.J.G., Mushegian, A., and Gurevich, V.V.G. (2012). Protein-coupled receptor kinases: more than just kinases and not only for GPCRs. *Pharmacol. Ther.* 133: 40–69.

35 Maeda, T., Maeda, A., Maruyama, I. et al. (2001). Mechanisms of photoreceptor cell death in cancer-associated retinopathy. *Investig. Ophthalmol. Vis. Sci.* 42: 705–712.

36 Zhou, L., Wang, M.Y., Liang, Z.Y. et al. (2016). G-protein-coupled receptor kinase 2 in pancreatic cancer: clinicopathologic and prognostic significance. *Hum. Pathol.* 56: 171–177.

37 Métayé, T., Menet, E., Guilhot, J., and Kraimps, J.L. (2002). Expression and activity of G protein-coupled receptor kinases in differentiated thyroid carcinoma. *J. Clin. Endocrinol. Metab.* 87: 3279–3286.

38 Wei, Z., Hurtt, R., Gu, T. et al. (2013). GRK2 negatively regulates IGF-1R signaling pathway and cyclins' expression in HepG2 cells. *J. Cell. Physiol.* 228: 1897–1901.

39 Hu, M., Wang, C., Li, W. et al. (2015). A KSHV microRNA directly targets G protein-coupled receptor kinase 2 to promote the migration and invasion of endothelial cells by inducing CXCR2 and activating AKT signaling. *PLoS Pathog.* 11: 1–27.

40 Nogués, L., Reglero, C., Rivas, V. et al. (2016). G protein-coupled receptor kinase 2 (GRK2) promotes breast tumorigenesis through a HDAC6-Pin1 axis. *eBioMedicine* 13: 132–145.

41 Nakata, H., Kinoshita, Y., Kishi, K. et al. (1996). Involvement of β-adrenergic receptor kinase-1 in homologous desensitization of histamine H_2 receptors in human gastric carcinoma cell line MKN-45. *Digestion* 57: 406–410.

42 Fitzhugh, D.J., McGinnis, M.W., Timoshchenko, R.G. et al. (2011). G protein coupled receptor kinase 3 (GRK3) negatively regulates CXCL12/CXCR4 signaling and tumor migration in breast cancer. *J. Allergy Clin. Immunol.* 127: AB230.

43 Li, W., Ai, N., Wang, S. et al. (2014). GRK3 is essential for metastatic cells and promotes prostate tumor progression. *Proc. Natl. Acad. Sci. U.S.A.* 111: 1521–1526.

44 Dautzenberg, F.M., Braun, S., and Hauger, R.L. (2001). GRK3 mediates desensitization of CRF_1 receptors: a potential mechanism regulating stress adaptation. *Am. J. Physiol. Regul. Integr. Comp. Physiol.* 280: R935–R946.

45 Brenninkmeijer, C.B.A.P., Price, S.A., López Bernal, A., and Phaneuf, S. (1999). Expression of G-protein-coupled receptor kinases in pregnant term and non-pregnant human myometrium. *J. Endocrinol.* 162: 401–408.

46 King, D.W., Steinmetz, R., Wagoner, H.A. et al. (2003). Differential expression of GRK isoforms in nonmalignant and malignant human granulosa cells. *Endocrine* 22: 135–141.

47 Kaur, G., Kim, J., Kaur, R. et al. (2013). G-protein coupled receptor kinase (GRK)-5 regulates proliferation of glioblastoma-derived stem cells. *J. Clin. Neurosci.* 20: 1014–1018.

48 Malbon, C.C. (2004). Frizzleds: new members of the superfamily of G-protein-coupled receptors. *Front. Biosci.* 9: 1048–1058.

49 Li, Y.P. (2013). GRK6 expression in patients with hepatocellular carcinoma. *Asian Pac. J. Trop. Med.* 6: 220–223.

50 Behrens, J. and Lustig, B. (2004). The Wnt connection to tumorigenesis. *Int. J. Dev. Biol.* 48: 477–487.

51 Zhan, T., Rindtorff, N., and Boutros, M. (2017). Wnt signaling in cancer. *Oncogene* 36: 1461–1473.

52 Anastas, J.N. and Moon, R.T. (2013). Wnt signalling pathways as therapeutic targets in cancer. *Nat. Rev. Cancer* 13: 11–26.

53 Bar-Shavit, R., Maoz, M., Kancharla, A. et al. (2016). G protein-coupled receptors in cancer. *Int. J. Mol. Sci.* 17: 1320.

54 Nusse, R. (2005). Wnt signaling in disease and in development. *Cell Res.* 15: 28–32.

55 Korinek, V., Barker, N., Morin, P.J. et al. (1997). Constitutive transcriptional activation by a β-catenin-Tcf complex in APC(−/−) colon carcinoma. *Science* 275: 1784–1787.

56 Gurney, A., Axelrod, F., Bond, C.J. et al. Wnt pathway inhibition via the targeting of frizzled receptors results in decreased growth and tumorigenicity of human tumors. *Proc. Natl. Acad. Sci. U.S.A.*

57 Praskova, M., Xia, F., and Avruch, J. (2008). MOBKL1A/MOBKL1B phosphorylation by MST1 and MST2 inhibits cell proliferation. *Curr. Biol.* 18: 311–321.

58 Wu, S., Huang, J., Dong, J., and Pan, D. (2003). hippo encodes a Ste-20 family protein kinase that restricts cell proliferation and promotes apoptosis in conjunction with salvador and warts. *Cell* 114: 445–456.

59 Mo, J.-S., Park, H.W., and Guan, K.-L. (2014). The hippo signaling pathway and cancer. *EMBO Rep.* 15: 642–656.

60 Zeng, Q. and Hong, W. (2008). The emerging role of the hippo pathway in cell contact inhibition, organ size control, and cancer development in mammals. *Cancer Cell* 13: 188–192.

61 Zanconato, F., Cordenonsi, M., and Piccolo, S. (2016). YAP/TAZ at the roots of cancer. *Cancer Cell* 29: 783–803.

62 Bao, Y., Nakagawa, K., Yang, Z. et al. (2011). A cell-based assay to screen stimulators of the Hippo pathway reveals the inhibitory effect of dobutamine on the YAP-dependent gene transcription. *J. Biochem.* 150: 199–208.

63 Mantovani, A., Allavena, P., Sica, A., and Balkwill, F. (2008). Cancer-related inflammation. *Nature* 454: 436–444.

64 Balkwill, F. and Mantovani, A. (2001). Inflammation and cancer: back to Virchow? *Lancet* 357: 539–545.

65 Balkwill, F. (2004). Cancer and the chemokine network. *Nat. Rev. Cancer* 4: 540–550.

66 Barbieri, F., Bajetto, A., and Florio, T. (2010). Role of chemokine network in the development and progression of ovarian cancer: a potential novel pharmacological target. *J. Oncol.* 2010: 426956.

67 Rollins, B.J. (2006). Inflammatory chemokines in cancer growth and progression. *Eur. J. Cancer* 42: 760–767.

68 Aldinucci, D. and Colombatti, A. (2014). The inflammatory chemokine CCL5 and cancer progression. *Mediators Inflammation* 2014: 1–12.

69 Kershaw, M.H., Westwood, J.A., and Darcy, P.K. (2013). Gene-engineered T cells for cancer therapy. *Nat. Rev. Cancer* 13: 525–541.

70 Brown, J.R. and DuBois, R.N. (2005). COX-2: a molecular target for colorectal cancer prevention. *J. Clin. Oncol.* 23: 2840–2855.

71 Gupta, R.A. and Dubois, R.N. (2001). Colorectal cancer prevention of cyclooxygenase-2. *Nat. Rev. Cancer* 1: 11–21.

72 Kim, T.H., Xiong, H., Zhang, Z., and Ren, B. (2005). β-Catenin activates the growth factor endothelin-1 in colon cancer cells. *Oncogene* 24: 597–604.

73 Mazhar, D., Ang, R., and Waxman, J. (2006). COX inhibitors and breast cancer. *Br. J. Cancer* 94: 346–350.

74 Lin, D.T., Subbaramaiah, K., Shah, J.P. et al. (2002). Cyclooxygenase-2: a novel molecular target for the prevention and treatment of head and neck cancer. *Head Neck* 24: 792–799.

75 Bresalier, R.S., Sandler, R.S., Quan, H. et al. (2005). Cardiovascular events associated with rofecoxib in a colorectal adenoma chemoprevention trial: commentary. *Dis. Colon Rectum* 48: 1330–1331.

76 Hull, M.A., Ko, S.C.W., and Hawcroft, G. (2004). Prostaglandin EP receptors: targets for treatment and prevention of colorectal cancer? *Mol. Cancer Ther.* 3: 1031–1039.

77 Hansen-Petrik, M.B., McEntee, M.F., Jull, B. et al. (2002). Prostaglandin E2 protects intestinal tumors from nonsteroidal anti-inflammatory drug-induced regression in Apcmin/+ mice. *Cancer Res.* 62: 403–408.

78 Sonoshita, M., Takaku, K., Sasaki, N. et al. (2001). Acceleration of intestinal polyposis through prostaglandin receptor EP2 in ApcΔ716 knockout mice. *Nat. Med.* 7: 1048–1051.

79 Watanabe, K., Kawamori, T., Nakatsugi, S. et al. (2000). Inhibitory effect of a prostaglandin E receptor subtype EP_1 selective antagonist, ONO-8713, on development of azoxymethane-induced aberrant crypt foci in mice. *Cancer Lett.* 156: 57–61.

80 Kim, J.I., Lakshmikanthan, V., Frilot, N., and Daaka, Y. (2010). Prostaglandin E2 promotes lung cancer cell migration via EP4-βArrestin1-c-Src signalsome. *Mol. Cancer Res.* 8: 569–577.

81 Hida, T., Yatabe, Y., Achiwa, H. et al. (1998). Increased expression of cyclooxygenase 2 occurs frequently in human lung cancers, specifically in adenocarcinomas. *Cancer Res.* 58: 3761–3764.

82 Chang, S.H., Ai, Y., Breyer, R.M. et al. (2005). The prostaglandin E2 receptor EP_2 is required for cyclooxygenase 2-mediated mammary hyperplasia. *Cancer Res.* 65: 4496–4499.

83 Liu, C.H., Chang, S.H., Narko, K. et al. (2001). Overexpression of cyclooxygenase-2 is sufficient to induce tumorigenesis in transgenic mice. *J. Biol. Chem.* 276: 18563–18569.

84 Kikuchi, A. (2003). Tumor formation by genetic mutations in the components of the Wnt signaling pathway. *Cancer Sci.* 94: 225–229.

85 Castellone, M.D., Teramoto, H., Williams, B.O. et al. (2005). Prostaglandin E2 promotes colon cancer cell growth through a Gs-axin-β-catenin signaling axis. *Science* 310: 1504–1510.

86 Shao, J., Jung, C., Liu, C., and Sheng, H. (2005). Prostaglandin E2 stimulates the β-catenin/T cell factor-dependent transcription in colon cancer. *J. Biol. Chem.* 280: 26565–26572.

87 Kamiyama, M., Pozzi, A., Yang, L. et al. (2006). EP_2, a receptor for PGE2, regulates tumor angiogenesis through direct effects on endothelial cell motility and survival. *Oncogene* 25: 7019–7028.

88 Taniguchi, T., Fujino, H., Israel, D.D. et al. (2008). Human EP3 prostanoid receptor induces VEGF and VEGF receptor-1 mRNA expression. *Biochem. Biophys. Res. Commun.* 377: 1173–1178.

89 Kubo, H., Hosono, K., Suzuki, T. et al. (2010). Host prostaglandin EP_3 receptor signaling relevant to tumor-associated lymphangiogenesis. *Biomed. Pharmacother.* 64: 101–106.

90 Amano, H., Ito, Y., Suzuki, T. et al. (2009). Roles of a prostaglandin E-type receptor, EP_3, in upregulation of matrix metalloproteinase-9 and vascular endothelial growth factor during enhancement of tumor metastasis. *Cancer Sci.* 100: 2318–2324.

91 Even-Ram, S., Uziely, B., Cohen, P. et al. (1998). Thrombin receptor overexpression in malignant and physiological invasion processes. *Nat. Med.* 4: 909–914.

92 Daaka, Y. (2004). G proteins in cancer: the prostate cancer paradigm. *Sci. STKE* 2004: 1–11.

93 Darash-Yahana, M., Pikarsky, E., Abramovitch, R. et al. (2004). Role of high expression levels of CXCR4 in tumor growth, vascularization, and metastasis. *FASEB J.* 18: 1240–1242.

94 Müller, A., Homey, B., Soto, H. et al. (2001). Involvement of chemokine receptors in breast cancer metastasis. *Nature* 410: 50–56.

95 Dhawan, P. and Richmond, A. (2002). Role of CXCL1 in tumorigenesis of melanoma. *J. Leukocyte Biol.* 72: 9–18.

96 Balkwill, F. (2004). The significance of cancer cell expression of the chemokine receptor CXCR4. *Semin. Cancer Biol.* 14: 171–179.

97 Kakinuma, T. (2006). Chemokines, chemokine receptors, and cancer metastasis. *J. Leukocyte Biol.* 79: 639–651.

98 Mills, G.B. and Moolenaar, W.H. (2003). The emerging role of lysophosphatidic acid in cancer. *Nat. Rev. Cancer* 3: 582–591.

99 Takuwa, Y. (2002). Subtype-specific differential regulation of Rho family G proteins and cell migration by the Edg family sphingosine-1-phosphate receptors. *Biochim. Biophys. Acta, Mol. Cell. Biol. Lipids* 1582: 112–120.

100 Balthasar, S., Samulin, J., Ahlgren, H. et al. (2006). Sphingosine 1-phosphate receptor expression profile and regulation of migration in human thyroid cancer cells. *Biochem. J.* 398: 547–556.

101 Akao, Y., Banno, Y., Nakagawa, Y. et al. (2006). High expression of sphingosine kinase 1 and S1P receptors in chemotherapy-resistant prostate cancer

PC3 cells and their camptothecin-induced up-regulation. *Biochem. Biophys. Res. Commun.* 342: 1284–1290.

102 Yamashita, H., Kitayama, J., Shida, D. et al. (2006). Sphingosine 1-phosphate receptor expression profile in human gastric cancer cells: differential regulation on the migration and proliferation. *J. Surg. Res.* 130: 80–87.

103 Chambers, A.F., Groom, A.C., and MacDonald, I.C. (2002). Dissemination and growth of cancer cells in metastatic sites. *Nat. Rev. Cancer* 2: 563–572.

104 Coughlin, S.R. (2000). Thrombin signalling and protease-activated receptors. *Nature* 407: 258–264.

105 Morris, D.R., Ding, Y., Ricks, T.K. et al. (2006). Protease-activated receptor-2 is essential for factor VIIa and Xa-induced signaling, migration, and invasion of breast cancer cells. *Cancer Res.* 66: 307–314.

106 Versteeg, H.H., Schaffner, F., Kerver, M. et al. (2008). Inhibition of tissue factor signaling suppresses tumor growth. *Blood* 111: 190–199.

107 Kato, M., Kitayama, J., Kazama, S., and Nagawa, H. (2003). Expression pattern of CXC chemokine receptor-4 is correlated with lymph node metastasis in human invasive ductal carcinoma. *Breast Cancer Res.* 5.

108 Cardones, A.R., Murakami, T., and Hwang, S.T. (2003). CXCR4 enhances adhesion of B16 tumor cells to endothelial cells in vitro and in vivo via β1 integrin. *Cancer Res.* 63: 6751–6757.

109 Kleeff, J., Kusama, T., Rossi, D.L. et al. (1999). Detection and localization of MIP-3α/LARC/Exodus, a macrophage proinflammatory chemokine, and its CCR6 receptor in human pancreatic cancer. *Int. J. Cancer* 81: 650–657.

110 Lee, Z., Swaby, R.F., Liang, Y. et al. (2006). Lysophosphatidic acid is a major regulator of growth-regulated oncogene α in ovarian cancer. *Cancer Res.* 66: 2740–2748.

111 Slinger, E., Langemeijer, E., Siderius, M. et al. (2011). Herpesvirus-encoded GPCRs rewire cellular signaling. *Mol. Cell. Endocrinol.* 331: 179–184.

112 White, M.K., Gorrill, T.S., and Khalili, K. (2006). Reciprocal transactivation between HIV-1 and other human viruses. *Virology* 352: 1–13.

113 Jenkins, F.J., Rowe, D.T., and Rinaldo, C.R. (2003). Herpesvirus infections in organ transplant recipients. *Clin. Diagn. Lab Immunol.* 10: 1–7.

114 Vischer, H.F., Siderius, M., Leurs, R., and Smit, M.J. (2014). Herpesvirus-encoded GPCRs: neglected players in inflammatory and proliferative diseases? *Nat. Rev. Drug Discovery* 13: 123–139.

115 Zhang, J., Feng, H., Xu, S., and Feng, P. (2016). Hijacking GPCRs by viral pathogens and tumor. *Biochem. Pharmacol.* 114: 69–81.

116 Feng, H., Shuda, M., Chang, Y., and Moore, P.S. (2008). Clonal integration of a polyomavirus in human Merkel cell carcinoma. *Science* 319: 1096–1100.

117 Martin, D. and Gutkind, J.S. (2008). Human tumor-associated viruses and new insights into the molecular mechanisms of cancer. *Oncogene* 27: S31–S42.

118 Montaner, S., Kufareva, I., Abagyan, R., and Gutkind, J.S. (2013). Molecular mechanisms deployed by virally encoded G protein–coupled receptors in human diseases. *Annu. Rev. Pharmacol. Toxicol.* 53: 331–354.

119 Arvanitakls, L., Geras-raakat, E., Varmat, A. et al. (1997). Human herpesvirus KSHV active G-protein-coupled proliferation. *Nature* 385: 347–350.

120 Sodhi, A., Montaner, S., and Gutkind, J.S. (2004). Viral hijacking of G-protein-coupled-receptor signalling networks. *Nat. Rev. Mol. Cell Biol.* 5: 998–1012.

121 Mitra, D., Luo, X., Morgan, A. et al. (2012). An ultraviolet-radiation-independent pathway to melanoma carcinogenesis in the red hair/fair skin background. *Nature* 491: 449–453.

122 Velasco, G., Sánchez, C., and Guzmán, M. (2012). Towards the use of cannabinoids as antitumour agents. *Nat. Rev. Cancer* 12: 436–444.

123 Green, J.A., Suzuki, K., Cho, B. et al. (2011). The sphingosine 1-phosphate receptor S1P2 maintains the homeostasis of germinal center B cells and promotes niche confinement. *Nat. Immunol.* 12: 672–680.

124 Lee, J.H., Miele, M.E., Hicks, D.J. et al. (1996). *KiSS-1*, a novel human malignant melanoma metastasis-suppressor gene. *J. Natl. Cancer Inst.* 88: 1731–1737.

125 O'Hayre, M., Vázquez-Prado, J., Kufareva, I. et al. (2013). The emerging mutational landscape of G proteins and G-protein-coupled receptors in cancer. *Nat. Rev. Cancer* 13: 412–424.

126 Kan, Z., Jaiswal, B.S., Stinson, J. et al. (2010). Diverse somatic mutation patterns and pathway alterations in human cancers. *Nature* 466: 869–873.

127 Broad Institute TCGA Genome Data Analysis Center (2016). Analysis-ready standardized TCGA data from Broad GDAC Firehose stddata_2015_08_21 run. Broad Institute of MIT and Harvard.

128 Jensen, M.A., Ferretti, V., Grossman, R.L., and Staudt, L.M. (2017). The NCI genomic data commons as an engine for precision medicine. *Blood* 130: 453–459.

129 Finch, A.M., Sarramegna, V., and Graham, R.M. (2006). Ligand binding, activation, and agonist trafficking. In: *The Adrenergic Receptors* (ed. D. Perez), 25–85. Totowa, NJ: Humana Press.

130 Peeters, M.C., van Westen, G.J.P., Guo, D. et al. (2011). GPCR structure and activation: an essential role for the first extracellular loop in activating the adenosine A_{2B} receptor. *FASEB J.* 25: 632–643.

131 Salon, J., Lodowski, D.T., and Palczewski, K. (2011). The significance of G protein-coupled receptor. *Pharmacol. Rev.* 63: 901–937.

132 Chang, S.D. and Bruchas, M.R. (2014). Functional selectivity at GPCRs: new opportunities in psychiatric drug discovery. *Neuropsychopharmacology* 39: 248–249.

133 Leff, P. (1995). The two-state model of receptor activation. *Trends Pharmacol. Sci.* 16: 89–97.

134 Seifert, R. and Wenzel-Seifert, K. (2002). Constitutive activity of G-protein-coupled receptors: cause of disease and common property of wild-type receptors. *Naunyn-Schmiedeberg's Arch. Pharmacol.* 366: 381–416.

135 Parnot, C., Miserey-Lenkei, S., Bardin, S. et al. (2002). Lessons from constitutively active mutants of G protein-coupled receptors. *Trends Endocrinol. Metab.* 13: 336–343.

136 Vassart, G., Pardo, L., and Costagliola, S. (2006). A molecular dissection of the glycoprotein hormone receptors. In: *Insights into Recept.* Funct. New Drug Dev. Targets (ed. M. Conn, C. Kordon and Y. Christen), 151–166. Berlin, Heidelberg: Springer Berlin Heidelberg.

137 Lebon, G., Warne, T., and Tate, C.G. (2012). Agonist-bound structures of G protein-coupled receptors. *Curr. Opin. Struct. Biol.* 22: 482–490.

138 Parma, J., Duprez, L., Van Sande, J. et al. (1993). Somatic mutations in the thyrotropin receptor gene cause hyperfunctioning thyroid adenomas. *Nature* 365: 649–651.

139 Tao, Y.X. and Segaloff, D.L. (2005). Functional analyses of melanocortin-4 receptor mutations identified from patients with binge eating disorder and nonobese or obese subjects. *J. Clin. Endocrinol. Metab.* 90: 5632–5638.

140 Gromoll, J., Simoni, M., Nordhoff, V. et al. (1996). Functional and clinical consequences of mutations in the FSH receptor. *Mol. Cell. Endocrinol.* 125: 177–182.

141 Laue, L., Wu, S.M., Kudo, M. et al. (1996). Heterogeneity of activating mutations of the human luteinizing hormone receptor in male-limited precocious puberty. *Biochem. Mol. Med.* 58: 192–198.

142 Rasmussen, S.G.F., Choi, H.J., Fung, J.J. et al. (2011). Structure of a nanobody-stabilized active state of the β_2 adrenoceptor. *Nature* 469: 175–181.

143 Wonerow, P., Chey, S., Führer, D. et al. (2000). Functional characterization of five constitutively activating thyrotrophin receptor mutations. *Clin. Endocrinol. (Oxf)* 53: 461–468.

144 Parma, J., Duprez, L., Van Sande, J. et al. (1997). Diversity and prevalence of somatic mutations in the thyrotropin receptor and $G_{s\alpha}$ genes as a cause of toxic thyroid adenomas. *J. Clin. Endocrinol. Metab.* 82: 2695–2701.

145 Stoy, H. and Gurevich, V.V. (2015). How genetic errors in GPCRs affect their function: possible therapeutic strategies. *Genes Dis.* 2: 108–132.

146 Tao, Y.X. (2006). Inactivating mutations of G protein-coupled receptors and diseases: structure-function insights and therapeutic implications. *Pharmacol. Ther.* 111: 949–973.

147 Ward, N.A., Hirst, S., Williams, J., and Findlay, J.B.C. (2012). Pharmacological chaperones increase the cell-surface expression of intracellularly retained mutants of the melanocortin 4 receptor with unique rescuing efficacy profiles. *Biochem. Soc. Trans.* 40: 717–720.

148 Bichet, D.G. (2009). Chapter 2 V2R mutations and nephrogenic diabetes insipidus. *Prog. Mol. Biol. Transl. Sci.* 89: 15–29.

149 Venkatakrishnan, A.J., Deupi, X., Lebon, G. et al. (2013). Molecular signatures of G-protein-coupled receptors. *Nature* 494: 185–194.

150 Ulloa-Aguirre, A., Dias, J.A., Bousfield, G. et al. (2013). *Trafficking of the follitropin receptor*, 1e, vol. 521. Elsevier.

151 Jespers, W., Schiedel, A.C., Heitman, L.H. et al. (2017). Structural mapping of adenosine receptor mutations: ligand binding and signaling mechanisms. *Trends Pharmacol. Sci.* 39: 75–89.

152 Rubin, L.L. and de Sauvage, F.J. (2006). Targeting the Hedgehog pathway in cancer. *Nat. Rev. Drug Discovery* 5: 1026–1033.

153 Epstein, E.H. (2008). Basal cell carcinomas: attack of the hedgehog. *Nat. Rev. Cancer* 8: 743–754.

154 Xie, J., Murone, M., Luoh, S. et al. (1998). Mutations in sporadic basal-cell carcinoma. *Nature* 391: 90–92.

155 Lum, L. and Beachy, P.A. (2004). The hedgehog response network: sensors, switches, and routers. *Science* 304: 1755–1759.

156 Xie, J., Murone, M., Luoh, S.M. et al. (1998). Activating smoothened mutations in sporadic basal-cell carcinoma. *Nature* 391: 90–92.

157 Clark, V.E., Zeynep Erson-Omay, E., Serin, A. et al. (2013). Genomic analysis of non-NF2 meningiomas reveals mutations in TRAF7, KLF4, AKT1, and SMO. *Science* 339: 1077–1081.

158 Brastianos, P.K., Horowitz, P.M., Santagata, S. et al. (2013). Genomic sequencing of meningiomas identifies oncogenic SMO and AKT1 mutations. *Nat. Genet.* 45: 285–289.

159 Scales, S.J. and de Sauvage, F.J. (2009). Mechanisms of Hedgehog pathway activation in cancer and implications for therapy. *Trends Pharmacol. Sci.* 30: 303–312.

160 Prickett, T.D., Wei, X., Cardenas-Navia, I. et al. (2011). Exon capture analysis of G protein-coupled receptors identifies activating mutations in GRM3 in melanoma. *Nat. Genet.* 43: 1119–1126.

161 Teh, J.L.F. and Chen, S. (2011). Glutamatergic signaling in cellular transformation. *Pigm. Cell Melanoma Res.* 25: 331–342.

162 Paavola, K.J. and Hall, R.A. (2012). Adhesion G protein-coupled receptors: signaling, pharmacology, and mechanisms of activation. *Mol. Pharmacol.* 82: 777–783.

163 Schuijers, J. and Clevers, H. (2012). Adult mammalian stem cells: the role of Wnt, Lgr5 and R-spondins. *EMBO J.* 31: 2685–2696.

164 Gessi, S., Merighi, S., Sacchetto, V. et al. (2011). Adenosine receptors and cancer. *Biochim. Biophys. Acta, Biomembr.* 1808: 1400–1412.

165 Jockers, R., Linder, M.E., Hohenegger, M. et al. (1994). Species difference in the G protein selectivity of the human and bovine A_1-adenosine receptor. *J. Biol. Chem.* 269: 32077–32084.

166 Palmer, T.M., Gettys, T.W., and Stiles, G.L. (1995). Differential interaction with and regulation of multiple G-proteins by the rat A_3 adenosine receptor. *J. Biol. Chem.* 270: 16895–16902.

167 Freund, S., Ungerer, M., and Lohse, M.J. (1994). A_1 adenosine receptors expressed in CHO-cells couple to adenylyl cyclase and to phospholipase C. *Naunyn-Schmiedeberg's Arch. Pharmacol.* 350: 49–56.

168 Zhou, Q.-Y., Li, C., Olah, M.E. et al. (1992). Molecular cloning and characterization of an adenosine receptor: the A_3 adenosine receptor. *Proc. Natl. Acad. Sci. U.S.A.* 89: 7432–7436.

169 Schulte, G. and Fredholm, B.B. (2003). The G_s-coupled adenosine A_{2B} receptor recruits divergent pathways to regulate ERK1/2 and p38. *Exp. Cell. Res.* 290: 168–176.

170 Hirano, D., Yoshiko, A., Ogasawara, H. et al. (1996). Functional coupling of adenosine A_{2a} receptor to inhibition of the mitogen-activated protein kinase cascade in Chinese hamster ovary cells. *Biochem. J.* 316: 81–86.

171 Linden, J., Thai, T., Figler, H. et al. (1999). Characterization of human A_{2B} adenosine receptors: radioligand binding, western blotting, and coupling to Gqin human embryonic kidney 293 cells and HMC-1 mast cells. *Mol. Pharmacol.* 56: 705–713.

172 Fredholm, B.B., Irenius, E., Kull, B., and Schulte, G. (2001). Comparison of the potency of adenosine as an agonist at human adenosine receptors expressed in Chinese hamster ovary cells. *Biochem. Pharmacol.* 61: 443–448.

173 Yaar, R., Jones, M.R., Chen, J.-F., and Ravid, K. (2005). Animal models for the study of adenosine receptor function. *J. Cell. Physiol.* 202: 9–20.

174 Palmer, T.M. and Stiles, G.L. (1995). Adenosine receptors. *Neuropharmacology* 34: 683–694.

175 Fredholm, B.B., IJzerman, A.P., Jacobson, K. et al. (2001). International Union of Pharmacology. XXV. Nomenclature and classification of adenosine receptors. *Pharmacol. Rev.* 53: 527–552.

176 Glukhova, A., Thal, D.M., Nguyen, A.T. et al. (2017). Structure of the adenosine A_1 receptor reveals the basis for subtype selectivity. *Cell* 168: 867–877.e13.

177 Nascimento, F.P., Macedo-Júnior, S.J., Pamplona, F.A. et al. (2015). Adenosine A_1 receptor-dependent antinociception induced by inosine in mice: pharmacological, genetic and biochemical aspects. *Mol. Neurobiol.* 51: 1368–1378.

178 Porkka-Heiskanen, T., Strecker, R.E., Thakkar, M. et al. (1997). Adenosine: a mediator of the sleep-inducing effects of prolonged wakefulness. *Science* 276: 1265–1267.

179 Elliott, K.J., Weber, E.T., and Rea, M.A. (2001). Adenosine A_1 receptors regulate the response of the hamster circadian clock to light. *Eur. J. Pharmacol.* 414: 45–53.

180 De Mendonça, A. and Ribeiro, J.A. (1997). Influence of metabotropic glutamate receptor agonists on the inhibitory effects of adenosine A_1 receptor activation in the rat hippocampus. *Br. J. Pharmacol.* 121: 1541–1548.

181 Jacobson, K.A. and Gao, Z.-G. (2006). Adenosine receptors as therapeutic targets. *Nat. Rev. Drug Discovery* 5: 247–264.

182 Brown, R.M. and Short, J.L. (2008). Adenosine A_{2A} receptors and their role in drug addiction. *J. Pharm. Pharmacol.* 60: 1409–1430.

183 Poon, A. and Sawynok, J. (1999). Antinociceptive and anti-inflammatory properties of an adenosine kinase inhibitor and an adenosine deaminase inhibitor. *Eur. J. Pharmacol.* 384: 123–138.

184 Gao, Y. and Phillis, J.W. (1994). CGS 15943, an adenosine A_1 receptor antagonist, reduces cerebral ischemic injury in the Mongolian gerbil. *Life Sci.* 55: 61–65.

185 Mori, A. and Shindou, T. (2003). Modulation of GABAergic transmission in the striatopallidal system by adenosine A_{2A} receptors. *Neurology* 61:S44 LP-S48.

186 El Yacoubi, M., Costentin, J., and Vaugeois, J.-M. (2003). Adenosine A_{2A} receptors and depression. *Neurology* 61:S82 LP-S87.

187 Klotz, K.N. (2000). Adenosine receptors and their ligands. *Naunyn-Schmiedeberg's Arch. Pharmacol.* 362: 382–391.

188 Franco, R., Ciruela, F., Casadó, V. et al. (2005). Partners for adenosine A_1 receptors. *J. Mol. Neurosci.* 26: 221–231.

189 Young, H.W.J., Molina, J.G., Dimina, D. et al. (2004). A_3 adenosine receptor signaling contributes to airway inflammation and mucus production in adenosine deaminase-deficient mice. *J. Immunol.* 173: 1380–1389.

190 Zhao, Z., Yaar, R., Ladd, D. et al. (2002). Overexpression of A_3 adenosine receptors in smooth, cardiac, and skeletal muscle is lethal to embryos. *Microvasc. Res.* 63: 61–69.

191 Lukashev, D., Ohta, A., and Sitkovsky, M. (2004). Targeting hypoxia-A_{2A} adenosine receptor-mediated mechanisms of tissue protection. *Drug Discovery Today* 9: 403–409.

192 Peyot, M., Gadeau, A., Dandré, F. et al. (2000). Extracellular adenosine induces apoptosis of human. *Circ. Res.* 76–85.

193 Rivkees, S.A., Thevananther, S., and Hao, H. (2000). Are A_3 adenosine receptors expressed in the brain? *Neuroreport* 11.

194 Kazemi, M.H., Raoofi Mohseni, S., Hojjat-Farsangi, M. et al. (2018). Adenosine and adenosine receptors in the immunopathogenesis and treatment of cancer. *J. Cell. Physiol.* 233: 2032–2057.

195 Mirza, A., Basso, A., Black, S. et al. (2005). RNA interference targeting of A_1 receptor-overexpressing breast carcinoma cells leads to diminished rates of cell proliferation and induction of apoptosis. *Cancer Biol. Ther.* 4: 1355–1360.

196 Dastjerdi, N.M., Valiani, A., Mardani, M., and Ra, M.Z. (2016). Adenosine A_1 receptor modifies P53 expression and apoptosis in breast cancer cell line MCF-7. *Bratislava Med. J.* 116: 242–246.

197 Woodhouse, E.C., Amanatullah, D.F., Schetz, J.A. et al. (1998). Adenosine receptor mediates motility in human melanoma cells. *Biochem. Biophys. Res. Commun.* 246: 888–894.

198 Sai, K., Yang, D., Yamamoto, H. et al. (2006). A(1) adenosine receptor signal and AMPK involving caspase-9/-3 activation are responsible for adenosine-induced RCR-1 astrocytoma cell death. *Neurotoxicology* 27: 458–467.

199 Saito, T. and Sadoshima, J. (2015). Molecular mechanisms of mitochondrial autophagy/mitophagy in the heart. *Circ. Res.* 116: 1477–1490.

200 D'Ancona, S., Ragazzi, E., Fassina, G. et al. (n.d.). Effect of dipyridamole, 5′-(*N*-ethyl)-carboxamidoadenosine and 1,3-dipropyl-8-(2-amino-4-chlorophenyl)-xanthine on LOVO cell growth and morphology. *Anticancer Res.* 14: 93–97.

201 Synowitz, M., Glass, R., Färber, K. et al. (2006). A_1 adenosine receptors in microglia control glioblastoma-host interaction. *Cancer Res.* 66: 8550–8557.

202 Etique, N., Grillier-Vuissoz, I., Lecomte, J., and Flament, S. (2009). Crosstalk between adenosine receptor (A_{2A} isoform) and ERα mediates ethanol action in MCF-7 breast cancer cells. *Oncol. Rep.* 21: 977–981.

203 Gessi, S., Merighi, S., Fazzi, D. et al. (2011). Adenosine receptor targeting in health and disease. *Expert Opin. Investig. Drugs* 20: 1591–1609.

204 Sexl, V., Mancusi, G., Höller, C. et al. (1997). Stimulation of the mitogen-activated protein kinase via the A_{2A}-adenosine receptor in primary human endothelial cells. *J. Biol. Chem.* 272: 5792–5799.

205 Lutty, G.A. and McLeod, D.S. (2003). Retinal vascular development and oxygen-induced retinopathy: a role for adenosine. *Prog. Retin Eye Res.* 22: 95–111.

206 Raskovalova, T., Huang, X., Sitkovsky, M. et al. (2005). G_s protein-coupled adenosine receptor signaling and lytic function of activated NK cells. *J. Immunol.* 175: 4383–4391.

207 Merighi, S., Mirandola, P., Milani, D. et al. (2002). Adenosine receptors as mediators of both cell proliferation and cell death of cultured human melanoma cells. *J. Invest. Dermatol.* 119: 923–933.

208 Fisher, J.W. and Brookins, J. (2001). Adenosine A_{2A} and A_{2B} receptor activation of erythropoietin production. *Am. J. Physiol. Physiol.* 281: F826–F832.

209 Nagashima, K. and Karasawa, A. (1996). Modulation of erythropoietin production by selective adenosine agonists and antagonists in normal and anemic rats. *Life Sci.* 59: 761–771.

210 Yasuda, Y., Saito, M., Yamamura, T. et al. (2009). Extracellular adenosine induces apoptosis in Caco-2 human colonic cancer cells by activating caspase-9/-3 via A_{2a} adenosine receptors. *J. Gastroenterol.* 44: 56–65.

211 Fredholm, B.B. (2010). Adenosine receptors as drug targets. *Exp. Cell. Res.* 316: 1284–1288.

212 Ciruela, F., Albergaria, C., Soriano, A. et al. (2010). Adenosine receptors interacting proteins (ARIPs): behind the biology of adenosine signaling. *Biochim. Biophys. Acta, Biomembr.* 1798: 9–20.

213 Igor, F., Goldstein, A.E., Sergey, R. et al. (2002). Differential expression of adenosine receptors in human endothelial cells. *Circ. Res.* 90: 531–538.

214 Merighi, S., Benini, A., Mirandola, P. et al. (2007). Caffeine inhibits adenosine-induced accumulation of hypoxia-inducible factor-1α, vascular endothelial growth factor, and interleukin-8 expression in hypoxic human colon cancer cells. *Mol. Pharmacol.* 72: 395–406.

215 Fernandez-Gallardo, M., González-Ramírez, R., Sandoval, A. et al. (2016). Adenosine stimulate proliferation and migration in triple negative breast cancer cells. *PLoS One* 11: e0167445.

216 Ryzhov, S., Novitskiy, S.V., Zaynagetdinov, R. et al. (2008). Host A_{2B} receptors promote carcinoma growth. *Neoplasia* 10: 987–995.

217 Mittal, D., Sinha, D., Barkauskas, D. et al. (2016). Adenosine 2B receptor expression on cancer cells promotes metastasis. *Cancer Res.* 76: 4372–4382.

218 Beavis, P.A., Milenkovski, N., Henderson, M.A. et al. (2015). Adenosine receptor 2A blockade increases the efficacy of anti-PD-1 through enhanced antitumor T-cell responses. *Cancer Immunol. Res.* 3: 506–517.

219 Cekic, C., Sag, D., Li, Y. et al. (2012). Adenosine A_{2B} receptor blockade slows growth of bladder and breast tumors. *J. Immunol.* 188: 198–205.

220 Dhillon, A.S., Hagan, S., Rath, O., and Kolch, W. (2007). MAP kinase signalling pathways in cancer. *Oncogene* 26: 3279–3290.

221 D'Alimonte, I., Nargi, E., Zuccarini, M. et al. (2015). Potentiation of temozolomide antitumor effect by purine receptor ligands able to restrain the in vitro growth of human glioblastoma stem cells. *Purinergic Signal* 11: 331–346.

222 Aghaei, M., Panjehpour, M., Karami-Tehrani, F., and Salami, S. (2011). Molecular mechanisms of A_3 adenosine receptor-induced G1 cell cycle arrest and apoptosis in androgen-dependent and independent prostate cancer cell lines: involvement of intrinsic pathway. *J. Cancer Res. Clin. Oncol.* 137: 1511–1523.

223 Jacobson, K.A. (1998). Adenosine A_3 receptors: novel ligands and paradoxical effects. *Trends Pharmacol. Sci.* 19: 184–191.

224 Feoktistov, I., Ryzhov, S., Goldstein, A.E., and Biaggioni, I. (2003). Mast cell-mediated stimulation of angiogenesis: cooperative interaction between A_{2B} and A_3 adenosine receptors. *Circ. Res.* 92: 485–492.

225 Merighi, S., Benini, A., Mirandola, P. et al. (2005). A_3 adenosine receptor activation inhibits cell proliferation via phosphatidylinositol 3-kinase/Akt-dependent inhibition of the extracellular signal-regulated kinase 1/2 phosphorylation in A375 human melanoma cells. *J. Biol. Chem.* 280: 19516–19526.

226 Torres, A., Vargas, Y., Uribe, D. et al. (2016). Adenosine A_3 receptor elicits chemoresistance mediated by multiple resistance-associated protein-1 in human glioblastoma stem-like cells. *Oncotarget* 7: 67373–67386.

227 Fishman, P., Bar-Yehuda, S., Liang, B.T., and Jacobson, K.A. (2012). Pharmacological and therapeutic effects of A_3 adenosine receptor agonists. *Drug Discovery Today* 17: 359–366.

228 Haskó, G., Antonioli, L., and Cronstein, B.N. (2018). Adenosine metabolism, immunity and joint health. *Biochem. Pharmacol.* 151: 307–313.

229 Antonioli, L., Fornai, M., Blandizzi, C. et al. (2019). Adenosine signaling and the immune system: when a lot could be too much. *Immunol. Lett.* 205: 9–15.

230 Palmer, T.M. and Trevethick, M.A. (2008). Suppression of inflammatory and immune responses by the A_{2A} adenosine receptor: an introduction. *Br. J. Pharmacol.* 153: S27–S34.

231 Hatfield, S.M., Kjaergaard, J., Lukashev, D. et al. (2015). Immunological mechanisms of the antitumor effects of supplemental oxygenation. *Sci. Transl. Med.* 7: 1–13.

232 Wang, Y.J., Fletcher, R., Yu, J., and Zhang, L. (2018). Immunogenic effects of chemotherapy-induced tumor cell death. *Genes Dis.* 5: 194–203.

233 Di Virgilio, F., Sarti, A.C., Falzoni, S. et al. (2018). Extracellular ATP and P2 purinergic signalling in the tumour microenvironment. *Nat. Rev. Cancer* 18: 601–618.

234 Antonioli, L., Csóka, B., Fornai, M. et al. (2014). Adenosine and inflammation: what's new on the horizon? *Drug Discovery Today* 19: 1051–1068.

235 Horenstein, A.L., Chillemi, A., Zaccarello, G. et al. (2013). A CD38/CD203A/CD73 ectoenzymatic pathway independent of CD39 drives a novel adenosinergic loop in human T lymphocytes. *Oncoimmunology* 2: 1–14.

236 Morandi, F., Morandi, B., Horenstein, A.L. et al. (2015). A non-canonical adenosinergic pathway led by CD38 in human melanoma cells induces suppression of T cell proliferation. *Oncotarget* 6: 25602–25618.

237 Ludwig, N., Yerneni, S.S., Azambuja, J.H. et al. (2020). Tumor-derived exosomes promote angiogenesis via adenosine A_{2B} receptor signaling. *Angiogenesis.*

238 Sek, K., Mølck, C., Stewart, G. et al. (2018). Targeting adenosine receptor signaling in cancer immunotherapy. *Int. J. Mol. Sci.* 19: 3837.

239 Young, A., Mittal, D., Stagg, J., and Smyth, M.J. (2014). Targeting cancer-derived adenosine: new therapeutic approaches. *Cancer Discov.* 4: 879–888.

240 Liu, H., Kuang, X., Zhang, Y. et al. (2020). ADORA1 inhibition promotes tumor immune evasion by regulating the ATF3-PD-L1 axis. *Cancer Cell* 37: 324–339.e8.

241 Boussiotis, V.A. (2016). Molecular and biochemical aspects of the PD-1 checkpoint pathway. *N. Engl. J. Med.* 375: 1767–1778.

242 Zou, W., Wolchok, J.D., and Chen, L. (2016). PD-L1 (B7-H1) and PD-1 pathway blockade for cancer therapy: mechanisms, response biomarkers, and combinations. *Sci. Transl. Med.* 8.

243 Mandapathil, M., Hilldorfer, B., Szczepanski, M.J. et al. (2010). Generation and accumulation of immunosuppressive adenosine by human $CD4^{+}CD25^{high}FOXP3^{+}$ regulatory T cells. *J. Biol. Chem.* 285: 7176–7186.

244 Deaglio, S., Dwyer, K.M., Gao, W. et al. (2007). Adenosine generation catalyzed by CD39 and CD73 expressed on regulatory T cells mediates immune suppression. *J. Exp. Med.* 204: 1257–1265.

245 Koshiba, M., Kojima, H., Huang, S. et al. (1997). Memory of extracellular adenosine A(2a) purinergic receptor-mediated signaling in murine T cells. *J. Biol. Chem.* 272: 25881–25889.

246 Ohta, A., Kjaergaard, J., Sharma, S. et al. (2009). In vitro induction of T cells that are resistant to A2 adenosine receptor-mediated immunosuppression. *Br. J. Pharmacol.* 156: 297–306.

247 Csóka, B., Selmeczy, Z., Koscsó, B. et al. (2012). Adenosine promotes alternative macrophage activation via A_{2A} and A_{2B} receptors. *FASEB J.* 26: 376–386.

248 Volpini, R., Costanzi, S., Vittori, S. et al. (2003). Medicinal chemistry and pharmacology of A_{2B} adenosine receptors. *Curr. Top. Med. Chem.* 3: 427–443.

249 Vaupel, P., Kallinowski, F., and Okunieff, P. (1989). Blood flow, oxygen and nutrient supply, and metabolic microenvironment of human tumors: a review. *Cancer Res.* 49: 6449–6465.

250 Fozard, J.R., Pfannkuche, H.J., and Schuurman, H.J. (1996). Mast cell degranulation following adenosine A_3 receptor activation in rats. *Eur. J. Pharmacol.* 298: 293–297.

251 Mediavilla-Varela, M., Castro, J., Chiappori, A. et al. (2017). A novel antagonist of the immune checkpoint protein adenosine A_{2a} receptor restores tumor-infiltrating lymphocyte activity in the context of the tumor microenvironment. *Neoplasia (United States)* 19: 530–536.

252 Willingham, S.B., Ho, P.Y., Hotson, A. et al. (2018). $A_{2A}R$ antagonism with CPI-444 induces antitumor responses and augments efficacy to anti-PD-(L)1 and anti-CTLA-4 in preclinical models. *Cancer Immunol. Res.* 6: 1136–1149.

253 Kim, D.G. and Bynoe, M.S. (2016). A_{2A} adenosine receptor modulates drug efflux transporter P-glycoprotein at the blood-brain barrier. *J. Clin. Invest.* 126: 1717–1733.

254 Jackson, S., Weingart, J., Nduom, E.K. et al. (2018). The effect of an adenosine A_{2A} agonist on intra-tumoral concentrations of temozolomide in patients with recurrent glioblastoma. *Fluids Barriers CNS* 15: 1–9.

255 Hatfield, S.M. and Sitkovsky, M. (2016). A_{2A} adenosine receptor antagonists to weaken the hypoxia-HIF-1α driven immunosuppression and improve immunotherapies of cancer. *Curr. Opin. Pharmacol.* 29: 90–96.

256 Wei, Q., Costanzi, S., Balasubramanian, R. et al. (2013). A_{2B} adenosine receptor blockade inhibits growth of prostate cancer cells. *Purinergic Signal* 9: 271–280.

257 Iannone, R., Miele, L., Maiolino, P. et al. (2013). Blockade of A_{2b} adenosine receptor reduces tumor growth and immune suppression mediated by myeloid-derived suppressor cells in a mouse model of melanoma. *Neoplasia (United States)* 15: 1400–1409.

258 Ohta, A., Gorelik, E., Prasad, S.J. et al. (2006). A_{2A} adenosine receptor protects tumors from antitumor T cells. *Proc. Natl. Acad. Sci. U.S.A.* 103: 13132–13137.

259 Waickman, A.T., Alme, A., Senaldi, L. et al. (2012). Enhancement of tumor immunotherapy by deletion of the A_{2A} adenosine receptor. *Cancer Immunol. Immunother.* 61: 917–926.

260 Iannone, R., Miele, L., Maiolino, P. et al. (2014). Adenosine limits the therapeutic effectiveness of anti-CTLA4 mAb in a mouse melanoma model. *Am. J. Cancer Res.* 4: 172–181.

261 Allard, B., Pommey, S., Smyth, M.J., and Stagg, J. (2013). Targeting CD73 enhances the antitumor activity of anti-PD-1 and anti-CTLA-4 mAbs. *Clin. Cancer Res.* 19: 5626–5635.

262 Barbhaiya, H., McClain, R., and IJzerman AP, Rivkees SA. (1996). Site-directed mutagenesis of the human A_1 adenosine receptor: influences of acidic and hydroxy residues in the first four transmembrane domains on ligand binding. *Mol. Pharmacol.* 50: 1635–1642.

263 Kim, J., Wess, J., van Rhee, A.M. et al. (1995). Site-directed mutagenesis identifies residues involved in ligand recognition in the human A_{2a} adenosine receptor. *J. Biol. Chem.* 270: 13987–13997.

264 Jiang, Q., Van Rhee, A.M., Kim, J. et al. (1996). Hydrophilic side chains in the third and seventh transmembrane helical domains of human A_{2A} adenosine receptors are required for ligand recognition. *Mol. Pharmacol.* 50: 512–521.

265 Kim, J., Jiang, Q., Glashofer, M. et al. (1996). Glutamate residues in the second extracellular loop of the human A_{2a} adenosine receptor are required for ligand recognition. *Mol. Pharmacol.* 49: 683–691.

266 Beukers, M.W., van Oppenraaij, J., van der Hoorn, P.P.W. et al. (2004). Random mutagenesis of the human adenosine A_{2B} receptor followed by growth selection in yeast. Identification of constitutively active and gain of function mutations. *Mol. Pharmacol.* 65: 702–710.

267 Thimm, D., Schiedel, A.C., Sherbiny, F.F. et al. (2013). Ligand-specific binding and activation of the human adenosine A_{2B} receptor. *Biochemistry* 52: 726–740.

268 IJzerman AP, Van Galen, P.J., and Jacobson, K.A. (1992). Molecular modeling of adenosine receptors. I. The ligand binding site on the A_1 receptor. *Drug Des. Discovery* 9: 49–67.

269 Dudley, M.W., Peet, N.P., Demeter, D.A. et al. (1993). Adenosine A_1 receptor and ligand molecular modeling. *Drug Dev. Res.* 28: 237–243.

270 Gray, V.E., Hause, R.J., and Fowler, D.M. (2017). Analysis of large-scale mutagenesis data to assess the impact of single amino acid substitutions. *Genetics* 207: 53–61.

271 Bromberg, Y. and Rost, B. (2008). Comprehensive in silico mutagenesis highlights functionally important residues in proteins. *Bioinformatics* 24: i207–i212.

272 Wang, X., Jespers, W., Bongers, B.J. et al. (2020). Characterization of cancer-related somatic mutations in the adenosine A_{2B} receptor. *Eur. J. Pharmacol.* 880: 173126.

19

Dopamine Receptors: Neurotherapeutic Targets for Substance Use Disorders

Ashley N. Nilson[2,3], *Daniel E. Felsing*[1,3], *and John A. Allen*[1,2,3]

[1] *Department of Pharmacology and Toxicology, University of Texas Medical Branch, Galveston, TX, USA*
[2] *Department of Neuroscience and Cell Biology, University of Texas Medical Branch, Galveston, TX, USA*
[3] *Center for Addiction Research, University of Texas Medical Branch, Galveston, TX, USA*

19.1 Introduction

The discovery of dopamine as a neurotransmitter by Arvid Carlsson [1] over 60 years ago opened an entire field of basic and translational neuroscience to follow. Decades of basic brain research have revealed that dopamine is a crucial neurotransmitter with essential physiological roles in motor control, memory, attention, motivation, and reward [2–13]. Early research included the identification and cloning of the five dopamine receptors, which have emerged as canonical, neuromodulatory G protein-coupled receptors [14–16]. The pivotal molecular cloning of the receptors further catalyzed efforts to advance dopamine receptor modulation as a neurotherapeutic strategy to treat both neurological and psychiatric disorders. While there has been great clinical success for targeting dopamine receptors for the treatment of psychosis (D2 receptor antagonists are clinically effective antipsychotics) and for Parkinson's disease (nonselective dopamine agonists improve motor function), many additional central nervous system disorders will likely benefit from pharmacotherapy that selectively targets subtypes or combinations of dopamine receptors. These include various cognitive disorders [4, 9] and substance use disorders that share the common pathology of altered dopamine signaling in response to drugs of abuse [2, 17–19].

In this chapter, a detailed review of dopamine receptor signaling and physiology is presented along with evidence that specific dopamine receptors control brain reward systems and behavioral responsiveness to substances of abuse. A basic review of dopamine receptor ligands, their structures, and pharmacology are also presented focusing on both preclinical models and clinical findings in which

GPCRs as Therapeutic Targets, Volume 2, First Edition. Edited by Annette Gilchrist.

dopamine receptor targeted drugs have been evaluated for therapeutic value in substance use disorders.

19.2 Substance Use Disorders: A Crisis of Unmet Clinical Need

Drug addiction, clinically defined as a substance use disorder, is a disease that profoundly affects the brain and behavior leading to the inability to control the use of legal or illegal drugs or medications. In the United States alone, abuse of alcohol, tobacco, and illicit drugs costs more than $740 billion annually due to crime, lost work productivity, and health care [20]. Tragically, in just 2019, more than 70 000 Americans died from a drug-induced overdose from illicit drugs and prescription opioids (source: CDC Wonder, https://wonder.cdc.gov). One example of this crisis is stimulant use disorder, which includes abuse and addiction to drugs such as cocaine, methamphetamine, ecstasy, and prescription drug stimulants. Approximately, 6 million Americans misused prescription stimulants in 2016, while nearly 5 million Americans reported current cocaine use in 2016, which together is close to 4% of the population [21]. The mortality rates from all psychostimulants has been increasing since 2010, and in 2017, overdose deaths involving cocaine increased by more than 34%, with almost 14 000 Americans dying from an overdose involving cocaine [21]. A number of dopamine-receptor targeted agents that modulate functional output of brain dopamine systems have been reported as potential first-line treatments for stimulant use disorders [22]. Researchers are also currently testing several medications that have been used to treat other disorders, including disulfiram (used to treat alcoholism), and buprenorphine (used for opioid addiction). Despite these drug therapy studies, there are no Food and Drug Administration (FDA)-approved pharmacotherapies available that reduce psychostimulant use, and the available psychosocial interventions have varied, often weak overall effects [23]. There is clearly a large unmet medical need to address this crisis and this article reviews dopamine receptors as promising neurotherapeutic targets to treat substance use disorders.

19.3 The Dopamine Hypothesis of Addiction

Perhaps the greatest empirical evidence for the involvement of brain dopamine in psychiatry emanates from research on substance use disorders [24]. It has also been proposed that a crucial mechanism for the development of substance use disorders is drug-induced activation of dopamine neurotransmission in the mesolimbic dopamine pathway, also referred to as the "dopamine hypothesis of addiction" [25, 26]. During acute drug taking, nearly all psychoactive drugs of abuse (for example, alcohol, amphetamines, cocaine, cannabinoids, nicotine, or opiates) induce alterations in the transmission of dopamine within the mesolimbic dopamine pathway, with most of these drugs increasing extracellular concentrations of dopamine resulting in a hyperactivated state during drug taking [27–29]. Experimental evidence also suggests that the mesolimbic dopamine system undergoes adaptation and may become hypofunctional in the addicted brain with decreased dopamine and/or dopamine receptor function in addicted subjects [30], resulting in decreased interest for nondrug-related stimuli [31]. The dopamine hypothesis of addiction also predicts that pharmacologically modulating dopamine function through targeting receptors or the dopamine transporter (DAT) may normalize function and prove therapeutic for various substance use disorders.

19.4 Overview of Dopaminergic Brain Pathways

There are four major dopaminergic circuit tracts in the mammalian brain: nigrostriatal, mesocortical, mesolimbic, and tuberoinfundibular pathways (Figure 19.1). The nigrostriatal pathway stretches from the substantia nigra (SN) to the striatum where it modulates voluntary movement and motor control pathways. Of particular clinical relevance, degeneration of the dopamine producing cells in the nigrostriatal pathway of the SN causes Parkinson's disease and a progressive loss of various motor control functions [32]. Correcting nigrostriatal pathway dopamine by either increasing dopamine synthesis (by levodopa), or by activating striatal dopamine receptors (by agonists) are successful therapeutic strategies for treating Parkinson's disease. A second major dopaminergic brain pathway is the mesocortical system which extends from the ventral tegmental area (VTA) to the prefrontal cortex where this pathway modulates various cognitive processes including working memory and attention [4, 9–11, 33–35]. Modulation of prefrontal dopamine neurotransmission may be useful to correct cognitive deficits in a range of neuropsychiatric disorders [9, 11]. The third major dopaminergic pathway is the mesolimbic dopamine pathway which is of particular importance for this chapter. This pathway extends from the VTA to the amygdala and nucleus accumbens and is important for processing both natural and synthetic rewards as well as motivation. Numerous substances of abuse alter the mesolimbic pathway including cocaine, morphine, and alcohol [36–38]. The rewarding effects of drugs of abuse derive in large part from increases in extracellular dopamine in mesolimbic pathways, particularly within the nucleus accumbens, during drug use [28, 39]. Therefore, modulating dopamine signaling in the mesolimbic pathway

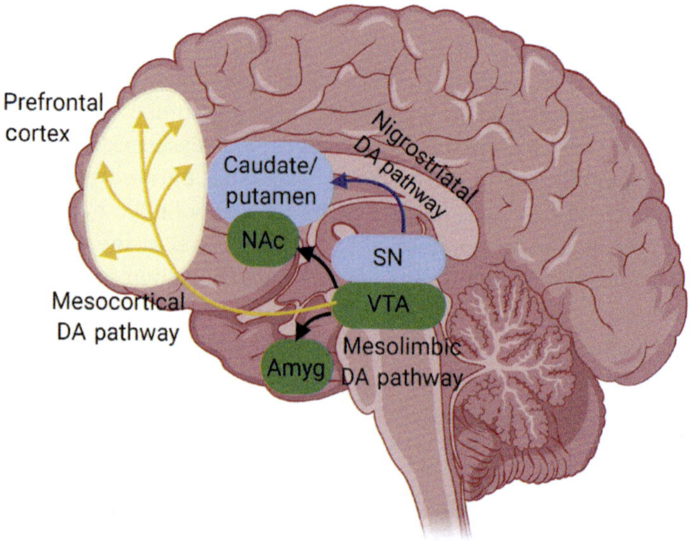

Figure 19.1 Major dopaminergic (DA) brain pathways. The nigrostriatal dopaminergic pathway extends from the substantia nigra (SN) to the caudate/putamen and is involved in voluntary motor control. The mesolimbic DA pathway extends from the ventral tegmental area (VTA) to the nucleus accumbens (NAc) and amygdala (Amyg) and is involved in the reward and emotional aspects of substance use disorders. The mesocortical DA pathway extends from the VTA to the prefrontal cortex and is involved is cognitive processes.

may be a therapeutic strategy for treating substance use disorders. The fourth major dopaminergic pathway is the tuberoinfundibular pathway that extends from dopamine neurons originating in the arcuate nucleus of the hypothalamus to the infundibular region (median eminence). The tuberoinfundibular pathway primarily expresses dopamine D2 receptors that control prolactin release from the pituitary gland [40]. For more extensive explanation of the dopamine neurocircuitry and addiction, we direct readers to these additional reviews and references (2, 41–43). Taken together, these four pathways provide essential dopamine neurotransmission in the brain and control a multitude of functions.

19.5 Dopamine Neurotransmission

The two main dopamine-producing regions of the brain are the VTA and the SN. Dopamine signaling and neurotransmission between neurons begins with an action potential generated in VTA or SN neurons which results in the presynaptic release of dopamine (Figure 19.2). This synaptic release of dopamine activates dopamine receptors on responsive neurons, such as the medium spiny

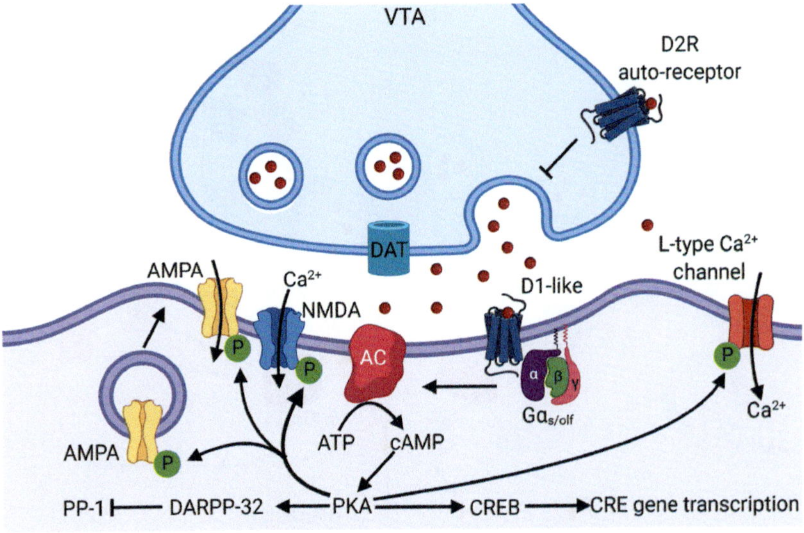

(a) Dopamine D1-like receptor signaling

Figure 19.2 Dopamine receptor signaling pathways. (a) Dopamine D1-like receptor signaling. D1-like receptors (D1R) stimulate cAMP production through adenylyl cyclase (AC). cAMP activates protein kinase A (PKA) which phosphorylates downstream targets including dopamine and cAMP regulated phosphoprotein-32 (DARPP-32), which prevents dephosphorylation of PKA targets by inhibiting protein phosphatase-1 (PP-1). PKA also phosphorylates cAMP response element-binding protein (CREB) leading to changes in gene transcription. Phosphorylation of multiple ion channel receptors (AMPA, NMDA) increases neuronal excitability and AMPA receptor insertion into the membrane. Phosphorylation of L-type voltage-dependent Ca^{2+} channels further increases neuronal excitability. (b) Dopamine D2-like receptor signaling. D2-like receptors (D2R) decrease cAMP level by inhibiting AC, which decreases the activity of PKA and increases the activity of PP-1. The D2R interacts with NMDA receptors decreasing phosphorylation and directly decreasing the activity of the NMDA receptor. The $G\beta\gamma$ facilitates interactions of the D2R with G protein-gated inwardly rectifying K^+ (GIRK) channels resulting in K^+ efflux, further decreasing neuronal excitability. The D2R also decreases the activity of L-type Ca^{2+} channels. β-arrestin recruitment to the D2R activates protein phosphatase 2A (PP2A) inhibiting Akt and leading to an increase in activity of glycogen synthase kinase 3 (GSK3). On the presynaptic side, D2R auto-receptors can decrease the release of dopamine by increasing the activity of GIRK channels. The dopamine transporter (DAT) reuptakes dopamine from the synapse. (a) Source: Based on Greengard [44].

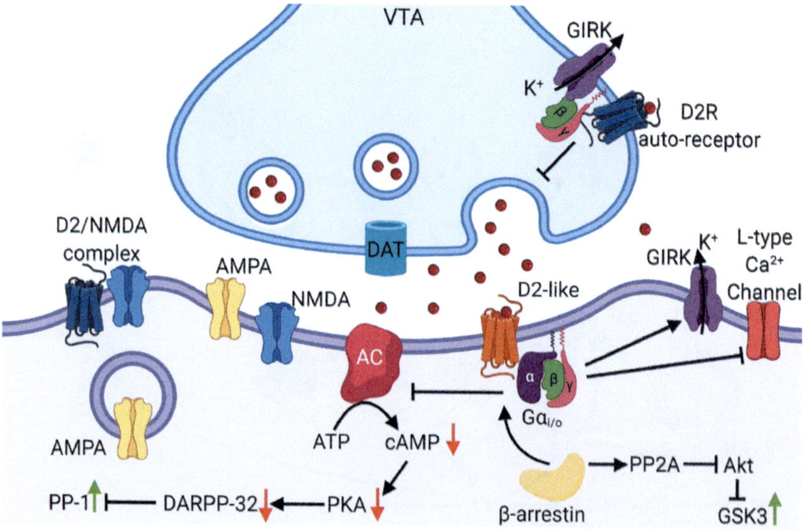

(b) Dopamine D2-like receptor signaling

Figure 19.2 (*Continued*)

neurons within the nucleus accumbens. The receptors are part of the G protein-coupled receptor family, and upon binding the transmitter activate heterotrimeric G proteins (G_{olf}/G_s or $G_{i/o}$). The G proteins in turn regulate the activity of effector proteins such as ion channels, or the enzyme adenylyl cyclase that produces the second messenger cyclic adenosine monophosphate (cAMP), resulting in neuromodulation and alterations in neural activity [2, 8]. The neural response to dopamine is terminated by reuptake of dopamine into the presynaptic neuron terminals, which is primarily controlled by the DAT. In addition, the enzymes monoamine oxidase and catechol-*o*-methyltransferase can terminate signaling by breaking down dopamine to the metabolites homovanillic acid or 3-methoxytyramine, respectively.

19.6 Alterations of Dopamine Signaling by Drugs of Abuse

The nucleus accumbens is essential for reward processing and increases in extrasynaptic dopamine in the nucleus accumbens have an important role in addiction [28, 39, 43]. Primarily using preclinical rodent models, numerous studies have determined most drugs of abuse, including stimulants, alcohol, opiates, cannabinoids, nicotine, and hallucinogens, facilitate the neurotransmission and release of dopamine within the mesolimbic dopamine pathway, with most of these agents increasing extracellular concentrations of dopamine [2]. The firing of VTA dopamine neurons that accompanies drug use encodes a number of reward properties, including reward expectancy [45], reward learning, and contextual memories [46]. All these neurobiological processes are believed to contribute to the intense motivation to attain drugs of abuse. Modulation of the strength of neurotransmission and synaptic plasticity induced by the excess dopamine is also thought to underlie the neuroadaptations in brain reward systems, resulting in drug addiction and substance use disorders [2, 41, 42]. The mesolimbic dopamine pathway is also important for the emotional processing associated with drugs of abuse. In addition, there are learning, memory, and attention processes involved, which are primarily mediated through the mesocortical dopamine pathway [43, 47, 48]. The neurobiology and circuitry of addiction are complex and modern techniques that specifically activate or inhibit dopamine release in specific neuron populations in a region or pathway are also refining our understanding [49, 50].

An early neuroadaptation common to all drugs of abuse and by example, observed after a single injection of cocaine, is increased excitability of the mesolimbic dopamine system reflected in long-term potentiation-dependent changes in both dopamine and glutamate activity [19]. Subsequently, the elevated dopamine contributes to increased excitability of medium spiny neurons in the ventral striatum of the nucleus accumbens with decreased glutamatergic activity during withdrawal and increased glutamatergic activity during drug-primed and cue-induced drug-seeking [2, 19, 51]. As drug abuse evolves into full-blown addiction, loss-of-function occurs in the frontal cortex that controls executive functions, contributing to poor decision-making and gain-of-function in the stress systems contributing to incentive salience for drugs over natural reinforcers [19].

19.7 Dopamine Receptors and Their Signaling

Dopamine exerts its physiological functions through five distinct G protein-coupled receptors (D1–D5), classified by gene sequence and biochemical signaling [8, 52, 53]. All five dopamine receptor subtypes have the characteristic seven transmembrane (TM) spanning α-helical domains, which interact with dopamine and orthosteric ligands. Receptor intracellular loops interact with various effector systems including G proteins and β-arrestins. The D1 and D5 belong to the D1-like subfamily (D1R) that share high-sequence homology (80% TM identity) and activate heterotrimeric G_s or G_{olf} G proteins. Activation of adenylyl cyclase increases cAMP production and thus promotes neuronal activity. The remaining three subtypes: D2, D3, and D4, belong to the D2-like receptor subfamily (D2R) which inhibit adenylyl cyclase to decrease cAMP, and activate K^+ channels, to reduce neuronal activity. Additionally, all dopamine receptors have been shown to interact with G protein-coupled receptor kinases to recruit the multifunctioning adaptor protein β-arrestin to promote receptor endocytosis, desensitization, and possibly a second wave of biochemical signaling [54]. While there are similarities between receptor signaling, a discussion of each receptor is overviewed in the following section and further illustrated in Figure 19.2.

19.8 Dopamine D1 Receptors

The D1R was the first dopamine receptor discovered and cloned in 1990 [15, 55]. The D1R couples to the G_s and G_{olf} G proteins that activate adenylyl cyclase to produce cAMP, initiating an intracellular signaling cascade. D1R signaling is further illustrated in Figure 19.2a. The G_s G protein is expressed throughout the body while the G_{olf} G protein has more restricted expression [56]. The G_{olf} G protein is highly expressed in the olfactory bulb (hence the name G_{olf}) as well as the striatum [57]. Interestingly, the G_s G protein has very low expression in the striatum where G_{olf} is highly expressed. This is important to note that as the D1R is also highly expressed in the striatum where G_{olf} may be the predominant Gα subunit that stimulates adenylyl cyclase. Furthermore, G_{olf} knockout mice do not increase cAMP production through adenylyl cyclase in the striatum after D1R agonism, but cAMP production is unaffected in the prefrontal cortex where G_s is expressed [58]. In addition, the G_{olf} knockout mice do not have the typical hyperlocomotor response to D1R agonists or cocaine, which elevates dopamine levels in the synapse, indicating that the D1R may primarily couple to G_{olf} in the striatum [57].

As mentioned, the D1R stimulates the production of the second messenger cAMP by activating adenylyl cyclase with the $G_{s/olf}$ G proteins [8]. Increases in cAMP in turn activate protein kinase A (PKA) that can phosphorylate a multitude of target proteins. PKA is a highly studied protein kinase that regulates a plethora of protein targets including dopamine- and cAMP-regulated phosphoprotein, 32 kDa (DARPP-32). DARPP-32 enhances PKA signaling by preventing the dephosphorylation of the downstream targets of PKA [44] (see Figure 19.2a). When DARPP-32 is phosphorylated at Thr34 by PKA, DARPP-32 binds to the active site of protein phosphatase-1 (PP-1), thereby inhibiting PP-1 from dephosphorylating its substrates and further enhancing PKA-mediated signaling [59]. Interestingly, DARPP-32 is highly enriched in medium spiny neurons in the striatum [44, 60, 61]. Furthermore, DARPP-32 knockout mice have deficits in dopaminergic signaling, altered electrophysiological responses to dopamine, and reduced locomotion induced by cocaine [62] indicating that DARPP-32 has an important function in D1R signaling.

PKA can also influence neuronal excitability by phosphorylating ion channels altering their conductance. D1R-mediated $G_{s/olf}$ /cAMP/PKA signaling increases the conductance of α-amino-3-hydroxy-5-methylisoxazole-4-propionic acid (AMPA) receptors by increasing the phosphorylation of the GluR1 subunit of AMPA [63–65]. The activation of DARPP-32 and the inhibition of PP-1 are important for the increases in GluR1 subunit phosphorylation. Indeed, GluR1 phosphorylation was reduced in DARPP-32 knockout mice after treatment with dopamine and psychostimulants [63]. Additionally, PKA increases AMPA channel insertion into the membrane to further increase neuronal excitability [66]. Directly phosphorylating AMPA receptors as well as increasing the insertion of AMPA channels into the membrane increases synaptic strength and may be part of the mechanism behind the synaptic plasticity that occurs with drugs of abuse. In addition to AMPA channels, PKA/DARPP-32 also inhibits γ-aminobutric acid (GABA) $GABA_A$ receptors by increasing the phosphorylation of the beta1/beta3 subunits of the $GABA_A$ receptor [67]. Together, the increase in excitatory AMPA receptor and decrease in inhibitory $GABA_A$ receptor currents increase the excitability of medium spiny neurons in the striatum.

D1R stimulation can also increase currents through the *N*-methyl-D-aspartate (NMDA) receptor by PKA-dependent and independent mechanisms. To begin, PKA/DARPP-32 activity increases phosphorylation of the NR1 subunit of the NMDA receptor increasing Ca^{2+} influx into the neuron when the NMDA receptor is open [62, 68, 69]. Interestingly, amphetamines increase NR1 subunit phosphorylation in striatal medium spiny neurons [70]. The PKA-independent regulation of NMDA involves the Fyn kinase, which is activated by Src [71]. Fyn phosphorylates the NR2 subunit to enhance translocation of the NMDA receptor to the postsynaptic compartment and enhances channel activity [71, 72]. Both Ca^{2+} and cAMP induce the phosphorylation and activation of the Ca^{2+}/cAMP response element binding protein (CREB). Thus, D1R stimulation increases the transcription of genes that have CRE promoters by increasing both cAMP (directly) and Ca^{2+} through the NMDA receptor. Moreover, the activation of both the D1R and NMDA receptors may synergistically activate extracellular signal regulated kinases 1 and 2 (ERK1/2), which are important kinases involved in synaptic plasticity in medium spiny neurons [73].

D1R stimulation also regulates the activity of other ion channels, primarily promoting excitability. In addition to interactions with the NMDA receptor leading to increases in Ca^{2+}, D1R stimulation also increases the activity of L-type Ca^{2+} channels by a PKA/DARPP-32 dependent mechanism in the striatum [74]. Furthermore, the ability of the D1R to enhance NMDA receptor long-term potentiation was attenuated by blocking L-type Ca^{2+} channels and DARPP-32 knockout mice have reduced long-term potentiation in the corticostriatal pathway [75–77]. D1R stimulation leads to the phosphorylation of Na^+ channels by PKA that reduces Na^+ channel activity [78]. Additionally, D1R agonists decrease the activity of inwardly rectifying K^+ channels, N-, and P/Q-type Ca^{2+} channels through PKA-mediated phosphorylation [78].

The D1R recruits β-arrestins and undergoes endocytosis after agonist stimulation, and this is thought to desensitize further D1R G protein signaling. In the classic model for G protein-coupled receptor desensitization, the G protein-coupled receptor recruits β-arrestin sterically hindering further G protein activation [79–83]. β-arrestins facilitate the endocytosis of G protein-coupled receptors; then, the receptor can be dephosphorylated and recycled back to the membrane to resensitize signaling or be degraded in the lysosome [84–88]. Agonist stimulation of the D1R induces the recruitment of β-arrestin1 or β-arrestin2 to the D1R, desensitizing G protein signaling [13, 79]. The D1R can also be internalized following agonist stimulation and is rapidly recycled back to the membrane [89–91]. Interestingly, the D1R may be able to undergo dephosphorylation at the membrane since endocytosis inhibitors did not prevent the dephosphorylation of the D1R [92]. Finally, D1R endocytosis may facilitate further G protein signaling from the endosome. The D1R colocalized in endosomes with G_s and adenylyl cyclase [93]. Furthermore, D1R cAMP production was attenuated in cells treated with clathrin-mediated endocytosis inhibitors but not when recycling was blocked [93]. Together, these studies indicate that the D1R recruits β-arrestins and undergoes desensitization and endocytosis. However, the D1R diverges from the classical model as dephosphorylation can occur at the membrane, rather than the endosome, and the D1R may stimulate cAMP production after endocytosis.

19.9 Dopamine D5 Receptors

The dopamine D5 receptor (D5R) is discussed here because it falls into the D1-like subclass of dopamine receptors and signaling is overviewed in Figure 19.2a. The D5R couples to $G_{s/olf}$ G proteins to stimulate cAMP production and initiate similar downstream cascades to the D1R (PKA/DARPP-32) [94, 95]. The D5R has 10-fold higher affinity for dopamine than the D1R but is otherwise near pharmacologically indistinguishable from the D1R [94]. Interestingly, the D5R has constitutive, agonist-independent effects on basal cAMP. Cells transfected with the D5R have higher basal levels of cAMP than cells transfected with similar levels of the D1R [96]. The D5R also interacts directly with the $GABA_A$ receptor and agonists applied to either receptor ($GABA_A$ or D5R) inhibit the activity of the other receptor [97]. This regulation is achieved by a protein–protein interaction between the C-terminal tail of the D5R and the second intracellular loop of the $GABA_A$ receptor [97]. Additionally, the D5R modulates the NMDA receptor in a similar manner to that of the D1R using a PKA-dependent mechanism to increase the insertion of NMDA receptors into the postsynaptic membrane [98]. Indeed, D5R knockout mice had significantly decreased NMDA receptor NR2B subunit in the hippocampus and had impaired spatial and recognition memory [99]. The D5R can recruit β-arrestin in an agonist-dependent manner [100, 101]. Furthermore, the D5R can undergo agonist-independent endocytosis through a mechanism involving protein kinase C [101]. Finally, there is evidence that the D5R regulates brain-derived neurotrophic factor and Akt signaling in the prefrontal cortex. Agonist stimulation increased brain-derived neurotrophic factor expression and signaling in D1R knockout mice [102]. Overall, less is known specifically about the D5R, and future studies are needed to fully characterize the D5R and distinguish it from the D1R. The development of a truly selective D5R agonist or antagonist would greatly advance our knowledge of D5R function *in vitro* and *in vivo* and allow comparative study of the D5R and the D1R.

19.10 Dopamine D2 Receptors

The D2-like dopamine receptors all exert an inhibitory effect on cAMP production and downstream signaling (see Figure 19.2b). The most extensively studied receptor in the D2-like subclass is the D2R. The D2R is alternatively spliced into two isoforms that differ by 29 amino acids [103, 104]. The D2 long receptor (D2LR) contains 29 amino acids in the intracellular loop 3 that the D2 short receptor (D2SR) does not [103, 104]. The D2LR has higher expression levels in the brain and is primarily localized to the postsynaptic compartment, whereas the D2SR is primarily a presynaptic auto-receptor that regulates dopamine release in nigrostriatal neurons [105–107]. The D2R inhibits adenylyl cyclase by coupling to $G_{i/o}$ G proteins that are pertussis toxin-sensitive [108]. Intracellular loops 2 and 3 are important regions for G protein coupling [109, 110] and the isoforms couple to G_{i1}, G_{i2}, and G_{i3} differently [110, 111]. There are a few reports of the D2R coupling to the G_z G protein to inhibit adenylyl cyclase in a pertussis toxin-insensitive manner [112]. In addition, in G_z knockout mice, quinpirole does not suppress locomotor activity to the same extent as wildtype mice suggesting that G_z G protein coupling to the D2R has a physiological function [113]. All the $G_{i/o}$ and G_z G proteins inhibit adenylyl cyclase and cAMP production.

The D2R inhibits cAMP production and the subsequent signaling events mediated by PKA/DARPP-32. DARPP-32 phosphorylation at Ser34 is decreased by D2R agonists but increased by antagonists [114–116]. Decreased DARPP-32 phosphorylation prevents it from inhibiting PP-1, which is a phosphatase that counteracts PKA. In addition, the D2R has opposing effects on AMPA and NMDA receptors compared to the D1R. D2R agonists decrease the phosphorylation of AMPA receptors [117]. Furthermore, the D2R agonist, quinpirole decreased NMDA receptor activity [118]. D2R antagonists increase the phosphorylation of NMDA receptors adding further evidence suggesting that the D2R negatively regulates NMDA currents [119]. Reduced NMDA Ca^{2+} currents decrease synaptic plasticity by reducing CREB activation and subsequent gene expression. Together, these studies indicate that D1R stimulation increases cAMP/PKA/DARPP-32 activity to promote neuronal excitability and promote synaptic plasticity while the D2R opposes these effects by inhibiting adenylyl cyclase and cAMP.

In addition to decreasing cAMP/PKA-mediated NMDA receptor phosphorylation, the D2R may also inhibit the NMDA receptor through a protein–protein interaction. The D2R can interact with the NR2B subunit of the NMDA receptor, which prevents the association of the NMDA receptor with Ca^{2+}/calmodulin-dependent protein kinase II (CaMKII) [120]. The reduction in CaMKII interaction decreases NMDA receptor phosphorylation at Ser1303 and currents in striatal neurons [120]. Interestingly, inhibiting the interaction of the D2R and NMDA receptor decreases cocaine-induced locomotor activity [120]. D2R-mediated inhibition of NMDA receptors is accomplished through multiple mechanisms (PKA-dependent and direct protein–protein interaction) that result in decreased phosphorylation of the NMDA receptor at multiple sites shown to influence NMDA receptor currents.

The D2R also reduces neuronal excitability through the regulation of multiple ion channels through the $G\beta\gamma$ subunit of the heterotrimeric G protein. The $G\beta\gamma$ subunit initiates a series of signaling events leading to decreased L-type Ca^{2+} channel activation in medium spiny neurons and opposes D1R stimulation of the L-type Ca^{2+} channel. The $G\beta\gamma$ subunit can increase the activity of phospholipase C and inositol-1,4,5-trisphosphate [121]. Inositol-1,4,5-trisphosphate activates Ca^{2+} channels on the endoplasmic reticulum releasing Ca^{2+} that can activate calcineurin, which dephosphorylates the L-type Ca^{2+} channel, decreasing its activity [121]. In neostriatal neurons, D2R $G\beta\gamma$ subunit can inhibit N-type Ca^{2+} channels in cholinergic interneurons [122]. However, the inhibition of the N-type Ca^{2+} channel does not depend on phospholipase C activation [122].

Stimulation of the D2R also increases the activation of G protein-gated inwardly rectifying K^+ channels (GIRK) [123, 124]. The D2R modulates GIRK activity through direct protein–protein interaction with the $G\beta\gamma$ subunit being critical to the formation of the D2R/GIRK complex [125]. The increased activity of the GIRK channel hyperpolarizes the neuron decreasing its excitability. Selective GIRK knockout in dopaminergic VTA neurons increased locomotor activity following cocaine administration and increased cocaine intake during self-administration tests, suggesting GIRK channels may function in D2R auto-receptor feedback inhibition [126, 127]. Together, these studies provide evidence for the important

D2R-mediated regulation of GIRK channels by D2R/GIRK association and Gβγ subunit signaling.

Another intracellular signaling pathway regulated by the D2R is the Akt/glycogen synthase kinase 3 (GSK3) pathway. Stimulation of the D2R and the D3R decreases the activity of Akt through dephosphorylation [128, 129]. The regulation of the Akt pathway is independent of the G protein and requires the recruitment of β-arrestin 2 to the D2R [129–131]. β-arrestin 2 then acts as a scaffold recruiting Akt and protein phosphatase-2A [131]. Akt is a kinase that phosphorylates GSK3α and β inactivating them [129]. Thus, D2R stimulation increases the activity of GSK3 by preventing its phosphorylation by Akt. Reducing the activity of the D2R/Akt/PP2A/GSK3 pathway also reduces locomotor activity induced by psychostimulants such as amphetamine. β-arrestin 2 knockout mice and genetic and pharmacological inhibition of GSK3 all significantly reduce psychostimulant-induced locomotor activity [129, 131].

Regulator of G protein signaling 9 (RGS9) has been implicated in reducing D2R signaling activity in the striatum. RGS9 is a protein that accelerates the GTPase activity of G proteins increasing GTP hydrolysis and Gα subunit inactivation [132]. RGS9 is expressed selectively in striatal medium spiny neurons [133]. The D2R and RGS9 colocalize in striatal neurons and in transfected mammalian cells [132]. In RGS9 knockout mice, cocaine-induced locomotor activity and rewarding properties are increased [134]. Furthermore, overexpression of RGS9 in the nucleus accumbens reduces cocaine and D2R agonist-induced increases in locomotion, while D1R agonist locomotion was not affected [134]. Taken together, these studies indicate that RGS9 has an important function in attenuating D2R signaling.

The D2R undergoes desensitization and endocytosis after which it can be recycled back to the membrane to resensitize signaling or be degraded in the lysosome. D2R phosphorylation may play a role in the recruitment of β-arrestin to the D2R and initiate receptor endocytosis. Specifically, protein kinase C can phosphorylate the D2R decreasing the ability of the D2R to couple to $G_{i/o}$ and inhibit cAMP production [135]. Protein kinase C phosphorylation increased D2R desensitization and endocytosis [135]. In addition, G protein-coupled receptor kinase (GRK) 2 phosphorylates the D2R and increases receptor endocytosis and recycling [136–138]. The function of the four nonvisual GRKs expressed in the striatum on dopamine receptor signaling is reviewed in the following reference (139). GRK2 and GRK6 appear to have important functions *in vivo* for D2R-mediated signaling and differentially regulate psychostimulant-induced behavior [139]. Interestingly, GRK6 knockout mice have enhanced locomotor responses to cocaine and amphetamine and D2R G protein signaling is augmented in the striatum [140]. D2R phosphorylation has an important function in regulating receptor signaling, desensitization, and trafficking.

19.11 Dopamine D3 Receptors

The D3R is a member of the D2-like subtype and is expressed in many of the same brain regions as the D2R. The D3R is expressed in the nucleus accumbens and expression is generally limited to limbic regions in rats [141] but may be more widely expressed in human brains [142, 143]. Due to the crossover in expression patterns, most studies focusing specifically on the D3R have been conducted in cell lines to avoid convoluting the results with D2R signaling. The D3R couples to $G_{i/o}$ G proteins, and there are rare reports that it may couple to the $G_{q/11}$ G proteins as well [144]. Interestingly, the D3R exhibits weaker inhibition of adenylyl cyclase than the D2R when it is expressed at similar or much higher levels than the D2R [144, 145]. In addition, the D3R G protein-mediated inhibition of cAMP appears to be selectively accomplished through adenylyl cyclase 5 activation rather than other adenylyl cyclase isoforms [146]. In addition to the effects on the second messenger cAMP, the D3R can modulate multiple ion channels by similar mechanisms as the D2R. Stimulation of the D3R increases the activation of GIRK channels, hyperpolarizing neurons [123]. GIRK activation is dependent on the Gβγ subunit of the heterotrimeric G protein and is pertussis toxin-sensitive [123, 147]. There are also several reports that the D3R modulates voltage-dependent K^+ channels through a $G_{i/o}$-mediated mechanism [148, 149]. The D3R modulates Ca^{2+} currents through N/Q-type Ca^{2+} channels decreasing Ca^{2+} influx into cells. Interestingly, the modulation of the Ca^{2+} channels was found to be $G_{i/o}$-dependent as pertussis toxin prevented D3R-mediated decreases in inward Ca^{2+} currents [150, 151]. Finally, the D3R has been implicated in the regulation of $GABA_A$ receptors through increasing the endocytosis of the $GABA_A$ receptor, likely by preventing PKA-induced phosphorylation of the $GABA_A$ receptor [152].

The D3R appears to be regulated by GRKs and β-arrestins in a different manner than the D2R. Compared to the D2R, the D3R underwent little receptor phosphorylation, β-arrestin recruitment, and subsequent endocytosis [153]. The authors of this study went on to further demonstrate that intracellular loop 3 residues were essential to the control of β-arrestin recruitment and endocytosis using chimera receptors that swapped intracellular loop 3 residues [153]. Interestingly, D3R desensitization is rapidly achieved through a β-arrestin–dependent mechanism that does not involve receptor endocytosis and was GRK2 independent [154]. Instead, β-arrestin recruitment leads to the sequestration of the D3R in hydrophobic lipid rafts on the plasma membrane [154]. Furthermore, D3R desensitization may involve the interaction of β-arrestin with the Gβγ subunit associated with the D3R heterotrimeric G protein [154, 155]. Regulation of the D3R is unique for dopamine receptors and appears not to require receptor phosphorylation or internalization for desensitization and resensitization.

19.12 Dopamine D4 Receptors

The final dopamine receptor discussed here is also in the D2-like subtype and inhibits cAMP production through adenylyl cyclase (see Figure 19.2b). Comparatively little is known about the D4R specifically. The D4R is widely expressed in the brain and is highly expressed in the prefrontal cortex, amygdala, hippocampus, hypothalamus, and pituitary [156, 157]. Unlike the other dopamine receptors, little D4R mRNA is found in the striatum [156]. Interestingly, the D4R is most abundant in the retina [158]. The D4R couples to $G_{i/o}$ and inhibits adenylyl cyclase production of cAMP via pertussis toxin-sensitive G proteins [159, 160]. In addition, the D4R inhibits L-type Ca^{2+} channels similar to the D2R in multiple cell lines [161, 162]. The D4R also increases the activation of GIRK channels via a Gβγ-dependent mechanism like the D2R [147, 163]. In contrast to the D2R and D3R, the D4R reduces voltage-dependent outward K^+ currents [148]. In the prefrontal cortex, the D4R decreases $GABA_A$ receptor currents by decreasing PKA activity and increasing PP-1 activity, likely decreasing $GABA_A$ receptor phosphorylation [164]. Interestingly, infusion of D4R agonists or psychostimulants decreases locomotor activity in rats when injected directly into the reticular nucleus of the thalamus or SN, likely due to decreasing the activity of GABAergic neurons [165]. Finally, the D4R uniquely has a proline-rich region in intracellular loop 3 that has been shown to interact with multiple proteins containing SH3-binding domains [166].

The D4R does not interact with β-arrestin and undergo agonist-dependent endocytosis in the same manner as the D2R. The D4R exhibits decreased phosphorylation, endocytosis, and degradation in CHO, HeLa, HEK293, HT22, and primary rat hippocampal cells [167]. Intriguingly, there is constitutive interaction between the D4R and β-arrestin 2 that has not been observed with the D2R [167, 168]. The D4R also interacts with a complex that induces D4R ubiquitination and that complex includes β-arrestins [168, 169]. However, the functional significance of D4R ubiquitination or constitutive β-arrestin engagement is not known especially since the D4R does not readily undergo endocytosis and D4R ubiquitination does not induce receptor degradation [170]. Another study reported that the D4R undergoes endocytosis but only when both visual and β-arrestins were transfected into the HEK293 cells, which may be functionally relevant for the retinal D4Rs but not for the D4R expressed in the brain [171].

19.13 Dopamine Receptors in Substance Use Disorders and Drug Taking: Preclinical Models

Several animal models have been created to investigate different aspects of drug addiction [27, 172, 173]. The groundbreaking animal research by Olds and Milner [174] provided a key foundation for our modern understanding of brain mechanisms in reward. In these seminal experiments, rats were provided the ability to self-administer electrical stimulation to different brain regions including the mesolimbic dopamine pathway. The rats repeatedly and often persistently would choose to stimulate the mesolimbic dopamine pathway, often to the exclusion of other behaviors. Behavioral studies in rodents also clearly indicate that dopamine is essential for the self-administration of drugs of abuse for which the mesolimbic dopamine pathway is a crucial substrate [39, 175]. Drug self-administration is a "gold standard" for animal models of drug abuse [27, 176]. A typical drug self-administration procedure involves animals working to obtain a drug by performing a simple behavior (such as pressing a lever), and animals will readily self-administer the same drugs that are abused by humans [176]. Therefore, models that utilize self-administration of drugs are thought to best capture the human condition where drugs that are self-administered by animals correspond well with those that have abuse potential in humans. Additional models include behaviors related to exposure to cues and contexts previously paired with drug self-administration and measurements of impulsive behavior (for recent comprehensive reviews see: [173, 177]).

19.13.1 Psychostimulants

The importance of the D1R in psychostimulant substance use has been known for more than 30 years. D1-like antagonist pretreatment dose-dependently increases cocaine fixed ratio self-administration in rats but decreases progressive ratio self-administration [178–180]. Interestingly, similar results were also obtained with D2-like receptor antagonists where fixed ratio, but not progressive ratio, self-administration of cocaine or amphetamine, was increased [179–181]. Inhibiting the D3R and D4R through genetic and pharmacological approaches produced no increase in cocaine fixed ratio self-administration even though D2R antagonism increased it [182–185]. However, D3R antagonists decrease cocaine self-administration on progressive ratio schedules suggesting that D3Rs are important in reward but are not involved in feedback regulation [184–186]. The early studies using mainly pharmacological approaches determined that D1Rs and D2Rs are essential for the rewarding effects of psychostimulants.

Stimulating the D1-like or D2-like receptors with full agonists reduces cocaine fixed ratio self-administration, which is the opposite effect of antagonists that increase cocaine fixed ratio self-administration. Cocaine self-administration follows a dose-dependent inverted U curve where low doses and high doses of cocaine do not effectively induce self-administration. Pretreatment with a full D1-like agonist shifted the inverted U dose–response down, deceasing self-administration across all doses tested [187]. In contrast, D2-like receptor agonists decreased cocaine self-administration by shifting the inverted U to the left (higher responses at lower cocaine doses) and increasing the intervals between cocaine infusions [187]. Similar results were also observed in rhesus monkeys trained to self-administer cocaine [188]. Thus, both D1-like and D2-like receptor agonists decrease cocaine self-administration but in different ways.

Knocking out the D1R or the D2R in mice further confirmed their essential roles in psychostimulant self-administration. D1R knockout mice do not readily acquire cocaine self-administration [189]. Furthermore, D1R knockout but not D5R knockout blocks the locomotor response to acute cocaine administration and cocaine sensitization after repeated injections [190, 191]. However, D2R knockout mice acquire cocaine self-administration and even display increased fixed ratio self-administration schedules, like antagonist treatments [182]. Interestingly, the D2-like antagonist, eticlopride, does not increase fixed ratio self-administration in D2R knockout mice indicating that it is D2R antagonism not D3R or D4R that mediate the increased cocaine intake [182]. It is unclear if the D3R plays a significant role in psychostimulant self-administration. There are reports that D3R knockout does not affect cocaine self-administration in fixed or progressive ratio compared to wildtype mice [192]. On the other hand, there are also reports that D3R knockout enhances both fixed and progressive ratio cocaine self-administration [193]. Further studies are required to determine the function of the D3R on psychostimulant self-administration and to resolve the conflicting reports. Consistent with the antagonist studies of the D4R, D4R knockout did not affect fixed ratio responding for cocaine [194, 195]. Taken together, these knockout studies demonstrate that the D1R is essential for the rewarding effects of cocaine, and the D2R is important for negative feedback decreasing psychostimulant self-administration.

Agonist and antagonist injections into specific brain regions have further clarified the function of D1-like and D2-like receptors. In the nucleus accumbens shell, injection of antagonists for the D1-like and D2-like receptors increases psychostimulant fixed ratio self-administration, like systemic antagonist treatment [196–199]. However, infusion of D1-like or D2-like receptor agonists into the nucleus accumbens shell did not decrease cocaine self-administration unlike systemic administration [198]. The authors suggested that the D1-like and D2-like receptors in the nucleus accumbens shell are super-saturated with dopamine and thus, stimulating them further has no affect [198]. Alternatively, the reduction in cocaine intake may require the activation of both D1-like and D2-like receptors or dopamine receptors in other brain regions [198, 200]. The nucleus accumbens shell also drives motivation for cocaine in progressive ratio self-administration schedules. Infusing D1-like or D2-like antagonists into the nucleus accumbens shell decreased cocaine self-administration on progressive ratio schedules [201, 202]. Interestingly, the effects of the D1-like and D2-like receptor antagonist injection into the nucleus accumbens shell was selective for cocaine and did not affect food reward [201]. Together, these results are consistent with systemic administration of D1-like and D2-like antagonist pretreatment.

Infusion of the D1-like receptor antagonist SCH-23390 indicates that psychostimulant intake and motivation is modulated by multiple brain regions. SCH-23390 increases cocaine fixed ratio self-administration when injected into the prefrontal cortex [203], insular cortex [204], and amygdala [199, 202, 204]. When injected into the prefrontal cortex, SCH-23390 also reduces progressive ratio responding for cocaine suggesting that D1-like receptors in the prefrontal cortex are involved in the motivation and intake of cocaine [203]. Together, these studies indicate that the D1-like receptors in the nucleus accumbens shell, prefrontal cortex, insular cortex, and amygdala all modulate psychostimulant self-administration.

19.13.2 Opioids

Opiates such as morphine, heroin, and others stimulate the mesolimbic dopaminergic pathways by increasing dopamine release. However, the mechanism by which dopamine is increased differs from that of psychostimulants. Opiates increase dopamine release from VTA neurons by inhibiting GABAergic neurons, disinhibiting the dopaminergic neurons [205, 206]. While dopamine has a major role in the rewarding effects of opiates, opiates also have dopamine-independent mechanisms for opiate reward [207, 208]. In rats that were trained to self-administer heroin, lesioning the dopaminergic neurons in the nucleus accumbens reduced heroin self-administration to 76% of prelesions levels but reduced cocaine self-administration to 30% prelesion levels indicating that there are dopamine-independent mechanisms for heroin reward [207]. It is well established that opiates elevate mesolimbic dopamine levels, but due to the indirect mechanism (disinhibition), there are fewer dopamine receptor subtype specific studies on opiate substance use disorders.

The role of the D1R in opioid reward is still under investigation. Systemic administration of the D1R antagonist SCH-23390 significantly reduced the acquisition of heroin self-administration [209]. However, local intracranial injection of SCH-23390 into the nucleus accumbens did not affect heroin intake in fixed ratio self-administration suggesting that D1-like receptors in other brain regions may be involved [209]. Mice can also be trained to self-administer morphine directly into the lateral septum, and this was blocked by systemic administration of the D1-like receptor antagonist SCH-23390 [210]. In addition, D1Rs in the ventromedial prefrontal cortex may be involved in opiate reward. Coinjection of the mu-opioid receptor agonist DAMGO and SCH-23390 into the ventromedial prefrontal cortex reduced progressive ratio self-administration compared to injecting DAMGO alone suggesting that D1R blockade in this brain region reduced motivation for opiate reward [211]. In contrast to systemic pharmacological D1-like receptor blockade, D1R knockout mice self-administer opioid receptor agonists like wildtype controls [189]. Little is known about the role of the D5R in opiate reward and the involvement of the D5R may explain some of the discrepancies between the pharmacological and D1R knockout studies. Finally, D1-like receptor agonism with SKF-81297 or SKF-82958 enhanced the rewarding effects of heroin by shifting the dose response curve leftward in rhesus monkeys [212]. Together, these studies indicate that D1-like receptors are important for the rewarding effects of opiates, but there are D1-like receptor-independent mechanisms as well.

In contrast to the results above, conditioned place preference studies indicate the D1-like receptors in the nucleus accumbens have a role in opiate reward. Injection of SCH-23390 into the nucleus accumbens decreased the development and maintenance of conditioned place preference to morphine suggesting that the D1R may be important for the rewarding properties of opiates [213, 214]. Similarly, 6-OHDA lesions of dopaminergic neurons in the nucleus accumbens also decreased the development of morphine conditioned place preference; however, this type of lesion reduces dopamine input into the area, affecting all dopamine receptor subtypes [214]. Additionally, injection of SCH-23390 into the nucleus accumbens reduced reinstatement of morphine conditioned place preference [215]. Finally, systemic D1R agonist administration during extinction training increased conditioned place preference and dendritic arborization in the nucleus accumbens core but not shell [216]. Together, the studies on opiate conditioned place preference indicate that D1-like receptors in the nucleus accumbens are important for opiate reward.

The D2R appears to be critical for opiate reward mechanisms. Wildtype mice will lever press for morphine in fixed ratio and progressive ratio schedules, but D2R knockout mice did not respond for either fixed ratio or progressive ratio self-administration [217]. When the D2R antagonist sulpiride (also a D3R antagonist) was injected into the VTA, morphine self-administration was reduced suggesting that D2R presynaptic auto-receptors are involved in morphine reward [218]. When sulpiride was injected into the nucleus accumbens, D2/D3R blockade had no effect on morphine self-administration adding further evidence that D2/D3R auto-receptors are involved in opiate reward rather than postsynaptic receptors [218]. Additionally, in rhesus monkeys trained to self-administer heroin, D2-like receptor agonist co-administration with heroin shifted the dose–response curve to the right which may reflect activation of the D2R auto-receptors [212]. The D2R auto-receptor activation may counteract the effects of heroin to disinhibit dopaminergic neurons. Interestingly, mice also self-administer morphine into the lateral septum. Self-administration into the lateral septum was dependent on D2/D3 receptors as systemic sulpiride blocked the acquisition of morphine self-administration [210].

Additionally, a recent study found that morphine increased the activity of excitatory D1R medium spiny neurons in the nucleus accumbens after forced abstinence but reduced the activity of D2R medium spiny neurons. Reversing the effects of morphine on the activity of D1R and D2R medium spiny neurons using optogentic stimulation prevented morphine-conditioned place preference [219]. Furthermore, morphine can reduce long-term potentiation and even induce long-term depression in GABAergic interneurons in the VTA and D2/D3R antagonists prevent the reduction in GABAergic interneuron activity [220].

The D3R appears to limit opioid self-administration since D3R knockout mice acquire heroin self-administration faster and had higher heroin intake during both acquisition and maintenance [221]. In contrast, D3R selective antagonists reduce the self-administration of heroin [222].

19.13.3 Alcohol

Alcohol reward processes are usually studied by giving the animal a choice between alcohol and water or sucrose rather than the self-administration/lever pressing as discussed for other substances of abuse. The D1-like receptor antagonist SCH-23390 and the D2-like receptor antagonist raclopride did not influence ethanol consumption [223]. Another D2-like antagonist, remoxipride, also had no effect on ethanol intake in a self-administration procedure requiring responses to obtain the reward (similar to self-administration of other substances) [224]. In contrast, the D1R partial agonist SKF-38393 reduced ethanol intake and preference in a dose-dependent manner [223]. The D2/D3R agonist 7-OH-DPAT increased ethanol consumption and preference but only at low doses [223]. In another study, SCH-23390 dose dependently decreased alcohol intake in alcohol preferring female rats and SKF-38393 also decreased alcohol intake in these rats [225]. However, it is important to note that SCH-23390 can also reduce water consumption indicating the effect may be nonspecific [223]. Alternatively, SKF-38393 is a partial agonist and reduces D1R activity under high dopamine tone making it appear as an antagonist under those conditions. Furthermore, SCH-39166, a D1-like receptor antagonist, reduced ethanol intake without reducing water or food ingestion at moderate doses indicating that D1-like receptors have an important function in alcohol reward [226]. Taken together, these early studies on alcohol reward indicate that D1-like receptors play a major role in alcohol reward, whereas the D2-like receptors may not be necessary.

D2R auto-receptors oppose the disinhibition of the dopamine neurons induced by alcohol. While D2-like receptor antagonists had no effect on alcohol intake, the

D2-like receptor agonist quinpirole reduces alcohol intake [225]. However, this contrasts with the results using low doses of 7-OH-DPAT [223]. To further clarify the role of the D2R auto-receptors, quinpirole co-injection with alcohol into the VTA reduced the acquisition and maintenance of alcohol self-administration [227]. This study confirmed that D2R auto-receptors in the VTA oppose the rewarding effects of alcohol. On the other end of the circuit in the nucleus accumbens, D1R antagonists also reduce alcohol self-administration [228]. In the bed nucleus of the stria terminalis, D1R antagonism but not D2R antagonism reduced alcohol self-administration [229]. These studies implicate both D1-like and D2-like receptors in alcohol reward although through different brain regions. Both D1R and D2R knockout mice exhibit an aversion to alcohol providing further support for the involvement of both receptors in alcohol reward [230, 231].

Studies that specifically investigate the role of the D3R in alcohol reward indicate that it plays an important role. Administration of the D3R antagonist U99191A did not affect alcohol preference in mice, but high doses of SB-277011A reduces alcohol intake in mice and rats [232–234]. Moreover, D3R knockout mice consumed significantly less alcohol in two-bottle choice and binge-like paradigms [235]. Additionally, viral-mediated knockdown of the D3R in the nucleus accumbens reduced alcohol intake in the two-bottle choice test while overexpression increased alcohol intake [236]. While additional studies are needed to determine the specific role of the D3R versus the D2R in alcohol reward, these studies indicate the D3R is important for alcohol intake. There are few studies looking at D4R-specific effects on alcohol consumption and reward due to the lack of selective D4R agonists and antagonists. Recently, new ligands targeting the D4R were used to demonstrate that D4R antagonism reduces alcohol intake, but agonists had no effect [237]. Intriguingly, D4R knockout reduced alcohol intake but only in male mice [238]. The role of the D4R in alcohol reward and consumption requires further investigation and is complicated by the polymorphism within the D4R gene, which may differentially affect alcohol reward or behavior [239].

19.14 Dopamine Receptor Pharmacology for Substance Use Disorders

19.14.1 Nonselective Ligands

Since the discovery of dopamine in 1910 by Wellcome labs [240], and its receptors 60 years later, numerous nonselective dopamine agonists have been created with activity at multiple dopamine and other biogenic amine receptors. These agents, such as apomorphine or bromocriptine, have been used to investigate dopamine

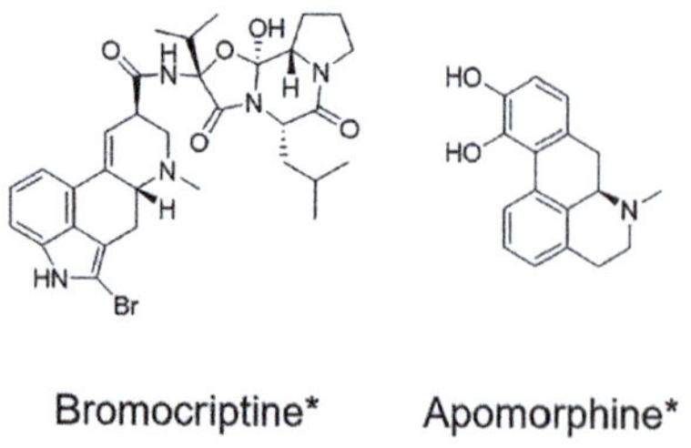

Figure 19.3 Structures of nonselective dopamine receptor agonists. The asterisk indicates compounds that have entered clinical trials for substance use disorders.

receptor activation as a therapeutic approach to substance use disorders. Apomorphine (Figure 19.3) is a partial agonist at all dopamine receptors, with highest affinity for the D4R. Bromocriptine (Figure 19.3) is a potent dopamine agonist with high affinity at D2R and D3R, but also serotonin (5-HT) and α-adrenergic receptors. There have been several clinical studies of nonselective dopamine agents for use in substance use disorders.

Apomorphine (Figure 19.3) was investigated several decades ago in clinical trials for alcohol use disorder, with mixed reviews. In one case series in 1952, ~40% of 500 alcohol-dependent patients reported abstinence after five years of apomorphine [241]. In 1972, 90% of 51 alcohol-dependent war veterans reported drug-craving relief [242]. In 1977, apomorphine was given to intoxicated patients and significantly increased sobriety on the following day, but it did not alter long-term abstinence [243]. However, many of these studies were not properly regulated or did not use the proper controls thus producing unreliable results to translate to modern-day positive clinical outcomes. This combined with the common side effect of emesis led to the discontinuation of apomorphine as a clinically viable candidate.

Bromocriptine (Figure 19.3) was also investigated in the clinic to reduce drinking in alcoholics with conflicting outcomes. One small study of 52 alcohol-dependent subjects showed bromocriptine was effective in reducing craving and anxiety scores [244]. However, a larger study of 366 participants showed bromocriptine was not effective at preventing relapse or reducing the amount of alcohol intake [245]. Interestingly, some evidence exists that bromocriptine is reinforcing in its own right as it is self-administered in rhesus monkeys [246]. Nonselective dopamine agonists also exhibit drawbacks in that there is a narrow therapeutic window and numerous deleterious side effects, such as emesis, hypotension, headaches, and worsening of psychotic symptoms, which limit their therapeutic potential. These combined challenges led to the abandonment of bromocriptine and other nonselective dopamine agents as pharmacotherapeutics for substance use disorders and refocused efforts to identify and evaluate receptor subtype selective compounds.

19.15 Dopamine D1 Receptor Subfamily Ligands

19.15.1 Agonists

The D1R has several selective agonists developed over the last several decades, the most studied of which are the benzazepines (e.g. SKF-81297, see Figure 19.4) and the tetrahydroisoquinolines (e.g. dihydrexidine, see Figure 19.4). Both pharmacophores mimic the structure of dopamine, with a dihydroxyphenyl (catechol) group attached to a conformationally restrained basic nitrogen (phenylethylamine). Molecules from both these pharmacophores exhibit moderate-to-excellent selectivity over D2R (10–1000-fold). However, the catechol moiety associated with these compounds significantly inhibit the pharmacokinetic (PK) profile, in that it causes negligible oral bioavailability, rapid metabolism, and poor central nervous system penetration [250]. Recently, Pfizer has developed a series of novel noncatechol agonists (e.g. PF-6142, see Figure 19.4) that

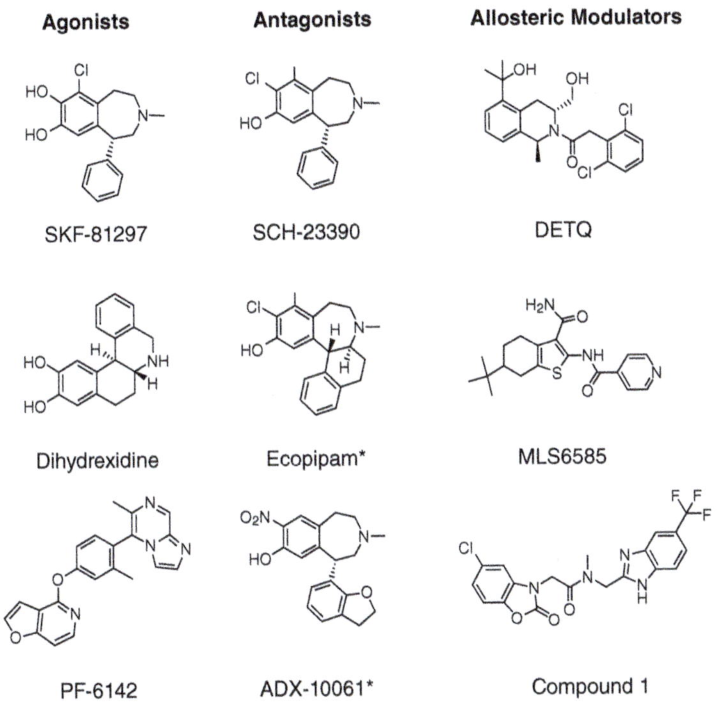

Figure 19.4 Structures of selective dopamine D1-like receptor family ligands. The asterisk indicates compounds that have entered clinical trials for substance use disorders. Source: Based on Hyttel [247], Iorio et al. [248], and Heidbreder [249].

have circumvented the poor PK profile, while maintaining moderate selectivity over D2R [13].

Benzazepines were some of the first D1R selective agonists discovered (e.g. SKF-81297 [Figure 19.4], SKF-38393) and remain as some of the best tools to probe D1R agonism *in vitro*. Numerous structure activity relationship (SAR) studies have been completed to interrogate necessary interactions for the pharmacophore [251–254]. Notably, these molecules display excellent affinity and potency for the D1R and D5R with ~100-fold selectivity over the D2R but are commonly partial agonists. Interestingly, select molecules from this pharmacophore show biased agonism for G_s signaling over β-arrestin recruitment, although the structural determinants governing biased responses are inconclusive [100]. Even after numerous decades of research and optimization, benzazepines suffer from poor PK, limiting the ability to sufficiently interrogate the D1R *in vivo*, and thus have not been useful clinically for substance use disorders.

Tetrahydroisoquinolines (e.g. dihydrexidine [Figure 19.4], dinapsoline) have also been studied extensively as D1R selective agonists, which display high affinity for D1R, with moderate selectivity over D2R (~10-fold). One advantage of the tetrahydroisoquinolines is that they were the first full D1R agonists discovered [255, 256], while the benzazepines are typically partial agonists at the D1R. Tetrahydroisoquinolines display slightly improved PK over benzazepines, but they still suffer from PK deficits and exhibit dose-limiting hypotension clinically.

The newly developed noncatechol D1R agonists from Pfizer have represented a new breakthrough for pharmacotherapeutic potential of D1R agonists. Using a high-throughput screen of ~3 million molecules, and meticulous medicinal chemistry efforts, a series of orthosteric agonists with high potency (low nanomolar), and a range of efficacies, from low partial to full agonism were developed (e.g. PF-6142 [Figure 19.4], PF-2334, PF-1119). Interestingly, these molecules also display varying degrees of G_s biased signaling from fully biased to balanced agonism [13, 257]. Although the molecular determinants of these interactions remain elusive, they represent an opportunity to interrogate how D1R signaling bias may affect therapeutic potential. Due to the removal of the catechol group, these molecules exhibit excellent PK properties including slow metabolism, high central nervous system penetration, and oral bioavailability, these molecules offer a new opportunity to interrogate D1R agonism *in vivo*. Recently, an exemplar molecule (PF-9751) from this series was determined to be safe and pharmacokinetically viable in phase I clinical trials [258]. While the primary implication of these molecules was for Parkinson's disease, they may also represent a unique opportunity for substance use disorders. Notably these molecules were recently licensed to Cerevel for clinical advancement (www.cerevel.com). One area largely unexplored is the testing and validation of D1R partial agonists in substance use disorder models. Since partial agonists have the unique activity of only weak or moderate agonism, such ligands might normalize D1R responses providing receptor activation when dopamine levels are low or decreased (e.g. during drug withdrawal and abstinence) while reducing D1R activation when dopamine levels are in excess, such as during drug taking.

19.15.2 Antagonists

Molecules from the benzazepine pharmacophore can also be highly selective antagonists for the D1R, with sub-nanomolar affinity and excellent selectivity over the D2R (>100-fold). The main determinant of this pharmacophore for agonism versus antagonism at the D1R is the halogen substitution at C7 [259]. The first discovered exemplar molecule from this pharmacophore was SCH-23390 (Figure 19.4) [247, 248]. SCH-23390 shows excellent affinity for the D1R, enantioselectivity (with the R-(−) enantiomer being active), but has activity at serotonin (5-HT) receptors. SCH-23390 remains a useful tool for interrogation of the D1R, due to high affinity, and is commonly used to radiolabel the D1R. Several SAR studies have also been completed to optimize the molecule for selectivity and PK [260, 261].

One of the results from these SAR studies was SCH-39166, or ecopipam (Figure 19.4). While affinity is similar to SCH-23390, the additional ethylene bridge and conformational restraint yielded increased selectivity over 5-HT receptors, and duration of action likely due to PK. Due to ecopipam's affinity, favorable PK, and selectivity, it has been used in several clinical trials, including for cocaine-use disorder. One study of 10 cocaine-dependent subjects showed it acutely blocked the euphoric effects of cocaine, but ultimately increased self-administration of cocaine [262]. Subsequent studies showed that repeated administration of ecopipam failed to alter the direct effects or desire for cocaine [263]. These results, coupled with the side effects of increased depression, anxiety, and suicidal thoughts in a large phase III clinical trial for obesity resulted in abandonment of ecopipam for substance use disorders [264]. Similarly, Addex therapeutics established a clinical trial with ADX-10061 [249] (Figure 19.4), a similar selective D1R antagonist for smoking cessation; however, it did not meet the primary efficacy endpoint compared to placebo in 145 subjects, and the compound was abandoned as a pharmacotherapy. Taken together, it appears that selective D1R antagonists may not be viable pharmacotherapies for substance use disorders.

19.15.3 Allosteric Modulators

Allosteric modulators are a recent addition to the pharmacological repertoire targeting the D1R. Since these molecules bind to nonevolutionarily conserved binding sites, they are not limited to the typical constraints of previously described D1R molecules, such as limited PK and vast off-target side effects. Within the last few years, several structurally diverse, and pharmacologically distinct, molecules were identified and optimized as D1R positive allosteric modulators (PAMs).

The first to be described was Eli Lilly's DETQ (Figure 19.4) [2-(2,6-dichlorophenyl)-1-((1S,3R)-3-(hydroxymethyl)-5-(2-hydroxypropan-2-yl)-1-methyl-3,4dihydroisoquinolin-2(1H)-yl)ethan-1-one] [265, 266]. DETQ induces a 21-fold potency shift of dopamine at the D1R but is inactive at the D5R, which represents a unique pharmacological aspect of distinguishing between the activation of the D1R and D5R, which had not been seen with previously described orthosteric agonists. DETQ displays agonist activity on its own, with a potency of 5 nM, suggesting that it is an ago-PAM. DETQ also increases locomotor activity in mice, with the human D1R knocked in, without causing stereotypy or tachyphylaxis [266]. Additionally, it causes an increase in spontaneous eye-blink rate in rhesus monkey, a common behavioral marker of D1R activation *in vivo* [267]. DETQ has been in several clinical trials designed around cognition in mild-to-moderate Parkinson's disease dementia (NCT03616795, NCT02562768, NCT02365571, NCT02603861, and NCT03305809), but could also be repurposed for stimulant use disorders.

Another pharmacologically unique D1R PAM is MLS6585 (Figure 19.4), discovered in a high-throughput screen using a small-molecule library from the NIH Molecular Libraries program [268]. This compound does not display any agonist activity on its own, suggesting it is a pure PAM. MLS6585 increased dopamine's potency by 8-fold, increasing its efficacy by 34%, while displaying low micromolar affinity. MLS6585 has been well characterized *in vitro*; however, future studies are needed to validate this tool for *in vivo* activity and therapeutic potential.

Similar to Eli Lilly, Astellas has recently disclosed structures of a PAM in a patent [269]. Several molecules based around the structure of compound 1 (Figure 19.4), were described that show moderate potency at the human D1R, and ~7-fold increase in the potency of dopamine at the D1R. Astellas does have a compound, ASP4345, in phase II clinical trials for cognitive impairment associated with schizophrenia, but the structure of this PAM has not been disclosed [270]. These are just some of the D1R PAMs that have recently been disclosed in the last five years, and they represent a unique and innovative strategy to investigate the impact of D1R activation in substance use disorders.

19.16 Dopamine D2 Receptor Subfamily Ligands

19.16.1 Agonists

Selective agonists of the D2R family have been around for 40 years with Eli Lilly's discovery of quinpirole (Figure 19.5) [271, 272]. However, selective D2R full

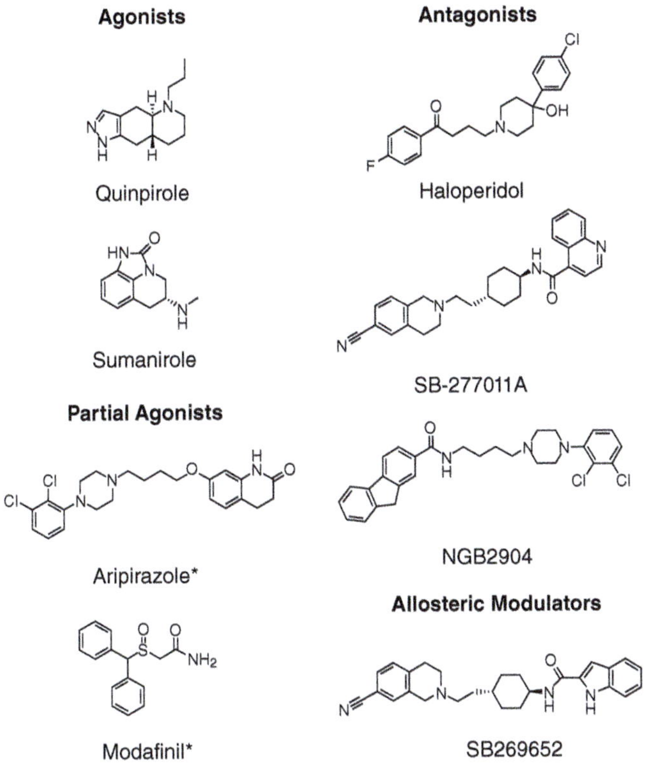

Figure 19.5 Structures of selective dopamine D2-like receptor family ligands. The asterisk indicates compounds that have entered clinical trials for substance use disorders. Source: Based on Bach et al. [271] and Tsuruta et al. [272].

agonists are commonly psychoactive and may be self-administered [273, 274], which would not be useful tools for substance use disorders. Interestingly, recent evidence suggests partial agonists, such as aripiprazole (Figure 19.5), at the D2R, act as dopamine normalizers and may be clinically useful tools for substance use disorders [275].

Quinpirole (Figure 19.5) is one of the most widespread tools used for studying the D2R both *in vitro* and *in vivo*. It displays potent full agonist D2R-mediated activation of G_i (3 nM) and β-arrestin recruitment (2 nM) [276]. Data do suggest that quinpirole is slightly selective for D3R over D2R *in vitro* [277, 278], but *in vivo* it seems to act as a dual agonist [279]. However, sumanirole (Figure 19.5), also a D2R full agonist, shows greater selectivity of D2R over D3R activation [280, 281] and may be able to separate activation of these receptor subtypes *in vivo* [273, 279].

Partial agonists of the D2R, on the other hand, have been heavily investigated for use clinically, as both antipsychotics and as a treatment for substance use.

The exemplar molecule for this series of molecules is aripiprazole (Figure 19.5), which is primarily used as an antipsychotic [282], but has been heavily studied in preclinical and clinical models of substance use disorders. Aripiprazole has not only potent activation of the D2R (38 nM) and partial agonist activity of G_i mediated signaling (51%) but also shows potent (2 nM) and efficacious (71%) D2R-mediated β-arrestin recruitment and is suggested to be β-arrestin biased [276, 283]. Depending on the context, due to partial agonist activity, aripiprazole should antagonize dopamine when tone is high, whereas it should increase dopamine transmission when basic tone is low, and thus represents a unique treatment for substance use disorders. The most encouraging evidence for aripiprazole being a pharmacotherapy for substance use disorders has been with alcohol-dependent subjects. In small clinical trials, aripiprazole reduces overall drinking [284], especially in subjects that struggle with impulse control [285]. A larger 12 week double-blind, placebo-controlled study of 295 subjects showed that it initially decreased heavy drinking days but was not significant when the larger target dose (30 mg) was reached. However, a separate study showed a lower dose was just as effective as naltrexone at reducing drinking and relapse [286]. Due to its primary indication as an antipsychotic, aripiprazole has also been investigated in patients with bipolar or schizoaffective disorders with comorbid substance use disorders and shown positive effects [287, 288]. Aripiprazole is commonly well tolerated with the primary side effects of nausea, tiredness, and headaches and represents one of the most-studied dopaminergic drugs for substance use disorders.

Modafinil (Figure 19.5) is another D2R partial agonist, which has relevance for substance use disorders. Modafinil displays a 16 nM affinity with a 120 nM potency and 48% efficacy vs. dopamine for the D2R [289], and decent selectivity over the D1R. Modafinil displays a complicated polypharmacologic profile, through several other possible mechanisms of actions, including activation of adrenergic α_1 receptors as well as modulation of glutamate, GABA, histamine, and hypcretin [290]. The primary indication of modafinil is as a eugeroic for the treatment of narcolepsy, shift work sleep disorder, and sleep apnea. However, there have been clinical studies of modafinil for the treatment of cocaine-use disorder [291, 292]. In these small clinical studies, modafinil (400 mg/day) was shown to significantly decrease positive urine samples of cocaine metabolite benzoylecgonine and significantly increase periods of cocaine abstinence. There has been recent scrutiny of abuse potential of modafinil due to athletes using it as a stimulant to train longer [293]. Modafinil has subsequently been banned by the World Anti-Doping Agency and classified as a Schedule IV controlled substance by the FDA for its low but significant addiction potential. However, in a study of chronic cocaine users, they were able to fully discriminate between modafinil and cocaine and reported no euphorigenic activity [294, 295]. Taken together, evidence from aripiprazole and modafinil studies suggest that D2R partial agonists represent a therapeutic avenue for substance use disorders.

19.16.2 Antagonists

Selective D2R subfamily antagonists (e.g. haloperidol, see Figure 19.5) are classified as typical antipsychotics and commonly used to treat psychosis associated with schizophrenia, Tourette syndrome, and bipolar disorder [296]. Interestingly, typical antipsychotics such as haloperidol have been used to treat delirium associated with alcohol withdrawal [297], but do not decrease consumption of abused substances. Most typical antipsychotics not only display low nanomolar to subnanomolar affinity for D2R-like receptors but also display relevant affinity for α1-adrenergic, 5-HT, and histamine receptors, and commonly yield adverse effects of extrapyramidal side effects, hypotension, somnolence, and weight gain. Atypical antipsychotics such as risperidone or aripiprazole commonly antagonize or partially activate the D2R-like receptors and have distinct selectivity profiles at other receptors compared to the typical antipsychotics. Atypical antipsychotics have not shown treatment efficacy for cocaine dependence, although results come from only 14 trials, which in some cases had small sample sizes [298]. At present, there is no convincing or consistent clinical evidence that existing D2R antagonists will be useful in the treatment of substance use disorders, but a truly selective D2R antagonist that does not have activity at D3R or D4R has been challenging to identify. Recently, ML321 was identified as a highly D2R subtype selective antagonist and is a promising compound leading toward D2R subtype selective *in vivo* studies [299]. Such highly selective D2R antagonists might allow revisitation and testing in substance use disorder models and may show differentiated activity from the typical and atypical antipsychotics.

Recently, significant evidence has been found that selective blockade of D3R over D2R and D4R receptors may represent a new pharmacotherapeutic approach to treat substance use disorders [300–302]. In 2000, SB-277011A (Figure 19.5) was the first discovered highly selective D3R antagonist showing 11 nM affinity, while showing 1 μM affinity for the D2R (100-fold selectivity) [303, 304]. Since then, SB-277011A has been shown in numerous preclinical animal models of substance use disorders to dose-dependently reduce intracranial self-stimulation of cocaine [305], nicotine [306], and methamphetamine [307]. Additionally, SB-277011A significantly decreased alcohol [233], cocaine [184], and nicotine [308] self-administration under a progressive ratio schedule as well as conditioned place preference [305, 309] and a host of other models (see [300]).

Several other selective D3R antagonists have been developed, through medicinal chemistry efforts, such as NGB 2904 (Figure 19.5) with higher affinity (0.9 nM) and selectivity vs. D2R (~300-fold) [310]. Similar to SB-277011A, NGB

2904 also reduces intracranial self-stimulation of cocaine as well as cocaine self-administration under a progressive ratio schedule [185, 311]. Taken together, SB-277011A and NGB 2904 provide evidence that selective blockade of the D3R is a therapeutically viable target for substance use disorders. However, high lipophilicity and poor aqueous solubility may limit these compound's clinical viability and may require further PK improvements for an optimal therapeutic drug. Notably, Glaxo Smith Kline developed GSK598809, a D3R antagonist that is 100-fold selective for D3R versus D2R. In clinical studies, GSK598809 reduced craving in nicotine-dependence [312] and attentional bias for food cues in overweight subjects [313]. GSK598809 also normalized ventral striatal reward responses and enhanced response in the D3R-rich regions of the human ventral pallidum and SN, suggesting that GSK598809 may remediate reward deficits in substance dependence [314]. D3R antagonists are among the most promising investigational drugs for treating substance use disorders, and further studies are warranted and underway [302, 315].

19.16.3 Allosteric Modulators

Discovery of allosteric modulators of the D2R family have been limited. Recently, a group discovered SB-269652 (Figure 19.5) displayed negative allosteric behavior at both D2R and D3R [316]. Interestingly, this compound was developed from optimization of SB-277011A (Figure 19.5), the selective D3R antagonist, and displays orthosteric effects at low concentrations of dopamine or [^{3}H] spiperone but negative allosteric effects at higher concentrations of dopamine or [^{3}H] spiperone, suggesting that this compound is an atypical negative allosteric modulator [317]. Due to the preclinical evidence of D3R selective antagonists in substance use disorder models, negative allosteric modulators of D3R represent a unique and understudied, but innovative opportunity as a therapeutic modality for substance use disorders.

In summary, numerous structurally diverse and pharmacologically distinct ligands for the dopamine receptor family have been elucidated since the discovery of the dopamine receptor subtypes. Selective and nonselective agonists, antagonists, and more recently allosteric modulators for both the D1R and D2R have been identified and are well characterized *in vitro*. From these tools, several molecules have been useful in both preclinical and clinical models of substance use disorders. The most promising have been the partial agonists of the D2R (aripiprazole and modafinil, see Figure 19.4) and selective antagonists of the D3R (SB-277011A and NGB 2904, see Figure 19.5). Further optimization and evidence are needed to validate and advance and these molecules as bona fide therapeutics for use in substance use disorders.

Abbreviations

5-HT	serotonin
AC	adenylyl cyclase
AMPA	α-amino-3-hydroxy-5-methylisoxazole-4-propionic acid
Amyg	amygdala
CaMKII	Ca^{2+}/calmodulin-dependent protein kinase II
cAMP	cyclic adenosine monophosphate
CDC	Center for Disease Control
CREB	Ca^{2+}/cAMP response element binding protein
D1R	dopamine D1 receptor
D2R	dopamine D2 receptor
D2LR	dopamine D2 long receptor
D2SR	dopamine D2 short receptor
D3R	dopamine D3 receptor
D4R	dopamine D4 receptor
D5R	dopamine D5 receptor
DA	dopamine
DARPP-32	dopamine- and cAMP-regulated phosphoprotein, 32 kDA
DAT	dopamine transporter
ERK	extracellular signal regulated kinases 1 and 2
FDA	Food and Drug Administration
GABA	γ-aminobutyric acid
GIRK	G protein-gated inwardly rectifying K^{+} channels
GRK	G protein-coupled receptor kinase
GSK3	Akt/glycogen synthase kinase 3
NAc	nucleus accumbens
NMDA	N-methyl-D-aspartate
PAM	positive allosteric modulator
PK	pharmacokinetic
PKA	protein kinase A
PP-1	protein phosphatase-1
PP2A	protein phosphatase 2A
RGS9	regulator of G protein signaling 9
SAR	structure activity relationship
SN	substantia nigra
TM	transmembrane
VTA	Ventral tegmental area.

Acknowledgments

This work is supported by the UTMB Center for Addiction Research, United States Public Health Service Grants U18 DA052543 and DA047643 and by a PhRMA Foundation research starter grant in Pharmacology and Toxicology RSGPT18.

References

1 Carlsson, A., Lindqvist, M., and Magnusson, T. (1957). 3,4-Dihydroxyphenylalanine and 5-hydroxytryptophan as reserpine antagonists. *Nature* 180 (4596): 1200.

2 Dichter, G.S., Damiano, C.A., and Allen, J.A. (2012). Reward circuitry dysfunction in psychiatric and neurodevelopmental disorders and genetic syndromes: animal models and clinical findings. *J. Neurodevelop. Disord.* 4 (1): 19.

3 Iversen, S.D. and Iversen, L.L. (2007). Dopamine: 50 years in perspective. *Trends Neurosci.* 30 (5): 188–193.

4 Arnsten, A.F. (2013). The neurobiology of thought: the groundbreaking discoveries of Patricia Goldman-Rakic 1937–2003. *Cereb. Cortex* 23 (10): 2269–2281.

5 Nutt, D.J., Lingford-Hughes, A., Erritzoe, D., and Stokes, P.R. (2015). The dopamine theory of addiction: 40 years of highs and lows. *Nat. Rev. Neurosci.* 16 (5): 305–312.

6 Buchanan, R.W., Freedman, R., Javitt, D.C. et al. (2007). Recent advances in the development of novel pharmacological agents for the treatment of cognitive impairments in schizophrenia. *Schizophr. Bull.* 33 (5): 1120–1130.

7 Girgis, R.R., Van Snellenberg, J.X., Glass, A. et al. (2016). A proof-of-concept, randomized controlled trial of DAR-0100A, a dopamine-1 receptor agonist, for cognitive enhancement in schizophrenia. *J. Psychopharmacol. (Oxford, England)* 30 (5): 428–435.

8 Beaulieu, J.M. and Gainetdinov, R.R. (2011). The physiology, signaling, and pharmacology of dopamine receptors. *Pharmacol. Rev.* 63 (1): 182–217.

9 Arnsten, A.F., Girgis, R.R., Gray, D.L., and Mailman, R.B. (2017). Novel dopamine therapeutics for cognitive deficits in schizophrenia. *Biol. Psychiatry* 81 (1): 67–77.

10 Sawaguchi, T. and Goldman-Rakic, P.S. (1991). D1 dopamine receptors in prefrontal cortex: involvement in working memory. *Science* 251 (4996): 947–950.

11 Goldman-Rakic, P.S., Castner, S.A., Svensson, T.H. et al. (2004). Targeting the dopamine D1 receptor in schizophrenia: insights for cognitive dysfunction. *Psychopharmacology (Berl).* 174 (1): 3–16.

12 Berke, J.D. (2018). What does dopamine mean? *Nat. Neurosci.* 21 (6): 787–793.

13 Gray, D.L., Allen, J.A., Mente, S. et al. (2018). Impaired beta-arrestin recruitment and reduced desensitization by non-catechol agonists of the D1 dopamine receptor. *Nat. Commun.* 9 (1): 674.

14 Seeman, P., Lee, T., Chau-Wong, M., and Wong, K. (1976). Antipsychotic drug doses and neuroleptic/dopamine receptors. *Nature* 261 (5562): 717–719.

15 Dearry, A., Gingrich, J.A., Falardeau, P. et al. (1990). Molecular cloning and expression of the gene for a human D1 dopamine receptor. *Nature* 347 (6288): 72–76.

16 Creese, I., Burt, D.R., and Snyder, S.H. (1976). Dopamine receptor binding predicts clinical and pharmacological potencies of antischizophrenic drugs. *Science* 192 (4238): 481–483.

17 Koob, G.F. and Volkow, N.D. (2010). Neurocircuitry of addiction. *Neuropsychopharmacology* 35 (1): 217–238.

18 Diana, M. (2011). The dopamine hypothesis of drug addiction and its potential therapeutic value. *Front. Psychiatry* 2: 64.

19 Koob, G.F. and Volkow, N.D. (2016). Neurobiology of addiction: a neurocircuitry analysis. *Lancet Psychiatry* 3 (8): 760–773.

20 Justice NDICUSDo (2011). *National Drug Threat Assessment*. Washington, DC: United States Department of Justice.

21 CDC. 2018 Annual Surveillance Report of Drug-Related Risks and Outcomes — United States, Surveillance Special Report 2. Centers for Disease Control and Prevention, US Department of Health and Human Services. 2018.

22 Karila, L., Weinstein, A., Aubin, H.J. et al. (2010). Pharmacological approaches to methamphetamine dependence: a focused review. *Br. J. Clin. Pharmacol.* 69 (6): 578–592.

23 Farrell, M., Martin, N.K., Stockings, E. et al. (2019). Responding to global stimulant use: challenges and opportunities. *Lancet* 394 (10209): 1652–1667.

24 Kalivas, P.W. and Volkow, N.D. (2005). The neural basis of addiction: a pathology of motivation and choice. *Am. J. Psychiatry* 162 (8): 1403–1413.

25 Wise, R.A. and Bozarth, M.A. (1987). A psychomotor stimulant theory of addiction. *Psychol. Rev.* 94 (4): 469–492.

26 Koob, G.F. and Le Moal, M. (2008). Addiction and the brain antireward system. *Annu. Rev. Psychol.* 59: 29–53.

27 Willuhn, I., Wanat, M.J., Clark, J.J., and Phillips, P.E. (2010). Dopamine signaling in the nucleus accumbens of animals self-administering drugs of abuse. *Curr. Top Behav. Neurosci.* 3: 29–71.

28 Di Chiara, G. and Imperato, A. (1988). Drugs abused by humans preferentially increase synaptic dopamine concentrations in the mesolimbic system of freely moving rats. *Proc. Natl. Acad. Sci. U.S.A.* 85 (14): 5274–5278.

29 Wise, R.A. (1980). Action of drugs of abuse on brain reward systems. *Pharmacol. Biochem. Behav.* 13 (Suppl 1): 213–223.

30 Volkow, N.D., Fowler, J.S., and Wang, G.J. (2002). Role of dopamine in drug reinforcement and addiction in humans: results from imaging studies. *Behav. Pharmacol.* 13 (5–6): 355–366.

31 Sanna, A., Fattore, L., Badas, P. et al. (2020). The hypodopaminergic state ten years after: transcranial magnetic stimulation as a tool to test the dopamine hypothesis of drug addiction. *Curr. Opin. Pharmacol.* 56: 61–67.

32 Kalia, L.V. and Lang, A.E. (2015). Parkinson's disease. *Lancet* 386 (9996): 896–912.

33 Castner, S.A., Williams, G.V., and Goldman-Rakic, P.S. (2000). Reversal of antipsychotic-induced working memory deficits by short-term dopamine D1 receptor stimulation. *Science* 287 (5460): 2020–2022.

34 Chudasama, Y. and Robbins, T.W. (2006). Functions of frontostriatal systems in cognition: comparative neuropsychopharmacological studies in rats, monkeys and humans. *Biol. Psychol.* 73 (1): 19–38.

35 Chudasama, Y. and Robbins, T.W. (2004). Dopaminergic modulation of visual attention and working memory in the rodent prefrontal cortex. *Neuropsychopharmacology* 29 (9): 1628–1636.

36 Corrigall, W.A., Coen, K.M., and Adamson, K.L. (1994). Self-administered nicotine activates the mesolimbic dopamine system through the ventral tegmental area. *Brain Res.* 653 (1): 278–284.

37 Corrigall, W.A., Franklin, K.B., Coen, K.M., and Clarke, P.B. (1992). The mesolimbic dopaminergic system is implicated in the reinforcing effects of nicotine. *Psychopharmacology (Berl)* 107 (2–3): 285–289.

38 Pierce, R.C. and Kumaresan, V. (2006). The mesolimbic dopamine system: the final common pathway for the reinforcing effect of drugs of abuse? *Neurosci. Biobehav. Rev.* 30 (2): 215–238.

39 Koob, G.F. and Bloom, F.E. (1988). Cellular and molecular mechanisms of drug dependence. *Science* 242 (4879): 715–723.

40 Grattan, D.R. (2015). 60 Years of neuroendocrinology: the hypothalamo-prolactin axis. *J. Endocrinol.* 226 (2): T101–T122.

41 Volkow, N.D., Wise, R.A., and Baler, R. (2017). The dopamine motive system: implications for drug and food addiction. *Nat. Rev. Neurosci.* 18 (12): 741–752.

42 Baik, J.-H. (2013). Dopamine Signaling in reward-related behaviors. *Front. Neural Circuits* 7 (152).

43 Hyman, S.E., Malenka, R.C., and Nestler, E.J. (2006). Neural mechanisms of addiction: the role of reward-related learning and memory. *Annu. Rev. Neurosci.* 29: 565–598.

44 Greengard, P. (2001). The neurobiology of slow synaptic transmission. *Science* 294 (5544): 1024–1030.

45 Volkow, N.D., Wang, G.J., Ma, Y. et al. (2003). Expectation enhances the regional brain metabolic and the reinforcing effects of stimulants in cocaine abusers. *J. Neurosci.* 23 (36): 11461–11468.

46 Waelti, P., Dickinson, A., and Schultz, W. (2001). Dopamine responses comply with basic assumptions of formal learning theory. *Nature* 412 (6842): 43–48.

47 Buchta, W.C. and Riegel, A.C. (2015). Chronic cocaine disrupts mesocortical learning mechanisms. *Brain Res.* 1628 (Pt A): 88–103.

48 Reynolds, L.M., Makowski, C.S., Yogendran, S.V. et al. (2015). Amphetamine in adolescence disrupts the development of medial prefrontal cortex dopamine connectivity in a DCC-dependent manner. *Neuropsychopharmacology* 40 (5): 1101–1112.

49 Carvalho Poyraz, F., Holzner, E., Bailey, M.R. et al. (2016). Decreasing striatopallidal pathway function enhances motivation by energizing the initiation of goal-directed action. *J. Neurosci.* 36 (22): 5988–6001.

50 Ferguson, S.M., Eskenazi, D., Ishikawa, M. et al. (2011). Transient neuronal inhibition reveals opposing roles of indirect and direct pathways in sensitization. *Nat. Neurosci.* 14 (1): 22–24.

51 Koob, G.F. (1992). Drugs of abuse: anatomy, pharmacology and function of reward pathways. *Trends Pharmacol. Sci.* 13 (5): 177–184.

52 Missale, C., Nash, S.R., Robinson, S.W. et al. (1998). Dopamine receptors: from structure to function. *Physiol. Rev.* 78 (1): 189–225.

53 Sibley, D.R. and Monsma, F.J. Jr. (1992). Molecular biology of dopamine receptors. *Trends Pharmacol. Sci.* 13 (2): 61–69.

54 Del'guidice, T., Lemasson, M., and Beaulieu, J.M. (2011). Role of Beta-arrestin 2 downstream of dopamine receptors in the Basal Ganglia. *Front. Neuroanat.* 5: 58.

55 Monsma, F.J. Jr., Mahan, L.C., McVittie, L.D. et al. (1990). Molecular cloning and expression of a D1 dopamine receptor linked to adenylyl cyclase activation. *Proc. Natl. Acad. Sci. U.S.A.* 87 (17): 6723–6727.

56 Milligan, G. and Kostenis, E. (2006). Heterotrimeric G-proteins: a short history. *Br. J. Pharmacol.* 147 (Suppl 1): S46–S55.

57 Zhuang, X., Belluscio, L., and Hen, R. (2000). G(olf)alpha mediates dopamine D1 receptor signaling. *J. Neurosci.* 20 (16): Rc91.

58 Corvol, J.C., Studler, J.M., Schonn, J.S. et al. (2001). Galpha(olf) is necessary for coupling D1 and A2a receptors to adenylyl cyclase in the striatum. *J. Neurochem.* 76 (5): 1585–1588.

59 Kwon, Y.G., Huang, H.B., Desdouits, F. et al. (1997). Characterization of the interaction between DARPP-32 and protein phosphatase 1 (PP-1): DARPP-32 peptides antagonize the interaction of PP-1 with binding proteins. *Proc. Natl. Acad. Sci. U.S.A.* 94 (8): 3536–3541.

60 Ouimet, C.C., Langley-Gullion, K.C., and Greengard, P. (1998). Quantitative immunocytochemistry of DARPP-32-expressing neurons in the rat caudatoputamen. *Brain Res.* 808 (1): 8–12.

61 Ouimet, C.C., Miller, P.E., Hemmings, H.C. Jr. et al. (1984). DARPP-32, a dopamine- and adenosine 3′:5′-monophosphate-regulated phosphoprotein enriched in dopamine-innervated brain regions. III. Immunocytochemical localization. *J. Neurosci.* 4 (1): 111–124.

62 Fienberg, A.A., Hiroi, N., Mermelstein, P.G. et al. (1998). DARPP-32: regulator of the efficacy of dopaminergic neurotransmission. *Science* 281 (5378): 838–842.

63 Snyder, G.L., Allen, P.B., Fienberg, A.A. et al. (2000). Regulation of phosphorylation of the GluR1 AMPA receptor in the neostriatum by dopamine and psychostimulants in vivo. *J. Neurosci.* 20 (12): 4480–4488.

64 Banke, T.G., Bowie, D., Lee, H. et al. (2000). Control of GluR1 AMPA receptor function by cAMP-dependent protein kinase. *J. Neurosci.* 20 (1): 89–102.

65 Roche, K.W., O'Brien, R.J., Mammen, A.L. et al. (1996). Characterization of multiple phosphorylation sites on the AMPA receptor GluR1 subunit. *Neuron* 16 (6): 1179–1188.

66 Mangiavacchi, S. and Wolf, M.E. (2004). D1 dopamine receptor stimulation increases the rate of AMPA receptor insertion onto the surface of cultured nucleus accumbens neurons through a pathway dependent on protein kinase A. *J. Neurochem.* 88 (5): 1261–1271.

67 Flores-Hernandez, J., Hernandez, S., Snyder, G.L. et al. (2000). D(1) dopamine receptor activation reduces GABA(A) receptor currents in neostriatal neurons through a PKA/DARPP-32/PP1 signaling cascade. *J. Neurophysiol.* 83 (5): 2996–3004.

68 Snyder, G.L., Fienberg, A.A., Huganir, R.L., and Greengard, P. (1998). A dopamine/D1 receptor/protein kinase A/dopamine- and cAMP-regulated phosphoprotein (Mr 32 kDa)/protein phosphatase-1 pathway regulates dephosphorylation of the NMDA receptor. *J. Neurosci.* 18 (24): 10297–10303.

69 Blank, T., Nijholt, I., Teichert, U. et al. (1997). The phosphoprotein DARPP-32 mediates cAMP-dependent potentiation of striatal *N*-methyl-D-aspartate responses. *Proc. Natl. Acad. Sci. U.S.A.* 94 (26): 14859–14864.

70 Liu, Z., Mao, L., Parelkar, N.K. et al. (2004). Distinct expression of phosphorylated *N*-methyl-D-aspartate receptor NR1 subunits by projection neurons and interneurons in the striatum of normal and amphetamine-treated rats. *J. Comp. Neurol.* 474 (3): 393–406.

71 Dunah, A.W., Sirianni, A.C., Fienberg, A.A. et al. (2004). Dopamine D1-dependent trafficking of striatal *N*-methyl-D-aspartate glutamate receptors requires Fyn protein tyrosine kinase but not DARPP-32. *Mol. Pharmacol.* 65 (1): 121–129.

72 Suzuki, T. and Okumura-Noji, K. (1995). NMDA receptor subunits epsilon 1 (NR2A) and epsilon 2 (NR2B) are substrates for Fyn in the postsynaptic density fraction isolated from the rat brain. *Biochem. Biophys. Res. Commun.* 216 (2): 582–588.

73 Valjent, E., Pascoli, V., Svenningsson, P. et al. (2005). Regulation of a protein phosphatase cascade allows convergent dopamine and glutamate signals to activate ERK in the striatum. *Proc. Natl. Acad. Sci. U.S.A.* 102 (2): 491–496.

74 Surmeier, D.J., Bargas, J., Hemmings, H.C. Jr. et al. (1995). Modulation of calcium currents by a D1 dopaminergic protein kinase/phosphatase cascade in rat neostriatal neurons. *Neuron* 14 (2): 385–397.

75 Cepeda, C., Colwell, C.S., Itri, J.N. et al. (1998). Dopaminergic modulation of NMDA-induced whole cell currents in neostriatal neurons in slices: contribution of calcium conductances. *J. Neurophysiol.* 79 (1): 82–94.

76 Liu, J.C., DeFazio, R.A., Espinosa-Jeffrey, A. et al. (2004). Calcium modulates dopamine potentiation of *N*-methyl-D-aspartate responses: electrophysiological and imaging evidence. *J. Neurosci. Res.* 76 (3): 315–322.

77 Calabresi, P., Gubellini, P., Centonze, D. et al. (2000). Dopamine and cAMP-regulated phosphoprotein 32 kDa controls both striatal long-term depression and long-term potentiation, opposing forms of synaptic plasticity. *J. Neurosci.* 20 (22): 8443–8451.

78 Neve, K.A., Seamans, J.K., and Trantham-Davidson, H. (2004). Dopamine receptor signaling. *J. Recept. Signal Transduct. Res.* 24 (3): 165–205.

79 Kim, O.J., Gardner, B.R., Williams, D.B. et al. (2004). The role of phosphorylation in D1 dopamine receptor desensitization: evidence for a novel mechanism of arrestin association. *J. Biol. Chem.* 279 (9): 7999–8010.

80 Dratz, E.A., Furstenau, J.E., Lambert, C.G. et al. (1993). NMR structure of a receptor-bound G-protein peptide. *Nature* 363 (6426): 276–281.

81 Konig, B. and Gratzel, M. (1994). Site of dopamine D1 receptor binding to Gs protein mapped with synthetic peptides. *Biochim. Biophys. Acta* 1223 (2): 261–266.

82 Kang, Y., Zhou, X.E., Gao, X. et al. (2015). Crystal structure of rhodopsin bound to arrestin by femtosecond X-ray laser. *Nature* 523 (7562): 561–567.

83 Rasmussen, S.G., DeVree, B.T., Zou, Y. et al. (2011). Crystal structure of the beta2 adrenergic receptor-Gs protein complex. *Nature* 477 (7366): 549–555.

84 Ferguson, S.S., Downey, W.E. 3rd, Colapietro, A.M. et al. (1996). Role of beta-arrestin in mediating agonist-promoted G protein-coupled receptor internalization. *Science* 271 (5247): 363–366.

85 Ahn, S., Wei, H., Garrison, T.R., and Lefkowitz, R.J. (2004). Reciprocal regulation of angiotensin receptor-activated extracellular signal-regulated kinases by beta-arrestins 1 and 2. *J. Biol. Chem.* 279 (9): 7807–7811.

86 Kohout, T.A., Lin, F.S., Perry, S.J. et al. (2001). Beta-Arrestin 1 and 2 differentially regulate heptahelical receptor signaling and trafficking. *Proc. Natl. Acad. Sci. U.S.A.* 98 (4): 1601–1606.

87 Mohan, M.L., Vasudevan, N.T., Gupta, M.K. et al. (2012). G-protein coupled receptor resensitization-appreciating the balancing act of receptor function. *Curr. Mol. Pharmacol.*

88 Kliewer, A., Reinscheid, R.K., and Schulz, S. (2017). Emerging paradigms of G protein-coupled receptor dephosphorylation. *Trends Pharmacol. Sci.* 38 (7): 621–636.

89 Vargas, G.A. and Von Zastrow, M. (2004). Identification of a novel endocytic recycling signal in the D1 dopamine receptor. *J. Biol. Chem.* 279 (36): 37461–37469.

90 Vickery, R.G. and von Zastrow, M. (1999). Distinct dynamin-dependent and -independent mechanisms target structurally homologous dopamine receptors to different endocytic membranes. *J. Cell. Biol.* 144 (1): 31–43.

91 Kong, M.M., Hasbi, A., Mattocks, M. et al. (2007). Regulation of D1 dopamine receptor trafficking and signaling by caveolin-1. *Mol. Pharmacol.* 72 (5): 1157–1170.

92 Gardner, B., Liu, Z.F., Jiang, D., and Sibley, D.R. (2001). The role of phosphorylation/dephosphorylation in agonist-induced desensitization of D1 dopamine receptor function: evidence for a novel pathway for receptor dephosphorylation. *Mol. Pharmacol.* 59 (2): 310–321.

93 Kotowski, S.J., Hopf, F.W., Seif, T. et al. (2011). Endocytosis promotes rapid dopaminergic signaling. *Neuron* 71 (2): 278–290.

94 Sunahara, R.K., Guan, H.C., O'Dowd, B.F. et al. (1991). Cloning of the gene for a human dopamine D5 receptor with higher affinity for dopamine than D1. *Nature* 350 (6319): 614–619.

95 Weinshank, R.L., Adham, N., Macchi, M. et al. (1991). Molecular cloning and characterization of a high affinity dopamine receptor (D1 beta) and its pseudogene. *J. Biol. Chem.* 266 (33): 22427–22435.

96 Tiberi, M. and Caron, M.G. (1994). High agonist-independent activity is a distinguishing feature of the dopamine D1B receptor subtype. *J. Biol. Chem.* 269 (45): 27925–27931.

97 Liu, F., Wan, Q., Pristupa, Z.B. et al. (2000). Direct protein-protein coupling enables cross-talk between dopamine D5 and gamma-aminobutyric acid A receptors. *Nature* 403 (6767): 274–280.

98 Schilström, B., Yaka, R., Argilli, E. et al. (2006). Cocaine enhances NMDA receptor-mediated currents in ventral tegmental area cells via dopamine D5

receptor-dependent redistribution of NMDA receptors. *J. Neurosci.* 26 (33): 8549–8558.

99 Moraga-Amaro, R., González, H., Ugalde, V. et al. (2016). Dopamine receptor D5 deficiency results in a selective reduction of hippocampal NMDA receptor subunit NR2B expression and impaired memory. *Neuropharmacology* 103: 222–235.

100 Conroy, J.L., Free, R.B., and Sibley, D.R. (2015). Identification of G protein-biased agonists that fail to recruit beta-arrestin or promote internalization of the D1 dopamine receptor. *ACS Chem. Neurosci.* 6 (4): 681–692.

101 Thompson, D. and Whistler, J.L. (2011). Trafficking properties of the D5 dopamine receptor. *Traffic* 12 (5): 644–656.

102 Perreault, M.L., Jones-Tabah, J., O'Dowd, B.F., and George, S.R. (2013). A physiological role for the dopamine D5 receptor as a regulator of BDNF and Akt signalling in rodent prefrontal cortex. *Int. J. Neuropsychopharmacol.* 16 (2): 477–483.

103 Monsma, F.J. Jr., McVittie, L.D., Gerfen, C.R. et al. (1989). Multiple D2 dopamine receptors produced by alternative RNA splicing. *Nature* 342 (6252): 926–929.

104 Dal Toso, R., Sommer, B., Ewert, M. et al. (1989). The dopamine D2 receptor: two molecular forms generated by alternative splicing. *EMBO J.* 8 (13): 4025–4034.

105 Centonze, D., Usiello, A., Gubellini, P. et al. (2002). Dopamine D2 receptor-mediated inhibition of dopaminergic neurons in mice lacking D2L receptors. *Neuropsychopharmacology* 27 (5): 723–726.

106 Lindgren, N., Usiello, A., Goiny, M. et al. (2003). Distinct roles of dopamine D2L and D2S receptor isoforms in the regulation of protein phosphorylation at presynaptic and postsynaptic sites. *Proc. Natl. Acad. Sci. U.S.A.* 100 (7): 4305–4309.

107 Usiello, A., Baik, J.H., Rougé-Pont, F. et al. (2000). Distinct functions of the two isoforms of dopamine D2 receptors. *Nature* 408 (6809): 199–203.

108 Neve, K.A., Henningsen, R.A., Bunzow, J.R., and Civelli, O. (1989). Functional characterization of a rat dopamine D-2 receptor cDNA expressed in a mammalian cell line. *Mol. Pharmacol.* 36 (3): 446–451.

109 Kozell, L.B., Machida, C.A., Neve, R.L., and Neve, K.A. (1994). Chimeric D1/D2 dopamine receptors. Distinct determinants of selective efficacy, potency, and signal transduction. *J. Biol. Chem.* 269 (48): 30299–30306.

110 Senogles, S.E., Heimert, T.L., Odife, E.R., and Quasney, M.W. (2004). A region of the third intracellular loop of the short form of the D2 dopamine receptor dictates Gi coupling specificity. *J. Biol. Chem.* 279 (3): 1601–1606.

111 Montmayeur, J.P., Guiramand, J., and Borrelli, E. (1993). Preferential coupling between dopamine D2 receptors and G-proteins. *Mol. Endocrinol.* 7 (2): 161–170.

112 Obadiah, J., Avidor-Reiss, T., Fishburn, C.S. et al. (1999). Adenylyl cyclase interaction with the D2 dopamine receptor family; differential coupling to Gi, Gz, and Gs. *Cell Mol. Neurobiol.* 19 (5): 653–664.

113 Leck, K.J., Blaha, C.D., Matthaei, K.I. et al. (2006). Gz proteins are functionally coupled to dopamine D2-like receptors in vivo. *Neuropharmacology* 51 (3): 597–605.

114 Bateup, H.S., Svenningsson, P., Kuroiwa, M. et al. (2008). Cell type-specific regulation of DARPP-32 phosphorylation by psychostimulant and antipsychotic drugs. *Nat. Neurosci.* 11 (8): 932–939.

115 Nishi, A., Snyder, G.L., and Greengard, P. (1997). Bidirectional regulation of DARPP-32 phosphorylation by dopamine. *J. Neurosci.* 17 (21): 8147–8155.

116 Svenningsson, P., Lindskog, M., Ledent, C. et al. (2000). Regulation of the phosphorylation of the dopamine- and cAMP-regulated phosphoprotein of 32 kDa in vivo by dopamine D1, dopamine D2, and adenosine A2A receptors. *Proc. Natl. Acad. Sci. U.S.A.* 97 (4): 1856–1860.

117 Hakansson, K., Galdi, S., Hendrick, J. et al. (2006). Regulation of phosphorylation of the GluR1 AMPA receptor by dopamine D2 receptors. *J. Neurochem.* 96 (2): 482–488.

118 Cepeda, C., Buchwald, N.A., and Levine, M.S. (1993). Neuromodulatory actions of dopamine in the neostriatum are dependent upon the excitatory amino acid receptor subtypes activated. *Proc. Natl. Acad. Sci. U.S.A.* 90 (20): 9576–9580.

119 Leveque, J.C., Macías, W., Rajadhyaksha, A. et al. (2000). Intracellular modulation of NMDA receptor function by antipsychotic drugs. *J. Neurosci.* 20 (11): 4011–4020.

120 Liu, X.Y., Chu, X.P., Mao, L.M. et al. (2006). Modulation of D2R-NR2B interactions in response to cocaine. *Neuron* 52 (5): 897–909.

121 Hernandez-Lopez, S., Tkatch, T., Perez-Garci, E. et al. (2000). D2 dopamine receptors in striatal medium spiny neurons reduce L-type Ca2+ currents and excitability via a novel PLC[beta]1-IP3-calcineurin-signaling cascade. *J. Neurosci.* 20 (24): 8987–8995.

122 Yan, Z., Song, W.J., and Surmeier, J. (1997). D2 dopamine receptors reduce N-type Ca2+ currents in rat neostriatal cholinergic interneurons through a membrane-delimited, protein-kinase-C-insensitive pathway. *J. Neurophysiol.* 77 (2): 1003–1015.

123 Kuzhikandathil, E.V., Yu, W., and Oxford, G.S. (1998). Human dopamine D3 and D2L receptors couple to inward rectifier potassium channels in mammalian cell lines. *Mol. Cell. Neurosci.* 12 (6): 390–402.

124 Huang, R., Griffin, S.A., Taylor, M. et al. (2013). The effect of SV 293, a D2 dopamine receptor-selective antagonist, on D2 receptor-mediated GIRK channel activation and adenylyl cyclase inhibition. *Pharmacology* 92 (1–2): 84–89.

125 Lavine, N., Ethier, N., Oak, J.N. et al. (2002). G protein-coupled receptors form stable complexes with inwardly rectifying potassium channels and adenylyl cyclase. *J. Biol. Chem.* 277 (48): 46010–46019.

126 McCall, N.M., Kotecki, L., Dominguez-Lopez, S. et al. (2017). Selective ablation of GIRK channels in dopamine neurons alters behavioral effects of cocaine in mice. *Neuropsychopharmacology* 42 (3): 707–715.

127 Rifkin, R.A., Huyghe, D., Li, X. et al. (2018). GIRK currents in VTA dopamine neurons control the sensitivity of mice to cocaine-induced locomotor sensitization. *Proc. Natl. Acad. Sci. U.S.A.* 115 (40): E9479–e88.

128 Beaulieu, J.M., Tirotta, E., Sotnikova, T.D. et al. (2007). Regulation of Akt signaling by D2 and D3 dopamine receptors in vivo. *J. Neurosci.* 27 (4): 881–885.

129 Beaulieu, J.-M., Sotnikova, T.D., Yao, W.-D. et al. (2004). Lithium antagonizes dopamine-dependent behaviors mediated by an AKT/glycogen synthase kinase 3 signaling cascade. *Proc. Natl. Acad. Sci. U.S.A.* 101 (14): 5099.

130 Beaulieu, J.M., Marion, S., Rodriguiz, R.M. et al. (2008). A beta-arrestin 2 signaling complex mediates lithium action on behavior. *Cell* 132 (1): 125–136.

131 Beaulieu, J.M., Sotnikova, T.D., Marion, S. et al. (2005). An Akt/beta-arrestin 2/PP2A signaling complex mediates dopaminergic neurotransmission and behavior. *Cell* 122 (2): 261–273.

132 Kovoor, A., Seyffarth, P., Ebert, J. et al. (2005). D2 dopamine receptors colocalize regulator of G-protein signaling 9-2 (RGS9-2) via the RGS9 DEP domain, and RGS9 knock-out mice develop dyskinesias associated with dopamine pathways. *J. Neurosci.* 25 (8): 2157–2165.

133 Thomas, E.A., Danielson, P.E., and Sutcliffe, J.G. (1998). RGS9: a regulator of G-protein signalling with specific expression in rat and mouse striatum. *J. Neurosci. Res.* 52 (1): 118–124.

134 Rahman, Z., Schwarz, J., Gold, S.J. et al. (2003). RGS9 modulates dopamine signaling in the basal ganglia. *Neuron* 38 (6): 941–952.

135 Namkung, Y. and Sibley, D.R. (2004). Protein kinase C mediates phosphorylation, desensitization, and trafficking of the D2 dopamine receptor. *J. Biol. Chem.* 279 (47): 49533–49541.

136 Ito, K., Haga, T., Lameh, J., and Sadée, W. (1999). Sequestration of dopamine D2 receptors depends on coexpression of G-protein-coupled receptor kinases 2 or 5. *Eur. J. Biochem.* 260 (1): 112–119.

137 Pack, T.F., Orlen, M.I., Ray, C. et al. (2018). The dopamine D2 receptor can directly recruit and activate GRK2 without G protein activation. *J. Biol. Chem.* 293 (16): 6161–6171.

138 Namkung, Y., Dipace, C., Javitch, J.A., and Sibley, D.R. (2009). G protein-coupled receptor kinase-mediated phosphorylation regulates post-endocytic trafficking of the D2 dopamine receptor. *J. Biol. Chem.* 284 (22): 15038–15051.

139 Gurevich, E.V., Gainetdinov, R.R., and Gurevich, V.V. (2016). G protein-coupled receptor kinases as regulators of dopamine receptor functions. *Pharmacol. Res.* 111: 1–16.

140 Gainetdinov, R.R., Bohn, L.M., Sotnikova, T.D. et al. (2003). Dopaminergic supersensitivity in G protein-coupled receptor kinase 6-deficient mice. *Neuron* 38 (2): 291–303.

141 Bouthenet, M.L., Souil, E., Martres, M.P. et al. (1991). Localization of dopamine D3 receptor mRNA in the rat brain using in situ hybridization histochemistry: comparison with dopamine D2 receptor mRNA. *Brain Res.* 564 (2): 203–219.

142 Gurevich, E.V. and Joyce, J.N. (1999). Distribution of dopamine D3 receptor expressing neurons in the human forebrain: comparison with D2 receptor expressing neurons. *Neuropsychopharmacology* 20 (1): 60–80.

143 Suzuki, M., Hurd, Y.L., Sokoloff, P. et al. (1998). D3 dopamine receptor mRNA is widely expressed in the human brain. *Brain Res.* 779 (1–2): 58–74.

144 Newman-Tancredi, A., Cussac, D., Audinot, V. et al. (1999). G protein activation by human dopamine D3 receptors in high-expressing Chinese hamster ovary cells: a guanosine-5′-O-(3-[35S]thio)- triphosphate binding and antibody study. *Mol. Pharmacol.* 55 (3): 564–574.

145 Chio, C.L., Lajiness, M.E., and Huff, R.M. (1994). Activation of heterologously expressed D3 dopamine receptors: comparison with D2 dopamine receptors. *Mol. Pharmacol.* 45 (1): 51–60.

146 Robinson, S.W. and Caron, M.G. (1997). Selective inhibition of adenylyl cyclase type V by the dopamine D3 receptor. *Mol. Pharmacol.* 52 (3): 508–514.

147 Werner, P., Hussy, N., Buell, G. et al. (1996). D2, D3, and D4 dopamine receptors couple to G protein-regulated potassium channels in Xenopus oocytes. *Mol. Pharmacol.* 49 (4): 656–661.

148 Liu, L.X., Burgess, L.H., Gonzalez, A.M. et al. (1999). D2S, D2L, D3, and D4 dopamine receptors couple to a voltage-dependent potassium current in N18TG2 x mesencephalon hybrid cell (MES-23.5) via distinct G proteins. *Synapse* 31 (2): 108–118.

149 Liu, L.X., Monsma, F.J. Jr., Sibley, D.R., and Chiodo, L.A. (1996). D2L, D2S, and D3 dopamine receptors stably transfected into NG108-15 cells couple to a voltage-dependent potassium current via distinct G protein mechanisms. *Synapse* 24 (2): 156–164.

150 Kuzhikandathil, E.V. and Oxford, G.S. (1999). Activation of human D3 dopamine receptor inhibits P/Q-type calcium channels and secretory activity in AtT-20 cells. *J. Neurosci.* 19 (5): 1698–1707.

151 Seabrook, G.R., Kemp, J.A., Freedman, S.B. et al. (1994). Functional expression of human D3 dopamine receptors in differentiated neuroblastoma x glioma NG108-15 cells. *Br. J. Pharmacol.* 111 (2): 391–393.

152 Chen, G., Kittler, J.T., Moss, S.J., and Yan, Z. (2006). Dopamine D3 receptors regulate GABAA receptor function through a phospho-dependent endocytosis mechanism in nucleus accumbens. *J. Neurosci.* 26 (9): 2513–2521.

153 Kim, K.M., Valenzano, K.J., Robinson, S.R. et al. (2001). Differential regulation of the dopamine D2 and D3 receptors by G protein-coupled receptor kinases and beta-arrestins. *J. Biol. Chem.* 276 (40): 37409–37414.

154 Min, C., Zheng, M., Zhang, X. et al. (2013). Novel roles for β-arrestins in the regulation of pharmacological sequestration to predict agonist-induced desensitization of dopamine D3 receptors. *Br. J. Pharmacol.* 170 (5): 1112–1129.

155 Zheng, M., Zhang, X., Sun, N. et al. (2020). A novel molecular mechanism responsible for phosphorylation-independent desensitization of G protein-coupled receptors exemplified by the dopamine D(3) receptor. *Biochem. Biophys. Res. Commun.* 528 (3): 432–439.

156 Meador-Woodruff, J.H., Damask, S.P., Wang, J. et al. (1996). Dopamine receptor mRNA expression in human striatum and neocortex. *Neuropsychopharmacology* 15 (1): 17–29.

157 Xiang, L., Szebeni, K., Szebeni, A. et al. (2008). Dopamine receptor gene expression in human amygdaloid nuclei: elevated D4 receptor mRNA in major depression. *Brain Res.* 1207: 214–224.

158 Matsumoto, M., Hidaka, K., Tada, S. et al. (1995). Full-length cDNA cloning and distribution of human dopamine D4 receptor. *Brain Res. Mol. Brain Res.* 29 (1): 157–162.

159 Oak, J.N., Oldenhof, J., and Van Tol, H.H. (2000). The dopamine D(4) receptor: one decade of research. *Eur. J. Pharmacol.* 405 (1–3): 303–327.

160 Nir, I., Harrison, J.M., Haque, R. et al. (2002). Dysfunctional light-evoked regulation of cAMP in photoreceptors and abnormal retinal adaptation in mice lacking dopamine D4 receptors. *J. Neurosci.* 22 (6): 2063–2073.

161 Kazmi, M.A., Snyder, L.A., Cypess, A.M. et al. (2000). Selective reconstitution of human D4 dopamine receptor variants with Gi alpha subtypes. *Biochemistry* 39 (13): 3734–3744.

162 Seabrook, G.R., Knowles, M., Brown, N. et al. (1994). Pharmacology of high-threshold calcium currents in GH4C1 pituitary cells and their regulation by activation of human D2 and D4 dopamine receptors. *Br. J. Pharmacol.* 112 (3): 728–734.

163 Wedemeyer, C., Goutman, J.D., Avale, M.E. et al. (2007). Functional activation by central monoamines of human dopamine D(4) receptor polymorphic variants coupled to GIRK channels in Xenopus oocytes. *Eur. J. Pharmacol.* 562 (3): 165–173.

164 Wang, X., Zhong, P., and Yan, Z. (2002). Dopamine D4 receptors modulate GABAergic signaling in pyramidal neurons of prefrontal cortex. *J. Neurosci.* 22 (21): 9185–9193.

165 Erlij, D., Acosta-García, J., Rojas-Márquez, M. et al. (2012). Dopamine D4 receptor stimulation in GABAergic projections of the globus pallidus to the reticular thalamic nucleus and the substantia nigra reticulata of the rat decreases locomotor activity. *Neuropharmacology* 62 (2): 1111–1118.

166 Oldenhof, J., Vickery, R., Anafi, M. et al. (1998). SH3 binding domains in the dopamine D4 receptor. *Biochemistry* 37 (45): 15726–15736.

167 Spooren, A., Rondou, P., Debowska, K. et al. (2010). Resistance of the dopamine D4 receptor to agonist-induced internalization and degradation. *Cell. Signalling* 22 (4): 600–609.

168 Skieterska, K., Shen, A., Clarisse, D. et al. (2016). Characterization of the interaction between the dopamine D4 receptor, KLHL12 and β-arrestins. *Cell. Signalling* 28 (8): 1001–1014.

169 Rondou, P., Haegeman, G., Vanhoenacker, P., and Van Craenenbroeck, K. (2008). BTB Protein KLHL12 targets the dopamine D4 receptor for ubiquitination by a Cul3-based E3 ligase. *J. Biol. Chem.* 283 (17): 11083–11096.

170 Rondou, P., Skieterska, K., Packeu, A. et al. (2010). KLHL12-mediated ubiquitination of the dopamine D4 receptor does not target the receptor for degradation. *Cell. Signalling* 22 (6): 900–913.

171 Deming, J.D., Shin, J.A., Lim, K. et al. (2015). Dopamine receptor D4 internalization requires a beta-arrestin and a visual arrestin. *Cell. Signalling* 27 (10): 2002–2013.

172 Nestler, E.J. and Hyman, S.E. (2010). Animal models of neuropsychiatric disorders. *Nat. Neurosci.* 13 (10): 1161–1169.

173 Cunningham, K.A. and Anastasio, N.C. (2014). Serotonin at the nexus of impulsivity and cue reactivity in cocaine addiction. *Neuropharmacology* 76 (Pt B): 460–478.

174 Olds, J. and Milner, P. (1954). Positive reinforcement produced by electrical stimulation of septal area and other regions of rat brain. *J. Comp. Physiol. Psychol.* 47 (6): 419–427.

175 Ritz, M.C., Lamb, R.J., Goldberg, S.R., and Kuhar, M.J. (1987). Cocaine receptors on dopamine transporters are related to self-administration of cocaine. *Science* 237 (4819): 1219–1223.

176 Panlilio, L.V. and Goldberg, S.R. (2007). Self-administration of drugs in animals and humans as a model and an investigative tool. *Addiction* 102 (12): 1863–1870.

177 Kuhn, B.N., Kalivas, P.W., and Bobadilla, A.C. (2019). Understanding addiction using animal models. *Front. Behav. Neurosci.* 13: 262.

178 Koob, G.F., Le, H.T., and Creese, I. (1987). The D1 dopamine receptor antagonist SCH 23390 increases cocaine self-administration in the rat. *Neurosci. Lett.* 79 (3): 315–320.

179 Britton, D.R., Curzon, P., Mackenzie, R.G. et al. (1991). Evidence for involvement of both D1 and D2 receptors in maintaining cocaine self-administration. *Pharmacol. Biochem. Behav.* 39 (4): 911–915.

180 Hubner, C.B. and Moreton, J.E. (1991). Effects of selective D1 and D2 dopamine antagonists on cocaine self-administration in the rat. *Psychopharmacology (Berl).* 105 (2): 151–156.

181 Izzo, E., Orsini, C., Koob, G.F., and Pulvirenti, L. (2001). A dopamine partial agonist and antagonist block amphetamine self-administration in a progressive ratio schedule. *Pharmacol. Biochem. Behav.* 68 (4): 701–708.

182 Caine, S.B., Negus, S.S., Mello, N.K. et al. (2002). Role of dopamine D2-like receptors in cocaine self-administration: studies with D2 receptor mutant mice and novel D2 receptor antagonists. *J. Neurosci.* 22 (7): 2977–2988.

183 Gál, K. and Gyertyán, I. (2003). Targeting the dopamine D3 receptor cannot influence continuous reinforcement cocaine self-administration in rats. *Brain Res. Bull.* 61 (6): 595–601.

184 Xi, Z.X., Gilbert, J.G., Pak, A.C. et al. (2005). Selective dopamine D3 receptor antagonism by SB-277011A attenuates cocaine reinforcement as assessed by progressive-ratio and variable-cost-variable-payoff fixed-ratio cocaine self-administration in rats. *Eur. J. Neurosci.* 21 (12): 3427–3438.

185 Xi, Z.X., Newman, A.H., Gilbert, J.G. et al. (2006). The novel dopamine D3 receptor antagonist NGB 2904 inhibits cocaine's rewarding effects and cocaine-induced reinstatement of drug-seeking behavior in rats. *Neuropsychopharmacology* 31 (7): 1393–1405.

186 Song, R., Yang, R.F., Wu, N. et al. (2012). YQA14: a novel dopamine D3 receptor antagonist that inhibits cocaine self-administration in rats and mice, but not in D3 receptor-knockout mice. *Addict. Biol.* 17 (2): 259–273.

187 Caine, S.B., Negus, S.S., Mello, N.K., and Bergman, J. (1999). Effects of dopamine D(1-like) and D(2-like) agonists in rats that self-administer cocaine. *J. Pharmacol. Exp. Ther.* 291 (1): 353–360.

188 Caine, S.B., Negus, S.S., and Mello, N.K. (2000). Effects of dopamine D(1-like) and D(2-like) agonists on cocaine self-administration in rhesus monkeys: rapid assessment of cocaine dose-effect functions. *Psychopharmacology (Berl).* 148 (1): 41–51.

189 Caine, S.B., Thomsen, M., Gabriel, K.I. et al. (2007). Lack of self-administration of cocaine in dopamine D1 receptor knock-out mice. *J. Neurosci.* 27 (48): 13140–13150.

190 Karlsson, R.-M., Hefner, K.R., Sibley, D.R., and Holmes, A. (2008). Comparison of dopamine D1 and D5 receptor knockout mice for cocaine locomotor sensitization. *Psychopharmacology* 200 (1): 117–127.

191 Xu, M., Guo, Y., Vorhees, C.V., and Zhang, J. (2000). Behavioral responses to cocaine and amphetamine administration in mice lacking the dopamine D1 receptor. *Brain Res.* 852 (1): 198–207.

192 Caine, S.B., Thomsen, M., Barrett, A.C. et al. (2012). Cocaine self-administration in dopamine D_3 receptor knockout mice. *Exp. Clin. Psychopharmacol.* 20 (5): 352–363.

193 Song, R., Zhang, H.Y., Li, X. et al. (2012). Increased vulnerability to cocaine in mice lacking dopamine D3 receptors. *Proc. Natl. Acad. Sci. U.S.A.* 109 (43): 17675–17680.

194 Thanos, P.K., Habibi, R., Michaelides, M. et al. (2010). Dopamine D4 receptor (D4R) deletion in mice does not affect operant responding for food or cocaine. *Behav. Brain Res.* 207 (2): 508–511.

195 Di Ciano, P., Grandy, D.K., and Le Foll, B. (2014). Dopamine D4 receptors in psychostimulant addiction. *Adv. Pharmacol.* 69: 301–321.

196 Maldonado, R., Robledo, P., Chover, A.J. et al. (1993). D1 dopamine receptors in the nucleus accumbens modulate cocaine self-administration in the rat. *Pharmacol. Biochem. Behav.* 45 (1): 239–242.

197 Phillips, G.D., Robbins, T.W., and Everitt, B.J. (1994). Bilateral intra-accumbens self-administration of d-amphetamine: antagonism with intra-accumbens SCH-23390 and sulpiride. *Psychopharmacology (Berl).* 114 (3): 477–485.

198 Bachtell, R.K., Whisler, K., Karanian, D., and Self, D.W. (2005). Effects of intra-nucleus accumbens shell administration of dopamine agonists and antagonists on cocaine-taking and cocaine-seeking behaviors in the rat. *Psychopharmacology (Berl).* 183 (1): 41–53.

199 Caine, S.B., Heinrichs, S.C., Coffin, V.L., and Koob, G.F. (1995). Effects of the dopamine D-1 antagonist SCH 23390 microinjected into the accumbens, amygdala or striatum on cocaine self-administration in the rat. *Brain Res.* 692 (1–2): 47–56.

200 Hurd, Y.L. and Pontén, M. (2000). Cocaine self-administration behavior can be reduced or potentiated by the addition of specific dopamine concentrations in the nucleus accumbens and amygdala using in vivo microdialysis. *Behav. Brain Res.* 116 (2): 177–186.

201 Bari, A.A. and Pierce, R.C. (2005). D1-like and D2 dopamine receptor antagonists administered into the shell subregion of the rat nucleus accumbens decrease cocaine, but not food, reinforcement. *Neuroscience* 135 (3): 959–968.

202 McGregor, A. and Roberts, D.C. (1993). Dopaminergic antagonism within the nucleus accumbens or the amygdala produces differential effects on intravenous cocaine self-administration under fixed and progressive ratio schedules of reinforcement. *Brain Res.* 624 (1–2): 245–252.

203 McGregor, A. and Roberts, D.C. (1995). Effect of medial prefrontal cortex injections of SCH 23390 on intravenous cocaine self-administration under both a fixed and progressive ratio schedule of reinforcement. *Behav. Brain Res.* 67 (1): 75–80.

204 Alleweireldt, A.T., Hobbs, R.J., Taylor, A.R., and Neisewander, J.L. (2006). Effects of SCH-23390 infused into the amygdala or adjacent cortex and basal ganglia on cocaine seeking and self-administration in rats. *Neuropsychopharmacology* 31 (2): 363–374.

205 Langlois, L.D. and Nugent, F.S. (2017). Opiates and plasticity in the ventral tegmental area. *ACS Chem. Neurosci.* 8 (9): 1830–1838.

206 Johnson, S.W. and North, R.A. (1992). Opioids excite dopamine neurons by hyperpolarization of local interneurons. *J. Neurosci.* 12 (2): 483–488.

207 Pettit, H.O., Ettenberg, A., Bloom, F.E., and Koob, G.F. (1984). Destruction of dopamine in the nucleus accumbens selectively attenuates cocaine but not heroin self-administration in rats. *Psychopharmacology (Berl).* 84 (2): 167–173.

208 Ettenberg, A., Pettit, H.O., Bloom, F.E., and Koob, G.F. (1982). Heroin and cocaine intravenous self-administration in rats: mediation by separate neural systems. *Psychopharmacology (Berl).* 78 (3): 204–209.

209 Gerrits, M.A., Ramsey, N.F., Wolterink, G., and van Ree, J.M. (1994). Lack of evidence for an involvement of nucleus accumbens dopamine D1 receptors in the initiation of heroin self-administration in the rat. *Psychopharmacology (Berl).* 114 (3): 486–494.

210 Le Merrer, J., Gavello-Baudy, S., Galey, D., and Cazala, P. (2007). Morphine self-administration into the lateral septum depends on dopaminergic mechanisms: evidence from pharmacology and Fos neuroimaging. *Behav. Brain Res.* 180 (2): 203–217.

211 Selleck, R.A., Giacomini, J., Buchholtz, B.D. et al. (2018). Modulation of appetitive motivation by prefrontal cortical mu-opioid receptors is dependent upon local dopamine D1 receptor signaling. *Neuropharmacology* 140: 302–309.

212 Rowlett, J.K., Platt, D.M., Yao, W.D., and Spealman, R.D. (2007). Modulation of heroin and cocaine self-administration by dopamine D1- and D2-like receptor agonists in rhesus monkeys. *J. Pharmacol. Exp. Ther.* 321 (3): 1135–1143.

213 Namvar, P., Zarrabian, S., Nazari-Serenjeh, F. et al. (2019). Involvement of D1- and D2-like dopamine receptors within the rat nucleus accumbens in the maintenance of morphine rewarding properties in the rats. *Behav. Neurosci.* 133 (6): 556–562.

214 Shippenberg, T.S., Bals-Kubik, R., and Herz, A. (1993). Examination of the neurochemical substrates mediating the motivational effects of opioids: role of the mesolimbic dopamine system and D-1 vs. D-2 dopamine receptors. *J. Pharmacol. Exp. Ther.* 265 (1): 53–59.

215 Sadeghzadeh, F., Babapour, V., and Haghparast, A. (2015). Role of dopamine D1-like receptor within the nucleus accumbens in acute food deprivation- and drug priming-induced reinstatement of morphine seeking in rats. *Behav. Brain Res.* 287: 172–181.

216 Kobrin, K.L., Arena, D.T., Heinrichs, S.C. et al. (2017). Dopamine D1 receptor agonist treatment attenuates extinction of morphine conditioned place preference while increasing dendritic complexity in the nucleus accumbens core. *Behav. Brain Res.* 322 (Pt A): 18–28.

217 Elmer, G.I., Pieper, J.O., Rubinstein, M. et al. (2002). Failure of intravenous morphine to serve as an effective instrumental reinforcer in dopamine D2 receptor knock-out mice. *J. Neurosci.* 22 (10): Rc224.

218 David, V., Durkin, T.P., and Cazala, P. (2002). Differential effects of the dopamine D2/D3 receptor antagonist sulpiride on self-administration of morphine into the ventral tegmental area or the nucleus accumbens. *Psychopharmacology (Berl).* 160 (3): 307–317.

219 Hearing, M.C., Jedynak, J., Ebner, S.R. et al. (2016). Reversal of morphine-induced cell-type-specific synaptic plasticity in the nucleus accumbens shell blocks reinstatement. *Proc. Natl. Acad. Sci. U.S.A.* 113 (3): 757–762.

220 Dacher, M. and Nugent, F.S. (2011). Morphine-induced modulation of LTD at GABAergic synapses in the ventral tegmental area. *Neuropharmacology* 61 (7): 1166–1171.

221 Zhan, J., Jordan, C.J., Bi, G.H. et al. (2018). Genetic deletion of the dopamine D3 receptor increases vulnerability to heroin in mice. *Neuropharmacology* 141: 11–20.

222 Boateng, C.A., Bakare, O.M., Zhan, J. et al. (2015). High affinity dopamine D3 receptor (D3R)-selective antagonists attenuate heroin self-administration in wild-type but not D3R knockout mice. *J. Med. Chem.* 58 (15): 6195–6213.

223 Silvestre, J.S., O'Neill, M.F., Fernandez, A.G., and Palacios, J.M. (1996). Effects of a range of dopamine receptor agonists and antagonists on ethanol intake in the rat. *Eur. J. Pharmacol.* 318 (2–3): 257–265.

224 Czachowski, C.L., Santini, L.A., Legg, B.H., and Samson, H.H. (2002). Separate measures of ethanol seeking and drinking in the rat: effects of remoxipride. *Alcohol* 28 (1): 39–46.

225 Dyr, W., McBride, W.J., Lumeng, L. et al. (1993). Effects of D1 and D2 dopamine receptor agents on ethanol consumption in the high-alcohol-drinking (HAD) line of rats. *Alcohol* 10 (3): 207–212.

226 Panocka, I., Ciccocioppo, R., Mosca, M. et al. (1995). Effects of the dopamine D1 receptor antagonist SCH 39166 on the ingestive behaviour of alcohol-preferring rats. *Psychopharmacology (Berl).* 120 (2): 227–235.

227 Rodd, Z.A., Melendez, R.I., Bell, R.L. et al. (2004). Intracranial self-administration of ethanol within the ventral tegmental area of male Wistar rats: evidence for involvement of dopamine neurons. *J. Neurosci.* 24 (5): 1050–1057.

228 Hodge, C.W., Samson, H.H., and Chappelle, A.M. (1997). Alcohol self-administration: further examination of the role of dopamine receptors in the nucleus accumbens. *Alcohol Clin. Exp. Res.* 21 (6): 1083–1091.

229 Eiler, W.J. 2nd, Seyoum, R., Foster, K.L. et al. (2003). D1 dopamine receptor regulates alcohol-motivated behaviors in the bed nucleus of the stria terminalis in alcohol-preferring (P) rats. *Synapse* 48 (1): 45–56.

230 El-Ghundi, M., George, S.R., Drago, J. et al. (1998). Disruption of dopamine D1 receptor gene expression attenuates alcohol-seeking behavior. *Eur. J. Pharmacol.* 353 (2–3): 149–158.

231 Phillips, T.J., Brown, K.J., Burkhart-Kasch, S. et al. (1998). Alcohol preference and sensitivity are markedly reduced in mice lacking dopamine D2 receptors. *Nat. Neurosci.* 1 (7): 610–615.

232 Boyce, J.M. and Risinger, F.O. (2002). Dopamine D3 receptor antagonist effects on the motivational effects of ethanol. *Alcohol* 28 (1): 47–55.

233 Thanos, P.K., Katana, J.M., Ashby, C.R. Jr. et al. (2005). The selective dopamine D3 receptor antagonist SB-277011-A attenuates ethanol consumption in ethanol preferring (P) and non-preferring (NP) rats. *Pharmacol. Biochem. Behav.* 81 (1): 190–197.

234 Heidbreder, C.A., Andreoli, M., Marcon, C. et al. (2007). Evidence for the role of dopamine D3 receptors in oral operant alcohol self-administration and reinstatement of alcohol-seeking behavior in mice. *Addict. Biol.* 12 (1): 35–50.

235 Leggio, G.M., Camillieri, G., Platania, C.B. et al. (2014). Dopamine D3 receptor is necessary for ethanol consumption: an approach with buspirone. *Neuropsychopharmacology* 39 (8): 2017–2028.

236 Bahi, A. and Dreyer, J.L. (2014). Lentiviral vector-mediated dopamine d3 receptor modulation in the rat brain impairs alcohol intake and ethanol-induced conditioned place preference. *Alcohol Clin. Exp. Res.* 38 (9): 2369–2376.

237 Kim, A., Di Ciano, P., Pushparaj, A. et al. (2020). The effects of dopamine D4 receptor ligands on operant alcohol self-administration and cue- and stress-induced reinstatement in rats. *Eur. J. Pharmacol.* 867: 172838.

238 Thanos, P.K., Roushdy, K., Sarwar, Z. et al. (2015). The effect of dopamine D4 receptor density on novelty seeking, activity, social interaction, and alcohol binge drinking in adult mice. *Synapse* 69 (7): 356–364.

239 Daurio, A.M., Deschaine, S.L., Modabbernia, A., and Leggio, L. (2020). Parsing out the role of dopamine D4 receptor gene (DRD4) on alcohol-related phenotypes: a meta-analysis and systematic review. *Addict. Biol.* 25 (3): e12770.

240 Barger, G. and Dale, H.H. (1910). Chemical structure and sympathomimetic action of amines. *J. Phys.* 41 (1–2): 19–59.

241 De Morsier, G. and Feldmann, H. (1952). Apomorphine therapy of alcoholism; report of 500 cases. *Schweiz Arch. Neurol. Psychiatr.* 70 (2): 434–440.

242 Schlatter, E.K. and Lal, S. (1972). Treatment of alcoholism with Dent's oral apomorphine method. *Q. J. Stud. Alcohol* 33 (2): 430–436.

243 Jensen, S.B., Christoffersen, C.B., and Noerregaard, A. (1977). Apomorphine in outpatient treatment of alcohol intoxication and abstinence: a double-blind study. *Br. J. Addict Alcohol Other Drugs* 72 (4): 325–330.

244 Lawford, B.R., Young, R.M., Rowell, J.A. et al. (1995). Bromocriptine in the treatment of alcoholics with the D2 dopamine receptor A1 allele. *Nat. Med.* 1 (4): 337–341.

245 Naranjo, C.A., Dongier, M., and Bremner, K.E. (1997). Long-acting injectable bromocriptine does not reduce relapse in alcoholics. *Addiction* 92 (8): 969–978.

246 Woolverton, W.L., Goldberg, L.I., and Ginos, J.Z. (1984). Intravenous self-administration of dopamine receptor agonists by rhesus monkeys. *J. Pharmacol. Exp. Ther.* 230 (3): 678–683.

247 Hyttel, J. (1983). SCH 23390 – the first selective dopamine D-1 antagonist. *Eur. J. Pharmacol.* 91 (1): 153–154.

248 Iorio, L.C., Barnett, A., Leitz, F.H. et al. (1983). SCH 23390, a potential benzazepine antipsychotic with unique interactions on dopaminergic systems. *J. Pharmacol. Exp. Ther.* 226 (2): 462–468.

249 Heidbreder, C. (2005). Novel pharmacotherapeutic targets for the management of drug addiction. *Eur. J. Pharmacol.* 526 (1–3): 101–112.

250 Meanwell, N.A. (2011). Synopsis of some recent tactical application of bioisosteres in drug design. *J. Med. Chem.* 54 (8): 2529–2591.

251 Neumeyer, J.L., Baindur, N., Niznik, H.B. et al. (1991). (+/−)-3-Allyl-6-bromo-7,8-dihydroxy-1-phenyl-2,3,4,5-tetrahydro-1H-3- benzazepin, a new high-affinity D1 dopamine receptor ligand: synthesis and structure-activity relationship. *J. Med. Chem.* 34 (12): 3366–3371.

252 Ross, S.T., Franz, R.G., Wilson, J.W. et al. (1986). Dopamine receptor agonists: 3-allyl-6-chloro-2,3,4,5-tetrahydro- 1-(4-hydroxyphenyl)-1H-3-benzazepine-7,8-diol and a series of related 3-benzazepines. *J. Med. Chem.* 29 (5): 733–740.

253 Neumeyer, J.L., Kula, N.S., Bergman, J., and Baldessarini, R.J. (2003). Receptor affinities of dopamine D1 receptor-selective novel phenylbenzazepines. *Eur. J. Pharmacol.* 474 (2–3): 137–140.

254 O'Boyle, K.M., Gaitanopoulos, D.E., Brenner, M., and Waddington, J.L. (1989). Agonist and antagonist properties of benzazepine and thienopyridine derivatives at the D1 dopamine receptor. *Neuropharmacology* 28 (4): 401–405.

255 Brewster, W.K., Nichols, D.E., Riggs, R.M. et al. (1990). Trans-10,11-dihydroxy-5,6,6a,7,8,12b-hexahydrobenzo[a]phenanthridine: a highly potent selective dopamine D1 full agonist. *J. Med. Chem.* 33 (6): 1756–1764.

256 Salmi, P., Isacson, R., and Kull, B. (2004). Dihydrexidine--the first full dopamine D1 receptor agonist. *CNS Drug Rev.* 10 (3): 230–242.

257 Wang, P., Felsing, D.E., Chen, H. et al. (2019). Synthesis and pharmacological evaluation of noncatechol G protein biased and unbiased dopamine D1 receptor agonists. *ACS Med. Chem. Lett.* 10 (5): 792–799.

258 Sohur, U.S., Gray, D.L., Duvvuri, S. et al. (2018). Phase 1 Parkinson's disease studies show the dopamine D1/D5 agonist PF-06649751 is safe and well tolerated. *Neurol. Ther.* 7 (2): 307–319.

259 Seiler, M.P. and Markstein, R. (1982). Further characterization of structural requirements for agonists at the striatal dopamine D-1 receptor. Studies with a series of monohydroxyaminotetralins on dopamine-sensitive adenylate cyclase and a comparison with dopamine receptor binding. *Mol. Pharmacol.* 22 (2): 281–289.

260 Wu, W.L., Burnett, D.A., Spring, R. et al. (2005). Dopamine D1/D5 receptor antagonists with improved pharmacokinetics: design, synthesis, and biological evaluation of phenol bioisosteric analogues of benzazepine D1/D5 antagonists. *J. Med. Chem.* 48 (3): 680–693.

261 Burnett DA, Greenlee WJ, Mckirtrick B, et al. (2009). Selective D1/D5 receptor antagonists for the treatment of obesity and CNS disorders. US Patent 20050075325A1, filed May 20, 2004, issued Mar. 17, 2009.

262 Haney, M., Ward, A.S., Foltin, R.W., and Fischman, M.W. (2001). Effects of ecopipam, a selective dopamine D1 antagonist, on smoked cocaine self-administration by humans. *Psychopharmacology (Berl).* 155 (4): 330–337.

263 Nann-Vernotica, E., Donny, E.C., Bigelow, G.E., and Walsh, S.L. (2001). Repeated administration of the D1/5 antagonist ecopipam fails to attenuate the subjective effects of cocaine. *Psychopharmacology (Berl).* 155 (4): 338–347.

264 Astrup, A., Greenway, F.L., Ling, W. et al. (2007). Randomized controlled trials of the D1/D5 antagonist ecopipam for weight loss in obese subjects. *Obesity (Silver Spring)* 15 (7): 1717–1731.

265 Bruns, R.F., Mitchell, S.N., Wafford, K.A. et al. (2018). Preclinical profile of a dopamine D1 potentiator suggests therapeutic utility in neurological and psychiatric disorders. *Neuropharmacology* 128: 351–365.

266 Svensson, K.A., Heinz, B.A., Schaus, J.M. et al. (2017). An allosteric potentiator of the dopamine D1 receptor increases locomotor activity in human D1 knock-in mice without causing stereotypy or tachyphylaxis. *J. Pharmacol. Exp. Ther.* 360 (1): 117–128.

267 Jutkiewicz, E.M. and Bergman, J. (2004). Effects of dopamine D1 ligands on eye blinking in monkeys: efficacy, antagonism, and D1/D2 interactions. *J. Pharmacol. Exp. Ther.* 311 (3): 1008–1015.

268 Luderman, K.D., Conroy, J.L., Free, R.B. et al. (2018). Identification of positive allosteric modulators of the D1 dopamine receptor that act at diverse binding sites. *Mol. Pharmacol.* 94 (4): 1197–1209.

269 Shiraki R, Tobe T, Kawakami S, et al. (2015) Heterocyclic Acetamide Compound. US Patent 8,937,087 B2, filed Apr. 18, 2013 and issued Jan. 20, 2015.

270 Hall, A., Provins, L., and Valade, A. (2018). Novel strategies to activate the dopamine D1 receptor: recent advances in orthosteric agonism and positive allosteric modulation. *J. Med. Chem.* 62 (1): 128–140.

271 Bach, N.J., Kornfeld, E.C., Jones, N.D. et al. (1980). Bicyclic and tricyclic ergoline partial structures. Rigid 3-(2-aminoethyl)pyrroles and 3- and 4-(2-aminoethyl)pyrazoles as dopamine agonists. *J. Med. Chem.* 23 (5): 481–491.

272 Tsuruta, K., Frey, E.A., Grewe, C.W. et al. (1981). Evidence that LY-141865 specifically stimulates the D-2 dopamine receptor. *Nature* 292 (5822): 463–465.

273 Koffarnus, M.N., Collins, G.T., Rice, K.C. et al. (2012). Self-administration of agonists selective for dopamine D2, D3, and D4 receptors by rhesus monkeys. *Behav. Pharmacol.* 23 (4): 331–338.

274 Collins, G.T. and Woods, J.H. (2007). Drug and reinforcement history as determinants of the response-maintaining effects of quinpirole in the rat. *J. Pharmacol. Exp. Ther.* 323 (2): 599–605.

275 Brunetti, M., Di Tizio, L., Dezi, S. et al. (2012). Aripiprazole, alcohol and substance abuse: a review. *Eur. Rev. Med. Pharmacol. Sci.* 16 (10): 1346–1354.

276 Allen, J.A., Yost, J.M., Setola, V. et al. (2011). Discovery of beta-arrestin-biased dopamine D2 ligands for probing signal transduction pathways essential for antipsychotic efficacy. *Proc. Natl. Acad. Sci. U.S.A.* 108 (45): 18488–18493.

277 Sokoloff, P., Giros, B., Martres, M.P. et al. (1990). Molecular cloning and characterization of a novel dopamine receptor (D3) as a target for neuroleptics. *Nature* 347 (6289): 146–151.

278 Freedman, S.B., Patel, S., Marwood, R. et al. (1994). Expression and pharmacological characterization of the human D3 dopamine receptor. *J. Pharmacol. Exp. Ther.* 268 (1): 417–426.

279 Collins, G.T., Newman, A.H., Grundt, P. et al. (2007). Yawning and hypothermia in rats: effects of dopamine D3 and D2 agonists and antagonists. *Psychopharmacology (Berl).* 193 (2): 159–170.

280 McCall, R.B., Lookingland, K.J., Bedard, P.J., and Huff, R.M. (2005). Sumanirole, a highly dopamine D2-selective receptor agonist: in vitro and in vivo pharmacological characterization and efficacy in animal models of Parkinson's disease. *J. Pharmacol. Exp. Ther.* 314 (3): 1248–1256.

281 de Paulis, T. (2003). Sumanirole Pharmacia. *Curr. Opin. Invest. Drugs* 4 (1): 77–82.

282 Casey, A.B. and Canal, C.E. (2017). Classics in chemical neuroscience: aripiprazole. *ACS Chem. Neurosci.* 8 (6): 1135–1146.

283 Urban, J.D., Vargas, G.A., von Zastrow, M., and Mailman, R.B. (2007). Aripiprazole has functionally selective actions at dopamine D2 receptor-mediated signaling pathways. *Neuropsychopharmacology* 32 (1): 67–77.

284 Kranzler, H.R., Covault, J., Pierucci-Lagha, A. et al. (2008). Effects of aripiprazole on subjective and physiological responses to alcohol. *Alcohol Clin. Exp. Res.* 32 (4): 573–579.

285 Voronin, K., Randall, P., Myrick, H., and Anton, R. (2008). Aripiprazole effects on alcohol consumption and subjective reports in a clinical laboratory paradigm--possible influence of self-control. *Alcohol Clin. Exp. Res.* 32 (11): 1954–1961.

286 Martinotti, G., Di Nicola, M., Di Giannantonio, M., and Janiri, L. (2009). Aripiprazole in the treatment of patients with alcohol dependence: a double-blind, comparison trial vs. naltrexone. *J. Psychopharmacol. (Oxford, England)* 23 (2): 123–129.

287 Brown, E.S., Jeffress, J., Liggin, J.D. et al. (2005). Switching outpatients with bipolar or schizoaffective disorders and substance abuse from their current antipsychotic to aripiprazole. *J. Clin. Psychiatry* 66 (6): 756–760.

288 Cuomo, I., Kotzalidis, G.D., de Persis, S. et al. (2018). Head-to-head comparison of 1-year aripiprazole long-acting injectable (LAI) versus paliperidone LAI in comorbid psychosis and substance use disorder: impact on clinical status, substance craving, and quality of life. *Neuropsychiatr. Dis. Treat.* 14: 1645–1656.

289 Seeman, P., Guan, H.C., and Hirbec, H. (2009). Dopamine D2High receptors stimulated by phencyclidines, lysergic acid diethylamide, salvinorin A, and modafinil. *Synapse* 63 (8): 698–704.

290 Ballon, J.S. and Feifel, D. (2006). A systematic review of modafinil: potential clinical uses and mechanisms of action. *J. Clin. Psychiatry* 67 (4): 554–566.

291 Dackis, C. and O'Brien, C. (2003). Glutamatergic agents for cocaine dependence. *Ann. N.Y. Acad. Sci.* 1003: 328–345.

292 Dackis, C.A., Kampman, K.M., Lynch, K.G. et al. (2005). A double-blind, placebo-controlled trial of modafinil for cocaine dependence. *Neuropsychopharmacology* 30 (1): 205–211.

293 Kaufman, K.R. (2005). Modafinil in sports: ethical considerations. *Br. J. Sports Med.* 39 (4): 241–244. discussion-4.

294 Rush, C.R., Kelly, T.H., Hays, L.R., and Wooten, A.F. (2002). Discriminative-stimulus effects of modafinil in cocaine-trained humans. *Drug Alcohol Depend.* 67 (3): 311–322.

295 Rush, C.R., Kelly, T.H., Hays, L.R. et al. (2002). Acute behavioral and physiological effects of modafinil in drug abusers. *Behav. Pharmacol.* 13 (2): 105–115.

296 Tyler, M.W., Zaldivar-Diez, J., and Haggarty, S.J. (2017). Classics in chemical neuroscience: haloperidol. *ACS Chem. Neurosci.* 8 (3): 444–453.

297 Schuckit, M.A. (2014). Recognition and management of withdrawal delirium (delirium tremens). *N. Engl. J. Med.* 371 (22): 2109–2113.

298 Indave, B.I., Minozzi, S., Pani, P.P., and Amato, L. (2016). Antipsychotic medications for cocaine dependence. *Cochrane Database Syst. Rev.* 3: CD006306.

299 Xiao, J., Free, R.B., Barnaeva, E. et al. (2014). Discovery, optimization, and characterization of novel D2 dopamine receptor selective antagonists. *J. Med. Chem.* 57 (8): 3450–3463.

300 Heidbreder, C.A. and Newman, A.H. (2010). Current perspectives on selective dopamine D(3) receptor antagonists as pharmacotherapeutics for addictions and related disorders. *Ann. N.Y. Acad. Sci.* 1187: 4–34.

301 Heidbreder, C.A., Gardner, E.L., Xi, Z.X. et al. (2005). The role of central dopamine D3 receptors in drug addiction: a review of pharmacological evidence. *Brain Res. Brain Res. Rev.* 49 (1): 77–105.

302 Keck, T.M., John, W.S., Czoty, P.W. et al. (2015). Identifying medication targets for psychostimulant addiction: unraveling the dopamine D3 receptor hypothesis. *J. Med. Chem.* 58 (14): 5361–5380.

303 Stemp, G., Ashmeade, T., Branch, C.L. et al. (2000). Design and synthesis of trans-N-[4-[2-(6-cyano-1,2,3,4-tetrahydroisoquinolin-2-yl)ethyl]cyclohexyl]-4-quinolinecarboxamide (SB-277011): a potent and selective dopamine D(3) receptor antagonist with high oral bioavailability and CNS penetration in the rat. *J. Med. Chem.* 43 (9): 1878–1885.

304 Reavill, C., Taylor, S.G., Wood, M.D. et al. (2000). Pharmacological actions of a novel, high-affinity, and selective human dopamine D(3) receptor antagonist, SB-277011-A. *J. Pharmacol. Exp. Ther.* 294 (3): 1154–1165.

305 Vorel, S.R., Ashby, C.R. Jr., Paul, M. et al. (2002). Dopamine D3 receptor antagonism inhibits cocaine-seeking and cocaine-enhanced brain reward in rats. *J. Neurosci.* 22 (21): 9595–9603.

306 Pak, A.C., Ashby, C.R. Jr., Heidbreder, C.A. et al. (2006). The selective dopamine D3 receptor antagonist SB-277011A reduces nicotine-enhanced brain reward and nicotine-paired environmental cue functions. *Int. J. Neuropsychopharmacology* 9 (5): 585–602.

307 Spiller, K., Xi, Z.X., Peng, X.Q. et al. (2008). The selective dopamine D3 receptor antagonists SB-277011A and NGB 2904 and the putative partial D3 receptor agonist BP-897 attenuate methamphetamine-enhanced brain stimulation reward in rats. *Psychopharmacology (Berl).* 196 (4): 533–542.

308 Ross, J.T., Corrigall, W.A., Heidbreder, C.A., and LeSage, M.G. (2007). Effects of the selective dopamine D3 receptor antagonist SB-277011A on the reinforcing effects of nicotine as measured by a progressive-ratio schedule in rats. *Eur. J. Pharmacol.* 559 (2–3): 173–179.

309 Cervo, L., Burbassi, S., Colovic, M., and Caccia, S. (2005). Selective antagonist at D3 receptors, but not non-selective partial agonists, influences the expression of cocaine-induced conditioned place preference in free-feeding rats. *Pharmacol. Biochem. Behav.* 82 (4): 727–734.

310 Yuan, J., Chen, X., Brodbeck, R. et al. (1998). NGB 2904 and NGB 2849: two highly selective dopamine D3 receptor antagonists. *Bioorg. Med. Chem. Lett.* 8 (19): 2715–2718.

311 Xi, Z.X. and Gardner, E.L. (2007). Pharmacological actions of NGB 2904, a selective dopamine D3 receptor antagonist, in animal models of drug addiction. *CNS Drug Rev.* 13 (2): 240–259.

312 Mugnaini, M., Iavarone, L., Cavallini, P. et al. (2013). Occupancy of brain dopamine D3 receptors and drug craving: a translational approach. *Neuropsychopharmacology* 38 (2): 302–312.

313 Nathan, P.J., O'Neill, B.V., Mogg, K. et al. (2012). The effects of the dopamine D(3) receptor antagonist GSK598809 on attentional bias to palatable food cues in overweight and obese subjects. *Int. J. Neuropsychopharmacol.* 15 (2): 149–161.

314 Murphy, A., Nestor, L.J., McGonigle, J. et al. (2017). Acute D3 antagonist GSK598809 selectively enhances neural response during monetary reward anticipation in drug and alcohol dependence. *Neuropsychopharmacology* 42 (7): 1559.

315 Galaj, E., Newman, A.H., and Xi, Z.X. (2020). Dopamine D3 receptor-based medication development for the treatment of opioid use disorder: rationale, progress, and challenges. *Neurosci. Biobehav. Rev.* 114: 38–52.

316 Silvano, E., Millan, M.J., Mannoury la Cour, C. et al. (2010). The tetrahydroisoquinoline derivative SB269,652 is an allosteric antagonist at dopamine D3 and D2 receptors. *Mol. Pharmacol.* 78 (5): 925–934.

317 Rossi, M., Fasciani, I., Marampon, F. et al. (2017). The first negative allosteric modulator for dopamine D2 and D3 receptors, SB269652 may lead to a new generation of antipsychotic drugs. *Mol. Pharmacol.* 91 (6): 586–594.

20

PTHR1 in Bone

Carole Le Henaff and Nicola C. Partridge

Department of Molecular Pathobiology, New York University College of Dentistry, New York, NY, USA

20.1 Introduction

This chapter focuses on the parathyroid receptor (PTHR1), its structure, regulation, action induced by its different ligands (parathyroid hormone (PTH), PTHrP, and related peptide, abaloparatide), and finally as a therapeutic target in osteoporosis and cancer.

20.2 PTHRs

20.2.1 PTHR1 or PTH/PTHrP Receptor 1

The receptor for parathyroid and related hormones, or PTHR1, was first successfully cloned in 1991 [1–3]. PTH has diverse actions and multiple target tissues but only a single G protein-coupled receptor, now referred to as the common PTHR1 [4].

20.2.1.1 PTH/PTHrP Receptor 1 Expression

The human *PTHR1* gene is located on chromosome 3 and consists of 14 exons. With its long amino terminal extracellular domain, this receptor belongs to the class B G-protein coupled receptors and shares the same basic structure with the ~800 other members of the GPCR superfamily of proteins (i.e. seven transmembrane domains [4]). The PTHR1 80 000-MW membrane glycoprotein is not exclusively located at the plasma membrane; it can also localize to the endosome and cell nucleus; however, its function in the nucleus remains unclear [5–8].

Recently, the structure of the extracellular domain of human PTHR1 has been analyzed by X-ray crystallography and shows an N-terminal α-helix, four

GPCRs as Therapeutic Targets, Volume 2, First Edition. Edited by Annette Gilchrist.

β-strands, a C-terminal α-helix, and three conserved disulfide bonds. The transmembrane domains are connected by three extracellular and three intracellular loops, and a C-terminal tail of about 130 amino acids extends intracellularly [9, 10]. The extracellular domain forms dimers and a hydrophobic α-β-β-α binding pocket for PTH. When PTH binds the receptor, it disrupts this oligomerization [11, 12].

PTHR1 expression is very important in bone, kidney, and cartilage. Indeed, it plays a key role in the regulation of Ca^{2+} concentrations in blood and of endochondral bone formation [13]. It is also expressed at lower levels in other tissues including the vasculature and certain developing organs, where it mediates the autocrine and paracrine effects of parathyroid hormone-related protein (PTHrP) at various times throughout development [14].

PTHR1 induces several intracellular signaling pathways. Activation of PTHR1 in different cell types initiates tissue-specific biochemical and cellular responses. For example, PTHR1 modulates the proliferation and apoptosis of osteoblasts and chondrocytes, as well as the production of many signaling factors involved in bone and cartilage metabolism [15, 16]. Activation of PTHR1 in renal tubular cells regulates the expression of proteins involved in mineral ion transport. Through autocrine mechanisms, PTHR1 regulates several molecular cascades: mechanisms of receptor desensitization [17, 18], feedback to control hormone release [19], and mechanisms for catabolism and removal of the hormone-ligand from the circulation [20, 21].

20.3 PTHR1 Ligands

20.3.1 Parathyroid Hormone or PTH

The human *PTH* gene is localized on the short arm of chromosome 11 at 11p15 [22, 23] and encodes a single-copy gene that consists of three exons and two introns [24].

PTH is secreted by the chief cells of the parathyroid glands. These four pea-sized parathyroid glands embedded at the back of the thyroid gland in the neck are part of the endocrine system. PTH is synthesized as a pre-pro-peptide [25]. After stimulation, the secretion of pro-hormone can be within minutes or hours. Cleavage of the prosequence 6-amino acids allows the mature full-length hormone, PTH (1–84), to be produced and stored in secretory vesicles [20, 26]. Further cleavage occurs when PTH is released into the circulation. PTH (1–34) is biologically active. The hormone is metabolized to amino-terminal and carboxyl-terminal fragments primarily in the liver, as well as in the kidney, the parathyroid glands, and the blood. The carboxyl-terminal fragments are cleared by glomerular filtration. Full-length PTH has a circulating half-life of less than five minutes. PTH is secreted in three distinct ways: tonic secretion, in a bimodal diurnal rhythm (with a primary peak in the early morning, and a secondary peak in the late afternoon, and nadirs in the morning and evening), and a pulsatility that appears to be stochastic (occurring unpredictably, 10 or more times a day) [27].

Various other PTH fragments, such as C-terminal PTH fragments, can be found in the body. These have been reported as a large NH_2-terminal truncated PTH (7–84) fragment, and (7–34) fragments which are 20% of the circulating PTH and 50% in kidney disease [28–30]. In general, PTH (7–84) and PTH (7–34) antagonize the effect of PTH: inhibit PTH release, inhibit kidney $1,25(OH)_2D_3$ production, and decrease calcium release by blocking PTH action or osteoclast formation in bone. The C-terminal fragments also have been reported to not always require PTHR1 for action [31–35].

PTH is highly regulated by serum calcium, phosphate, magnesium, and fluoride. For example, when the calcium level in the blood is high, calcitonin from the thyroid glands is secreted, and the calcium level is lowered by depositing calcium in bones and through urinary excretion. Inversely, when the level of calcium in the blood is slightly lowered, PTH is released to increase the level of serum calcium by reabsorption of calcium in the kidney, release of calcium by resorption from bone, and by stimulating increased synthesis of kidney $1,25(OH)_2D_3$ to increase uptake of calcium across the intestine. PTH production is also inhibited by $1,25(OH)_2D_3$ or FGF23 [36, 37].

20.3.2 Parathyroid Hormone-Related Protein or PTHrP

PTHrP is normally produced in many tissues and is recognized for its endocrine, paracrine, autocrine, and intracrine effects [38].

The gene coding for PTHrP is located in a similar position as PTH on sibling chromosomes 11 and 12, showing that they were likely duplicated from the same ancestral gene [38].

The human *PTHrP* gene (*PTHLH*) is composed of nine exons, transcribed from three functionally distinct promoters (two TATA boxes and a GC-rich region). The three induced transcripts are alternatively spliced to give rise to three mature proteins (139, 141, or 173 amino acids). But all of the peptides have identical sequences within the first 139 amino acids [39, 40]. Most PTHrP mRNA has a short half-life of 90–120 minutes, whereas the PTHrP (1–173) isoform has a longer half-life, around four hours [38]. The *PTHLH* gene is regulated at a transcriptional level by noncoding RNAs, miRNAs [41], and by multiple transcription factors, growth factors, cytokines, and hormones (cAMP response element binding protein (CREB), $1,25(OH)_2D_3$/vitamin D receptor (VDR), Est, Tx1, Sp-1, transforming growth factor β (TGFβ), epidermal growth factor (EGF)) [39, 40, 42, 43].

PTHrP protein itself is liable to extensive proteolytic processing. PTHrP can act as a cytokine and have a paracrine role, an autocrine function, or act as a circulating hormone [38, 44, 45].

First, a pre-pro-peptide with a 24-amino acid presequence is synthesized. After cleavage of the presequence, the 12-amino acid prosequence is important for intracellular trafficking and secretion of the protein. Posttranslational processing of PTHrP yields at least three mature fragments with distinct biological functions [38, 46]. The N-terminal 1–36 is structurally similar to PTH (1–34) and acts through the PTH1R with the same affinity [38, 47]. A histidine at position 5 of PTHrP contributes to the ability of PTHR2 to discriminate between PTH and PTHrP. The middle-region 36–139 is encoded by all three isoforms of PTHrP mRNA. It can allow nuclear localization, and the phosphorylation of Threonine 85 induces its exit from the nucleus [48]. The 35–84 nuclear sequence has a role in regulating placental calcium transport for fetal skeletal development [49]. In contrast with other fragments, the C-terminal (107–139 or 107–111) has been named osteostatin because of its ability to inhibit osteoclastic bone resorption [50, 51]. The peptides were also found to be anti-proliferative, to stimulate intracellular calcium, and activate PKC (Protein kinase C) [52, 53]. Surprisingly, PTHrP (107–139) interacts with vascular endothelial growth factor (VEGF) and enhances osteoblastic cell differentiation [54]. Thus, this polyhormone has several bioactivities produced by its three fragments.

PTHrP (1–36) has been most studied in bone metabolism because of its homology to PTH (1–34) and because it binds the same receptor (PTHR1). PTHrP (1–36) mimics PTH (1–34) in inducing bone resorption, renal phosphate wasting, and is implicated in hypercalcemia of malignancy [38].

20.4 Biochemical Reactions

20.4.1 Interaction of Hormone/Receptor

Structurally, a number of ligands can bind PTHR1: PTH, PTHrP, and their analogs. However, these occur with distinct, high-affinity receptor conformations like the G-protein-coupled (RG, GTPγS-sensitive) and uncoupled (R^0). The receptor conformation is more "closed" in the RG conformation than with R^0 [55]. These different conformations induce different signaling responses with different duration, longer for R^0 than for RG [56]. R^0 is an intermediary between the classical R and RG states of the two-state or ternary complex models of GPCR action. R^0 is a pre-active state that is primed to interact efficiently with a G protein as it encounters one. Thus, the ligand-R^0 complex can isomerize to ligand-RG and become signaling competent. The affinity with which a ligand binds to R^0 will be a determinant, in part, of the overall signaling response capacity of that ligand [57].

Different PTHR1 ligands can have different capacities to stabilize the R^0 conformation, caused by their mode of interaction with the N (extracellular) and/or J (juxtamembrane – extracellular loop/transmembrane) domains of the receptor. The C-terminal peptide (14–34) of PTH or PTHrP interacts with the extracellular N-terminal receptor region (N interaction) [55]. The N-terminal peptide (1–14) of PTH and PTHrP is sufficient for activation of the PTHR1 by their binding to the J domain of the receptor (J interaction) [9, 58]. The J interaction plays a major role in inducing the conformational changes in the receptor that leads to G protein coupling [55, 58, 59]. PTHR1 couples with the heterotrimeric G protein containing $G_{\alpha S}$. For the RG state, agonist binding is pseudo-irreversible, suggesting that the ligand is trapped within the ligand-receptor-G-protein complex [55].

PTH (1–34) and PTHrP (1–36) bind similarly to RG, but PTH (1–34) has a greater capacity to bind to R^0. This is thought to produce cumulatively greater signaling responses with greater duration of cAMP levels (via $R^0 \rightarrow RG$ isomerization) compared with PTHrP. In addition, PTHrP (1–36) dissociates more rapidly than PTH (1–34) from PTHR1. This induces the same increase in cAMP, but this is followed by a faster decline with PTHrP (1–36) than PTH (1–34). This explains the prolonged signal with PTH (1–34) compared with PTHrP (1–36) [38, 57, 60].

20.4.2 Signaling Pathways Activated by PTHR1 in Bone

After ligand coupling, PTHR1 activates four different intracellular signaling cascades:

- GαS-adenylyl cyclase-cAMP-protein kinase A (PKA) [61]. This pathway is followed by increased gene expression of several growth factors, and factors controlling bone formation and resorption [62]. PKA is formed of two regulatory subunits and two catalytic subunits. When cAMP is produced, it binds the regulatory subunits and allows the release of the active catalytic subunits. Mutations in the type 1 regulatory subunit of PKA or phosphodiesterase 4 (PDE4D) cause acrodysostosis. In contrast, Prkar1a (R1α regulatory subunit of PKA) mutations result in the release of the catalytic subunits, and overactivated PKA are associated with osteosarcoma [63, 64]. Constitutive inhibition of PKA (by preventing the binding of cAMP) induces acrodysostosis with renal resistance to PTH and chondrodysplasia [65]. In bone, the GαS-adenylyl cyclase-cAMP-PKA pathway signals through Runx2, HDAC4/5, Mef2c, sirtuin-1, and SIKs-CRTCs. By these pathways, PTH or PTHrP regulates gene expression such as for cFos, MMP13, Sclerostin, or RANKL [65–75].
- Gαq-phospholipase C (PLC) β-inositol triphosphate-cytoplasmic Ca^{2+}-protein kinase C (PKC). PKC signaling is not required for PTH effects in bone and may be inhibitory to the osteoanabolic actions of PTH [4, 21].
- $G\alpha_{12/13}$-phospholipase D-transforming protein RhoA.

- β-arrestin-extracellular signal-regulated kinase 1/2 (ERK1/2). When PTH binds PTHR1, the receptor can recruit β-arrestins to the plasma membrane, which allows the stimulation of ERK1/2 phosphorylation [76]. With intermittent PTH administration, this pathway limits osteoclastogenesis and contributes to increased bone mineral density [77, 78]. In osteoblastic cell lines, PTH or PTHrP seem to have some opposite effects with activation or inhibition of this pathway depending on the proliferation or differentiated state of the cells [79–82]. Indeed, by endocytosis of the receptor, β-arrestin not only reduces the availability of the receptor at the cell surface [18, 83] but also stabilizes the persistent ternary PTH–PTHR1–β-arrestin complex and prolongs signaling responses via adenylyl cyclases and cAMP. This process is associated with prolonged physiological calcemic and phosphate responses observed for long-acting PTH analogs in animals [9, 76, 84, 85]. PTHrP also binds PTHR1, but the signal is very transient, and the PTHrP/PTHR1 complex does not appear to be internalized [86].

PTH (1–34) can also influence the Wnt pathway by forming a ternary PTH/PTHR1/Lrp6 complex and stimulating bone formation. This signaling pathway plays an important role in bone cell function in skeletal modeling and remodeling [87–91]. Several studies revealed that PTHR1 regulates Wnt signaling in a G protein-independent manner in bone. Indeed, PTH and PTHrP upregulate the expression of LRP6, frizzled-1 (FZD-1), and β-catenin and decrease LRP5 and Dkk-1 (Dikkopf-1) [92, 93]. PTH can act indirectly through sclerostin, an inhibitor of the canonical Wnt pathway [94], by inhibiting its expression, allowing increased bone formation.

20.5 Physiological Function of PTHR1 in Bone

20.5.1 Physiological Function of PTH

20.5.1.1 PTH in Calcium/Vitamin D Homeostasis

The primary function of PTH is to maintain the extracellular fluid Ca^{2+} concentration. The hormone acts directly on bone and kidney and indirectly on the intestine. For the latter, PTH induces the synthesis of $1,25(OH)_2D_3$ that acts on the small intestine to increase absorption of Ca^{2+}. PTH production is closely regulated by serum Ca^{2+}. This feedback regulation allows fine control of Ca^{2+} homeostasis. PTH-PTHR1 binding facilitates the increased renal reabsorption of Ca^{2+} and stimulates the activity of the enzyme CYP27B1, which converts inactive 25-hydroxyvitamin D3 to its active form (1,25-dihydroxyvitamin D3 or calcitriol) [95, 96]. This active form is the most powerful physiological agent to stimulate active transport of calcium, and to a lesser degree phosphorus and magnesium, across the small intestine. Calcitriol has effects predominantly through the VDR to increase intestinal Ca^{2+} absorption [97].

In bone, binding of PTH to PTHR1 increases the rate of dissolution of bone mineral, increasing the Ca^{2+} flow from the bone to blood. The PTH action on bone can be seen within minutes, leading to an increase in bone remodeling. Continuous exposure to elevated PTH induces increased osteoclast activity and bone breakdown. However, intermittent PTH administration with elevated hormone levels for one to two hours each day, leads to increased bone formation.

20.5.2 Physiological Function of PTHrP

Produced by many tissues, PTHrP has autocrine and paracrine roles in postnatal physiology. These include tooth eruption, branching morphogenesis, regulation of β-cell proliferation and insulin production, regulation of vascular smooth muscle, and regulation of keratinocyte differentiation [38].

During pregnancy, PTHrP is produced by the fetus and regulates calcium transport between the mother and the fetus [49, 98]. In lactation, PTHrP (1–36) is produced by the breast, circulates in the body, and is highly secreted into milk [99–101]. During lactation, the increased PTHrP causes bone loss for provision of calcium for milk [102, 103].

During skeletal development, PTHrP is important in cartilage and endochondral bone formation [38, 104]. Indeed, in the absence of PTHrP, mice showed skeletal abnormalities with defective endochondral bone development and cartilage with more hypertrophic chondrocytes [105–107]. PTHrP inhibits chondrocyte differentiation and stimulates their proliferation regulated by Ihh feedback [38].

PTHrP is also produced by cells of the osteoblast lineage [108, 109]. It stimulates bone formation by promoting the differentiation of osteoblast precursors and by inhibiting apoptosis of mature osteoblasts and osteocytes [38, 105, 108]. Indeed, mice which are haploinsufficient for *Pthlh* or have osteoblast-specific deletion of PTHrP showed a low bone mass after three months of age. This is caused by decreased bone formation induced by lesser recruitment of bone marrow precursors coupled with increased osteoblast apoptosis [38, 105, 110].

20.5.3 PTH and Bone Remodeling

Bone is a dynamic organ: trabecular bone (20% of skeleton) is metabolically more active than cortical bone (80% of skeleton), with an annual turnover (remodeling) of approximately 20–30% for the trabeculae and 3–10% for cortical bone.

Throughout life, this process replaces old or damaged bone which is resorbed by osteoclasts, followed by macrophages, to leave a site where osteoblasts form new bone [111]. The process differs from one bone to another, between cortical and trabecular bone [38]. The key regulators of this process are RANK (Receptor activator of nuclear factor-κB, expressed by osteoclasts), RANKL (Receptor activator of nuclear factor-κB ligand, produced by osteoblasts to stimulate osteoclastogenesis), and OPG (Osteoprotegerin, a decoy receptor to RANKL produced by osteoblasts which inhibits osteoclastogenesis).

In vivo, both continuous and intermittent administrations of PTH induce an increased bone turnover. However, unlike intermittent administration, continuous administration results in a net increased osteoclast differentiation and activity, and thus, enhanced resorption [112–114].

20.5.3.1 Anabolic Action of PTH

PTHR1 is found on cells of the osteoblast lineage, including osteocytes [115, 116]. PTH binds to PTH1R on these cells.

Intermittent PTH (1–34) administration causes an increase in bone formation through an increase in osteoblast number and activity. This leads to increased mineralized matrix deposition and reactivates quiescent lining cells [15, 16, 117]. In the osteoblastic lineage, PTH (1–34) acts by inhibition of osteoblast apoptosis, by the conversion of quiescent bone lining cells into active osteoblasts [117], by the increase in preosteoblast proliferation [118, 119], and by the induction of osteoblast differentiation [120]. All of these lead to an increase in bone formation [121]. PTH also inhibits adipocyte differentiation from very early stage stem cells in the osteoblast lineage [122].

In vivo, bone formation has been demonstrated to occur within the first six to nine hours following PTH administration. Daily intermittent PTH (1–34) administration repeats these cycles of upregulation, thus leading to an overall net anabolic effect [90, 123]. In rats, the number of Colony Forming Units (CFU) doubled [124]. Thus, PTH increases the number of osteoblast precursors in the bone marrow in rodents [121]. In osteocytes, PTH regulates Notch, Wnt, bone morphogenetic protein (BMP) pathways, and genes associated with osteocyte differentiation and action. PTH also decreases the expression of sclerostin, a negative regulator of bone formation and an inhibitor of the Wnt pathway [125, 126].

In vitro, the anabolic effects of PTH remain incompletely understood, difficult to replicate, and depend on a number of parameters such as cell density, cell-cycle stage, concentration of PTH, and mode of PTH treatment (intermittent or continuous) [75, 127]. In most papers, PTH promotes osteoblastic lineage commitment from bone marrow-derived cells, primary calvarial cells, and periosteal cells. PTH promotes osteoblast differentiation and RANKL expression in osteoblastic cells which, *in vivo*, induces osteoclastogenesis [128–130].

20.5.3.2 Catabolic Action of PTH

In hyperparathyroidism or with continuous PTH administration, the hormone causes a net increase in bone resorption by indirectly increasing osteoclastogenesis [93, 131–133]. There have been reports of a direct resorptive effect of PTH by activation of a PTHR1 receptor identified on osteoclasts [134, 135]. However, most findings demonstrate that PTH acts indirectly via osteoblasts to stimulate osteoclastogenesis.

This process is induced by the OPG/RANKL balance. The two factors play a critical role in bone resorption. PTH stimulates RANKL expression and inhibits osteoprotegerin (OPG, a decoy receptor of RANK) expression in osteoblasts and osteocytes. By this mechanism, RANKL binds RANK on the surface of hematopoietic precursors of osteoclasts and induces osteoclastogenesis and osteoclast survival [136–139]. A short and transient exposure of PTH (1–34) allows a favorable response with less bone resorption than bone formation [140]. In contrast, continuous infusion of PTH leads to a sustained, constant increase in *Rankl* mRNA and decreased *Opg* mRNA in osteoblasts and osteocytes [126, 141, 142]. This results in enhanced osteoclastogenesis and consequently, net bone resorption [93, 143–145]. Mice without RANKL in osteocytes develop increased bone mass associated with reduced osteoclast number. Moreover, these mice have reduced bone loss caused by secondary hyperparathyroidism induced by dietary calcium deficiency [135]. In mice with conditional deletion of the PTHR1 in osteocytes, PTH administration fails to increase RANKL expression [142].

In patients with mild hyperparathyroidism, circulating levels of RANKL are increased and positively correlated with bone resorption markers and rates of bone loss at the total femur [146]. One year after parathyroidectomy, these patients show a decreased *Rankl*/*Opg* ratio in iliac crest bone biopsies [147, 148].

RANKL exists in both transmembrane and soluble forms [149]. *In vitro*, matrix metalloproteinase 14 (MMP14) causes proteolytic cleavage of transmembrane RANKL to generate the soluble form (sRANKL) in osteoblastic cells [149, 150]. Using *in vitro*, *ex vivo*, and *in vivo* approaches, it was shown that PTH upregulates the expression of MMP14 in osteocytes. This enzyme is required to increase numbers of osteoclasts, bone resorption, and consequent bone formation [150].

PTH affects the production of macrophage colony-stimulating factor (M-CSF), a cytokine involved in the regulation of both cell proliferation and differentiation of hematopoietic stem cells from the bone marrow niche. PTH causes its release from osteoblasts at the same time as RANKL and subsequent effects on osteoclasts, *in vivo* [134, 151]. These dual effects have an unclear role in overall bone anabolism [90].

Monocyte chemoattractant protein-1 (MCP-1) is a chemokine highly upregulated by daily hPTH (1–34) injections *in vivo* in rats [152, 153]. Only 2 h after PTH (1–34) injection, serum MCP1 is increased and facilitates osteoclast recruitment and differentiation [153, 154]. MCP-1−/− mice showed that MCP-1 appears to be necessary for PTH's anabolic and catabolic effects [154]. In fact, continuous infusion of PTH in rats induces moderate, constantly increased MCP-1 expression in bone which enhances RANKL-mediated osteoclastic bone resorption [153, 155]. In primary hyperparathyroidism, a positive association was found in women between serum levels of MCP-1 and PTH [156]. After parathyroidectomy, the elevated MCP-1 levels immediately decrease [157]. These results show a link between PTH and MCP-1 regulation.

20.6 PTHR1 as a Therapeutic Target in Osteoporosis

In humans, bone mass is acquired up to the fourth decade, with a rapid phase during adolescent growth. Most peak bone mass is genetically determined. Women have approximately 30% less peak bone mass than men and experience an accelerated loss after menopause. Both genders experience age-related loss of bone mass [158].

Osteoporosis is a common skeletal disorder characterized by a reduced bone mass and an architectural deterioration caused by an imbalance between osteoblasts and osteoclasts. This high bone resorption associated with a low bone formation leads to reduced bone strength and increased risk of fracture often occurring with minimal trauma, in particular at the hip and spine, the most common sites of osteoporotic fractures [159, 160].

Osteoporosis affects more than 200 million patients worldwide and is a major growing global health problem [161]. Prior to the mid-twentieth century, osteoporosis was largely untreatable and discovered after one or more spontaneous or pathological fractures, mostly in the spine or hip [4]. With the development of noninvasive methods of measuring bone mass, diagnosis improved and treatment could begin earlier [162]. The second great advance has been the development of effective anti-resorptive therapies which do not enhance bone formation but stop osteoclast activity and the corresponding bone resorption [4, 163]: Bisphosphonates (Alendronate, Risendronate, Ibandronate, and Zoledronic acid), receptor activator of nuclear factor κ-B ligand antibody (RANKL-antibody or denosumab), estrogen, selective estrogen modulators (SERMs, Raloxifene), and calcitonin.

A coveted but elusive goal in osteoporosis therapy is to replace the bone lost in this disease and likewise reduce fracture risk [164]. For stimulation of new bone formation, teriparatide, or PTH (1–34), was the first anabolic treatment approved in the USA to treat osteoporosis [159]. Later, abaloparatide, an analog of PTHrP, was approved by the Food and Drug Administration (FDA) in April 2017 as a new anabolic treatment for osteoporosis [165]. Another anabolic treatment has recently been approved by the FDA: the antibody to sclerostin or Romosozumab.

The antibody binds sclerostin and prevents its negative effects on osteoblast differentiation and bone remodeling [166, 167].

20.6.1 PTH as an Anabolic Treatment for Osteoporosis

PTH was the first anabolic agent developed for osteoporosis that became the most effective therapy. Full-length PTH is not required to activate PTH1R. The NH_2-terminal domain of PTH mediates most of the physiological effects of the hormone: PTH (1–34) is used as a PTH analog with identical biological activity [95]. It stimulates osteoblast activity and thus, bone formation.

However, a paradox still exists: constant infusion/elevation of PTH and hyperparathyroidism cause bone loss, whereas intermittent injection/elevation of PTH in hypoparathyroidism and osteoporosis causes an increase in bone volume [4, 168, 169]. This anabolic effect of PTH was first reported in the 1930s, but was largely neglected due to this paradox [170]. It was only in the 1970s that clinical trials with PTH began in osteoporosis patients. The first clinical trial was published in 2001 [171], but it was stopped after 18 months because of the concurrent appearance of osteosarcomas in Fischer rats treated long term with PTH (1–34) [108, 172]. In mice and rats, they showed that intermittent injection of PTH is anabolic for bone, while continuous elevation is catabolic [4, 120, 140, 168].

Over the years and over a number of trials, studies showed that PTH (1–34) strikingly reduced fracture incidence and new nonvertebral fractures, increased trabecular bone, bone mass, and connectivity [169, 171, 173]. In 2002, the FDA in the Unites States approved teriparatide, a recombinant human PTH (1–34), for the treatment of osteoporosis in postmenopausal women and men who are at high risk for fracture (20 μg daily injection for less than two years) [174]. Intact human PTH (1–84, Allelix) was approved and is available in Europe. This was studied because of the thought that intact PTH (1–84) is the major PTH in the circulation and may be more efficient than the already active PTH (1–34). The first clinical trial on the osteoanabolic effect of PTH was done with PTH (1–84) which showed a great improvement in the quality of life of osteoporotic patients [175]. The therapeutic use of PTH is limited to two years because of the cancers that occurred in rats treated long term. Since the animals were treated from weaning and the osteosarcomas developed near the end of the lifetime of the animals, it does not seem to be relevant to humans. No osteosarcomas have been found in humans with long-term treatment with PTH because of hypoparathyroidism [171, 174].

When teriparatide therapy is started, circulating bone formation markers increase rapidly, with a delayed appearance of circulating bone resorption markers, giving rise to the concept of the "anabolic window." This shows bone formation is the first response to PTH (1–34), but followed and superseded by subsequent increased bone resorption. This theory is only based on serum bone markers [108, 176, 177]. The delay in resorption marker increase after starting PTH treatment can be explained by the principles of basic multicellular unit (BMU)-based remodeling [178–180]. BMUs remodeling bone are divided into several phases: resorptive (two to three weeks, activation of osteoclasts and then osteoclast resorption), reversal (two to three weeks, cleaning the pits by mononuclear cells and osteoblast activation and differentiation), or formative phases (three to four months, osteoblast activity with collagen deposition and mineralization). BMUs are initiated asynchronously throughout the skeleton, geographically and chronologically separated from each other. Because the duration of the formation phase is longer than the resorption phase, at any time, there are more BMUs in their formation phase at various locations than there are BMUs in their resorption phase at other locations [181]. Teriparatide promotes differentiation and activity of osteoblasts and inhibits osteoblast apoptosis. It acts on BMUs in their reversal or formation phases shown by an increase in P1NP (osteoblast formation marker) [16, 120, 182, 183]. However, teriparatide induces an increase in RANKL expression in osteoblasts and osteocytes, with resultant osteoclast activation. Therefore, teriparatide also acts on BMUs in their resorptive phase and generates new BMUs. When enough matrices have been resorbed, resorption markers such as CTX (collagen crosslink degradation product, osteoclast activity marker) reach a measurable level in the circulation. This can explain the delay in the increase in resorption markers relative to formation markers after the start of teriparatide treatment and the "anabolic window" [108, 182, 184]. At cortical sites, PTH typically does not increase bone density, but tends to increase porosity. PTH increases periosteal bone mass and enlarges the cross-sectional area, strengthening cortical bone [185, 186].

To avoid bone resorption, some teams have combined PTH (1–34) treatment concurrently with anti-resorptive treatments such as with a bisphosphonate (Alendronate). However, this did not yield a better-expected effect due to the coupling that seems necessary between osteoblast/osteoclast or bone formation/bone resorption. As a result, it led to the conclusion that in intermittent PTH treatment, bone resorption is necessary for inducing bone formation [187–190]. While concurrent treatment with PTH and biphosphonate did not appear to be additive, bisphosphonate therapy initiated immediately upon completion of PTH treatment is beneficial with a continual gain or maintenance of bone mineral density (BMD). This combination avoids bone loss after cessation of PTH (1–34) treatment [191–193]. In the Denosumab and Teriparatide Administration (DATA-trial), a combination of concurrent daily teriparatide (20 μg) + denosumab (60 mg) injection every six months showed a better increase in spine and hipbone density than monotherapies [163, 194]. However, the long-term effect of this combination remains to be determined and results on fractures are unknown. Another study showed the advantage of the sequential therapy with one year of

PTH (1–84) treatment (50, 75, or 100 μg/day) followed by one year of alendronate treatment (10 mg/day) in comparison with PTH alone or alendronate alone [193].

20.6.2 PTHrP as an Anabolic Treatment for Osteoporosis

Because PTHrP shares the same receptor with PTH, PTHrP (1–36) has been investigated as a potential anabolic agent to treat osteoporosis [195]. The anabolic action of daily injection of PTHrP (1–36) in human subjects has been investigated extensively by Stewart and colleagues. They suggested a very low resorptive effect determined by measuring bone formation and resorption markers. In normal human subjects, infused PTHrP (1–34) was less potent than PTH (1–34) [196]. Single injections of several doses of PTHrP (1–36) (maximum 3.20 μg/kg) resulted in increased phosphate excretion and nephrogenous cAMP and increased plasma calcitriol [197]. The highest PTHrP (1–36) dose (600 μg/day) resulted in a decrease in resorption marker excretion [198]. However, the dose of PTHrP used in order to have the same bone marker levels was much higher (400 μg/day) than with PTH (1–34) (20 μg/day) [199–201]. Unlike PTH (1–34), PTHrP (1–36) is degraded more rapidly after injection and has a lower volume of distribution to activate BMUs. However, if less agonist is available to the receptor, then this should lead to less anabolic and resorptive effects [108]. The use of PTHrP and the results remain unclear.

20.6.3 Abaloparatide as an Anabolic Treatment for Osteoporosis

A working hypothesis has emerged concerning the interaction of different regions of the biologically active amino terminal 34-aminoacid region of the peptide with the receptor (cf PTHR1 conformation; [4, 55]). To increase the effect through this receptor, a new peptide was generated: abaloparatide, or Tymlos. It is a 34 amino acid synthetic analog of PTHrP and has 41% homology to PTH (1–34) and 76% homology to PTHrP (1–34) [202]. Abaloparatide is identical to PTHrP in its first 20 amino acids, while over half of the remaining amino acids are different [203]. Five of the 13 residues between 22 and 34 of PTHrP (1–34) have been altered to increase the interaction between abaloparatide and PTHR1. Like PTH and PTHrP, abaloparatide works as an anabolic agent for bone, through selective activation of the PTH1R. Abaloparatide appears to exhibit greater selectivity for the RG receptor conformation than PTH (1–34) and considerably lower affinity for R^0. It is associated with more transient cAMP signaling responses and activation period compared with PTH.

In April 2017, subcutaneous abaloparatide (Tymlos™) received its approval for the treatment of osteoporosis in postmenopausal women, in the USA. Radius Health is also developing a transdermal formulation of abaloparatide, with administration via a microneedle patch [204].

In a short-term phase II clinical trial, Leder et al. tested three doses of abaloparatide (20, 40, and 80 μg/day). At 24 weeks, they saw a dose-dependent increase in bone mineral density similar or greater than PTH (1–34) at 20 μg/day [205]. The same results were observed in the Abaloparatide Comparator Trial in Vertebrae End-points (ACTIVE) double blind phase III trial, after 18 months with patients treated with abaloparatide at 80 μg/day or PTH (1–34) at 20 μg/day [203]. After 18 months, in the ACTIVExtend trial, patients with abaloparatide or placebo received alendronate treatment for 6 months [206]. This last anti-resorptive treatment confirms a better anabolic action of abaloparatide compared with PTH (1–34).

Most preclinical studies of the anabolic effect of abaloparatide were done in aged osteopenic, ovariectomized, or orchidectomized rats. In these models, rats were treated with two or three doses of abaloparatide (1, 5, and 25 μg/kg) and showed similar dose–responsive increases in bone mineral density and bone formation without an effect on bone resorption compared with PTH (1–34) [207–209]. Abaloparatide has an anabolic effect with dose-dependent bone formation. This is characterized by an increase in trabecular, endocortical, and periosteal surfaces and bone formation markers. Abaloparatide shows an absence or less increased, eroded surfaces, osteoclast numbers, or bone resorption markers compared with PTH (1–34) [108]. However, wildtype male adult mice treated with abaloparatide (80 μg/kg/day) at the same dose as PTH (1–34) show the same anabolic and catabolic effects compared with PTH (1–34), but not the same molecular and cellular mechanisms [72, 210]. *In vitro*, abaloparatide and PTH (1–34) do not have the same molecular mechanisms with different activation of PTHR1. These induce different cAMP production and regulation of osteoblastic gene expression, observed in primary calvarial cells and in mice [210, 211]. Two years daily treatment with several doses of abaloparatide in rats results in dose and time-dependent formation of osteosarcomas similar to PTH (1–34) at a similar exposure [212]. In ovariectomized monkeys, abaloparatide administration was associated with increases in bone formation, bone mass, and bone strength without increases in serum calcium or bone resorption parameters [213].

20.6.4 LA-PTH as an Anabolic Treatment for Osteoporosis

Recent structure–activity studies have identified PTH analogs that exhibit markedly prolonged cAMP signaling actions in PTHR1-expressing cells, as well as prolonged functional responses in animals [214, 215]. These analogs emerged from efforts initially aimed at optimizing the N-terminal region of PTH (1–34). It is formed by the optimized N-terminal PTH (1–14) sequence and the

C-terminal segment is derived from the (15–36) region of PTHrP and is called long-acting parathyroid hormone (LA-PTH). Compared with PTH (1–34), this hybrid binds with severalfold higher affinity to the R^0 conformation of PTHR1, which maintains high affinity for the ligand through multiple rounds of G protein coupling and mediates more prolonged cAMP responses [58, 215].

When injected into wildtype and thyroparathyroidectomized animals, LA-PTH causes calcemic and phosphate responses that persist for several hours longer than that induced by PTH (1–34) without exhibiting a longer half-life in the circulation or hypercalciuria or increasing bone resorption markers. The peptides disappear from the blood within several minutes. Injection of LA-PTH resulted in rapid and robust increases in cAMP in both blood and urine of mice [214, 215]. This strong dissociation between pharmacokinetic and pharmacodynamic properties suggests that the analogs are cleared rapidly from the circulation, in part by binding efficiently and stably to PTH receptors in target cells, in which they mediate extended downstream action [214]. LA-PTH is thought to induce prolonged responses *in vivo* via a pseudo-irreversible binding to the PTH receptor in target cells, and hence persistent signaling responses, possibly from within internalized cell compartments [216, 217]. In any case, the responsible mechanisms bring about a therapeutically favorable but discordant pharmacokinetic/pharmacodynamic relationship – a serum half-life of less than eight minutes, as measured in rodents, versus biological effects on calcium and phosphate blood levels lasting 24 hours or more, depending on the dose. The prolonged biological actions of these compounds are associated with an extended pharmacokinetic profile, which is thought to arise, in part, from enhanced resistance to serum proteases [214, 216, 218].

20.7 PTHR1: PTH as Treatment for Other Bone Diseases

Teriparatide has been used to treat osteonecrosis of the jaw in animal studies and several clinical case reports. However, after the development of this disease, teriparatide did not show any beneficial effect [219].

Teriparatide shows a positive effect on implant fixation in total joint arthroplasty, in rat and canine models, and in human cases [220–222]. However, the results remain controversial: some retrospective studies showed a bony ingrowth at the bone-prothesis surface or a significant reduction in the rate of femoral stem subsidence after a cementless hemiarthroplasty for femoral neck fractures or no effects [223–225].

20.8 PTHR1 in Cancer

20.8.1 PTHrP as a Diagnostic for Cancer

PTHrP (1–36) is implicated as a hormone in cancer and produced in excess by certain cancers [226, 227]. Hypercalcemia of malignancy is usually due to increased bone resorption. That can be caused by skeletal metastases of the tumors or by the tumoral production of PTHrP [228]. Most of the time, tumor cells stimulate neighboring osteoclasts and bone resorption. The most common factor involved seems to be PTHrP, especially in solid tumors where abnormal PTHrP expression is implicated in up to 80% of the patients. PTHrP expression is noted to be common not only in squamous cell cancers but also in breast, lung, and prostate cancer. Primary tumors, such as breast and lung cancers, that produce PTHrP are more likely to metastasize to bone. For prostate cancer, PTHrP production does not usually cause hypercalcemia, perhaps because this type of tumor processes the polypeptide to a nonhypercalcemic peptide [158].

Strangely, in breast cancer, tumors positive for PTHrP at surgery were independently predictive of improved patient survival with reduced metastases at all sites, including bone [229, 230]. Such a mechanism is distinct from the bone resorbing action later in the disease. The role of PTHrP as a diagnostic for cancer remains unclear: in mouse breast cancer models, PTHrP protects against mammary tumor emergence and infiltration, whereas in another study, PTHrP promotes the initiation and progression of primary tumors. At the same time, PTHrP has an important role in early mammary gland development [44] and therefore might play a part in early stages of breast cancer development [38].

20.8.2 PTHrP as a Target in Osteosarcoma

Osteosarcomas are the most common primary tumor of bone, occurring mainly in the second decade of life. They arise from malignant osteoblastic mesenchymal cells that produce an osteoid matrix that mineralizes in the case of the osteoblastic subtype of osteosarcoma (60%). The less mineralized fibroblastic and the chondroblastic subtypes comprise ~30% [38].

Some rodent models, which overexpress PTHR1 or PTHrP, were developed and found to be like human osteosarcomas [231]. *In vitro* studies showed an autocrine production of PTHrP in human osteosarcomas and in UMR 106 rat osteosarcoma cells [232, 233]. In all mouse osteosarcoma models, the tumors produce PTHrP that acts through the PTHR1 and can induce osteosarcoma proliferation. The absence of PTHR1 or PTHrP in osteosarcoma cells decreases the tumor growth and invasion [234]. These deletions, *in vivo*, increase apoptosis and cause profound growth inhibition of transplanted tumors [234]. On the other hand, normal murine osteoblasts survived depletion of PTHrP. Therefore, for osteosarcoma an autocrine circuit was identified: the PTHrP-PTHR1-PKA-CREB axis as a critical proliferation and survival pathway. Nevertheless, PTHrP production by cancer cells acts generally upon the skeleton to promote excessive bone resorption and to facilitate bone metastasis formation and expansion [38].

20.8.3 PTHR1: PKA as a Target in Osteosarcoma

To understand the role of cAMP-dependent PKA-CREB activation in osteosarcoma, a mouse model was developed with deletion of the α regulatory subunit of PKA type I or PRKAR1A in osteoblasts using the α1-collagen promoter-Cre recombinase. The functional consequence of reduced PRKAR1A is enhanced PKA activity. The *Prkar1a*$^{ob/ob}$ mouse model shows a highly invasive osteosarcoma. This leads to the conclusion that PRKAR1A is an osteosarcoma tumor suppressor [64]. In osteosarcoma, an amplification of *Prkaca* RNA was also identified in tumors. This codes for the catalytic component of PKA. Thus, the activation of the PKA/CREB cascade contributes to osteosarcoma invasion and growth [235].

20.9 Conclusions and Future directions

PTHR1 has important physiological roles in bone development as well as bone remodeling. It is now also involved in treating osteoporosis. Only two osteoanabolic peptides are in use, which are PTH and abaloparatide and bind PTHR1. Gaps in our understanding that need to be pursued include further investigation of the functioning of PTHR1 in bone. The understanding of pathological situations involving PTHR1 could lead to possibilities of drug therapies.

References

1 Juppner, H. and Hesch, R.D. (1991). Biochemical characterization of cellular hormone receptors. *Curr. Top. Pathol.* 83: 53–69.

2 Juppner, H. (2000). Role of parathyroid hormone-related peptide and Indian hedgehog in skeletal development. *Pediatr. Nephrol.* 14 (7): 606–611.

3 Abou-Samra, A.B., Juppner, H., Force, T. et al. (1992). Expression cloning of a common receptor for parathyroid hormone and parathyroid hormone-related peptide from rat osteoblast-like cells: a single receptor stimulates intracellular accumulation of both cAMP and inositol trisphosphates and increases intracellular free calcium. *Proc. Natl. Acad. Sci. U.S.A.* 89 (7): 2732–2736.

4 Potts, J.T. (2005). Parathyroid hormone: past and present. *J. Endocrinol.* 187 (3): 311–325.

5 Watson, P.H., Fraher, L.J., Natale, B.V. et al. (2000). Nuclear localization of the type 1 parathyroid hormone/parathyroid hormone-related peptide receptor in MC3T3-E1 cells: association with serum-induced cell proliferation. *Bone* 26 (3): 221–225.

6 Pickard, B.W., Hodsman, A.B., Fraher, L.J., and Watson, P.H. (2006). Type 1 parathyroid hormone receptor (PTH1R) nuclear trafficking: association of PTH1R with importin α1 and β. *Endocrinology* 147 (7): 3326–3332.

7 Pickard, B.W., Hodsman, A.B., Fraher, L.J., and Watson, P.H. (2007). Type 1 parathyroid hormone receptor (PTH1R) nuclear trafficking: regulation of PTH1R nuclear-cytoplasmic shuttling by importin- α/β and chromosomal region maintenance 1/exportin 1. *Endocrinology* 148 (5): 2282–2289.

8 Patterson, E.K., Hodsman, A.B., Hendy, G.N. et al. (2010). Functional analysis of a type 1 parathyroid hormone receptor intracellular tail mutant [KRK(484-6)AAA]: effects on second messenger generation and cellular targeting. *Bone* 46 (4): 1180–1187.

9 Gensure, R.C., Gardella, T.J., and Juppner, H. (2005). Parathyroid hormone and parathyroid hormone-related peptide, and their receptors. *Biochem. Biophys. Res. Commun.* 328 (3): 666–678.

10 Gensure, R. and Juppner, H. (2005). Parathyroid hormone without parathyroid glands. *Endocrinology* 146 (2): 544–546.

11 Pioszak, A.A., Harikumar, K.G., Parker, N.R. et al. (2010). Dimeric arrangement of the parathyroid hormone receptor and a structural mechanism for ligand-induced dissociation. *J. Biol. Chem.* 285 (16): 12435–12444.

12 Pioszak, A.A. and Xu, H.E. (2008). Molecular recognition of parathyroid hormone by its G protein-coupled receptor. *Proc. Natl. Acad. Sci. U.S.A.* 105 (13): 5034–5039.

13 Urena, P., Kong, X.F., Abou-Samra, A.B. et al. (1993). Parathyroid hormone (PTH)/PTH-related peptide receptor messenger ribonucleic acids are widely distributed in rat tissues. *Endocrinology* 133 (2): 617–623.

14 Clemens, T.L., Cormier, S., Eichinger, A. et al. (2001). Parathyroid hormone-related protein and its receptors: nuclear functions and roles in the renal and cardiovascular systems, the placental trophoblasts and the pancreatic islets. *Br. J. Pharmacol.* 134 (6): 1113–1136.

15 Kousteni, S. and Bilezikian, J.P. (2008). The cell biology of parathyroid hormone in osteoblasts. *Curr. Osteoporos. Rep.* 6 (2): 72–76.

16 Jilka, R.L. (2007). Molecular and cellular mechanisms of the anabolic effect of intermittent PTH. *Bone* 40 (6): 1434–1446.

17 Tawfeek, H.A., Qian, F., and Abou-Samra, A.B. (2002). Phosphorylation of the receptor for PTH and PTHrP is required for internalization and regulates receptor signaling. *Mol. Endocrinol.* 16 (1): 1–13.

18 Lohse, M.J. (1993). Molecular mechanisms of membrane receptor desensitization. *Biochim. Biophys. Acta* 1179 (2): 171–188.

19 Brown, E.M. (1991). Extracellular Ca^{2+} sensing, regulation of parathyroid cell function, and role of Ca^{2+} and other ions as extracellular (first) messengers. *Physiol. Rev.* 71 (2): 371–411.

20 Habener, J.F., Rosenblatt, M., and Potts, J.T. Jr. (1984). Parathyroid hormone: biochemical aspects of biosynthesis, secretion, action, and metabolism. *Physiol. Rev.* 64 (3): 985–1053.

21 Yavropoulou, M.P., Michopoulos, A., and Yovos, J.G. (2017). PTH and PTHR1 in osteocytes. New insights into old partners. *Hormones (Athens)* 16 (2): 150–160.

22 Antonarakis, S.E., Phillips, J.A. 3rd, Mallonee, R.L. et al. (1983). Beta-globin locus is linked to the parathyroid hormone (PTH) locus and lies between the insulin and PTH loci in man. *Proc. Natl. Acad. Sci. U.S.A.* 80 (21): 6615–6619.

23 Zabel, B.U., Kronenberg, H.M., Bell, G.I., and Shows, T.B. (1985). Chromosome mapping of genes on the short arm of human chromosome 11: parathyroid hormone gene is at 11p15 together with the genes for insulin, c-Harvey-ras 1, and β-hemoglobin. *Cytogenet. Cell Genet.* 39 (3): 200–205.

24 Kronenberg, H.M., Igarashi, T., Freeman, M.W. et al. (1986). Structure and expression of the human parathyroid hormone gene. *Recent. Prog. Horm. Res.* 42: 641–663.

25 Habener, J.F., Kemper, B.W., Rich, A., and Potts, J.T. Jr. (1976). Biosynthesis of parathyroid hormone. *Recent. Prog. Horm. Res.* 33: 249–308.

26 Kemper, B., Habener, J.F., Rich, A., and Potts, J.T. Jr. (1975). Microtubules and the intracellular conversion of proparathyroid hormone to parathyroid hormone. *Endocrinology* 96 (4): 903–912.

27 el-Hajj Fuleihan, G., Klerman, E.B., Brown, E.N. et al. (1997). The parathyroid hormone circadian rhythm is truly endogenous – a general clinical research center study. *J. Clin. Endocrinol. Metab.* 82 (1): 281–286.

28 Brossard, J.H., Cloutier, M., Roy, L. et al. (1996). Accumulation of a non-(1–84) molecular form of parathyroid hormone (PTH) detected by intact PTH assay in renal failure: importance in the interpretation of PTH values. *J. Clin. Endocrinol. Metab.* 81 (11): 3923–3929.

29 Brossard, J.H., Whittom, S., Lepage, R., and D'Amour, P. (1993). Carboxyl-terminal fragments of parathyroid hormone are not secreted preferentially in primary hyperparathyroidism as they are in other hypercalcemic conditions. *J. Clin. Endocrinol. Metab.* 77 (2): 413–419.

30 Lepage, R., Roy, L., Brossard, J.H. et al. (1998). A non-(1–84) circulating parathyroid hormone (PTH) fragment interferes significantly with intact PTH commercial assay measurements in uremic samples. *Clin. Chem.* 44 (4): 805–809.

31 Nguyen-Yamamoto, L., Rousseau, L., Brossard, J.H. et al. (2001). Synthetic carboxyl-terminal fragments of parathyroid hormone (PTH) decrease ionized calcium concentration in rats by acting on a receptor different from the PTH/PTH-related peptide receptor. *Endocrinology* 142 (4): 1386–1392.

32 Divieti, P., Inomata, N., Chapin, K. et al. (2001). Receptors for the carboxyl-terminal region of PTH(1–84) are highly expressed in osteocytic cells. *Endocrinology* 142 (2): 916–925.

33 Divieti, P., John, M.R., Juppner, H., and Bringhurst, F.R. (2002). Human PTH-(7–84) inhibits bone resorption in vitro via actions independent of the type 1 PTH/PTHrP receptor. *Endocrinology* 143 (1): 171–176.

34 Divieti, P.P. (2005). PTH and osteocytes. *J. Musculoskelet. Neuronal. Interact.* 5 (4): 328–330.

35 Murray, T.M., Rao, L.G., Divieti, P., and Bringhurst, F.R. (2005). Parathyroid hormone secretion and action: evidence for discrete receptors for the carboxyl-terminal region and related biological actions of carboxyl-terminal ligands. *Endocr. Rev.* 26 (1): 78–113.

36 Liberman, U.A. (2000). Endotext [Internet]. In: *Disorders in Vitamin D Action* (ed. L.J. De Groot, G. Chrousos, K. Dungan, et al.). South Dartmouth, MA: MDText.com, Inc.

37 Drueke, T.B. (2000). Endotext [Internet]. In: *Hyperparathyroidism in Chronic Kidney Disease* (ed. L.J. De Groot, G. Chrousos, K. Dungan, et al.). South Dartmouth, MA: MDText.com, Inc.

38 Martin, T.J. (2016). Parathyroid hormone-related protein, its regulation of cartilage and bone development, and role in treating bone diseases. *Physiol. Rev.* 96 (3): 831–871.

39 Nishishita, T., Okazaki, T., Ishikawa, T. et al. (1998). A negative vitamin D response DNA element in the human parathyroid hormone-related peptide gene binds to vitamin D receptor along with Ku antigen to mediate negative gene regulation by vitamin D. *J. Biol. Chem.* 273 (18): 10901–10907.

40 Dittmer, J., Pise-Masison, C.A., Clemens, K.E. et al. (1997). Interaction of human T-cell lymphotropic virus type I Tax, Ets1, and Sp1 in transactivation of the PTHrP P2 promoter. *J. Biol. Chem.* 272 (8): 4953–4958.

41 Betel, D., Koppal, A., Agius, P. et al. (2010). Comprehensive modeling of microRNA targets predicts functional non-conserved and non-canonical sites. *Genome Biol.* 11 (8): R90.

42 Karperien, M., Farih-Sips, H., Lowik, C.W. et al. (1997). Expression of the parathyroid hormone-related peptide gene in retinoic acid-induced differentiation: involvement of ETS and Sp1. *Mol. Endocrinol.* 11 (10): 1435–1448.

43 Chilco, P.J., Leopold, V., and Zajac, J.D. (1998). Differential regulation of the parathyroid hormone-related protein gene P1 and P3 promoters by cAMP. *Mol. Cell. Endocrinol.* 138 (1, 2): 173–184.

44 Wysolmerski, J.J., Cormier, S., Philbrick, W.M. et al. (2001). Absence of functional type 1 parathyroid hormone (PTH)/PTH-related protein receptors in humans is associated with abnormal breast development and tooth impaction. *J. Clin. Endocrinol. Metab.* 86 (4): 1788–1794.

45 Wysolmerski, J.J. (2010). Interactions between breast, bone, and brain regulate mineral and skeletal metabolism during lactation. *Ann. N.Y. Acad. Sci.* 1192: 161–169.

46 Philbrick, W.M., Wysolmerski, J.J., Galbraith, S. et al. (1996). Defining the roles of parathyroid hormone-related protein in normal physiology. *Physiol. Rev.* 76 (1): 127–173.

47 Tregear, G.W. and Potts, J.T. Jr. (1975). Synthetic analogues of residues 1–34 of human parathyroid hormone: influence of residue number 1 on biological potency in vitro. *Endocr. Res. Commun.* 2 (8): 561–570.

48 Jans, D.A., Thomas, R.J., and Gillespie, M.T. (2003). Parathyroid hormone-related protein (PTHrP): a nucleocytoplasmic shuttling protein with distinct paracrine and intracrine roles. *Vitam. Horm.* 66: 345–384.

49 Kovacs, C.S., Lanske, B., Hunzelman, J.L. et al. (1996). Parathyroid hormone-related peptide (PTHrP) regulates fetal-placental calcium transport through a receptor distinct from the PTH/PTHrP receptor. *Proc. Natl. Acad. Sci. U.S.A.* 93 (26): 15233–15238.

50 Fenton, A.J., Kemp, B.E., Hammonds, R.G. Jr. et al. (1991). A potent inhibitor of osteoclastic bone resorption within a highly conserved pentapeptide region of parathyroid hormone-related protein; PTHrP[107–111]. *Endocrinology* 129 (6): 3424–3426.

51 Fenton, A.J., Martin, T.J., and Nicholson, G.C. (1994). Carboxyl-terminal parathyroid hormone-related protein inhibits bone resorption by isolated chicken osteoclasts. *J. Bone Miner. Res.* 9 (4): 515–519.

52 Valin, A., Garcia-Ocana, A., De Miguel, F. et al. (1997). Antiproliferative effect of the C-terminal fragments of parathyroid hormone-related protein, PTHrP-(107–111) and (107–139), on osteoblastic osteosarcoma cells. *J. Cell. Physiol.* 170 (2): 209–215.

53 Valin, A., Guillen, C., and Esbrit, P. (2001). C-terminal parathyroid hormone-related protein (PTHrP) (107–139) stimulates intracellular Ca^{2+} through a receptor different from the type 1 PTH/PTHrP receptor in osteoblastic osteosarcoma UMR 106 cells. *Endocrinology* 142 (7): 2752–2759.

54 de Gortazar, A.R., Alonso, V., Alvarez-Arroyo, M.V., and Esbrit, P. (2006). Transient exposure to PTHrP (107–139) exerts anabolic effects through vascular endothelial growth factor receptor 2 in human osteoblastic cells in vitro. *Calcif. Tissue. Int.* 79 (5): 360–369.

55 Hoare, S.R., Gardella, T.J., and Usdin, T.B. (2001). Evaluating the signal transduction mechanism of the parathyroid hormone 1 receptor. Effect of receptor-G-protein interaction on the ligand binding mechanism and receptor conformation. *J. Biol. Chem.* 276 (11): 7741–7753.

56 Hattersley, G., Dean, T., Corbin, B.A. et al. (2016). Binding selectivity of abaloparatide for PTH-type-1-receptor conformations and effects on downstream signaling. *Endocrinology* 157 (1): 141–149.

57 Dean, T., Vilardaga, J.P., Potts, J.T. Jr., and Gardella, T.J. (2008). Altered selectivity of parathyroid hormone (PTH) and PTH-related protein (PTHrP) for distinct conformations of the PTH/PTHrP receptor. *Mol. Endocrinol.* 22 (1): 156–166.

58 Dean, T., Linglart, A., Mahon, M.J. et al. (2006). Mechanisms of ligand binding to the parathyroid hormone (PTH)/PTH-related protein receptor: selectivity of a modified PTH(1–15) radioligand for GαS-coupled receptor conformations. *Mol. Endocrinol.* 20 (4): 931–943.

59 Gensure, R.C., Ponugoti, B., Gunes, Y. et al. (2004). Identification and characterization of two parathyroid hormone-like molecules in zebrafish. *Endocrinology* 145 (4): 1634–1639.

60 Vilardaga, J.P., Gardella, T.J., Wehbi, V.L., and Feinstein, T.N. (2012). Non-canonical signaling of the PTH receptor. *Trends Pharmacol. Sci.* 33 (8): 423–431.

61 Gardella, B., Porru, D., Allegri, M. et al. (2014). Pharmacokinetic considerations for therapies used to treat interstitial cystitis. *Expert Opin. Drug Metab. Toxicol.* 10 (5): 673–684.

62 Kramer, I., Loots, G.G., Studer, A. et al. (2010). Parathyroid hormone (PTH)-induced bone gain is blunted in SOST overexpressing and deficient mice. *J. Bone Miner. Res.* 25 (2): 178–189.

63 Amieux, P.S., Howe, D.G., Knickerbocker, H. et al. (2002). Increased basal cAMP-dependent protein kinase activity inhibits the formation of mesoderm-derived structures in the developing mouse embryo. *J. Biol. Chem.* 277 (30): 27294–27304.

64 Molyneux, S.D., Di Grappa, M.A., Beristain, A.G. et al. (2010). Prkarla is an osteosarcoma tumor suppressor that defines a molecular subclass in mice. *J. Clin. Invest.* 120 (9): 3310–3325.

65 Le Stunff, C., Tilotta, F., Sadoine, J. et al. (2017). Knock-in of the recurrent R368X mutation of PRKAR1A that represses cAMP-dependent protein kinase A activation: a model of type 1 acrodysostosis. *J. Bone Miner. Res.* 32 (2): 333–346.

66 D'Alonzo, R.C., Selvamurugan, N., Karsenty, G., and Partridge, N.C. (2002). Physical interaction of the activator protein-1 factors c-Fos and c-Jun with Cbfa1 for collagenase-3 promoter activation. *J. Biol. Chem.* 277 (1): 816–822.

67 Fei, Y., Shimizu, E., McBurney, M.W., and Partridge, N.C. (2015). Sirtuin 1 is a negative regulator of parathyroid hormone stimulation of matrix

metalloproteinase 13 expression in osteoblastic cells: role of sirtuin 1 in the action of PTH on osteoblasts. *J. Biol. Chem.* 290 (13): 8373–8382.
68 Nakatani, T., Chen, T., Johnson, J. et al. (2018). The deletion of Hdac4 in mouse osteoblasts influences both catabolic and anabolic effects in bone. *J. Bone Miner. Res.* 33 (7): 1362–1375.
69 Nakatani, T. and Partridge, N.C. (2017). MEF2C interacts with c-FOS in PTH-stimulated Mmp13 gene expression in osteoblastic cells. *Endocrinology* 158 (11): 3778–3791.
70 Nakatani, T., Chen, T., and Partridge, N.C. (2016). MMP-13 is one of the critical mediators of the effect of HDAC4 deletion on the skeleton. *Bone* 90: 142–151.
71 Wein, M.N., Liang, Y., Goransson, O. et al. (2016). SIKs control osteocyte responses to parathyroid hormone. *Nat. Commun.* 7: 13176.
72 Ricarte, F.R., Le Henaff, C., Kolupaeva, V.G. et al. (2018). Parathyroid hormone (1–34) and its analogs differentially modulate osteoblastic RANKL expression via PKA/PP1/PP2A and SIK2/SIK3-CRTC3 signaling. *J. Biol. Chem.* 293: 20200–20213.
73 Shimizu, E., Nakatani, T., He, Z., and Partridge, N.C. (2014). Parathyroid hormone regulates histone deacetylase (HDAC) 4 through protein kinase A-mediated phosphorylation and dephosphorylation in osteoblastic cells. *J. Biol. Chem.* 289 (31): 21340–21350.
74 Shimizu, E., Selvamurugan, N., Westendorf, J.J. et al. (2010). HDAC4 represses matrix metalloproteinase-13 transcription in osteoblastic cells, and parathyroid hormone controls this repression. *J. Biol. Chem.* 285 (13): 9616–9626.
75 Wein, M.N. and Kronenberg, H.M. (2018). Regulation of bone remodeling by parathyroid hormone. *Cold Spring Harb. Perspect. Med.* 8 (8): a031237.
76 Gesty-Palmer, D., Chen, M., Reiter, E. et al. (2006). Distinct beta-arrestin- and G protein-dependent pathways for parathyroid hormone receptor-stimulated ERK1/2 activation. *J. Biol. Chem.* 281 (16): 10856–10864.
77 Gesty-Palmer, D., Flannery, P., Yuan, L. et al. (2009). A beta-arrestin-biased agonist of the parathyroid hormone receptor (PTH1R) promotes bone formation independent of G protein activation. *Sci. Transl. Med.* 1 (1): 1ra.
78 Ferrari, S.L., Pierroz, D.D., Glatt, V. et al. (2005). Bone response to intermittent parathyroid hormone is altered in mice null for {beta}-arrestin2. *Endocrinology* 146 (4): 1854–1862.
79 Verheijen, M.H. and Defize, L.H. (1995). Parathyroid hormone inhibits mitogen-activated protein kinase activation in osteosarcoma cells via a protein kinase A-dependent pathway. *Endocrinology* 136 (8): 3331–3337.
80 Chen, C.H., Paing, M.M., and Trejo, J. (2004). Termination of protease-activated receptor-1 signaling by beta-arrestins is independent of receptor phosphorylation. *J. Biol. Chem.* 279 (11): 10020–10031.

81 Datta, N.S., Chen, C., Berry, J.E., and McCauley, L.K. (2005). PTHrP signaling targets cyclin D_1 and induces osteoblastic cell growth arrest. *J. Bone Miner. Res.* 20 (6): 1051–1064.

82 Datta, N.S., Pettway, G.J., Chen, C. et al. (2007). Cyclin D_1 as a target for the proliferative effects of PTH and PTHrP in early osteoblastic cells. *J. Bone Miner. Res.* 22 (7): 951–964.

83 Wehbi, V.L., Stevenson, H.P., Feinstein, T.N. et al. (2013). Noncanonical GPCR signaling arising from a PTH receptor-arrestin-Gbetagamma complex. *Proc. Natl. Acad. Sci. U.S.A.* 110 (4): 1530–1535.

84 Ferrandon, S., Feinstein, T.N., Castro, M. et al. (2009). Sustained cyclic AMP production by parathyroid hormone receptor endocytosis. *Nat. Chem. Biol.* 5 (10): 734–742.

85 Feinstein, T.N., Wehbi, V.L., Ardura, J.A. et al. (2011). Retromer terminates the generation of cAMP by internalized PTH receptors. *Nat. Chem. Biol.* 7 (5): 278–284.

86 Gardella, T.J. and Vilardaga, J.P. (2015). International Union of Basic and Clinical Pharmacology. XCIII. The parathyroid hormone receptors--family B G protein-coupled receptors. *Pharmacol. Rev.* 67 (2): 310–337.

87 Revollo, L., Kading, J., Jeong, S.Y. et al. (2015). N-cadherin restrains PTH activation of Lrp6/β-catenin signaling and osteoanabolic action. *J. Bone Miner. Res.* 30 (2): 274–285.

88 Datta, N.S. and Abou-Samra, A.B. (2009). PTH and PTHrP signaling in osteoblasts. *Cell. Signalling* 21 (8): 1245–1254.

89 Wan, M., Yang, C., Li, J. et al. (2008). Parathyroid hormone signaling through low-density lipoprotein-related protein 6. *Genes Dev.* 22 (21): 2968–2979.

90 Osagie-Clouard, L., Sanghani, A., Coathup, M. et al. (2017). Parathyroid hormone 1–34 and skeletal anabolic action: the use of parathyroid hormone in bone formation. *Bone Joint Res.* 6 (1): 14–21.

91 Yu, B., Zhao, X., Yang, C. et al. (2012). Parathyroid hormone induces differentiation of mesenchymal stromal/stem cells by enhancing bone morphogenetic protein signaling. *J. Bone Miner. Res.* 27 (9): 2001–2014.

92 Kulkarni, N.H., Halladay, D.L., Miles, R.R. et al. (2005). Effects of parathyroid hormone on Wnt signaling pathway in bone. *J. Cell. Biochem.* 95 (6): 1178–1190.

93 Silva, B.C. and Bilezikian, J.P. (2015). Parathyroid hormone: anabolic and catabolic actions on the skeleton. *Curr. Opin. Pharmacol.* 22: 41–50.

94 Rhee, Y., Allen, M.R., Condon, K. et al. (2011). PTH receptor signaling in osteocytes governs periosteal bone formation and intracortical remodeling. *J. Bone Miner. Res.* 26 (5): 1035–1046.

95 Kopic, S. and Geibel, J.P. (2013). Gastric acid, calcium absorption, and their impact on bone health. *Physiol. Rev.* 93 (1): 189–268.

96 Brenza, H.L. and DeLuca, H.F. (2000). Regulation of 25-hydroxyvitamin D_3 1α-hydroxylase gene expression by parathyroid hormone and 1,25-dihydroxyvitamin D_3. *Arch. Biochem. Biophys.* 381 (1): 143–152.

97 Goltzman, D. and Hendy, G.N. (2015). The calcium-sensing receptor in bone–mechanistic and therapeutic insights. *Nat. Rev. Endocrinol.* 11 (5): 298–307.

98 Rodda, C.P., Kubota, M., Heath, J.A. et al. (1988). Evidence for a novel parathyroid hormone-related protein in fetal lamb parathyroid glands and sheep placenta: comparisons with a similar protein implicated in humoral hypercalcaemia of malignancy. *J. Endocrinol.* 117 (2): 261–271.

99 Grill, V., Hillary, J., Ho, P.M. et al. (1992). Parathyroid hormone-related protein: a possible endocrine function in lactation. *Clin. Endocrinol. (Oxf).* 37 (5): 405–410.

100 Wojcik, S.F., Schanbacher, F.L., McCauley, L.K. et al. (1998). Cloning of bovine parathyroid hormone-related protein (PTHrP) cDNA and expression of PTHrP mRNA in the bovine mammary gland. *J. Mol. Endocrinol.* 20 (2): 271–280.

101 Rakopoulos, M., Vargas, S.J., Gillespie, M.T. et al. (1992). Production of parathyroid hormone-related protein by the rat mammary gland in pregnancy and lactation. *Am. J. Phys* 263 (6): E1077–E1085.

102 Brommage, R. and DeLuca, H.F. (1985). Regulation of bone mineral loss during lactation. *Am. J. Phys* 248 (2 Pt 1): E182–E187.

103 Kent, G.N., Price, R.I., Gutteridge, D.H. et al. (1990). Human lactation: forearm trabecular bone loss, increased bone turnover, and renal conservation of calcium and inorganic phosphate with recovery of bone mass following weaning. *J. Bone Miner. Res.* 5 (4): 361–369.

104 Kronenberg, H.M. (2003). Developmental regulation of the growth plate. *Nature* 423 (6937): 332–336.

105 Miao, D., He, B., Jiang, Y. et al. (2005). Osteoblast-derived PTHrP is a potent endogenous bone anabolic agent that modifies the therapeutic efficacy of administered PTH 1–34. *J. Clin. Invest.* 115 (9): 2402–2411.

106 Karaplis, A.C., Luz, A., Glowacki, J. et al. (1994). Lethal skeletal dysplasia from targeted disruption of the parathyroid hormone-related peptide gene. *Genes Dev.* 8 (3): 277–289.

107 Karaplis, A.C. and Kronenberg, H.M. (1996). Physiological roles for parathyroid hormone-related protein: lessons from gene knockout mice. *Vitam. Horm.* 52: 177–193.

108 Martin, T.J. and Seeman, E. (2017). Abaloparatide is an anabolic, but does it spare resorption? *J. Bone Miner. Res.* 32 (1): 11–16.

109 Moseley, J.M., Hayman, J.A., Danks, J.A. et al. (1991). Immunohistochemical detection of parathyroid hormone-related protein in human fetal epithelia. *J. Clin. Endocrinol. Metab.* 73 (3): 478–484.

110 Amizuka, N., Karaplis, A.C., Henderson, J.E. et al. (1996). Haploinsufficiency of parathyroid hormone-related peptide (PTHrP) results in abnormal postnatal bone development. *Dev. Biol.* 175 (1): 166–176.

111 Eriksen, E.F. (1986). Normal and pathological remodeling of human trabecular bone: three dimensional reconstruction of the remodeling sequence in normals and in metabolic bone disease. *Endocr. Rev.* 7 (4): 379–408.

112 Grosso, M.J., Courtland, H.W., Yang, X. et al. (2015). Intermittent PTH administration and mechanical loading are anabolic for periprosthetic cancellous bone. *J. Orthop. Res.* 33 (2): 163–173.

113 Amugongo, S.K., Yao, W., Jia, J. et al. (2014). Effects of sequential osteoporosis treatments on trabecular bone in adult rats with low bone mass. *Osteoporos. Int.* 25 (6): 1735–1750.

114 Zhou, H., Iida-Klein, A., Lu, S.S. et al. (2003). Anabolic action of parathyroid hormone on cortical and cancellous bone differs between axial and appendicular skeletal sites in mice. *Bone* 32 (5): 513–520.

115 Calvi, L.M., Sims, N.A., Hunzelman, J.L. et al. (2001). Activated parathyroid hormone/parathyroid hormone-related protein receptor in osteoblastic cells differentially affects cortical and trabecular bone. *J. Clin. Invest.* 107 (3): 277–286.

116 Powell, W.F. Jr., Barry, K.J., Tulum, I. et al. (2011). Targeted ablation of the PTH/PTHrP receptor in osteocytes impairs bone structure and homeostatic calcemic responses. *J. Endocrinol.* 209 (1): 21–32.

117 Kim, S.W., Pajevic, P.D., Selig, M. et al. (2012). Intermittent parathyroid hormone administration converts quiescent lining cells to active osteoblasts. *J. Bone Miner. Res.* 27 (10): 2075–2084.

118 Jilka, R.L., Weinstein, R.S., Bellido, T. et al. (1999). Increased bone formation by prevention of osteoblast apoptosis with parathyroid hormone. *J. Clin. Invest.* 104 (4): 439–446.

119 Wang, Y., Nishida, S., Boudignon, B.M. et al. (2007). IGF-I receptor is required for the anabolic actions of parathyroid hormone on bone. *J. Bone Miner. Res.* 22 (9): 1329–1337.

120 Dempster, D.W., Cosman, F., Parisien, M. et al. (1993). Anabolic actions of parathyroid hormone on bone. *Endocr. Rev.* 14 (6): 690–709.

121 Balani, D.H., Ono, N., and Kronenberg, H.M. (2017). Parathyroid hormone regulates fates of murine osteoblast precursors in vivo. *J. Clin. Invest.* 127 (9): 3327–3338.

122 Fan, Y., Hanai, J.I., Le, P.T. et al. (2017). Parathyroid hormone directs bone marrow mesenchymal cell fate. *Cell Metab.* 25 (3): 661–672.

123 Brunner, S., Theiss, H.D., Murr, A. et al. (2007). Primary hyperparathyroidism is associated with increased circulating bone marrow-derived progenitor cells. *Am. J. Physiol. Endocrinol. Metab.* 293 (6): E1670–E1675.

124 Nishida, S., Yamaguchi, A., Tanizawa, T. et al. (1994). Increased bone formation by intermittent parathyroid hormone administration is due to the stimulation of proliferation and differentiation of osteoprogenitor cells in bone marrow. *Bone* 15 (6): 717–723.

125 Keller, H. and Kneissel, M. (2005). SOST is a target gene for PTH in bone. *Bone* 37 (2): 148–158.

126 Bellido, T., Saini, V., and Pajevic, P.D. (2013). Effects of PTH on osteocyte function. *Bone* 54 (2): 250–257.

127 Swarthout, J.T., D'Alonzo, R.C., Selvamurugan, N., and Partridge, N.C. (2002). Parathyroid hormone-dependent signaling pathways regulating genes in bone cells. *Gene* 282 (1, 2): 1–17.

128 Ishizuya, T., Yokose, S., Hori, M. et al. (1997). Parathyroid hormone exerts disparate effects on osteoblast differentiation depending on exposure time in rat osteoblastic cells. *J. Clin. Invest.* 99 (12): 2961–2970.

129 Locklin, R.M., Khosla, S., Turner, R.T., and Riggs, B.L. (2003). Mediators of the biphasic responses of bone to intermittent and continuously administered parathyroid hormone. *J. Cell. Biochem.* 89 (1): 180–190.

130 Ogita, M., Rached, M.T., Dworakowski, E. et al. (2008). Differentiation and proliferation of periosteal osteoblast progenitors are differentially regulated by estrogens and intermittent parathyroid hormone administration. *Endocrinology* 149 (11): 5713–5723.

131 Chambers, T.J., Fuller, K., McSheehy, P.M., and Pringle, J.A. (1985). The effects of calcium regulating hormones on bone resorption by isolated human osteoclastoma cells. *J. Pathol.* 145 (4): 297–305.

132 McSheehy, P.M. and Chambers, T.J. (1986). Osteoblastic cells mediate osteoclastic responsiveness to parathyroid hormone. *Endocrinology* 118 (2): 824–828.

133 Xiong, J. and O'Brien, C.A. (2012). Osteocyte RANKL: new insights into the control of bone remodeling. *J. Bone Miner. Res.* 27 (3): 499–505.

134 Jacome-Galarza, C.E., Lee, S.K., Lorenzo, J.A., and Aguila, H.L. (2011). Parathyroid hormone regulates the distribution and osteoclastogenic potential of hematopoietic progenitors in the bone marrow. *J. Bone Miner. Res.* 26 (6): 1207–1216.

135 Xiong, J., Piemontese, M., Thostenson, J.D. et al. (2014). Osteocyte-derived RANKL is a critical mediator of the increased bone resorption caused by dietary calcium deficiency. *Bone* 66: 146–154.

136 Simonet, W.S., Lacey, D.L., Dunstan, C.R. et al. (1997). Osteoprotegerin: a novel secreted protein involved in the regulation of bone density. *Cell* 89 (2): 309–319.

137 Fuller, K., Owens, J.M., and Chambers, T.J. (1998). Induction of osteoclast formation by parathyroid hormone depends on an action on stromal cells. *J. Endocrinol.* 158 (3): 341–350.

138 Khosla, S. (2001). Minireview: the OPG/RANKL/RANK system. *Endocrinology* 142 (12): 5050–5055.

139 Kearns, A.E., Khosla, S., and Kostenuik, P.J. (2008). Receptor activator of nuclear factor kappaB ligand and osteoprotegerin regulation of bone remodeling in health and disease. *Endocr. Rev.* 29 (2): 155–192.

140 Canalis, E., Giustina, A., and Bilezikian, J.P. (2007). Mechanisms of anabolic therapies for osteoporosis. *N. Engl. J. Med.* 357 (9): 905–916.

141 O'Brien, C.A., Nakashima, T., and Takayanagi, H. (2013). Osteocyte control of osteoclastogenesis. *Bone* 54 (2): 258–263.

142 Saini, V., Marengi, D.A., Barry, K.J. et al. (2013). Parathyroid hormone (PTH)/PTH-related peptide type 1 receptor (PPR) signaling in osteocytes regulates anabolic and catabolic skeletal responses to PTH. *J. Biol. Chem.* 288 (28): 20122–20134.

143 Lee, S.K. and Lorenzo, J.A. (1999). Parathyroid hormone stimulates TRANCE and inhibits osteoprotegerin messenger ribonucleic acid expression in murine bone marrow cultures: correlation with osteoclast-like cell formation. *Endocrinology* 140 (8): 3552–3561.

144 Lee, S.K. and Lorenzo, J.A. (2002). Regulation of receptor activator of nuclear factor-kappa B ligand and osteoprotegerin mRNA expression by parathyroid hormone is predominantly mediated by the protein kinase a pathway in murine bone marrow cultures. *Bone* 31 (1): 252–259.

145 Huang, J.C., Sakata, T., Pfleger, L.L. et al. (2004). PTH differentially regulates expression of RANKL and OPG. *J. Bone Miner. Res.* 19 (2): 235–244.

146 Nakchbandi, I.A., Lang, R., Kinder, B., and Insogna, K.L. (2008). The role of the receptor activator of nuclear factor-kappaB ligand/osteoprotegerin cytokine system in primary hyperparathyroidism. *J. Clin. Endocrinol. Metab.* 93 (3): 967–973.

147 Stilgren, L.S., Rettmer, E., Eriksen, E.F. et al. (2004). Skeletal changes in osteoprotegerin and receptor activator of nuclear factor-kappab ligand mRNA levels in primary hyperparathyroidism: effect of parathyroidectomy and association with bone metabolism. *Bone* 35 (1): 256–265.

148 Szymczak, J. and Bohdanowicz-Pawlak, A. (2013). Osteoprotegerin, RANKL, and bone turnover in primary hyperparathyroidism: the effect of parathyroidectomy and treatment with alendronate. *Horm. Metab. Res.* 45 (10): 759–764.

149 Sabbota, A.L., Kim, H.R., Zhe, X. et al. (2010). Shedding of RANKL by tumor-associated MT1-MMP activates Src-dependent prostate cancer cell migration. *Cancer Res.* 70 (13): 5558–5566.

150 Delgado-Calle, J., Tu, X., Pacheco-Costa, R. et al. (2017). Control of bone anabolism in response to mechanical loading and PTH by distinct mechanisms downstream of the PTH receptor. *J. Bone Miner. Res.* 32 (3): 522–535.

151 Jacquin, C., Koczon-Jaremko, B., Aguila, H.L. et al. (2009). Macrophage migration inhibitory factor inhibits osteoclastogenesis. *Bone* 45 (4): 640–649.

152 Li, X., Liu, H., Qin, L. et al. (2007). Determination of dual effects of parathyroid hormone on skeletal gene expression in vivo by microarray and network analysis. *J. Biol. Chem.* 282 (45): 33086–33097.

153 Li, X., Qin, L., Bergenstock, M. et al. (2007). Parathyroid hormone stimulates osteoblastic expression of MCP-1 to recruit and increase the fusion of pre/osteoclasts. *J. Biol. Chem.* 282 (45): 33098–33106.

154 Tamasi, J.A., Vasilov, A., Shimizu, E. et al. (2013). Monocyte chemoattractant protein-1 is a mediator of the anabolic action of parathyroid hormone on bone. *J. Bone Miner. Res.* 28 (9): 1975–1986.

155 Siddiqui, J.A. and Partridge, N.C. (2017). CCL2/monocyte chemoattractant protein 1 and parathyroid hormone action on bone. *Front. Endocrinol. (Lausanne).* 8: 49.

156 Sukumar, D., Partridge, N.C., Wang, X., and Shapses, S.A. (2011). The high serum monocyte chemoattractant protein-1 in obesity is influenced by high parathyroid hormone and not adiposity. *J. Clin. Endocrinol. Metab.* 96 (6): 1852–1858.

157 Patel, H., Trooskin, S., Shapses, S. et al. (2014). Serum monocyte chemokine protein-1 levels before and after parathyroidectomy in patients with primary hyperparathyroidism. *Endocr. Pract.* 20 (11): 1165–1169.

158 Shaker, J.L. and Deftos, L. (2000). Endotext [Internet]. In: *Calcium and Phosphate Homeostasis* (ed. L.J. De Groot, G. Chrousos, K. Dungan, et al.). South Dartmouth, MA: MDText.com, Inc.

159 Ensrud, K.E. and Crandall, C.J. (2017). Osteoporosis. *Ann. Intern. Med.* 167 (3): ITC17–ITC32.

160 Kannegaard, P.N., van der Mark, S., Eiken, P., and Abrahamsen, B. (2010). Excess mortality in men compared with women following a hip fracture. National analysis of comedications, comorbidity and survival. *Age Ageing* 39 (2): 203–209.

161 Sozen, T., Ozisik, L., and Basaran, N.C. (2017). An overview and management of osteoporosis. *Eur. J. Rheumatol.* 4 (1): 46–56.

162 Hans, D., Alekxandrova, I., Njeh, C. et al. (2005). Appropriateness of internal digital phantoms for monitoring the stability of the UBIS 5000 quantitative ultrasound device in clinical trials. *Osteoporos. Int.* 16 (4): 435–445.

163 Harslof, T. and Langdahl, B.L. (2016). New horizons in osteoporosis therapies. *Curr. Opin. Pharmacol.* 28: 38–42.

164 Shao, Y., Hernandez-Buquer, S., Childress, P. et al. (2017). Improving combination osteoporosis therapy in a preclinical model of heightened osteoanabolism. *Endocrinology* 158 (9): 2722–2740.

165 Chew, C.K. and Clarke, B.L. (2017). Abaloparatide: recombinant human PTHrP (1–34) anabolic therapy for osteoporosis. *Maturitas* 97: 53–60.

166 Langdahl, B.L., Libanati, C., Crittenden, D.B. et al. (2017). Romosozumab (sclerostin monoclonal antibody) versus teriparatide in postmenopausal women with osteoporosis transitioning from oral bisphosphonate therapy: a randomised, open-label, phase 3 trial. *Lancet* 390 (10102): 1585–1594.

167 Liu, M., Kurimoto, P., Zhang, J. et al. (2018). Sclerostin and DKK1 inhibition preserves and augments alveolar bone volume and architecture in rats with alveolar bone loss. *J. Dent. Res.* 97 (9): 1031–1038.

168 Tam, C.S., Heersche, J.N., Murray, T.M., and Parsons, J.A. (1982). Parathyroid hormone stimulates the bone apposition rate independently of its resorptive action: differential effects of intermittent and continuous administration. *Endocrinology* 110 (2): 506–512.

169 Dempster, D.W., Cosman, F., Kurland, E.S. et al. (2001). Effects of daily treatment with parathyroid hormone on bone microarchitecture and turnover in patients with osteoporosis: a paired biopsy study. *J. Bone Miner. Res.* 16 (10): 1846–1853.

170 Albright, F., Bauer, W., Ropes, M., and Aub, J.C. (1929). Studies of calcium and phosphorus metabolism: IV. The effect of the parathyroid hormone. *J. Clin. Invest.* 7 (1): 139–181.

171 Neer, R.M., Arnaud, C.D., Zanchetta, J.R. et al. (2001). Effect of parathyroid hormone (1–34) on fractures and bone mineral density in postmenopausal women with osteoporosis. *N. Engl. J. Med.* 344 (19): 1434–1441.

172 Vahle, J.L., Sato, M., Long, G.G. et al. (2002). Skeletal changes in rats given daily subcutaneous injections of recombinant human parathyroid hormone (1–34) for 2 years and relevance to human safety. *Toxicol. Pathol.* 30 (3): 312–321.

173 Jiang, Y., Zhao, J.J., Mitlak, B.H. et al. (2003). Recombinant human parathyroid hormone (1–34) [teriparatide] improves both cortical and cancellous bone structure. *J. Bone Miner. Res.* 18 (11): 1932–1941.

174 Cipriani, C., Irani, D., and Bilezikian, J.P. (2012). Safety of osteoanabolic therapy: a decade of experience. *J. Bone Miner. Res.* 27 (12): 2419–2428.

175 Moricke, R., Rettig, K., and Bethke, T.D. (2011). Use of recombinant human parathyroid hormone(1–84) in patients with postmenopausal osteoporosis: a prospective, open-label, single-arm, multicentre, observational cohort study of the effects of treatment on quality of life and pain--the PROPOSE study. *Clin. Drug Investig.* 31 (2): 87–99.

176 Bilezikian, J.P., Rubin, M.R., and Finkelstein, J.S. (2005). Parathyroid hormone as an anabolic therapy for women and men. *J. Endocrinol. Invest.* 28 (8 Suppl): 41–49.

177 Rubin, M.R. and Bilezikian, J.P. (2005). Parathyroid hormone as an anabolic skeletal therapy. *Drugs* 65 (17): 2481–2498.

178 Baron, R., Tross, R., and Vignery, A. (1984). Evidence of sequential remodeling in rat trabecular bone: morphology, dynamic histomorphometry, and changes during skeletal maturation. *Anat. Rec.* 208 (1): 137–145.

179 Parfitt, A.M. (1994). Osteonal and hemi-osteonal remodeling: the spatial and temporal framework for signal traffic in adult human bone. *J. Cell. Biochem.* 55 (3): 273–286.

180 Sims, N.A. and Martin, T.J. (2015). Coupling signals between the osteoclast and osteoblast: how are messages transmitted between these temporary visitors to the bone surface? *Front. Endocrinol. (Lausanne)* 6: 41.

181 Parfitt, A.M. (1983). Assessment of trabecular bone status. *Henry Ford Hosp. Med. J.* 31 (4): 196–198.

182 Dempster, D.W., Zhou, H., Recker, R.R. et al. (2016). A longitudinal study of skeletal histomorphometry at 6 and 24 months across four bone envelopes in postmenopausal women with osteoporosis receiving teriparatide or zoledronic acid in the SHOTZ trial. *J. Bone Miner. Res.* 31 (7): 1429–1439.

183 Compston, J. (2007). Does parathyroid hormone treatment affect fracture risk or bone mineral density in patients with osteoporosis? *Nat. Clin. Pract. Rheumatol.* 3 (6): 324–325.

184 Lindsay, R., Cosman, F., Zhou, H. et al. (2006). A novel tetracycline labeling schedule for longitudinal evaluation of the short-term effects of anabolic therapy with a single iliac crest bone biopsy: early actions of teriparatide. *J. Bone Miner. Res.* 21 (3): 366–373.

185 Parfitt, A.M. (2002). Parathyroid hormone and periosteal bone expansion. *J. Bone Miner. Res.* 17 (10): 1741–1743.

186 Burr, D.B., Hirano, T., Turner, C.H. et al. (2001). Intermittently administered human parathyroid hormone(1–34) treatment increases intracortical bone turnover and porosity without reducing bone strength in the humerus of ovariectomized cynomolgus monkeys. *J. Bone Miner. Res.* 16 (1): 157–165.

187 Finkelstein, J.S., Hayes, A., Hunzelman, J.L. et al. (2003). The effects of parathyroid hormone, alendronate, or both in men with osteoporosis. *N. Engl. J. Med.* 349 (13): 1216–1226.

188 Finkelstein, J.S., Wyland, J.J., Lee, H., and Neer, R.M. (2010). Effects of teriparatide, alendronate, or both in women with postmenopausal osteoporosis. *J. Clin. Endocrinol. Metab.* 95 (4): 1838–1845.

189 Black, D.M., Greenspan, S.L., Ensrud, K.E. et al. (2003). The effects of parathyroid hormone and alendronate alone or in combination in postmenopausal osteoporosis. *N. Engl. J. Med.* 349 (13): 1207–1215.

190 Rodan, G.A. and Martin, T.J. (2000). Therapeutic approaches to bone diseases. *Science* 289 (5484): 1508–1514.

191 Kurland, E.S., Heller, S.L., Diamond, B. et al. (2004). The importance of bisphosphonate therapy in maintaining bone mass in men after therapy with teriparatide [human parathyroid hormone(1–34)]. *Osteoporos. Int.* 15 (12): 992–997.

192 Black, D.M., Bilezikian, J.P., Ensrud, K.E. et al. (2005). One year of alendronate after one year of parathyroid hormone (1–84) for osteoporosis. *N. Engl. J. Med.* 353 (6): 555–565.

193 Rittmaster, R.S., Bolognese, M., Ettinger, M.P. et al. (2000). Enhancement of bone mass in osteoporotic women with parathyroid hormone followed by alendronate. *J. Clin. Endocrinol. Metab.* 85 (6): 2129–2134.

194 Tsai, J.N., Lee, H., David, N.L. et al. (2019). Combination denosumab and high dose teriparatide for postmenopausal osteoporosis (DATA-HD): a randomised, controlled phase 4 trial. *Lancet Diabetes Endocrinol.* 7 (10): 767–775.

195 Juppner, H., Abou-Samra, A.B., Freeman, M. et al. (1991). A G protein-linked receptor for parathyroid hormone and parathyroid hormone-related peptide. *Science* 254 (5034): 1024–1026.

196 Fraher, L.J., Hodsman, A.B., Jonas, K. et al. (1992). A comparison of the in vivo biochemical responses to exogenous parathyroid hormone-(1–34) [PTH-(1–34)] and PTH-related peptide-(1–34) in man. *J. Clin. Endocrinol. Metab.* 75 (2): 417–423.

197 Henry, J.G., Mitnick, M., Dann, P.R., and Stewart, A.F. (1997). Parathyroid hormone-related protein-(1–36) is biologically active when administered subcutaneously to humans. *J. Clin. Endocrinol. Metab.* 82 (3): 900–906.

198 Plotkin, H., Gundberg, C., Mitnick, M., and Stewart, A.F. (1998). Dissociation of bone formation from resorption during 2-week treatment with human parathyroid hormone-related peptide-(1–36) in humans: potential as an anabolic therapy for osteoporosis. *J. Clin. Endocrinol. Metab.* 83 (8): 2786–2791.

199 Horwitz, M.J., Tedesco, M.B., Garcia-Ocana, A. et al. (2010). Parathyroid hormone-related protein for the treatment of postmenopausal osteoporosis: defining the maximal tolerable dose. *J. Clin. Endocrinol. Metab.* 95 (3): 1279–1287.

200 Horwitz, M.J., Augustine, M., Khan, L. et al. (2013). A comparison of parathyroid hormone-related protein (1–36) and parathyroid hormone (1–34) on markers of bone turnover and bone density in postmenopausal women: the PrOP study. *J. Bone Miner. Res.* 28 (11): 2266–2276.

201 Augustine, M. and Horwitz, M.J. (2013). Parathyroid hormone and parathyroid hormone-related protein analogs as therapies for osteoporosis. *Curr. Osteoporos. Rep.* 11 (4): 400–406.

202 Tella, S.H., Kommalapati, A., and Correa, R. (2017). Profile of abaloparatide and its potential in the treatment of postmenopausal osteoporosis. *Cureus* 9 (5): e1300.

203 Gonnelli, S. and Caffarelli, C. (2016). Abaloparatide. *Clin. Cases Miner. Bone Metab.* 13 (2): 106–109.

204 Shirley, M. (2017). Abaloparatide: first global approval. *Drugs* 77 (12): 1363–1368.

205 Leder, B.Z., O'Dea, L.S., Zanchetta, J.R. et al. (2015). Effects of abaloparatide, a human parathyroid hormone-related peptide analog, on bone mineral density in postmenopausal women with osteoporosis. *J. Clin. Endocrinol. Metab.* 100 (2): 697–706.

206 Cosman, F., Hattersley, G., Hu, M.Y. et al. (2017). Effects of abaloparatide-SC on fractures and bone mineral density in subgroups of postmenopausal women with osteoporosis and varying baseline risk factors. *J. Bone Miner. Res.* 32 (1): 17–23.

207 Besschetnova, T., Brooks, D.J., Hu, D. et al. (2019). Abaloparatide improves cortical geometry and trabecular microarchitecture and increases vertebral and femoral neck strength in a rat model of male osteoporosis. *Bone* 124: 148–157.

208 Chandler, H., Lanske, B., Varela, A. et al. (2019). Abaloparatide, a novel osteoanabolic PTHrP analog, increases cortical and trabecular bone mass and architecture in orchiectomized rats by increasing bone formation without increasing bone resorption. *Bone* 120: 148–155.

209 Varela, A., Chouinard, L., Lesage, E. et al. (2017). One year of abaloparatide, a selective peptide activator of the PTH1 receptor, increased bone mass and strength in ovariectomized rats. *Bone* 95: 143–150.

210 Le Henaff, C., Ricarte, F., Finnie, B. et al. (2019). Abaloparatide at the same dose has the same effects on bone as PTH (1–34) in mice. *J. Bone Miner. Res.*

211 Ricarte, F.R., Le Henaff, C., Kolupaeva, V.G. et al. (2018). Parathyroid hormone(1–34) and its analogs differentially modulate osteoblastic Rankl expression via PKA/SIK2/SIK3 and PP1/PP2A-CRTC3 signaling. *J. Biol. Chem.* 293 (52): 20200–20213.

212 Jolette, J., Attalla, B., Varela, A. et al. (2017). Comparing the incidence of bone tumors in rats chronically exposed to the selective PTH type 1 receptor agonist abaloparatide or PTH(1–34). *Regul. Toxicol. Pharm.* 86: 356–365.

213 Doyle, N., Varela, A., Haile, S. et al. (2018). Abaloparatide, a novel PTH receptor agonist, increased bone mass and strength in ovariectomized cynomolgus monkeys by increasing bone formation without increasing bone resorption. *Osteoporos. Int.* 29 (3): 685–697.

214 Shimizu, M., Joyashiki, E., Noda, H. et al. (2016). Pharmacodynamic actions of a long-acting PTH Analog (LA-PTH) in thyroparathyroidectomized (TPTX) rats and normal monkeys. *J. Bone Miner. Res.* 31 (7): 1405–1412.

215 Maeda, A., Okazaki, M., Baron, D.M. et al. (2013). Critical role of parathyroid hormone (PTH) receptor-1 phosphorylation in regulating acute responses to PTH. *Proc. Natl. Acad. Sci. U.S.A.* 110 (15): 5864–5869.

216 Cheloha, R.W., Gellman, S.H., Vilardaga, J.P., and Gardella, T.J. (2015). PTH receptor-1 signalling-mechanistic insights and therapeutic prospects. *Nat. Rev. Endocrinol.* 11 (12): 712–724.

217 Vilardaga, J.P., Jean-Alphonse, F.G., and Gardella, T.J. (2014). Endosomal generation of cAMP in GPCR signaling. *Nat. Chem. Biol.* 10 (9): 700–706.

218 Cheloha, R.W., Maeda, A., Dean, T. et al. (2014). Backbone modification of a polypeptide drug alters duration of action in vivo. *Nat. Biotechnol.* 32 (7): 653–655.

219 Keskinruzgar, A., Bozdag, Z., Aras, M.H. et al. (2016). Histopathological effects of teriparatide in medication-related osteonecrosis of the jaw: an animal study. *J. Oral. Maxillofac. Surg.* 74 (1): 68–78.

220 Daugaard, H. (2011). The influence of parathyroid hormone treatment on implant fixation. *Dan. Med. Bull.* 58 (9): B4317.

221 Zati, A., Sarti, D., Malaguti, M.C., and Pratelli, L. (2011). Teriparatide in the treatment of a loose hip prosthesis. *J. Rheumatol.* 38 (4): 778–780.

222 Suzuki, T., Sukezaki, F., Shibuki, T. et al. (2018). Teriparatide administration increases periprosthetic bone mineral density after total knee arthroplasty: a prospective study. *J. Arthroplasty.* 33 (1): 79–85.

223 Kaneko, T., Otani, T., Kono, N. et al. (2016). Weekly injection of teriparatide for bone ingrowth after cementless total knee arthroplasty. *J. Orthop. Surg. (Hong Kong).* 24 (1): 16–21.

224 Kobayashi, N., Inaba, Y., Uchiyama, M. et al. (2016). Teriparatide versus alendronate for the preservation of bone mineral density after total hip arthroplasty – a randomized controlled trial. *J. Arthroplasty.* 31 (1): 333–338.

225 Ledin, H., Good, L., Johansson, T., and Aspenberg, P. (2017). No effect of teriparatide on migration in total knee replacement. *Acta Orthop.* 88 (3): 259–262.

226 Burtis, W.J., Brady, T.G., Orloff, J.J. et al. (1990). Immunochemical characterization of circulating parathyroid hormone-related protein in patients with humoral hypercalcemia of cancer. *N. Engl. J. Med.* 322 (16): 1106–1112.

227 Grill, V., Ho, P., Body, J.J. et al. (1991). Parathyroid hormone-related protein: elevated levels in both humoral hypercalcemia of malignancy and hypercalcemia complicating metastatic breast cancer. *J. Clin. Endocrinol. Metab.* 73 (6): 1309–1315.

228 Deftos, L.J. (2002). Hypercalcemia in malignant and inflammatory diseases. *Endocrinol. Metab. Clin. North Am.* 31 (1): 141–158.

229 Henderson, A., Andreyev, H.J., Stephens, R., and Dearnaley, D. (2006). Patient and physician reporting of symptoms and health-related quality of life in trials of treatment for early prostate cancer: considerations for future studies. *Clin. Oncol. (R Coll Radiol).* 18 (10): 735–743.

230 Henderson, M.A., Burt, J.D., Jenner, D. et al. (2001). Radical surgery with omental flap for uncontrolled locally recurrent breast cancer. *ANZ. J. Surg.* 71 (11): 675–679.

231 Yang, D., Singh, R., Divieti, P. et al. (2007). Contributions of parathyroid hormone (PTH)/PTH-related peptide receptor signaling pathways to the anabolic effect of PTH on bone. *Bone* 40 (6): 1453–1461.

232 Partridge, N.C., Alcorn, D., Michelangeli, V.P. et al. (1983). Morphological and biochemical characterization of four clonal osteogenic sarcoma cell lines of rat origin. *Cancer Res.* 43 (9): 4308–4314.

233 Pasquini, G.M., Davey, R.A., Ho, P.W. et al. (2002). Local secretion of parathyroid hormone-related protein by an osteoblastic osteosarcoma (UMR 106-01) cell line results in growth inhibition. *Bone* 31 (5): 598–605.

234 Ho, P.W., Goradia, A., Russell, M.R. et al. (2015). Knockdown of PTHR1 in osteosarcoma cells decreases invasion and growth and increases tumor differentiation in vivo. *Oncogene* 34 (22): 2922–2933.

235 Tsang, K.M., Starost, M.F., Nesterova, M. et al. (2010). Alternate protein kinase A activity identifies a unique population of stromal cells in adult bone. *Proc. Natl. Acad. Sci. U.S.A.* 107 (19): 8683–8688.

21

Activators of G-Protein Signaling in the Normal and Diseased Kidney

Frank Park

Department of Pharmaceutical Sciences, University of Tennessee Health Science Center, Memphis, TN, USA

21.1 Introduction

The kidney is a fundamentally important organ found in many types of multicellular invertebrates and higher-order mammals. A primary function of the kidney is to ensure that the internal homeostasis remains constant in the face of a continual influx of nutrients, fluids, and electrolytes by appropriating the necessary energy for sufficient reabsorption of the filtered load while maintaining proper efflux of metabolic waste. In addition, the kidney is involved in fine tuning variations in pH, maintenance of arterial blood pressure, production of red blood cells, and regulation of bone density through the synthesis or activation of hormones.

In many of these processes, heterotrimeric guanine nucleotide-binding proteins (hereafter called G-proteins) act as intermediaries to activate intracellular signaling pathways by transmitting biochemical signals initiated by G-protein coupled receptors (GPCRs) at the plasma membrane due to changes in the extracellular milieu [1–3]. GPCRs function as prototypical guanine exchange factors (GEFs) to control the activation and inactivation state of G-proteins [4]. Aberrant activity of G-proteins due to modest-to-severe perturbations in GPCR activity can result in pathological changes in cellular architecture and renal function. Other than interacting with GPCRs as classical Gαβγ heterotrimers, Gα and Gβγ subunits can also exhibit alternate modes of action by complexing with accessory proteins to affect the magnitude, duration, and in some cases, spatial location away from the plasma membrane [5, 6]. A major family of accessory proteins known as Activators of G-protein Signaling (AGS), which is the focus of this chapter, can regulate G-protein signaling independent of GPCR activation by directly influencing either nucleotide exchange with specific Gα subunits or other novel modes of interaction with the G-protein Gα or Gβγ subunits [7–13].

GPCRs as Therapeutic Targets, Volume 2, First Edition. Edited by Annette Gilchrist.

In this chapter, the rationale and the methodology used to isolate the AGS proteins will be provided. Each AGS member and their associated binding partners (i.e. heterotrimeric G-protein α and $\beta\gamma$ subunits) will be described with regards to any known information on their expression profile, localization, and putative signaling function during normal and diseased states of the kidney.

21.2 Heterotrimeric G-Protein Subunits in the Kidney

G-proteins are evolutionarily conserved from simple invertebrates to higher-order mammals. Currently, there are four families of Gα subunits, which include Gα_s, G$\alpha_{i/o}$, G$\alpha_{q/11}$, and G$\alpha_{12/13}$, along with 5 Gβ and 12 Gγ subunits [14]. The localization of G-protein subunits within the kidney is important as they must be present in the same cells as their AGS protein partners for signal modulation to occur. A detailed summary of the Gα and G$\beta\gamma$ subunits in the kidney has been described elsewhere [15].

In brief, the specific cell types in the kidney that express one or more G-protein subunits has been studied using molecular and pharmacological techniques. In the glomerulus and preglomerular vasculature, all of the Gα subunits were detected, including Gα_s [16–21], Gα_i [16–21], G$\alpha_{q/11}$ [16, 18, 20, 21], and G$\alpha_{12/13}$ [22, 23]. In the nephron, however, the localization of specific Gα subunits is segment-specific where the proximal tubule expressed Gα_s [24–26], Gα_{i3} [24], Gα_q [27, 28], and G$\alpha_{12/13}$ [22, 23] using molecular, cellular, or pharmacological approaches. In the thick ascending limb of Henle's loop, most of the Gα subunits were expressed except for Gα_{11} [19], Gα_o, and Gα_{13} [23]. On the other hand, all of the classes of Gα were detected in the collecting duct epithelial cells [19, 22–24], other than Gα_q and some of the G$\alpha_{i/o}$ (i.e. Gα_{i1}, α_{i3}, and Gα_o) subunits [19].

For the Gβ subunits, all five were detected in the kidney using various species of animals [29–32], whereas only 6 of the 12 Gγ subunits were detected in the adult mammalian kidney [33, 34]. To date, the specific cell types that express individual subtypes of G$\beta\gamma$ remains to be fully elucidated.

21.3 Identification of AGS Proteins

The budding yeast *Saccharomyces cerevisiae* was chosen as the platform to generate a high throughput functional screen to identify potential modulators of heterotrimeric G-protein activity [8–11, 35]. To achieve this goal, the mating pheromone–response pathway, which is governed by a single GPCR, was targeted for assay development. First, the endogenous pheromone receptor was removed from the yeast [8]. Second, the native yeast Gα protein was replaced with a modified N-terminal version of yeast Gα fused to either human or rat Gα isoforms, which are more efficient in activating the pheromone–response pathway following transformation with various sources of mammalian cDNA libraries [8]. The activation of the pathway leads to Gβγ bioactivity that induces changes in the growth of the yeast, which is the phenotype that is used to identify potential G-protein activators. Further epistatic testing of the isolated activators was examined in similarly modified yeast with additional deletions in other points of the mating pathway to identify proteins that directly interact with the G-protein subunits. Immunoprecipitation assays using purified mammalian Gα and Gβγ subunits were subsequently performed to confirm the legitimacy of the interaction.

Using these approaches, 13 heterotrimeric G-protein activators were isolated and classified as AGS proteins [8–11, 35]. Protein analysis led to homology matches that tentatively added an extra six proteins to this family by virtue of their similarities in protein domains and potential G-protein function [36, 37]. Of these

Table 21.1 List of AGS proteins and alternate names.

AGS group number	AGS number	Alternate common names	G-protein subunit interaction
1	AGS1	DexRas1/RasD1	Gαi
1		GIV/Girdin	Gαi
1		Ric-8A/B	Gαi, Gαq, and Gα12/13 (Ric-8A); Gαs and Gαq (Ric-8B)
1		Rhes/RasD2	Gαi
2	AGS3	GPSM1	4 × GDP-Gαi/o
2	AGS4	GPSM3/G18.1b	3 × GDP-Gαi/o
2	AGS5	GPSM2/LGN	4 × GDP-Gαi/o
2	AGS6	RGS12	1 × GDP-Gαi/o
2		RGS14	1 × GDP-Gαi/o
2		Rap1GAP	1 × GDP-Gαi/o
2	AGS3-SHORT		3 × GDP-Gαi/o
2		GPSM4/L7/Pcp-2	1 × GDP-Gαi/o
3	AGS2	Tctex-1	Gβγ
3	AGS7	TRIP13	Gβγ
3	AGS8	FNDC1	Gβγ
3	AGS9	Rpn10/S5a	Gβγ
3	AGS10	Gαo	Gβγ
4	AGS11	TFE3	Gα16
4	AGS12	TFEB	Gα16
4	AGS13	MiTF	Gα16

Source: Based on Refs [39–46].

20 proteins considered to be part of the AGS family, there are several proteins already categorized into other protein families. For example, AGS6 is known as Regulator of G-protein Signaling 12 (RGS12), which is a member of the larger RGS family [9], while AGS7 is also known as thyroid receptor interacting protein 13 (TRIP13), which is a member of the AAA^{+} ATPase family [9, 38]. A list of the AGS proteins, their alternate names and associated G-protein subunits are summarized in Table 21.1.

Even though the homology to other existing proteins helps our basic understanding of its role in some organ systems, there still remains a paucity of information as to the role of these proteins in terms of its G-protein regulation. To date, AGS proteins fall into four distinct functional classes (Figure 21.1) and their modes

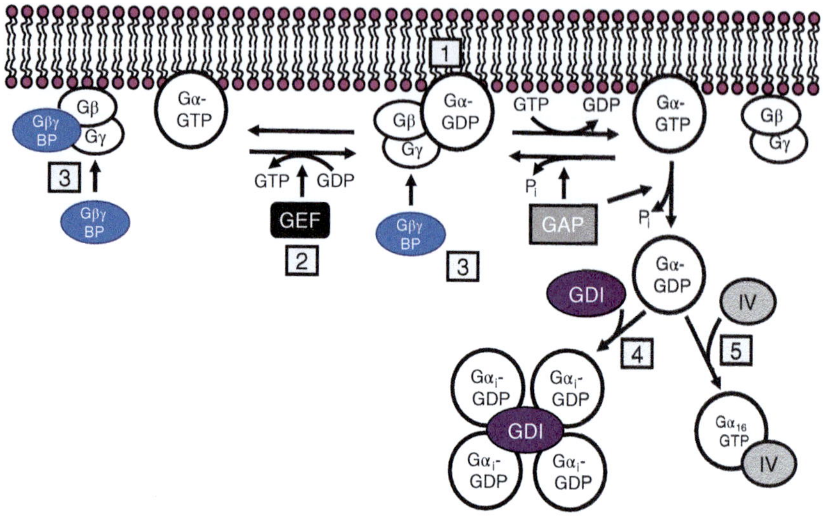

Figure 21.1 Proposed mechanisms of action by the different groups of AGS proteins. The numbers in the figure is to highlight the specific modes of action by which AGS proteins are proposed to function in the cells. (1) Heterotrimeric G-proteins are in their inactive GDP-bound Gαβγ state. (2) Perturbations to the cell results in the conformation changes to the heterotrimeric G-proteins where Group I AGS proteins (GEF) facilitate the exchange of GDP with GTP on the Gα subunit to activate the Gα and Gβγ subunits. (3) The Gβγ that is bound or unbound with Gα-GDP can also interact with Group III AGS proteins (Gβγ BP) to regulate downstream signaling. (4) Upon inactivation by GTPase-activating proteins (GAP), GDP-bound Gα subunits can either form heterotrimers with its natural Gβγ partner or Group II AGS proteins (GDI) can intercept up to four inactive GDP-Gα subunits to form a multiprotein complex to activate signaling pathways. There is evidence for Group II AGS proteins controlling GPCR-G protein signaling at the plasma membrane or at other site within the cells, such as the spindle poles or other intracellular organelles. (5) Alternatively, GDP-bound Gα16 subunits may be bound by Group IV AGS proteins (IV) for transport to the nuclei to regulate transcriptional output of specific genes, such as connexin-43 (CX-43). Source: Based on Refs [9, 11, 37, 39, 47–50].

of action is illustrated in Figure 21.1: (1) direct Gα activators (Group 1); (2) modulators of Gα–Gβγ interaction by binding to Gα (Group 2); and (3) modulators of Gα–Gβγ interaction by binding to Gβγ (Group 3); and (4) interaction with Gα16 (Group 4). This chapter will highlight our knowledge of each individual AGS protein with respect to their expression profile and their putative role, if known, in the kidney.

21.4 Activators of G-Protein Signaling in the Kidney

21.4.1 Group I AGS Proteins

This group of accessory proteins act as receptor-independent GEFs, which facilitates the switching of guanine dinucleotide phosphate (GDP) and guanine trinucleotide phosphate (GTP) in the Gα subunits (Figure 21.1). Direct binding by the Group I AGS protein to the Gα subunit was dependent upon the specific Gα subunit used in the functional screen [8]. In addition to AGS1 [8, 12, 13], there are three other members classified as Group I AGS proteins were based upon their similarity in sequence homology and/or mechanism of action: (i) Gα-interacting, Vesicle-associated protein (GIV)/Girders of actin filaments (Girdin)/Akt phosphorylation enhancer (APE) [5]; (ii) resistance to inhibitors of cholinesterase 8 (Ric-8) [51–53]; and (iii) RasD2/Ras homolog enriched in striatum (Rhes)/tumor endothelial marker 2 (TEM2) [54].

21.4.1.1 AGS1/Dexras1/RasD1

The first isolated AGS protein from the functional yeast screen using a human liver cDNA library was aptly named AGS1 [8], and its sequence identity was homologous to dexamethasone-inducible Ras protein (Dexras1/RasD1) [8]. Dexras1/RasD1 is a member of the Ras superfamily of small GTPases and is a protein comprised of 281 amino acids with an expected mass of ~32 kD [55]. AGS1/Dexras1 mRNA expression can be induced in many organs throughout the body, including the kidney, by exposure to dexamethasone [55–58]. Structurally, AGS1 acts as a GEF through specific protein domains to facilitate transfer of GTP binding to monomeric $G\alpha_{i/o}$ subunits for downstream signal activation (Figure 21.2). This was confirmed in experiments where AGS1 signaling was negatively impacted when a variant form of $G\alpha_{i2}$ (G204A) incapable of GTP binding, or over-expression of Regulator of G-protein Signaling 4 (RGS), which accelerates GTP hydrolysis on $G\alpha_{i/o}$ was used in combination with AGS1 [8]. AGS1 did not interact with alternate $G\alpha_s$ and $G\alpha_{16}$ subunits or Gβγ dimers [8, 12, 13, 57]. There is a subsequent study showing that AGS1 may indirectly activate downstream ERK1/2 signaling by Gβγ [63]. The activation of ERK1/2 could be blocked by pertussis toxin [63], which is known to only inhibit $G\alpha_{i/o}$ subunits ribosylated as

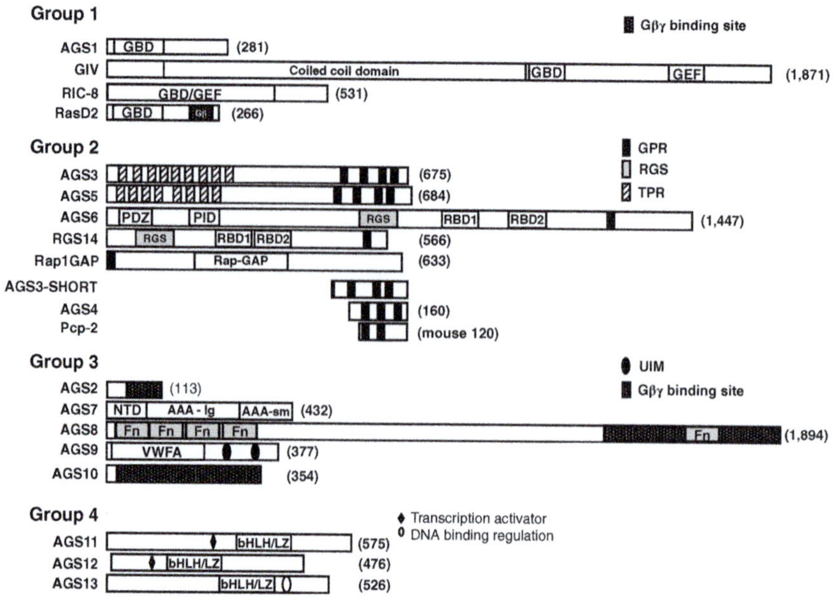

Figure 21.2 Protein structure of AGS proteins. AGS proteins are listed in their respective groups, and the specific protein domains or regions are labeled with abbreviations, which are described as follows for each group: Group 1: GBD/GEF = G-protein binding domain/ guanine exchange factor site; Gβ = Gβ binding site. Group 2: GPR = G-protein regulatory motifs; RGS = Regulatory of G-protein Signaling domain (grey box); PDZ = PSD95/Discs Large/ZO-1; PID = phosphotyrosine interaction domain; RBD1/2 = Ras binding domain 1/2. Group 3: NTD = N-terminal domain; AAA-lg/-sm = AAA$^+$ ATPase large or small domain; FN = fibronectin-binding site; VWFA = von Willebrand factor A binding site; Ubiquitin-interacting motif (UIM; ⬮). The Gβγ binding region is marked with vertical lines for AGS2, AGS8, and AGS10, but the sites are not known for AGS7 and AGS9. Group 4: TA = transcription activation site; bHLH/LZ = basic Helix–Loop–Helix and leucine zipper sites. The amino acid length for the human AGS protein is provided in parentheses. The location of the Gα16 binding site in Group 4 remains to be determined, but does not require the C-terminal 27 amino acids. Source: Based on Refs [59–62].

a heterotrimeric complex with Gβγ. [64]. This effect provides some evidence that AGS1 can also modulate GPCR-dependent signaling at the plasma membrane by affecting channel activity through the actions of Gβγ [65]. In this respect, AGS1 may intercept a select pool of Gαi subunits that normally target muscarinic receptors to reduce their ability to activate GIRK channels [65]. Using cultured cells, AGS1 exhibited heterologous sensitization of adenylyl cyclase 1 (AC1) through the release of Gβγ dimers following stimulation of the dopamine D2 receptor [64]. This effect was not dependent upon the GEF function of AGS1, since this was observed using a nucleotide binding defective G31V variant. These data suggest that Dexras1 selectively regulates receptor-mediated Gβγ signaling pathways [64]. This may implicate another alternate role for AGS1/Dexras1 in the kidney by possibly regulating Na^+–K^+ ATPase activity in a cAMP-dependent mechanism through the dopamine D2 receptor [66]. Other biological roles for AGS1 have been described that could either impact renal function through a direct action in tubular epithelial cells or indirectly through its actions on organs associated with the cardiovascular system. Alternative roles for AGS1 have been shown in the kidney to control cardiovascular function through its actions on renin transcription [67] or nitric oxide synthase activity [68], which are well established to play a crucial role in long-term sodium reabsorption and blood pressure regulation. Therefore, the function of AGS1 may be more complex and demonstrate multiple functions within the cell to control heterotrimeric G-protein function as a monomer as well as a trimer.

The expression pattern of AGS1/Dexras1 in the kidney is developmentally regulated with levels increasing from the fetus toward adulthood [69], but those levels are still far lower than other organs in the mouse [55] or human [57, 69]. Under normal conditions, AGS1 was primarily localized to the proximal tubules in the cortical and outer medullary region in the mouse kidney regardless of the type of renal injury [48] (Figure 21.3). Ischemia–reperfusion injury (IRI), which is a common form of acute kidney injury (AKI), led to an induction of AGS1 mRNA after 24 hours and remained elevated for the duration of the experimental period of seven days with no change in the localization pattern of AGS1 outside of the

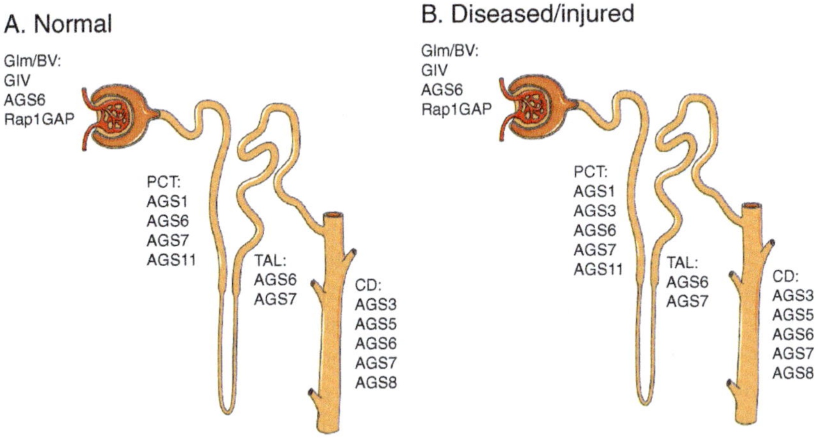

Figure 21.3 Localization of AGS proteins in different segments of the kidney. This figure illustrates the known sites for some of the AGS proteins in the kidney using either mRNA or protein detection methods during (A) normal and (B) diseased/injured states. Other than AGS6/RGS12, the expression of AGS proteins in the kidney have been detected exclusively in either the tubular epithelial cells or in the glomerulus. Glm = glomerulus; BV = blood vessels; PCT = proximal convoluted tubule; TAL = thick ascending limb of Henle's loop; CD = collecting duct. Source: The images of the nephron/Servier Medical Art/CC BY-3.0.

proximal tubules [48]. Similarly, AGS1 protein was not observed in cystic epithelial cells, but only in the proximal tubules from a mouse model of polycystic kidney disease (PKD). Dexamethasone promoted a prosurvival effect following renal IRI [70], but exacerbated a loss in renal function by accelerating cystic disease progression [71, 72]. These findings suggest that if AGS1 does, indeed, have a role in the kidney, the function would likely be context-dependent so their mode of action would need further elucidation.

21.4.1.2 GIV/Girdin/APE

GIV/Girdin is a large 1871 amino acid protein that acts as a central hub to control various cell processes within the cell, such as cell migration [73, 74], autophagy [75], and metastasis [73]. GIV/Girdin achieves its multiplicity of function by its wide array of binding partners in the C-terminal end of the protein, including receptor tyrosine kinases, Akt, and G-protein subunits [5]. The mechanism by which GIV/Girdin regulates G-protein function is unique and appears to be compartment-dependent in the cell. GIV/Girdin has a conserved Gα binding and activation (GBA) motif that promotes both guanine nucleotide exchange (GEF) using Gαi subunits at the plasma membrane, but can also serve as a guanine nucleotide dissociation inhibitor (GDI) by interacting with Gαs-GDP at the endosomes [59] (Figure 21.2). This dual role can promote cell migration at the plasma membrane while initiating cell proliferation within the cell. Moreover, the GEF function in GIV provides an opportunity to transfer G-protein α subunits from other AGS proteins, namely Group II AGS proteins, to regulate pathway activity. As an example, GIV can intercept inactive GDP-bound Gαi3-AGS3, a Group II AGS protein, to switch autophagy from active to inactive progression at the lysosome [76].

Little is known about the role of GIV/Girdin in the kidney, due in large part to its discrete localization to only the podocytes of the glomerulus and not other cell types within the nephron, vasculature, or interstitial cells [77] (Figure 21.3). This would be consistent with the inability to detect GIV mRNA or protein using whole kidneys [78]. Induction of GIV/Girdin expression in the podocytes was observed following administration of puromycin aminonucleoside, which is a known podocyte toxin that produces a reproducible model of nephrotic syndrome [77]. Functionally, the increased production of podocyte GIV/Girdin provided protection by activating phosphoinositide-3 kinase (PI3K)/Akt prosurvival pathways through the formation of a multiprotein complex consisting of Gαi and vascular endothelial growth factor receptor 2 [77].

21.4.1.3 Ric-8

In mammals, there are two isoforms, Ric-8A (531 amino acids) and Ric-8B (520 amino acids), that act as a chaperone to unfold Gα subunits and exert GEF activity

[51–53]. Mechanistically, Ric-8A/B interacts with only GDP-bound Gα proteins to stimulate the release of GDP and form a stable nucleotide-free transition state complex with the Gα protein. Upon binding of GTP to the Gα subunit, the Ric-8 complex dissociates from the GTP-bound Gα subunit [51]. Recent data have shown that the minimal Gα binding domain and catalytic activity for Ric-8 is found within the first 452 amino acids (Figure 21.2), but that additional sites from amino acid 453 to 491 are required to provide tighter binding to Gα subunits following phosphorylation of two specific Ser/Thr residues [79]. This phosphorylation step was found to be an essential part of promoting both Gα subunit folding and guanine nucleotide exchange [80]. Both of the Ric-8 isoforms have preferential selectivity for Gα subunits where Ric-8A interaction is more promiscuous by binding to Gαi, Gαq, and Gα12/13, and Ric-8B interacts with only Gαs and Gαq subunits (Table 21.1).

Functionally, Ric-8 stabilizes Gα subunits to enable interaction with Gβγ at the plasma membrane. In the absence of Ric-8, a marked reduction (>95%) is observed in the steady-state Gα subunit levels compared to wildtype cells, which can negatively impact G-protein signaling [81]. At present, the only evidence for a potential role for Ric-8 in tubular epithelial cells in the kidney was recently published by Chishiki et al. [82]. In this study, a three-dimensional renal tubular epithelial cell *in vitro* system demonstrated that Ric-8A could regulate tight junction assembly and abnormal epithelial cell orientation through Gαi switching with other class II AGS proteins, AGS5/LGN and AGS3/GPSM1 [82]. This is similar to the transfer of the Gαi subunits between AGS3 and GIV [76]. Reduced or complete loss in the expression of Ric-8A or Gαi delayed the cortical delivery of Gαi, which was associated with an increased formation of intercellular lumens surrounded by membranes enriched with Gαi3 and gp135, an apical membrane protein [82]. Although this is an encouraging finding to demonstrate the relationship of accessory protein function within the AGS family to orient mitotic spindles and control cyst formation *in vitro*, there is no published study that has identified a biological role for Ric-8 in the kidney, due in large part to a lack of molecular evidence for renal Ric-8 transcription in lower invertebrates and mice [83–85], except adult frogs [86]. Studies will need to be performed to validate the expression and function of Ric-8 in normal and diseased states of the kidney using animal models and human tissue analyses.

21.4.1.4 RasD2/Rhes/TEM2

RasD2 has ~62% sequence homology to AGS1, but it did not exert a Gαi activating function in yeast [54]. Other studies evaluating modes of action for RasD2 have not yet shown whether GEF activity is present to control guanine nucleotide turnover in Gα subunits [87–90]. In nonrenal cells, there is evidence that RasD2 may exhibit an alternative mode of action by interacting with Gαi bound to Gβ [91].

This suggests that a complex of RasD2 with Gαiβγ regulates plasma membrane-G protein signaling [92–95], which was confirmed by the ability of pertussis toxin to block Gαi-dependent accumulation of cAMP following dopamine D1 receptor activation. In the whole kidney, RasD2 mRNA was detected at an extremely low expression levels [96], and a subsequent study detected transcripts in microdissected distal tubules [97]. It remains to be determined whether the minimal transcription of RasD2 has any physiological role in the kidney.

21.4.2 Group II AGS Proteins

This group of proteins was initially categorized together based on the presence of one or more G-protein regulatory (GPR)/GoLoco motifs of ~20 amino acids [98–100]. These motifs selectively interacts with inactive GDP-bound Gαi/o family of G-proteins [37, 47] (Figure 21.1), which enables the Group II AGS proteins to function as GDI. At present, there are seven members in this class that are further stratified into three subtypes depending on their protein structure. The first type consists of AGS3 and AGS5 (LGN), which has four GPR/GoLoco motifs with tetratricopeptide repeat (TPR) domains [11, 101]. The second type contains RGS12 (AGS6), RGS14, and Rap1GAP, which has only the GPR/GoLoco motif plus other defined domains (RGS) that accelerate Gα–GTP hydrolysis. The third type includes a truncated form of AGS3 (known as AGS3-SHORT) [102], AGS4 (GPSM3) [35, 103, 104], and PCP2/L7 (GPSM4), which contains multiple GPR/GoLoco motifs without any other clearly defined regulatory protein domains.

21.4.2.1 Activator of G-Protein Signaling 3 (AGS3)

AGS3 or G-protein Signaling Modulator 1 (GPSM1) is the first member of the Group II AGS proteins [8, 11, 101]. The murine AGS3 is 673 amino acids, and 675 amino acids in humans with an expected size ~72 kD. The GPR/GoLoco motifs in the C-terminal part of AGS interact preferentially to Gαi subunits with little to no binding to Gαs, Gαq, or Gαo subunits [11, 98, 99, 105]. The biological role of AGS3 in the kidney has garnered much interest over the past decade in a number of experimental and genetic models of renal injury or diseases [48–50, 106, 107]. Under basal conditions, AGS3 deficient mice have no observable differences in renal architecture or function compared to their wildtype littermates [48–50, 107]. The expression profile of AGS3 mRNA [102, 108] and protein [48–50, 101, 102, 107, 108] is primarily dormant in both adult mouse and rat kidneys with selective localization to the distal tubular epithelial cells in the nephron [48–50] (Figure 21.1).

In multiple animal models of experimental kidney injury and genetic disease, AGS3 expression was induced or aberrantly elevated [50, 107]. Following IRI, which is a common form of AKI [109–111], AGS3 was temporally induced and associated with markers of proliferation during the recovery phase of the tubular epithelial cells [48, 50]. The inability to induce AGS3 during this phase of recovery exacerbated the extent of tubular epithelial cell injury due to IRI [50]. In PKD, which is the most common genetic disease in the kidney, AGS3 expression was abnormally elevated in kidneys from multiple rodent models of PKD [48, 49, 107]. Moreover, AGS3 was localized to the cystic collecting duct epithelia [48, 107]. In addition, AGS3 protein was highly detected in human kidney tissue and urinary exosomes from PKD patients, but not patients with chronic kidney disease (CKD) unrelated to PKD [49, 106]. Genetic removal of AGS expression on the background of a polycystic kidney disease 1 (*Pkd1*) variant mouse, which simulates autosomal dominant polycystic kidney disease (ADPKD), cystic disease progression was accelerated [107].

From these studies, it would appear that AGS3 may act as a stress–response protein that is induced by cellular injury to repair damaged tubular epithelia during states of active proliferation. In the case of AKI, the damaged tubular epithelia induced AGS3 to initiate repair that facilitated recovery, which subsequently returned AGS3 levels back to normal quiescent levels. In PKD, germline and somatic mutations lead to a constant state of increased proliferation and cyst formation [2], such that this disease is considered to be a model of futile cellular repair [112, 113]. Therefore, the continued increase in AGS3 production may be a compensatory effect where the precystic epithelial cells attempt to thwart the deleterious cell altering effects by the variant cystic genes. Interestingly, not all types of renal injury will lead to a change in the AGS3 levels. Compensatory hypertrophy attributed to unilateral nephrectomy did not either induced AGS3 expression nor alter the rate and magnitude of the hypertrophic effect on the kidney [48].

On a cellular level, AGS3 has been implicated in several signaling pathways associated with cell survival [50, 114], cAMP production [49, 115], mitotic spindle orientation [116], ion channel activity [107], and autophagy [73, 117–120]. In tubular epithelial cells *in vitro*, AGS3 overexpression or knockdown increased or decreased cell number, respectively [50, 114]. Intuitively, the ability of AGS3 to scavenge multiple Gαi subunits by the GPR/GoLoco motifs should influence cAMP production [115], but this was not the case in cystic tubular epithelial cells [49]. Another potential role for AGS3 to promote survival is to activate autophagic pathways by interacting with Gαi subunits [73, 117–119], but this does not occur in all cell types [120]. Instead, the increased cell survival was Gβγ-dependent on the activation of MAPK pathways [50, 114]. Consistent with this effect, Gβγ signaling promoted tubular epithelial cell recovery following IRI *in vivo* [121].

Two other Gβγ-dependent processes associated with AGS3 that could affect tubular cell pathogenesis are ion channel activity and mitotic spindle orientation. AGS3 increased heteromeric polycystin-1/-2 (PC1/PC2) ion channel activity by Gβγ activity and was not blocked by pertussis toxin [107]. This was suggestive that PC1/PC2 activity was not dependent upon the GDP form of Gαi bound as a heterotrimer with Gβγ, and that there may be a separate pool of inactive Gαi-GDP that was intercepted by AGS3 to prolong Gβγ. Alternatively, AGS3 has been implicated in mitotic spindle orientation through Gβγ-dependent pathways in neurons [116]. There is debate whether this process of oriented cell division is integral to cystic disease initiation [122–125], and there are no current data showing that AGS3 is even involved in this particular function in the kidney. Other protein–protein interactions with AGS3, such as liver kinase B1 (LKB1) [126], may have the opportunity to influence signal transduction pathways. LKB1 is a serine/threonine kinase known to regulate AMP-activated protein kinase signaling during states of energetic stress [7]. In PKD, dysregulation of LKB1 can alter the progression of cyst formation [93, 127], but as yet there is no evidence that dysregulation expression of AGS3 is involved in this process. Further studies are needed to fully elucidate the specific mechanisms of action by AGS3 within various cell types and how they impact disease pathogenesis in the kidney.

21.4.2.2 AGS5/GPSM2/LGN

AGS5 along with AGS3/GPSM1 is considered a mammalian homolog of the *Drosophila* protein, Partner of Inscuteable (Pins), which is a key protein that is involved in cell polarity and spindle positioning [128]. AGS5 has other common names, LGN (Leu–Gly–Asn repeat-enriched protein) and G protein signaling modulator 2 (GPSM2), with a length of 679 and 684 amino acids in mice and humans, respectively. Similar to AGS3, AGS5 has N-terminal TPR domains that interact with a multitude of proteins, which in many cases, are distinct from the proteins that bind to AGS3. In mouse, rat, and human tubular epithelial cell lines *in vitro* and whole kidneys *in vivo*, AGS5 expression is quite high [48–50, 101, 107]. Similar to AGS3, the localization of AGS5 was limited to the distal tubules from the thick ascending limb to the collecting ducts [48] (Figure 21.3). Although AGS3 is induced during genetic disease or experimental perturbations, AGS5 expression remains largely unchanged [48–50, 107]. Considering that the protein structure of AGS5 is similar to AGS3, there may be functional overlap, but the role in the kidney remains to be determined.

There is a mouse model with homozygous C-terminal deletion of the GPR motifs in AGS5/LGN ($LGN^{\Delta C/\Delta C}$), which did not affect the postnatal viability, but the effects on the kidney were not evaluated in this study [129]. The $LGN^{\Delta C/\Delta C}$ mice did, however, exhibit defective spindle orientation in neurons during apical divisions [129]. As mentioned for AGS3, this process is known to be regulated by AGS5/LGN in many cell types [130–133], including kidney cells [134, 135]. Reduced AGS5 levels resulted in asymmetric spindle orientation and cyst formation *in vitro* [134, 135], but further studies will need to confirm whether this occurs in animal models. Alternatively, AGS5 has been shown to activate other pathways, such as soluble guanylyl cyclase [136], which could play a role in the dysfunctional vascular and tubular function of the kidney during PKD [137]. Continued studies are needed to clearly define the important of not only AGS5 but also AGS3 to understand their pivotal role in kidney disease, and whether compensatory mechanisms, such as upregulation of AGS3 and vice versa, may provide a sufficient buffer to protect the cells from greater damage in the absence of one of these proteins.

21.4.2.3 AGS6/RGS12

AGS6 is part of the second type of Group II AGS proteins where the protein structure includes a single GPR motif to bind inactive Gαi–GDP subunits with other protein regulatory domains that are involved in accelerated Gα–GTP hydrolysis [60] (Figure 21.2). More specifically, AGS6/RGS12 is an 85 kD protein that interacts directly with Gαi/o subunits at the N-terminal RGS domain to accelerate their intrinsic GTPase activity to inactivate their G protein [138]. On the C-terminal end, all isoforms of inactive Gαi-GDP subunits are bound by a single GoLoco/GPR motif [139]. Localization of AGS6/RGS12 in the rat kidney was relatively ubiquitous with low detection observed throughout the nephron and endothelial cells in the blood vessels [48] (Figure 21.3). AGS6 expression in the kidney may be species- or isoform-specific, since human kidneys expressed high levels of *Rgs12* mRNA with multiple isoforms [140], but the kidneys from mice had undetectable levels [141]. Currently, there is no study to show that AGS6 has a direct role in regulating renal function, but there are hypertrophic effects by RGS12 in the heart that could exert a long-term effect on renal biology [142]. Moreover, there are electrophysiological studies documenting a key role in RGS12-dependent regulation of ion channels, such as N-type [141, 143], $Ca_{V2.2}$ calcium channels [144–146], and transient receptor potential C4 channels (TRPC4) [147]. In the latter study, both the RGS domain and GPR motif in RGS12 played an additive role in blocking the receptor-dependent TRPC4 activation [147]. Site-directed variants in either the GPR and RGS domains only partially restored TRPC4 channel activity, but when both sites were simultaneously defective in Gαi binding, the channel activity was fully restored. These studies provide new information as to the potential interaction of RGS domain along with the GPR motif to provide intricate fine-tuning of ion channel activity by a single protein, which could impact normal and pathological states in the kidney.

21.4.2.4 RGS14

RGS14 is a 544 amino acid protein with a predicted size of ~60 kDa [140], and has moderately similar homology (63%) with AGS6/RGS12 in the GAIP or GOS homology (GH) subdomain, which is common to all RGS proteins. RGS14 has a N-terminal canonical RGS domain, a tandem Ras/Rap binding domain (R1 and

R2 RBD), and a single C-terminal GoLoco/GPR motif (Figure 21.2). Similar to AGS6/RGS12, RGS14 selectively interacts with Gαi/o subunits in the RGS box [139, 148], whereas the GPR/GoLoco motif selectively binds to all of the Gαi subunits, but has a weaker interaction with $G\alpha_{i2}$ [139] compared to the other Gαi1 and Gαi3 subunits [149].

Under normal conditions, the transcript level of *Rgs14* mRNA was minimal in the rat kidney compared to other organs in the body [140, 150]. A recent study, however, detected sufficient *Rgs14* mRNA in normal mouse kidneys [151], which could suggest species differences in gene expression profiles. The role of RGS14 has begun to gain interest as a potential regulator of renal function over the past three years. Multiple genome-wide association studies (GWASs) have identified variants in the *RGS14* gene with linkage to biological parameters suggest a putative role for this gene to have a crucial involvement in the pathogenesis of CKD [151] or control phosphate reabsorption in the nephron [152, 153]. A variant in the *SLC41A* gene was associated with decreasing values for glomerular filtration rate (GFR), which was used as a proxy for CKD determination in patients [151]. This led to molecular studies identifying *Rgs14* as a site of transcriptional activity in renal tubular epithelial cells, which were subsequently linked to an expression quantitative trait loci (eQTL) using nonrenal organs. The potential impact of RGS14 on renal function was confirmed using the 129S6 mouse, which is salt-sensitive and is predisposed to glomerulosclerosis. *Rgs14* mRNA expression was decreased in the kidney following administration of a high 6% sodium chloride diet and following progression to CKD using a model of subtotal nephrectomy [151]. In the other two GWAS studies, another variant (rs4074995) in the *RGS14* gene, which is near the *SLC34A1* gene, was examined to exhibit a strong association with perturbations in circulating FGF23 [152] and PTH concentrations [153]. Due to the proximity of Rgs14 to SLC34A1, there was speculation that this would be a strong candidate in the regulation of phosphate handling in the kidneys, particularly at the proximal tubules. These association studies provide the rationale for further pursuit into the mechanisms that may be involved by RGS14 in the kidney regarding either phosphate handling or its role in salt balance during the progression of CKD.

21.4.2.5 Rap1GAP

The role of Rap1GAP in the kidney is relatively limited. The protein structure has the GPR motif at the extreme N-terminus end with the GTPase-activating activity in the middle of the 663 amino acid protein [61] (Figure 21.2). Like other members in the Group II AGS protein, the expression of Rap1GAP is relatively quiescent throughout the body, but can be detected under normal conditions in the kidney [61, 154]. However, there is evidence that Rap1GAP can be induced in renal glomerular podocytes from HIV-1 transgenic mice and in human kidney biopsies of focal and segmental glomerulosclerosis [155], which was believed to contribute to the dysfunction observed in the podocytes due to an altered β1 integrin adhesion process. On the other hand, renal cancer cells have shown reduced [156] or increased levels of Rap1GAP [61]. In the tumor cells with reduced Rap1GAP, there was an increased ability for cellular invasion [156]. The reason for the differences in Rap1GAP levels is not known, but may be due to differences in the culture conditions versus *in vivo* conditions.

The third type of group II AGS proteins consists of three proteins, which contain multiple C-terminal GPR domains, but no other defined regulatory protein binding sites: (i) a truncated form of AGS3 (AGS3-SHORT) [102], (ii) AGS4/GPSM3/G18.1b [35], and (iii) Pcp-2/L7/GPSM4 [157]. To date, there is no study that has detected physiologically relevant expression of AGS3-SHORT [102], AGS4/GPSM3/G18.1b [35, 158], or Pcp-2/L7/GPSM4 [157, 159] in the kidney, so their biological roles remain to be defined in the kidneys.

21.4.3 Group III AGS Proteins

Similar to Group II AGS proteins, Group III proteins were capable of activating signaling pathways in a functional yeast screen expressing either a Giα_2 variant, G204A, or the GTPase-activating protein RGS4. Instead of binding to the inactive GDP-bound Gαi subunit, Group III proteins were shown to interact directly with G$\beta\gamma$ subunits complexed with or without the Gα subunits [9, 11, 39] (Figure 21.1). As shown in Table 21.1, there are five proteins that were isolated in this group, which includes AGS2/Tctex1 [39, 40], AGS7/TRIP13 [41, 42], AGS8/FNDC1 [43], AGS9/Rpn10/S5a/Psmd4 [44], and AGS10/GNAO [45, 46]. To date, there is no obvious consensus motif that specifically directs G$\beta\gamma$ binding to AGS proteins. The role of these proteins is largely undefined in the kidney, but some of their relevant activity in other nonrenal cell types may be applicable to the kidney.

21.4.3.1 AGS2/Tctex1

AGS2 was the second protein isolated from the pheromone receptor-independent yeast screen using the NG-10815 cDNA library [8], and was determined to have identical sequence to mouse Tctex-1. AGS2 activity did not require a specific Gα subunit (i.e. functioned in the yeast screen with either Gαi2/i3, Gαs, or Gα16). It was determined that AGS2 could directly bind to G$\beta\gamma$ [12, 39].

AGS2/Tctex-1 is a small (113 amino acids) cytoplasmic dynein light chain that plays an accessory role in the assembly of the dynein complex [40], which is a microtubule-based molecular motor that can regulate a number of fundamental cellular activities. In the cytoplasm, dyneins power the retrograde trafficking of many components, including endosomes and lysosomes [160], Golgi maintenance [161], breakdown of the nuclear envelope [162], cell cycle regulation, and cilia dynamics [163–165]. Recent studies using neural progenitor cells have shown that

AGS2/Tctex1 is recruited to the ciliary transition zone prior to the DNA synthesis phase of the cell cycle [163], and is modulated by interaction with AGS3/GPSM1 [165]. AGS2 may play a central role in this process by intercepting unbound Gβγ subunits following receptor stimulation by insulin growth factor 1 (IGF-1). The cycling Gαi–GTP to Gαi–GDP is subsequently bound to AGS3 leading to dysregulation in the activation–inactivation cycle of the G-protein system. Tctex1 can also be phosphorylated to regulate actin polymerization and periciliary endocytosis during the initial phase of ciliary resorption [164]. AGS2/Tctex1 may have a similar role in the kidney, which has many types of tubular epithelial cells with cilia, especially since the kidney is one of the major organs expressing copious amounts of Tctex1 mRNA and protein during development to maturation in mice and humans [40, 166]. Abnormalities in the tubular epithelial cell architecture due to cilia defects, such as cystic disease, which already showed abnormal changes in AGS3, could be an important site of action for AGS2 during disease initiation and its progression.

In addition, Ismail et al. [167] discovered a previously undescribed function for AGS2 in tubular epithelial cells *in vitro*. AGS2/Tctex-1 was found to interact with KIM-1, which is a phosphatidylserine receptor that can be specifically induced following AKI. KIM-1 reduces tissue damage by promoting efferocytosis, which is a process where cells undergoing apoptosis are cleared by phagocytosis. Upon injury, KIM-1 expression prevented the phosphorylation of AGS2 at Threonine 94, which can disrupt the binding of AGS2 to dynein on the microtubules and is the same amino acid involved in cilia dynamics. Loss of AGS2/Tctex-1 significantly inhibited efferocytosis. These studies demonstrate that AGS2/Tctex-1 has potential functions in the kidney, but the exact role by which Gβγ signaling is disrupted to affect AGS2 activity remains to be determined.

21.4.3.2 AGS7

AGS7 is a 432 amino acid protein with an expected size of ~49 kD (Figure 21.2). TRIP13 was previously classified as a member of the evolutionarily conserved AAA$^+$ ATPase enzyme family, and its functions has been described in many facets of DNA damage and cell cycle control, such as promoting double-stranded DNA repair [168, 169], mitotic checkpoint control [170–177], and meiotic recombination [41, 42]. Although TRIP13 was isolated from two different modified yeast systems either binding to bait proteins tagged with the thyroid receptor [38] or components of the heterotrimeric G-proteins [178], there remains no published data that demonstrate that these proteins are involved in regulating cellular function through these binding partners.

The expression pattern of AGS7/TRIP13 in the kidney appears to be species dependent where the basal expression is quite low in the rat kidney with relatively selective localization to the principal cells of the collecting duct [179]. In the mouse kidney, AGS7/TRIP13 is weakly detected throughout the nephron, but not the glomeruli or vasculature, which is consistent with the minimal detection of *Trip13* mRNA in the whole kidney [41] (Figure 21.3). However, there is a recent study clearly demonstrating a crucial role for TRIP13 to promote tubular epithelial cell survival following IRI, which is a common type of AKI [179]. The temporal recovery of the kidney following IRI was markedly attenuated in $Trip13^{Gt/Gt}$ hypomorph mice compared to wildtype $Trip13^{+/+}$ littermates [179]. Mechanistically, reduced expression of TRIP13 was associated with increased DNA damage leading to persistent tubular epithelial cell injury due to repressed activation of p53. This is consistent with findings in murine oocytes where cell numbers were markedly reduced through increased apoptotic signaling due to dysregulated TP53 [169].

AGS7/TRIP13 may function as an adaptor protein by providing a platform to form larger protein complexes that facilitate and control signaling pathways. TRIP13 directly interacts with $p31^{comet}$ [170–177], which regulates the conformational change of Mad2 and controls whether the spindle assembly checkpoint is active or inhibited for progression from metaphase to anaphase during mitosis. In double-stranded DNA repair, TRIP13 binds to DNA protein kinase (DNA-PKcs) to increase its activity to promote tumorigenesis in head and neck cancer [168]. AGS7 was identified as a Gβγ binding protein [9, 12], but to date, there is no direct evidence of this interaction in the kidney and whether this interaction is involved in the canonical modes for TRIP13 or other pathways. Recalling AGS3 control of the mitotic spindle using Gβγ, it may also be interesting to speculate that AGS7 could be involved in this type of process, considering the timing in which TRIP13 is generally active and aids in the surveillance of intact spindle at the kinetochore. Further studies are needed to better understand the role of AGS7/TRIP13 in the mammalian kidney.

21.4.3.3 AGS8/FNDC1/KIAA1866

AGS8 was isolated from the AGS yeast screen using a cDNA library created from ischemic injured rat hearts [43]. Functionally, AGS8 was identified as an ischemia-inducible protein from rat hearts, in which the rat AGS8 protein consisted of 1730 amino acids. AGS8 was matched to mouse (1732 amino acids) and human (1894 amino acids) Fibronectin type III containing I (FNDC1) [62] (Figure 21.2). An alignment of other mammalian sequences of FNDC1, even different strains of rats and mice, demonstrated a fairly wide range of protein sizes where the N-terminal end can vary up to 200 amino acids (*personal observation*). However, the C-terminal region of AGS8 is highly conserved across species between mice, rats, and humans and is known to bind with Gβ1γ2 to activate Gβγ signaling as shown by Sato et al. [9]. Epistasis experiments demonstrated that AGS8 interacted selectively with Gβγ, but not Gαs or any of the Gαi/o subunits [9].

The role for AGS8 in the kidney remains largely undescribed, but there is a recent GWAS suggesting a link between AGS8/FNDC1 and urinary albumin-to-creatinine ratio [180]. However, their study using morpholino knockdown of AGS8/FNDC1 in zebrafish did not support a role for this protein during renal development or dysfunction [180]. It is possible that the zebrafish is not ideal to definitively confirm the effects of AGS8 with the G-protein signaling pathway. In another genetic study using congenic rats, variants in AGS8 were identified in the promoter and coding region, which were associated with elevated blood pressure measurements [181]. In the kidneys with the variants in the AGS8 locus, gene expression was highly elevated in the heart and kidney, but the renal induction was ~100-fold less than the heart. It is not known from this study whether the kidney had any role in mediating any dysregulation in blood pressure.

Outside of the kidney, however, there are additional potential pathways that may have relevance on the control of renal function. In cardiomyocytes, the AGS8-Gβγ complex activated connexin-43 (CX-43), which is a gap junction protein, to control apoptotic cell death pathways [9, 182, 183]. Since CX-43 is expressed in the normal kidney [9, 184] with an expression pattern that includes the collecting ducts [184, 185], which is the predominant site of AGS8 expression [186] (Figure 21.3), it is possible to speculate a similar role in the kidney. Another set of studies have implicated AGS8 to regulate vascular endothelial growth factor receptor 2 (VEGFR2) [138, 156], but this is not likely to occur in the kidney due to the exclusive expression of VEGFR2 in peritubular and glomerular capillaries [187], which is not a site that coincides with AGS8 expression [186]. A first step in understanding the potential role for AGS8 in the kidney, surgical bilateral renal IRI was induced leading to a marked reduction in AGS8/Fdnc1 mRNA after 24 hours, which returned back toward normal after 168 hours [186]. Since the collecting ducts are normally insensitive to the damaging effects of IRI, it remains unclear as to how the changes in AGS8 transcript levels would be involved in the injury and recovery of the kidney. In any event, these studies demonstrate that the expression of distal tubular AGS8 has the potential to have some involvement in the regulation of renal function, but further studies are clearly warranted to elucidate these pathways to expand our knowledge of AGS8 as a regulator of normal and diseased states in the kidney.

21.4.3.4 AGS9

AGS9 was isolated from two different cDNA libraries generated from human heart and rats with repetitive cardiac ischemia [9], and demonstrated sequence homology to a previously identified protein known as the 26S proteosome regulatory subunit Rpn10 or S5a. Rpn10 plays a role in the ubiquitin-26S proteosome system, which is an ATP-dependent protein degradation system in eukaryotic cells [188]. Rpn10/S5a is a 268 amino acid protein that has a N-terminal binding site for von

Willebrand factor A and 2 ubiquitin interacting motifs (UIM) at the C-terminus (Figure 21.2). Rpn10 functions as an ubiquitin receptor that binds to ubiquitinated substrates, including Gγ subunits [189–191], that are recruited to the large polyprotein 26S proteosome for catabolic removal.

The kidney is one of the organs that showed the highest level of Rpn10 expression [192], which can be detected as five variant forms, Rpn10a–Rpn10e [193]. A complete loss of Rpn10 expression in mice was embryonic lethal [193], but selective expression of a specific isoform, Rpn10a, allowed the mice to survive. This demonstrated that Rpn10 is essential component of the proteosome, but whether the interaction with Gβγ is direct or indirect through ubiquitylation needs further investigation.

21.4.3.5 AGS10/Gαo

AGS10 is also known as the guanine nucleotide-binding Gαo subunit, which is a natural binding partner for Gβγ, and is encoded by the *GNAO* gene. This particular Gα subunit was originally isolated as a 39 kDa protein from the brain, liver, and heart of rats and cows [45, 46]. Activated forms of Gαo has been shown to function similarly as its closely related Gαi subunits using kidney derived cell lines [194], but there is little evidence for the expression of GNAO in the kidneys from zebrafish [195] and rats [19, 196]. Considering that the GNAO knockout mice exhibit a lethal phenotype, future studies would require organ-specific deletion or expression of GNAO to determine whether there is any function role for this particular Gα subunit to regulate renal function.

21.4.4 Group IV AGS Proteins

The last group of AGS proteins was isolated from the functional yeast screen using cDNA libraries generated from mice that underwent cardiac hypertrophy by either a surgical transverse aortic constriction or chronic infusion of isoproterenol. In this group, the three AGS proteins were identified as members of the microphthalmia-associated transcription (MiT) subfamily of transcription factors, which included transcription factor E3 (TFE3 or AGS11), transcription factor EB (TFEB or AGS12), and microphthalmia transcription factor (MiTF or AGS13) [10]. Sato et al. [10] demonstrated that AGS11 (TFE3) could bind selectively to Gα16 within the Gαq family. Little to no interaction was observed with either Gαs or Gαi3, respectively, by any of the Group IV AGS proteins. This was confirmed by the lack of interaction using co-immunoprecipitation assays and the lack of bioactivity for G-protein signaling using Gαs or Gαi3. In addition, the interaction of TFE3 with Gα16 was likely dependent upon a free inactive GDP-bound form as determined by the lack of TFE3 binding to GTP-bound variants of Gα16 [10]. Bioactivity for G-protein signaling as measured by β-galactosidase activity was strongest for AGS11 (TFE3), with moderate to no effect by AGS13 (MiTF) and AGS12 (TFEB), respectively. C-terminal deletion of the final 27 amino acids in AGS11, which was 67% identical to both AGS12 and AGS13, did not prevent either binding to Gα16 or the preferential colocalization to the nucleus in the cardiomyocytes. Instead, the loss of the C-terminal end eliminated the induced ability for yeast to grow and blunted the expression of claudin-14, a cell junction protein, in mammalian cells.

Although this finding was not performed in renal cells, there may be a similar role for AGS11/TFE3 due to the widespread pattern of expression for claudins, including claudin-14, in the kidney to regulate paracellular fluid and ion transport [197]. Consistent with this logic, TFE3 has been shown to bind to a specific sequence in the inorganic phosphate response element (5′-CACGTG-3′) to activate the expression of the NPT2a co-transporter during periods of dietary phosphate deprivation [198, 199]. MiTF (AGS13) has also been shown to induce gene expression for hypoxia inducible factor 1a [200], which is abundantly expressed in many cell types in the kidney [201]. In these latter two studies, the potential interactive role with Gα subunits was not investigated. However, the role of AGS11–AGS13 could be limited in the normal kidney due to the relatively low expression levels for both the MiTF/TFE family of transcription factors [202, 203] and Gα15/α16 subunits [204]. Moreover, the localization of MiTF/TFE is primarily in the cytoplasm or weakly in the nucleus [205].

During diseased states, such as translocation renal cell carcinoma, there may be a greater possibility for these Group IV AGS proteins in the kidney as there has been evidence that there is predominant nuclear localization of the MiTF/TFE transcription factors, particularly AGS11/TFE3 and to a lesser degree, AGS12/TFEB, from the cytoplasm [205, 206]. This cellular transformation is attributed to aberrant chromosomal translocation or inversion of genomic DNA containing either TFE3 [207–209] or TFEB [203] into other genes [210–212], but further investigations are needed to determine whether part of the mechanism driving the localization to the nucleus involves G-protein interaction with the Group IV AGS proteins. AGS13/MiTF has also been discussed as an oncogene in renal cell carcinoma, but its biological function is more prevalent in the cancer of melanocytes [213, 214]. There is evidence that MiTF/TFE transcription factors have an inter-relationship between each other through redundant oncogenic function whereby TFE3 knockdown decreased viability of TFE3-translocated papillary renal carcinoma cells, but could be rescued by expression of MiTF [215]. Further studies are needed to confirm the G-protein component in the pathogenesis of cancer.

In addition to renal oncogenesis, there is increasing evidence that TFE3 plays a role in the pathogenesis of renal cystic disease. Increased chimeric TFE3 protein led to dedifferentiation of renal proximal tubule cells as determined by a loss of cilia formation and key functional transporter proteins [210]. Other studies have shown that folliculin, a crucial negative regulator of TFE3 nuclear translocation, promoted renal cell carcinoma and produced the phenotypic appearance of PKD [205, 216]. In folliculin-deficient mice, renal tubular epithelial cells transformed into a cystic phenotype that increased nuclear localization of TFE3 compared with the noncystic kidneys [216]. Moreover, folliculin has recently been shown to interact with multiple Rab proteins to facilitate the loading of active GTP-bound Rab proteins onto their effector target [217, 218], or mediate degradation of membrane receptors either as a GTPase activating protein (GAP) [219] or a GEF [220]. It remains to be determined whether GTP–GDP dynamics on Gα16 through the actions of Group IV AGS proteins with or without other accessory proteins, such as folliculin, play any role in these and other processes in the kidney.

21.5 Summary and Perspective

Manipulation of cell surface GPCRs using molecular and pharmacologic interventions remains one of the hallmark methods to understand heterotrimeric G-protein signaling in the kidney. In the past decade, however, there appears to be a greater appreciation by which G-proteins are regulated by atypical intracellular accessory proteins depending upon the specific renal cell types, including the glomerulus, vasculature, and tubular epithelial cells [5]. Not only can these accessory proteins modulate the magnitude and duration of G-protein activity but also their spatial location within and even away from the plasma membrane, such as the mitotic spindle poles during cell division or within the nucleus. Our current knowledge by which AGS proteins are involved in the normal and pathological states of the kidney are only beginning to be discovered, due in large part that these proteins were identified using cDNA libraries derived from organs outside of the kidney. Even with this limitation, AGS proteins have been shown to have a wide range of activity from fine tuning signaling output to playing a more central role in many facets of renal function, such as the regulation of fluid and solute transport, DNA repair mechanisms, cell survival pathways, and cystic disease progression. Further progress in this area by screening new kidney disease-specific cDNA libraries may either confirm the importance of existing AGS proteins that were not previously studied in the kidney or possibly identify previously unidentified proteins to open new doors of G-protein regulation. At this point, our understanding of G-protein regulation by unconventional modes of action is growing by leaps and bounds in the kidney, and further discoveries in this field may bear fruit by providing new information about the basic physiology of the kidney and potentially identify new therapeutic targets to treat renal diseases through our improved knowledge about G-protein regulation.

References

1 Cazorla-Vazquez, S. and Engel, F.B. (2018). Adhesion GPCRs in kidney development and disease. *Front. Cell Dev. Biol.* 6: 9.

2 Hama, T. and Park, F. (2016). Heterotrimeric G protein signaling in polycystic kidney disease. *Physiol. Genomics* 48 (7): 429–445.

3 Weiss, R.H. (1998). G protein-coupled receptor signalling in the kidney. *Cell Signal.* 10 (5): 313–320.

4 Syrovatkina, V., Alegre, K.O., Dey, R., and Huang, X.Y. (2016). Regulation, signaling, and physiological functions of G-proteins. *J. Mol. Biol.* 428 (19): 3850–3868.

5 Park, F. (2015). Accessory proteins for heterotrimeric G-proteins in the kidney. *Front. Physiol.* 6: 219.

6 Sjogren, B. (2017). The evolution of regulators of G protein signalling proteins as drug targets – 20 years in the making: IUPHAR review 21. *Br. J. Pharmacol.* 174 (6): 427–437.

7 Alexander, A. and Walker, C.L. (2011). The role of LKB1 and AMPK in cellular responses to stress and damage. *FEBS Lett.* 585 (7): 952–957.

8 Cismowski, M.J., Takesono, A., Ma, C. et al. (1999). Genetic screens in yeast to identify mammalian nonreceptor modulators of G-protein signaling. *Nat. Biotechnol.* 17 (9): 878–883.

9 Sato, M., Cismowski, M.J., Toyota, E. et al. (2006). Identification of a receptor-independent activator of G protein signaling (AGS8) in ischemic heart and its interaction with Gβγ. *Proc. Natl. Acad. Sci. U.S.A.* 103 (3): 797–802.

10 Sato, M., Hiraoka, M., Suzuki, H. et al. (2011). Identification of transcription factor E3 (TFE3) as a receptor-independent activator of Gα16: gene regulation by nuclear Gα subunit and its activator. *J. Biol. Chem.* 286 (20): 17766–17776.

11 Takesono, A., Cismowski, M.J., Ribas, C. et al. (1999). Receptor-independent activators of heterotrimeric G-protein signaling pathways. *J. Biol. Chem.* 274 (47): 33202–33205.

12 Cismowski, M.J., Takesono, A., Bernard, M.L. et al. (2001). Receptor-independent activators of heterotrimeric G-proteins. *Life Sci.* 68 (19, 20): 2301–2308.

13 Cismowski, M.J., Ma, C., Ribas, C. et al. (2000). Activation of heterotrimeric G-protein signaling by a ras-related protein. Implications for signal integration. *J. Biol. Chem.* 275 (31): 23421–23424.

14 Wettschureck, N. and Offermanns, S. (2005). Mammalian G proteins and their cell type specific functions. *Physiol. Rev.* 85 (4): 1159–1204.

15 Park, F. (2015). Activators of G protein signaling in the kidney. *J. Pharmacol. Exp. Therap.* 353 (2): 235–245.

16 Brunskill, N., Bastani, B., Hayes, C. et al. (1991). Localization and polar distribution of several G-protein subunits along nephron segments. *Kidney Int.* 40 (6): 997–1006.

17 Hansen, P.B., Castrop, H., Briggs, J., and Schnermann, J. (2003). Adenosine induces vasoconstriction through Gi-dependent activation of phospholipase C in isolated perfused afferent arterioles of mice. *J. Am. Soc. Nephrol.* 14 (10): 2457–2465.

18 Ruan, X., Chatziantoniou, C., and Arendshorst, W.J. (1999). Impaired prostaglandin E_2/prostaglandin I_2 receptor-G_s protein interactions in isolated renal resistance arterioles of spontaneously hypertensive rats. *Hypertension* 34 (5): 1134–1140.

19 Senkfor, S.I., Johnson, G.L., and Berl, T. (1993). A molecular map of G protein α chains in microdissected rat nephron segments. *J. Clin. Invest.* 92 (2): 786–790.

20 Yanagisawa, H., Kurihara, N., Klahr, S. et al. (1994). Regional characterization of G-protein subunits in glomeruli, cortices and medullas of the rat kidney. *Nephron* 66 (4): 447–452.

21 Yanagisawa, H., Morrissey, J., and Klahr, S. (1993). Bilateral ureteral obstruction alters levels of the G-protein subunits $G_{\alpha s}$ and $G_{\alpha q/11}$. *Kidney Int.* 43 (4): 865–871.

22 Boucher, I., Yu, W., and Beaudry, S. (2012). Gα12 activation in podocytes leads to cumulative changes in glomerular collagen expression, proteinuria and glomerulosclerosis. *Lab Invest.* 92 (5): 662–675.

23 Zheng, S., Yu, P., and Zeng, C. (2003). Gα12- and Gα13-protein subunit linkage of D5 dopamine receptors in the nephron. *Hypertension* 41 (3): 604–610.

24 Stow, J.L., Sabolic, I., and Brown, D. (1991). Heterogeneous localization of G protein α-subunits in rat kidney. *Am. J. Physiol.* 261 (5 Pt 2): F831–F840.

25 Williamson, C.M., Turner, M.D., Ball, S.T. et al. (1996). Glomerular-specific imprinting of the mouse Gsα gene: how does this relate to hormone resistance in albright hereditary osteodystrophy? *Genomics* 36 (2): 280–287.

26 Yu, S., Yu, D., Lee, E. et al. (1998). Variable and tissue-specific hormone resistance in heterotrimeric Gs protein α-subunit (Gsα) knockout mice is due to tissue-specific imprinting of the gsα gene. *Proc. Natl. Acad. Sci. U.S.A.* 95 (15): 8715–8720.

27 Derrickson, B.H. and Mandel, L.J. (1997). Parathyroid hormone inhibits Na(+)-K(+)-ATPase through Gq/G11 and the calcium-independent phospholipase A2. *Am. J. Physiol.* 272 (6 Pt 2): F781–F788.

28 He, Q., Zhu, Y., Corbin, B.A. et al. (2015). The G protein α subunit variant $XL\alpha_s$ promotes inositol 1,4,5-trisphosphate signaling and mediates the renal actions of parathyroid hormone in vivo. *Sci. Signal.* 8 (391): ra84.

29 Fong, H.K., Amatruda, T.T. 3rd, Birren, B.W., and Simon, M.I. (1987). Distinct forms of the β subunit of GTP-binding regulatory proteins identified by molecular cloning. *Proc. Natl. Acad. Sci. U.S.A.* 84 (11): 3792–3796.

30 Snow, B.E., Betts, L., Mangion, J. et al. (1999). Fidelity of G protein β-subunit association by the G protein γ-subunit-like domains of RGS6, RGS7, and RGS11. *Proc. Natl. Acad. Sci. U.S.A.* 96 (11): 6489–6494.

31 Tummala, H., Fleming, S., Hocking, P.M. et al. (2011). The D153del mutation in GNB3 gene causes tissue specific signalling patterns and an abnormal renal morphology in Rge chickens. *PLoS One* 6 (8): e21156.

32 von Weizsacker, E., Strathmann, M.P., and Simon, M.I. (1992). Diversity among the β subunits of heterotrimeric GTP-binding proteins: characterization of a novel β-subunit cDNA. *Biochem. Biophys. Res. Commun.* 183 (1): 350–356.

33 Cali, J.J., Balcueva, E.A., Rybalkin, I., and Robishaw, J.D. (1992). Selective tissue distribution of G protein γ subunits, including a new form of the γ subunits identified by cDNA cloning. *J. Biol. Chem.* 267 (33): 24023–24027.

34 Ray, K., Kunsch, C., Bonner, L.M., and Robishaw, J.D. (1995). Isolation of cDNA clones encoding eight different human G protein γ subunits, including three novel forms designated the γ_4, γ_{10}, and γ_{11} subunits. *J. Biol. Chem.* 270 (37): 21765–21771.

35 Cao, X., Cismowski, M.J., Sato, M. et al. (2004). Identification and characterization of AGS4: a protein containing three G-protein regulatory motifs that regulate the activation state of Giα. *J. Biol. Chem.* 279 (26): 27567–27574.

36 Blumer, J.B., Smrcka, A.V., and Lanier, S.M. (2007). Mechanistic pathways and biological roles for receptor-independent activators of G-protein signaling. *Pharmacol. Therap.* 113 (3): 488–506.

37 Sato, M., Blumer, J.B., Simon, V., and Lanier, S.M. (2006). Accessory proteins for G proteins: partners in signaling. *Annu. Rev. Pharmacol. Toxicol.* 46: 151–187.

38 Lee, J.W., Choi, H.S., Gyuris, J. et al. (1995). Two classes of proteins dependent on either the presence or absence of thyroid hormone for interaction with the thyroid hormone receptor. *Mol. Endocrinol.* 9 (2): 243–254.

39 Sachdev, P., Menon, S., and Kastner, D.B. (2007). G protein βγ subunit interaction with the dynein light-chain component Tctex-1 regulates neurite outgrowth. *EMBO J.* 26 (11): 2621–2632.

40 DiBella, L.M., Benashski, S.E., Tedford, H.W. et al. (2001). The Tctex1/Tctex2 class of dynein light chains. Dimerization, differential expression, and interaction with the LC8 protein family. *J. Biol. Chem.* 276 (17): 14366–14373.

41 Li, X.C. and Schimenti, J.C. (2007). Mouse pachytene checkpoint 2 (trip13) is required for completing meiotic recombination but not synapsis. *PLoS Genet.* 3 (8): e130.

42 Roig, I., Dowdle, J.A., and Toth, A. (2010). Mouse TRIP13/PCH2 is required for recombination and normal higher-order chromosome structure during meiosis. *PLoS Genet.* 6 (8).

43 Nielsen, S., DiGiovanni, S.R., Christensen, E.I. et al. (1993). Cellular and subcellular immunolocalization of vasopressin-regulated water channel in rat kidney. *Proc. Natl. Acad. Sci. U.S.A.* 90 (24): 11663–11667.

44 Deveraux, Q., van Nocker, S., Mahaffey, D. et al. (1995). Inhibition of ubiquitin-mediated proteolysis by the Arabidopsis 26 S protease subunit S5a. *J. Biol. Chem.* 270 (50): 29660–29663.

45 Neer, E.J., Lok, J.M., and Wolf, L.G. (1984). Purification and properties of the inhibitory guanine nucleotide regulatory unit of brain adenylate cyclase. *J. Biol. Chem.* 259 (22): 14222–14229.

46 Sternweis, P.C. and Robishaw, J.D. (1984). Isolation of two proteins with high affinity for guanine nucleotides from membranes of bovine brain. *J. Biol. Chem.* 259 (22): 13806–13813.

47 Willard, F.S., Kimple, R.J., and Siderovski, D.P. (2004). Return of the GDI: the GoLoco motif in cell division. *Annu. Rev. Biochem.* 73: 925–951.

48 Lenarczyk, M., Pressly, J.D., Arnett, J. et al. (2015). Localization and expression profile of group I and II activators of G-protein signaling in the kidney. *J. Mol. Histol.* 46 (2): 123–136.

49 Nadella, R., Blumer, J.B., Jia, G. et al. (2010). Activator of G protein signaling 3 promotes epithelial cell proliferation in PKD. *J. Am. Soc. Nephrol.* 21 (8): 1275–1280.

50 Regner, K.R., Nozu, K., Lanier, S.M. et al. (2011). Loss of activator of G-protein signaling 3 impairs renal tubular regeneration following acute kidney injury in rodents. *FASEB J.* 25 (6): 1844–1855.

51 Chan, P., Gabay, M., Wright, F.A., and Tall, G.G. (2011). Ric-8B is a GTP-dependent G protein αs guanine nucleotide exchange factor. *J. Biol. Chem.* 286 (22): 19932–19942.

52 Tall, G.G., Krumins, A.M., and Gilman, A.G. (2003). Mammalian Ric-8A (synembryn) is a heterotrimeric Gα protein guanine nucleotide exchange factor. *J. Biol. Chem.* 278 (10): 8356–8362.

53 Thomas, C.J., Briknarová, K., and Hilmer, J.K. (2011). The nucleotide exchange factor Ric-8A is a chaperone for the conformationally dynamic nucleotide-free state of Gαi1. *PLoS One* 6 (8): e23197.

54 Cismowski, M.J. (2006). Non-receptor activators of heterotrimeric G-protein signaling (AGS proteins). *Semin. Cell Dev. Biol.* 17 (3): 334–344.

55 Kemppainen, R.J. and Behrend, E.N. (1998). Dexamethasone rapidly induces a novel ras superfamily member-related gene in AtT-20 cells. *J. Biol. Chem.* 273 (6): 3129–3131.

56 Brogan, M.D., Behrend, E.N., and Kemppainen, R.J. (2001). Regulation of Dexras1 expression by endogenous steroids. *Neuroendocrinology* 74 (4): 244–250.

57 Tu, Y. and Wu, C. (1999). Cloning, expression and characterization of a novel human Ras-related protein that is regulated by glucocorticoid hormone. *Biochim. Biophys. Acta* 1489 (2, 3): 452–456.

58 Vaidyanathan, G., Cismowski, M.J., Wang, G. et al. (2004). The Ras-related protein AGS1/RASD1 suppresses cell growth. *Oncogene* 23 (34): 5858–5863.

59 Gupta, V., Bhandari, D., Leyme, A. et al. (2016). GIV/Girdin activates Gαi and inhibits Gαs via the same motif. *Proc. Natl. Acad. Sci. U.S.A.* 113 (39): E5721–E5730.

60 Blumer, J.B. and Lanier, S.M. (2014). Activators of G protein signaling exhibit broad functionality and define a distinct core signaling triad. *Mol. Pharmacol.* 85 (3): 388–396.

61 Rubinfeld, B., Munemitsu, S., Clark, R. et al. (1991). Molecular cloning of a GTPase activating protein specific for the Krev-1 protein $p21^{rap1}$. *Cell* 65 (6): 1033–1042.

62 Gao, M., Craig, D., Lequin, O. et al. (2003). Structure and functional significance of mechanically unfolded fibronectin type III_1 intermediates. *Proc. Natl. Acad. Sci. U.S.A.* 100 (25): 14784–14789.

63 Graham, T.E., Prossnitz, E.R., and Dorin, R.I. (2002). Dexras1/AGS-1 inhibits signal transduction from the Gi-coupled formyl peptide receptor to Erk-1/2 MAP kinases. *J. Biol. Chem.* 277 (13): 10876–10882.

64 Nguyen, C.H. and Watts, V.J. (2005). Dexras1 blocks receptor-mediated heterologous sensitization of adenylyl cyclase 1. *Biochem. Biophys. Res. Commun.* 332 (3): 913–920.

65 Takesono, A., Nowak, M.W., Cismowski, M. et al. (2002). Activator of G-protein signaling 1 blocks GIRK channel activation by a G-protein-coupled receptor: apparent disruption of receptor signaling complexes. *J. Biol. Chem.* 277 (16): 13827–13830.

66 Bertorello, A. and Aperia, A. (1990). Short-term regulation of Na^+,K^+-ATPase activity by dopamine. *Am. J. Hypertens.* 3 (6 Pt 2): 51S–54S.

67 Tan, J.J., Ong, S.A., and Chen, K.S. (2011). Rasd1 interacts with Ear2 (Nr2f6) to regulate renin transcription. *BMC Mol. Biol.* 12: 4.

68 Fang, M., Jaffrey, S.R., Sawa, A. et al. (2000). Dexras1: a G protein specifically coupled to neuronal nitric oxide synthase via CAPON. *Neuron* 28 (1): 183–193.

69 Kemppainen, R.J., Cox, E., Behrend, E.N. et al. (2003). Identification of a glucocorticoid response element in the 3′-flanking region of the human Dexras1 gene. *Biochim. Biophys. Acta* 1627 (2, 3): 85–89.

70 Rusai, K., Prokai, A., Juanxing, C. et al. (2013). Dexamethasone protects from renal ischemia/reperfusion injury: a possible association with SGK-1. *Acta Physiol. Hung.* 100 (2): 173–185.

71 Wilson, P.D., Du, J., and Norman, J.T. (1993). Autocrine, endocrine and paracrine regulation of growth abnormalities in autosomal dominant polycystic kidney disease. *Eur. J. Cell Biol.* 61 (1): 131–138.

72 McDonald, A.T., Crocker, J.F., Digout, S.C. et al. (1990). Glucocorticoid-induced polycystic kidney disease – a threshold trait. *Kidney Int.* 37 (3): 901–908.

73 Ghosh, P., Beas, A.O., Bornheimer, S.J. et al. (2010). A Gαi-GIV molecular complex binds epidermal growth factor receptor and determines whether cells migrate or proliferate. *Mol. Biol. Cell* 21 (13): 2338–2354.

74 Ghosh, P., Garcia-Marcos, M., Bornheimer, S.J., and Farquhar, M.G. (2008). Activation of Gαi3 triggers cell migration via regulation of GIV. *J. Cell Biol.* 182 (2): 381–393.

75 Garcia-Marcos, M., Jung, B.H., Ear, J. et al. (2011). Expression of GIV/Girdin, a metastasis-related protein, predicts patient survival in colon cancer. *FASEB J.* 25 (2): 590–599.

76 Garcia-Marcos, M., Ear, J., Farquhar, M.G., and Ghosh, P. (2011). A GDI (AGS3) and a GEF (GIV) regulate autophagy by balancing G protein activity and growth factor signals. *Mol. Biol. Cell* 22 (5): 673–686.

77 Wang, H., Misaki, T., Taupin, V. et al. (2015). GIV/girdin links vascular endothelial growth factor signaling to Akt survival signaling in podocytes independent of nephrin. *J. Am. Soc. Nephrol.* 26 (2): 314–327.

78 Anai, M., Shojima, N., Katagiri, H. et al. (2005). A novel protein kinase B (PKB)/AKT-binding protein enhances PKB kinase activity and regulates DNA synthesis. *J. Biol. Chem.* 280 (18): 18525–18535.

79 Zeng, B., Mou, T.C., Doukov, T.I. et al. (2019). Structure, function, and dynamics of the Gα binding domain of Ric-8A. *Structure* 27 (7): 1137–1147.e5.

80 Papasergi-Scott MM, Stoveken HM, MacConnachie L, *et al.* (2018) Dual phosphorylation of Ric-8A enhances its ability to mediate G protein α subunit folding and to stimulate guanine nucleotide exchange. *Sci. Signal.* 11(532).

81 Gabay, M., Pinter, M.E., Wright, F.A. et al. (2011). Ric-8 proteins are molecular chaperones that direct nascent G protein α subunit membrane association. *Sci. Signal.* 4 (200): ra79.

82 Chishiki, K., Kamakura, S., Hayase, J., and Sumimoto, H. (2017). Ric-8A, an activator protein of Gαi, controls mammalian epithelial cell polarity for tight junction assembly and cystogenesis. *Genes Cells* 22 (3): 293–309.

83 Miller, K.G., Emerson, M.D., McManus, J.R., and Rand, J.B. (2000). RIC-8 (Synembryn): a novel conserved protein that is required for G(q)α signaling in the *C. elegans* nervous system. *Neuron* 27 (2): 289–299.

84 Tonissoo, T., Kõks, S., Meier, R. et al. (2006). Heterozygous mice with Ric-8 mutation exhibit impaired spatial memory and decreased anxiety. *Behav. Brain Res.* 167 (1): 42–48.

85 Tonissoo, T., Meier, R., Talts, K. et al. (2003). Expression of ric-8 (synembryn) gene in the nervous system of developing and adult mouse. *Gene Expr. Patterns* 3 (5): 591–594.

86 Maldonado-Agurto, R., Toro, G., Fuentealba, J. et al. (2011). Cloning and spatiotemporal expression of RIC-8 in *Xenopus* embryogenesis. *Gene Expr. Patterns* 11 (7): 401–408.

87 Falk, J.D., Vargiu, P., Foye, P.E. et al. (1999). Rhes: a striatal-specific Ras homolog related to Dexras1. *J. Neurosci. Res.* 57 (6): 782–788.

88 Harrison, L.M. and He, Y. (2011). Rhes and AGS1/Dexras1 affect signaling by dopamine D1 receptors through adenylyl cyclase. *J. Neurosci. Res.* 89 (6): 874–882.

89 Thapliyal, A., Bannister, R.A., Hanks, C., and Adams, B.A. (2008). The monomeric G proteins AGS1 and Rhes selectively influence Gαi-dependent signaling to modulate N-type ($Ca_{V2.2}$) calcium channels. *Am. J. Physiol. Cell Physiol.* 295 (5): C1417–C1426.

90 Vargiu, P., De Abajo, R., Garcia-Ranea, J.A. et al. (2004). The small GTP-binding protein, Rhes, regulates signal transduction from G protein-coupled receptors. *Oncogene* 23 (2): 559–568.

91 Hill, C., Goddard, A., Ladds, G., and Davey, J. (2009). The cationic region of Rhes mediates its interactions with specific Gβ subunits. *Cell. Physiol. Biochem.* 23 (1–3): 1–8.

92 Jockers, R., Linder, M.E., Hohenegger, M. et al. (1994). Species difference in the G protein selectivity of the human and bovine A1-adenosine receptor. *J. Biol. Chem.* 269 (51): 32077–32084.

93 Viau, A., Bienaimé, F., Lukas, K. et al. (2018). Cilia-localized LKB1 regulates chemokine signaling, macrophage recruitment, and tissue homeostasis in the kidney. *EMBO J.* 37 (15): e98615.

94 Fung, B.K. (1983). Characterization of transducin from bovine retinal rod outer segments. I. Separation and reconstitution of the subunits. *J. Biol. Chem.* 258 (17): 10495–10502.

95 Fung, B.K. and Nash, C.R. (1983). Characterization of transducin from bovine retinal rod outer segments. II. Evidence for distinct binding sites and conformational changes revealed by limited proteolysis with trypsin. *J. Biol. Chem.* 258 (17): 10503–10510.

96 Spano, D., Branchi, I., Rosica, A. et al. (2004). Rhes is involved in striatal function. *Mol. Cell. Biol.* 24 (13): 5788–5796.

97 Pradervand, S., Zuber Mercier, A., Centeno, G. et al. (2010). A comprehensive analysis of gene expression profiles in distal parts of the mouse renal tubule. *Pflugers Arch.* 460 (6): 925–952.

98 Bernard, M.L., Peterson, Y.K., Chung, P. et al. (2001). Selective interaction of AGS3 with G-proteins and the influence of AGS3 on the activation state of G-proteins. *J. Biol. Chem.* 276 (2): 1585–1593.

99 Peterson, Y.K., Bernard, M.L., Ma, H. et al. (2000). Stabilization of the GDP-bound conformation of Giα by a peptide derived from the G-protein regulatory motif of AGS3. *J. Biol. Chem.* 275 (43): 33193–33196.

100 Siderovski, D.P., Diverse-Pierluissi, M., and De Vries, L. (1999). The GoLoco motif: a Gαi/o binding motif and potential guanine-nucleotide exchange factor. *Trends Biochem. Sci* 24 (9): 340–341.

101 Blumer, J.B., Chandler, L.J., and Lanier, S.M. (2002). Expression analysis and subcellular distribution of the two G-protein regulators AGS3 and LGN indicate distinct functionality. Localization of LGN to the midbody during cytokinesis. *J. Biol. Chem.* 277 (18): 15897–15903.

102 Pizzinat, N., Takesono, A., and Lanier, S.M. (2001). Identification of a truncated form of the G-protein regulator AGS3 in heart that lacks the tetratricopeptide repeat domains. *J. Biol. Chem.* 276 (20): 16601–16610.

103 Robichaux, W.G. 3rd, Branham-O'Connor, M., Hwang, I.Y. et al. (2017). Regulation of chemokine signal integration by activator of G-protein signaling 4 (AGS4). *J. Pharmacol. Exp. Therap.* 360 (3): 424–433.

104 Giguere, P.M., Laroche, G., Oestreich, E.A. et al. (2012). Regulation of the subcellular localization of the G-protein subunit regulator GPSM3 through direct association with 14-3-3 protein. *J. Biol. Chem.* 287 (37): 31270–31279.

105 Natochin, M., Lester, B., Peterson, Y.K. et al. (2000). AGS3 inhibits GDP dissociation from Gα subunits of the Gi family and rhodopsin-dependent activation of transducin. *J. Biol. Chem.* 275 (52): 40981–40985.

106 Keri, K.C., Regner, K.R., Dall, A.T., and Park, F. (2018). Urinary exosomal expression of activator of G protein signaling 3 in polycystic kidney disease. *BMC Res. Notes* 11 (1): 359.

107 Kwon, M., Pavlov, T.S., Nozu, K. et al. (2012). G-protein signaling modulator 1 deficiency accelerates cystic disease in an orthologous mouse model of autosomal dominant polycystic kidney disease. *Proc. Natl. Acad. Sci. U.S.A.* 109 (52): 21462–21467.

108 De Vries, L., Fischer, T., Tronchère, H. et al. (2000). Activator of G protein signaling 3 is a guanine dissociation inhibitor for $G\alpha_i$ subunits. *Proc. Natl. Acad. Sci. U.S.A.* 97 (26): 14364–14369.

109 Basile, D.P., Anderson, M.D., and Sutton, T.A. (2012). Pathophysiology of acute kidney injury. *Compr. Physiol.* 2 (2): 1303–1353.

110 Bonventre, J.V. and Yang, L. (2011). Cellular pathophysiology of ischemic acute kidney injury. *J. Clin. Invest.* 121 (11): 4210–4221.

111 Zuk, A. and Bonventre, J.V. (2016). Acute kidney injury. *Annu. Rev. Med.* 67: 293–307.

112 Weimbs, T. (2006). Regulation of mTOR by polycystin-1: is polycystic kidney disease a case of futile repair? *Cell Cycle* 5 (21): 2425–2429.

113 Weimbs, T. (2007). Polycystic kidney disease and renal injury repair: common pathways, fluid flow, and the function of polycystin-1. *Am. J. Physiol. Renal Physiol.* 293 (5): F1423–F1432.

114 Rasmussen, S.A., Kwon, M., Pessly, J.D. et al. (2015). Activator of G-protein signaling 3 controls renal epithelial cell survival and ERK5 activation. *J. Mol. Signal.* 10: 6.

115 Fan, P., Jiang, Z., Diamond, I., and Yao, L. (2009). Up-regulation of AGS3 during morphine withdrawal promotes cAMP superactivation via adenylyl cyclase 5 and 7 in rat nucleus accumbens/striatal neurons. *Mol. Pharmacol.* 76 (3): 526–533.

116 Sanada, K. and Tsai, L.H. (2005). G protein βγ subunits and AGS3 control spindle orientation and asymmetric cell fate of cerebral cortical progenitors. *Cell* 122 (1): 119–131.

117 Groves, B., Abrahamsen, H., Clingan, H. et al. (2010). An inhibitory role of the G-protein regulator AGS3 in mTOR-dependent macroautophagy. *PLoS One* 5 (1): e8877.

118 Pattingre, S., De Vries, L., Bauvy, C. et al. (2003). The G-protein regulator AGS3 controls an early event during macroautophagy in human intestinal HT-29 cells. *J. Biol. Chem.* 278 (23): 20995–21002.

119 Pattingre, S., Petiot, A., and Codogno, P. (2004). Analyses of Gα-interacting protein and activator of G-protein-signaling-3 functions in macroautophagy. *Methods Enzymol.* 390: 17–31.

120 Vural, A., TJ, M.Q., Blumer, J.B. et al. (2013). Normal autophagic activity in macrophages from mice lacking $G\alpha_{i3}$, AGS3, or RGS19. *PLoS One* 8 (11): e81886.

121 White, S.M., North, L.N., Haines, E. et al. (2014). G-protein βγ subunit dimers modulate kidney repair after ischemia-reperfusion injury in rats. *Mol. Pharmacol.* 86 (4): 369–377.

122 Fischer, E., Legue, E., Doyen, A. et al. (2006). Defective planar cell polarity in polycystic kidney disease. *Nat. Genet.* 38 (1): 21–23.

123 Nishio, S., Tian, X., Gallagher, A.R. et al. (2010). Loss of oriented cell division does not initiate cyst formation. *J. Am. Soc. Nephrol.* 21 (2): 295–302.

124 Patel, V., Li, L., Cobo-Stark, P. et al. (2008). Acute kidney injury and aberrant planar cell polarity induce cyst formation in mice lacking renal cilia. *Hum. Mol. Genet.* 17 (11): 1578–1590.

125 Saburi, S., Hester, I., Fischer, E. et al. (2008). Loss of Fat4 disrupts PCP signaling and oriented cell division and leads to cystic kidney disease. *Nat. Genet.* 40 (8): 1010–1015.

126 Blumer, J.B., Bernard, M.L., Peterson, Y.K. et al. (2003). Interaction of activator of G-protein signaling 3 (AGS3) with LKB1, a serine/threonine kinase involved in cell polarity and cell cycle progression: phosphorylation of the G-protein regulatory (GPR) motif as a regulatory mechanism for the interaction of GPR motifs with Giα. *J. Biol. Chem.* 278 (26): 23217–23220.

127 Flowers, E.M., Sudderth, J., Zacharias, L. et al. (2018). Lkb1 deficiency confers glutamine dependency in polycystic kidney disease. *Nat. Commun.* 9 (1): 814.

128 Yu, F., Morin, X., Cai, Y. et al. (2000). Analysis of partner of inscuteable, a novel player of *Drosophila* asymmetric divisions, reveals two distinct steps in inscuteable apical localization. *Cell* 100 (4): 399–409.

129 Konno, D., Shioi, G., Shitamukai, A. et al. (2008). Neuroepithelial progenitors undergo LGN-dependent planar divisions to maintain self-renewability during mammalian neurogenesis. *Nat. Cell Biol.* 10 (1): 93–101.

130 Culurgioni, S., Alfieri, A., Pendolino, V. et al. (2011). Inscuteable and NuMA proteins bind competitively to Leu-Gly-Asn repeat-enriched protein (LGN) during asymmetric cell divisions. *Proc. Natl. Acad. Sci. U.S.A.* 108 (52): 20998–21003.

131 Fuja, T.J., Schwartz, P.H., Darcy, D., and Bryant, P.J. (2004). Asymmetric localization of LGN but not AGS3, two homologs of *Drosophila* pins, in dividing human neural progenitor cells. *J. Neurosci. Res.* 75 (6): 782–793.

132 Lechler, T. and Fuchs, E. (2005). Asymmetric cell divisions promote stratification and differentiation of mammalian skin. *Nature* 437 (7056): 275–280.

133 Zhu, J., Wen, W., Zheng, Z. et al. (2011). LGN/mInsc and LGN/NuMA complex structures suggest distinct functions in asymmetric cell division for the Par3/mInsc/LGN and Gαi/LGN/NuMA pathways. *Mol. Cell* 43 (3): 418–431.

134 Xiao, Z., Wan, Q., Du, Q., and Zheng, Z. (2012). Gα/LGN-mediated asymmetric spindle positioning does not lead to unequal cleavage of the mother cell in 3-D cultured MDCK cells. *Biochem. Biophys. Res. Commun.* 420 (4): 888–894.

135 Zheng, Z., Zhu, H., Wan, Q. et al. (2010). LGN regulates mitotic spindle orientation during epithelial morphogenesis. *J. Cell Biol.* 189 (2): 275–288.

136 Chauhan, S., Jelen, F., Sharina, I., and Martin, E. (2012). The G-protein regulator LGN modulates the activity of the NO receptor soluble guanylate cyclase. *Biochem. J.* 446 (3): 445–453.

137 Raptis, V., Loutradis, C., and Sarafidis, P.A. (2018). Renal injury progression in autosomal dominant polycystic kidney disease: a look beyond the cysts. *Nephrol. Dialysis Transplant.* 33 (11): 1887–1895.

138 Hayashi, H., Mamun, A.A., Takeyama, M. et al. (2019). Activator of G-protein signaling 8 is involved in VEGF-induced choroidal neovascularization. *Sci. Rep.* 9 (1): 1560.

139 Kimple, R.J., De Vries, L., Tronchère, H. et al. (2001). RGS12 and RGS14 GoLoco motifs are $G\alpha_i$ interaction sites with guanine nucleotide dissociation inhibitor activity. *J. Biol. Chem.* 276 (31): 29275–29281.

140 Snow, B.E., Antonio, L., Suggs, S. et al. (1997). Molecular cloning and expression analysis of rat Rgs12 and Rgs14. *Biochem. Biophys. Res. Commun.* 233 (3): 770–777.

141 Yang, S. and Li, Y.P. (2007). RGS12 is essential for RANKL-evoked signaling for terminal differentiation of osteoclasts in vitro. *J. Bone Miner. Res.* 22 (1): 45–54.

142 Huang, J., Chen, L., Yao, Y. et al. (2016). Pivotal role of regulator of G-protein signaling 12 in pathological cardiac hypertrophy. *Hypertension* 67 (6): 1228–1236.

143 Schiff, M.L., Siderovski, D.P., Jordan, D.P. et al. (2000). Tyrosine-kinase-dependent recruitment of RGS12 to the N-type calcium channel. *Nature* 408 (6813): 723–727.

144 Anantharam, A. and Diverse-Pierluissi, M.A. (2002). Biochemical approaches to study interaction of calcium channels with RGS12 in primary neuronal cultures. *Methods Enzymol.* 345: 60–70.

145 Richman, R.W. and Diverse-Pierluissi, M.A. (2004). Mapping of RGS12-$Ca_{v2.2}$ channel interaction. *Methods Enzymol.* 390: 224–239.

146 Richman, R.W., Strock, J., Hains, M.D. et al. (2005). RGS12 interacts with the SNARE-binding region of the $Ca_{v2.2}$ calcium channel. *J. Biol. Chem.* 280 (2): 1521–1528.

147 Jeon, J.P., Thakur, D.P., Tian, J.B. et al. (2016). Regulator of G-protein signalling and GoLoco proteins suppress TRPC4 channel function via acting at $G\alpha_{i/o}$. *Biochem. J.* 473 (10): 1379–1390.

148 Traver, S., Bidot, C., Spassky, N. et al. (2000). RGS14 is a novel Rap effector that preferentially regulates the GTPase activity of Gαo. *Biochem. J.* 350 (Pt 1): 19–29.

149 Mittal, V. and Linder, M.E. (2004). The RGS14 GoLoco domain discriminates among $G\alpha_i$ isoforms. *J. Biol. Chem.* 279 (45): 46772–46778.

150 Reif, K. and Cyster, J.G. (2000). RGS molecule expression in murine B lymphocytes and ability to down-regulate chemotaxis to lymphoid chemokines. *J. Immunol.* 164 (9): 4720–4729.

151 Mahajan, A., Rodan, A.R., Le, T.H. et al. (2016). Trans-ethnic fine mapping highlights kidney-function genes linked to salt sensitivity. *Am. J. Hum. Genet.* 99 (3): 636–646.

152 Robinson-Cohen, C., Bartz, T.M., Lai, D. et al. (2018). Genetic variants associated with circulating fibroblast growth factor 23. *J. Am. Soc. Nephrol.* 29 (10): 2583–2592.

153 Robinson-Cohen, C., Lutsey, P.L., Kleber, M.E. et al. (2017). Genetic variants associated with circulating parathyroid hormone. *J. Am. Soc. Nephrol.* 28 (5): 1553–1565.

154 Kurachi, H., Wada, Y., Tsukamoto, N. et al. (1997). Human *SPA-1* gene product selectively expressed in lymphoid tissues is a specific GTPase-activating protein for Rap1 and Rap2. Segregate expression profiles from a *rap1GAP* gene product. *J. Biol. Chem.* 272 (44): 28081–28088.

155 Potla, U., Ni, J., Vadaparampil, J. et al. (2014). Podocyte-specific RAP1GAP expression contributes to focal segmental glomerulosclerosis-associated glomerular injury. *J. Clin. Invest.* 124 (4): 1757–1769.

156 Hayashi, H., Al Mamun, A., Sakima, M., and Sato, M. (2016). Activator of G-protein signaling 8 is involved in VEGF-mediated signal processing during angiogenesis. *J. Cell Sci.* 129 (6): 1210–1222.

157 Nordquist, D.T., Kozak, C.A., and Orr, H.T. (1988). cDNA cloning and characterization of three genes uniquely expressed in cerebellum by Purkinje neurons. *J. Neurosci.* 8 (12): 4780–4789.

158 Zhao, P., Nguyen, C.H., and Chidiac, P. (2010). The proline-rich N-terminal domain of G18 exhibits a novel G protein regulatory function. *J. Biol. Chem.* 285 (12): 9008–9017.

159 Saito, H., Tsumura, H., Otake, S. et al. (2005). L7/Pcp-2-specific expression of Cre recombinase using knock-in approach. *Biochem. Biophys. Res. Commun.* 331 (4): 1216–1221.

160 Suikkanen, S., Sääjärvi, K., Hirsimäki, J. et al. (2002). Role of recycling endosomes and lysosomes in dynein-dependent entry of canine parvovirus. *J. Virol.* 76 (9): 4401–4411.

161 Barr, F.A. and Egerer, J. (2005). Golgi positioning: are we looking at the right MAP? *J. Cell Biol.* 168 (7): 993–998.

162 Salina, D., Bodoor, K., Eckley, D.M. et al. (2002). Cytoplasmic dynein as a facilitator of nuclear envelope breakdown. *Cell* 108 (1): 97–107.

163 Li, A., Saito, M., Chuang, J.Z. et al. (2011). Ciliary transition zone activation of phosphorylated Tctex-1 controls ciliary resorption, S-phase entry and fate of neural progenitors. *Nat. Cell Biol.* 13 (4): 402–411.

164 Saito, M., Otsu, W., Hsu, K.S. et al. (2017). Tctex-1 controls ciliary resorption by regulating branched actin polymerization and endocytosis. *EMBO Rep.* 18 (8): 1460–1472.

165 Yeh, C., Li, A., Chuang, J.Z. et al. (2013). IGF-1 activates a cilium-localized noncanonical Gβγ signaling pathway that regulates cell-cycle progression. *Dev. Cell* 26 (4): 358–368.

166 King, S.M., Barbarese, E., Dillman, J.F. et al. (1998). Cytoplasmic dynein contains a family of differentially expressed light chains. *Biochemistry* 37 (43): 15033–15041.

167 Ismail, O.Z., Sriranganathan, S., Zhang, X. et al. (2018). Tctex-1, a novel interaction partner of Kidney Injury Molecule-1, is required for efferocytosis. *J. Cell. Physiol.* 233 (10): 6877–6895.

168 Banerjee, R., Russo, N., Liu, M. et al. (2014). TRIP13 promotes error-prone nonhomologous end joining and induces chemoresistance in head and neck cancer. *Nat. Commun.* 5: 4527.

169 Bolcun-Filas, E., Rinaldi, V.D., White, M.E., and Schimenti, J.C. (2014). Reversal of female infertility by Chk2 ablation reveals the oocyte DNA damage checkpoint pathway. *Science* 343 (6170): 533–536.

170 Eytan, E., Wang, K., Miniowitz-Shemtov, S. et al. (2014). Disassembly of mitotic checkpoint complexes by the joint action of the AAA-ATPase TRIP13 and p31(comet). *Proc. Natl. Acad. Sci. U.S.A.* 111 (33): 12019–12024.

171 Ma, H.T. and Poon, R.Y.C. (2016). TRIP13 regulates both the activation and inactivation of the spindle-assembly checkpoint. *Cell Rep.* 14 (5): 1086–1099.

172 Ma, H.T. and Poon, R.Y.C. (2018). TRIP13 functions in the establishment of the spindle assembly checkpoint by replenishing O-MAD2. *Cell Rep.* 22 (6): 1439–1450.

173 Miniowitz-Shemtov, S., Eytan, E., Kaisari, S. et al. (2015). Mode of interaction of TRIP13 AAA-ATPase with the Mad2-binding protein p31comet and with mitotic checkpoint complexes. *Proc. Natl. Acad. Sci. U.S.A.* 112 (37): 11536–11540.

174 Tipton, A.R., Wang, K., Oladimeji, P. et al. (2012). Identification of novel mitosis regulators through data mining with human centromere/kinetochore proteins as group queries. *BMC Cell Biol.* 13: 15.

175 Wang, K., Sturt-Gillespie, B., Hittle, J.C. et al. (2014). Thyroid hormone receptor interacting protein 13 (TRIP13) AAA-ATPase is a novel mitotic checkpoint-silencing protein. *J. Biol. Chem.* 289 (34): 23928–23937.

176 Ye, Q., Kim, D.H., Dereli, I. et al. (2017). The AAA+ ATPase TRIP13 remodels HORMA domains through N-terminal engagement and unfolding. *EMBO J.* 36 (16): 2419–2434.

177 Ye, Q., Rosenberg, S.C., Moeller, A. et al. (2015). TRIP13 is a protein-remodeling AAA+ ATPase that catalyzes MAD2 conformation switching. *Elife* 28 (4): e07367.

178 Cismowski, M.J. and Lanier, S.M. (2005). Activation of heterotrimeric G-proteins independent of a G-protein coupled receptor and the implications for signal processing. *Rev. Physiol. Biochem. Pharmacol.* 155: 57–80.

179 Pressly, J.D., Hama, T., Brien, S.O. et al. (2017). TRIP13-deficient tubular epithelial cells are susceptible to apoptosis following acute kidney injury. *Sci. Rep.* 7: 43196.

180 Liu, C.T., Garnaas, M.K., Tin, A. et al. (2011). Genetic association for renal traits among participants of African ancestry reveals new loci for renal function. *PLoS Genet.* 7 (9): e1002264.

181 Deng, A.Y., Chauvet, C., and Menard, A. (2016). Alterations in fibronectin type III domain containing 1 protein gene are associated with hypertension. *PLoS One* 11 (4): e0151399.

182 Sato, M., Hiraoka, M., Suzuki, H. et al. (2014). Protection of cardiomyocytes from the hypoxia-mediated injury by a peptide targeting the activator of G-protein signaling 8. *PLoS One* 9 (3): e91980.

183 Sato, M., Jiao, Q., Honda, T. et al. (2009). Activator of G protein signaling 8 (AGS8) is required for hypoxia-induced apoptosis of cardiomyocytes: role of Gβγ and connexin 43 (CX43). *J. Biol. Chem.* 284 (45): 31431–31440.

184 Guo, R., Liu, L., and Barajas, L. (1998). RT-PCR study of the distribution of connexin 43 mRNA in the glomerulus and renal tubular segments. *Am. J. Physiol.* 275 (2): R439–R447.

185 Barajas, L., Liu, L., and Tucker, M. (1994). Localization of connexin43 in rat kidney. *Kidney Int.* 46 (3): 621–626.

186 Lenarczyk, M., Regner, K.R., and Park, F. (2014). Localization and expression profile of activator of G-protein signaling proteins in the kidney (690.6). *FASEB J.* 28 (Suppl 1): 690–696.

187 Dimke, H., Sparks, M.A., Thomson, B.R. et al. (2015). Tubulovascular cross-talk by vascular endothelial growth factor a maintains peritubular microvasculature in kidney. *J. Am. Soc. Nephrol.* 26 (5): 1027–1038.

188 Ciechanover, A. and Stanhill, A. (2014). The complexity of recognition of ubiquitinated substrates by the 26S proteasome. *Biochim. Biophys. Acta* 1843 (1): 86–96.

189 Hamilton, M.H., Cook, L.A., McRackan, T.R. et al. (2003). γ2 subunit of G protein heterotrimer is an N-end rule ubiquitylation substrate. *Proc. Natl. Acad. Sci. U.S.A.* 100 (9): 5081–5086.

190 Obin, M., Lee, B.Y., Meinke, G. et al. (2002). Ubiquitylation of the transducin βγ subunit complex. Regulation by phosducin. *J. Biol. Chem.* 277 (46): 44566–44575.

191 Obin, M., Nowell, T., and Taylor, A. (1994). The photoreceptor G-protein transducin (Gt) is a substrate for ubiquitin-dependent proteolysis. *Biochem. Biophys. Res. Commun.* 200 (3): 1169–1176.

192 Pusch, W., Jahner, D., and Ivell, R. (1998). Molecular cloning and testicular expression of the gene transcripts encoding the murine multiubiquitin-chain-binding protein (Mcb1). *Gene* 207 (1): 19–24.

193 Hamazaki, J., Sasaki, K., Kawahara, H. et al. (2007). Rpn10-mediated degradation of ubiquitinated proteins is essential for mouse development. *Mol. Cell. Biol.* 27 (19): 6629–6638.

194 Faivre, S., Régnauld, K., Bruyneel, E. et al. (2001). Suppression of cellular invasion by activated G-protein subunits Gαo, Gαi1, Gαi2, and Gαi3 and sequestration of Gβγ. *Mol. Pharmacol.* 60 (2): 363–372.

195 Zhang, B., Tran, U., and Wessely, O. (2018). Polycystin 1 loss of function is directly linked to an imbalance in G-protein signaling in the kidney. *Development* 145 (6): dev158931. https://doi.org/10.1242/dev.158931.

196 Brann, M.R., Collins, R.M., and Spiegel, A. (1987). Localization of mRNAs encoding the α-subunits of signal-transducing G-proteins within rat brain and among peripheral tissues. *FEBS Lett.* 222 (1): 191–198.

197 Yu, A.S. (2015). Claudins and the kidney. *J. Am. Soc. Nephrol.* 26 (1): 11–19.

198 Miyamoto, K.I. and Itho, M. (2001). Transcriptional regulation of the NPT2 gene by dietary phosphate. *Kidney Int.* 60 (2): 412–415.

199 Kido, S., Miyamoto, K., Mizobuchi, H. et al. (1999). Identification of regulatory sequences and binding proteins in the type II sodium/phosphate cotransporter *NPT2* gene responsive to dietary phosphate. *J. Biol. Chem.* 274 (40): 28256–28263.

200 Busca, R., Berra, E., Gaggioli, C. et al. (2005). Hypoxia-inducible factor 1α is a new target of microphthalmia-associated transcription factor (MITF) in melanoma cells. *J. Cell Biol.* 170 (1): 49–59.

201 Haase, V.H. (2006). Hypoxia-inducible factors in the kidney. *Am. J. Physiol. Renal Physiol.* 291 (2): F271–F281.

202 Kuiper, R.P., Schepens, M., Thijssen, J. et al. (2004). Regulation of the MiTF/TFE bHLH-LZ transcription factors through restricted spatial expression and alternative splicing of functional domains. *Nucleic Acids Res.* 32 (8): 2315–2322.

203 Kuiper, R.P., Schepens, M., Thijssen, J. et al. (2003). Upregulation of the transcription factor TFEB in t(6;11)(p21;q13)-positive renal cell carcinomas due to promoter substitution. *Hum. Mol. Genet.* 12 (14): 1661–1669.

204 Wilkie, T.M., Scherle, P.A., Strathmann, M.P. et al. (1991). Characterization of G-protein α subunits in the Gq class: expression in murine tissues and in stromal and hematopoietic cell lines. *Proc. Natl. Acad. Sci. U.S.A.* 88 (22): 10049–10053.

205 Hong, S.B., Oh, H., Valera, V.A. et al. (2010). Inactivation of the FLCN tumor suppressor gene induces TFE3 transcriptional activity by increasing its nuclear localization. *PLoS One* 5 (12): e15793.

206 Armah, H.B. and Parwani, A.V. (2010). Xp11.2 translocation renal cell carcinoma. *Arch. Pathol. Lab Med.* 134 (1): 124–129.

207 Meloni, A.M., Dobbs, R.M., Pontes, J.E., and Sandberg, A.A. (1993). Translocation (X;1) in papillary renal cell carcinoma. A new cytogenetic subtype. *Cancer Genet. Cytogenet.* 65 (1): 1–6.

208 Weterman, M.A., Wilbrink, M., and Geurts van Kessel, A. (1996). Fusion of the transcription factor TFE3 gene to a novel gene, PRCC, in t(X;1)(p11;q21)-positive papillary renal cell carcinomas. *Proc. Natl. Acad. Sci. U.S.A.* 93 (26): 15294–15298.

209 Weterman, M.A., Wilbrink, M., Janssen, I. et al. (1996). Molecular cloning of the papillary renal cell carcinoma-associated translocation (X;1)(p11;q21) breakpoint. *Cytogenet. Cell Genet.* 75 (1): 2–6.

210 Mathur, M. and Samuels, H.H. (2007). Role of PSF-TFE3 oncoprotein in the development of papillary renal cell carcinomas. *Oncogene* 26 (2): 277–283.

211 Weterman, M.A., van Groningen, J.J., den Hartog, A., and Geurts van Kessel, A. (2001). Transformation capacities of the papillary renal cell carcinoma-associated PRCCTFE3 and TFE3PRCC fusion genes. *Oncogene* 20 (12): 1414–1424.

212 Weterman, M.J., van Groningen, J.J., Jansen, A., and van Kessel, A.G. (2000). Nuclear localization and transactivating capacities of the papillary renal cell carcinoma-associated TFE3 and PRCC (fusion) proteins. *Oncogene* 19 (1): 69–74.

213 Bertolotto, C., Lesueur, F., Giuliano, S. et al. (2011). A SUMOylation-defective MITF germline mutation predisposes to melanoma and renal carcinoma. *Nature* 480 (7375): 94–98.

214 Granter, S.R., Weilbaecher, K.N., Quigley, C., and Fisher, D.E. (2002). Role for microphthalmia transcription factor in the diagnosis of metastatic malignant melanoma. *Appl. Immunohistochem. Mol. Morphol.* 10 (1): 47–51.

215 Davis, I.J., Kim, J.J., Ozsolak, F. et al. (2006). Oncogenic MITF dysregulation in clear cell sarcoma: defining the MiT family of human cancers. *Cancer Cell* 9 (6): 473–484.

216 Chen, J., Futami, K., Petillo, D. et al. (2008). Deficiency of FLCN in mouse kidney led to development of polycystic kidneys and renal neoplasia. *PLoS One* 3 (10): e3581.

217 Starling, G.P., Yip, Y.Y., Sanger, A. et al. (2016). Folliculin directs the formation of a Rab34-RILP complex to control the nutrient-dependent dynamic distribution of lysosomes. *EMBO Rep.* 17 (6): 823–841.

218 Zhao, L., Ji, X., Zhang, X. et al. (2018). FLCN is a novel Rab11A-interacting protein that is involved in the Rab11A-mediated recycling transport. *J. Cell Sci.* 131 (24).

219 Laviolette, L.A., Mermoud, J., Calvo, I.A. et al. (2017). Negative regulation of EGFR signalling by the human folliculin tumour suppressor protein. *Nat. Commun.* 8: 15866.

220 Zheng, J., Duan, B., Sun, S. et al. (2017). Folliculin interacts with Rab35 to regulate EGF-induced EGFR degradation. *Front. Pharmacol.* 8: 688.

Part IV

Novel Approaches

22

Screening and Characterizing of GPCR–Ligand Interactions with Mass Spectrometry-Based Technologies

Shanshan Qin[1], *Yan Lu*[1,2], *and Wenqing Shui*[1,2]

[1] *iHuman Institute, ShanghaiTech University, Shanghai, China*
[2] *School of Life Science and Technology, ShanghaiTech University, Shanghai, China*

22.1 Introduction

The superfamily of G-protein-coupled receptors (GPCRs) represents the largest class of cell surface receptors which play central roles in a variety of pathophysiological conditions [1]. GPCRs constitute a prominent family of therapeutic targets which account for -over 40% of marketed drugs for the treatment of various diseases [2]. Considerable efforts have been made in both industry and academia to screen and characterize novel ligands that can act on specific GPCR targets with high efficacy and selectivity.

A vast array of techniques suitable for assaying protein–ligand interactions have been directly applied or tailored to ligand discovery and/or characterization for GPCR targets, which can be classified into four main categories (Table 22.1). Affinity-based assays monitor the physical interactions between a GPCR protein (usually in a purified recombinant form) and individual test compounds. Signaling-based assays measure downstream effectors (e.g. cAMP, Ca^{2+}, and IP1/IP3) of specific signaling pathways known to be coupled to the GPCR target, which reflect functional consequence of ligand binding. Stability-based assays assess the variation of thermal stability for a purified GPCR protein when treated by test compounds. Finally, structure-based approaches aim to elucidate the interaction mode of a given ligand associated with the receptor or the ligand-induced conformational changes of the receptor. These different techniques vary in the throughput of ligand screening and the characteristics of ligand binding that can determined (Table 22.1). In the early-phase lead discovery pipeline for various protein drug targets, affinity-based and signaling/activity-based assays are the most often implemented in a parallel or sequential manner as the multipronged

GPCRs as Therapeutic Targets, Volume 2, First Edition. Edited by Annette Gilchrist.

use of complementary techniques would reduce the overall false positive and false negative rates [3–6].

Due to the high sensitivity and selectivity of modern mass spectrometry (MS), especially high-resolution mass spectrometry (HRMS), for both protein and small molecule analysis, versatile MS-based technologies have been developed in the last two decades for screening ligands to a given protein target or characterization of ligand-binding properties (Table 22.1). These MS-based methods were originally developed for the analysis of ligand interactions with soluble proteins, and recently they have been adapted to ligand discovery for the much more challenging GPCR family members. The majority of MS-based technologies are known collectively as affinity mass spectrometry (affinity MS) or affinity selection–mass spectrometry (AS–MS) which detect ligands to a purified receptor from a compound mixture. The most widely used approaches include automated ligand identification system (ALIS), ultrafiltration–mass spectrometry (UF/MS), frontal affinity chromatography mass spectrometry (FAC–MS), and microbead-based affinity MS. Both an industrial team from Edelris and other companies as well as Breemen group have provided extensive reviews on affinity MS for ligand screening against various protein targets [7, 8]. Alternatively, a competitive MS binding assay evaluates ligand binding based on a marker ligand response measured by targeted MS. Native MS monitors the entire ligand-bound protein complexes as a measure of ligand detection. In addition, hydrogen-deuterium exchange mass spectrometry (HDX–MS) enables the analysis of structural dynamics of GPCR–ligand complexes. In general, MS-based methods for ligand detection demonstrate several major advantages over the receptor functional assays or other binding-based assays. These include the following: (i) The unbiased measurement of direct receptor–ligand binding facilitates identification of both orthosteric and allosteric ligands; (ii) Confirmation of ligand identity with accurate mass measurement; (iii) No chemical labeling of test compounds; (iv) The receptor target can be either in a purified form or embedded in the native cell membrane; (v) Ability to simultaneously distinguish multiple compounds from a complex mixture; (vi) Quantitative MS analysis enables ranking of ligand affinity or determination of ligand-binding characteristics. It is noteworthy that these MS-based techniques when hyphenated to different biochemical front-ends or liquid chromatography can achieve varying throughput, sensitivity, or quantification accuracy, thus making them more suited to either ligand screening or ligand characterization. Herein, we review the method principle, advantages and challenges, as well as applications of specific MS-based technologies to either high-throughput GPCR ligand screening or characterization of binding properties of ligands to a GPCR target.

Table 22.1 A variety of technologies applied to screening or characterizing of GPCR–ligand interactions.

Class	Method	Principle	Readout	Ligand-binding characterization	Screening Throughput
Affinity-based assays	Radioligand binding	Detect binding of a radioisotope-labeled ligand to a target in competition with a test compound	Radioactivity	K_i, k_{on}, k_{off}	Medium
	DEL	DNA-encoded compounds bound to a target are affinity selected, and their structures revealed by DNA sequencing	DNA sequence	Affinity ranking	Ultra-high
	SPR	Detect changes in the refractive index of the gold film surface when ligands interact with a target immobilized on the chip surface	Refractive index (mass on surface)	K_d, stoichiometry (n), K_{on}, K_{off}	Medium
	MST	Detect directed movement of molecules through a temperature gradient using covalently attached or intrinsic fluorophores	Fluorescence intensity	K_d, stoichiometry (n)	Medium
	TR–FRET	Detect fluorescence resonance energy transfer caused by interaction between target and ligand labeled with specific fluorophores	Fluorescence intensity	K_i	Medium
	ALIS	Target-ligand complexes are separated by fast SEC, and dissociated ligands are identified by LC–MS	m/z and MS intensity	Affinity ranking, K_d, ACE_{50}, k_{off}	High/Ultra-high

(Continued)

Table 22.1 (Continued)

Class	Method	Principle	Readout	Ligand-binding characterization	Screening Throughput
	UF–LC/MS	Target-ligand complexes are separated from unbound compounds by ultrafiltration prior to ligand identification by LC–MS	m/z and MS intensity	Affinity ranking, K_d	High
	FAC–MS	Detect ligands flowing through a protein-immobilized column based on breakthrough curves determined by MS	m/z and MS intensity	Affinity ranking, K_d, B_t	Medium
	Microbead-based affinity MS	Ligands bound to a protein target immobilized on microbeads are released and detected by LC–MS	m/z and MS intensity	Affinity ranking	High/Ultra-high
	Membrane-based affinity MS	Target-ligand complexes in the cell membranes are separated from unbound compounds by filtration prior to ligand identification by LC–MS	m/z and MS intensity	Affinity ranking	High
	Competitive MS binding	Detect binding of a non-radioactive ligand to a target in competition with a test compound by MRM-based MS analysis	m/z and MS intensity	K_d, B_{max}, K_i, k_{on}, k_{off}	Low
	Native MS	Detect intact protein-ligand complexes in the gas phase by MS	m/z and MS intensity	K_d, stoichiometry (n)	Low

Class	Method	Principle	Readout	Ligand binding characterization	Screening Throughput
Stability-based assay	DSF	Detect changes in protein fluorescence over a temperature gradient	Fluorescence intensity	T_m	Medium
	DLS	Measure changes in the protein aggregate size based on static light scattering properties over a temperature gradient	Light scattering intensity	T_{agg}	Medium
Signaling-based assay	GTPγS	Detect 35S-GTPγS binding to GPCR-expressing cell membranes as a result of receptor activation	Radioactivity	EC_{50}, IC_{50}	High
	cAMP	Detect cellular levels of cAMP coupled to Gαs or Gαi activation	Luminescence	EC_{50}, IC_{50}	High
	Ca^{2+}	Detect cellular levels of Ca^{2+} coupled to Gαq/11 or Gα15/16 activation	Fluorescence	EC_{50}, IC_{50}	High
	IP3/IP1	Detect cellular levels of IP3 coupled to Gαq or Gαi activation	Fluorescence	EC_{50}, IC_{50}	High
	Luciferase reporter gene	Measure reporter gene transcriptional activity downstream of receptor activation	Luminescence	EC_{50}, IC_{50}	High
	β-arrestin recruitment	Detect interaction of cellular β-arrestin with receptor	Luminescence	EC_{50}, IC_{50}	Medium--High

(Continued)

Table 22.1 (Continued)

Class	Method	Principle	Readout	Ligand-binding characterization	Screening Throughput
Structure-based approaches	NMR	Provide structural information based on interactions between nuclei and magnetic field	Magnetic resonance	Protein-ligand complex structure; binding mode	Low
	X-ray	Generate high-resolution structures of protein-ligand complexes based on X-ray diffraction from the crystal	Electron density maps	Protein-ligand complex structure; binding mode	Low
	Cryo-EM	Generate high-resolution images of large protein complexes *in situ*	3D EM maps	Protein-ligand complex structure; binding mode	Low
	Virtual screen	Ligand identification by docking analysis using protein 3D structures	Docking scores and poses	Protein–ligand binding mode	Ultrahigh
	HDX–MS	Exchange of hydrogen atoms in the protein backbone with deuterium atoms in solvent	H/D exchange rate	Protein structural dynamics	Low

22.2 High-Throughput GPCR Ligand Screening with Affinity MS

The generation of new chemical leads as a starting point for drug development is a critical step in pharmaceutical drug discovery. To realize the full potential of combinatorial chemistry-based drug discovery, high-throughput screening (HTS) has become an integral part of the lead discovery process [9, 10]. A routine HTS campaign is usually based on cell signaling assays for a specific GPCR target and typically involves screening a library of several million compounds against a target of interest. Recently, the traditional HTS platforms have become enhanced by the use of diverse affinity-based techniques to address challenging targets with no bioactivity assays adaptable to HTS.

22.2.1 Automated Ligand Identification System (ALIS)

ALIS, originally developed at NeoGenesis Pharmaceuticals and later acquired and refined by Schering-Plough and Merck, is currently the most common affinity MS-based technique employed in pharmaceutical companies for high-throughput ligand screening from large-scale compound libraries [6, 11, 12]. This system incorporates four consecutive modules: (i) an affinity selection module, where a protein binds to ligand mixtures; (ii) SEC step that isolates protein–ligand complexes from unbound compounds; (iii) reverse-phase chromatography (RPC) step that dissociates the bound ligands from the complex; (iv) an MS platform for identification and quantitation of the dissociated ligands. The initial platform was built to screen mass-encoded, 2500-member combinatorial libraries, leading to the discovery of a novel, bioactive ligand for the anti-infective target *Escherichia coli* dihydrofolate reductase (DHFR) [13]. Successful application of ALIS to screen therapeutically important proteins against combinatorial libraries at the million-compound scale has yielded novel ligands to a number of targets, including ERK inhibitors [14] and non-ATP-competitive MK2 inhibitors [15]. Further extension of ALIS application to drug discovery is the combination of ALIS with nanoscale synthesis to accelerate the SAR analysis of reaction space [5]. When coupled to nanomole-scale synthesis with each reaction performed at a 0.1-M concentration and consuming less than 0.05 mg of substrate, the ALIS platform of superior sensitivity allowed determination of the affinity of the reaction products to target proteins. This led to rapid identification of the bioactive molecules and conditions for their synthesis with minimal consumption of starting materials.

The application of ALIS to ligand screening for membrane receptors substantially lagged behind soluble proteins due to the difficulty of obtaining membrane proteins of sufficient purity, activity, and stability. The earliest case of using ALIS for GPCR ligand screening was described more than a decade ago by Whitehurst et al. (2006) for the muscarinic M_2 acetylcholine receptor (AChR), which opened a new avenue for GPCR drug discovery with affinity-based assays. In this experiment, purified porcine M_2 protein was incubated with different compound mixtures consisting of 1500 members, followed by SEC to separate protein–ligand complexes from nonbinding ligands. It is noted that relatively low-affinity ligands ($K_d < 10\,\mu M$) with moderate dissociate rates ($k_{off} < 0.1\,s^{-1}$) were not missed due to the fast SEC separation. After ligand detection with the RPC–MS platform, 48 primary hits were identified through screening about a total of 350 K compound pools which consumed 250 μg purified pM2 AChR. Two hits were further validated to be an orthosteric antagonist or an allosteric modulator of the receptor. A similar strategy was implemented by a research team from Merck and other pharmaceutical companies to screen chemical ligands for the human chemokine receptor CXCR4, an established drug target for anti-viral and cancer therapies [12]. An extremely large library consisting of 2.75 million compounds was screened by ALIS toward the purified CXCR4 protein. Each reaction only consumed 2.5 μM protein which was incubated with a mixture of approximately 100 (from the 48K library) or 2000 (from the NeoMorph Library) compounds. Hits were selected and characterized by ALIS-based ligand saturation binding assays (ALIS-K_d) and competition binding assays (ACE_{50}). A total of 362 primary hits were identified from the library screen and 34 of them were found to be new antagonists of CXCR4.

22.2.2 UF–MS

In addition to SEC, ultrafiltration (UF) using a semipermeable membrane with a low-molecular weight cut-off is another common method for separating protein complexes from free ligands in solution [16, 17]. A combination of UF with affinity MS analysis, also termed UF–MS, was initially developed by Richard Breemen et al. to screen active ingredients from combinatorial libraries that may act on serum albumin or adenosine deaminase [18]. Since then, UF–MS has been widely employed for ligand screening from compound libraries or natural product extracts toward specific enzyme targets [19–22]. The maximal screening throughput could reach 2000–3000 compounds per mixture [19, 20]. The Shui group has adapted the UF–MS approach to ligand detection or screening toward a number of protein targets that are critical for anti-bacterial or anti-viral drug development [21–25]. They first compared side-by-side UF–MS with nanoESI–MS in ligand detection. UF–MS showed better reproducibility, more tolerance of interference, and ability to do mixture screening [21]. Next, Chen et al. integrated UF–MS into a fragment-based ligand discovery (FBLD) program targeting the HCV RNA polymerase NS5B [22]. FBLD plays an important role in lead discovery and candidate optimization, as most fragments possess simple scaffolds convenient for chemical modification, but they usually have very low affinity to the protein target [26]. Their UF–MS analysis resulted in the discovery of 10 hits from a small fragment library through two independent screens of complex cocktails and a follow-up validation assay. Moreover, this MS-based approach enabled quantitative measurement of very weak binding affinities (high μM ~ low mM) which was in general consistent with SPR measurement of individual fragments. Five fragment hits were further co-crystallized with the protein target to lay a foundation for structure-based design of anti-viral inhibitors. This UF–MS assay has emerged as a valuable addition to the repertoire of current fragment screening techniques.

In contrast to ligand screening for soluble protein targets, UF–MS has rarely been applied to GPCR ligand screening given that detergents needed for membrane protein solubilization may affect the recovery of receptor-ligand complexes from UF membranes. Lu and coworkers used engineered nanodiscs to reconstitute a GPCR protein, adenosine A_{2A} receptor ($A_{2A}R$), free of detergent for ligand detection with UF–MS [27]. Nanodiscs are composed of self-assembled disc-like phospholipid bilayers that are encircled by two amphipathic helical membrane scaffold proteins (MSPs). They can maintain GPCRs at a nearly native state with better thermostability and homogeneity in detergent-free buffer. In this study, they first prepared a 15-compound mixture containing 8 known agonists or antagonists and 7 unrelated compounds to be incubated with $A_{2A}R$ nanodisc or empty nanodisc. After UF–MS analysis, all known $A_{2A}R$ ligands were selected based on their binding index (BI), while the unrelated compounds were negatives. Notably, the ranking of ligand affinity generally agreed with the BI value determined for each compound, which indicated the feasibility of this nanodisc-aided UF–MS workflow for GPCR ligand screening in a large scale.

Plants have been a rich source of compounds for drug discovery especially as anticancer and antimicrobial agents [28, 29]. There is a sustaining interest in herbal medicine, especially Traditional Chinese Medicine (TCM), for generation of lead compounds, as they possess more diverse chemical scaffolds and more therapeutically useful properties than synthetic compounds [30]. Peng and coworkers employed a thermal stability assay [31] to identify several bioactive ligands for 5-HT_{2c} receptor from an alkaloid library of ~300 TCM-derived compounds [32]. However, it is labor-intensive and cost-prohibitive to build such a library of pure compounds isolated from various natural product extracts. Thus, this type of ligand screening with isolated natural compound libraries is most often conducted in a small scale at a low throughput. Owing to the capability of LC–HRMS in separation and identification of diverse chemical constituents in a complex mixture, UF–MS has become an efficient and powerful technique in the discovery of small molecule ligands from the crude extracts of natural products [33–39]. Since plant extracts frequently consist of a large number of secondary metabolites, Fu et al. combined the metabolomics data processing pipeline with UF–MS analysis for efficient discovery of new chemical ligands from TCM extracts for the nucleocapsid proteins of Ebola viruses and Marburg viruses [24]. Compared to most other binding capacity studies of the most abundant compounds found in plants by HPLC analysis [37–40], this metabolomics-augmented workflow considerably increased the throughput and sensitivity by simultaneously screening hundreds to thousands of components from plant sources. Later on, Wang et al. developed another strategy that integrated virtual screen, UF–MS, and targeted metabolomics for discovery of plant-derived ligands toward a specific protein target site [25]. This newly integrated approach was showcased by identification of chemical ligands that target the hydrophobic pocket of the Ebola viral nucleoprotein for which no small molecule ligands have been reported. Taken together, these studies have demonstrated the great potential of UF–MS for discovery of natural product-derived bioactive ligands especially toward nonenzyme protein targets which could include GPCRs in future investigations.

22.2.3 FAC–MS

Heterogeneous binding-based approaches with proteins immobilized on solid phase have gained increased popularity in the protein–ligand interaction analysis. FAC–MS represents a classical method that relies on immobilized proteins to capture interacting ligands from a flowing system [41]. Protein targets are usually immobilized on the stationary phase surface of a chromatographic column. When compounds flow through the column, those ligands with interactions to the protein are retained on the column to different extent and elute with different profiles, generating characteristic breakthrough curves. In this experiment, a single point determination of the breakthrough volume of each detectable compound from a pooled mixture provides a measurement of the fraction of compound bound, which relates to the rank order of ligand affinity. Such an online FAC–MS system running at a single concentration has been established to screen pools of tens to hundreds of compounds for ligand discovery toward antibodies [42, 43], lectins [44, 45], and enzymes space [46, 47].

Ng et al. modified the FAC–MS system by insertion of an LC purification step to fractionate the FAC effluent prior to MS analysis, which increased the sensitivity in ligand identification from complex mixtures [48]. They applied this system to screening a mixture of ~1000 compounds targeting an enzyme *N*-acetylglucosaminyltransferase V (GnT-V) immobilized on the FAC column. Four quality ligands were identified from the compound pool with K_ds in the 0.6–1.5 mM range (determined in separate assays) and served as potential enzyme inhibitors. Ligand detection for GPCR targets by FAC–MS has been reported in several studies [49–51]. For example, Calleri et al. developed an FAC–MS approach to identify ligands to immobilized GPR17 from simple mixtures of nucleotide derivatives [51]. Mixture A containing 11 compounds and B containing 7 compounds were prepared separately with three known GPR17 ligands with nM to μM potency. The ranking order of these three reference ligands revealed by breakthrough curves was in agreement with their cellular activity. This allowed the determination of the affinity of other test compounds, were found to be low, medium, and high affinity ligands for GPR17. One test compound exhibited high agonist potency at GPR17 in a follow-up signaling-based assay. Furthermore, FAC–MS results were combined with steered molecular dynamics simulations to elaborate the structure details of receptor–ligand interaction.

These proof-of-concept experiments demonstrate the potential of FAC–MS in the screening of synthetic combinatorial libraries or natural product extracts. However, time and effort needed to pack and maintain a column, especially to keep immobilized proteins in an active state, remains a primary limitation of FAC–MS and may be the reason why this technique has not been widely explored. This could be particularly challenging for GPCRs that tend to aggregate or change conformations when exposed to solid support in a crowded environment, and the immobilization conditions are likely to be receptor-dependent.

22.2.4 Microbead-Based Affinity MS

Compared to LC column packing, affinity capture to microbeads offers a more convenient way of protein immobilization for ligand detection. Such an microbead-based affinity MS approach requires covalent or non-covalent attachment of the protein target to solid support such as magnetic beads, and subsequent incubation with a compound mixture in a suitable binding buffer. Separation of ligand–receptor complexes is carried out by washing away the unbound compounds, while the beads are retained in a magnetic field. The ligands are then released from the receptors using an organic or acidic solvent and analyzed by UPLC–MS analysis.

Originally developed by Breemen laboratory, microbead-based affinity MS has been successfully employed to screen chemical ligands against a variety of protein targets such as estrogen receptor and 15-lipoxygenase from either combinatorial libraries or natural product extracts [52, 53]. Notably, Zhang and coworkers pioneered in extending this technique to GPCR ligand screening from herbal extracts [54]. In this work, the receptor construct was designed to drive the heterogeneous protein population towards active conformations, and affinity MS was integrated with the metabolomics data mining workflow to identify ligands towards a serotonin receptor 5-HT_{2C}. After screening a panel of herbal extracts with agonist activity on 5-HT_{2C}, the authors rapidly discovered a naturally occurring aporphine compound that displayed strong selectivity for activating 5-HT_{2C} without activating the 5-HT_{2A} or 5-HT_{2B} receptors. Moreover, this novel ligand exhibited exclusive bias towards G protein signaling and showed in vivo efficacy for food intake suppression and weight loss. Thus, this study established an efficient approach to discovering novel GPCR ligands by exploring the largely untapped chemical space of natural products [54].

In the early phase of lead discovery, DNA-encoded library (DEL) and affinity MS assays are two binding-based techniques that allow high-throughput ligand screening and complement the classical activity-based HTS assays (Table 22.1). Both techniques adopt a mixture-based screening format in which a pool of compounds is incubated with the target of interest and multiple ligands are selected simultaneously due to their affinity to the target. DEL has afforded unprecedented screening throughput, given that 1–100 million compounds are typically synthesized and screened in one pool toward a specific target [55–57]. The high sensitivity and selectivity of a DEL screen is enabled by amplification of the distinct DNA barcode attached to each compound [55]. By contrast, various affinity selection MS-based methods as described above permit ligand screening from pools of typically 400–2000 compounds [14, 22, 58, 59]. Therefore, far more protein is needed when a large-scale library has to be subdivided into hundreds of less complex cocktails compared to screening the entire library as a whole.

In order to accelerate the throughput of microbead-based affinity MS towards GPCR targets, Lu et al. developed a new strategy that enabled screening of 20 000 compounds in one pool [60]. This new approach was applied to ligand screening for the A_{2A} adenosine receptor ($A_{2A}R$), an attractive GPCR target for immune-oncology, inflammation, and central nervous system disorders. Specifically, they modified the workflow by performing iterative rounds of affinity selection of compounds associated with the protein target (Figure 22.1a). After each selection round, the associated compounds were eluted from $A_{2A}R$ by chemical denaturation of the receptor and reincubated with freshly prepared $A_{2A}R$. Three sequential selection rounds were carried out before the compounds were analyzed by UPLC–MS. Similar to the previously described single-round affinity MS screening assay, quantitative measurement of binding index for each test compound identified of high-affinity ligands in this experiment. By comparing the selection of 16 benchmark $A_{2A}R$ ligands from screening compound pools of 480-mix, 2400-mix, 4800-mix, and 20K-mix, the authors concluded that the accelerated affinity MS screening approach, using either the purified receptor or receptor-expressing cell membranes, enables detection of most of the high-affinity $A_{2A}R$ ligands in compound pools of increased complexity ($K_d/K_i < 5\,\mu M$) (Figure 22.1b,c).

This screening experiment resulted in identification of three hit molecules that are new antagonists for $A_{2A}R$. Using the iterative selection procedure developed in this study, the authors envision that the throughput of affinity MS screening can be further increased to assay hundreds of thousands or even 1 million compounds

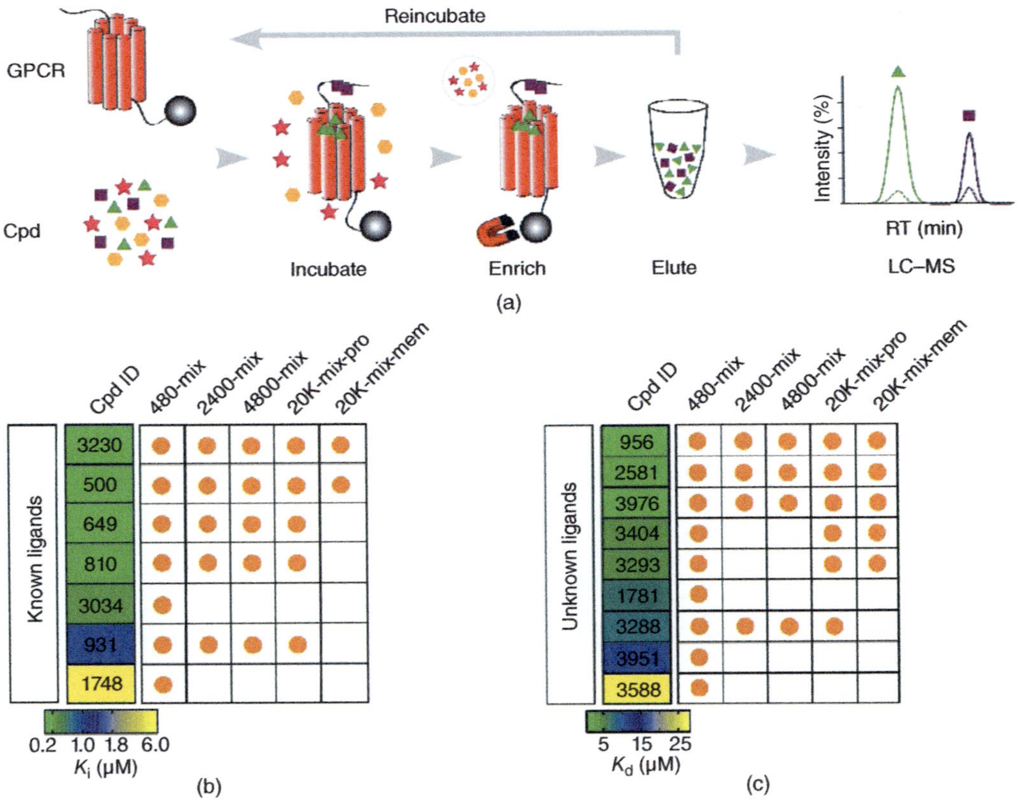

Figure 22.1 Outline of accelerated affinity MS with iterative selection. (a) Modified workflow integrating iterative incubation of eluted compounds with the target. Selection of seven known ligands (b) and nine unknown ligands (c) for $A_{2A}R$ in the benchmark ligand set by different affinity MS screens. The range of binding affinity of known ligands from literature (K_i) and unknown ligands measured by our SPR analysis (K_d) is color coded. Cpd, compound; Pro, protein; mem, membrane; RT, retention time. Source: Lu et al. [60]/with permission of American Chemical Society.

in one pool, approaching the throughput of the DEL screen [55, 61]. Indeed, microbead-based affinity MS has been utilized to screen combinatorial peptide libraries consisting of 1 million members in a single assay in order to identify high-affinity binders that inhibit protein–protein interactions [62]. However, it is worth mentioning that in such extremely complex mixtures, multi-ligand competition would become more severe and impair the reproducibility in ligand selection.

22.2.5 Membrane-Based Affinity MS

Although the aforementioned affinity MS techniques have been showcased to be feasible for GPCR ligand screening against certain receptors, almost all methods require the preparation of purified and stable GPCR proteins. Up till now, it is still very challenging to implement these affinity MS assays to a wider range of GPCR targets, given that many of them are unstable and at low yield when isolated out of the cell membrane. In fact, tremendous efforts were needed to optimize the receptor expression and purification conditions for a GPCR target subject to affinity MS-based screening [12, 63]. Thus, binding-based assays such as isothermal titration calorimetry (ITC), surface plasmon resonance (SPR), as well as different affinity MS techniques are most successful with purified proteins yet less amenable to many GPCRs that are attractive drug targets.

To address this challenge, Qin et al. established a membrane-based affinity MS technique that enables ligand screening toward wildtype active GPCRs embedded in the cell membranes [64]. With this new approach, high throughput, label-free, and unbiased ligand screening was achieved with two GPCR targets (5-HT_{2C}R [65], GLP-1R [66]). This membrane-based affinity MS approach features isolation of membrane fractions from cells expressing a GPCR target at high yield and incubation of the membranes with a cocktail of compounds. After washing, compounds associated with the receptor-expressing membranes were then released and subjected to liquid chromatography coupled to high-resolution mass spectrometry (LC–HRMS) analysis (Figure 22.2a). Primary hits were selected based on a binding index (BI) defined as the ratio of MS response of the compound detected in the target versus the control incubations, which was used to distinguish putative ligands from nonspecific binders. Screening a small compound library with this approach led to the rapid discovery of an antagonist for the 5-HT_{2c} receptor and four positive allosteric modulators for GLP-1 receptor that are not reported before (Figure 22.2b–c). Compared with other binding-based methods for ligand detection, this unique affinity MS workflow minimally modifies the receptor sequence, retains the receptor native conformation, and eliminates the needs of laborious receptor purification.

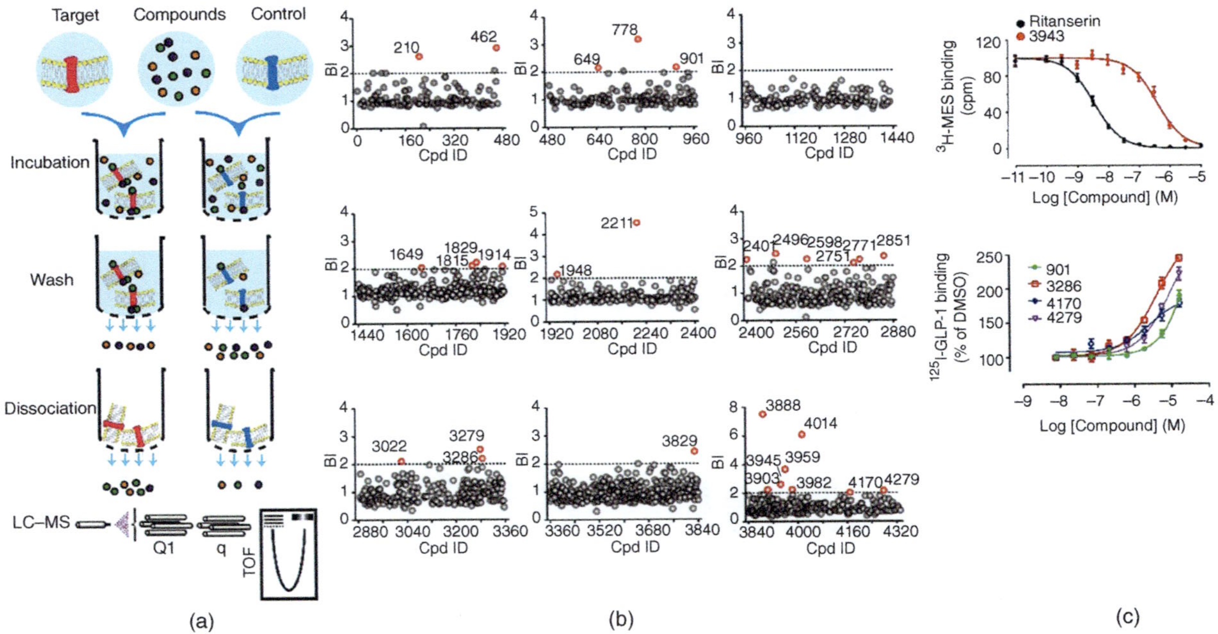

Figure 22.2 Outline of Membrane-based affinity MS. (a) Experimental workflow. (b) Affinity MS screening of a 4333-member compound library split into nine cocktails against GLP-1R. Initial hits are indicated by red dots, while gray dots represent negatives. (c) Binding of one new ligand to $5HT_{2C}R$ (upper) and four new ligands to GLP-1R (lower) were validated by radioligand binding assays. TOF, time of flight; cpm, counts per minute. Source: Qin et al. [64]/Royal Society of Chemistry/CC BY-3.0.

22.3 Characterization of GPCR–Ligand Interactions with MS-Based Techniques

For biophysical characterization of the ligand interaction with a given protein target, the most important properties are its affinity, stoichiometry, binding site, and association/dissociation rates. Diverse binding-based assays listed in Table 22.1 play an increasingly vital role in the drug development process as they can independently confirm specific binding of an HTS hit to the target and the binding characteristics measured on a lead compound provide crucial guidance to its downstream structural optimization. In GPCR ligand discovery, direct binding assays are more valuable for those less characterized targets for which no radioligands are available and no signaling pathways are decoded to construct a functional assay to evaluate potential ligands. Many of the MS-based techniques established for GPCR ligand screening from compound mixtures can be modified to determine thermodynamic and kinetic parameters of individual ligand binding to the receptor. Native MS and HDX–MS are two powerful technologies to gain structural insights into protein complex assembly and dynamics, and they have been successfully applied to GPCR structural characterization.

22.3.1 ALIS

The ALIS system has been configured to measure ligand equilibrium binding constant (K_d), ligand dissociation rate (k_{off}), relative affinity ranking, and classification of the ligand binding site (orthosteric or allosteric) [6, 11, 67, 68].

An ALIS titration assay is developed to assess specific binding affinity of a test compound [1, 3, 68]. K_d is determined based on the equation $K_d = \frac{([P]_0-[PL])([L]_0-[PL])}{[PL]}$ assuming single-site binding. Here, $[P]_0$ and $[L]_0$ represent the total receptor and total ligand concentrations, respectively. $[PL]$ refers to the concentration of the protein–ligand complex, which corresponds to the calibrated MS signal measured for the ligand dissociated from the complex. In this assay, plotting the ALIS MS response from a titration series of the ligand versus the total ligand concentration would yield a saturation-binding curve and the K_d of the test ligand [6]. When the ALIS screening platform was employed to discover the new antagonist NGD-3350 for M_2 muscarinic acetylcholine receptor, its affinity was determined using ALIS titration assay of the purified receptor treated by increasing concentration of the ligand (Figure 22.3a). Of note, K_d determined for this antagonist with the ALIS titration assay ($0.7 \pm 0.1\ \mu M$) agreed very well with K_i from the classical radioligand binding assay ($0.2 \pm 0.6\ \mu M$) [6].

Accurate measurement of the dissociation kinetics for protein–ligand interactions is critical in the drug discovery process as a necessary measure of ligand's potential efficacy. The ALIS dissociation assay which mimics the radioligand

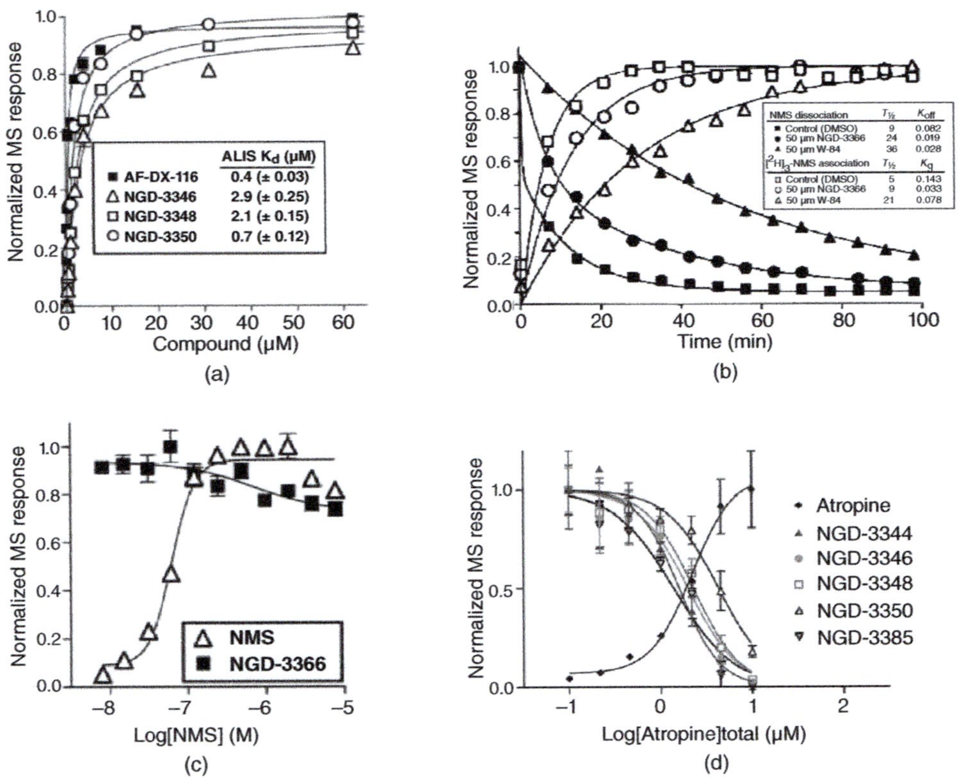

Figure 22.3 Ligand binding characterization by ALIS-based assays: (a) K_d of ligands to M2AChR are determined by the ALIS titration assay. (b) Simultaneous measure of the effect of NGD-3366 or W-84 on NMS dissociation (k_{off}) from M2AChR by the ALIS dissociation assay. (c) Nonlinear correlation of NGD-3366 binding with the increase of NMS concentration as shown in the ALIS competition assay. (d) Competition curves from an ALIS ACE_{50} experiment indicate NGD-3350 has the highest affinity to M2AChR than the other compounds in the mixture. Source: (a–c) Whitehurst et al. [6]/with permission of SAGE Publications and (d) from Annis et al. [11]/with permission of American Chemical Society.

quench method enables determination of dissociation constants for single ligands or ligands in a mixture. By addition of an excess competitor with equal or higher affinity of the test compound, the concentration of the ligand released from the protein–ligand complex will diminish with time, which can be measured by ALIS. For example, after NGD-3366 was validated as an allosteric binder of M_2 receptor, its binding kinetics was evaluated with the ALIS dissociation assay [6]. The equilibrated mixture of M_2 receptor plus *N*-methyl-scopolamine (NMS) was quenched by the isotopic $[^2H]_3$-NMS in the presence of the test ligand NGD-3366, and the released ligands were sampled every seven minutes for ALIS analysis. Through curve fitting, NGD-3366 decreased the dissociation rate of NMS from M_2 receptor, proving the allostery of NGD-3366 and its positive allosteric effect on NMS binding (Figure 22.3b).

Binding site characterization for ligands is vital for drug lead optimization and mechanistic study of protein–ligand interactions [69]. Although ligand-binding sites can be revealed at atomic resolution by NMR or X-ray crystallography-based structural analysis, these techniques are not efficient and require a large amount of proteins. The ALIS competition assay classifies the binding sites in a more efficient and low-cost manner though with limited resolution. In this assay, affinity selection is performed at a fixed concentration of the test ligand(s) incubated with serially increasing concentrations of a competitor ligand. If the test ligand binds the same site with the competitor, the ratio of the two compounds will be linear with the increasing amount of the competitor. If not, it may indicate different sites for two compounds and allostery between them. Positive cooperativity or negative coordination can be inferred, depending on whether one ligand binding promotes or reduces the binding of another. For example, the M_2 ligand NGD-3350 is displaced by the orthosteric titrant NMS and resulted in a linear correlation of NGD-3350 and NMS, indicating the orthosteric site occupied by NGD-3350. Another screening hit, NGD-3366, displayed a nonlinear relationship with NMS, which indicates the allosteric site by NGD-3366 (Figure 22.3c).

Another format of ALIS competition assay enables accurate ranking or determination of binding affinity for multiple ligands based on a specialized parameter ACE_{50}, which describes the concentration of a competitor ligand required to compete out 50% of the test ligand. It follows that a higher ACE_{50} can identify high-affinity ligands which need high concentrations of the competitor to be displaced. A group of hits arising from the ALIS screening experiment toward the M_2 receptor was evaluated by the ACE_{50} method [11]. The known antagonist atropine was used as the titrant against M_2 receptor in the presence of the compound mixture, and ACE_{50} measurement indicated ligand NGD-3350 had the highest affinity among all components in the mixture (Figure 22.3d). Moreover, if compounds with known K_d are included as internal calibrants, K_ds for other mixture components can be derived from a calibration curve based on the calibrants' ACE_{50} and K_d

values. K_d determined with this calibrated ACE_{50} method for several CDK2 leads [67] are consistent with those determined by the ALIS titration assay [68] and a fluorescence polarization assay [70].

Taken together, the ALIS-based methods for characterizing protein–ligand interactions are generally applicable to a broad range of protein families including GPCRs, and they are readily adaptable to other designs of affinity MS assays coupled to different forms of protein complex isolation as described in the ligand screening session.

22.3.2 UF–MS

Although different formats of ALIS binding assays allow for accurate and flexible determination of thermodynamic and kinetic parameters for ligand characterization, they usually require known ligands serving as internal calibrants, or a titration experiment at varying concentrations of a given test ligand [67]. Thus, it would be less feasible to apply these assays when no ligands are known or available for a newly discovered target, or sample materials are quite limited for the serial titration experiment.

While UF–MS has been applied to ligand identification for diverse protein targets, most studies only reported relative ranking of ligand affinities rather than measuring the equilibrium dissociation constant (K_d) that reflects the absolute binding strengths [71, 72]. Qin et al. first evaluated the accuracy of K_d determination for five inhibitors of the human carbonic anhydrase I (hCAI) through UF–LC/MS analysis, which were then compared with classical ITC measurement of individual inhibitors [23]. Compounds of varying binding strengths were added to the mixture to examine the effect of competitive binding in affinity measurement by UF–LC/MS. When they compared saturation binding curve analysis and single-point calculation of K_d, they found that K_d measurement at a single-point concentration of the compound mixture with high P:L mixing ratios (to minimize competitive binding between ligands) was more reliable and in accordance with standard ITC measurement of single compounds.

Next, the single-point K_d calculation method was established in a fragment mixture screen against the HCV RNA polymerase NS5B. Considering the weak binding affinity of fragments, the authors developed a workflow of unbound fraction analysis (UFA) which monitors the unbound compounds fraction to increase the sensitivity for ligand detection. In the UFA assay, K_d of each ligand in a mixture under a specific protein–ligand incubation condition is calculated using the following equation:

$$Kd_{\mathrm{m}} = \frac{\left([P_0] - \sum_{i=1}^{n}[PL_{\mathrm{i}}]\right)([L_{\mathrm{m},0}] - [\mathrm{PL}_{\mathrm{m}}])}{[PL_{\mathrm{m}}]} \tag{22.1}$$

whereas Kd_m is K_d of ligand m within a mixture of ligands (1, 2,..., n) competitively bound to the protein, $[P_0]$ is the initial protein concentration, $[L_{m,0}]$ is the initial concentration of ligand m, $[PL_i]$, and $[PL_m]$ are derived from the following equations:

$$[PL_i] = [L'_i]_b - [L'_i]_a \tag{22.2}$$

$$[PL_m] = [L'_m]_b - [L'_m]_a \tag{22.3}$$

where $[L'_i]_b$ and $[L'_m]_b$ are the concentrations of inhibitor i and inhibitor m present in the filtrate of the control and detected by LC–MS, $[L'_i]_a$ and $[L'_m]_a$ are the concentrations of inhibitor i and inhibitor m in the filtrate from the protein incubation and detected by LC–MS.

It turned out that this new UFA approach increased the sensitivity in identification of lower-affinity ligands (high μM to mM) as well as simplified the sample preparation process. More importantly, the binding affinity measurement for each component in the fragment mixture by UF–LC/MS were in good accordance with single ligand evaluation by conventional SPR analysis. In conclusion, this study demonstrates the reliability of affinity measurement for multiple ligands including very weak fragment binders to a specific target through UF–LC/MS, which is adaptable to various other protein targets including GPCRs.

22.3.3 Competitive MS Binding Assay

The radioligand binding assay using cell membrane fractions or intact cells expressing a specific receptor serves as a gold standard for GPCR ligand characterization [73]. But this type of assay has inherent drawbacks such as safety precautions, restriction for handling and disposing radioisotopes, and expensive reagent production. The competitive MS binding assay initially developed by the Wanner group [74–76] employs an unlabeled marker ligand to measure binding of test compounds, which closely mimics the radioligand binding assay yet avoids the use of radioisotopes. In this assay, a selected marker liberated from the protein target is quantified using an optimized SRM or MRM-based MS method of high sensitivity and selectivity for ligand binding equilibrium or kinetics measurement.

Analogous to the radioligand binding assay, this competitive MS binding assay can be performed in multiple formats: (i) saturation, association, and dissociation assays to determine the affinity and kinetic constants of the marker ligands and (ii) marker displacement or competition assays to determine the affinity and kinetic constants of test compounds competing with the marker [77]. It is also noteworthy that the target sources of this assay are cell membranes overexpressing a particular membrane protein, which maintains the protein–ligand interactions in a native environment and is especially useful for GPCR ligand characterization [74, 75].

The competitive MS binding assays have been established for a number of membrane transporters such as the neuronal subtype of the γ-aminobutyric acid (GABA) transporters (GAT1) and the serotonin transporter (SERT) as well as the nicotinic acetylcholine receptor which is an ion channel [78–81]. They are equally effective in addressing GPCR targets as recently demonstrated with the adenosine receptors A_1/A_{2A} and dopamine receptors $D_1/D_2/D_5$ [77, 82, 83]. In both cases, the authors demonstrated that the use of unlabeled marker ligands could substitute for their radiolabeled counterparts in all kinds of marker-based MS binding experiments, including saturation, displacement, dissociation, and competition association assays. Impressively, all standard parameters commonly determined with radioligand binding assays (i.e. K_d, B_{max}, K_i, k_{on}, k_{off}) are in excellent accordance with those determined with the new marker-based MS binding assays. The exception was the association rate determination, which was accurate by direct measurement on the marker ligands yet not on the test compounds [77, 82, 83]. Recently, the Wanner group proved the concept of assessing the binding of test compounds to more than one target in the cell membranes by simultaneously monitoring multiple marker ligands of high selectivity for different dopamine receptor subtypes [82]. This again exploits the unique advantage of MS analysis of ligand mixtures, which enables label-free MS binding assays to substitute for or even surpass the conventional radioligand binding assays in ligand characterization for various GPCR targets.

22.3.4 FAC–MS

In the FAC–MS binding assay, GPCRs immobilized in the column with sufficient and lasting activity is critical to ensure data quality, yet it is a very challenging step to optimize considering the structural vulnerability of GPCR proteins in a nonnative chemical environment. A useful technique is immobilized artificial membrane (IAM) originally developed by Pidgeon et al. which is a monolayer of phospholipid analogs with functional headgroups covalently attached to silica beads [84]. This IAM-based stationary phase was then exploited in FAC analysis for immobilization of membrane proteins including quite a few GPCRs like opioid GPCRs [85], β_2-AR [86], P2Y_1 [87], GPR17 [49], and A_{2A}R [50]. This IAM method mimics the native environment to retain the activity and stability of GPCR, but nonspecific binding remains an intrinsic limit to certain degree [50]. Another method for receptor immobilization without IAM which relies on noncovalent recognition of biomolecules can partially circumvent the problem [50], but still has the risk of receptor dissociation. More recently, Zeng et al. developed a covalent site-specific immobilization method to minimize receptor dissociation from the stationary phase [88]. In this work, they engineered an haloalkane dehalogenase (Halo) tag to be installed on the C-terminus of three receptors

(β_2-AR, AT_1, and AT_2) so that they can be covalently immobilized on the silica gel modified with Halo-tag substrates.

With continuous improvement in the immobilization technique, experimental process, and mathematical modeling, FAC–MS has emerged as an alternative method for equilibrium binding measurement of GPCR–ligand interactions [50, 89, 90]. Temporini et al. first described the use of immobilized A_{2A}R-expressing membranes for determination of K_d and the number of binding sites for several known A_{2A}R ligands [50]. For the high affinity ligand ANR152, its K_d determined by FAC–MS (2.4 nM) was in good agreement with reported data (3.7 nM) [91]. By immobilizing A_{2A}R onto the inner surface of the silica capillary through avidin–biotin coupling to increase specific interaction, they were able to rank the affinity of multiple compounds for A_{2A}R in a mixture experiment.

To improve the speed and accuracy of GPCR–ligand interaction analysis through FAC–MS, Zhao et al. developed a mathematical model using the relationship between the molar amount of an injected solute and its capacity factor to measure the interactions between β_2-AR and five known ligands [89]. K_d values measured with this modified model correlated well to the radioligand binding data. Li et al. used adsorption energy distribution (AED) calculations to improve the adsorption models for the binding of three different ligands to immobilized β_2-AR, which resulted in more accurate K_d determination [90]. Collectively, these results have shown promises of using FAC–MS in characterization of specific GPCR–ligand interaction.

22.3.5 Native MS and HDX–MS

Native mass spectrometry (Native MS) and HDX–MS are two MS technologies of unique strengths in structural characterization of protein–ligand complexes as a whole. They provide structural information complementary to the classical structural approaches by probing the assembly and dynamics of macromolecular protein complexes in gas phase or in solution (Table 22.1).

The Robinson group has played a leading role in developing native MS techniques for the detection of intact membrane protein complexes in the gas phase [92–94]. Membrane proteins are usually encapsulated in optimized detergent micelles that preserve noncovalent protein–ligand interactions and the protein tertiary structure during transfer into the gas phase (Figure 22.4a). Native MS analysis facilitates measurements of not only the stoichiometry of macromolecular complexes but also the affinity and kinetics of ligand binding when conducted as a titration assay [96, 97]. Because of the low yield and instability after extraction of GPCRs from cell membranes, this membrane protein family had remained almost intractable until breakthroughs were made recently [97, 98]. Yen et al. established conditions to preserve noncovalent ligand binding to a GPCR protein, namely the human purinergic receptor $P2Y_1$ [95]. They observed the direct

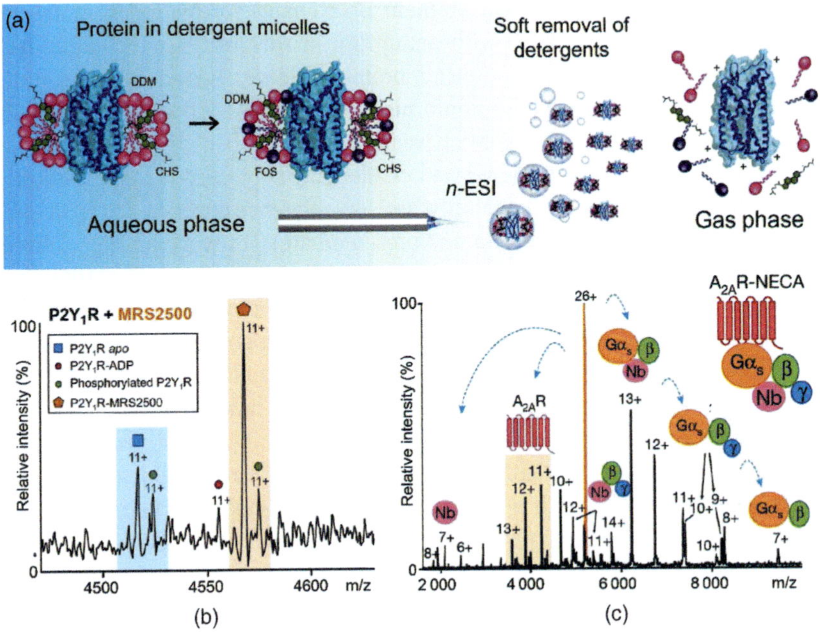

Figure 22.4 Native MS for intact GPCR complex analysis: (a) Illustration of gas phase activation of membrane proteins encapsulated in micelles, and transfer from solution to gas phase while retaining the ligand-binding site. (b) Mass spectrum recorded on $P2Y_1$ receptor following addition of the drug MRS2500. (c) Mass spectrum of $A_{2A}R$ receptor coupled to a trimeric G-protein complex in the presence of the stabilizing nanobody (Nb35). DDM: *n*-dodecyl β-D-maltoside; CHS, cholesteryl hemisuccinate; FOS, foscholine. Source: (a and b) Yen et al. [95]/AAAS/CC BY-4.0 and (c) Yen et al. [96]/with permission of Springer Nature.

binding of the endogenous ligands ADP and ATP, as well as an antagonist drug MRS2500 to $P2Y_1R$ in the gas phase (Figure 22.4b). Furthermore, they found phosphorylation of the receptor was suppressed in the presence of the drug by native MS analysis, which agrees with the role of the drug in inhibition of downstream signaling. More recently, Yen et al. utilized native MS to characterize three different class A GPCRs in complex with trimeric G protein assembly [96] (Figure 22.4c). Most interestingly, they found preferential binding of a particular phospholipid PIP2 over related lipids to receptors and binding of two PIP2 molecules could stabilize the complex formed between the β_1-adrenergic receptor (β_1-AR) and an engineered Gα subunit (mini-$G\alpha_s$), indicating the potential role of PIP2 in mediating coupling of GPCR and G_s protein. These new findings of GPCR interaction with endogenous lipids by native MS analysis could shed new light on receptor function and regulation, G protein coupling selectivity, as well as design of GPCR-targeting drugs [94, 99].

HDX–MS is based on the measurement of exchange of hydrogen atoms in the backbone of a biomolecule with deuterium in solvent. This technique has been extensively used for monitoring protein folding, ligand binding sites, protein structure dynamics, and conformational changes [100–103]. Although experimental conditions of HDX–MS have been well established for many soluble proteins, GPCRs are still considered very challenging subjects due to their notorious properties of insolubility, instability, and low sequence coverage for HDX–MS study [104–106]. The Griffin group first presented an HDX workflow specialized for charactering the β_2 adrenergic receptor bound to an inverse agonist [106]. The sequence coverage of protein dynamic regions was significantly increased under optimized conditions, which paved the way for probing conformational changes of β_2-AR receptor induced by binding of different types of ligands [106–108].

Additional H/D exchange-based techniques are available to determine protein–ligand binding affinities, such as stability of unpurified proteins from rates of H/D exchange (SUPREX) [109]. SUPREX monitors the global unfolding equilibrium of proteins with deuterium exchange to measure their thermodynamic stability. Epitope mapping is another application of the HDX toolbox to GPCR structural biology. Zhang et al. studied the effect of antibody binding on GPCR activation [110]. This study reveals that one antibody mAb1 behaves as an antagonist by blocking the orthosteric ligand access while the other one mAb23 not only blocks glucagon binding but also reduces basal receptor activity, acting as an inverse agonist.

Recently, a new emerging technique called Time-resolved hydrogen-deuterium exchange mass spectrometry (TRHDX–MS) has been applied to characterize the transient dynamic changes of GPCR complexes that cannot be captured by conventional HDX–MS at equilibrium states. Du et al. performed TRHDX analysis in millisecond resolution to monitor the structural dynamics of G_s protein upon coupling with the β_2-AR [111]. TRHDX also allows measurement of the kinetics parameters, i.e. fast k_{on} and k_{off} of protein–ligand interaction. Of course, in all types of HDX-MS experiments, changes in the HDX rate in certain protein regions may reflect direct interactions occurring in the ligand-binding site or indirect allosteric effects [102]. Therefore, the results from HDX–MS or native MS analysis to reveal new structural features or functional implications of GPCR–ligand interactions usually need to be corroborated by other orthogonal approaches.

22.4 Conclusion

In this chapter, we reviewed the fundamentals and recent applications of various MS-based technologies in detection, screening, and characterization of chemical ligands bound to specific GPCR targets. We also discussed technical strengths and limitations of each technique especially in the context of lead compound identification in the early-phase drug discovery pipeline. Although the majority of applications showcased here lead to discovery of small molecule ligands, they can be easily adapted to screening or detection of peptide ligands when conjugated with proteomics techniques. Owing to the remarkable sensitivity, specificity and high-throughput of current LC–MS/MS platforms, we anticipate that combination of affinity selection techniques with cutting-edge proteomics or metabolomics workflows would accelerate GPCR ligand discovery from diverse chemical resources including natural products and animal tissues. Unknown endogenous ligands to certain orphan receptors may be identified directly from crude tissue extracts using such unbiased and efficient screening technologies rather than by tracing down the bioactive compound through extensive fractionation.

To maintain the native conformation and full activity of the GPCR target is vital in any ligand screening or characterization experiment, which remains a great challenge for purified recombinant receptors. This can be resolved, at least in part, by developing more on-membrane or in-cell screening methods with high-resolution mass spectrometry analysis as a readout with high selectivity. Compared to DEL, current MS-based platforms still have large room to improve its throughput and sensitivity. Yet we should recognize the label-free advantage of affinity MS screen which avoids chemical functionalization-induced interference of protein–ligand interaction. We envision the MS-based toolkit will be continuously improved and expanded to facilitate GPCR ligand discovery and structural biology in an indispensable manner.

Acknowledgments

We thank ShanghaiTech University, the National Natural Science Foundation of China (No. 31971362), and National Key Research and Development Program of China (No. 2018YFA0507004) for funding.

Conflict of Interest

None declared.

References

1 Lagerstrom, M.C. and Schioth, H.B. (2008). Structural diversity of G protein-coupled receptors and significance for drug discovery. *Nat. Rev. Drug Discovery* 7 (4): 339–357.

2 Santos, R., Ursu, O., Gaulton, A. et al. (2017). A comprehensive map of molecular drug targets. *Nat. Rev. Drug Discovery* 16 (1): 19–34.

3 Ahn, S., Kahsai, A.W., Pani, B. et al. (2017). Allosteric "beta-blocker" isolated from a DNA-encoded small molecule library. *Proc. Natl. Acad. Sci. U.S.A.* 114 (7): 1708–1713.

4 Chen, D., Errey, J.C., Heitman, L.H. et al. (2012). Fragment screening of GPCRs using biophysical methods: identification of ligands of the adenosine A_{2A} receptor with novel biological activity. *ACS Chem. Biol.* 7 (12): 2064–2073.

5 Gesmundo, N.J., Sauvagnat, B., Curran, P.J. et al. (2018). Nanoscale synthesis and affinity ranking. *Nature* 557 (7704): 228–232.

6 Whitehurst, C.E., Nazef, N., Annis, D.A. et al. (2006). Discovery and characterization of orthosteric and allosteric muscarinic M_2 acetylcholine receptor ligands by affinity selection-mass spectrometry. *J. Biomol. Screening* 11 (2): 194–207.

7 Prudent, R., Annis, D.A., Dandliker, P.J. et al. (2021). Exploring new targets and chemical space with affinity selection–mass spectrometry. *Nat. Rev. Chem.* 5: 62–71.

8 Muchiri, R.N. and van Breemen, R.B. (2021). Affinity selection–mass spectrometry for the discovery of pharmacologically active compounds from combinatorial libraries and natural products. *J. Mass Spectrom.* 56 (5): e4647.

9 Chen, L., Jin, L., and Zhou, N. (2012). An update of novel screening methods for GPCR in drug discovery. *Expert Opin. Drug Discovery* 7 (9): 791–806.

10 Soudijn, W., Van Wijngaarden, I., and AP IJ. (2004). Allosteric modulation of G protein-coupled receptors: perspectives and recent developments. *Drug Discovery Today* 9 (17): 752–758.

11 Annis, D.A., Nazef, N., Chuang, C.C. et al. (2004). A general technique to rank protein-ligand binding affinities and determine allosteric versus direct binding site competition in compound mixtures. *J. Am. Chem. Soc.* 126 (47): 15495–15503.

12 Whitehurst, C.E., Yao, Z., Murphy, D. et al. (2012). Application of affinity selection-mass spectrometry assays to purification and affinity-based screening of the chemokine receptor CXCR4. *Comb. Chem. High Throughput Screening* 15 (6): 473–485.

13 Annis, D.A., Athanasopoulos, J., Curran, P.J. et al. (2004). An affinity selection-mass spectrometry method for the identification of small molecule ligands from self-encoded combinatorial libraries – discovery of a novel antagonist of *E coli* dihydrofolate reductase. *Int. J. Mass Spectrom.* 238 (2): 77–83.

14 Deng, Y., Shipps, G.W. Jr., Cooper, A. et al. (2014). Discovery of novel, dual mechanism ERK inhibitors by affinity selection screening of an inactive kinase. *J. Med. Chem.* 57 (21): 8817–8826.

15 Huang, X., Shipps, G.W. Jr., Cheng, C.C. et al. (2011). Discovery and hit-to-lead optimization of non-ATP competitive MK2 (MAPKAPK2) inhibitors. *ACS Med. Chem. Lett.* 2 (8): 632–637.

16 Luque-Garcia, J.L. and Neubert, T.A. (2007). Sample preparation for serum/plasma profiling and biomarker identification by mass spectrometry. *J. Chromatogr. A* 1153 (1, 2): 259–276.

17 Greening, D.W. and Simpson, R.J. (2010). A centrifugal ultrafiltration strategy for isolating the low-molecular weight (<or=25K) component of human plasma proteome. *J. Proteomics* 73 (3): 637–648.

18 van Breemen, R.B., Huang, C.R., Nikolic, D. et al. (1997). Pulsed ultrafiltration mass spectrometry: a new method for screening combinatorial libraries. *Anal. Chem.* 69 (11): 2159–2164.

19 Comess, K.M., Schurdak, M.E., Voorbach, M.J. et al. (2006). An ultraefficient affinity-based high-throughout screening process: application to bacterial cell wall biosynthesis enzyme MurF. *J. Biomol. Screening* 11 (7): 743–754.

20 Comess, K.M., Trumbull, J.D., Park, C. et al. (2006). Kinase drug discovery by affinity selection/mass spectrometry (ASMS): application to DNA damage checkpoint kinase Chk1. *J. Biomol. Screening* 11 (7): 755–764.

21 Chen, X., Li, L., Chen, S. et al. (2013). Identification of inhibitors of the antibiotic-resistance target New Delhi metallo-β-lactamase 1 by both nano-electrospray ionization mass spectrometry and ultrafiltration liquid chromatography/mass spectrometry approaches. *Anal. Chem.* 85 (16): 7957–7965.

22 Chen, X., Qin, S., Chen, S. et al. (2015). A ligand-observed mass spectrometry approach integrated into the fragment based lead discovery pipeline. *Sci. Rep.* 5: 8361.

23 Qin, S., Ren, Y., Fu, X. et al. (2015). Multiple ligand detection and affinity measurement by ultrafiltration and mass spectrometry analysis applied to fragment mixture screening. *Anal. Chim. Acta* 886: 98–106.

24 Fu, X., Wang, Z., Li, L. et al. (2016). Novel chemical ligands to Ebola virus and Marburg virus nucleoproteins identified by combining affinity mass spectrometry and metabolomics approaches. *Sci. Rep.* 6: 29680.

25 Wang, Z.H., Liang, H., Cao, H.J. et al. (2019). Efficient ligand discovery from natural herbs by integrating virtual screening, affinity mass spectrometry and targeted metabolomics. *Analyst* 144 (9): 2881–2890.

26 Chessari, G. and Woodhead, A.J. (2009). From fragment to clinical candidate – a historical perspective. *Drug Discovery Today* 14 (13, 14): 668–675.

27 Ma, J., Lu, Y., Wu, D. et al. (2017). Ligand identification of the adenosine A_{2A} receptor in self-assembled nanodiscs by affinity mass spectrometry. *Anal. Methods* 9 (40): 5851–5858.

28 Newman, D.J. and Cragg, G.M. (2012). Natural products as sources of new drugs over the 30 years from 1981 to 2010. *J. Nat. Prod.* 75 (3): 311–335.

29 Harvey, A.L., Edrada-Ebel, R., and Quinn, R.J. (2015). The re-emergence of natural products for drug discovery in the genomics era. *Nat. Rev. Drug Discovery* 14 (2): 111–129.

30 Rosen, J., Gottfries, J., Muresan, S. et al. (2009). Novel chemical space exploration via natural products. *J. Med. Chem.* 52 (7): 1953–1962.

31 Alexandrov, A.I., Mileni, M., Chien, E.Y.T. et al. (2008). Microscale fluorescent thermal stability assay for membrane proteins. *Structure* 16 (3): 351–359.

32 Peng, Y., Zhao, S., Wu, Y. et al. (2018). Identification of natural products as novel ligands for the human 5-HT_{2C} receptor. *Biophys. Rep.* 4 (1): 50–61.

33 Choi, Y., Jermihov, K., Nam, S.J. et al. (2011). Screening natural products for inhibitors of quinone reductase-2 using ultrafiltration LC–MS. *Anal. Chem.* 83 (3): 1048–1052.

34 Mulabagal, V. and Calderon, A.I. (2010). Development of binding assays to screen ligands for *Plasmodium falciparum* thioredoxin and glutathione reductases by ultrafiltration and liquid chromatography/mass spectrometry. *J. Chromatogr. B* 878 (13, 14): 987–993.

35 Chen, G.L., Wu, J.L., Li, N. et al. (2018). Screening for anti-proliferative and anti-inflammatory components from *Rhamnus davurica* Pall. using bio-affinity ultrafiltration with multiple drug targets. *Anal. Bioanal.Chem.* 410 (15): 3587–3595.

36 Wang, L., Liu, Y., Luo, Y. et al. (2018). Quickly screening for potential α-glucosidase inhibitors from guava leaves tea by bioaffinity ultrafiltration coupled with HPLC–ESI–TOF/MS method. *J. Agric. Food Chem.* 66 (6): 1576–1582.

37 Yang, Z.Z., Zhang, Y.F., Sun, L.J. et al. (2012). An ultrafiltration high-performance liquid chromatography coupled with diode array detector and mass spectrometry approach for screening and characterising tyrosinase inhibitors from mulberry leaves. *Anal. Chim. Acta* 719: 87–95.

38 Song, H.P., Zhang, H., Fu, Y. et al. (2014). Screening for selective inhibitors of xanthine oxidase from Flos Chrysanthemum using ultrafiltration LC–MS combined with enzyme channel blocking. *J. Chromatogr. B* 961: 56–61.

39 Song, H.P., Chen, J., Hong, J.Y. et al. (2015). A strategy for screening of high-quality enzyme inhibitors from herbal medicines based on ultrafiltration LC–MS and in silico molecular docking. *Chem. Commun.* 51 (8): 1494–1497.

40 Song, H.P., Wu, S.Q., Qi, L.W. et al. (2016). A strategy for screening active lead compounds and functional compound combinations from herbal medicines based on pharmacophore filtering and knockout/knockin chromatography. *J. Chromatogr. A* 1456: 176–186.

41 Schriemer, D.C., Bundle, D.R., Li, L. et al. (1998). Micro-scale frontal affinity chromatography with mass spectrometric detection: a new method for the screening of compound libraries. *Angew. Chem. Int. Ed.* 37 (24): 3383–3387.

42 Zhu, L.L., Chen, L.R., Luo, H.P. et al. (2003). Frontal affinity chromatography combined on-line with mass spectrometry: a tool for the binding study of different epidermal growth factor receptor inhibitors. *Anal. Chem.* 75 (23): 6388–6393.

43 Luo, H.P., Chen, L.R., Li, Z.Q. et al. (2003). Frontal immunoaffinity chromatography with mass spectrometric detection: a method for finding active compounds from traditional Chinese herbs. *Anal. Chem.* 75 (16): 3994–3998.

44 Zhang, B.Y., Palcic, M.M., Mo, H.Q. et al. (2001). Rapid determination of the binding affinity and specificity of the mushroom *Polyporus squamosus* lectin using frontal affinity chromatography coupled to electrospray mass spectrometry. *Glycobiology* 11 (2): 141–147.

45 Fort, S., Kim, H.S., and Hindsgaul, O. (2006). Screening for galectin-3 inhibitors from synthetic lacto-*N*-biose libraries using microscale affinity chromatography coupled to mass spectrometry. *J. Org. Chem.* 71 (19): 7146–7154.

46 Toledo-Sherman, L., Deretey, E., Slon-Usakiewicz, J.J. et al. (2005). Frontal affinity chromatography with MS detection of EphB2 tyrosine kinase receptor. 2. Identification of small-molecule inhibitors via coupling with virtual screening. *J. Med. Chem.* 48 (9): 3221–3230.

47 Zhang, B., Palcic, M.M., Schriemer, D.C. et al. (2001). Frontal affinity chromatography coupled to mass spectrometry for screening mixtures of enzyme inhibitors. *Anal. Biochem.* 299 (2): 173–182.

48 Ng, E.S.M., Yang, F., Kameyama, A. et al. (2005). High-throughput screening for enzyme inhibitors using frontal affinity chromatography with liquid chromatography and mass spectrometry. *Anal. Chem.* 77 (19): 6125–6133.

49 Temporini, C., Ceruti, S., Calleri, E. et al. (2009). Development of an immobilized GPR17 receptor stationary phase for binding determination using frontal affinity chromatography coupled to mass spectrometry. *Anal. Biochem.* 384 (1): 123–129.

50 Temporini, C., Massolini, G., Marucci, G. et al. (2013). Development of new chromatographic tools based on A_{2A} adenosine receptor subtype for ligand characterization and screening by FAC–MS. *Anal. Bioanal.Chem.* 405 (2, 3): 837–845.

51 Calleri, E., Ceruti, S., Cristalli, G. et al. (2010). Frontal affinity chromatography-mass spectrometry useful for characterization of new ligands for GPR17 receptor. *J. Med. Chem.* 53 (9): 3489–3501.

52 van Breemen, R.B. and Choi, Y. (2008). Development of a screening assay for ligands to the estrogen receptor based on magnetic microparticles and LC–MS. *Comb. Chem. High Throughput Screen* 11 (1): 1–6.

53 Rush, M.D., Walker, E.M., Burton, T., and van Breemen, R.B. (2016). Magnetic microbead affinity selection screening (MagMASS) of botanical extracts for inhibitors of 15-lipoxygenase. *J. Nat. Prod.* 79 (11): 2898–2902.

54 Zhang, B., Zhao, S., Yang, D. et al. (2020). A novel G protein-biased and subtype-selective agonist for a G protein-coupled receptor discovered from screening herbal extracts. *ACS Cent. Sci.* 6 (2): 213–225.

55 Goodnow, R.A., Dumelin, C.E., and Keefe, A.D. (2017). DNA-encoded chemistry: enabling the deeper sampling of chemical space. *Nat. Rev. Drug Discovery* 16 (2): 131–147.

56 Yang, H.F., Medeiros, P.F., Raha, K. et al. (2015). Discovery of a potent class of PI3K α inhibitors with unique binding mode via encoded library technology (ELT). *ACS Med. Chem. Lett.* 6 (5): 531–536.

57 Deng, H.F., Zhou, J.Y., Sundersingh, F.S. et al. (2015). Discovery, SAR, and X-ray binding mode study of BCATm inhibitors from a novel DNA-encoded library. *ACS Med. Chem. Lett.* 6 (8): 919–924.

58 O'Connell, T.N., Ramsay, J., Rieth, S.F. et al. (2014). Solution-based indirect affinity selection mass spectrometry – a general tool for high-throughput screening of pharmaceutical compound libraries. *Anal. Chem.* 86 (15): 7413–7420.

59 Bergsdorf, C. and Ottl, J. (2010). Affinity-based screening techniques: their impact and benefit to increase the number of high quality leads. *Expert Opin. Drug Discovery* 5 (11): 1095–1107.

60 Lu, Y., Qin, S.S., Zhang, B.J. et al. (2019). Accelerating the throughput of affinity mass spectrometry-based ligand screening toward a G protein-coupled receptor. *Anal. Chem.* 91 (13): 8162–8169.

61 Yang, H., Medeiros, P.F., Raha, K. et al. (2015). Discovery of a potent class of PI3Kα inhibitors with unique binding mode via encoded library technology (ELT). *ACS Med. Chem. Lett.* 6 (5): 531–536.

62 Touti, F., Gates, Z.P., Bandyopadhyay, A. et al. (2019). In-solution enrichment identifies peptide inhibitors of protein–protein interactions. *Nat. Chem. Biol.* 15 (4): 410–418.

63 Rich, R.L., Errey, J., Marshall, F. et al. (2011). Biacore analysis with stabilized G-protein-coupled receptors. *Anal. Biochem.* 409 (2): 267–272.

64 Qin, S., Meng, M., Yang, D. et al. (2018). High-throughput identification of G protein-coupled receptor modulators through affinity mass spectrometry screening. *Chem. Sci.* 9 (12): 3192–3199.

65 Narayanaswami, V. and Dwoskin, L.P. (2017). Obesity: current and potential pharmacotherapeutics and targets. *Pharmacol. Ther.* 170: 116–147.

66 Graaf, C., Donnelly, D., Wootten, D. et al. (2016). Glucagon-like peptide-1 and its class B G protein-coupled receptors: a long march to therapeutic successes. *Pharmacol. Rev.* 68 (4): 954–1013.

67 Annis, D.A., Shipps, G.W. Jr., Deng, Y. et al. (2007). Method for quantitative protein–ligand affinity measurements in compound mixtures. *Anal. Chem.* 79 (12): 4538–4542.

68 Annis, D.A., Chuang, C.C., and Nazef, N. (2007). ALIS: an affinity selection–mass spectrometry system for the discovery and characterization of protein–ligand interactions. *Mass Spectrom. Med. Chem.* 121–156.

69 Robertson, J.G. (2005). Mechanistic basis of enzyme-targeted drugs. *Biochemistry* 44 (15): 5561–5571.

70 Zhang, R.M., Mayhood, T., Lipari, P. et al. (2004). Fluorescence polarization assay and inhibitor design for MDM2/p53 interaction. *Anal. Biochem.* 331 (1): 138–146.

71 Liu, D.T., Guo, J., Luo, Y. et al. (2007). Screening for ligands of human retinoid X receptor-α. Using ultrafiltration mass spectrometry. *Anal. Chem.* 79 (24): 9398–9402.

72 Cheng, X. and van Breemen, R.B. (2005). Mass spectrometry-based screening for inhibitors of β-amyloid protein aggregation. *Anal. Chem.* 77 (21): 7012–7015.

73 Noel, F., Mendonca-Silva, D.L., and Quintas, L.E.M. (2001). Radioligand binding assays in the drug discovery process: potential pitfalls of high throughput screenings. *Arzneim.-Forsch.* 51 (2): 169–173.

74 Hofner, G. and Wanner, K.T. (2003). Competitive binding assays made easy with a native marker and mass spectrometric quantification. *Angew. Chem. Int. Ed.* 42 (42): 5235–5237.

75 Niessen, K.V., Hofner, G., and Wanner, K.T. (2005). Competitive MS binding assays for dopamine D_2 receptors employing spiperone as a native marker. *Chembiochem* 6 (10): 1769–1775.

76 Zepperitz, C., Hofner, G., and Wanner, K.T. (2006). MS-binding assays: kinetic, saturation, and competitive experiments based on quantitation of bound marker as exemplified by the GABA transporter $mGAT_1$. *ChemMedChem* 1 (2): 208–217.

77 Massink, A., Holzheimer, M., Holscher, A. et al. (2015). Mass spectrometry-based ligand binding assays on adenosine A_1 and A_{2A} receptors. *Purinergic Signal.* 11 (4): 581–594.

78 Grimm, S.H., Hofner, G., and Wanner, K.T. (2015). MS binding assays for the three monoamine transporters using the triple reuptake inhibitor (1*R*,3*S*)-indatraline as native marker. *ChemMedChem* 10 (6): 1027–1039.

79 Schmitt, S., Hofner, G., and Wanner, K.T. (2015). Application of MS transport assays to the four human γ-aminobutyric acid transporters. *ChemMedChem* 10 (9): 1498–1510.

80 Kern, F.T. and Wanner, K.T. (2015). Generation and screening of oxime libraries addressing the neuronal GABA transporter GAT_1. *ChemMedChem* 10 (2): 396–410.

81 Sichler, S., Hofner, G., Rappengluck, S. et al. (2018). Development of MS binding assays targeting the binding site of MB327 at the nicotinic acetylcholine receptor. *Toxicol. Lett.* 293: 172–183.

82 Schuller, M., Hofner, G., and Wanner, K.T. (2017). Simultaneous multiple MS binding assays addressing D_1 and D_2 dopamine receptors. *ChemMedChem* 12 (19): 1585–1594.

83 Neiens, P., Hofner, G., and Wanner, K.T. (2015). MS binding assays for D_1 and D_5 dopamine receptors. *ChemMedChem* 10 (11): 1924–1931.

84 Pidgeon, C. and Venkataram, U.V. (1989). Immobilized artificial membrane chromatography: supports composed of membrane lipids. *Anal. Biochem.* 176 (1): 36–47.

85 Beigi, F. and Wainer, I.W. (2003). Syntheses of immobilized G protein-coupled receptor chromatographic stationary phases: characterization of immobilized μ and κ opioid receptors. *Anal. Chem.* 75 (17): 4480–4485.

86 Beigi, F., Chakir, K., Xiao, R.P. et al. (2004). G-protein-coupled receptor chromatographic stationary phases. 2. Ligand-induced conformational mobility in an immobilized β_2-adrenergic receptor. *Anal. Chem.* 76 (24): 7187–7193.

87 Moaddel, R., Calleri, E., Massolini, G. et al. (2007). The synthesis and initial characterization of an immobilized purinergic receptor ($P2Y_1$) liquid chromatography stationary phase for online screening. *Anal. Biochem.* 364 (2): 216–218.

88 Zeng, K.Z., Li, Q., Wang, J. et al. (2018). One-step methodology for the direct covalent capture of GPCRs from complex matrices onto solid surfaces based on the bioorthogonal reaction between haloalkane dehalogenase and chloroalkanes. *Chem. Sci.* 9 (2): 446–456.

89 Zhao, X., Li, Q., Chen, J. et al. (2014). Exploring drug–protein interactions using the relationship between injection volume and capacity factor. *J. Chromatogr. A* 1339: 137–144.

90 Li, Q., Ning, X.H., An, Y.X. et al. (2018). Reliable analysis of the interaction between specific ligands and immobilized β_2-adrenoceptor by adsorption energy distribution. *Anal. Chem.* 90 (13): 7903–7911.

91 Volpini, R., Dal Ben, D., Lambertucci, C. et al. (2009). Adenosine A_{2A} receptor antagonists: new 8-substituted 9-ethyladenines as tools for in vivo rat models of Parkinson's disease. *ChemMedChem* 4 (6): 1010–1019.

92 Chorev, D.S., Baker, L.A., Wu, D. et al. (2018). Protein assemblies ejected directly from native membranes yield complexes for mass spectrometry. *Science* 362 (6416): 829–834.

93 Liko, I., Hopper, J.T.S., Allison, T.M. et al. (2016). Negative ions enhance survival of membrane protein complexes. *J. Am. Soc. Mass. Spectrom.* 27 (6): 1099–1104.

94 Robinson, C.V. (2019). Mass spectrometry: from plasma proteins to mitochondrial membranes. *Proc. Natl. Acad. Sci. U.S.A.* 116 (8): 2814–2820.

95 Yen, H.Y., Hopper, J.T.S., Liko, I. et al. (2017). Ligand binding to a G protein-coupled receptor captured in a mass spectrometer. *Sci. Adv.* 3 (6).

96 Yen, H.Y., Hoi, K.K., Liko, I. et al. (2018). PtdIns(4,5)P-2 stabilizes active states of GPCRs and enhances selectivity of G-protein coupling. *Nature* 559 (7714): 423–427.

97 Gault, J., Donlan, J.A.C., Liko, I. et al. (2016). High-resolution mass spectrometry of small molecules bound to membrane proteins. *Nat. Methods* 13 (4): 333–336.

98 Chun, E., Thompson, A.A., Liu, W. et al. (2012). Fusion partner toolchest for the stabilization and crystallization of G protein-coupled receptors. *Structure* 20 (6): 967–976.

99 Ambrose, S., Housden, N.G., Gupta, K. et al. (2017). Native desorption electrospray ionization liberates soluble and membrane protein complexes from surfaces. *Angew. Chem. Int. Ed.* 56 (46): 14463–14468.

100 Konermann, L., Pan, J.X., and Liu, Y.H. (2011). Hydrogen exchange mass spectrometry for studying protein structure and dynamics. *Chem. Soc. Rev.* 40 (3): 1224–1234.

101 Zheng, J., Strutzenberg, T., Pascal, B.D. et al. (2019). Protein dynamics and conformational changes explored by hydrogen/deuterium exchange mass spectrometry. *Curr. Opin. Struct. Biol.* 58: 305–313.

102 Chalmers, M.J., Busby, S.A., Pascal, B.D. et al. (2011). Differential hydrogen/deuterium exchange mass spectrometry analysis of protein–ligand interactions. *Expert. Rev. Proteomic.* 8 (1): 43–59.

103 Chandramohan, A., Krishnamurthy, S., Larsson, A. et al. (2016). Predicting allosteric effects from orthosteric binding in Hsp90-ligand interactions: implications for fragment-based drug design. *PLoS Comput. Biol.* 12 (6): e1004840.

104 Li, S., Lee, S.Y., and Chung, K.Y. (2015). Conformational analysis of G protein-coupled receptor signaling by hydrogen/deuterium exchange mass spectrometry. *Methods Enzymol.* 557: 261–278.

105 Zhang, X. (2017). Seven perspectives on GPCR H/D-exchange proteomics methods. *F1000Res.* 6: 89.

106 Zhang, X., Chien, E.Y.T., Chalmers, M.J. et al. (2010). Dynamics of the β_2-adrenergic G-protein coupled receptor revealed by hydrogen-deuterium exchange. *Anal. Chem.* 82 (3): 1100–1108.

107 Duc, N.M., Du, Y., Thorsen, T.S. et al. (2015). Effective application of bicelles for conformational analysis of G protein-coupled receptors by hydrogen/deuterium exchange mass spectrometry. *J. Am. Soc. Mass. Spectrom.* 26 (5): 808–817.

108 West, G.M., Chien, E.Y.T., Katritch, V. et al. (2011). Ligand-dependent perturbation of the conformational ensemble for the GPCR β_2 adrenergic receptor revealed by HDX. *Structure* 19 (10): 1424–1432.

109 Hopper, E.D., Roulhac, P.L., Campa, M.J. et al. (2008). Throughput and efficiency of a mass spectrometry-based screening assay for protein–ligand binding detection. *J. Am. Soc. Mass. Spectrom.* 19 (9): 1303–1311.

110 Zhang, H.N., Qiao, A.N., Yang, D.H. et al. (2017). Structure of the full-length glucagon class B G-protein-coupled receptor. *Nature* 546 (7657): 259–264.

111 Du, Y., Duc, N.M., Rasmussen, S.G.F. et al. (2019). Assembly of a GPCR-G protein complex. *Cell* 177 (5): 1232–1242.

23

Bioluminescence Resonance Energy Transfer (BRET) Technologies to Study GPCRs

Natasha C. Dale[1,2,3], *Carl W. White*[1,2,3], *Elizabeth K.M. Johnstone*[1,2,3], *and Kevin D.G. Pfleger*[1,2,3,4]

[1] *Molecular Endocrinology and Pharmacology, Harry Perkins Institute of Medical Research, QEII Medical Centre, Nedlands, Western Australia, Australia*
[2] *Centre for Medical Research, The University of Western Australia, Crawley, Western Australia, Australia*
[3] *National Centre, Australian Research Council Centre for Personalised Therapeutics Technologies, Australia*
[4] *Dimerix Limited, Nedlands, Western Australia, Australia*

23.1 Introduction

Bioluminescence Resonance Energy Transfer (BRET) technologies are widely used to study G protein-coupled receptor (GPCR) pharmacology, with the distinct advantages of being able to do so in live cells and in real-time [1–3]. Using BRET techniques, proximity can theoretically be monitored between any proteins, peptides, or other small molecules of interest, as long as BRET tags can be successfully appended without adversely affecting function.

BRET relies on the biophysical principle of nonradiative energy transfer from an energy donor to an energy acceptor (the BRET tags) when the complementary donor–acceptor pair are in close proximity (<10 nm) [4]. This phenomenon can be utilized to study cellular interactions and proximity by appending the BRET tags to small molecules, peptides, and/or proteins of interest via conjugation or genetic fusion.

BRET utilizes a luciferase enzyme along with a complementary luciferin substrate as the energy donor. Upon oxidation of the substrate, the luciferase will bioluminesce, producing an intrinsic energy source for subsequent energy transfer to occur. Forms of Renilla luciferase (Rluc), particularly the variant Renilla luciferase 8 (Rluc8), have historically been the preferred luciferases utilized for BRET assays [5–7]. More recently, an engineered luciferase called Nanoluciferase (Nanoluc; Nluc) was developed and has quickly gained traction within the GPCR

GPCRs as Therapeutic Targets, Volume 2, First Edition. Edited by Annette Gilchrist.

pharmacology field, as well as for wider biomedical applications. This rapid and broad adoption is due to its small size (19 kDa) and bright luminescence properties that can be used for a variety of improved and novel applications [8, 9].

With the possible applications of BRET rapidly expanding, the technique is known as the "undisputed workhorse for studying protein–protein interactions relevant to GPCR signaling" [10]. This chapter will discuss emerging applications of BRET to study GPCR pharmacology with rapidly increasing precision and depth.

23.2 BRET Overview: Advantages and Limitations

Figure 23.1 summarizes the principle of using BRET to study GPCR pharmacology. BRET offers many advantages over alternative techniques, making it an attractive experimental option. It enables time-dependent dynamics to be easily studied because of the ability to investigate interactions and/or proximity in live, intact cells in real time [1]. This provides a major advantage over assays that require cell lysis or solubilization, such as immunoassays, which can only provide

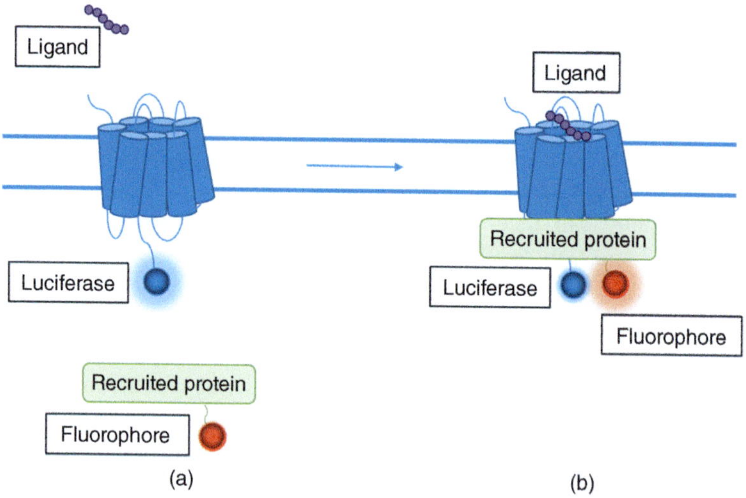

Figure 23.1 Schematic diagram of BRET between C′-terminally luciferase-tagged GPCR and a fluorophore-tagged intracellular protein. (a) Without agonist-stimulated recruitment of the intracellular protein to the GPCR, the luciferase and fluorophore tags are not in close proximity, as such no appreciable BRET signal is produced. (b) Upon agonist binding, the fluorophore-tagged intracellular protein is recruited to the GPCR bringing the luciferase and fluorophore tags in close proximity. Resonance energy transfer occurs from the luciferase to the fluorophore, producing a measurable BRET signal that denotes the protein has been recruited to the GPCR.

end-point readings. The scalability of BRET also makes the technique amenable to higher throughput applications. As such, generation of pharmacological parameters (such as affinity and potency) for a large set of receptors and/or ligands is achievable with relative ease and efficiency. BRET also side-steps experimental difficulties seen with fluorescence resonance energy transfer (FRET) such as photobleaching [11, 12]. Additionally, BRET has proven to be an assay platform that is readily adaptable to new technological advances. The integration of recent developments in engineered luciferases [13], biosensors [14], and genome-editing techniques [15] allows for the production of high powered and broadly applicable BRET-based assays. Indeed, substantial strides have been made in developing novel ways of using BRET to study an ever-widening range of facets of GPCRs with greater precision.

BRET is a versatile and valuable tool, but as with any approach, it is important to be aware of its potential limitations for a given purpose. As the luciferase enzyme requires an oxidizable complementary substrate to produce luminescence, this can be a source of limitation for some experimental conditions. Rluc8 and Nluc's substrates, coelenterazine-h and furimazine, respectively, limit BRET detection to approximately one hour (with variation depending on experimental conditions such as cell number and luciferase expression) as both substrates are oxidized by the luciferase relatively rapidly. This limitation has been overcome by the development of caged substrate derivations, EnduRen (caged coelenterazine-h for Renilla luciferases) and Endurazine (caged furimazine for Nluc). These substrates enable prolonged detection of BRET (>6 hours) as metabolism of the caged substrates by intracellular esterases results in sustained release of bioavailable substrate for oxidation. This in turn produces sustained luminescence, enabling BRET to be detected for extended periods of time [16–19]. Due to the requirement for an external substrate, additional consideration is needed when adding biological agents like ascorbic acid [20] to ensure there is no interference with the luciferase–substrate reaction. It may also be important to consider the effect of the oxidation reaction and substrate addition on oxidation-sensitive applications.

Perhaps the largest limitation of BRET is the need to create fusion constructs for at least the luciferase-tagged protein of interest. However, recent advances mean that not all of the luciferase needs to be fused to the protein of interest (see below). In some respects, this presents an experimental constraint by requiring generation of specific constructs before BRET experiments can be conducted. However, cellular expression can be more straightforward than having to purify protein. More importantly, fusing to a BRET tag has the potential to perturb the function of the molecule of interest. This effect can be partially mitigated by thorough validation of fusion constructs prior to use for BRET and through comparison to results produced using additional adjunct methods [1]. While it is important to be aware of the possibility of this effect, it is not unique to BRET. Fusion constructs have become widely used within the biomedical field, particularly with microscopy applications. In addition, as the efficiency of BRET is dependent on the proximity and orientation of the donor and acceptor species, the placement of the BRET tags needs to be considered for BRET to be successfully observed as well as for optimal assay sensitivity and dynamic range to be achieved. While for many proteins placement of BRET tag(s) on the N- or C-terminus (e.g. GPCRs and β-arrestins) is sufficient to monitor interaction, for other proteins such as G proteins or intramolecular BRET biosensors this is not always the case. Without structural information to guide the placement of the tags, inefficient BRET configurations may be generated. Indeed, many first-generation G protein biosensors that assess G protein activation by monitoring changes in BRET following heterotrimeric G protein conformational change have suboptimal signals, which were only recently improved following extensive optimization [21].

Additional advantages and limitations that are specific to certain BRET applications are identified and addressed throughout this chapter as appropriate.

23.3 Emerging BRET Techniques

23.3.1 Trafficking Assays

While receptor trafficking through cellular compartments following biosynthesis and internalization is an integral mechanism in the regulation of GPCR signaling [22, 23], monitoring the cellular localization of receptors and other proteins using BRET has traditionally been limited [1].

A bystander BRET-based assay to monitor receptor trafficking was first described by Lan et al. [24]. This assay utilizes Venus-tagged K-Ras and Rab proteins as markers of specific subcellular compartments, enabling identification of the location of luciferase-tagged receptors within the same subcellular compartments. The specificity of this assay technique comes from the well-characterized localizations of K-Ras and Rab proteins that show restricted, highly specific expression in different subcellular compartments. Rab proteins indicate different endosomal and subcellular compartments and K-Ras is localized to the plasma membrane [25, 26] as indicated in Figure 23.2.

In Lan et al. [24], the trafficking of β_2AR-Rluc8 was monitored through proximity to the C′-terminal fragment of K-Ras, which acts as a plasma membrane marker when conjugated with Venus (Venus-K-Ras), and the early endosome marker Rab5 (Venus-Rab5). Confocal microscopy confirmed the restricted expression of both markers to the appropriate compartments. Agonist stimulation resulted in decreased BRET with Venus-K-Ras and a concomitant increase in BRET signal with Venus-Rab5. The BRET response was agonist

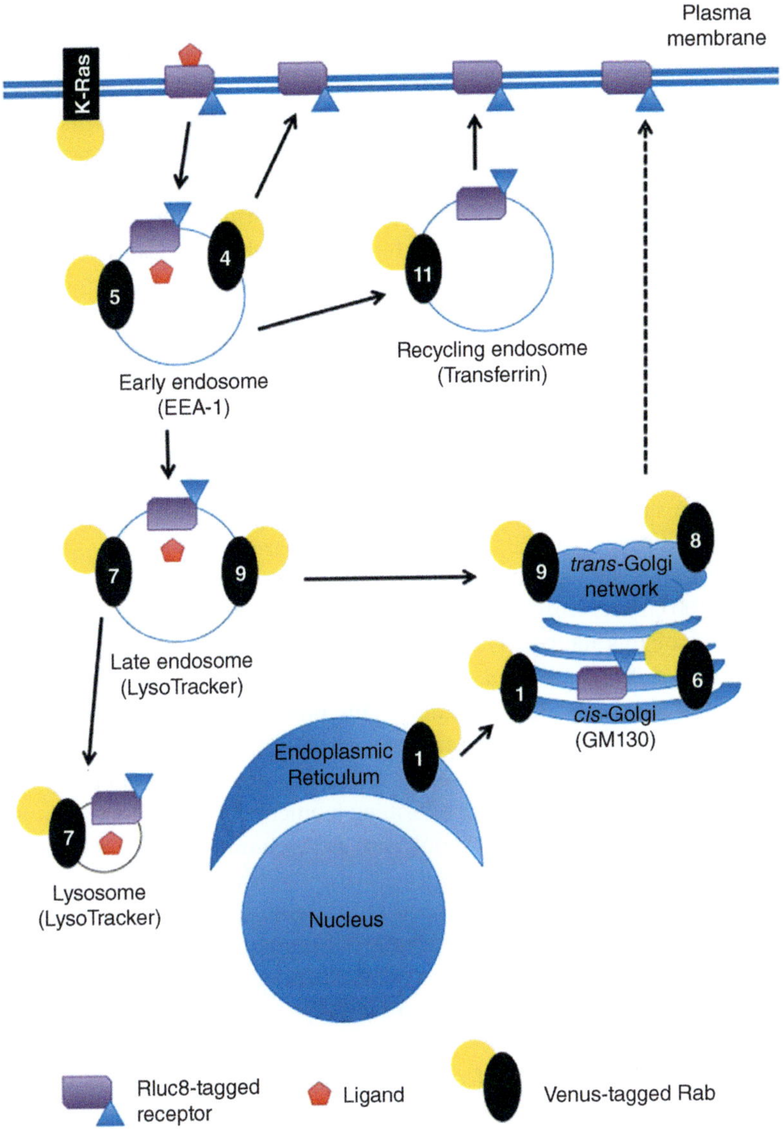

Figure 23.2 A simplified schematic representation of subcellular marker localization and receptor trafficking. Ligand-induced trafficking as well as constitutive localization was monitored using Rluc8-tagged wildtype or mutant V_2R by measuring proximity via extended BRET (eBRET) with the plasma membrane marker Venus/K-Ras, or the subcellular compartment markers Rabs: Venus/Rab5a (5) for early endosomes; Venus/Rab4 (4) for early endosome recycling; Venus/Rab11 (11) for recycling endosomes; Venus/Rab7a (7) for late endosomes/lysosomes; Venus/Rab9 (9) for late endosome trafficking to the *trans*-Golgi network; Venus/Rab1 (1) for endoplasmic reticulum trafficking to the *cis*-Golgi; Venus/Rab6 (6) for Golgi apparatus and *trans*-Golgi network; or Venus/Rab8 (8) for *trans*-Golgi network to plasma membrane. The subcellular markers used for the confocal microscopy validation are also included for comparison. Source: Tiulpakov et al. [19]/with permission of Oxford University Press.

concentration-dependent, with various inhibitors of internalization machinery giving evidence the BRET response was internalization-dependent. A subsequent study used a similar technique to produce markers for intracellular organelles using organelle membrane-restricted proteins as compartment markers [27]. This study also emphasized the potential advantages of monitoring subcellular localization by BRET compared to techniques like confocal microscopy. The high proximity-dependence of BRET allows for the determination of subcellular location with greater resolution than can be observed with confocal microscopy. Additionally, the relatively low intervention needed for this assay lends itself well to larger-scale investigations.

Endosomal trafficking and localization of receptors using Rab proteins was further expanded upon by Tiulpakov et al. [19], who described and validated additional Venus-Rab constructs to expand the BRET endosomal compartment marker panel (Figure 23.2). Using this expanded assay, differences in the trafficking of Vasopressin receptor 2 (V_2R) due to clinically relevant mutations was successfully observed [19]. A similar trafficking assay approach using RlucII and recombinant green fluorescent protein (rGFP) as the donor/acceptor pair has also been developed [28]. This approach was subsequently used to investigate G protein-coupling of a panel of 100 GPCRs with their prototypical ligands [29].

A limitation of this method is the need for an appropriate high density of acceptor-labeled compartment markers to be expressed in order to sufficiently detect luciferase-tagged protein proximity. This presents the issue of potential perturbation of function that accompanies overexpression; however, the required expression level of acceptor appears to be well tolerated. The possibility of the acceptor-tagged compartment markers forming a functional interaction with the luciferase-tagged protein of interest should also be taken into account, as this will impact the analysis of a BRET signal as a passive proximity reading [27].

BRET trafficking assays have been used successfully in a variety of contexts to investigate internalization and trafficking of a number of receptors [15, 19, 30–36]. In addition, it is now recognized that GPCR signaling occurs not only at the plasma membrane but also from intracellular compartments [37]. Using the same bystander BRET framework developed to monitor receptor internalization and trafficking, subcellular receptor signaling can also be investigated. Martin and Lambert [35] used bystander BRET approaches to monitor localization and trafficking of $G\alpha_s$-Rluc8 in different cellular compartments following β_2-adrenoceptor (β_2AR) activation. Similarly, ligand-mediated recruitment of mini-G (mG) proteins (discussed below) to adenosine A_1 receptors (A_1Rs) or β_2AR located in endosomal compartments was investigated by monitoring changes in BRET between Nluc on the mG protein and Venus-tagged intracellular markers [38]. The widespread use of these assays demonstrates their versatility as a tool to better understand how internalization and trafficking mechanisms may influence, or be influenced by, other aspects of a receptor's characteristics or function.

23.3.2 Intramolecular Conformational BRET Biosensors

In 2016, Lee et al. [39] developed intramolecular fluorescein arsenical hairpin (FlAsH) BRET biosensors to monitor conformational changes in β-arrestin2 (β-arr2) following recruitment to GPCRs. FlAsH is a fluorescent tag system in which a FlasH recognition motif (CCPGCC) is introduced into the protein sequence. A fluorescent detection reagent (fluorescein) will subsequently bind to the motif upon addition [40], forming the fluorescent acceptor. These BRET-FlAsH-based biosensors used both Rluc (energy donor) and FlAsH (energy acceptor) fused to β-arr2, with changes to intramolecular BRET reporting conformational changes in β-arr2.

Lee et al. [39] used these biosensors to study receptor-specific conformations of β-arr2 and investigated how different conformational states may affect β-arr2 function. By introducing the FlAsH binding motif at different points within the β-arr2 sequence and keeping Rluc's position constant, different β-arr2 conformations can be monitored using BRET. Using these biosensors, evidence of receptor-specific β-arr2 conformational states were observed. Additionally, similar BRET response profiles were observed for receptors thought to have similar β-arr2 interactions. Similar biosensors were subsequently used to probe for biased agonism at Galanin receptor 2 (GAL_2R) [41]. This study replaced Rluc with Nluc and subsequently used the biosensors to give evidence of ligand-specific β-arr2 conformations at GAL_2R, suggesting the differences seen in β-arr2 conformation induced by different ligands may play a role in the biased agonism observed at this receptor.

A similar approach utilizing Nluc and the fluorescent protein CyOFP1 fused to different forms of β-arr2 (wildtype, a R170E mutant that detects for partially phosphorylation-independent conformations and a finger-loop region deletion mutant that detects C′-terminal GPCR interactions) has since been employed [42]. This method has the advantage of producing clear, conformational readouts as opposed to the relative conformational profile given by the BRET-FlAsH biosensors.

While FlAsH-based biosensors require stringent technical procedure adherence once established [43], BRET-based intramolecular FlAsH biosensors show potential as powerful tools to screen for the relative effects of a range of ligands and receptors on β-arr2 conformation. This intramolecular BRET technique also has potential to be expanded to other signaling proteins to further probe the importance of effector conformation to GPCR signaling. Indeed, these BRET-FlasH biosensors have been used to investigate conformational changes within GPCRs upon activation. Without a ligand, GPCRs adopt various conformations. Binding of ligand to receptor increases the chance of adopting a distinct conformation. A number of BRET-FlAsH intramolecular-based biosensors have now been developed to investigate ligand-induced GPCR conformational changes [44]. These biosensors typically consist of a luciferase on the C′-terminus of the receptor and the FlAsH-tag within an intracellular loop. Unlike larger fluorescent acceptors, the small size of the tetracysteine recognition motif (CCPGCC) means receptor function is less likely to be disrupted. However, placement still needs to be carefully optimized and GPCRs with small intracellular loops may not tolerate these small insertions. Bourque et al. [45] described the generation of FlAsH-BRET biosensors to examine agonist-induced conformational changes in β_2AR, Angiotensin II receptor type I (AT_1R) and Prostaglandin $F_{2\alpha}$ receptor (FP). The same authors then used the same FlAsH-BRET approach to detect allosteric changes in conformation in an AT_1R-FP heterodimer [46]. Further demonstrating the applicability of the approach, Powlowski et al. [47] investigated ligand-induced conformational changes in the Serotonin 5-HT_{2A} receptor in both the model HEK293 cell line as well as the more physiologically relevant neuro-2A cell line. This is an important consideration as the cellular proteome and thus GPCR-interactome may result in different conformations depending on the cellular and subcellular contexts.

23.3.3 Mini-G Proteins

mG proteins are engineered GTPase domains of the Gα subunits from heterotrimeric G proteins. They form stable complexes with receptors, differing from the binding of wildtype G proteins in which the interaction is transient. Initially developed for use in structural studies such as crystallization and Cryo-EM to stabilize receptors within an active-state [48–51], they were subsequently optimized for use as biosensors of GPCR coupling and activation, with implications for the development of further novel assays using the mG protein biosensors.

mGs was the first mG biosensor to be developed, following isolation of the GTPase domain of $G\alpha_s$ and subsequent mutagenesis to improve thermostability of mGs, in particular the receptor-bound mGs conformation [52]. Of particular functional importance, these changes caused loss of plasma membrane and Gβγ association. This resulted in the mG protein existing in, and being recruited from, the cytosol. Changes to the GTPase also resulted in the uncoupling of nucleotide release upon binding to a GPCR, leading to stable complex formation between GPCR and mG [38, 52].

As the mutations introduced into mGs were in conserved regions of the Gα structure, it was hypothesized that these same mutations would be transferable to other Gα subunits to produce mG proteins for all Gα families [53]. From this initial experiment, mGolf, mGo1, and mG12 were successfully developed. In contrast, incorporation of the mutations used to produce mGs were not successful in producing mGz, mGt, mG16, mGq, and mGi1 proteins. Further attempts focused on producing an mG protein with $G\alpha_q$ characteristics. It was suggested that the mGq produced by introducing the mutations used to create mGs may not fold correctly, causing it to be unstable. As such, the strategy for producing an mG protein with $G\alpha_q$ characteristics was changed, now focusing on the possibility of creating a chimeric mG protein from mGs that had the functional specificity of $G\alpha_q$. This was achieved through mutation of amino acids in mGs to those found within $G\alpha_q$ to produce the chimeric mGsq. Following the success of this method, a similar method was employed to produce the chimeric mG protein mGsi [53]. From this work, a panel of mG proteins for multiple Gα families was produced (mGs, mGolf, mGo1, mG12, mGsq, and mGsi), which show promise not only for structural studies but also for probing GPCR function.

The potential to use mG proteins as biosensors to study GPCR pharmacology when used with BRET and other biophysical techniques was explored by Wan et al. [38]. The mG proteins mGs, mGsi, mGsq, and mG12 were N′-terminally Venus tagged and also tagged with an N-terminal nuclear export sequence (NES) to restrict Venus-mG expression to the cytosol only. When NES-Venus-mGs and β_2AR-Rluc8 were co-transfected into HEK293 cells, isoprotenerol concentration-response curves were successfully constructed for proximity between NES-Venus-mGs and β_2AR-Rluc8, indicating successful recruitment and the formation of a GPCR-mG complex. This complex was observed to be very stable, with noradrenaline-induced proximity remaining intact despite addition of the β_2AR inverse agonist ICI 118551, with dissociation observed at 15 minutes post-addition. In contrast, acetylcholine-induced proximity between Rluc8-tagged muscarinic acetylcholine receptors (M_3R and M_4R) and either NES-Venus-mGsq or NES-Venus-mGsi exhibited rapid dissociation upon treatment with atropine, indicating variation in the stability of GPCR-mG complexes. Additionally, this BRET-mG recruitment assay was observed to accurately report relative ligand potencies at β_2AR.

Using the panel of mG proteins to screen for receptor coupling may also prove to be an effective tool to determine the coupling partners of receptors. Prominent and minor coupling partners appear to be accurately reported by the mG proteins; however, this screen should be used as a tool for probing potential coupling for further research. As shown in Wan et al. [38], recruitment of multiple mG proteins can often be observed for one receptor, indicating the need for further research into the coupling partners of some receptors.

Important considerations for the use of mG proteins in functional assays include taking into account the effect mG-induced stabilization of the receptor may have on the perceived potencies of ligands. Furthermore, basal BRET signal appears to be at least in part a result of mG complex formation in the absence of ligand, evidenced by a decrease in BRET upon addition of a known antagonist [38]. While the slow dissociation kinetics exhibited by mGs has beneficial applications, the ability for this to negatively impact results, especially those investigating kinetics, is also an important consideration. Nevertheless, Wan et al. [38] clearly showed the power of using mG biosensors combined with BRET-based assays to investigate ligand efficacy, bias, and GPCR-G protein coupling with the potential for further novel applications.

mG proteins have since seen rapid uptake in their use to probe the pharmacology of a range of receptor targets, with BRET-based assays proving a popular assay technique to utilize these biosensors. mG recruitment to wildtype and mutant Class F receptors was monitored using BRET [54]. Using this assay format, evidence for mutations within a proposed molecular switch region of Class F receptors leading to the stabilization of the active receptor conformation was observed, as mutant receptors showed higher potency recruitment of mG proteins compared to wildtype receptors. This assay format has also been used to observe potentiation of AngII-induced mGsq recruitment at AT_1R by LVV-Hemorphin-7 [55], to characterize Smoothened receptor (SMO) agonist SAG1.3 as a Frizzled-6 receptor (FZD_6) partial agonist [56] and to characterize novel opioid agonists [57], among other applications [58].

Pritchard et al. [36] in particular highlights the power of BRET-based mG recruitment assays, using mG proteins to investigate the functional implications of a novel missense variant in the gene encoding Endothelin A receptor (ET_A). In combination with additional cutting-edge BRET assays such as the trafficking assay, which was used to determine the effect of the variant on receptor trafficking and cell surface expression, the mG protein recruitment assay was used to characterize the G protein coupling capabilities of the mutant ET_A receptor. Using this method, a decrease in mG recruitment to the mutant receptor was observed, giving evidence for the mechanism underlying the variant's loss of function. This study gives an example of the broad reach of these technologies that we are likely to see further developed in the future, alongside non-BRET-based techniques utilizing mG proteins [59].

23.3.4 Signaling Pathway Biosensor Panels

A real-time assay technique that is capable of providing robust, rapid results on the signaling pathways of receptors and ligands is highly desirable for understanding the signaling capacity of GPCRs as well as for drug development, particularly if its throughput is good relative to other profiling technologies. When understanding GPCR function, it is important to consider that for receptors to generate a signaling response they must couple to a subset of the possible G protein subunits present within a particular cellular environment. While G protein coupling can be promiscuous, many GPCRs show some level of preference for particular G proteins. However, for many GPCRs the mechanisms and selectivity of G protein coupling are poorly understood. Two recent studies have aimed to provide such an assay platform to investigate GPCR-G protein coupling using BRET-based technologies.

Avet et al. [29] utilize an effector membrane translocation assay (EMTA) that uses a rGFP-tagged plasma membrane marker to detect translocation of RlucII-tagged G protein- or β-arrestin-specific effector proteins to the plasma membrane upon receptor activation. Using this method, the panel of pathway-specific effector proteins can be used to probe the signaling pathways of receptors and the effect of biased ligands. Indeed, Avet et al. [29] demonstrate the successful use of the panel to characterize the coupling of 100 receptors with their prototypical ligand. An advantage of this strategy in particular is the use of a consistent method to measure effector recruitment. This allows for direct comparison between BRET outputs, given the appropriate normalizations have been conducted. In addition, a "quasi-universal" biosensor is described with the potential to be used as a rapid receptor activation biosensor for high-throughput ligand screens. This has exciting potential for use to detect off-target receptor activation that could see applications in safety screening for therapeutics, among other applications.

In comparison, Olsen et al. [21] describe a complete panel of 14 optimized Gα-Rluc8/Gβγ-GFP2 probes to screen for activation of specific G protein signaling cascades. Of note, these biosensors were reported to detect the activation of G proteins that are not reported by nonoptimized G protein dissociation assays, and as such represent a more sensitive assay for screening G protein activation than previously used constructs.

Both these studies give evidence using their developed assay panels that many GPCRs show diverse and often promiscuous coupling to signaling cascades, with both studies identifying previously unreported pathway coupling. Such assays could play an important role in future research aiming to uncover the large-scale signaling properties of GPCRs and as tools to screen for biased ligands with therapeutic potential. Moreover, neither approach requires modification of the receptor making them applicable to study endogenous receptors. This is an important consideration, as both studies were performed in model cells lines with exogenously expressed proteins, and GPCR-G protein coupling in a native cellular environment will be dependent on the available proteome. As such, these approaches performed in physiologically relevant cell lines with endogenous receptors or coupled with genome-editing to maintain native protein expression levels may allow for better prediction of GPCR signaling *in vivo*.

23.3.5 BRET Assays and Nanobodies

Nanobodies are single domain antibody fragments generated from heavy chain only antibodies found in camelids or cartilaginous fish. Nanobodies have been widely used in the GPCR field due to their ability to recognize and stabilize specific receptor/effector conformations as well as act as allosteric modulators of GPCR function. In addition, nanobodies expressed intracellularly, known as intrabodies, can be used to investigate GPCR signaling, particularly from intracellular compartments [60]. Due to their versatility, a number of recent studies have demonstrated coupling of nanobodies with BRET-based approaches to detect active/inactive GPCR states in live cell assays.

As part of their work determining the active structure of the Kappa opioid receptor (κOR), Che et al. [61] used a BRET approach to confirm that nanobody (Nb6) recognized the κOR inactive state. In this study, they observed high basal BRET between Nb6-YFP (yellow fluorescent protein) and κOR-Rluc8 that was reduced by a selective κOR agonist. Using this BRET approach, they also found that Nb39, which was originally described as recognizing the active state of the Mu opioid receptor (μOR), could be recruited to and stabilize the κOR active state. Demonstrating the utility of these approaches in understanding GPCR function, the authors found that $G\alpha_{i1}$ inhibited κOR/Nb39 interactions while β-arr2 modestly promoted the κOR/Nb39 interaction, which the authors use to support previous indications that β-arrestins and G proteins recognize different receptor conformations. The authors subsequently used the same techniques to fully characterize Nb6 as acting allosterically to stabilize κOR in a ligand-dependent inactive state [58].

Using Nb80, which detects the active state of β_2AR, Kilpatrick et al. [62] established a live cell NanoBRET approach to investigate agonist-induced receptor activation. In cells expressing β_2AR-Nluc and Nb80-GFP, application of isoprenaline resulted in a concentration-dependent increase in BRET that was completely inhibited by the β_2AR-selective antagonist ICI 118551. Similarly, in a recent study investigating biased agonism at the μOR, Gillis et al. [57] used a nanobody-BRET-based approach to investigate the ability of ligands to induce the active conformation of μOR. As part of their systematic evaluation of μOR ligands, BRET between Nb33-Venus (which recognizes the active conformation of μOR) and μOR-Nluc was monitored as part of a panel of differentially amplified assays. As this BRET signal can be assumed to represent a direct interaction, it helped demonstrate that low intrinsic efficacy may explain the improved therapeutic profile of so-called "biased μOR ligands." In addition to developing BRET-based assays where the nanobody binds to intracellular domains, a recent example has demonstrated a BRET and luciferase complementation assay using a nanobody that recognizes an extracellular domain. By fusing a small 11 amino acid fragment

(HiBiT) of Nluc (both Nluc and fragments are discussed in detail below) to the nanobody VUN400 that recognizes the second extracellular loop of the human CXCR4 chemokine receptor, Soave et al. [63] developed NanoBRET and Nluc complementation approaches to monitor specific CXCR4 conformations. This approach could be used to detect binding of orthosteric and allosteric ligands. It could also be used in both model cell systems, i.e. HEK293 cells overexpressing CXCR4, as well as in an immortalized Jurkat T-cell line natively expressing CXCR4.

While coupling of nanobodies with BRET-based approaches is still in its infancy, the available toolbox of nanobodies against GPCRs and/or their effectors is being continuously expanded. Due to their small size, stability, and ability to detect specific conformational states, it is likely that these approaches will become powerful tools to investigate GPCR function.

23.4 Novel NanoBRET Assays

NanoBRET is the name given to BRET assays that utilize Nluc as the luciferase [18, 64]. Optimized substrates have been developed and utilized for Nluc, namely furimazine and its caged derivative [19] now known as Endurazine. Most if not all BRET assays can be translated to the NanoBRET platform, but NanoBRET affords the opportunity for novel assay development through the distinct features of Nluc compared to other luciferases used in BRET. Specifically, Nluc's relatively small size (19 kDa compared to 36 kDa for Rluc8), bright luminescence output and stability in various environmental conditions allow for a widened range of BRET applications through NanoBRET [9].

23.4.1 Ligand Binding

A major advantage of NanoBRET is the ability to monitor ligand binding to GPCRs using an N′-terminally Nluc-tagged receptor and a fluorophore-tagged ligand. Nluc's derivation from a secreted protein enables the fusion construct to overcome the apparent limitation of other luciferases that prevents the N′-terminally tagged fusion construct from trafficking correctly to the cell surface, remaining trapped within the cell [13]. Additionally, Nluc's smaller size compared to Rluc8 reduces the chance of the luciferase tag sterically hindering the ligand-binding pocket [65]. The NanoBRET ligand-binding assay is a major development in the ability to observe ligand binding to GPCRs with many advantages over traditional methods. It overcomes limitations of other methods such as safety concerns that limit use of high concentrations of radiolabeled ligands when using radioligand binding assays [65] and high background signal with FRET-based ligand binding assays [66]. The assay exhibits high specificity and sensitivity due to the inherent distance dependence of BRET, which greatly reduces background signal, and the high luminescence yield of Nluc contributes to assay sensitivity. As such, this technique has seen swift uptake to study ligand binding for a range of receptors and ligands.

Both direct binding of a fluorescently tagged ligand and competitive binding of untagged ligands in the presence of a saturating amount of fluorescent ligand have become widely used methods to study ligand binding using NanoBRET [56, 62, 67–75]. In addition to studying direct, equilibrium-based binding parameters, these assays have also been used as a way to probe other aspects of ligand binding such as allosteric modulation and ligand-binding kinetics [13]. Examples of saturation, competition, and kinetic ligand-binding assays are shown in Figure 23.3.

Allosteric modulation and the role of multiple binding sites can be investigated using competition binding. Differences in the reported affinity of unlabeled ligands depending on the fluorescent ligand used, a phenomenon known as probe dependence, can indicate potential allosterism [65, 76]. Potential allosterism has been identified using this method for histamine H_1 and adenosine A_3 receptors [13, 77]. The effect of known allosteric ligands on orthosteric ligand binding can also be investigated [78, 79]. For example, Boursier et al. [79] monitored the effect of an allosteric modulator on the affinities of a panel of β_2AR ligands. Using competition binding of unlabeled ligands to HiBiT-β_2AR conjugated to LgBiT in the presence of fluorescently labeled-propranolol (propranolol-NB590), changes to the affinity of each ligand in the presence of a known positive allosteric modulator (PAM) were observed. The use of HiBiT–LgBiT complementation to study BRET ligand binding is further discussed under NanoBiT complementation.

Monitoring the kinetics of ligand binding has been dramatically simplified with NanoBRET, as measurements can be taken in live cells in real-time, as first demonstrated by Stoddart et al. [13]. From these kinetic studies, association (K_{on}) and dissociation (K_{off}) rate constants can be determined, and from these, ligand affinity (K_D) can be derived. Studies that have used this technique have shown similar K_D values determined from saturation and kinetic analyses, further validating this method to study single-step binding modes [13, 67, 77, 80]. Multistep binding can be indicated by discrepancy between the K_D derived from saturation binding and kinetic data [70]. Alcobia et al. [68] used kinetic ligand-binding analysis to show that propranolol-BODIPY630/650 rapidly bound to Nluc-β_2AR but dissociated slowly as determined by the association and dissociation rate constants. Stoddart et al. [77] developed fluorescent histamine H_1 receptor antagonists from mepyramine and VUF13816. Kinetic analysis showed VUF13816-based fluorescent ligands exhibited approximately fourfold longer target residence times compared to mepyramine-based fluorescent ligands despite the ligands showing similar K_D values in saturation binding experiments.

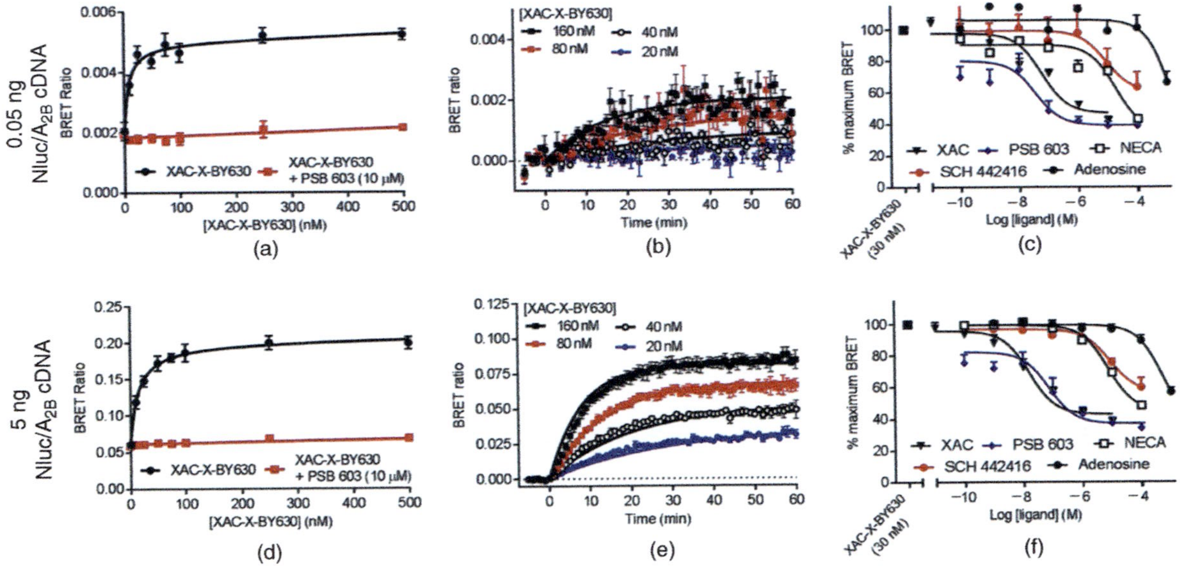

Figure 23.3 NanoBRET ligand binding at over-expressed Nluc/A_{2B} receptors. Ligand binding in HEK293 cells transiently transfected with 0.05 ng (a–c) or 5 ng (d–f) plasmid DNA coding for Nluc/A_{2B} per well of a six-well plate. (a and d) saturable ligand binding could be observed by incubating cells with increasing concentrations of xanthine amine congener (XAC)-X-BY630 in the absence (black circles) and presence (red squares) of 10 μM PSB 603. (b and e) Kinetic ligand binding in cells transiently transfected with Nluc/A_{2B} receptors was observed by treatment of cells with 20, 40, 80, and 160 nM XAC-X-BY630 (blue circles, open circles, red squares, and closed black squares, respectively) and monitoring BRET between Nluc and the fluorescent ligand for 60 minutes. (c and f) Displacement of 30 nM XAC-X-BY630 binding in cells transiently transfected with Nluc/A_{2B} receptors by XAC (10 pM–10 μM, triangles), PSB 603 (100 pM–100 μM, blue diamonds), NECA (100 pM–100 μM, open squares), SCH 442416 (100 pM–100 μM, red circles), and adenosine (1 nM–1 mM, closed black circles). Corrected BRET ratio was calculated as described in White et al. [72]. (c and f) plotted as % of maximum BRET observed for binding of 30 nM XAC-X-BY630 in the presence of vehicle only. Saturation, competition, and kinetic curves were fitted as described in White et al. [72] with points representing the mean ± S.E.M. of three independent experiments. Source: White et al. [72]/with permission of Elsevier.

The binding kinetics of unlabeled ligands can also be elucidated using competition NanoBRET ligand binding, given the kinetics of the fluorescent ligand are known [77, 81–83]. Kinetic parameters of the unlabeled ligands can subsequently be derived using the model outlined in Motulsky and Mahan [84]. The ability to study the kinetics of unlabeled ligands using this technique greatly expands the applicability of NanoBRET-based ligand kinetic analysis. Kinetic parameters of ligand binding are an important, but often underappreciated facet of ligand–receptor interactions [85, 86]. These NanoBRET-based kinetic methods have the potential to make kinetic analysis of ligand binding more widespread, potentially leading to a greater understanding and appreciation for the role kinetics play in ligand properties [87].

In addition to *in vitro* studies, Alcobia et al. [68] successfully monitored the binding of a red fluorescent propranolol-BODIPY630/650 to Nluc-β_2AR *in vivo*. Human breast cancer cells stably expressing Nluc-β_2AR were injected into the mammary fat pad of mice and Nluc-β_2AR expression within the resulting tumor was confirmed by bioluminescence imaging. Specific binding of intratumorally injected propranolol-BODIPY630/650 to Nluc-β_2AR *in vivo* was subsequently observed. The BRET response was successfully antagonized by a β_2AR-selective antagonist and target engagement monitored over 72 hours. This demonstrated the ability of the method to monitor binding of both labeled and unlabeled ligands, as well as report target residence time *in vivo*, giving exciting potential applications for drug discovery. This application of NanoBRET ligand binding demonstrates the potential for this technique beyond cell-based assays, offering applications for monitoring target occupancy and ligand properties within higher-order, more physiologically relevant systems.

23.4.2 Other Novel NanoBRET Uses

Nluc's brightness can be exploited to increase the dynamic range and sensitivity of assays that utilize Nluc as an energy donor. This is particularly important to assays aiming to monitor low-signal output interactions/proximity that classical BRET assays may lack sufficient sensitivity to detect. This applies to interactions that show low BRET signal due to low prevalence and also assays with more physiologically relevant expression levels [88, 89], such as when BRET-tagged proteins are expressed under endogenous promotion [15]. This is discussed further under genome-editing and bioluminescent techniques.

23.4.3 NanoBiT Complementation

NanoBiT is the name given to a split-luciferase reporter system derived from Nluc. This system was developed to offer a protein-fragment complementation assay (PCA) that would give sensitive detection due to the high luminescence output of Nluc. The split-luciferase was optimized to give an improved stability profile through the 18 kDa fragment termed LgBiT with low intrinsic affinity (K_D = 190 μM) for the smaller 1.3 kDa fragment, termed SmBiT [90]. This NanoBiT complementation system has shown great success as a tool to study protein–protein interactions and has been successfully used as a PCA assay to study GPCR pharmacology [38, 90–96]. An example of SmBiT complementation used in a BRET assay is Smith et al. [97], which tagged β-arrestin, Gα and either extracellular signal-regulated kinase (ERK) or the C′-terminus of a GPCR with SmBiT, LgBiT, and the fluorophore monomeric Kusabira Orange (mKO) in various conformations to investigate the formation of multiprotein complexes following receptor stimulation (Figure 23.4a).

An alternative SmBiT fragment termed HiBiT has since been described that exhibits a very high affinity for LgBiT (K_D = 700 pM) [90, 98]. While SmBiT complementation is useful for BRET assays aiming to study interactions involving more than two proteins and/or small molecules (Figure 23.4a), HiBiT complementation has a wider breadth of BRET-based applications. Due to the high affinity between HiBiT and LgBiT, HiBiT can be used in place of Nluc as a fusion construct tag to conduct BRET assays, with exogenously administered LgBiT reconstituting with HiBiT creating the functional luciferase tag (Figure 23.4b). As LgBiT is cell-impermeable, the application of this method to study intracellular interactions is limited. However, using HiBiT complementation to conduct BRET presents some advantages and novel applications over Nluc.

The HiBiT–LgBiT complementation system was first used to conduct NanoBRET in a non-GPCR study investigating hydroxylation of hypoxia-inducible factor (HIF1α) [98]. This study utilized clustered regularly interspaced short palindromic repeats (CRISPR)–Cas9 to endogenously express HIF1α-HiBiT, allowing for BRET to be conducted at more physiologically relevant levels. This endogenously expressed construct was used along with a fluorophore-conjugated antibody targeted toward hydroxylated HIF1α and exogenously added LgBiT to conduct BRET. Accumulation of HIF1α resulted in complementation of HiBiT and LgBiT, resulting in a luminescence signal. Hydroxylation of HIF1α resulted in targeting of the fluorophore-tagged antibody toward the hydroxylated HIF1α, resulting in measurable BRET between the HiBiT–LgBiT complex and the fluorophore-antibody conjugate.

A similar method was employed in another non-GPCR study to investigate proteolysis-targeting chimera (PROTAC)-mediated degradation of bromodomain and extra-terminal (BET) family proteins [99]. Luminescence produced by endogenously expressed HiBiT-tagged BET proteins conjugated with stably expressed LgBiT was used as a measure of protein degradation. NanoBRET signal between the HiBiT–LgBiT conjugated BET proteins and HaloTag-tagged

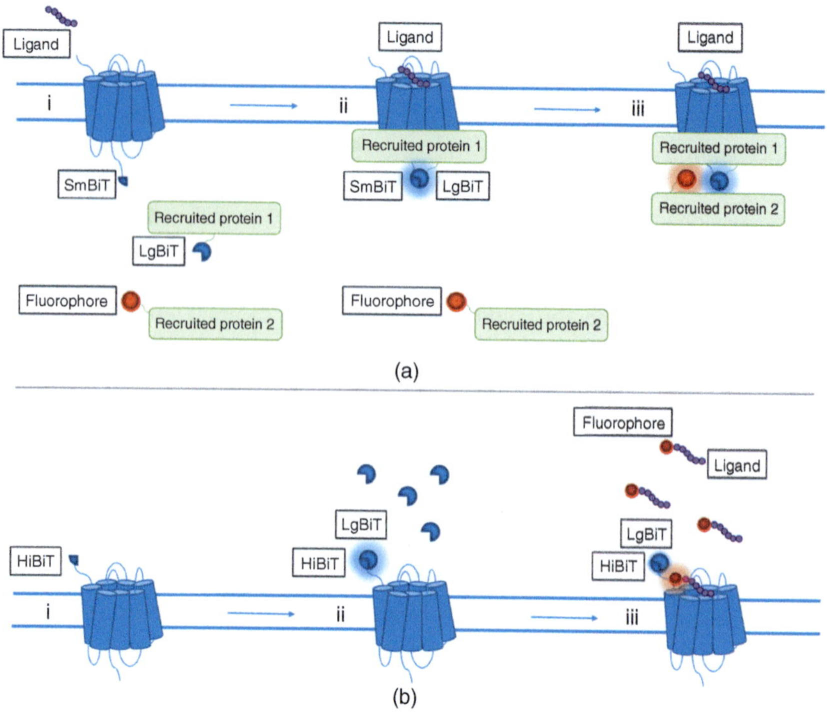

Figure 23.4 Examples of using NanoBiT for BRET assays. (a) Using SmBiT-LgBiT complementation and BRET to investigate multiprotein complexes. (ai) GPCR C′-terminally tagged with SmBiT expressed alongside an interacting protein tagged with LgBiT and a second interacting protein tagged with a fluorophore. (aii) Upon binding of an agonist, the LgBiT-tagged protein is recruited to the GPCR and this interaction allows SmBiT and LgBiT to reconstitute, producing a functional luciferase that emits luminescence. (aiii) The fluorophore-tagged protein is also recruited to the complex, bringing the fluorophore and luciferase in close proximity. Energy transfer from the reconstituted luciferase to the fluorophore occurs, producing a measurable BRET response. (b) Using HiBiT–LgBiT complementation and BRET to investigate ligand binding. (bi) GPCR N′ terminally tagged with HiBiT is expressed, (bii) LgBiT is exogenously added, resulting in reconstitution of HiBiT and LgBiT to produce a functional luciferase that emits luminescence. (biii) A fluorescent ligand is subsequently added and binds to the receptor. The close proximity of the reconstituted HiBiT–LgBiT luciferase and the fluorescent ligand produces a measurable BRET signal.

E3 ligase components, or ubiquitin, was used to further characterize the protein degradation mechanisms.

Reyes-Alcaraz et al. [41], Soave et al. [100], and White et al. [71] describe a novel GPCR internalization assay using HiBiT–LgBiT complementation. The NanoBiT internalization assay described exploits the cell-impermeability of LgBiT. As such, N′-terminally HiBiT-tagged receptors will reconstitute with LgBiT when on the plasma membrane but not if internalized. Thus, by measuring the decrease in luminescence caused by agonist stimulation, receptor internalization can be monitored. This assay was conducted with Galanin receptor 2 (GAL_2R) [41] and adenosine A_1R [100], as well as CXCR4 and β_2AR [71]. While subsequent studies demonstrate that this assay should translate well to multiple different GPCRs [79], it has been noted that this approach does not strictly report internalization. White et al. [71] observed an unexpected increase in luminescence output following application of antagonists to cells expressing HiBiT-tagged CXCR4 or agonists to cells expressing HiBiT-tagged ACKR3, implying noninternalization events were being observed. Subsequent investigations found that fusion of HiBiT to the N′-terminus of CXCR4 drastically reduced the affinity of HiBiT–LgBiT complementation, but that ligand binding to CXCR4 increased affinity of this complementation, and therefore luminescence output due to extracellular conformational changes. While changes in the affinity of NanoBiT complementation following fusion of HiBiT to a receptor is a clear confounder of the NanoBiT internalization approach, such effects following fusion to a protein are not unexpected and can readily be determined empirically. Furthermore, as seen for CXCR4, such changes in complementation affinity opens up the possibility of investigating ligand-induced conformational changes by a simple luminescence assay.

Soave et al. [100] additionally demonstrated the use of the HiBiT–LgBiT system for NanoBRET ligand binding as an alternative to an N′-terminally Nluc-tagged receptor. Of importance, this study established that complementation of a HiBiT tag N′-terminally fused to the adenosine A_1R with LgBiT did not appear to interfere with ligand binding and results were comparable to previous studies using an N′-terminally Nluc-tagged adenosine A_1R [13, 78].

Boursier et al. [79] also used N′-terminally HiBiT tagged β-adrenoceptors to study binding of unlabeled ligands through competition with a fluorescently labeled propranolol (propranolol-NB590). It was observed that this assay format translated well between cell types (HEK293 and PC3), as well as transient and endogenous expression systems. They also propose the reduced background luminescence from internalized receptors when using HiBiT is a feature that may be well suited to the investigation of low affinity or low prevalence ligand binding, such as binding to endogenously expressed receptors. This is evidenced by the difficulty in monitoring the relatively low affinity binding of propranolol-NB590 to β_3-adrenoceptor (β_3AR) when tagged with Nluc that was overcome with the use of HiBiT-tagged β_3AR.

Hoare et al. [101] exploited the cell-impermeability of LgBiT in a novel assay to investigate homomerization of relaxin family peptide 1 receptor (RXFP1) on the cell surface. Previous reports suggested RXFP1 forms homomers at the cell surface and that this may be important to the receptor's activation [102, 103]. However, this study observed transiently overexpressed RXFP1 expressed poorly on the plasma membrane, likely sequestering within internal cellular compartments [101]. As such, when using overexpression-based BRET to detect RXFP1 homomer formation, BRET signal due to tagged RXFP1 being in close proximity within subcellular compartments is likely, and the effect of this is difficult to dissect from a genuine signal at the plasma membrane. Given this, Hoare et al. [101] utilized N′-terminally HiBiT-tagged RXFP1 as well as N′-terminally mCitrine-tagged RXFP1 to develop a NanoBRET-based assay to monitor RXFP1 homomerization specifically at the plasma membrane. This assay indicated no relaxin-mediated change to RXFP1 homomerization at the plasma membrane, indicating that potential homomer formation does not appear to be relaxin-induced. This study also observed that RXFP1–RXFP1 proximity at the cell surface did not appear to be stable, with BRET signal increasing in correlation to increased receptor expression at the cell surface over time. These results give evidence that homomerization of RXFP1 may not be critical to receptor activation. This novel use of HiBiT is an example of the potential this assay format has to monitoring a range of plasma membrane-restricted interactions with high sensitivity.

23.5 Genome-Editing and Bioluminescent Techniques

While BRET has been widely used to investigate GPCR pharmacology, it is not without its limitations. Principle among these caveats is that BRET requires fusion of a donor luciferase to a protein of interest, traditionally with subsequent exogenous expression of the fusion construct. Exogenous expression can be readily achieved in a variety of cell backgrounds by current transfection, electroporation, or viral transduction methods; however, these approaches commonly result in high levels of protein over-expression that may disrupt protein function. At the very least, over-expression disrupts the normal cellular environment and poorly replicates the "native" stoichiometry between interacting proteins [104]. These are important considerations for GPCRs since many proteins are allosteric modulators of GPCRs. Therefore, the absence or presence and/or relative stoichiometry of these interactions may regulate GPCR function/signaling.

A classic example where the need for exogenous expression can confound interpretation is using BRET assays to understand GPCR oligomerization. BRET can establish proximity between two GPCRs tagged with a luminescent donor and fluorescent acceptor due to the high distance dependence of energy transfer. However, to rule out over-expression "forcing" complex formation, careful analysis and controls need to be performed. A number of BRET-based techniques and analyses have been developed in an attempt to limit these confounders [105–108]; however, even with such analyses, for many GPCR-oligomers it is unclear if complex formation occurs at native expression levels [109]. Properly replicating or maintaining the native cellular stoichiometry is therefore important to understand the molecular interactions modulating receptor function.

With the advance of genome-editing techniques such as transcription activator-like effector nucleases (TALENs) and CRISPR/Cas9 reporters, BRET tags can now be directly inserted into the endogenous genome (Figure 23.5a). CRISPR/Cas9 genome-editing approaches are now relatively accessible and have been widely adopted in the GPCR field to produce cell lines deficient of GPCR signaling and scaffolding proteins [110]. In contrast, the use of these approaches to generate models with proteins of interest expressed under endogenous promotion fused to a luciferase [98, 111] or fluorescent protein [112, 113] is largely still in its infancy.

The first successful report of BRET using a genome-edited protein under endogenous promotion was described by White et al. [15]. In CRISPR/Cas9-edited HEK293 cells, NanoBRET was used to investigate ligand-induced recruitment of genome-edited β-arr2 fused to Nluc (β-arr2-Nluc) to exogenously expressed GPCRs fused to the fluorescent acceptor Venus (GPCR-Venus), as well as exogenous β-arr2-Venus to genome-edited CXCR4-Nluc. In line with the importance of cellular context and stoichiometry on receptor function, depending on the relative levels of GPCR or β-arr2 expression, differences in the kinetic profiles of recruitment were observed. Additionally, it was possible to observe trafficking of genome-edited CXCR4-Nluc using the bystander BRET trafficking approach.

Endogenously expressed proteins are frequently expressed in orders of magnitude lower than that routinely achieved in overexpression-based assays; therefore, the sensitivity and measurable luminescence output is an important consideration when using BRET to investigate endogenous proteins (Figure 23.5b). Due to its brightness, Nluc (or Nluc fragments) have been most commonly used to investigate genome-edited proteins. This was highlighted in White et al. [15], which found greater assay sensitivity when using genome-edited proteins fused to Nluc compared to Rluc8. In addition to the sensitivity of the luciferase, a specific luciferase's suitability for use in genome-editing should also be considered. CRISPR/Cas9-mediated homology-directed repair is an inefficient process for large insertions [114]. The smaller size of Nluc compared to other luciferases confers the advantage of increasing editing rates, with further efficiencies and simplification of the editing process achieved using the small (11 amino acid) Nluc fragments HiBiT and SmBiT.

Demonstrating the ability to detect lowly expressed proteins, White et al. [72] used CRISPR/Cas9 to generate HEK293 cells expressing adenosine A_{2B} receptor genome-edited to express Nluc on the N′-terminus (Nluc-A_{2B}). Here, very low levels of Nluc-A_{2B} expression were observed, approximately 70-fold lower than the levels of genome-edited CXCR4-Nluc observed previously (Figure 23.5) [15].

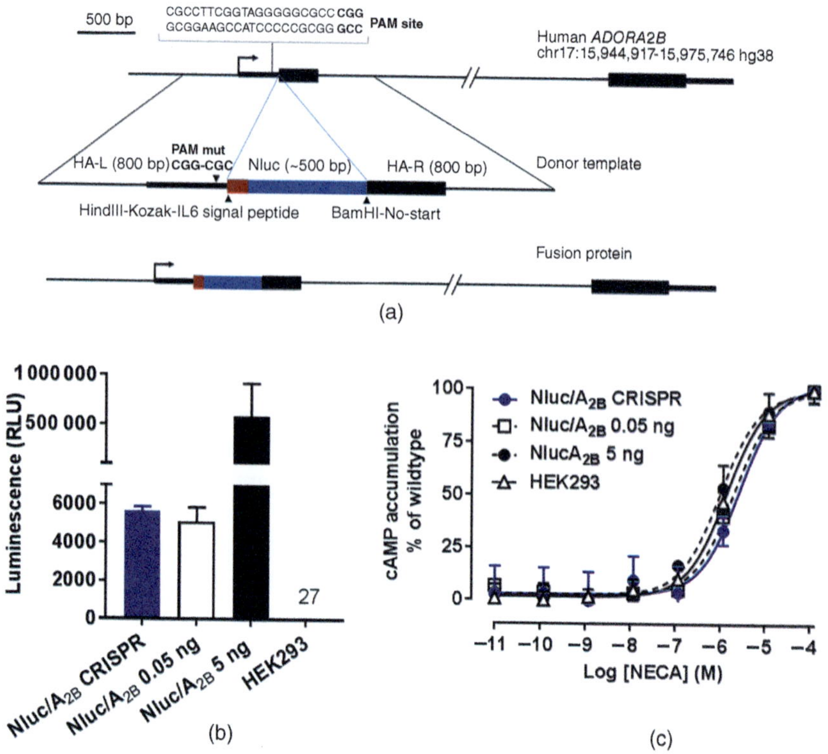

Figure 23.5 Design and validation of gene-edited adenosine A_{2B} receptors. (a) Gene-editing strategy and features used to fuse Nluc onto the N-terminus of adenosine A_{2B} receptors expressed in HEK293 cells via CRISPR/Cas9 mediated homology-directed repair. (b) Comparison of luminescence generated by cells expressing genome-edited Nluc/A_{2B} (blue bars) or from cells transiently transfected with cDNA coding for Nluc/A_{2B} at 0.05, 5, or 0 ng per well of a six-well plate (open, black, and no [RLU 27] bar, respectively). (c) Cyclic adenosine monophosphate (cAMP) accumulation mediated by NECA (10 pM–100 μM) in cells expressing genome-edited Nluc/A_{2B} (blue points) or in cells transiently transfected with cDNA coding for Nluc/A_{2B} at 0.05, 5, or 0 ng per well of a six-well plate (open squares, black circles, and open triangles, respectively). Points or bars represent mean ± S.E.M. of three (c) or six (b) independent experiments. Source: White et al. [72]/with permission of Elsevier.

Despite these exceptionally low levels of expression, fluorescent ligand binding could readily be observed in NanoBRET ligand-binding assays. This genome-editing approach was subsequently used to investigate fluorescent ligand (CXC12-AF647) binding to native CXCR4 or ACKR3 expressed in HEK293 and HeLa cells by NanoBRET [71]. In these studies, full length Nluc was inserted into the native genome using CRISPR/Cas9 genome-editing to create N′-terminally tagged receptors. In addition, NanoBRET ligand binding to genome-edited

CXCR4 receptors tagged with the small high-affinity Nluc fragment HiBiT was also reported. As discussed above, since internalization of receptors tagged on the N′-terminus with HiBiT can be monitored, the authors also investigated ligand-induced internalization of HiBiT-tagged CXCR4 and β_2AR expressed under endogenous promotion in HEK293 cells. The broad applicability of using Nluc/fragments and genome-editing to investigate the function of GPCRs was further confirmed by Boursier et al. [79], who investigated fluorescent ligand binding and internalization of natively expressed β_2AR N′-terminally tagged with HiBiT.

CRISPR/Cas9 genome-editing approaches appear particularly suited to investigate ligand binding by NanoBRET at natively expressed receptors. In this approach, the fluorescent acceptor is attached to the ligand that is applied to the cells to investigate binding; however, as of yet BRET between two genome-edited proteins has not been reported. In White et al. [15], while β-arr2 recruitment, trafficking and receptor oligomerization were investigated using natively expressed donor-tagged proteins, the protein-tagged with the fluorescent acceptor was still exogenously expressed. Similarly in Kilpatrick et al. [62] where the formation of VEGFR2 (vascular endothelial growth factor receptor 2)-β_2AR complexes were investigated by NanoBRET in HEK293 cells expressing genome-edited Nluc-β_2AR, fluorescently labeled VEGFR2 was exogenously expressed. This is perhaps not surprising since BRET assays performed with exogenously expressed proteins are routinely optimized for optimal BRET ratios. Typically, this results in assay configurations where the acceptor-fused species is in excess of the donor to maximize signal-to-background ratio. Clearly, such optimization is not possible with proteins under endogenous promotion whose expression is dictated by cellular requirements. As highlighted in White et al. [15], the relative levels of receptor or effector expression within the cell can drastically change the kinetics of β-arr2 recruitment to CXCR4. As such, systems where both partners are endogenously expressed are highly desirable.

By turning to Nluc complementation instead of BRET, it appears that these apparent limitations can be overcome. White et al. [71] probed interactions between natively expressed β-arr2 and CXCR4 in HEK293 using low-affinity Nluc complementation. Here, CRISPR/Cas9 genome-editing was used to generate HEK293 cells that expressed both CXCR4-LgBiT and β-arr2-SmBiT under endogenous promotion and ligand-induced recruitment of β-arr2-SmBiT to CXCR4-LgBiT was monitored by changes in luminescence. As expected, differences in the kinetic profile of recruitment were observed in the line expressing both partners under endogenous promotion compared to HEK293 cells expressing only CXCR4-LgBiT or β-arr2-SmBiT and transfected with the other partner, or HEK293 cells over-expressing both CXCR4-LgBiT and β-arr2-SmBiT.

Approaches using CRISPR/Cas9 genome-editing and BRET have been developed to eliminate the requirement for over-expression and allow maintenance of the normal endogenous expression levels; however, genome-editing is not a scarless procedure. Culture procedures to produce cell lines with genome-edited proteins may result in genetic-wiring and changes in protein levels [115]. Similarly, off-target genome-editing may produce changes in normal gene-expression coalescing in a change in the levels of a protein of interest. Furthermore, fusion of a tag to a natively expressed protein of interest can change expression levels, as seen previously for fluorescently tagged CXCR4 [112]. While previous studies have only observed minimal changes in expression at the mRNA level following insertion of Nluc or Nluc fragments into the genome [71], Nluc was initially engineered to increase stability [9], and differences in expression should therefore be considered and tested. However, any changes in expression are likely still much closer to native expression compared to the levels achieved using exogenously expressed proteins, which are commonly orders of magnitude higher.

The combination of CRISPR/Cas9 and NanoBRET or Nluc complementation has the potential to allow for investigation of the molecular aspects of GPCR pharmacology in native systems without over-expression. Probing receptor oligomerization and receptor signaling are likely areas of GPCR pharmacology that will particularly benefit from these approaches. Indeed, G proteins, β-arrestins and other receptors can act as allosteric modulators of GPCRs; therefore, maintaining the native cellular stoichiometry better models receptor function.

23.6 Summary

Technological development within the biomedical field is moving at a staggering rate. The development of BRET-based technologies to study GPCR pharmacology is no exception, with recent advancements improving our ability to study ligand binding, GPCR and effector protein conformations, effector recruitment, and many more aspects of GPCR function with impressive clarity. The technologies discussed have already been successfully implemented to deepen our understanding of GPCR function and will undoubtedly be indispensable to future research.

References

1 Pfleger, K.D. and Eidne, K.A. (2006). Illuminating insights into protein-protein interactions using bioluminescence resonance energy transfer (BRET). *Nat. Methods* 3 (3): 165–174.

2 Lohse, M.J., Nuber, S., and Hoffmann, C. (2012). Fluorescence/bioluminescence resonance energy transfer techniques to study G-protein-coupled receptor activation and signaling. *Pharmacol. Rev.* 64 (2): 299–336.

3 Milligan, G. (2004). Applications of bioluminescence-and fluorescence resonance energy transfer to drug discovery at G protein-coupled receptors. *Eur. J. Pharm. Sci.* 21 (4): 397–405.

4 Dacres, H., Michie, M., Wang, J. et al. (2012). Effect of enhanced Renilla luciferase and fluorescent protein variants on the Förster distance of bioluminescence resonance energy transfer (BRET). *Biochem. Biophys. Res. Commun.* 425 (3): 625–629.

5 Lorenz, W., Cormier, M., O'Kane, D. et al. (1996). Expression of the Renilla reniformis luciferase gene in mammalian cells. *J. Biolumin. Chemilumin.* 11 (1): 31–37.

6 Fraga, H. (2008). Firefly luminescence: a historical perspective and recent developments. *Photochem. Photobiol. Sci.* 7 (2): 146–158.

7 Kocan, M., See, H.B., Seeber, R.M. et al. (2008). Demonstration of improvements to the bioluminescence resonance energy transfer (BRET) technology for the monitoring of G protein–coupled receptors in live cells. *J. Biomol. Screening* 13 (9): 888–898.

8 England, C.G., Ehlerding, E.B., and Cai, W. (2016). NanoLuc: a small luciferase is brightening up the field of bioluminescence. *Bioconjugate Chem.* 27 (5): 1175–1187.

9 Hall, M.P., Unch, J., Binkowski, B.F. et al. (2012). Engineered luciferase reporter from a deep sea shrimp utilizing a novel imidazopyrazinone substrate. *ACS Chem. Biol.* 7 (11): 1848–1857.

10 Horioka, M., Huber, T., and Sakmar, T.P. (2020). Playing tag with your favorite GPCR using CRISPR. *Cell Chem. Biol.* 27 (6): 642–644.

11 Algar, W.R., Hildebrandt, N., Vogel, S.S. et al. (2019). FRET as a biomolecular research tool – understanding its potential while avoiding pitfalls. *Nat. Methods* 16 (9): 815–829.

12 Pfleger, K.D. and Eidne, K.A. (2005). Monitoring the formation of dynamic G-protein-coupled receptor–protein complexes in living cells. *Biochem. J.* 385 (3): 625–637.

13 Stoddart, L.A., Johnstone, E.K., Wheal, A.J. et al. (2015). Application of BRET to monitor ligand binding to GPCRs. *Nat. Methods* 12 (7): 661–663.

14 Namkung, Y., LeGouill, C., Kumar, S. et al. (2018). Functional selectivity profiling of the angiotensin II type 1 receptor using pathway-wide BRET signaling sensors. *Sci. Signal.* 11 (559): eaat1631.

15 White, C.W., Vanyai, H.K., See, H.B. et al. (2017). Using nanoBRET and CRISPR/Cas9 to monitor proximity to a genome-edited protein in real-time. *Sci. Rep.* 7 (1): 1–14.

16 Pfleger, K.D., Dromey, J.R., Dalrymple, M.B. et al. (2006). Extended bioluminescence resonance energy transfer (eBRET) for monitoring prolonged protein–protein interactions in live cells. *Cell Signal.* 18 (10): 1664–1670.

17 Pfleger, K.D., Seeber, R.M., and Eidne, K.A. (2006). Bioluminescence resonance energy transfer (BRET) for the real-time detection of protein–protein interactions. *Nat. Protoc.* 1 (1): 337–345.

18 Dale, N., Johnstone, E.K.M., White, C. et al. (2019). NanoBRET: the bright future of proximity-based assays. *Front. Bioeng. Biotechnol.* 7: 56.

19 Tiulpakov, A., White, C.W., Abhayawardana, R.S. et al. (2016). Mutations of vasopressin receptor 2 including novel L312S have differential effects on trafficking. *Mol. Endocrinol.* 30 (8): 889–904.

20 Pfleger, K.D. (2009). Analysis of protein–protein interactions using bioluminescence resonance energy transfer. In: *Bioluminescence* (ed. P.B. Rich and C. Douillet), 173–183. Totowa, NJ: Humana Press.

21 Olsen, R.H., DiBerto, J.F., English, J.G. et al. (2020). TRUPATH, an open-source biosensor platform for interrogating the GPCR transducerome. *Nat. Chem. Biol.* 16: 841–849.

22 Pavlos, N.J. and Friedman, P.A. (2017). GPCR signaling and trafficking: the long and short of it. *Trends Endocrinol. Metab.* 28 (3): 213–226.

23 Lobingier, B.T. and von Zastrow, M. (2019). When trafficking and signaling mix: how subcellular location shapes G protein-coupled receptor activation of heterotrimeric G proteins. *Traffic* 20 (2): 130–136.

24 Lan, T.-H., Kuravi, S., and Lambert, N.A. (2011). Internalization dissociates β_2-adrenergic receptors. *PLoS One* 6 (2): e17361.

25 Simpson, J.C. and Jones, A.T. (2005). Early endocytic Rabs: functional prediction to functional characterization. In: *Biochemical Society Symposia*, 99–108. London: Portland on behalf of The Biochemical Society.

26 Rodman, J.S. and Wandinger-Ness, A. (2000). Rab GTPases coordinate endocytosis. *J. Cell Sci.* 113 (2): 183–192.

27 Lan, T.H., Liu, Q., Li, C. et al. (2012). Sensitive and high resolution localization and tracking of membrane proteins in live cells with BRET. *Traffic* 13 (11): 1450–1456.

28 Namkung, Y., Le Gouill, C., Lukashova, V. et al. (2016). Monitoring G protein-coupled receptor and β-arrestin trafficking in live cells using enhanced bystander BRET. *Nat. Commun.* 7 (1): 1–12.

29 Avet, C., Mancini, A., Breton, B. et al. (2022). Effector membrane translocation biosensors reveal G protein and βarrestin coupling profiles of 100 therapeutically relevant GPCRs. *Elife* 11.

30 Pickering, R.J., Tikellis, C., Rosado, C.J. et al. (2019). Transactivation of RAGE mediates angiotensin-induced inflammation and atherogenesis. *J. Clin. Invest.* 129 (1): 406–421.

31 Jensen, D.D., Lieu, T., Halls, M.L. et al. (2017). Neurokinin 1 receptor signaling in endosomes mediates sustained nociception and is a viable therapeutic target for prolonged pain relief. *Sci. Transl. Med.* 9 (392): eaal3447.

32 Halls, M.L., Yeatman, H.R., Nowell, C.J. et al. (2016). Plasma membrane localization of the μ-opioid receptor controls spatiotemporal signaling. *Sci. Signal.* 9 (414): ra16.

33 Yeatman, H.R., Lane, J.R., Choy, K.H.C. et al. (2014). Allosteric modulation of M_1 muscarinic acetylcholine receptor internalization and subcellular trafficking. *J. Biol. Chem.* 289 (22): 15856–15866.

34 Jimenez-Vargas, N.N., Pattison, L.A., Zhao, P. et al. (2018). Protease-activated receptor-2 in endosomes signals persistent pain of irritable bowel syndrome. *Proc. Natl. Acad. Sci. U.S.A.* 115 (31): E7438–E7447.

35 Martin, B.R. and Lambert, N.A. (2016). Activated G protein $G\alpha_s$ samples multiple endomembrane compartments. *J. Biol. Chem.* 291 (39): 20295–20302.

36 Pritchard, A.B., Kanai, S.M., Krock, B. et al. (2020). Loss-of-function of Endothelin receptor type A results in Oro-Oto-Cardiac syndrome. *Am. J. Med. Genet. Part A* 182 (5): 1104–1116.

37 Irannejad, R., Pessino, V., Mika, D. et al. (2017). Functional selectivity of GPCR-directed drug action through location bias. *Nat. Chem. Biol.* 13 (7): 799–806.

38 Wan, Q., Okashah, N., Inoue, A. et al. (2018). Mini G protein probes for active G protein-coupled receptors (GPCRs) in live cells. *J. Biol. Chem.* 293 (19): 7466–7473.

39 Lee, M.-H., Appleton, K.M., Strungs, E.G. et al. (2016). The conformational signature of β-arrestin2 predicts its trafficking and signalling functions. *Nature* 531 (7596): 665–668.

40 Adams, S.R. and Tsien, R.Y. (2008). Preparation of the membrane-permeant biarsenicals FlAsH-EDT_2 and ReAsH-EDT_2 for fluorescent labeling of tetracysteine-tagged proteins. *Nat. Protoc.* 3 (9): 1527–1534.

41 Reyes-Alcaraz, A., Lee, Y.-N., Yun, S. et al. (2018). Conformational signatures in β-arrestin2 reveal natural biased agonism at a G-protein-coupled receptor. *Commun. Biol.* 1 (1): 1–12.

42 Oishi, A., Dam, J., and Jockers, R. (2019). β-arrestin-2 BRET biosensors detect different β-arrestin-2 conformations in interaction with GPCRs. *ACS Sens.* 5 (1): 57–64.

43 Hoffmann, C., Gaietta, G., Bünemann, M. et al. (2005). A FlAsH-based FRET approach to determine G protein–coupled receptor activation in living cells. *Nat. Methods* 2 (3): 171–176.

44 Kauk, M. and Hoffmann, C. (2018). Intramolecular and intermolecular FRET sensors for GPCRs–monitoring conformational changes and beyond. *Trends Pharmacol. Sci.* 39 (2): 123–135.

45 Bourque, K., Pétrin, D., Sleno, R. et al. (2017). Distinct conformational dynamics of three G protein-coupled receptors measured using FlAsH-BRET biosensors. *Front. Endocrinol. (Lausanne).* 8: 61.

46 Sleno, R., Devost, D., Pétrin, D. et al. (2017). Conformational biosensors reveal allosteric interactions between heterodimeric AT_1 angiotensin and prostaglandin $F_{2\alpha}$ receptors. *J Biol Chem.* 292 (29): 12139–12152.

47 Powlowski, P., Bourque, K., Jones-Tabah, J. et al. (2018). Conformational profiling of the 5-HT_{2A} receptor using FlAsH BRET. In: *Receptor–Receptor Interactions in the Central Nervous System* (ed. K. Fuxe and D. Borroto-Escuela), 265–282. New York, NY: Humana Press.

48 Carpenter, B., Nehmé, R., Warne, T. et al. (2016). Structure of the adenosine A_{2A} receptor bound to an engineered G protein. *Nature* 536 (7614): 104–107.

49 Tsai, C.-J., Pamula, F., Nehmé, R. et al. (2018). Crystal structure of rhodopsin in complex with a mini-Go sheds light on the principles of G protein selectivity. *Sci. Adv.* 4 (9): eaat7052.

50 García-Nafría, J., Lee, Y., Bai, X. et al. (2018). Cryo-EM structure of the adenosine A_{2A} receptor coupled to an engineered heterotrimeric G protein. *Elife* 7: e35946.

51 Garcia-Nafria, J., Nehme, R., Edwards, P.C. et al. (2018). Cryo-EM structure of the serotonin 5-HT_{1B} receptor coupled to heterotrimeric G_o. *Nature* 558 (7711): 620–623.

52 Carpenter, B. and Tate, C.G. (2016). Engineering a minimal G protein to facilitate crystallisation of G protein-coupled receptors in their active conformation. *Protein Eng. Des. Sel.* 29 (12): 583–594.

53 Nehme, R., Carpenter, B., Singhal, A. et al. (2017). Mini-G proteins: novel tools for studying GPCRs in their active conformation. *PLoS One* 12 (4): e0175642.

54 Wright, S.C., Kozielewicz, P., Kowalski-Jahn, M. et al. (2019). A conserved molecular switch in Class F receptors regulates receptor activation and pathway selection. *Nat. Commun.* 10 (1): 1–12.

55 Ali, A., Palakkott, A., Ashraf, A. et al. (2019). Positive modulation of angiotensin II type 1 receptor–mediated signaling by LVV–Hemorphin-7. *Front. Pharmacol.* 10: 1258.

56 Kozielewicz, P., Turku, A., Bowin, C.-F. et al. (2020). Structural insight into small molecule action on Frizzleds. *Nat. Commun.* 11 (1): 1–16.

57 Gillis, A., Gondin, A.B., Kliewer, A. et al. (2020). Low intrinsic efficacy for G protein activation can explain the improved side effect profiles of new opioid agonists. *Sci. Signal.* 13 (625): eaaz3140.

58 Che, T., English, J., Krumm, B.E. et al. (2020). Nanobody-enabled monitoring of kappa opioid receptor states. *Nat. Commun.* 11 (1): 1–12.

59 Carpenter, B. (2018). Current applications of mini G proteins to study the structure and function of G protein-coupled receptors. *AIMS Bioeng.* 5 (4): 209–225.

60 Irannejad, R., Tomshine, J.C., Tomshine, J.R. et al. (2013). Conformational biosensors reveal GPCR signalling from endosomes. *Nature* 495 (7442): 534–538.

61 Che, T., Majumdar, S., Zaidi, S.A. et al. (2018). Structure of the nanobody-stabilized active state of the kappa opioid receptor. *Cell* 172 (1,2): 55–67.e15.

62 Kilpatrick, L.E., Alcobia, D.C., White, C.W. et al. (2019). Complex Formation between VEGFR2 and the β_2-adrenoceptor. *Cell Chem. Biol.* 26 (6): 830–41.e9.

63 Soave, M., Heukers, R., Kellam, B. et al. (2020). Monitoring allosteric Interactions with CXCR4 using NanoBiT conjugated nanobodies. *Cell Chem. Biol.* 27 (10): 1250–1261.

64 Machleidt, T., Woodroofe, C.C., Schwinn, M.K. et al. (2015). NanoBRET – a novel BRET platform for the analysis of protein–protein interactions. *ACS Chem. Biol.* 10 (8): 1797–1804.

65 Stoddart, L.A., Kilpatrick, L.E., and Hill, S.J. (2018). NanoBRET approaches to study ligand binding to GPCRs and RTKs. *Trends Pharmacol. Sci.* 39 (2): 136–147.

66 Stoddart, L.A., White, C.W., Nguyen, K. et al. (2016 Oct). Fluorescence- and bioluminescence-based approaches to study GPCR ligand binding. *Br. J. Pharmacol.* 173 (20): 3028–3037.

67 Kilpatrick, L.E., Friedman-Ohana, R., Alcobia, D.C. et al. (2017). Real-time analysis of the binding of fluorescent VEGF165a to VEGFR2 in living cells: effect of receptor tyrosine kinase inhibitors and fate of internalized agonist-receptor complexes. *Biochem. Pharmacol.* (136): 62–75.

68 Alcobia, D.C., Ziegler, A.I., Kondrashov, A. et al. (2018). Visualizing ligand binding to a GPCR in vivo using NanoBRET. *iScience* 6: 280–288.

69 Conroy, S., Kindon, N.D., Glenn, J. et al. (2018). Synthesis and evaluation of the first fluorescent antagonists of the human $P2Y_2$ receptor based on AR-C118925. *J. Med. Chem.* 61 (7): 3089–3113.

70 Hoare, B.L., Bruell, S., Sethi, A. et al. (2019). Multi-component mechanism of H_2 relaxin binding to RXFP1 through NanoBRET kinetic analysis. *iScience* 11: 93–113.

71 White, C.W., Caspar, B., Vanyai, H.K. et al. (2020). CRISPR-mediated protein tagging with NanoLuciferase to investigate native chemokine receptor function and conformational changes. *Cell Chem. Biol.* 27 (5): 499–510.

72 White, C.W., Johnstone, E.K., See, H.B. et al. (2019). NanoBRET ligand binding at a GPCR under endogenous promotion facilitated by CRISPR/Cas9 genome editing. *Cell Signal.* 54: 27–34.

73 Sakyiamah, M.M., Kobayakawa, T., Fujino, M. et al. (2019). Design, synthesis and biological evaluation of low molecular weight CXCR4 ligands. *Bioorg. Med. Chem.* 27 (6): 1130–1138.

74 Sakyiamah, M.M., Nomura, W., Kobayakawa, T. et al. (2019). Development of a NanoBRET-based sensitive screening method for CXCR4 ligands. *Bioconjugate Chem.* 30 (5): 1442–1450.

75 Kozielewicz, P., Bowin, C.-F., Turku, A. et al. (2020). A NanoBRET-based binding assay for smoothened allows real-time analysis of ligand binding and distinction of two binding sites for BODIPY-cyclopamine. *Mol. Pharmacol.* 97 (1): 23–34.

76 Soave, M., Stoddart, L.A., Brown, A. et al. (2016 Oct). Use of a new proximity assay (NanoBRET) to investigate the ligand-binding characteristics of three fluorescent ligands to the human β_1-adrenoceptor expressed in HEK-293 cells. *Pharmacol. Res. Perspect.* 4 (5): e00250.

77 Stoddart, L.A., Vernall, A.J., Bouzo-Lorenzo, M. et al. (2018). Development of novel fluorescent histamine H_1-receptor antagonists to study ligand-binding kinetics in living cells. *Sci. Rep.* 8 (1): 1572.

78 Cooper, S.L., Soave, M., Jörg, M. et al. (2019 Apr). Probe dependence of allosteric enhancers on the binding affinity of adenosine A_1-receptor agonists at rat and human A_1-receptors measured using Nano BRET. *Br. J. Pharmacol.* 176 (7): 864–878.

79 Boursier, M.E., Levin, S., Zimmerman, K. et al. (2020). The luminescent HiBiT peptide enables selective quantitation of G protein–coupled receptor ligand engagement and internalization in living cells. *J. Biol. Chem.* 295 (15): 5124–5135.

80 Peach, C.J., Kilpatrick, L.E., Friedman-Ohana, R. et al. (2018). Real-time ligand binding of fluorescent VEGF-A isoforms that discriminate between VEGFR2 and NRP1 in living cells. *Cell Chem. Biol.* 25 (10): 1208–18.e5.

81 Mocking, T.A., Verweij, E.W., Vischer, H.F. et al. (2018). Homogeneous, real-time NanoBRET binding assays for the histamine H_3 and H_4 receptors on living cells. *Mol. Pharmacol.* 94 (6): 1371–1381.

82 Bouzo-Lorenzo, M., Stoddart, L.A., Xia, L. et al. (2019). A live cell NanoBRET binding assay allows the study of ligand-binding kinetics to the adenosine A_3 receptor. *Purinergic Signal.* 15 (2): 139–153.

83 Barkan, K., Lagarias, P., Stampelou, M. et al. (2020). Pharmacological characterisation of novel adenosine A_3 receptor antagonists. *Sci. Rep.* 10 (1): 1–21.

84 Motulsky, H.J. and Mahan, L. (1984). The kinetics of competitive radioligand binding predicted by the law of mass action. *Mol. Pharmacol.* 25 (1): 1–9.

85 Schuetz, D.A., de Witte, W.E.A., Wong, Y.C. et al. (2017). Kinetics for drug discovery: an industry-driven effort to target drug residence time. *Drug Discovery Today* 22 (6): 896–911.

86 Hoffmann, C., Castro, M., Rinken, A. et al. (2015). Ligand residence time at G-protein–coupled receptors – why we should take our time to study it. *Mol. Pharmacol.* 88 (3): 552–560.

87 Sykes, D.A., Stoddart, L.A., Kilpatrick, L.E. et al. (2019). Binding kinetics of ligands acting at GPCRs. *Mol. Cell. Endocrinol.* 485: 9–19.

88 Mo, X.-L., Luo, Y., Ivanov, A.A. et al. (2016). Enabling systematic interrogation of protein–protein interactions in live cells with a versatile ultra-high-throughput biosensor platform. *J. Mol. Cell. Biol.* 8 (3): 271–281.

89 Sampaio, N.G., Kocan, M., Schofield, L. et al. (2018). Investigation of interactions between TLR_2, MyD88 and TIRAP by bioluminescence resonance energy transfer is hampered by artefacts of protein overexpression. *PLoS One* 13 (8): e0202408.

90 Dixon, A.S., Schwinn, M.K., Hall, M.P. et al. (2016). NanoLuc complementation reporter optimized for accurate measurement of protein interactions in cells. *ACS Chem. Biol.* 11 (2): 400–408.

91 Laschet, C., Dupuis, N., and Hanson, J. (2019). A dynamic and screening-compatible nanoluciferase-based complementation assay enables profiling of individual GPCR–G protein interactions. *J. Biol. Chem.* 294 (11): 4079–4090.

92 Storme, J., Cannaert, A., Van Craenenbroeck, K. et al. (2018). Molecular dissection of the human A_3 adenosine receptor coupling with β-arrestin2. *Biochem. Pharmacol.* 148: 298–307.

93 Cannaert, A., Storme, J., Franz, F. et al. (2016). Detection and activity profiling of synthetic cannabinoids and their metabolites with a newly developed bioassay. *Anal. Chem.* 88 (23): 11476–11485.

94 Szpakowska, M., Nevins, A.M., Meyrath, M. et al. (2018). Different contributions of chemokine N-terminal features attest to a different ligand binding mode and a bias towards activation of ACKR3/CXCR7 compared with CXCR4 and CXCR3. *Br. J. Pharmacol.* 175 (9): 1419–1438.

95 Dupuis, N., Laschet, C., Franssen, D. et al. (2017). Activation of the orphan G protein–coupled receptor GPR27 by surrogate ligands promotes β-arrestin2 recruitment. *Mol. Pharmacol.* 91 (6): 595–608.

96 Bodle, C.R., Hayes, M.P., O'Brien, J.B. et al. (2017). Development of a bimolecular luminescence complementation assay for RGS: G protein interactions in cells. *Anal. Biochem.* 522: 10–17.

97 Smith, J.S., Pack, T.F., Inoue, A. et al. (2021). Noncanonical scaffolding of Gαi and β-arrestin by G protein–coupled receptors. *Science* 371 (6534): eaay1833.

98 Schwinn, M.K., Machleidt, T., Zimmerman, K. et al. (2018). CRISPR-mediated tagging of endogenous proteins with a luminescent peptide. *ACS Chem. Biol.* 13 (2): 467–474.

99 Riching, K.M., Mahan, S., Corona, C.R. et al. (2018). Quantitative live-cell kinetic degradation and mechanistic profiling of PROTAC mode of action. *ACS Chem. Biol.* 13 (9): 2758–2770.

100 Soave, M., Kellam, B., Woolard, J. et al. (2020). NanoBiT complementation to monitor agonist-induced adenosine A_1 receptor internalization. *SLAS Discov.* 25 (2): 186–194.

101 Hoare, B.L., Kocan, M., Bruell, S. et al. (2019). Using the novel HiBiT tag to label cell surface relaxin receptors for BRET proximity analysis. *Pharmacol. Res. Perspect.* 7 (4): e00513.

102 Svendsen, A.M., Zalesko, A., Kønig, J. et al. (2008). Negative cooperativity in H_2 relaxin binding to a dimeric relaxin family peptide receptor 1. *Mol. Cell. Endocrinol.* 296 (1, 2): 10–17.

103 Hartley, B.J., Scott, D.J., Callander, G.E. et al. (2009 Apr). Resolving the unconventional mechanisms underlying RXFP1 and RXFP2 receptor function. *Ann. N.Y. Acad. Sci.* 1160 (1): 67–73.

104 Gibson, T.J., Seiler, M., and Veitia, R.A. (2013). The transience of transient overexpression. *Nat. Methods* 10 (8): 715–721.

105 Guo, H., An, S., Ward, R. et al. (2017). Methods used to study the oligomeric structure of G-protein-coupled receptors. *Biosci. Rep.* 37 (2): BSR20160547.

106 Felce, J.H., MacRae, A., and Davis, S.J. (2019). Constraints on GPCR heterodimerization revealed by the type-4 induced-association BRET assay. *Biophys. J.* 116 (1): 31–41.

107 Felce, J.H., Latty, S.L., Knox, R.G. et al. (2017). Receptor quaternary organization explains G protein-coupled receptor family structure. *Cell Rep.* 20 (11): 2654–2665.

108 Johnstone, E.K.M. and Pfleger, K. (2012). Receptor-Heteromer Investigation Technology and its application using BRET. *Front. Endocrinol. (Lausanne)* 3: 101.

109 Gomes, I., Ayoub, M.A., Fujita, W. et al. (2016). G protein–coupled receptor heteromers. *Annu. Rev. Pharmacol. Toxicol.* 56: 403–425.

110 Milligan, G. and Inoue, A. (2018). Genome editing provides new insights into receptor-controlled signalling pathways. *Trends Pharmacol. Sci.* 39 (5): 481–493.

111 Oh-hashi, K., Furuta, E., Fujimura, K. et al. (2017). Application of a novel HiBiT peptide tag for monitoring ATF4 protein expression in Neuro2a cells. *Biochem. Biophys. Rep.* 12: 40–45.

112 Khan, A.O., White, C.W., Pike, J.A. et al. (2019). Optimised insert design for improved single-molecule imaging and quantification through CRISPR-Cas9 mediated knock-in. *Sci. Rep.* 9 (1): 1–13.

113 Leonetti, M.D., Sekine, S., Kamiyama, D. et al. (2016). A scalable strategy for high-throughput GFP tagging of endogenous human proteins. *Proc. Natl. Acad. Sci. U.S.A.* 113 (25): E3501–E3508.

114 Ran, F.A., Hsu, P.D., Wright, J. et al. (2013). Genome engineering using the CRISPR-Cas9 system. *Nat. Protoc.* 8 (11): 2281–2308.

115 Luttrell, L.M., Wang, J., Plouffe, B. et al. (2018). Manifold roles of β-arrestins in GPCR signaling elucidated with siRNA and CRISPR/Cas9. *Sci. Signal.* 11 (549): eaat7650.

24

The Application of ^{19}F NMR to Studies of Protein Function and Drug Screening

Geordi Frere[1], *Aditya Pandey*[1,2], *Jerome Gould*[1], *Advait Hasabnis*[1], *Patrick T. Gunning*[1], *and Robert S. Prosser*[1,2]

[1] *Department of Chemistry, Chemical and Physical Sciences, University of Toronto, Mississauga, ON, Canada*
[2] *Department of Biochemistry, University of Toronto, Toronto, ON Canada*

24.1 Introduction

The vast majority of proteins relies on both structure and dynamic properties to accomplish function, such as catalysis, transport, allosteric regulation, or signaling. For the most part, dynamics are enabled simply by thermal energy and result in cooperative transitions between functional states which define a reaction pathway. For example, in the case of enzymes, the actual chemical steps are preceded by specific functional states which might include a stable "apo" ground state, a conformation primed to capture substrate(s), and the Michaelis–Menten intermediate. These states in turn exist in a dynamic equilibrium and define the conformational ensemble of the protein. Ligands, substrate(s), or allosteric factors will influence this ensemble and enable sampling of successive conformations along the reaction pathway. A classic example is represented in Figure 24.1, where we consider the role of G protein coupled receptors (GPRCs) in signaling. GPCRs are an ubiquitous family of 7-transmembrane receptors that when activated by exogenous ligands, may engage specific heterotrimeric G proteins (Gαβγ), initiating nucleotide exchange in the Gα subunit, dissociation of the activated Gα subunit and Gβγ subunits, and downstream signaling. Cryo-electron microscopy (cryoEM), and X-ray crystallography have been used to capture detailed structural information regarding many aspects of this complex free energy landscape. NMR holds the promise of adding to this knowledge from the perspective of dynamics and the conformational ensemble.

In this review, we consider current approaches to the study of conformational equilibria and dynamics of proteins and protein complexes using fluorine

GPCRs as Therapeutic Targets, Volume 2, First Edition. Edited by Annette Gilchrist.

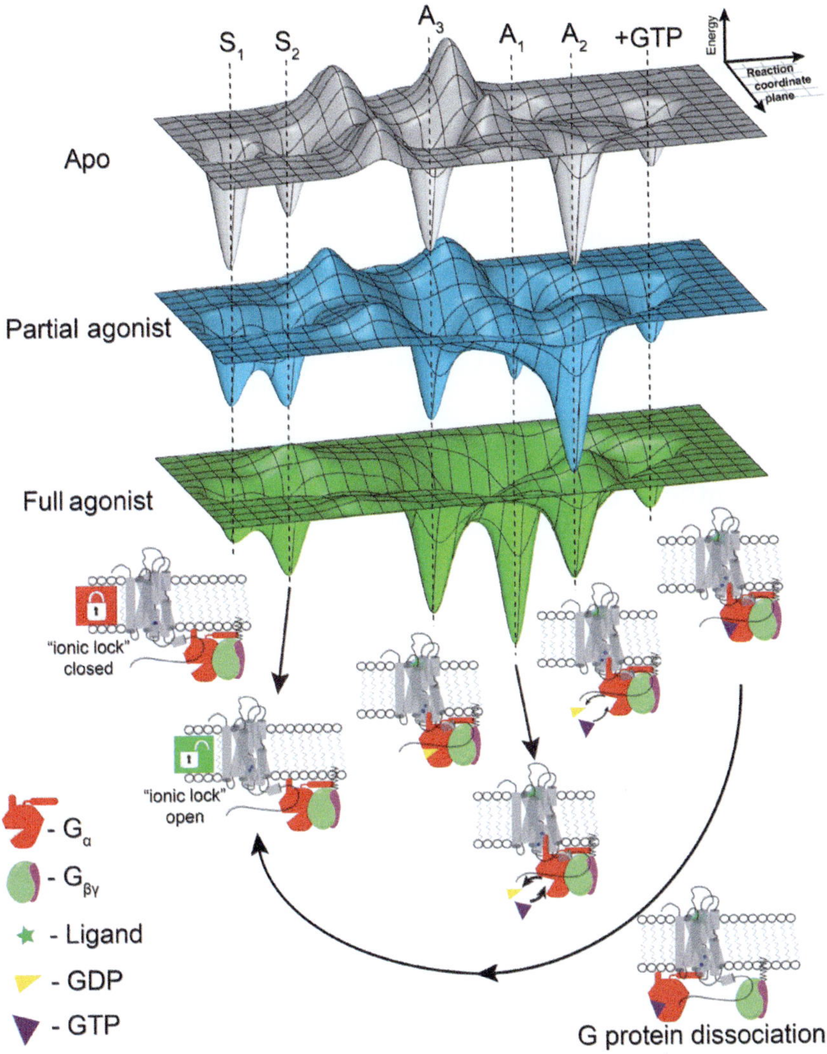

Figure 24.1 The free energy landscape. GPCRs are envisioned to sample a diverse conformational ensemble consisting of distinct inactive conformers (S_1, S_2), specific activation intermediate states (A_1, A_2), GTP bound state (+GTP) and a precoupled state (A_3) shown to be associated with the G protein. Binding by an agonist results in an allosteric response which engages the G protein, allowing sampling of subsequent steps involving release of GDP, binding of GTP to the Gα subunit, and dissociation of the activated Gα and Gβγ subunits. Ligands and adjuvants greatly influence the relative free energies and the exchange rates of the functional states in the ensemble. Techniques such a single molecule spectroscopy and ^{19}F NMR can identify functional states in the ensemble and measure their exchange rates. Source: Figure adapted from unpublished work.

(^{19}F) NMR. The associated spectroscopy and relaxation experiments have been described elsewhere and are repeated here for completeness. However, the labeling chemistry is always evolving and some next generation tags and labeling approaches are therefore discussed. At the same time, it is important to understand the NMR-derived states in the context of crystal- and cryoEM structures. Thus, additional ^{19}F NMR experiments which assess state-specific topologies or which measure inter-residue distances are discussed from the perspective of validating structural models and assigning ^{19}F NMR resonances to specific conformers. Finally, we briefly review the application of ^{19}F NMR to ligand interactions and drug discovery. Many of the lessons learned in protein ^{19}F NMR and labeling chemistry can be applied to ligand spectroscopy with the hope of improved delineation of small molecule interactions with proteins.

24.2 Fluorinated Amino Acid Analogs Used in Biosynthetic Labeling Approaches

Like ^{1}H,^{13}C, and ^{15}N, the ^{19}F nucleus is a spin $I = 1/2$ species and is thus straightforward to apply in the context of bio-NMR. Its gyromagnetic ratio is such that it exhibits relatively high sensitivity, second only to ^{1}H of stable nuclei in the periodic table. At the same time, assuming the NMR probe is Teflon-free and optimized for ^{19}F NMR, biological samples typically exhibit no background signal. Given the considerable interest in fluorinated compounds by the pharmaceutical community, low background room temperature, and cryogenic NMR probes are commercially available which can accommodate (^{1}H-decoupled) ^{19}F NMR applications. These probes suffice equally well for ^{19}F NMR studies of proteins, although as discussed below there is additional advantage to dual resonance ^{19}F,^{13}C probes which incorporate ^{13}C detection.

^{19}F NMR has proven to be exceptionally useful in quantifying conformational equilibria in ensembles. The real utility of ^{19}F NMR in studies of protein conformational dynamics arises from the sensitivity of the chemical shift to local van der Waals, electrostatic, and solvent environments. Due largely to the fluorine lone pair electrons, ^{19}F reporters exhibit very large chemical shift dispersions [1–3] which translates into the direct resolution of distinct functional states. This expected improvement in resolution over conventional ^{1}H, ^{13}C, and ^{15}N NMR is even more important in the discrimination of dynamics and exchange between states. For example, in cases where two states, separated by a frequency $\Delta\nu$, undergo fast exchange, resulting in a single average resonance, Carr Purcell Meiboom Gill (CPMG) relaxation dispersions depend on the square of the separation (i.e. $\Delta\nu^2$) meaning that ^{19}F CPMG relaxation dispersion measurements and other types of ^{19}F NMR exchange spectroscopy are typically more sensitive to exchange dynamics and the delineation of states over conventional spectroscopic approaches [4].

Fluoroaromatics are readily incorporated into proteins biosynthetically without the need for auxotrophic strains. Most commonly, 4-, 5-, 6-, or 7-fluorotryptophan can be added to cell cultures at induction and are efficiently incorporated, after first suppressing aromatic amino acid synthesis through the addition of glyphosate. Tryptophan derivatives can also be conveniently incorporated in tryptophan auxotrophs or induced auxotrophs via ^{19}F-labeled indole [5]. Similarly, 2-,3-, or 4-fluorophenylalanine can be readily incorporated and often serve as useful markers for protein folding [6, 7]. Finally, 2- and 3-fluorotyrosine readily incorporate biosynthetically, allowing characterization of both structure and function in proteins [8–11]. In general, monofluorinated aromatic amino acids are weakly perturbing to protein structure and function. This is not surprising since the substitution of a fluorine atom for a proton on an aromatic ring amounts to a 25% increase in the van der Waals radius although a trifluoromethyl group has a comparable volume to an isopropyl group. Cumulative perturbations from fluoroaromatics may nonetheless arise from fluorophobic effects [12] or clustering of aromatic residues in the hydrophobic protein interior [13]. Fluorotyrosine also gives rise to a shift in the side chain pK_a and/or hydrogen bonding capacity of the amino acid side chain [14]. For example, 2- and 3-fluorotyrosine analogs typically reduce the hydroxyl pK_a by 1 and 2 pH units, respectively [15].

One of the most challenging limitations in ^{19}F NMR studies of high molecular weight complexes using fluoroaromatics as reporters involves their large chemical shift anisotropy (CSA) [16]. This is evident in the choice of labeling sites for tyrosine, phenylalanine, and tryptophan analogs. For example, a comparison of 4,5,6, and 7-fluoro tryptophan labeling in a membrane peptide revealed that 5-fluorotryptophan gave rise to superior T_2-relaxation, while both 7-fluoro and 5-fluorotryptophan exhibited maximal chemical shift dispersion. In principle, at appropriate magnetic field strengths, it is possible to play relaxation effects associated with ^{19}F–^{13}C dipole-dipole and the ^{19}F CSA against each other to improve line widths [17]. This requires first enriching the fluoroaromatic with ^{13}C and then implementing a two-dimensional ^{13}C,^{19}F transverse relaxation optimized spectroscopy (TROSY) NMR experiment [16]. Building on earlier work that introduced the synthesis of ^{13}C-enriched 3-fluorotyrosine, and associated ^{13}C,^{19}F two-dimensional NMR [18], it was shown that a significant improvement in overall resolution was possible through ^{13}C-detected ^{13}C,^{19}F TROSY. Most fluoroaromatics exhibit 1-bond couplings between 250 and 300 Hz, allowing for efficient transfer of magnetization between the ^{19}F and ^{13}C nuclei. Thus, while the TROSY approach would in principle provide access to larger proteins via ^{19}F,^{13}C NMR of fluoroaromatic amino acid analogs, the current challenge lies in efficient routes to the synthesis of ^{13}C-enriched fluoroaromatic amino acid analogs.

A wide variety of fluorinated alkyl amino acid analogs have been synthesized for purposes of peptide and protein engineering [19, 20]. Of note, 2-fluorohistidine proved amenable to biosynthetic labeling, with the added caveat that the imidazole pK_a is lowered by 4.8 units [21]. Consequently, in situations where protein function may be pH-dependent or where catalysis directly employs histidine residues, 2-fluorohistidine represents a useful control [22]. Similarly, methionine is a convenient target for biosynthetic labeling via fluorinated analogs, given its relatively low abundance (<2%) which minimizes cumulative structural perturbations and spectral overlap. Methionine residues also often play a key role in nonpolar interactions. Using auxotrophic *Escherichia coli* strains, both trifluoro- and difluoromethionine have been successfully incorporated in proteins [23–25]. It has also been demonstrated that proteins can be biosynthetically labeled with monofluorinated proline residues [26]. In this case, fluorine greatly disrupts cis/trans isomerizations in proline and the mono-fluorinated analogs are therefore potentially useful controls in studies focusing on the structural and dynamic roles of proline residues in proteins. Trifluoromethylated versions of leucine, isoleucine, and valine have also been synthesized and shown to incorporate in *E. coli* auxotrophs [14]. More recently, cell-free synthesis approaches have helped to improve incorporation levels of fluorinated aliphatic amino acid analogs [27]. While useful in bioengineering from the perspective of modulating protein fold stability through protein packing and fluorous effects, it is prohibitively difficult to utilize these fluorinated amino acids as NMR reporters on protein conformational dynamics and state equilibria.

The challenges associated with spectral overlap and cumulative perturbations, which frequently arise in biosynthetic labeling, can of course be avoided by the use of engineered tRNA/amino-acyl-tRNA synthetase pairs designed to recognize and facilitate site-specific incorporation of unnatural amino acids at alternate stop codon sites in prokaryotes and eukaryotes. This field of synthetic biology has advanced considerably in recent years as has the possibility of introducing specific trifluoromethylated reporters [28, 29]. While a wide variety of unnatural amino acids have been developed [30, 31], 4-(trifluoromethoxy)phenylalanine and 4-(trifluoromethyl)phenylalanine hold great promise as reporters for ^{19}F NMR [28, 32, 33].

24.3 An Overview of Chemical Tagging and Orthogonal Labeling

While the repertoire of fluorinated amino acid analogs is extensive, biosynthetic labeling is nevertheless challenging in large proteins due to anticipated overlap of multiple resonances, and challenges in expression yield. In cases where single cysteine mutants (or a small number of solvent-exposed cysteines) are tolerated, chemical tagging postexpression is one of the most useful ways of achieving site-specific labeling and thus, site-specific insights into conformational equilibria and dynamics. Moreover, with the possible exception of fluorophenylalanine and fluorotryptophan, the majority of biosynthetic labeling approaches via fluorinated amino acid precursors have been demonstrated in *E. coli* or *E. coli* auxotrophs, leaving chemical tagging by fluorinated reporters a necessity in cases where the protein of interest is heterologously expressed via other means. A variety of thiol-specific fluorinated tags are commercially available and have been used in ^{19}F NMR applications. Notably 3-bromo-1,1,1-trifluoroacetone (BTFA), can be used to covalently tag solvent-exposed cysteine residues with a trifluoromethyl acetone moiety in one step under physiological pH conditions. The resulting ^{19}F NMR signal is conveniently boosted by a lack of observable ^{1}H couplings, a modest CSA, and narrow line widths through fast methyl rotation [34, 35]. 2,2,2-trifluoroethanethiol (TFET), which is also commonly used in ^{19}F NMR applications, provides a small trifluoromethyl tag via a disulfide bond in two steps [36–38]. The resulting ^{19}F NMR spectra tend to be well dispersed although one must contend with a modest ^{1}H coupling and the possibility of loss of the tag over time through weak reducing conditions. A third thiol-specific moiety used recently in our lab is 2-bromo-*N*-[4-(trifluoromethyl)phenyl]acetamide (BTFMA) [39]. While sterically bulky, the electron-withdrawing environment of the phenyl moiety was suggested to leave the fluorine shielding tensor more sensitive to electrostatic environments [40]. Consequently, we typically observe a greater chemical shift dispersion and thus a greater resolving capacity for this trifluoromethyl tag. This observation suggests that there may be other thiol-specific tags with the capacity to "amplify" chemical shift differences between states. Figure 24.2a depicts several thiol-specific fluorinated tags in addition to the more recent (trifluoromethylphenyl)acetanilide moiety and a hexafluorinated derivative, currently under investigation in our lab. In the latter case, the equivalent trifluoromethyl moieties are meant to boost sensitivity while the hydroxyl group is installed for purposes of chemical shift sensitivity to solvent and electrostatics. Chemical shift sensitivity can in principle be assessed by installing the fluorine tag on a small peptide, as shown in Figure 24.2b, and recording ^{19}F NMR chemical shifts as a function of solvent polarity. Varying solvent polarities can be attained by resorting to mixtures of water and methanol. The ^{19}F chemical shift response for three tags are shown in Figure 24.2b, for 95% MeOH/5% D_2O, 95% H_2O/5% D_2O, and 100% D_2O. The results reveal comparable chemical shift sensitivity and T_1 for all tags including a newly synthesized 4-(1,1,1,3,3,3-hexafluoro-2-hydroxyisopropyl)-2-iodoacetanilide (HFPA) tag. Recently, a unique approach to labeling cysteines was devised in which the alpha proton of the cysteine of interest is used to direct a specific conjugating moiety [41]. While not yet adapted for ^{19}F NMR, such an approach could be used

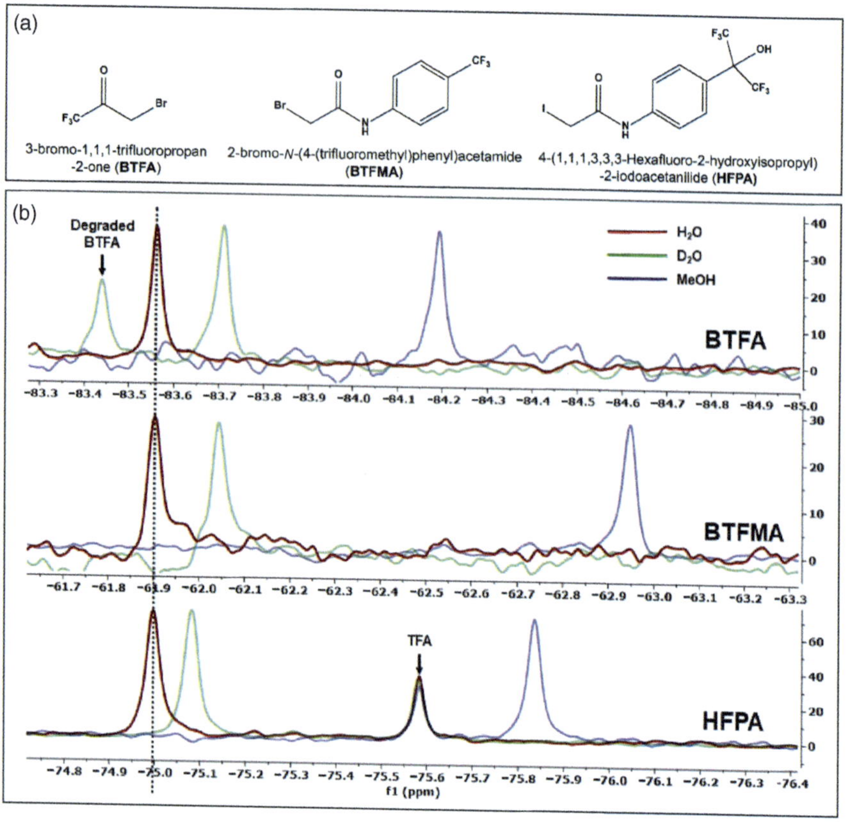

Figure 24.2 (a) Current thiol-specific ^{19}F NMR tags include BTFA, BTFMA, and HFPA. (b) ^{19}F NMR spectra of BTFA, BTFMA, and HFPA as a function of solvent (i.e. 95% MeOH/5% D_2O, 95% H_2O/5% D_2O, and 100% D_2O). Chemical shift differences between those in water and methanol are a useful predictor of chemical shift sensitivity of the tag of interest.

to selectively target a specific cysteine residue or selectively modify a cysteine residue with a paramagnetic tag. It is also possible that disulfides might be similarly used to install a CF_3 reporter, borrowing from chemistry that makes use of disulfide cross-bridging under nondenaturing conditions [42, 43]. The challenge would arise in situations where multiple disulfide linkages existed.

While cysteine mutagenesis and the associated ^{19}F-labeling is one of the most robust means to obtain site-specific information on protein conformational dynamics, there are many other possibilities for direct chemical tagging. For example, there are several means to directly tag lysine residues [44, 45] without altering charge at the ε-amino group. Transglutaminase has also been shown to append trifluoroethyl moieties on glutamine side chains [46]. The efficiency of both lysine and glutamine tagging depends on local solvent exposure. Consequently, labeling results in mixtures of species, although this can in principle be an advantage since peak intensity becomes a marker for solvent exposure and local dynamics. Both glutamine and lysine propensities are high in most proteins (~4% and 7%, respectively) making uniform labeling problematic from the perspective of spectral overlap and protein stability. One solution might be to make use of enzymes which utilize short recognition epitopes to append a reactive moiety. For example, formylglycine generating enzymes will react with cysteine residues in CXPXR motifs, thereby possibly allowing for site-specific conjugation of paramagnetic tags or fluorinated reporters [47].

24.4 Orthogonal Methods for Protein Labeling with ^{19}F NMR Probes

The utility of ^{19}F NMR for monitoring the conformational changes and interactions of biomolecules has motivated the development of chemical methods to label proteins with fluorinated species in a site-specific manner. While cysteine modification with conventional electrophilic reagents such as maleimides and α-halocarbonyls remains the standard for attaching synthetic molecules into target proteins without the need for genetic code expansion, this approach is limited by the need for pre-engineered protein constructs containing a single solvent-exposed cysteine and require highly reactive electrophilic reagents that may react promiscuously with various proteinogenic nucleophiles. To overcome these limitations, a suite of orthogonal protein labeling methodologies has emerged in recent years that target unique functionalities in target proteins and thus preclude the possibility for promiscuous labeling.

A particularly successful subset of these orthogonal methods makes use of bivalent labeling reagents. These reagents share a similar basic structure and are comprised of: (i) A reversibly interacting functionality F_{rev}, (ii) A linker region of appropriate length, and (iii) A reactive electrophilic functionality F_{irrev} that establishes an irreversible covalent interaction with a target residue in a chemoselective and site-specific manner, as shown in Figure 24.3. Early examples of this methodology within the context of ^{19}F NMR protein labeling involved ligand-directed tosyl (LDT) chemistry, whereby F_{rev} represents a modified ligand that binds to the target protein with high affinity and orients an electrophilic tosyl group for irreversible alkylation of a residue on the surface of the protein with a ^{19}F NMR probe [48, 49]. While an effective strategy, the dependence on a pre-existing high-affinity ligand for the target protein and the high intrinsic reactivity of arylsulfonate ester electrophiles limit the utility of this technology.

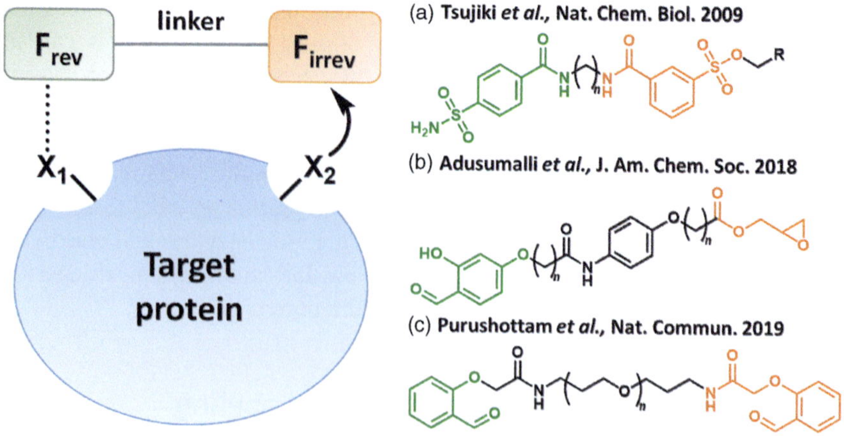

Figure 24.3 Orthogonal labeling methods achieve site-selective labeling by relying on a specific pair of amino acids in the protein, X_1 and X_2. X_1 serves as a site for (fast) reversible chemical labeling, to which an irreversible ligand with slower reaction chemistry is appended. The irreversible ligand is then conjugated uniquely to X_2 to due to its specific proximity, whereupon the reversible ligand is then removed.

A further advancement in site-specific ^{19}F protein labeling with bivalent molecules known as linchpin-directed modification (LDM) relies on global, reversible labeling of lysine residues with a benzaldehyde-derived moiety followed by chemoselective histidine alkylation with a mild alkyl epoxide electrophile [50]. Despite global labeling with the lysine-targeted "linchpin," only that histidine which is appropriately spaced from a lysine residue as dictated by the length of the spacer benefits from an acceleration of the irreversible reaction kinetics through a proximity-driven increase in the local concentration of the electrophile. Once the site-selective alkylation has taken place, all nonproductive linchpin molecules are washed away and the covalently linked protein may be functionalized with ^{19}F probes via subsequent oxime formation. A modification of this chemical technology has also been adapted to single-site labeling of N-terminal glycine residues in proteins, whereby a 2-formylphenoxyacetyl moiety mediates the formation of a latent enol nucleophile which reacts with another benzaldehyde electrophile to form a stable 1,2-aminoalcohol adduct [51]. This methodology offers many advantages over other bioconjugation strategies in that it targets a unique functionality in the target protein (i.e. an N-terminal glycine) and it generates the reactive nucleophile *in situ*, strongly disfavoring any possibility for promiscuous labeling. Once the adduct has been formed, the benzaldehyde at the other end of the molecule may be functionalized with ^{19}F probes via oxime formation in a similar fashion to LDM. Since these methods allow for labeling in addition to thiol-specific chemistry they are particularly attractive as a means for both site-directed mutagenesis and the (orthogonal) installation of a paramagnetic shift reagent or relaxation agent. This would greatly enhance the resulting spectroscopy as discussed below since specific resonances could additionally be distinguished by distance measurements to a specific (paramagnetic) marker.

24.5 Current Studies of Conformational Dynamics of Proteins

Subtle structural changes of proteins such as conformational rearrangements upon ligand binding, conformational exchange of active sites to shuttle substrates along reaction coordinates, and cooperative structural changes transmitted along allosteric pathways are crucial and ubiquitous facets of protein function [52, 53]. While advances in X-ray crystallography and cryoEM have greatly informed our molecular perspective of the interplay between protein structure and function [54], they fall short of providing a holistic understanding of the conformational landscape of the protein ensemble [55]. Indeed, such structural techniques generate significant bias toward stable, long-lived protein conformations over the transient, weakly populated states that often chiefly orchestrate the function of the protein [56]. In this regard, solution NMR spectroscopy is uniquely poised to study these dynamic processes with high temporal and spatial resolution in a native-like environment.

Owing to its exquisite sensitivity to chemical environment and virtual absence in biological systems, fluorine has been employed extensively as an NMR reporter moiety for probing the dynamics and chemical exchange of complex biological equilibria such as protein folding [57, 58], conformational selection [59, 60], and enzymatic catalysis [4, 61]. Monofluorinated amino acids and their biosynthetic precursors readily incorporate into proteins to furnish fractionally or uniformly fluorinated protein samples suitable for NMR conformational studies [13]. In our experience, these fluorine reporters are nonperturbing and highly responsive to subtle differences in their van der Waals and electrostatic environments associated with conformational transitions, enabling detailed mechanistic, and kinetic analysis of protein conformational dynamics [4].

Recently, Ruben et al. exploited 5-fluorotryptophan labeling of the trypsin-like clotting protease thrombin to investigate the conformational properties of the free enzyme and its zymogen precursor prethrombin-2 [62]. Both thrombin and its inactive zymogen precursor can exist in two distinct states, an "open" form (E) and a "closed" form (E^*) which restricts access to the primary specificity pocket [63]. While crystallographic [64, 65] and rapid kinetics [66, 67] studies suggest that thrombin and prethrombin-2 exist in conformational equilibrium between these states from which the ligand selects the optimal fit (i.e. conformational selection), contrasting evidence maintains that free thrombin is zymogen-like and switches to the mature conformation upon ligand binding by induced fit [68, 69]. By fluorine labeling all nine tryptophan residues in thrombin and prethrombin-2, the authors provide direct evidence for conformational plasticity of these proteins in the absence of ligands and show that the conformational ensemble of free thrombin is significantly different from that of prethrombin-2 and more similar to the active Na^+-bound form of the protease.

The ^{19}F NMR spectrum of free thrombin shows seven well-dispersed resonances spanning nearly 8 ppm while the spectrum of prethrombin-2 shows only four broad resonances (Figure 24.4), suggesting that the tryptophan residues in the zymogen exchange between multiple conformations leading to broad, overlapping resonances. Thus, the authors concluded that thrombin samples a smaller conformational space than prethrombin-2 and is intrinsically more rigid. Furthermore, residue-specific dynamics identified a tryptophan residue (W51) located 33 Å away from the Na^+-binding site with a CPMG relaxation dispersion profile consistent with fast conformational exchange for thrombin and prethrombin-2. Intriguingly, this exchange is entirely abrogated in thrombin upon the addition of Na^+, suggesting that Na^+ binding rigidifies this distal residue through a hitherto unidentified allosteric pathway. Another interesting feature in the ^{19}F spectrum of prethrombin-2 is the existence of two widely separated resonances that both correspond to W215, a tryptophan residue responsible for keeping the active site open (i.e. reduce the rate of E→E* conversion) through a

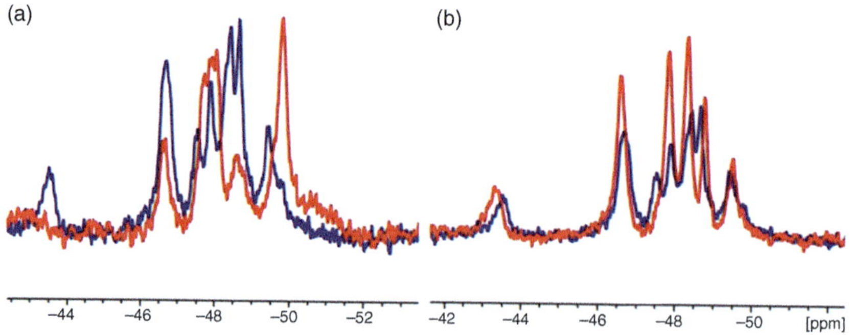

Figure 24.4 Overlay of 1D NMR spectra comparing a zymogen precursor prethrombin-2 and the active enzyme, thrombin (a) and between thrombin free and bound to Na^+ (b). The results reveal that free thrombin is more similar to its Na^+-bound form than the zymogen precursor prethrombin-2, and speak to a conformational selection model for activation in which prothrombin adopts an equilibrium between and open (E) and closed state (E*). Here, ligand binding takes place through the E state, resulting in an equilibrium shift and enzyme activation. Source: Figure obtained from [62] under CC-BY open access license.

π-stacking interaction with F227. The presence of these two peaks indicate that W215 adopts two distinct conformations that exchange very slowly or not at all. Upon conversion to thrombin, W215 becomes a single peak consistent with fast exchange between two alternative conformations.

24.6 Enhancing ^{19}F NMR Spectroscopy with Topology and Distance Measurements

While ^{19}F NMR chemical shift dispersions are large and therefore often provide delineation of multiple resonances in spectroscopic studies of biosynthetically fluorinated proteins, it is challenging to interpret residue-specific shifts in terms of specific structures. Prior ^{19}F NMR studies of 3-fluorotyrosine enriched BRD4, a bromodomain protein involved in histone acetylation, resolved all seven residue-specific resonances [70]. Via single site tyrosine/tryptophan mutagenesis, these resonances were then assigned [70]. By making use of Density Functional Theory (DFT)-based simulations from crystallographic and molecular dynamics (MD)-derived structures, these authors then attempted to recapitulate the observed ^{19}F NMR shifts [11]. The results provided a "first-order" correlation between the observed and simulated shifts, although 3-fluorotyrosine is arguably one of the most difficult fluoroaromatics to simulate given the solvation dynamics and orientation states of the adjacent hydroxyl group. In principle, if fluoroaromatic shifts were to improve to within a fraction of the chemical shift dispersion, assignments could be largely estimated from crystal structures and MD simulations, facilitating studies of conformational dynamics.

Empirically, ^{19}F NMR chemical shifts tend to correlate roughly with solvent exposure and are therefore useful in understanding protein topology [8]. The contributions to the chemical shift due to solvent exposure can be more precisely studied by measuring differences in observed chemical shifts in D_2O and H_2O. Such shifts reach a maximum on the order of 0.25 ppm, depending on residue-specific solvent exposure. Typically, these isotope shifts also correlate with ^{19}F spin-lattice relaxation times (T_1) in that greater solvent exposure results in larger T_1 [71]. Solvent isotope shifts may be complemented by paramagnetic shifts resulting from dissolved dioxygen. Dioxygen is a well-known paramagnetic species and its short electronic spin-lattice relaxation time result in convenient paramagnetic (contact) shifts and spin-lattice relaxation enhancements of ^{19}F nuclei. Typically, after allowing for equilibration of dissolved oxygen by monitoring ^{2}H lock signal, spin-lattice relaxation rates of roughly 10 Hz and paramagnetic shifts on the order of 0.25 ppm are observed for solvent exposed fluorinated probes, using oxygen partial pressures of 20 atm [72, 73]. Such pressures are attainable through commercially available medium- and thick-walled NMR tubes and have the added benefit of drastically reducing experimental time through quenching T_1. In most systems, accessibility of either water or dioxygen will depend on both diffusional accessibility and chemical potential of the species of interest. Water, for example will tend to weakly penetrate hydrophobic interiors, depending on both details of diffusional access and the local microenvironment. A fluorinated tag may reside in the protein interior as part of an accessible hydrogen-bonded water network. Such networks would be expected to give rise to both solvent isotope shifts and paramagnetic effects from oxygen. The combination of both measurements would help provide a more realistic measure of solvent accessibility, exposed surface area, and local hydrophobicity. The complementary chemical potentials of water and dioxygen have also been utilized previously to attain reliable measures of immersion depth of fluorinated labeling sites in detergents and lipid bilayers [58, 74–76]. In membrane proteins, for example precise measures of both solvent exposure and immersion depth (to within ~2–3 Å) can be simultaneously determined to provide a comprehensive assessment of topology [35].

As discussed above, ^{19}F NMR studies of proteins promise to build on high-resolution structures determined largely by cryoEM and X-ray crystallography. These high-resolution structural methods tend to bias the ensemble to single "ground state" conformers. Thus, ^{19}F NMR spectroscopy can be used to both validate specific conformers and interpret these in terms of the overall conformational ensemble. At present, it is technically challenging to validate structures through ^{19}F NMR chemical shifts alone, although topology measurements through solvent isotope shifts and O_2-derived paramagnetic effects improve the possibility of connecting spectra to structural models. In principle distance measurements between fluorinated reporters and paramagnetic sites on the protein could greatly add to the information content of a given series of spectra and the possibility of assigning spectra to specific conformers. This was recently demonstrated in a ^{19}F NMR study of a homotrimeric membrane transporter, GltPh, in detergent micelles [77]. In this study, the authors made use of biosynthetic labeling via 5-fluorotryptophan in combination with a di-histidine motif which upon chelation to Ni(II), provided distance measurements to site-specific trifluoromethyl tags via T_1. Conveniently, the paramagnetic properties of Ni(II) are such that no (paramagnetic) perturbations in linewidth were observed upon addition of the contrast agent, while the dipole-mediated electron spin-nuclear spin interactions could be reliably measured through T_1 to determine distances extending to ~12 Å. The results allowed the authors to distinguish three conformers in the ensemble associated with an inward facing and two outward facing states of the aspartate/sodium symporter (Figure 24.5). In the future, improved capacity to simulate ^{19}F NMR spectra (chemical shifts) from structural models, precise topology measurements from solvent isotope shifts and O_2-derived paramagnetic shifts, and distance measurements from paramagnetic centers may all be used to assign specific conformers in ^{19}F NMR spectra and elaborate on ensemble properties.

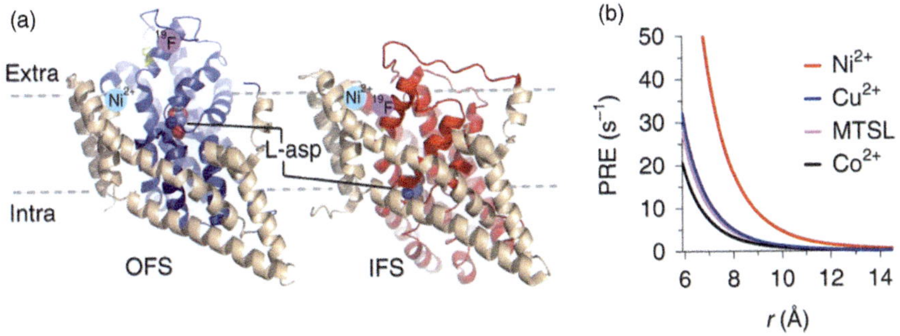

Figure 24.5 (a) Structures of *L*-Asp bound protomer in the outward facing and inward facing states (OFS and IFS) from the GltPh homotrimer. Here, the dynamic transport domain is highlighted in blue and red in the OFS and IFS, respectively. (b) Predicted dependence on the longitudinal R_1 ($1/T_1$) paramagnetic relaxation enhancement as a function of various paramagnetic centers. Source: Figure obtained from [77] with permission.

24.7 Studies of Ligand Interactions and Drug Discovery by ^{19}F NMR

The pronounced sensitivity of the ^{19}F NMR chemical shift to environment suggests that ^{19}F NMR might be successfully trained on the problem of drug discovery and in particular fragment-based drug discovery. Indeed, over 200 known pharmaceuticals contain fluorine while at the same time diverse fluorinated fragments continue to be developed, making the prospect of direct observation of ligand interactions very practical [78–80]. Many methodologies have since been refined that enable direct detection of fluorinated ligands, fragments, and drugs to protein targets, or enhanced detection via competition assays that depend on differentiating the bound and free states through either chemical shift and/or T_2 and T_1 [81–87].

The majority of ^{19}F NMR drug screening experiments rely on differences in either chemical shift or CSA-driven T_2 relaxation of the bound versus ligand-free states, although multiple quantum states, ^{1}H-^{19}F relaxation, and relative diffusion rates have also been used to delineate drug binding. As illustrated in Figure 24.6, the Fluorine Chemical Shift Anisotropy and Exchange for Screening (FAXS) experiment relies upon the fact that the bound fluorinated ligand is characterized by a unique chemical shift signature and line-broadening [88]. In this case, we imagine a so-called “spy” molecule, designated molecule S in Figure 24.6, which exhibits a broad and shifted bound state signature in the presence of the receptor. If the spy molecule is added in large excess to that of the receptor, then the average amount of exchange broadening associated with the bound state is reduced, although the effects of exchange broadening with the receptor are nonetheless easily detected after employing a spin-lock or CPMG filter after the initial

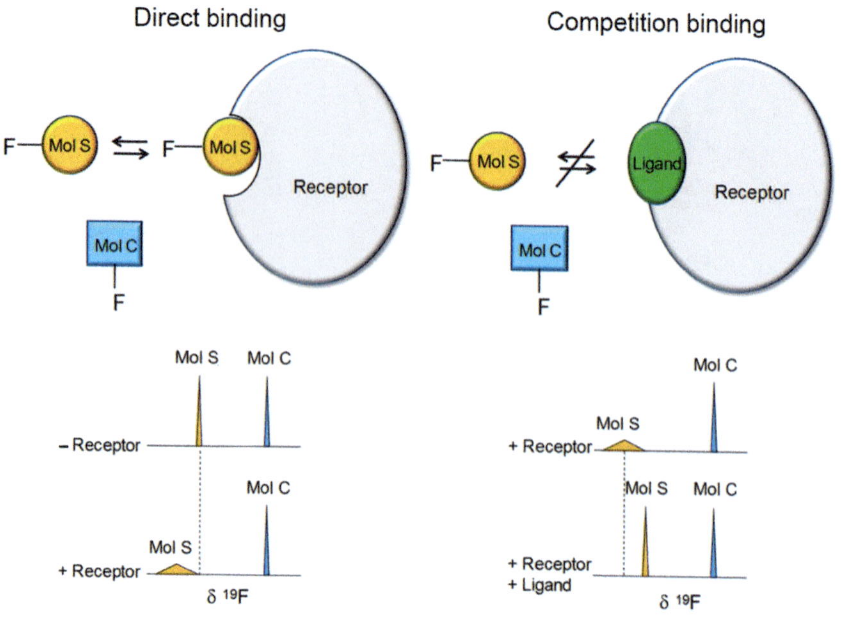

Figure 24.6 The FAXS experiment performed in direct (left) and competition (right) format. The ^{19}F NMR signal of the "spy" ligand, S, is distinguished by a drop intensity in the presence of receptor, while the signal of the control molecule C that does not bind to the receptor remains sharp and does not change its chemical shift (left). In the competition experiment (right), the broad signal of the spy (reporter) molecule S becomes sharp and returns to the chemical shift of the free state due to its displacement from the receptor in the presence of a competitive ligand. Typically, the spy molecule would be used in large excess to improve signal to noise in the experiment, and a CPMG or spin-lock (T_2-based) filter would be employed to reliably detect intensity changes of molecule S due to exchange between bound and free states. Source: Figure obtained from [85] with permission.

excitation pulse. Thus, ligands which exhibit binding to the receptor will exhibit a reduced intensity in accordance with the extent of binding, while a large ligand excess (~20–30×) ensures that the experiment is performed with low receptor concentrations and in a short period of time. As shown in Figure 24.6, this experiment can also be applied to the study of receptor binding by unlabeled ligands, in a classic competition assay. In this case, if the added ligand outcompetes the spy molecule, the exchange broadening on the part of the spy ligand is reduced.

Invariably, binding studies such as those described above, attempt to identify lead candidates that bind to a target receptor with sub-millimolar dissociation constants. Indeed, such "low affinity" hits are commonplace in fragment-based drug discovery. As binding affinity improves, the bound state signature becomes more pronounced and exchange rates between bound and free states drop. In this case, it becomes necessary to rely more on the ^{19}F T_1, rather than T_2, and perform CEST-type experiments tailored for ^{19}F NMR. Here, the bound-state resonance is first pre-saturated using a train of selective pulses over a period of seconds, whereupon an excitation pulse reveals the free ligand signature(s). In a different experiment using on- and off-resonant saturation pulse trains, the free spy molecule signature will exhibit and intensity change resulting from ligand exchange with the bound state during the pre-saturation period [83].

24.8 Final Comments

Advances in fluorine chemistry and progress with cryogenic probes have enabled protein and ligand-based ^{19}F NMR spectroscopy to flourish. Next generation probes will exhibit greater sensitivity to structure and topology, enabling greater detail in the study of conformational dynamics of ensembles and understanding protein function. New approaches to orthogonal labeling are now at hand which should enable the combination of distinct ^{19}F-tags or the paramagnetic tags in conjunction with ^{19}F NMR reporters. This will improve prospects for routine distance measurements in conjunction with spectroscopic studies. There are now a large number of papers with ^{19}F NMR chemical shifts of well-known fluoroaromatics. It may be possible in the near future to bring together DFT, MD simulations, and libraries of ^{19}F NMR shifts of fluoroaromatics in proteins to assemble quasi-empirical algorithms for predicting and validating chemical shifts. By combining such measurements with results from studies of solvent exposure and topology via solvent isotope shifts and O_2-derived paramagnetic effects, alongside distance measurements via paramagnetic tags, it should be possible to significantly improve structure validation and study of conformational ensembles by ^{19}F NMF. Similarly, fragment-based drug discovery may also see improvements if such methodology is routinely combined with next generation labels and appropriate paramagnetic tags for site-specific detection.

Acknowledgments

We wish to thank the Natural Sciences and Engineering Research Council (NSERC) of Canada for supporting this research.

References

1 Danielson, M.A. and Falke, J.J. (1996). Use of 19F NMR to probe protein structure and conformational changes. *Annu. Rev. Biophys. Biomol. Struct.* 25 (1): 163–195.

2 Lau, E.Y. and Gerig, J.T. (2000 May). Origins of fluorine NMR chemical shifts in fluorine-containing proteins †. *J. Am. Chem. Soc.* 122 (18): 4408–4417.

3 Pearson, J.G., Oldfield, E., Lee, F.S., and Warshel, A. (1993). Chemical-shifts in proteins – a shielding trajectory analysis of the fluorine nuclear-magnetic-resonance spectrum of the *Escherichia coli* galactose binding-protein using a multipole shielding polarizability local reaction field molecular-dynamics approach. *J. Am. Chem. Soc.* 115 (15): 6851–6862.

4 Kim, T.H., Mehrabi, P., Ren, Z. et al. (2017). The role of dimer asymmetry and protomer dynamics in enzyme catalysis. *Science* 355 (6322): eaag2355-39.

5 Broos, J., Gabellieri, E., Biemans-Oldehinkel, E., and Strambini, G.B. (2003). Efficient biosynthetic incorporation of tryptophan and indole analogs in an integral membrane protein. *Protein Sci.* 12 (9): 1991–2000.

6 Li, H. and Frieden, C. (2005). NMR studies of 4-19F-phenylalanine-labeled intestinal fatty acid binding protein: evidence for conformational heterogeneity in the native state. *Biochemistry* 44 (7): 2369–2377.

7 Kitevski-LeBlanc, J.L., Evanics, F., and Prosser, R.S. (2010). Approaches to the assignment of (19)F resonances from 3-fluorophenylalanine labeled calmodulin using solution state NMR. *J. Biomol. NMR* 47 (2): 113–123.

8 Sykes, B.D., Weingarten, H.I., and Schlesinger, M.J. (1974). Fluorotyrosine alkaline phosphatase from *Escherichia coli*: preparation, properties, and fluorine-19 nuclear magnetic resonance spectrum. *Proc. Natl. Acad. Sci.* 71 (2): 469–473.

9 Oyala, P.H., Ravichandran, K.R., Funk, M.A. et al. (2016). Biophysical characterization of fluorotyrosine probes site-specifically incorporated into enzymes: *E. coli* ribonucleotide reductase as an example. *J. Am. Chem. Soc.* 138 (25): 7951–7964.

10 Arntson, K.E. and Pomerantz, W.C.K. (2016). Protein-observed fluorine NMR: a bioorthogonal approach for small molecule discovery. *J. Med. Chem.* 59 (11): 5158–5171. http://pubs.acs.org.myaccess.library.utoronto.ca/doi/abs/10.1021/acs.jmedchem.5b01447.

11 Isley, W.C., Urick, A.K., Pomerantz, W.C.K., and Cramer, C.J. (2016). Prediction of (19)F NMR chemical shifts in labeled proteins: computational protocol and case study. *Mol. Pharmaceutics* 13 (7): 2376–2386.

12 Berger, R., Resnati, G., Metrangolo, P. et al. (2011). Organic fluorine compounds: a great opportunity for enhanced materials properties. *Chem. Soc. Rev.* 40 (7): 3496–3508.

13 Kitevski-LeBlanc, J.L., Evanics, F., and Prosser, R.S. (2010). Optimizing ^{19}F NMR protein spectroscopy by fractional biosynthetic labeling. *J. Biomol. NMR* 48 (2): 113–121.

14 Merkel, L. and Budisa, N. (2012). Organic fluorine as a polypeptide building element: in vivo expression of fluorinated peptides, proteins and proteomes. *Org. Biomol. Chem.* 10 (36): 7241–7261.

15 Ycas, P.D., Wagner, N., Olsen, N.M. et al. (2020). 2-Fluorotyrosine is a valuable but understudied amino acid for protein-observed 19F NMR. *J. Biomol. NMR* 74 (1): 61–69.

16 Boeszoermenyi, A., Chhabra, S., Dubey, A. et al. (2019). Aromatic 19F-13C TROSY: a background-free approach to probe biomolecular structure, function, and dynamics. *Nat. Methods* 16 (4): 1–12.

17 Pervushin, K., Riek, R., Wider, G., and Wüthrich, K. (1998). Transverse relaxation-optimized spectroscopy (TROSY) for NMR studies of aromatic spin systems in 13C-labeled proteins. *J. Am. Chem. Soc.* 120 (25): 6394–6400.

18 Kitevski-LeBlanc, J.L., Al-Abdul-Wahid, M.S., and Prosser, R.S. (2009). A mutagenesis-free approach to assignment of (19)F NMR resonances in biosynthetically labeled proteins. *J. Am. Chem. Soc.* 131 (6): 2054–2055.

19 Berger, A.A., Völler, J.-S., Budisa, N., and Koksch, B. (2017). Deciphering the fluorine code-the many hats fluorine wears in a protein environment. *Acc. Chem. Res.* 50 (9): 2093–2103.

20 Moschner, J., Stulberg, V., Fernandes, R. et al. (2019). Approaches to obtaining fluorinated α-amino acids. *Chem. Rev.* 119 (18): 10718–10801.

21 Klein, D.C., Weller, J.L., Kirk, K.L., and Hartley, R.W. (1977). Incorporation of 2-fluoro-L-histidine into cellular protein. *Mol. Pharmacol.* 13 (6): 1105–1110.

22 Kasireddy, C., Ellis, J.M., Bann, J.G., and Mitchell-Koch, K.R. (2017). The biophysical probes 2-fluorohistidine and 4-fluorohistidine: spectroscopic signatures and molecular properties. *Sci. Rep.* 15 (7): 42651.

23 Duewel, H., Daub, E., Robinson, V., and Honek, J.F. (1997). Incorporation of trifluoromethionine into a phage lysozyme: implications and a new marker for use in protein 19F NMR †. *Biochemistry* 36 (11): 3404–3416. https://pubs-acs-org.myaccess.library.utoronto.ca/doi/abs/10.1021/bi9617973.

24 Vaughan, M.D., Cleve, P., Robinson, V. et al. (1999). Difluoromethionine as a novel F-19 NMR structural probe for internal amino acid packing in proteins. *J. Am. Chem. Soc.* 121 (37): 8475–8478.

25 Duewel, H.S., Daub, E., Robinson, V., and Honek, J.F. (2001). Elucidation of solvent exposure, side-chain reactivity, and steric demands of the trifluoromethionine residue in a recombinant protein †. *Biochemistry* 40 (44): 13167–13176. https://pubs-acs-org.myaccess.library.utoronto.ca/doi/abs/10.1021/bi011381b.

26 Kim, W., George, A., Evans, M., and Conticello, V.P. (2004). Cotranslational incorporation of a structurally diverse series of proline analogues in an *Escherichia coli* expression system. *ChemBioChem* 5 (7): 928–936.

27 Arthur, I.N., Hennessy, J.E., Padmakshan, D. et al. (2013). In situ deprotection and incorporation of unnatural amino acids during cell-free protein synthesis. *Chemistry* (Weinheim an der Bergstrasse, Germany). 19 (21): 6824–6830.

28 Cellitti, S.E., Jones, D.H., Lagpacan, L. et al. (2008). In vivo incorporation of unnatural amino acids to probe structure, dynamics, and ligand binding in a large protein by nuclear magnetic resonance spectroscopy. *J. Am. Chem. Soc.* 130 (29): 9268–9281.

29 Jones, D.H., Cellitti, S.E., Hao, X. et al. (2010). Site-specific labeling of proteins with NMR-active unnatural amino acids. *J. Biomol. NMR* 46 (1): 89–100.

30 Miyake-Stoner, S.J., Refakis, C.A., Hammill, J.T. et al. (2010). Generating permissive site-specific unnatural aminoacyl-tRNA synthetases. *Biochemistry* 49 (8): 1667–1677.

31 Yang, F., Yu, X., Liu, C. et al. (2015). Phospho-selective mechanisms of arrestin conformations and functions revealed by unnatural amino acid incorporation and 19F-NMR. *Nat. Commun.* 1–15.

32 Jackson, J.C., Hammill, J.T., and Mehl, R.A. (2007). Site-specific incorporation of a (19)F-amino acid into proteins as an NMR probe for characterizing protein structure and reactivity. *J. Am. Chem. Soc.* 129 (5): 1160–1166.

33 Shi, P., Li, D., Chen, H. et al. (2012). In situ 19F NMR studies of an *E. coli* membrane protein. *Protein Sci.* 21 (4): 596–600.

34 Huestis, W.H. and Raftery, M.A. (1978). Bromotrifluoroacetone alkylates hemoglobin at cysteine beta93. *Biochem. Biophys. Res. Commun.* 81 (3): 892–899.

35 Luchette, P.A., Prosser, R.S., and Sanders, C.R. (2002). Oxygen as a paramagnetic probe of membrane protein structure by cysteine mutagenesis and (19)F NMR spectroscopy. *J. Am. Chem. Soc.* 124 (8): 1778–1781.

36 Klein-Seetharaman, J., Getmanova, E.V., Loewen, M.C. et al. (1999). NMR spectroscopy in studies of light-induced structural changes in mammalian rhodopsin: applicability of solution (19)F NMR. *Proc. Natl. Acad. Sci. U.S.A.* 96 (24): 13744–13749.

37 Loewen, M.C., Klein-Seetharaman, J., Getmanova, E.V. et al. (2001). Solution 19F nuclear overhauser effects in structural studies of the cytoplasmic domain of mammalian rhodopsin. *Proc. Natl. Acad. Sci.* 98 (9): 4888–4892.

38 Liu, J.J., Horst, R., Katritch, V. et al. (2012). Biased signaling pathways in β2-adrenergic receptor characterized by 19F-NMR. *Science* 335 (6072): 1106–1110.

39 Ye, L., Eps, N.V., Zimmer, M. et al. (2016). Activation of the A2A adenosine G-protein-coupled receptor by conformational selection. *Nature* 533 (7602): 265. https://doi.org/10.1038/nature17668.

40 Ye, L., Larda, S.T., Li, Y.F.F. et al. (2015). A comparison of chemical shift sensitivity of trifluoromethyl tags: optimizing resolution in 19F NMR studies of proteins. *J. Biomol. NMR* 62 (1): 97–103.

41 Nathani, R.I., Moody, P.R., Chudasama, V. et al. (2013). A novel approach to the site-selective dual labelling of a protein via chemoselective cysteine modification. *Chem. Sci.*

42 Agrawalla, B.K., Wang, T., Riegger, A. et al. (2018). Chemoselective dual labeling of native and recombinant proteins. *Bioconjugate Chem.* 29 (1): 29–34.

43 Marculescu, C., Kossen, H., Morgan, R.E. et al. (2014). Aryloxymaleimides for cysteine modification, disulfide bridging and the dual functionalization of disulfide bonds. *Chem. Commun.* 50 (54): 7139.

44 Larda, S.T., Pichugin, D., and Prosser, R.S. (2015). Site-specific labeling of protein lysine residues and N-terminal amino groups with indoles and indole-derivatives. *Bioconjugate Chem.* 26 (12): 2376–2383.

45 Chen, H., Huang, R., Li, Z. et al. (2017). Selective lysine modification of native peptides via aza-Michael addition. *Org. Biomol. Chem.* 15 (35): 7339–7345.

46 Hattori, Y., Heidenreich, D., Ono, Y. et al. (2017). Protein 19F-labeling using transglutaminase for the NMR study of intermolecular interactions. *J. Biomol. NMR* 68 (4): 271–279.

47 Appel, M.J. and Bertozzi, C.R. (2015). Formylglycine, a post-translationally generated residue with unique catalytic capabilities and biotechnology applications. *ACS Chem. Biol.* 10 (1): 72–84.

48 Tsukiji, S., Miyagawa, M., Takaoka, Y. et al. (2009). Ligand-directed tosyl chemistry for protein labeling in vivo. *Nat. Chem. Biol.* 5 (5): 341–343.

49 Takaoka, Y., Sun, Y., Tsukiji, S., and Hamachi, I. (2011). Mechanisms of chemical protein 19F-labeling and NMR-based biosensor construction in vitro and in cells using self-assembling ligand-directed tosylate compounds. *Chem. Sci.* 2 (3): 511–520.

50 Adusumalli, S.R., Rawale, D.G., Singh, U. et al. (2018). Single-site labeling of native proteins enabled by a chemoselective and site-selective chemical technology. *J. Am. Chem. Soc.* 140 (44): 15114–15123. https://doi.org/10.1021/jacs.8b10490.

51 Purushottam, L., Adusumalli, S.R., Singh, U. et al. (2019). Single-site glycine-specific labeling of proteins. *Nat. Commun.* 10 (1): 1–9.

52 Motlagh, H.N., Wrabl, J.O., Li, J., and Hilser, V.J. (2014). The ensemble nature of allostery. *Nature* 508 (7496): 331–339.

53 Kay, L.E. (2016). New views of functionally dynamic proteins by solution NMR spectroscopy. *J. Mol. Biol.* 428 (2): 323–331.

54 Congreve, M., de Graaf, C., Swain, N.A., and Tate, C.G. (2020). Impact of GPCR structures on drug discovery. *Cell* 181 (1): 81–91.

55 Pietrantonio, C.D., Pandey, A., Gould, J. et al. (2019). *Understanding Protein Function Through an Ensemble Description: Characterization of Functional States by 19F NMR*, 1e. Elsevier Inc. (G Protein Coupled Receptors; vol. 615).

56 Manglik, A., Kim, T.H., Masureel, M. et al. (2015). Structural insights into the dynamic process of β2-adrenergic receptor signaling. *Cell* 161 (5): 1101–1111. http://www.sciencedirect.com/science/article/pii/S0092867415004997.

57 Ropson, I.J. and Frieden, C. (1992). Dynamic NMR spectral analysis and protein folding: identification of a highly populated folding intermediate of rat intestinal fatty acid-binding protein by 19F NMR. *Proc. Natl. Acad. Sci. U.S.A.* 89 (15): 7222–7226.

58 Evanics, F., Bezsonova, I., Marsh, J. et al. (2006). Tryptophan solvent exposure in folded and unfolded states of an SH3 domain by 19F and 1H NMR. *Biochemistry* 45 (47): 14120–14128.

59 Chrisman, I.M., Nemetchek, M.D., de IMS, V. et al. (2018). Defining a conformational ensemble that directs activation of PPARγ. *Nat. Commun.* 9 (1): 1794.

60 Frei, J.N., Broadhurst, R.W., Bostock, M.J. et al. (2020). Conformational plasticity of ligand-bound and ternary GPCR complexes studied by 19F NMR of the β1-adrenergic receptor. *Nat. Commun.* 11 (1): 669.

61 Mehrabi, P., Pietrantonio, C.D., Kim, T.H. et al. (2019). Substrate-based allosteric regulation of a homodimeric enzyme. *J. Am. Chem. Soc.* 141 (29): 11540–11556. https://pubs-acs-org.myaccess.library.utoronto.ca/doi/10.1021/jacs.9b03703.

62 Ruben, E.A., Gandhi, P.S., Chen, Z. et al. (2020). 19F NMR reveals the conformational properties of free thrombin and its zymogen precursor prethrombin-2. *J. Biol. Chem.* 295 (24): 8227–8235.

63 Chinnaraj, M., Chen, Z., Pelc, L.A. et al. (2018). Structure of prothrombin in the closed form reveals new details on the mechanism of activation. *Sci. Rep.* 8 (1): 2945.

64 Chen, Z., Pelc, L.A., and Cera, E.D. (2010). Crystal structure of prethrombin-1. *Proc. Natl. Acad. Sci. U.S.A.* 107 (45): 19278–19283.

65 Pozzi, N., Chen, Z., Gohara, D.W. et al. (2013). Crystal structure of prothrombin reveals conformational flexibility and mechanism of activation. *J. Biol. Chem.* 288 (31): 22734–22744.

66 Vogt, A.D., Chakraborty, P., and Cera, E.D. (2015). Kinetic dissection of the pre-existing conformational equilibrium in the trypsin fold. *J. Biol. Chem.* 290 (37): 22435–22445.

67 Chakraborty, P., Acquasaliente, L., Pelc, L.A., and Cera, E.D. (2018). Interplay between conformational selection and zymogen activation. *Sci. Rep.* 8 (1): 4080.

68 Huntington, J.A. (2009). Slow thrombin is zymogen-like. *J. Thromb. Haemost.* 7 (SUPPL. 1): 159–164.

69 Kamath, P., Huntington, J.A., and Krishnaswamy, S. (2010). Ligand binding shuttles thrombin along a continuum of zymogen- and proteinase-like states. *J. Biol. Chem.* 285 (37): 28651–28658.

70 Mishra, N.K., Urick, A.K., Ember, S.W.J. et al. (2014). Fluorinated aromatic amino acids are sensitive 19F NMR probes for bromodomain-ligand interactions. *ACS Chem. Biol.* 9 (12): 2755–2760. https://pubs-acs-org.myaccess.library.utoronto.ca/doi/10.1021/cb5007344.

71 Hull, W.E. and Sykes, B.D. (1976). Fluorine-19 nuclear magnetic resonance study of fluorotyrosine alkaline phosphatase: the influence of zinc on protein structure and a conformational change induced by phosphate binding. *Biochemistry* 15 (7): 1535–1546.

72 Prosser, R.S., Luchette, P.A., and Westerman, P.W. (2000). Using O2 to probe membrane immersion depth by 19F NMR. *Proc. Natl. Acad. Sci. U.S.A.* 97 (18): 9967–9971.

73 Prosser, R.S., Luchette, P.A., Westerman, P.W. et al. (2001). Determination of membrane immersion depth with O2: a high-pressure 19F NMR study. *Biophys. J.* 80 (3): 1406–1416.

74 Evanics, F. (2005). Discriminating binding and positioning of amphiphiles to lipid bilayers by 1H NMR. *Anal. Chim. Acta* 534 (1): 21–29. http://www.sciencedirect.com/science/article/pii/S0003267004007986.

75 Evanics, F., Hwang, P.M., Cheng, Y. et al. (2006). Topology of an outer-membrane enzyme: measuring oxygen and water contacts in solution NMR studies of PagP. *J. Am. Chem. Soc.* 128 (25): 8256–8264.

76 Al-Abdul-Wahid, M.S., Yu, C.-H., Batruch, I. et al. (2006). A combined NMR and molecular dynamics study of the transmembrane solubility and diffusion rate profile of dioxygen in lipid bilayers †. *Biochemistry* 45 (35): 10719–10728. https://pubs.acs.org/doi/10.1021/bi060270f.

77 Huang, Y., Wang, X., Lv, G. et al. (2020). Use of paramagnetic 19F NMR to monitor domain movement in a glutamate transporter homolog. *Nat. Chem. Biol.* 29: 1–16.

78 Vulpetti, A. and Dalvit, C. (2013). Design and generation of highly diverse fluorinated fragment libraries and their efficient screening with improved 19F NMR methodology. *ChemMedChem* 8 (12): 2057–2069.

79 Wang, J., Sánchez-Roselló, M., Aceña, J.L. et al. (2014). Fluorine in pharmaceutical industry: fluorine-containing drugs introduced to the market in the last decade (2001-2011). *Chem. Rev.* 114 (4): 2432–2506.

80 Troelsen, N., Shanina, E., Gonzalez-Romero, D. et al. (2019). The 3F library: fluorinated Fsp3-rich fragments for expeditious 19F-NMR-based screening. *Angew. Chem. Int. Ed.* 1–8.

81 Dalvit, C. (2007). Ligand- and substrate-based 19F NMR screening: principles and applications to drug discovery. *Prog. Nucl. Magn. Reson. Spectrosc.* 51 (4): 243–271.

82 Norton, R., Leung, E., Chandrashekaran, I., and MacRaild, C. (2016). Applications of 19F-NMR in fragment-based drug discovery. *Molecules* 21 (7): 860–813. https://www.mdpi.com/1420-3049/21/7/860/htm.

83 Dalvit, C. and Piotto, M. (2017). 19F NMR transverse and longitudinal relaxation filter experiments for screening: a theoretical and experimental analysis. *Magn. Reson. Chem.* 55 (2): 106–114.

84 Sugiki, T., Furuita, K., Fujiwara, T., and Kojima, C. (2018). Current NMR techniques for structure-based drug discovery. *Molecules* 23 (1): 148–127. Available from: http://www.mdpi.com/1420-3049/23/1.

85 Dalvit, C. and Vulpetti, A. (2019). Ligand-based fluorine NMR screening: principles and applications in drug discovery projects. *J. Med. Chem.* 62 (5): 2218–2244.

86 Rüdisser, S.H., Goldberg, N., Ebert, M.-O. et al. (2020). Efficient affinity ranking of fluorinated ligands by 19F NMR: CSAR and FastCSAR. *J. Biomol. NMR* 7: 215.

87 Stadmiller, S.S., Aguilar, J.S., Waudby, C.A., and Pielak, G.J. (2020). Rapid quantification of protein-ligand binding via 19F NMR lineshape analysis. *Biophys. J.* 118 (10): 2537–2548.

88 Dalvit, C., Fagerness, P.E., Hadden, D.T.A. et al. (2003). Fluorine-NMR experiments for high-throughput screening: theoretical aspects, practical considerations, and range of applicability. *J. Am. Chem. Soc.* 125 (25): 7696–7703.

25

Optical Approaches for Dissecting GPCR Signaling

Patrick R. O'Neill[1], Bryan A. Copits[2], and Michael R. Bruchas[3,4]

[1] *Hatos Center for Neuropharmacology, Department of Psychiatry and Biobehavioral Sciences, University of California Los Angeles, Los Angeles, CA, USA*
[2] *Pain Center, Department of Anesthesiology, Washington University School of Medicine, St. Louis, MO, USA*
[3] *Center of Excellence in the Neurobiology of Addiction, Pain, and Emotion, Department of Anesthesiology, University of Washington, Seattle, WA, USA*
[4] *Department of Pharmacology, University of Washington, Seattle, WA, USA*

25.1 Introduction

Cells communicate over distances ranging from the cellular scale to the entire organism using a diverse collection of peptides, hormones, and secreted small molecules that bind to GPCRs and activate intracellular signaling cascades. There has been a growing appreciation over the past decade that the biological information communicated through GPCR signaling is encoded not only in the specific biochemical cascades activated by a given receptor but also in the spatial and temporal patterns of signaling activity [1]. Varied patterns of agonist concentration, such as temporal ramps, pulses, and spatial gradients, carry information that can direct a cell to proliferate, differentiate, or migrate in a specific direction. Intracellular networks of signaling proteins, lipids, and second messengers convert these signals into appropriate cellular responses. Notably, patterns of intracellular signaling can be completely different from those of the stimulus input [2]. For example, a sustained stepwise increase in receptor activity can produce pulsatile or transient intracellular signaling responses, and cells can generate switch-like responses to temporally graded stimuli [3, 4]. Additionally, GPCR trafficking plays an active role in shaping cellular responses by controlling the timing and location of receptor-mediated signaling [5, 6].

An in-depth understanding of GPCR function requires detailed information on the spatiotemporal dynamics of both the stimuli (endogenous agonists) and the intracellular responses, and how these exert changes in cellular behaviors that

GPCRs as Therapeutic Targets, Volume 2, First Edition. Edited by Annette Gilchrist.

control physiological processes. Genetically encoded sensors have made it possible to observe subcellular dynamics of many of the signals generated following GPCR activation, as well as dynamic changes in the concentrations of various endogenous GPCR agonists [7–13]. Imaging-based methods have been crucial for discovering unanticipated patterns of signaling, such as pulses of kinase activity in individual cells that are averaged out in population based assays, or the presence of active GPCRs and G proteins at numerous intracellular organelles. These approaches are useful for identifying correlations between GPCR activation, intracellular signaling dynamics, and cellular responses, but lack some specificity since establishing causal mechanisms requires dynamic perturbation of different components of the signaling network.

Optically controllable perturbations, using optogenetic or photopharmacological approaches, are particularly useful in this regard because optical stimuli can be applied rapidly, reversibly, and with spatial control on scales ranging from anatomical to subcellular. These tools fall broadly into two main categories: optical control of GPCR stimulation, either by photosensitive ligands or GPCR opsins, and optical control of intracellular signaling proteins or second messengers. The current GPCR tool-kit enables optical activation of a diverse collection of signaling modules, as well as inhibition or allosteric modulation of several different endogenous signaling proteins. This chapter surveys these tools and discusses how they have been used in combination with imaging-based readouts, physiological measurements, or behavioral assays to identify causal links between signaling dynamics and molecular, cellular, physiologic, and behavioral responses.

25.2 Optical Control of GPCRs

Optical control of GPCRs has been achieved through two main approaches: optogenetics and photopharmacology [14–16]. Optogenetic approaches use naturally occurring or genetically modified retinal-binding opsin GPCRs to achieve light-dependent signaling. Photopharmacology uses photosensitive ligands to activate, inhibit, or modulate otherwise light-insensitive GPCRs. This section discusses both approaches, along with assays that have proven useful in screening new tools that are designed for optical control of GPCR signaling.

25.2.1 Retinal Binding Opsin GPCRs

The term "opsin" has been widely used to refer to proteins that covalently bind to a vitamin A-based retinaldehyde (retinal) chromophore, resulting in light-dependent activity. Among the various known types of photoreceptive proteins, only opsins are integral-membrane proteins. Animal opsins typically function as GPCRs, whereas microbial opsins are usually ion channels or pumps. Here we focus on animal opsin GPCRs, which have been the basis for a growing collection of optogenetic tools that use exogenous expression of naturally occurring opsins or engineered chimeric forms to control G protein and/or arrestin signaling. The main advantages of opsins compared to photopharmacological approaches are that they are fully genetically encoded, enabling selective cell type perturbations, and can also be engineered to have a wide variety of signaling properties. These benefits are also shared by widely used chemogenetic Designer Receptors Exclusively Activated by Designer Drugs (DREADDs), yet opsins are uniquely advantageous for biological questions requiring signaling perturbations with rapid onset, reversibility, or tight spatial localization.

Animal opsins serve well-studied visual functions in the retina, but they are also important for nonvisual photoreception in other tissues [17]. Most animals express multiple opsin genes, and there are over 2000 known animal opsins. These have been classified into eight groups based on phylogeny (see Table 25.1), which mostly parallels their different signaling properties: G_t-coupled opsin (vertebrate visual and non-visual opsins), G_q-coupled opsin (invertebrate visual opsin and melanopsin), invertebrate G_o-coupled opsin, $G_{i/o}$-coupled Opn3 (encephalopsin and teleost multiple tissue [TMT] opsin), G_i-coupled Opn5 (neuropsin), peropsin, cnidarian G_s-coupled opsin, and retinochromes [31]. Opsins in most of these groups are known to function as GPCRs, with the exception of retinochromes, which are considered to be retinal isomerases that generate 11-cis-retinal and do not activate G proteins. Peropsins were originally thought to function solely as isomerases as well, but recent studies have suggested that at least some peropsins also function as GPCRs [34]. Most of these different groups contain specific opsins that have either been directly used in optogenetic experiments, or have been characterized in mammalian cell lines with an eye toward future optogenetics applications.

25.2.1.1 Opsin Photoactivation Cycles

Animal opsin-based pigments, consisting of an opsin protein and retinal chromophore, mostly follow one of two different photoactivation cycles, with important practical implications for their use as optogenetic tools (Figure 25.1). G_t-coupled vertebrate visual opsins form photobleaching pigments, whereas most other animal opsins are now thought to form bistable pigments [31]. Parapinopsin, a G_t-coupled vertebrate non-visual opsin, forms a bistable pigment and is thought to be an evolutionary intermediate between bistable and bleaching pigments [24].

Most bleaching and bistable pigments share the same photoactivation mechanism, but differ in how they return to the dark (inactive) state. In both cases, the opsin first binds 11-cis-retinal through a Schiff base linkage to a lysine residue in

Table 25.1 Opsin GPCRs used as optogenetic tools.

Phylogenic group	G protein coupling	Examples in optogenetics	Photocycle	Chimeras	References
Vertebrate visual	Gt (G_i, G_o)	Rhodopsin	Bleaching	Opto-α1AR (G_q)	[18]
				Opto-β2AR (G_s)	[19]
				Opto-β2AR-SS (G_s biased)	[20]
				Opto-β2AR-LYY (arrestin biased)	[20]
				Opto-D1	[21]
				Opto-MOR ($G_{i/o}$)	[22]
		Blue opsin (human)	Bleaching	CrBlue (G_s)	[23]
Vertebrate non-visual	Gt (G_i, G_o)	Parapinopsin (lamprey)	Bistable		[24]
Opn3 (encephalopsin)	$G_{i/o}$	Opn3	Bistable		[25]
Invertebrate visual	G_q	Honeybee blue	Bistable		[26]
Opn4 (melanopsin)	G_q, possibly G_i	Melanopsin	Bistable	Melanopsin-$5HT_{2A}$	[27–30]
Invertebrate Go	G_o	None so far			[31]
Cnidarian G_s	G_s	Jellyfish G_s opsin	Atypical		[32]
Opn5/Neuropsin	G_i	Opn5m	Bistable		[17]
		Opn5L1 (chicken)	Reverse photoreceptor		[17, 33]
Peropsin	Unknown	Spider peropsin	Reverse photoreceptor		[34]
Retinal photoisomerases	None	None so far	Regenerates 11cis retinal		[31]

This table illustrates the scope of naturally occurring and engineered opsin GPCRs, and is not meant to be an exhaustive list. Several additional examples are discussed in the text. Opsin GPCRs can be classified in phylogenic groups, which also share signaling properties. Vertebrate visual and non-visual opsins couple natively to Gt, but can activate $G_{i/o}$ in heterologous systems. Jellyfish opsin has an atypical photocycle [32]. Like bistable opsins, it generates a stable photoproduct follow photoactivation, but unlike bistable opsins subsequent light absorption does not convert it back to its dark state.
Source: Table is adapted from Koyanagi and Terakita [31].

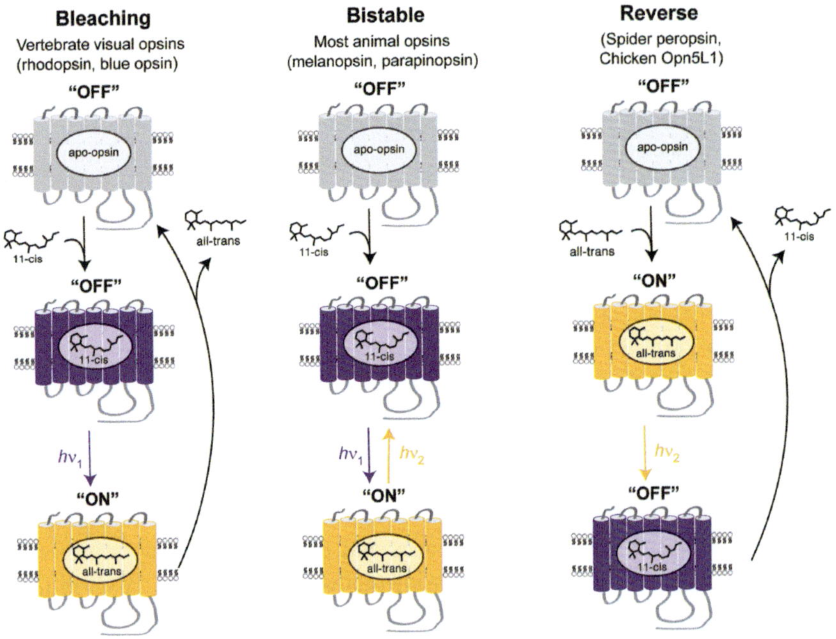

Figure 25.1 Opsin GPCR photocycles. Bleaching pigments release all-trans retinal through a thermal-mediated process following photoactivation. Binding of a new 11-cis retinal molecule and subsequent photoactivation are required for continued signaling. Bistable pigments retain the bound retinal and can be photoswitched between active and inactive signaling states. Reverse photoreceptors bind to all-trans retinal and exhibit light-independent activation. They can however, be deactivated by illumination with light of appropriate wavelength. Source: Adapted from Yamashita [17] and Koyanagi and Terakita [31].

the seventh transmembrane helix. Light absorption triggers isomerization of the chromophore from 11-cis to all-trans, inducing a conformational change in the receptor that activates signaling (Figure 25.1). In photobleaching pigments, like the prototypical GPCR rhodopsin, all-trans-retinal is released by the opsin and a new molecule of 11-cis-retinal must bind to the opsin before a new photoactivation cycle can be initiated. In bistable pigments like parapinopsin, the all-trans state is stable, and the opsin remains active until subsequent absorption of a second photon reverses the pigment back to the 11-cis bound inactive state [24, 35]. Experiments comparing two G_t-coupled opsins, bovine rhodopsin and lamprey parapinopsin, showed that rhodopsin undergoes a larger conformational change upon photoactivation, and is ~20-fold more effective at activating G proteins [36]. It has been proposed that this may reflect a general difference between bistable and bleaching opsins [36].

It is hypothesized that bistable opsins evolved first, minimizing the required supply of 11-cis-retinal, and that vertebrate visual opsins evolved later, in parallel with tissue-specific mechanisms for actively regenerating 11-cis retinal, allowing for optimized visual responses to low light levels. For most bistable opsins, the chromophore-free opsin ("apo-opsin") can only bind 11-cis retinal, such that activation can only be achieved through light absorption and not by direct binding of all-trans retinal. Exceptions to this rule exist, exemplified by recent studies of spider peropsin and chicken Opn5L1 [33, 34]. All-trans-retinal can directly bind and activate these receptors in the dark, analogous to a GPCR agonist [33, 34]. Light absorption with λ_{max} of 540 nm (peropsin) or 510 nm (Opn5L1) results in photoisomerization to the 11-cis state and receptor deactivation. These two opsins are therefore considered examples of reverse photoreceptors. The physiological functions of these opsins are unknown, but it has been suggested that their unique photocycles might be harnessed for optogenetics applications requiring chronic activation that can be rapidly turned off at any desired time [34].

Bleaching and bistable opsins can each have their own advantages depending on the desired optogenetic application. On one hand, bistable opsins may be preferred for *in vivo* optogenetics in tissues with low endogenous levels of 11-cis-retinal, because they do not require a continual supply of the chromophore. For certain *in vitro* applications, where a surplus of exogenous retinal is available or can be supplied and maximal signaling output is desired, photobleaching opsins may be advantageous.

25.2.1.2 Opsin Spectral Properties

Nature has achieved spectral tuning of opsin pigments with maximum absorption wavelengths (λ_{max}) ranging from UV to red light [37] (Table 25.2). Spectral tuning occurs mainly through changes in opsin amino acids that either alter bond rotations in the retinal chromophore or stabilize or extend the delocalization of the positive charge in the Schiff base [63]. The absence of naturally occurring opsins with λ_{max} in the infrared has been attributed to biophysical constraints related to noise from thermal activation, a notion that is supported by recent quantum chemical modeling of bovine rhodopsin revealing that the same electronic transition state mediates thermal and photoactivation [64, 65].

Spectral response is particularly important for multiplexing experiments where the investigator may attempt to combine fluorescence imaging with optogenetics. Ideally, it is possible to excite various fluorophores at wavelengths that do not cause photoactivation of the optogenetic construct, so that spatial and temporal parameters of photoactivation can be controlled independent of imaging. Although many bovine rhodopsin based chimeras have been used to control signaling *in vitro* and *in vivo* when imaging was not required [18, 20, 21], rhodopsin based constructs are, in general, poor options for experiments also requiring live cell imaging due

Table 25.2 List of tools for optical control of intracellular signaling.

Target	Construct	Effect	Mechanism	References
		heterotrimeric G proteins		
Gαi Gαq	CRY2-RGS4	Inhibition	CRY2/CIBN based membrane recruitment of GAP	[38]
Gαq	CRY2-RGS2	Inhibition	CRY2/CIBN based membrane recruitment of GAP	[39]
Gβγ	CRY2-GRK2ct	Inhibition	CRY2/CIBN based membrane recruitment of sequestering domain	[40]
Gαq	CRY2-Gαq (mutant)	Activation	CRY2/CIBN based membrane recruitment of palmitoylation-deficient, GTPase deficient mutant	[41]
		Arrestins		
β–Arrestin2	CRY2-arrestin	Activation	CRY2/CIBN based recruitment of arrestin to GPCR	[38]
		Second messengers		
cAMP	bPAC	Increase	Light activated adenylyl cyclase	[42]
cAMP	CaRhAc	Increase	Light activated adenylyl cyclase	[43]
cAMP	LAPD	Decrease	Light activated phosphodiesterase	[44, 45]
Diacylglycerol	PhoDAGs/ OptoDArG	Increase	Photoswitchable chemical	[46, 47]
Ca^{2+}	OptoSTIM1	Increase	Activation of endogenous calcium channels by induced aggregation	[48, 49]
		Small GTPases		
Rac1	Lov-Rac	Activation	Light-induced unmasking of constitutively active mutant	[50]

(continued)

Table 25.2 (Continued)

Target	Construct	Effect	Mechanism	References
Rac1	PIF3-Tiam1	Activation	PhyB/PIF based membrane recruitment of a Rac1 GEF	[51]
Rac1	Tiam1-SspB	Activation	iLID based membrane recruitment of a Rac1 GEF	[52]
Rac1	BcLOV4-Rac	Activation	Membrane recruitment of wildtype Rac using a plasma membrane binding LOV domain	[53]
Cdc42	ITSN-SspB	Activation	iLID based membrane recruitment of a Cdc42 GEF	[52]
RhoA	LARG-SspB	Activation	iLID based membrane recruitment of a RhoA GEF	[54]
Ras	PIF6-SOS	Activation	PhyB/PIF based membrane recruitment of a Ras GEF	[55]
Ran	RanTRAP	Activation	Light-induced release of sequestered Ran using LOVTRAP	[56]
Rab5, Rab11	IM-LARIAT	Inhibition	Light-induced aggregation using CRY2/CIBN	[57]
Rab6	CIBN-Rab6a	Inhibition	Light-induced aggregation using CRY2/CIBN	[58]
		MAP kinase cascades		
C-raf	C-raf-CRY2	Activation	CRY2 based dimerization	[59, 60]
MEK-1	psMEK	Activation	Photoswitching based on pdDronpa	[61]
p38	Optop38i	Inhibition	Light-induced unmasking of inhibitory peptide	[62]
JNK	OptoJNKi	Inhibition	Light-induced unmasking of inhibitory peptide	[62]

to spectral overlap with many commonly used fluorescent proteins. Given the exquisite sensitivity of rhodopsin, considerable activation is detected across the visible light spectrum, even for wavelengths well separated from λ_{max} and using imaging modalities that require relatively low illumination intensities like spinning disc confocal microscopy [23].

The spectral limitations of rhodopsin led to the exploration of color opsins (cone opsins), involved in color vision, as alternative optogenetic tools [23]. The three human color opsins absorb maximally at 414 nm (blue), 533 nm (green), and 560 nm (red), and these G_t coupled opsins have all been shown capable of activating endogenous $G_{i/o}$ signaling in mammalian cells [23]. For imaging based experiments, blue opsin has emerged as the most useful color opsin and is compatible with imaging of yellow and red fluorophores when exogenously expressed in a variety of cell types, enabling subcellular control of $G_{i/o}$ signaling while simultaneously imaging dynamic cellular responses like immune cell migration or neurite outgrowth [23, 66]. Color opsins also deactivate and recover more rapidly than rhodopsin, enabling better temporal control over signaling [67]. Additionally, unlike rhodopsin, color opsins can be repeatedly photoactivated without a noticeable desensitization of the response [68].

For optogenetics using bistable opsins, the amount of spectral overlap between the inactive and activate states has practical importance. Some bistable opsins have nearly identical absorption spectra in the inactive and active states. For example, G_q coupled spider opsin absorbs maximally around 535 nm in both states [69]. In contrast, lamprey parapinopsin exhibits a large spectral shift of λ_{max} from 370 to 515 nm between inactive and active states [24]. As an optogenetic tool, this gives parapinopsin the advantage that it can be selectively activated and deactivated using different wavelengths of light [35], which enables precise temporal control over the switching between active and inactive signaling states [70].

25.2.1.3 Opsin Signaling Selectivity and Chimeric Opsins

Photoactivation of select G protein signaling pathways ($G_{i/o}$, G_s, G_q) has been achieved using naturally occurring opsins or engineered chimeric opsins. Chimeric opsins typically consist of changes to the intracellular loops and C-terminal domain of a naturally occurring opsin, resulting in altered intracellular signaling while retaining photosensitivity. The term "opto-XRs" has commonly been used when discussing chimeras based on mammalian rhodopsin [21]. The ability to generate chimeric receptors, and to mutate specific residues involved with G protein or arrestin signaling, opens additional doors for dissecting the consequences of different GPCR mediated signaling pathways. Chimeric opsins have also provided insights into the G protein coupling selectivity of orphan GPCRs [71].

$G_{i/o}$ activation by naturally occurring opsins has been achieved using various vertebrate visual opsins including rhodopsin and color opsins. Whereas these opsins couple to G_t in their native cell types, they can also efficiently activate $G_{i/o}$ signaling in heterologous systems. This capability is not surprising given the homology of $G\alpha_t$ with the $G\alpha_{i/o}$ subunits, although several natively $G_{i/o}$ coupled opsins fail to activate G_t. Lamprey parapinopsin, a bistable G_t coupled opsin, also activates $G_{i/o}$ signaling in a variety of cell types [24, 54, 70].

G_s activation has been achieved using G_s-coupled jellyfish opsin, as well as a rhodopsin-β2 adrenergic receptor (β2AR) chimera. Jellyfish opsin generated higher amplitude cAMP responses, and repeatedly produced cAMP responses with less fatigue than the rhodopsin-β2 adrenergic receptor chimera [72]. A chimera containing blue opsin with intracellular portions of jellyfish opsin, termed CrBlue, was used to shift the spectral response from green to blue, making it better suited for experiments requiring live cell imaging [23]. CrBlue has thus enabled orthogonal photoactivation and imaging of cAMP production using a green/red fluorescence resonance energy transfer (FRET) sensor, which was not possible using jellyfish opsin [23].

G_q signaling activation has been achieved using melanopsin as well as a rhodopsin-α1 adrenergic receptor chimera [18, 27]. Bistable melanopsin (Opn4) has been used for rapid, reversible control over signaling in a variety of cell types, and has been shown to be compatible with both one and two photon imaging [27–30]. Users should note that it has been reported to activate $G_{i/o}$ signaling in addition to G_q in several cell types, suggesting this may be a general feature of Opn4 [26, 29]. Alternatively, invertebrate G_q-coupled pigments have also been shown to function in cultured mammalian cells and exhibit diverse spectral sensitivity ranging from UV to green [69, 73, 74].

So far, no opsins have been reported that selectively activate G_{12} signaling, and no chimeric opsins have been engineered to control this family of Gα subunits. A chemogenetic DREADD receptor has recently been designed that specifically couples to G_{12}, suggesting that it may also be possible to engineer a chimeric G_{12} coupled opsin [75]. New, forward genetics and high-throughput testing should aid on this front. Given that native GPCRs that couple to $G_{12/13}$ signaling generally also couple to other G protein subtypes [76], optogenetic and chemogenetic tools exhibiting selective coupling to G_{12}, G_{13}, or both subunit types will likely be useful in dissecting $G_{12/13}$ signaling. Such tools can help interrogate $G_{12/13}$ signaling in processes such as cellular dynamics, embryonic development, and metastatic invasion, where its role remains poorly defined.

In addition to heterotrimeric G proteins, GPCRs interact with a number of other proteins, and these interactions may also be tailored using opsin chimeras [77, 78]. For example, mutagenesis of a rhodopsin-β2AR chimera generated G protein-biased (opto-β2AR^{SS}) and arrestin-biased (opto-β2AR^{LYY}) mutants with stereotypical differences in G protein- and arrestin-mediated cAMP responses and receptor trafficking following photostimulation (Figure 25.2) [20]. Different GPCRs can exhibit unique trafficking properties, and function in distinct subcellular domains. Chimeric opsins provide an opportunity to optically control signaling in specific subcellular microdomains or compartments. For example, melanopsin exhibited distinct localization patterns in neurons, depending on whether it was expressed as a chimera with the serotonin receptor $5HT_{2A}$ [81]. Notably, the chimeric and non-chimeric versions generated distinct photoactivated Ca^{2+} responses, however, only the chimeric receptor resembled endogenous $5HT_{2A}$ signaling. The extent to which exogenously expressed chimeras can truly replicate the trafficking and subcellular distribution of endogenous receptors is not known. Some of the photopharmacological approaches described later may be better suited for addressing questions where the expression level, subcellular localization, or specific protein couplings of the native GPCR under study are particularly important.

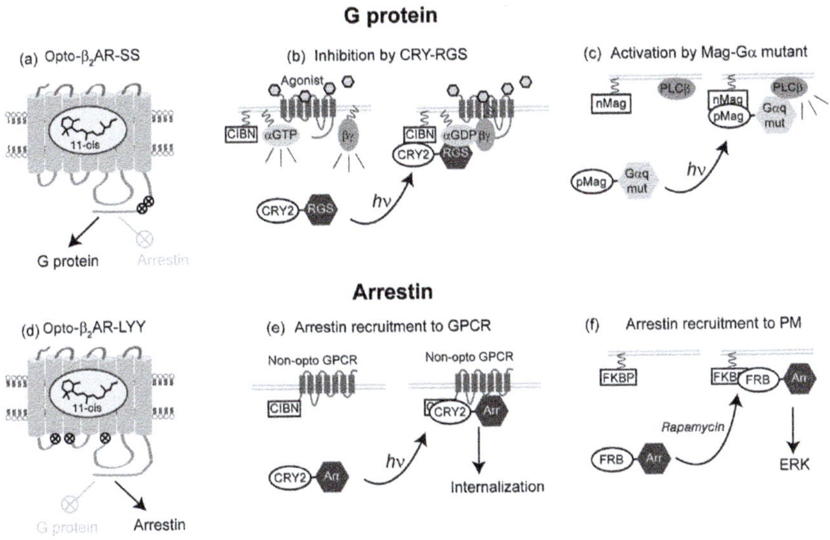

Figure 25.2 Control of G protein or arrestin signaling. (a) G protein biased opsin. Source: Siuda et al. [20]/Springer Nature/CC BY 4.0. (b) Optically controlled plasma membrane recruitment of RGS proteins (RGS4, RGS2) using the CRY2-CIBN optogenetic dimerization scheme. Source: Based on O'Neill and Gautam [38]. (c) Optically controlled plasma membrane recruitment of a palmitoylation-deficient, GTPase-deficient Gαq subunit using the Magnets optogenetic dimerization scheme. Source: Based on Yu et al. [79].
(d) Arrestin biased opsin. Source: Siuda et al. [20]/Springer Nature/CC BY 4.0. (e) Optical control of arrestin–GPCR interactions and receptor trafficking in the absence of receptor activation, using the CRY2-CIBN scheme. Source: Takenouchi et al. [80]/Springer Nature/CC BY 4.0. (f) Chemogenetic control of arrestin-mediated signaling by inducible cytosol-to-plasma membrane translocation. Source: Based on Terrillon et al. [42].

25.2.2 Photopharmacology: Optically Controlled Ligands

Photopharmacological control of GPCRs has been achieved using diffusible photocaged and photoswitchable ligands, or tethered photoswitchable ligands attached either to specific GPCRs or to diffusible transmembrane domains at the plasma membrane. Tethered ligands offer the fastest activation onsets, and have been used to target genetically defined subsets of cells, but their design faces more complex constraints that have limited their use to a much smaller subset of GPCRs compared to diffusible ligands. Collectively, these approaches can be used to control different classes of ligands including agonists, antagonists, and allosteric modulators. All of these approaches, with the exception of GPCR-tethered ligands, enable optical control of endogenous GPCRs. Photocaged versions of glutamate, GABA, dopamine, and serotonin are commercially available, but more generally, the need for in-house chemical synthesis of many photosensitive ligands requires more specialized expertise than approaches that are fully genetically encoded, like those based on opsins. This section covers basic concepts and illustrative examples in photopharmacological control of GPCR signaling. Detailed discussions, including comprehensive lists of ligands and photochemical mechanisms, can be found in recent reviews of GPCR photopharmacology [15, 16].

25.2.2.1 Photocaged Ligands

Photocaged ligands are functionally inactive until the application of a photostimulus, typically a brief pulse of UV light or 2-photon stimulation by near infrared light, releases the active ligand from a molecular caging moiety such as 4-methoxy-7-nitroindolinyl (MNI). These ligands can be bath applied *in vitro*, or systemically applied *in vivo*, and rapidly uncaged in specific anatomical regions or even subcellular locations such as dendritic spines. Photocaging has been applied to a diverse array of GPCR ligands including glutamate [82], a negative allosteric modulator (NAM) of $mGlu_5$ [83], adrenergic receptor agonists [84], dopamine and the dopamine receptor antagonist sulpiride [85], endothelin receptor agonists and antagonists [86], angiotensin-II [87], opioid peptides and the opioid receptor antagonist naloxone [88, 89], and chemotactic peptides targeting the formyl peptide receptor [90, 91].

Anatomical control of photolysis has been performed, for example, in the peripheral or central nervous system *in vivo* [83]. Localized photolysis has been performed *in vitro* to generate well-controlled chemoattractant gradients [91]. Additionally, photolysis of intracellular agonists has been used to selectively study signaling by GPCRs localized on internal membranes by ligands such as glutamate, endothelin, and angiotensin-II [86, 87, 92, 93].

25.2.2.2 Diffusible Photoswitchable Ligands

Diffusible photoswitchable ligands, also referred to as photochromic ligands (PCLs), switch between two photoisomers with different pharmacological properties. The most common photoswitch is a UV sensitive azobenzene, although red shifted versions have also been developed [94, 95]. Photoisomerization is typically achieved using light-sensitive moieties that either switch between cis and trans isomers (azobenzenes, stilbenes, and hemi-thioindigos), or between open and closed forms (spiropyrans, diarllethenes, and fulgides) [16]. Typically, one wavelength of light will convert the switch from the more thermodynamically stable isomer to the other, which can return to the original isomer either by absorption of a different wavelength of light [96], or by thermal relaxation in the dark [97]. The ability to switch reversibly between active or inactive photoisomers using different wavelengths of light presents a distinct advantage over photocleavable ligands. For example, a photoswitchable M2 muscarinic receptor has been used to reversibly control cardiac function in translucent frog tadpoles [98].

Some PCLs transition between inert and active photoisomers, whereas other switch between two different pharmacologically active states. For example, the nonselective photoswitchable adenosine receptor ligand MRS5543 behaves as a full A_3R agonist in both photoisomerization states, but switches from a partial $A_{2A}R$ agonist in the dark to a full $A_{2A}R$ antagonist upon photoisomerization with 460 nm light [99]. Meanwhile, azobenzene-containing ligands for the peptidergic GPCR CXCR3 demonstrated photoswitching from antagonism to partial agonism and even full agonism [100]. Large photo-induced efficacy changes from antagonism to full agonism can provide more robust control over GPCR signaling than light-induced changes in ligand binding affinity.

25.2.2.3 Tethered Photoswitchable Ligands

Unlike optogenetic methods, photo-pharmacological approaches based on diffusible ligands lack the ability to target genetically defined subpopulations of cells. To overcome this limitation, hybrid approaches have been developed that involve the use of a genetically encoded component that is chemically modified with a photoswitchable ligand, such that light switches the ability of the ligand to bind the receptor with high affinity. This has been achieved by linking the ligand to the receptor [101], or to a transmembrane targeting domain [102]. Tethering to the GPCR provides receptor subtype specificity, which has enabled studies of specific metabotropic glutamate receptors despite the lack of drugs that selectively bind to just one member of the family [101]. In principle, it can also be used to study signaling by different receptor mutants or splice variants. On the other hand, tethering to a transmembrane domain enables optical control of unmodified endogenous GPCRs at their native expression levels [102]. Indeed, this is the only approach currently available for optical control of endogenous GPCRs with cell-type selectivity. However, thus far it has only been applied so far to mGluRs, class C GPCRs which contain a large clamshell-like ligand binding domain, and it is not yet clear whether this approach will be applicable to studying other classes of GPCRs.

Direct tethering to the GPCR of interest has been achieved primarily through engineered cysteines or a genetically encoded, N-terminal self-labeling tag (i.e. SNAP) [101, 103]. It is even possible to use bioorthogonal enzyme-based self-labelling by SNAP or CLIP protein tags to achieve independent optical control of two different GPCRs in the same experiment, when used with photoswitchable ligands with different optical properties [104]. SNAP labeling also simplifies the use of tethered photoswitchable ligands for *in vivo* application. For example, one recent application targeted SNAP-mGluR2 expression to cells that natively expressed mGluR2 in mice [105]. A branched photoswitchable tethered ligand was then attached to the SNAP-mGluR2 receptors by infusion into the prefrontal cortex (PFC). Photoactivation of mGluR2 in this population of cells inhibited working-memory-related behaviors, and reversing the photoisomerization eliminated this effect. Overall, the use of hybrid genetic-photopharmacological control has lagged behind the use of opsin GPCRs for cell type-selective applications *in vivo*. However, unique advantages such as the opportunity to control GPCRs by allosteric modulation are likely to inspire future *in vivo* applications of the hybrid approach.

25.2.3 Assays for Testing and Validation of GPCR-Based Optogenetic Tools

Optogenetic tools that target GPCR or G protein signaling can be tested using assays that directly measure G protein activation or downstream signaling responses like second messenger production. Initial tests are often performed in easy to transfect cell lines like HEK293 or HeLa, and constructs are then tested in the specific system or cell type of interest for a given application. Assays vary in terms of their signaling specificity, sensitivity, temporal resolution and reversibility, spectral overlap with optogenetic tools, and their ability to report activity in populations of cells or single cells down to the subcellular level.

Plate reader based assays offer advantages in terms of throughput, sensitivity, and temporal resolution in testing new optogenetic tools. Biosensors based on bioluminescence or fluorescence resonance energy transfer (BRET or FRET) have been used extensively to study GPCR, G protein, and arrestin interactions. When applied to heterotrimeric Gα and Gβγ subunits, these methods can measure rapid kinetics of G protein activation by measuring dissociation or reorganization of the subunits that accompanies activation. In general, bioluminescence resonance energy transfer (BRET) approaches are suited for testing a wider array of optogenetic constructs than FRET, because they do not require excitation of the donor and therefore avoid unwanted photoactivation of the optogenetic construct. Notably, though, FRET measurements have been used with UV-light triggered uncaging of agonists to provide temporal resolution in the submillisecond range [106]. These assays can directly measure light induced changes in G protein activation, and reversal upon subsequent photoactivation at different wavelengths or removal of the light stimulus. Highly optimized BRET biosensors for measuring heterotrimeric G protein activation are now available [107]. Additionally, unimolecular BRET sensors that detect either GαGTP or free Gβγ, rather than heterotrimer dissociation or rearrangement, can detect activation of endogenous G proteins [108]. Whereas most BRET studies have been performed in cell lines like HEK293 due to the need to express multiple exogenous constructs, these unimolecular sensors have proven useful for detecting G protein activation by endogenous GPCRs in primary neurons following lentiviral transduction [108]. An alternative assay uses engineered domains of Gα subunits, called mini G proteins. These probes translocate from cytosol to the plasma membrane upon receptor activation, which can be measured by BRET using a tagged receptor, or by live cell imaging [109]. Unlike the aforementioned approaches, the mini G protein assay directly measures receptor interaction with different Gα subtypes rather than activation-associated changes in heterotrimeric G proteins.

Bioluminescence assays can also be used to monitor second messenger production. One popular approach has been to use genetically encoded sensors that convert binding of a target molecule into a change in luciferase luminescence activity. For example, the GloSensor cAMP assay was used to show reversible activation and deactivation of lamprey parapinopsin in cell lines [35]. This approach is limited to measuring activation of G_s mediated increases in cAMP, or $G_{i/o}$ mediated inhibition of forskolin-induced cAMP. However, it can be extended to other G proteins using the recently developed G_sX assay [110]. This assay includes the additional expression of chimeric Gα subunits in which different Gα C-terminal domains are fused to the $G\alpha_s$ subunit. These chimeras exhibit GPCR-G protein coupling selectivity determined by the C-terminal domain, but they all generate cAMP increases reflective of G_s signaling. Using this assay, it is possible to generate a G protein coupling profile for a receptor using a single readout. The profile is considered relative rather than absolute, but can be quantitatively compared between different receptors. Additionally, it can distinguish between members of the G_i family, including G_o, G_i, G_t, and G_z. The G_sX assay has been used to screen different $G_{i/o}$ coupled opsins for specific coupling to G_o versus G_i [110].

Three live cell imaging methods have been used to directly monitor G protein activation and deactivation with subcellular resolution: G protein βγ subunit translocation, conformationally selective nanobodies that bind activated Gα subunits, and FRET based imaging of heterotrimer dissociation. Of these, only Gβγ translocation has been used so far to validate subcellular optogenetic control. The Gβγ translocation assay involves imaging a fluorescently tagged Gγ subunit, or a Gβγ complex in which the β and γ subunits jointly express a split fluorescent protein [111–115]. Agonist induced GPCR activation generates translocation of the fluorescent Gβγ from the plasma membrane to intracellular membranes, with kinetics that vary by more than 10-fold depending on the Gγ subunit type [112, 113]. Using fast translocating subunits like Gγ9, this assay has a temporal resolution on the order of several seconds. The main advantages of Gβγ translocation for testing optogenetic tools are that (i) it reverses upon G protein deactivation; (ii) it produces a robust response that can report localized activation in a region of the plasma membrane; and (iii) it can be used with a single fluorescent protein, allowing imaging and photoactivation to be performed in distinct spectral channels. Although the assay has shown the capacity to detect activation of overexpressed G_s, and G_q, it is best suited to the study of $G_{i/o}$ activation and typically does not require exogenous $G\alpha_{i/o}$.

Additionally, several indirect measures of G protein activation by live cell imaging have been used for initial opsin characterization. These include the use of genetically encoded biosensors for Ca^{2+} responses, PIP_2 hydrolysis, cAMP, $PI(3,4,5)P_3$ production and a myriad of other intracellular responses [23, 66]. These live cell imaging based assays provide subcellular resolution over opsin-mediated responses that is not possible in high throughput plate reader assays. In some cases, the use of these sensors has revealed unexpected consequences of photoactivation. For example, in cells supplied with exogenous retinal, but lacking any opsin expression, photoactivation generated robust translocation of a biosensor typically used to study G_q mediated PIP_2 hydrolysis, the PH domain of PLCδ [116]. This surprising finding led to the discovery that retinal mediated photochemistry can alter membrane lipid composition, highlighting the importance of parameters like retinal concentration and photoactivation power and wavelength for performing optogenetic experiments without spurious light-induced cellular responses [116].

Electrophysiological assays have also played an important role in opsin characterization. Although lower throughput than both plate reader and microscopy based assays, electrophysiological measurements benefit from excellent temporal resolution, and provide a direct measure of cellular responses that are important for applications in excitable cells like neurons or cardiomyocytes. For example, measurement of G protein coupled inwardly rectifying potassium (GIRK) channel activity using patch clamp electrophysiology has been a popular approach for testing $G_{i/o}$ activation by tethered ligands and opsins [40, 68].

25.3 Optical Control of Signaling Downstream of GPCRs

Whereas optically controlled GPCRs provide the ability to determine the consequences of receptor activation with great spatiotemporal precision, additional tools are needed in order to further dissect GPCR-triggered signaling cascades inside the cell. For some questions, opto-GPCRs can be combined with traditional genetic or pharmacological perturbations that target downstream signaling proteins. However, in many cases it is desirable to have direct optical control of the downstream signaling proteins.

In contrast to GPCRs, most families of downstream signaling proteins lack naturally occurring photosensitive members that can serve as templates for engineering new optogenetic tools. Nonetheless, several light sensitive protein domains, genetically isolated from plants, bacteria, and fungi, have been used to engineer genetically encoded constructs that enable optical control of a growing collection of signaling proteins. The modern intracellular optogenetic toolkit enables control of different nodes in GPCR signaling networks, including heterotrimeric G proteins, arrestins, second messengers, and downstream signaling proteins.

25.3.1 Light-Sensing Protein Domains and Strategies for Optically Controlling Signaling Proteins

Naturally occurring modular light-sensing and signaling domains can be used to achieve control over signaling functions when engineered into fusion proteins. Both naturally occurring and engineered light-inducible dimers have been widely used to control signaling protein function through manipulation of subcellular location, most often by recruiting a signaling protein domain from the cytosol to the plasma membrane. Optically induced unmasking of peptides has been used to activate or inhibit signaling by endogenous proteins. Light-induced conformational changes in a photoreceptor domain have also been leveraged to induce conformational changes in a signaling domain to regulate its function.

The major families of photoreceptor proteins used in optogenetic control of cell signaling include flavoproteins and phytochromes. Blue-light sensing flavoproteins use either flavin adenine dinucleotide (FAD) or flavin mononucleotide (FMN) chromophores and can be used in mammalian cells, *in vitro* and *in vivo*, without the need to add exogenous chromophore. Flavoprotein based optogenetic tools have used photosensing domains from cryptochromes, light-oxygen-voltage sensing (LOV) domains, and blue light using flavin adenine dinucleotide (BLUF) domains [14].

The cryptochrome CRY2 from *Arabidopsis thaliana* has been used in light-inducible dimerization schemes with its binding partner CIB1 or its N-terminal domain CIBN [117]. This system has been used to control a wide range of signaling proteins, including heterotrimeric G proteins and arrestins [38, 39, 80]. CRY2 can also form light-induced protein clusters in the absence of CIB1, a feature that has been used to control various exogenously expressed signaling proteins as well as endogenous receptor tyrosine kinases (RTKs) [118, 119]. CRY2 clustering has since been optimized for enhanced clustering and faster assembly and disassembly of clusters upon application and removal of blue light [50, 120].

AsLOV2, a photosensor domain from *Avena sativa* phototropin1, has been used in diverse approaches to optically control signaling. Fusing a constitutively active Rac1 mutant to the C-terminus of the AsLOV2 Jα helix enabled photoswitchable control over signaling by sterically hindering Rac interactions with other proteins in the dark state [121]. The conformational change in the C-terminal Jα helix has also been used for light-inducible unmasking of various peptides, including inhibitory peptides that modulate endogenous kinase activities [122]. It has also been used to engineer optical control over the high-affinity interaction between a bacterial peptide (ssrA) and its binding partner (SspB) to generate a rapidly reversible light inducible dimerization system called iLID [52]. Additionally, an engineered protein called Zdark selectively binds to the dark state of asLOV2, and dissociates upon photoactivation. This "LOVTRAP" system has been used to control signaling by exogenously expressed proteins by reversibly localizing them to mitochondria [123].

Systematic studies of blue-light induced heterodimers have compared the CRY2/CIB1 and iLID systems with a third dimerization system called "magnets" engineered from the fungal photoreceptor Vivid from *Neurospora crassa* [51]. The iLID and magnets systems both offered advantages for achieving the tightest spatial control over protein localization due to their faster dissociation kinetics, although the CRY2 system remained useful for applications requiring the maximal number of photoinduced dimers [124].

Red and near-infrared sensing phytochromes offer the ability to control cell signaling using longer wavelengths than tools based on flavoproteins, which can be advantageous due to greater tissue penetration of the photoactivation light. Light-inducible dimerization schemes based on plant phytochrome PhyB and its interacting partner PIF6 have been used for cytosol-to-plasma membrane recruitment of various guanine nucleotide exchange factor (GEF) domains to activate endogenous Rho family G proteins [41]. PhyB undergoes a bistable photocycle, and can be activated by 650 nm light and deactivated using >750 nm light, enabling rapid photoswitching. An important limitation of plant phytochromes, especially for *in vivo* applications, is the requirement of a phycocyanobilin chromophore not produced in mammalian cells. In contrast, bacterial phytochromes use a biliverdin chromophore that is endogenously produced in mammalian cells. The interaction between a bacterial phytochrome BphP1 and its partner PpsR2 from *Rhodopseudomonas palustris* upon photoactivation with near infrared light at 740–780 nm has been introduced as a promising alternative to the PhyB-PIF6 system, although so far it has not been as widely adopted [125].

25.3.2 Optogenetic Control of Heterotrimeric G Proteins

Heterotrimeric G proteins are central transducers of GPCR signaling. Non-GPCR-based optogenetic tools have now been developed that enable direct control of heterotrimeric G proteins (Figure 25.2). These include tools for light-induced inhibition of endogenous Gα and/or Gβγ subunits [38, 39], or optical activation of exogenously expressed mutant Gα subunits [79]. Both approaches provide an ability to dissect discrete Gα and Gβγ signaling pathways.

G protein activity is controlled by a guanine nucleotide cycle. In the inactive state, Gα-GDP binds Gβγ subunits with high affinity. GPCRs act as guanine nucleotide exchange factors (GEFs), inducing release of GDP from the binding pocket in the Gα subunit, which is subsequently occupied by GTP present at high concentrations in the cytoplasm. Gα-GTP then dissociates from Gβγ. In some cases this is a complete physical dissociation; in other circumstances it is a functional dissociation. The dissociated Gα-GTP and Gβγ both interact with downstream signaling proteins [126]. Gα subunits slowly hydrolyze GTP to GDP, a process that is accelerated by regulators of G protein signaling (RGS) proteins, which function as GTPase-activating proteins (GAPs), resulting in reformation of the inactive heterotrimer [127].

Multiple labs have reported optogenetic tools based on RGS proteins. The net concentration of activated (GTP-coupled) Gα subunits depends on the competing GEF activity of GPCRs, and presumably other non-receptor, cytosolic GEFs which can activate free GαGDP independent of Gβγ, as well as GAP activity of RGS proteins and native GAP activity of the Gα subunits. At sufficiently high RGS concentrations, it is possible to drive the active Gα concentration down to very low levels. It is worth noting though that native RGS proteins can control the kinetics of downstream responses without altering their magnitude or function, by limiting the spatial range of G protein–effector interactions such that the G protein only activates effectors present close to the activated receptors [128]. The opto-RGS constructs described later have been designed for G protein inhibition, but one can also imagine future applications of opto-RGS tools that achieve more nuanced control of G protein signaling akin to endogenous RGS proteins.

The first RGS-based optogenetic tool used the CRY2-CIBN system to achieve optical control over the plasma membrane localization of RGS4 [38]. Rationale for this approach was inspired by previous studies showing that plasma membrane localization is tightly coupled with the ability of RGS4 to inhibit the pheromone response pathway in the yeast *Saccharomyces cerevisiae* [129]. CRY2-mCh-RGS4 was generated by fusing CRY2 and mCherry to the N-terminus of a truncated version of RGS4 lacking its native N-terminal palmitoylated membrane-targeting domain. This resulted in a cytosolic fusion construct that could be optically recruited to the plasma membrane with subcellular control when used with a membrane-targeted CIBN.

Optical inhibition of G protein signaling by CRY-RGS4 was demonstrated in two distinct assays: Gβγ translocation and cell migration. In HeLa cells, pronounced translocation of mCherry-Gγ9 from the plasma membrane to intracellular membranes was observed upon global activation of endogenous CXCR4 receptors by bath application of stromal derived factor (SDF-1α). Subsequent blue light photoactivation of the plasma membrane at one side of the cell resulted in reverse translocation, but only to the photoactivated region, consistent with light-induced G protein deactivation. In macrophage-like RAW 264.7 cells, application of agonists for either CXCR4 or C5a receptors, combined with blue light photoactivation at one side of the cell, resulted in cell migration in the opposite direction, consistent with the establishment of an intracellular gradient of G protein signaling.

More recently, an optogenetic construct based on RGS2 was created that utilizes the CRY2-CIBN system [39]. Whereas RGS4 acts on both $G\alpha_{i/o}$ and $G\alpha_q$ family subunits, RGS2 is selective for $G\alpha_q$. Wild-type RGS2 contains an amphipathic N-terminal domain that facilitates membrane interaction. A CRY2-RGS2 construct lacking this N-terminal domain was optically recruited to the plasma membrane through its interaction with CIBN fused to the RGS2 N-terminal domain to inhibit G_q signaling. This system enabled rapid light-triggered suppression of calcium dynamics without inducing compensatory mechanisms that have been observed in some types of genetic knockouts. This rapid control was used to establish a role for RGS2 in negative feedback control over the stochasticity of G_q-mediated calcium spike timing.

Interestingly, a bioinformatics study that mined for candidate LOV domains identified 77 protein sequences, primarily in fungi and protists, corresponding to predicted LOV-RGS fusions [53]. An in-depth study of one of these photoreceptors, BcLOV4, from the fungus *Botrytis cinerea*, revealed blue-light induced translocation from the cytosol to the plasma membrane when heterologously expressed in mammalian (HEK293) or fungal (*S. cerevisiae*) cells [130]. Translocation results from an electrostatic interaction between anionic plasma membrane phospholipids and a polybasic amphipathic helix at the LOV interface with the presumed RGS domain. RGS functionality within this family of photoreceptors has yet to be demonstrated, but it is tempting to speculate that organisms lacking opsin GPCRs may have evolved a unique mechanism for photo-regulation (albeit inhibitory regulation) of G protein signaling based on LOV-RGS fusion proteins. LOV-RGS fusions from fungi and protists might not exhibit functional control over mammalian G proteins, but one can imagine engineering modified versions that incorporate mammalian RGS proteins or their GAP domains. These could be advantageous over the CRY-RGS constructs, because they only require transfection of a single transgene. Indeed, the LOV domain of BcLOV4 was recently fused with the monomeric G protein Rac1 to achieve optical control over its signaling, suggesting that this LOV domain could be widely used in place of dimerization schemes to achieve optical control of plasma membrane targeting [131, 132].

The opto-RGS approach has two main limitations. First, it cannot be used to manipulate all G protein subtypes. Twenty mammalian RGS proteins act as GAPs for $G\alpha_{i/o}$, $G\alpha_q$ or both, but there are currently no known RGS proteins that selectively target $G\alpha_s$ or $G\alpha_{12/13}$ family subunits [133]. Second, the approach cannot be used to manipulate Gα signaling independent of Gβγ. Since RGS proteins inhibit Gα signaling by shifting the dynamic equilibrium toward the GαGDP state, Gβγ signaling is also perturbed due to the high affinity interaction between GαGDP and Gβγ.

An alternative approach has been developed that can manipulate Gα subunit signaling independent of Gβγ. It involves direct activation of Gα subunits by light-induced cytosol-to-plasma membrane translocation of constitutively active mutants lacking their normal palmitoylation sites [79]. A $G\alpha_q$ Q209L mutant lacks GTPase activity, exhibits constitutively active signaling, and has low affinity for Gβγ. Further mutation of palmitoylated residues in the N-terminal domain (C9S, C10S) prevents interaction with the plasma membrane and $G\alpha_q$ effectors like PLCβ located there. Optically recruiting this mutant $G\alpha_q$ to the plasma membrane enables inducible $G\alpha_q$ signaling, validated by the generation of calcium responses. Like several optogenetic tools using optical dimerizers, this work was inspired by earlier studies using chemically inducible dimerization to control Gα subunits [134]. It is worth noting that in the chemically inducible system, it was demonstrated that resting inositol trisphosphate (IP_3) levels were not altered by expression of the inducible $G\alpha_q$ components, suggesting that inducible $G\alpha_q$ does not exhibit basal activity. A similar approach was used to optically control $G\alpha_s$ mediated cAMP signaling [79]. These inducible Gα subunits could in principle be targeted to many other locations besides the plasma membrane, which could aide greatly in high resolution dissection of G protein signaling at organelles.

Optical control over the subcellular location of a Gβγ sequestering protein domain has been used for targeted inhibition of Gβγ signaling [38]. The C-terminal domain of GRK2 (GRK2ct) has been widely used for selective inhibition of Gβγ signaling. GRK2ct has sufficiently high affinity to compete with effector proteins for binding to Gβγ, enabling it to efficiently block Gβγ signaling. However, Gβγ has a higher affinity for GαGDP that GRK2ct, which enables the G protein activation cycle to continue in the presence of GRK2ct. As a result, Gβγ mediated signaling is inhibited while Gα signaling remains intact [135].

By performing photostimulation at one side of a macrophage-like cell stimulated globally with a chemoattractant receptor agonist, localized Gβγ inhibition by CRY2-GRK2ct was shown to be capable of directing phosphatidylinositol (3,4,5)-trisphosphate ($PI(3,4,5)P_3$) production and lamellipodia formation in the opposite direction [38]. This work demonstrated the ability to direct Gβγ inhibition to a subcellular location, but it did not test whether the construct also globally inhibits Gβγ to some extent in the dark state. Since a lack of dark-state inhibition would be important for many applications, it is worth noting that Gβγ inhibition has also been performed using chemical dimerization to recruit a GRK2ct K567E/R578E mutant (GRKct-KERE) to either the plasma membrane or the Golgi membranes [136]. These mutations disrupt binding of GRK2 to phospholipid headgroups. Unlike GRK2ct, cytoplasmic GRKct-KERE did not inhibit Gβγ. The KERE mutant only inhibited Gβγ signaling after it was recruited to membranes using the FRB/FKBP heterodimerization system. Future optogenetic constructs based on this mutant could enable rapid, reversible, subcellular control of Gβγ inhibition with minimal dark-state perturbation. It is worth noting that mammals express 5 Gβ and 12 Gγ subunit types, resulting in a very large number of possible Gβγ combinations. GRK2ct based tools are thought to generally inhibit all Gβγ combinations, although this has not been tested exhaustively. Signaling diversity among different Gβγ subunit types remains understudied and would greatly benefit from the future development of optogenetic tools that can selectively perturb specific Gβγ subtypes.

Most of the tools described earlier for perturbing heterotrimeric G protein subunits are based on CRY2-CIBN dimerization. Light-induced translocation of CRY2 constructs from the cytosol to the plasma membrane is typically observed within a few seconds of photoactivation. Downstream signaling effects are observed in less than a minute. A more precise assessment of the perturbation kinetics will require the use of higher temporal resolution readouts such as the BRET or electrophysiological assays discussed earlier. The somewhat slow dissociation of CRY2-CIBN dimers upon removing the light stimulus means that reversing the signaling effects takes several minutes. In principle, one can imagine reengineering these constructs using the iLID or Magnets dimerization systems, which could greatly increase the reversal kinetics from about 5 minutes to about 30 seconds.

25.3.3 Optogenetic Control of Arrestins

Arrestins regulate GPCR desensitization and internalization, and can initiate G protein-independent signaling. Optical control of arrestin function has been achieved through two main approaches (Figure 25.2). The first approach uses mutant GPCR opsins that exhibit biased signaling toward arrestins [20, 137]. The second approach uses light-inducible dimerization to directly control arrestin–receptor or arrestin–membrane interactions [80]. Both approaches have been used to study arrestin function downstream of the β2AR, a widely used receptor for studies of GPCR trafficking and arrestin interactions. The opsin-based approach is well suited for dissecting the consequences of receptor-mediated arrestin responses, whereas the dimerization approach is uniquely suited for addressing questions related to the role of receptor–arrestin interaction dynamics in controlling receptor trafficking.

Mutation of specific residues involved in G protein coupling has enabled the creation of arrestin-biased opsins. Specifically, mutations of the canonical DRY G protein-coupling motif were applied to a rhodopsin-β2AR chimera to achieve biased signaling [20, 138]. Notably, an arrestin biased mutant of G_s coupled jellyfish opsin has been used to dissect circadian signaling [137]. This general approach can preserve G protein-independent interactions between the activated GPCR and arrestins. Arrestin binding typically requires phosphorylation of specific residues on the GPCR, mediated by G protein receptor kinases (GRKs) or other kinases that recognize the activated conformation of the receptor. Some kinases (GRK2) are recruited to the plasma membrane by interaction with Gβγ subunits, which requires G protein activation. However, other kinases (GRK5, GRK6) can phosphorylate activated GPCRs independent of G protein signaling [138]. Arrestin-biased opsins will only trigger arrestin responses that are mediated by this later set of kinases, and so their ability to generate arrestin signaling will depend on the GRK expression profile in a given cell type.

Direct optical control over GPCR–arrestin interactions has been performed using both chemogenetic and optogenetic dimerization schemes [42, 43, 80]. In these schemes, one component of the inducible dimer (i.e. FRB or CRY2) is fused to the arrestin subtype of interest, and the other component of the dimer (i.e. FKBP or CIBN) is fused to the GPCR of interest. Alternatively, the FKBP or CIBN can be fused to a generic plasma membrane targeting domain to help dissect the importance of specific receptor–arrestin interactions versus non-specific arrestin recruitment to the plasma membrane [42]. Recruitment of β-Arrestin2 to β2AR using either chemical or optical dimerization schemes generated robust receptor internalization. Reversibility of the optical dimerization scheme enabled experiments to dissect how the duration of arrestin–receptor interactions determined whether internalized receptors were recycled to the plasma membrane or trafficked for lysosomal degradation [80].

25.3.4 Optical Control of Second Messengers

Second messengers are small molecules or ions that bind various effector proteins and alter their functions. Following GPCR activation, cells can rapidly generate large changes in the concentrations of second messengers such as cAMP, diacylglycerol (DAG), IP_3, Ca^{2+}, and PIP_3. The rapid kinetics and diffusion of second messengers enables coordinated responses by many effector proteins. One might expect that second messengers change more or less uniformly throughout a cell given their rapid diffusion, but it has become widely appreciated that they can exhibit complex spatiotemporal dynamics at the subcellular level. Together with live cell imaging based approaches, optical control of second messengers can help dissect the cellular consequences of these dynamics.

Different Gα subunits famously produce distinct changes in second messengers, so one route to controlling second messengers is by optical activation of the appropriate Gα subunit. However, complex feedback loops regulate second messenger levels, and they often do not directly correlate in time with GPCR activation levels. For example, persistent G protein activation can yield transient PIP_3 increases, or oscillating Ca^{2+} dynamics. Determining the consequences of different spatiotemporal patterns of second messengers therefore requires a unique series of optogenetic tools that bypass GPCRs and G proteins and exert more direct control over second messengers themselves.

Optical control of cAMP production has been achieved in a variety of cell types using heterologous expression of photoactivatable adenylyl cyclases. A cytosolic light activated adenylyl cyclase from the bacterium *Beggiatoa* (bPAC) has been used as an optogenetic tool to control cAMP [139]. In bPAC, an adenylyl cyclase is directly fused to a BLUF domain, resulting in a 300-fold increase in cyclase activity on photoactivation. Alternatively, an engineered green-light activated transmembrane adenylyl cyclase, CaRhAC, was generated by mutation of a rhodopsin-guanylyl cyclase from *Catenaria anguillulae* [44]. Both constructs have shown the ability to rapidly increase cAMP levels in primary cultures of mammalian neurons. CaRhAC offers 10-fold faster reversal of cAMP production upon removal of the photoactivation light. On the other hand, the cytosolic nature bPAC has enabled its targeting to subcellular regions by fusion with appropriate organelle-targeting domains [45].

Intracellular cAMP concentrations reflect a dynamic competition between cAMP-producing adenylyl cyclases and cAMP-degrading phosphodiesterases. Tight optogenetic regulation of cAMP can therefore benefit from orthogonal optical control of adenylyl cyclases and phosphodiesterases. This has been achieved by the introduction of a red light activated phosphodiesterase (LAPD), and through subsequent engineering of improved LAPDs [46, 47]. Orthogonal control of cilium-targeted bPAC and LAPD using blue and red light enabled systematic studies of cAMP control of cilia length [45].

Canonical G_q signaling involves PLC-mediated hydrolysis of the lipid $PI(4,5)P_2$ to generate membrane-bound DAG and soluble IP_3. DAG regulates ion channels and recruits specific proteins like PKC to the plasma membrane, whereas IP_3 activates receptors on the endoplasmic reticulum to release intracellular Ca^{2+} stores. Photoswitchable diacylglycerols (PhoDAGs and OptoDArG) have been synthesized that allow DAG-mediated signaling to be studied independent of IP_3 production or other G_q pathways [140, 141]. These tools have helped to determine the mechanisms by which DAG regulates transient receptor potential (TRP) channels [141, 142].

Dynamic Ca^{2+} signals arise from coordinated actions of ion channels, pumps and receptors at the plasma membrane and endoplasmic reticulum. They control diverse cellular responses such as fertilization, contraction, differentiation, proliferation, and neurotransmission [143]. Optical control of Ca^{2+} signals can help determine how Ca^{2+} patterns with varying amplitude, frequency, and spatial localization produce distinct cellular outcomes. Numerous optogenetic tools have now been generated to dynamically control Ca^{2+} signals, including various approaches to activate endogenous calcium release activated channels (CRACs) that mediate Ca^{2+} influx at the plasma membrane [48, 49, 144, 145]. For example, OptoSTIM1 uses CRY2-based aggregation of stromal interaction molecule 1 (STIM-1) to activate endogenous CRAC, and has been modified for optimized two-photon activation [49, 145]. These tools have been used to study Ca^{2+} mediated cellular responses ranging from transcription factor translocation to stem cell differentiation and T cell migration in lymph nodes [49, 145].

25.3.5 Optogenetic Control of Small GTPases and Mitogen Activated Protein Kinases

The past several years have seen considerable growth in the number of intracellular signaling proteins targeted by optogenetic constructs, and a comprehensive list is beyond the scope of this chapter [55, 146–149]. In this section we focus on optogenetic control of two families of downstream signaling proteins with important roles in GPCR signaling: monomeric GTPases of the Ras superfamily, and mitogen activated protein kinases (MAPKs). Small GTPases control diverse cellular processes such as cell division, migration, growth, and differentiation through their actions on cytoskeletal dynamics, membrane trafficking, and downstream regulators of gene expression. They also relay upstream signals from GPCRs and

RTKs to downstream activation of MAPKs, which in turn regulate diverse cellular responses ranging from cell fate to synaptic plasticity [150].

The Ras superfamily is divided into five main families of small GTPases: Ras, Rho, Ran, Rab and Arf. Similar to the heterotrimeric Gα subunits, these monomeric G proteins function as molecular switches, controlled by guanine nucleotide exchange and hydrolysis and regulated by auxiliary proteins such as GEFs and GAPs. Optogenetic approaches have been extensively applied to members of the Ras and Rho families [41, 52, 54, 151], and some applications have also targeted members of the Ran and Rab families [56–58].

Optogenetic studies of the small GTPase Rac1 illustrate how multiple approaches can be used for optical control of the same protein, each with different advantages. The first approach used a constitutively active Rac1 mutant, fused with asLOV2 such that its signaling was masked in the dark and unmasked by photoactivation with blue light [121]. This scheme allowed optical control of different Rac1 mutants, including a dominant negative Rac1. However, extending the approach to closely related signaling proteins has proven difficult. The second approach used light-inducible dimerization schemes to optically recruit a Rac1 GEF, Tiam1, to the plasma membrane [41, 52, 54]. This approach activates endogenous Rac1, and has been readily extended to other small GTPases through the use of GEFs selective for Ras, Cdc42, and RhoA [52, 151]. The third approach used a fusion of wild type Rac1 to the C-terminal region of BcLOV4, an LOV domain that rapidly binds the plasma membrane upon photoactivation [131]. In this case, plasma membrane recruitment of wild type Rac1 was sufficient to generate its activation by endogenous GEFs. Rac1 is activated by multiple GEFs, which in addition to catalyzing the GTP switch on the G protein, can also function as scaffolds that control the proximity of different effector proteins to the activated G protein [152]. Thus an advantage of this third approach is that it leverages activation by endogenous GEFs and may replicate some aspects of native Rac1 signaling that could be lost in the other two approaches. Several specific applications using optical control of Rac1, Cdc42 and RhoA are discussed in Sections 25.4.1–25.4.3, 25.4.4.1, and 25.4.4.2.

GPCRs regulate all of the major MAPK families, including ERK-1/2, JNK, p38, ERK5, and ERK6, through a variety of pathways involving different Gα subunits, as well as Gβγ and arrestins [153]. Typical MAPK activation pathways involve a multi-kinase cascade. In a classic example, Ras activates C-Raf (MAP3K), which activates MEK-1/2 (MAP2K), which activates ERK-1/2 (MAPK). Several optogenetic tools have been developed that target different steps of this cascade. Ras activation has been achieved using plasma membrane recruitment of a Ras GEF, Sos, as discussed earlier for Rac1 [151]. C-Raf and related kinases have been activated by light-induced dimerization and clustering based on CRY2, mimicking the native activation mechanism involving Ras-induced C-Raf dimerization [59, 60]. A photoswitchable MEK-1 construct has been created using an engineered dimeric protein called pdDronpa, which dissociates in cyan (500nm) light and re-associates in violet (400 nm) light. Attaching two pdDronpa domains at select locations in the MEK1 kinase domain enabled photoreversible caging of MEK1's active site [61]. Additional mutations to the MEK sequence in this construct resulted in improved light-induced kinase activity [154]. Meanwhile, optogenetic inhibition of two other MAPKs, p38 and JNK, has been achieved by photo-triggered unmasking of inhibitory peptides using LOV based fusions [62].

25.4 Experimental Applications and Biological Insights

Optical control of cell signaling has enabled new insights into diverse cellular and physiological processes. In this section, we focus the discussion on a subset of recent applications that exemplify emerging themes in how these new approaches are being used and the types of questions they can help address. In addition to the examples in the following text, it is worth noting that there is also now a rich collection of research in the area of developmental biology that leverages optical control of cell signaling, although it is more focused on RTK signaling pathways and not covered in detail here [155].

25.4.1 Subcellular Signaling: Signaling Gradients and Organelle-Targeted Signaling

Optogenetic approaches can achieve subcellular control either by applying a precisely localized light stimulus to part of a cell or by engineering a subcellular targeting domain into the genetically encoded construct (Figure 25.3). Photoactivation can be rapidly repositioned, for example, from one end of a spatially polarizing cell to the other, or in neurons from one neurite tip to another, or even one dendritic spine to another. For optical control at specific organelles, genetically encoded organelle-targeting motifs can be used with either global or subcellular photoactivation to control organelle-specific signaling. This approach can also be used to achieve organelle specific control in a population of cells with widefield photoactivation.

Spatially confined photostimulation has been used extensively to study the generation, maintenance, and reversal of spatially polarized signaling underlying cell migration. Many migratory cell types respond to chemoattractants that activate $G_{i/o}$ coupled receptors, and both human blue opsin and lamprey parapinopsin have been used to optically control immune cell migration *in vitro* [54, 66].

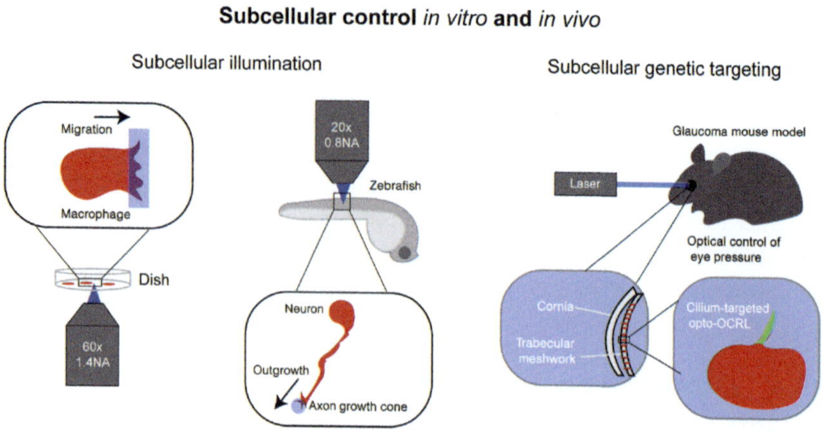

Figure 25.3 Subcellular optogenetics. Subcellular control of cell signaling can be achieved using either subcellular illumination or subcellular genetic targeting. Examples using subcellular illumination include optical control of immune cell migration *in vitro* [66], and control of axonal outgrowth *in vivo* in zebrafish [156]. Alternatively, whole-field illumination can be combined with genetic approaches for subcellular targeting. One example is the use of ciliary-targeting domains in optogenetic constructs for the control of an inositol phosphatase (OCRL) to control ciliary signaling in trabecular meshwork cells of the eye to optically regulate eye pressure in a glaucoma mouse model [157].

These studies have shown that the mechanisms that drive cell migration arise simply from $G_{i/o}$ signaling and do not require other specific pathways unique to chemoattractant receptors.

GPCRs control cellular dynamics, at least in part, by downstream activation of Rho-family G proteins that regulate the cytoskeleton. Roles for these signaling pathways in cell migration have been established using pharmacological and genetic perturbations. However, before the availability of optogenetic perturbations, the sufficiency of different Rho family G proteins to drive cell polarization and migration was unclear, as was the causal relationship between different signaling gradients. The ability to bypass the chemoattractant GPCR and directly control downstream signaling proteins has been leveraged, for example, to demonstrate that localized activation of either Rac1 or Cdc42 is sufficient to initiate, sustain, and reverse directional cell migration in a variety of cell types [121, 158]. Moreover, localized Cdc42 activation at the leading edge of a macrophage-like cell was sufficient to trigger both Rac1 activity at the front of the cell and RhoA-dependent myosin contraction at the rear [158]. Detailed studies using spatially varying light intensity gradients, rather than localized photoactivation pulses, uncovered mechanisms responsible for generating distinct spatial profiles of Rac1 and Cdc42 signaling in migratory cells, and dissected how the steepness, amplitude, and extent of different signaling gradients affect migration speed and directional control [159].

Optogenetics can also uncover links between subcellular signaling and physiological responses, exemplified by a recent study of the role of primary cilia in eye pressure regulation [157]. This work used the CRY2/CIBN dimerization system with CRY2 fused to an inositol 5-phosphatase domain, and CIBN fused to the third intracellular loop of the GPCR somatostatin receptor 3, to dynamically control phosphoinositide signaling in the primary cilia of trabecular meshwork cells of the eye. *In vivo* experiments in mice showed that optical stimulation of the ciliary inositol 5-phosphatase signaling increases fluid outflow, reducing intraocular pressure, and can even rescue defective intraocular pressure in a mouse model of congenital glaucoma.

25.4.2 Molecular Signaling Network Motifs: Feedback and Feed-Forward Loops

Whereas early models portrayed GPCR signaling as linear pathways, it is now widely appreciated that GPCRs control dynamic signaling networks involving many branched pathways, feedback and feed-forward loops, and intricate time-dependent properties that can generate temporal bias in downstream signaling [2, 160, 161]. Optogenetic approaches present two advantages for dissecting these dynamic signaling networks. First, unlike traditional genetic approaches, they can perturb signaling on time scales that are short compared to the dynamics of the system. This is important because it can reveal dynamic responses that are masked in genetic knock-outs, while avoiding compensatory mechanisms that arise with chronic perturbations. Second, the ability to optically control a growing number of signaling proteins enables direct activation of different submodules within a larger signaling network. This can help uncover functional connections that would otherwise be masked by feedback and redundancy within the larger network [162]. Importantly, the subcellular precision of optogenetic perturbations enables interrogation of the interplay between location-specific signaling and feedback or feed-forward loops [38].

Identifying feedback loops is essential for understanding the molecular basis by which cell signaling networks achieve functions such as bistability, oscillations, polarization, and robustness [163]. Optogenetic approaches have been used to uncover both positive and negative feedback loops within signaling networks. For example, in fission yeast, optogenetic activation of Cdc42 resulted in co-recruitment of a Cdc42 GEF to the plasma membrane, revealing that Cdc42 activates its own GEF in a positive feedback loop [164]. In migrating HeLa cells, optical activation of Rac1 was found to recruit β2-chimaerin, a Rac1 GAP, demonstrating the presence of a negative feedback loop [159]. In the

Raf–MEK–ERK pathway, the ability to generate brief pulses of Raf activity using a pdDronpa-based photoswitchable Raf helped to uncover negative feedback from ERK to MEK [61].

GPCR-mediated chemotaxis, cell migration along a chemical gradient, has been a popular model system for dissecting how different dynamic signaling motifs are combined to control cellular responses [165]. Migrating cells are able to sense shallow gradients over a large range of ligand concentrations. This is critical to immune system function, and requires cells to "adapt" to different levels of GPCR activation without becoming desensitized. This can be achieved by generating transient downstream signaling responses that return to basal levels (i.e. "adapt") even when the upstream components remain active. Subcellular optogenetic inhibition of G proteins was applied to cells following adaptation of a downstream PIP_3 response to generate direct experimental evidence for a "local excitation, global inhibition" model that explains how cells can generate transient and spatially uniform responses to a global stimulus, or sustained directional responses to a gradient stimulus [38, 166]. This work focused on how $G_{i/o}$ signaling controls adaptive PIP_3 responses, but other optogenetic tools have been applied to uncover additional adaptive subcircuits further downstream. Optogenetically driven PIP_3 production, independent of upstream GPCRs or heterotrimeric G proteins, showed that sustained PIP_3 signaling generates transient Rac1 responses by triggering competing parallel activation and inhibition pathways that act on Rac1 [162]. Collectively, these results showed that adaptation, in the context of immune cell chemotaxis, is governed by multiple, layered signaling motifs that cannot be deconvolved using conventional pharmacologic or genetic perturbations.

25.4.3 Cellular Dynamics: Interplay of Signaling and Biomechanics

Optogenetic control of cell signaling, especially by manipulating proteins that regulate cytoskeletal or adhesive dynamics, has emerged as a valuable approach for probing cellular dynamics and biomechanics in single and multicellular systems. For example, optical control of RhoA, a known GPCR effector that regulates actomyosin contraction, has been used to gain new insights into the physical properties of stress fibers [167], mechanisms of adhesion-independent cell migration [54], the role of collective actomyosin dynamics in controlling multicellular behaviors in embryonic development [168], and how signaling at different subcellular locations triggers distinct mechanotransduction pathways [169].

One approach for dissecting the interplay between signaling and cellular dynamics has been to control a single protein using different spatial patterns of activation. For example, in macrophage-like RAW264.7 cells, different spatial patterns of RhoA-induced actomyosin contractility caused either directional migration or formation of a cell division furrow [54, 170]. Detailed studies of these responses uncovered shared features of membrane flow and vesicular trafficking underlying these distinct cellular dynamics [54, 170]. In confluent MDCK epithelial cell monolayers, optical activation of RhoA at the plasma membrane or mitochondria resulted in opposing changes in cellular traction, intercellular tension, and tissue compaction [169]. Alternatively, optical control of different signaling proteins has been used to control distinct modes of a given cell behavior. For example, localized photoactivation of either parapinopsin or RhoA generates directional migration of macrophage-like cells from the front and rear, respectively [54]. Optical control over these two distinct migration modes enabled experiments comparing their ability to propel cells under adhesive or non-adhesive conditions, and uncovered new insights into the biophysical mechanisms underlying adhesion-independent cell migration.

25.4.4 Neuroscience

25.4.4.1 Cellular and Molecular Neuroscience

Optogenetic control of signaling at neurite growth cones, using dynamically positioned photoactivation, has helped advance our mechanistic understanding of neurite outgrowth *in vitro* and axon guidance *in vivo*. In primary cultures of rat hippocampal neurons, localized photoactivation of human blue opsin initiated neurite formation and extension through a process involving $G_{i/o}$, PIP_3, and Rac signaling [23]. Meanwhile, optically controlled production of PIP_3 at neurite tips in hippocampal neurons, using a light-activated PI3K, produced growth cones with dynamic filopodia and lamellipodia, but failed to induce neurite elongation [171]. Collectively these results showed that PIP_3 is necessary, but not sufficient to mediate $G_{i/o}$ induced neurite elongation. Other studies have addressed how organelle positioning influences axonal outgrowth. Light-induced dimerization between cytoskeletal motor proteins and Rab11 was used to enrich or deplete recycling endosomes in axonal growth cones and uncover their role in axon elongation [172]. More distal signaling within neurons has also been linked to axon growth and pathfinding through the use of optogenetic constructs targeted to primary cilia for control of ciliary cAMP, PIP_3, and AKT signaling [173]. *In vivo* applications have taken advantage of the transparent tissues of zebrafish to apply localized photoactivation of Rac1 for long-range control of axon guidance in an intact organism [156]. Remarkably, this approach was able to outcompete endogenous repulsive axon guidance barriers and rescue genetic axon guidance defects, demonstrating potential as a new model system for systematic studies of the development, function, and regeneration of neural networks.

Optogenetic control of signaling proteins also provided new insights regarding timing of specific molecular events that control synaptic plasticity. Structural plasticity at the level of individual dendritic spines was studied using a light-controlled inhibitor of CaMKII consisting of a potent inhibitory peptide, AIP2, fused to the C-terminal Jα helix of asLOV2 [174]. Two-photon glutamate uncaging in rat hippocampal brain slices induced spine enlargement associated with long-term potentiation (referred to as structural long term potentiation or sLTP). Optogenetic inhibition of CaMKII, before, during, or after glutamate uncaging revealed that transient and sustained phases of sLTP have distinct temporal requirements for CaMKII activity. This work also determined the temporal requirements for CaMKII signaling in electrical stimulation-induced LTP, as well as inhibitory avoidance learning. Similarly, an optogenetic inhibitor of JNK helped to clarify its role in spine dynamics in cultured hippocampal rat neurons [175]. Localized optogenetic inhibition demonstrated that JNK acts locally within the spine-head to exert rapid control over spine motility, and that JNK controls ionotropic glutamate receptor removal and regression of spines in response to corticosterone stress.

25.4.4.2 Systems Neuroscience

The primary strength of optogenetics for systems neuroscience applications lies in the ability to express genetically encoded light sensitive proteins to manipulate distinct types of neurons that underlie specific behavioral processes. Signaling optogenetics complements the use of microbial opsins in studying the function of neural circuits in intact organisms. Whereas channelrhodopsin and other ion-conducting microbial opsins have been useful for driving collective firing of a subset of neurons at a specific frequency to drive behaviors, this does not always represent normal circuit dynamics. Additionally, direct manipulations of membrane potential may not be ideal for non-excitable cells like astrocytes and microglia. Manipulations that more naturally mimic physiological processes might provide more detail into how the brain processes and integrates different types of information. Signaling optogenetics can determine how distinct biochemical events at the cellular or subcellular level affect neural circuit dynamics and behavioral responses. For example, microbial opsins have shown that activation of specific subsets of neurons can generate addiction related behaviors in rodents, but many questions remain about the underlying cellular and synaptic changes that drive those behaviors. Classic approaches using local knockout of Rac1, or overexpression of a dominant negative Rac1 mutant, showed that this protein is essential for cocaine-induced structural plasticity in nucleus accumbens neurons in mice [176]. Interestingly, repeated cocaine exposure induces transient decreases in Rac1 signaling, and compensating for this transient decrease using temporally controlled Rac1 photoactivation prevented the formation of cocaine-induced place preference, a measure of the rewarding effects of the drug [176]. Whereas the knockout approach alone might have been interpreted as meaning that Rac1 signaling drives the cocaine-induced behavioral changes, direct optogenetic manipulations of these signaling cascades revealed that the actual requirement is for a combination of basal Rac1 signaling with cocaine-induced transient suppression in Rac1 activity. Optical control of Rac1 has also been used to link dendritic spine dynamics to memory formation. In the cortex, motor learning tasks in mice produced synaptic remodeling and new spine growth. Photoactivation of Rac1 after learning led to spine shrinkage and erased these newly acquired memories [177]. There are many areas where direct optical control of various signaling cascades described earlier might enable new questions or reveal significant insights into how the brain is wired.

Neuronal GPCRs largely function in a modulatory capacity, adjusting cellular and synaptic excitability depending on the downstream effectors. The nervous system employs a tremendous variety of neuromodulatory transmitters and peptides that alter the internal state and behavioral output of nearly all circuits. Many of these roles have been established in traditional behavioral pharmacology studies. These studies are limited by compound availability and selectivity, and do not allow for the study of neuromodulators in discrete cell types.

Optical control of GPCRs can bypass many of these limitations and help dissect how neuromodulation by different receptors alters circuit dynamics and behaviors in various animal models. In systems neuroscience, these approaches have largely focused on the use of chimeric opsins or "opto-XRs" to mimic the signaling of distinct neuromodulators. Importantly, the opto-XR can be selectively targeted to only those neurons that natively express the original GPCR using a combination of transgenic animals and viral transduction methods [21]. These opto-XRs now allow for optical control of adenosine, adrenergic, dopamine, glutamate, opioid, and serotonin GPCR signaling, and have helped to more definitively link different behavioral states to activation of GPCRs in specific types of neurons [18–22, 68, 178–181]. For example, these opto-XRs have demonstrated the sufficiency of α1AR activation in the nucleus accumbens (NAc) for reward-related behaviors, while mimicking D1 dopamine receptor activation in the NAc increased social interactions [18, 21].

A chimera of rat rhodopsin and mu opioid receptor ("opto-MOR") revealed the cells underlying both reward and aversion behaviors attributed to drugs of abuse, like morphine [22]. This chimera was shown to exhibit signaling, desensitization, and internalization properties that parallel MOR. Using cell-specific expression, light stimulation of opto-MOR receptors expressed in inhibitory GABAergic neurons of the rostromedial tegmental nucleus (RMTg) produced preference behaviors, while activating MOR signaling in GABAergic neurons of the ventral pallidum produced aversion. While these two brain regions were previously linked to these opposing behavior responses to opioids, classical behavioral pharmacology experiments were unable to dissociate the specific cell types mediating these effects, highlighting a major strength of optogenetic approaches. Similar strategies have determined the cell types and specific intracellular signaling cascades underlying the known anxiolytic effects of β2-AR antagonists in the amygdala [20].

Chemogenetic approaches based on DREADDs have been widely used to "activate" or "inhibit" neurons with engineered G_q- or G_i-coupled GPCRs, respectively. These relatively simple approaches have been widely adopted and have proven very successful for relatively straightforward circuit manipulations. The use of opsin GPCRs as generalized tools in systems neuroscience has lagged behind DREADDs, partly because DREADD activation does not require any customized equipment for light delivery into the brain. However, many applications will benefit from the rapid activation and reversibility of opsin-GPCRs.

Several new modifications of these genetically encoded tools have increased their utility for systems neuroscience. For example, an axon-targeted G_i/o coupled DREADD, hM4D-neurexin, was engineered for projection-selective synaptic silencing [182]. However, this approach is still temporally limited by delivery of an exogenous drug. G_i/o-coupled opsins, on the other hand, are particularly promising tools for optical inhibition of synaptic transmission at axon terminals, which was recently demonstrated using mosquito Opn3 and lamprey parapinopsin [183, 184]. These are both bistable G_i-coupled GPCRs that were shown to couple to endogenous inhibitory signaling cascades to reduce neuronal excitability and reversibly inhibit transmission at synaptic terminals. The generation of photoactivatable neurotoxins represents another intriguing avenue for inducible suppression of transmission [185]. While light-induced cleavage of synaptic vesicle proteins is well-suited for long-term silencing, this approach lacks the temporal precision and reversibility achieved by opsins. Overall, optical control of endogenous signaling pathways by opsin-GPCRs or their disruption by photoactivatable neurotoxins represent superior alternatives to inhibitory ion-based microbial opsins, which can produce unintended effects at synaptic projections, including direct depolarization and rebound spiking [186–190].

Photopharmacological approaches have also been employed in systems neuroscience, where they can provide unique capabilities such as allosteric modulation of endogenous receptors. For example, NAMs have been used to study different mGluR subtypes, for which highly selective orthosteric ligands are not available. A diffusible, photoswitchable NAM of mGlu4, termed OptoGluNAM4.1, exerted effects on mechanical hypersensitivity in a mouse model of inflammatory chronic pain [97]. A photocaged $mGlu_5$, JF-NP-26, generated light-dependent analgesia in both neuropathic and inflammatory pain models [83]. As mentioned earlier, tethered photoswitchable ligands have also been used *in vivo*, where they enable control over signaling in specific cell types [105]. Of course, the targeted receptors in this case are no longer endogenous, but their trafficking and signaling properties may more closely mimic endogenous receptors compared to chimeric opsins.

25.5 Future Directions

25.5.1 G Protein Signaling at Organelles

Heterotrimeric G protein signaling has classically been thought to occur at the plasma membrane, but the last decade has seen work from several labs indicating that G protein signaling can also occur at various organelles including the Golgi apparatus, endosomes, mitochondria, and the nucleus [191]. Imaging of agonist induced G protein subunit translocation between the plasma membrane and intracellular membranes, and conformationally selective nanobodies have provided evidence that active GPCRs and G protein subunits are not exclusively localized to the plasma membrane [111–113, 192–194]. However, selective perturbation of G protein signaling at organelles has been somewhat limited. One approach has been to use cell permeable versus impermeable agonists or antagonists, which can provide information on functions of cell-surface versus intracellular GPCR signaling [195, 196]. Agonists that activate quantitatively different amounts of G protein signaling from endosomes have also been used. Additionally, knockdown of rapidly translocating Gγ subunit types, or overexpression of slow translocating Gγ subunits has been used as a dominant negative to study Gβγ signaling from organelle membranes [197].

Optogenetics can complement these approaches by enabling selective activation or inhibition of G protein signaling at organelles. GPCRs localized to endosomes or the trans Golgi network appear to be oriented with the N-terminus (classically extracellular) facing the organelle lumen, and the C-terminus (classically intracellular) facing the cytosol. Tools like the CRY-RGS and CRY-GRK2ct constructs described earlier could be re-engineered such that light activation induces translocation from the cytosol to Golgi or endosomal membranes instead of the plasma membrane. Indeed, an analogous chemogenetic approach based on inducible recruitment of a GRK2ct mutant to the Golgi apparatus has been used with great success to show that Golgi localized Gβγ signaling regulates basolateral, but not apical, cargo transport out of the trans-Golgi network in polarized cells [136]. The faster kinetics and reversibility of optogenetic-based schemes may be even better suited to a number of organelle-specific applications. Alternatively, optogenetic tools based on non-receptor GEFs, such as the recently reported LOV2GIVe construct, could enable optical activation of different G protein subtypes in an organelle-targeted manner [198].

25.5.2 Nanobody-Based Optogenetics

Optical control of endogenous GPCRs is currently limited to approaches using diffusible or tethered photoswitchable ligands. An alternative approach is to use genetically encoded constructs that interact with the intracellular portions of GPCRs in a light-dependent manner to activate or inhibit signaling. For example, an optogenetic tool was recently reported based on Nb60, a nanobody that binds an inactive form of the β2AR [199]. A split version of the nanobody was generated, resulting in a functional nanobody only after the split fragments were brought together using light-inducible dimers. This approach demonstrated the ability to optically suppress isoprenaline-induced β2AR signaling and internalization in HEK293 cells. Slow kinetics (~1 hour) and irreversibility will limit the utility of this particular approach, but alternative approaches for generating nanobody-based optogenetic tools promise to provide much faster kinetics and reversibility [200, 201]. Given the recent development of *in vitro* assays for rapid screening of conformationally selective nanobodies, one can imagine applying nanobody-based optogenetics to control a large number of different endogenous GPCRs [202].

25.6 Concluding Remarks

The unique combination of rapid kinetics, reversibility, and subcellular control provided by optogenetic and photopharmacological approaches makes them ideally suited for interrogating dynamic GPCR signaling networks. The opto-GPCR toolkit is currently in a stage of rapid growth, but already offers the ability to apply targeted perturbations to many different signaling proteins within a GPCR signaling cascade. The applications discussed here highlight the remarkable breadth of new biological insights these approaches have already uncovered. Most of these optogenetic tools could not have been developed without support for basic research outside of the biomedical sciences, including fundamental studies of opsin-GPCRs across diverse species, and studies of the plants and fungi from which many of the light-sensing protein domains have been derived. Going forward, the opto-GPCR field is poised to provide diverse opportunities for scientists, such as the development of minimally invasive methods to deliver light stimuli *in vivo* [203, 204], molecular engineering of new optogenetic constructs, or the construction of theoretical models that can provide a framework for interpreting the molecular, cellular, and physiological responses generated by spatiotemporally targeted signaling perturbations.

References

1 Lohse, M.J. and Hofmann, K.P. (2015). Spatial and temporal aspects of signaling by G-protein-coupled receptors. *Mol. Pharmacol.* 88 (3): 572–578.

2 Kholodenko, B.N., Hancock, J.F., and Kolch, W. (2010). Signalling ballet in space and time. *Nat. Rev. Mol. Cell Biol.* 11 (6): 414–426.

3 Devreotes, P.N. and Zigmond, S.H. (1988). Chemotaxis in eukaryotic cells: a focus on leukocytes and *Dictyostelium*. *Annu. Rev. Cell Biol.* 4: 649–686.

4 Shah, N.A. and Sarkar, C.A. (2011). Robust network topologies for generating switch-like cellular responses. *PLoS Comput. Biol.* 7 (6): e1002085.

5 Eichel, K. and von Zastrow, M. (2018). Subcellular organization of GPCR signaling. *Trends Pharmacol. Sci.* 39 (2): 200–208.

6 Lobingier, B.T. and von Zastrow, M. (2019). When trafficking and signaling mix: how subcellular location shapes G protein-coupled receptor activation of heterotrimeric G proteins. *Traffic* 20 (2): 130–136.

7 Patriarchi, T. et al. (2018). Ultrafast neuronal imaging of dopamine dynamics with designed genetically encoded sensors. *Science* 360 (6396): eaat4422.

8 Marvin, J.S. et al. (2019). A genetically encoded fluorescent sensor for in vivo imaging of GABA. *Nat. Methods* 16 (8): 763–770.

9 Marvin, J.S. et al. (2013). An optimized fluorescent probe for visualizing glutamate neurotransmission. *Nat. Methods* 10 (2): 162–170.

10 Sun, F. et al. (2018). A genetically encoded fluorescent sensor enables rapid and specific detection of dopamine in flies, fish, and mice. *Cell* 174 (2): 481–496.e19.

11 Feng, J. et al. (2019). A genetically encoded fluorescent sensor for rapid and specific in vivo detection of norepinephrine. *Neuron* 102 (4): 745–761.e8.

12 Halls, M.L. and Canals, M. (2018). Genetically encoded FRET biosensors to illuminate compartmentalised GPCR signalling. *Trends Pharmacol. Sci.* 39 (2): 148–157.

13 Ni, Q., Mehta, S., and Zhang, J. (2018). Live-cell imaging of cell signaling using genetically encoded fluorescent reporters. *FEBS J.* 285 (2): 203–219.

14 Spangler, S.M. and Bruchas, M.R. (2017). Optogenetic approaches for dissecting neuromodulation and GPCR signaling in neural circuits. *Curr. Opin. Pharmacol.* 32: 56–70.

15 Ricart-Ortega, M., Font, J., and Llebaria, A. (2019). GPCR photopharmacology. *Mol. Cell. Endocrinol.* 488: 36–51.

16 Berizzi, A.E. and Goudet, C. (2020). Strategies and considerations of G-protein-coupled receptor photopharmacology. *Adv. Pharmacol.* 88: 143–172.

17 Yamashita, T. (2020). Unexpected molecular diversity of vertebrate nonvisual opsin Opn5. *Biophys. Rev.* 12 (2): 333–338.

18 Airan, R.D. et al. (2009). Temporally precise in vivo control of intracellular signalling. *Nature* 458 (7241): 1025–1029.

19 Kim, J.M. et al. (2005). Light-driven activation of β_2-adrenergic receptor signaling by a chimeric rhodopsin containing the β_2-adrenergic receptor cytoplasmic loops. *Biochemistry* 44 (7): 2284–2292.

20 Siuda, E.R. et al. (2015). Optodynamic simulation of β-adrenergic receptor signalling. *Nat. Commun.* 6: 8480.

21 Gunaydin, L.A. et al. (2014). Natural neural projection dynamics underlying social behavior. *Cell* 157 (7): 1535–1551.

22 Siuda, E.R. et al. (2015). Spatiotemporal control of opioid signaling and behavior. *Neuron* 86 (4): 923–935.

23 Karunarathne, W.K. et al. (2013). Optically triggering spatiotemporally confined GPCR activity in a cell and programming neurite initiation and extension. *Proc. Natl. Acad. Sci. U.S.A.* 110 (17): E1565–E1574.

24 Koyanagi, M. et al. (2004). Bistable UV pigment in the lamprey pineal. *Proc. Natl. Acad. Sci. U.S.A.* 101 (17): 6687–6691.

25 Koyanagi, M. et al. (2013). Homologs of vertebrate Opn3 potentially serve as a light sensor in nonphotoreceptive tissue. *Proc. Natl. Acad. Sci. U.S.A.* 110 (13): 4998–5003.

26 Bailes, H.J. and Lucas, R.J. (2013). Human melanopsin forms a pigment maximally sensitive to blue light ($\lambda_{max} \approx 479$ nm) supporting activation of $G_{q/11}$ and $G_{i/o}$ signalling cascades. *Proc. Biol. Sci.* 280 (1759): 20122987.

27 Melyan, Z. et al. (2005). Addition of human melanopsin renders mammalian cells photoresponsive. *Nature* 433 (7027): 741–745.

28 McGregor, K.M. et al. (2016). Using melanopsin to study G protein signaling in cortical neurons. *J. Neurophysiol.* 116 (3): 1082–1092.

29 Kankanamge, D. et al. (2018). Melanopsin (Opn4) utilizes $G\alpha_i$ and $G\beta\gamma$ as major signal transducers. *J. Cell Sci.* 131 (11): jcs212910.

30 Mederos, S. et al. (2019). Melanopsin for precise optogenetic activation of astrocyte-neuron networks. *Glia* 67 (5): 915–934.

31 Koyanagi, M. and Terakita, A. (2014). Diversity of animal opsin-based pigments and their optogenetic potential. *Biochim. Biophys. Acta* 1837 (5): 710–716.

32 Koyanagi, M. et al. (2008). Jellyfish vision starts with cAMP signaling mediated by opsin-G_s cascade. *Proc. Natl. Acad. Sci. U.S.A.* 105 (40): 15576–15580.

33 Sato, K. et al. (2018). Opn5L1 is a retinal receptor that behaves as a reverse and self-regenerating photoreceptor. *Nat. Commun.* 9 (1): 1255.

34 Nagata, T. et al. (2018). An all-trans-retinal-binding opsin peropsin as a potential dark-active and light-inactivated G protein-coupled receptor. *Sci. Rep.* 8 (1): 3535.

35 Kawano-Yamashita, E. et al. (2015). Activation of transducin by bistable pigment parapinopsin in the pineal organ of lower vertebrates. *PLoS One* 10 (10): e0141280.

36 Tsukamoto, H. et al. (2009). The magnitude of the light-induced conformational change in different rhodopsins correlates with their ability to activate G proteins. *J. Biol. Chem.* 284 (31): 20676–20683.

37 Saito, T. et al. (2019). Spectral tuning mediated by helix III in butterfly long wavelength-sensitive visual opsins revealed by heterologous action spectroscopy. *Zoological Lett.* 5: 35.

38 O'Neill, P.R. and Gautam, N. (2014). Subcellular optogenetic inhibition of G proteins generates signaling gradients and cell migration. *Mol. Biol. Cell* 25 (15): 2305–2314.

39 Hannanta-Anan, P. and Chow, B.Y. (2018). Optogenetic inhibition of Gα_q protein signaling reduces calcium oscillation stochasticity. *ACS Synth. Biol.* 7 (6): 1488–1495.

40 Acosta-Ruiz, A., Broichhagen, J., and Levitz, J. (2019). Optical regulation of class C GPCRs by photoswitchable orthogonal remotely tethered ligands. *Methods Mol. Biol.* 1947: 103–136.

41 Levskaya, A. et al. (2009). Spatiotemporal control of cell signalling using a light-switchable protein interaction. *Nature* 461 (7266): 997–1001.

42 Terrillon, S. and Bouvier, M. (2004). Receptor activity-independent recruitment of βarrestin2 reveals specific signalling modes. *EMBO J.* 23 (20): 3950–3961.

43 Liebick, M. et al. (2017). Functional consequences of chemically-induced β-arrestin binding to chemokine receptors CXCR4 and CCR5 in the absence of ligand stimulation. *Cell. Signal.* 38: 201–211.

44 Scheib, U. et al. (2018). Rhodopsin-cyclases for photocontrol of cGMP/cAMP and 2.3 Å structure of the adenylyl cyclase domain. *Nat. Commun.* 9 (1): 2046.

45 Hansen, J.N. et al. (2020). Nanobody-directed targeting of optogenetic tools to study signaling in the primary cilium. *Elife* 9: e57907.

46 Gasser, C. et al. (2014). Engineering of a red-light-activated human cAMP/cGMP-specific phosphodiesterase. *Proc. Natl. Acad. Sci. U.S.A.*

47 Stabel, R. et al. (2019). Revisiting and redesigning light-activated cyclic-mononucleotide phosphodiesterases. *J. Mol. Biol.* 431 (17): 3029–3045.

48 Pham, E., Mills, E., and Truong, K. (2011). A synthetic photoactivated protein to generate local or global Ca^{2+} signals. *Chem. Biol.* 18 (7): 880–890.

49 Kyung, T. et al. (2015). Optogenetic control of endogenous Ca^{2+} channels in vivo. *Nat. Biotechnol.* 33 (10): 1092–1096.

50 Taslimi, A. et al. (2014). An optimized optogenetic clustering tool for probing protein interaction and function. *Nat. Commun.* 5: 4925.

51 Kawano, F. et al. (2015). Engineered pairs of distinct photoswitches for optogenetic control of cellular proteins. *Nat. Commun.* 6: 6256.

52 Guntas, G. et al. (2015). Engineering an improved light-induced dimer (iLID) for controlling the localization and activity of signaling proteins. *Proc. Natl. Acad. Sci. U.S.A.* 112 (1): 112–117.

53 Glantz, S.T. et al. (2016). Functional and topological diversity of LOV domain photoreceptors. *Proc. Natl. Acad. Sci. U.S.A.* 113 (11): E1442–E1451.

54 O'Neill, P.R. et al. (2018). Membrane flow drives an adhesion-independent amoeboid cell migration mode. *Dev. Cell* 46 (1): 9–22 e4.

55 Leopold, A.V., Chernov, K.G., and Verkhusha, V.V. (2018). Optogenetically controlled protein kinases for regulation of cellular signaling. *Chem. Soc. Rev.* 47 (7): 2454–2484.

56 Huang, Y.A. et al. (2020). Actin waves transport RanGTP to the neurite tip to regulate non-centrosomal microtubules in neurons. *J. Cell Sci.* 133 (9): jcs241992.

57 Nguyen, M.K. et al. (2016). Optogenetic oligomerization of Rab GTPases regulates intracellular membrane trafficking. *Nat. Chem. Biol.* 12: 431–436, 6.

58 Cornejo, V.H. et al. (2020). Non-conventional axonal organelles control TRPM8 ion channel trafficking and peripheral cold sensing. *Cell Rep.* 30 (13): 4505–4517 e5.

59 Wend, S. et al. (2014). Optogenetic control of protein kinase activity in mammalian cells. *ACS Synth. Biol.* 3 (5): 280–285.

60 Chatelle, C.V. et al. (2016). Optogenetically controlled RAF to characterize BRAF and CRAF protein kinase inhibitors. *Sci. Rep.* 6: 23713.

61 Zhou, X.X. et al. (2017). Optical control of cell signaling by single-chain photoswitchable kinases. *Science* 355 (6327): 836–842.

62 Melero-Fernandez de Mera, R.M. et al. (2017). A simple optogenetic MAPK inhibitor design reveals resonance between transcription-regulating circuitry and temporally-encoded inputs. *Nat. Commun.* 8: 15017.

63 Wang, W., Geiger, J.H., and Borhan, B. (2014). The photochemical determinants of color vision: revealing how opsins tune their chromophore's absorption wavelength. *Bioessays* 36 (1): 65–74.

64 Luo, D.G. et al. (2011). Activation of visual pigments by light and heat. *Science* 332 (6035): 1307–1312.

65 Gozem, S. et al. (2012). The molecular mechanism of thermal noise in rod photoreceptors. *Science* 337 (6099): 1225–1228.

66 Karunarathne, W.K. et al. (2013). Optical control demonstrates switch-like PIP_3 dynamics underlying the initiation of immune cell migration. *Proc. Natl. Acad. Sci. U.S.A.* 110 (17): E1575–E1583.

67 Imamoto, Y. and Shichida, Y. (2014). Cone visual pigments. *Biochim. Biophys. Acta* 1837 (5): 664–673.

68 Masseck, O.A. et al. (2014). Vertebrate cone opsins enable sustained and highly sensitive rapid control of $G_{i/o}$ signaling in anxiety circuitry. *Neuron* 81 (6): 1263–1273.

69 Nagata, T. et al. (2012). Depth perception from image defocus in a jumping spider. *Science* 335 (6067): 469–471.

70 Eickelbeck, D. et al. (2020). Lamprey parapinopsin ("UVLamP"): a bistable UV-sensitive optogenetic switch for ultrafast control of GPCR pathways. *ChemBioChem* 21 (5): 612–617.

71 Morri, M. et al. (2018). Optical functionalization of human Class A orphan G-protein-coupled receptors. *Nat. Commun.* 9 (1): 1950.

72 Bailes, H.J., Zhuang, L.Y., and Lucas, R.J. (2012). Reproducible and sustained regulation of Gαs signalling using a metazoan opsin as an optogenetic tool. *PLoS One* 7 (1): e30774.

73 Terakita, A. et al. (2008). Expression and comparative characterization of G_q-coupled invertebrate visual pigments and melanopsin. *J. Neurochem.* 105 (3): 883–890.

74 Wakakuwa, M. et al. (2010). Evolution and mechanism of spectral tuning of blue-absorbing visual pigments in butterflies. *PLoS One* 5 (11): e15015.

75 Inoue, A. et al. (2019). Illuminating G-protein-coupling selectivity of GPCRs. *Cell* 177 (7): 1933–1947 e25.

76 Riobo, N.A. and Manning, D.R. (2005). Receptors coupled to heterotrimeric G proteins of the G_{12} family. *Trends Pharmacol. Sci.* 26 (3): 146–154.

77 Bockaert, J. et al. (2010). GPCR interacting proteins (GIPs) in the nervous system: roles in physiology and pathologies. *Annu. Rev. Pharmacol. Toxicol.* 50: 89–109.

78 Magalhaes, A.C., Dunn, H., and Ferguson, S.S. (2012). Regulation of GPCR activity, trafficking and localization by GPCR-interacting proteins. *Br. J. Pharmacol.* 165 (6): 1717–1736.

79 Yu, G. et al. (2016). Optical manipulation of the alpha subunits of heterotrimeric G proteins using photoswitchable dimerization systems. *Sci. Rep.* 6: 35777.

80 Takenouchi, O., Yoshimura, H., and Ozawa, T. (2018). Unique roles of β-arrestin in GPCR trafficking revealed by photoinducible dimerizers. *Sci. Rep.* 8 (1): 677.

81 Eickelbeck, D. et al. (2019). CaMello-XR enables visualization and optogenetic control of $G_{q/11}$ signals and receptor trafficking in GPCR-specific domains. *Commun. Biol.* 2: 60.

82 Callaway, E.M. and Katz, L.C. (1993). Photostimulation using caged glutamate reveals functional circuitry in living brain slices. *Proc. Natl. Acad. Sci. U.S.A.* 90 (16): 7661–7665.

83 Font, J. et al. (2017). Optical control of pain in vivo with a photoactive $mGlu_5$ receptor negative allosteric modulator. *Elife* 6: e23545.

84 Muralidharan, S. and Nerbonne, J.M. (1995). Photolabile "caged" adrenergic receptor agonists and related model compounds. *J. Photochem. Photobiol. B* 27 (2): 123–137.

85 Asad, N. et al. (2020). Photoactivatable dopamine and sulpiride to explore the function of dopaminergic neurons and circuits. *ACS Chem. Neurosci.* 11 (6): 939–951.

86 Merlen, C. et al. (2013). Intracrine endothelin signaling evokes IP_3-dependent increases in nucleoplasmic Ca^{2+} in adult cardiac myocytes. *J. Mol. Cell. Cardiol.* 62: 189–202.

87 Tadevosyan, A. et al. (2016). Caged ligands to study the role of intracellular GPCRs. *Methods* 92: 72–77.

88 Banghart, M.R. and Sabatini, B.L. (2012). Photoactivatable neuropeptides for spatiotemporally precise delivery of opioids in neural tissue. *Neuron* 73 (2): 249–259.

89 Banghart, M.R. et al. (2013). Caged naloxone reveals opioid signaling deactivation kinetics. *Mol. Pharmacol.* 84 (5): 687–695.

90 Pirrung, M.C. et al. (2000). Caged chemotactic peptides. *Bioconjugate Chem.* 11 (5): 679–681.

91 Collins, S.R. et al. (2015). Using light to shape chemical gradients for parallel and automated analysis of chemotaxis. *Mol. Syst. Biol.* 11 (4): 804.

92 Purgert, C.A. et al. (2014). Intracellular mGluR5 can mediate synaptic plasticity in the hippocampus. *J. Neurosci.* 34 (13): 4589–4598.

93 Jong, Y.I. and O'Malley, K.L. (2017). Mechanisms associated with activation of intracellular metabotropic glutamate receptor, mGluR5. *Neurochem. Res.* 42 (1): 166–172.

94 Dong, M. et al. (2015). Red-shifting azobenzene photoswitches for in vivo use. *Acc. Chem. Res.* 48 (10): 2662–2670.

95 Konrad, D.B., Frank, J.A., and Trauner, D. (2016). Synthesis of redshifted azobenzene photoswitches by late-stage functionalization. *Chemistry* 22 (13): 4364–4368.

96 Pittolo, S. et al. (2014). An allosteric modulator to control endogenous G protein-coupled receptors with light. *Nat. Chem. Biol.* 10 (10): 813–815.

97 Rovira, X. et al. (2016). OptoGluNAM4.1, a photoswitchable allosteric antagonist for real-time control of $mGlu_4$ receptor activity. *Cell Chem. Biol.* 23 (8): 929–934.

98 Riefolo, F. et al. (2019). Optical control of cardiac function with a photoswitchable muscarinic agonist. *J. Am. Chem. Soc.* 141 (18): 7628–7636.

99 Bahamonde, M.I. et al. (2014). Photomodulation of G protein-coupled adenosine receptors by a novel light-switchable ligand. *Bioconjugate Chem.* 25 (10): 1847–1854.

100 Gómez-Santacana, X. et al. (2018). Photoswitching the efficacy of a small-molecule ligand for a peptidergic GPCR: from antagonism to agonism. *Angew. Chem. Int. Ed. Engl.* 57 (36): 11608–11612.

101 Levitz, J. et al. (2013). Optical control of metabotropic glutamate receptors. *Nat. Neurosci.* 16 (4): 507–516.

102 Donthamsetti, P.C. et al. (2019). Genetically targeted optical control of an endogenous G protein-coupled receptor. *J. Am. Chem. Soc.* 141 (29): 11522–11530.

103 Broichhagen, J. et al. (2015). Orthogonal optical control of a G protein-coupled receptor with a SNAP-tethered photochromic ligand. *ACS Cent. Sci.* 1 (7): 383–393.

104 Levitz, J. et al. (2017). Dual optical control and mechanistic insights into photoswitchable group II and III metabotropic glutamate receptors. *Proc. Natl. Acad. Sci. U.S.A.* 114 (17): E3546–E3554.

105 Acosta-Ruiz, A. et al. (2020). Branched photoswitchable tethered ligands enable ultra-efficient optical control and detection of G protein-coupled receptors in vivo. *Neuron* 105 (3): 446–463.e13.

106 Grushevskyi, E.O. et al. (2019). Stepwise activation of a class C GPCR begins with millisecond dimer rearrangement. *Proc. Natl. Acad. Sci. U.S.A.* 116 (20): 10150–10155.

107 Olsen, R.H.J. et al. (2020). TRUPATH, an open-source biosensor platform for interrogating the GPCR transducerome. *Nat. Chem. Biol.*

108 Maziarz, M. et al. (2020). Revealing the activity of trimeric G-proteins in live cells with a versatile biosensor design. *Cell* 182 (3): 770–785 e16.

109 Wan, Q. et al. (2018). Mini G protein probes for active G protein-coupled receptors (GPCRs) in live cells. *J. Biol. Chem.* 293 (19): 7466–7473.

110 Ballister, E.R. et al. (2018). A live cell assay of GPCR coupling allows identification of optogenetic tools for controlling G_o and G_i signaling. *BMC Biol.* 16 (1): 10.

111 Akgoz, M., Kalyanaraman, V., and Gautam, N. (2004). Receptor-mediated reversible translocation of the G protein βγ complex from the plasma membrane to the Golgi complex. *J. Biol. Chem.* 279 (49): 51541–51544.

112 Ajith Karunarathne, W.K. et al. (2012). All G protein βγ complexes are capable of translocation on receptor activation. *Biochem. Biophys. Res. Commun.* 421 (3): 605–611.

113 O'Neill, P.R. et al. (2012). G-protein signaling leverages subunit-dependent membrane affinity to differentially control beta gamma translocation to intracellular membranes. *Proc. Natl. Acad. Sci. U.S.A.* 109 (51): E3568–E3577.

114 Ratnayake, K. et al. (2017). Measurement of GPCR-G protein activity in living cells. *Methods Cell Biol.* 142: 1–25.

115 Senarath, K. et al. (2016). Reversible G protein βγ9 distribution-based assay reveals molecular underpinnings in subcellular, single-cell, and multicellular GPCR and G protein activity. *Anal. Chem.* 88 (23): 11450–11459.

116 Ratnayake, K. et al. (2018). Blue light excited retinal intercepts cellular signaling. *Sci. Rep.* 8 (1): 10207.

117 Kennedy, M.J. et al. (2010). Rapid blue-light-mediated induction of protein interactions in living cells. *Nat. Methods* 7 (12): 973–U48.

118 Bugaj, L.J. et al. (2013). Optogenetic protein clustering and signaling activation in mammalian cells. *Nat. Methods* 10 (3): 249–252.

119 Bugaj, L.J. et al. (2015). Regulation of endogenous transmembrane receptors through optogenetic Cry2 clustering. *Nat. Commun.* 6: 6898.

120 Park, H. et al. (2017). Optogenetic protein clustering through fluorescent protein tagging and extension of CRY2. *Nat. Commun.* 8 (1): 30.

121 Wu, Y.I. et al. (2009). A genetically encoded photoactivatable Rac controls the motility of living cells. *Nature* 461 (7260): 104–U111.

122 Yi, J.J. et al. (2014). Manipulation of endogenous kinase activity in living cells using photoswitchable inhibitory peptides. *ACS Synth. Biol.* 3 (11): 788–795.

123 Wang, H. et al. (2016). LOVTRAP: an optogenetic system for photoinduced protein dissociation. *Nat. Methods* 13 (9): 755–758.

124 Benedetti, L. et al. (2018). Light-activated protein interaction with high spatial subcellular confinement. *Proc. Natl. Acad. Sci. U.S.A.* 115 (10): E2238–E2245.

125 Kaberniuk, A.A., Shemetov, A.A., and Verkhusha, V.V. (2016). A bacterial phytochrome-based optogenetic system controllable with near-infrared light. *Nat. Methods* 13 (7): 591–597.

126 Bondar, A. and Lazar, J. (2014). Dissociated GαGTP and Gβγ protein subunits are the major activated form of heterotrimeric $G_{i/o}$ proteins. *J. Biol. Chem.* 289 (3): 1271–1281.

127 Ross, E.M. and Wilkie, T.M. (2000). GTPase-activating proteins for heterotrimeric G proteins: regulators of G protein signaling (RGS) and RGS-like proteins. *Annu. Rev. Biochem.* 69: 795–827.

128 Zhong, H. et al. (2003). A spatial focusing model for G protein signals. Regulator of G protein signaling (RGS) protein-mediated kinetic scaffolding. *J. Biol. Chem.* 278 (9): 7278–7284.

129 Srinivasa, S.P. et al. (1998). Plasma membrane localization is required for RGS4 function in *Saccharomyces cerevisiae*. *Proc. Natl. Acad. Sci. U.S.A.* 95 (10): 5584–5589.

130 Glantz, S.T. et al. (2018). Directly light-regulated binding of RGS-LOV photoreceptors to anionic membrane phospholipids. *Proc. Natl. Acad. Sci. U.S.A.* 115 (33): E7720–E7727.

131 Berlew, E.E. et al. (2020). Optogenetic Rac1 engineered from membrane lipid-binding RGS-LOV for inducible lamellipodia formation. *Photochem. Photobiol. Sci.* 19 (3): 353–361.

132 Hannanta-Anan, P., Glantz, S.T., and Chow, B.Y. (2019). Optically inducible membrane recruitment and signaling systems. *Curr. Opin. Struct. Biol.* 57: 84–92.

133 Stewart, A. and Fisher, R.A. (2015). Introduction: G protein-coupled receptors and RGS proteins. *Prog. Mol. Biol. Transl. Sci.* 133: 1–11.

134 Putyrski, M. and Schultz, C. (2011). Switching heterotrimeric G protein subunits with a chemical dimerizer. *Chem. Biol.* 18 (9): 1126–1133.

135 Koch, W.J. et al. (1994). Cellular expression of the carboxyl terminus of a G protein-coupled receptor kinase attenuates $G_{\beta\gamma}$-mediated signaling. *J. Biol. Chem.* 269 (8): 6193–6197.

136 Klayman, L.M. and Wedegaertner, P.B. (2017). Inducible inhibition of Gβγ reveals localization-dependent functions at the plasma membrane and Golgi. *J. Biol. Chem.* 292 (5): 1773–1784.

137 Bailes, H.J. et al. (2017). Optogenetic interrogation reveals separable G-protein-dependent and -independent signalling linking G-protein-coupled receptors to the circadian oscillator. *BMC Biol.* 15 (1): 40.

138 Shenoy, S.K. et al. (2006). β-Arrestin-dependent, G protein-independent ERK1/2 activation by the beta2 adrenergic receptor. *J. Biol. Chem.* 281 (2): 1261–1273.

139 Stierl, M. et al. (2011). Light modulation of cellular cAMP by a small bacterial photoactivated adenylyl cyclase, bPAC, of the soil bacterium *Beggiatoa*. *J. Biol. Chem.* 286 (2): 1181–1188.

140 Frank, J.A. et al. (2016). Photoswitchable diacylglycerols enable optical control of protein kinase C. *Nat. Chem. Biol.* 12 (9): 755–762.

141 Lichtenegger, M. et al. (2018). An optically controlled probe identifies lipid-gating fenestrations within the TRPC3 channel. *Nat. Chem. Biol.* 14 (4): 396–404.

142 Leinders-Zufall, T. et al. (2018). PhoDAGs enable optical control of diacylglycerol-sensitive transient receptor potential channels. *Cell Chem. Biol.* 25 (2): 215–223 e3.

143 Smedler, E. and Uhlen, P. (2014). Frequency decoding of calcium oscillations. *Biochim. Biophys. Acta* 1840 (3): 964–969.

144 Fukuda, N., Matsuda, T., and Nagai, T. (2014). Optical control of the Ca^{2+} concentration in a live specimen with a genetically encoded Ca^{2+}-releasing molecular tool. *ACS Chem. Biol.* 9 (5): 1197–1203.

145 Bohineust, A. et al. (2020). Optogenetic manipulation of calcium signals in single T cells in vivo. *Nat. Commun.* 11 (1): 1143.

146 Toettcher, J.E. et al. (2011). The promise of optogenetics in cell biology: interrogating molecular circuits in space and time. *Nat. Methods* 8 (1): 35–38.
147 Karunarathne, W.K., O'Neill, P.R., and Gautam, N. (2015). Subcellular optogenetics – controlling signaling and single-cell behavior. *J. Cell Sci.* 128 (1): 15–25.
148 Zhang, K. and Cui, B. (2015). Optogenetic control of intracellular signaling pathways. *Trends Biotechnol.* 33 (2): 92–100.
149 Rost, B.R. et al. (2017). Optogenetic tools for subcellular applications in neuroscience. *Neuron* 96 (3): 572–603.
150 Chiariello, M. et al. (2010). Activation of Ras and Rho GTPases and MAP kinases by G-protein-coupled receptors. *Methods Mol. Biol.* 661: 137–150.
151 Toettcher, J.E., Weiner, O.D., and Lim, W.A. (2013). Using optogenetics to interrogate the dynamic control of signal transmission by the Ras/Erk module. *Cell* 155 (6): 1422–1434.
152 Marei, H. and Malliri, A. (2017). GEFs: dual regulation of Rac1 signaling. *Small GTPases* 8 (2): 90–99.
153 Goldsmith, Z.G. and Dhanasekaran, D.N. (2007). G protein regulation of MAPK networks. *Oncogene* 26 (22): 3122–3142.
154 Patel, A.L. et al. (2019). Optimizing photoswitchable MEK. *Proc. Natl. Acad. Sci. U.S.A.* 116 (51): 25756–25763.
155 Johnson, H.E. and Toettcher, J.E. (2018). Illuminating developmental biology with cellular optogenetics. *Curr. Opin. Biotechnol.* 52: 42–48.
156 Harris, J.M. et al. (2020). Long-range optogenetic control of axon guidance overcomes developmental boundaries and defects. *Dev. Cell* 53 (5): 577–588.e7.
157 Prosseda, P.P. et al. (2020). Optogenetic stimulation of phosphoinositides reveals a critical role of primary cilia in eye pressure regulation. *Sci. Adv.* 6 (18): eaay8699.
158 O'Neill, P.R., Kalyanaraman, V., and Gautam, N. (2016). Subcellular optogenetic activation of Cdc42 controls local and distal signaling to drive immune cell migration. *Mol. Biol. Cell* 27 (9): 1442–1450.
159 de Beco, S. et al. (2018). Optogenetic dissection of Rac1 and Cdc42 gradient shaping. *Nat. Commun.* 9 (1): 4816.
160 O'Neill, P.R. et al. (2014). The structure of dynamic GPCR signaling networks. *Wiley Interdiscip. Rev. Syst. Biol. Med.* 6 (1): 115–123.
161 Grundmann, M. and Kostenis, E. (2017). Temporal bias: time-encoded dynamic GPCR signaling. *Trends Pharmacol. Sci.* 38 (12): 1110–1124.
162 Graziano, B.R. et al. (2017). A module for Rac temporal signal integration revealed with optogenetics. *J. Cell Biol.* 216 (8): 2515–2531.
163 Brandman, O. and Meyer, T. (2008). Feedback loops shape cellular signals in space and time. *Science* 322 (5900): 390–395.

164 Lamas, I. et al. (2020). Optogenetics reveals Cdc42 local activation by scaffold-mediated positive feedback and Ras GTPase. *PLoS Biol.* 18 (1): e3000600.

165 Iglesias, P.A. and Devreotes, P.N. (2012). Biased excitable networks: how cells direct motion in response to gradients. *Curr. Opin. Cell Biol.* 24 (2): 245–253.

166 Xiong, Y.A. et al. (2010). Cells navigate with a local-excitation, global-inhibition-biased excitable network. *Proc. Natl. Acad. Sci. U.S.A.* 107 (40): 17079–17086.

167 Oakes, P.W. et al. (2017). Optogenetic control of RhoA reveals zyxin-mediated elasticity of stress fibres. *Nat. Commun.* 8: 15817.

168 Shellard, A. et al. (2018). Supracellular contraction at the rear of neural crest cell groups drives collective chemotaxis. *Science* 362 (6412): 339–343.

169 Valon, L. et al. (2017). Optogenetic control of cellular forces and mechanotransduction. *Nat. Commun.* 8: 14396.

170 Castillo-Badillo, J.A. et al. (2020). SRRF-stream imaging of optogenetically controlled furrow formation shows localized and coordinated endocytosis and exocytosis mediating membrane remodeling. *ACS Synth. Biol.* 9 (4): 902–919.

171 Kakumoto, T. and Nakata, T. (2013). Optogenetic control of PIP_3: PIP_3 is sufficient to induce the actin-based active part of growth cones and is regulated via endocytosis. *PLoS One* 8 (8): e70861.

172 van Bergeijk, P. et al. (2015). Optogenetic control of organelle transport and positioning. *Nature* 518 (7537): 111–114.

173 Guo, J. et al. (2019). Primary cilia signaling promotes axonal tract development and is disrupted in Joubert syndrome-related disorders models. *Dev. Cell* 51 (6): 759–774 e5.

174 Murakoshi, H. et al. (2017). Kinetics of endogenous CaMKII required for synaptic plasticity revealed by optogenetic kinase inhibitor. *Neuron* 94 (1): 37–47.e5.

175 Hollos, P. et al. (2020). Optogenetic control of spine-head JNK reveals a role in dendritic spine regression. *eNeuro* 7 (1): ENEURO.0303-19.2019.

176 Dietz, D.M. et al. (2012). Rac1 is essential in cocaine-induced structural plasticity of nucleus accumbens neurons. *Nat. Neurosci.* 15 (6): 891–896.

177 Hayashi-Takagi, A. et al. (2015). Labelling and optical erasure of synaptic memory traces in the motor cortex. *Nature* 525 (7569): 333–338.

178 Oh, E. et al. (2010). Substitution of 5-HT_{1A} receptor signaling by a light-activated G protein-coupled receptor. *J. Biol. Chem.* 285 (40): 30825–30836.

179 Li, P. et al. (2015). Optogenetic activation of intracellular adenosine A_{2A} receptor signaling in the hippocampus is sufficient to trigger CREB phosphorylation and impair memory. *Mol. Psychiatry* 20 (11): 1339–1349.

180 Barish, P.A. et al. (2013). Design and functional evaluation of an optically active μ-opioid receptor. *Eur. J. Pharmacol.* 705 (1–3): 42–48.

181 van Wyk, M. et al. (2015). Restoring the ON switch in blind retinas: Opto-mGluR6, a next-generation, cell-tailored optogenetic tool. *PLoS Biol.* 13 (5): e1002143.

182 Stachniak, T.J., Ghosh, A., and Sternson, S.M. (2014). Chemogenetic synaptic silencing of neural circuits localizes a hypothalamus→midbrain pathway for feeding behavior. *Neuron* 82 (4): 797–808.

183 Mahn, M. et al. (2021). Efficient optogenetic silencing of neurotransmitter release with a mosquito rhodopsin. *Neuron* 109 (10): 1621–1635.e8.

184 Copits, B.A. et al. (2021). A photoswitchable GPCR-based opsin for presynaptic inhibition. *Neuron* 109 (11): 1791–1809.e11.

185 Liu, Q. et al. (2019). A photoactivatable botulinum neurotoxin for inducible control of neurotransmission. *Neuron* 101 (5): 863–875.e6.

186 Mahn, M. et al. (2016). Biophysical constraints of optogenetic inhibition at presynaptic terminals. *Nat. Neurosci.* 19 (4): 554–556.

187 Mahn, M. et al. (2018). High-efficiency optogenetic silencing with soma-targeted anion-conducting channelrhodopsins. *Nat. Commun.* 9 (1): 4125.

188 Messier, J.E. et al. (2018). Targeting light-gated chloride channels to neuronal somatodendritic domain reduces their excitatory effect in the axon. *Elife* 7: e38506.

189 Raimondo, J.V. et al. (2012). Optogenetic silencing strategies differ in their effects on inhibitory synaptic transmission. *Nat. Neurosci.* 15 (8): 1102–1104.

190 Wiegert, J.S. et al. (2017). Silencing neurons: tools, applications, and experimental constraints. *Neuron* 95 (3): 504–529.

191 Plouffe, B., Thomsen, A.R.B., and Irannejad, R. (2020). Emerging role of compartmentalized G protein-coupled receptor signaling in the cardiovascular field. *ACS Pharmacol. Transl. Sci.* 3 (2): 221–236.

192 Azpiazu, I. et al. (2006). G protein βγ11 complex translocation is induced by G_i, G_q and G_s coupling receptors and is regulated by the α subunit type. *Cell. Signal.* 18 (8): 1190–1200.

193 Irannejad, R. et al. (2017). Functional selectivity of GPCR-directed drug action through location bias. *Nat. Chem. Biol.* 13 (7): 799–806.

194 Stoeber, M. et al. (2018). A genetically encoded biosensor reveals location bias of opioid drug action. *Neuron* 98 (5): 963–976 e5.

195 Jong, Y.J., Kumar, V., and O'Malley, K.L. (2009). Intracellular metabotropic glutamate receptor 5 (mGluR5) activates signaling cascades distinct from cell surface counterparts. *J. Biol. Chem.* 284 (51): 35827–35838.

196 Nash, C.A. et al. (2019). Golgi localized β_1-adrenergic receptors stimulate Golgi PI4P hydrolysis by PLCε to regulate cardiac hypertrophy. *Elife* 8: e48167.

197 Cho, J.H. et al. (2011). Alteration of Golgi structure in senescent cells and its regulation by a G protein γ subunit. *Cell. Signal.* 23 (5): 785–793.

198 Garcia-Marcos, M. et al. (2020). Optogenetic activation of heterotrimeric G_i proteins by LOV2GIVe—a rationally engineered modular protein. *bioRxiv* 2020.08.17.253781.

199 Yu, D. et al. (2019). Optogenetic activation of intracellular antibodies for direct modulation of endogenous proteins. *Nat. Methods* 16 (11): 1095–1100.

200 Gil, A.A. et al. (2020). Optogenetic control of protein binding using light-switchable nanobodies. *Nat. Commun.* 11 (1): 4044.

201 Redchuk, T.A. et al. (2020). Optogenetic regulation of endogenous proteins. *Nat. Commun.* 11 (1): 605.

202 McMahon, C. et al. (2018). Yeast surface display platform for rapid discovery of conformationally selective nanobodies. *Nat. Struct. Mol. Biol.* 25 (3): 289–296.

203 Kim, T.-i. et al. (2013). Injectable, cellular-scale optoelectronics with applications for wireless optogenetics. *Science* 340 (6129): 211–216.

204 Park, S.I. et al. (2015). Soft, stretchable, fully implantable miniaturized optoelectronic systems for wireless optogenetics. *Nat. Biotechnol.* 33 (12): 1280–1286.

26

GPCR Signaling in Nanodomains: Lessons from Single-Molecule Microscopy

Davide Calebiro[1,2], Jak Grimes[1,2], Emma Tripp[1,2], and Ravi Mistry[1,2]

[1] *College of Medical and Dental Sciences, Institute of Metabolism and Systems Research, University of Birmingham, Birmingham, UK*
[2] *Centre of Membrane Proteins and Receptors (COMPARE), Universities of Nottingham and Birmingham, UK*

26.1 Introduction

G protein-coupled receptors (GPCRs) are the largest family of membrane receptors in the human genome and are encoded by approximately 800 different genes [1]. These seven-transmembrane domain receptors act as signal transducers to a wide range of stimuli, including light, odors, hormones, and neurotransmitters [1, 2]. As GPCRs are fundamental for physiological homeostasis and are implicated in several pathophysiological processes, it comes as no surprise that they are targeted by at least a third of all FDA approved drugs [2].

26.2 The Basic Mechanisms of GPCR Signaling

GPCR signaling has been heavily researched, initially using classical biochemical and pharmacological approaches. Pioneering studies initiated by Earl Sutherland, Ted Rall, Alfred G. Gilman, and Martin Rodbell, among others, were instrumental in delineating the main events involved in GPCR signaling at the plasma membrane [3–6]. This research formed the basis of what is now known as the central mechanism of GPCR signaling, which involves an agonist, a receptor, a heterotrimeric G protein (composed of Gα, Gβ, and Gγ subunits), and an effector.

Upon binding of an agonist to a GPCR, the receptor adopts a conformation which has a higher affinity for heterotrimeric G proteins, leading to G protein recruitment. The agonist-occupied receptor then acts as a guanine nucleotide exchange factor (GEF), promoting the exchange of a GDP for a GTP molecule at

GPCRs as Therapeutic Targets, Volume 2, First Edition. Edited by Annette Gilchrist.

the Gα subunit, thereby activating the G protein. This is followed by dissociation of the GTP-bound Gα subunit from the receptor and its separation from the Gβγ dimer, with both the Gα and Gβγ units interacting with various downstream effectors [7]. The nature of the activated effector mainly depends on the identity of the Gα subunit of which there are at least 18 isoforms expressed in mammalian cells [8]. These isoforms can be categorized into four families: $G\alpha_s$, which activates adenylyl cyclases; $G\alpha_{i/o}$, which inhibit the activity of adenylyl cyclases; $G\alpha_{q/11}$, which activates the β-isoforms of phospholipase C (PLC-β); and $G\alpha_{12/13}$, which modulates the activity of RhoGEFs [8]. Although multiple isoforms exist for the Gβ and Gγ subunits, 5 and 12, respectively, these different isoforms appear to have similar functions once assembled into the Gβγ dimer, which modulates the activity of G protein-dependent inwardly rectifying K^+ (GIRK) channels, voltage-gated Ca^{2+} channels, adenylyl cyclases, and PLC-β [8].

Since the discovery of the central components of the GPCR signaling machinery, various hypotheses and models have been formulated to rationalize how these proteins interact to mediate their effects [9–12]. This work led to the development of the ternary complex model [13, 14]. This model posits that receptors exist in equilibrium between an active conformation, which is stabilized by agonists, and an inactive conformation, which is stabilized by inverse agonists. In its extended formulation, the ternary complex model predicts that both agonist-bound and free receptors can interact with G proteins, albeit with differing probabilities [14]. This suggests the possibility for receptor and G protein to form complexes in the absence of agonist, i.e. to be "pre-coupled." Although the development of the ternary complex model marked a major achievement, this model theoretically allows for very different scenarios depending on the frequency and stability of receptor-G protein interactions in a given system. These scenarios range from stable, precoupled receptor-G protein complexes, potentially capable of rapid and localized signals, to highly transient receptor-G protein complexes, which would maximize signal amplification. Moreover, the ternary complex and similar models do not take into account the complex organization of the plasma membrane and its impact on receptor-G protein signaling [15].

Alongside the early studies into receptor-G protein interactions, research was carried out to understand how GPCR-mediated signaling is regulated. Work from Hermann Kühn's and Robert Lefkowitz's groups revealed the fundamental mechanism behind GPCR homologous desensitization [16–20]. After prolonged stimulation, the active, agonist-bound GPCRs are phosphorylated by specialized GPCR kinases (GRKs), increasing the affinity of the receptors for arrestins [21]. Visual GPCRs interact with visual arrestins (arrestin-1 and 4), while all other GPCRs interact with β-arrestins (arrestin-2 and 3, also known as β-arrestin-1 and 2) [21]. Binding of an arrestin molecule to the core of an active GPCR prevents G protein binding to the same receptor, therefore, inhibiting the activation of downstream effectors [21].

Subsequent studies showed that, besides their involvement in signal desensitization, β-arrestins also play fundamental roles in receptor internalization and G protein-independent signaling [22, 23]. The role of β-arrestins in receptor internalization is mediated via their interactions with clathrin and the adaptor protein complex AP2, which promote receptor recruitment to clathrin-coated pits [24]. The receptor-containing pits are then internalized to endosomal compartments, from where the receptors are either recycled back to the plasma membrane in a dephosphorylated state, or trafficked to lysosomes for degradation [24]. The role of β-arrestins in G protein-independent signaling involves their interactions with several components of mitogen-associated protein kinase (MAPK) cascades [24]. By acting as scaffolds for MAPK components, arrestins promote MAPK signaling independently of G protein activation.

26.3 The Structural Basis for GPCR Signaling

In the past two decades, there has been tremendous progress toward understanding the molecular mechanisms of GPCR activation thanks to the attainment of high resolution, three-dimensional structures of GPCRs in different conformations [25, 26]. These structures include free receptors [27], those bound to agonists [28–30] or antagonists [31–33], and receptors in complex with G proteins [34–42], arrestins [43–46] and even receptor-G protein-arrestin supercomplexes [47].

Comparisons of the active and inactive structures of GPCRs have revealed the main conformational changes that are associated with receptor activation. Upon agonist binding, conformational changes within the transmembrane segments (TMs) cause an opening of the intracellular portion of the receptor, revealing a pocket for G proteins and arrestins to bind [7]. These changes include an outward movement up to 14 Å of the intracellular end of TM6 (Figure 26.1).

Additionally, recent structural data have revealed the conformational changes that occur in both G proteins and arrestins upon their interaction with an agonist-bound receptor. After being recruited to an active receptor, a major rotation of the helical domain relative to the GTPase domain occurs in G proteins [34, 40, 41], which is associated with their activation (Figure 26.1). The binding of arrestins to agonist occupied receptors appears to be more complex, possibly occurring in a multistep process. Arrestins are hypothesized to initially interact with the plasma membrane and the phosphorylated C terminus of the receptor. These interactions are then believed to trigger a series of conformational changes in the arrestin molecule [45, 46]. These conformational changes promote arrestin

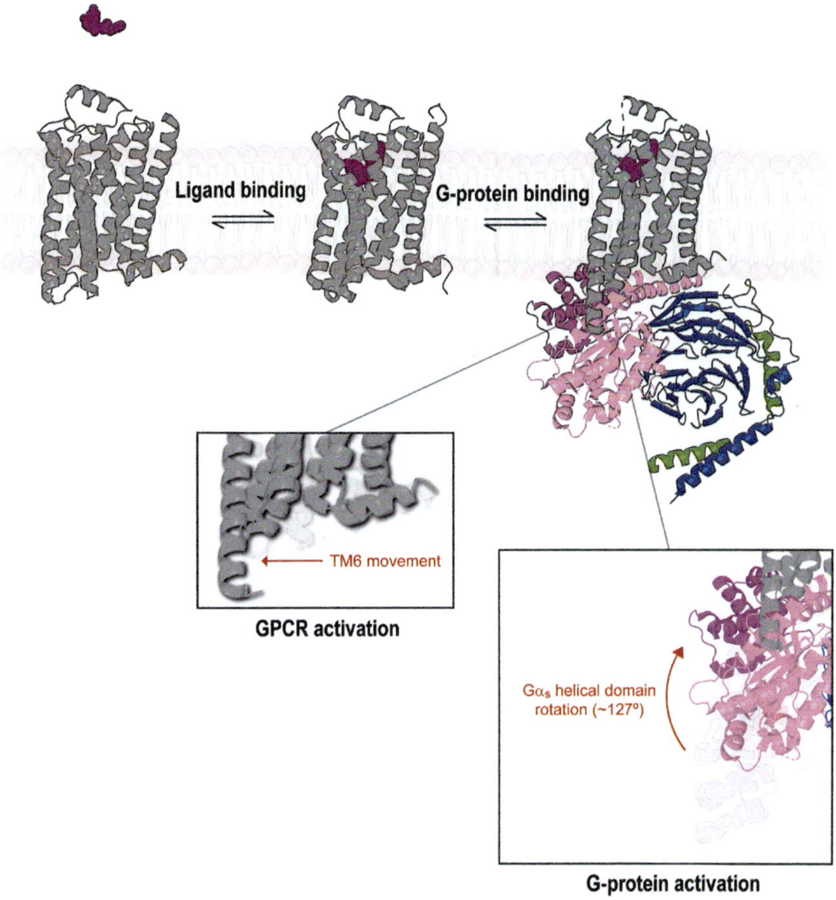

Figure 26.1 Conformational changes during the formation of an active receptor-G protein complex. GPCRs exist in dynamic equilibrium, transitioning among multiple conformations prior to ligand binding. Upon binding of a ligand to a GPCR, a series of conformational changes shift this equilibrium toward a partially active state, revealing an intracellular binding pocket for G proteins. The agonist-bound conformation represents an intermediate state in the GPCR activation process, with full activation requiring both agonist and G protein binding. The fully active state is characterized by a major, outward movement of transmembrane-domain 6 (TM6) in the receptor, and a major rotation in the helical domain of the Gα subunit in the G protein.

interactions with the receptor core, ultimately resulting in the formation of a fully engaged receptor-arrestin complex [45, 46].

Interestingly, comparison of these structures has revealed that agonist-bound receptors represent an intermediate conformation between the inactive and active state, with full receptor activation requiring both agonist and G protein or arrestin binding [28]. This view is further supported by biophysical studies on purified proteins [48–50]. These studies revealed that GPCRs adopt multiple conformations that are in dynamic equilibrium and are selectively stabilized by ligand binding and G protein coupling [7].

26.4 Emerging Concepts in GPCR Signaling

Although GPCRs have been extensively researched for nearly 70 years, recent studies have presented evidence that challenge previously widely accepted concepts in GPCR signaling. These studies owe part of their success to the introduction of innovative biophysical techniques that allow researchers to monitor GPCR signaling in living cells. Two such methods are fluorescence resonance energy transfer (FRET) and bioluminescence resonance energy transfer (BRET) [51].

One concept that has emerged from these studies is biased signaling [52]. The idea behind biased signaling is that different ligands for the same receptor might differentially activate or inhibit certain signaling pathways over others. This phenomenon has been linked to the ability of GPCRs to assume multiple conformations, each preferentially stabilized by a specific ligand and coupled to a distinct downstream signaling pathway. This model is supported by the results of biophysical and structural studies conducted on purified proteins and living cells [53–59].

Another intriguing concept is that of signal compartmentalization. The hypothesis of signal compartmentalization dates back at least to the late 1970s, when researchers compared the response of heart tissue to β-adrenergic or prostaglandin E1 (PGE1) receptor stimulation [60, 61]. Although both treatments were shown to cause similar increases in the concentration of intracellular cAMP, it was observed that only β-adrenergic stimulation activates glycogen phosphorylase, increases heart contractility, and induces troponin I phosphorylation [60, 61]. Later, β-adrenergic stimulation was found to selectively increase cAMP and protein kinase A (PKA) activity in the particulate fraction of rabbit heart lysates [62]. Based on these findings, Buxton and Brunton proposed that β-adrenergic and PGE1 receptors might induce cAMP/PKA signaling in separate subcellular compartments within cardiomyocytes, thus leading to distinct biological responses [63]. However, direct evidence for the occurrence of signal compartmentalization has only been obtained some 20 years later with the introduction of FRET biosensors to monitor cAMP levels and PKA activity in living cells [64, 65]. These approaches have revealed the presence of cAMP nanodomains in both neurons [66, 67] and cardiomyocytes [68–70].

More recently, ensemble FRET/BRET methods have been instrumental in showing that GPCRs can remain active after internalization from the plasma membrane and are able to signal via G protein-dependent pathways from intracellular sites, such as the Golgi/trans-Golgi network or the endosomal compartment [71–74]. Intriguingly, recent evidence suggests that endosomal GPCR signaling may involve the formation of a super-complex containing a receptor, a β-arrestin, and a heterotrimeric G protein [47, 75]. In addition to these direct observations of active receptor signaling at endosomes and the Golgi/trans-Golgi network, there is growing evidence to suggest that GPCRs can signal at other intracellular sites, including the nucleus [76–78], mitochondria [79–81], and the endoplasmic reticulum [82].

Taken together, these new concepts have profoundly changed our understanding of GPCR signaling. GPCRs should no longer be viewed as simple on-off switches signaling exclusively at the plasma membrane, but, rather, as complex and highly dynamic molecular machines capable of signaling in distinct subcellular compartments. While our knowledge of these fundamental processes expands, it has also become increasingly apparent that new and better techniques are required to further investigate GPCR signaling in living cells. In this context, single-molecule microscopy methods have recently emerged as a powerful approach to study GPCRs in intact cells with unprecedented spatiotemporal resolution.

26.5 Single-Molecule Microscopy

In a notable 1959 lecture, Richard Feynman proposed the visionary idea of directly observing individual molecules [83]. However, it was not until the work of Thomas Hirschfield in 1976 that initial steps toward the detection of single particles were taken [84]. By attaching approximately 80–100 molecules of a fluorescent dye to single γ globulin molecules, Hirschfield was able to both visually observe and photometrically detect individual γ globulin molecules in solution [84], proving that visualizing single molecules was indeed possible. A major breakthrough took place in 1989, when William Moerner and Lothar Kador were able to detect the absorption spectrum of single dopant molecules of pentacene in *p*-terphenyl crystals [85]. This was expanded upon a year later by Michel Orrit and Jacky Bernard, who used a similar approach to detect single pentacene molecules using fluorescence excitation [86]. The same year marked the first case of efficient detection of single fluorophore molecules in solution [87], while the first image of individual fluorophore molecules was achieved soon after using near-field scanning optical microscopy [88].

Over the last two decades, the development of more sensitive cameras, the synthesis of brighter fluorophores, and progress in image analysis have made visualizing and manipulating individual molecules accessible to an expanding research community. Thanks to these improvements, researchers can now investigate individual receptors and other biologically relevant proteins in solution, reconstituted systems and, more importantly, in living cells or tissues [89].

Several labeling strategies are commonly used for single-molecule microscopy, including the incubation of proteins of interest with fluorescently labeled antibodies, binding of fluorescent ligands, or fusion with genetically encoded fluorescent proteins. A major advance in the field has been the introduction of so-called "self-labeling tags" such as SNAP or CLIP, which can be genetically fused to a protein of interest and react specifically with small, organic fluorescent substrates. SNAP is a 20 kDa protein tag derived from the human DNA repair enzyme O^6-alkylguanine-DNA alkyltransferase and forms irreversible adducts with fluorophore-conjugated benzylguanine derivatives [90]. CLIP, a further engineered tag derived from SNAP, can instead be labeled with a range of fluorophore-conjugated benzylcytosine derivatives [91]. Self-labeling tags offer multiple benefits over fluorescent proteins, including the flexibility, brightness, and photostability resulting from the use of organic fluorescent substrates.

In a typical single-molecule experiment, the molecules of interest are fluorescently labeled and imaged on a fast, highly sensitive camera. In the case of membrane molecules, this is usually achieved via total internal reflection fluorescence (TIRF) illumination. This method was originally developed by E. J. Ambrose [92] and was further expanded by D. Axelrod, who demonstrated its use to selectively observe fluorescently labeled molecules on the plasma membrane of cells [93]. TIRF exploits the evanescent excitation field that is generated when light is reflected at the interface between two media that have different refractive indices. This is, for instance, the case of a cell growing on a glass coverslip. Since the intensity of the evanescent field decreases exponentially with the distance from the interface, fluorophores located on the cell surface are selectively illuminated without interference from the rest of the cell. The resulting low background and favorable signal-to-noise ratio compared with epifluorescence facilitate the single-molecule visualization of membrane molecules (Figure 26.2).

The obtained TIRF images contain diffraction-limited spots, each corresponding to a single fluorescently labeled molecule. Thereafter, automated computer algorithms are used to detect the diffraction-limited spots and determine their position. Although the diffraction-limited spots are usually much larger (200–300 nm) than the actual size of the molecules under investigation (2–4 nm for a typical membrane protein), their position can be determined with nanometer precision by localizing their center of mass. This is generally achieved by fitting the pixel intensities of each diffraction-limited spot with a two-dimensional Gaussian distribution (Figure 26.2). The precision of this localization procedure is influenced by multiple factors. These include the wavelength of the emitted light and, importantly, the number of photons emitted and detected for a single fluorophore. While there is no theoretical limit to the

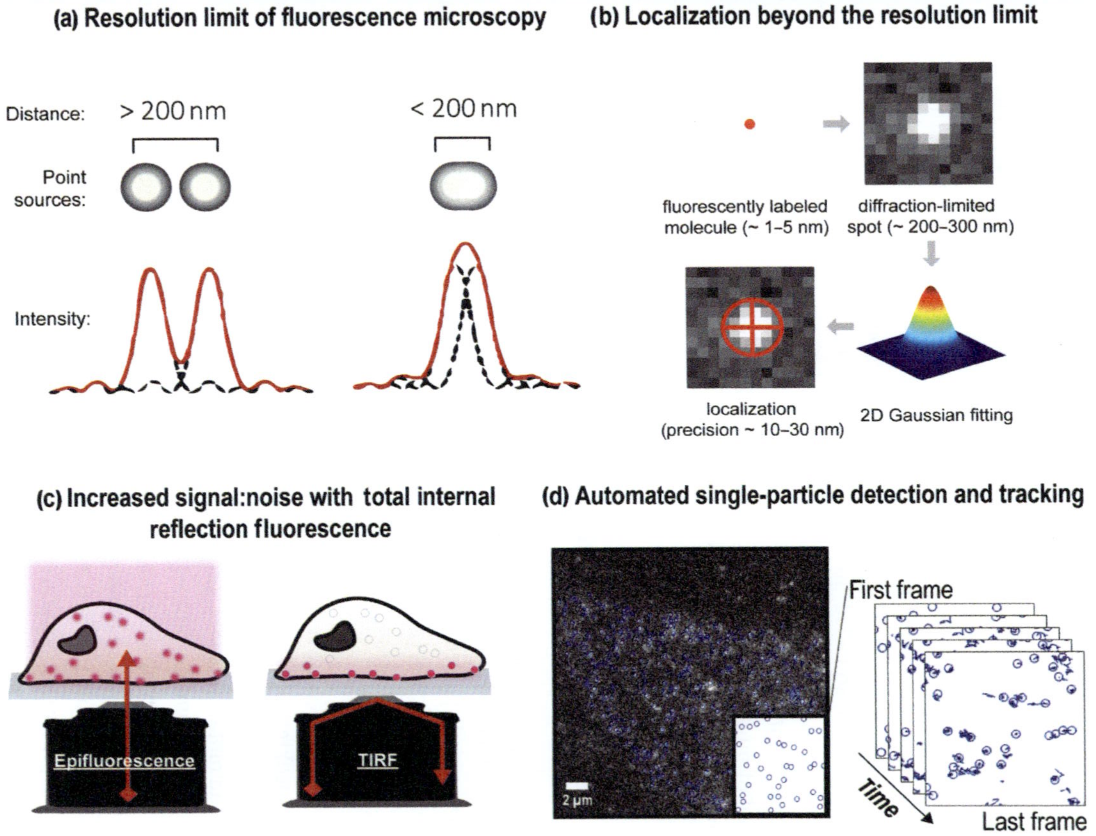

Figure 26.2 Studying membrane proteins by single-molecule microscopy. (a) The resolution limit of conventional fluorescence microscopy. Two diffraction-limited objects, such as two fluorophores, cannot be separated when they are within approximately 200–300 nm of each other. (b) Resolution improvement through single-molecule localization. Individual molecules can be localized with a precision of approximately 10–30 nm. (c) Comparison between epifluorescence and TIRF. (d) Single particle tracking. Shown are individual GPCRs on the surface of a living cell and their corresponding trajectories in blue.

localization precision of single-molecule microscopy, values of 10–30 nm are typically attained with currently available fluorophores and technology [94, 95]. This is approximately ten times higher than the resolution limit of conventional fluorescence microscopy methods, such as confocal microscopy, and approaches the typical size of membrane proteins. A limitation of single-molecule microscopy is that it can only analyze fluorescent molecules present at low concentrations or densities, typically less than 1 molecule/μm^2 for membrane proteins, as higher concentrations/densities impede their correct localization [15, 94, 96].

In the case of live cell imaging, automated single-particle tracking algorithms are then used to link the particles detected in an image sequence and thus reconstruct their trajectories over time (Figure 26.2). A computationally expensive but powerful approach is so-called "multiple-hypothesis" tracking [97]. This approach involves assigning a cost to all possible events a particle can undertake at a given frame. These events include particle appearance, disappearance, and the merging or splitting of multiple particles. By minimizing the sum of all costs, the global solution with the highest probability is selected.

26.6 Applications of Single-Particle Tracking

An important application of single-particle tracking is the study of the diffusion of fluorescently labeled molecules. A widely used approach relies on calculating the mean square displacement (*MSD*) of the tracked particles, which is related to their diffusion coefficient (*D*) via the following equation:

$$MSD = 4Dt^{\alpha}$$

where t denotes time, and α is the anomalous diffusion exponent.

Broadly speaking, the MSD describes how effectively a given molecule explores space. In the case of normal diffusion, as in the case of molecules undergoing simple Brownian motion, the MSD grows linearly with the time between two observation points ($\alpha = 1$) and the diffusion coefficient can be estimated directly from the slope of the MSD. Deviations from this linear relationship are suggestive of either superdiffusion ($\alpha > 1$), as in the case of directional motion, or subdiffusion ($\alpha < 1$), as observed for confined molecules [98, 99].

Another important application of single-particle tracking is the study of the stoichiometry of protein complexes. Two main approaches are commonly used for this purpose, the first of which is based on analyzing the distribution of the intensities of the detected particles. In the case of a mixture of monomers, dimers, and/or oligomers, the particle intensities are expected to have a distribution corresponding to the sum of Gaussians with increasing mean values. The relative abundance of the individual components can therefore be estimated by performing a mixed

Gaussian fitting. Alternatively, the number of fluorophores, and hence molecules, present in a given spot can be estimated by following their photobleaching over time. Since the photobleaching of each individual fluorophore produces a characteristic step in the intensity profiles, the initial number of fluorescent molecules can be estimated by counting the number of bleaching steps. Appropriate controls, including individual fluorophores as well as monomeric and dimeric proteins, are usually used to calibrate and validate these analyses [94].

In addition, single-particle tracking is emerging as a powerful method to precisely characterize dynamic protein–protein interactions, such as those occurring among receptors and other membrane proteins. Since this approach allows researchers to simultaneously estimate the density of the labeled molecules and both the frequency and duration of their interactions, it can be used to derive microscopic association and dissociation rate constants [100]. A possible complication of this approach is the simultaneous occurrence of random colocalizations. However, it has been shown that a method based on the deconvolution of the observed colocalization durations with those due to random colocalization can be used to correct for this factor and efficiently estimate the durations of the underlying interactions [95].

26.7 Single-Molecule Localization Super-Resolution Microscopy Methods

Instead of separating molecules in space, super-resolution microscopy techniques, such as direct stochastic optical reconstruction microscopy (*d*STORM) and photoactivated localization microscopy (PALM), rely on the temporal separation of fluorophores obtained by stochastically activating only a subset of them at a time.

*d*STORM exploits the spontaneous blinking of suitable organic fluorophores [101–104]. Under favorable environmental conditions, these fluorophores randomly switch between a dark and a fluorescent state. Since only a small fraction of fluorophores are visible at any given frame of an image sequence, their position can be determined with high accuracy by 2D Gaussian fitting. This process is repeated thousands of times, resulting in the reconstruction of high-resolution images. This makes *d*STORM particularly useful for imaging densely packed cellular structures such as the cytoskeleton. Since image acquisition typically takes several minutes, *d*STORM and similar methods have only limited temporal resolution and, as a result, are often applied to fixed samples. Additionally, the stochastic nature of *d*STORM often results in molecules being counted multiple times, complicating quantitative analyses. *d*STORM has been successfully employed to investigate the organization of GPCRs in cells and tissues. For instance, Szalai et al. employed *d*STORM to investigate corticotropin-releasing hormone type 1 receptors in the mouse hippocampus [105]. More recently, our group applied *d*STORM to study the nanoscale organization of metabotropic glutamate receptors in the mouse cerebellum [106]. This approach revealed that $mGluR_4$s are localized in small nanodomains within presynaptic active zones, where they are found in close proximity to voltage-dependent $Ca_v2.1$ channels and Munc18-1, a component of the secretory machinery. It is tempting to speculate that this tight spatial organization might be implicated in the fast inhibition of glutamate release by $mGluR_4$.

PALM is based on a concept similar to that of *d*STORM but utilizes photoswitchable or photo-convertible probes instead. These probes can be activated by excitation with low-intensity near-UV light [107]. The molecules carrying the probes are then localized and rapidly brought to a dark state, usually via photobleaching [107]. PALM has been employed to analyze GPCR clustering on the plasma membrane in a study by Scarselli et al. [108]. This study showed that β_2ARs form preassociated clusters on the membrane of cardiac myocytes with the participation of the actin cytoskeleton [108].

26.8 Single-Molecule FRET

FRET – often described as a spectroscopic ruler – is a physical phenomenon consisting in the nonradiative transfer of energy between a donor and an acceptor fluorophore located at close distance, generally less than 10 nm [51, 109]. Several additional conditions must be met for FRET to occur, including a favorable relative orientation of the donor and acceptor fluorophores and a sufficient overlap of their emission and excitation spectrum, respectively. The use of FRET to study single-molecules was first reported in 1996 by Ha et al. who proposed its application to study conformational, rotational, and distance changes between structures within a single molecule [110]. Since then, single-molecule FRET (smFRET) has been applied to study a variety of biological systems with purified and fluorescently labeled proteins *in vitro* [89]. In a typical smFRET experiment, purified receptors or other proteins of interest are labeled at two distinct sites via insertion of a suitable pair of small organic fluorophores. The proteins are subsequently immobilized on a glass coverslip, for instance by taking advantage of the avidin-biotin system. The conformational dynamics within individual proteins can then be followed in real time by monitoring FRET changes between the two fluorophores. For a comprehensive review see [111].

26.9 Fluorescence Correlation Spectroscopy

Fluorescence correlation spectroscopy (FCS) is a method complementary to single-molecule microscopy. FCS is based on the measurement of intensity fluctuations that occur as fluorescently labeled molecules diffuse across a small confocal detection volume. The obtained data are then used to perform a time-dependent autocorrelation analysis, which provides quantitative information about the average number and dwell time of the fluorescent molecules in the detection volume. This allows quantification of the concentration, stoichiometry, and diffusion of heterogeneous mixtures of fluorescent molecules. An extension of FCS is fluorescence cross-correlation spectroscopy (FCCS), which measures the cross correlation between two or more emission channels. FCCS finds an important application in the study of protein–protein interactions. These methods have been employed to investigate various aspects of GPCR signaling, including receptor diffusion and ligand-receptor interactions [112]. Several GPCRs including the β_2AR, adenosine A_1 and A_3 receptors, and the histamine H_1 receptor have been investigated by FCS [113–116]. These studies have revealed a high heterogeneity in the diffusion of receptors on the plasma membrane.

26.10 Single-Molecule Microscopy Versus Ensemble Methods

Single-molecule microscopy offers several advantages over ensemble methods such as FRET and BRET. An important limitation of ensemble FRET and BRET approaches is that they often require overexpression of the investigated molecules, which can potentially alter their properties [117]. Second, ensemble methods measure the average behavior of thousands or millions of molecules. Since these molecules are usually not synchronized, this precludes direct estimation of kinetic rates [100]. In contrast, single-molecule methods can directly visualize and investigate individual fluorescent molecules at physiological concentrations. This allows estimation of kinetic parameters, such as association (k_{on}) and dissociation (k_{off}) rate constants [95]. Moreover, it enables the analysis of rare events that are typically hidden in ensemble measurements. Finally, single-molecule microscopy has a spatial resolution of tens of nanometers, allowing researchers to directly study the nanoscale organization of the investigated molecules. However, single-molecule methods currently suffer from some drawbacks too, such as the requirement for complex and time-consuming computational analyses. Additionally, single-particle tracking methods are not well suited to investigate structures that contain high concentrations of fluorescently labeled molecules, as the localization of overlapping particles cannot be correctly determined.

26.11 Early Single-Molecule Studies

The first application of single-molecule microscopy to a biological question was the measurement of ATP turnover by individual myosin molecules in aqueous solution [118]. The researchers used fluorescently labeled ATP to observe its binding to immobilized myosin molecules that had been labeled with a second fluorophore [118]. Soon after, TIRF microscopy was utilized to study signaling events in living cells. Through the use of fluorescently labeled ligands and antibodies, Sako and colleagues were able to delineate the mechanisms behind epidermal growth factor (EGF) receptor (EGFR) activation [119]. Single-particle tracking revealed that upon binding of a single EGF molecule to a single EGFR, EGFR dimerization occurs and a resulting EGF-$(EGFR)_2$ complex is formed. This complex has a higher affinity for EGF than monomeric EGFR, and, therefore, a second EGF molecule binds preferentially to this complex as opposed to the EGFR monomer. These observations, alongside those showing that EGFR dimerization is concomitant with EGFR autophosphorylation, uncovered the early events of EGFR signal transduction [119].

The use of single-molecule microscopy for studying GPCRs was pioneered by work of Akihiro Kusumi's group. In what has now become a historic study, they successfully imaged individual μ-opioid receptors labeled with gold nanoparticles as they diffuse on the surface of living cells [120]. By imaging individual μ-opioid receptors at the impressive speed of one frame every 25 μs, the researchers found that the receptors exhibit rapid hop diffusion as they jump among small membrane compartments defined by the actin cytoskeleton, in which they are transiently confined. Along with similar results, these seminal studies led to the formulation of the fence-and-picket model of the plasma membrane [120, 121]. This model proposes that the plasma membrane is partitioned into small domains with a size of approximately 30–400 nm, defined by the subcortical cytoskeleton and integral membrane proteins bound to it, in which membrane proteins and lipids are transiently trapped.

26.12 Lessons from Single-molecule Microscopy *In Vitro*

Over the last decade, single-molecule methods have been successfully applied to investigate the conformational dynamics of solubilized GPCRs *in vitro*. One of the first studies investigated the β_2AR, which was site-specifically labeled at the cytoplasmic end of TM6 with a small organic fluorophore [122]. By analyzing photon-burst histograms, this study provided evidence that the β_2AR was in a dynamic equilibrium between at least two conformational states that increased to three upon agonist addition [122]. This idea was expanded upon by Bockenhauer et al. who used single-molecule measurements of individual β_2ARs immobilized in an anti-Brownian electrokinetic (ABEL) trap to provide further evidence that

GPCRs exist in a dynamic equilibrium among multiple, discrete states [123]. The exchange between these discrete states was further investigated by single molecule fluorescence spectroscopy on β_2ARs reconstituted in phospholipid nanodiscs and labeled with an environmentally sensitive fluorophore (Cy3) at the cytoplasmic end of TM6 [124]. In the absence of G proteins and ligands, receptors were found to mainly occupy an inactive state and spontaneously transition to and from an active state. Stimulation with a full agonist or an inverse agonist shifted the equilibrium toward the active or the inactive state, respectively. An analysis of the dwell times in the two states revealed complex transition kinetics, with faster and slower components [124].

Subsequently, Gregorio et al. used smFRET between two fluorophores placed at the cytoplasmic ends of TM4 and TM6 of the β_2AR to investigate its conformational dynamics and, importantly, how this is affected by ligand or G_s protein binding. In the absence of G protein, agonists with increasing efficacy were found to cause a progressive reduction of FRET efficiency, from which they estimated an average outward movement of the cytoplasmic end of TM6 of 4 Å caused by the full agonist epinephrine. The addition of G_s protein stabilized the receptor in a state characterized by an additional reduction of FRET efficiency, further supporting the view that G protein binding is required to stabilize the receptor into the fully active conformation observed in the crystallographic β_2AR-G_s protein complex [125].

In another interesting study, Cao et al. applied smFRET to investigate the conformational dynamics of the platelet activating factor receptor (PAFR) [126]. The PAFR is a widely expressed receptor that mediates the effects of platelet activating factor (PAF), a biologically active phospholipid that acts as a pro-inflammatory mediator and stimulates platelet aggregation. Because of this, PAFR is a potential target for the treatment of inflammation, asthma, and cardiovascular diseases [127]. By combining X-ray crystallography and smFRET on purified receptors, this study showed that PAFR adopts different conformational states depending on whether it is bound to an inverse agonist (ABT-491) or an antagonist (SR27417). The crystal structure of PAFR bound to SR271417 revealed an unusual orientation of TM2 and TM4, whereas the structure of PAFR bound to ABT-491 was closer to that of other inactive GPCR structures. The existence of distinct PAFR conformational states induced by the two ligands was further supported by the results of smFRET measurements [126].

In addition, smFRET has been used to investigate conformational rearrangements within family C GPCR dimers. Class C GPCRs are known to exist as dimers, a phenomenon essential for their correct trafficking, cell surface expression, and function [128–131]. This receptor subfamily includes metabotropic glutamate receptors (mGluRs), the calcium sensing receptor (Ca-SR), and the γ-aminobutyric acid B receptor ($GABA_BR$), which is an obligate heterodimer of $GABA_{B1}$, responsible for ligand binding, and $GABA_{B2}$, responsible for G protein coupling [129, 130]. mGluRs are involved in the modulation of neuronal excitability and synaptic plasticity, and are promising drug targets for neurological disorders such as epilepsy and depression [132, 133]. They are characterized by the presence of a large extracellular ligand-binding domain (LBD) that resembles a Venus Fly Trap. Vafabakhsh et al. investigated the conformational dynamics within the LBDs of individual $mGluR_2$ and $mGluR_3$ receptor dimers that were solubilized and immobilized on a glass substrate. This was achieved by monitoring smFRET between donor and acceptor fluorophores placed at the N-terminus of each LBD of the dimer. In the absence of glutamate, a high-FRET state was observed, consistent with the two N-termini being in close proximity. Agonist stimulation caused a concentration-dependent shift toward a low FRET state as well as the appearance of an intermediate state. Based on these results, the authors concluded that the LBDs of $mGluR_2$ and $mGluR_3$ exist in a dynamic equilibrium between an inactive, an active, and an intermediate, short-lived inactive conformation [134]. The same group then went on to show, again using smFRET, that mGluR homo/heterodimerization is mainly due to noncovalent interactions between the LBDs, rather than being largely mediated by a disulfide bridge between the LBDs, as previously thought. Moreover, these interactions between the LBDs were shown to prevent LBD closure in the absence of glutamate and, thus, mediate cooperativity between the two LBDs [135].

More recently, smFRET has been used to investigate the dimerization of family A GPCRs. Unlike family C GPCRs, these receptors were long believed to function as monomers. However, this view has been challenged by growing evidence obtained with biochemical methods and biophysical approaches in living cells. In an elegant study, Dijkman et al. used smFRET to investigate the homodimerization of the neurotensin-1 receptor (NST_1R) [136]. In this study, purified NST_1Rs labeled with either Cy3 (donor) or Cy5 (acceptor) fluorophores at the intracellular end of TM4, a region that has previously been suggested to be the dimer interface of the NST_1R [137], were reconstituted into a droplet interface bilayer. The receptors were then followed by single-particle tracking and smFRET was used to quantify receptor interactions. smFRET intensity distributions were found to be multimodal, indicating that the NST_1R assembled into multiple dimeric states. The observed interactions between NST_1Rs were transient, with an average half-life of 1.2 seconds, consistent with the results of previous single-molecule experiments in living cells on other GPCRs [94, 138, 139]. Based on these findings and those of double electron–electron resonance measurements, the authors concluded that NST_1Rs rapidly associate and dissociate into dimeric complexes which exist in multiple conformations.

26.13 Lessons from Single-molecule Microscopy in Living Cells

Besides *in vitro* studies with solubilized proteins, single-molecule microscopy has been successfully applied to investigate GPCR dimerization in intact cells [94, 138–141].

An initial study by Hern et al. investigated M_1 receptors expressed in CHO cells and labeled with a high affinity fluorescent antagonist (Cy3B-telenzepine). Using single-molecule microscopy and single-particle tracking in living cells, they found that M_1 receptors were apparently freely diffusing on the surface of CHO cells. Moreover, they concluded that approximately 30% of the receptors formed transient and reversible dimers with an average lifetime of 0.5 seconds at 23 °C. They found no evidence for receptor trimers or higher-order oligomers as no more than two photobleaching steps were observed [138].

Using a similar approach, Kasai et al. showed that the N-formyl peptide (FP) receptor forms transient dimers in living cells. In this study, which was the first to provide a full characterization of a GPCR monomer-dimer dynamic equilibrium, approximately 40% of the FP receptors were found to be present as dimers. On average, FP receptors were estimated to form dimers every 150 ms and separate every 90 ms at 37 °C, indicative of fast association and dissociation kinetics [139].

Fluorescent agonists and antagonists can be used to visualize endogenous receptors, but they do not allow for the investigation of unliganded receptors. Moreover, selective fluorescent ligands are available only for a subset of GPCRs. To circumvent these limitations, our group took advantage of direct labeling with small organic fluorophores via fusion of a SNAP tag to the receptor N-terminus. This approach allowed us to compare three prototypical GPCRs, the β_1AR, β_2AR, and $GABA_B$, by single-molecule microscopy [94]. When expressed in CHO cells, both β_1- and β_2AR were found to rapidly associate and dissociate to form transient dimers, and, to a lesser extent, higher-order oligomers. The fraction of receptors in the dimeric state increased with the density of receptors on the plasma membrane, consistent with receptors being in a dynamic monomer–dimer equilibrium. At comparable densities, the β_2AR had a higher propensity to dimerize than the β_1AR, and agonist stimulation did not affect the monomer–dimer equilibrium for either receptor. Consistent with previous observations, $GABA_B$ receptors were found to form larger supramolecular complexes, mainly consisting of two to four $GABA_{B1}$–$GABA_{B2}$ heterodimers. Whereas β_1- and β_2ARs were largely diffusing across the plasma membrane, $GABA_B$ receptors were mainly immobile and arranged into ordered arrays through interaction with the actin cytoskeleton [94].

Altogether, these studies shed important light on the nature of receptor dimerization by providing direct evidence for a highly dynamic nature of receptor–receptor interactions. However, the functional consequences of the observed transient receptor interactions remain poorly understood. In an attempt to address these open questions, Ge et al. recently applied single-molecule microscopy to investigate the consequences of G protein coupling on the dimerization of the chemokine receptor type 4 (CXCR4) [142]. In an experiment using pertussis toxin (PTX) to interfere with CXCR4 coupling to $G\alpha_{i/o}$ proteins, the group showed that PTX reduces agonist-dependent CXCR4 dimerization/oligomerization. Moreover, blocking receptor internalization with chlorpromazine increased the relative abundance of dimers and oligomers on the plasma membrane. Based on these results, the authors concluded that CXCR4 dimerization is influenced by G protein coupling and has an impact on CXCR4 internalization and, possibly, function [142].

Single-molecule methods have also been applied to fixed cells. While fixation can potentially introduce artifacts, it also simplifies the analysis by removing the requirement for single-particle tracking. Moreover, it allows the experiments to be performed under conditions that are not compatible with living cells. This has found an interesting application in cryogenic single-molecule microscopy [143]. By performing imaging at temperatures as low as that of liquid helium (4.3 K), it is possible to substantially improve the photochemistry of the employed fluorophores, which translates into a further improvement in spatial resolution. Using this approach, Tabor et al. interrogated the dimerization of SNAP-tagged dopamine D_2 receptors in fixed CHO cells, achieving a spatial resolution of less than 10 nm [109]. Consequently, D_2 receptors were imaged with a high enough resolution to measure the distance between the centers of mass of the two receptors in a dimer, which was estimated to be approximately 9 nm – a distance consistent with a physical interaction. D_2 receptor dimerization was further investigated in living cells, providing evidence that it is transient in nature [144], similar to what has been observed for other family A GPCRs.

26.14 Hot Spots for Receptor-G protein Interactions

Our group recently used fast, multicolor TIRF microscopy to image individual GPCRs and G proteins as they diffuse and interact on the surface of living cells [95]. In this study, adrenergic receptors (α_{2A}AR and β_2AR) and their cognate G proteins, G_i and G_s respectively, were imaged simultaneously with a spatiotemporal resolution of approximately 20 nm and 30 ms. The constructs were fused to SNAP and CLIP tags and subsequently labeled with a pair of bright, organic fluorophores. Experiments were performed in both CHO cells and human primary endothelial cells, where α_{2A}AR and β_2AR exert important physiological roles.

The study revealed a high heterogeneity and complexity in the diffusion of both receptors and G proteins on the plasma membrane, with both molecules frequently switching among phases of immobility, confinement, and fast diffusion [95]. The phases of virtual immobility and slow diffusion could be explained by transient trapping of receptors and G proteins in small areas of the plasma membrane. Simultaneous imaging of the cytoskeleton underneath the plasma membrane revealed that this phenomenon was at least partly due to barriers provided by the actin cytoskeleton, in agreement with the fence-and-picket model of the plasma membrane [120, 121]. Importantly, this phenomenon was shown to lead to the accumulation of receptors and G proteins in "hot spots" on the plasma membrane, where they preferentially interact [95] (Figure 26.3). By using a nanobody-based biosensor (Nb37) [72] that binds selectively to nucleotide-free, active Gα_s subunits, it was possible to show that G protein activation preferentially occurs at these hot spots, rather than homogenously across the plasma membrane [95]. Since the location and composition of these hot spots may vary among cell types and receptors, this might provide an important mechanism to diversify GPCR responses. In particular, it might help to explain the high specificity observed in GPCR signaling despite the fact that the hundreds of different GPCRs expressed in a typical cell converge onto a relatively small number of common signaling pathways.

Furthermore, the same study reported a novel mathematical approach based on deconvolution of the apparent single-molecule colocalization times to estimate the kinetic rates of protein–protein interactions. This allowed the estimation of the microscopic association (k_{on}) and dissociation (k_{off}) rate constants of receptor-G protein interactions. In the absence of agonists, a low level of basal interactions

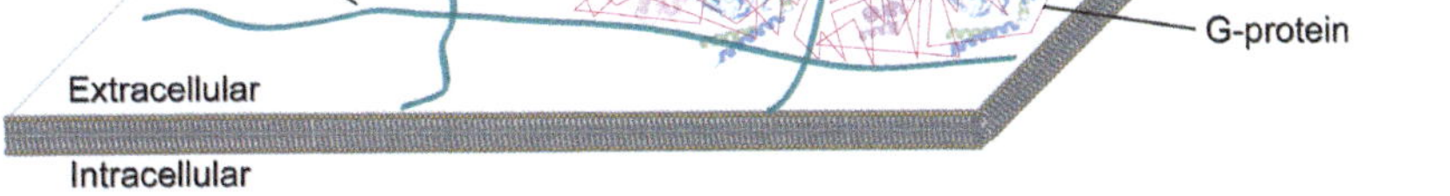

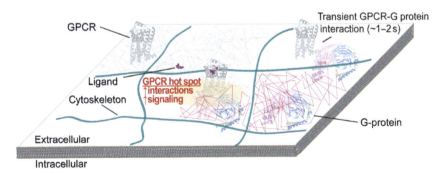

Figure 26.3 Schematic representation of GPCR signaling hot spots at the plasma membrane. Hot spots are areas of the plasma membrane where GPCRs and G proteins preferentially interact and signal. These hot spots arise from the presence of cytoskeletal elements with their associated membrane proteins and possibly other structural components, which provide barriers to the lateral diffusion of GPCRs and G proteins. Source: Based on Sungkaworn et al. [95].

were observed between receptors and G proteins ($k_{on} \approx 0.015$ μm^2 molecule^{-1}s^{-1} for α_{2A}AR-Gα_i interactions), which lead to transient interactions lasting approximately 1–2 seconds ($k_{off} \approx 0.5 - 1$ s^{-1}). Agonists were observed to modulate receptor-G protein interactions by increasing the association rate approximately 10-fold. Interestingly, this rate was approximately 10-times higher for α_{2A}AR-Gα_i than for β_2AR-Gα_s interactions, which is consistent with the view that receptor-G protein coupling might be more efficient for G_i- than for G_s-coupled receptors. Additionally, through the comparison of a panel of agonists for the α_{2A}AR, a positive correlation was found between agonist efficacy and the rate of receptor-G protein association, with only minor changes to the duration of the interactions [95]. This relationship between agonist efficacy and the kinetics of G protein coupling is similar to what was observed by Gregorio et al. on purified β_2ARs by smFRET [125]. Together, these results suggest that agonist efficacy operates, at least to some extent, at the level of receptor-G protein interactions. Moreover, the kinetic differences observed between the two investigated receptor systems might help explain the conundrum of why, although both G_i- and G_s-coupled receptors can theoretically trigger Gβγ release [145], only G_i-coupled receptors are capable of efficiently activating GIRK channels. This hypothesis is further supported by the results of a recent study by Touhara and Mackkinnon (2018) [146].

Single-molecule investigations of receptor-G protein interactions have also shed new light on the long-debated question of whether GPCRs and G proteins are precoupled in the absence of agonists [147, 148]. A low frequency of basal interactions between both α_{2A}AR and β_2AR and their cognate G proteins was observed by single-molecule microscopy [95]. However, these interactions were transient, lasting on average approximately 1–2 seconds, ruling out the occurrence of a relevant fraction of stable, pre-formed complexes between the tested receptors and G proteins. Moreover, the observed receptor-G protein interactions appear to be due to receptor constitutive activity and require a functional G protein, since they can be largely prevented by treatment with an inverse agonist or inactivation of the Gα_i protein with PTX [95]. It is conceivable that the transient interactions and the diffusional barriers provided by the cytoskeleton contribute to maintaining the receptor and G protein in close proximity, leading to rapid and efficient responses to agonist-induced receptor activation. At the same time, the occurrence of transient rather than stable interactions allows a single receptor to catalytically activate multiple G proteins, leading to signal amplification.

Interestingly, the duration of receptor-G protein interactions estimated by single-particle tracking [95], albeit transient in nature, is significantly longer than the time required to trigger rapid responses like those mediated by GIRK channels, which can occur within 100 ms from agonist stimulation [149]. Further studies are required to understand the implications of these findings. One hypothesis is that G proteins might be able to regulate some of their effectors while still bound to their activating receptors. This hypothesis is consistent with previous evidence that G proteins might not fully dissociate into Gα and Gβγ subunits upon activation [150, 151] and that receptors, G proteins and effectors, including both GIRK channels and adenylyl cyclases, might form multimolecular complexes [152, 153].

26.15 Summary and Future Perspectives

Single-molecule studies have begun to unravel the complexity of GPCR signaling at the plasma membrane. Among other lessons, these studies have taught us that GPCRs are highly dynamic and can engage in transient interactions among themselves, with other membrane proteins, and with structural elements of the plasma membrane to form short-lived complexes and functional nanodomains on the plasma membrane. Although a lot remains to be understood about the spatiotemporal dynamics of these nanodomains and their consequence for GPCR signaling, single-molecule microscopy methods are likely to play a fundamental role in further elucidating their organization and function. These efforts will likely benefit from an integrated approach, combining single-molecule microscopy and other biophysical methods in living cells and reconstituted systems, structural biology, and advanced computational methods.

Further technological developments are likely to play an important role in this effort. One important direction is the synthesis of novel organic fluorophores with improved characteristics. The time a single fluorophore can be visualized, and the precision with which it can be localized, largely depend on the number of photons that the fluorophore emits before being irreversibly photobleached. Therefore, efforts are underway to improve the photostability and the quantum yield of fluorophores, i.e. their efficiency in emitting photons. Among various strategies to improve fluorophore photostability, an intriguing approach is their conjugation with triple state quenchers that prevent photobleaching [154].

Another important area concerns the development of new strategies to achieve less-invasive labeling of proteins in living cells. One promising approach is to use genetic code expansion (GCE) in combination with biocompatible chemical reactions to insert small and bright organic fluorophores directly onto a protein of interest [155]. In GCE-based labeling, an unnatural amino acid (UAA) containing a functional group is incorporated site-specifically onto the target protein. The protein is then fluorescently labeled via a highly selective bioorthogonal reaction between the functional group of the UAA and a suitable fluorescent compound. Since the attached fluorophores are much smaller than a traditional protein tag (~0.5 versus ~5 nm), they are less likely to perturb the function of the target protein. Moreover, they can offer important advantages in smFRET studies as their use substantially reduces the uncertainty in fluorophore position that is associated with large protein tags. For instance, König et al. determined the plasma membrane distribution and diffusion of the EGFR with nanoscale precision using GCE-based labeling combined with single-molecule localization super-resolution microscopy and single-particle tracking in living cells [155].

Besides multicolor single-molecule studies in living cells, another area expected to yield further developments is that of single-molecule studies with purified proteins in reconstituted systems. With the explosion of structural information about the main conformations adopted by GPCRs and their signaling partners, there is a growing demand to understand their dynamics. In this respect, smFRET studies with purified receptors and other signaling proteins are likely to play an important role. Improved reconstitution and nano-encapsulation approaches that preserve the natural lipid environment of the protein of interest are likely going to be critical for the success of these studies. Besides lipid nanodiscs [156], a new approach is offered by poly-styrene co-maleic acid lipid particles (SMALPs) [157]. For example, Grime et al. recently demonstrated the use of SMALPs to extract adenosine A_{2A} receptors from the plasma membrane together with their surrounding native lipids and monitor ligand binding to individual SMALP-receptor complexes in solution by FCS [158].

While we are beginning to see the early fruits of single-molecule methods applied to GPCR signaling, the combination of single-molecule microscopy with complementary approaches in a multidisciplinary effort promises to further untangle the complex organization of GPCR signaling at the plasma membrane, as well as at other subcellular structures. Not only is this expected to help answer some of the still open questions about how our cells process extracellular stimuli, but might also provide a new basis for the development of innovative treatments for common diseases such as Alzheimer's, diabetes, or heart failure.

References

1 Pierce, K.L., Premont, R.T., and Lefkowitz, R.J. (2002). Seven-transmembrane receptors. *Nat. Rev. Mol. Cell Biol.* 3 (9): 639–650.

2 Hauser, A.S., Attwood, M.M., Rask-Andersen, M. et al. (2017). Trends in GPCR drug discovery: new agents, targets and indications. *Nat. Rev. Drug Discovery* 16 (12): 829–842.

3 Rall, T.W., Sutherland, E.W., and Wosilait, W.D. (1956). The relationship of epinephrine and glucagon to liver phosphorylase. III. Reactivation of liver phosphorylase in slices and in extracts. *J. Biol. Chem.* 218 (1): 483–495.

4 Berthet, J., Rall, T.W., and Sutherland, E.W. (1957). The relationship of epinephrine and glucagon to liver phosphorylase. IV. Effect of epinephrine

and glucagon on the reactivation of phosphorylase in liver homogenates. *J. Biol. Chem.* 224 (1): 463–475.

5 Rodbell, M. (1980). The role of hormone receptors and GTP-regulatory proteins in membrane transduction. *Nature* 284 (5751): 17–22.

6 Gilman, A.G. (1987). G proteins: transducers of receptor-generated signals. *Annu. Rev. Biochem.* 56: 615–649.

7 Weis, W.I. and Kobilka, B.K. (2018). The molecular basis of G protein-coupled receptor activation. *Annu. Rev. Biochem.* 87: 897–919.

8 Syrovatkina, V., Alegre, K.O., Dey, R., and Huang, X.Y. (2016). Regulation, signaling, and physiological functions of G-proteins. *J. Mol. Biol.* 428 (19): 3850–3868.

9 Katz, B. and Thesleff, S. (1957). A study of the desensitization produced by acetylcholine at the motor end-plate. *J. Physiol.* 138 (1): 63–80.

10 Fletcher, J.E., Spector, A.A., and Ashbrook, J.D. (1970). Analysis of macromolecule-ligand binding by determination of stepwise equilibrium constants. *Biochemistry* 9 (23): 4580–4587.

11 Tolkovsky, A.M. and Levitzki, A. (1978). Mode of coupling between the β-adrenergic receptor and adenylate cyclase in turkey erythrocytes. *Biochemistry* 17 (18): 3795.

12 Kent, R.S., De Lean, A., and Lefkowitz, R.J. (1980). A quantitative analysis of β-adrenergic receptor interactions: resolution of high and low affinity states of the receptor by computer modeling of ligand binding data. *Mol. Pharmacol.* 17 (1): 14–23.

13 De Lean, A., Stadel, J.M., and Lefkowitz, R.J. (1980). A ternary complex model explains the agonist-specific binding properties of the adenylate cyclase-coupled β-adrenergic receptor. *J. Biol. Chem.* 255 (15): 7108–7117.

14 Samama, P., Cotecchia, S., Costa, T., and Lefkowitz, R.J. (1993). A mutation-induced activated state of the β_2-adrenergic receptor. Extending the ternary complex model. *J. Biol. Chem.* 268 (7): 4625–4636.

15 Calebiro, D. and Koszegi, Z. (2019). The subcellular dynamics of GPCR signaling. *Mol. Cell. Endocrinol.* 483: 24–30.

16 Kühn, H. and Dreyer, W.J. (1972). Light dependent phosphorylation of rhodopsin by ATP. *FEBS Lett.* 20 (1): 1–6.

17 Kühn, H. (1978). Light-regulated binding of rhodopsin kinase and other proteins to cattle photoreceptor membranes. *Biochemistry* 17 (21): 4389–4395.

18 Wilden, U., Hall, S.W., and Kühn, H. (1986). Phosphodiesterase activation by photoexcited rhodopsin is quenched when rhodopsin is phosphorylated and binds the intrinsic 48-kDa protein of rod outer segments. *Proc. Natl. Acad. Sci. U.S.A.* 83 (5): 1174–1178.

19 Stadel, J.M., Nambi, P., Shorr, R.G. et al. (1983). Catecholamine-induced desensitization of turkey erythrocyte adenylate cyclase is associated with

phosphorylation of the β-adrenergic receptor. *Proc. Natl. Acad. Sci. U.S.A.* 80 (11): 3173–3177.

20 Benovic, J.L., Kühn, H., Weyand, I. et al. (1987). Functional desensitization of the isolated β-adrenergic receptor by the β-adrenergic receptor kinase: potential role of an analog of the retinal protein arrestin (48-kDa protein). *Proc. Natl. Acad. Sci. U.S.A.* 84 (24): 8879–8882.

21 Gurevich, V.V. and Gurevich, E.V. (2019). GPCR signaling regulation: the role of GRKs and arrestins. *Front. Pharmacol.* 10: 125.

22 Goodman, O.B. Jr., Krupnick, J.G., Santini, F. et al. (1996). β-arrestin acts as a clathrin adaptor in endocytosis of the β_2-adrenergic receptor. *Nature* 383 (6599): 447–450.

23 Luttrell, L.M., Ferguson, S.S., Daaka, Y. et al. (1999). β-arrestin-dependent formation of β_2 adrenergic receptor-Src protein kinase complexes. *Science* 283 (5402): 655–661.

24 Pierce, K.L. and Lefkowitz, R.J. (2001). Classical and new roles of β-arrestins in the regulation of G-protein-coupled receptors. *Nat. Rev. Neurosci.* 2 (10): 727–733.

25 Ghosh, E., Kumari, P., Jaiman, D., and Shukla, A.K. (2015). Methodological advances: the unsung heroes of the GPCR structural revolution. *Nat. Rev. Mol. Cell Biol.* 16 (2): 69–81.

26 Safdari, H.A., Pandey, S., Shukla, A.K., and Dutta, S. (2018). Illuminating GPCR Signaling by Cryo-EM. *Trends Cell Biol.* 28 (8): 591–594.

27 Huang, J., Chen, S., Zhang, J.J., and Huang, X.Y. (2013). Crystal structure of oligomeric β_1-adrenergic G protein-coupled receptors in ligand-free basal state. *Nat. Rev. Struct. Mol. Biol.* 20 (4): 419–425.

28 Lebon, G., Warne, T., Edwards, P.C. et al. (2011). Agonist-bound adenosine A_{2A} receptor structures reveal common features of GPCR activation. *Nature* 474 (7352): 521–525.

29 Rasmussen, S.G., Choi, H.J., Fung, J.J. et al. (2011). Structure of a nanobody-stabilized active state of the β_2 adrenoceptor. *Nature* 469 (7329): 175–180.

30 Rosenbaum, D.M., Zhang, C., Lyons, J.A. et al. (2011). Structure and function of an irreversible agonist-β_2 adrenoceptor complex. *Nature* 469 (7329): 236–240.

31 Cherezov, V., Rosenbaum, D.M., Hanson, M.A. et al. (2007). High-resolution crystal structure of an engineered human β_2-adrenergic G protein-coupled receptor. *Science* 318 (5854): 1258–1265.

32 Rasmussen, S.G., Choi, H.J., Rosenbaum, D.M. et al. (2007). Crystal structure of the human β_2 adrenergic G-protein-coupled receptor. *Nature* 450 (7168): 383–387.

33 Warne, T., Serrano-Vega, M.J., Baker, J.G. et al. (2008). Structure of a β_1-adrenergic G-protein-coupled receptor. *Nature* 454 (7203): 486–491.

34 Rasmussen, S.G., DeVree, B.T., Zou, Y. et al. (2011). Crystal structure of the β_2 adrenergic receptor-G_s protein complex. *Nature* 477 (7366): 549–555.

35 Carpenter, B., Nehme, R., Warne, T. et al. (2016). Structure of the adenosine A_{2A} receptor bound to an engineered G protein. *Nature* 536 (7614): 104–107.

36 Liang, Y.L., Khoshouei, M., Radjainia, M. et al. (2017). Phase-plate cryo-EM structure of a class B GPCR-G-protein complex. *Nature* 546 (7656): 118–123.

37 Zhang, Y., Sun, B., Feng, D. et al. (2017). Cryo-EM structure of the activated GLP-1 receptor in complex with a G protein. *Nature* 546 (7657): 248–253.

38 Draper-Joyce, C.J., Khoshouei, M., Thal, D.M. et al. (2018). Structure of the adenosine-bound human adenosine A_1 receptor-G_i complex. *Nature* 558 (7711): 559–563.

39 Garcia-Nafria, J., Nehme, R., Edwards, P.C., and Tate, C.G. (2018). Cryo-EM structure of the serotonin 5-HT_{1B} receptor coupled to heterotrimeric G_o. *Nature* 558 (7711): 620–623.

40 Kang, Y., Kuybeda, O., de Waal, P.W. et al. (2018). Cryo-EM structure of human rhodopsin bound to an inhibitory G protein. *Nature* 558 (7711): 553–558.

41 Koehl, A., Hu, H., Maeda, S. et al. (2018). Structure of the μ-opioid receptor–G_i protein complex. *Nature* 558 (7711): 547–552.

42 Liang, Y.L., Khoshouei, M., Deganutti, G. et al. (2018). Cryo-EM structure of the active, G_s-protein complexed, human CGRP receptor. *Nature* 561 (7724): 492–497.

43 Kang, Y., Zhou, X.E., Gao, X. et al. (2015). Crystal structure of rhodopsin bound to arrestin by femtosecond X-ray laser. *Nature* 523 (7562): 561–567.

44 Zhou, X.E., He, Y., de Waal, P.W. et al. (2017). Identification of phosphorylation codes for arrestin recruitment by G protein-coupled receptors. *Cell* 170 (3): 457–469 e413.

45 Staus, D.P., Hu, H., Robertson, M.J. et al. (2020). Structure of the M_2 muscarinic receptor-β-arrestin complex in a lipid nanodisc. *Nature* 579 (7798): 297–302.

46 Huang, W., Masureel, M., Qu, Q. et al. (2020). Structure of the neurotensin receptor 1 in complex with β-arrestin 1. *Nature* 579 (7798): 303–308.

47 Nguyen, A.H., Thomsen, A.R.B., Cahill, T.J. 3rd et al. (2019). Structure of an endosomal signaling GPCR-G protein-β-arrestin megacomplex. *Nat. Rev. Struct. Mol. Biol.* 26 (12): 1123–1131.

48 Nygaard, R., Zou, Y., Dror, R.O. et al. (2013). The dynamic process of β_2-adrenergic receptor activation. *Cell* 152 (3): 532–542.

49 Manglik, A., Kim, T.H., Masureel, M. et al. (2015). Structural insights into the dynamic process of β_2-adrenergic receptor signaling. *Cell* 161 (5): 1101–1111.

50 Ye, L., Van Eps, N., Zimmer, M. et al. (2016). Activation of the A_{2A} adenosine G-protein-coupled receptor by conformational selection. *Nature* 533 (7602): 265–268.

51 Calebiro, D., Sungkaworn, T., and Maiellaro, I. (2014). Real-time monitoring of GPCR/cAMP signalling by FRET and single-molecule microscopy. *Hormone Metab. Res.* 46 (12): 827–832.

52 Smith, J.S., Lefkowitz, R.J., and Rajagopal, S. (2018). Biased signalling: from simple switches to allosteric microprocessors. *Nat. Rev. Drug Discovery* 17 (4): 243–260.

53 Vilardaga, J.P., Steinmeyer, R., Harms, G.S., and Lohse, M.J. (2005). Molecular basis of inverse agonism in a G protein-coupled receptor. *Nat. Chem. Biol.* 1 (1): 25–28.

54 Zürn, A., Zabel, U., Vilardaga, J.P. et al. (2009). Fluorescence resonance energy transfer analysis of α_{2A}-adrenergic receptor activation reveals distinct agonist-specific conformational changes. *Mol. Pharmacol.* 75 (3): 534–541.

55 Reiner, S., Ambrosio, M., Hoffmann, C., and Lohse, M.J. (2010). Differential signaling of the endogenous agonists at the β_2-adrenergic receptor. *J. Biol. Chem.* 285 (46): 36188–36198.

56 Liu, J.J., Horst, R., Katritch, V. et al. (2012). Biased signaling pathways in β_2-adrenergic receptor characterized by 19F-NMR. *Science* 335 (6072): 1106–1110.

57 Rahmeh, R., Damian, M., Cottet, M. et al. (2012). Structural insights into biased G protein-coupled receptor signaling revealed by fluorescence spectroscopy. *Proc. Natl. Acad. Sci. U.S.A.* 109 (17): 6733–6738.

58 Wacker, D., Wang, C., Katritch, V. et al. (2013). Structural features for functional selectivity at serotonin receptors. *Science* 340 (6132): 615–619.

59 Wingler, L.M., Elgeti, M., Hilger, D. et al. (2019). Angiotensin analogs with divergent bias stabilize distinct receptor conformations. *Cell* 176 (3): 468–478.e411.

60 Brunton, L.L., Hayes, J.S., and Mayer, S.E. (1979). Hormonally specific phosphorylation of cardiac troponin I and activation of glycogen phosphorylase. *Nature* 280 (5717): 78–80.

61 Keely, S.L. (1979). Prostaglandin E_1 activation of heart cAMP-dependent protein kinase: apparent dissociation of protein kinase activation from increases in phosphorylase activity and contractile force. *Mol. Pharmacol.* 15 (2): 235–245.

62 Hayes, J.S., Brunton, L.L., and Mayer, S.E. (1980). Selective activation of particulate cAMP-dependent protein kinase by isoproterenol and prostaglandin E_1. *J. Biol. Chem.* 255 (11): 5113–5119.

63 Buxton, I.L. and Brunton, L.L. (1983). Compartments of cyclic AMP and protein kinase in mammalian cardiomyocytes. *J. Biol. Chem.* 258 (17): 10233–10239.

64 Zhang, J., Ma, Y., Taylor, S.S., and Tsien, R.Y. (2001). Genetically encoded reporters of protein kinase A activity reveal impact of substrate tethering. *Proc. Natl. Acad. Sci. U.S.A.* 98 (26): 14997–15002.

65 Nikolaev, V.O., Bünemann, M., Hein, L. et al. (2004). Novel single chain cAMP sensors for receptor-induced signal propagation. *J. Biol. Chem.* 279 (36): 37215–37218.

66 Castro, L.R., Gervasi, N., Guiot, E. et al. (2010). Type 4 phosphodiesterase plays different integrating roles in different cellular domains in pyramidal cortical neurons. *J. Neurosci.* 30 (17): 6143–6151.

67 Maiellaro, I., Lohse, M.J., Kittel, R.J., and Calebiro, D. (2016). cAMP signals in Drosophila motor neurons are confined to single synaptic boutons. *Cell Rep.* 17 (5): 1238–1246.

68 Zaccolo, M. and Pozzan, T. (2002). Discrete microdomains with high concentration of cAMP in stimulated rat neonatal cardiac myocytes. *Science* 295 (5560): 1711–1715.

69 Nikolaev, V.O., Bünemann, M., Schmitteckert, E. et al. (2006). Cyclic AMP imaging in adult cardiac myocytes reveals far-reaching β_1-adrenergic but locally confined β_2-adrenergic receptor-mediated signaling. *Circ. Res.* 99 (10): 1084–1091.

70 Nikolaev, V.O., Moshkov, A., Lyon, A.R. et al. (2010). β_2-adrenergic receptor redistribution in heart failure changes cAMP compartmentation. *Science* 327 (5973): 1653–1657.

71 Calebiro, D., Nikolaev, V.O., Gagliani, M.C. et al. (2009). Persistent cAMP-signals triggered by internalized G-protein-coupled receptors. *PLoS Biol.* 7 (8): e1000172.

72 Irannejad, R., Tomshine, J.C., Tomshine, J.R. et al. (2013). Conformational biosensors reveal GPCR signalling from endosomes. *Nature* 495 (7442): 534–538.

73 Tsvetanova, N.G. and von Zastrow, M. (2014). Spatial encoding of cyclic AMP signaling specificity by GPCR endocytosis. *Nat. Chem. Biol.* 10 (12): 1061–1065.

74 Godbole, A., Lyga, S., Lohse, M.J., and Calebiro, D. (2017). Internalized TSH receptors en route to the TGN induce local G_s-protein signaling and gene transcription. *Nat. Commun.* 8 (1): 443.

75 Thomsen, A.R.B., Plouffe, B., Cahill, T.J. 3rd et al. (2016). GPCR-G protein-β-arrestin super-complex mediates sustained G protein signaling. *Cell* 166 (4): 907–919.

76 Di Benedetto, A., Sun, L., Zambonin, C.G. et al. (2014). Osteoblast regulation via ligand-activated nuclear trafficking of the oxytocin receptor. *Proc. Natl. Acad. Sci. U.S.A.* 111 (46): 16502–16507.

77 Joyal, J.S., Nim, S., Zhu, T. et al. (2014). Subcellular localization of coagulation factor II receptor-like 1 in neurons governs angiogenesis. *Nat. Med.* 20 (10): 1165–1173.

78 Kinsey, C.G., Bussolati, G., Bosco, M. et al. (2007). Constitutive and ligand-induced nuclear localization of oxytocin receptor. *J. Cell. Mol. Med.* 11 (1): 96–110.

79 Benard, G., Massa, F., Puente, N. et al. (2012). Mitochondrial CB_1 receptors regulate neuronal energy metabolism. *Nat. Neurosci.* 15 (4): 558–564.

80 Suofu, Y., Li, W., Jean-Alphonse, F.G. et al. (2017). Dual role of mitochondria in producing melatonin and driving GPCR signaling to block cytochrome c release. *Proc. Natl. Acad. Sci. U.S.A.* 114 (38): E7997–E8006.

81 Wang, Q., Zhang, H., Xu, H. et al. (2016). 5-HTR_3 and 5-HTR_4 located on the mitochondrial membrane and functionally regulated mitochondrial functions. *Sci. Rep.* 6: 37336.

82 Revankar, C.M., Cimino, D.F., Sklar, L.A. et al. (2005). A transmembrane intracellular estrogen receptor mediates rapid cell signaling. *Science* 307 (5715): 1625–1630.

83 Feynman, R.P. (1961). There's plenty of room at the bottom. *Miniaturization* 282–296.

84 Hirschfeld, T. (1976). Optical microscopic observation of single small molecules. *Appl. Opt.* 15 (12): 2965–2966.

85 Moerner, W.E. and Kador, L. (1989). Optical detection and spectroscopy of single molecules in a solid. *Phys. Rev. Lett.* 62 (21): 2535–2538.

86 Orrit, M. and Bernard, J. (1990). Single pentacene molecules detected by fluorescence excitation in a p-terphenyl crystal. *Phys. Rev. Lett.* 65 (21): 2716–2719.

87 Shera, B.E., Seitzinger, N.K., Davis, L.M. et al. (1990). Detection of single fluorescent molecules. *Chem. Phys. Lett.* 174 (6): 553–557.

88 Betzig, E. and Chichester, R.J. (1993). Single molecules observed by near-field scanning optical microscopy. *Science* 262 (5138): 1422–1425.

89 Walter, N.G., Huang, C.Y., Manzo, A.J., and Sobhy, M.A. (2008). Do-it-yourself guide: how to use the modern single-molecule toolkit. *Nat. Methods* 5 (6): 475–489.

90 Keppler, A., Gendreizig, S., Gronemeyer, T. et al. (2003). A general method for the covalent labeling of fusion proteins with small molecules in vivo. *Nat. Biotechnol.* 21 (1): 86–89.

91 Gautier, A., Juillerat, A., Heinis, C. et al. (2008). An engineered protein tag for multiprotein labeling in living cells. *Chem. Biol.* 15 (2): 128–136.

92 Ambrose, E.J. (1956). A surface contact microscope for the study of cell movements. *Nature* 178 (4543): 1194.

93 Axelrod, D. (1981). Cell-substrate contacts illuminated by total internal reflection fluorescence. *J. Biol. Chem.* 89 (1): 141–145.

94 Calebiro, D., Rieken, F., Wagner, J. et al. (2013). Single-molecule analysis of fluorescently labeled G-protein-coupled receptors reveals complexes with distinct dynamics and organization. *Proc. Natl. Acad. Sci. U.S.A.* 110 (2): 743–748.

95 Sungkaworn, T., Jobin, M.L., Burnecki, K. et al. (2017). Single-molecule imaging reveals receptor-G protein interactions at cell surface hot spots. *Nature* 550 (7677): 543–547.

96 Calebiro, D. and Jobin, M.L. (2018). Hot spots for GPCR signaling: lessons from single-molecule microscopy. *Curr. Opin. Cell Biol.* 57: 57–63.

97 Jaqaman, K., Loerke, D., Mettlen, M. et al. (2008). Robust single-particle tracking in live-cell time-lapse sequences. *Nat. Methods* 5 (8): 695–702.

98 Lanoiselée, Y. and Grebenkov, D.S. (2016). Revealing nonergodic dynamics in living cells from a single particle trajectory. *Phys. Rev. E* 93 (5): 052146.

99 Michalet, X. (2010). Mean square displacement analysis of single-particle trajectories with localization error: Brownian motion in an isotropic medium. *Phys. Rev. E* 82: 041914.

100 Calebiro, D. and Sungkaworn, T. (2018). Single-molecule imaging of GPCR interactions. *Trends Pharmacol. Sci.* 39 (2): 109–122.

101 Wolter, S., Loschberger, A., Holm, T. et al. (2012). rapidSTORM: accurate, fast open-source software for localization microscopy. *Nat. Methods* 9 (11): 1040–1041.

102 Whelan, D.R. and Bell, T.D. (2015). Image artifacts in single molecule localization microscopy: why optimization of sample preparation protocols matters. *Sci. Rep.* 5: 7924.

103 Klein, T., Loschberger, A., Proppert, S. et al. (2011). Live-cell dSTORM with SNAP-tag fusion proteins. *Nat. Methods* 8 (1): 7–9.

104 Sauer, M. and Heilemann, M. (2017). Single-molecule localization microscopy in eukaryotes. *Chem. Rev.* 117 (11): 7478–7509.

105 Szalai, A.M., Armando, N.G., Barabas, F.M. et al. (2018). A fluorescence nanoscopy marker for corticotropin-releasing hormone type 1 receptor: computer design, synthesis, signaling effects, super-resolved fluorescence imaging, and in situ affinity constant in cells. *Phys. Chem. Chem. Phys.* 20 (46): 29212–29220.

106 Siddig, S., Aufmkolk, S., Doose, S. et al. (2020). Super-resolution imaging reveals the nanoscale organization of metabotropic glutamate receptors at presynaptic active zones. *Sci. Adv.* 6 (16): 7193.

107 Betzig, E., Patterson, G.H., Sougrat, R. et al. (2006). Imaging intracellular fluorescent proteins at nanometer resolution. *Science* 313 (5793): 1642–1645.

108 Scarselli, M., Annibale, P., and Radenovic, A. (2012). Cell type-specific β_2-adrenergic receptor clusters identified using photoactivated localization microscopy are not lipid raft related, but depend on actin cytoskeleton integrity. *J. Biol. Chem.* 287 (20): 16768–16780.

109 Tabor, A., Weisenburger, S., Banerjee, A. et al. (2016). Visualization and ligand-induced modulation of dopamine receptor dimerization at the single molecule level. *Sci. Rep.* 6: 33233.

110 Ha, T., Enderle, T., Ogletree, D.F. et al. (1996). Probing the interaction between two single molecules: fluorescence resonance energy transfer between a single donor and a single acceptor. *Proc. Natl. Acad. Sci. U.S.A.* 93 (13): 6264–6268.

111 Roy, R., Hohng, S., and Ha, T. (2008). A practical guide to single-molecule FRET. *Nat. Methods* 5 (6): 507–516.

112 Briddon, S.J., Kilpatrick, L.E., and Hill, S.J. (2018). Studying GPCR pharmacology in membrane microdomains: fluorescence correlation spectroscopy comes of age. *Trends Pharmacol. Sci.* 39 (2): 158–174.

113 Briddon, S.J., Middleton, R.J., Cordeaux, Y. et al. (2004). Quantitative analysis of the formation and diffusion of A_1-adenosine receptor-antagonist complexes in single living cells. *Proc. Natl. Acad. Sci. U.S.A.* 101 (13): 4673–4678.

114 Cordeaux, Y., Briddon, S.J., Alexander, S.P. et al. (2008). Agonist-occupied A_3 adenosine receptors exist within heterogeneous complexes in membrane microdomains of individual living cells. *FASEB J.* 22 (3): 850–860.

115 Corriden, R., Kilpatrick, L.E., Kellam, B. et al. (2014). Kinetic analysis of antagonist-occupied adenosine-A_3 receptors within membrane microdomains of individual cells provides evidence of receptor dimerization and allosterism. *FASEB J.* 28 (10): 4211–4222.

116 Hegener, O., Prenner, L., Runkel, F. et al. (2004). Dynamics of β_2-adrenergic receptor-ligand complexes on living cells. *Biochemistry* 43 (20): 6190–6199.

117 Lohse, M.J., Nuber, S., and Hoffmann, C. (2012). Fluorescence/bioluminescence resonance energy transfer techniques to study G-protein-coupled receptor activation and signaling. *Pharmacol. Rev.* 64 (2): 299–336.

118 Funatsu, T., Harada, Y., Tokunaga, M. et al. (1995). Imaging of single fluorescent molecules and individual ATP turnovers by single myosin molecules in aqueous solution. *Nature* 374 (6522): 555–559.

119 Sako, Y., Minoghchi, S., and Yanagida, T. (2000). Single-molecule imaging of EGFR signalling on the surface of living cells. *Nat. Cell Biol.* 2 (3): 168–172.

120 Suzuki, K., Ritchie, K., Kajikawa, E. et al. (2005). Rapid hop diffusion of a G-protein-coupled receptor in the plasma membrane as revealed by single-molecule techniques. *Biophys. J.* 88 (5): 3659–3680.

121 Murase, K., Fujiwara, T., Umemura, Y. et al. (2004). Ultrafine membrane compartments for molecular diffusion as revealed by single molecule techniques. *Biophys. J.* 86 (6): 4075–4093.

122 Peleg, G., Ghanouni, P., Kobilka, B.K., and Zare, R.N. (2001). Single-molecule spectroscopy of the β_2 adrenergic receptor: observation of conformational substates in a membrane protein. *Proc. Natl. Acad. Sci. U.S.A.* 98 (15): 8469–8474.

123 Bockenhauer, S., Furstenberg, A., Yao, X.J. et al. (2011). Conformational dynamics of single G protein-coupled receptors in solution. *J. Phys. Chem. B* 115 (45): 13328–13338.

124 Lamichhane, R., Liu, J.J., Pljevaljcic, G. et al. (2015). Single-molecule view of basal activity and activation mechanisms of the G protein-coupled receptor β_2AR. *Proc. Natl. Acad. Sci. U.S.A.* 112 (46): 14254–14259.

125 Gregorio, G.G., Masureel, M., Hilger, D. et al. (2017). Single-molecule analysis of ligand efficacy in β_2AR-G-protein activation. *Nature* 547 (7661): 68–73.

126 Cao, C., Tan, Q., Xu, C. et al. (2018). Structural basis for signal recognition and transduction by platelet-activating-factor receptor. *Nat. Rev. Struct. Mol. Biol.* 25 (6): 488–495.

127 van der Meijden, P.E.J. and Heemskerk, J.W.M. (2019). Platelet biology and functions: new concepts and clinical perspectives. *Nat. Rev. Cardiol.* 16 (3): 166–179.

128 Romano, C., Yang, W.L., and O'Malley, K.L. (1996). Metabotropic glutamate receptor 5 is a disulfide-linked dimer. *J. Biol. Chem.* 271 (45): 28612–28616.

129 Jones, K.A., Borowsky, B., Tamm, J.A. et al. (1998). $GABA_B$ receptors function as a heteromeric assembly of the subunits $GABA_B$R1 and $GABA_B$R2. *Nature* 396 (6712): 674–679.

130 White, J.H., Wise, A., Main, M.J. et al. (1998). Heterodimerization is required for the formation of a functional $GABA_B$ receptor. *Nature* 396 (6712): 679–682.

131 Pin, J.P., Kniazeff, J., Liu, J. et al. (2005). Allosteric functioning of dimeric class C G-protein-coupled receptors. *FEBS J.* 272 (12): 2947–2955.

132 Celli, R., Santolini, I., Van Luijtelaar, G. et al. (2019). Targeting metabotropic glutamate receptors in the treatment of epilepsy: rationale and current status. *Expert Opin. Therap. Targets* 23 (4): 341–351.

133 Potter, L.E., Zanos, P., and Gould, T.D. (2020). Antidepressant effects and mechanisms of group II mGlu receptor-specific negative allosteric modulators. *Neuron* 105 (1): 1–3.

134 Vafabakhsh, R., Levitz, J., and Isacoff, E.Y. (2015). Conformational dynamics of a class C G-protein-coupled receptor. *Nature* 524 (7566): 497–501.

135 Levitz, J., Habrian, C., Bharill, S. et al. (2016). Mechanism of assembly and cooperativity of homomeric and heteromeric metabotropic glutamate receptors. *Neuron* 92 (1): 143–159.

136 Dijkman, P.M., Castell, O.K., Goddard, A.D. et al. (2018). Dynamic tuneable G protein-coupled receptor monomer-dimer populations. *Nat. Commun.* 9 (1): 1710.

137 Casciari, D., Dell'Orco, D., and Fanelli, F. (2008). Homodimerization of neurotensin 1 receptor involves helices 1, 2, and 4: insights from quaternary structure predictions and dimerization free energy estimations. *J. Chem. Inf. Model.* 48 (8): 1669–1678.

138 Hern, J.A., Baig, A.H., Mashanov, G.I. et al. (2010). Formation and dissociation of M_1 muscarinic receptor dimers seen by total internal reflection fluorescence imaging of single molecules. *Proc. Natl. Acad. Sci. U.S.A.* 107 (6): 2693–2698.

139 Kasai, R.S., Suzuki, K.G., Prossnitz, E.R. et al. (2011). Full characterization of GPCR monomer-dimer dynamic equilibrium by single molecule imaging. *J. Cell Biol.* 192 (3): 463–480.

140 Gentzsch, C., Seier, K., Drakopoulos, A. et al. (2020). Selective and wash-resistant fluorescent dihydrocodeinone derivatives allow single-molecule imaging of mu-opioid receptor dimerization. *Angew. Chem. Int. Ed.* 59 (15): 5958–5964.

141 Drakopoulos, A., Koszegi, Z., Lanoiselee, Y. et al. (2020). Investigation of inactive state κ opioid receptor homodimerization via single molecule microscopy using new antagonistic fluorescent probes. *J. Med. Chem.* 3596–3609.

142 Ge, B., Lao, J., Li, J. et al. (2017). Single-molecule imaging reveals dimerization/oligomerization of $CXCR_4$ on plasma membrane closely related to its function. *Sci. Rep.* 7 (1): 16873.

143 Weisenburger, S., Jing, B., Renn, A., and Sandoghdar, V. (2013). Cryogenic localization of single molecules with angstrom precision. *SPIE* 8815.

144 Kasai, R.S., Ito, S.V., Awane, R.M. et al. (2018). The class-A GPCR dopamine D_2 receptor forms transient dimers stabilized by agonists: detection by single-molecule tracking. *Cell Biochem. Biophys.* 76 (1-2): 29–37.

145 Dascal, N. and Kahanovitch, U. (2015). The roles of Gβγ and Gα in gating and regulation of GIRK channels. *Int. Rev. Neurobiol.* 123: 27–85.

146 Touhara, K.K. and MacKinnon, R. (2018). Molecular basis of signaling specificity between GIRK channels and GPCRs. *eLife* 7 (e42908).

147 Neubig, R.R., Gantzos, R.D., and Thomsen, W.J. (1988). Mechanism of agonist and antagonist binding to α_2 adrenergic receptors: evidence for a precoupled receptor-guanine nucleotide protein complex. *Biochemistry* 27 (7): 2374–2384.

148 Gether, U. and Kobilka, B.K. (1998). G protein-coupled receptors. II. Mechanism of agonist activation. *J. Biol. Chem.* 273 (29): 17979–17982.

149 Hein, P., Frank, M., Hoffmann, C. et al. (2005). Dynamics of receptor/G protein coupling in living cells. *EMBO J.* 24 (23): 4106–4114.

150 Bünemann, M., Frank, M., and Lohse, M.J. (2003). G_i protein activation in intact cells involves subunit rearrangement rather than dissociation. *Proc. Natl. Acad. Sci. U.S.A.* 100 (26): 16077–16082.

151 Galés, C., Van Durm, J.J., Schaak, S. et al. (2006). Probing the activation-promoted structural rearrangements in preassembled receptor-G protein complexes. *Nat. Rev. Struct. Mol. Biol.* 13 (9): 778–786.

152 Richard-Lalonde, M., Nagi, K., Audet, N. et al. (2013). Conformational dynamics of Kir3.1/Kir3.2 channel activation via δ-opioid receptors. *Mol. Pharmacol.* 83 (2): 416–428.

153 Nagi, K. and Pineyro, G. (2014). Kir3 channel signaling complexes: focus on opioid receptor signaling. *Front. Cell. Neurosci.* 8: 186.

154 Altman, R.B., Terry, D.S., Zhou, Z. et al. (2011). Cyanine fluorophore derivatives with enhanced photostability. *Nat. Methods* 9 (1): 68–71.

155 Konig, A.I., Sorkin, R., Alon, A. et al. (2020). Live cell single molecule tracking and localization microscopy of bioorthogonally labeled plasma membrane proteins. *Nanoscale* 12 (5): 3236–3248.

156 Denisov, I.G. and Sligar, S.G. (2017). Nanodiscs in membrane biochemistry and biophysics. *Chem. Rev.* 117 (6): 4669–4713.

157 Knowles, T.J., Finka, R., Smith, C. et al. (2009). Membrane proteins solubilized intact in lipid containing nanoparticles bounded by styrene maleic acid copolymer. *J. Am. Chem. Soc.* 131 (22): 7484–7485.

158 Grime, R.L., Goulding, J., Uddin, R. et al. (2020). Single molecule binding of a ligand to a G-protein-coupled receptor in real time using fluorescence correlation spectroscopy, rendered possible by nano-encapsulation in styrene maleic acid lipid particles. *Nanoscale* 12: 11518.

Index

GPCRs as Therapeutic Targets, Volume 2, First Edition. Edited by Annette Gilchrist.

h

i

j

k

l